Modern Radar Systems

Second Edition

For a listing of recent titles in the *Artech House Radar Series*,
turn to the back of this book.

Modern Radar Systems

Second Edition

Hamish Meikle

ARTECH
HOUSE

BOSTON | LONDON
artechhouse.com

Library of Congress Cataloging-in-Publication Data
A catalog record for this book is available from the U.S. Library of Congress.

British Library Cataloguing in Publication Data
A catalogue record for this book is available from the British Library.

ISBN-13: 978-1-59693-242-5

Cover design by Yekaterina Ratner

685 Canton Street
Norwood, MA 02062

10 9 8 7 6 5 4 3 2 1

To my wife, Monika

Contents

Foreward xvii
Preface to the second edition xix
Preface to the first edition xxi

Chapter 1 The radar and its ground environment 1
1.1 Primary and secondary radar 1
1.1.1 Other types of radar 3
1.2 Coordinate systems and range 5
1.3 Main monostatic radar components 7
1.3.1 Transmitter 8
1.3.2 Waveguide or transmission line system 8
1.3.3 Diplexer 8
1.3.4 Antenna 8
1.3.5 Factors external to the radar 8
1.3.6 Receiver 8
1.3.7 Matched and matching filters 9
1.3.8 Detector 9
1.3.9 Analogue-to-digital converter 9
1.3.10 Signal processor 9
1.3.11 Threshold 9
1.3.12 Determination of position 9
1.3.13 Common components and timing 9
1.4 Basic quantities, maximum range 12
1.5 Secondary radar 17
1.6 Radars with separately located transmitters and receivers 17
1.6.1 Elliptical coordinates 18
1.6.2 Bistatic radar maximum range 19
1.7 Performance 21
1.7.1 Effects on range 22
1.7.2 Resolution 23
1.7.3 Accuracy 23
1.7.4 Stability 24
References 24

Chapter 2 Usual and unusual concepts 25
2.1 An example of three-dimensional representation: the Wien bridge oscillator 25
2.2 Vector representation 27
2.3 Order of linear processing 28
2.4 Polyphase modulation and demodulation 28
2.5 Symmetrical components in polyphase circuits 33
2.6 Harmonics in balanced polyphase circuits 36
2.7 Polyphase, or bottle-brush, noise 37
2.8 Time and spectral domains, helical spectra 41
2.8.1 Convolution and correlation 45
2.9 Gaussian pulses, spectra, and beam shapes 47
2.9.1 Gaussian pulses and spectra 47
2.9.2 Gaussian beam shapes 49
2.9.3 Gaussian illumination functions 49
2.10 Use of brackets and other symbols 50

	References	50
Chapter 3	Transmitters	51
	3.1 Transmitter power	51
	3.2 Power output stage	51
	3.2.1 Semiconductor transmitters	52
	3.3 Spectrum and sidebands	53
	3.3.1 Trapezoidal edges	54
	3.3.2 Cosine and cosine squared edges	56
	3.3.3 Extra modulator power needed for shaping	57
	3.4 Pulse compression	58
	3.4.1 Linear frequency modulation	59
	3.4.2 Simple phase modulation	62
	3.4.3 Other types of modulation and their spectra	65
	3.5 Harmonics from the transmitter	72
	3.6 Figures affecting radar performance	72
	3.6.1 Range	72
	3.6.2 Resolution	73
	3.6.3 Accuracy	73
	3.6.4 Stability	73
	3.6.5 Interference to neighboring systems	79
	References	79
Chapter 4	Microwave waveguide and transmission line system	81
	4.1 Mismatch	84
	4.2 Components	86
	4.2.1 Coaxial cables	86
	4.2.2 Waveguides	86
	4.2.3 Strip or microstrip lines	89
	4.2.4 Microwave passive components	90
	4.3 Monitoring	95
	4.3.1 Power level	95
	4.3.2 Voltage standing wave ratio (VSWR)	95
	4.4 Effect on radar performance	95
	4.4.1 Effects on maximum range	95
	4.4.2 Effects on stability	96
	References	96
Chapter 5	Antennas	97
	5.1 Linear and rectangular radiators	98
	5.1.1 Tapering the illumination function to reduce sidelobes	107
	5.1.2 Uniform, trapezoidal, and triangular illumination tapering	109
	5.1.3 Simply tapered illumination functions	109
	5.1.4 Low-sidelobe tapering functions	110
	5.1.5 General rules for tapering	115
	5.2 Radiation from circular apertures	115
	5.2.1 Simply shaped circular tapering functions	118
	5.2.2 Circular Taylor low-sidelobe tapering function	118
	5.3 Monopulse radar antennas	119
	5.3.1 Tapering functions for monopulse antennas with low sidelobes	121
	5.4 Arrays of discrete radiators	129
	5.4.1 Tapered illumination functions	131
	5.4.2 Ways of driving discrete elements	131
	5.4.3 Grating effects	133
	5.4.4 Beam-steering quantization effects	134

5.5 Creating shaped beams 138
5.5.1 Inverse Fourier transform method 139
5.5.2 The Woodward-Lawson method 140
5.6 Circular polarization 147
5.6.1 Circular polarizer for horn feeds 147
5.6.2 Reflecting polarizers 148
5.6.3 Transmission polarizers 148
5.6.4 Phased array polarization 149
5.6.5 Engineers' and physicists' conventions 151
5.6.6 Ellipticity or the quality of circular polarization 151
5.6.7 Rain echo suppression 152
5.7 Antenna hardware losses 152
5.7.1 Illumination function loss 153
5.7.2 Blocking loss 153
5.7.3 Spillover loss 153
5.7.4 Surface tolerance loss 155
5.7.5 Losses in power dividers, phase shifters, and other beam-forming network components 155
5.7.6 Other effects giving losses 156
5.8 Beam shape loss 156
5.8.1 Coherent integration 156
5.8.2 Noncoherent integration 158
5.8.3 Small numbers of pulses 159
5.9 Scanning loss 160
5.10 The equivalence of different signal combining systems 162
5.11 Noise received from an antenna 164
5.12 Sidelobe cancelers and adaptive beam forming 168
5.13 Antennas mounted on aircraft 169
5.13.1 Synthetic apertures 170
5.13.2 Mapping 176
5.13.3 Radars on satellites 176
5.13.4 Other considerations 177
5.14 Figures affecting radar performance 177
5.14.1 Range 177
5.14.2 Resolution 178
5.14.3 Accuracy 178
5.14.4 Stability 179
References 179
Appendix 5A Mathematical appendix 180
5A.1 Taylor distribution 180
5A.2 Zolotarëv distribution 180
5A.3 Bayliss distribution 182

Chapter 6 Factors outside the radar: propagation, scattering, and clutter 185
6.1 Amplitude and phase of the echo 185
6.1.1 Amplitude of the echo 185
6.1.2 Phases of the echoes and Doppler frequency 186
6.2 Effects of the atmosphere 187
6.2.1 Exponential atmosphere models 188
6.2.2 Constant *k* atmosphere model 190
6.2.3 Range-height paper 192
6.2.4 Lens loss 192
6.2.5 Anomalous propagation and superrefraction 194
6.2.6 Attenuation in the atmosphere 195
6.3 Scattering 200
6.3.1 Scattering without fading 201
6.3.2 Scattering with fluctuation 205

	6.4 Second-time-around effect	217
	6.5 Ground reflections and multiple paths	217
	6.5.1 Flat or nearly flat earth	217
	6.5.2 Factors that reduce reflections and lobing	227
	6.6 Scenario to simulate a "typical" radar environment	231
	6.6.1 Echoes of interest, the reference echo	232
	6.6.2 A land clutter model	235
	6.6.3 The Weibull clutter model	241
	6.6.4 Land clutter spectrum	242
	6.6.5 Sea clutter	244
	6.6.6 Volume clutter, rain or snow clutter, and chaff or window	246
	6.6.7 Rain and chaff spectra	248
	6.6.8 Total signal at the input of the receiver	250
	6.7 Figures affecting performance	250
	6.7.1 Range	250
	6.7.2 Accuracy	251
	6.7.3 Stability	251
	References	251
	Appendix 6A Range-height paper	253
Chapter 7	Receivers	273
	7.1 Dynamic range, control of gain, and sensitivity time control	274
	7.2 Radio frequency section	277
	7.2.1 Radio frequency amplifier	277
	7.2.2 Radio frequency filter	279
	7.2.3 Mixer	280
	7.3 Intermediate frequency amplifier and filter	281
	7.4 Limiters	282
	7.4.1 Effects on amplitude	283
	7.4.2 Effects on spectrum	283
	7.5 Receiver characteristics	285
	7.5.1 Minimum range	285
	7.5.2 Gain	285
	7.5.3 Bandwidth, filtering, ringing, and the ability to reject interference	285
	7.5.4 Dynamic range before and after sensitivity time control (STC)	286
	7.5.5 Ability to withstand transmitter pulse spikes	286
	7.5.6 Constant signal or noise output	286
	7.6 Figures affecting radar performance	286
	7.6.1 Range budget	286
	7.6.2 Resolution	287
	7.6.3 Accuracy budget	287
	7.6.4 Stability budget	287
	References	290
Chapter 8	Matched and matching filters	291
	8.1 Uncompressed pulses	292
	8.1.1 Rectangular transmitter pulse and rectangular filter	292
	8.1.2 Rectangular transmitter pulse and Gaussian filter	295
	8.1.3 Doppler frequency shift, detuning	298
	8.1.4 Filtering after limiting, Dicke-fix receiver	298
	8.2 Pulse compression using frequency modulation	301
	8.2.1 Linear frequency modulation	301
	8.2.2 Nonlinear frequency modulation	309
	8.2.3 The effects of limiting before the pulse compression filter	311
	8.2.4 General correlator	312
	8.3 Discrete phase shift modulated pulse compression	312

8.3.1 Binary codes 315
8.3.2 Polyphase codes 321
8.4 Other forms of modulation 326
8.5 Negative phase sequence signals 326
8.6 Output signals 326
8.7 Figures affecting radar performance 327
8.7.1 Range budget 327
8.7.2 Accuracy and resolution budget 327
8.7.3 Stability budget 327
References 327

Chapter 9 Detectors 329
9.1 Incoherent detectors 331
9.1.1 Logarithmic intermediate frequency amplifiers 333
9.2 Coherent detectors 334
9.2.1 Ring modulator or demodulator 336
9.3 Vector detectors 338
9.3.1 Polar detectors 339
9.3.2 Cartesian or two-phase detector 339
9.4 Sampling waveforms and spectra 341
9.4.1 Simple or single phase sampling 341
9.4.2 Complex or two-phase sampling 343
9.4.3 Sampling at intermediate frequency 344
9.5 Measurement of noise 344
9.5.1 Gaussian noise 345
9.5.2 Rayleigh noise 346
9.6 Figures affecting radar performance 348
9.6.1 Range budget 348
References 348

Chapter 10 Analogue-to-digital conversion 349
10.1 Principle 349
10.2 Dynamic range 349
10.3 Nature and treatment of errors 350
10.3.1 Types of errors 350
10.3.2 Measurement of errors 352
10.3.3 Correction of errors 354
10.3.4 Analogue-to-digital conversion using intermediate frequency signals 355
10.4 Figures affecting radar performance 358
10.4.1 Range budget 358
10.4.2 Accuracy and resolution budget 360
10.4.3 Stability budget 360
References 361

Chapter 11 Signal processing 363
11.1 Altering the form of the video during one sweep 365
11.1.1 Limiting 366
11.1.2 Differentiation of the video: fast (or short) time constant 366
11.1.3 Pulse length (or width) discriminator 366
11.1.4 Logarithmic video and log FTC 366
11.1.5 Bandwidth or stretching of the video signals for display 369
11.1.6 Noise clipping 369
11.1.7 Constant false alarm rate processing by cell averaging 370
11.1.8 Gating the video 375
11.1.9 Combining the videos from several beams 377

11.2 Signal processing over a number of sweeps 378
11.2.1 Video integration 384
11.2.2 Binary integration 385
11.2.3 Rejection of echoes that have unwanted Doppler frequencies, moving target indicator 385
11.2.4 Doppler frequency processing which selects the Doppler frequencies of interest 407
11.2.5 Comparison of moving target indicators and detectors 432
11.3 Processing over many scans and maps 432
11.3.1 Area moving target indication (AMTI) or clutter map 433
11.3.2 Maps for a moving target detector 435
11.3.3 Quality selection 437
11.4 Airborne moving target signal processing 438
11.4.1 The TACCAR moving target indicator 438
11.4.2 Displaced phase center antenna 439
11.4.3 Pulse Doppler radars 442
11.4.4 Sideways-looking radars 443
11.5 Figures affecting radar performance 443
11.5.1 Range budget 443
11.5.2 Accuracy and resolution budget 444
11.5.3 Stability budget 444
References 444
Appendix 11A An approximation to solve for thresholds above clutter 445

Chapter 12 Threshold and detection 447
12.1 Dwell time and the number of echoes 447
12.2 False alarm probabilities, times, and thresholds 448
12.2.1 False alarm time 448
12.2.2 False alarm number 448
12.2.3 False alarm probability 449
12.2.4 Changing the threshold levels 452
12.3 Probability of detection 452
12.3.1 Marcum case: no fluctuation 453
12.3.2 Swerling case I: slow fluctuation 458
12.3.3 Swerling case II: fast fluctuation 464
12.3.4 Swerling case III: slow chi-squared fluctuation 468
12.3.5 Swerling case IV: fast chi-squared fluctuation 472
12.4 Comparison of probability of detection cases 475
12.4.1 Coherent integration: slow fluctuation 476
12.4.2 Noncoherent integration: slow fluctuation 476
12.4.3 Noncoherent integration: fast fluctuation 477
12.5 Joint probabilities of detection 478
12.5.1 Other forms of integration after the threshold 478
12.5.2 Frequency diversity radars 483
12.6 Useful approximations 485
12.6.1 Albersheim's approximation for the Marcum case 485
12.6.2 Snidman's approximation for the Marcum and Swerling cases 486
12.7 Figures affecting radar performance 487
References 487

Chapter 13 Determination of position 489
13.1 Fire control radars 489
13.1.1 Conical scanning 490
13.1.2 Amplitude monopulse receivers 490
13.1.3 Phase monopulse receivers 492
13.1.4 Measurement of range 493
13.1.5 Extracting the Doppler frequency 494

13.2 Sector scan radars 494
13.3 Fast scanning radars 494
13.4 Surveillance radars 495
13.4.1 Binary detection decisions 495
13.4.2 Maximum signal 496
13.4.3 Center of gravity or centroid 496
13.4.4 Monopulse angle measurement with search radars 497
13.5 Accuracy 497
13.5.1 Angular accuracies and root mean square aperture 500
13.5.2 Time accuracies and root mean square bandwidth 501
13.5.3 Frequency accuracy and root mean square signal duration 502
13.6 The display of position 502
13.6.1 Displays used to measure range 502
13.6.2 Displays for surveillance 502
13.6.3 Airborne displays 504
13.6.4 Displays for aiming weapons 504
13.6.5 Displays to indicate interference or jamming 506
13.6.6 Signals used for displays 506
13.7 Figures affecting performance 507
References 508

Chapter 14 Performance 509
14.1 Range 509
14.2 Accuracy 511
14.2.1 Bias errors 512
14.2.2 Random errors 512
14.3 Resolution 515
14.3.1 Resolution in range 517
14.3.2 Resolution in azimuth angle 517
14.3.3 Resolution in elevation angle 518
14.3.4 Resolution in Doppler frequency 518
14.4 Stability, the cancellation ratio 518
14.5 Interfering or jamming signals 519
14.5.1 Chaff or window 519
14.5.2 Active jamming 519
14.5.3 Deception jamming 520
14.6 Tables 521
14.6.1 Basic radar requirements 521
14.6.2 Derived characteristics 523
14.6.3 Factors for calculating range 523
14.6.4 Resolution 524
14.6.5 Accuracy 524
14.6.6 Stability 524
References 525

Chapter 15 Statistics 527
15.1 Terms 527
15.1.1 Mean and expected values 528
15.1.2 Variance 528
15.1.3 Standard deviation 529
15.1.4 Histogram and probability distribution 529
15.1.5 Cumulative distribution function 529
15.1.6 Percentiles and quartiles 530
15.1.7 Moment generating functions 530
15.1.8 Fourier transform 531

15.2 Families of distributions 534
15.2.1 Gaussian or normal distribution: two sided from -∞ to +∞ 534
15.2.2 Log-normal distribution 537
15.2.3 Rayleigh distribution 538
15.2.4 Ricean distribution 541
15.3 Gamma distribution family: bounded on one side from 0 to ∞ 544
15.3.1 Erlangian distribution 546
15.3.2 Chi-squared distribution 547
15.3.3 Chi distribution 548
15.3.4 Negative exponential distribution 549
15.4 Other distributions bounded on one side 549
15.4.1 Weibull distribution 550
15.5 Discrete distribution: binomial distribution 551
15.6 Random numbers 552
References 553

Chapter 16 Transforms 555
16.1 Conventions for the Fourier transform 555
16.2 Some polyphase and single phase Fourier transforms 556
16.2.1 Single phase cosine wave 558
16.2.2 Single phase sine wave 559
16.2.3 Rectangular pulse 560
16.2.4 Time shift 561
16.2.5 Phase shift 563
16.2.6 Examples of Fourier transforms in the complex plane 563
16.2.7 Addition and subtraction 564
16.2.8 Differentiation 564
16.2.9 Convolution, the multiplication of Fourier transforms 565
16.2.10 Cross-correlation, multiplication with complex conjugates 568
16.2.11 Autocorrelation, multiplication with its own complex conjugate 569
16.2.12 Energy and power 570
16.3 Discrete Fourier transform 570
16.3.1 Differences between continuous and discrete transform functions 574
16.3.2 Fast transforms 576
16.4 Summary of properties of the Fourier transform 576
16.5 Tapering 577
16.5.1 Gains and losses 579
16.5.2 Spectral leakage or two-tone characteristics 584
16.5.3 Resolution 585
16.5.4 Example: von Hann and Hamming tapering functions 585
16.6 Relationships to other transforms 590
16.6.1 The *z* transform 590
16.7 The use of Fourier transforms for finite impulse response filters 592
16.7.1 Single-phase signal filtering 593
16.7.2 Video integration in a finite impulse response filter 597
16.7.3 Noncoherent moving target indicator finite impulse response filter 597
16.7.4 Polyphase signal filtering 599
References 600
Appendix 16A Complete correlation 601

Appendix A Language and glossary 603
A.1 Unified terminology 603
A.1.1 Military words that have been avoided 603
A.1.2 Unfortunate words 603
A.1.3 Technical words 604
A.1.4 Words retained 604

A.1.5 New words 605
A.1.6 Russian names 605
A.2 Glossary 605
A.3 Displays 614
A.3.1 Surveillance radars 615
A.3.2 Anti-aircraft (AA) artillery radars 615
A.3.3 Airborne intercept (AI) radars 615
A.3.4 Display types 615
A.3.5 Displays for showing the effects of jamming (military) 616
A.4 Symbols 617
References 617

Appendix B Tapering functions 619
B.1 Conventions and normalization 619
B.1.1 Scaloping loss and worst-case processing loss 621
B.1.2 Spectral leakage or two-tone characteristic 621
B.1.3 Other names used for tapering functions 621
B.2 Tapering functions 622
B.2.1 Trapezoidal tapering 622
B.2.2 $(1-4p^2)^n$ tapering. 626
B.2.3 Cosine to the power n tapering 629
B.2.4 Cosine on a pedestal tapering 633
B.2.5 Hamming, Blackman, and Blackman-Harris tapering 636
B.2.6 Truncated Gaussian tapering 646
B.2.7 Even Taylor tapering 649
B.3 Tapering with discrete elements 653
B.3.1 Dolph-Chebyshëv tapering for a discrete 10 element system 653
B.4 Tapering or illumination functions for circular antennas 657
B.4.1 Circular $(1-4r^2)^n$ tapering 657
B.4.2 Circular truncated Gaussian tapering 660
B.4.3 Circular Taylor tapering 662
B.5 Odd tapering functions 664
B.5.1 Odd rectangular tapering 664
B.5.2 Odd triangular tapering 667
B.5.3 Odd cosine to the power n tapering 669
B.5.4 Odd truncated Rayleigh tapering 671
B.5.5 Odd Taylor derivative tapering 673
B.5.6 Bayliss tapering 675
B.5.7 Zolotarëv tapering 679
References 681

Appendix C Frequency band letters 683

List of symbols 685

About the author 693

Index 695

Foreword

The publication of the first edition of *Modern Radar Systems*, in 2001, made available to the radar community a valuable resource for understanding the principles of radar systems and their several subsystems. Radar theory was explained in words, equations, diagrams, and plots, notably including three-dimensional representations of complex signals that provide visualization of otherwise obscure concepts. Accurate, closed-form expressions were included to permit the radar engineer to calculate and plot the functions that affect radar performance.

Now, in an extensively revised and updated second edition, Hamish Meikle has refined the original discussions and added significant new material to cover areas of importance to all those working in radar design and analysis, including the statistical domain. Much of the new material reflects the evolution of radar, independent of the technology used, from early designs to more modern digital and solid-state circuits and subsystems.

Chapter 1 presents the fundamentals of radar, including both primary radar that receives echoes reflected from the target and secondary radar that relies on a transponder to generate the response. Monostatic and bistatic radar systems are defined and described. Before going into details of radar circuits and subsystems, Chapter 2 introduces the fundamental methods of vector presentation, modulation, polyphase signals and noise, and the probability distributions that are used to represent them. A new section covers the mathematical processes of convolution and correlation, essential for analysis of radar performance and increasingly used in modern digital signal processors.

The discussion of radar subsystems starts with Chapter 3 on transmitters, where new material has been included on solid-state devices that have begun to replace vacuum tubes in both modulators and RF power sources. The discussion of pulse compression has been updated to reflect current design practice, including waveforms such as the NRL P-codes, and references to recent publications have been added. Chapter 4 on waveguides and transmission lines now includes material on stripline technology, and the description of the many types of microwave components has been updated.

An expanded Chapter 5 describes antennas, including a discussion and plots of low-sidelobe monopulse patterns based on the Bayliss and Zolotarev distributions. A discussion on electronically scanned antennas using horn-fed lens or reflector arrays, as used in the U.S. Patriot system and many Russian radars, has been added. As in the first edition, a thorough discussion of losses associated with antennas has been included. New sections on sidelobe cancelers and adaptive arrays now appear, along with a section on aircraft antennas, including synthetic aperture systems. Chapter 6 covers factors outside the radar: propagation, scattering, and clutter. The mathematical basis of the Swerling target models is now described more thoroughly.

Continuing with radar subsystems, Chapter 7 is dedicated to the receiver, followed by Chapter 8 on matched and matching filters. A section has been added on the Dicke-fix receiver, an important counter to electronic jamming. In the discussion of pulse compression, the nonlinear FM technique is now covered as well as the linear FM technique, the effects of limiting before pulse compression are described, and the correlator processing method is presented. Additional data on discrete phase codes, including the polyphase codes, has been added to the material of the first edition.

Chapter 9 covers detectors, and Chapter 10 discusses analogue-to-digital conversion. Here, material on the techniques of conversion at intermediate frequency has been added. Signal processing is the subject of Chapter 11. New material includes detailed mathematics of the log-FTC circuit response to clutter and on the operation of range-cell-averaging constant-false-alarm-rate (CFAR) processing. A section on airborne MTI has been added, covering time-averaged coherent clutter airborne radar (TACCAR) and the displace-phase-center antenna (DPCA).

Threshold and detection issues are discussed in Chapter 12, with plots and graphs covering the steady target and the four Swerling models. Chapter 13 covers the determination of position, with equations for the errors for different measuring procedures. A section on the display of position has been added. Chapter 14 discusses radar performance, including the range equation, accuracy, resolution, stability considerations, and operation in jamming. Finally, chapters on statistics and transforms explain in more detail the mathematical techniques needed to perform radar system analysis. A section on moment generating functions has been added to Chapter 13. The transform discussion now includes material on convolution, the multiplication of the Fourier transform, supporting the earlier discussion in Chapter 2.

Those who have used the first edition in their work will welcome the new material included in this new edition. Those who have not previously discovered the unique advantages of Meikle's presentation are guaranteed new insight into the analysis of radar signals and circuits provided by the second edition of *Modern Radar Systems*.

David K. Barton
Hanover, New Hampshire
February 2008

Preface to the second edition

As with the first edition, the chapters are arranged in the order that the signals pass through the radar hardware and, as far as possible, the section numbers have not been changed. This leads to pulse compression being divided between two chapters, Chapters 3 and 8. In reworking, a number of errors from the first edition have been corrected.

An important domain in radar is the statistical domain. This edition extends the Fourier transform treatment introduced in the first edition into the statistical domain for the calculation of probabilities of detection for wanted echo signals surrounded by thermal noise or clutter.

The second edition continues to use simple, civil language, using military terms only when appropriate and very few of the abbreviations that are beloved by the military community.

Again I would like to thank the staff of Artech House and their reviewers for their encouragement, support, and suggestions for the improvement of this book.

Hamish D. Meikle
Holzfeld, near 56154 Boppard am Rhein, Germany
Meiklehd@t-online.de
February 2008

Preface to the first edition

A radar is placed among its surroundings in space. It sends, receives, and processes signals in time. These signals are also described by their spectra and statistical distributions. The radar engineer must be able to think in all four of the domains, know their relationships, and be able to move freely between them using the shortest and most dependable routes.

The creative talents of engineers are primarily to look into the future, grab an idea, and create the three-dimensional drawings so that production can start today. This book is an attempt to support thinking in many domains and dimensions.

The main theme is to convey the shapes of the phenomena so that an engineer can place numbers on them and apply them to radar. The Maple V program has been used as the calculating engine to present the values and shapes in three dimensions for a better, more complete understanding, especially with vectors. To this end, in addition to amplitude and phase or x and y (I and Q) graphs, three-dimensional space curves are shown which emphasize the vector relationships involved. In practice purely real or purely imaginary simplifications seldom occur.

Radar has attracted electrical engineers, physicists, mathematicians, and later software specialists, each of whom brought their own terminology. Many have forgotten whence the techniques came: electrical engineering. The unity of thought and terminology throughout electrical engineering fell apart as engineers started to specialize; and worse still, electrical engineering training has followed this trend. The growing value of software in products is tending to change software for systems into systems for software (Appendix A).

Radar literature is peppered with abbreviations which often leads to woolly thinking and are roundly condemned by psychologists as a method of distinguishing those within the privileged, knowing circle from those without [1]. The author has tried to avoid them here, but they are included in parentheses (round brackets) for those who are more familiar with the abbreviations alone.

Radar technology grew up during the Second World War (1939-1945) and was refined during the Cold War (1946-1989). The language that developed is full of belligerent terms, and those who use them today are mostly ignorant of their origins. This is especially the case for those with a limited knowledge of English who do not know the principal English meaning. I was horrified once to learn that the surveillance radars at a civilian airport had to be "boresighted" and that the monopulse azimuth accuracy enhancement gave an "off-boresight-error" in civil secondary radar. Boresights are cross-wire sights mounted by hand in the bores or barrels of guns to provide the fundamental sighting reference. In field guns they are used to realign the main sights should they become displaced. In anti-aircraft gunnery, the boresight mounted in the barrel of the master gun is pointed at the alignment telescope of the fire control radar which is the reference. The gun azimuth is then set to the back bearing of the radar azimuth, and the elevations, set by spirit levels, are checked. I thought of the approach controller at the destination airport shepherding a passenger aircraft full of happy, carefree holiday-makers home using the representation of the "target" by his radar. At this time, the boresights would have been removed to allow the "target" to be engaged accurately. In the case of a miss, there was always the "execute" button on the console. To avoid such misunderstandings, demilitarized language is used in this book.

Having used W. A. Skillman's book on radar calculations [2] later, the version for personal computers, I thought that a more transparent book could be written using Maple V on a computer as the calculating engine. The graphics available in Maple bring the phenomena to life and give them shape.

I started to write useful Maple V worksheets with the "book of the disk" containing explanations. This was to give the look and feel on the computer display for the phenomena described using stereo images. I was advised that the market would be too small because too few people use the Maple V program. Here the Maple V program has carried out hard mathematical work and provided the diagrams in three dimensions. If demand calls for it, the "disk of the book" containing the Maple worksheets could be published. That would give hands-on experience in the handling of the shapes and sizes of the phenomena described in parallel to using published results.

When writing such a book, limits must be set. For that reason no techniques using Monte Carlo methods, covariance matrices, or adaptive characteristics are described.

I would like to thank Mr. David Barton for his many useful comments and suggestions including the name for the title and Mr. Robert Schmidt for his advice in the beginning. I would also like to thank Waterloo Maple Incorporated who kindly sent me an update of their program.

Hamish D. Meikle
Holzfeld, near 56154 Boppard am Rhein, Germany
February 2001

REFERENCES

1. Dixon, N.F., *On the Psychology of Military Incompetence*, London: Johnathon Cape Limited, 1976.
2. Skillman, W.A., *Radar Calculations Using the TI-59 Programmable Calculator*, Dedham, Massachusetts: Artech House, 1983.

Chapter 1

The radar and its ground environment

The early radars worked by sending out a continuous wave and the reflected energy showed the presence of an object, such as a ship or an aircraft. Separate transmitting and receiving antennas were used. Today, police radars, designed for speed traps, use this principle as do intruder alarms, and simple radars used as proximity fuses mounted in the noses of shells. Frequency modulated continuous wave radars, in addition to those used in radar altimeters and in the HAWK anti-aircraft missile system, are starting to be used in vehicles. Current pulse radars switch the same antenna between the transmitter and receiver to allow transmission and reception (monostatic) and are capable of measuring the range of the reflecting or scattering object, its azimuth angle, elevation angle, and radial velocity.

1.1 PRIMARY AND SECONDARY RADAR

A primary radar illuminates its surroundings, like a searchlight, and picks up part of the energy scattered by the objects it illuminates. Monostatic radars, the majority of all radars, use the same antenna for transmission, or illumination, and reception. Such a radar is shown in Figure 1.1.

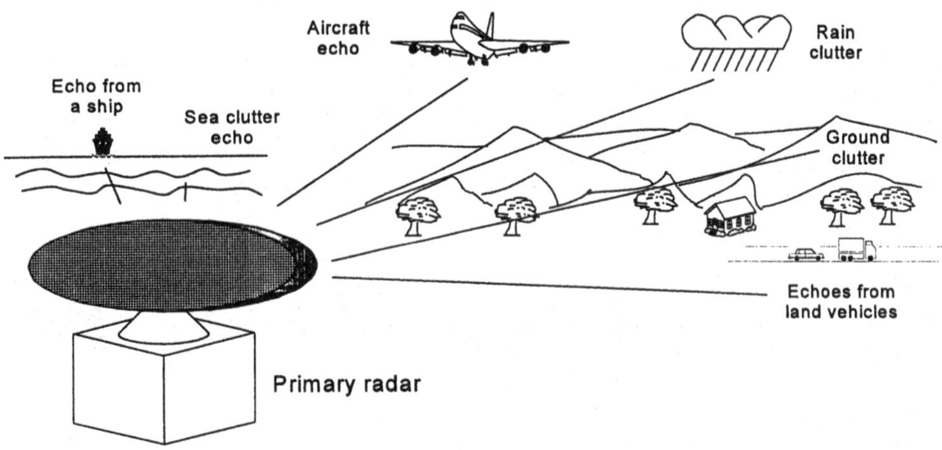

Figure 1.1 A primary ground radar and its environment.

If the transmitter sends short pulses, the scattered energy from the objects that form the environment are copies of the transmitted pulses with a time delay proportional to the distance traveled, as with a sound echo. The frequency of the echoes, as with sound, depends on the rate of change of range or radial velocity. Thus the echoes from the objects in Figure 1.1 may be classed from fastest, probably the aircraft echoes, through ground vehicle and rain clutter to the waving of the trees.

When directing aircraft using radar during the Second World War (1939-1945), a need appeared to be able to differentiate between one's own aircraft and encroaching aircraft, which were their targets. This led to the development of radar repeaters which were carried on friendly aircraft and doubled the "blip" on the display. This system is called identification friend or foe (IFF). All the repeaters in the aircraft had to be tuned to the frequency of the radar that observed it and this led to the development of separate systems, initially the IFF Mark III in the band 157 MHz to 187 MHz. Radar systems that require the object being observed to cooperate by carrying a receiver-transmitter, transponder, or beacon are called secondary radar systems.

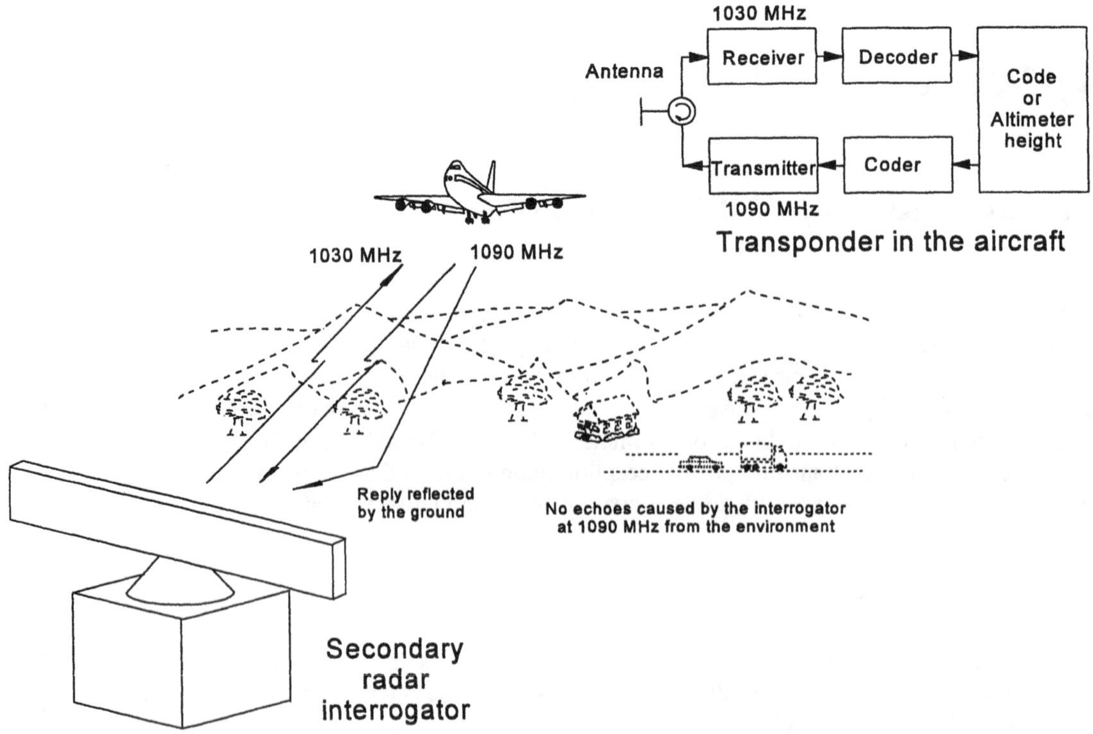

Figure 1.2 A secondary radar interrogator and its environment.

Currently, frequencies of 1 030 MHz are used for the interrogation path from ground to air path and 1 090 MHz for the reply path from air to ground, as shown in Figure 1.2. The system, originally known as the military Mark X (Roman ten) was opened for civil use with mode 3 and A common to both. The interrogator transmitter transmits two 0.8 μs pulses with a power between 21 dBW and 27 dBW (about 125 W to 500 W) spaced according to the mode of interrogation in the main beam. The STANAG and ICAO pulse separations are listed in Table 1.1.

Table 1.1
A selection of STANAG and ICAO pulse separations for secondary radar

	Military mode	Civil mode	Pulse spacing (μs)
	1		3
	2		5
Military and civil (shared)	3	A	8
		B	17
Height reply		C	21
		D	25

Aircraft that use this system carry a combination of receiver and transmitter called a transponder. The pulses from the interrogator are decoded, and the code, if any, for the mode is sent to the coder. The coder sends two 0.45 μs bracket pulses with 20.3 μs spacing with up to 12 pulses carrying the code in between. The codes are generally set by the pilots on the instruction of the air controllers and are used to identify the aircraft. Mode C is reserved for the standard altimeter height, which obviates the need for civil three-dimensional radars.

There are extra high security ciphered modes for military use, and the civil modes have been extended to mode S. Aircraft equipped with mode S reply normally to modes 1 to 3/A to D. Mode S has extra modes that are used to interrogate aircraft individually and transfer data [1].

The receiver listens on 1 090 MHz and the echoes from the environment return at a frequency near 1 030 MHz; thus, no echoes are received to be confused with wanted replies. The aircraft antennas for the transponders are quasi-omnidirectional, so there will be asynchronous replies in the receiver that come from interrogations from other sites.

Other types of special transponder are used on objects, such as space vehicles, to extend the range of radars.

1.1.1 Other types of radar

This book is principally concerned with pulse radars where the antennas turn continuously to provide signals or data for surveillance as at airports, for example. At the end of the Second World War two types of continuous wave radar were in production: proximity fuses and radar altimeters, and other types of radar followed.

The proximity fuse is mounted in the noses of anti-aircraft shells and has to be cheap, robust, and simple. When the shell is fired, a glass ampule containing electrolyte in the battery is shattered, and the battery provides power to a radio frequency oscillator. When the shell arrives near a large object, such as the ground or its target, the returning radio frequency waves have a Doppler frequency shift. The returning signals beat with signals from the oscillator to give a signal in the audio frequency range, separated out from direct voltage biasses by a capacitor as in Figure 1.3. When the signal is large enough, the shell is detonated.

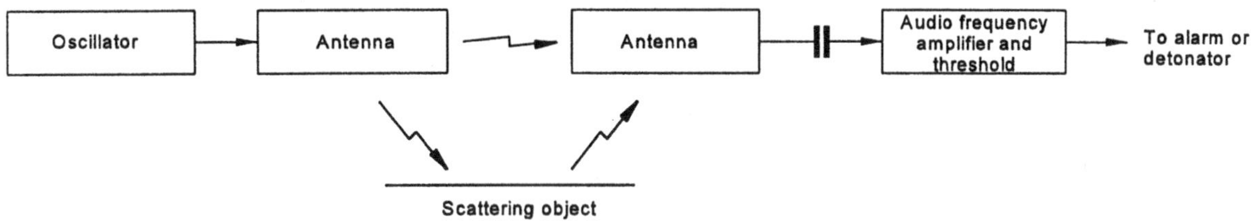

Figure 1.3 A primitive continuous wave radar.

Inside buildings intruder alarms fill the rooms with radio frequency energy and give an alarm when any signals exhibiting Doppler frequency are returned. Similar radars are used to open doors and to indicate when vehicles approach traffic lights.

The police use radars for the measurement of vehicle speed in order to enforce speed limits. A counter counts the audio frequency for a short time and the speed is displayed directly. Examples of Doppler frequency are shown in Table 1.2.

Table 1.2

Doppler frequencies for common speed limits

Frequency band	Doppler frequency, Hz	
	30 mph	50 kmph
3 GHz	267.67	277.78
10 GHz	811.11	841.75
30 GHz	2 676.67	2 777.78

Other examples of continuous wave radars that measure velocity are those that measure the speed of cricket balls or baseballs, the muzzle velocities of guns, and as well as speedometers mounted under vehicles.

The signal from continuous wave radars must be frequency modulated in order to measure range as in Figure 1.4.

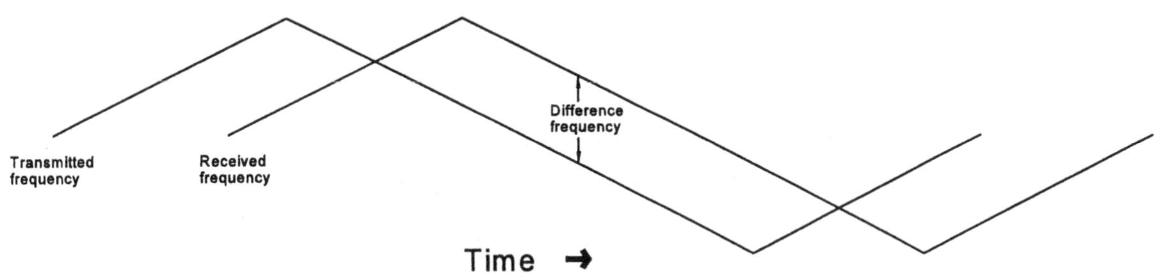

Figure 1.4 A linear frequency modulation for a radar altimeter.

A radar or radio altimeter is dimensioned to have a good signal from the ground at a height of up to, say, 15 km (49 200 feet) and is able to change the range of heights that it measures. A block diagram is shown in Figure 1.5.

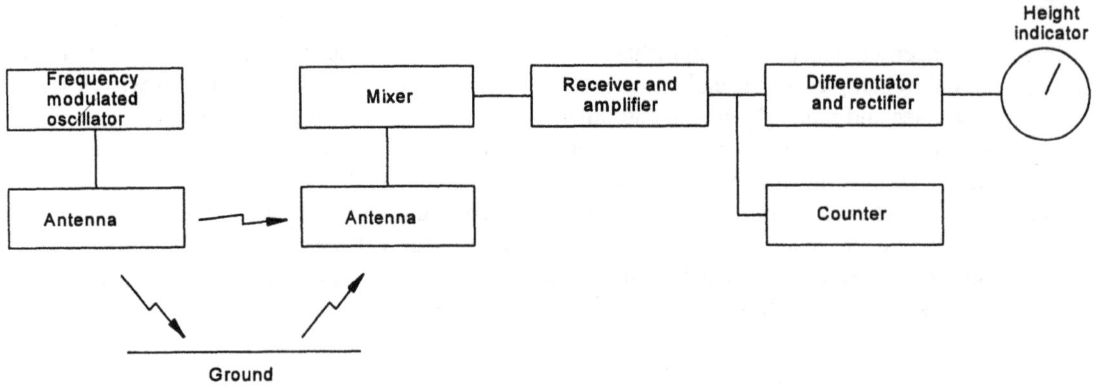

Figure 1.5 Block diagram of a radar altimeter.

The HAWK radar system uses a search radar that uses continuous wave and frequency modulated continuous wave modes on alternate scans to indicate velocity and range.

Frequency modulated continuous wave radars with semiconductor transmitters are starting to be used with road vehicles. There is much research and development to find the best form of frequency modulation, often stepped, to give the best range determination and resolution, to find the best antenna design for azimuth accuracy and resolution, and to manage the display of the information to increase safety on the roads.

Radar signals do not propagate well under ground nor is the velocity constant. Ground penetrating radars [2] have been developed to look into the first 5 m, say, to find buried pipes and cables, voids in the ground that may contain corpses, voids in buildings (secret chambers), archeological sites that have not been excavated, and ores around holes that have been drilled. The common way of using them is to mount the radar on a small cart or sledge, drag it across the area to be surveyed while recording the data, and then analyze the data later: the objects observed do not move. Distances are short so pulsewidths must be short, leading to the use of ultra wide band techniques. A simple example is shown in Figure 1.6.

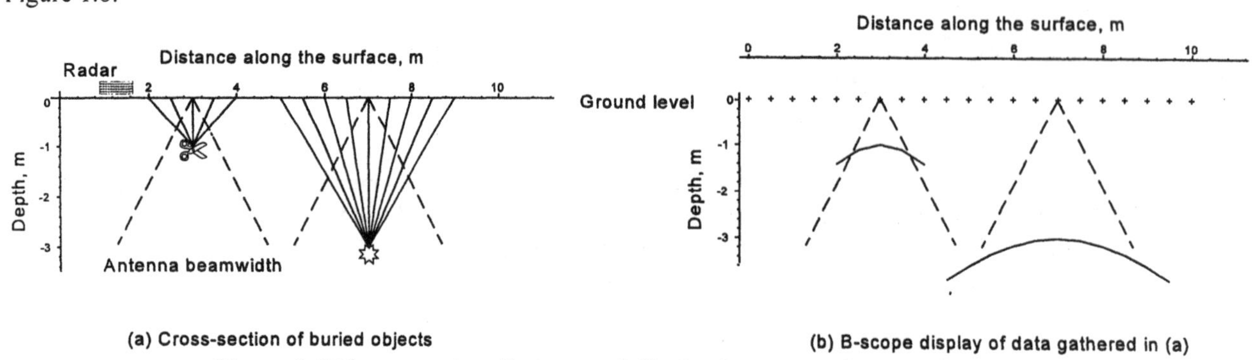

Figure 1.6 The geometry of sensor and display for a ground penetrating radar.

Commonly the antenna beamwidths are wider than with other radars and the positions where the radar sees the object are shown in Figure 1.6(a). The radar pictures obtained are often murky and need experience to be interpreted, and typically ranges, or distances between the radar and the buried object, have a wider variation so that the B-scope evaluation of the data shows the objects as hyperbolae in Figure 1.6(b).

1.2 COORDINATE SYSTEMS AND RANGE

A radar can be made to measure the position of an object in three dimensions and its rate of change of range, radial velocity. "Radar" coordinates are used throughout this book and, to avoid confusion, these are compared to physical polar coordinates in Figure 1.7.

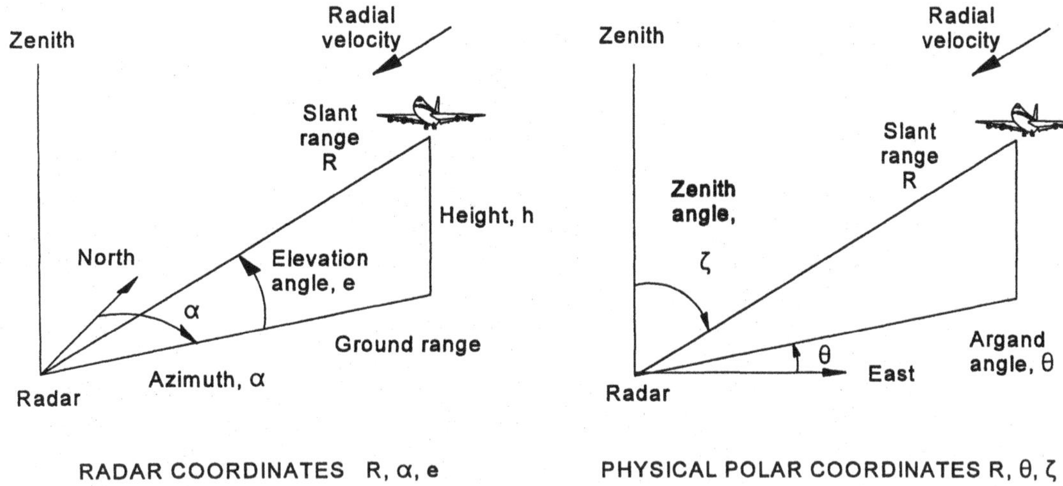

Figure 1.7 Radar coordinates and their physical polar equivalents.

Principal causes of confusion are the different starting points and opposite senses of the angle coordinates. The main units used are listed in Table 1.3.

Table 1.3
Units used for radar coordinates

Type of coordinate	Range	Angle (number in a circle)	Doppler frequency	Comments
Fundamental	Meter	Radian (2π)	Hertz	
Metric	Meter, Kilometer		Meters per second	
Navigation	Nautical miles	Degrees (360) ACP (2^n)	Knots	ACP is the azimuth change pulse with (2^n) in a circle. These vary from 256 (8 bits) to 16 384 (14 bits) ACP to a circle. The nautical mile was defined as the distance of one minute of angle measured at the center of the Earth on the earth's surface. The United States nautical mile is now 1 852 m. The metric equivalent is the angle subtended by 1 km, one centigrad. The grad angle unit gives 400 grad to a circle and is used by metric surveyors. Always check angle measuring devices to see which units are used.
Land surveillance	Meters, yards, miles	Degrees (360) grad (400)	Meters/sec mph	
Artillery	Meters, yards	Mils (6400)	Meters/sec Feet/sec	Warsaw Pact equipment uses 6 000 mils to a circle.

The transmission has to be modulated to measure range. The normal way is to emit pulses of radio frequency energy which are reflected, and the time to the reflecting object and back is measured. The time is proportional to range. The modulation is usually pulse modulation but frequency modulation, phase modulation, and combinations have been used. If light travels at 300 m/s, then each microsecond of time between transmission and reception (there and back) represents 150 m.

$$Range = \frac{3.10^8}{2} \times time\ (s)\ \text{m} \\ = 150 \times time\ (\mu s)\ \text{m} \tag{1.1}$$

In other units, instead of the constant 3 × 10⁸/2 m/µs, the range is given by:

- Kilometers 0.15 km/µs or 6.667 µs/km;
- Statute miles 0.0934 statute miles/µs or 10.729 µs/statute mile;
- Nautical miles 0.0899 nautical mile/µs or 12.347 µs/nautical mile.

Azimuth and elevation angles are derived from the antenna characteristics, and sometimes the elevation angle determination uses reflections from the ground. The ability to produce fine antenna beams using smaller antennae has favored the use of shorter wavelengths, commonly between 3 mm (100 GHz) and 23 cm (1.25 GHz). There are, however, radars that operate outside these frequency ranges.

Radial velocity is measured using the Doppler effect, when the reflected signal undergoes a change in frequency. Typically this is 20 Hz/m/s at 3 GHz (wavelength 10 cm).

Radars are designed for a particular purpose and to detect and find the position of particular classes of objects. These objects are normally in the atmosphere (including space) and part of the energy is reflected by the surroundings (land or rain clouds) that clutter the signal on the display. The unwanted signals are thus called clutter.

Pulse radars send out a pulse and wait until all echoes have returned before sending out the next pulse. The frequency of the pulses is called the pulse repetition frequency, and the time between two pulses, the pulse repetition time. A pulse repetition frequency of 1 000 Hz gives a listening time of 1 000 µs or a maximum range of 150 km. Echoes beyond 150 km will return after the second pulse has been transmitted and will be displayed at short ranges. These are called second time around echoes.

Early radars used a magnetron in the transmitter. The magnetron was driven by a high-voltage pulse from the modulator, and the radio frequency was determined by the size of its cavities. The phase of the output is random and uniformly distributed. This made for simple, robust radars, many of which are still in service. When moving target indicators appeared, the random phase had to be compensated with an agile coherent oscillator as a reference for the phase detector.

The pulse repetition frequency of pulse radars limits the range of Doppler frequencies that can be identified uniquely (see Chapter 9, Detectors). The pulses are a form of sampling and this limits the Doppler frequency determination to a span of the pulse repetition frequency (see Chapter 11, Signal processing). If, for example, a pulse repetition frequency of 1 000 Hz is used in S-band (3 GHz), this represents a velocity span of 50 m/s or 180 km/h. This is not enough to give the correct speed on a German autobahn let alone a flying aircraft. Thus there is a compromise between choosing the pulse repetition frequency to give a maximum range and being able to estimate the Doppler frequency simply.

Table 1.4 shows the major division between pulse radars that measure range without ambiguities and pulse Doppler radars.

Table 1.4

Ambiguous range and Doppler frequency

	Ambiguous range	Ambiguous Doppler frequency
High pulse repetition frequency	Yes	No
Medium pulse repetition frequency	Yes	Yes
Low pulse repetition frequency	No	Yes

Ambiguities can be resolved by the observation of the object of interest using a number of radio frequencies, pulse repetition frequencies, or both. This is called radio or pulse repetition frequency diversity except when the pulse repetition frequency is changed from pulse to pulse in moving target indicator processors, then the term staggered pulse repetition frequency is used.

Radar detection and measurement are subject to statistics. As with fading in short wave radio, diversity allows separate possibilities of detection and measurement during a look. Older frequency diversity radars have two or more separate transmitter-receiver systems that use the same antenna to provide extra chances of detection and more accurate measurement. When one radar channel developed a fault, the other channel was still operating. Radars with wider band transmitters can transmit waveforms for frequency scan (see Section 5.4.2) or at two frequencies for frequency diversity, with a separate receiver for each frequency. Semiconductor transmitters have many modules working in parallel and, if one fails, the others provide enough power to continue operation. Standby receivers may be switched in quickly to provide continuity on the receiving side. Coherent radars often use trains of pulses at two pulse repetition frequencies during a look to give an additional opportunity to separate wanted echoes from clutter. This is in contrast to signal processors which suppress signals with near zero Doppler frequencies that use staggered triggering (giving staggered pulse repetition frequencies) to give better responses for echoes with higher Doppler frequencies (see Chapter 11, Signal processing).

1.3 MAIN MONOSTATIC RADAR COMPONENTS

A block diagram for a basic monostatic radar is shown in Figure 1.8. The radar timing is contained in the coherent oscillator and trigger generator box. The coherent oscillator provides the phase reference for the radar and the basic timing clock.

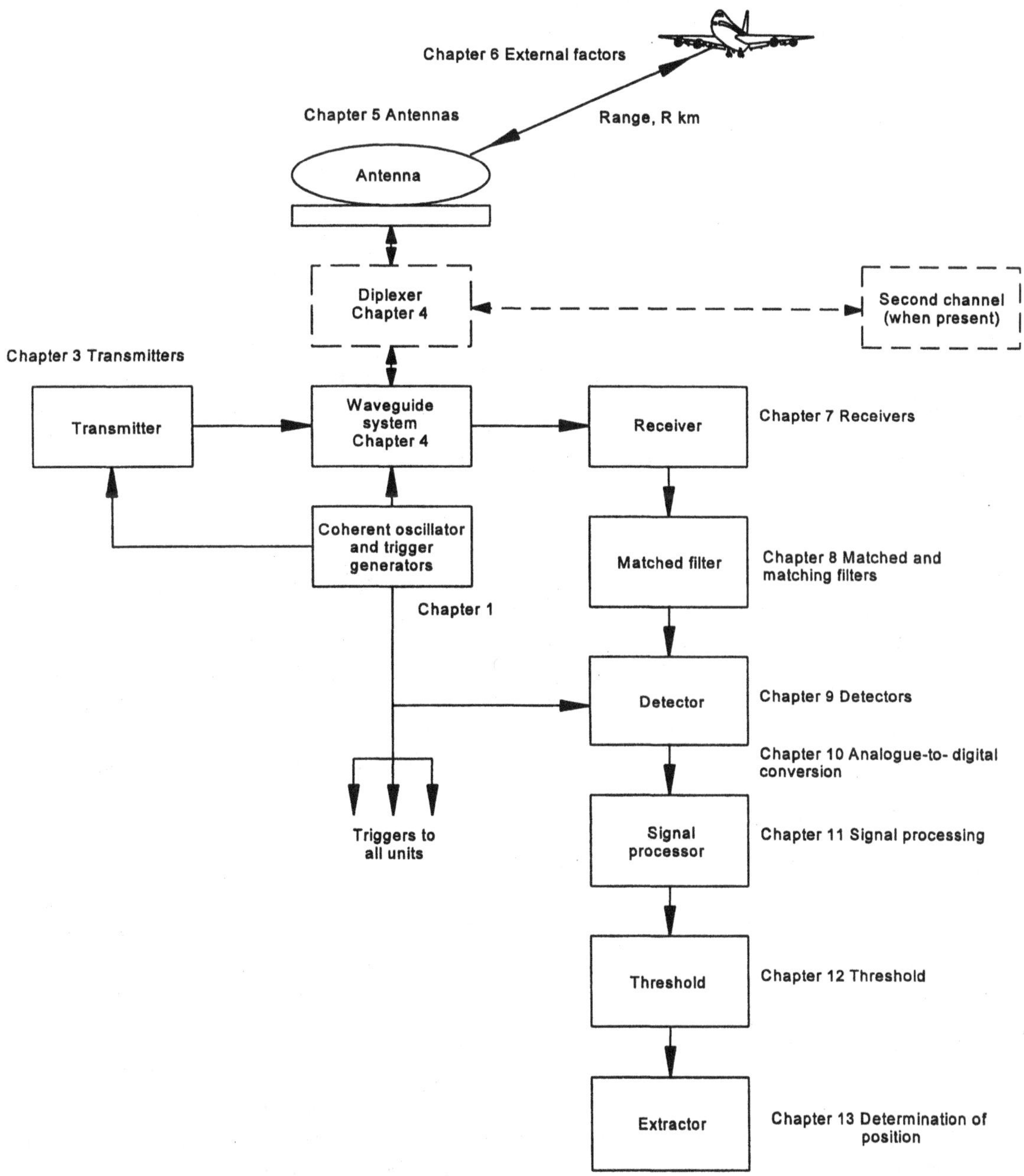

Figure 1.8 Block diagram of a basic monostatic radar.

Radar components cover a wide range of sizes and use a number of different special techniques so that they are normally developed by specialists in their own field. A number of common factors must be observed throughout the design of a radar, such as the avoidance of signal distortion, which requires a wide dynamic range, and the best arrangement of the stages using the components at hand.

Figure 1.8 shows a typical arrangement of stages for a modern radar. As long as no signals are distorted or suppressed, linear signal processing may take place in any order, and the processes used in analogue processing (mostly at intermediate frequency), and digital processing (digital representations of a two-phase video signal) are equivalent. Vector representations of the processes involved in both the time and the spectral domain help to give clarity and provide unity of treatment.

1.3.1 Transmitter

The transmitter converts mains power into radio frequency power, which leaves the radar by the antenna to illuminate its surroundings in the same way as a searchlight. Many texts describe transmitters [2, 3], and Chapter 3 is limited to describing the transmitter spectrum in vector form and the transmitter stability.

1.3.2 Waveguide or transmission line system

The waveguide or transmission line system connects the transmitters and receivers to a common antenna. Chapter 4 describes these with particular attention to using oversize waveguides.

1.3.3 Diplexer

The diplexer is only used for frequency diversity radars that operate on two or more frequencies simultaneously. The diplexer separates the signals according to frequency and conducts them to the appropriate transmitter-receiver. The transmitter pulses must be offset in time so that they do not overlap. Some older radars had a second channel as stand-by channel and a waveguide switch was installed in place of the diplexer.

1.3.4 Antenna

The antenna directs the transmitter energy into a beam that illuminates the radar's surroundings. The echoes of the illumination are gathered by the antenna for the receivers. The antenna is the principal filter to attenuate signals (interference or jamming) coming from other directions. Chapter 5 describes the shapes of antenna beams from linear and circular apertures with the accompanying side effects: sidelobes and grating lobes. Loss figures are given for rotating antennas owing to energy leaving and entering the antenna off the tip of the beam. The antenna is the first component in the receiving chain, the effects of the antenna on receiver noise are discussed in Chapter 5.

1.3.5 Factors external to the radar

Once the radio frequency energy leaves the antenna, it is subject to atmospheric refraction and attenuation. Chapter 6 describes common models, and Appendix 6A gives examples for range-height paper for these models. Conventionally, the radio frequency energy is scattered, and part of the energy returns to the radar. The shapes of a number of scattering models are illustrated and described.

1.3.6 Receiver

The receiver is normally a superheterodyne receiver with one or more intermediate frequencies. The first stage of the receiver is the principal component that defines the noise level and thus the sensitivity of the radar. The signals that arrive at the signal processor must not be distorted so they may be separated into the original components without the distortion components. The use of sensitivity time control (STC) reduces the necessity for a very wide dynamic in the receiver circuits, and diagrams show the effects of limiting.

1.3.7 Matched and matching filters

The matched filter has a shape and a bandwidth that give best selection of signal energy out of the mix of signals and noise coming from the receiver. These filters are often difficult to produce, so commonly simpler filters are used, called matching filters. Matching filters introduce losses that are discussed and illustrated in Chapter 8. The chapter also shows how the signal energy is compressed in time in pulse compression systems.

1.3.8 Detector

The intermediate frequency signals need to be detected before they can be displayed. Simpler detectors reject either phase or amplitude information and change the statistics of the thermal noise on a notional carrier from the receiver. Coherent signal processors, which follow detectors, require the full amplitude and phase information in the echo signals. This information is preserved in vector detectors. Detectors are described in Chapter 9.

1.3.9 Analogue-to-digital converter

Digital signal processors perform arithmetic on trains of binary numbers that represent the signals. Analogue to digital converters, described in Chapter 10, convert the signals, normally from a vector detector and more recently directly from the intermediate frequency signal, to digital form. The side effects are the limiting of the dynamic range, quantization noise, and distortion.

1.3.10 Signal processor

Chapter 11 describes the signal processing that may take place during one sweep, many sweeps, and many scans. Historically, first limiting of the video signals was performed to provide a better match to cathode ray tube phosphors and to reduce burn-in and glare, and this was augmented by using fast or short time constant shaping for the video. Over a number of sweeps, there are equivalent forms of signal processing for vector signals on a carrier, normally at intermediate frequency, and for detected vector signals. Mapping over a number of scans is also described.

1.3.11 Threshold

The bases for decisions whether the radar detects an object are described in Chapter 12. They are described in terms of a steady scatterer (Marcum model) and the four Swerling model scatterers.

1.3.12 Determination of position

Once the decision is made as to whether something is there, the best estimation must be made for its location. Methods and the accuracy obtained are described in Chapter 13.

1.3.13 Common components and timing

The components in a radar must operate together and in proper time, so every radar needs a number of control pulses or triggers. Typical trigger timings are shown in Figure 1.9, where the time is divided into receiving or listening time (when the echoes return) and the dead time in between. Activity starts long before the transmitter pulse with the basic pulse repetition frequency or system trigger. This is delayed to generate the other triggers. The names and numbers of triggers vary from radar to radar. Typical triggers are listed in Table 1.5.

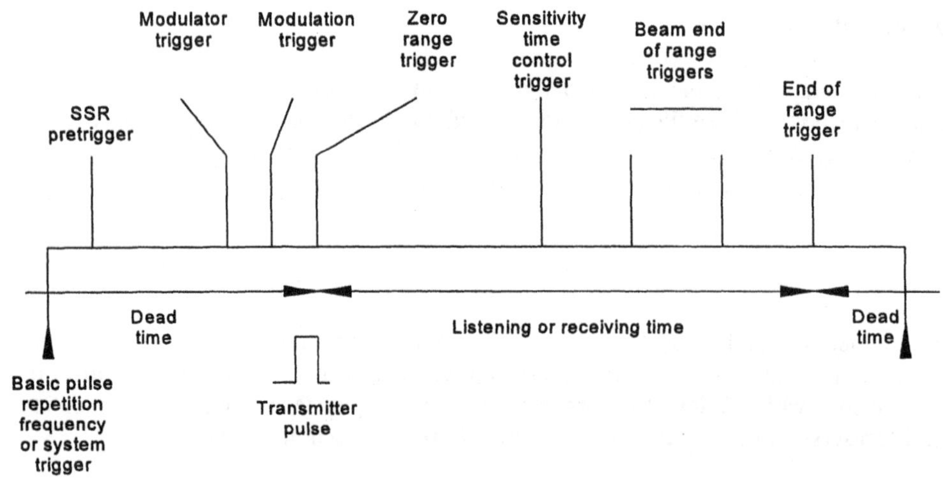

Figure 1.9 The principal triggers in a low pulse repetition frequency radar.

Table 1.5
Typical radar triggers

Time	Trigger	Purpose
End of dead time	Secondary surveillance radar (SSR) pretrigger	Sent to the SSR so that it has time to choose and code its interrogation pulses
	Modulator trigger	Sent to the modulator in the transmitter to start the modulator pulse which is sent to the output stage
	Modulation trigger	Sent to the stage which modulates the radio frequency input for the output stage
Receiving or listening time	Zero range trigger	Starts the sweeps on the displays and so on.
	Sensitivity time control (STC) trigger	Switches off the sensitivity time control
	End of beam triggers	Switch off the video outputs from the receivers to reduce collapsing loss
	End of range trigger	Switches off all the video outputs and starts the dead time
Start of dead time		This gives a pause until the end of dead time during which: • Echoes beyond maximum range return • The receiver is gated off to allow the use of test and calibration pulses in the system

Medium and high pulse repetition frequency radars send pulses during the listening times for longer ranges. During transmission the receivers are blanked or eclipsed, giving holes in the coverage. This reduces the probability of detection for the radar, and the increase of the signal-to-noise ratio necessary to bring back the probability of detection to the specified value is called the eclipsing loss.

Early and present day magnetron radars use a relatively simple trigger generation and distribution as is shown in Figure 1.10.

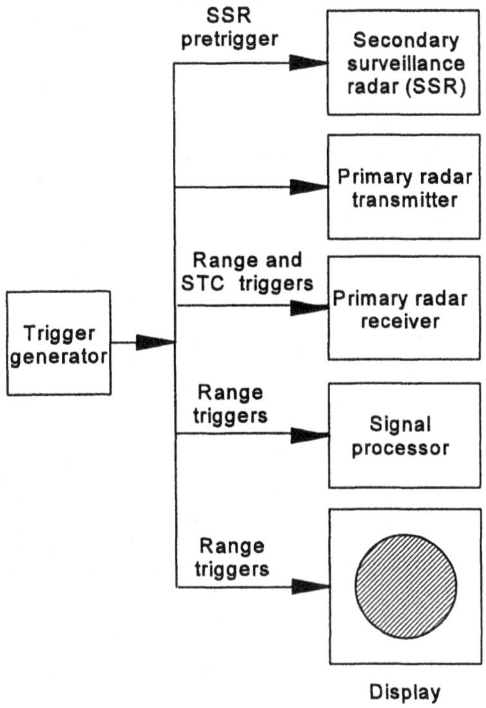

Figure 1.10 Trigger generation for a basic magnetron radar.

Modern radars with higher performance operate coherently. A central coherent oscillator (COHO), often 30 MHz or 60 MHz, provides a reference phase for the transmitted and received signals. The oscillator frequency is divided down to provide clocks to delay and synchronize all the triggers, the digital signal processors, and computers. This provides a much tighter control of the timing to reduce jitter losses for greater signal processing performance. Figure 1.11 shows the coherent oscillator and trigger generator block for the main block diagram for a coherent radar in Figure 1.8.

The timing accuracy of the triggers is critical for the performance of coherent signal processing (Chapter 11). The primary frequency reference is the coherent oscillator, the sine wave from it is counted to generate the triggers, and the triggers are distributed giving a propagation of timing jitter. Each logical circuit has its jitter random component (standard deviation) and, if the jitter for the kth stage is $\tau_{j,k}$, then the standard deviation of the jitter from n stages will be

$$Total\ jitter = \sqrt{\sum_{k=1}^{n} \tau_{j,k}^2} \tag{1.2}$$

The same applies for the jitter at the output of a counter which counts up to n except that the values of τ_j are equal so that the jitter is $\tau_j\sqrt{n}$. An example is given in Section 3.6.4.

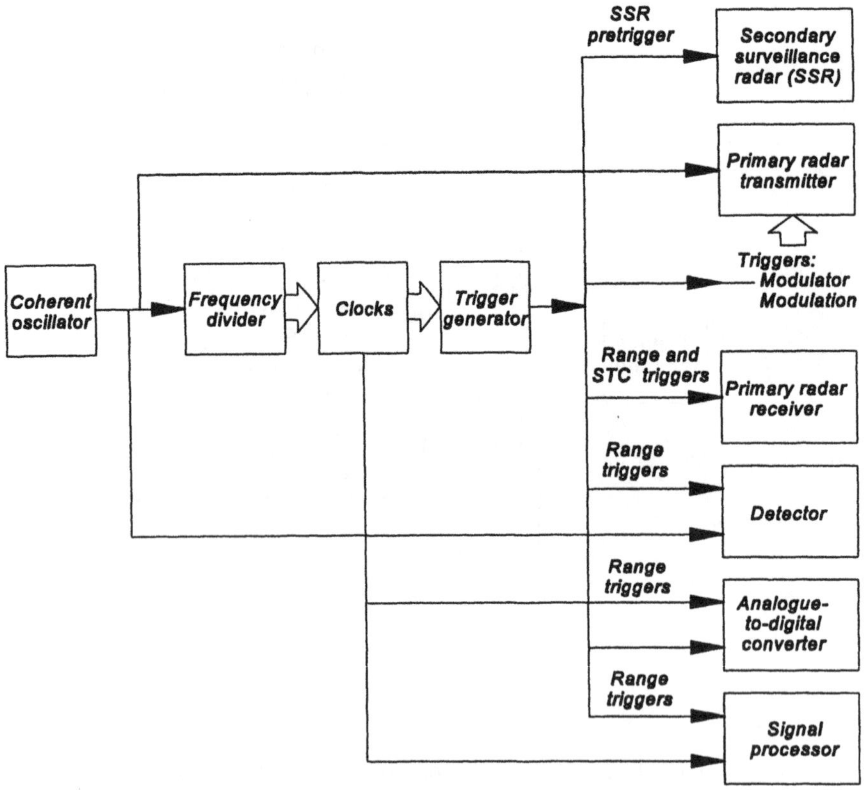

Figure 1.11 Timing, trigger generation, and distribution.

1.4 BASIC QUANTITIES, MAXIMUM RANGE

The power relationships used in the initial models are based on a single pulse and do not include any losses. The losses are considered in each chapter. Integration of a number of these pulses is considered in the signal processing and threshold chapters (Chapters 11 and 12).

The transmitter transmits a large pulse with a peak power of P_t watts, a width of τ seconds, and a wavelength of λ meters. The pulse is radiated by the antenna and the pulse travels to the object at a range, R meters. The power density reaching the object is (compare with (6.2))

$$\textit{Power density at distant object} = \frac{P_t G_t}{4\pi R^2} \text{ W/m}^2 \tag{1.3}$$

The object has a scattering area (often called the reflecting area) of σ square meters. The power is scattered equally in all directions so that the power density returning to the radar is

$$\textit{Power density returning to radar} = \frac{P_t G_t}{4\pi R^2} \frac{\sigma}{4\pi R^2} \text{ W/m}^2 \tag{1.4}$$

The antenna gain is proportional to its area measured in wavelengths. This is from (5.13):

$$\textit{Antenna gain, } G = \frac{4\pi \textit{ Antenna area}}{\lambda^2} \tag{1.5}$$

Thus, the effective antenna area is

$$\textit{Effective antenna area} = \frac{G\lambda^2}{4\pi} \quad \mathrm{m}^2 \tag{1.6}$$

The echo signal power intercepted by the antenna is from (6.4):

$$\textit{Power intercepted by the antenna} = \frac{P_t G_t}{4\pi R^2} \frac{\sigma}{4\pi R^2} \frac{G_r \lambda^2}{4\pi} \quad \mathrm{W} \tag{1.7}$$

The critical power ratio is the signal-to-noise ratio. Most of the noise is generated by the first stage of the receiver and depends on the equivalent receiver noise temperature T K.

$$\textit{Receiver noise power} = kTB \quad \mathrm{W} \tag{1.8}$$

where k is Boltzmann's constant 1.38×10^{-23} J/K;
T is the noise temperature K;
B is the receiver bandwidth Hz.

The receiver bandwidth is taken to be the reciprocal of the transmitter pulse width, τ seconds. Thus, the signal-to-noise ratio at the receiver is taken to be (no losses)

$$\begin{aligned}\textit{Signal-to-noise ratio} &= \frac{P_t G_t}{4\pi R^2} \frac{\sigma}{4\pi R^2} \frac{G_r \lambda^2}{4\pi} \frac{\tau}{kT} \\ &= \frac{P_t \tau G_t G_r \lambda^2 \sigma}{(4\pi)^3 R^4 kT}\end{aligned} \tag{1.9}$$

Losses in signal-to-noise ratio are caused by resistive losses in the system and by design. Sometimes it is better to sacrifice gain in the system to obtain better side lobes or better clutter suppression, so (1.9) is extremely dangerous. If the product of the losses is L, then (1.9) becomes

$$\begin{aligned}\textit{Signal-to-noise ratio} &= \frac{P_t G_t}{4\pi R^2} \frac{\sigma}{4\pi R^2} \frac{G_r \lambda^2}{4\pi} \frac{\tau}{kT} \frac{1}{L} \\ &= \frac{P_t \tau G_t G_r \lambda^2 \sigma}{(4\pi)^3 R^4 kTL}\end{aligned} \tag{1.10}$$

To illustrate the calculation, a hypothetical radar is assumed that has the characteristics shown in Table 1.6.

Table 1.6

Characteristics of a hypothetical radar with an assumed receiver noise figure of 2 dB

Characteristic	Original units	MKS units	Value for calculation	dB
Transmitter peak power, P_t	1 MW	1 000 000 W	1 000 000	60
Transmitter pulse width, τ_t	1 μs	10^{-6} s	10^{-6}	-60
Wavelength, λ	10 cm	0.1 m	0.01	-20
Antenna gain, transmitting, G_t	35 dB	3 162.278	3 162.278	35
Antenna gain, receiving, G_r	35 dB	3 162.278	3 162.278	35
Subtotal numerator in (1.10)		1 000 000	100 000	50
$(4\ a)^3$	1 984.402		1 984.402	32.976
Boltzmann's constant, k	$1.381.10^{-23}$J/K		$1.381.10^{-23}$	-228.599
Receiver temperature, a noise	460 K		460	26.628
Product of the losses	10 dB	10	10	10
Subtotal denominator in (1.10)			$1.260\ 10^{-18}$	-178.995
Constant, C			$7.935\ 10^{+20}$	208.995

Table 1.6 allows the fixed radar parameters to be combined into a constant, C. These calculations have a wide arithmetic range, so that in the past decibels had to be used for hand calculation as in the example in Table 1.6. The signal-to-noise equation reduces to

$$Signal\text{-}to\text{-}noise\ ratio \quad = \quad C\ \frac{\sigma}{R^4} \tag{1.11}$$

where C is the constant from Table 1.6.

An example of the signal-to-noise ratio for a one square meter area scatterer is shown in Figure 1.12. If the signal-to-noise ratio for the given probability of detection for a given false alarm rate in Chapter 12 is 10 dB, then the radar will have a range of 94.4 km. The dynamic range of the signal above the noise level is 100 dB here; for longer range radars, the dynamic range is greater. The dynamic range for signal processing can be reduced using sensitivity time control (STC) (see Chapter 7, Receivers) and, as in the past, limiting.

For normal air surveillance radars, Figure 1.12 shows in addition the mean clutter expected if the radar is situated near the ground (clutter models are discussed in Section 6.6). Whereas the aircraft cross-section does not vary with range, the clutter cross-section is proportional to range up to the radar horizon (assumed to be at 30 km) and decreases beyond the horizon. Clutter is rarely solid, and the log-normal model is used (see Section 6.6.2). A typical simulated clutter scenario is added in Figure 1.13.

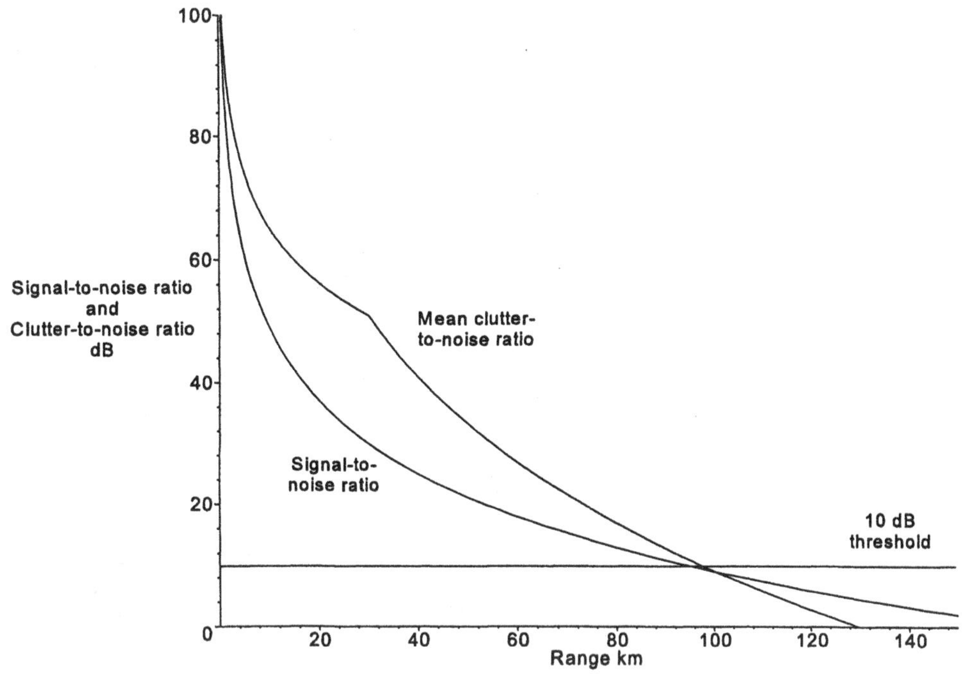

Figure 1.12 Typical signal-to-noise and clutter-to-noise ratios for the radar in Table 1.6.

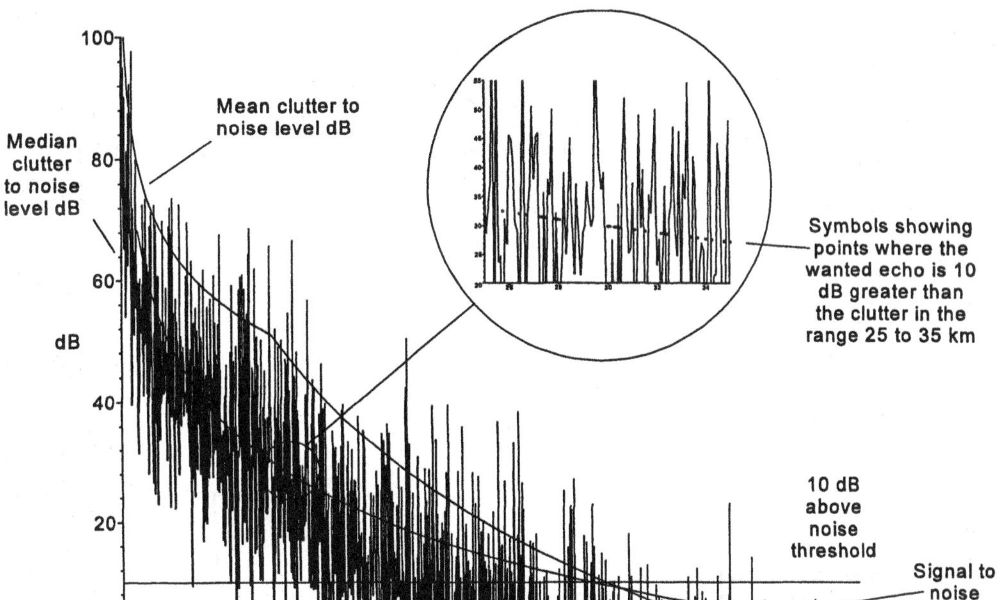

Figure 1.13 Signal-to-noise ratio with simulated log-normal clutter signals.

With a mean-to-median ratio of 20 dB, the mean level of the clutter is determined by the clutter spikes. The enlargement shows the situation between 25 km and 35 km and that there is a possibility of detection with a signal-to-clutter level of 10 dB between the spikes as shown by the dots in the enlargement circle in Figure 1.13. Figure 1.14 shows the density of these dots over the full range, and Figure 1.15 shows the probability of detection per kilometer within the radar's range.

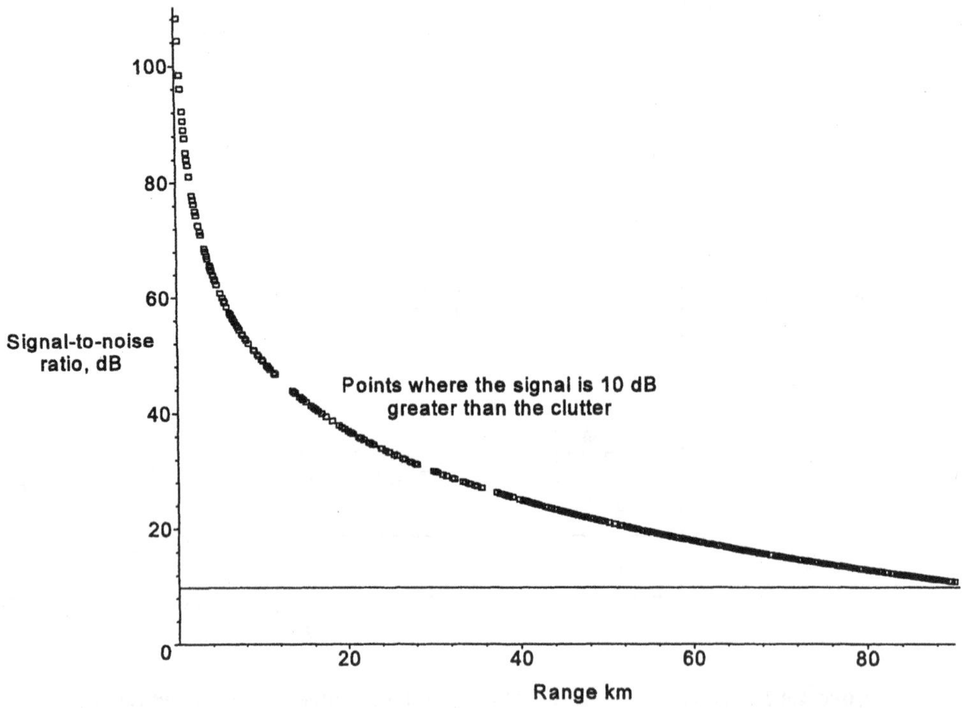

Figure 1.14 Signal-to-noise ratio of a standard scatterer when the signal-to-clutter ratio exceeds 10 dB.

The probability of detection is not sufficient for normal operation; typically between 80% and 95% is required. In practice, the clutter is not independent from cell to cell but follows the relief of the land around the radar, that is, there will be groups of cells with bad probabilities of detection. Either the radar must be used only for larger scatterers, for example for airliners, the clutter level may be reduced by reducing the resolution cell in range using a shorter transmitter pulse width or antenna beamwidth, or the clutter level reduced by signal processing (Chapter 11, Signal processing). In practice, it is difficult to see between bright clutter points on a cathode ray tube, and the radar picture requires a large amount of memory to reduce the amount of clutter displayed.

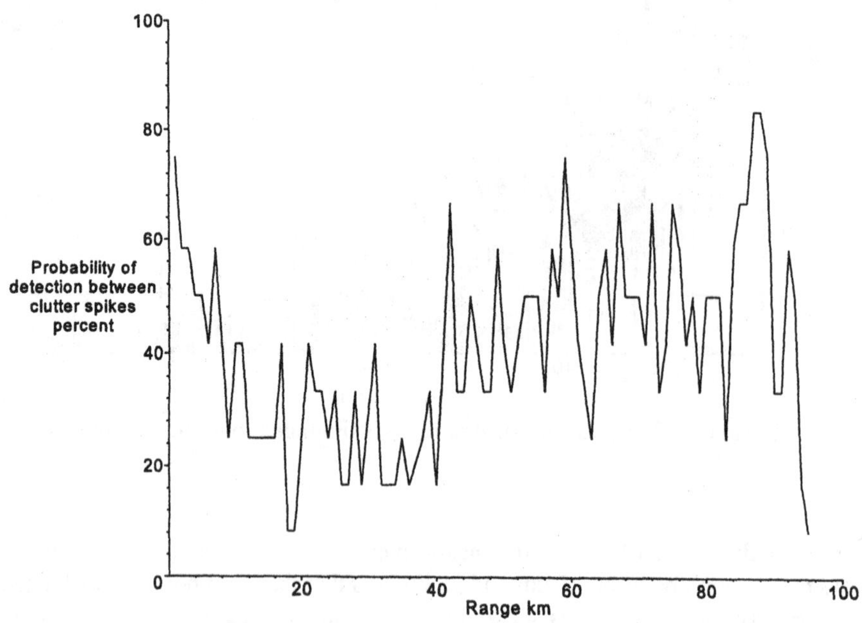

Figure 1.15 The probabilities of seeing a standard scatterer between clutter.

1.5 SECONDARY RADAR

In contrast to primary radar, secondary radar consists of two separate paths, each with its own range equation:

$$R_I = \frac{\lambda_I}{2\pi}\sqrt{\frac{P_I\, G_I\, G_R}{P_{RR}\, L_I}} \qquad R_R = \frac{\lambda_R}{2\pi}\sqrt{\frac{P_R\, G_I\, G_R}{P_{IR}\, L_R}} \tag{1.12}$$

where R_I is the range from the interrogator to the transponder;
R_R is the range from the transponder to the interrogator;
λ_I is the interrogator wavelength for 1 030 MHz, 29.13 cm;
λ_R is the interrogator wavelength for 1 090 MHz, 27.52 cm;
P_I is the interrogator transmitter peak power;
P_R is the transponder transmitter peak power;
G_I is the interrogator antenna gain;
G_R is the transponder antenna gain;
P_{IR} is the interrogator receiver sensitivity for the required probability of decoding a reply;
P_{RR} is the transponder receiver sensitivity for replying to an interrogation;
L_I and L_R are the total losses on the interrogation and reply paths, respectively.

The International Civil Air Organization (ICAO) regulations for the operation of secondary radar require that the minimum interrogator power be used for the range of the secondary radar. The aircraft transponder sensitivity limits are given as between -69 dBm and -77 dBm [1] at the antenna end of the cable leading to the transponder for a 90% probability of reply. The losses in the ground-to-air case (uplink) include the cable between the interrogator and its antenna on the ground (gain in the region of 25 dB) and the one-way atmospheric loss (see Sections 6.2.4 to 6.2.6). Note that the losses in the cable between the transponder and antenna are included in the sensitivity figure. The transponder antenna is quasi-omnidirectional and the gain is assumed to be near 0 dB. On the reply side (air to ground or down-link), the losses are the same.

Beacons in spacecraft are not standard; most operate with only one type of radar and must be treated on a case by case basis.

1.6 RADARS WITH SEPARATELY LOCATED TRANSMITTERS AND RECEIVERS

There are principally two reasons for placing a radar transmitter with its transmitting antenna and the receiver with its antenna in two separate places or buildings:

- *Lack of suitable transmit-receive switch devices with sufficient isolation.* Early radars, the British GL Mark II, were split into transmitting and receiving cabins which were placed 100 yards apart. Later, microwave radars, for example, the British Radar AA No. 3 Mk II, used separate transmitting antennas on the same mount, and currently, higher power continuous wave radars.
- *Radar transmitters are expensive to develop, purchase, and operate.* It has long been an aim to use echoes from another transmitter and antenna to build up a radar picture of the surroundings.

The use of echoes from a distant transmitter was employed by the Germans (*Parasit*) using the Chain Home radar stations as illuminators during the Second World War. A conventional radar is like a high-power searchlight at night, and antiradiation missiles (ARM) have been developed to seek and destroy such radiating sources. There is a current military interest in passive radar systems using illuminators "from the other side" which are not likely to be jammed. Further examples are the use of frequency modulated radio transmitters in VHF band II as illuminators.

Civil uses are to use a local receiving-only radar to give better coverage in relatively small but important areas where the coverage of a primary radar is not adequate.

When the transmitting site and the receiving site or sites are widely separated, the radar system is called bistatic and a typical arrangement for a bistatic radar system is shown in Figure 1.16. Notice that the bearing angles are referenced to the base line, the line between the transmitting and receiving sites.

Part of the pulse from the transmitting site follows the direct path to the receiving site. The signals giving echoes will have traveled from the transmitting site to the scatterer and then to the receiving site, which takes longer. Bistatic radars measure range by measuring the time that the echoes arrive after the direct pulse, so that lines of constant range are ellipses. No range measurements are possible along the base line.

The first process is to change the range sum, $\overline{R_T+R_R}$, into radar radial range, normally the slant range to the receiver, R_R. For this the azimuth of the echoes at the receiving antenna must be available. The slant range is [4, p. 25.9]

$$R_R = \frac{\overline{R_T + R_R}^2 - L_B^2}{2(\overline{R_T + R_R} + L_B \sin \alpha_R)} \tag{1.13}$$

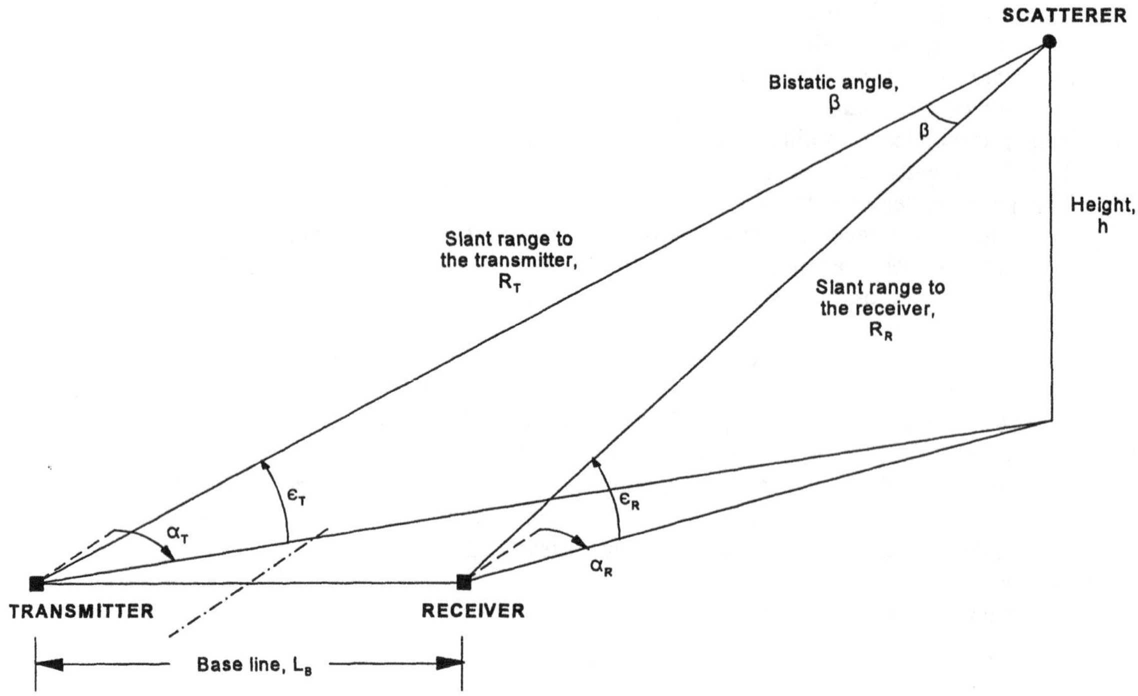

Figure 1.16 The geometry of a bistatic radar.

When the elevation angle of arrival is available, then the height and the ground range may be calculated.

1.6.1 Elliptical coordinates

Ellipses with the major axis along the x-axis are commonly defined by two of the following

- Major axis, $2a$;
- Minor axis, $2b$;
- Distance between the foci, $2c$, or the base line length, L_B.

The ellipticity is the ratio of the distance between the foci to the length of the major axis, or c/a.

Classical elliptical coordinates [5, p. 529] are defined in terms of u and v. The transformation into Cartesian coordinates is given by

$$\begin{aligned} x &= c \cosh u \cos v \\ y &= c \sinh u \sin u \end{aligned} \tag{1.14}$$

where $2c$ is the distance between the foci or the length of the base line;
$1/\cosh u$ is the ellipticity;
$\tan v$ is the slope of the line for large values of u.

The range sum is the major axis of the ellipse, or $2c \cosh u$. Elliptical coordinates for a base line length of unity (foci at ±½) and a number of values of u and v are shown in Figure 1.17. The ellipses are lines of constant range sum, and the hyperbolae are the courses for scatterers that give a constant Doppler frequency.

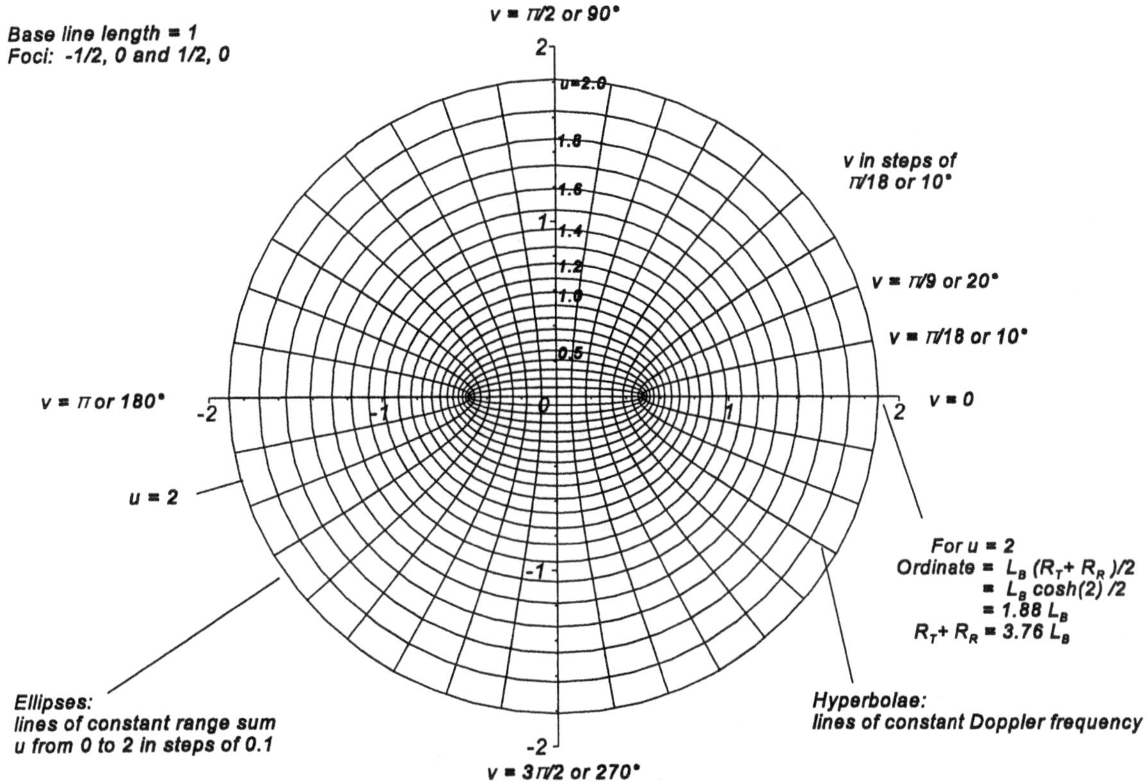

Figure 1.17 A grid of elliptical coordinates.

The radar measures time delay or range in terms of the range sum, $R_T + R_R$, which is related to the coordinate u by

$$\frac{R_T + R_R}{2\,L_B} = \cosh u \tag{1.15}$$

1.6.2 Bistatic radar maximum range

Chapter 6 deals with the scattering of energy back to a monopulse radar only. With a bistatic system, the energy is not scattered back directly but at an angle, β, that changes the scattering characteristics. Military aircraft that see themselves as targets for anti-aircraft defenses have reduced scattering characteristics for monopulse radars and this extends over a wide range of the bistatic angle, β, except where β approaches 180 degrees. The range, R, in the monostatic radar equation, (14.1), is replaced by the geometric mean, $\sqrt{R_T\,R_R}$ in the bistatic radar equation:

$$\sqrt{R_T\,R_R} = \left(\frac{P_t\,G_T G_R\,\sigma_B\,F_T^2 F_R^2}{(4\pi)^3\,k\,T\,B\,D_s\,L}\,\frac{c^2}{f^2}\right)^{\frac{1}{4}} \tag{1.16}$$

where R_T is the distance from the transmitter to the scatterer;
R_R is the distance from the scatterer to the receiver;
G_T is the gain of the transmitting antenna;
G_R is the gain of the receiving antenna;
σ_B is the bistatic radar cross-section;

F_T is the propagation factor for transmission;
F_R is the propagation factor for reception;
c is the velocity of light;
k is Boltzmann's constant;
T is the system noise temperature, K;
D_S is the signal-to-noise ratio required for detection;
f is the frequency of the radar;
L is the product of the losses.

The radar cross-section for scattering depends on the angle to the transmitting and receiving sites. For military aircraft, the cross-section is minimized towards the direction of the transmitter site, so the bistatic cross-section may be greater.

The low-altitude coverage of monostatic radars is limited by the shadow formed by the Earth's curvature. With bistatic radar, there are shadows in the coverages of both transmitting and receiving sites. Echoes occur only in the volume of common coverage, and airborne elements may help in reducing these shadows. The clutter echoes come from cells defined not only by the pulse length and the antenna beam widths but also from the angle between the two antenna beams.

Cassini's ovals. Figure 1.18 shows the curves for a constant value of the geometric mean divided by the base line length, $\sqrt{R_T\ R_R}\ /\ L_B$, for a base line distance of unity. For values of 1.5 and greater, the quasi-monostatic radar case is

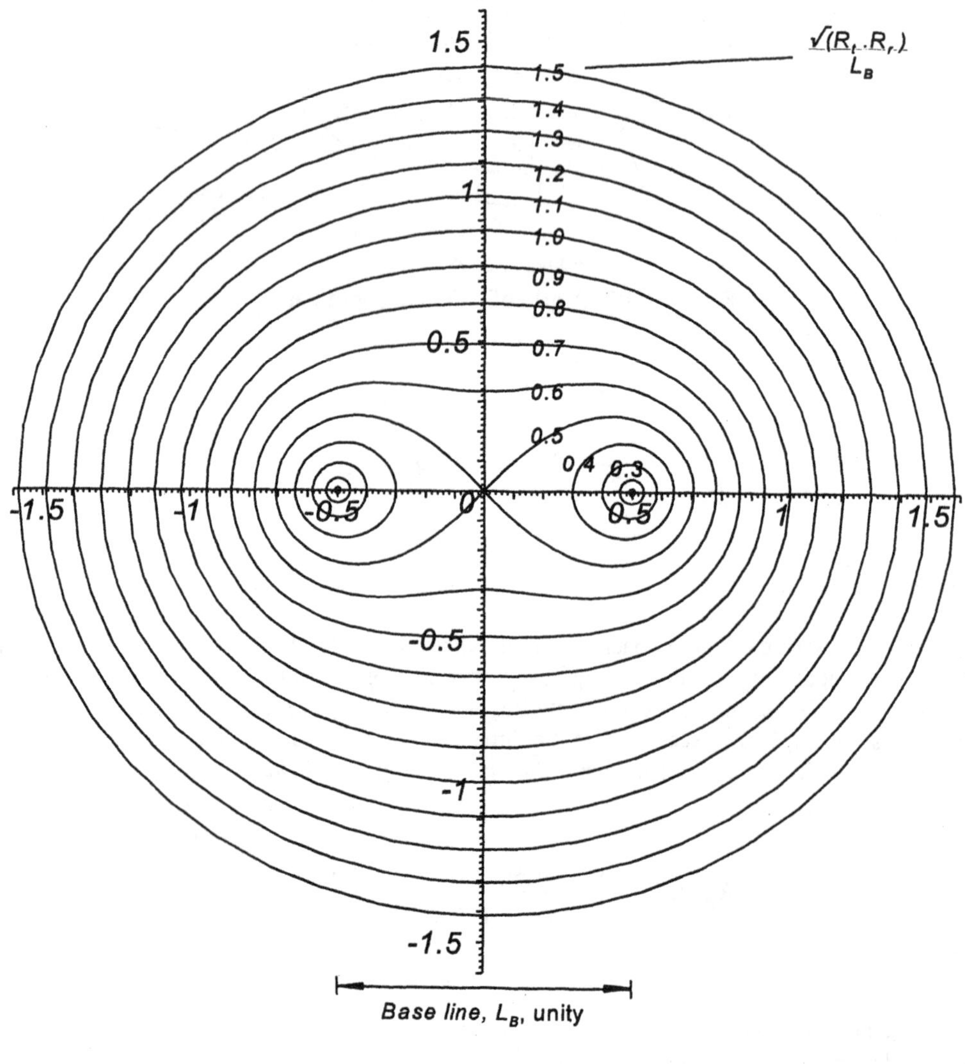

Figure 1.18 Curves showing lines of equal bistatic signal-to-noise ratio in terms of √(R_t R_r)/L.

approached with similar ranges and signal-to-noise ratios in directions along and across the base line. Below this value, the coverage elongates and finally splits into two, giving only local coverage around the two sites. Notice that this also applies to the elevation coverage. This has the effect of changing the traditional coordinates as shown in Table 1.7.

Table 1.7
Comparison of monostatic and bistatic radar characteristics

	Monostatic radar	Bistatic radar	
		Measured	Coverage
Range	Slant range, R	Slant range, $R_T + R_R$	$\sqrt{R_T \, R_R}$
		Elliptical coordinates	Cassinian coordinates
		Ranges to transmitter or receiver must be calculated	
Radial velocity	Doppler frequency	Hyperbolic Doppler frequency contours	
Azimuth and elevation	Radar antenna	Usually measured by the receiving antenna	

1.7 PERFORMANCE

A radar is designed for a specific range of tasks that determine the technical performance. This book is written around a large class of radars on the ground that detect moving aircraft for air traffic control. The principles may be applied to radars for other purposes. A radar "sees" objects of a certain size at a distance so that one of the primary quantities is a definition of distance. For example, for:

- Scatterer cross-section, σ 1 m²;
- Range, R_{max} 60 km.

The above defines the range for scatterers of the defined size, but larger sizes may be seen out to longer ranges, the maximum instrumented range. This sets the time required for the pulse to reach the maximum instrumented range, R_{inst}, and return. The reciprocal of this time is the highest pulse repetition frequency. Other pulse repetition frequencies are chosen to fulfill the need to switch to other maximum instrumented ranges, to provide staggering for moving target indicator signal processing (Section 11.2.3.5) or pulse repetition frequency diversity for moving target detector signal processing (Section 11.2.4.5). Pulse repetition frequency staggering and diversity also may be used to resolve range ambiguities in medium and high pulse repetition frequency radars.

The minimum range must be defined for pulse radars. With short pulses, the transmit-receive switch sets the lower limit for range (Section 4.2.4.3). It is sometimes necessary for radars that use long pulses to use a separate short pulse short range mode.

The amount of azimuth coverage determines whether continuous or sector scanning is used in the search mode. The elevation dimension may be either scanned in the search mode or covered using a broad beam in the elevation plane, as with civil air traffic control radars (aircraft report their height using the replies to mode C secondary radar interrogation). These broad beams in elevation are shaped to give coverage up to a specific height above ground (see Section 5.5).

Radar detection takes a definite time and is described statistically. Often the number of false alarms is fixed to reduce false detections to an acceptable level which gives the probability of false alarm P_{fa}. This allows a threshold to be set above the thermal noise in the system: signals plus noise exceeding this threshold are reported as detections only P_d percent of the time, or probability of detection. An example of a common way of defining these is

- Probability of false alarm, P_{fa}, 10^{-6};
- Probability of detection, P_d, 80%.

The diagrams in Chapter 12 give the signal-to-noise ratio required for detection with one or more echoes from the scatterer, called hits. Surveillance radars, generally speaking, do not observe their environment by staring at it. Instead they probe in the darkness as we would, handlamp in hand. The time that the object is illuminated times the pulse repetition frequency gives the number of hits, which is important in radars that use integration.

A radar must be able to distinguish between the different objects that return echo signals. Unlike the human eye, radar is able to measure range by measuring the echo time. The effective pulse length must be short enough so that one returning echo does not mask its neighbor. Angular resolution is obtained by narrow azimuth and elevation antenna beam widths.

The fourth radar dimension is the radial velocity of the scatterer, which gives rise to the Doppler frequency of the echo. In surveillance radars, this is used mainly to reject or sort out echoes of near zero Doppler frequency, which represent land clutter. Moving target detector radars are able to estimate an ambiguous Doppler frequency, that is above the range of the pulse repetition frequency.

The choice of radar frequency is limited, for civil radars, to the bands allocated to radars. The gain available from a given size of antenna depends on its dimensions in wavelengths; hence, portable radars and small radars tend to use higher frequencies. Radars mounted on vehicles and ships can support larger antennas and the largest are used by fixed ground installations. Radars mounted on vehicles or ships generally must operate in all types of weather, and the higher frequency radars, above S-band, are particularly susceptible to rain clutter. Examples are *en route* radars for air traffic control (range 150 nautical miles, or 278 km) which tend to use L-band; airfield approach radars S-band (range 60 nautical miles or 111 km); and those that cover the runways out to 10 km use X or a higher band, as do small ships' radars.

The accuracy of radar measurement depends on the signal-to-noise ratio (see Chapter 13). The signal-to-noise ratio for accurate measurement is normally greater than that necessary for detection so that the range or signal-to-noise ratio for measuring the radar accuracy must be defined separately. The following are examples:

- Range accuracy, standard deviation 10 m;
- Azimuth accuracy, standard deviation 0.1 degree;
- Elevation accuracy, standard deviation 0.1 degree;
- Doppler frequency accuracy, standard deviation 10 Hz.

(All at a range of 30 km.)

A knowledge of the radar site supported by map studies allows the amplitude and extent of the radar clutter to be estimated to find how great the dynamic range or the receiving system must be (including sensitivity time control) and how much clutter cancellation is required. Coherent signal processing allows precise sorting or suppression of signals according to their Doppler frequency. In addition to not allowing large clutter signals to swamp wanted echo signals using tapering, the radar itself must not generate any interfering signal components that may be mistaken for wanted echo signals. This gives rise to the concept of stability, namely that interfering signals and noise caused by spurious instabilities in gain and timing must be a small fraction, often 60 dB to 80 dB, of the useful echo signal.

These figures give bases for the range, resolution, accuracy, and stability budgets. The range budget determines the sizes of the transmitter, antenna, and receiver sensitivity. Performance margins must be estimated to compensate for the losses in the system, and most of the factors listed under performance reduce the signal-to-noise ratio, thus reducing maximum range and accuracy. During the development of a radar, the goals for the characteristics of some components may not be met while others may be exceeded. The most convenient way of managing the daily state of the performance budgets is to use a computer spreadsheet.

The remaining chapters in this book end with the effects of the stages in the chapter on radar performance, for Chapter 1 these are the effects of the common components. These figures are gathered in Chapter 14 (Performance). Four measures of performance are treated: the effects on range, resolution, accuracy, and stability.

1.7.1 Effects on range

The choice of pulse repetition frequencies determines the maximum instrumented range and the maximum unambiguous Doppler frequency. The probability of having an echo to measure is reduced in medium and high pulse repetition frequency radars that have parts of the listening time blanked for transmitter pulses. To cover the gaps, such radars often operate over a number of pulse repetition frequencies, and to have the same sensitivity as a radar without such holes, or eclipses, the signal-to-noise ratio has to be increased. This increase is called eclipsing loss.

Bringing a radar into operation from development that does not have expensive excess performance is a constant struggle to limit the losses to a minimum. Examples of losses are:

- Radio frequency losses:
  - Waveguide or transmission loss from the transmitter to the duplexer;
  - Duplexer losses from input port to the port to the antenna;
  - Waveguide or transmission loss from the duplexer to the antenna;
  - For frequency diversity radars, the diplexer losses;
  - Internal losses in the antenna;
  - Atmospheric loss between the antenna and the scatterer of interest;
  - Losses caused by the scatterer not being at an optimum aspect, fluctuation losses;
  - Atmospheric loss between the scatterer and the antenna;
  - Losses in the receiving antenna, in stacked beam radars there is a different radio frequency path for transmission and reception;
  - Losses in the waveguide or transmission line between the antenna and the duplexer;
  - For frequency diversity radars, the diplexer losses;
  - Losses in the receiver protector and sensitivity time control (STC) attenuator;

- Receiver and signal processor losses:
  - Loss caused mismatch in the matching filter or tapering losses in the pulse compression stage;
  - Quantization loss in the analogue-to-digital converter;
  - Losses in signal processing caused by logarithmic video, moving target indicator or detector processing;
  - Losses during the recognition (extraction) of valid echo signals.

1.7.2 Resolution

The resolutions in range, angle, and Doppler frequency are part of the basic design of the radar and depend on the effective transmitter pulse width, antenna beamwidth, and signal processor, respectively.

During the path of the signals from the transmitter back to the point where signals are extracted or displayed, the pulse becomes stretched in range and angle (azimuth and elevation), and the principal stages are shown in Table 1.8.

Table 1.8
Places where waveforms are stretched

Cause	Range	Angle
Large scatterers	Over one pulse length radially	Over one beamwidth
Limited receiver bandwidth	Stretches echo pulse	
Matching or pulse compression stage	Stretched in range	
Moving target indicator or detector		Shifted and stretched in angle

1.7.3 Accuracy

The accuracy of measurement is based on the Woodward equation [6, p. 105 (27)] which is general and applies to measurements in range, angle, and Doppler frequency. The standard deviation of the error, σ, is given by

$$\sigma = k \frac{\zeta}{\sqrt{R}} \tag{1.17}$$

where σ is the standard deviation of the error for the dimension of ζ, namely:
for range, the range resolution cell width (time or distance);
for angle, the angular resolution;
for Doppler frequency, the frequency resolution (frequency or radial velocity);
k is a constant;
R is the signal-to-noise ratio.

The point of measurement for τ and R is where the information is used in either the extractor or the display. Table 1.8 shows where τ is stretched in time and Section 1.7.1 shows where the signal to noise ratio, R, is reduced.

An equation similar to (1.17) is used for angle accuracy in azimuth and elevation with similar dilutions of accuracy for signal-to-noise ratio and antenna beamwidth. Moving target indicators and the tapering used to increase the suppression of clutter in moving target detectors widen the effective antenna beamwidth at the extractor. The estimation of Doppler frequency depends on the effective bandwidth and signal to noise ratio at the point of estimation.

Most of the statistical distributions describing accuracy are Gaussian although angles derived from monopulse radars should have, theoretically, a Cauchy distribution. In trials of three-dimensional stacked beam radars with an aircraft flying at a constant height, the resulting distribution was found experimentally to be Gaussian.

The timing error (jitter) is normally small, compared with other factors when all pulses are derived from the same, stable frequency source, generally the stable oscillator (STALO).

1.7.4 Stability

The trigger generation and distribution system is the start of the propagation of synchronization jitter (see Section 3.6.4), which is critical so that radar instability does not limit the quality of the signal processing.

Modern moving target detector signal processors demand a rock stable clock and stable trigger system.

REFERENCES

1. Stevens, M. C., *Secondary Surveillance Radar*, Norwood, Massachusetts: Artech House, 1988.
2. Daniels, D., ed., *Ground Penetrating Radar*, 2nd ed., Stevenage: I.E.E., 2005.
3. Skolnik, M. I., *Radar Handbook,* New York: McGraw-Hill, 1970.
4. Skolnik, M. I., *Radar Handbook,* 2nd ed., New York: McGraw-Hill, 1990.
5. Weisstein, E. W., ed., *CRC Concise Encyclopaedia of Mathematics,* Boca Raton, Florida: CRC Press, 1999.
6. Woodward, P. M., *Probability and Information Theory with Applications to Radar*, 2nd ed., Oxford: Pergamon Press, 1964 and reprinted by Artech House in 1980.

Chapter 2

Usual and unusual concepts

This chapter introduces a number of other ideas and conventions used in this book. Fuller explanations are given in the mathematics chapters. This is an introduction to the three-dimensional representation of complex values that vary with distance, time, or frequency. The types of diagrams described are:

- Three-dimensional locus or working point diagrams;
- Polyphase modulation, demodulation, and waveforms;
- Symmetrical components;
- Polyphase or bottlebrush noise;
- Simultaneous representation of waveforms and spectra;
- Gaussian pulses, spectra, and beam shapes.

These diagrams are chosen to communicate the shapes of the phenomena, to allow engineers to put figures on the shapes to allow calculation, and to apply them to radar. To start, the working point of a Wien bridge oscillator is described.

2.1 AN EXAMPLE OF THREE-DIMENSIONAL REPRESENTATION: THE WIEN BRIDGE OSCILLATOR

The Wien bridge oscillator in Figure 2.1 is well known.

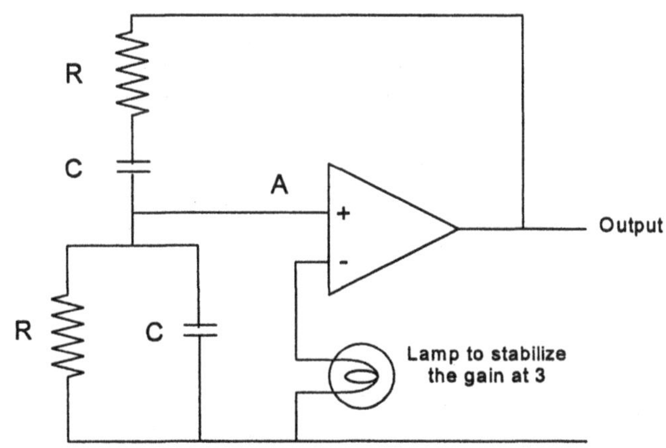

Figure 2.1 Block diagram of a Wien bridge oscillator.

The working point at the point A, referenced to zero by subtracting the network attenuation factor, is given by

$$Working\ point = \frac{R}{(j2\pi fCR + 1)\left(R + \dfrac{1}{j2\pi fC} + \dfrac{R}{j2\pi fCR+1}\right)} - \frac{1}{3} \tag{2.1}$$

and the frequency of oscillation as

$$\textit{Working frequency} = \frac{1}{2\pi CR} \text{ Hz} \tag{2.2}$$

In this example R is 1 000 Ω, and C is 1 μF, so that the oscillating frequency, f, is 159.1549 Hz.

Traditionally Bode diagrams, showing the gain and phase shift around the loop at point A, are drawn as in Figure 2.2. The diagrams show discontinuities at the frequency of oscillation, which, by instinct, nature abhors.

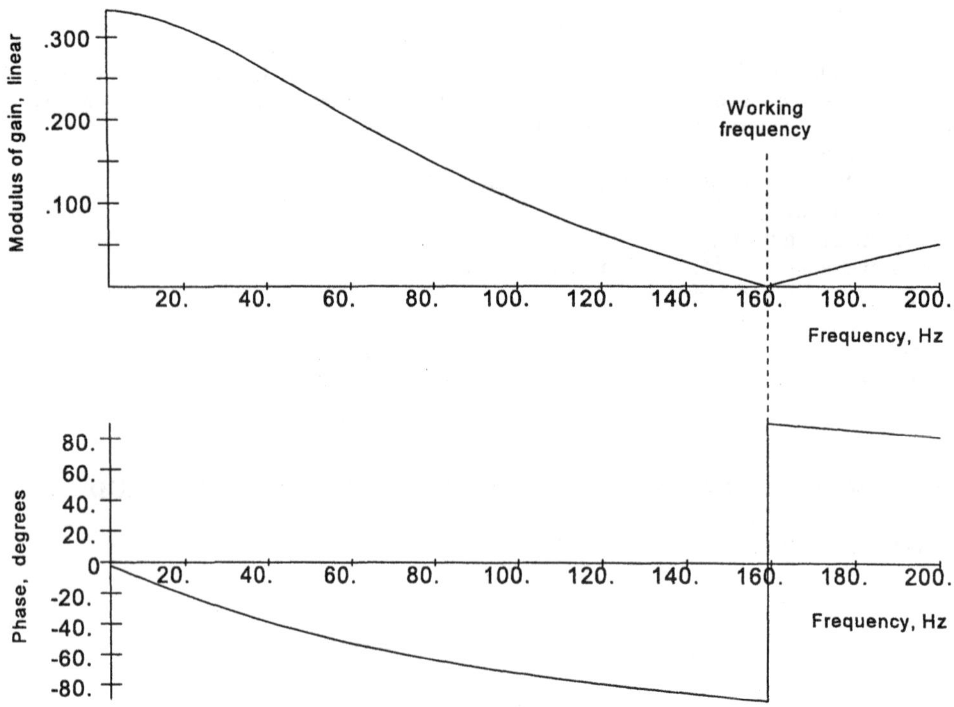

Figure 2.2 Modulus of the gain and phase of the working point of the Wien bridge oscillator.

Modern mathematics programs allow the plotting of gain and phase together. The locus diagram in gain or amplitude and phase for the Wien bridge oscillator is shown in Figure 2.3.

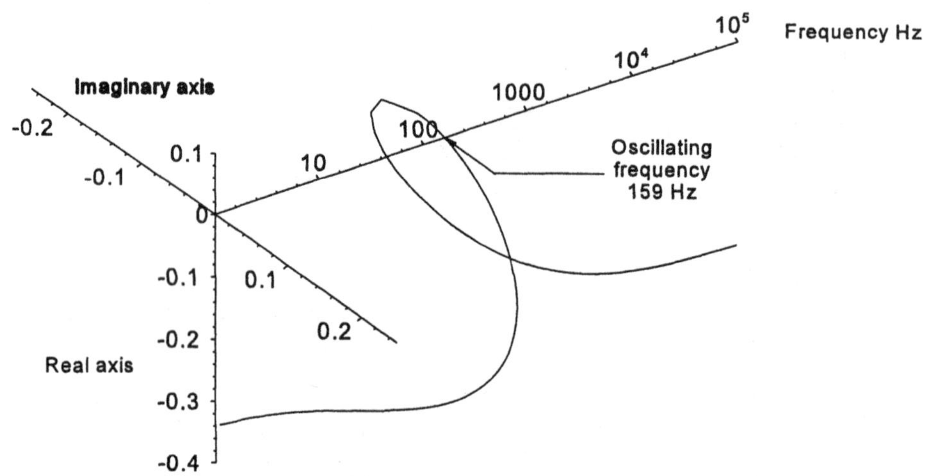

Figure 2.3 Three-dimensional representation of the working point of a Wien bridge oscillator.

In this representation of complex working point versus frequency, there is no discontinuity and nature feels much better. There is one unbroken curve, which loops once only. Here it is obvious that there is only one frequency of oscillation where the working point cuts the frequency axis, as shown in the enlarged views. Logarithmic scales widen the dynamic range of such diagrams. This type of diagram is used to represent complex time waveforms and spectra in the remainder of this book.

2.2 VECTOR REPRESENTATION

The calculation in the complex plane follows the rules of vector arithmetic. Addition is represented by a parallelogram, as in Figure 2.4. Diagrams are often represented by half of the parallelogram. The complex algebra representation is

$$\begin{aligned} Sum &= a+jb + p+jq \\ &= (a+p) + j(p+q) \end{aligned} \tag{2.3}$$

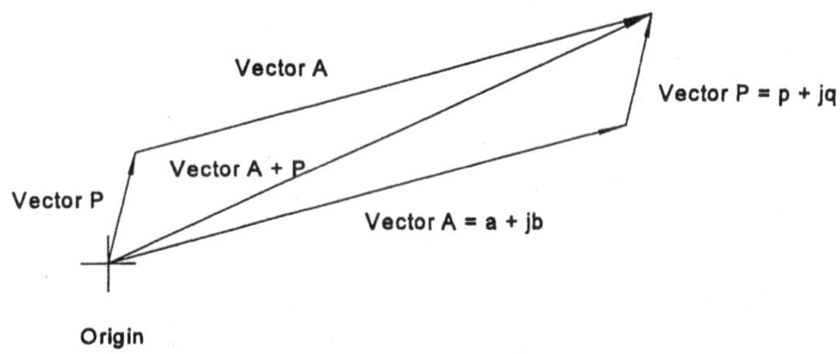

Figure 2.4 The sum of two vectors.

Subtraction is an extension of addition where one vector is drawn in the negative sense.

Multiplication is defined in polar coordinates by the product of the lengths of the vectors at an angle that is the sum of the angles for the directions of the vectors. This is shown in Figure 2.5. In polar coordinates, this is

$$Product = A\angle\alpha \; P\angle\beta = AP\angle(\alpha+\beta) \tag{2.4}$$

Commonly, Cartesian coordinates are used so that

$$Product = (a+jb)(p+jq) = ap-bq + j(aq+bp) \tag{2.5}$$

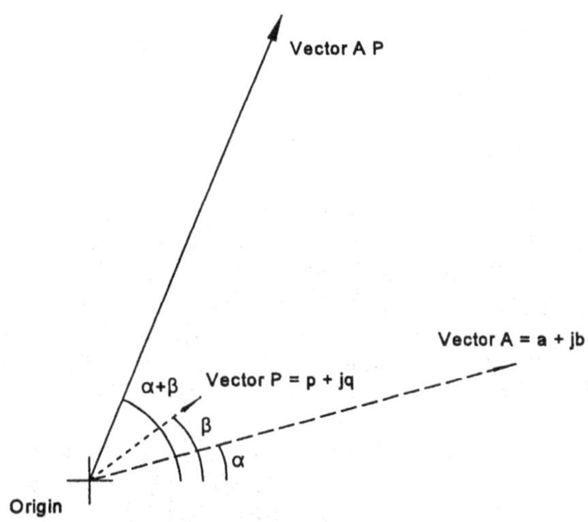

Figure 2.5 Vector multiplication.

Division is defined in polar coordinates only:

$$Quotient = \frac{A\angle\alpha}{P\angle\beta} = \frac{A}{P}\angle(\alpha - \beta) \tag{2.6}$$

Powers of complex numbers are an extension of multiplication:

$$n^{th}\ power = \left(A\angle\alpha\right)^n = A^n\angle n\alpha \tag{2.7}$$

The calculation of power is carried out by multiplying the voltage by the complex conjugate of the current. The conjugate of a complex quantity is the quantity with the sign of the imaginary part reversed.

$$\begin{aligned} Voltage &= a + jb \\ Current &= p + jq \qquad \text{the complex conjugate is } p - jq \\ Power &= (a + jb)(p - jq) = ap + bq - j(aq - bp) \end{aligned} \tag{2.8}$$

The calculation of power is also used for the correlation of two waveforms in Chapter 8, Matched and matching filters, and Chapter 11, Signal processing.

2.3 ORDER OF LINEAR PROCESSING

Just as $x + y = y + x$ and $xy = yx$,additive and multiplicative processes may take place in any order so long as they are linear. In contrast, subtraction and selection can destroy information that may no longer be recovered.

2.4 POLYPHASE MODULATION AND DEMODULATION

Simple amplitude modulation stays in the alternating current domain, namely,

$$\begin{aligned} & \qquad\qquad \text{Carrier wave} \qquad\qquad \text{Modulating wave} \\ Modulated\ wave &= A\sin(2\pi f_c t + \phi_c)\ \ B\sin(2\pi f_m t + \phi_m) \\ &= \frac{AB}{2}\left[\cos\left(2\pi(f_c - f_m)t + \phi_c - \phi_m\right) - \cos\left(2\pi(f_c + f_m)t + \phi_c + \phi_m\right)\right] \end{aligned} \tag{2.9}$$

where A is the carrier frequency amplitude;
f_c is the carrier frequency;
ϕ_c is the carrier phase;
B is the modulation frequency amplitude;
f_m is the modulation frequency;
ϕ_m is the modulation phase;
t is the time in seconds.

The modulated wave consists of two sidebands at $f_c - f_m$ and $f_c + f_m$. Notice that the phases of the modulation of these sidebands have opposite signs. Thus, one is the complex conjugate of the other.

A signal returned from a scatterer contains the original modified by the movement of the scatterer. This modification is the change of phase per second between the transmitter and the receiver. The phase of the transmitted pulse changes at 2π radians along each wavelength, λ. If the scatterer moves at v m/s towards the radar, then the change of distance to and from the scatterer per second is $2v$, and the change of phase is $2v\ 2\pi/\lambda$ radians per second or $2v/\lambda$ Hz. For an S-band radar, wavelength 0.1 m, a transmitted pulse at 3 GHz is returned at 3 000 001 000 Hz by a scatterer moving at 50 m/s towards the radar.

In practice, the echoes are a mixture of fixed and moving echoes. This is equivalent to single-sideband modulation in communications. In Figure 2.6, the fixed echo vector or clutter is drawn, by convention, vertically. The moving echo changes its phase with time so that the combined vector is amplitude and phase modulated. Mathematically, this may be represented by

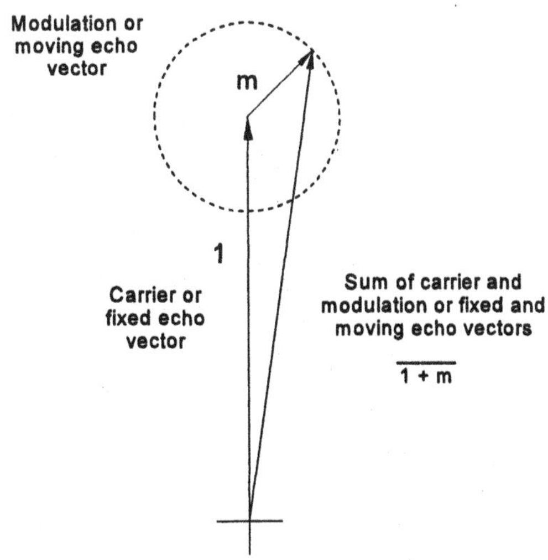

Figure 2.6 Vectors representing fixed and moving echoes.

$$Resultant\ vector = 1 + m\ \exp(j\phi) \tag{2.10}$$

where $\phi = 2\pi f_{Doppler.}$

A representation of how this looks on an oscillograph with a greatly reduced ratio of carrier to modulation frequency is shown in Figure 2.7.

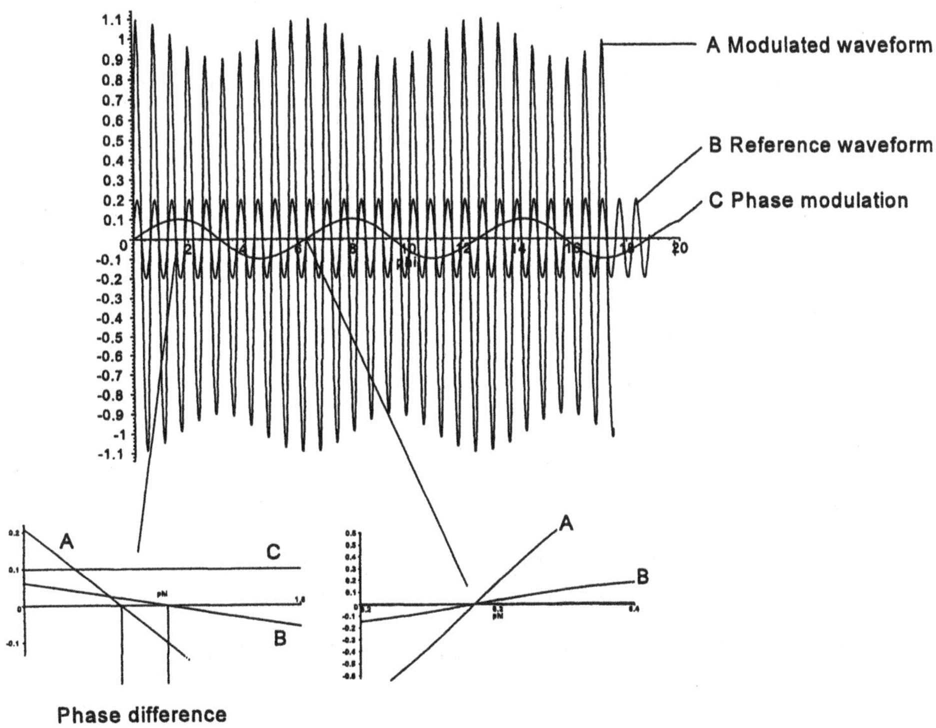

Figure 2.7 A single-sideband modulated carrier.

The phase differences in Figure 2.7 are too small to be seen without magnification. Figure 2.8 shows the same waveform but with the phase angle transferred to the complex plane so as to look like the vectors in Figure 2.6. In Figure

2.8 the axes have been rotated to give the best view of the waveform.

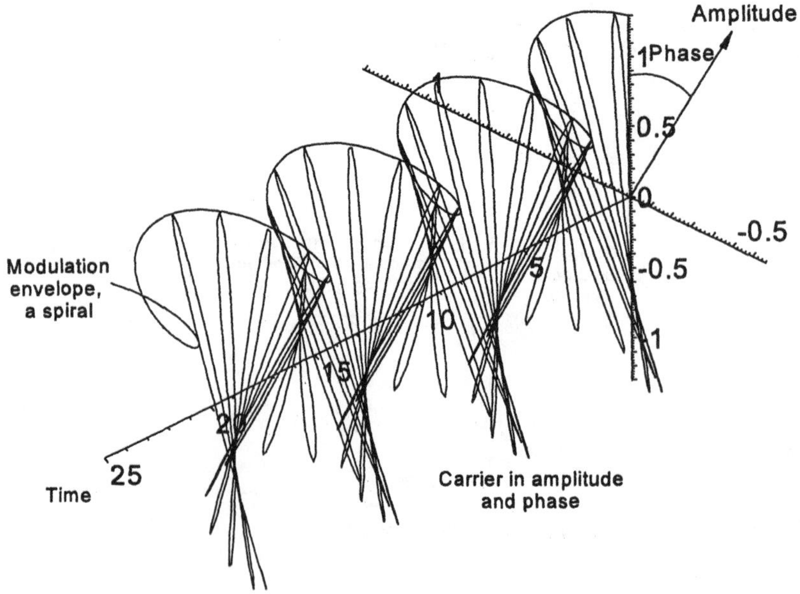

Figure 2.8 A single-sideband modulated carrier with the phase shown in the complex plane.

Ideally a vector detector would remove the carrier and leave the helix as a signal for further processing. Because it is impossible to pass ideal vector videos on a single wire, such components do not exist and another solution must be found. This is the same as the voltage in a three-phase power circuit or a circuit with any number of phases. Three-phase and two-phase systems are shown in Figure 2.9. The two-phase system has Cartesian reference axes.

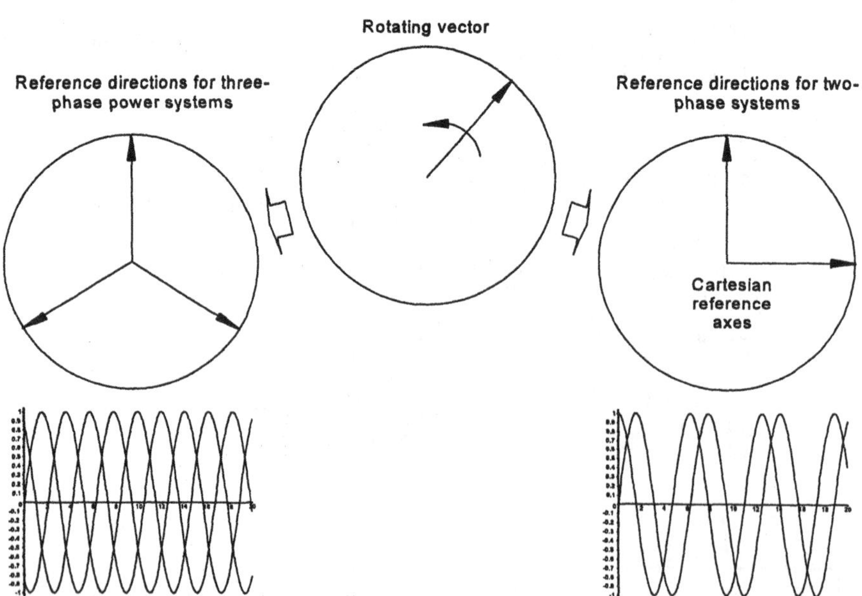

Figure 2.9 Three-phase and two-phase representations of a rotating vector.

The number of phases for power systems was settled early in the 20th century, although a number of two-phase cables were still in use many, many years later. Le Blanc transformers perform the transformation for voltage and the number of phases. Counting the return wires and the neutral, four wires are required for transmission in a three-phase system and five for a two-phase system. For balanced circuits, three and four conductors are required. This leaves the three-phase power distribution system as that which transmits polyphase power with the least number of conductors.

Mathematically, for three phases [1],

$$\begin{aligned} Polyphase\ voltage &= e^{j\,2\pi ft} \\ phase1 &= V\sin(2\pi ft) &= V_1 \\ phase2 &= V\sin(2\pi ft + 120°) &= \lambda\, V_1 \\ phase3 &= V\sin(2\pi ft + 240°) &= \lambda^2 V_1 \end{aligned} \tag{2.11}$$

where $\lambda = \exp\left(j\dfrac{2\pi}{3}\right) = -\dfrac{1}{2} + j\dfrac{\sqrt{3}}{2}$

The line voltage for a three-phase system is the voltage between the vector arrow tips. That is $\sqrt{3}\ V_{phase}$, where V_{phase} is the phase voltage or the length of the arrowed vector. These voltages are commonly as follows:

Phase voltage	110	117	220	240
Three-phase line voltage	191	203	381	416

For two-phase or Cartesian representation this is

$$\begin{aligned} Polyphase\ voltage &= V\exp(\,j2\pi ft) \\ &= V\,(\cos 2\pi ft + j\sin 2\pi ft) \\ &= Re(V) + Im(V) \end{aligned} \tag{2.12}$$

This is shown in Figure 2.10.

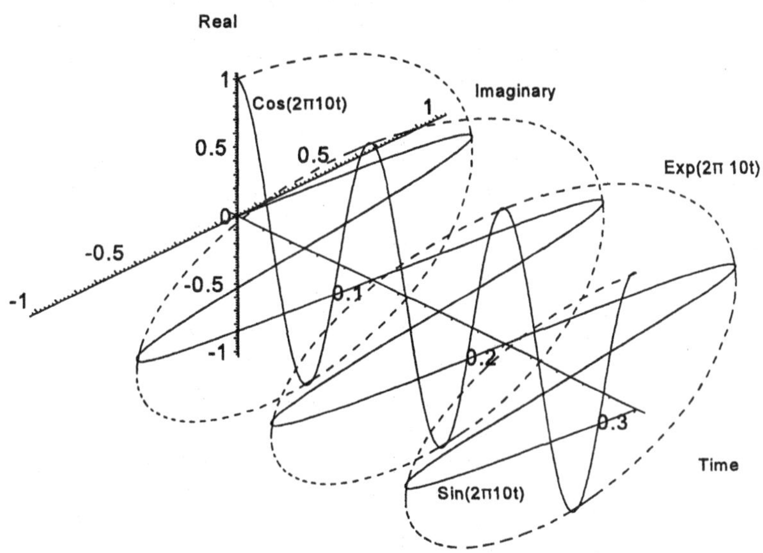

Figure 2.10 A 10 Hz polyphase voltage and its two-phase components.

The power delivered by a three-phase power circuit is the sum of the powers in each of the individual phases. In alternating current and polyphase systems, the power is the voltage times the conjugate of the current. Statistically, this is the correlation function which, for time limited waveforms, one waveform is dragged by the other and the integral of their product is plotted. This is shown in Figure 2.11.

Polyphase power may be converted into large numbers of phases using transformers with star and zig-zag windings. This is useful for rectifying polyphase alternating currents. Each of the phases is fed to a separate rectifier so that the ripple amplitude is decreased and the ripple frequency is increased. Smaller capacitors are required for smoothing.

The energy in the correlation function is represented by its peak. With constant impedance, the current is proportional to the voltage so that power is proportional to the voltage multiplied by its complex conjugate. The peak of the autocorrelation function is always real. For nonperiodic waveforms, the envelope tends to a steady line. The energy per second gives the power.

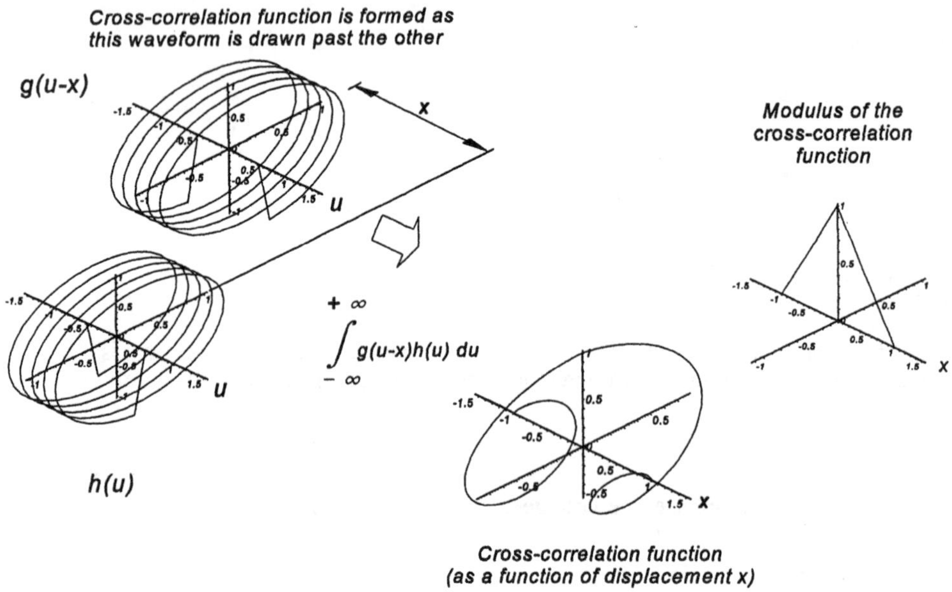

Figure 2.11 Polyphase power representation.

In order to remove the carrier and preserve all the amplitude and phase information, the real and imaginary components of the modulation must be determined. The reference phase of the carrier is provided by the coherent oscillator (COHO) of the radar. Synchronous detectors[1] give the Cartesian component with respect to the reference phase so that two separate synchronous detectors are required for each of the two Cartesian component outputs. The block diagram of such a detector is shown in Figure 2.12.

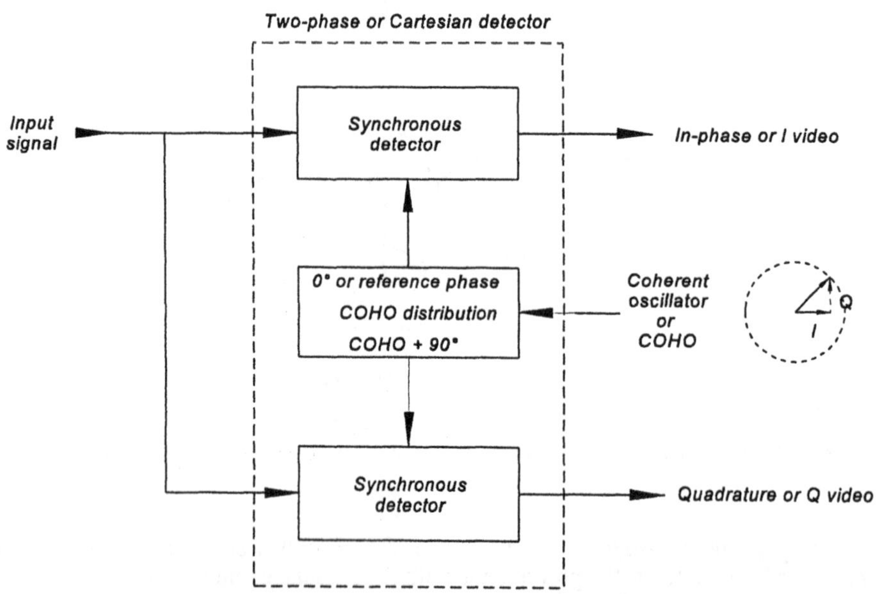

Figure 2.12 A two-phase or Cartesian detector.

The rotating vector on the right of Figure 2.12 may rotate to the right or the left depending on whether the Doppler

---

[1] Synchronous detectors have been in use since the late 1930s as channel detectors in multichannel frequency division multiplex (FDM) telephone transmission systems and, by extension, for the detection of single-sideband (SSB) radio signals. Before the Second World War the normally 60 kHz pilot signal for the modulators and demodulators was distributed from one central oscillator. This is the case with the coherent oscillator (COHO) in a radar.

frequency of the echo is above or below the carrier. This affects the order, or sequence, of the peaks appearing in the in-phase and quadrature videos, which are called positive and negative sequences.

Theoretically, polar detection is possible. Historically, videos formed from first the modulus (linear detection) or the square of the modulus (square law detection) were used for display on A-scopes and plan position indicators (PPI). Much theoretical work has been carried out on these signals. Later, moving target indication (MTI) stages were developed using the phase information only from a phase detector. So far, the integrated circuit components used for signal processing are made for Cartesian components, so polar detectors are a curiosity. Figure 2.13 is a block diagram of such a detector.

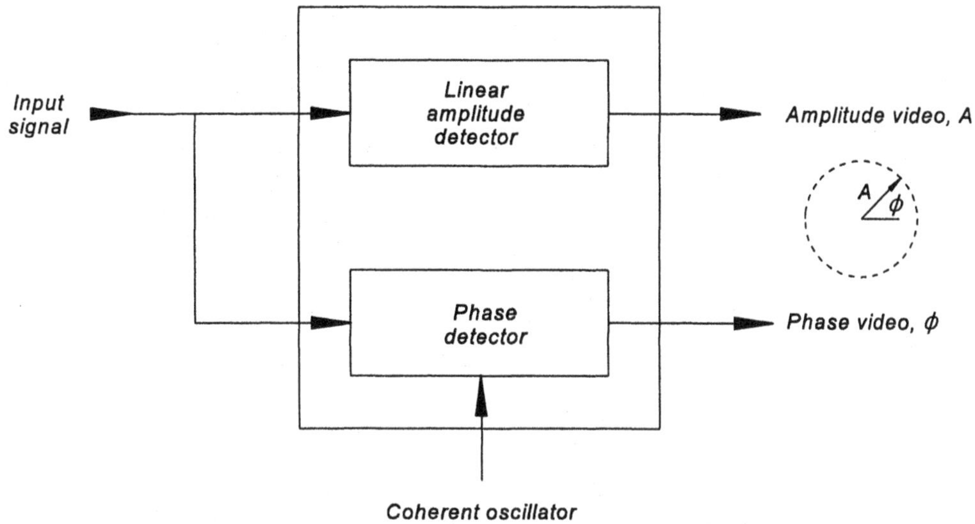

Figure 2.13 Block diagram of a polar detector.

The term "second detector" comes from superheterodyne radio receivers and this is the stage where the modulation is recovered (see Chapter 9, Detectors).

2.5 SYMMETRICAL COMPONENTS IN POLYPHASE CIRCUITS

Other than synchros, motors that drive fans, and power transformers, there are physically no polyphase components used in radar. Power circuits are rarely fully balanced, that is, all the phase vectors are of the same length. Power engineers simplify the calculations by resolving the unbalanced quantity into symmetrical components:

- Positive phase sequence component, which is the working component;
- Negative phase sequence component, which is caused by unbalance and is represented by a vector rotating in the opposite direction to the positive sequence component;
- Zero phase sequence component, which is the direct voltage component, caused by unbalance and a vector representing a constant offset of the origin.

In radar, a Cartesian detector is used to split a vector signal into its x (I) and y (Q) components. Errors are caused by differences in the gain or scaling in the x and y dimensions and when the angle between the x-axis and y-axis is not exactly 90 degrees.

Figure 2.14 shows the effects of the sum of positive, negative, and zero phase sequences on the original offset circle represented by the positive phase sequence component. The arrows on the vectors in the left diagram show the positive phase sequence component rotating anticlockwise, the negative phase sequence component rotating clockwise, and the zero phase sequence component as a constant vector.

Expressed mathematically, the positive and negative phase sequence components are given by

$$\begin{aligned} \textit{Positive phase sequence component} &= V_+ \exp(+j(2\pi ft + \theta)); \\ \textit{Negative phase sequence component} &= V_- \exp(-j(2\pi ft + \theta)) \end{aligned} \tag{2.13}$$

where V_+ is the phase voltage of the positive phase sequence component;
V_- is the phase voltage of the negative phase sequence component;
θ is the phase angle between the positive sequence vector at 0° and the negative sequence vector.

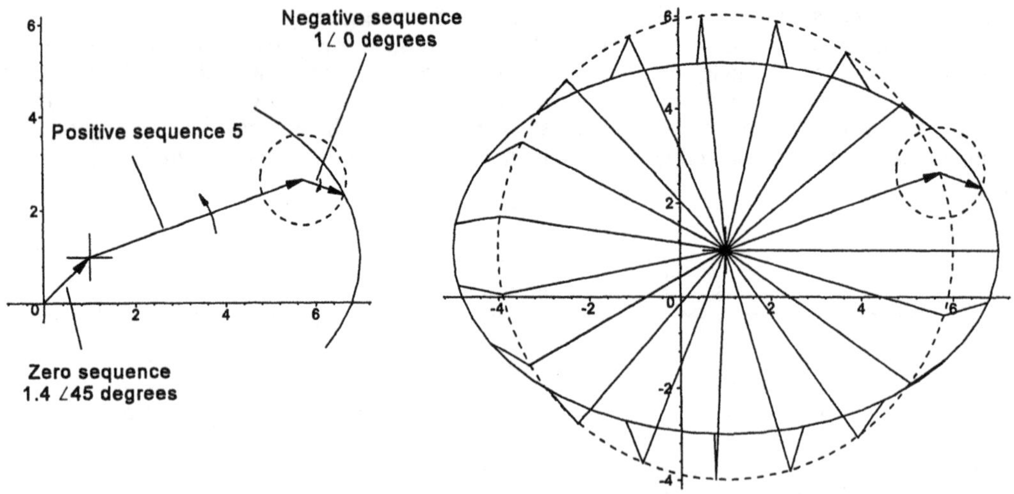

Figure 2.14 The effects of negative and zero sequences on polyphase symmetry.

The zero phase sequence component is a constant vector, say, $a + jb$. The direction where the egg points is given by the phase angle, θ, of the negative component, as is illustrated in Figure 2.15.

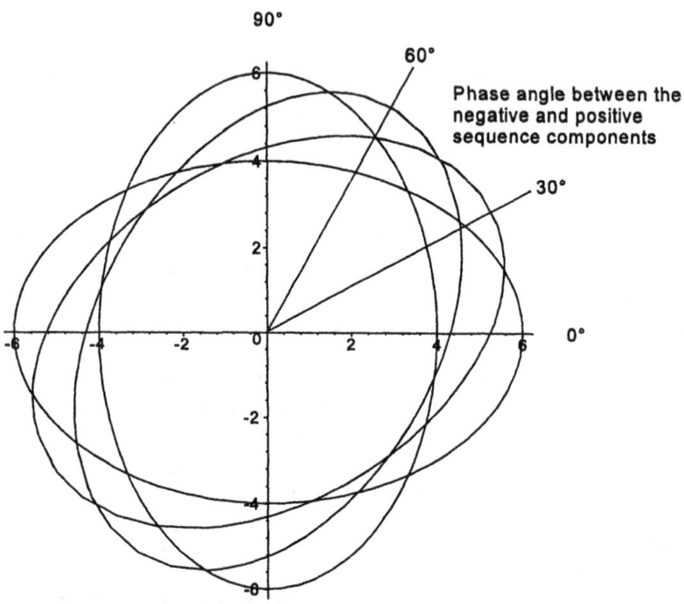

Figure 2.15 Examples of unbalance caused errors in the axes of the two phases.

An ellipse has a major axis radius *a* and a minor radius *b*, so that

$$\begin{aligned} a &= V_+ + V_- \\ b &= V_+ - V_- \end{aligned} \tag{2.14}$$

or, conversely,

$$\begin{aligned} \textit{Positive phase sequence component, } V_+ &= \frac{a+b}{2} \\ \textit{Negative phase sequence component, } V_- &= \frac{a-b}{2} \end{aligned} \qquad (2.15)$$

The shape of the egg is given by the geometrical eccentricity and is given by

$$\textit{Eccentricity, } \epsilon = \sqrt{1 - \left(\frac{b}{a}\right)^2} = \sqrt{1 - \left(\frac{(V_+ - V_-)}{(V_+ + V_-)}\right)^2} \qquad (2.16)$$

where a is the major radius;
b is the minor radius;
V_+ is the positive sequence component;
V_- is the negative sequence component.

The negative and zero phase sequence components are interfering byproducts and are discussed in the description of the errors from Cartesian detectors. Another method of describing the egg shape used in radar is the ellipticity. This is defined as

$$\textit{Ellipticity} = \frac{a}{b} \text{ or } 20 \log_{10}\left(\frac{a}{b}\right) \text{ dB} \qquad (2.17)$$

Equation (2.17) is used for the calculation of the suppression of rain echoes in Chapter 5, Antennas. The negative phase sequence component represents the rain echo signal that has not been canceled.

Extreme examples of elliptic polyphase waveforms are when the positive and negative phase sequence voltages are equal. This gives a single phase waveform. Mathematically from (2.13),

$$\begin{aligned} \textit{Positive + negative phase sequence component} &= \frac{V_+}{2}\exp(j(2\pi ft+\theta)) + \frac{V_-}{2}\exp(-j(2\pi ft+\theta)) = V\cos(2\pi ft+\theta) \\ \textit{Positive − negative phase sequence component} &= \frac{V_+}{2}\exp(j(2\pi ft+\theta)) - \frac{V_-}{2}\exp(-j(2\pi ft+\theta)) = jV\sin(2\pi ft+\theta) \end{aligned} \qquad (2.18)$$

These are illustrated in Figures 2.16 and 2.17. The peaks of the cosine curve occur when the contrarotating helices cross, and those for the sine curve when the points for the same value are diametrically opposite.

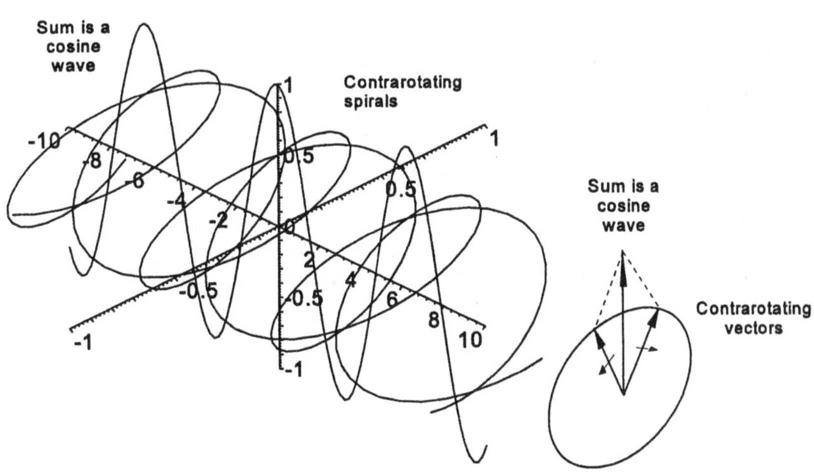

Figure 2.16 Formation of a cosine wave from two contrarotating helices or vectors.

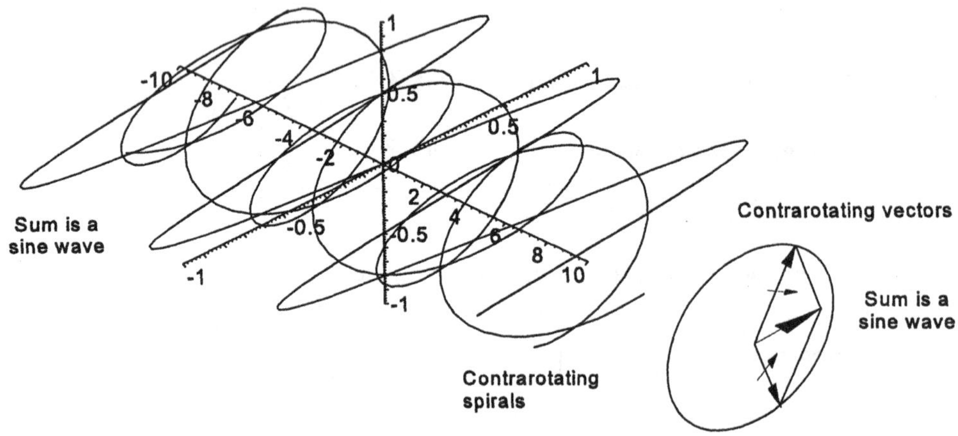

Figure 2.17 Formation of a sine wave from two contrarotating helices or vectors.

2.6 HARMONICS IN BALANCED POLYPHASE CIRCUITS

In addition to phase, polyphase waveforms have an extra property: phase sequence. The most common form of polyphase electricity is three-phase power distribution which transfers the maximum of polyphase power with the minimum number of conductors. Harmonics occur when there are nonlinear loads, for example controlled rectifiers. Table 2.1 shows the phase angles of the odd harmonics.

Table 2.1
Phase angles of odd harmonics in a three-phase system

Harmonic	Phase angles of			Comment
	Phase A	Phase B	Phase C	
Fundamental	0 degree	120 degrees	240 degrees	
3rd harmonic	0 degree	0 degree	0 degree	Zero phase sequence component
5th harmonic	0 degree	240 degrees	120 degrees	Negative phase sequence component
7th harmonic	0 degree	120 degrees	240 degrees	Positive phase sequence component

The third harmonic produces a zero frequency component, which is short-circuited in delta-connected transformers. The fifth harmonic has a negative phase sequence. The seventh harmonic has a positive phase sequence and can cause creeping when induction motors are started.

Return wires are necessary in two-phase power circuits, so that four phase angles are shown in Table 2.2.

Table 2.2
Phase angles of odd harmonics in a two-phase system

Harmonic	Phase angles of				Comment
	I phase	Q phase	I phase return	Q phase return	
Fundamental	0 degree	90 degrees	180 degrees	270 degrees	
3rd harmonic	0 degree	270 degrees	180 degrees	90 degrees	Negative phase sequence
5th harmonic	0 degree	90 degrees	180 degrees	270 degrees	Positive phase sequence

Conventional power circuits operate at a fixed frequency, commonly 16.67 Hz, 50 Hz, 60 Hz, 400 Hz, and so on, so that lowpass filters are able to remove them. Polyphase signals occurring in radar exist from zero frequency to the full video bandwidth, so that filtering is not a good option because these harmonics appear in the later processed or filtered signals.

2.7 POLYPHASE, OR BOTTLE-BRUSH, NOISE

Cosmic noise comes from outer space into the antenna. A greater amount of noise is generated by the first active component in the receiver. Formerly, these components were crystal diodes, but currently they are mostly transistors. This noise is filtered by the waveguide between the antenna and the receiver and is equally distributed in the band. The noise power in a resistor is given by the Nyquist expression:

$$\textit{Noise power} = kTB \quad \text{W} \tag{2.19}$$

where k is Boltzmann's constant, 1.380622 10^{-23} J/K [2, pp. 3-16];
T is the absolute temperature of the resistor in Kelvin;
B is the bandwidth in Hz.

The filtering at radio frequency gives a band of noise around a central radio frequency extending from half the bandwidth on one side to half the bandwidth on the other side. There is no carrier frequency. In the receiver, the noise is block converted to intermediate frequency, filtered again, and then detected. The bandwidth defining the noise power sent for detection is determined by this final filter, as is shown in Figure 2.18.

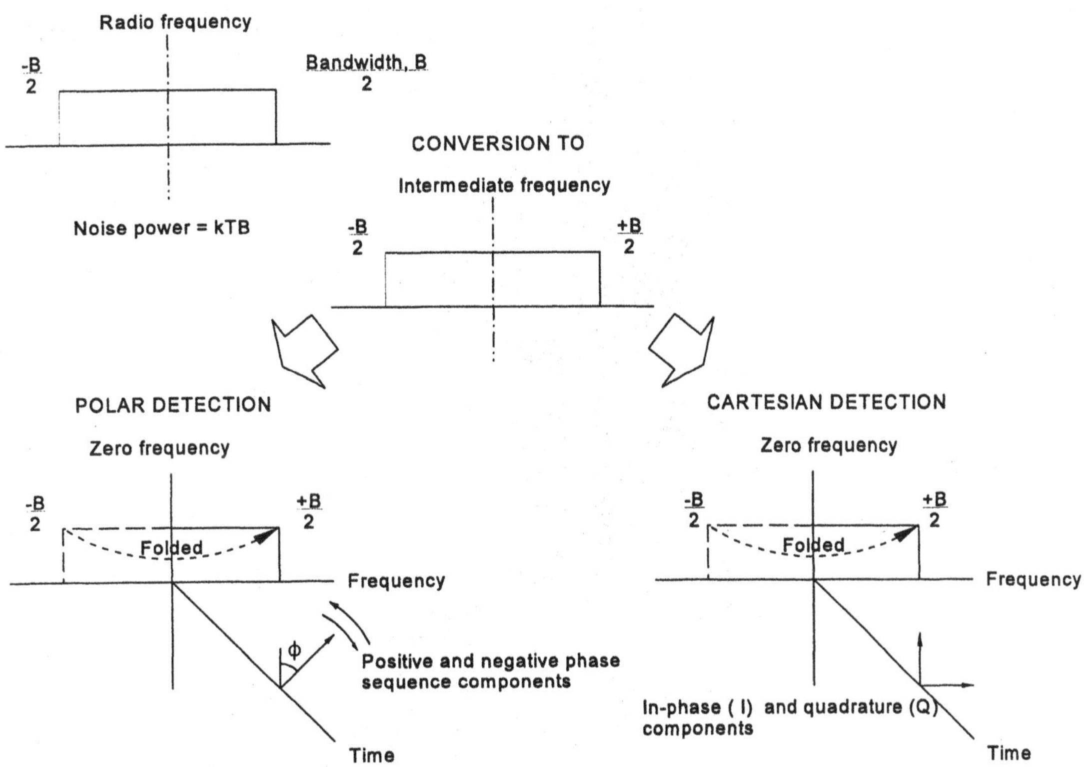

Figure 2.18 Conversion of noise from radio to zero frequency.

With polar detection zero frequency components convert to stationary vectors. The components with a frequency of up to $B/2$ convert to positive phase sequence components, and those with a notional "negative" frequency convert to negative phase sequence components. Frequency is defined in cycles per second so that negative frequencies do not exist, and the negative portion, shown by a dotted line in Figure 2.18, is said to be folded into the positive domain. A linear detector gives the voltage of the vector, and a square law detector gives the power of the vector.

Signals that undergo Cartesian detection are resolved into two orthogonal components that represent the polar vector. These components vary with time if the polar vector rotates. When the noise is detected by a Cartesian or polar converter, the complex waveform is represented as in Figure 2.18. Remember, a constant polyphase voltage, as in electrical power systems, is represented by a helix. In contrast, the noise has the appearance of a shaggy, stochastic bottlebrush, as shown in Figure 2.19. This is a bivariate Gaussian distribution with zero mean, and the standard deviation corresponds to the root mean square noise phase voltage. If the noise passes through a Cartesian detector, then the noise is resolved into the in-phase component and the quadrature component as shown in Figure 2.20.

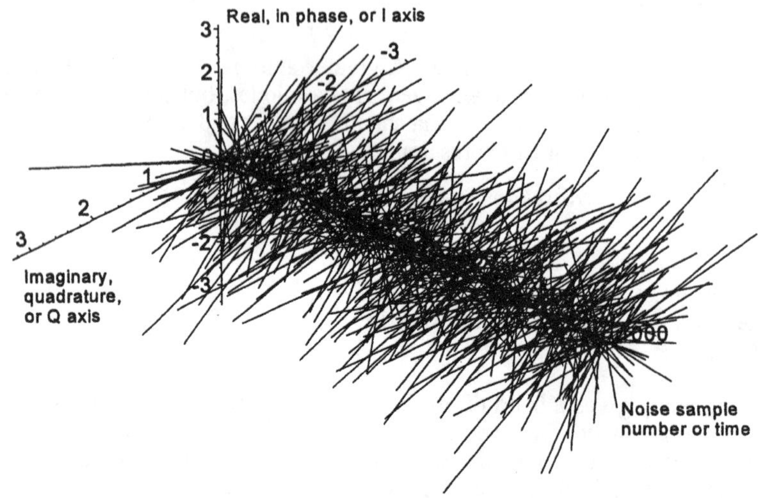

Figure 2.19 Bottle brush or polyphase noise.

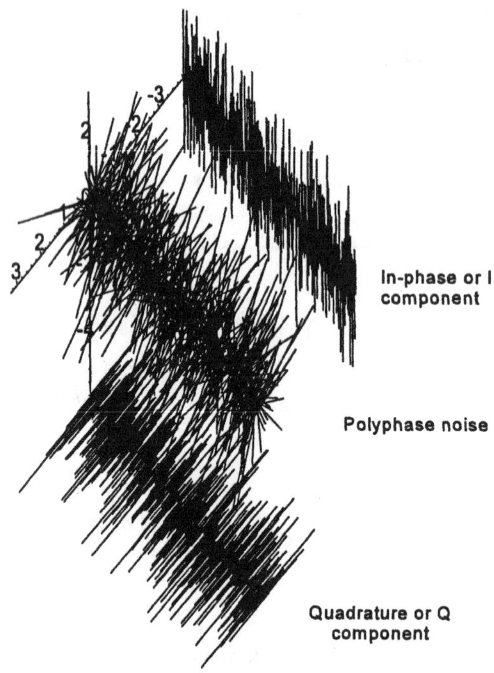

Figure 2.20 Bottle-brush, or polyphase, noise and its resolution into two single-phase components.

Figure 2.21 shows the scatter diagram of the noise, or the tips of the bottle-brush, as seen from the end. In the x and y directions, the components are part of a bivariate Gaussian distribution in that the samples are bunched towards the origin. The noise samples are uniformly distributed in phase. This is shown in three dimensions in Figure 2.22. The amplitude output of a polar detector is the radius, r, of the noise sample in Figure 2.22. A representation of amplitude with time, as on an A-scope, is shown in Figure 2.23. This probability distribution of the noise around the origin is, in x and y, the product of the two Gaussian distributions

$$p(x,\ y) = \frac{\exp\left(-\dfrac{x^2}{2\sigma_{xy}^2}\right)}{\sqrt{2\pi\sigma_{xy}^2}} \frac{\exp\left(-\dfrac{y^2}{2\sigma_{xy}^2}\right)}{\sqrt{2\pi\sigma_{xy}^2}} = \frac{\exp\left(-\dfrac{x^2+y^2}{2\sigma_{xy}^2}\right)}{2\pi\sigma_{xy}^2} \tag{2.20}$$

where σ_{xy} is the root mean square (rms) value of the x or y component of the noise voltage. The radial measure for σ is the sum of the equal x and y components or $\sqrt{2}\,\sigma_{xy}$.

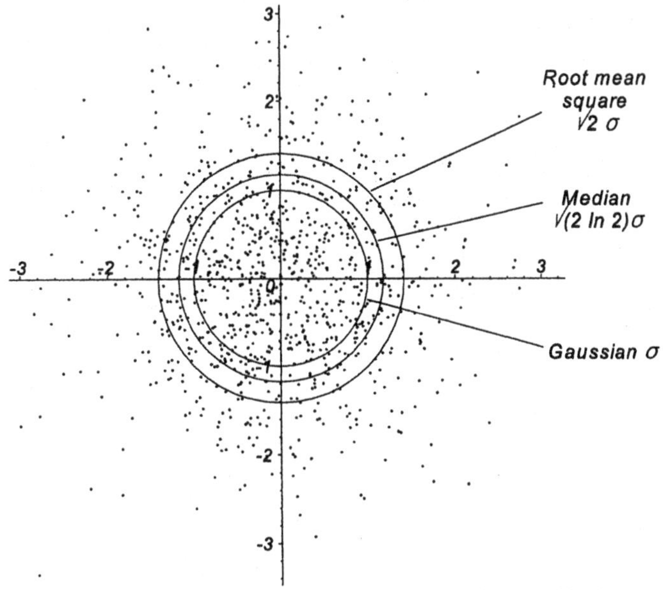

Figure 2.21 A scatter diagram showing the amplitudes and phases of the bottle-brush noise samples.

Substituting for x and y:

$$r^2 = x^2 + y^2$$

$$\textit{Height of hump} = \frac{\exp\left(-\dfrac{r^2}{2\sigma^2}\right)}{2\pi\sigma^2} \qquad (2.21)$$

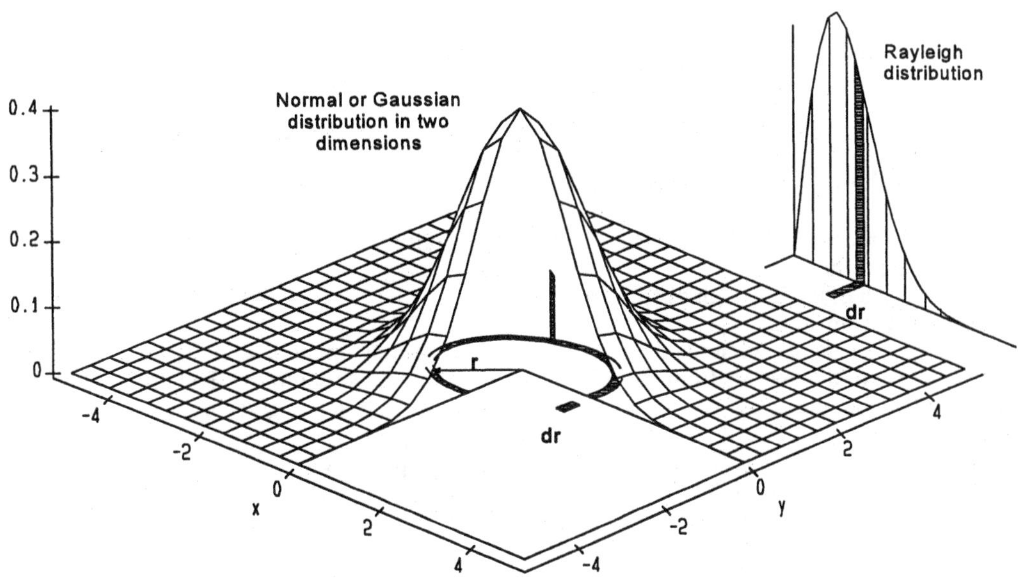

Figure 2.22 Bivariate Gaussian distribution of noise.

The probability at the radius, r, is

$$p(r) = 2\pi r\,dr \times Height\ of\ hump$$

$$= \frac{r \exp\left(-\frac{r^2}{2\sigma^2}\right)}{\sigma^2} \tag{2.22}$$

This is the Rayleigh distribution and is the probability distribution of the detected noise, as shown on the right of Figure 2.22.

The following quantities are defined for the Rayleigh distribution:

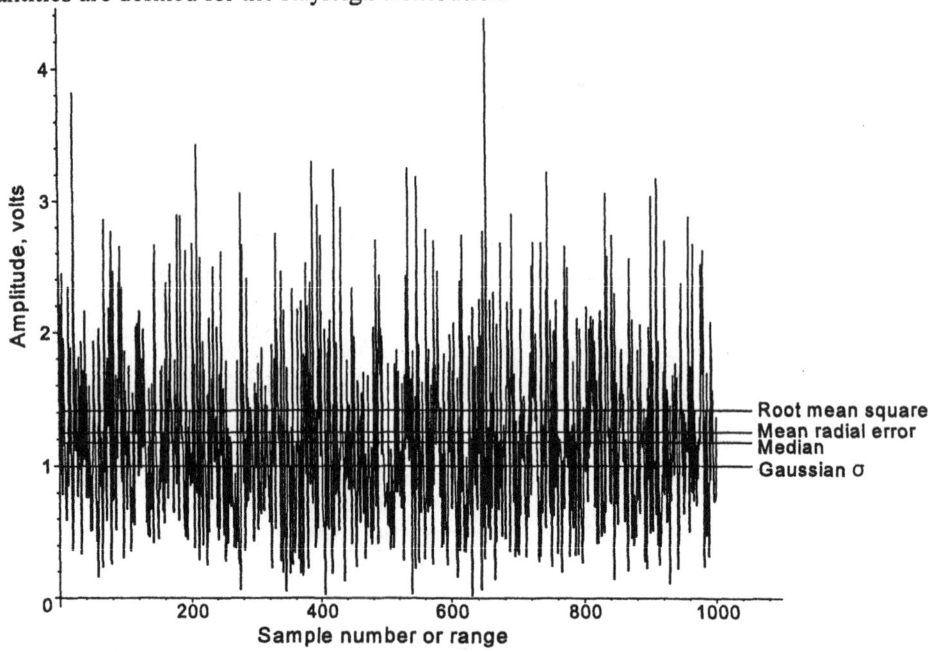

Figure 2.23 Rayleigh noise, or the plot of the modulus of bottle-brush noise.

- σ_{xy}, the standard deviation of each of the two parts of the bivariate Gaussian distribution;
- The mean of the Rayleigh distribution $\sqrt{\frac{\pi}{2}}\sigma_{xy} = 1.2533\ \sigma_{xy}$, measured by a direct voltage voltmeter;
- The root mean square value of the Rayleigh distribution $= \sqrt{2}\sigma_{xy} = 1.4142\ \sigma_{xy}$, measured by a root mean square voltmeter;
- The median of the Rayleigh distribution $\sqrt{2\ \ln\ 2}\sigma_{xy} = 1.1774\ \sigma_{xy}$ (half the points are inside this radius, and half are outside).

Traditionally, the sum of a vector and noise is shown as a washing up mop, as in Figure 2.24. The individual signal plus noise samples are the vectors from the origin to each of the noise samples. The probability distribution for the distance of a point from the origin is a Ricean distribution [3, p. 325] and is given by

$$p(r) = \frac{r}{\sigma^2} \exp\left(-\frac{r^2+S^2}{2\sigma^2}\right) I_0\left(\frac{rS}{\sigma^2}\right) \tag{2.23}$$

where r is the variable to a noise sample;
S is the value of the steady vector or signal;
σ is the standard deviation of the bivariate Gaussian distribution;
I_0 is a Bessel function of zero order and imaginary argument.

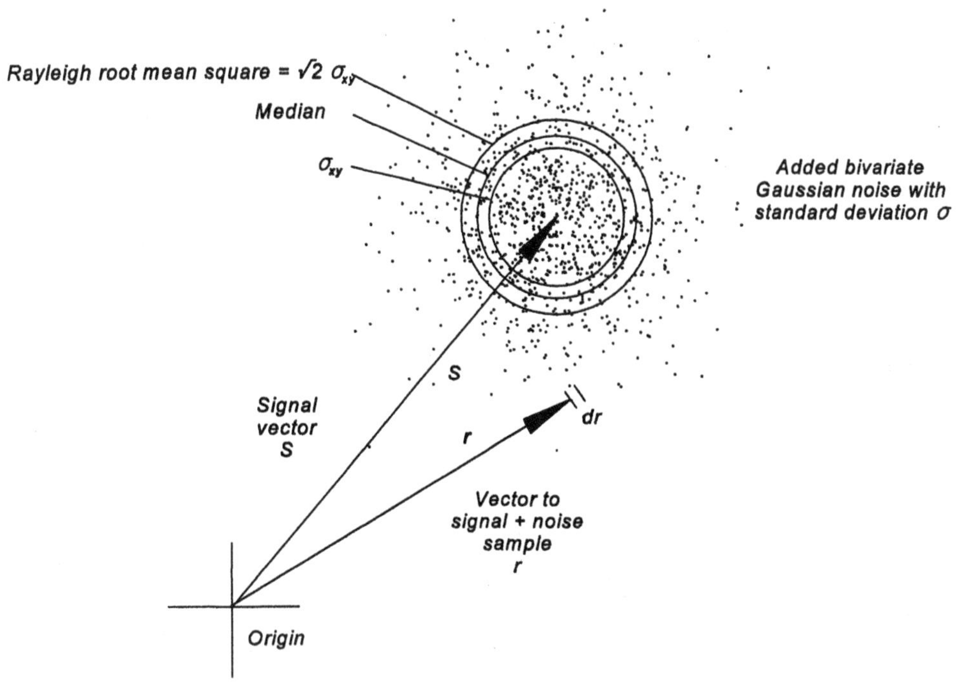

Figure 2.24 Bivariate Gaussian noise added to a signal vector.

Figure 2.25 shows the Ricean distribution for $\sigma = 1$ for various values of S. When $S = 0$, this is a Rayleigh distribution, which is used to determine the probability of detection for echoes that do not fade.

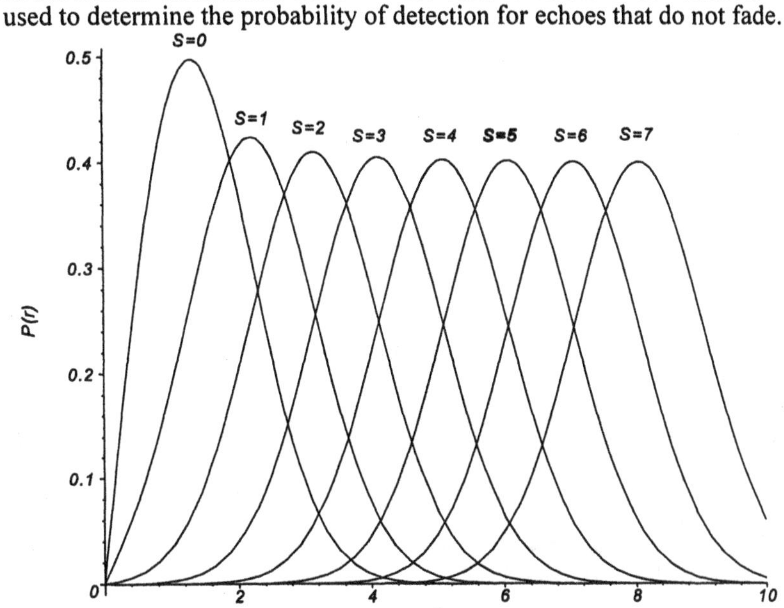

Figure 2.25 Ricean distribution for various values of S.

2.8 TIME AND SPECTRAL DOMAINS, HELICAL SPECTRA

The engineer must hold in view simultaneously the time, spectral, and statistical domains. A way of bringing the time and spectral domains, or their equivalents, onto the same page is to plot them on separate orthogonal axes. This can be illustrated by a single rectangular pulse.

$$\begin{aligned} \textit{Rectangular pulse} &= 1 \quad \text{for} \quad -\tau/2 < t < \tau/2 \\ &= 0 \quad \text{otherwise} \end{aligned} \tag{2.24}$$

The spectrum for the pulse with a constant height of 1 (unity) along its length, centered on zero is

$$\text{Spectrum} = \int_{-\frac{\tau}{2}}^{+\frac{\tau}{2}} 1 \, e^{-j\,2\pi f t}\, dt = \frac{\sin \pi f \tau}{\pi f \tau} \tag{2.25}$$

Figure 2.26 shows the combination of a time waveform and its spectrum into a single diagram. For an example the familiar rectangular pulse of width T and its sin x/x spectrum have been chosen.

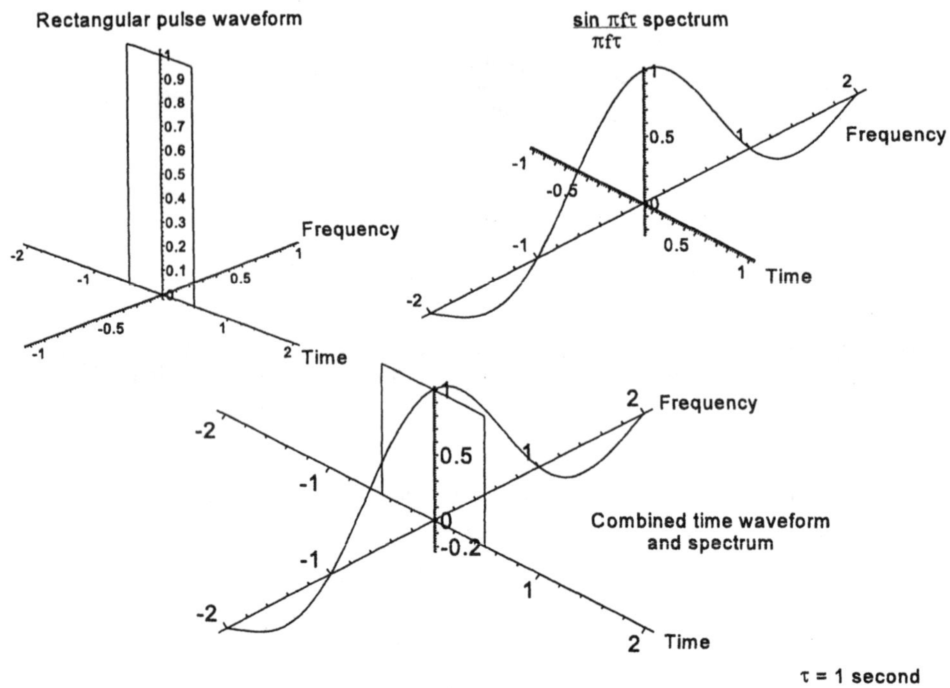

Figure 2.26 The assembly of a time waveform, width 1 s, and spectrum onto a single diagram.

Both the time waveform and its spectrum are real and can be easily represented on a flat piece of paper. The orientation and axes have been chosen such that:

- The common reference vector is along the real axis in time and frequency. This common direction is up as in polyphase vector diagrams;
- The diagrams represent the standard two-dimensional real form when one is looking from the front and the imaginary form when looking from above.

As the time pulse moves away from being centered on zero time, the spectrum becomes complex. This can be accomplished by changing the limits for integration from $-\tau/2..+\tau/2$ to $-\tau/2 + t_1$ to $\tau/2 + t_1$, where the pulse occurs t_1 seconds later. Thus, the spectrum of the delayed pulse is

$$\text{Spectrum of the pulse delayed by } t_1 = \frac{\sin \pi f \tau}{\pi f \tau}\, e^{-j\,2\pi f t_1} \tag{2.26}$$

The exponential is a helix with a pitch depending on t_1, which allows the time waveform to be reconstructed uniquely from the spectrum. The greater t_1, the tighter the helix in the spectrum, as in Figure 2.27, hence the expression helical spectrum. The exponential term in (2.26) is in angle only, a helix of radius unity, so the modulus of the spectrum remains the same.

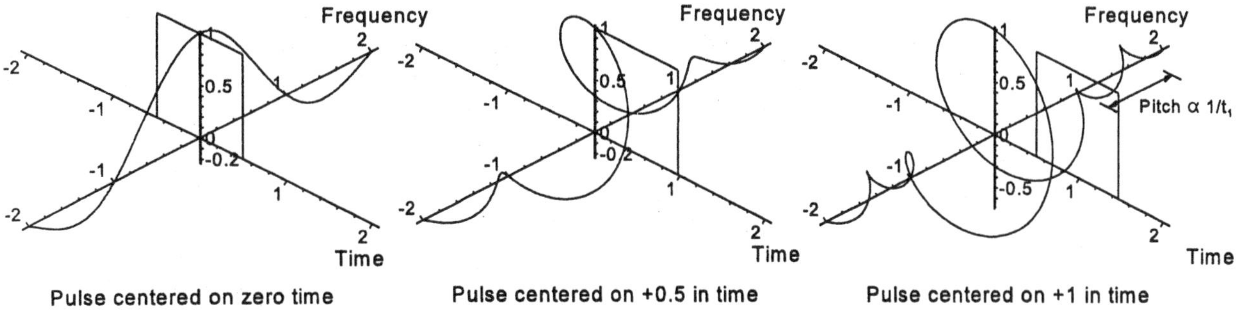

Figure 2.27 The effects of the pulse not being centered on zero.

The convention in Figure 2.27 and the direction of the axes are common throughout this book.

In contrast to standard textbook descriptions, most practical spectra are complex. It is often useful to plot the complex spectrum of a time series in three dimensions. The phenomena and transients causing the unwanted spectral components tend to separate into the real and imaginary planes. This eases the conceptual analysis. Modern mathematical software makes this presentation possible.

The general expression for the Fourier transform used here is

$$\textit{Spectrum},\ H(f) = \int_{-\infty}^{+\infty} h(t)\ \exp(-j\ 2\pi f t)\ dt \tag{2.27}$$

and its inverse is

$$\textit{Waveform},\ h(t) = \int_{-\infty}^{+\infty} H(f)\ \exp(+j\ 2\pi f t)\ df \tag{2.28}$$

Notice that the $-f$, or frequency, convention is used [4, p. 381] so the $1/2\pi$ factor is not necessary for the inverse transform. The exponential terms represent helices. A polyphase signal is itself a helix, say, $h(t) = \exp(j\ 2\pi\ f_1\ t)$, as shown in Figure 2.28. The multiplier in the Fourier transform in (2.27) is $\exp(-j\ 2\pi\ f\ t)$. During its time of existence, the function $H(f)$ will have a value only when the frequencies of the helices are the same and their sense or phase sequences are opposite.

The helical component is effectively unraveled or stopped. With a polyphase signal of frequency f_1 Hz, this will occur at $+f_1$ Hz for a positive phase sequence signal and $-f_1$ Hz for a negative phase sequence. The spectrum in Figure 2.29, in contrast to that in Figure 2.27, is a fine line, or delta function (zero width, undetermined height, area unity), at only the "positive" frequency. This makes delta functions difficult to draw, so that their height is conventionally their area or unity [5].

For a cosine wave of frequency f Hz, which is composed of two polyphase waveforms of opposite phase sequences, there will be two spectral components at $+f$ Hz and $-f$ Hz. For noise of power P W measured over a bandwidth of zero to F Hz, the spectral density over all frequency space (from $-\infty$ to $+\infty$) is $P/2F$ W/Hz. This is discussed in Chapter 16, Transforms.

A delta function has a value only at time zero and has an area of unity. Thus,

$$\int_{-\infty}^{+\infty} \delta(t)\ \exp(-j\ 2\pi\ f\ t)\ dt = 1 \tag{2.29}$$

for all frequencies as shown in the center of Figure 2.29. If the delta function occurs at time a, then [5, p. 75]

$$\int_{-\infty}^{+\infty} \delta(t-a)\, f(t)\, dt = f(a) \tag{2.30}$$

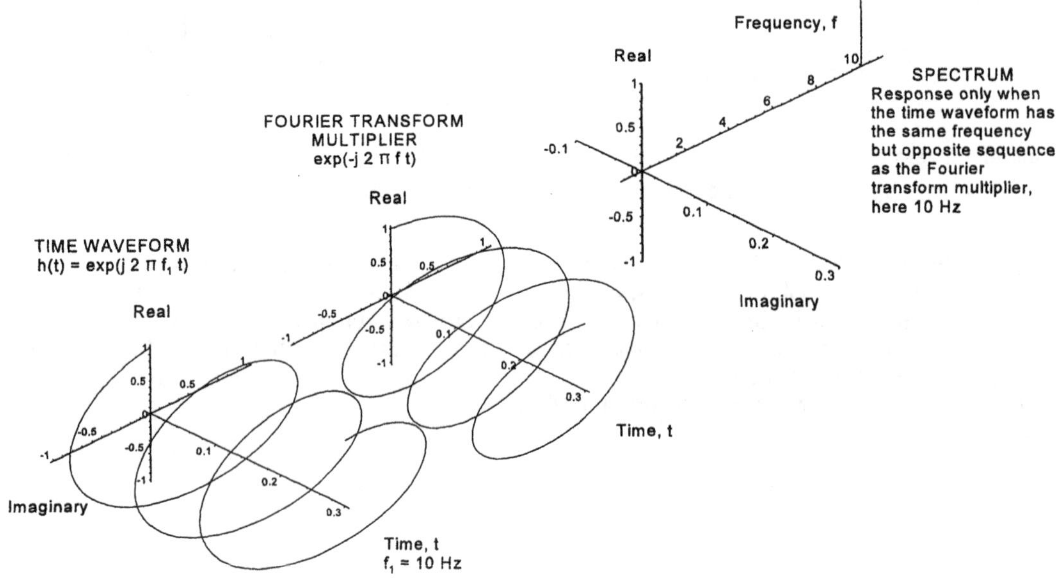

Figure 2.28 The spectrum of a 10 Hz polyphase waveform.

For the spectrum of a pulse occurring at a seconds, the Fourier transform is the helix

$$\int_{-\infty}^{+\infty} \delta(t-a)\, \exp(-j\, 2\pi\, f\, t)\, dt = \exp(-j\, 2\pi\, f\, t) \tag{2.31}$$

The outer portions of Figure 2.29 show the helices, $\exp(\pm j\ f)$, each with a pitch of 2 Hz, for displacements of ∓½ second.

The transformation of a two-dimensional real (balanced even) or imaginary (balanced odd) spectrum into a helical spectrum allows the time of occurrence of all its components to be expressed uniquely. For example, if in Figure 2.29 there are two unequal components, the spectrum will be similar to a "coiled coil" in incandescent lamps.

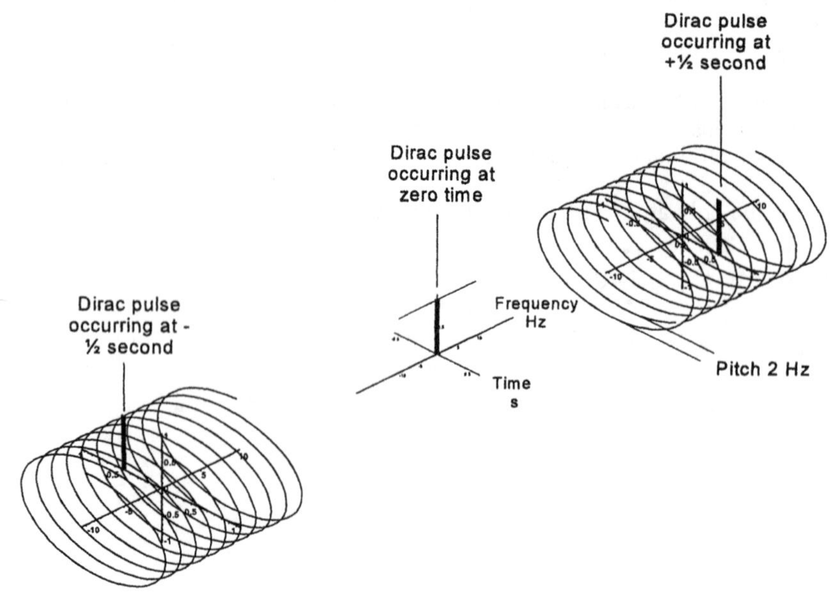

Figure 2.29 The helical spectra from Dirac pulses at various times.

2.8.1 Convolution and correlation

The Fourier transform, as well as the formula that links waveforms in time to their spectra, is also used in statistics. Complex Fourier transforms may be multiplied together either simply for convolution or by the complex conjugate, as in alternating current power calculations, called correlation.

The notation in this book used for statistics is matched to that in signal theory so that the Fourier transform or characteristic function in statistics, $C(\xi)$, for a probability distribution $p(x)$ is

$$C(\xi) = \int_a^b p(x)\, e^{-j2\pi x\xi}\, dx \quad \text{and} \quad p(x) = \int_{-\infty}^{+\infty} C(\xi)\, e^{j2\pi x\xi}\, d\xi \tag{2.32}$$

where the limits of integration, a and b, are over the full range of x.

The convention is different from the $+i\,x\,\xi$ found in the exponent in statistical texts.

2.8.1.1 Convolution

Convolution is used in this book to find the shapes of the sums from statistical distributions and their moments, for example, statistically described signals and noise. In Figure 2.31(a) two Gaussian curves with a variance of unity are shown and the Fourier transforms are to be seen in Figure 2.31(b) with 4 cycles per frequency unit and in Figure 2.31(c) with 6 cycles per frequency unit. The curve for the Fourier transform of the sum of samples from the two curves is found by multiplication: multiply the amplitudes and add the phase angles to give a narrower transform with 10 cycles per frequency unit — see Figure 2.31(e). The final curve in Figure 2.31(g) is found by inversion and has a mean of 10 and a variance of 2, found by adding the moments of the curves in Figure 2.31(a). This is the basis for summing the means to find bias and the variances for accuracy and the shape of the resulting Figure 2.31(g) may be used to find percentiles.

The probability distributions in Chapters 11 and 12 are not symmetric, so convolution must be used to find the shapes of the distributions of the sums.

Any number of Fourier transforms may be multiplied together to give a final distribution function and moments, as with the sum of noise samples, each with an exponential distribution in Figure 2.30. The product of an infinite number of convolutions inverts to a Gaussian distribution as stated in the central limit theorem [6, p. 219], and an example is shown in Figure 2.30.

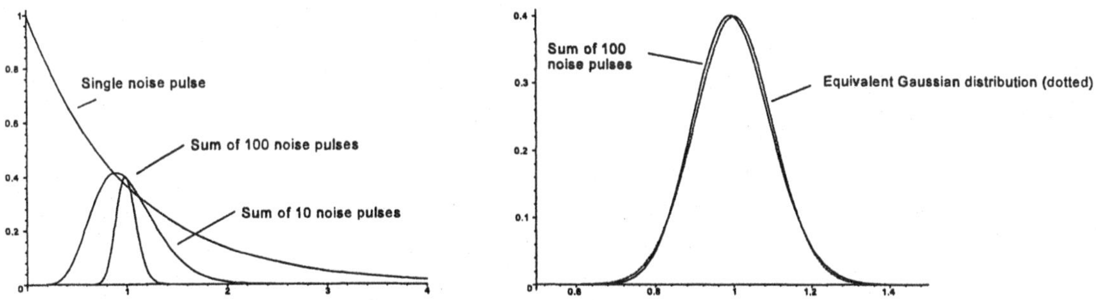

Figure 2.30 The normalized distributions of the sums of 1, 10, and 100 noise pulses and an equivalent Gaussian distribution.

2.8.1.2 Cross-correlation

Correlation is used to match a signal with a reference copy held in memory — see Chapter 8. In this case the phases of the signals are subtracted by reversing the sign of the Q phase of the reference signal before multiplication, as in Figure 2.31(d). After multiplication the curve in Figure 2.31(f) is found with a frequency of two cycles per frequency unit or the difference of the positions of the means in Figure 2.31(a). The inverse transform, Figure 2.31(h), gives a pulse with a lag of 2 and a variance of the sum of the variances of the original curves, 2 (standard deviation $\sqrt{2}$).

2.8.1.3 Autocorrelation

When a waveform is cross-correlated with itself, the tip of the function is the classical calculation of power, and the process is called autocorrelation. Autocorrelation is also used to give the shape of an expanded pulse and its time sidelobes after pulse compression (see Chapter 8). The process can be imagined from Figure 2.31(a) if both Gaussian curves coincide so that the lag in Figure 2.31(h) is zero and the curve is symmetrical around the ordinate axis.

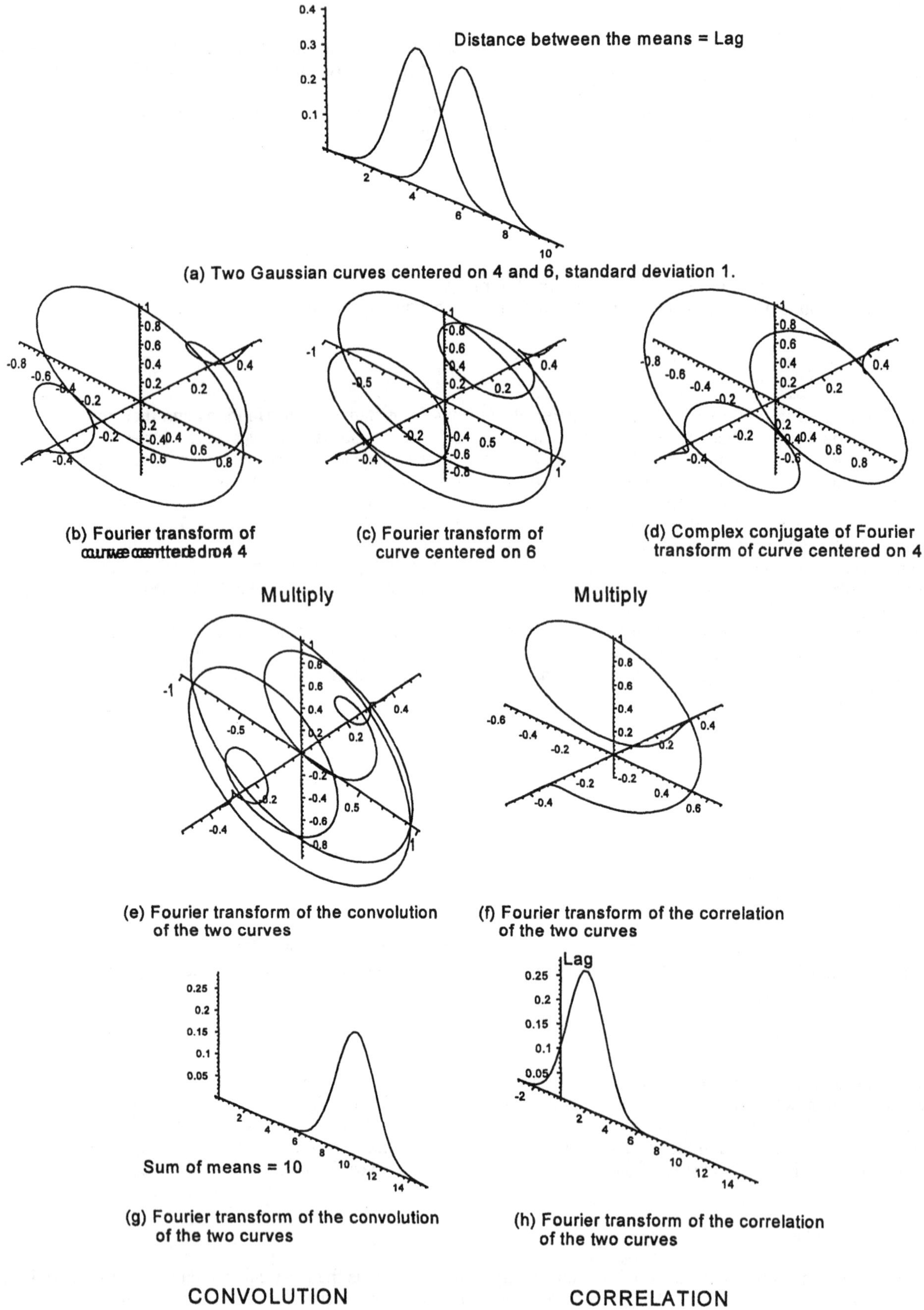

Figure 2.31 Convolution and correlation with Fourier transforms.

2.9 GAUSSIAN PULSES, SPECTRA, AND BEAM SHAPES

Gaussian pulses, spectra, and beam shapes are based on the curves for the Gaussian or normal distribution in statistics. The position and the shape of the curve are defined by the mean, μ, and the standard deviation, σ, respectively. The other measures used to define the widths are shown next.

2.9.1 Gaussian pulses and spectra

A Gaussian pulse is assumed to be centered on zero and has a duration or width in time of τ seconds between the 50% points. The equations are as follows:

Gaussian or normal distribution

$$p(x) = \frac{\exp\left(\frac{-x^2}{2\sigma^2}\right)}{\sqrt{2\pi}\sigma} \tag{2.33}$$

Gaussian pulse

$$f(t) = \exp\left(-4\ \ln(2)\left(\frac{t}{\tau}\right)^2\right) = \exp\left(-2.77259\left(\frac{t}{\tau}\right)^2\right) \tag{2.34}$$

These shapes are shown in Figure 2.32.

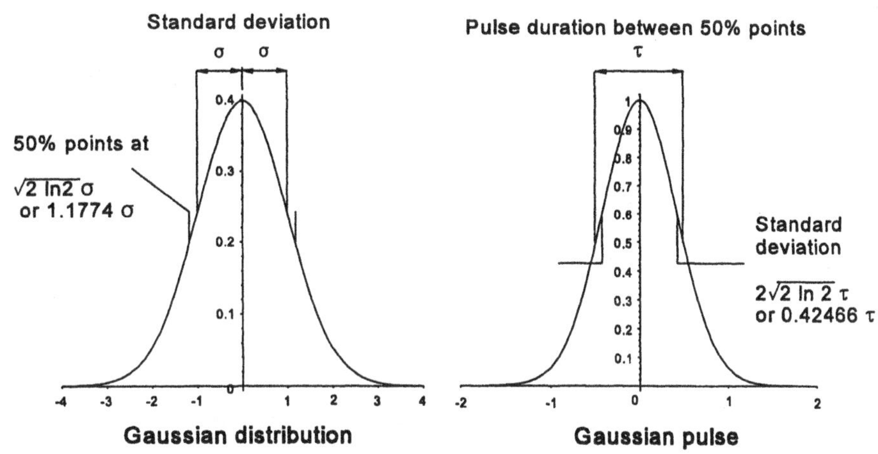

Figure 2.32 Gaussian or normal distribution and Gaussian pulse.

The positions of the 50% points are shown in Figure 2.32, and positions of others points are given in Table 2.3.

Table 2.3
Values for the ordinates for Gaussian distributions and pulses

Ordinate	Power ratio (dB)	Voltage ratio (dB)	Gaussian or normal distribution	Ratio	Gaussian pulse
$1/\sqrt{2}$	1.5	3.0	$\sqrt{\ln(2)}\sigma = 0.8325546112\sigma$		$\frac{1}{\sqrt{2}} = 0.7071067810$
1/2	3.0	6.0	$\sqrt{2\ln(2)}\sigma = 1.177410022\sigma$	$\sqrt{2}$	$\frac{1}{2} = 0.5$
1/4	6.0	12.0	$\sqrt{2\ln(4)}\sigma = 1.665109221\sigma$	$\sqrt{2}$	$\frac{1}{\sqrt{2}} = 0.7071067810$

The spectra are given by the Fourier transforms. The integrals may be evaluated using mathematical software or also

[7, Eq. 3.323.2; 8, p. A-12].

Fourier transform of Gaussian or normal distribution:

$$F(f) \quad \frac{1}{\sqrt{2\pi}\sigma}\int_{-\infty}^{\infty}\exp\left(-\frac{1}{2}\left(\frac{x}{\sigma}\right)^2\right)\exp(-j\ 2\ \pi\ f\ t)\ dx$$
$$= \exp(-2\ \sigma^2\ \pi^2\ f^2) \tag{2.35}$$

$$Standard\ deviation = \frac{1}{2\ \pi\ \sigma} = \frac{0.1596}{\sigma}$$
$$50\%\ points = \pm\frac{1}{\pi\ \sigma}\sqrt{\frac{\ln(2)}{2}} = \pm\frac{0.1874}{\sigma} \tag{2.36}$$

Fourier transform of Gaussian pulse:

$$F(f) = \int_{-\infty}^{\infty}\exp\left(-4\ln(2)\left(\frac{t}{\tau}\right)^2\right)\exp(-j\ 2\ \pi\ f\ t)\ dt$$
$$= \frac{\tau}{2}\sqrt{\frac{\pi}{\ln(2)}}\exp\left(-\frac{\pi^2\ \tau^2\ f^2}{4\ \ln(2)}\right) \tag{2.37}$$

$$Standard\ deviation = \frac{\sqrt{2\ln(2)}}{\pi\ \tau} = \frac{0.37478}{\tau}$$
$$Bandwidth = \frac{4\ \ln(2)}{\pi\ \tau} \tag{2.38}$$

The shapes of the spectra are shown in Figure 2.33.

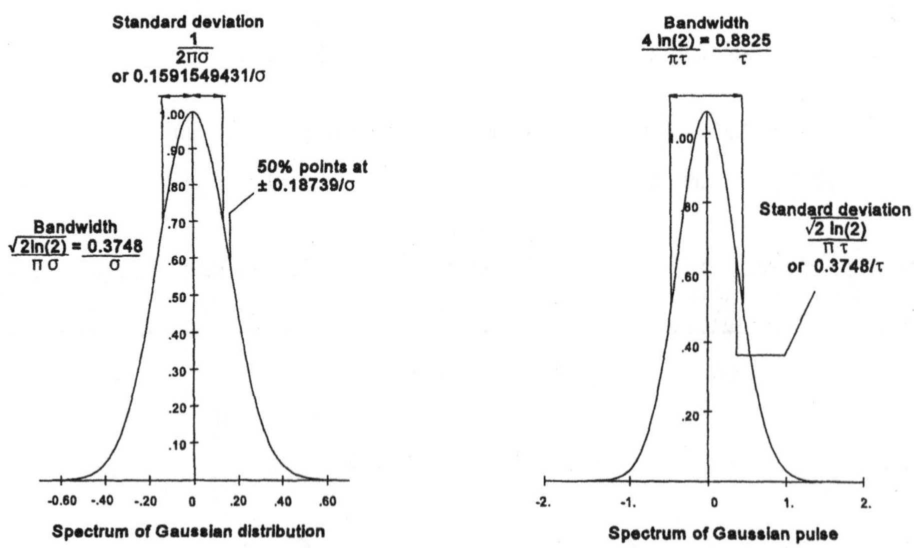

Figure 2.33 The spectrum of the Gaussian or normal distribution for a standard deviation of unity and the spectrum of a Gaussian pulse of length 1 s.

The spectrum of a Gaussian time pulse is itself a Gaussian or bell shape, which is the reason for its usefulness. When a Gaussian curve is plotted with the ordinate in decibels, the curve is a parabola open underneath.

2.9.2 Gaussian beam shapes

If the tip of the beam is exponential in shape, it has the form

$$\textit{Voltage pattern, one way} = \exp\left(-a\frac{x^2}{\theta_3^2}\right) \quad (2.39)$$

The power patterns are given by

$$\textit{Power pattern, one way} \quad \exp\left(-2a\frac{x^2}{\theta_3^2}\right) \quad (2.40)$$

$$\textit{Power pattern, two way} \quad \exp\left(-4a\frac{x^2}{\theta_3^2}\right) \quad (2.41)$$

where $a = 2\ \ln(2) = 1.38629$;
θ_3 is the half power beam width.

The exponential beam patterns for a half-power beam width of unity are shown in Figure 2.34. The two-way power pattern is also shown compared with the equivalent sin *x/x* pattern. The pattern decreases to zero at large angles without lobing as does the Gaussian distribution. Examples of these functions are given in Appendix B.

2.9.3 Gaussian illumination functions

The Gaussian waveform exists over all time and gives a Gaussian spectrum without lobing which exists over all frequencies. Gaussian illumination is finite and is cut off (truncated) at the sides of the reflector which gives sidelobes. The function is given by

$$g(x) = \exp\left(-a\ (n\ x)^2\right) \quad (2.42)$$

where $a = 2\ \ln(2) = 1.38629$;
n is a parameter.

The values of n determine how much of the central part of the Gaussian function is used to illuminate the aperture. In linear measure,

$$\begin{aligned} \textit{Half power length} &= \frac{1}{n} \\ \textit{Edge illumination ratio} &= 10\log_{10}(2)\ n^2 \approx 3n^2 \ \text{dB} \end{aligned} \quad (2.43)$$

Examples of these functions are given in Appendix B.

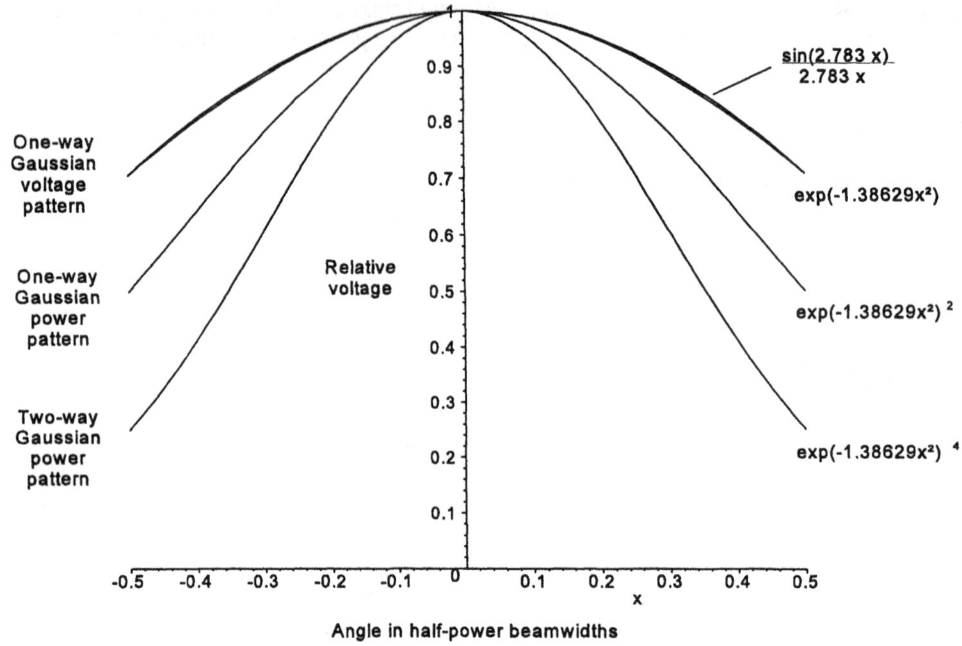

Figure 2.34 Exponential representations of the tip of a beam pattern.

2.10 USE OF BRACKETS AND OTHER SYMBOLS

The use of normal ( ), square [ ], or curly { } brackets in mathematical expressions partly follows the common convention in many programming languages and hand-held calculators (for example, the Hewlett-Packard 48).

- Round brackets ( ) are used generally in mathematical expressions;
- Square brackets [ ] are used to enclose mathematical expressions in round brackets and to index members of lists;
- Curly brackets { } are used for sets.

Other conventions used are:

- Two full-stops .. are used to indicate a range;
- A cross × is used for arithmetic multiplication;
- The International Standards Organisation (ISO) in Geneva recommends a space as a thousands separator. The space is used in this book, namely, 123 456.123 456 with long numbers where necessary.

REFERENCES

1. Say, M. G., *Electrotechnology*, 2nd ed., London: George Newnes, 1955.
2. Jordan, E. C., *Reference Data for Engineers*, 7th ed. Indianapolis: Sams, 1986.
3. Blake, L. V., *Radar Range-Performance Analysis,* Norwood, Massachusetts: Artech House, 1986.
4. Press, W. H., B. P. Flannery, S. A. Teukolsky, and W. T. Vetterling, *Numerical Recipes*, Cambridge: Cambridge University Press, 1986.
5. Bracewell, R. N., *The Fourier Transform and Its Applications*, 2nd ed., New York: McGraw-Hill, 1965.
6. Weisstein, E. W., *CRC Concise Encyclopedia of Mathematics*, Boca Raton, Florida: Chapman and Hall, 1999.
7. Gradshteyn, I. S., and I. M. Ryzhik, *Table of Integrals, Series, and Products*, New York: Academic Press, 1980.
8. Ludloff, A., *Handbuch Radar- und Radarsignalverarbeitung*, Braunschweig: Vieweg, 1993.

Chapter 3

Transmitters

This chapter discusses pulse shapes, with their influence on spectra, and stability. The wavelengths used vary from millimeters to meters. Initially radar designers were pleased to have all the power that their magnetron could give reliably to illuminate the space around them. Today, the radar bands are crowded so that sidebands must be held to a minimum, and modern signal processing requires very stable transmitters.

Although continuous wave radars are used by the police to measure vehicle speeds, in most cases the transmitter signal is modulated to be able to measure the echo time and thus range. This modulation may be:

- Continuous wave: Frequency modulated;
  Phase modulated;
- Pulse modulated: No modulation inside the pulse;
  Frequency modulation inside the pulse;
  Phase modulation inside the pulse.

The majority of radars use pulse modulated transmitters. Early radars used triode or tetrode transmitters, sometimes self- oscillating until they were superseded by magnetrons during the Second World War. Greater stability and bandwidth are provided by klystron and traveling wave tube output stages, which came later in the more expensive military radars.

3.1 TRANSMITTER POWER

The transmitter power is a major factor in the radar range equation. Just as with a searchlight, the greater the power, the greater the range. Power in the microwave bands is expensive and inefficient to generate, so it must not be wasted. The range equation uses the transmitter energy or power times the pulse width. This represents the energy in each pulse. With shaped pulses, either or both the peak power and time must be modified to give the correct product. This is normally the case when the time is measured between the 3 dB points of pulses with symmetrical edges.

Early microwave radars used magnetrons, and currently many basic radars still use magnetrons. The frequencies were fixed and the purity of the output was controlled by strapping and the rise time of the direct voltage driving pulse. Later power amplifiers such as klystrons and traveling wave tubes became available to give a greater control of the transmitter pulse and its shape. Such a transmitter is shown in Figure 3.1.

With the trend of eliminating all vacuum tubes, modern radars are increasingly using all solid-state transmitters, which are introduced in Section 3.2.1.

3.2 POWER OUTPUT STAGE

The radar pulse power is chosen to provide the maximum range and accuracy. The output tube in Figure 3.1 requires a high-voltage pulse up to 250 000 V, which is generally the practical limit. Another limit is the power that can be carried by the waveguide before voltage breakdown or arcing occurs. The input power is normally generated in a pulse modulator, which allows capacitors in delay lines to charge between radar pulses and then delivers a short pulse as required. These short pulses are generated at a lower voltage and transformed up to drive the cathode. Often the body and anode are at near ground potential. Large powers are generated and great effort is made to ensure that there is as little variation as possible in the voltages of each pulse (see Section 3.6.4). During each pulse, the direct voltage pulse is turned on, the current in the output tube stabilizes, the radio frequency drive pulse is applied, and then the modulator pulse is turned off.

The amplification and phase of output tubes depend on the cathode to anode voltage. This causes a distortion when

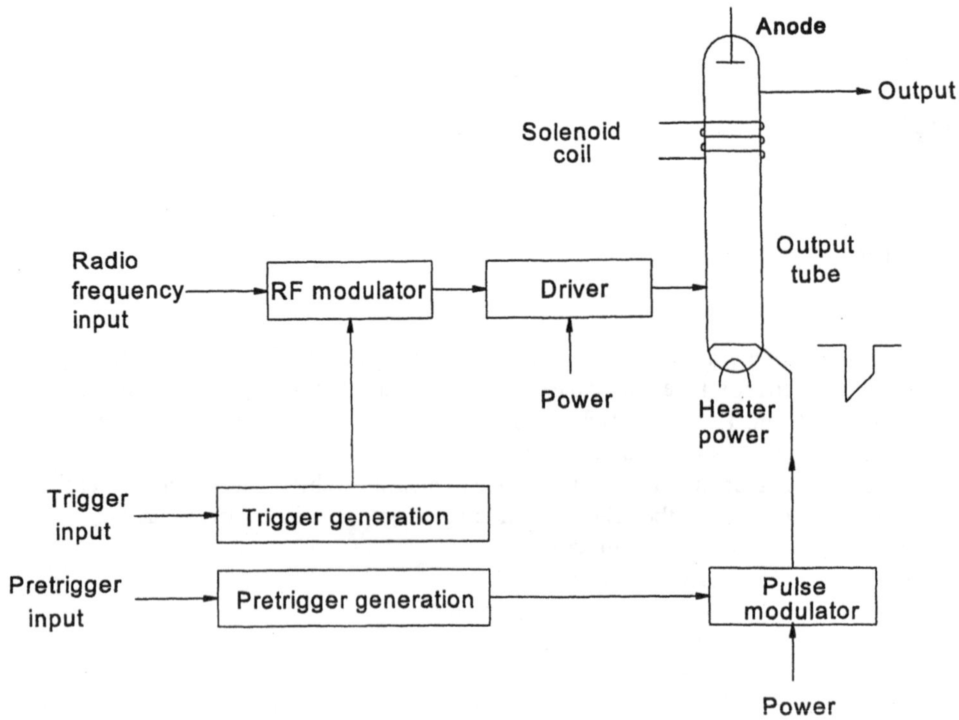

Figure 3.1 Simplified block diagram of a radar transmitter output stage.

the pulse is angle (frequency or phase) modulated for pulse compression. Nonlinearity produces harmonics. The drive pulse is at a relatively low power, and semiconductor amplifier blocks are commonly used. This pulse may be shaped and the shape determines the transmitted waveform in frequency, phase, and amplitude.

The transmitted pulse width determines the range (energy) and range resolution (range extent) of the radar. Increased pulse width is sometimes used to reduce the peak power in the waveguide to avoid arcing. Increased resolution may be obtained by frequency or phase modulating the radio frequency pulse. This is called pulse compression and has a different transmitter pulse spectrum.

3.2.1 Semiconductor transmitters

Modern transmitters use semiconductors instead of gas-filled switches in modulators and also to replace power tubes in the radio frequency amplifier stages, including the final power amplifiers.

In modulators a single thyratron and pulse forming network is often replaced by a number of printed circuit cards, each with its own pulse forming network and switching thyristor. With, typically 20, such modules the output impedance is of the order of fractions of ohms and the modulator pulse from each module needs to be carried away on one or a number of cables in parallel. Each modulator module often has a fuse and control circuitry monitors each module to report faults and, if necessary, to switch in reserve modules in case of single failures. The radar can often be designed to operate with power when less than the full number of modules is working, called graceful degradation. The modules can be designed so that the cards can be withdrawn and replaced while the transmitter is running at full power, and the radar is able to operate at full performance until the maintenance time and a second transmitter is no longer necessary. Semiconductors are very sensitive to voltage spikes so that additional protective circuitry is required to discharge the pulse forming networks at each pulse repetition time so that accumulations of charge in the pulse forming networks from pulse to pulse never happen.

The transistor amplifier is essentially a triode and smaller and smaller structures are required for higher frequencies leading to many amplifiers being connected in parallel. The gain per transistor can be relatively small, only 10 dB compared with 60 dB for a klystron, and care must be taken controlling the losses in the power splitters and combiners necessary to pass the power to a single output waveguide flange. Power recombination can be avoided by distributing the transmitter elements into separate transmit-receive modules in the antenna, each behind its separate element or line of elements, in a planar (or otherwise shaped) array. Distributing the transmitter often eases cooling problems. Smaller continuous wave radars for intruder alarms may use Gunn oscillators as transmitters, and Impatt oscillators are used in the millimeter region.

High power vacuum tube amplifiers use relatively large parts that use the thermal inertia to smooth over temperature changes from pulse to pulse. This does not apply to semiconductor amplifiers so that to achieve the same mean power longer pulses must be used, increasing the duty cycle. Pulse compression is essential with low and medium pulse repetition frequency radars and the transmitters in frequency modulated continuous wave radars use a special form of modulation. Often simple, short pulses are used to cover the volume that is hidden by the increased dead time of a stretched transmitter pulse.

The spectrum from the transmitter, or array of transmitters, depends on how it is modulated and pulse compression, described in Section 3.4, is often used to send enough power in the pulse.

3.3 SPECTRUM AND SIDEBANDS

Crowded radar bands require that the radar transmits only in its allocated band. As more radars are used, these bands become more restricted, and greater control over the spectrum is required. In addition to the radar carrier frequency, the (pulse) modulation of the radar transmitter creates sidebands around the carrier frequency in addition to any modulation sidebands caused by modulation within the pulse.

The sidebands from pulse radars are based on two structures. The dominant coarse structure is determined by the pulse width, τ_t seconds, and gives the frequency spectrum

$$\textit{Coarse spectrum} = \frac{\sin(\pi f \tau_t)}{\pi f \tau_t} \tag{3.1}$$

Equation (3.1) is the envelope for the individual spectral lines spaced at the pulse repetition frequency. If the time between the pulses is T_t seconds, then each line is a spectrum of the form

$$\textit{Fine spectrum} = \frac{\sin\left(\pi f \frac{\tau_t}{T_t}\right)}{\pi f \frac{\tau_t}{T_t}} \tag{3.2}$$

The envelopes of the peaks are given by $\frac{1}{\pi f \tau_t}$ and $\frac{T_t}{\pi f \tau_t}$. The factor $f\tau_t$ always appears together so the spectra may be plotted in terms of frequency times pulse width. This is shown in decibels for a transmitter with a pulse width of 1 μs and pulse repetition frequency of 10 000 Hz, $T = 100$ μs in Figure 3.2.

Radar transmitters produce peak powers often in the order of megawatts and the decibel scale with respect to 1 MW corresponds to Table 3.1.

Table 3.1
Attenuation required for screening against a 1 MW transmitter

Decibels	Power	Decibels	Power
0	1 MW	90	1 mW
30	1 kW	120	1 μW
60	1 W		

These large powers easily overload the amplifiers in nearby electronic equipment, so very good screening is required. Where the interference is caused by the signal from the antenna, the antenna gain, up to maybe 30 dB, must be taken into account. The great dynamic range between the peak transmitter power and receivers causes spectral components to be detectable at frequencies well away from the transmitter frequency. In contrast to antennas and filters, which normally have rounded peaks, the most efficient transmitter pulse is rectangular because the transmitter often runs in a saturated mode. The leading and trailing edges are the only area of freedom for control of the spectrum to reduce interference at neighboring frequencies.

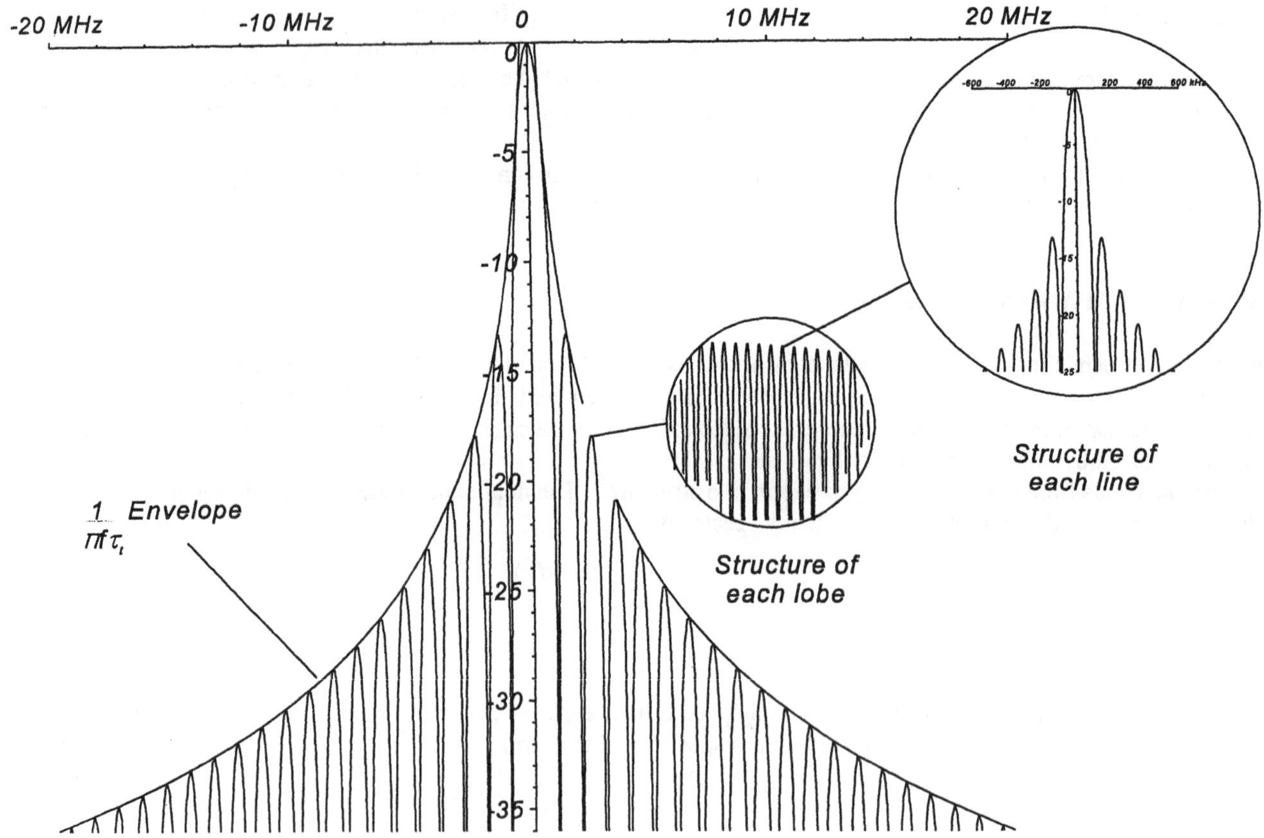

Figure 3.2 The levels of the structure of a pulse spectrum.

Commonly with amplifying output stages, a high-voltage pulse is applied for a time slightly longer than the transmitter pulse so the electron beam remains stable during the radio frequency drive pulse. This drive pulse is often large enough to saturate the output stage, but the leading and trailing edges may be shaped to control the output spectrum.

The spectra of the pulse waveforms may be constructed from the sum of the spectra of the components. Namely,

$$\begin{aligned} \text{Complete spectrum} = & \text{ Spectrum of leading edge} \\ & + \text{spectrum of constant center section} \\ & + \text{spectrum of trailing edge} \end{aligned} \tag{3.3}$$

It is assumed that the center section is symmetric about zero time to give a symmetrical, real spectrum. The waveforms of the edges are offset in time to fit the leading and trailing edges of the center section. Each of these edge spectra is complex, and the imaginary parts of these two spectra add to zero. Thus, the spectrum of the symmetrical composite pulse has real components only. In the following discussions, all the diagrams are normalized for a pulse width of unity. The spectra are plotted in terms of the frequency-time product, $f\tau_t$.

3.3.1 Trapezoidal edges

The first attempts to control the transmitter spectra used trapezoidal pulse waveforms which are shown in a normalized form in Figure 3.3. The width of the pulse, $\tau_{t\ av}$, at the 50% voltage point is unity and the widths of the edges, $\tau_{t\ edge}$, are expressed as a percentage.

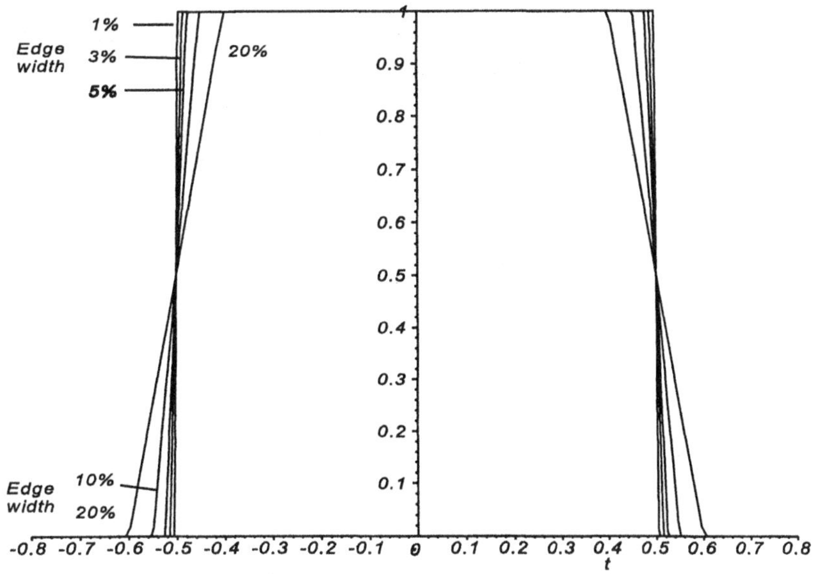

Figure 3.3 Normalized trapezoidal waveforms with edges from 1% to 20%.

The spectra from these shaped pulses are given by [1]

$$Trapezoidal\ spectrum = \frac{\sin(\pi f \tau_{t\ av})}{\pi f \tau_{t\ av}} \frac{\sin(\pi f \tau_{t\ edge})}{\pi f \tau_{t\ edge}} \tag{3.4}$$

Spectra for 1%, 3%, 5%, 10%, and 20% edges are shown in Figure 3.4.

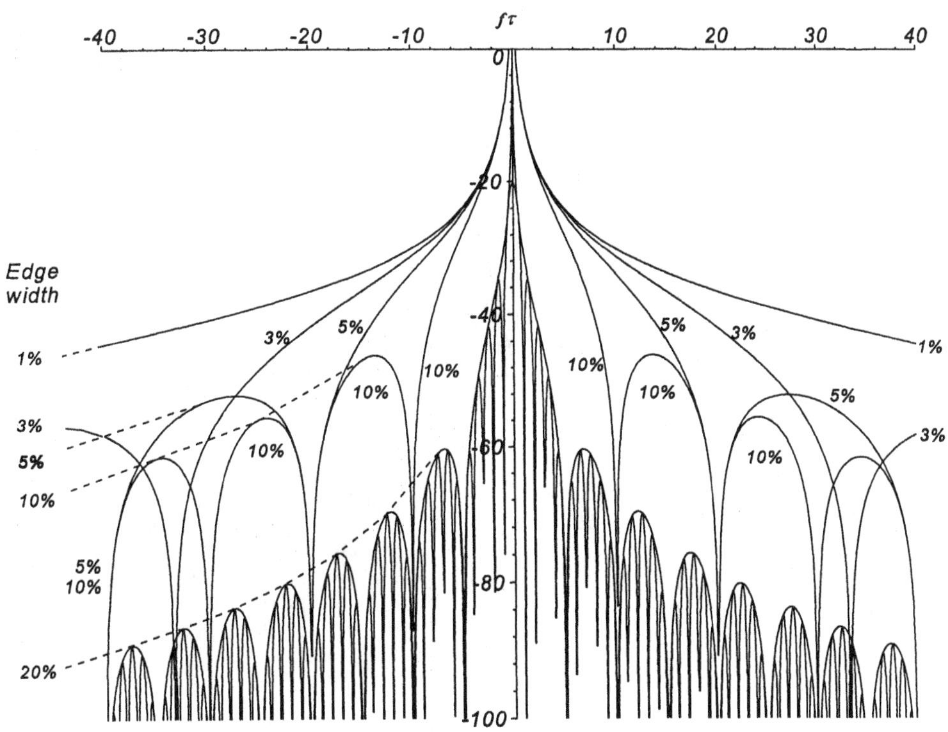

Figure 3.4 Normalized spectra of trapezoidal pulses with edges from 1% to 20%.

3.3.2 Cosine and cosine squared edges

Corners in any waveform produce sidebands. Cosine edges remove the corners at the top of the waveform and cosine squared edges also remove the corners at the bottom, as shown in Figure 3.5. These spectra have been obtained by performing the Fourier transform numerically. Higher cosine powers do not improve matters. The spectra with cosine edges for edge widths of 1%, 3%, 5%, 10%, and 20% are shown in Figure 3.6. The line above the 1% edge line is the spectrum without shaping. Figure 3.7 shows the spectra for cosine squared edges.

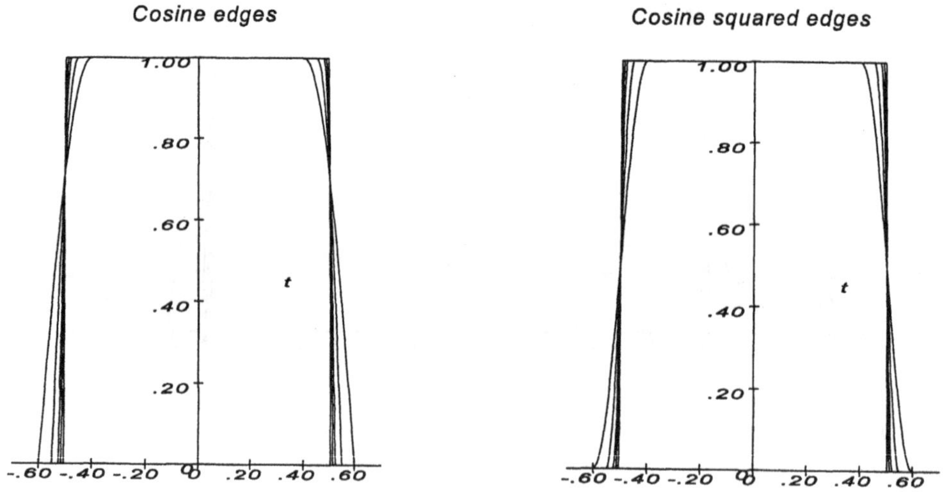

Figure 3.5 Pulses with normalized cosine (left) and cosine squared (right) edges of 1%, 3%, 5%, 10%, and 20%.

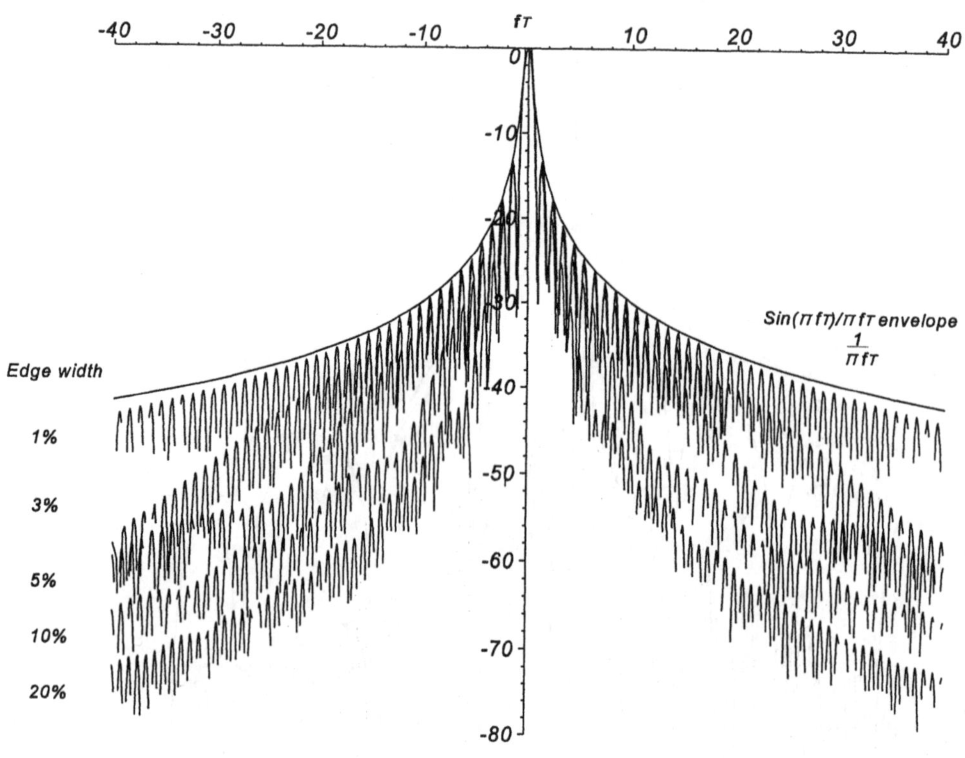

Figure 3.6 Normalized spectra for pulses with cosine edges, widths 1%, 3%, 5%, 10%, and 20%.

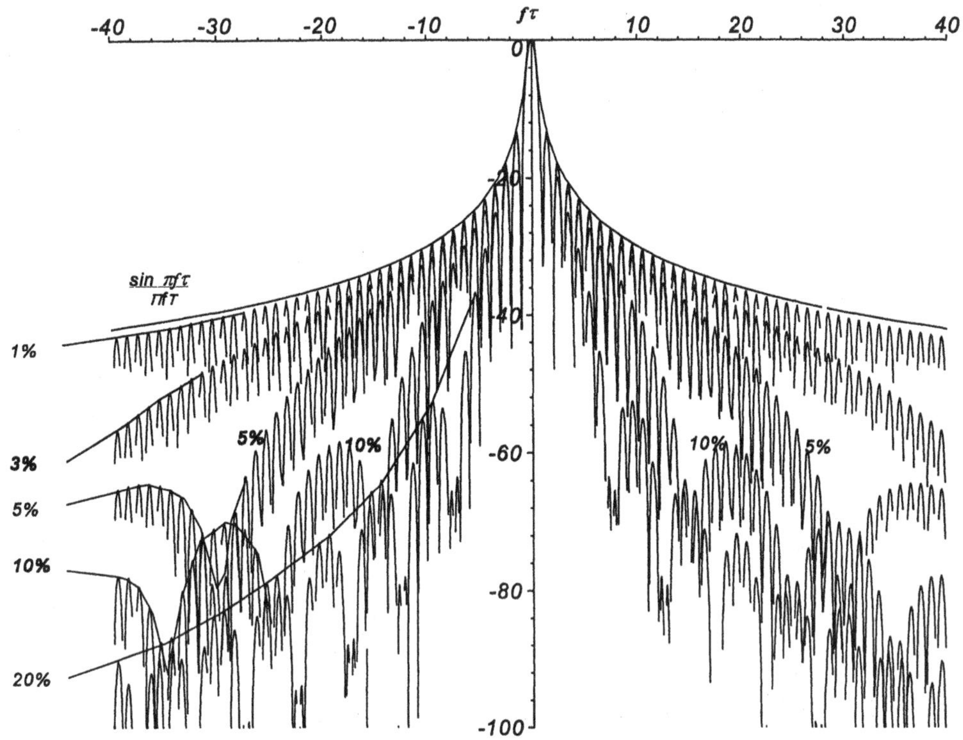

Figure 3.7 Normalized spectra for pulses with cosine squared edges, widths 1%, 3%, 5%, 10%, and 20%.

3.3.3 Extra modulator power needed for shaping

The beam in the output tube must be fully switched on and stabilized before the shaped radio frequency drive starts. Figure 3.3 and Figure 3.5 also show the extra time during which the modulator must deliver full power. During the pulse edges the modulator is delivering full power but the transmitter delivers only the edge power at its output.

The power output of the tube is proportional to the square of the drive voltage, and the power input is proportional to the time that the modulator is switched on.

Efficiency is defined in (3.5) and is plotted in Figure 3.8.

$$Efficiency = \frac{1}{\tau_{av} + \tau_{flank}} \int_{\frac{-\tau_{av} - \tau_{flank}}{2}}^{\frac{\tau_{av} + \tau_{flank}}{2}} (drive\ modulation\ voltage)^2\ dt \qquad (3.5)$$

Control of the spectrum by shaping the driving signal comes at an extra cost. The modulator consumes a large proportion of the power used by the larger radars. A modulator provisionally sized for 5 kW must be increased to 5.556 kW to compensate for a 10% loss. There is extra manufacturing cost, the extra 556 W is consumed throughout the life of the radar, and extra cooling is required.

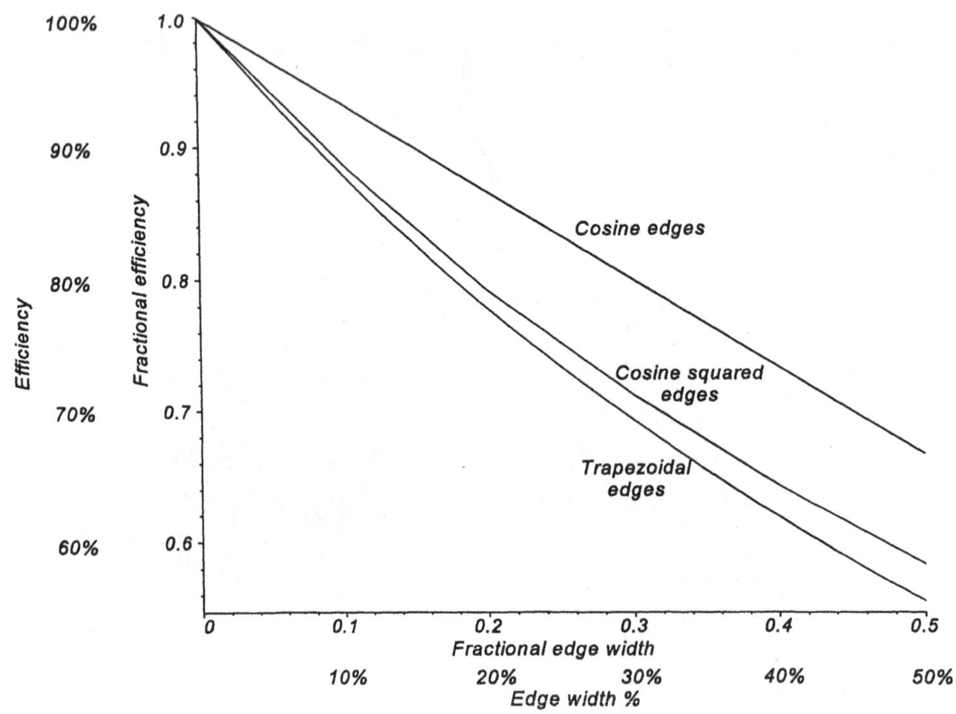

Figure 3.8 Output stage efficiency with shaping of pulse edges.

3.4 PULSE COMPRESSION

Pulse compression has gained importance since the practical limit for the voltage of an output klystron is about a quarter of a million volts, and today pulse compression is essential because cost-effective semiconductor output stages often do not produce the peak powers necessary to cover the specified range for the radar. Radars have been made and have used pulse compression to increase the ratio of wanted echoes to clutter and to exploit the gaps in the clutter (see Figures 1.9 and 1.10) as detection depends on the ratio of the wanted echo to clutter signal inside the volume of the transmitted pulse in space. At that time longer range radars used pulses of lengths 3 μs to 6 μs that bridged 450 m to 900 m gaps in the clutter, whereas a 50 ns pulse covers only 75 m in range, the length of a larger aircraft. Compared to a radar with a 6 μs pulse, there is a reduction of a factor of 120 or 20.8 dB in clutter or rain echo.

The radar pulse masks 150 m in range for each microsecond in length plus the time taken by the transmit-receive switch to switch. Before pulse compression large radars used pulses typically between 3 μs and 6 μs in length, giving minimum ranges of 450 m to 900 m plus the switching time and a similar range resolution. For an airport radar with a 70 μs pulse to achieve 60 nautical miles (111 km) range, the pulse covers 5.7 nautical miles (10.5 km) in range so that this inner volume must be covered by short pulses placed in between the long pulses (interleaving). The short pulses mostly have the same bandwidth as the compressed long range pulses so that the same hardware may be used in the receiver and signal processor. In modern radars pulse lengths of hundreds of microseconds are used for ranges outside clutter and may occupy up to 50% of the range so that shorter pulses with a higher pulse repetition frequency must be interleaved to cover the shorter ranges and to allow coherent clutter suppression (moving target indication or detection). The extra equipment has to be built into the transmitter (this chapter) and acts as the matching filter (Chapter 8) that reconstructs the original narrow pulse. A block diagram is shown in Figure 3.9.

On the transmission side the pulse is expanded at intermediate frequency, and frequency changing stages change the signal to the radar frequency. On the receiving side the long pulse may be compressed by the matching filter being placed after the receiver, for modulated pulses after analogue to digital conversion, or after a signal processor with a limited dynamic range. Theoretically with linear signal processing the boxes may be placed in any order. Pulse compression stages have a delay and the output has time sidelobes before and after the (delayed) main narrow peak that can lead to the masking of a weak echo by a strong echo within the long transmitter pulse time. Time sidelobes may be reduced by tapering (see Appendix B) which gives a loss and widens the compressed pulse or by predistorting the long pulse to have a larger bandwidth, so there is a price to be paid.

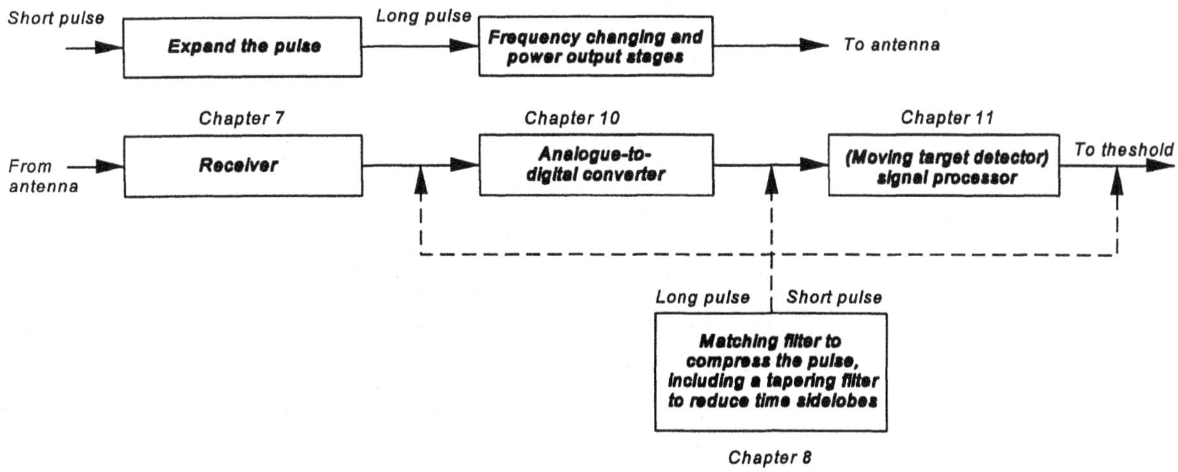

Figure 3.9 The components required for pulse compression.

Increased resolution allows an improved ability to explore objects in the radar beam and to improve the ability to see aircraft flying through weather echoes or chaff (window). Stretching and modulating the transmitter pulse gives additional advantages to military radars as radar reconnaissance or warning receivers in the past relied on the large peak transmitter pulse powers and their easy recognition to achieve their sensitivity. A long modulated pulse has much less amplitude and possibly unknown, changing, modulation so that it is difficult to find a matching filter to integrate the pulses and achieve full sensitivity. Other examples are the use of very wide bandwidths to keep the spectral density too small for reconnaissance receivers at longer ranges. Low probability of intercept (LPI) radars use these concepts leading to an extreme case where the radar transmits a long random coded pulse continuously and uses sophisticated processing to recognize echoes in range. Unlike data communications systems that can take time to resynchronize and can request a retransmission, each of the underlying short pulses in the received echoes must be reconstructed coherently and normally be coherent from pulse to pulse for the signal processing that follows. There are, though, types of radar that can quickly move their antenna beam to have another look (or backscan) or to verify what was seen in a volume of interest.

In order to keep the transmitter at full power during the pulse, frequency or phase modulation is used. The types of modulation and spectra are listed in this chapter and the compression or recognition and the losses involved are discussed in Chapter 8.

3.4.1 Linear frequency modulation

One of the first ways to make the range resolution shorter than the pulse width is to change the frequency during the time of the pulse, called linear frequency modulation. The frequency sweep extends over a bandwidth B that results in a range resolution $1/B$ where there is no tapering. The pulse width times the frequency sweep is called the time-bandwidth product, τB. If the time waveform is normalized to exist between times of $-1/2$ and $+1/2$ and the frequency f is a fraction of the time-bandwidth product, then, in exponential form [2],

$$\textit{Modulating waveform} = \exp(j\ \pi\tau B t^2) \qquad \text{for } -1/2 < t < +1/2, \text{ otherwise } 0 \tag{3.6}$$

The spectrum is given by the Fourier transform

$$\textit{spectrum} = \sqrt{\tau B} \int_{-\tau/2}^{+\tau/2} \exp(j\ \pi\tau B t^2)\ \exp(-j\ 2\pi f\tau B t)dt \tag{3.7}$$

The spectrum is complex. The normalized spectrum for a time-bandwidth product of 100 is shown in Figure 3.10. The spectrum exists between -50 and +50, giving a time resolution 1/100 of the pulse width.

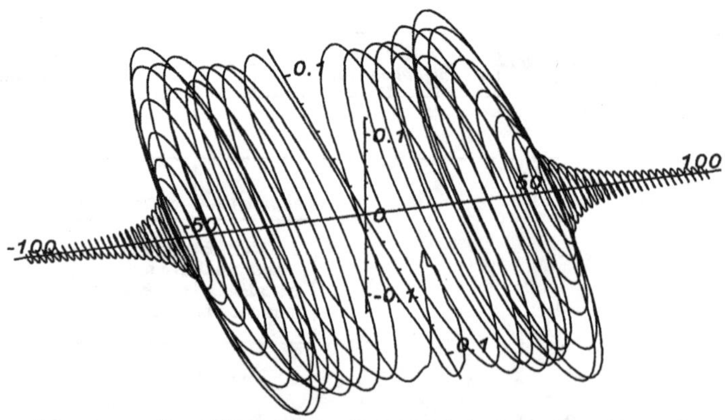

Figure 3.10 Normalized spectrum for a linear frequency modulated pulse centered on zero time with a time-bandwidth product of 100.

The spectrum is calculated for times between -½ and ½ or half in negative time. A spiral in a spectrum represents a time delay: the tighter the spiral, the greater the delay. Figure 3.10 shows a spiral that reverses at the center, alternatively the pulse energy is put forward and retarded to be grouped at zero. In practice, a filter can only retard energy, so with this time delay the spectrum becomes that shown in Figure 3.11. The modulus of the spectrum in Figure 3.10 is shown in Figure 3.12 for time-bandwidth products of 3, 10, 30, 100, and 1 000, with the highest time-bandwidth products giving the flattest spectra. The far sidebands which interfere with other equipment are shown in Figure 3.13.

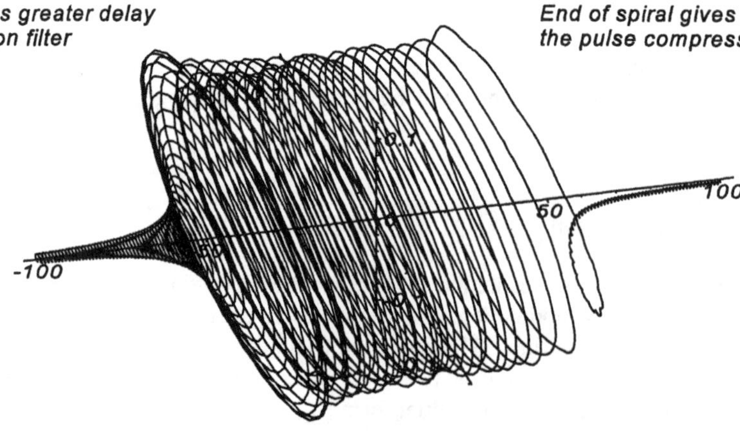

Figure 3.11 Normalized spectrum for a frequency modulated pulse with a time-bandwidth product of 100 with time delay.

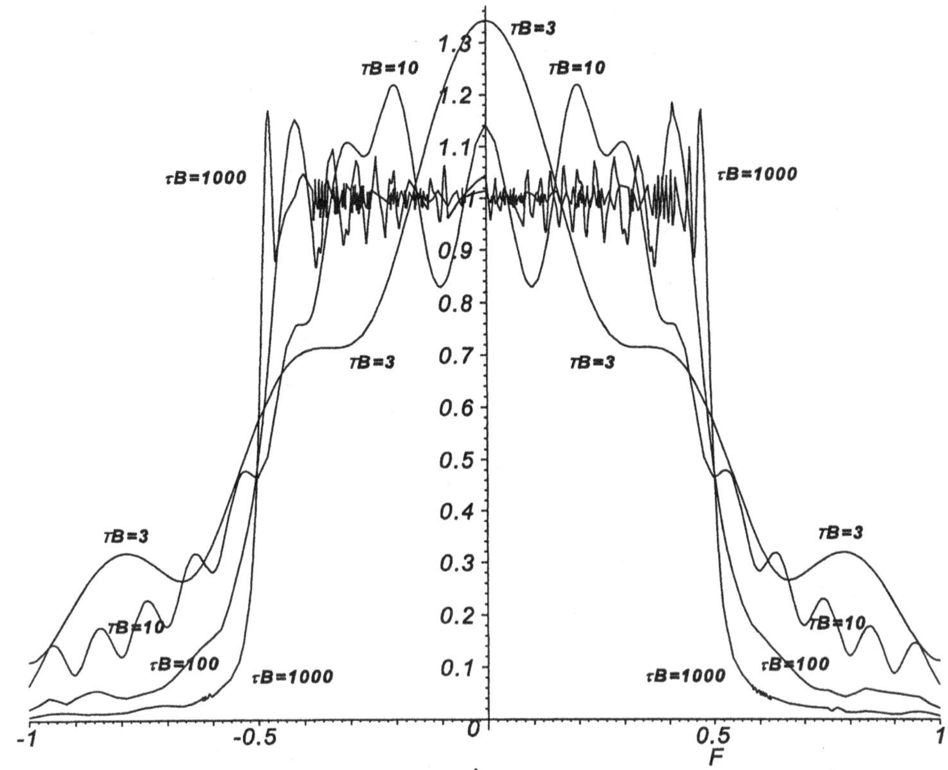

Figure 3.12 The near sidebands for time-bandwidth products of 1 000, 100, 10, and 3.

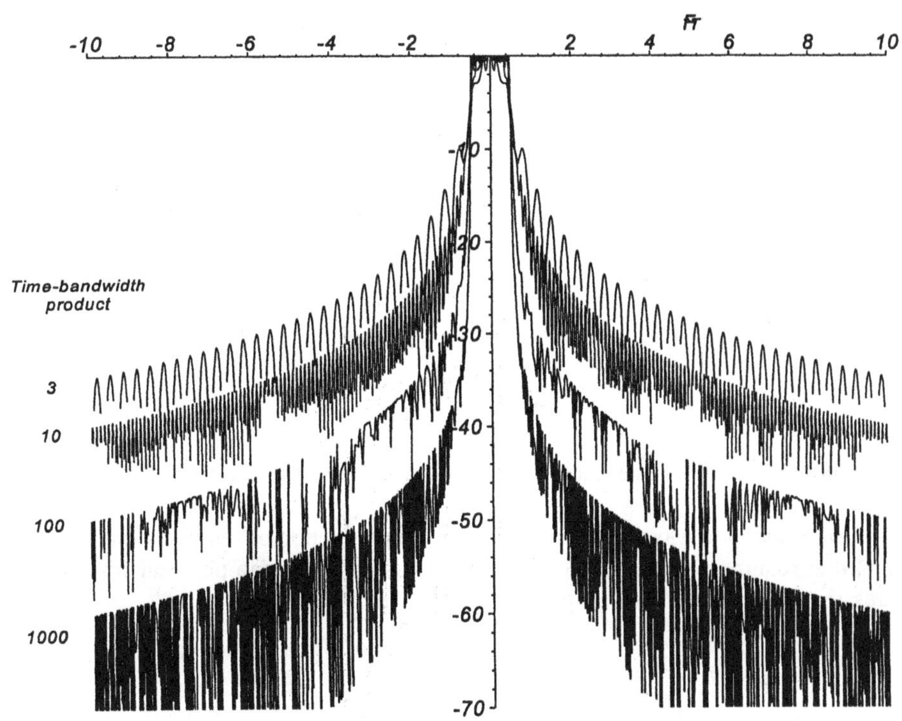

Figure 3.13 Far sidebands for time-bandwidth products of 1 000, 100, 10, and 3.

Dispersive networks for linear frequency modulation or chirp were expensive and difficult to adjust as they used inductors and capacitors working in the short wave band. A single network for both transmission and reception may be used in radars with a single transmitter and receiver but the complex conjugate of the normal echo signals is necessary for recombination. Figure 3.14 shows a scheme that takes advantage of the fact that the sidebands from a mixer are complex conjugates [1, p. 14].

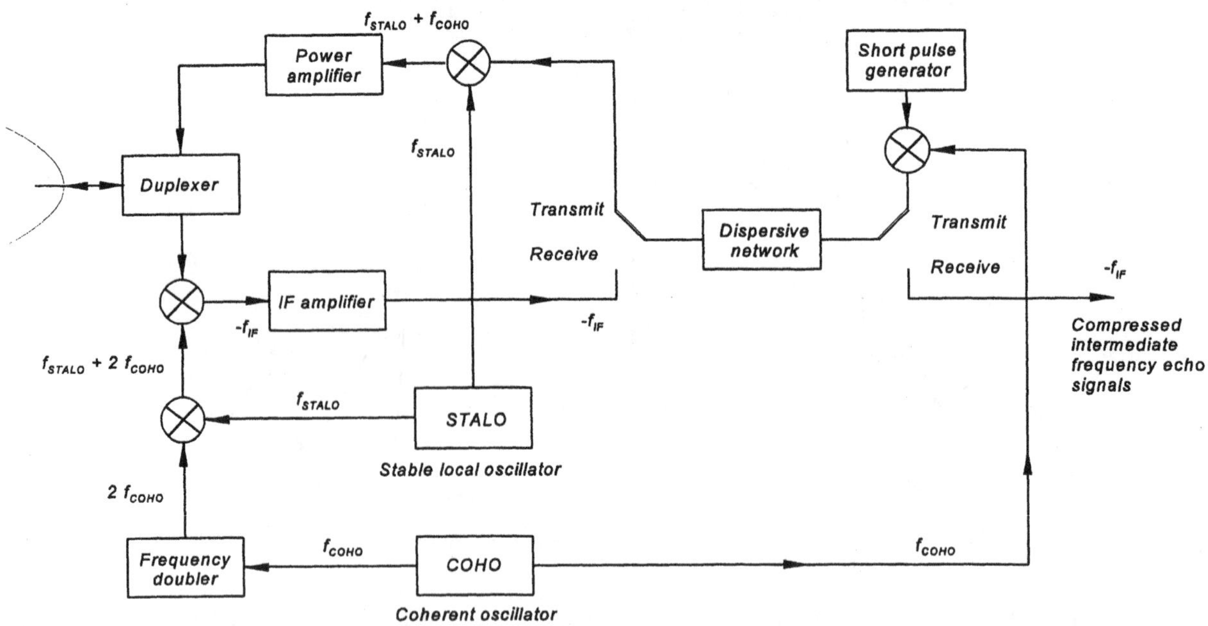

Figure 3.14 Frequency changing in a frequency modulated pulse compression radar with a common dispersive network.

The transmitter signal frequency is the coherent oscillator (COHO) signal plus the stable local oscillator (STALO) frequency:

$$\textit{Transmitter frequency} = \textit{Echo signal frequency} = f_{STALO} + f_{COHO} \tag{3.8}$$

The returning echoes signals are mixed with a frequency consisting of the local oscillator frequency plus twice the intermediate frequency, namely,

$$\begin{aligned}\textit{Echo signal intermediate frequency} &= (f_{STALO} + f_{COHO}) - (f_{STALO} + 2f_{COHO}) \\ &= -f_{COHO}\end{aligned} \tag{3.9}$$

The minus sign attached to the frequency indicates an opposite rotation or it is modulated by complex conjugate of the echo signal.

The development of surface acoustic wave filters has allowed the production of relatively cheap and reproducible passive dispersive filters, that require no maintenance, so that the use of the same physical filter for expansion on transmission and compression on reception is no longer necessary [2, p. 1073].

3.4.2 Simple phase modulation

A commonly used type of phase modulation is the Barker code. Phase shifts of 180 degrees are applied between code elements. The Barker codes are given in Table 3.2.

Table 3.2
Barker codes

Code length	Barker code
2	[+1, -1]
3	[+1, +1, -1]
4	[+1, +1, -1, +1]
5	[+1, +1, +1, -1, +1]
7	[+1, +1, +1, -1, -1, +1, -1]
11	[+1, +1, +1, -1, -1, -1, +1, -1, -1, +1, -1]
13	[+1, +1, +1, +1, +1, -1, -1, +1, +1, -1, +1, -1, +1]

The waveform for the phase modulation of a 13 element Barker code is shown by the dotted line in Figure 3.15 with the modulated signal. The spectrum of the modulation of the transmitted pulse with the pulse centered on zero time is the sum of the spectra of the individual elements displaced in time:

$$Spectrum(f) = \sum_{i=-\frac{L_{phase}-1}{2}}^{\frac{L_{phase}-1}{2}} \int_{i-1/2}^{i+1/2} \exp\left(j\,\frac{\pi}{2}\,L_{code}\left[i + \frac{L_{phase}-1}{2} + 1\right]\right) \exp(j\,2\pi ft)\,dt \tag{3.10}$$

where L_{phase} is the length of the phase code;
L_{code} is the list containing the phase shift;
I is the counter pointing to the phase code element;
τ is the pulse width of the chip, *s*;
f is the frequency variable.

The term $f\tau$ appears in the integral in (3.10) and the normalized abscissae of Figure 3.16 and Figure 3.17 are in terms of $f\tau$. The time waveform is not symmetrical, so the spectrum (3.10) is generally complex. The complex plot (amplitude and phase) of the center of the spectrum with the time waveform centered on zero is shown in Figure 3.16.

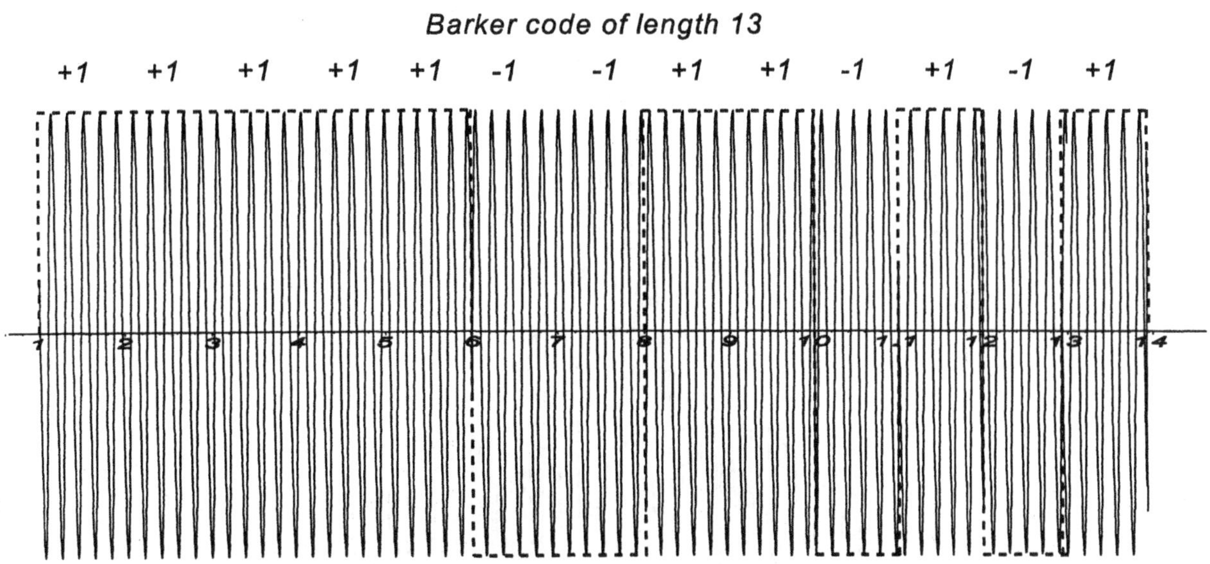

Figure 3.15 The 180 degree phase changes in a Barker coded pulse (dotted line) and a phase modulated sine wave.

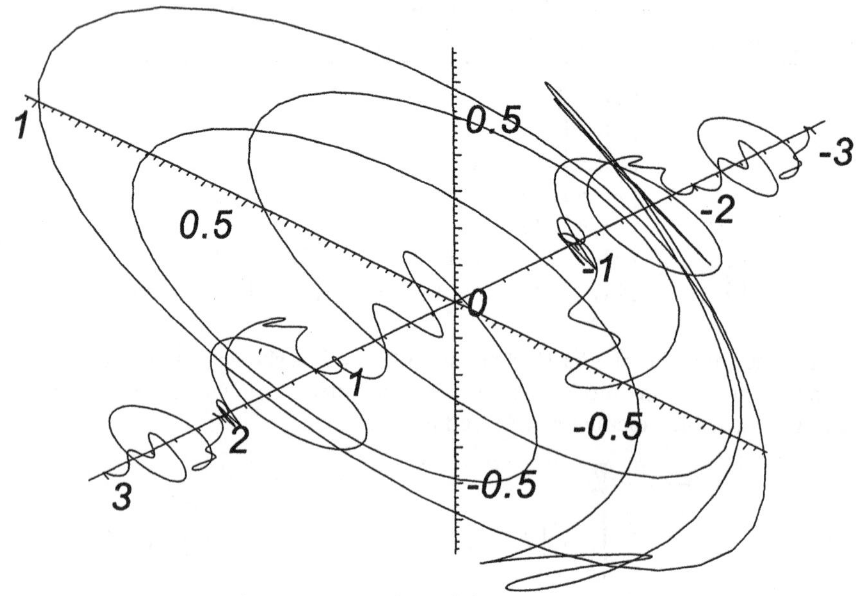

Figure 3.16 A complex representation of the spectrum of a Barker coded pulse, length 13.

Figure 3.17 shows the far sidebands with the $1/\pi f\tau$ line, as mentioned in [3]. This is almost the same as a rectangular pulse with the width the same as a phase element. The peaks of the lobes consist of two parts: a main lobe of width $1/f\tau$ and a tip of width $1/nf\tau$ where n is the number of elements. The tip is wider with lesser compression ratios.

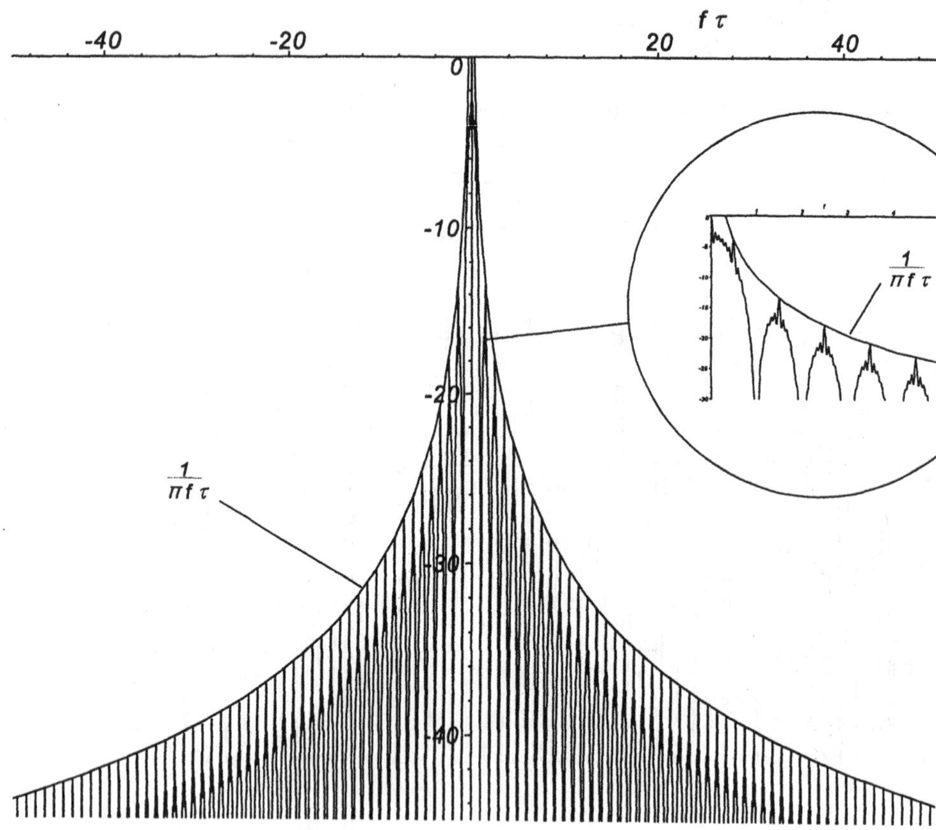

Figure 3.17 The far sidebands for a 13 element Barker coded pulse in terms of $f\tau$.

3.4.3 Other types of modulation and their spectra

Much effort has been expended to find codes that are as simple as the Barker codes, give a narrow spike with low range sidelobes on recompression, have little degradation for echo signals with Doppler frequency shift, and can be implemented using digital circuitry. Digital circuitry is reproducible using standard hardware with no adjustment necessary, and it is more flexible as changing the clock frequency changes the pulse compression ratio and switching the delays may be used to change the length.

In the search for codes that may be recombined to give narrow compressed pulses, a number of polyphase codes have been developed. These codes have low time sidelobes without resorting to tapering giving the resulting widening and losses.

In the sections that follow a number of codes are described that are used for pulse compression with examples with lengths of 16 and 100. For completeness the diagrams start with binary codes in Figure 3.18.

3.4.3.1 Binary code examples

Figure 3.18 shows a 13 element Barker code in comparison with a 100 element random binary code.

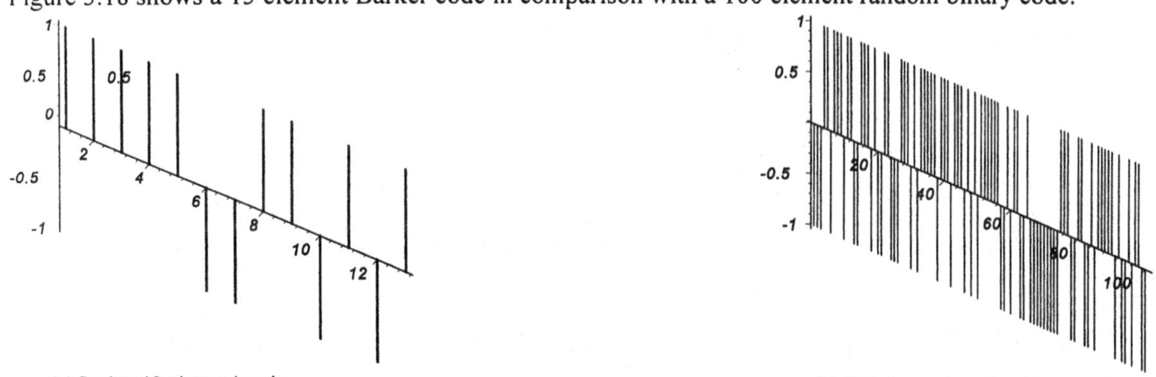

(a) Barker 13 element code *(b) 100 element random binary code*

Figure 3.18. Two examples of the modulation for binary codes.

Binary codes may be stored in read only memory (ROM). A number of long codes with acceptable peak sidelobes may be found in [3, p. 108] or be generated using shift registers and exclusive OR blocks as shown in the example in Figure 3.19 for a maximum length sequence, pseudo-random sequence or Galois code [4, p. 693; 5, p. 242] of length 15 (four steps, length $2^4 - 1$). The code in Figure 3.19 is

1, 1, 1, 1, 0, 0, 0, 1, 0, 0, 1, 1, 0, 1, 0

and phase shifts are $\pi, \pi, \pi, \pi, 0, 0, 0, \pi, 0, 0, \pi, \pi, 0, \pi, 0$ radians.

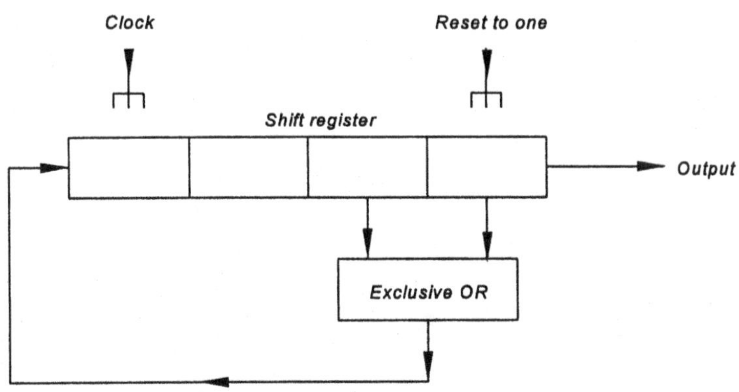

Figure 3.19 An example of the generation of maximum length, pseudo-random, or Galois codes.

Other binary codes can to be found in [6, p. 457].

A simple tapped transformer may be used to split the coherent oscillator signal into an in-phase and antiphase signal, and a semiconductor switch is used to choose the phase as in Figure 3.20.

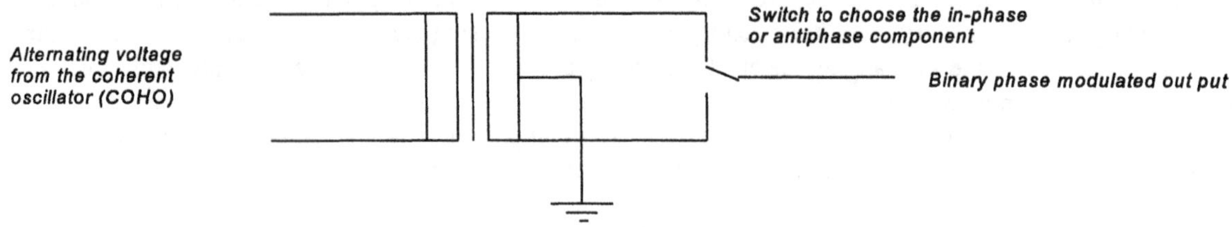

Figure 3.20 One way of modulating a binary coded signal.

Where more than one phase is used, the modulating signals are generated either by the systems shown below or are held in memory as two phase signals (I and Q). In both cases a vector modulator is used to generate the waveform for the transmitter as shown in Figure 3.21.

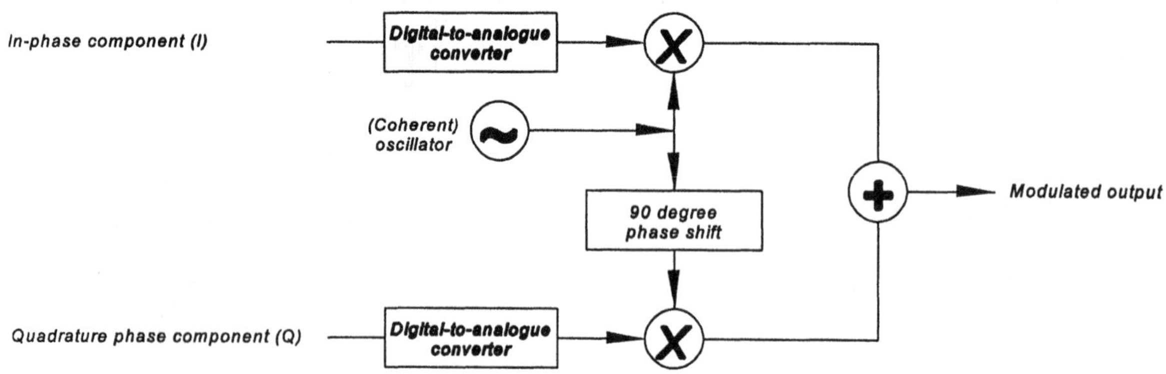

Figure 3.21 A two-phase vector modulator.

Binary codes are sensitive to Doppler frequency shifts and their ambiguity functions are discussed in Chapter 8. Polyphase codes described in the next sections bridge the gap between binary codes and linear frequency modulation and have been constructed so that no tapering is required in the compression stages (Chapter 8).

3.4.3.2 Frank code

The Frank code was the first code to give an accelerating phase using digital components. The phase of each element is given by

$$\phi_{Frank} = \frac{2\pi}{N}(i-1)(j-1) \tag{3.11}$$

where the indices i and j are valued from 1 to N. The compression ratio ρ is N^2.

For $N = 4$ the phases are

$$0,0,0,0, \quad 0, \pi/2, \pi, 3\pi/2, \quad 2\pi, 3\pi, 4\pi, 5\pi, \quad 6\pi, 8\pi, 10\pi, 12\pi$$

and it can be seen that these phases are added to the initial frequency f_0 to give blocks of accelerating phase shown in Figure 3.22.

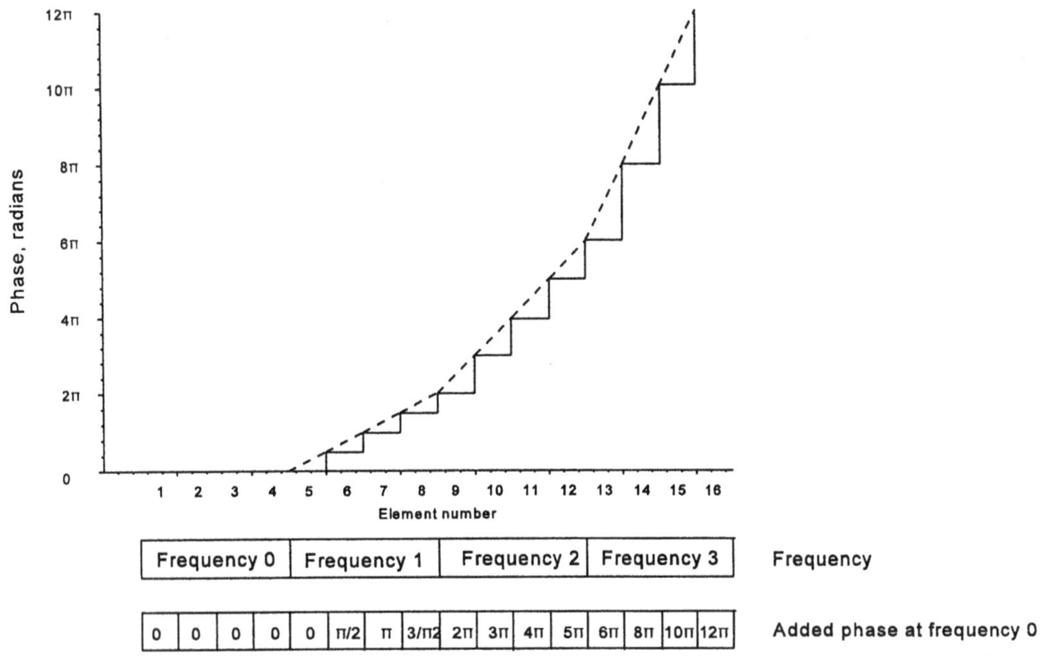

Figure 3.22 The accelerating phase and frequency blocks in a 16 element Frank code ($N = 4$).

The modulation for the 16 element (N =4) code in Figure 3.22 is shown in Figure 3.23(a). With greater pulse compression ratios, $\rho = 100$ and $N = 10$, in Figure 3.23(b) and (c). N phases are used.

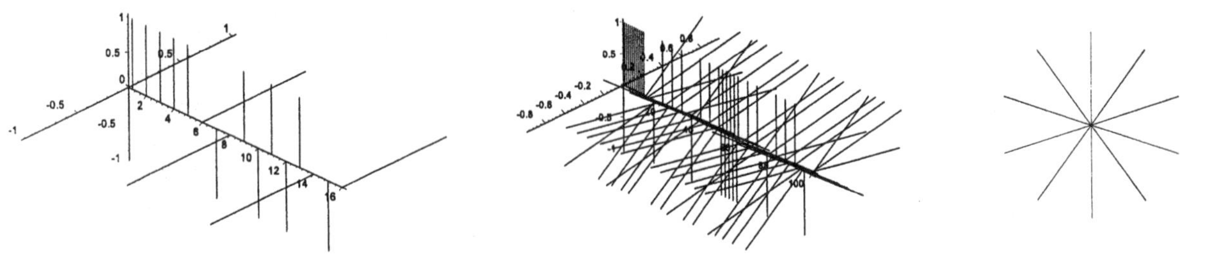

(a) 16 element Frank code $N = 4$ (b) 100 element Frank code $N = 10$ (c) Phases in a 100 element Frank code

Figure 3.23 Examples of the modulation used for Frank codes.

The Cartesian values (I and Q) may be stored in a read only memory (ROM) but historically digital circuit components used for the fast Fourier transform may be used to expand and compress pulses; an example is shown in Figure 3.24. The components are delay elements for one chip width (top) and four chip widths (bottom), multiplier and adder blocks denoted by x +, and adders, +. Where complex multiplication is used, denoted by j at the inputs, both the I and Q phases must be implemented: the conjugates of the echo signals are always complex. Synchronous logic is used and the clock frequency may be changed as necessary for each pulse length.

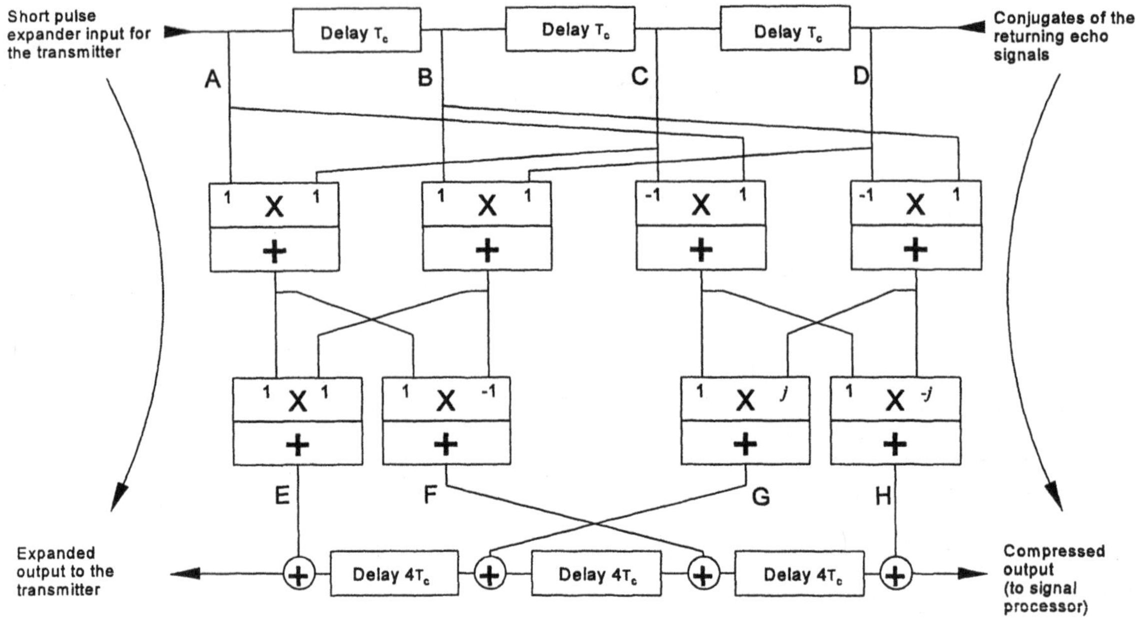

Figure 3.24 A circuit to expand and compress signals for a 16 element ($N = 4$) Frank code.

To expand the pulse a single pulse is clocked along the top row of delays implemented as a shift register. Before entering the first delay, the pulse is present at the input of the first and third multipliers from the left where they are both multiplied by the weighting factors +1 and are passed to the adder so that there are two pulses present simultaneously at the outputs of the first row of multipliers and adders. The second row of multipliers includes weights of j and $-j$ and the adders combine the weighted inputs to give an output for each position of the pulse in the top row shift register.

The bottom row consists of three shift registers that have a delay of four clock pulses so that for the first four clock pulses the outputs are 1 from the leftmost multiplier and adder block. The next group of four pulses come from the third multiplier-adder, as would be the case with a four point fast Fourier transform circuit and have the values $1, j, -1, -j$ or phases angles $0, \pi/2, \pi, 3\pi/2$ radians. Note that the add symbol means .OR. as there is only one input to each of the multiplier-adder blocks.

The complete run of 16 phases is thus

$$0, 0, 0, 0 \qquad 0, \pi/2, \pi, 3\pi/2 \qquad 0, \pi, 2\pi, 3\pi \qquad 0, 2\pi, 3\pi, 4\pi$$

The Frank code does not tolerate bandwidth clipping in the transmitter and receiver circuits [1, p. 15], and attempts to overcome this led to the development of the U.S. Naval Research Laboratory P-codes [1]. The first was the P1 code, see [1, p. 68].

3.4.3.3 P1 code

With the P1 code the frequency groups are rearranged in time to place the smallest phase jumps in the center of the pulse to give the phase run, giving

$$\phi_{P1} = -\frac{\pi}{N}\,(N - (2j - 1))\,(N(j - 1) + (i - 1)) \tag{3.12}$$

where the indices i and j are valued from 1 to N. The compression ratio ρ is N^2.

P1 code sequences for 16 and 100 elements are shown in Figure 3.25.

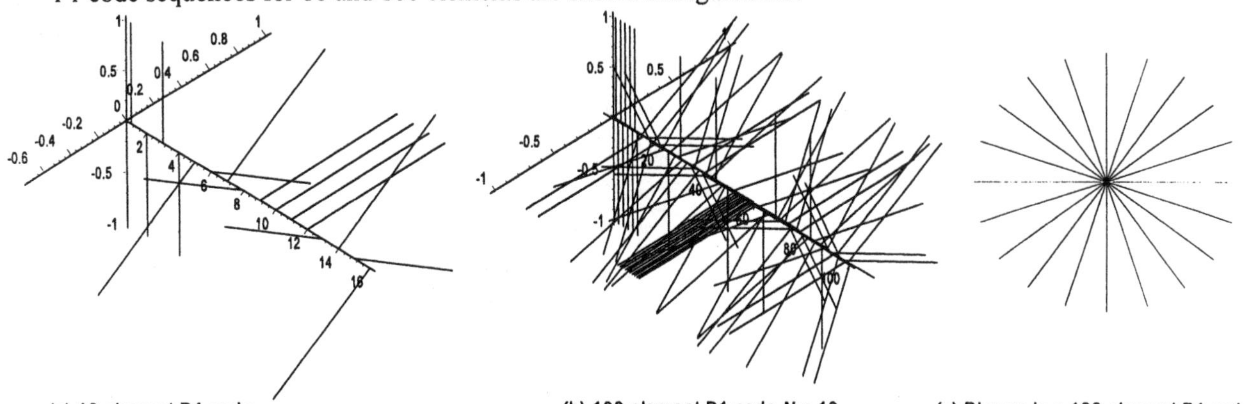

(a) 16 element P1 code (b) 100 element P1 code N = 10 (c) Phases in a 100 element P1 code

Figure 3.25 P1 code phase sequences for lengths of 16 and 100 elements.

3.4.3.4 P2 code

The P2 code is a modification of the P1 code using the weighting in a Butler matrix for antennas to give a symmetrical code and has the bandwidth tolerance of the P1 code. The phases of the individual elements are given by

$$\phi_{P2} = \left(\frac{\pi}{2}\, \frac{N-1}{N} - \frac{\pi}{N}\,(i-1) \right) (N+1-2j) \tag{3.13}$$

where the indices i and j are valued from 1 to N. The compression ratio ρ is N^2.

The phases of the modulation for 16 and 100 element P2 codes are shown in Figure 3.26.

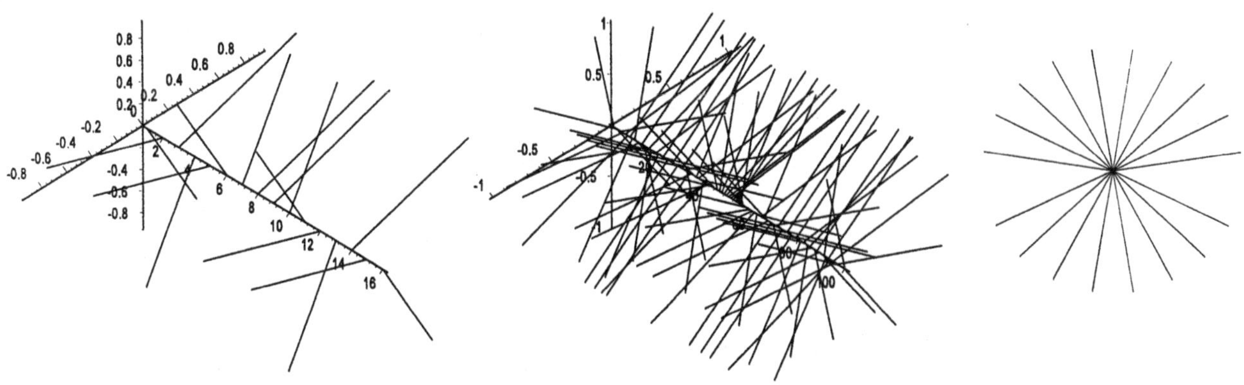

(a) 16 element P2 code (b) 100 element P2 code N = 10 (c) Phases in a 100 element P2 code

Figure 3.26 The modulation phases for 16 and 100 element P2 codes.

3.4.3.5 P3 code

Linear frequency modulation has much better Doppler frequency characteristics than stepped frequency modulation. The phases of the elements are then given by

$$\phi_{P3} = \frac{\pi\,(i-1)^2}{\rho} \tag{3.14}$$

where the index i is valued from 1 to the compression ratio ρ.

The accelerating phases are shown for 16 and 100 element codes in Figure 3.27.

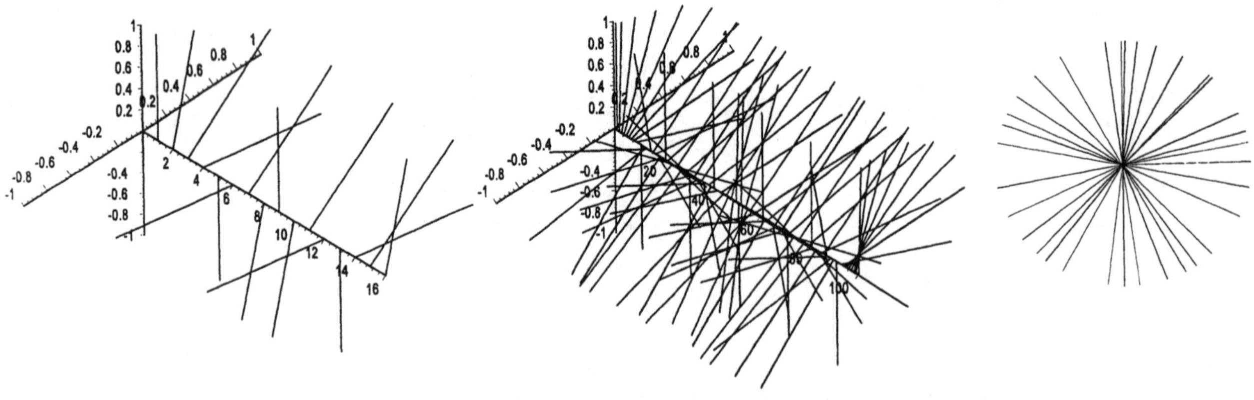

(a) 16 element P3 code (b) 100 element P3 code ρ = 100 (c) Phases in a 100 element P3 code

Figure 3.27 The phases for the modulation for 16 and 100 element P3 codes.

The run of phases can be better illustrated by a line joining the phase vectors as in Figure 3.28.

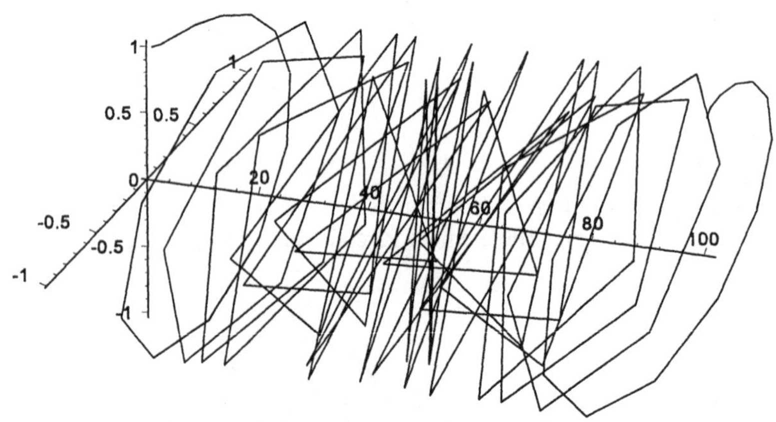

Figure 3.28 The run of phases in a 100 element P3 code.

3.4.3.6 P4 code

The P4 code has the same relationship to the P3 code as the P1 code has to the Frank code and is given by

$$\phi_{P4} = \frac{\pi\,(i-1)^2}{\rho} - \pi(i-1) \tag{3.15}$$

where the index i is valued from 1 to the compression ratio ρ.

The phase changes are greatest at the start and end of the expanded pulse, and phase vectors for 16 and 100 element P4 codes are given in Figure 3.29.

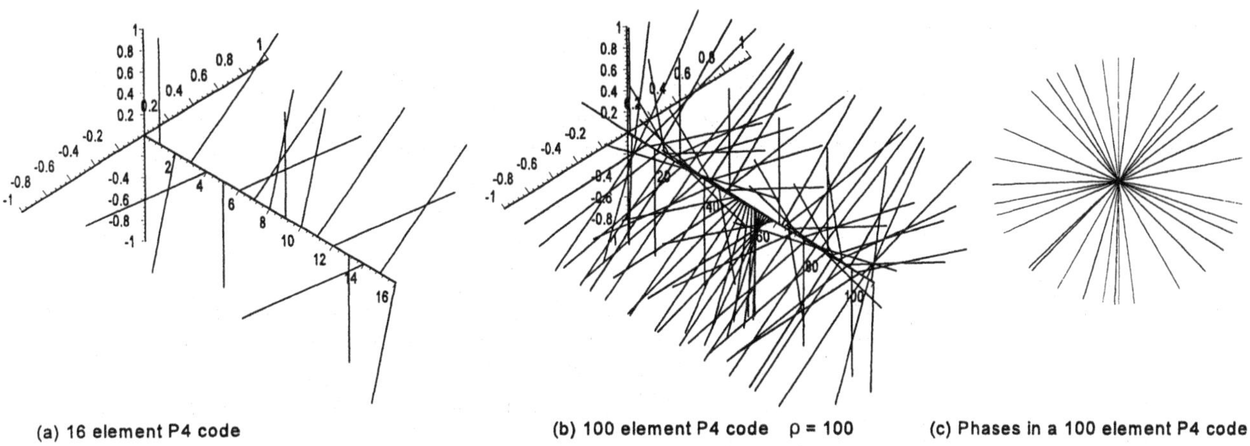

Figure 3.29 The phase vectors for 16 and 100 element P4 codes.

Joining the tips of the phase vectors shows its relation to the P3 code and is shown in Figure 3.30.

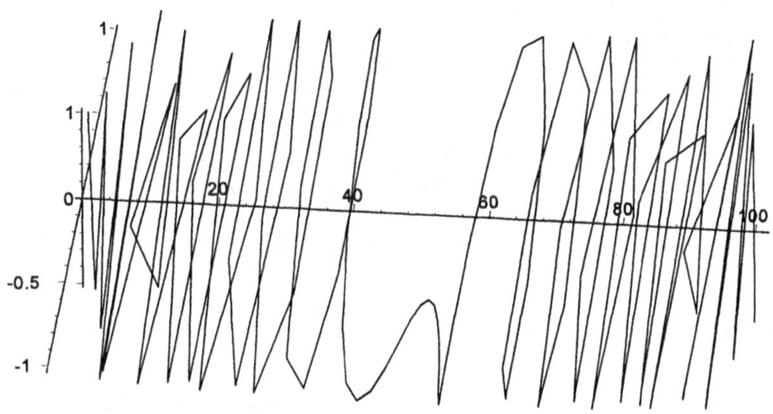

Figure 3.30 The run of phases in a 100 element P4 code.

The phase shifts necessary in Figure 2.24 are shown in Table 3.3.

Table 3.3
Phasing for the generation of the P codes

Code	A	B	C	D	E	F	G	H
Frank	0	0	0	0	0	0	0	0
P1	0	$5\pi/4$	$\pi/2$	$7\pi/4$	π	π	0	0
P2	0	$5\pi/4$	$\pi/2$	$7\pi/4$	$9\pi/8$	$-3\pi/8$	$3\pi/8$	$-9\pi/8$
P3	0	$\pi/16$	$4\pi/16$	$9\pi/16$	0	0	π	π

P4 The strapping from the lower row of adders is different and is shown below.

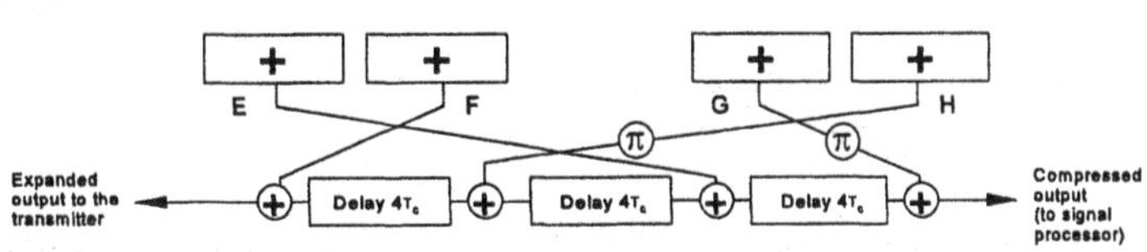

0	$\pi/16$	$4\pi/16$	$9\pi/16$	0	0	π	π

3.4.3.7 Other polyphase codes

Barker codes may be nested but they can have awkward peak sidelobes, whereas examples of polyphase Barker codes may be found in [3, p. 110]. Frank, P1, and P2 codes have a length of N^2 and these have been extended to include any length by Zadoff and Chu [3, p. 122]. The reference also describes other types of codes.

3.4.3.8 Costas codes

Costas codes are the result of rearranging the order of the frequency steps in stepped frequency linear frequency modulation [3] so that the parts may be recombined to give a thumbtack ambiguity function. The spectrum of the signal resembles linear frequency modulation with additional sidebands for the frequency switching function.

Other forms of phase modulation using less abrupt phase changes give narrower spectra. Examples are:

- Widest spectrum, sin*x*/*x* spectrum width between the first nulls $2/\tau_c$;
  - Barker codes, pseudo-random and random binary phase codes;
  - P1 to P4 codes;
  - Frank codes, stepped frequency modulation [4];
- Narrower spectra with relatively steep sides;
  - Nonlinear frequency modulation, Costas codes;
  - Linear frequency modulation (depends on time-bandwidth product).

The transmitter must amplify the full signal bandwidth with minimal distortion in amplitude and phase so that when the expanded pulses are passed through a compression filter, the original narrow pulse is restored with the resolution in range. The distorted components, such as amplitude and phase distortion, limited bandwidth, and analogue-to-digital converter errors, go to increasing the time sidelobes that decrease resolution and probability of detection (see Chapter 8, Matched and matching filters). This process takes time and gives rise to time lobes before and after the delayed restored pulse. These time lobes are reduced by windowing or weighting.

3.5 HARMONICS FROM THE TRANSMITTER

All transmitters produce harmonics. Modern slotted-wall waveguide harmonic filters absorb the harmonics, so the harmonic content is reduced to acceptable levels. These filters also absorb some of the expensively generated transmitter power.

3.6 FIGURES AFFECTING RADAR PERFORMANCE

The following figures are used in Chapter 14 to calculate radar performance.

3.6.1 Range

The range equation contains the expression $P_t \tau_t$, which is the energy in one transmitter pulse. More accurately, it is

$$Energy = \int_{\text{over the pulse width } \tau_t} P_t \, dt \tag{3.16}$$

where P_t is the peak transmitter power, W;
t is the time in seconds.

Transmitter power is generally measured as the average power, using a thermal power meter. Peak power is then calculated as $P_t = P_{av}/D_u$, where $D_u = \tau_t f_r$ is the duty cycle, estimated from the pulse width, τ_t, and pulse repetition frequency f_r. The parameters P_t and τ_t are used in Chapter 14.

3.6.2 Resolution

The transmitted bandwidth B is the primary value for range resolution. Weighting for sidelobe reduction, when pulse compression is used (see Chapter 8), decreases the effective value of B, broadening the output pulse. In the absence of pulse compression, $B \approx 1/\tau$.

3.6.3 Accuracy

The time jitter variance is added to the other variances in the calculation of range accuracy. Normally this is small because the timing jitter has to be held within tighter limits for signal processing (see Section 3.6.4).

3.6.4 Stability

The stability of the transmitted pulse depends on the reproducibility of its timing, amplitude, radio frequency, and phase. The requirement for stability depends on the signal processor used. The times in question are:

- From one pulse to the next pulse;
- During one coherent processing interval (CPI).

Each radar transmitter is different and only a typical master oscillator power amplifier (MOPA) transmitter is shown here. The analysis and summing of the various components are an exact bookkeeping operation and are best carried out using a spreadsheet. It is assumed in this section that the transmitter receives its triggers and transmitter frequency (pulse) signal from an outside unit. Any signals at the radar frequency in the transmitter frequency are best switched off until they are needed, as they represent the perfect jammer signal.

The transmitted pulse timing depends on the timing of the modulator, which forms the driving pulse in radars with a power amplifier or on the modulator pulse with magnetron radars. Errors in the timing are called jitter. The timing may be at exact or random intervals within a range, or jittered. The latter timing was used in older, simpler radars to reject repeater jamming. A typical block diagram for the power amplifier stages was shown in Figure 3.1. How this affects transmitter stability is shown in the simple example in Figure 3.31.

The sections that follow contain an example used to estimate the stability of a transmitter for a radar. There are a number of delays that contain an element of jitter that is different for each counter chain.

3.6.4.1 Trigger generation

The transmitter pretrigger is normally derived from and resynchronized with the coherent oscillator (COHO) (see Chapter 1). The trigger passes through a number of gates, each of which adds to the jitter in a random manner. The total standard deviation of the jitter is the sum of the squares of the individual components. For example, if there are four gates between the reference point and the radio frequency modulator and each has a jitter of 50 ps, the total jitter at A in Figure 3.31 is

$$\begin{aligned} \textit{Total timing jitter} &= \sqrt{50^2+50^2+50^2+50^2}\ \text{ps} \\ &= 100\ \text{ps} \end{aligned} \tag{3.17}$$

This neglects the jitter in the following linear amplifier stages. The limiting cancellation ratio (CR_{lim}) caused by timing is given by [7, p. 15.52]

$$CR_{timing} = 20\,\log_{10}\left(\frac{\tau}{\Delta t}\frac{1}{\sqrt{2B\tau}}\right)\ \text{dB} \tag{3.18}$$

where τ is the transmitter pulse length;
$\Delta\tau$ the standard deviation of the pulse jitter;
$B\tau$ is the time-bandwidth product (or pulse compression ratio) or unity where pulse compression is not used.

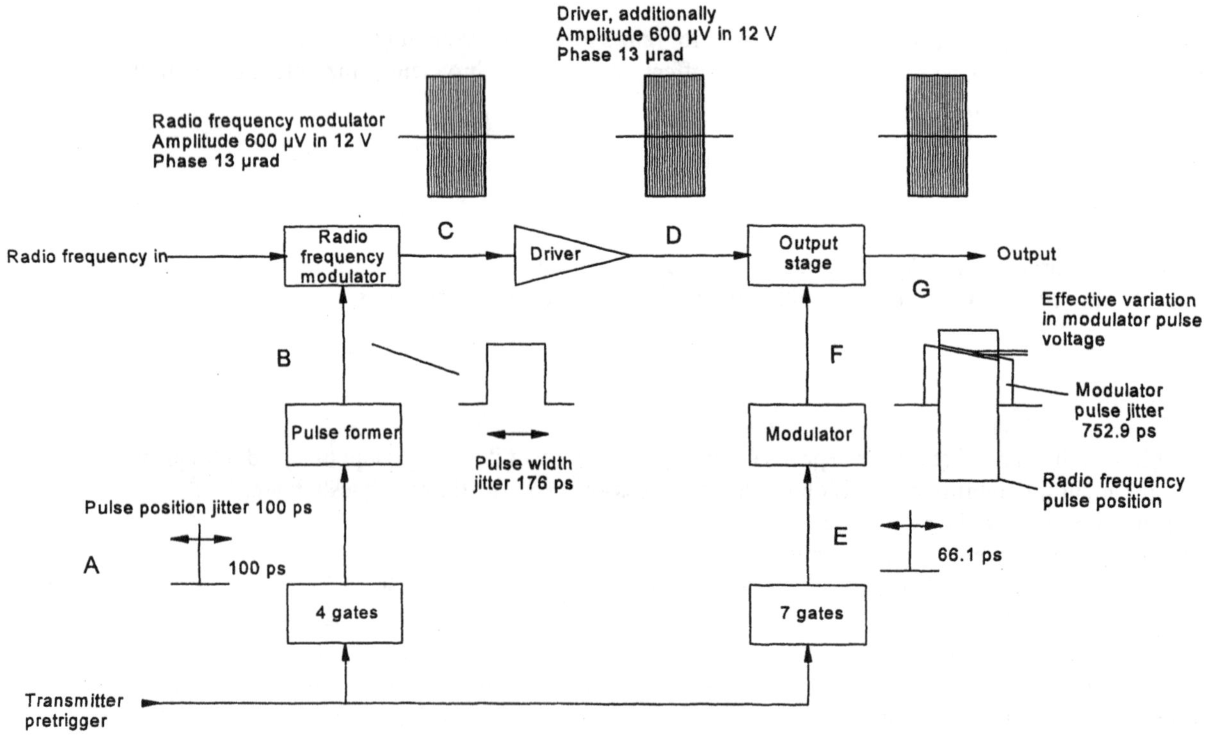

Figure 3.31 The components of transmitter stability.

The limiting cancellation ratio for a 1 µs pulse caused by 100 ps timing jitter is given by

$$\begin{aligned} CR_{timing} &= 20 \log_{10}\left(\frac{10^{-6}}{100 \times 10^{-12}} \frac{1}{\sqrt{2}}\right) \text{ dB} \\ &= 76.990 \text{ dB} \end{aligned} \tag{3.19}$$

3.6.4.2 Transmitter frequency

The change of transmitter frequency from pulse to pulse limits the cancellation ratio to

$$CR_{frequency} = 20 \log_{10} \frac{1}{\pi \, \Delta f_t \, \tau_t} \tag{3.20}$$

where Δf_t is the transmitter frequency change between pulses;
τ_t is the transmitter pulse width.

A frequency drift of 100 Hz between pulses gives

$$\begin{aligned} CR_{frequency} &= 20 \log_{10} \frac{1}{\pi \, 100 \, . \, 10^{-6}} \text{ dB} \\ &= 70.07 \text{ dB} \end{aligned} \tag{3.21}$$

3.6.4.3 Radio frequency modulator

The radio frequency (RF) modulator is responsible for forming the radio frequency pulse that will be amplified and transmitted. The leading edge is assumed to have the jitter of the pulse from the fourth gate of the preceding stage. A stop trigger defines the end of the pulse. If, for example, this is achieved with 50 counts, there will be a pulse width jitter (B in Figure 3.31) of

$$\begin{aligned} Jitter\ for\ 50\ counts &= \sqrt{50}\ .\ 25\ \text{ps} \\ &= 176.777\ \text{ps} \end{aligned} \tag{3.22}$$

This limits the cancellation ratio to

$$CR_{pulsewidth} = 20\ \log_{10}\left(\frac{\tau}{\Delta pulse\ width}\ \frac{1}{\sqrt{B\tau}} \right)\ \text{dB} \tag{3.23}$$

This case gives

$$\begin{aligned} CR_{pulsewidth} &= 20\ \log_{10}\left(\frac{10^{-6}}{176.777\ \times\ 10-12}\ \frac{1}{\sqrt{1}} \right)\ \text{dB} \\ &= 75.051\ \text{dB} \end{aligned} \tag{3.24}$$

At this stage, it is normal to form the leading and trailing edges of the pulses to reduce interfering sidebands as shown in Section 3.3.

The voltage from the stabilized power supplies in this block have a noise like voltage variation. It is assumed that each block is supplied from an independent power supply so that the power supply noise voltages add only in the root mean square (rms) sense. For example, an rms noise of 600 μV on a 12 V power supply gives a variation of

$$\begin{aligned} Amplitude\ variation &= \frac{600\ \times\ 10^{-6}}{12} = 0.005\ \% \\ CR_{rfmod\ amp} &= 20\log_{10}\frac{12}{600\ \times\ 10^{-6}}\ \text{dB} \\ &= 86.021\ \text{dB} \end{aligned} \tag{3.25}$$

There is also phase variation from the modulator block with supply voltage. If this is 7.5 degrees for a 3 dB change in output caused by low power supply voltage, the phase error is

$$\begin{aligned} Phase\ variation &= 7.5\ \times\ 2\ \times\ \frac{0.005}{100} = 0.00075\ \text{degrees} = 13.090\ \mu\text{radians} \\ CR_{rfmod\ phase} &= 20\log_{10}\frac{1}{13.090\ \times\ 10^{-6}}\ \text{dB} \\ &= 97.661\ \text{dB} \end{aligned} \tag{3.26}$$

3.6.4.4 Driver

The driver stages for the radio frequency pulses carry relatively low powers and are supplied by stable power supplies with a voltage variation similar to noise. As with the radio frequency modulator, the power supplies amplitude and phase modulate the output from the driver. The example assumes 600 μV in 12 V and 13 μradians in phase.

3.6.4.5 Modulator trigger generation

The pulse modulator stores electrical energy during the interpulse period and releases it as necessary for the output stage. This process takes time, so that pretriggers are required to time the pulse-forming operation in time for the transmitter pulse. In the example, there are seven gates using different components as in Section 3.6.4.3, each with 25 ps jitter, giving an rms total of 66.14 ps at point E in Figure 3.31.

3.6.4.6 Pulse modulator

The modulator output pulse switches on the output stage, so the direct current can stabilize before the radio frequency drive is applied. The radio frequency drive is applied and then ends. Soon after this point in time, the modulator has discharged or is switched off.

The variation or jitter in the switching times may be grouped as follows:

- Trigger amplification;
- Jitter in the main thyristors or other switches.

The effect of the jitter is to move the modulator pulse with respect to the radio frequency driver pulse. If the modulator pulse had a flat top, there would be no problem. Normally, modulator pulses droop from start to finish, so this movement represents a change of amplitude. The modulator also has amplitude and phase modulating effects on the output stage caused by amplitude and timing variations. In the example this jitter is taken to be 750 ps, which added to the jitter in Section 3.6.4.5 gives 752.9 ps at point F in Figure 3.31. These effects are discussed under output stage in the next section.

3.6.4.7 Output stage

The reproducibility of the amplitude of the transmitted pulse is determined primarily by the modulator of the output stage. The waveforms are often not very square but must be the same from pulse to pulse. The variation in the amplitude of the modulator pulses gives echoes of varying amplitudes which a signal processor may mistake for a Doppler frequency. The limit on the cancellation ratio is given by [7, p. 15.52]

$$CR_{output\ amp} = 20\log_{10}\frac{A}{\Delta A} \quad \text{dB} \tag{3.27}$$

where A is the pulse amplitude, V;
ΔA is the standard deviation of the pulse amplitude, V.

In a master oscillator power amplifier transmitter (MOPA) the reproducibility of the phase of the transmitted pulse is determined primarily by the sources of the radio frequency [7, p. 15.52]. These are the coherent and local oscillators (COHO and LO or STALO). Magnetron radars transmit with random phase, and the phase stability is determined by the accuracy of either a rephased coherent oscillator or the phase corrector. This phase error limits the cancellation ratio as

$$CR_{output\ phase} = 20\log_{10}\frac{1}{\Delta\phi} \quad \text{dB} \tag{3.28}$$

where $\Delta\phi$ is the standard deviation of the phase error in radians between pulses.

This example uses a klystron output stage. The main components of the amplitude and phase errors are as follows:

Heater voltage.

With the trend to smaller components the heater supply frequency may be in the order of many kilohertz. In indirectly heated tubes, this voltage couples capacitively with the cathode and adds to the anode voltage, giving a different voltage during each pulse. This may be reduced by rectifying the heater power or by switching it off completely during the transmitter pulse. In the example, the heater to cathode isolation is assumed to be 53 dB.

- Magnetic field.
As with the heater voltage, the trend is for small components running at higher frequencies to supply the current, normally many ampères, for the solenoids for the magnetic field (excluding triodes, pentodes, transistors, and so on). A change of magnetic field changes the phase at the output and may be fast enough for a change between pulses.
- Cathode voltage.
This is supplied from the modulator, often through a transformer. Each pulse from the modulator has a different amplitude, which is translated into a phase error. The transformer may not return to exactly the same rest state between pulses. This remaining random component of the magnetic field adds to the field from primary winding carrying the modulator pulse.
- Timing of modulator and radio frequency pulses, anode voltage droop.
In the example, the droop of the modulator pulse is taken to be 3% in 1 μs. When the modulator pulse jitter is 752 ps, the limitation in cancellation ratio reciprocal is 2.10^{-5}.

The output stages thus give amplitude and phase errors that limit the cancellation ratio of the radar. An example is given in Table 3.4.

3.6.4.8 Omitted components

Each transmitter is different. The timing uncertainty caused by cable-length tolerances, actual connector resistances, saturable reactors, and many other types of components has been omitted here, as have the magnetization variations and tolerances of the transformers.

3.6.4.9 Summing the components in Table 3.4

The components for stability are expressed as fractions that represent standard deviations or the reciprocal of the individual cancellation ratio limitations. Amplitude and phase variances are summed to give the final standard deviations, that assumes that they are not correlated. In the case of power supply ripple, the amplitude and phase errors must be added linearly if the components are fed from a common power supply. Switching power supplies normally operate at different frequencies, so the ripple of the power supplies is not synchronous.

Once the fractions for timing, pulse width, frequency, amplitude, and phase have been calculated, they can be added to give a final reciprocal of the cancellation ratio limitation.

The total limitation of the cancellation ratio for Chapter 14 for the transmitter, where there is no correlation between the various factors, is given by

$$\begin{aligned}\frac{1}{CR_{transmitter}} = {} & \frac{1}{CR_{timing}} + \frac{1}{CR_{frequency}} + \frac{1}{CR_{pulsewidth}} \\ & + \frac{1}{CR_{rfmod\ amp}} + \frac{1}{CR_{rfmod\ phase}} \\ & + \frac{1}{CR_{output\ amp}} + \frac{1}{CR_{output\ phase}}\end{aligned} \tag{3.29}$$

As an example, if each of the five components has the value 67 dB, then the combined result for Chapter 14 is 60 dB.

Table 3.4
The stability calculation in spreadsheet form

Section	Parameter			Timing, ps	Frequency, Hz	Amplitude	Phase, radians	Cancellation ratio limitation, dB
Factors affecting pulse timing								
3.6.4.1	Pulse timing		100 ps	0.0001				76.990
Factors affecting radio frequency								
3.6.4.2	Radio frequency change between pulses 100 Hz				0.0001			80.000
Factors affecting pulse width								
3.6.4.3	Pulse width variation 122.4 ps			0.0001				78.244
Factors affecting radio frequency, pulse amplitude, and phase								
3.6.4.3	Radio frequency Modulator							
		Amplitude				0		
		Phase					0	
3.6.4.4	Driver							
		Amplitude				0		
		Phase					0	
3.6.4.5	Modulator trigger							
	Amplifying gates		7 gates @ 25 ps = 66.144 ps					
3.6.4.6	Pulse modulator							
	Thyristors		750 ps					
	Total		752.91 ps					
3.6.4.7	Output stage							
	Heater voltage							
		Amplitude	7.07V 53dB isolation			0		
		Phase	45% for 80 kV				6.40e-09	
	Anode voltage							
		Amplitude	0.004% of 80 kV			0		
		Phase	45% for 80 kV				0	
	Anode voltage droop							
		Amplitude	3% in 1μs times modulator jitter (752.9 ps)			0		
		Phase	45% for 80 kV				0	
	Solenoid current							
		Amplitude						
		Phase	0.2° for 1% current				0	
			5 mA in 20 A					
					rms sum	0	0	
	Limits to cancellation ratio dB					81.484	80.366	

	Fraction		dB
Limit to cancellation ratio caused by pulse timing jitter	0.0001		76.990
Limit to cancellation ratio caused by pulse width variation	0.0001		78.244
Limit to cancellation ratio caused by pulse amplitude variation	0		81.484
Limit to cancellation ratio caused by pulse phase variation	0		80.366
Limit to cancellation ratio caused by radio frequency variation	0.0001		80.000
Total cancellation ratio	0.0005	66.021	66.021

3.6.5 Interference to neighboring systems

A radar has a powerful transmitter that is capable of interfering with all other receiving equipment or other equipment which uses amplifiers or solid-state logic nearby. To reduce out-of-band interference, the shape of the transmitter spectrum is often specified, for example, components of the spectrum x MHz from the radar must be attenuated by y dB.

REFERENCES

1 Lewis, B. L., F. F. Kretschmer, and W. W. Shelton, *Aspects of Radar Signal Processing*, Norwood, Massachusetts: Artech House, 1986.
2. Dorf, R. C., ed., *The Electrical Engineering Handbook*, Boca Raton, Florida: CRC Press, 1993.
3. Levanon, N., and E. Mozeson, *Radar Signals*, New York: Wiley, 2004.
4. Weisstein, E. W., *CRC Concise Encyclopedia of Mathematics*, Boca Raton, Florida: Chapman and Hall, 1999.
5. Galati, G., ed., *Advanced Radar Techniques and Systems*, Stevenage, Herts: Peter Peregrinus, 1993.
6. Nathanson, F. E., *Radar Design Principles*, New York: McGraw-Hill, 1991.
7. Skolnik, M. I., *Radar Handbook*, New York: McGraw-Hill, 1990.

Chapter 4

Microwave waveguide and transmission line system

This chapter describes the components that connect the transmitters, antennas, and receivers together. During the Second World War (1939-1945), before technicians understood what was happening, they derogatorily used the word "plumbing" for this amazing assembly of magic pipes. These connecting elements absorb and reflect the energy in the transmitter pulses and the returning echoes. These losses are costly because radio frequency power lost in heat is the most expensive type of electrical power in the radar and especially critical in the parts of the system common to transmission and reception.

Historically, radars had to have two separate antennas in order to protect the receivers from the large transmitter power. This is shown in Figure 4.1(a). This is often the case with continuous wave radars.

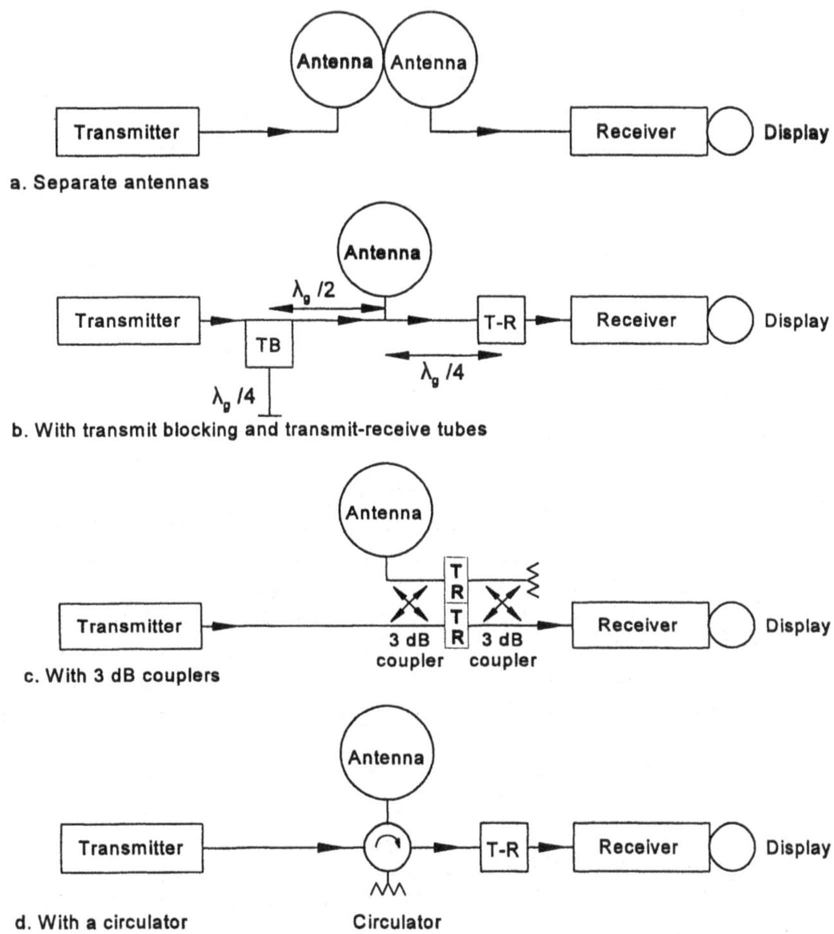

Figure 4.1 The history of transmitter and receiver switching.

Later, during the Second World War, ways were found to connect microwave transmitters and receivers to a common antenna. A gas discharge tube was placed at the input to the receiver to protect it during the transmitter pulses. Originally, the magnetron transmitter reflected the received echo signals into the receiver. Later, transmitter blocking (TB) tubes were introduced to give better reflection. They were at the mouth of a quarter wavelength shorted stub (or open circuit at the mouth) and fired at the start of the transmitter pulse to allow it to pass, as in Figure 4.1(b). Tuned line lengths restrict the bandwidth, and an untuned switch may be made using two 3 dB couplers and two tubes, as in Figure 4.1(c) [1, p. 8.32]. Active or gas discharge tubes have a life inversely proportional to the power they have to handle, so currently a circulator is placed between the components as in Figure 4.1(d). The circulator directs approximately 99% of the transmitter power into the antenna. The remaining 1% (-20 dB) passes onto the next exit and fires the transmit-receive (T-R) tube. The reflected energy is then circulated to the load. For the case of a full reflection in the waveguide to the antenna, the T-R tube must be dimensioned to reflect the full transmitter power for the short time until the transmitter is switched off. The transmitter and receiver switching allows duplex operation with the antenna[1] [2].

Later, it was found to be of advantage to connect a number of transmitter-receivers on separate frequencies to a common antenna to give frequency diversity operation, as shown in Figure 4.2. The diplexer is a frequency filter that gives independent ways to the antenna for the two frequencies. The path between the diplexer and the antenna is common to both systems, so the transmitter pulses are spaced by a small number of microseconds to avoid their voltages adding and thus causing arcing.

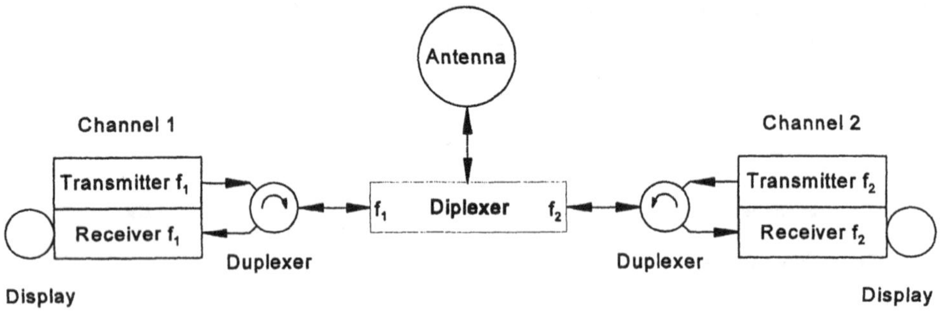

Figure 4.2 The combination of two transmitter-receivers for diversity operation.

The arrangement in Figure 4.2 gives the following advantages:

- Single-channel operation: Redundancy for reliability;
  One channel may be tuned while the other is used (military);
- Dual-channel operation: Diversity gain;
  Greater bandwidth (military).

Figure 4.3 shows a simplified arrangement for a two-channel radar. The transmitter pulse passes around the circulator to the next exit, where it leaves via a directional coupler (A) to a waveguide switch. This switch directs the transmitter pulse either to the diplexer, as shown, or to a high-power dummy load (1) capable of dissipating the full transmitter output power. This allows the maintenance and tuning of the transmitter without radiation. The directional coupler (A) is used to monitor the power from transmitter A and the reflected power in the channel A waveguide. The reflected power is a measure of the standing wave ratio and, in the case of a short circuit, all the transmitter power will be reflected. The detector attached to the reverse power coupler must act fast enough to switch off the transmitter before the next pulse.

The diplexer is a frequency filter that connects the A and B channels to the antenna. The common waveguide to the antenna has a directional coupler (X) built into it on its way to the rotary joint. The forward and reverse couplers allow the monitoring of the power from both transmitters and the standing waves in the common waveguide. On the revolving side of the rotary joint, the waveguide goes to the feeding system for the antenna.

The returning echo signals from both channels enter the antenna, pass through the rotary joint, and are separated by frequency by the diplexer for the two channels. They pass through the switch via the directional coupler, the circulator,

[1] There is often confusion between the words *duplex* and *diplex*, which come from communications. *Duplex* refers to the simultaneous transmission in both directions of information. *Diplex* refers to the simultaneous transmission and reception of two signals using a common medium. The diplexer filters separate the two channels in the frequency domain.

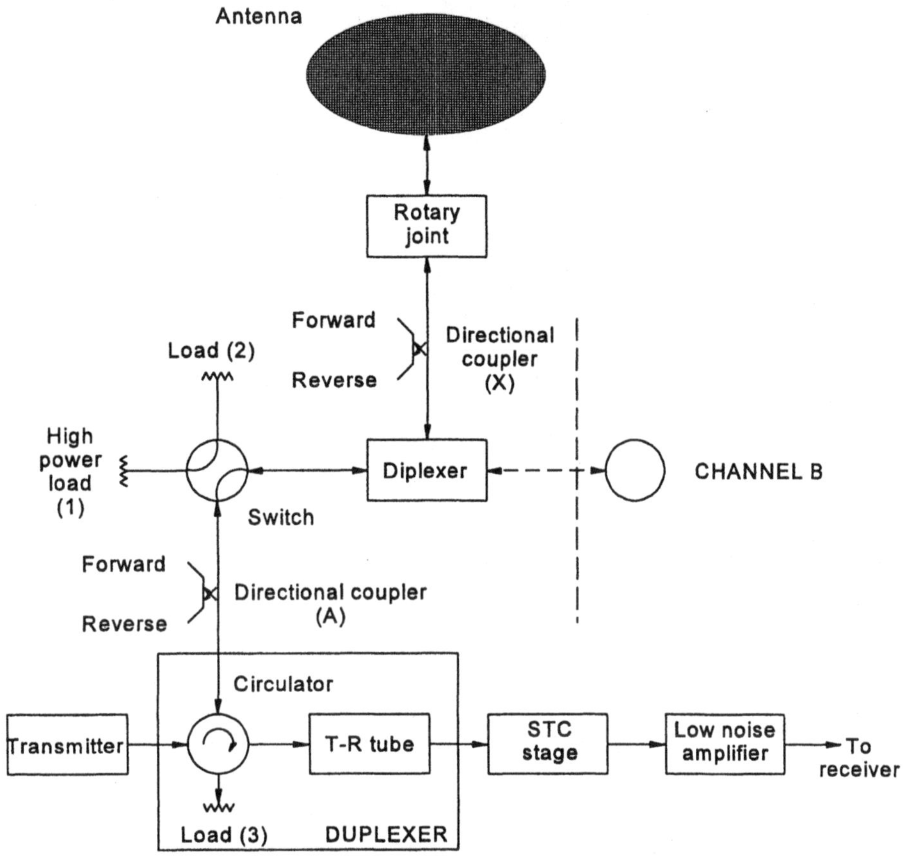

Figure 4.3 A typical arrangement for a two channel primary radar.

and the T-R tube to the receiver.

The diplexer is not a perfect filter. Part of the transmitter pulse from the B channel will leave the diplexer by the A channel waveguide and will then be directed towards the A channel receiver, fire the T-R tube, and be reflected into the dummy load (3). Commonly, bandstop filters tuned to reject signals from the other channel are used to reduce the effects of this type of leakage. When one channel is switched to the high-power dummy load (1), a second dummy load (2) is used to terminate the diplexer. The load must be dimensioned large enough to dissipate the full power from transmitter B until the fault is recognized (up to 30 seconds) and the transmitter is switched off. Load (3) must be dimensioned to absorb the full transmitter power until the transmitter can be switched off.

The forward power couplers may be used to inject test signals for the radar receiving system. A common form in older radars is a radio frequency test pulse injected at long range or in the dead time. The amplitude of the test signal is adjusted until the signal is just discernable on a display, the minimum discernable signal. Similarly, noise may be injected for receiver noise measurements and test signal trains for the signal processor and extractor.

Where there are several receiving channels through the rotary joint from the antenna, each channel has its separate diplexer to route the echo signals to the appropriate channel. The inputs to the receivers are protected by T-R tubes as with the main channel.

With secondary radar, only one interrogator-receiver at a time is switched to the antenna. Figure 4.4 shows two channels feeding a switch. In the past, there would have been only the Σ path for the P_1 and P_3 interrogation pulses and the returning replies, which is connected to the main beam of the antenna. Later, improvements have included the addition of an omnidirectional pattern to the antenna, which is used for the transmission of a P_2 pulse as part of the interrogator side lobe suppression (ISLS) system. Either a high-power switch is used to switch the interrogator transmitter pulse to the antenna input for the omnidirectional pattern or a separate transmitter is used, as in Figure 4.4. The omnidirectional pattern may also be used to suppress replies entering the side lobes of the antenna, that is receiver side lobe suppression (RSLS). A monopulse difference pattern, Δ, was introduced still later and is used to improve azimuth accuracy.

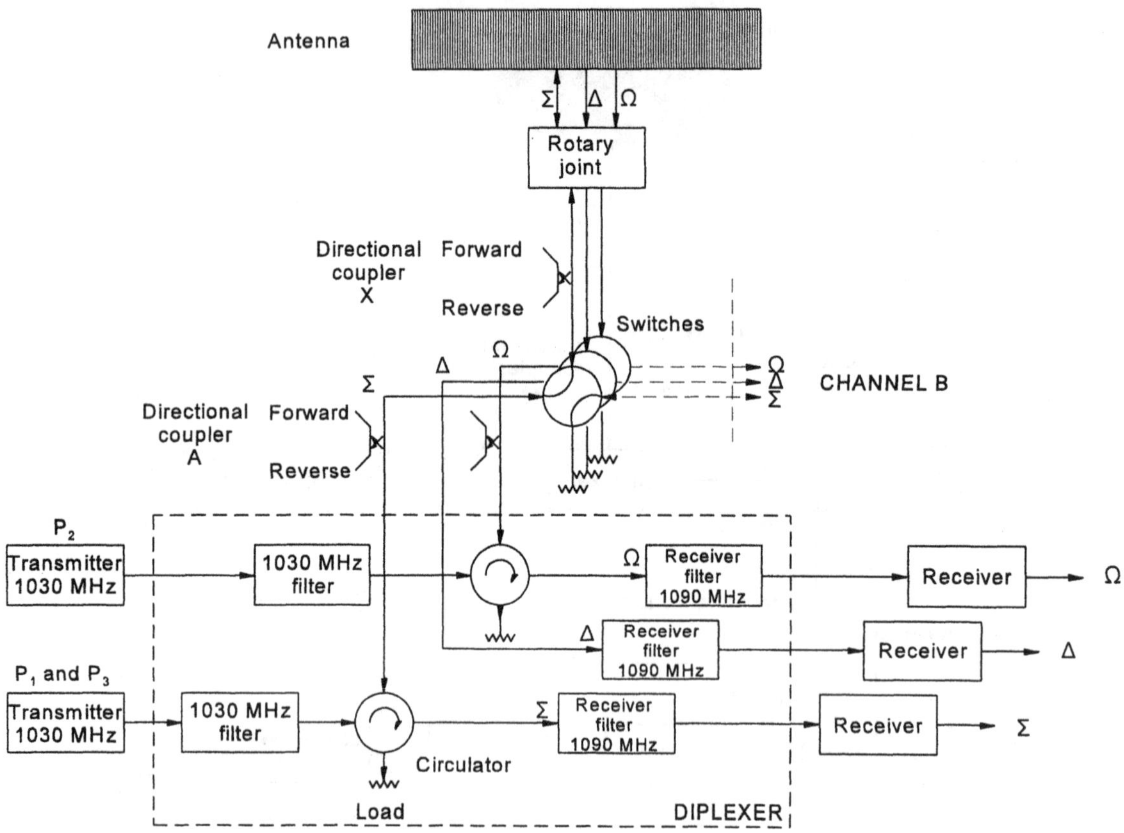

Figure 4.4 A typical arrangement for a secondary radar with two interrogator-receivers.

In the common form of secondary radar, the transmitter frequency is 1 030 MHz, and the receiver frequency is 1 090 MHz. With this arrangement a diplexer (frequency filter) is able to provide the connection between the interrogator (transmitter), the antenna, and the receiver. With the advent of circulators, a circulator is used to route approximately 99% of the power to give additional isolation and a diplexer containing a circulator is shown in Figure 4.4. The circulator load performs a useful function by absorbing any power reflected by the filters.

The design of feeder systems using coaxial lines or waveguides with minimum loss and the best matching is the domain of the microwave engineer [3-5]. The losses, which are then expressed in decibels, must be added together separately for the transmission and reception paths.

4.1 MISMATCH

Mismatch causes part of the power to be reflected and not to arrive at its proper destination. The voltages of the forward and reverse components interfere, giving standing waves. The mismatch is normally expressed as the voltage standing wave ratio (VSWR), which can be measured using a slotted line or waveguide. The amounts of power reflected at an interface depends on the voltage reflection coefficient, ρ:

$$\textit{Voltage reflection coefficient, } \rho = \frac{VSWR - 1}{VSWR + 1} \tag{4.1}$$

Normally the voltage standing wave ratio is a number greater than unity, but it is sometimes given as a number less than one. In this case, the numerator in (4.1) is 1 - VSWR and the denominator 1 + VSWR. The forward and reflected or reverse powers are given by

$$\begin{aligned} \textit{Fraction of power reflected} &= \rho^2 \\ \textit{Fraction of power transmitted} &= 1 - \rho^2 \end{aligned} \tag{4.2}$$

Graphs of (4.2) are shown in Figures 4.5 and 4.6.

In older, fixed-frequency radars, either phase shifters or tuners were used to tune out the remaining reactance in a waveguide run to reduce the voltage standing wave ratio to below 1.1:1, often 1.05:1. As an example, a 1 MW peak pulse has a reverse power component in a waveguide with a voltage standing wave ratio of 1.1:1 is 2 267.6 W, peak. This power has to be dissipated as it will start to cause damage to the weakest component in the line. The advent of high-power circulators where the load on the fourth arm absorbs the reverse power greatly alleviates the situation so that higher standing wave ratios may be tolerated. This is necessary for radars that operate over a large band, for example, frequency agile radars.

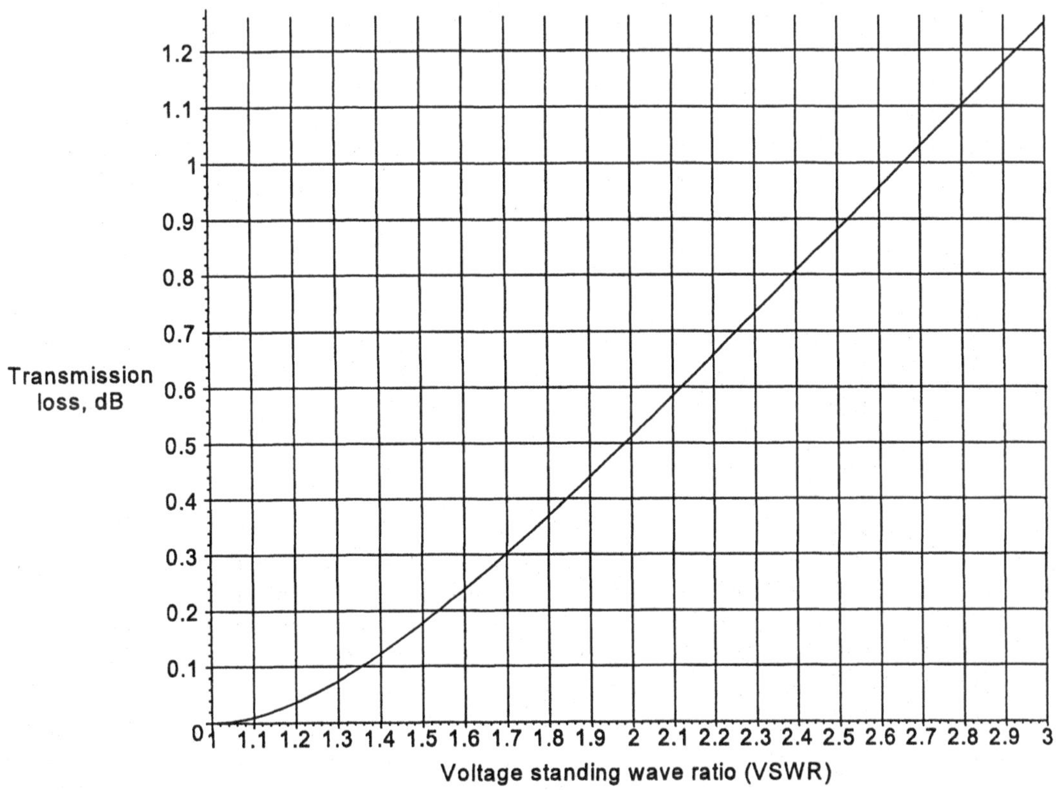

Figure 4.5 Transmission loss (dB) against voltage standing wave ratio (VSWR).

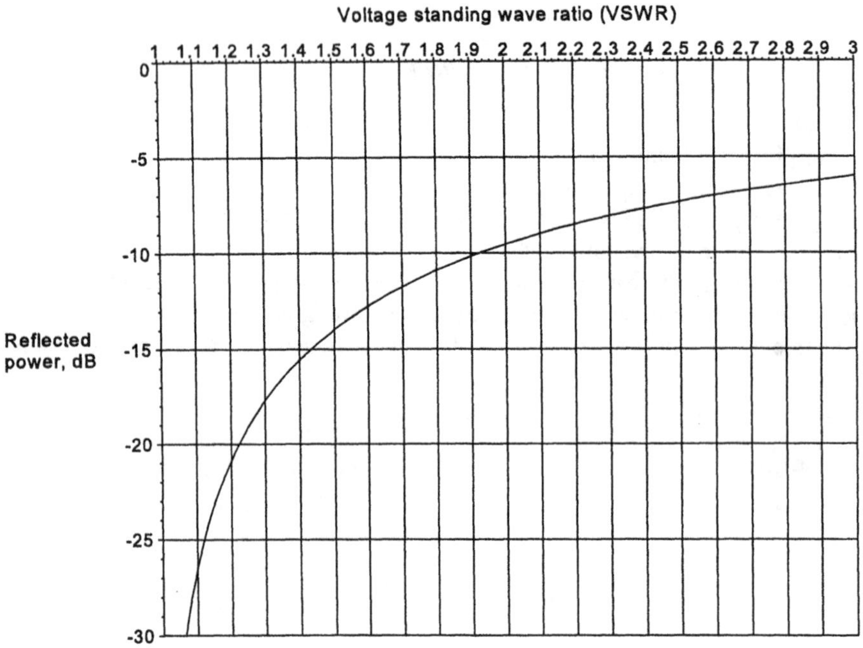

Figure 4.6 The fraction of power reflected (dB) in a mismatched waveguide against voltage standing wave ratio (VSWR).

4.2 COMPONENTS

As far as radar performance is concerned, this section contains components that absorb part of the energy in the transmitter pulses and the echo signals. The main components are chosen for their low losses, low voltage standing wave ratio (VSWR), and power handling capabilities in the frequency range used.

4.2.1 Coaxial cables

Coaxial cables are used for video, intermediate, and radio frequencies and exist in flexible, semirigid, and rigid forms. Generally, they are used for low-power signals up to the powers used in secondary radar, typically up to 1.5 kW peak. Coaxial cable losses are specified by the manufacturer per foot or per meter. To these losses the connector losses must be added. Longer lengths of cable are available to reduce connector and adaptor losses. Dry air or nitrogen is often used to keep moisture out of these cables and thus hold the losses as specified.

4.2.2 Waveguides

Waveguides are used for maximum power capacity and minimum losses. For each frequency band, there is a common size of waveguide that is the smallest reasonable size for the bandwidth involved. Oversized waveguides are used to reduce losses. The power carrying capacity of normal waveguides can be increased by filling them with desiccated compressed air and, if necessary, cooling them with water. Dry air under pressure (sometimes as low as 10 mbar) avoids condensation and corrosion that leads to additional losses. Often, even when air under pressure is not necessary, waveguides are filled with desiccated air to reduce corrosion and thus losses.

Waveguide losses are specified by the manufacturer per foot or per meter for the waveguide. To this must be added the flange losses. Runs with low loss use long waveguide sections to reduce the number of flanges and their losses. Waveguides must be firmly attached to the supporting structure to avoid vibration, which would have a negative effect on the stability of the radar.

Commonly decibels are used as a unit of loss. Decibels being logarithmic units, the total loss is the sum of the decibel values. A typical waveguide run for an air surveillance radar was shown in Figure 4.3 and connects the following:

- Transmitter output waveguide flange or connector.
  This is the interface point where the transmitter power, pulse width, and spectrum are measured and have been found to be within tolerance.
- Antenna waveguide flange or connector.
  This is the point where the antenna gain has been measured on the antenna measurement range. All components up to this point must be accounted for.
- Receiver input waveguide flange or connector.
  This is the point where the receiver noise figure or temperature has been tested and accepted in the factory.

Individual manufacturers publish the characteristics of the components they make. The waveguides that connect them may be treated generically in terms of loss and power handling. Power handling is divided into peak power, which is limited by voltage breakdown, and average power, which is limited by cooling.

Normally, one waveguide size is used to connect the components. Waveguides carry radio frequency energy above their cut-off frequency. The cut-off wavelength for the TE10 mode is twice the broad dimension:

$$\lambda_{cut\text{-}off} = 2\,a \qquad f_{cut\text{-}off} = \frac{c}{2\,a} \tag{4.3}$$

where a is the major dimension;
c is the velocity of light.

The useful frequency band is approximately 1.25 × $f_{cut\text{-}off}$ to 2 × $f_{cut\text{-}off}$. For the standard S-band WR284 waveguide, the figures are as follows:

- Type WR284 waveguide;
- Inner dimension, a = 2.84 inches = 7.2135 cm;
- Cut-off wavelength = 14.4272 cm giving a cut-off frequency of 2 079.406 MHz;
- Frequency range is approximately from 2 600 MHz to 4 160 MHz.

The losses in an air-filled waveguide are caused mainly by the skin effect resistance of the internal walls. This is, theoretically, after [4, p. 65, Eq. 2-192]

$$\alpha = \sqrt{\frac{\pi}{2\,\sigma}\sqrt{\frac{\epsilon_0}{\mu_0}}}\;\frac{1}{a^{3/2}}\;\frac{2\left(\dfrac{f}{f_c}\right)^{-1/2}+\left(\dfrac{a}{b}\right)\left(\dfrac{f}{f_c}\right)^{3/2}}{\sqrt{\left(\dfrac{f}{f_c}\right)^{2}-1}}\ \text{Np/m}$$

$$= \sqrt{\frac{\pi}{2\,\sigma}\sqrt{\frac{\epsilon_0}{\mu_0}}}\;\frac{1}{a^{3/2}}\;\frac{2\left(\dfrac{f}{f_c}\right)^{-1/2}+\left(\dfrac{a}{b}\right)\left(\dfrac{f}{f_c}\right)^{3/2}}{\sqrt{\left(\dfrac{f}{f_c}\right)^{2}-1}}\ 20\log_{10}(e)\ \text{dB/m} \tag{4.4}$$

where f is the working frequency, Hz;
f_c is the cutoff frequency, Hz;
a and b are the major and minor waveguide dimensions, m;
σ is the conductivity of the inner walls S/m;
ϵ_0 is the permittivity of free space;
μ_0 is the permeability of free space.

The theoretical losses are calculated in Figure 4.7 for WR284 (WG10) waveguide and two larger sizes (WR340 or WG9A) and WR439 (or WG8) for comparison. These are valid when the inner walls are very smooth.

The waveguide tables [6] give two sizes of waveguide for most frequencies. In order to save size and weight, the smaller size is normally used, which gives greater attenuation; see Figure 4.7. For long lengths of waveguide it is often an advantage to choose a size where the working frequency is on the minimum point on the attention curve. These larger waveguides must be connected to the normal waveguides using long funnel-shaped transition waveguides.

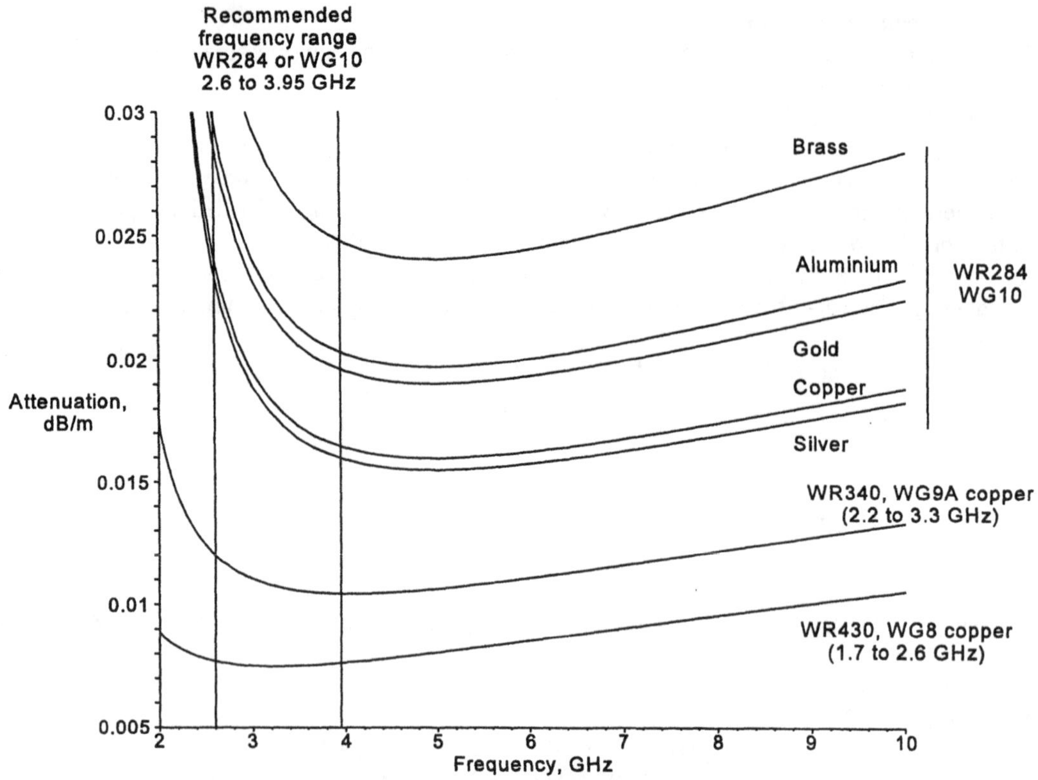

Figure 4.7 Waveguide attenuation in dB/m for WR284 waveguides with inner walls made of brass, aluminium, gold, copper, and silver and for copper WR340 and 430 waveguides.

The peak power in the waveguide is limited by the voltage breakdown between the broad sides. The normal breakdown voltage at sea level is approximately 30 kV/cm. The peak power in a waveguide in TE01 mode is with perfect matching [3, p. 179, Eq. 6.36]:

$$\textit{Maximum power} = \frac{E_{max}^2}{120\pi} \frac{\lambda}{\lambda_g} \frac{ab}{4} \text{ W} \tag{4.5}$$

where E_{max} is the maximum electric field between the broad faces of the waveguide, V/m;
λ is the working wavelength, m;
λ_g is the guide wavelength, m;
a and b are the major and minor internal dimensions of the waveguide, m.

If this voltage is limited to 15 kV/cm (a safety factor of two), then the peak power in WR284 waveguide is defined in Figure 4.8.

Mismatch in the waveguide increases the peak voltage in the waveguide by the voltage standing wave ratio (VSWR). The peak power able to be carried in the waveguide must be divided by this voltage ratio squared:

$$\textit{Peak maximum power} = \frac{\textit{Peak power with perfect match}}{(\textit{Voltage standing wave ratio})^2} \tag{4.6}$$

Normally, waveguides are filled with desiccated air at slightly above atmospheric pressure. The peak power carried by the waveguide may be increased if the pressure is increased. At a conservative estimate [7], the breakdown voltage is proportional to pressure and inversely proportional to the absolute temperature with respect to 30 KV/cm at 1 013.25 HPa (mb) and 25°C. This change is shown in Figure 4.8. A limit on the amount of air pressure that a waveguide can withstand is given by the manufacturer. For a time, high dielectric strength gases, such as sulfur hexafluoride (SF_6), were used, but most produce toxic by-products after arcing and gas leaks in the waveguides caused problems. The maximum mean power is limited by the temperature rise and must be found from the manufacturer's literature.

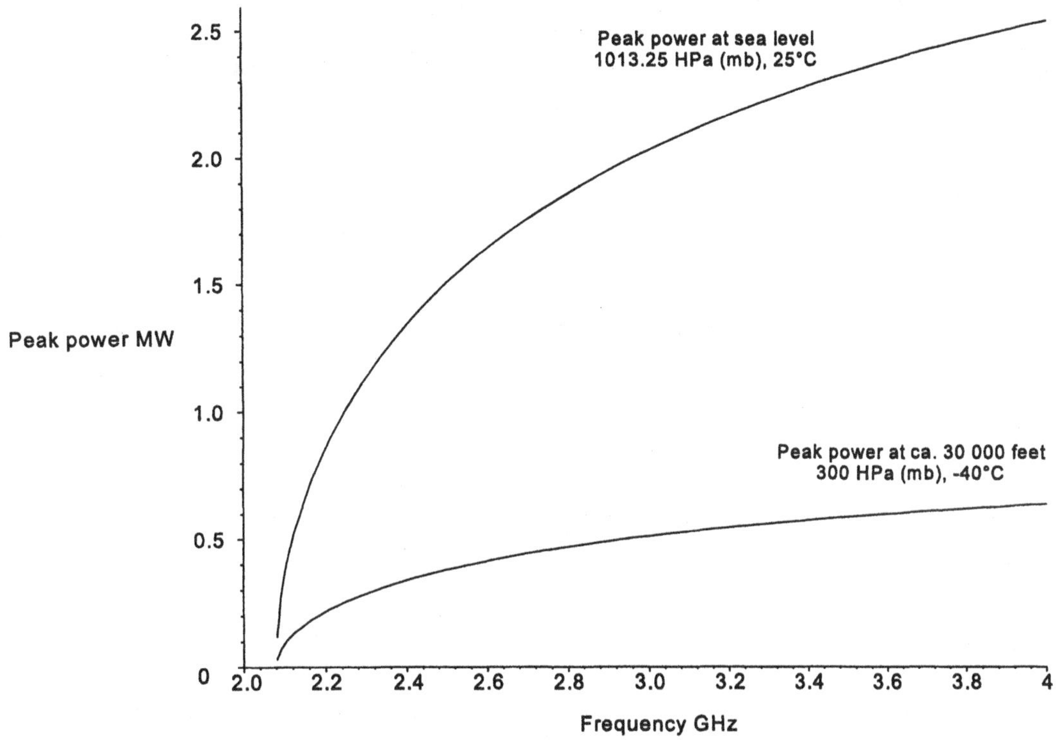

Figure 4.8 Maximum peak power in WR284 waveguide at sea level and at 30 000 feet.

Commonly standard rectangular waveguides are used in the lowest mode. Feeds for circularly polarized waves (see Section 5.6) must use waveguides that are symmetrical on both axes, namely, square or circular waveguides. Circular waveguides are also used as the central member of a rotating joint.

4.2.3 Striplines or microstrip lines

Waveguides and coaxial cables are, by today's ideas, complicated to work and reproduce so that a printed circuit form of radio frequency transmission for low powered signals has been developed. Microstrip lines are conductors machined or etched on one side of a printed circuit dielectric with a ground plane on the other, as shown in Figure 4.9(a). With striplines there is a second, upper ground plane, as shown in Figure 4.9(b). They are robust and relatively easy to manufacture and find their place in low power radio frequency modules, especially when large numbers are required.

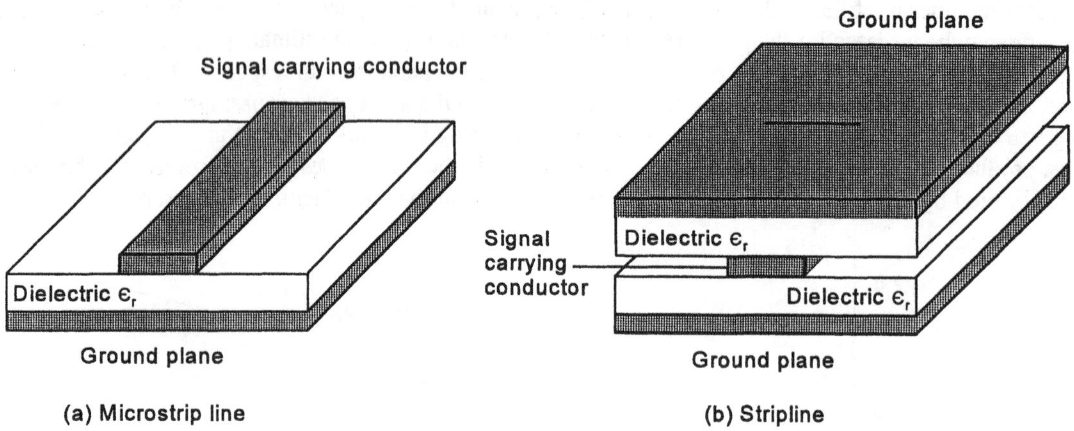

Figure 4.9 Microstrip line and stripline.

There are equivalent stripline and microstrip-line components for most waveguide and coaxial components.

4.2.4 Microwave passive components

Microwave passive components are used to connect the radio frequency generators and amplifiers inside the transmitter, the transmitter to the antenna, the antenna to the receiver, and the radio frequency signals inside the receiver. Other microwave components are used to direct and sample the signals for monitoring.

4.2.4.1 Circulators

Circulators are used to route the transmitter pulses to the antenna and the returning echo signals to the receiver. These are ferrite components and need cooling for the higher powers. The symbol for a circulator is shown in Figure 4.10 where signals entering one arm leave by the next arm.

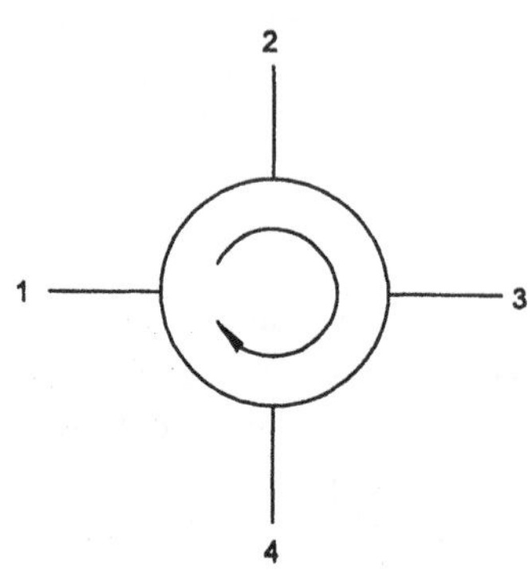

Figure 4.10. Symbol for a circulator.

They are specified by:

- The loss from one arm to the next, typically 0.5 dB;
- Voltage standing wave ratio;
- The power that does not leave by the next arm which may be typically 20 dB down (1/100) compared to the input power;
- Peak power;
- Mean power.

Circulators are realized as waveguide, strip, or microstrip components. Typically a four-port circulator is used and arm 4 would be connected to the transmitter, arm 1 to the antenna, and arm 2 to the receiver. Arm 3 is connected to a dummy load for the case that there is a short circuit or other blockage in the antenna waveguide and the transmitter pulse is reflected with full power. The TR tube or receiver protector in the waveguide to the receiver will fire to reflect the transmitter pulse back into the circulator to be passed into the dummy load. The dummy load must be capable of dissipating the full transmitter power until protective circuitry switches off the transmitter. The dummy load prevents the transmitter pulse entering the transmitter to cause damage.

4.2.4.2 Isolators

Isolators are used to absorb reflected energy in order to improve the voltage standing wave ratio (VSWR). These either absorb the returning power internally or are circulators which route the returning power to a separately cooled load. They are specified by:

- The forward loss and voltage standing wave ratio (VSWR);
- The loss for returning energy;
- Peak forward power;
- Mean forward power.

4.2.4.3 Transmit-receive tubes

Semiconductor limiters have not yet taken over from gas-filled tubes. Gas-filled tubes have a limited life and need replacing regularly when the losses increase. They are specified by:

- The loss when the tube is extinguished, which must be added to the receiver input losses;
- The amount of the energy spike passed to the receiver when the tube starts to fire;
- The attenuation when the tube has fired;
- The mean and peak power handling (reflection) capacity;
- The time taken to extinguish which determines the minimum range of the radar (see Section 1.6);
- Voltage standing wave ratio.

Though circulators have taken over the main burden of routing the transmitter pulse away from the receiver, transmit-receive tubes still have to block the transmitter pulse in the case of a fault in the antenna. If the tube does not fire, the first receiver stages will be destroyed.

4.2.4.4 Diplexers, triplexers, and so on

A number of transmitter-receivers using different frequencies may be connected to a single antenna. Two transmitter-receiver groups working on different frequencies allow two statistically independent looks at an aircraft, for example, each antenna scan, which increases the probability of detection and false alarms.

These components are specified mainly by:

- Losses and voltage standing wave ratio;
- Peak power;
- Mean power;
- Amount of power that leaks into the other channel.

Two types of diplexers are those that combine the two channels into one waveguide before the rotary joint and those that require two high power rotary joints and combine the signals in orthogonal polarizations in a square waveguide on their way to the horn in the antenna. Polarization diversity additionally gives a look at a different polarization. The radar transmitters normally transmit one after the other and do not overlap in order to hold the peak power requirement to that of one transmitter alone. The mean power is the sum of the mean powers of the transmitters.

Great care must be taken to keep the losses as small as possible as they can nullify the effects of the "diversity gain", or changing the effective scattering model from Swerling case I to Swerling case III (see Chapter 12).

4.2.4.5 Rotary joints

Rotary joints connect a rotating antenna to a fixed transmitter and receiver system. Before rotary joints became common, the transmitter and receivers were rotated with the antenna and this made maintenance more difficult. If the rotation is only occasional and less than 370 degrees, then (spiral) flexible cables may be used. Rotary joints often have a number of channels, each of which is specified mainly by:

- Frequency range;
- Losses and voltage standing wave ratio: average and variation about the average or wow;
- Peak power;
- Mean power.

The number of channels in the rotary joint depends on the radar. Figure 4.11 shows the arrangement for a three-dimensional radar with four receiving beams. The output from the transmitter is carried by the high-power rotating joint to the power divider for the formation of a single cosecant squared transmitter beam. The returning echoes are passed to the receivers by low-power joints. To these the rotating joints of the secondary radar channels must be added (see Figure 4.4).

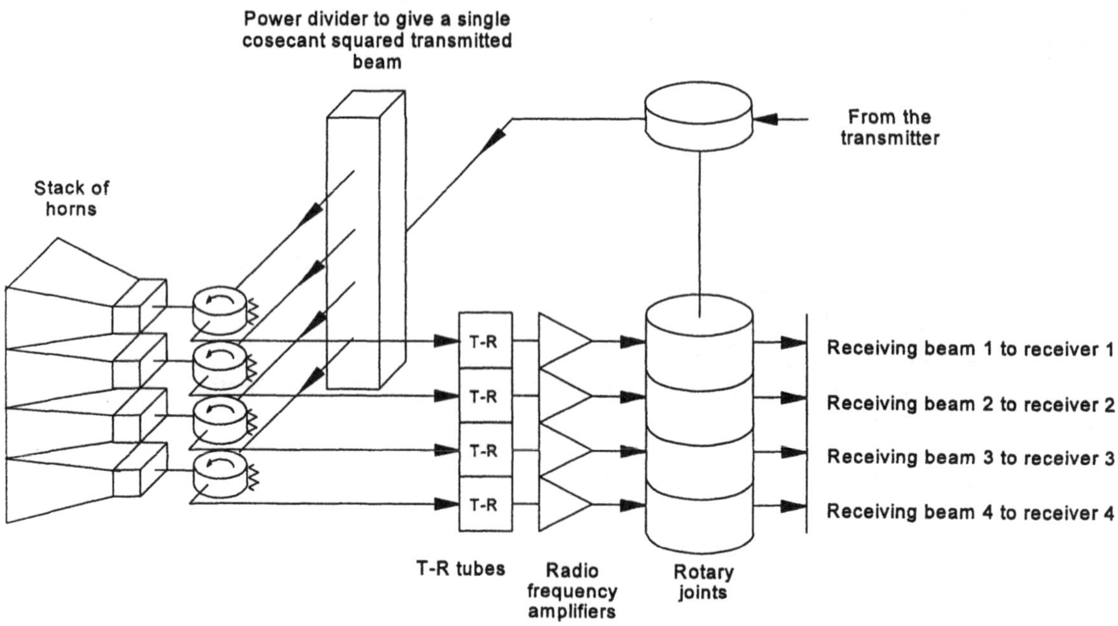

Figure 4.11 The arrangement of duplexers and rotary joints in a stacked beam radar.

4.2.4.6 Switches

Switches are used to switch the energy carried by the waveguide from the working direction to dummy loads or other components for maintenance. Normally, the switching takes place when the transmitter is switched off or muted by the removal of the radio frequency drive and a common form of waveguide switch is shown in Figure 4.12. The rotor consists of two waveguide bends to connect, on the left of Figure 4.12, port 1 with port 4 and port 2 with port 3. The rotor may be turned by hand or by electric motor but must be locked in one of the two positions before it starts to carry radio frequency power.

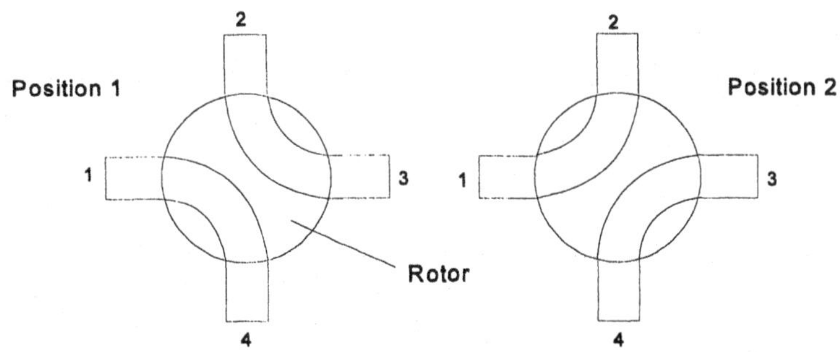

Figure 4.12 The two positions of a waveguide switch.

A typical switch for two transmitters would connect transmitter A (port 4) to the antenna (port 1) and transmitter B (port 2) to the dummy load (port 3). The waveguide switch may be switched so that transmitter may be tested at full power with the dummy load, and in the other position it is connected to the antenna.

Switches are specified mainly by:

- Losses and voltage standing wave ratio;
- Isolation in the unconnected ports;
- Peak power;
- Mean power;
- Switching time.

High-speed synchronous switches using quartz blades have been used to connect two pulse transmitters to a single antenna.

4.2.4.7 Dummy loads

Dummy loads are load resistors in coaxial or waveguide form. They are dimensioned to absorb the expected radio frequency power under both normal and, for a short time, under fault conditions. This short time is long enough to allow the fault detectors to switch off the transmitter. Dummy loads are specified mainly by:

- Peak power;
- Mean power;
- Continuous and short time power ratings;
- Voltage standing wave ratio.

A number of water cooled loads are equipped with thermometers and water flow meters to measure the power dissipated.

4.2.4.8 Directional couplers for injecting signals or measurement

Directional couplers take a sample energy from the main waveguide for transmitter and standing wave monitoring, or they feed test signals into the receiver waveguides and two couplers as shown in Figure 4.13. They are especially useful for monitoring the voltage standing wave ratio.

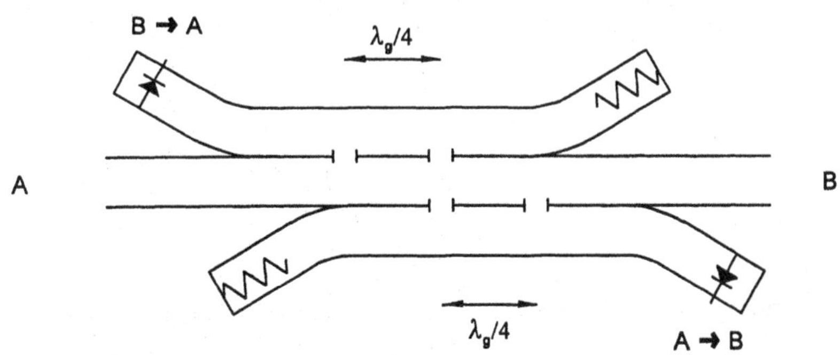

Figure 4.13 Directional couplers for monitoring forward and returning power.

These couplers are specified mainly by:

- Attenuation between the signal in the waveguide and the coupled signal in the same direction;
- Attenuation between the signal in the waveguide and the coupled signal in the reverse direction;
- Leakage between these two signals, or directivity;
- Peak power;
- Mean power.

Figure 4.13 shows two-hole couplers that have a rather narrow frequency range. Greater bandwidth can be obtained using more holes with diameters given by Pascal's triangle, namely 1,2,1 or 1, 3, 3, 1 and so on [8, p. 286]. The coupling holes are often circular for coupling factors of 1% (20 dB) or less, and the coupling depends on the sixth power of the diameter. For greater amounts of coupling, rectangular holes are used leading to the 3 dB hybrid couplers in Figure 4.1(c). There are other forms with angled or crossed waveguides.

4.2.4.9 Hybrid couplers

Hybrid transformers are used in communications circuits to divide a signal equally or reassemble them. In Figure 4.14 the input at A is divided equally between B and D, and the difference from any reflections from B and D exits at C. The hybrid transformers are used in a two-wire repeater to router the signals to the amplifiers and prevent positive feedback leading to oscillation. The loads on the C terminals dissipate any unbalanced signals.

A form of waveguide hybrid, the rat race, is shown on the right of Figure 4.14. Again there is no coupling between A and C additionally between B and D.

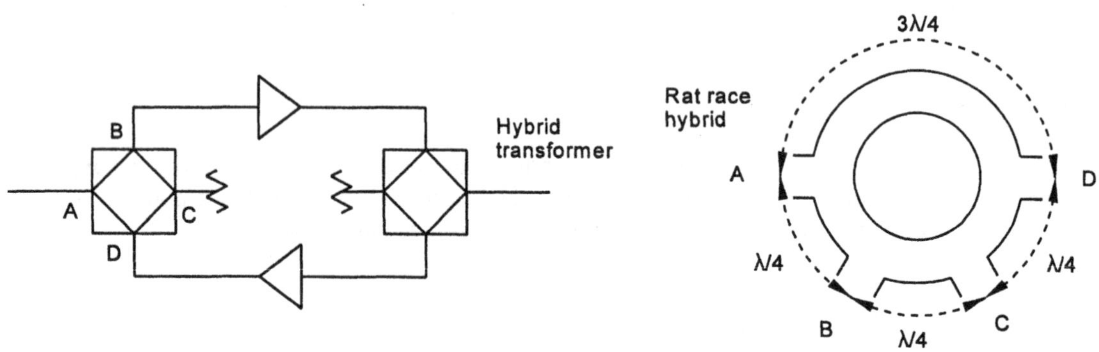

Figure 4.14 Hybrid transformers and the rat race waveguide hybrid.

The size of the rat race depends on wavelength, and a waveguide hybrid with a wider bandwidth is the hybrid or magic T shown in Figure 4.15 [3, p. 291]. The magic T is more compact and is often used with horns as a feed in monopulse antennas with the sum channel connected via the transmit-receive switch to the transmitter and the sum receiver. The difference output from the magic T is only used for reception and is fed to the difference receiver.

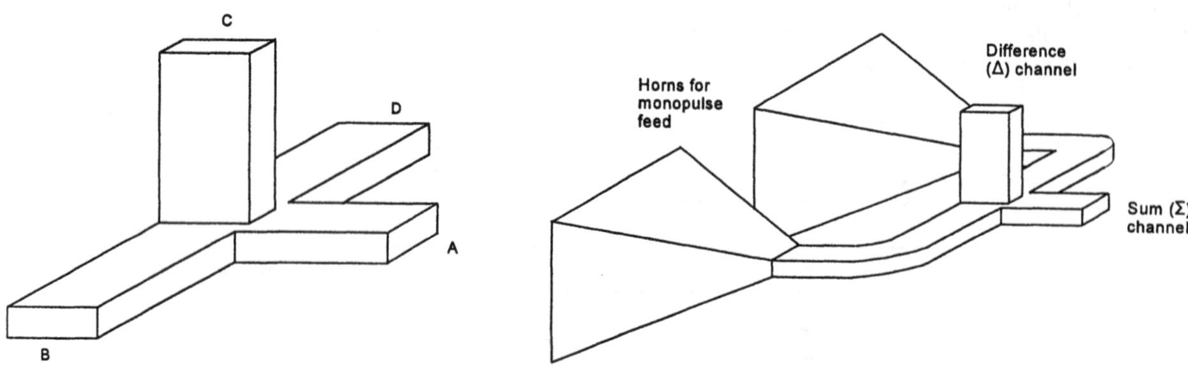

Figure 4.15 The hybrid or magic T hybrid and its use for monopulse sum and difference channels.

4.2.4.10 Tuners

In the past, waveguide tuners reflected the conjugate of the mismatch impedance into the waveguide, and voltage standing wave ratios of less than 1.1:1 were specified. Broadband radars and the advent of less frequency dependent components, circulators, and "clean" waveguide runs allowed the tuners to be eliminated. The maximum voltage standing wave ratio requirement was relaxed to 2:1. This gives a forward loss of 0.51152 dB, the returned power is 9.54242 dB down (about 1/9) compared with the forward power, but the voltage in the waveguide is doubled which divides the maximum power that can be carried by the waveguide by four.

4.3 MONITORING

Continuous monitoring is necessary to alert the operational and maintenance areas as to when the radar is not performing normally. For this section the transmitter power level and the voltage standing wave ratio are monitored continuously. Signals for receiver sensitivity, receiver noise figure, and test signals for the verification of the receiving system for test plots are injected using directional couplers. When the channel is not in use, switches can be used to connect the test equipment.

4.3.1 Power level

The transmitter power output can be measured with a power meter connected to one of the directional couplers. For accurate measurement, the frequency characteristic of the directional coupler must be taken into account. For continuous monitoring, a simple diode detector with a threshold for low power indication may be used.

4.3.2 Voltage standing wave ratio (VSWR)

Detectors connected to the reverse power couplers measure the transmitter power that is not passed to the antenna for radiation. Any fault in the waveguide system increases the amount of reflected power so that a threshold is set to give an alarm and to inhibit the next transmitter pulses.

4.4 EFFECT ON RADAR PERFORMANCE

Resistive losses and reflections reduce the maximum range of a radar. In extreme cases with continuous wave radars waveguide vibration can cause problems.

4.4.1 Effects on maximum range

As with all connecting wiring the waveguide or coaxial cable runs should be kept as short as possible to keep the losses as low as possible. For example, for a 1 dB extra loss, the full or 100% range is reduced as shown in Table 4.1.

Table 4.1
Effect of losses on maximum range

Connection	Proportion of maximum range
Transmitter to duplexer	97.163%
Duplexer to antenna	94.406%
Antenna to receiver	97.163%

The duplexer to antenna run is the more sensitive as both the transmitter pulses and the received echo signals pass the same way. The modern tendency is not to tune the waveguide system to particular frequencies, so it must be kept as clean and as simple as possible. The effect on performance is a meticulous accounting process best carried out using a spread-sheet. For each component, the following are required:

• Input connector or flange and gasket loss;
• Input cable or waveguide loss;
• Component loss;
• Output cable or waveguide loss;
• Output connector or flange and gasket loss.

The sums of the transmission and reception losses in decibels are used in Chapter 14, Performance. The waveguide system often has different losses in the transmitting and receiving directions. The losses for the range equation are

$$\begin{aligned} L_{t\ waveguide\ system} &= Sum\ of\ the\ losses\ in\ the\ transmittting\ direction \\ L_{r\ waveguide\ system} &= Sum\ of\ the\ losses\ in\ the\ receiving\ direction \end{aligned} \quad (4.7)$$

4.4.2 Effects on stability

The waveguide system is normally mechanically rigid enough so as not to deform enough to affect the transmitter pulse or the echo signals. In extreme cases with continuous wave (CW) radars, critical waveguide components are milled out of a solid block of metal. There must be no rotary joint wow components (variation with rotation as with record players) over the angle contained in one look in the Doppler frequency range of coherent radars.

REFERENCES

1. Skolnik, M. I., *Radar Handbook*, New York: McGraw-Hill, 1970, p. 8-32.
2. *IEEE Standard 100*, Piscataway, New Jersey: IEEE Press, 1993.
3. Glazier, E. V. D., and H. R. L. Lamont, *Services Textbook of Radio, Volume 5, Transmission and Propagation,* London: Her Majesty's Stationery Office, 1958.
4. Gardiol, F. E., *Introduction to Microwaves*, Dedham, Massachusetts: Artech House, 1984.
5. Collin, R. E., *Field Theory of Guided Waves*, Piscataway, New Jersey: IEEE Press, 1990.
6. Saad, T. S., *Microwave Engineers' Handbook*, Dedham, Massachusetts: Artech House, 1971.
7. Jordan, E. C., *Reference Data for Engineers*, 7th ed., Indianapolis: Howard Sams, 1986.
8. Alison, W. B. W., *A Handbook for the Mechanical Tolerancing of Waveguides*, London: Her Majesty's Stationery Office, 1972. Reprinted by Artech House, 1987.

Chapter 5

Antennas

Antennas are used to launch signals to free space or to recover them. This chapter discusses the main characteristics of antennas, including:

- Antenna gain and sidelobes for linear radiators, simple planar arrays, and circular arrays;
- Antenna patterns with low sidelobes, Chebyshëv, Zolotarëv, Taylor, and Bayliss distributions;
- Phased arrays with quantizing effects;
- Antenna losses, including beam shape and scanning losses;
- Synthesis of beams;
- Equivalence of active and passive antenna systems;
- Introduction to sidelobe cancellers and synthetic apertures.

Antennas radiate and bunch the transmitted energy in the correct direction and provide a gathering area for the returning echo energy. The most common form of antenna, currently, is a large parabolic reflector fed by a horn (or group of horns) shown diagrammatically in Figure 5.1. The larger reflector dimension is between tens and many hundreds of wavelengths long.

Here the horn is shown at the center of the reflector, which causes blockage. Offset feed systems have the horn placed outside the edge of the reflector so that little or no transmitted energy is reflected back to the horn. At the center of the beam, the electromagnetic waves are in the same phase as they would be if they came from the directrix. On transmission the energy from the horn is concentrated in one direction. During reception, the area of the reflector collects the echo signals and presents them in phase at the mouth of the horn.

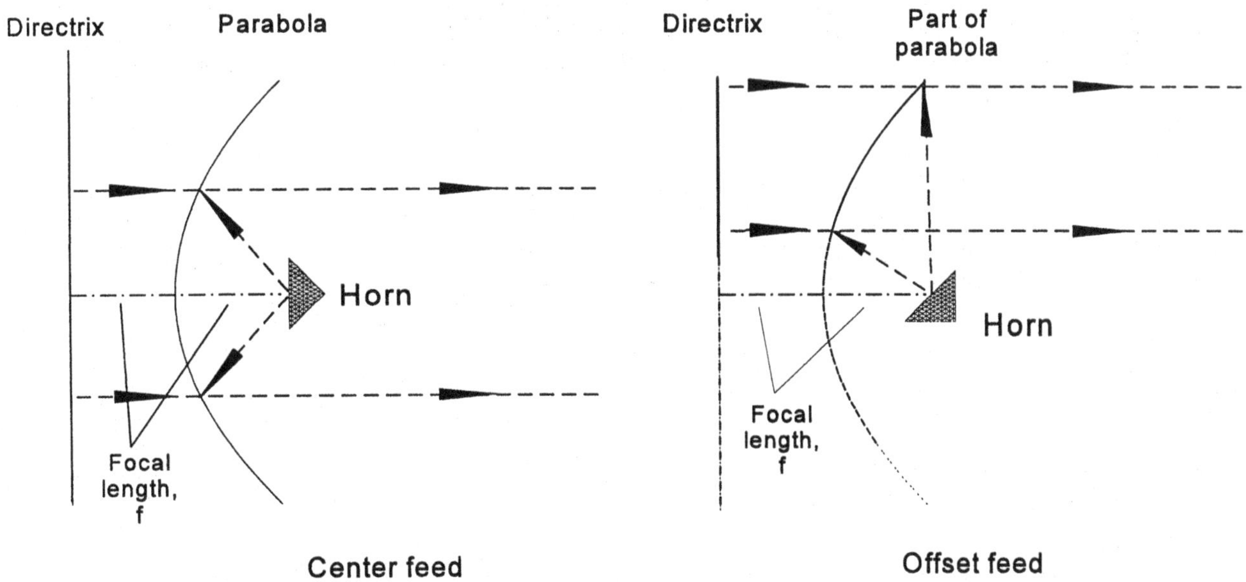

Figure 5.1 Reflector antennas.

The antennas are normally focused at infinity, that is, the horn is at the focus of the parabola to give the narrowest possible beam in the far field. The distance between the horn and the nearest point on the parabola is the focal length, *f*. An imaginary line or plane exists behind the parabola so that any point on the parabola is equidistant from the focus (the focal point in the horn) and the directrix. These antennas may be treated as if the directrix line or plane was a linear or planar array.

As in optics, a beam may be formed using a lens. In Figure 5.2, the lens is formed using waveguides and the focal plane, or an equivalent radiating plane exists through the focal point of the horn.

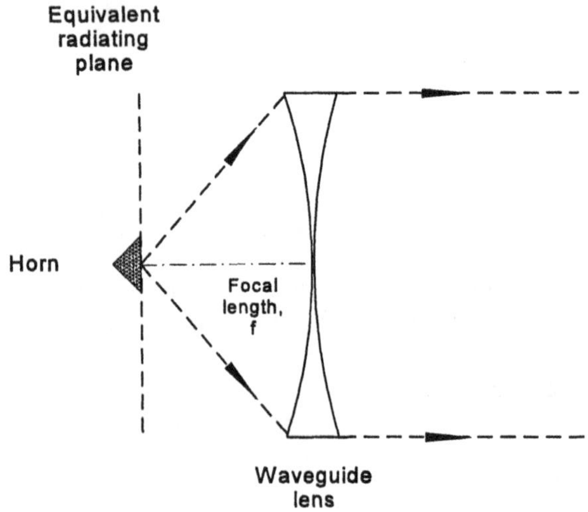

Figure 5.2 The equivalent radiating plane of a lens antenna.

In this chapter these antennas are treated by considering the radiation as if it came from the directrix or the focal plane.

The same effect may be achieved by transmitting pulses with a defined phase at defined points while traveling along the directrix and receiving the echoes at these points. The echo signals are stored and processed in phase as they would be by the reflector and horn. This is the basis of synthetic aperture radar.

The linear direction in Figure 5.1 may be extended to a plane in the same form as the equivalent radiating plane in Figure 5.2. The radiation pattern from a rectangular continuous surface, defined in two Cartesian dimensions, is the product of the patterns that would have been formed in each dimension (Figures 5.10 and 5.11 are examples). Circular surfaces require the separate treatment given in Section 5.2 with Figures 5.21 and 5.22 as examples. The surfaces may be composed of discrete radiators that have their own patterns, as described in Section 5.4. The radiation pattern is that for the surface times the pattern for the individual elements.

5.1 LINEAR AND RECTANGULAR RADIATORS

Equation (5.1) represents the radiation pattern from a uniformly excited line source. The field from a point source at a distance s radiating at a frequency f in a two-dimensional plane is given by [1, p. 428]

$$dE = \frac{K_1}{2\pi s} \exp\left(j\, 2\pi f \left(t - \frac{s}{c} \right) \right) dx \tag{5.1}$$

where K_1 is an amplitude constant;
$2\pi ft$ is the phase of the transmitted wave;
$c = f\lambda$ is the velocity of light;
$2\pi f \frac{s}{c} = 2\pi \frac{s}{\lambda}$ is the phase lag caused by the time delay between source and distant point.

Figure 5.3 shows a wavefront radiated in the direction θ_m from a line source or a notional directrix in an antenna using a reflector of length w [2, pp. 26, 174]. The currents driving the elements of the line are of normalized amplitude $g(x)$ and have ϕ radians phase difference per unit length along the line. Two elemental radiators A and B distance dx apart are shown with the elemental Huygens wavefronts. At a distant point, the components add, so that

$$E = K_1 \int_{-w/2}^{+w/2} \frac{g(x)}{2\pi s} \exp\left(j\, 2\pi f \left(t - \frac{s}{c}\right)\right) dx \tag{5.2}$$

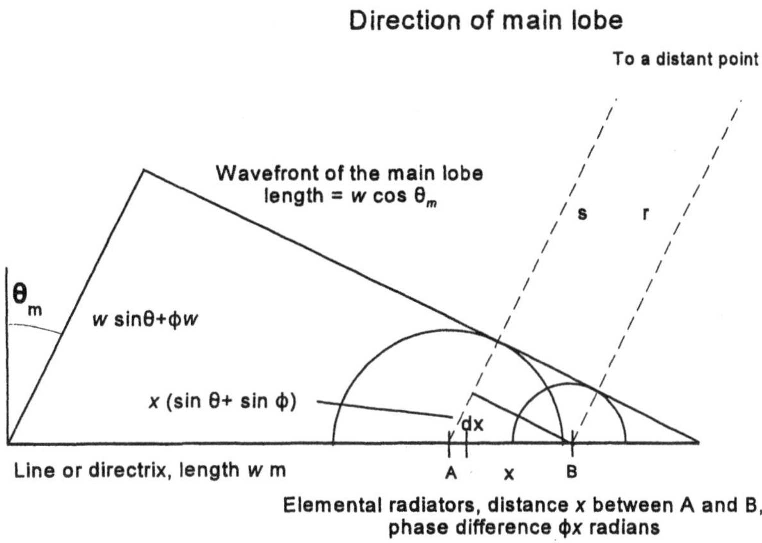

Figure 5.3 Geometry of the radiation from a line source.

If s is written in terms of r, that is, $s = r - x(\sin\theta + \sin\phi)$ and $\sin\phi = -\sin\theta_m$, then

$$\begin{aligned} E &= \frac{K_1}{2\pi(r - x\sin\theta)} \int_{-w/2}^{+w/2} g(x) \exp\left(j\, 2\pi f \left(t - \frac{r - x(\sin\theta - \sin\theta_m)}{c}\right)\right) dx \\ &= \frac{A \exp\left(j\, 2\pi\left(ft - \frac{r}{\lambda}\right)\right)}{2\pi(r - x\sin\theta)} \int_{-w/2}^{+w/2} g(x) \exp\left(j\, \frac{2\pi x(\sin\theta - \sin\theta_m)}{\lambda}\right) dx \end{aligned} \tag{5.3}$$

If $x \sin\theta$ is much smaller than r, then

$$E = \frac{A}{2\pi r} \exp\left(j\, 2\pi\left(ft - \frac{r}{\lambda}\right)\right) \int_{-w/2}^{+w/2} g(x) \exp\left(j\, \frac{2\pi x(\sin\theta - \sin\theta_m)}{\lambda}\right) dx \tag{5.4}$$

Equation (5.4) consists of two parts:

- To the left of the integral sign: Amplitude and phase caused by distance to the distant point;
- The integral: Amplitude and phase caused by angle off the axis, or antenna array factor.

The integral is the Fourier transform of $g(x)$ in terms of $(\sin\theta - \sin\theta_m)/\lambda$. If the illumination function is uniform, that is, $g(x) = 1$, then the normalized antenna pattern is given by

$$Antenna\ pattern = \frac{\sin\left(\frac{w\pi}{\lambda}(\sin\theta - \sin\theta_m)\right)}{\frac{\pi}{\lambda}(\sin\theta - \sin\theta_m)} \tag{5.5}$$

The function has the value w when $\theta - \theta_m = 0$. Often, this is divided by the length w to give the normalized function, which has a maximum value of unity.

$$Normalized\ antenna\ pattern = \frac{\sin\left(\frac{w\pi}{\lambda}(\sin\theta - \sin\theta_m)\right)}{\frac{w\pi}{\lambda}(\sin\theta - \sin\theta_m)} \tag{5.6}$$

For a more general treatment compatible with the other chapters, normalized functions representing angles are used, $u' = \frac{w}{\lambda}(\sin\theta - \sin\theta_m)$, then the function becomes

$$\frac{\sin(\pi u')}{\pi u'} \quad \text{or} \quad \operatorname{sinc} \pi u' \tag{5.7}$$

The distant point where the far field is fully formed is a matter of judgement, normally taken to be where the wave from the antenna approximates to a plane wave. Figure 5.4 shows a wavefront from the center of the directrix at a distance $D + x$. This wavefront is a distance D from the edges of the antenna, or the inaccuracy from a plane wavefront is x. The criteria for choosing x are $\lambda/16$ or, occasionally, $\lambda/32$.

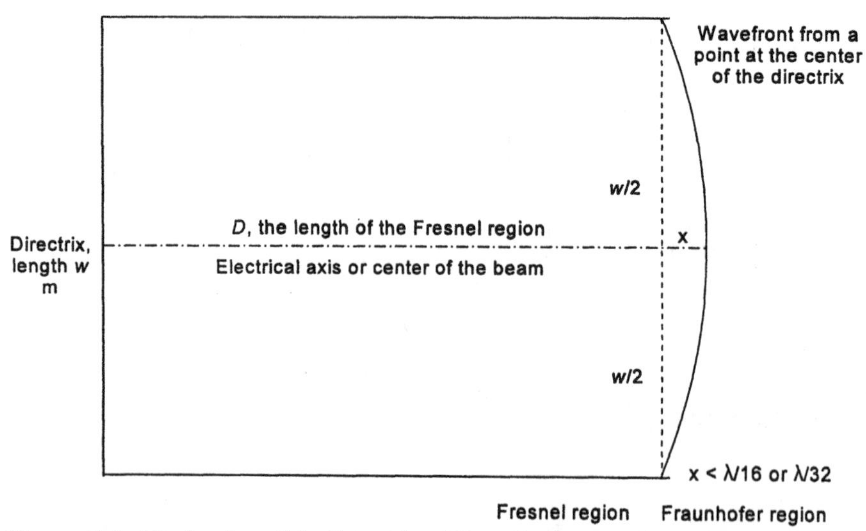

Figure 5.4 The border of the Fresnel and Fraunhofer regions, where the wavefront becomes almost flat.

Using the theorem for the product of intersecting chords in a circle

$$\begin{aligned} x(2D + x) &= \left(\frac{w}{2}\right)^2 \\ D &\approx \frac{w^2}{8x} = 2\frac{w^2}{\lambda} \text{ or } 4\frac{w^2}{\lambda} \text{ for greater precision} \end{aligned} \tag{5.8}$$

Conventionally, the region within the distance D is called the Fresnel region, and the region beyond D is called the Fraunhofer region. All considerations of beam shape conventionally apply only in the Fraunhofer region where the beam has been properly formed. The transmitting antennas used for antenna measurements must be in the Fraunhofer region. For an S-band radar, with an antenna 10 m in diameter, the Fraunhofer region starts between 2 km and 4 km from the antenna.

The beam thus formed sends most of the energy along the desired direction, but a measurable amount of energy is found to be sent at other angles, called sidelobes, as with the diffraction bands when light shines through a slit. The sin *x*/*x* pattern from a uniformly illuminated source gives the narrowest main beam and is shown in Figure 5.5.

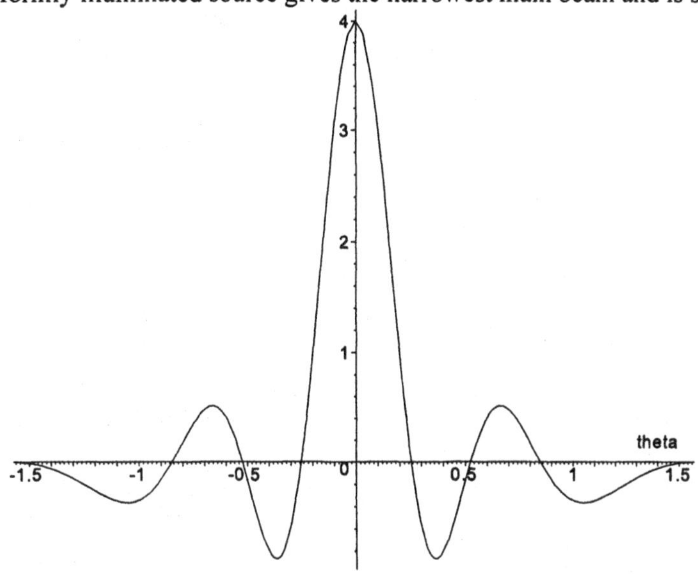

Figure 5.5 The voltage pattern for a 4λ line antenna.

Antenna diagrams are drawn conventionally with a linear angle scale on the *x*-axis. The example shown in Figure 5.5 is a 4λ line antenna with a voltage gain of 4. The phase angle between each adjacent (side)lobe is π radians, or 180 degrees. Normally, the pattern is plotted as a power pattern in decibels with the maximum power conventionally 0 dB. This allows the plotting of a greater dynamic range and allows the sidelobe levels to be read directly from the y scale. This is shown in Figure 5.6. Notice that the indication of phase shift from sidelobe to sidelobe is no longer shown. In real life, the antenna radiates according to the polar diagram, which is shown in Figure 5.7.

The width of the beam is conventionally taken to be the angle between points where the power measured by a probe is half the maximum power. This is the half-power or 3dB beamwidth, which can be obtained by a numerical solution of (5.6).

$$\begin{aligned} \textit{Half-power (3 dB) beamwidth} &= 0.88589 \frac{\lambda}{w} \text{ radians} \\ &= 50.75793 \frac{\lambda}{w} \text{ degrees} \end{aligned} \tag{5.9}$$

Other common measures are the standard beamwidth and the Rayleigh beamwidth, both given by arcsin(λ/w) or λ/w radians for small angles. The Rayleigh beamwidth is the angle from the peak to the first null.

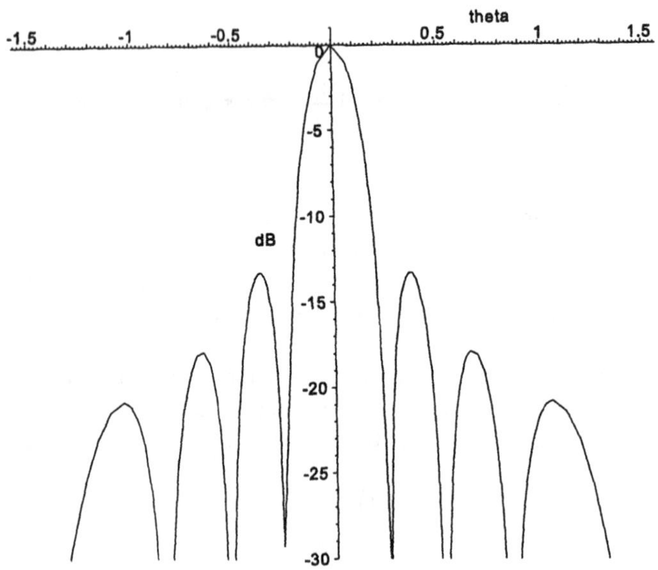

Figure 5.6 The antenna characteristic for a 4λ line antenna.

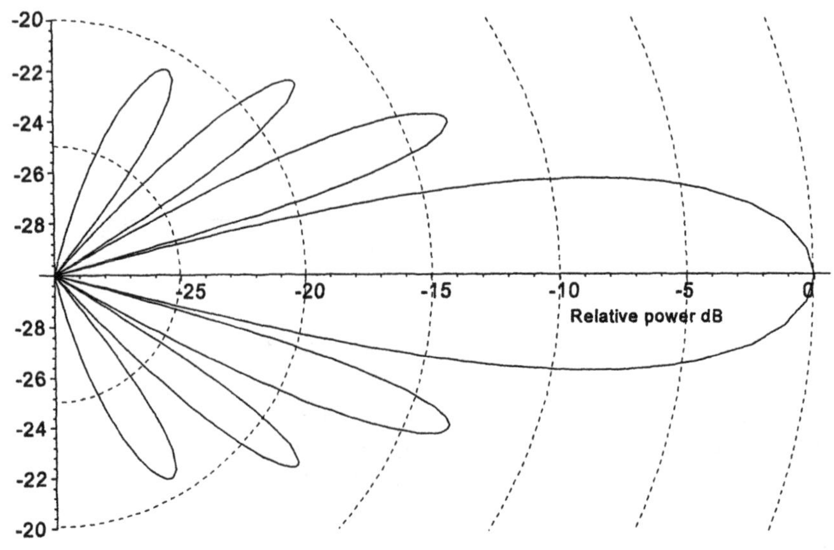

Figure 5.7 Polar diagram for a 4λ line antenna.

The amplitudes of the sidelobes in the example are as follows:

- First sidelobe amplitude: -13.261 dB;
- Second sidelobe amplitude: -17.830 dB;
- Third sidelobe amplitude: -20.788 dB.

A progressive phase shift along a 10λ line source moves the beam as shown in Figure 5.3, and the patterns arising from this are illustrated in Figure 5.8. Looking down the beam, the projected length of the antenna is $w \cos \theta$, which widens the beam by a factor of $1/\cos \theta$ and reduces the power gain by $\cos \theta$ or the voltage gain by $\sqrt{\cos \theta}$. Where the patterns of the elemental radiators are not quasi-isotropic, their patterns reduce the voltage gain by between $\cos^{1/2} \theta$ and $\cos^{3/4} \theta$.

The amplitude and the phase of the aperture function needed to give a broadside beam and a beam with 15 degrees or $\pi/12$ radians shift of the beam is shown in Figure 5.9.

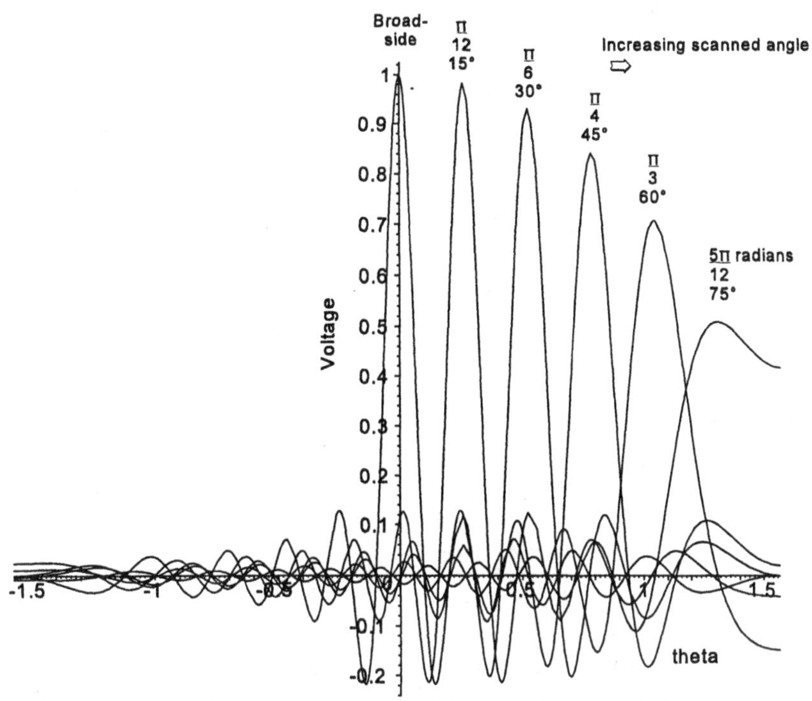

Figure 5.8 The effects of scanning on the beam from a 10λ line source.

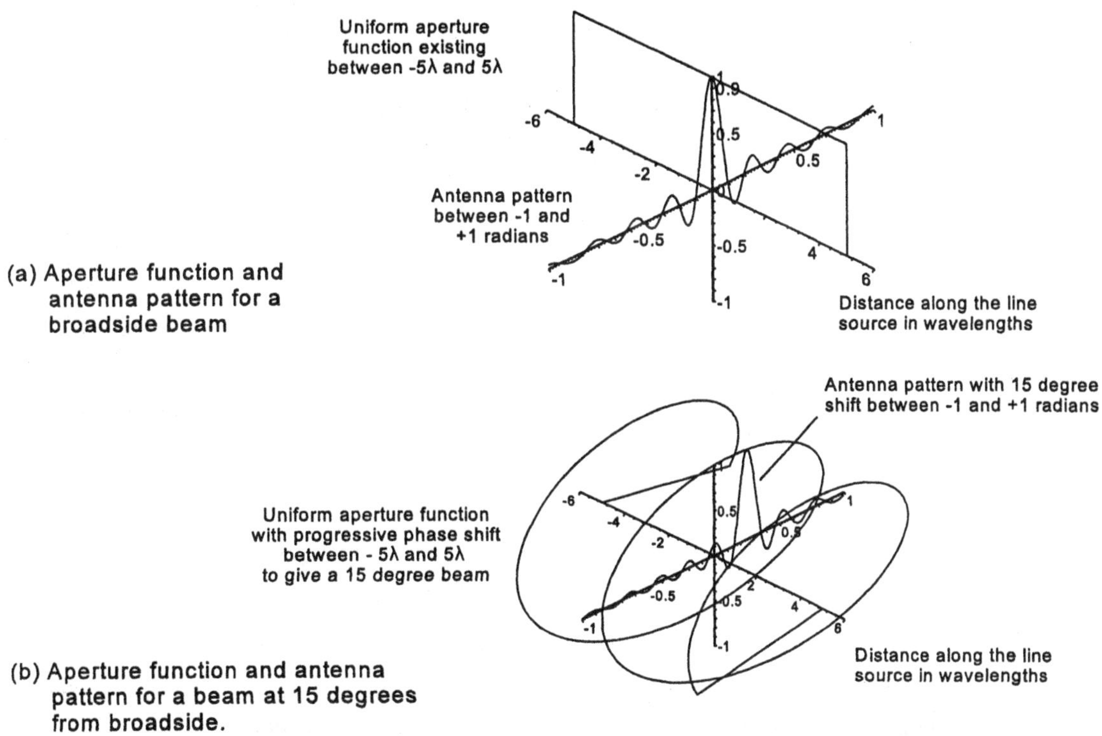

Figure 5.9 A three-dimensional representation of aperture function for a 10λ line source and antenna patterns without and with scanning.

The antenna is the main filter for sorting wanted from unwanted signals and echoes. For radar purposes, the transmission and reception through the sidelobes add clutter from these angles to the receiver noise, making detection more difficult. Interference, in times of military conflict jamming, may enter the antenna at angles where there are sidelobes. In most radars, the reflector is illuminated from a horn that sends most of its energy to the center, and the energy density is tapered towards the edges to give a broader main lobe with reduced sidelobes.

A number of radars have an antenna with a linear source illuminating a cylindrical reflector. The line source allows a close control of the beam shape and sidelobes in the azimuth dimension. The shape of the cylinder gives the elevation pattern.

Most antennas have large reflectors or planar radiating surfaces that shape the beam in the azimuth and elevation dimensions. The illumination function and the beam pattern are the product of the line patterns in the two dimensions (see Section 5.1.1). Such a pattern from a square aperture, traditionally plotted in Cartesian coordinates, is shown in Figure 5.10. Alternatively the plot may be made with scales in sin θ and sin ϕ each in the range ±1. This is called sine space. The major sidelobes occur in the cardinal planes with smaller intercardinal plane sidelobes between them.

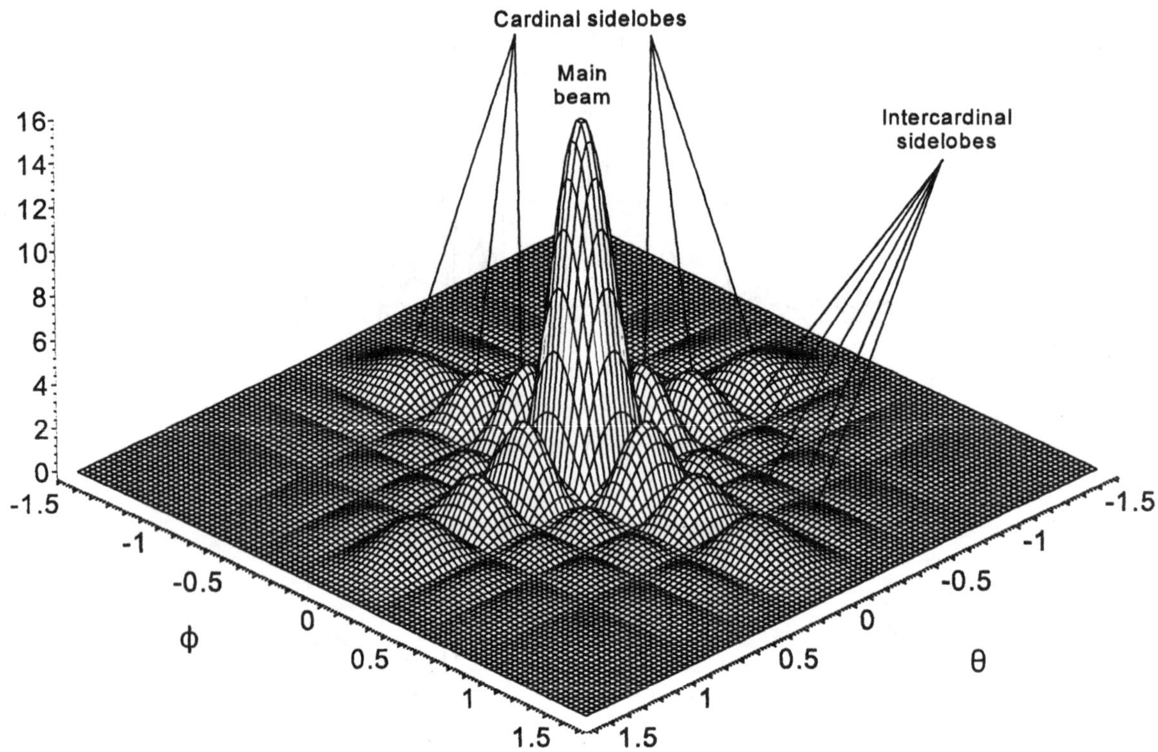

Figure 5.10 Two-dimensional antenna diagram from a square aperture with uniform illumination.

In real life, the azimuth and elevation dimensions are angles in polar coordinates. The antenna pattern using a logarithmic decibel scale is as in Figure 5.11.

The qualities of antennas are described in terms of gain, beamwidth, and sidelobes. The gain of a receiving antenna is defined at the maximum point as

$$Gain = \frac{Signal\ power\ delivered\ at\ the\ output\ terminals\ of\ the\ antenna}{Signal\ power\ delivered\ at\ the\ output\ terminals\ of\ an\ isotropic\ antenna} \tag{5.10}$$

A similar definition for transmitting gain substitutes the power density of the power density of the transmitted wave for signal power in (5.10). The gain includes the losses between the points where the waves excite the antenna and the output terminals. Where these losses are not taken into account the term *directivity* is used, namely,

$$Directivity = \frac{Signal\ power\ delivered\ at\ the\ output\ terminals\ of\ the\ antenna\ (neglecting\ internal\ losses)}{Signal\ power\ delivered\ at\ the\ output\ terminals\ of\ an\ isotropic\ antenna} \tag{5.11}$$

An isotropic antenna radiates or receives the same signal power in all directions. If in a chosen antenna all the

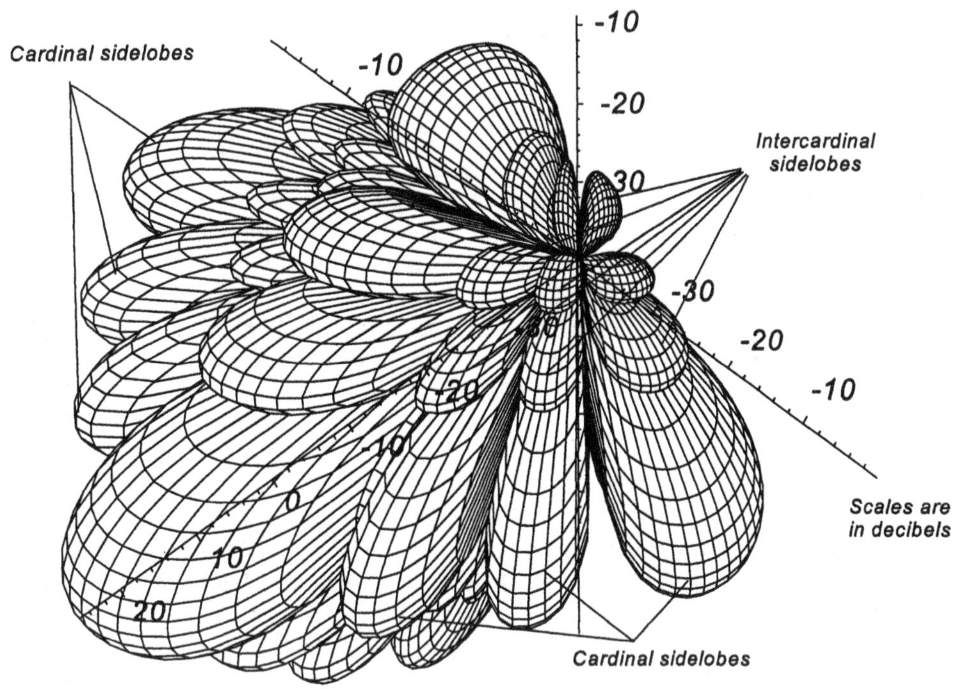

Figure 5.11 Three-dimensional plot of the antenna characteristic.

signals are radiated only in a solid angle Ω (Figure 5.12), then equating the volumes

$$\begin{aligned} \text{Volume of sphere of unity gain, } \frac{4}{3}\pi\, 1^3 &= \text{Volume of cone, } \frac{1}{3}\,\Omega\, D \\ \text{Directivity, } D &= \frac{4\,\pi}{\Omega} \end{aligned} \tag{5.12}$$

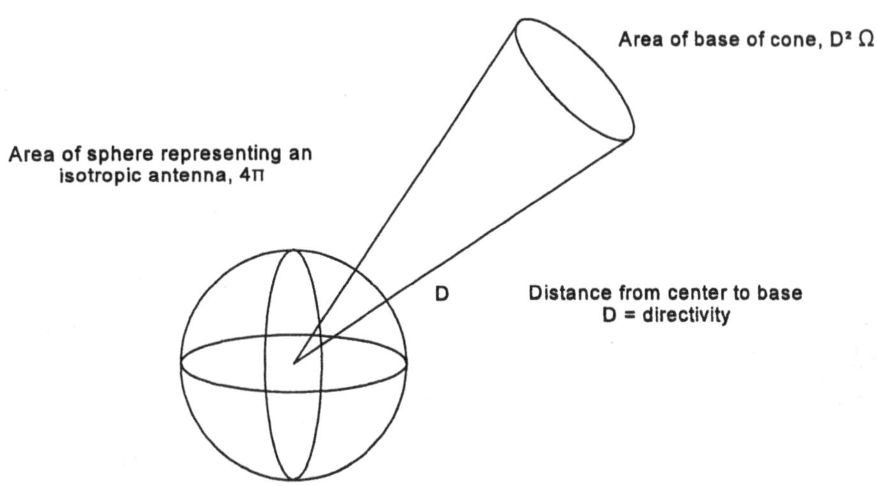

Figure 5.12 Comparison of an isotropic antenna with a directional antenna of directivity *D*.

Alternatively, the antenna in Figure 5.12 can be considered to have the area $D^2\,\Omega$ at the base of the cone. If it is compared with the area of an isotropic antenna $\lambda^2/4\pi$ [3, Section 2.14], then

$$\text{Directivity, } D = \frac{\text{Antenna area, } A}{\text{Isotropic antenna area}} = \frac{A}{\frac{\lambda^2}{4\pi}} = \frac{4\pi A}{\lambda^2} \tag{5.13}$$

For rectangular apertures of length w and height h, the area is wh, giving a directivity of $4\pi\ wh/\lambda^2$ [2, p. 390]. For line sources very many wavelengths long ($w \gg \lambda$) that are uniformly excited, the contribution from each half-wavelength add in the broadside direction. When compared to an isotropic source, it has a directivity of $2w/\lambda$ [2, p. 182, Eq. 4-21]. The standard beamwidth is λ/w radians.

In a practical antenna, the greater power in the solid angle Ω is achieved at the expense of radiation at other angles. Not all the signal power collected is present at the antenna terminals, reducing the electrical gain by the aperture efficiency, η. Directivity and gain are commonly given in decibels.

$$\text{Gain, } G = D\eta \tag{5.14}$$

The antenna is the first filter in the chain, and the main beam is used to direct and concentrate the energy from the transmitter and select the wanted returning echoes. Each sidelobe acts as a separate antenna to send energy and receive interfering signals or environmental noise. In the case of clutter, if the main beam is pointed away from clutter, the sidelobes often point towards the clutter, so this can become a problem. The elevation view in Figure 5.13 shows aircraft echoes in the main lobe with rain and clutter echoes entering through the elevation sidelobes. Rain and clutter echoes are often much stronger than the wanted aircraft echoes, so the smaller the response of the antenna in the direction of clutter and rain, the better the separation. In the case of marine and airborne radars, the clutter moves in opposite directions on either side of the ship or aircraft. Additionally with airborne radars, the ground under the radar gives a very strong echo, the altitude line. This sidelobe clutter widens the clutter spectrum and makes clutter filtering more difficult.

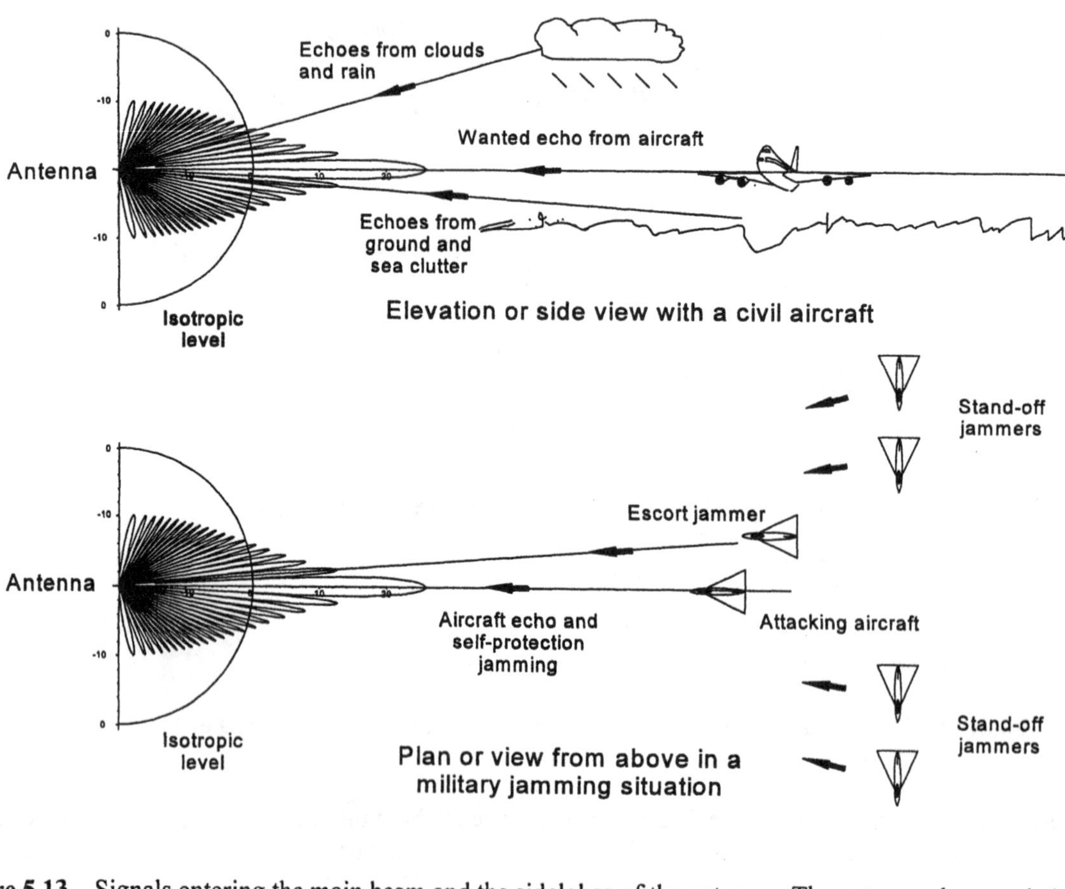

Figure 5.13 Signals entering the main beam and the sidelobes of the antenna. The antenna characteristic has been drawn using a logarithmic scale (dB).

In addition to the considerations of clutter and weather, military air surveillance radars are constructed to work in the presence of deliberately generated interference, called active jamming, and the antennas are designed accordingly. The plan view in Figure 5.13 shows a military situation with an attacking aircraft, an escort jamming aircraft, and barrage jammers. Attacking aircraft, which carry weapons, are the main danger. They may carry a small self-protection jammer to deflect anti-aircraft weapons used against them. If the jammer is switched on too soon, the jammer reveals the attacking aircraft's direction. When the attacking aircraft enter the surveillance range of the radar, escorting aircraft packed with electronic countermeasure (ECM) equipment try to mask the echoes from the attacking aircraft by transmitting a jamming signal. The jamming signal must be powerful enough to give an interfering signal around the formation when entering the antenna through the near sidelobes. Stand-off jammers that operate outside the range of the air defences produce a barrage of jamming signals using powerful jammers, often carried in transport aircraft, to penetrate the antenna sidelobes. A summary is given in Table 5.1.

Table 5.1
Types of military jammers and jamming

	Type of jammer	Comment
Active countermeasures	Self-protection or self-screening jammers	A light jammer for use against anti-aircraft weapons.
	Escort jammers	Heavier jammers carried in aircraft similar to the attacking aircraft. They need extra power to give the same effect when jamming through the near sidelobes.
	Barrage jammers	Heavy jammers often carried in transport aircraft. These need much greater power to be able to jam radars via the far sidelobes and for the extra distance so as to be outside the shooting range of the air defense systems.
Passive countermeasures	Decoys, chaff, window	Metal strips called window, chaff, or rope have the same confusion effect as rain.

Military antennas are designed to have as narrow a main beam as reasonable to give higher resolution and the lowest sidelobes possible to reject echoes and jamming outside the main beam. A narrow beam in the elevation dimension reduces the reception of active jamming at other elevation angles. The height extent for rain or passive countermeasures (chaff or window) that have the same horizontal velocity component as wind is also reduced. The wind speed varies with altitude, and the variation may be great enough to give echoes with Doppler frequencies that spread up to the pulse repetition frequency to fill the Doppler frequency domain completely.

Low azimuth sidelobes are not so critical for civil surveillance radars, typically a first sidelobe level of 25 dB is specified. Antennas are often specially developed with a low sidelobe level on the underside of the beam to reduce clutter.

5.1.1 Tapering the illumination function to reduce sidelobes

Radar was first developed for military purposes. The early surveillance radars often used a line feed and a cylindrical reflector. The radiation pattern from the line feed was shaped to provide lower sidelobes and a slightly broader main beam with slightly less gain. The elevation pattern was controlled by shaping the cylindrical reflector. A number of performance factors have been developed to rank these illumination functions in comparison with the uniform illumination function and are shown in Table 5.2.

Table 5.2
The effects of the antenna parameters

Antenna parameter		Use
Gain, efficiency, and pattern		Radar range equation
Beamwidth		Used to calculate angular resolution
Sidelobes	First or near sidelobes	Range in the presence of clutter and escort jammers
	Far sidelobes	Range in the presence of clutter and barrage jammers

The antenna patterns from rectangular apertures are found by multiplying the patterns in the two linear directions. It is therefore customary to consider a normalized linear aperture existing between -1/2 and +1/2. If the width of the aperture is w, the normalized distance variable x (between $-w/2$ and $+w/2$) becomes $x' = x/w$, which has values between -1/2 and +1/2. The notation is broadly in agreement with [4].

As with the early antennas, the pattern from a rectangular aperture with the dimensions w_1 by w_2 is the product of the patterns from two independent orthogonal linear arrays, each with a directivity of w_1/λ and w_2/λ, respectively. The radiation pattern is shaped by reducing the excitation of parts of the array using the normalized function $g(x')$ or normalized over the width $g(x')$. If the normalized length is unity, the effective normalized length is A compared with a uniformly excited array. Note that the symbol A is chosen to be consistent with Appendix B.

$$A = \int_{-\frac{1}{2}}^{+\frac{1}{2}} g(x')\, dx \qquad \text{no dimensions} \tag{5.15}$$

The normalized effective power radiated, C, is proportional to the square of the excitation currents $(g(x'))^2$. Thus,

$$C = \int_{-\frac{1}{2}}^{+\frac{1}{2}} (g(x'))^2\, dx \qquad \text{no dimensions} \tag{5.16}$$

The efficiency of the linear array compared with a uniformly excited array, η_x, is

$$\eta_x = \frac{(\textit{Effective aperture length})^2}{\textit{Effective power radiated}} = \frac{A^2}{C} \qquad \text{no dimensions} \tag{5.17}$$

The common ways of describing beamwidths are shown in Table 5.3.

Table 5.3
Ways of describing beamwidth

	Symbol	Comment
Half-power or 3 dB beamwidth	θ_3	Used commonly and is found by solving the equation for the pattern at the half-power points
Half-voltage or 6 dB beamwidth	θ_6	This is found by solving the equation for the pattern at the half-voltage points
Standard beamwidth	$\arcsin(\lambda/w)$	Simple expression, often used in theory
Rayleigh beamwidth		Angle between the peak and the first null
Noise beamwidth	θ_n	For uniform illumination this is equal to the standard beamwidth λ/w (radians) for uniform illumination

For tapered illumination the noise beamwidth is greater than the standard beamwidth or $\theta_n\lambda/w > 1$, and is given by

$$\theta_n = \frac{C}{A^2} \qquad \text{no dimensions} \tag{5.18}$$

The positions and amplitudes of first and far sidelobes are calculable and depend on the illumination function and blocking (see Section 5.7.2). The far sidelobes decay from the first sidelobe but a considerable contribution comes from diffraction around the reflector or planar array, signals entering the feed horn directly, and leakage through the reflector mesh. All sidelobes must be measured on a test range.

Typical even illumination functions for linear and rectangular apertures are as follows:

- Uniform, trapezoidal, and triangular;
- Simply curved illumination functions:
  - $[1 - 4x^2]^n$;
  - $\cos^n x$;
  - Cosine on pedestal;
  - Truncated Gaussian;
- Low sidelobe illumination functions:
  - Dolph-Chebyshëv, for discrete illumination functions only (see Section 5.4, Chapter 11, Signal processing, and Appendix B, Tapering functions);
  - Taylor.

Most radars have antennas that consist of one or more horns and a reflector. The horns normally have a curved radiation characteristic with a peak along their axis. They give radiation patterns different from those from uniform illumination, which may be approximated by the $[1 - 4x^2]^n$, $\cos^n x$, cosine on pedestal, and the truncated Gaussian illumination characteristics. In practice, all antenna characteristics are measured to verify the theoretical design. Early metric radars, which used arrays of dipoles, changed from using linear illumination to shaped illumination functions, and microwave radars using waveguide feeds and cylindrical reflectors often used Chebyshev illumination functions.

The graphs of the characteristics that follow are normalized in terms of w/λ. An antenna of width w handling signals of wavelength λ with uniform illumination has the following:

- One main lobe of width $\sin^{-1} 2\lambda/w$ radians between the zeroes of the characteristic;
- $2(w/\lambda - 1)$ complete sidelobes, which exist from zero to zero, of width λ/w between the zeroes. With values of w/λ which are not integers, there is part of a sidelobe at ±90 degrees, or $\pm\pi/2$ radians.

5.1.2 Uniform, trapezoidal, and triangular illumination tapering

The uniform illumination function is used when the narrowest beam regardless of sidelobes is required. The triangular function is the other extreme of linear shaping, which produces regular sidelobes with the same phase. Trapezoidal illumination, shown in Figure 5.14, is the difference of two triangular illumination functions.

$$Trapezoid = Large\ triangle - small\ triangle \tag{5.19}$$

The functions have periods depending on w, the full antenna width, for the large triangle and w_t, the flat part of the trapezium, for the small triangle. The sidelobes of different periods can be seen to be beating, producing an irregular pattern. The antenna pattern from trapezoidal illumination is similar to the spectrum of a trapezoidal transmitter pulse (see Chapter 3, Transmitters). The first sidelobe varies from 13.26 dB below the main lobe to 26.52 dB below the main lobe for triangular illumination. The efficiency is reduced from 100% to 75%, and the beamwidth with triangular illumination is 1.28 standard beamwidths (λ/w), or 1.445 times the beamwidth from uniform illumination.

5.1.3 Simply tapered illumination functions

Simple functions used to investigate horn-fed reflector antennas have been $[1 - 4x^2]^n$ and $\cos^n x$ illumination functions. In practice, the reflector illumination does not reduce to zero at the edges, which gives rise to the cosine on pedestal and truncated Gaussian illumination functions. In the illumination function,

$$g(x) = (1 - (2x)^2)^n \tag{5.20}$$

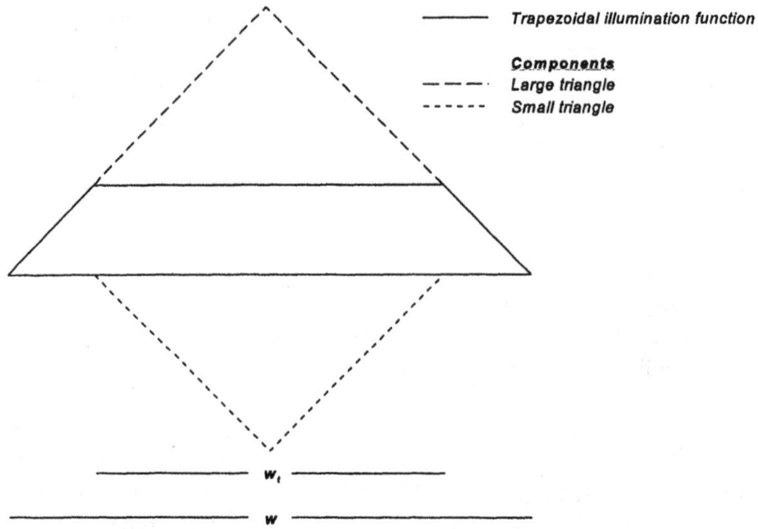

Figure 5.14 Two triangular components of trapezoidal illumination.

the value of n may be zero (for uniform illumination), unity (for parabolic illumination), to higher powers which give lower sidelobes. The function

$$g(x) = (\cos \pi x)^n \tag{5.21}$$

has lower sidelobes than the $(1 - 4x^2)^n$ function with corresponding lower efficiency.

The above functions assume that the illumination at the edges of the aperture is zero. If a horn illuminates a reflector, energy is radiated by the horn beyond the limits of the reflector. This energy, with the energy that leaks through the reflector, contributes to the backlobe. The cosine on pedestal and truncated Gaussian functions are attempts to model the shapes of beams formed by reflection from the reflector. The illumination functions of array antennas are able to be defined more exactly to use illumination functions that are not zero at the edges (Taylor, for example).

The cosine on pedestal illumination function is given by

$$g(x) = k + (1 - k)\cos(\pi x) \tag{5.22}$$

The Gaussian waveform gives a Gaussian spectrum without sidebands but exists over all time (see Sections 2.9.2 and 2.9.3). Gaussian illumination is finite and is cut off (truncated) at the sides of the reflector, which give sidelobes. The function is given by

$$g(x) = \exp(-1.386\ (n\ x)^2) \tag{5.23}$$

Examples of these functions are given in Appendix B.

5.1.4 Low-sidelobe tapering functions

Dolph used Chebyshëv polynomials to create an antenna pattern that were used in the late 1940s to produce antennas with much lower sidelobe levels, mainly with cylindrical reflector antennas fed by slotted waveguides using the slots as discrete radiating elements. The illumination function was derived from the antenna pattern and the sidelobe level attainable depended on how accurately the radiating slots could be cut into the waveguide.

Taylor used the Dolph-Chebyshëv pattern for use with continuous radiators and examples of uniform, Dolph-Chebyshëv, and Taylor patterns are shown in Figure 5.15.

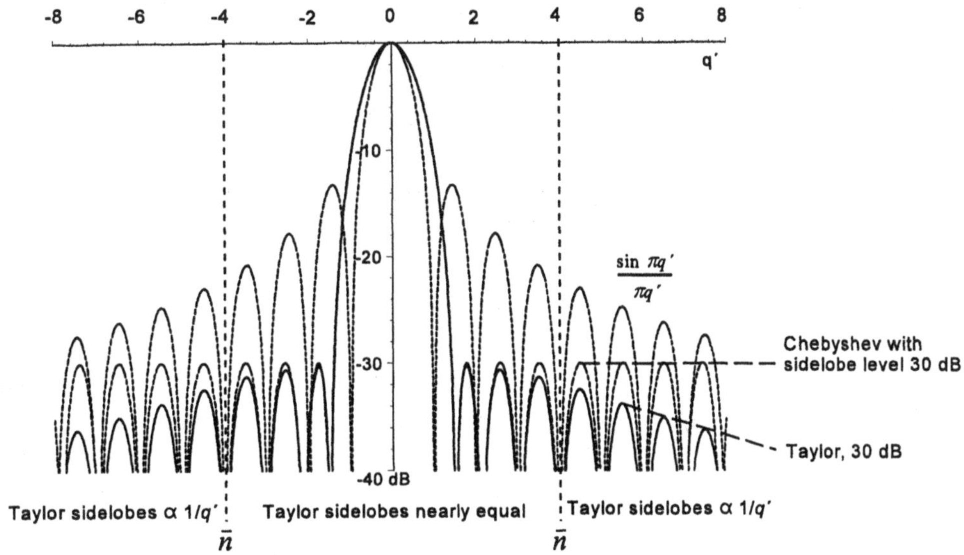

Figure 5.15 Comparison of antenna patterns with uniform, Chebyshëv, and Taylor distributions. [Source: Meikle, H. D., *A New Twist to Fourier Transforms*, Wiley-VCH, 2004.]

5.1.4.1 Dolph-Chebyshëv tapering function

The Dolph-Chebyshëv pattern has equal sidelobes outside the main beam together with minimum main beam spreading for a given sidelobe level. This function is defined for a line of discrete radiators, so strictly this topic belongs in Section 5.4 [5, p. 714]. Chebyshëv polynomials are used for economizing approximations in the domain ±1, in which the error, or ripple, oscillates between ±1. The first seven polynomials are shown in Figure 5.16.

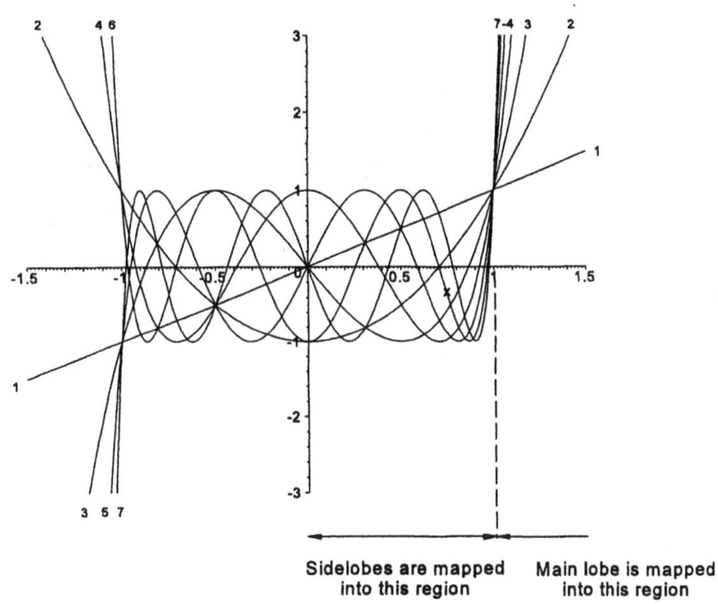

Figure 5.16 Plots of the first seven Chebyshëv polynomials.

Chebyshëv polynomials of degree n are given by [5, p. 775; 6, p. 774]

$$T_n(x) = \begin{cases} -1^n \cosh(n \operatorname{arccosh}|x|) & x < -1 \\ \cos(n \arccos x) & -1 < x < 1 \\ \cosh(n \operatorname{arccosh} x) & x > 1 \end{cases} \tag{5.24}$$

Dolph [7] saw that if the center (cosine) region could be mapped onto the sidelobes and the (hyperbolic cosine) region beyond +1 onto the main lobe, then an illumination function for a pattern with constant sidelobes could be calculated. An N element array has N - 1 zeroes in its antenna pattern. All curves pass through the point 1, 1. The Chebyshëv polynomial has n zeroes, so a polynomial of order n-1 is used for mapping.

The antenna pattern is of the type [5, p. 715]

$$\begin{aligned} G(u') &= T_{N-1}(x_0 \cos \pi u') \qquad \text{where } u' = \frac{w}{\lambda} \sin\theta \\ &= \cos\left((N-1)\arccos(x_0 \cos \pi u')\right) \end{aligned} \tag{5.25}$$

The antenna pattern $G(u)$ must be normalized by dividing it by the maximum value $T_{N-1}(x_0)$. The value u' has a range ±½ so that $\pi u'$ radians varies between ±90 degrees. The value N is the number of elements in the antenna, and x_0 is a scaling factor that defines the sidelobe levels. This is calculated from the Chebyshëv polynomial of degree n beyond +1.

$$T_n(x) = \cosh(n \operatorname{arccosh} x) \qquad x > 1 \tag{5.26}$$

$T_n(x)$ is equal to the desired voltage sidelobe level, SLR. Commonly, the sidelobe level is given in decibels, SLL dB, for example 20, 30, and 40 dB, so that

$$\text{Sidelobe ratio, } SLR = 10^{\frac{SLL}{20}} \quad \text{in the above example 10, 31.62, 100, voltage ratio} \tag{5.27}$$

Solving for x_0, the value of x giving the sidelobe ratio, SLR

$$x_0 = \cosh \frac{\operatorname{arccosh} SLR}{N-1} \tag{5.28}$$

The values of x_0 for a seven element array in the example are 1.127, 1.248, and 1.416.

Figure 5.17 shows the normalized voltage field antenna pattern for a seven element antenna. This has six zeroes. The Chebyshëv polynomial with six zeroes is the $T_6(x)$ polynomial. The antenna diagrams using the $T_6(x)$ polynomials have been plotted for 20, 30, or 40 dB sidelobes on the same diagram. The customary decibel plot is shown in Figure 5.18.

The weighting coefficients are obtained from the inverse Fourier transform of (5.25). A number of methods for the calculation have been developed, one of which uses a large number of whole numbers. They are suitable for calculation by mathematical programs [8, p. 260]. For the Kth weight for an antenna of N elements,

$$\begin{aligned} weight[K] &= \frac{N-1}{N-K} \sum_{S=0}^{K-2} \frac{(K-2)!\,(N-K)!\,\alpha^{S+1}}{S!\,(K-2-S)!\,(S+1)!\,(N-K-S-1)!} \\ \text{where } \alpha &= \left[\tanh \frac{\operatorname{arccosh}(SLR)}{N-1}\right]^2 \\ \text{and } weight[1] &= weight[N] = 1 \end{aligned} \tag{5.29}$$

These weights are all greater than 1 and must be divided by the center weight to give the normalized values which are less than unity.

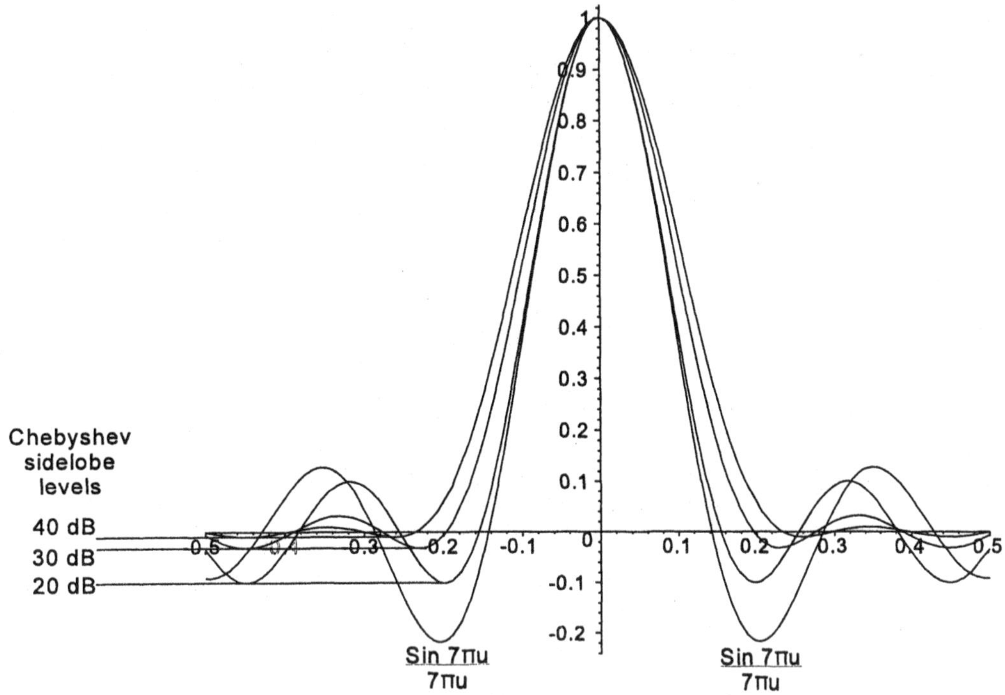

Figure 5.17 Antenna pattern with scaled Chebyshëv polynomials for a seven element array.

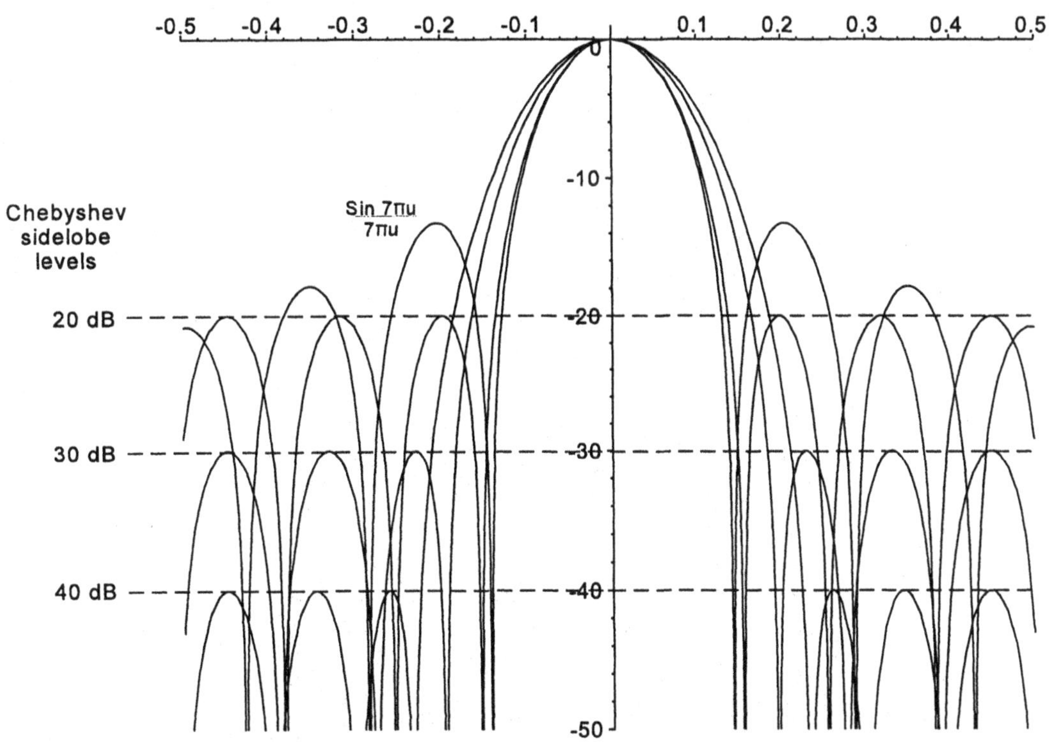

Figure 5.18 Decibel plot of the antenna pattern with scaled Chebyshëv polynomials for a seven element array.

5.1.4.2 Taylor tapering function

Taylor modified the Dolph-Chebyshëv illumination function for continuous illumination [5, p. 719]. As with the Dolph-Chebyshëv taper, first choose the level of the first sidelobe SLL dB. Then,

$$R = 10^{\frac{SLL}{20}} \quad \text{voltage ratio} \tag{5.30}$$

Calculate the value A_t :

$$A_t = \frac{\operatorname{arccosh}(R)}{\pi} \tag{5.31}$$

The antenna pattern for the center region is defined as [5, p. 719]

$$\begin{aligned} G(u) &= \cosh\left(\pi\sqrt{A_t^2 - u^2}\right) \quad \text{for } u < A \qquad \text{main beam} \\ \text{or } &\cos\left(\pi\sqrt{u^2 - A_t^2}\right) \quad \text{for } u > A \qquad \text{sidelobe} \end{aligned} \tag{5.32}$$

Equation (5.32) describes a central beam of amplitude A and sidelobes of amplitude 1. The zeroes of the sidelobes are at $\sqrt{A_t^2 + (n - 1/2)^2}$. Taylor chose to dilate the main and sidelobe widths and null positions inside the $\bar{n}$th sidelobe so that the amplitudes decay with a rate of $1/u$. A value of $\bar{n}$ for the transition point is chosen such that

$$\textit{minimum value of } \bar{n} \geq 2A_t^2 + 0.5 \tag{5.33}$$

The dilation factor, σ, is

$$\sigma = \frac{\bar{n}}{\sqrt{A_t^2 + (\bar{n} - 0.5)^2}} \tag{5.34}$$

which gives the zero locations for the pattern

$$\begin{aligned} z(n) &= \pm\sigma\sqrt{A_t^2 + \left(n - \frac{1}{2}\right)^2} \quad &&\text{for } 1 \leq n \leq \bar{n} \\ &= \pm n &&\text{for } \bar{n} \leq n \leq \infty \end{aligned} \tag{5.35}$$

Using the unit circle approach, the radiation pattern is given by

$$F(u') = \frac{\sin(\pi u')}{\pi u'} \prod_{n=1}^{\bar{n}-1} \frac{1 - \dfrac{u'^2}{z(n)^2}}{1 - \dfrac{u'^2}{n^2}} \tag{5.36}$$

The illumination function is found from the radiation pattern. Typically,

$$g(x') = 1 + 2\sum_{n=1}^{\bar{n}-1} F_n \cos(2\pi n x') \tag{5.37}$$

where the coefficients F_n are given by

$$F_n = \frac{\left((\bar{n} - 1)!\right)^2}{(\bar{n} - 1 + n)!\ (\bar{n} - 1 - n)!} \prod_{m=1}^{\bar{n}-1} 1 - \frac{n^2}{z_m^2} \tag{5.38}$$

A number of Taylor illumination functions and patterns are given in Appendix B. This is not the end of the design and [5, p. 739] lists a number of references for optimization procedures to improve the design.

5.1.5 General rules for tapering

A number of general rules have been derived by Taylor [9, p. 698] and others for tapering functions, namely:

- Symmetric distributions give lower sidelobes.
- $F(u')$ should be an even entire function of u'.
- A distribution with a pedestal produces a sidelobe envelope falling off as $1/u'$ and is more efficient.
- A distribution going linearly to zero produces a far sidelobe envelope falling off as $1/u'^2$.

5.2 RADIATION FROM CIRCULAR APERTURES

The early reflectors were circular or elliptical paraboloids, and this has not changed. They are basically two-dimensional, and the illumination of the reflector, or imaginary directrix, is defined in polar coordinates of the amplitude $g(r)$ at a radius r. The normalized pattern is given by [3, p. 193]

$$F(u', R) = 2 \pi R^2 \int_0^1 f(r)\ r'\ J_0(u' r')\ dr' \tag{5.39}$$

where $u' = 2\pi \frac{R}{\lambda} \sin \theta = \pi \frac{D}{\lambda} \sin \theta$;

r' is the normalized radius variable;

R is the radius of the antenna, D is the diameter;

$J_n(x)$ is a Bessel function of the first kind and order n.

Using the relationship [10, p. 683, eq. 6.5615]

$$\int_0^1 x^{\nu+1}\ J_\nu(a\ x)\ dx = \frac{J_{\nu+1}(a)}{a} \tag{5.40}$$

The beam pattern for uniform illumination ($g(r) = 1$) is

$$F(u', R) = 2 \pi R^2 \frac{J_1(u')}{u'} \tag{5.41}$$

which is illustrated in Figure 5.18. In comparison with the rectangular surface, circular illumination is slightly tapered in the corners, which gives lower sidelobes. The half-power beamwidth is wider than that for a rectangular aperture and is

$$\begin{aligned} Half\text{-}power\ (3\ \text{dB})\ beamwidth &= 1.02899\ \frac{\lambda}{D}\ \text{radians} \\ &= 58.957\ \frac{\lambda}{D}\ \text{degrees} \end{aligned} \tag{5.42}$$

The sidelobe levels for the first sidelobe with uniform illumination are:

- First: -17.570 dB;
- Second: -23.811 dB;
- Third: -27.957 dB.

In contrast with a linear array, all circular arrays are always two-dimensional, that is, the gain depends on the area. For a square antenna with 4λ sides, the directivity is, from (5.13), $\dfrac{4\pi\ 4\lambda\ 4\lambda}{\lambda^2} = 201.062$ or 23.033 dB. For a circular antenna 4λ in diameter or 2λ in radius, the gain is $\dfrac{4\pi\ (2\lambda)^2\pi}{\lambda^2} = 157.913$ *or* 21.984 dB, which is shown in Figure 5.19.

The normalized power characteristic in decibels is shown in Figure 5.20 with a linear angle scale, whereas the antenna pattern is in fact polar, and the polar plot is shown in Figure 5.21.

Circular apertures radiate in the azimuth and elevation dimensions. A plot in cylindrical coordinates (azimuth and elevation in Cartesian coordinates) is shown in Figure 5.22. A polar plot in three dimensions is shown in Figure 5.23.

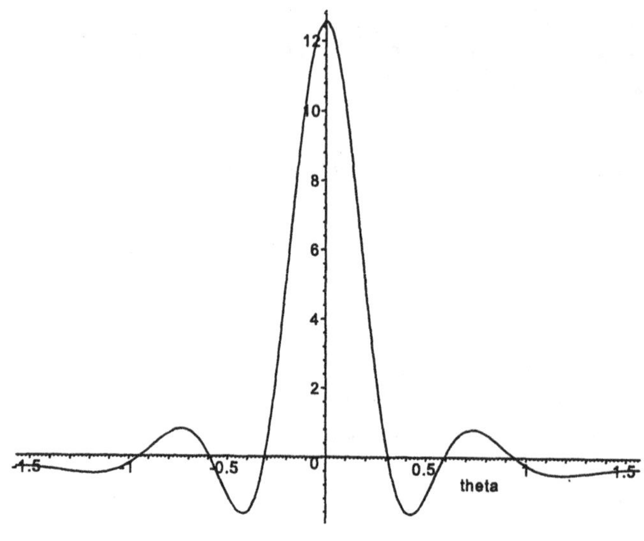

Figure 5.19 The voltage pattern of a circular antenna 4λ in diameter.

Uniformly illuminated circular apertures have lower sidelobes than square apertures of the same diameter. If a circle is cut out of a square and the excitation in the y direction is summed on to a diameter in the x direction, the function represents a tapered illumination of a line source in the x direction.

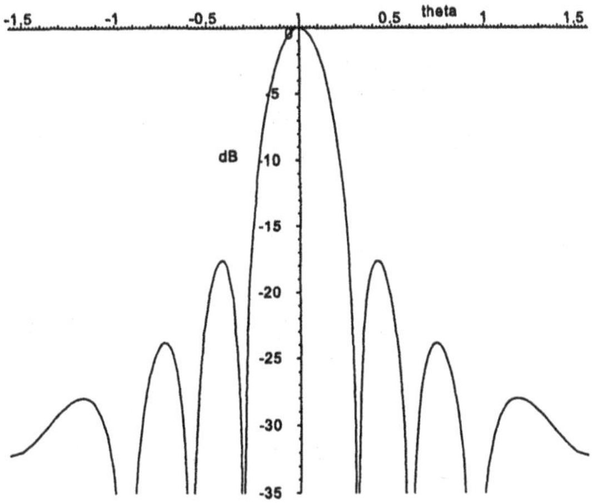

Figure 5.20 The power pattern of a circular antenna 4λ in diameter.

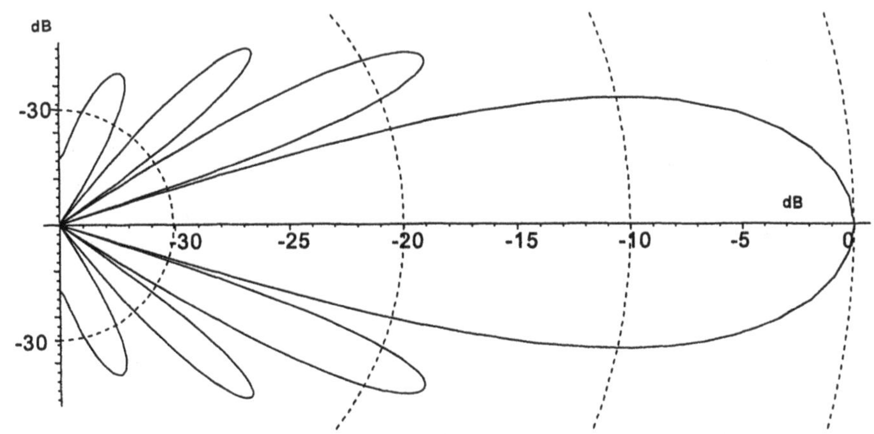

Figure 5.21 The power pattern of a circular antenna 4λ in diameter.

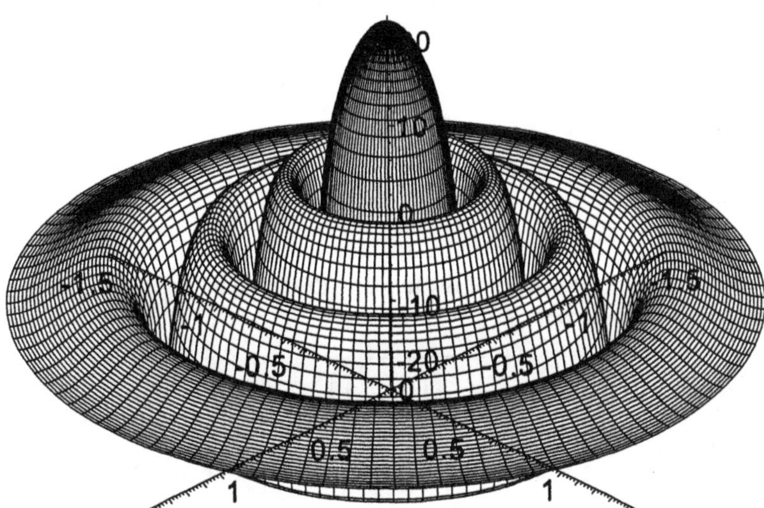

Figure 5.22 Cylindrical plot of the power characteristic of a uniformly illuminated circular antenna 4λ in diameter.

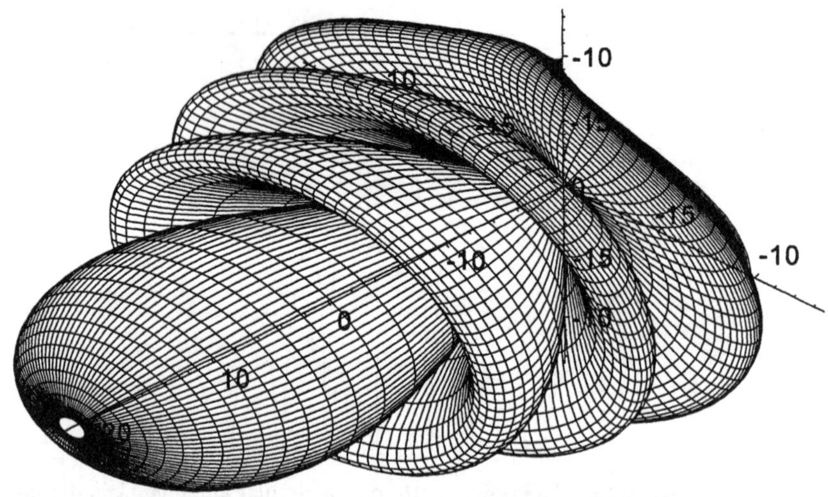

Figure 5.23 Polar plot of a uniformly illuminated circular antenna 4λ in diameter.

5.2.1 Simply shaped circular tapering functions

As with linear illumination functions, there are a number of illuminations defined by the distance from the center of a circular directrix. Those treated here are $[1-4r^2]^n$ and truncated Gaussian. The areas of circles may be compared to rectangles without their corners which gives an equivalent illumination function with even greater taper. Though lower sidelobes are found, the smaller area reduces the radiation efficiency. The characteristics are shown in the figures in Appendix B.

5.2.2 Circular Taylor low-sidelobe tapering function

The circular Taylor illumination function [5, p. 850] is derived from the linear function. First the desired sidelobe level, *SLL* dB, is chosen and the voltage sidelobe level, *R*, calculated.

$$R = 10^{\frac{SLL}{20}} \quad \text{voltage ratio} \tag{5.43}$$

The value of A_t is, again,

$$A_t = \frac{\operatorname{arccosh}(R)}{\pi} \tag{5.44}$$

The zero locations of the Bessel J function $\mu(n)$ of the uniform distribution are the solutions of $J_1(\pi x) = 0$ are replaced inside the transition $\bar{n}$ point by new zeroes

$$z(n) = \pm\sigma\sqrt{A_t^2 + \left(n - \frac{1}{2}\right)^2} \quad \text{for } 1 \le n \le \bar{n} \tag{5.45}$$

The transition point is chosen such that

$$\text{minimum value of} \quad \bar{n} \ge 2A_t^2 + 0.5 \tag{5.46}$$

and the dilation factor, σ, is

$$\sigma = \frac{\bar{n}}{\sqrt{A_t^2 + (\bar{n} - 0.5)^2}} \tag{5.47}$$

The antenna pattern is given by

$$F(u') = \frac{2J_1(\pi u')}{\pi u'} \prod_{n=1}^{\bar{n}-1} \left(\frac{1 - \dfrac{u'^2}{z(n)^2}}{1 - \dfrac{u'^2}{\mu_n^2}} \right) \tag{5.48}$$

and the aperture function is

$$g(r') = \frac{2}{\pi^2} \sum_{m=1}^{\bar{n}-1} \left(\frac{F_m J_0\left(\dfrac{2\pi r \mu_m}{D} \right)}{(J_0(\pi\mu_m))^2} \right) \tag{5.49}$$

where

$$F_0 = 1$$

$$F_m = -J_0(\pi\mu_m) \frac{\prod_{n=1}^{\bar{n}-1} \left(1 - \dfrac{\mu_m^2}{u_n^2} \right)}{\prod_{n=1, n \neq m}^{\bar{n}-1} \left(1 - \dfrac{\mu_m^2}{u_n^2} \right)} \tag{5.50}$$

5.3 MONOPULSE RADAR ANTENNAS

Radar was first developed for military purposes and this initial application has colored the language used today. In parallel to surveillance radars, gun laying (GL) radars were developed to measure the exact position of aircraft that were targets for anti-aircraft guns. Earlier radars used to track or follow aircraft first used manual beam switching to see how far off the electrical axis of the antenna the aircraft was flying. The antenna was turned by hand to bring the electrical axis of the antenna in line with the target, as with the sights of a rifle. Later this was automated using a rotating offset dipole which produced a conical scan around the electrical axis of the antenna. Any change in the signal during the rotation of the dipole showed a pointing error. This method of operation has a number of disadvantages:

- Variations in transmitter power during the conical scan lead to error.
- The aircraft is not in the center of the beam, which leads to a lower signal-to-noise ratio and greater error.
- The aircraft can send modulated jamming to mislead the radar's tracking circuits in order to throw the radar off the track.

Later antennas were developed with multiple feed horns. The feed horns are coupled, typically by magic T hybrids, to give the sum of the signals entering the horns and their difference, as shown in Figure 5.24. The sum channel is connected to a radar transmitter and receiver in the normal way and provides a detection channel. The difference channel is connected to a separate receiver.

Currently the secondary radar mode S uses a monopulse function of the antenna in the search mode to improve azimuth accuracy when there are missing interrogations. There are military radars that use monopulse reception to improve angle accuracy when small numbers of pulses are transmitted or they work in the single hit mode.

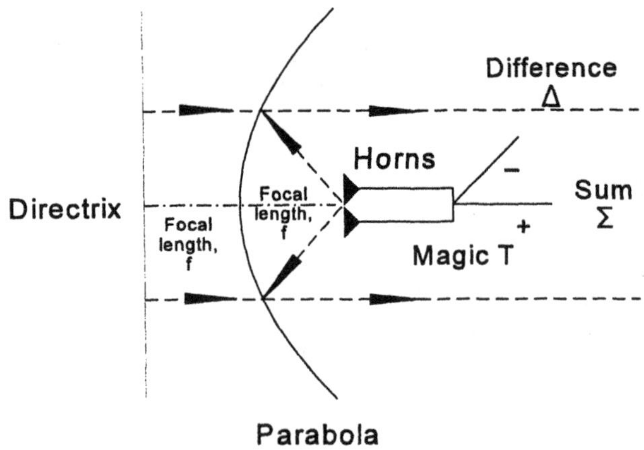

Figure 5.24 Elementary monopulse reflector antenna.

If the normalized illumination function of a pencil beam antenna, or sum pattern, is $g(x')$, then the pattern is

$$F(u') = \int_{-1/2}^{1/2} g(x') \exp(-j2\pi x' u') \, dx' \tag{5.51}$$

where $u' = w/\lambda \sin\theta$.

One way of obtaining a difference function to give a reasonable model pattern obtained from clusters of horns in the focus of a parabola or phased arrays is to change $g(x')$ to $x'g(x')$. Then the illumination function is no longer symmetrical and the difference pattern becomes

$$F(u') = \int_{-1/2}^{1/2} x' \, g(x') \exp(-j2\pi x' u') \, dx \tag{5.52}$$

where $u' = w/\lambda \sin\theta$.

Equation (5.52) is the differential of the sum pattern divided by 2π, as given by the rule for the differentiation of Fourier transforms. The combined representation shows the fact that an odd antenna pattern requires an imaginary aperture function which is shown in Figure 5.25.

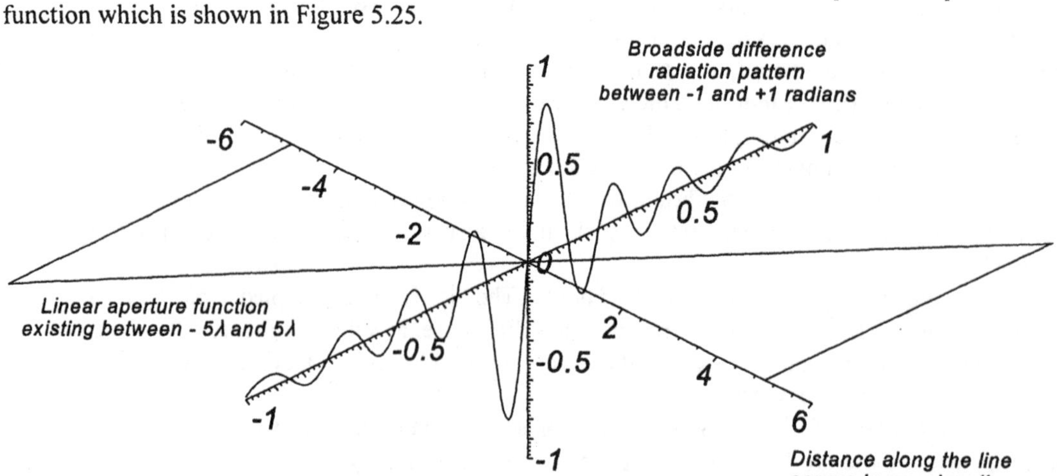

Figure 5.25 A three-dimensional representation of a linear odd antenna pattern and the imaginary aperture function.

Uniform illumination is the standard reference function for the sum pattern; linear illumination is the standard for the difference pattern. The figures used to compare the performance are as follows:

- The effective radiated power C, as with even illumination;
- The slope, K_r, compared with a linearly illuminated antenna;
- The peak of the difference diagram compared to the sum diagram, G_r;
- The value of the first sidelobe, normally in dB.

As with the sum patterns, the illumination function of the difference pattern is shaped to reduce the sidelobes but maintain the slope. Appendix B gives examples of typical odd illumination functions. They are:

- Rectangular, triangular;
- Simply shaped illumination functions:
  - $x \cos^n x$;
  - Truncated Rayleigh;
- Low-sidelobe illumination functions:
  - Taylor derivative, Zolotarëv, and Bayliss.

5.3.1 Tapering functions for monopulse antennas with low sidelobes

Low sidelobes for monopulse antennas give the same advantages as with simple antennas, namely reduction of clutter in the sidelobes and jamming entering through the sidelobes. Figure 5.26 shows a number of monopulse antenna patterns starting with the odd uniform distribution, and first attempts were made to achieve low sidelobes using the Taylor derivative pattern. Figure 5.26 shows what happens when the aim is 30 dB sidelobes. Bayliss modified the positions of the zeroes in the Taylor derivative pattern to give a difference pattern with really low sidelobes and the 30 dB aim is met. Later Zolotarëv polynomials were suggested as an analogue for antennas that have a Chebyshëv pattern.

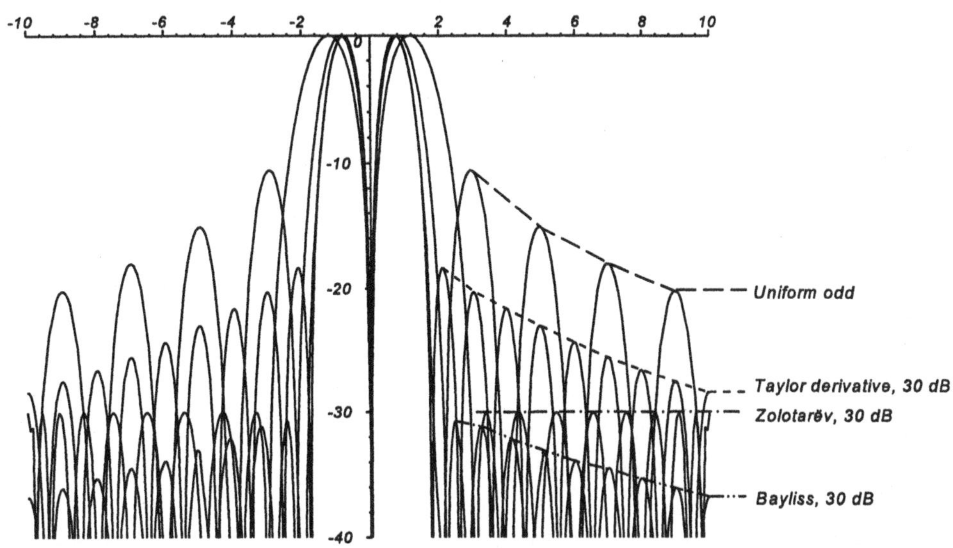

Figure 5.26. A number of odd distributions with low sidelobes.

5.3.1.1 Taylor derivative distribution

Many of the derivatives of sum patterns may be used to form difference patterns and the Taylor derivative [4, p. 300] looked interesting but does not live up to its promise and the aim of 30 dB sidelobes is not met as shown in Figures 5.27, 5.28, and 5.29.

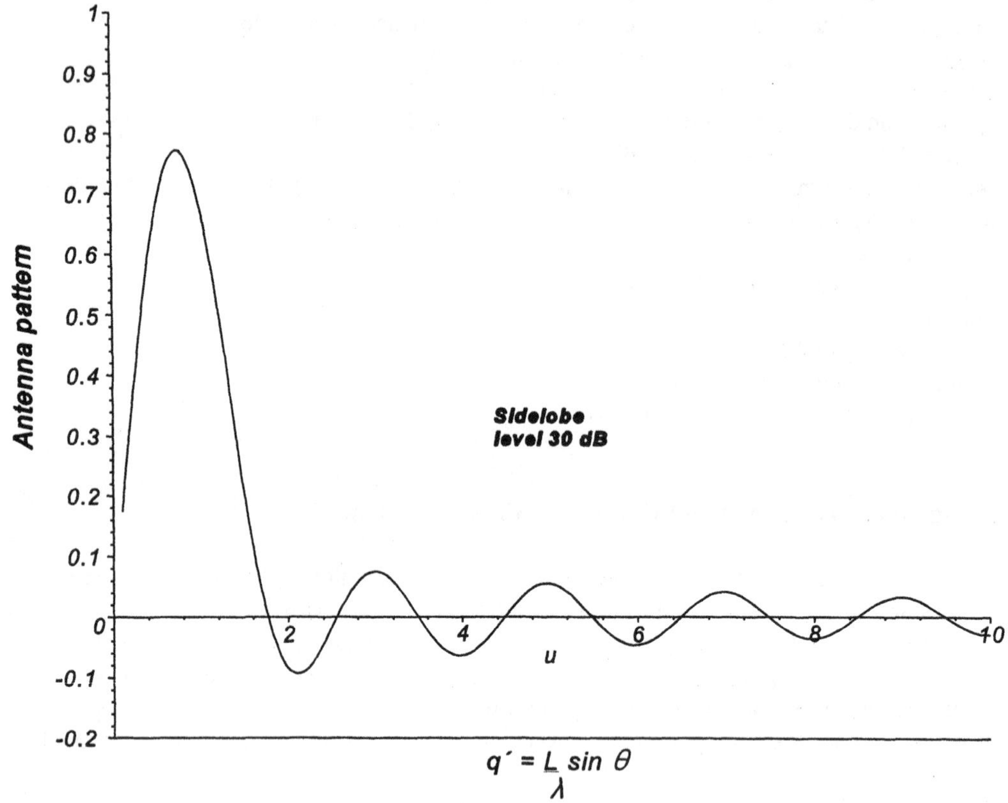

Figure 5.27 Taylor derivative characteristic.

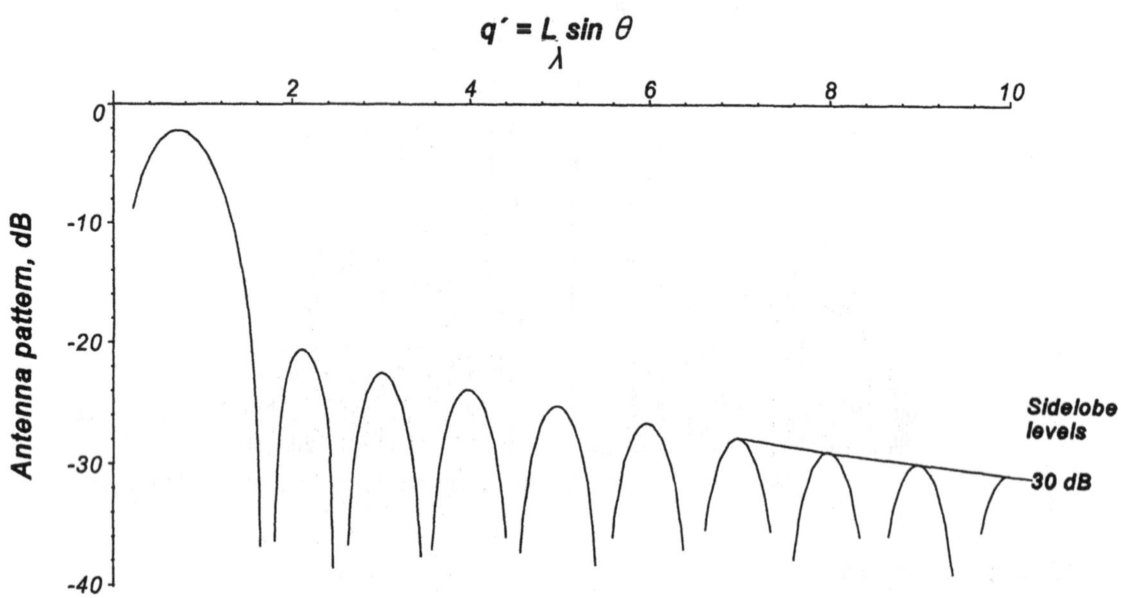

Figure 5.28 Taylor derivative antenna characteristic in decibels.

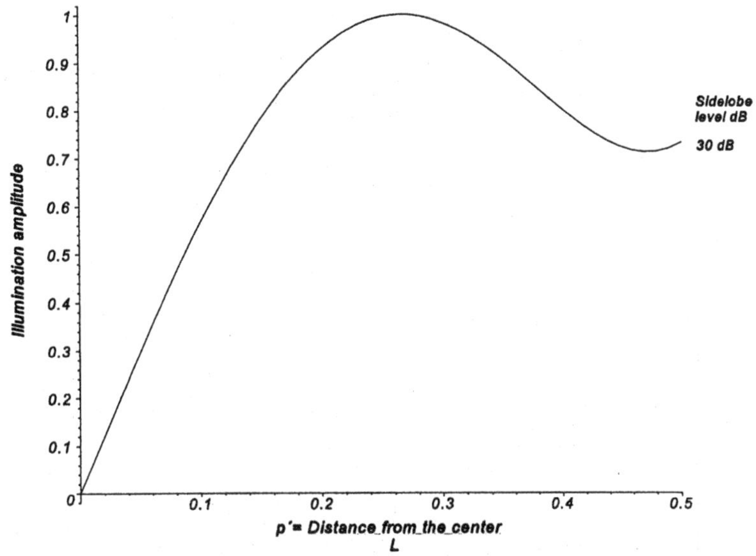

Figure 5.29 The illumination function for a Taylor derivative characteristic.

Bayliss [11] modified the Taylor derivative to give reduced near sidelobes of equal height and decaying far sidelobes.

5.3.1.2 Zolotarëv distribution

Yegor Zolotarëv was a pupil of Chebyshëv at the university of St. Petersburg and later took over from him. He developed polynomials that give a monopulse antenna pattern with low sidelobes for discrete elements described in [12, 13]. The function, like the Chebyshëv for discrete numbers of elements, has a single peak and sidelobes that alternate between ±1 and is defined parametrically and is of the form

$$Z_{2n+1}(x) = \cosh\left((n+\frac{1}{2}) \ln\left(\frac{\mathrm{H}(M+\nu,\ k)}{\mathrm{H}(M-\nu,\ k)} \right) \right) \tag{5.53}$$

where the number of elements is $2n + 1$;
$\mathrm{H}(\nu, k)$ is the Jacobi Eta function with modulus k [13, p. 577];

$$x = \frac{sn(M,\ k)\ cn(\nu,\ k)}{\sqrt{sn^2(M,\ k) - sn^2(\nu,\ k)}};$$

$sn(M, k)$, $cn(M, k)$, and $dn(M, k)$ are the Jacobi elliptic functions [6, p. 570; 14, p. 942];

$M = -\frac{K(k)}{2n + 1}$ $K(k)$ is the complete elliptic integral of the first kind [6, p. 590; 14, p. 537];

k_1 is the complementary modulus $\sqrt{1 - k^2}$.

Note: In the literature k', K', and q' are used for complementary functions that has lead to misprints, so that k_1, K_1, and q_1 are used here for clarity. The full procedure for calculating a Zolotarëv distribution, from [12], is given in the appendix at the end of this chapter.

The sidelobe ratio is determined by the value at $x = x_2$ and depends on n, and Figure 5.30 gives the values of k to achieve this.

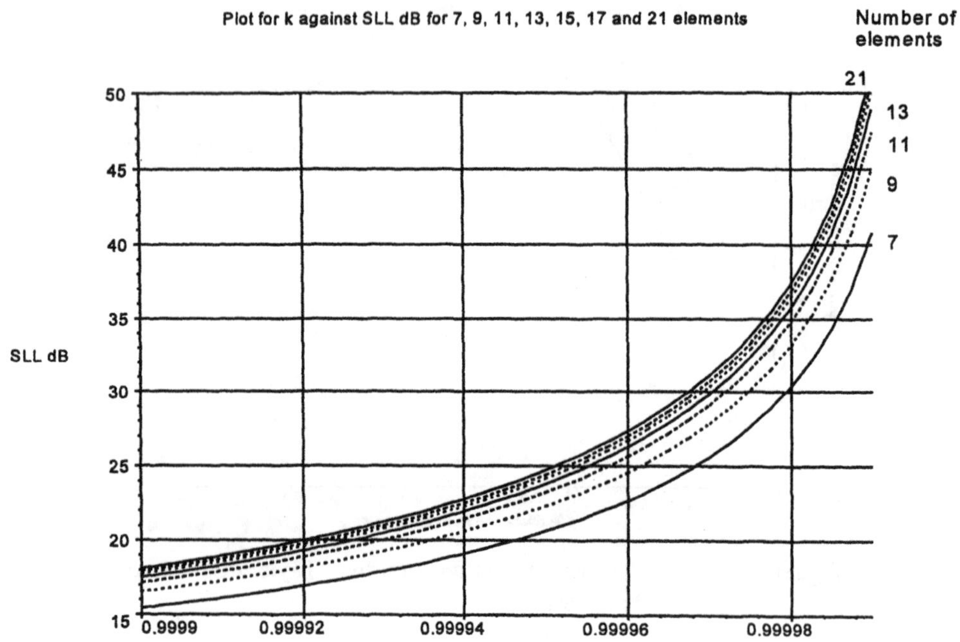

Figure 5.30 Values of k for sidelobe levels and numbers of elements.

Figures 5.31and 5.32 show the characteristics of an array of 20 elements (the center element is not driven) as the Zolatarëv polynomial and the plot of one side of the antenna difference pattern in decibels.

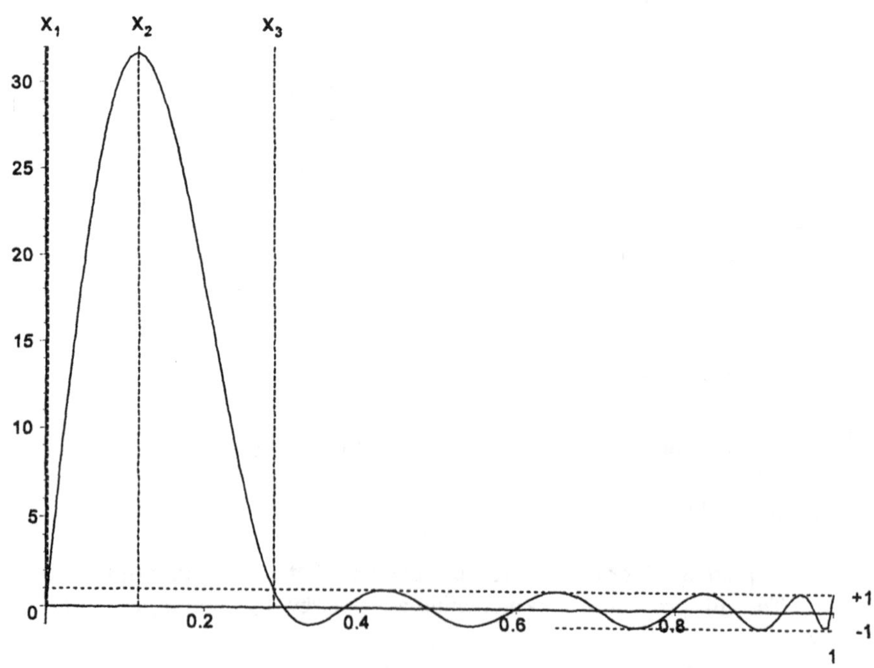

Figure 5.31 Zolotarëv characteristic with 30 dB sidelobes.

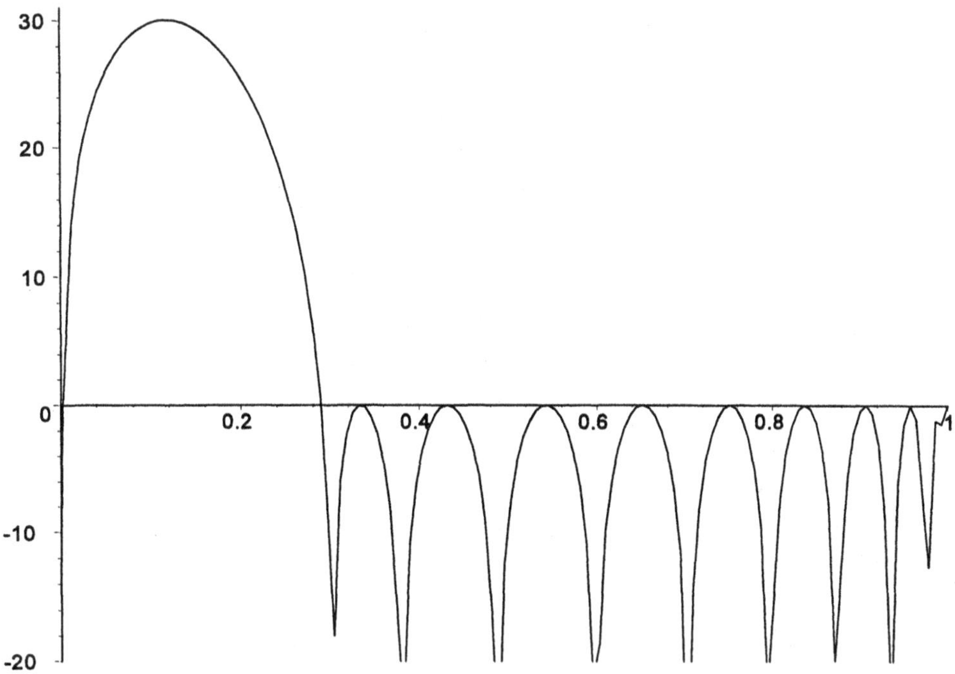

Figure 5.32 Zolotarëv antenna characteristic in decibels with 30 dB sidelobes.

Figure 5.33 shows the excitation derived numerically from the discrete Fourier transform. The left-hand 10 elements are driven with the same amplitude but in antiphase.

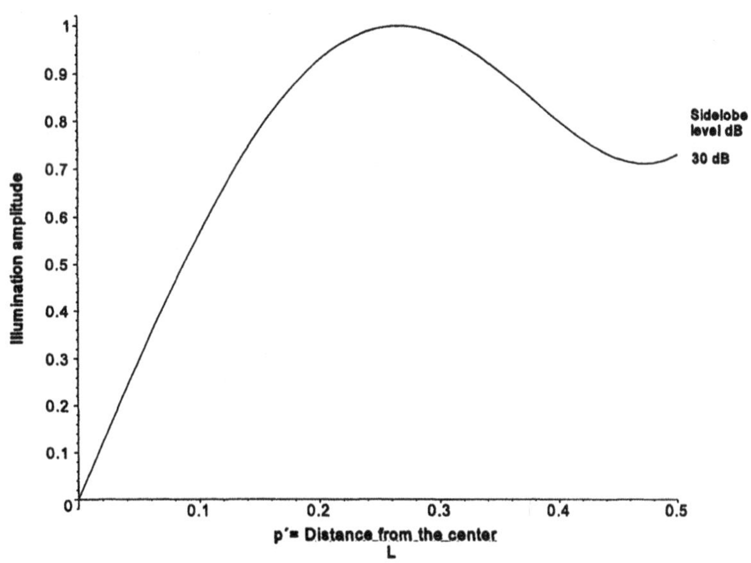

Figure 5.33 Zolotarëv antenna illumination function for the example.

5.3.1.3 Bayliss distribution

Bayliss [11] found the way to reduce the first four sidelobes of the Taylor derivative pattern by moving the first nulls. No closed function is available for the positions of the nulls and an iterative procedure was used to find them. The modified or corrected table [15] for finding the positions of the zeroes is reproduced in Appendix B, Tapering functions.

Similarly to the Taylor distribution the pattern is defined and the distribution is derived from the pattern. First for a given sidelobe level, *SLL* db, the factor A is found from the polynomial approximation given in Appendix 5A.

The pattern is given by

$$F(u) = u \cos \pi u \frac{\prod_{n=1}^{\bar{n}-1} \left(1 - \frac{u^2}{\sigma^2 z_n^2} \right)}{\prod_{n=0}^{\bar{n}-1} \left(1 - \frac{u^2}{(n + ½)^2} \right)} \tag{5.54}$$

where z_n is calculated from the polynomials in the Appendix for n less than $\bar{n}$ or $\pm \sqrt{A^2 + n^2}$ for n outside the center region and the dilation factor σ is given by $\sigma = \frac{\bar{n} + ½}{z_{\bar{n}}}$.

A typical pattern with 30 dB sidelobes is illustrated in Figure 5.34(a) or in decibels in Figure 5.34(b).

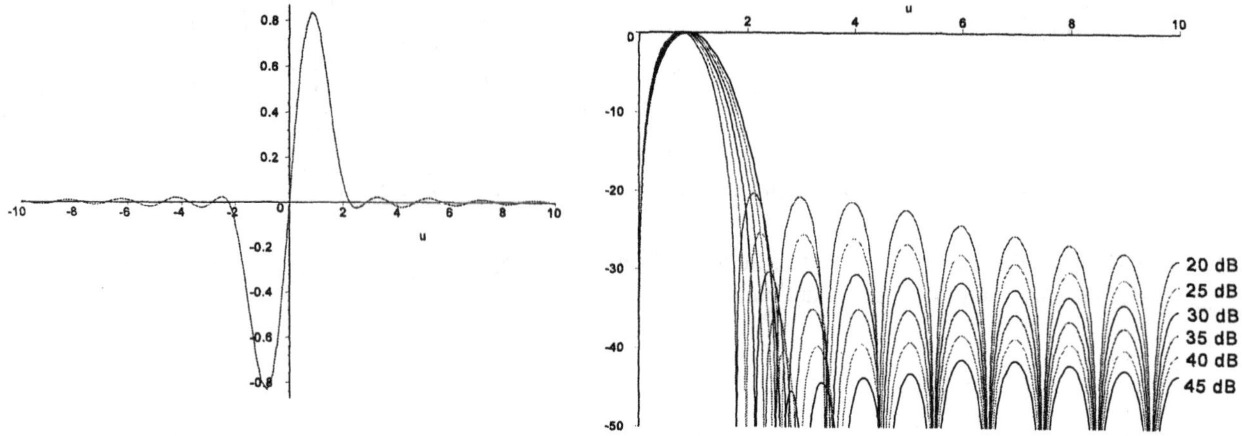

(a) 30 dB Bayliss difference pattern (b) Bayliss difference patterns with 20 dB to 45 dB sidelobes

Figure 5.34 A linear Bayliss pattern with 30 dB sidelobes and a family of patterns in decibels.

The excitation with p as the distance from the center (values from 0 to 1/2) is given by

$$g(p) = \sum_{n=0}^{\bar{n}-1} B_n \sin \frac{\pi p}{n + ½} \tag{5.55}$$

where

$$B_m = -(-1)^m (m + ½)^2 \frac{\prod_{n=1}^{\bar{n}-1} \left(1 - \frac{(m + ½)^2}{\sigma^2 z_n^2} \right)}{\prod_{n=0,\, n \neq m}^{\bar{n}-1} \left(1 - \frac{(m + ½)^2}{(n + ½)^2} \right)} \tag{5.56}$$

Examples are shown in Figure 5.35.

An alternative expression for the difference pattern using B_n is given by

$$F(u) = u \cos \pi u \sum_{n=0}^{\bar{n}-1} \frac{(-1)^n B_n}{\left(n + \frac{1}{2}\right)^2 - u^2} \tag{5.57}$$

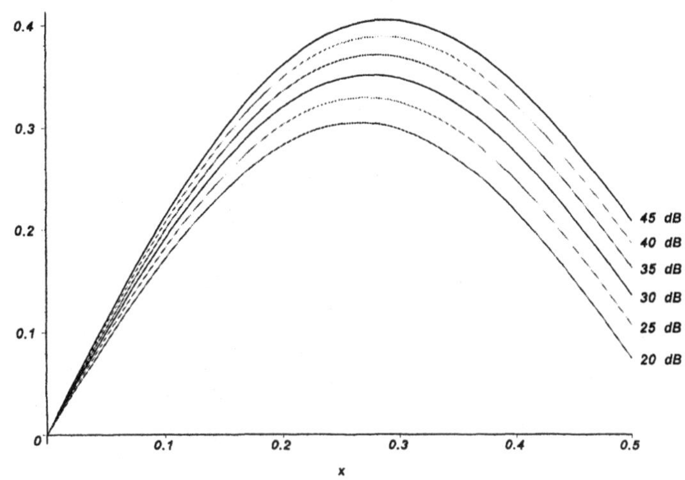

Figure 5.35 Examples of the illumination function for Bayliss antenna patterns.

Bayliss' article [11] started with his difference pattern for a circular aperture in the plane of the cross-section of the difference pattern. The calculation for A and the zeroes using the data from Table 5A.1 is the same and the pattern is given by (the cosϕ term for the other dimension has been omitted)

$$F(u) = u\, J'_1(\pi u) \frac{\prod_{n=1}^{\bar{n}-1} \left(1 - \frac{u^2}{\sigma^2 z^2_n}\right)}{\prod_{n=0}^{\bar{n}-1} \left(1 - \frac{u^2}{\mu_n^2}\right)} \tag{5.58}$$

where $J_1'(x)$ is the differential of the Bessel function of the first kind, order one $J_1(x)$;
μ_n is the nth zero of $J_1'(\pi x)$;
the dilation factor, σ, is given by $\frac{\mu_{\bar{n}}}{\sqrt{A^2 + \bar{n}^2}}$.

Typical Bayliss circular patterns are shown in Figure 5.36.

Alternately the pattern may be calculated from the expression

$$F(u) = u\, J'_1(\pi u) \sum_{n=0}^{\bar{n}-1} \frac{B_n\, J_1(\pi \mu_n)}{\mu_n^2 - u^2} \tag{5.59}$$

where

$$B_m = 2\mu_m^2 \frac{\prod_{n=1}^{\bar{n}-1}\left(1 - \frac{\mu_m^2}{\sigma^2 z_n^2}\right)}{J_1(\pi\mu_m)\prod_{n=0,\ n\neq m}^{\bar{n}-1}\left(1 - \frac{\mu_m^2}{\mu_n^2}\right)} \qquad m = 0,\ 1,\ 2,\ ..\ \bar{n} - 1 \tag{5.60}$$

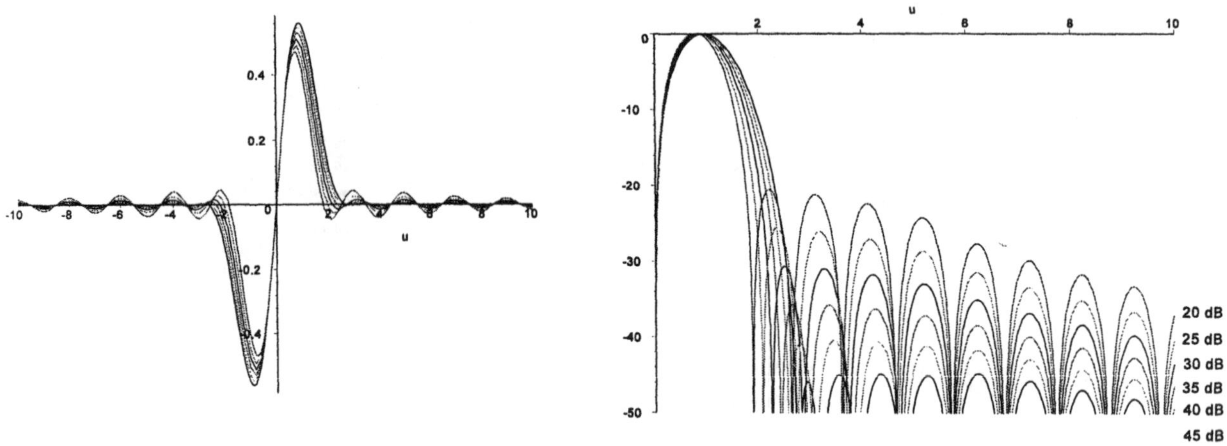

(a) Circular Bayliss patterns with sidelobe levels from 20 dB to 45 dB (b) Circular Bayliss patterns in dB w ith sidelobe levels from 20 dB to 45 dB

Figure 5.36 A family of circular Bayliss antenna patterns in the plane of the difference pattern.

The aperture distribution is given by

$$g(p) = \sum_{n=0}^{\bar{n}-1} J_1(\mu_n\ p) \tag{5.61}$$

Typical aperture distributions for circular apertures with Bayliss patterns are shown in Figure 5.37.

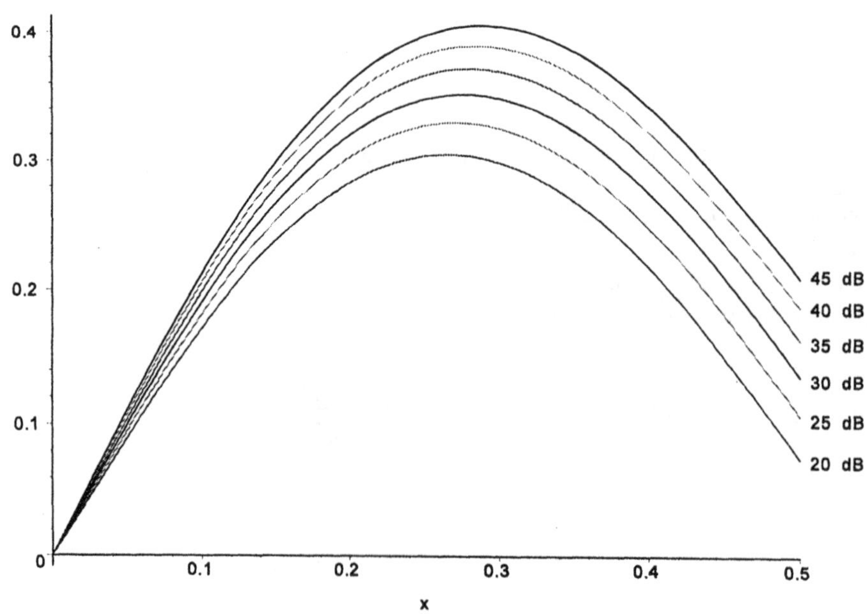

Figure 5.37 Aperture distribution for a number of linear Bayliss patterns.

5.4 ARRAYS OF DISCRETE RADIATORS

Some of the first radar antennas were lines or arrays of dipoles. If all the elements are fed in phase, then the beam is normal to the line (or broadside) as in Figure 5.38.

The contributions to a wavefront at a distant point normal to the antenna consist of an amplitude term and an exponential representing phase, namely,

$$\textit{Broadside wavefront field} = K \sum_{n=0}^{N-1} A_n \exp\left(j\, R_n \frac{2\pi}{\lambda}\right) \tag{5.62}$$

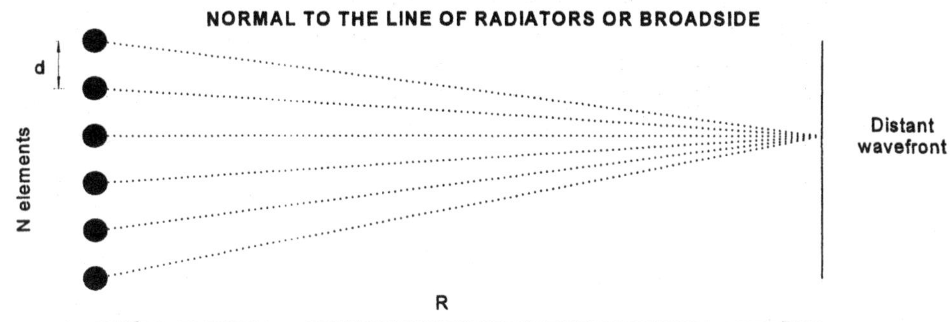

Figure 5.38 The contribution of individual antenna elements to a distant wavefront.

where K is a constant depending on the range to the radiator, R_n, m;
R_n is taken to be the same for all n at this range;
A_n is the excitation of the element;
λ is the wavelength;
$\frac{2\pi}{\lambda}$ is the number of radians of phase change per meter.

At other angles, the range differences from the elements change the phases of the components so that the field strength is less. This situation is shown in Figure 5.39.

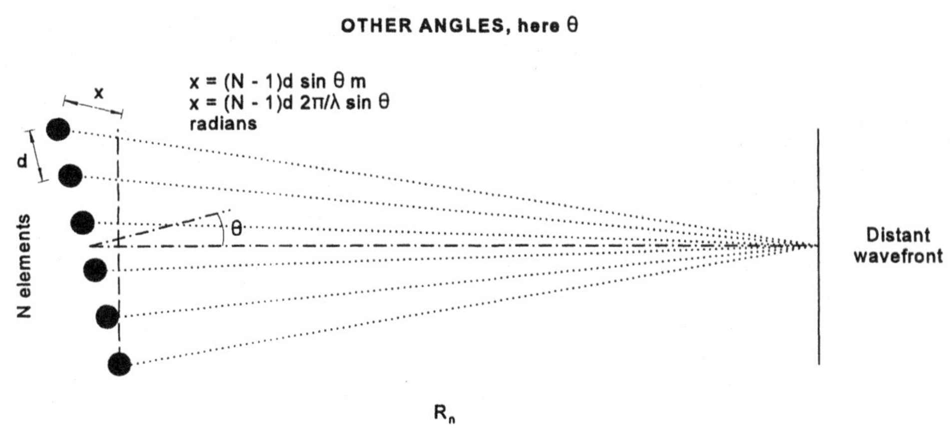

Figure 5.39 The contribution of individual antenna elements to a distant wavefront.

Equation (5.62) is rearranged to take account of linear change of distance of the elements to give

$$\textit{Wavefront field} = K \sum_{n=0}^{N-1} A_n \exp\left(j\, \frac{2\pi}{\lambda}(R + \overline{nd \sin\theta})\right) \tag{5.63}$$

where R is the range to the nearest element, m;
d is the distance between two elements, m;
n is the element number;
A_n is the excitation of the nth element.

Taking out the constant terms K and R gives the array factor

$$\textit{Relative wavefront field or array factor} = \sum_{n=0}^{N-1} A_n \exp\left(j\,\frac{2\pi}{\lambda}\,nd\,\sin\theta\right) \tag{5.64}$$

The individual elements may be driven individually with control over amplitude and phase. The tapering in amplitude is normally to shape the beam, in particular, to reduce sidelobes (see Sections 5.2.2 and 5.5). Control of the phase angles of the driving currents is used, principally to steer the beam away from broadside. Figure 5.40 shows the general case with part of such a line of N radiators equally spaced d meters apart.

To turn the beam from broadside to an angle θ_m, the phase shift of the currents driving the adjacent elements is made to be α radians. The components of the wavefront all have the same phase, so that the phase shift represented by the distance X in Figure 5.40 is

$$\frac{2\pi}{\lambda}\,d\,\sin\theta + \alpha = 2\,\pi\,k \qquad k = 0,\ 1,\ 2,\ \ldots \tag{5.65}$$

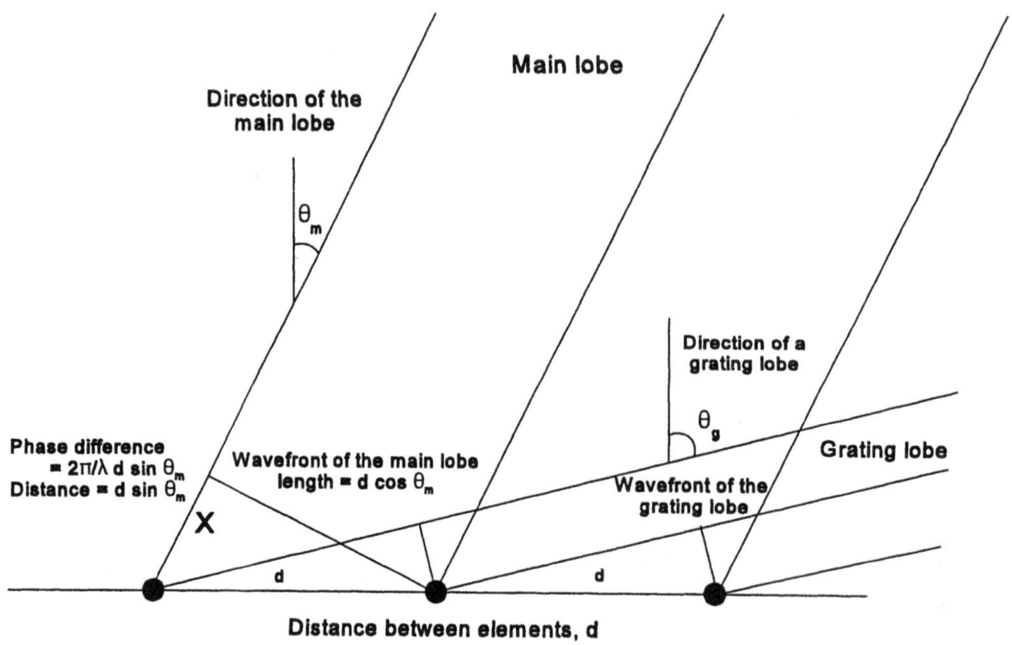

Figure 5.40 The formation of main and grating lobes by a linear array.

For the center of the beam, $\theta = 0$, so that θ_m is

$$\alpha = \frac{2\pi}{\lambda}\,d\,\sin\theta_m \tag{5.66}$$

The right hand side of (5.65) represents the fact that the phase angle repeats itself each wavelength. If the distance, d, between elements in Figure 5.40 is large enough to allow values of k greater than 0, then grating lobes occur (see Section 5.2.2).

$$\textit{Array factor} = \sum_{n=0}^{N-1} A_n \exp\left(j\, n\left(\frac{2\pi}{\lambda}\, d \sin\theta + \alpha\right)\right) = \sum_{n=0}^{N-1} A_n \exp(j\, n\psi) \tag{5.67}$$

where $\psi = \frac{2\pi}{\lambda} d \sin\theta + \alpha = \frac{2\pi}{\lambda} d(\sin\theta + \sin\theta_m)$.

When the $\exp(j\, n\psi)$ terms are replaced by $(\exp(j\,\psi))^n$, a polynomial is formed (Schelkunoff) [5, p. 696] or it may be represented by the z-transform (see Section 16.7). For uniform excitation, (5.68) is a geometric progression. Using the fact that

$$\sum_{n=0}^{N-1} q^n = \frac{q^n - 1}{q - 1}$$

$$\textit{Array factor for uniform illumination} = z^{N-1}\,\frac{z^N - 1}{z-1} = \exp(j\pi(N-1))\,\frac{\sin(N\frac{\psi}{2})}{N\sin\left(\frac{\psi}{2}\right)} \tag{5.68}$$

Reference [2, p. 125] gives a number of rules for the shapes of the broadside radiation patterns for radiators spaced at integer multiples of half a wavelength. There is one main lobe with a directivity NG_e, where G_e is the gain of the individual elements. It has a standard beamwidth $\arcsin \lambda/w = \arcsin(2/N)$ and sidelobes of standard width $\arcsin 1/N$. The sidelobe level decreases with a larger number of elements (see Section 16.3.1).

Steering the main beam away from broadside reduces the width of the antenna when viewed from the axis of the main beam. Then, for the N radiators,

$$\textit{Projected width} = Nd\cos\theta_m \tag{5.69}$$

The gain is reduced and the beamwidth increased when pointed to θ_m, (see Figure 5.9).

5.4.1 Tapered illumination functions

As with continuous aperture antennas, uniform illumination gives patterns with high sidelobes which may be reduced by tapering the illumination (see Sections 5.1.1 to 5.1.5). These illumination functions may be obtained by sampling the functions used for continuous apertures. The exception is the Chebyshëv illumination which is defined for discrete arrays only (see Appendix B).

5.4.2 Ways of driving discrete elements

Discrete radiators may be driven either in series or in parallel. Figure 5.41 illustrates these two main types. With a change of frequency, the value of d/λ changes to give a beam width that depends on wavelength.

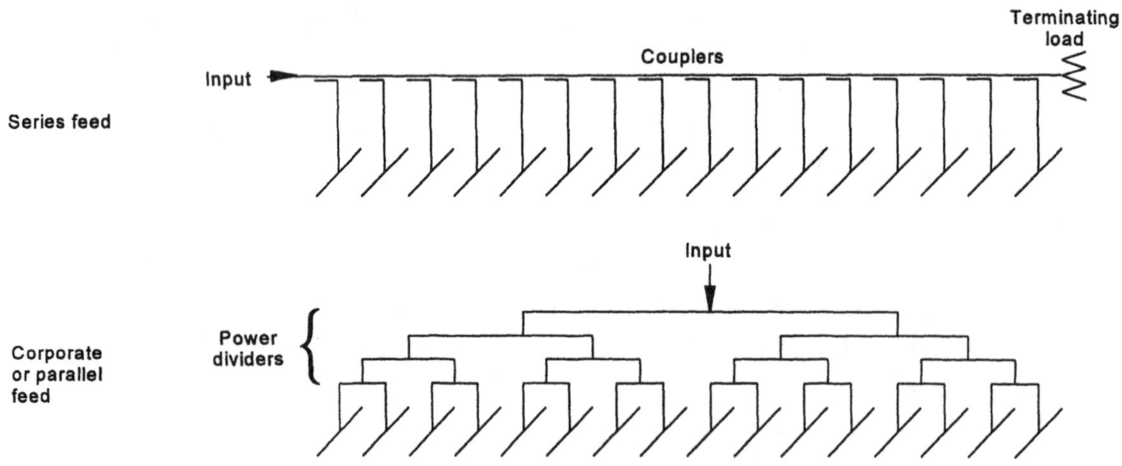

Figure 5.41 Series and parallel feeds.

Additional considerations are:

- Series feed.

  The phase shift between elements is proportional to d/λ so that a change of frequency causes such a change of phase and swings the main beam. This is called squint and must be taken into account when setting end-fed antennas. This property is used in antennas that scan by changing the frequency. The outgoing signal must fill the series feed completely before the antenna beam is formed as designed.

- Parallel feed.

  The path length to each element is the same so that the beam axis does not change with frequency variations.

For phased arrays the elements are driven with a progressively advancing or retarding phase as the beam is steered. Phase shifters (see Figure 5.42) can be controlled by beam-steering logic driven by a computer. The phase shifters may be situated either behind the elements (high power) or in the driving chain. The phase shifters change their characteristics with frequency and temperature, and these changes cause a rotation of the beam, or squint.

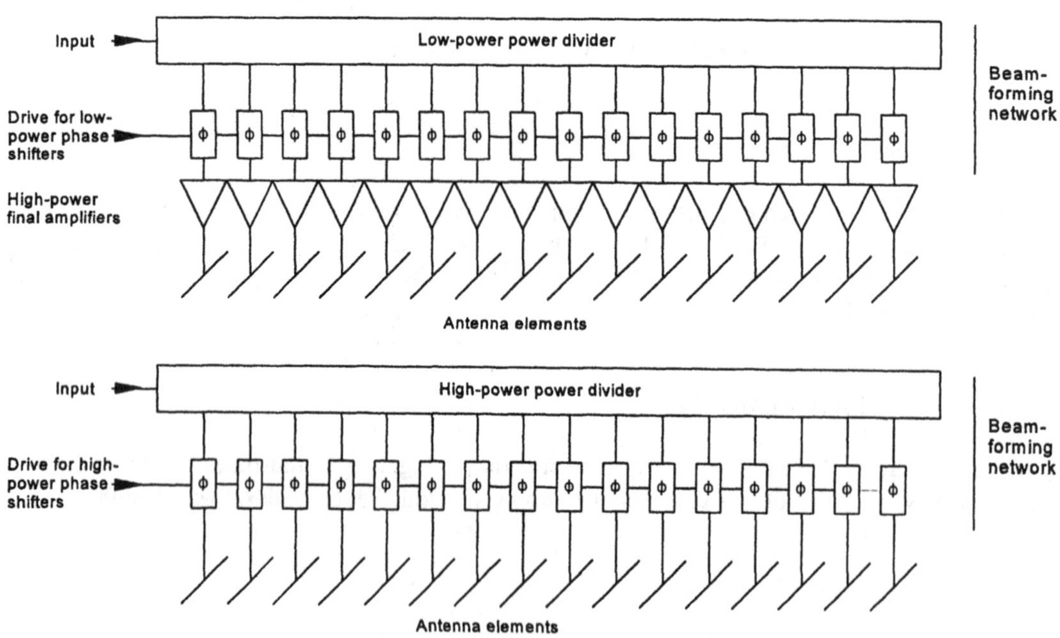

Figure 5.42 Beam steering for a phased array.

The phase shifters can be made to change their values in microseconds which allows fast beam steering for:

- Mechanically rotating three-dimensional surveillance radars that scan in elevation;
- Surveillance or tracking radars with no rotating parts.

Though mechanically directed antennas can follow the movements of mortar or artillery rounds, phased arrays are now preferred for this.

Instead of the use of expensive waveguides to distribute the transmitter signal to the antenna elements and recollect the echo signals in the receiving time, a horn or similar radiator can illuminate a surface containing phase shifters. The surface may represent a lens, as in the case of the Patriot radar (Figure 5.43(a)) or a reflector as shown in Figure 5.43(b).

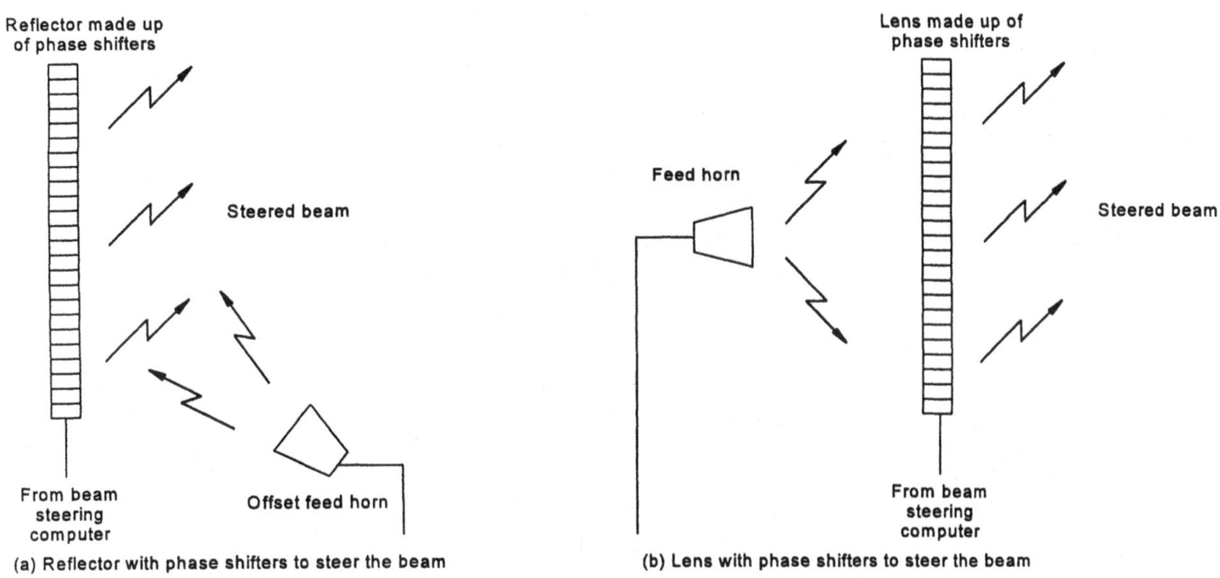

Figure 5.43 Space feeds for a reflector containing phase shifters or a lens containing phase shifters.

This configuration allows a horn system to generate sum and difference patterns in two planes and the beams can be moved quickly to search for and track fast moving objects.

The three-dimensional representations of the sampled aperture function in a 10λ line source with a broadside and a beam steered to 15 degrees offset are shown in Figure 5.44. For clarity, the samples are shown λ/4 apart.

5.4.3 Grating effects

An array of discrete radiators acts like an optical diffraction grating. Groups of radiators with an abrupt change of excitation between the groups act as a larger diffraction grating framing the groups of elements. The steering of the main beam away from broadside increases the occurrence of grating lobes as shown in Figure 5.39. With element spacings greater than λ/2 as the main beam is steered away from broadside, the first grating lobe ($n = 1$ in (5.71)) will appear in the opposite quadrant. With greater element spacings additional grating lobes may appear closer to the main lobe. The relationships between the phases of the elements and the angles of the main and grating lobes are [5, p. 704].

$$\begin{aligned} \frac{2\pi d}{\lambda} \sin \theta_m + \alpha &= 2\pi m \qquad m = 0, 1, 2, \ldots \\ \frac{2\pi d}{\lambda} \sin \theta_g + \alpha &= 2\pi p \qquad p = 0, 1, 2, \ldots \end{aligned} \tag{5.70}$$

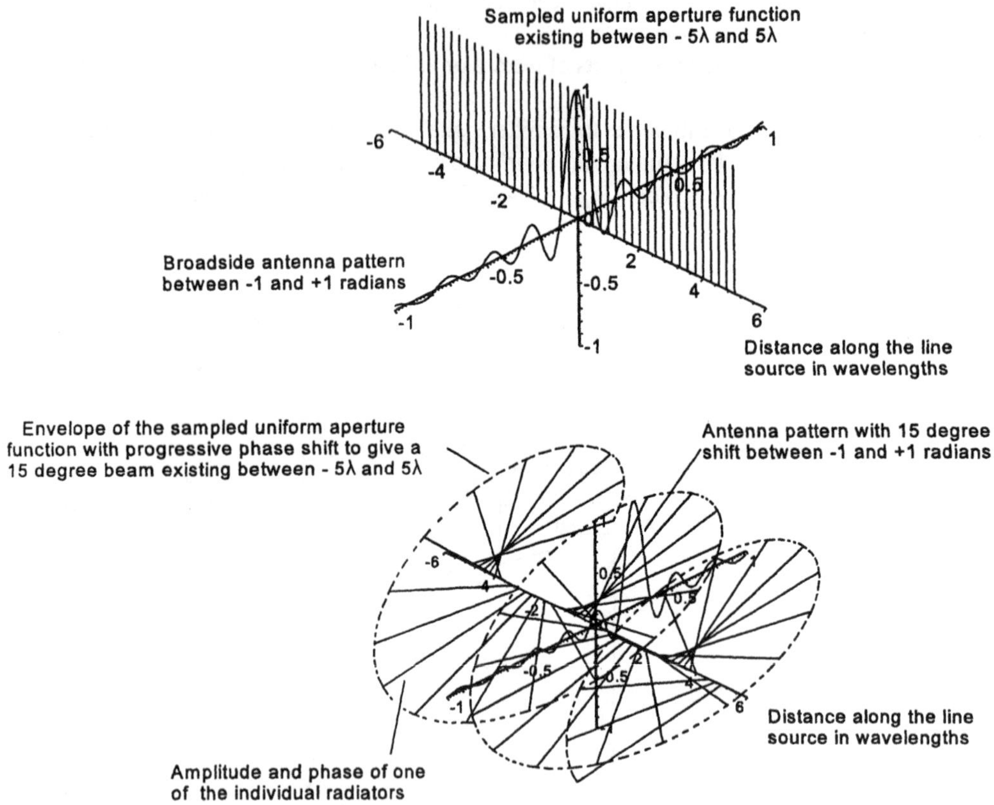

Figure 5.44 A three-dimensional representation of the aperture function for a 10λ line source for a broadside and a beam 15 degrees off broadside.

Subtracting to find the d/λ spacing for a steered main beam and grating lobe, we have

$$\frac{d}{\lambda} = \frac{n}{|\sin\theta_m - \sin\theta_g|} \qquad n = 1, 2, \ldots \tag{5.71}$$

where n is $m - p$ from (5.70).

This is shown in Figure 5.45 for $n = 1$. From this it can be seen that there are two rules of thumb:

- There are no grating lobes when the spacing is one half of a wavelength, or less.
- When the spacing is one wavelength, the broadside beam has a grating lobe at 90 degrees.

The phase shifts obtainable from commonly available phase shifters are binary fractions of 360 degrees, or 2π radians, and not the exact phase shift required by the equations. As individual phase shifters are expensive, their number may be reduced by feeding a number of elements from the same phase shifter, as discussed in the next section.

5.4.4 Beam-steering quantization effects

The phase shifters used to steer the main beam are normally controlled by a computer. Commonly, the steps are binary fractions of half a circle and this quantization leads to a small change in the direction of the main beam and extra sidelobes. A sawtooth error is caused when the elements are being switched to the nearest correct value of phase possible and either of the following occurs:

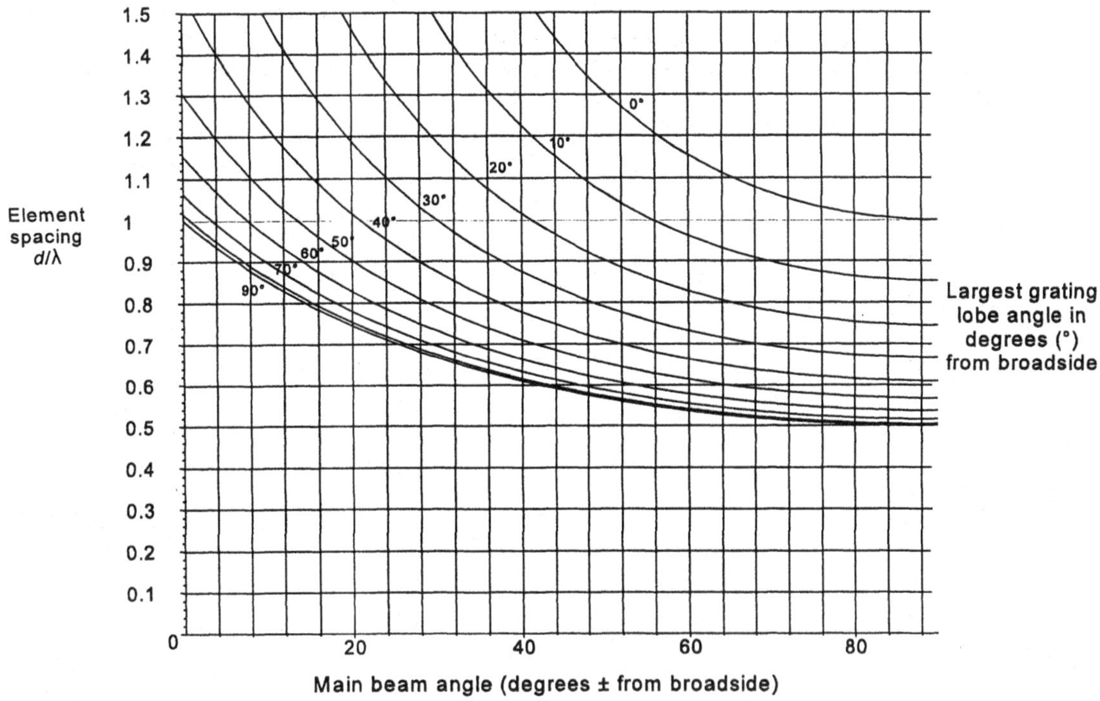

Figure 5.45 Element spacing plotted against main beam angle for largest grating lobe.

- There are more elements than twice the number of steps, that is, adjacent elements are switched to the same phase.
- The array elements are grouped and fed from a common phase shifter as in Figure 5.46.

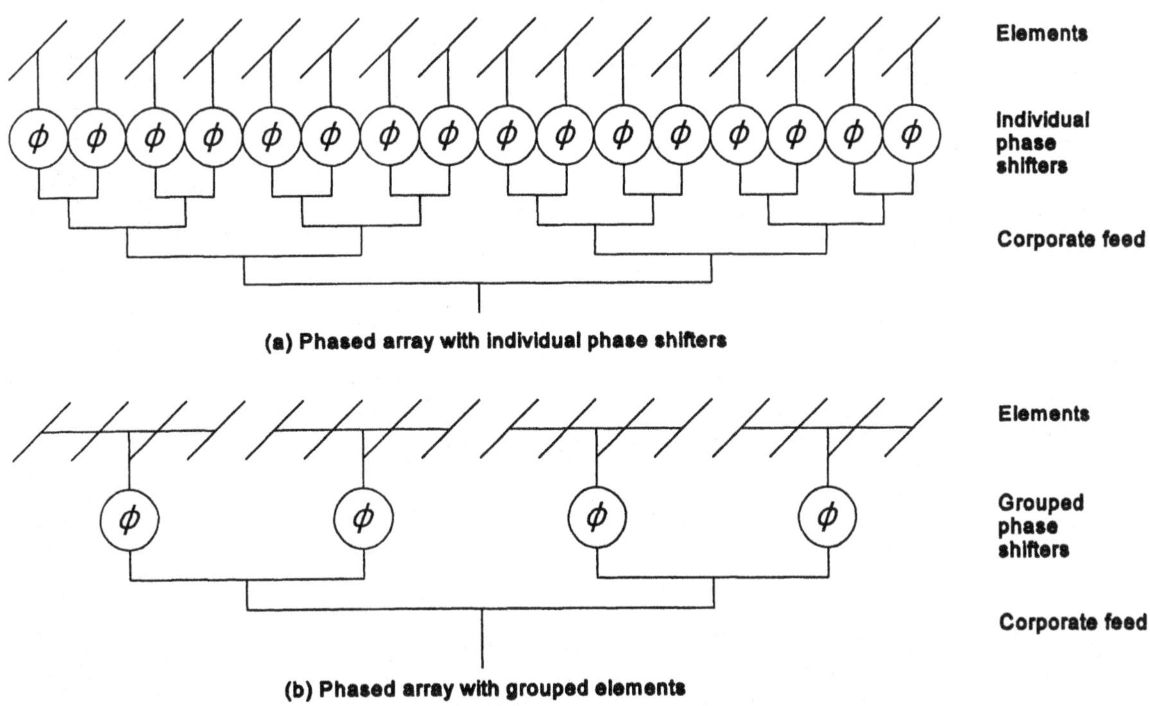

Figure 5.46 Phased array with individual phase shifters and grouped elements.

In Figure 5.47 a straight diagonal line shows the desired phase along the antenna aperture to swing the beam to a given angle. The staircase around the line shows the digital approximation actually used to steer the beam. The approximate minus the desired phase shift gives the error, which has a sawtooth form. The width of the steps is either the number of elements with the same phase or between jumps in the approximation. The height of the steps is the actual phase error between the elements which is always less than the continuous form shown in Figure 5.47.

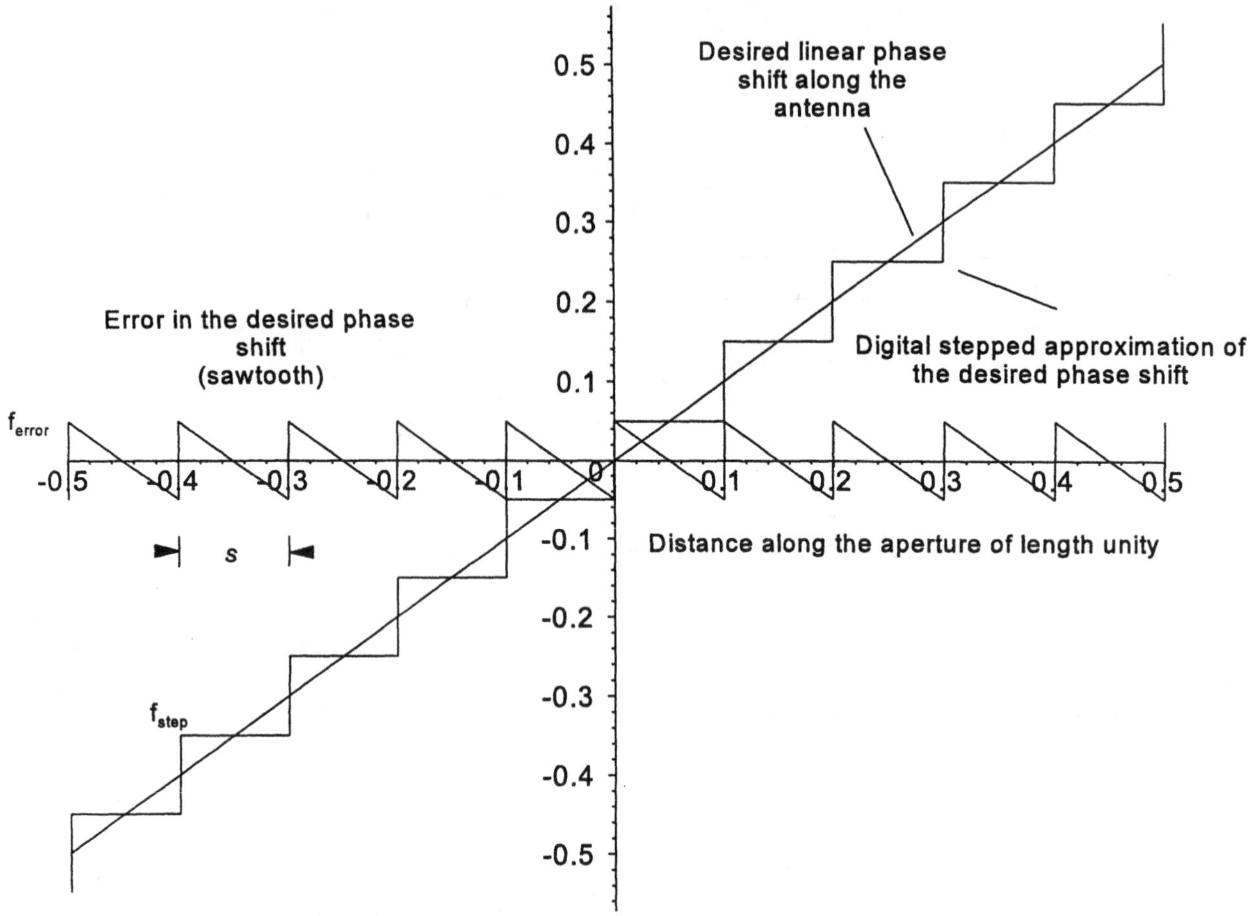

Figure 5.47 Phase shift, quantized phase shift, and phase shift errors.

These errors are periodic and may be subtracted from the desired phase shift function to give the function

$$f_{error}(x) \;=\; f(x) \;-\; f_{step}(x) \tag{5.72}$$

The number of steps varies from one, when each half of the aperture is switched to the same phase, up to the number of phase shifters. The length of the steps, s, may be from half the aperture length downwards. This component may cause grating lobes [16, p. 11-41]. The quantizing lobes occur at the angles given by

$$\theta_1 \;=\; \arcsin\left(\sin\,\theta_m \,\pm\, n\frac{\lambda}{s}\right) \qquad n \;=\; 1,\, 2,\, 3,\, \ldots \tag{5.73}$$

where θ_l is the angle of the lobe between ±90 degrees from broadside;
θ_m is the angle of the main lobe;
s is the width of the sawtooth error cycle for the particular main lobe position.

The maximum power gain relative to the mainlobe gain at broadside is given by

$$\textit{Lobe amplitude} = \frac{G_e(\theta_1)}{2^{2p}} \text{ with respect to the mainlobe power} \tag{5.74}$$

where p is the number of binary phase shifter bits;
$G_e(\theta)$ is the element pattern;
θ_1 is the position of the grating lobe.

An error pointing the main beam from its desired or commanded position is caused by the tilting of the teeth of the sawtooth error function, given by [16, p. 11-38, eqs. 38, 39; 17, pp. 7.44 and 7.45, Eqs. 7.32 and 7.33]

$$\sigma_\theta \approx \frac{\theta_s}{2^p\, N} \tag{5.75}$$

where p is the number of binary phase shifter bits;
N is the number phase changes over the aperture;
θ_s is the standard beam width.

To take an example, the quantizing sidelobe caused by varying levels of sawtooth error in an antenna 40λ wide is shown in Figure 5.34 [18]. In this example, the elements are half a wavelength apart fed by 4 bit phase shifters and the steps are eight elements wide. Thus, when

$d = \lambda/2$ and
$s = 8\,d = 4\,\lambda$
Number of phase steps $= 2^4 = 16$
Number of groups, $N = 10$

Then

Standard beamwidth, $\theta_s = 1/40$ radian $= 1.432$ degree
Phase step between groups, $\alpha = 2\pi/2^4 = 0.0156$ radian $= 0.895$ degree

Using (5.66) with one phase step, the main beam is offset by

$$\textit{Main lobe offset} = \arcsin\left(\alpha\frac{\lambda}{2\pi d}\right) = \arcsin\left(\frac{2\pi}{16}\,\frac{\lambda}{2\pi\; 4\lambda}\right) = 0.8953 \text{ degree} \tag{5.76}$$

From (5.75) when there is one phase step at the center

$$\textit{Mainlobe pointing error (standard deviation)} = \frac{\theta_s}{2^p\, N} = \frac{1/40}{2^4 \times 1} = 0.08952 \text{ degree} \tag{5.77}$$

The positions of the quantizing lobes are, from (5.73),

$$\textit{Quantizing sidelobe positions} = \arcsin\left(\sin\theta_m \pm \frac{n\,\lambda}{s}\right) = \arcsin\left(\frac{1}{64} \pm \frac{n}{4}\right) \quad n = 1, 2, 3, \ldots \tag{5.78}$$

The positions of the quantizing sidelobes are listed in Table 5.4.

Table 5.4
Positions of quantizing sidelobes

	At positive angles			At negative angles			
	$n = 1$	$n = 2$	$n = 3$	$n = 1$	$n = 2$	$n = 3$	$n = 4$
In radians	0.269	0.542	0.872	-0.2366	-0.5056	-0.8247	-1.3938
In degrees	15.404	31.039	49.963	-13.555	-28.972	-47.254	-79.858
In terms of u	10.625	20.625	30.625	-9.375	-19.375	-29.375	-39.375

The amplitudes of the first lobes are, assuming quasi-omnidirectional radiating elements,

$$\textit{Quantizing sidelobe voltage ratio} = \frac{1}{2^4}\sqrt{\frac{\cos\,\theta_1}{\cos\,\theta_m}} = \frac{1}{16}\sqrt{\frac{\cos\,15.404}{\cos\,0.8953}} = 24.241\ \text{dB}$$
$$\text{or}\quad \frac{1}{16}\sqrt{\frac{\cos\,13.555}{\cos\,0.8953}} = 24.205\ \text{dB} \qquad (5.79)$$

Figure 5.48 was plotted using the error sawtooth with uniform illumination. It shows the sin x/x sidelobe pattern for the uniform illumination together with the pattern with phase errors. The first quantizing lobes are at $u = 10.625$ and -9.375 and are clearly visible.

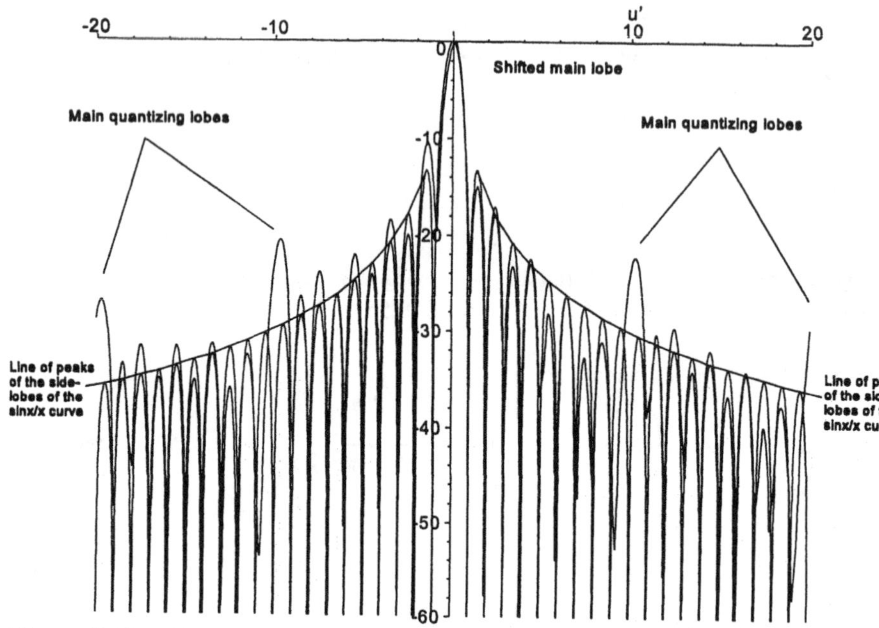

Figure 5.48 The sin x/x pattern for uniform illumination and the same pattern with phase errors caused by quantization.

5.5 CREATING SHAPED BEAMS

The discussion up to now has concentrated on producing a single narrow beam with acceptable sidelobes in two separable angular dimensions or circular symmetry. Air surveillance radars are manufactured to observe the air volume above an area of the earth's surface. Military surveillance radars often search up to 120 000 feet (36 576 m), and airfield radars up to 30 000 feet (9 144 m). The shape of such a beam is shown in Figure 5.49.

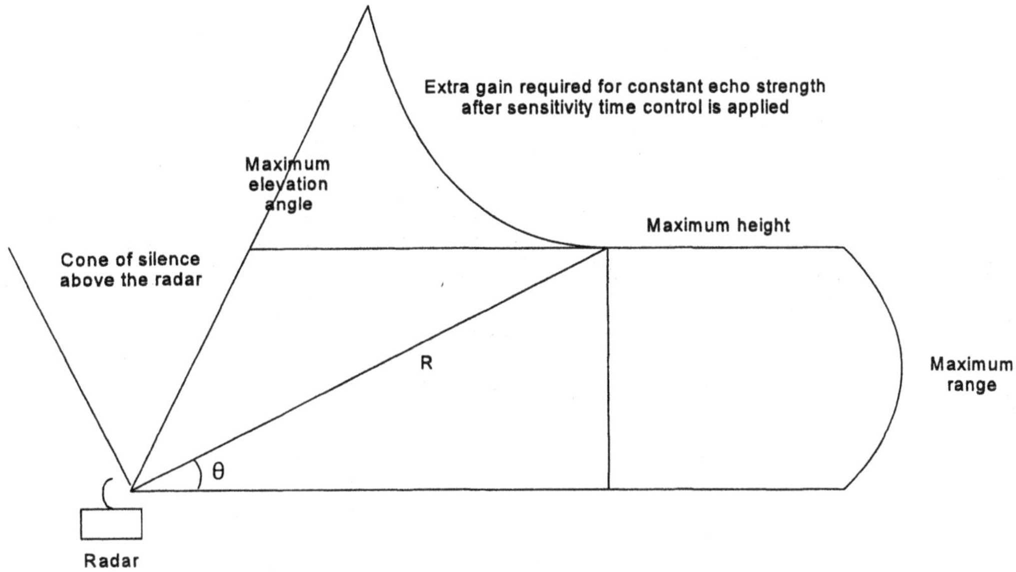

Figure 5.49 The constant height or cosecant squared pattern.

The echo signal strength from an aircraft flying along the maximum height line is proportional to $1/range^4$.

$$Echo\ signal\ power \ \propto \ \frac{(Antenna\ gain)^2}{Range^4} \tag{5.80}$$

From Figure 5.41,

$$Slant\ range,\ R \ = \ H_{max} \operatorname{cosec} \theta \tag{5.81}$$

So that there is a constant signal strength for an aircraft echo, the antenna power gain must be proportional to $\operatorname{cosec}^2 \theta$. Normally, the signal strength after sensitivity time control (STC) is held constant which requires the gain to be higher for close-in high aircraft to counter the effects of sensitivity time control.

There are a number of methods for finding the illumination function to give a chosen pattern, including:

- The inverse Fourier transform;
- The Woodward-Lawson sample value method.

5.5.1 Inverse Fourier transform method

This is the theoretical inverse Fourier transform [2, p. 524] of (5.72) and is easy to calculate for simple pattern, say, a sector, for a fan search beam from the horizon up to 60 degrees. Such an example is a linear antenna 10 wavelengths long having a design beam width of $2c_{end}$ = 2arcsin(c) or 60 degrees.

$$Beam\ shape\ in\ sine\ space \ = \ \begin{matrix} 1 \text{ for } -c_{end} < u < c_{end} + c_{end} = 1/2 \\ \text{otherwise } 0 \end{matrix} \tag{5.82}$$

The normalized antenna patterns are given in sine space terms where $u' = w \sin \theta / \lambda$ and x in terms of the antenna width, w, and the wavelength, λ. The sine space beamwidth Θ is 1. The current distribution of the radiators in the line or illumination function is

$$\text{Current distribution, } g(x) = \frac{\sin \pi \frac{w}{\lambda} x}{\pi \frac{w}{\lambda} x} \quad \text{for} \quad -\frac{w}{2} > x > +\frac{w}{2} \tag{5.83}$$

The inverse Fourier transform assumes that the illumination function exists between minus and plus infinity. The real antenna exists only between -*w*/2 and +*w*/2, so the illumination function is truncated. The Fourier transform of this truncated illumination function gives an antenna pattern with sidelobes and ripple in the main beam:

$$\text{Pattern from an antenna length } \frac{w}{\lambda} = \frac{1}{\pi}\left(\text{Si}\left(\frac{w}{\lambda}\pi(u' + c_{end})\right) - \text{Si}\left(\frac{w}{\lambda}\pi(u' - c_{end})\right) \right) \tag{5.84}$$

where Si is the sine integral $\text{Si}(z) = \int_0^z \frac{\sin t}{t}\, dt.$

The curves are shown in Figure 5.50.

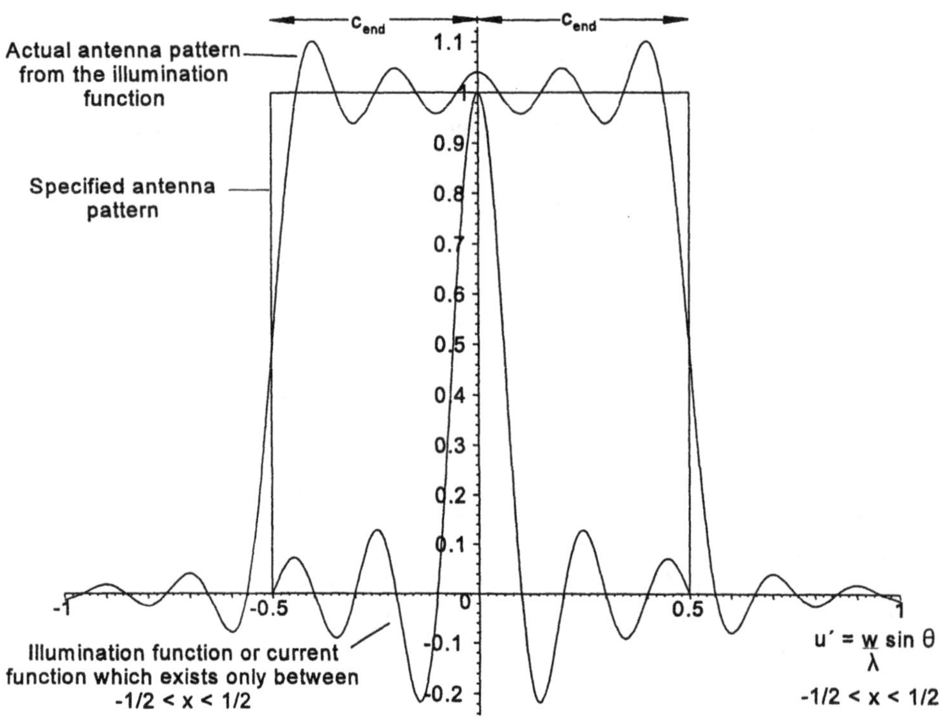

Figure 5.50 A specified sector pattern in sine space, synthesized pattern, and aperture function.

Closed forms for the Fourier transform are not available for arbitrary functions. For arbitrary patterns, such as the cosecant squared with sensitivity time control, the Woodward-Lawson method is often used.

5.5.2 The Woodward-Lawson method

Another method of finding the illumination pattern for a chosen beam shape is the Woodward-Lawson sample value method [2, p. 526; 5, p. 419]. This method is illustrated here in the example for a cosecant squared pattern.

The desired pattern is defined first in (elevation) angle as (see Figure 5.51):

- Elevation angle 2.87 degrees (arcsin 0.05): Gain 0 dB;
- Elevation angle 10 degrees: Gain 0 dB;
- Elevation angle 30 degrees : Gain -9.1 dB; following a cosecant squared characteristic from 10 degrees.

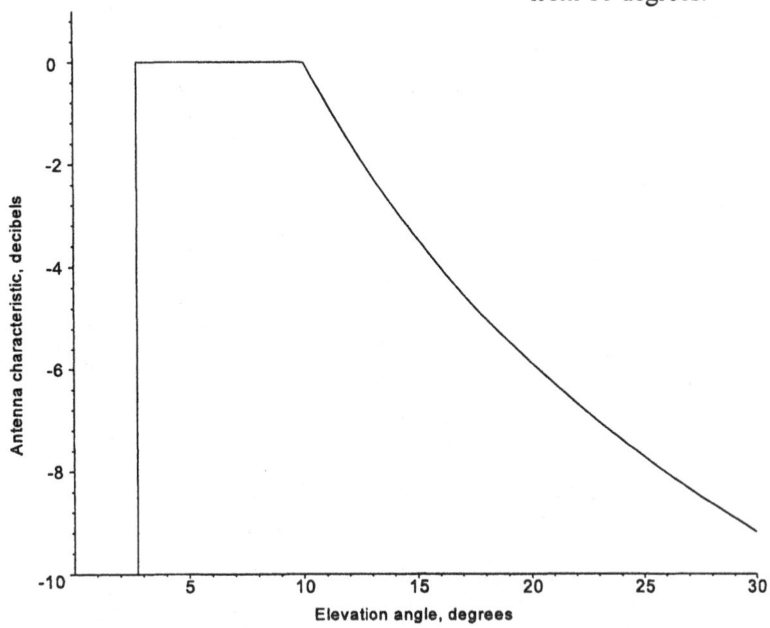

Figure 5.51 The specified cosecant squared pattern.

Equally spaced samples in sine space are chosen. The spacing is normally chosen to be around the standard beamwidth, λ/w, to avoid possible grating lobes at wide angles.

The antenna pattern is built up from these samples as the sum of displaced patterns of linear apertures of the same length. The radiation pattern of a uniform line source is

$$\text{Sample line source radiation pattern, } F_k(\theta) = a_k \frac{\sin\left(\pi \frac{w}{\lambda} \sin(\theta - k\delta)\right)}{\pi \frac{w}{\lambda} \sin(\theta - k\delta)} \tag{5.85}$$

where a_k is the scaling factor taken from Figure 5.51 at a point $\theta = k\delta$;
k is an integer between $k_{\max}$ and $k_{\min}$;
$k_{\max} - k_{\min} = K$ the number of samples;
δ is the spacing between the sampled patterns.

One such sample pattern is shown in Figure 5.52. The sin x/x function has two interesting properties:

$$\int_{-0.6132}^{0.6132} \frac{\sin \pi q}{\pi q} dq = 1$$

$$\int_{-\infty}^{-0.6132} \frac{\sin \pi q}{\pi q} dq = \int_{0.6132}^{\infty} \frac{\sin \pi q}{\pi q} dq = 0 \tag{5.86}$$

where $q = (\theta - \delta)/\lambda$.

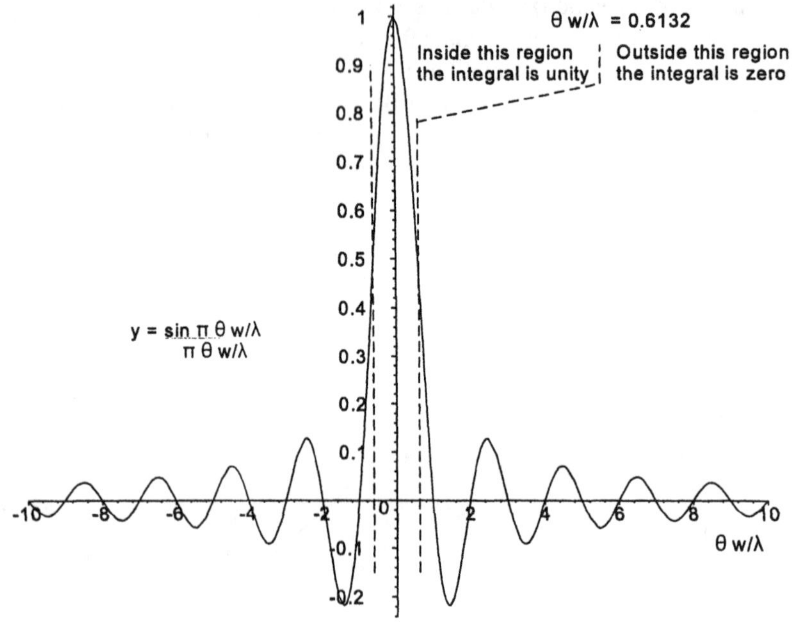

Figure 5.52 The sinc function showing the regions inside and outside $\pm\theta w/\lambda = 0.6132$.

Inside the bounds $\pm\theta w/\lambda = 0.6132$ the integral is unity, and outside these bounds there is no contribution to the inverse Fourier transform. The radiation patterns of these samples are shown in Figure 5.53, where it is easy to see that the contributions away from the peaks tend to zero. The synthesized antenna pattern is the sum of these elemental patterns and is shown in the traditional angle versus decibel form in Figure 5.54. The polar form is shown in Figure 5.55.

The aperture function, illumination function, or current distribution is the inverse Fourier transform of the antenna pattern [2, p. 534; 8, p. 24]. This is calculated from the inverse Fourier transforms of each sample, one of which is shown in Figure 5.57. The inverse transform of a displaced sin x/x function has two component rectangular functions equally spaced from the origin, and the inverse transform of all the samples is the envelope. This is shown as the illumination pattern in Figure 5.57. Sin x/x functions extend to infinity, so that the illumination function taken for a radiation pattern over a finite angle is

$$\textit{Aperture illumination function, } g(x) = \sum_{k=k_{min}}^{k_{max}} g_k(x) = \sum_{k=k_{min}}^{k_{max}} \sum_{n=-\frac{N}{2}}^{+\frac{N}{2}} F_k(\theta_n) \exp\left(j2\pi \frac{n\lambda}{2} \sin\theta_n\right) \tag{5.87}$$

where θ_n is one of N equally spaced pattern samples within $\pm 0.6132\ \lambda/w$ of the angle $k\delta$. Note that for an angle sample spacing of a standard beamwidth (λ/w), the aperture illumination function samples are spaced $\lambda/2$ apart.

The antenna pattern is not symmetrical. In the Woodward-Lawson synthesis, the amplitude function is an even function. The aiming and unbalance are provided by the phase characteristic and are given by

$$\begin{aligned} \textit{Amplitude} &= \left| \textit{Aperture function} \right| \\ \textit{Phase} &= \arctan\left(\frac{\textit{Imaginary part of aperture function}}{\textit{Real part of aperture function}} \right) \end{aligned} \tag{5.88}$$

The aperture functions are shown in complex form in Figure 5.57, amplitude in Figure 5.59, and phase in Figure 5.60.

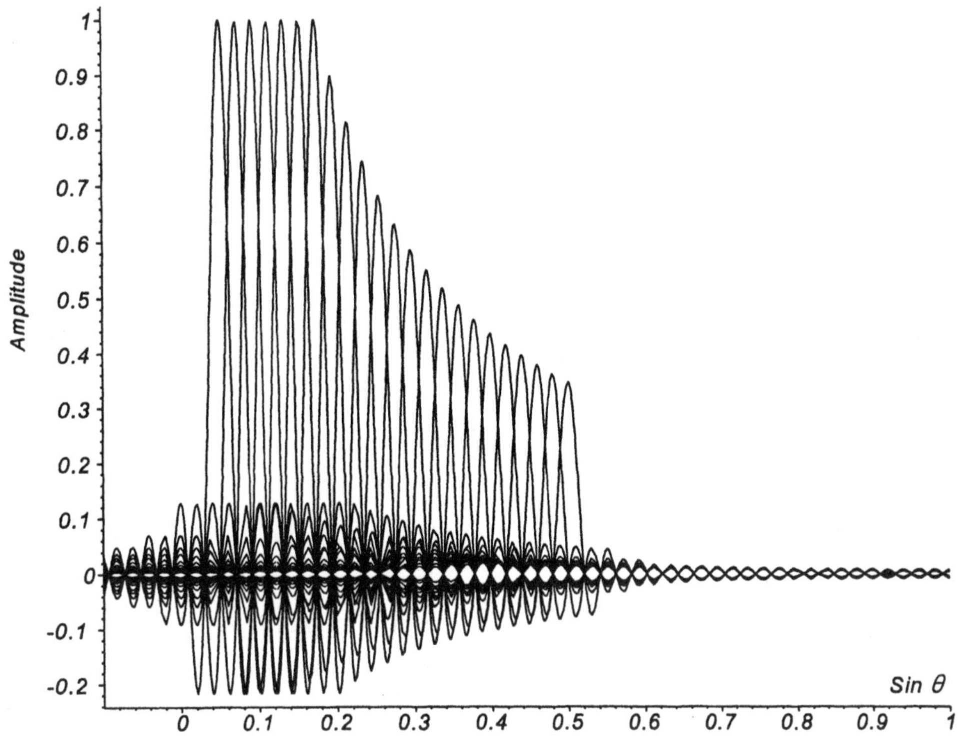

Figure 5.53 The construction of the synthesized pattern using sinc curves spaced at $\lambda/2$ along a length of 11.5λ.

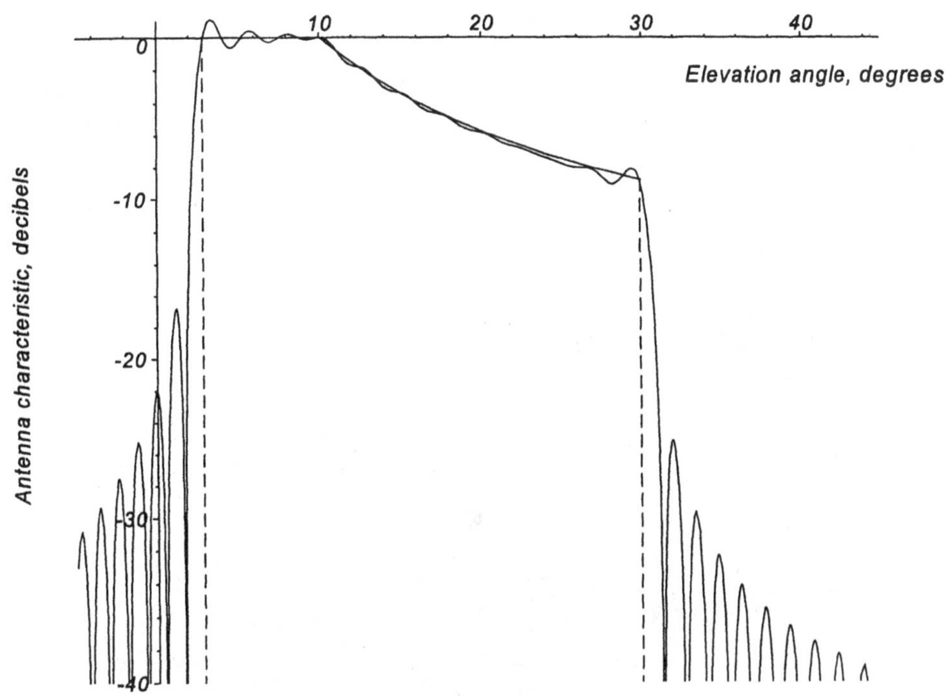

Figure 5.54 The desired pattern and its approximation in decibels using sinc curves spaced at $\lambda/2$ along a length of 11.5λ.

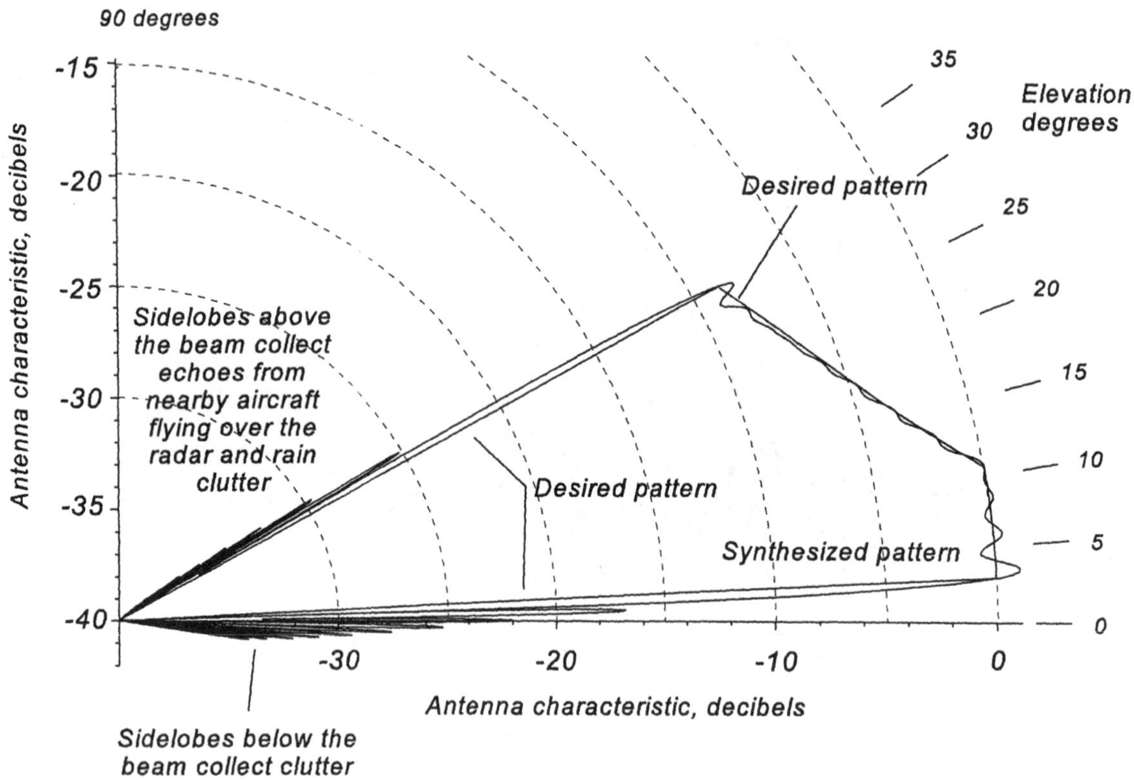

Figure 5.55 The polar diagram of the synthesized pattern for samples spaced at λ/2 along a length of 11.5λ.

The illumination function is the sum of the Fourier transforms of the individual sinc functions in Figure 5.53. The Fourier transform of a sinc function is a helix of limited length with a phase depending on the position of the sinc function. The cluster of sinc functions before summing is shown in Figure 5.56, their vector sum in Figure 5.57, the modulus in Figure 5.58, and the phase angle along the aperture in Figure 5.59.

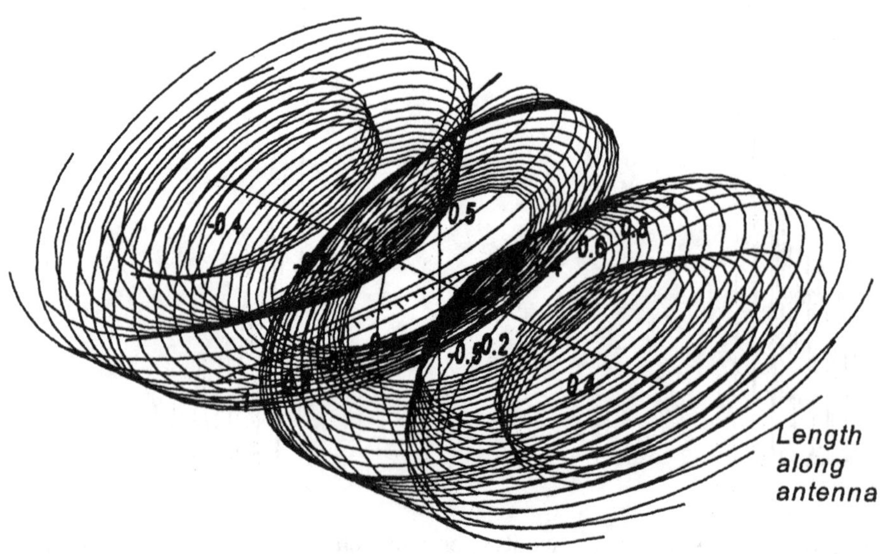

Figure 5.56 Inverse Fourier transforms of the individual sinc functions.

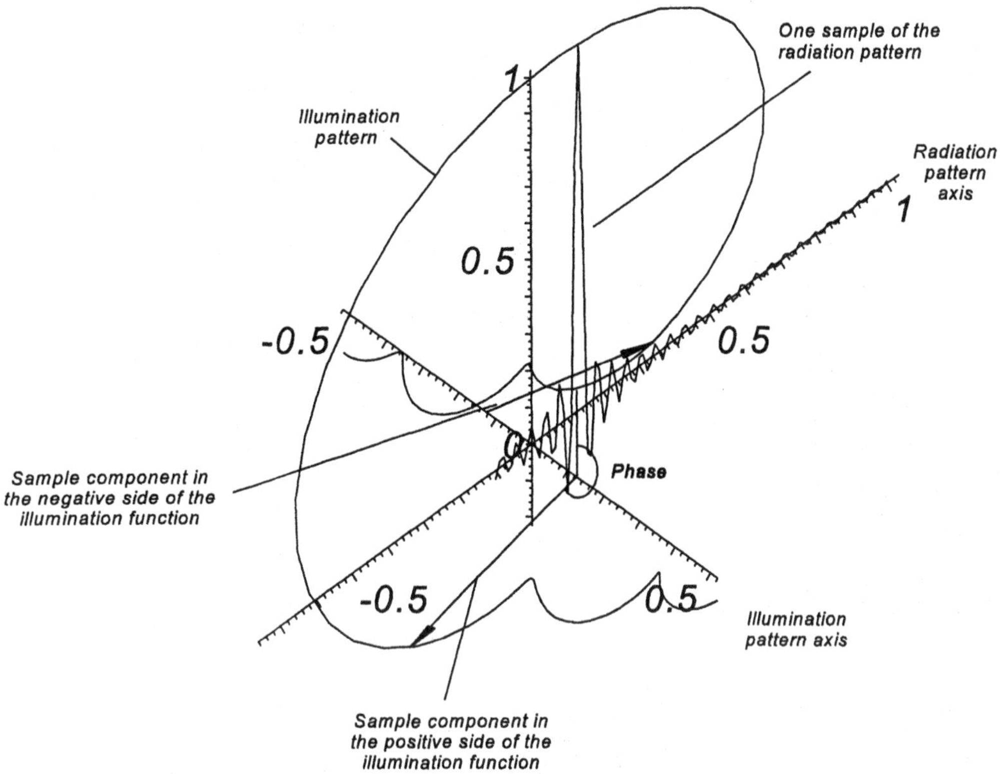

Figure 5.57 The complex aperture function for the cosecant squared antenna of length 11.5λ.

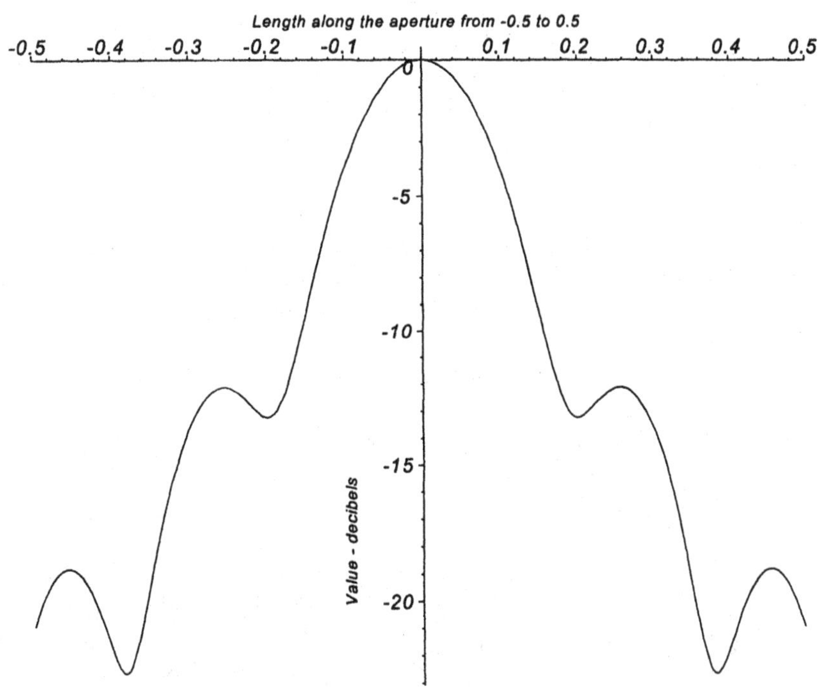

Figure 5.58 The amplitude of the aperture function for the cosecant squared antenna of length 11.5λ.

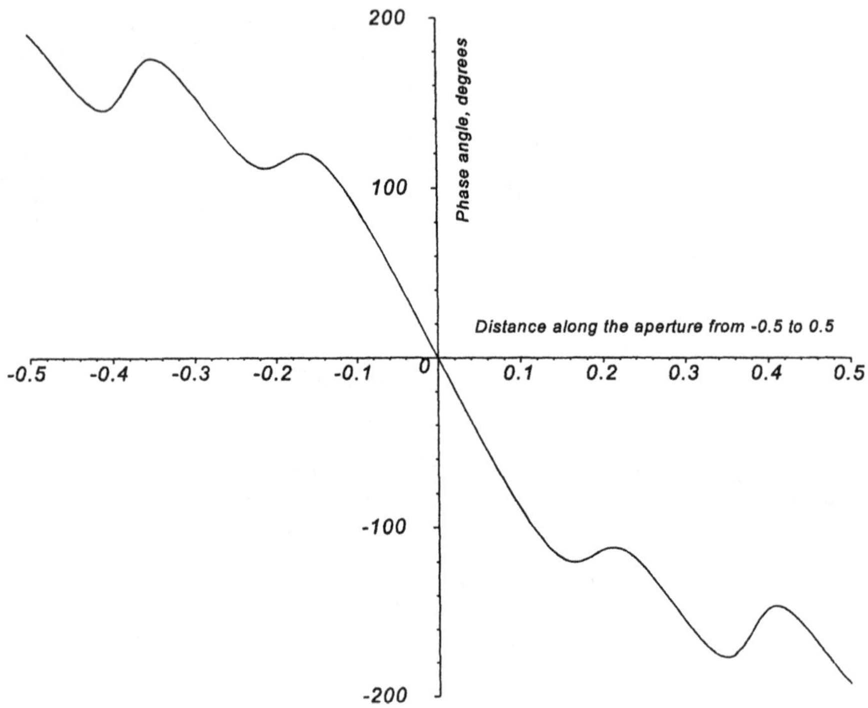

Figure 5.59 The phase angle of the aperture function for the cosecant squared antenna of length 11.5λ.

There are a number of points to note:

- The samples must be independent or orthogonal [5, p. 739].
- The pattern is the same as the desired pattern at the K sample points but is not controlled between them. If there are problems caused, more sample points must be used. The distance between the sampling points is normally one half a wavelength to avoid grating lobes. Thus more sampling points can be achieved only by increasing the size of the antenna [8, p. 24].
- Aperture function is symmetrical in amplitude and conjugate in phase [8, p. 24].
- The phase is tilted in the direction of the bulk of the beam [5, p. 739].
- This synthesis applies to linear antennas and planar arrays where the dimensions are separable [8, p. 24].

The pattern that has been formed will give a number of problems. The beam has ripple, which, in this case, will slightly alter the maximum ranges. Above and below the beam a number of sidelobes are present. With a surface radar, the sidelobe at 1.24 degrees 16.7 dB down and those below it will collect nearby clutter. The sidelobes above the beam, for example, at 30.3 degrees, 25.5 dB down, will give echoes from aircraft flying over the radar at short ranges.

After the basic design a number of numerical optimization techniques [6, p. 739] may be used to decrease the ripple and the sidelobes.

The aperture function can be approximated by specially developed horns. The phase characteristic may be controlled for a reflector antenna by changing the distance of the virtual directrix from the radiating horn. The shape of a parabolic reflector is often modified to simulate this shaped directrix.

Often, a reflector is fed by a number of horns (or similar feeds) stacked vertically to produce a cosecant-squared beam. In Figure 5.60, the transmit-receive devices are positioned between the individual horns and the power divider fed by the transmitter. The power divider gives the illumination function for a cosecant-squared transmitted beam. The returning echoes are switched to separate receivers which allow the measurement of elevation angle using monopulse evaluation in three-dimensional surveillance radars. It must be noted that the transmitting and receiving patterns of such antennas are not reciprocal.

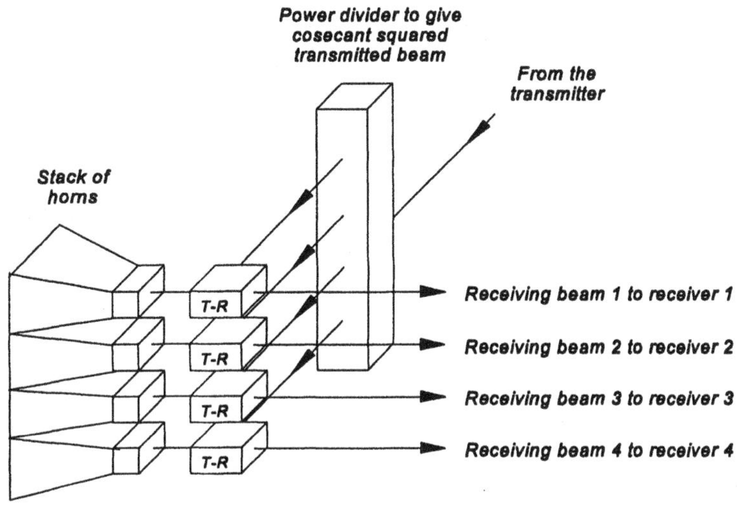

Figure 5.60 The arrangement of horns and transmit-receive switches to give a single transmitted beam and a number of stacked receiving beams.

5.6 CIRCULAR POLARIZATION

Circular polarization is often used as a primary measure against rain in spite of diminished scattering from normal objects of interest. To preserve the polarization, the antenna must work equally with horizontal and vertical polarization, that is, a symmetrical mesh or solid surface must be used on the reflector. Any deviation from symmetry will favor one of the orthogonal components to produce elliptical polarization. The negative rotation component of the ellipse (equivalent to the negative phase sequence component) will reflect rain echoes.

Waves with circular polarization may be radiated from arrays of circularly polarized elements or converted from linear polarization. Circular polarizers, which convert linearly to circularly polarized waves, may be built into waveguides, be part of special reflectors, or be held in front of radiating elements.

5.6.1 Circular polarizer for horn feeds

Waveguide polarizers are placed between a normal waveguide and the radiating element, normally a horn. The circular mode is symmetrical in both Cartesian coordinates so that all waveguide components are square or circular. Such a generic polarizer is shown in Figure 5.61.

Circular polarizers depend on resolving the TE_{10} or H_{10} wave in a square or circular waveguide into two diagonal components, as shown in Figure 5.62. These components undergo different phase shifts until they are 90 degrees out of phase. Then the voltage field vector rotates, giving circular polarization. Polarizers may take a number of forms. Generally they are used only when necessary and there is a requirement not to interrupt operation during the changeover. With some older polarizers, it was necessary to switch off the transmitter when the polarization was changed. For this reason, only polarizers built into circular waveguides are considered. They consist of a rectangular to circular waveguide tapered transition, a rotatable phase changing section, and a circular output waveguide. This is often followed by a circular to square transition for a rectangular horn.

The rotatable center section may be rotated such that the sides of the iris are horizontal or the post or blade is vertical. This gives no change in the polarization and is often the normal mode of operation that gives greater radar range. Rotating the center section either to the right or to the left by 45 degrees gives one or the other of the two senses of circular polarization which is used, when necessary, to reduce rain clutter.

The section that changes the polarization normally consists of a number of irises placed a quarter of a wavelength apart. This spacing suppresses reflections. The irises have narrower openings towards the center of the section, giving

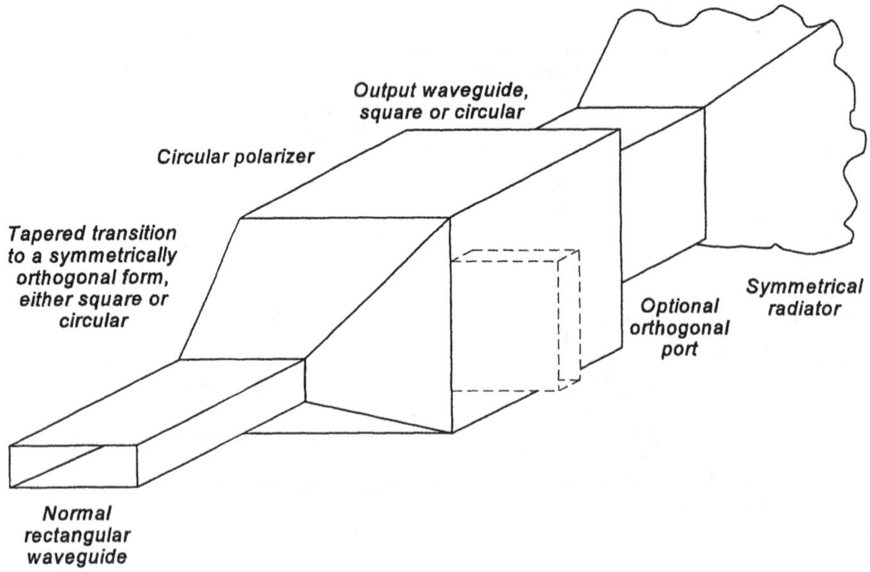

Figure 5.61 A generic waveguide polarizer for feed horns.

higher reactance in the middle [19]. The gradual transition reduces the standing waves and increases the bandwidth. The total reactance of the irises is chosen to give the orthogonal waves exactly 90 degrees phase difference. The waveguide containing the irises may be rotated through 45 degrees to be in line with the input and output waveguides. Although there is a phase shift in the main mode, there is no transition to circular polarization. The linear polarization mode is used in clear conditions to obtain maximum radar range.

Other types of circular polarizer use diagonal posts placed a quarter of a wavelength apart or dielectric vanes, which have arrow-shaped ends.

If the waves in the waveguide in Figure 5.62 are horizontally polarized, the delay occurs in the opposite order and right circular polarization is produced. The coupler shown dotted in Figure 5.61 couples the horizontal component into or out of the polarizer, which is used to extract the rain clutter signal for a rain or weather display in airport radars.

5.6.2 Reflecting polarizers

Reflecting polarizers are in the form of diagonal grids in reflectors [20, p. 23-26]. Such a polarizer is shown in Figure 5.63. Two diagonal orthogonal grids are placed an eighth of a wavelength apart, so there is an extra path length of a quarter wavelength or 90 degrees in the phase of one component. The example shown produces left circular polarization when looking in the direction of propagation. Note that illumination from a horizontally polarized wave would produce right circular polarization.

5.6.3 Transmission polarizers

Parallel plate, dielectric, or meander transmission polarizers [20, p. 23-26] are placed in front of the antenna or feed horn system and one is shown in Figure 5.64. The reference gives the following equations for the values of a and l:

$$\textit{Slat depth, } l = \frac{3\lambda_g}{4} \text{ or } \frac{\lambda_g}{4} \text{ according to the sense of polarization} \tag{5.89}$$

$$\textit{Slat separation, } a = 0.671\,\lambda_g \tag{5.90}$$

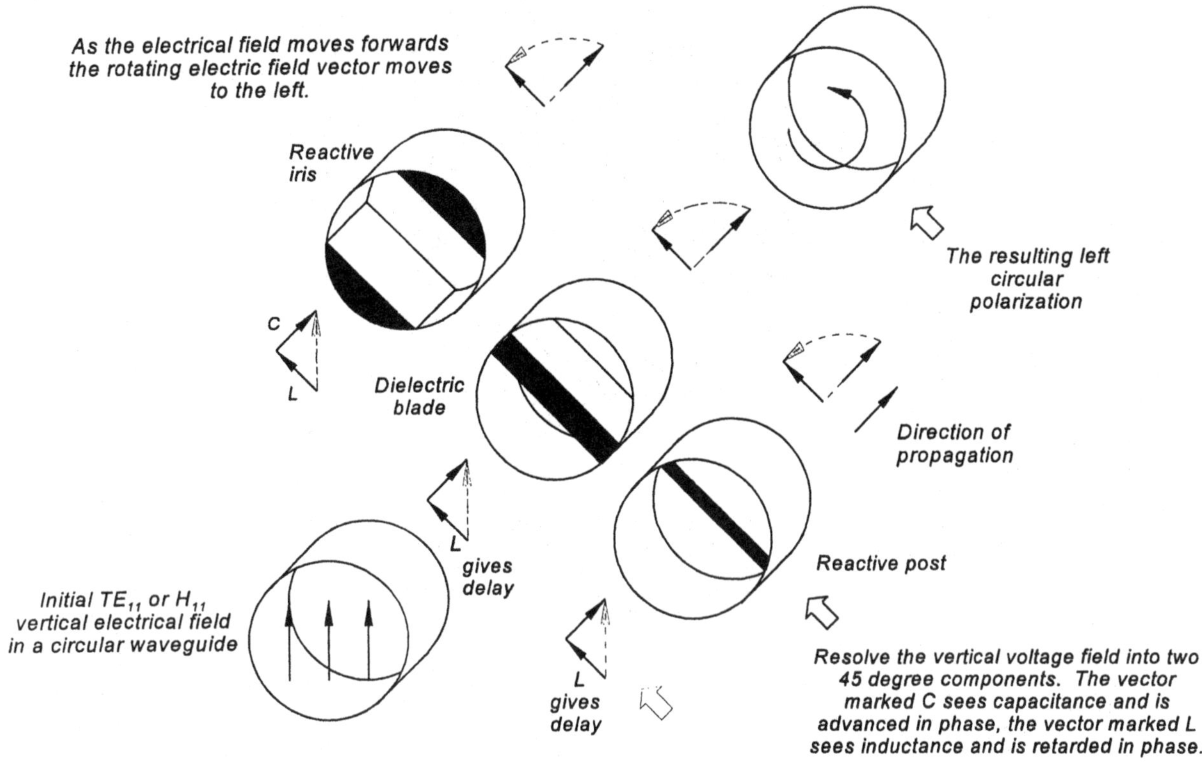

Figure 5.62 The principle of circular polarizers using irises, posts, and blades.

$$\text{Guide wavelength, } \lambda_g = \frac{\lambda_0}{\sqrt{1 - \left(\frac{\lambda_0}{2a}\right)^2}} \tag{5.91}$$

where λ_0 is the free space wavelength.

The slats are moved mechanically either in front of the feed system or to storage at the side.

5.6.4 Phased array polarization

Phased arrays may be designed for circular polarization either with an external polarizer (slats) or with each discrete element radiating in a circularly polarized mode. Such elements are quadrature fed crossed dipoles, axial mode helices, or circularly polarized patches. To switch between the types of polarization, each element must be switched to radiate in a different mode.

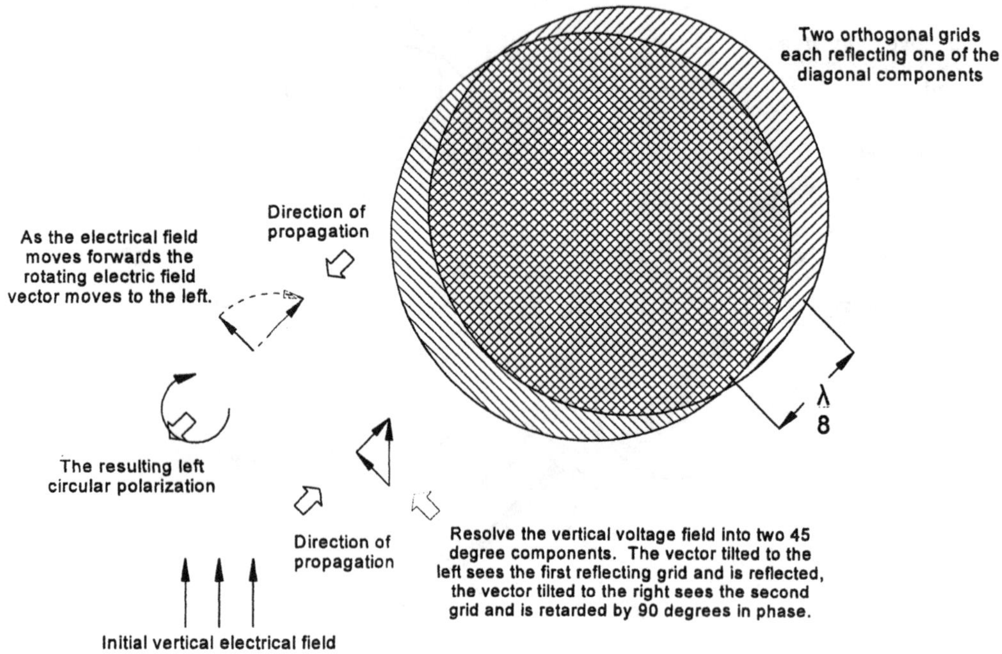

Figure 5.63 Principle of a reflector polarizer.

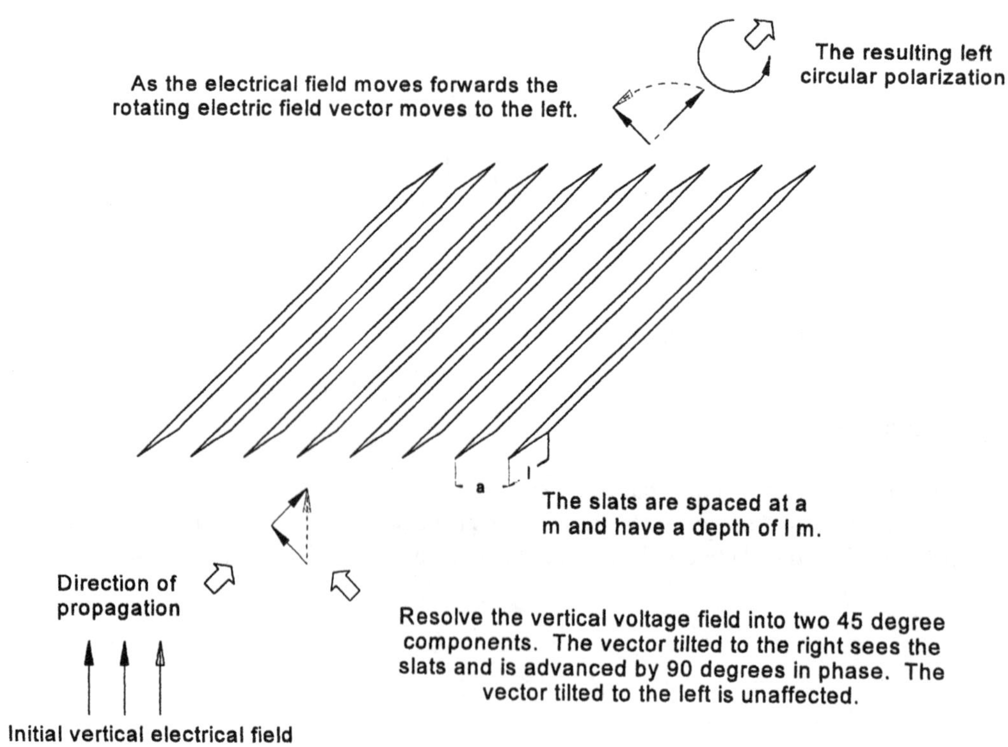

Figure 5.64 Circular polarization formed by diagonal slats.

5.6.5 Engineers' and physicists' conventions

There is a contrast as to how engineers and physicists define polarization, as shown in Table 5.5.

Table 5.5
Engineers' and physicists' conventions for circular polarization

Electrical engineering	Physics
Linear polarization The direction of linear polarization is given by the direction of the electric field normal to the direction of propagation (see IEEE Standard 100).	Linear polarization The definition is not so clear: older literature uses the direction of the magnetic field vector and newer literature the electric field vector [21, p. 6-4].
Circular polarization The sense is given by the direction of rotation of the electric field vector with time as the wave passes through a plane normal to the direction of propagation as seen by an observer at the source (see IEEE Standard 100).	Circular polarization This is given by the rotation of the electric field vector with time as the wave passes through a plane normal to the direction of propagation as seen by an observer looking towards the source [21, p. 6-4].

The different conventions caused problems and delays with the first transmissions using satellites between the United States and England.

5.6.6 Ellipticity or the quality of circular polarization

The quality of circular polarization is measured by using a rotating dipole as a transmitting antenna during antenna testing. If the polarization of the antenna on test is perfectly circular, it will deliver a constant signal. Otherwise, the signal will be modulated by the transmitting antenna rotation, as in Figure 5.65. If the peaks are a voltage units and the troughs b voltage units, the ellipticity is given by

$$\textit{ellipticity},\ e_{ell} = \frac{a}{b} \quad \text{or} \quad 20\ \log_{10}\frac{a}{b}\ \text{dB} \tag{5.92}$$

Note that eccentricity is defined as $\epsilon = \sqrt{1 - \dfrac{b^2}{a^2}}$

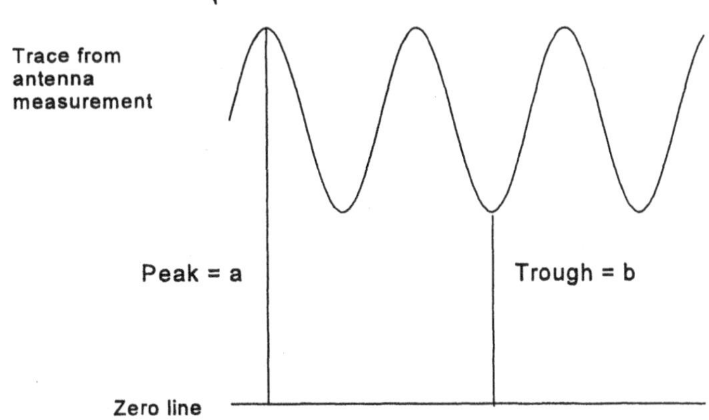

Figure 5.65 The wavy antenna characteristic trace with a rotating transmitting dipole.

It is customary to measure the ellipticity, e_{ell}, over the whole of the main beam, at least between the 10 dB points. The average used for finding the integrated cancellation ratio (ICR) is weighted by the two-way antenna voltage gain in that direction.

5.6.7 Rain echo suppression

The amount of rain echo reduction [17, p. 7.6] is given by the measure of suppression with e_{ell} as the average ellipticity

$$\textit{Measure of suppression} = \frac{e_{ell}^2 + 1}{e_{ell}^2 - 1} \text{ ratio or } 20 \log_{10} \frac{e_{ell}^2 + 1}{e_{ell}^2 - 1} \text{ dB} \tag{5.93}$$

If (5.93) is weighted by the two-way antenna pattern and integrated, it is called the integrated cancellation ratio, where e_{ell} is the average ellipticity weighted by the two-way antenna voltage characteristic.

Figure 5.66 shows this in graphical form.

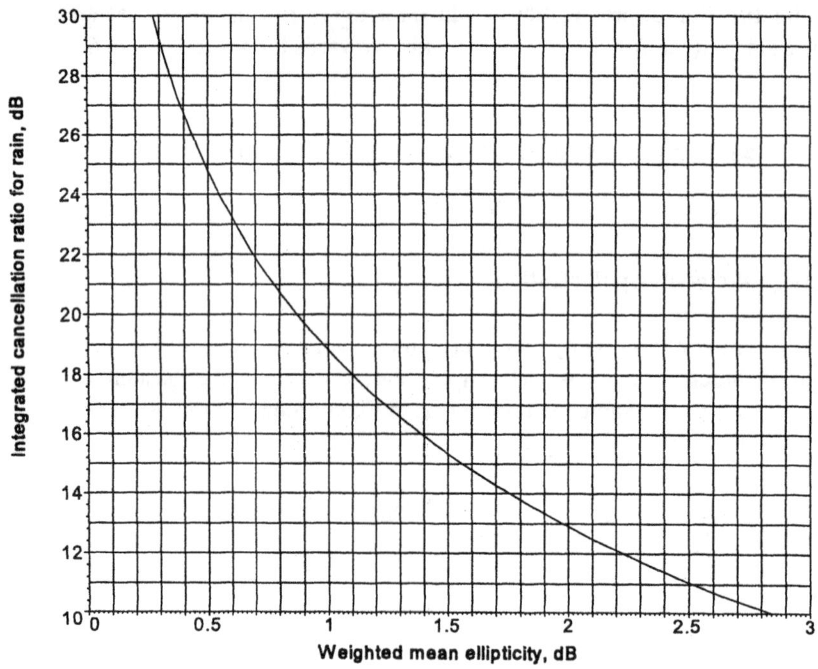

Figure 5.66 The effects of ellipticity on rain echo reduction.

5.7 ANTENNA HARDWARE LOSSES

The principal losses compared to a perfect aperture with uniform illumination are as follows:

- Illumination function loss;
- Blocking loss;
- Spillover loss;
- Surface loss;
- Losses in the power dividers, phase shifters, and other beam forming network components.

Antenna patterns are normally measured on a test range. The measured patterns include the losses mentioned in the preceding list.

5.7.1 Illumination function loss

This is a so-called loss because it is a reduction of the gain in a practical antenna compared with a uniformly illuminated antenna of the same size. Losses for aperture functions are given in Appendix B. The antenna gain can be restored to the desired value by increasing the size of the antenna or changing the illumination function.

5.7.2 Blocking loss

Any obstacle placed in front of the reflector affects the antenna pattern. Effectively, looking perpendicularly at the directrix plane, there is a hole in this plane. Using the fact that the Fourier transform is the sum of the Fourier transforms of the components, the pattern is the Fourier transform of the illumination of the directrix minus the Fourier transform of the illumination of the "hole". These effects are shown for one dimension of a rectangular antenna with cosine-squared illumination for 1% and 3% linear blockage in Figure 5.67. The sidelobe levels with and without blocking are shown in Figure 5.68. The equivalent sidelobe envelope for 1% blockage is nearly constant, while that for 3% shows less gain at wider angles. When the linear blocking pattern is subtracted from the linear antenna pattern, the odd sidelobes are increased, and the even sidelobes are decreased. The result in decibels is also shown.

There will be a loss of (coherent) gain caused by the blocking. The loss is $1 - w_b/w$ in amplitude, where w_b is the effective blocking width. There is a formula in [3, p. 101] for the increase in odd (1, 3, 5, ...) sidelobes:

$$\textit{blocked sidelobe ratio, } p' = \frac{p + \dfrac{w_b}{w}}{1 - \dfrac{w_b}{w}} \quad \text{voltage ratio} \tag{5.94}$$

where p is the unblocked sidelobe voltage ratio.

For structures that are short along the antenna axis, the area and the shape of the blocking structure may be used. For horns and similar radiators, there is a rule of thumb [5, p. 144; 22, p. 158] that states that this area should be multiplied by two. Longer structures must be treated as end-fire arrays with an appropriate, larger blocking area extending many times the physical width [17, p. 6.38]

$$\textit{Effective width} = \textit{width} + \sqrt{\frac{\textit{depth}\ \lambda}{2}} \tag{5.95}$$

The effects of increased blockage between 0.5% and 10% are shown in Figure 5.68.

5.7.3 Spillover loss

Not all the energy from the transmitter radiated by a horn hits the reflector. Some of the energy passes by around the edges and is called the spillover, causing extra clutter as the echoes enter the horn by the same path. In military radars, this path may be a way for active jamming to enter the antenna. The amount of spillover represents a loss.

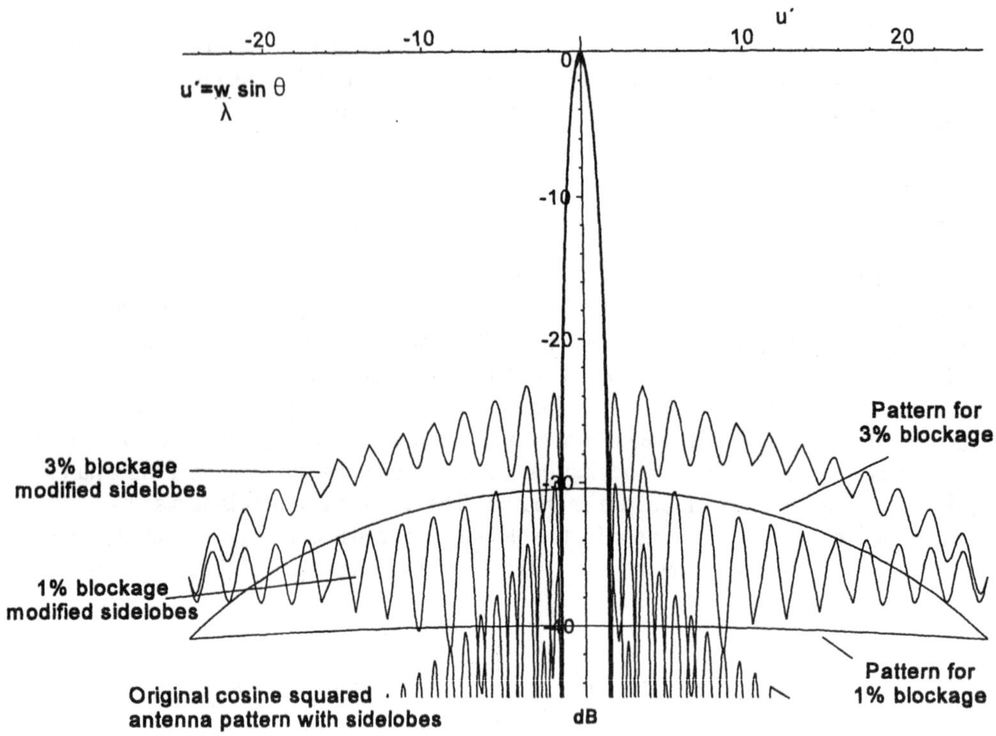

Figure 5.67 The effects of 1% and 3% linear blockage for an antenna with cosine-squared illumination.

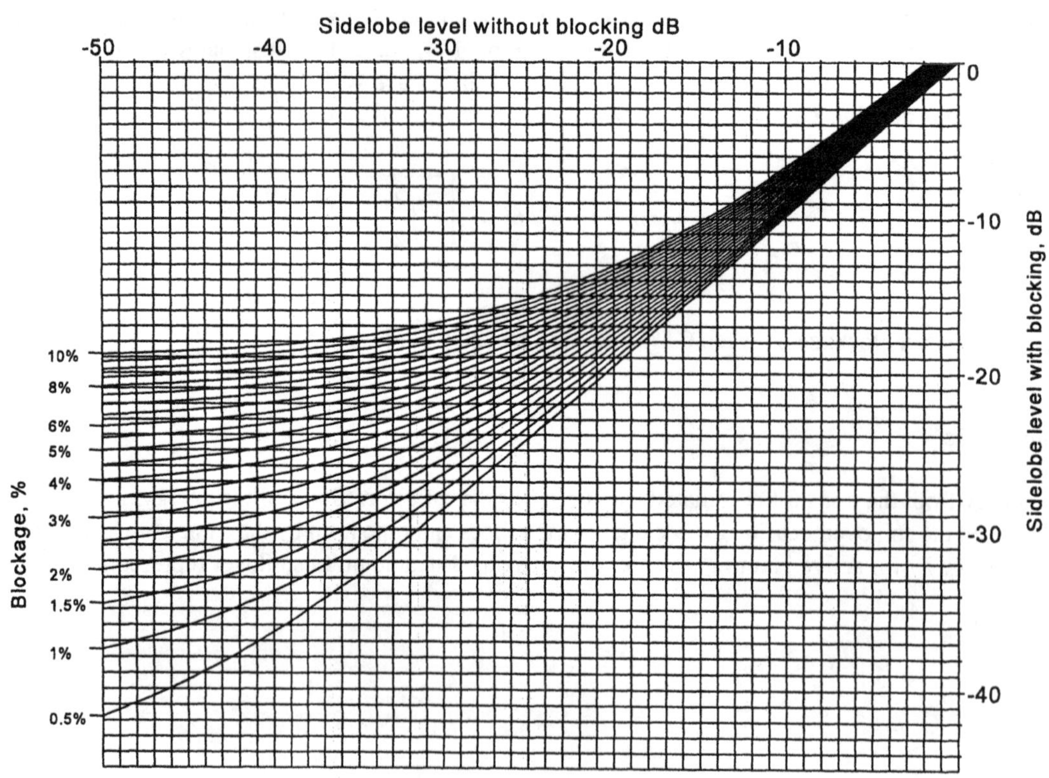

Figure 5.68 The effects of blocking on the antenna sidelobes.

5.7.4 Surface tolerance loss

The surfaces of a planar array or a reflector always have tolerances. If the standard deviation or root mean square (rms) tolerance of a planar radiator is δ, then the phase error standard deviation is $2\pi\delta/\lambda$ radians. Ruze in [17, p. 6-41] has developed the relationship for the loss when radiating from a surface:

$$\begin{aligned} Loss &= \exp(rms\ phase\ error\ in\ radians^2) \\ &= \exp\left(\frac{2\ \pi\ \delta}{\lambda}\right)^2 \\ &= 4.3429\left(2\pi\frac{\delta}{\lambda}\right)^2 = 171.4526\left(\frac{\delta}{\lambda}\right)^2 \quad \text{dB} \end{aligned} \tag{5.96}$$

Tolerances in reflectors have an effect on the incident and reflected portions of the energy, so the tolerance figure for δ must be doubled. These losses are shown in the graph in Figure 5.69.

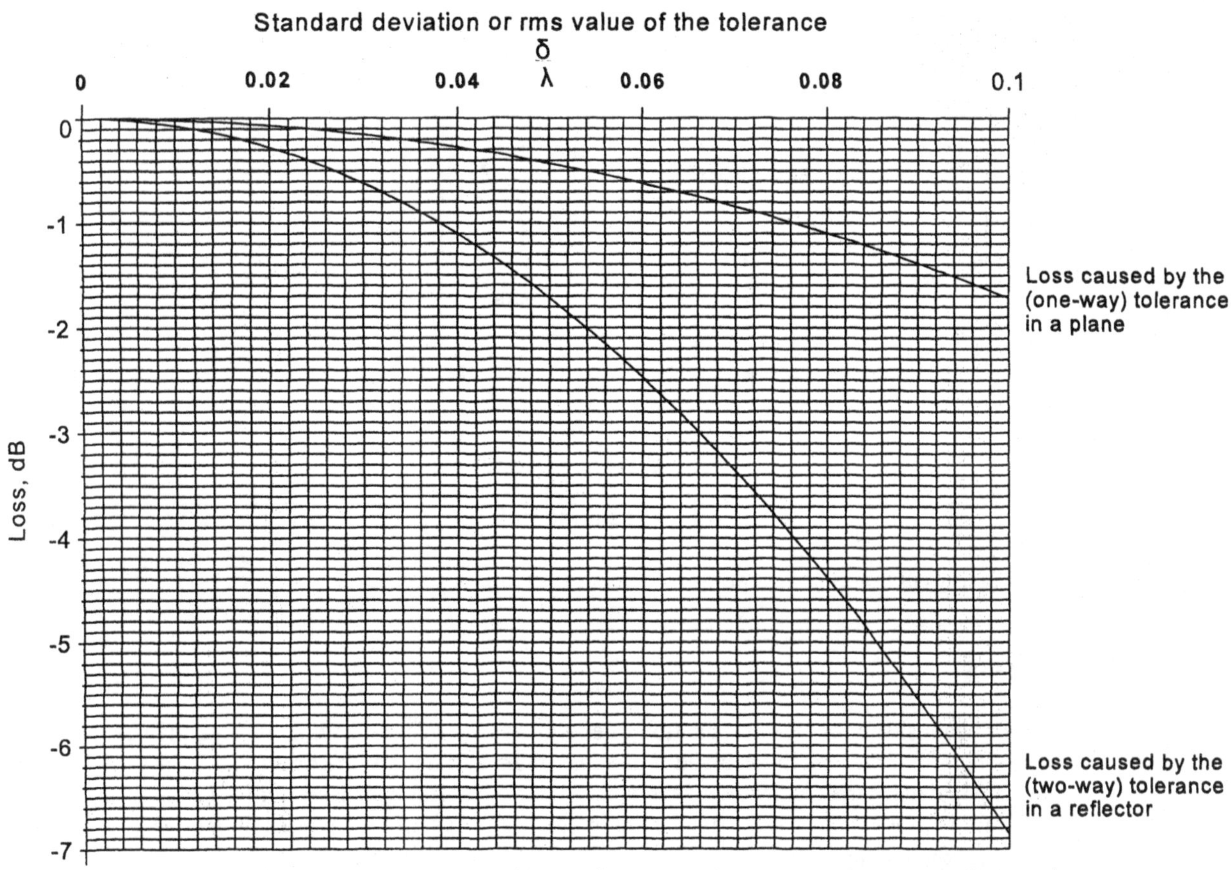

Figure 5.69 The loss caused by surface errors in an array plane or a reflector.

5.7.5 Losses in power dividers, phase shifters, and other beam-forming network components

The gain and patterns of an antenna are measured on a test range at a particular reference point or waveguide flange. There are losses between the points where the returning echoes land on the antenna and the point where the gain is measured, as is shown in Figure 5.70.

5.7.6 Other effects giving losses

Other effects that are recorded in the patterns during antenna testing are:

- Leakage through the mesh forming the reflector which gives a backlobe diffraction around:
  - The edges of a planar array;
  - The edges of a reflector;
  - The edges of a lens;
- Energy that avoids the reflector or lens and enters the feeder system directly, called spillover.

Leakage and diffraction account for the backlobe of the antenna. Nearby interference can enter the feed system directly which is shown by a small sidelobe at between 90 and 120 degrees. Antenna patterns at elevation angles below zero are rarely made, so screens to reduce interference entering the feed horn under the reflector are exceptional.

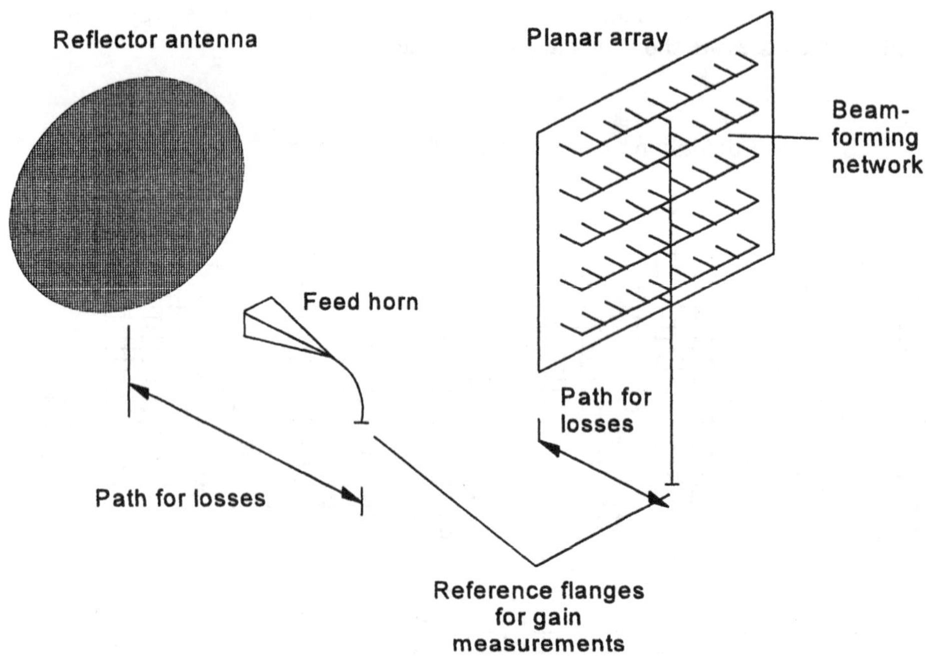

Figure 5.70 Losses up to the reference point for antenna gain measurement.

5.8 BEAM SHAPE LOSS

The scanning, pattern, or beam shape loss is the reduction of energy caused by the fact that the antenna beam is not a sector of a sphere. The strength of the signals received is at a maximum at the tip of the beam and less at the sides, in contrast to the constant strength using the sector pattern shown in Figure 5.71. Equivalence is for the signal-to-noise ratio before detection and is different for coherent and noncoherent integration. As before, the phenomenon is resolved into separate azimuth and elevation dimensions.

5.8.1 Coherent integration

Where coherent integration is used [23] the signal-to-noise ratio after integration is proportional to the number of pulses integrated, *n*. The useful signals, those with high signal-to-noise ratio, come from the tip of the beam whose shape may be represented by the Gaussian function (from Section 2.9.2):

$$G_{One\ way,\ voltage}(\theta) = \exp\left(-2\ \ln(2)\ \frac{\theta^2}{\theta_3^2}\right) \tag{5.97}$$

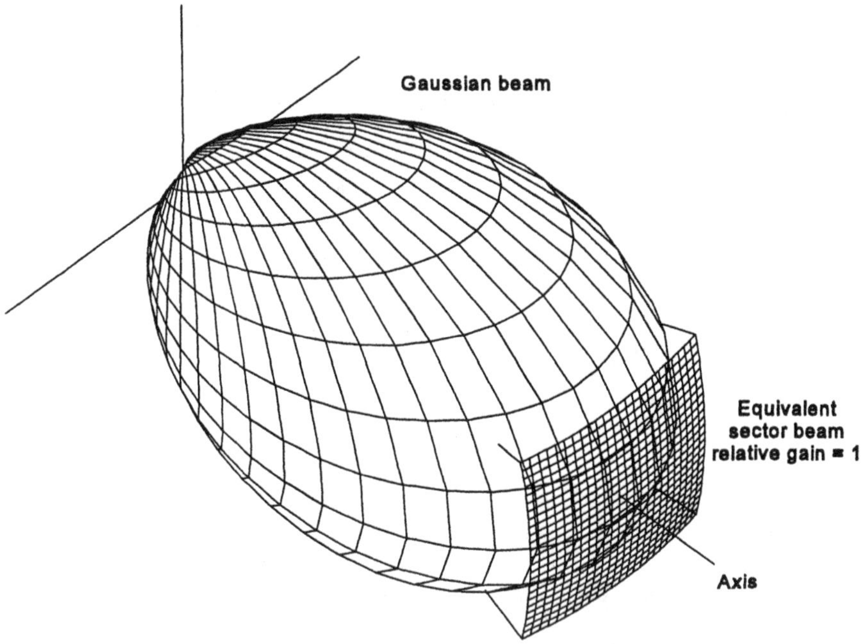

Figure 5.71 A Gaussian beam and its equivalent sector beam.

The one-way power pattern and the two-way voltage pattern are

$$G_{One\ way,\ power}(\theta) = G_{Two\ way,\ voltage}(\theta) = \exp\left(-4\ \ln(2)\ \frac{\theta^2}{\theta_3^2}\right) \tag{5.98}$$

The two-way power pattern is given by

$$G_{Two\ way,\ power}(\theta) = \exp\left(-8\ \ln(2)\ \frac{\theta^2}{\theta_3^2}\right) \tag{5.99}$$

If the radar transmits k pulses per radian from its antenna, which scans at ω radians/s, $k = \frac{f_{prf}}{\omega}$, the energy received is

$$\begin{aligned} Energy_{Two\ way,\ power} &= \int_{-\infty}^{\infty} \exp\left(-8\ \ln(2)\ \frac{\theta^2}{\theta_3^2}\right) d\theta \\ &= k\ \theta_3 \sqrt{\frac{\pi}{8\ \ln\ 2}} = 0.75269\ k\ \theta_3 \end{aligned} \tag{5.100}$$

The equivalent width of a sector beam is $0.75269\ \theta_3$, which receives 1.2338 dB less signal energy than a sector beamwidth of θ_3. The relationships are shown in Figure 5.72.

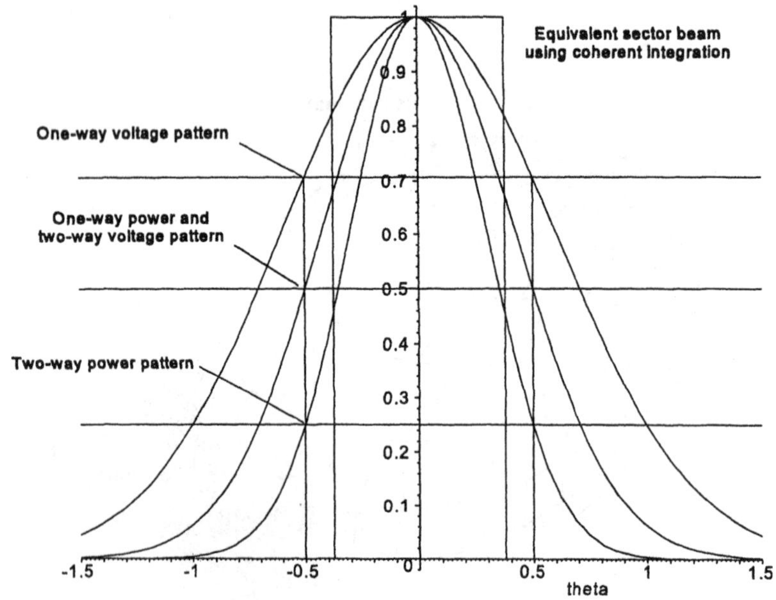

Figure 5.72 One-way and two-way Gaussian beam patterns with the equivalent sector beamwidth with coherent integration.

It should be noted that the equivalent sector beamwidth one way is 1.064 θ_3, giving a "beam shape gain" of 0.2713 dB.

5.8.2 Noncoherent integration

Blake [24] first investigated noncoherent integration as a reduction of the number of effective pulses. The noncoherently integrated signal-to-noise ratio for a peak signal value S_0, pulse density k pulses per radian from an antenna scanning at ω radians per second, received noise level σ_N volt over an angle θ_2 is given by

$$R = \frac{k \int_{-\theta_2}^{+\theta_2} S_0 \exp\left(-8 \ln(2) \frac{\theta^2}{\theta_3^2}\right) d\theta}{\sqrt{k \int_{-\theta_2}^{+\theta_2} \sigma_N^2 \, d\theta}} = R_0 \sqrt{\frac{2\,k}{\theta_2}} \int_0^{+\theta_2} \exp\left(-8 \ln(2) \frac{\theta^2}{\theta_3^2}\right) d\theta \qquad (5.101)$$

where $R_0 = S_0 / \sigma_N$.

The ratio R/R_0 is plotted against θ_2 in Figure 5.73 and shows a distinct maximum at $\theta_2 = 0.84078\ \theta_3$.

The equivalent sector beamwidth θ_{noncoh} of a sector beam antenna is given by equating the noncoherently integrated signals. The energy from a sector beam antenna is $\sqrt{k\ \theta_{noncoh}}$, thus,

$$\sqrt{k\ \theta_{noncoh}} = \sqrt{\frac{2\,k}{\theta_2}} \int_0^{+\theta_2} \exp\left(-8 \ln(2) \frac{\theta^2}{\theta_3^2}\right) d\theta \qquad (5.102)$$

Equation (5.102) gives the equivalent sector beamwidth of 0.47374 θ_3. When the signals are noncoherently integrated, the ratio becomes $\sqrt{0.47374\ \theta_3} = 0.68829\ \theta_3$. This represents a loss of 1.62229 dB compared with the signals integrated over the original half-power beamwidth. These relationships are shown in Figure 5.74.

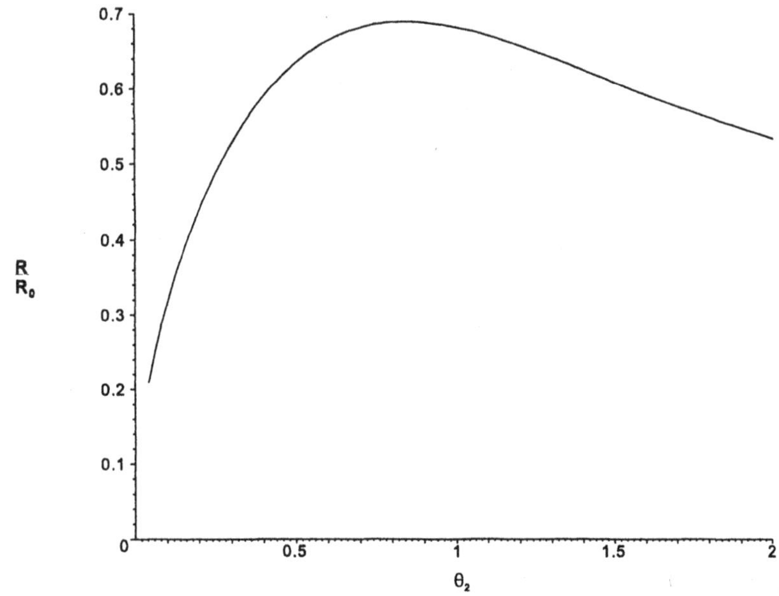

Figure 5.73 The signal-to-noise ratio with noncoherent integration plotted against the angle over which integration occurs.

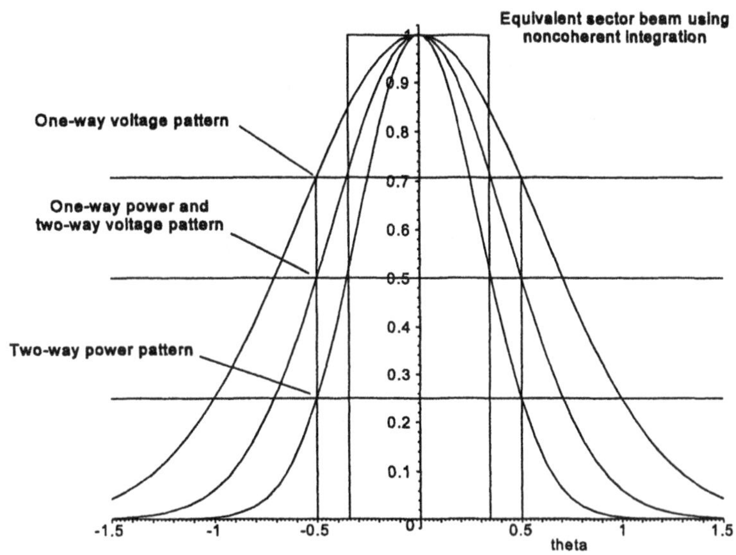

Figure 5.74 One-way and two-way Gaussian beam patterns with the equivalent sector beamwidth with noncoherent integration.

5.8.3 Small numbers of pulses

Both the estimations for coherent and noncoherent integration use integrals that assume many pulses between the half-power points of the beam pattern. The estimations assume that the noncoherent integration gain is proportional to the square root of the number of pulses integrated and are valid down to about four integrated pulses. Further, the distributions of the sums of small numbers of noise samples are gamma distributions instead of Gaussian distributions. For less than two pulses, the value will be greater than 1.23 dB for high detection probability.

5.9 SCANNING LOSS

The beam shape loss does not account for the movement of the antenna between the moments of transmission and reception. Figure 5.75 shows a reference two-way Gaussian beam, the positions of the transmitting and reception beams each offset a tenth of a beamwidth from the center, and the resulting two-way pattern.

The resulting transmission and reception beam with movement gives an effective loss of gain. When the combined beam is integrated over a beamwidth, we obtain with incoherent integration

$$\textit{Incoherent scan loss} = \frac{\displaystyle\int_{-1/2}^{1/2} \exp(-4ax^2)dx}{\displaystyle\int_{-1/2}^{1/2} \exp\left(-2a\left(x + \frac{\alpha}{2}\right)^2\right)\exp\left(-2a\left(x - \frac{\alpha}{2}\right)^2\right)dx} \text{ power ratio} \tag{5.103}$$

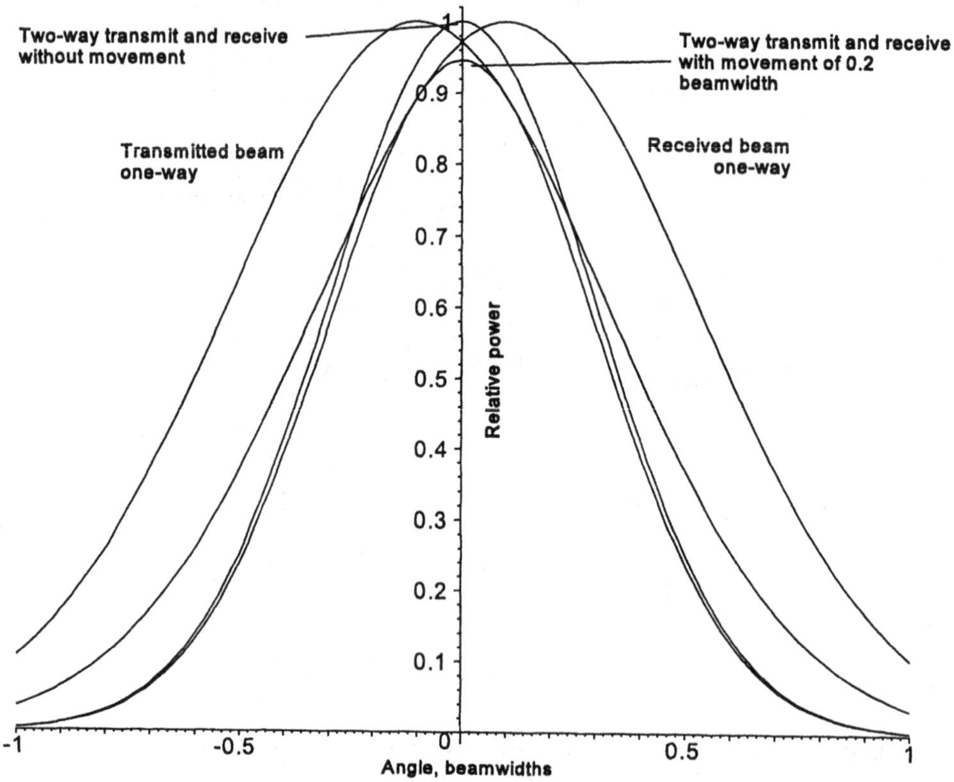

Figure 5.75 The effective one-way and two-way patterns when the antenna has rotated two tenths of a beamwidth between transmission and reception.

and for coherent integration

$$\textit{Coherent scan loss} = \frac{\left[\displaystyle\int_{-1/2}^{1/2} \exp(-2ax^2)dx\right]^2}{\left[\displaystyle\int_{-1/2}^{1/2} \exp\left(-a\left(x + \frac{\alpha}{2}\right)^2\right)\exp\left(-a\left(x - \frac{\alpha}{2}\right)^2\right)dx\right]^2} \text{ power ratio} \tag{5.104}$$

Figure 5.76 shows the scanning loss in decibels for from zero to half a beamwidth between transmission and reception.

Scanning loss, in practice, applies only to scanning antennas with narrow beams at long ranges. For example, if a surveillance radar rotates at six revolutions per minute and has a beamwidth of half a degree, then the angle between transmission and reception depends on range.

- Antenna rotation speed: 6 × 360/60 36 degrees/second
  36/0.5 72 beamwidths/second

Each kilometer in range needs 6.67 μs for the pulse to travel out and the echo to return which gives an angle difference that increases with range.

- Converting to range: beamwidths/second 1 000/150 = 0.00048 beamwidths/km

The result is shown in Figure 5.77 which is an extreme case. Commonly, beamwidths of 1 degree to 1.5 degrees are used which reduce the scanning loss to an almost negligible value.

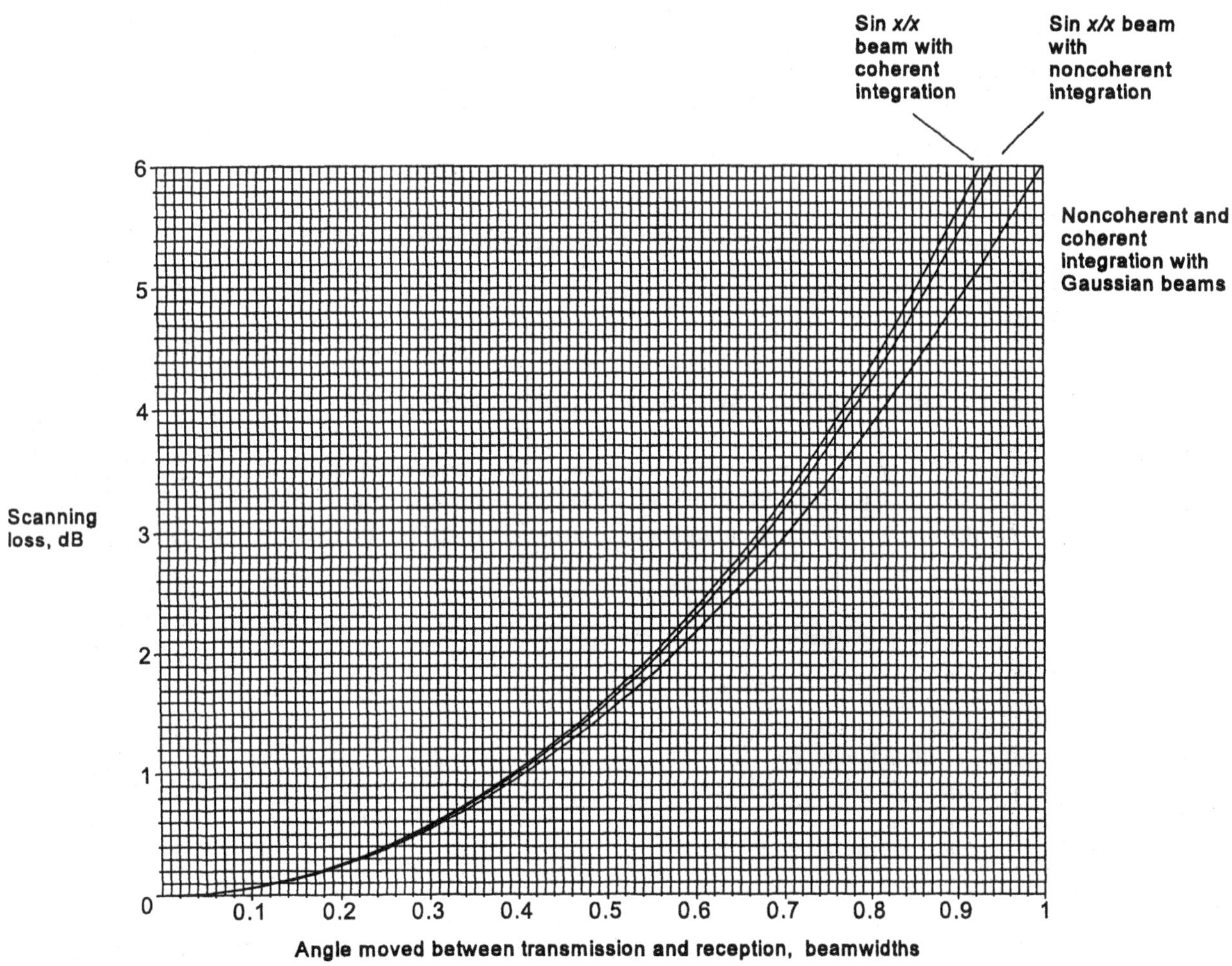

Figure 5.76 Scanning loss when the antenna rotates between transmission and reception.

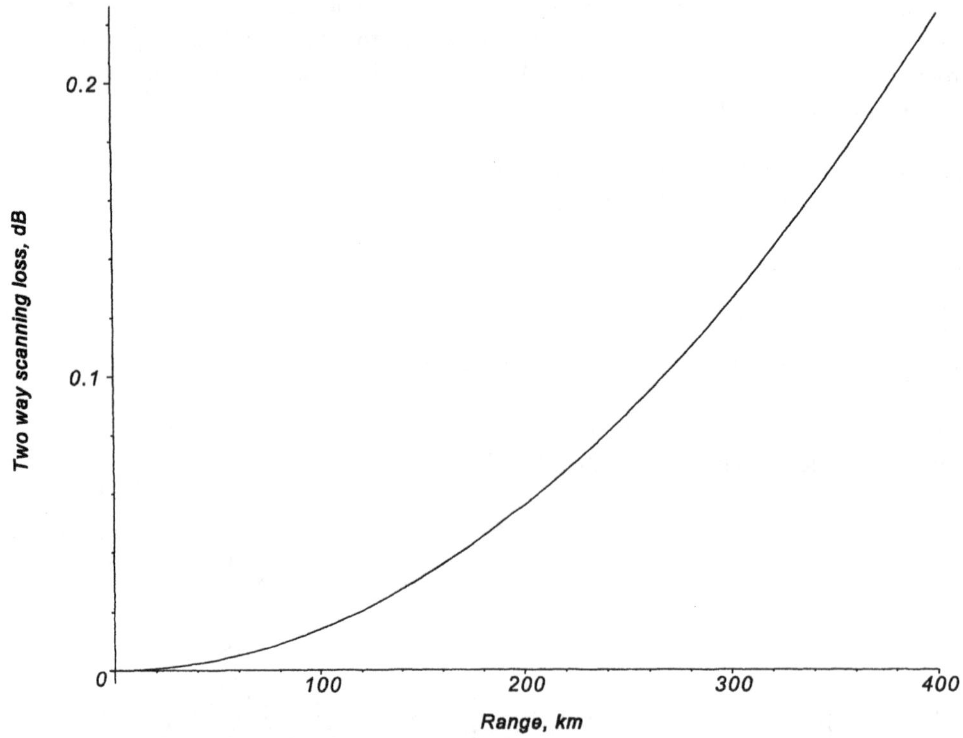

Figure 5.77 Two-way scanning loss for a surveillance radar with a beamwidth of 0.5 degrees rotating at 6 rpm.

5.10 THE EQUIVALENCE OF DIFFERENT SIGNAL COMBINING SYSTEMS

With a planar array with a radio frequency combiner (or beam former) or a parabolic reflector reflecting the signals into a horn there is a single output feeding the radio frequency amplifier. Neglecting the cosmic noise and antenna loss noise, the signal-to-noise ratio at the output of the radio frequency amplifier, gain A_{rf}, fed by the radio frequency combiner, is

$$\text{Output signal to noise ratio} = \frac{n\ S_{in}}{k\ T\ B}\ \frac{A_{rf}}{A_{rf}} \tag{5.105}$$

where n is the number of antenna elements;
S_{in} is the signal power delivered by each element;
k is Boltzmann's constant;
T is the receiver noise temperature in K;
B is the receiver bandwidth.

There are a number of applications where it is a great advantage in having a radio frequency amplifier behind each antenna element:

- A number of receiving beams are to be formed by dividing the output of the amplifiers between a number of beam forming networks. Without the amplifiers there would be a loss.
- A complete transmitter-receiver is mounted behind each element or a group of elements in an active array.

The combination of the signals from each element or group of elements may take place at radio frequency or intermediate frequency using combining networks using phase shifters to steer the beam and attenuators to shape the beam. At baseband, polyphase combination must be used and digital techniques are mostly used to achieve this, called digital beam forming. These three types of combinations are shown in Figure 5.78.

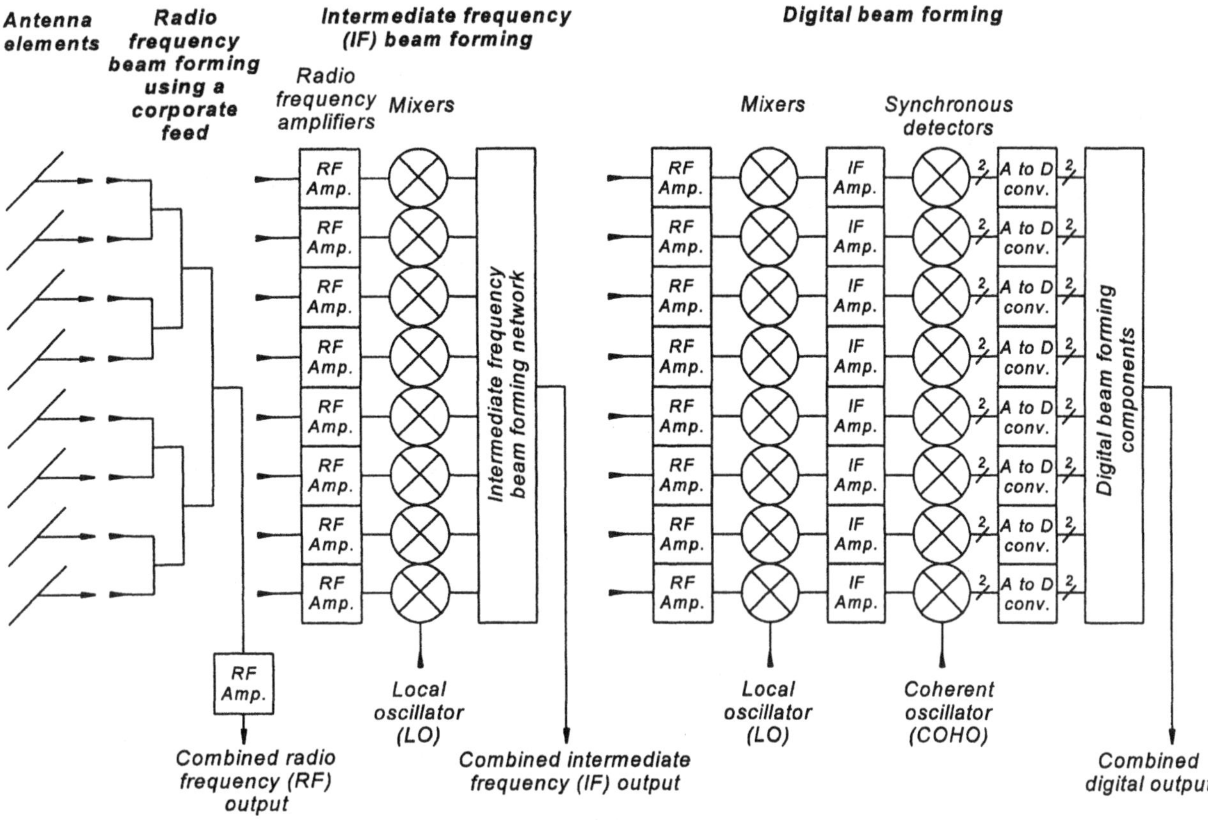

Figure 5.78 Radio frequency, intermediate frequency, and digital beam forming systems.

When each element feeds a radio frequency amplifier, the signal-to-noise ratio at the output of each amplifier is

$$\textit{Output signal to noise ratio, each amplifier} = \frac{S_{in}}{k\ T\ B}\ \frac{A_{rf}}{A_{rf}} \tag{5.106}$$

Each of the wanted signals ($\sqrt{S_{in}}$ volts) adds coherently and the sum is $n\sqrt{S_{in}}$ volts. The noise adds incoherently to give $\sqrt{nkTB}$ in voltage.

$$\textit{Output signal to noise ratio, all amplifiers} = \frac{n^2\ S_{in}}{n\ k\ T\ B}\ \frac{A_{rf}}{A_{rf}} = \frac{n\ S_{in}}{k\ T\ B} \tag{5.107}$$

This is the same as (5.106) above so that there is no difference in performance between passive and active radio frequency combination. The signal-to-noise ratio is not changed by shaping the beam as noise is attenuated at the tips of the aperture function as well as the signal.

Distributing the transmitter and receiver components among many parallel elements in an antenna gives a characteristic known as graceful degradation. That is, if one transmitter-receiver element fails, the radar can continue in operation with slightly reduced, but acceptable, performance. Low sidelobe antennas depend on very accurate illumination functions so that the sidelobe performance will be affected to a much greater degree.

5.11 NOISE RECEIVED FROM AN ANTENNA

The noise at the output port of an antenna is the cosmic noise from outer space, the noise from the sun or radio stars in certain cases, the noise from the warm atmosphere, and the noise from the resistive losses in the antenna itself [25, p. 415]. It is assumed that there are no amplifiers and that the system is completely passive. The noise components add and are then attenuated.

The behavior of an attenuator with a source and a load is illustrated in Figure 5.79. If the source temperature is 0 K then the noise power delivered to the load is totally generated in the attenuator. Referring the noise temperature to the input [1], the load thinks that the effective source temperature T_e is

$$T_e = T_T\,(L-1) \tag{5.108}$$

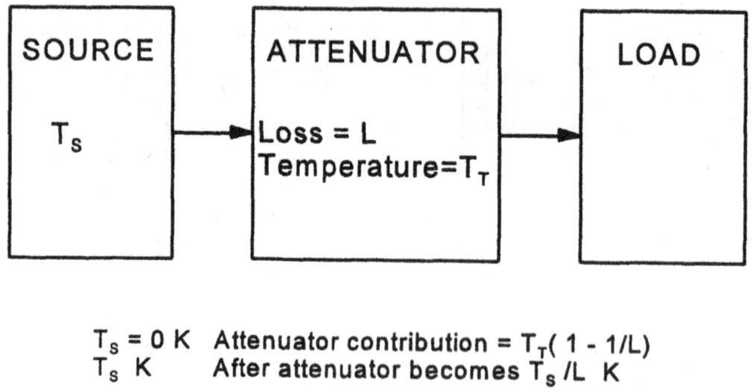

Figure 5.79 Arrangement of source, attenuator, and load.

The load sees an impedance which is partly that of the source with the remainder the resistive impedance of the attenuator. It is like looking through mist as the resistive component in the attenuator is itself a noise generator.

If the source has a temperature T_S, the noise power is kT_SB W. This is attenuated by the attenuator to give kT_SB/L W so that the source temperature as seen by the load is T_S/L K. The total temperature at the output is

$$T_{out} = \underbrace{\frac{T_S}{L}}_{\text{Source}} + \underbrace{T_T\left(1-\frac{1}{L}\right)}_{\text{Attenuator}} \quad \text{(Contributions)} \tag{5.109}$$

The effects of this are shown in the block diagram Figure 5.80. The input noise is attenuated and a component generated in the attenuator is added.

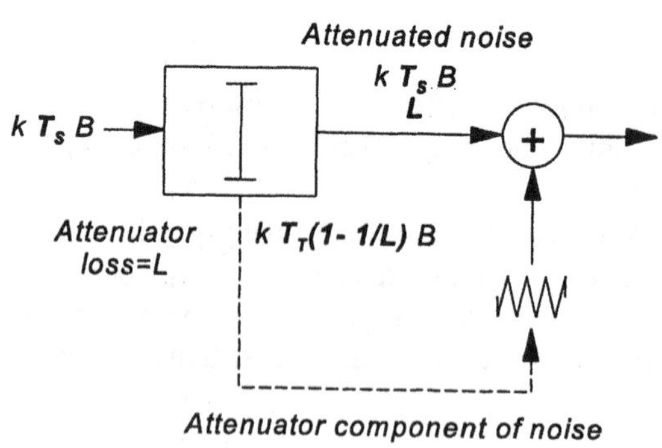

Figure 5.80 Additional noise from an attenuator.

Figure 5.81 shows the effects of an attenuator at 290 K, the IEEE standard temperature, on the effective noise temperature out. If the attenuator is warmer than the source, the attenuator adds noise. If the source is warmer, the attenuator reduces the noise temperature. At infinite attenuation the output sees a matched load.

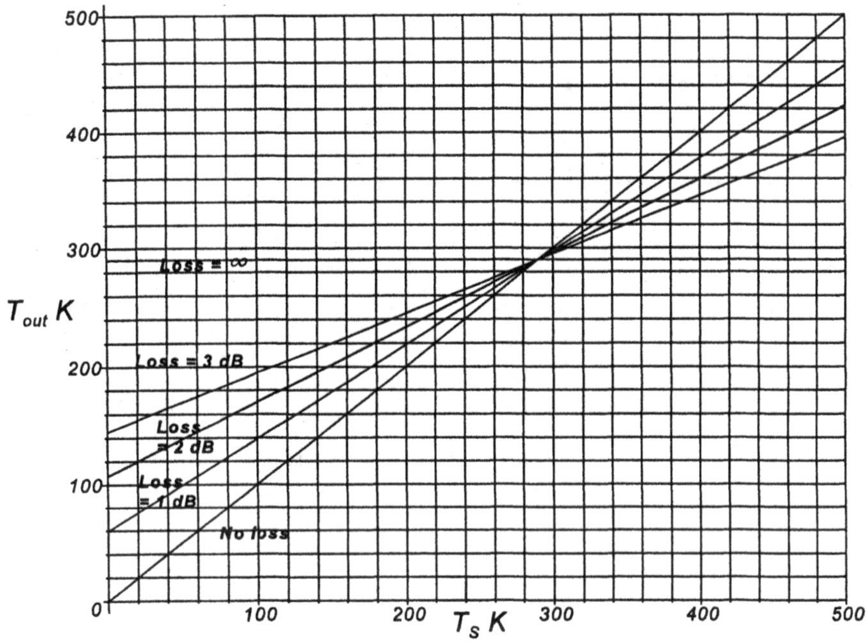

Figure 5.81 The effects of an attenuator at 290 K on the effective noise temperature at the output.

Figure 5.82 shows the additional noise temperature produced by a resistive attenuator at 290 K according to (5.109).

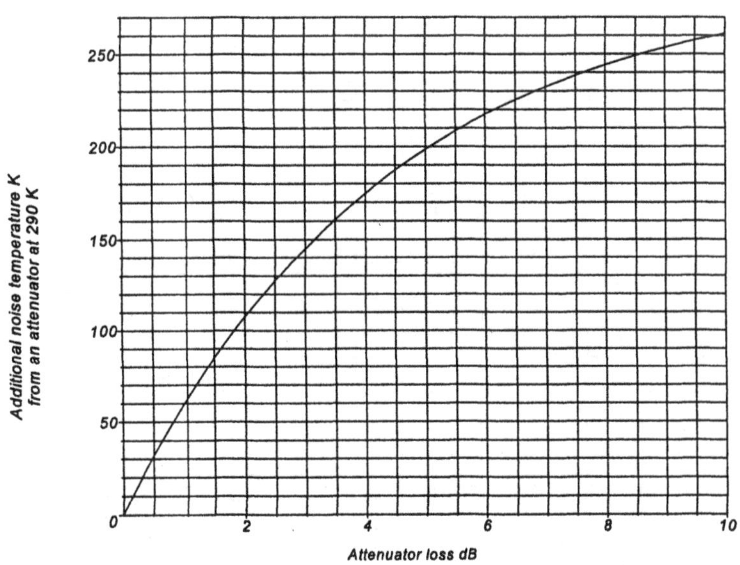

Figure 5.82 The additional temperature component from an attenuator at 290 K.

The noise contributions for a ground radar from cosmic noise, the sun, the ground, and the antenna are shown in Figure 5.83 [27]. Noise components from the sun and outer space [26] are first attenuated by the atmosphere.

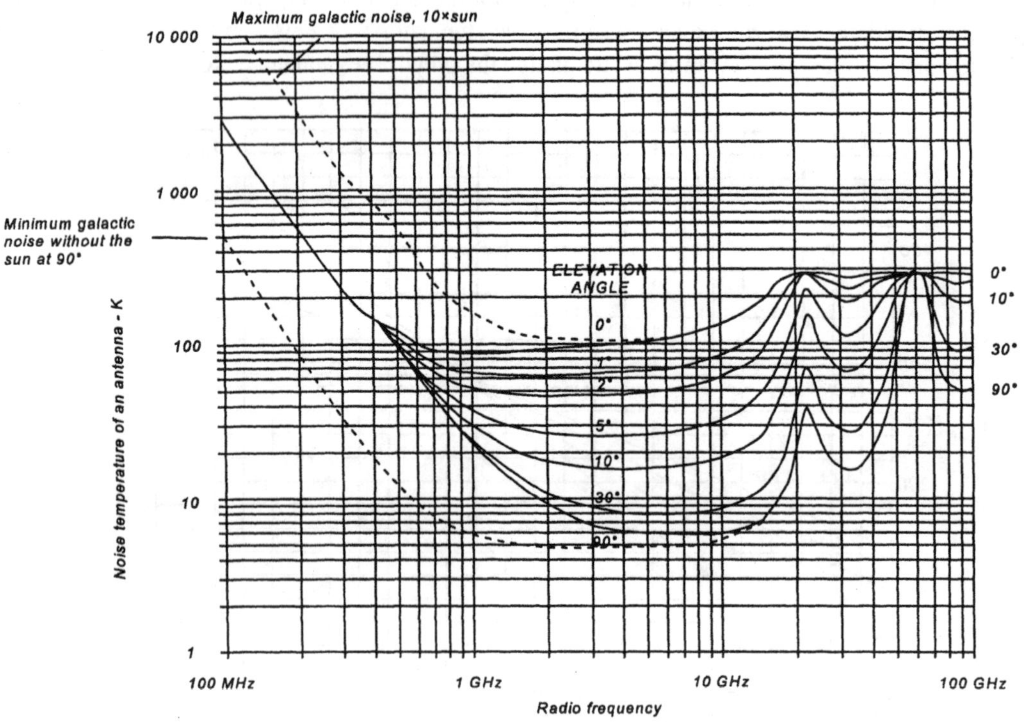

Figure 5.83 Noise temperature of an idealized antenna as a function of frequency and elevation angle. [Source: Blake L. V., *A Guide to Basic Pulse Radar Maximum Range Calculation*, U.S. Naval Research Laboratory Report, 23 December 1969, p. 126, Figure 11.]

This attenuation is shown in Figure 5.84.

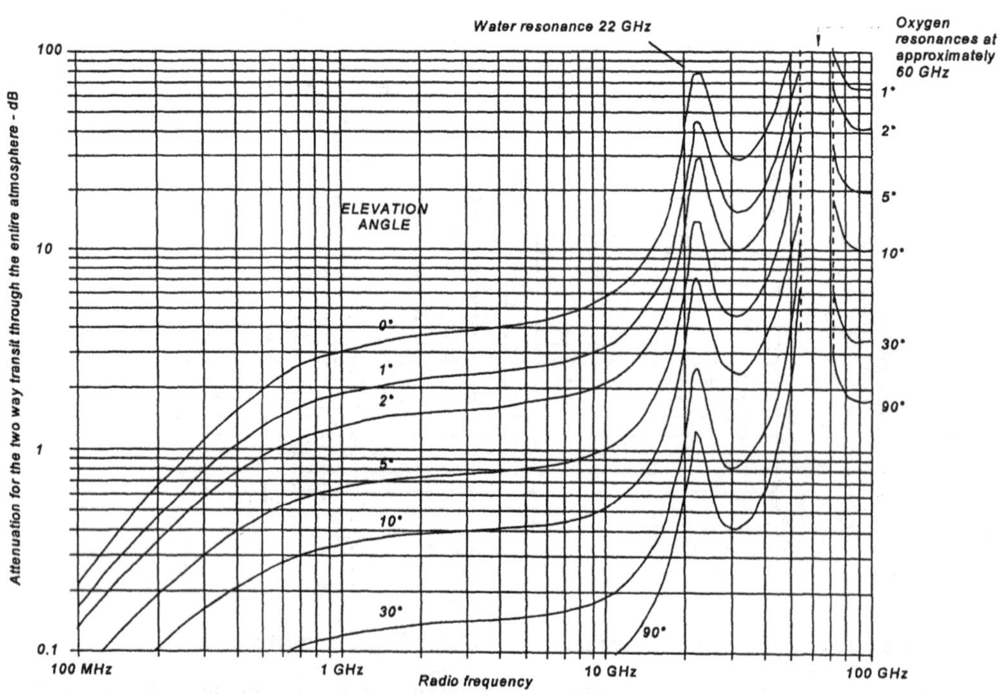

Figure 5.84 Two-way atmospheric attenuation for a transit of the entire troposphere. [Source: Blake L. V., *A Guide to Basic Pulse Radar Maximum Range Calculation*, U.S. Naval Research Laboratory Report, 23 December 1969, p. 73, Figure 21.]

The contributions made by the various noise sources [28, p. 152 et seq.] depend on the integrated antenna gain over the solid angle where the noise occurs. These are attenuated and added as shown in Figure 5.85 where the weighting factors α in Figure 5.86 represent the various gain factors.

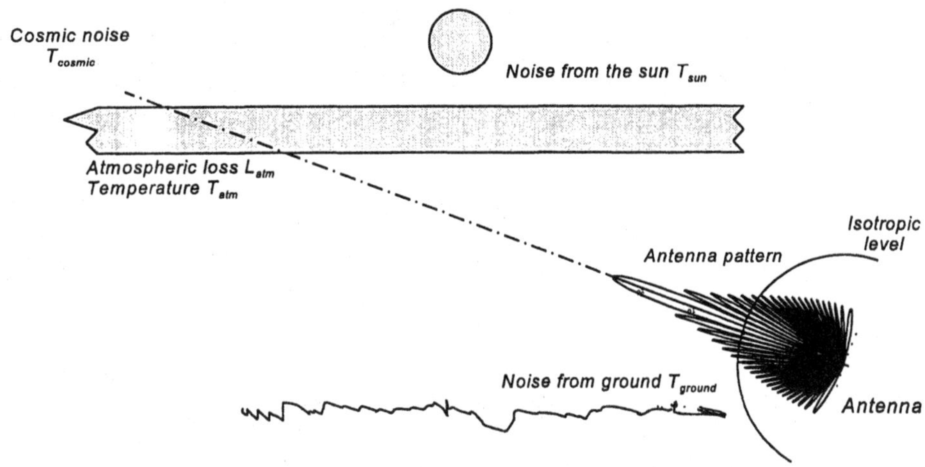

Figure 5.85 Noise components in the antenna.

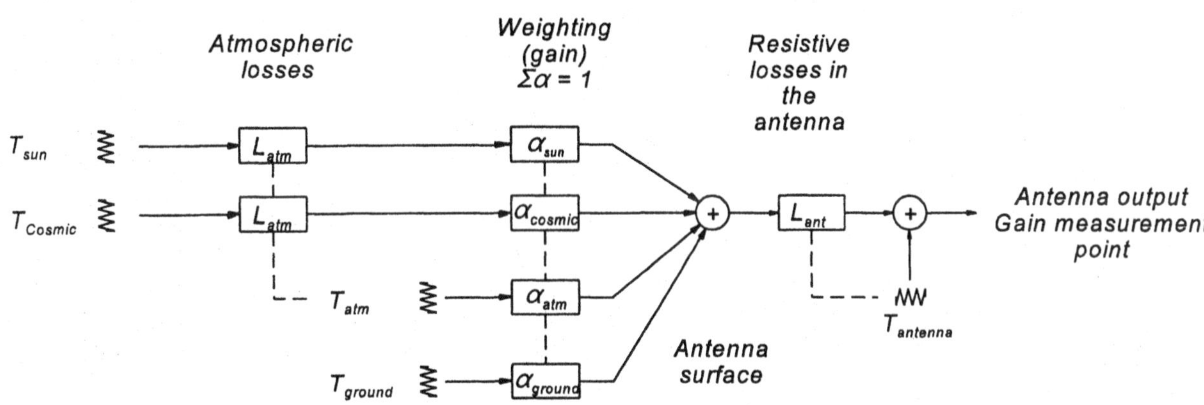

Figure 5.86 How the noise components sum at the antenna output.

- α_{sun} is the integrated sidelobe gain over the solid angle subtended by the sun;
- α_{cosmic} is the noise gain (integrated over the solid noise beamwidth) of the antenna;
- α_{ground} is the integrated sidelobe gain over the solid angle subtended by the ground.;
- α_{atm} is the integrated gain for atmospheric noise.

The noise arriving at the surface of the antenna is given by

$$T_{antenna\ surface} = \frac{\alpha_{cosmic}\,T_{cosmic}}{L_{atm}} + \frac{\alpha_{sun}\,T_{sun}}{L_{atm}} + \alpha_{atm}\,T_{atm} + \alpha_{ground}\,T_{ground} \tag{5.110}$$

$$\text{Note: } \alpha_{cosmic} + \alpha_{ground} = 1$$

Rule of thumb values for α assume that $\alpha_{cosmic} = \alpha_{atm} = 1 - \alpha_{ground}$. Values for the component of T_{ground} at 290 K for a number of values of α are [28, p. 171]:

α_{cosmic}	0.975	0.924	0.876
α_{ground}	0.025	0.076	0.124
Ratio dB	16.02	11.19	9.067
T_g	7.25	22.04	35.96 K

The noise which arrives at the output is attenuated by the resistive losses in the reflector, the feed system and beam forming networks, and the waveguide or coaxial line leading to the output flange where the antenna gain measurement was made. This is

$$T_{output} = \frac{T_{antenna\ surface}}{L_{antenna}} + T_{antenna}\left(1 - \frac{1}{L_{antenna}}\right) \tag{5.111}$$

5.12 SIDELOBE CANCELERS AND ADAPTIVE BEAM FORMING

Military radars have to contend with electronic countermeasures (ECM), when the adversary does not wish to be detected. One way is to scatter a cloud of small scatterers, typically aluminium foil, that has the same effect as rain (note that circular polarization has no effect on chaff) on a radar, and the other is to aim modulated radio frequency energy at the radar of interest, called active jamming.

The main element to filter out active jamming is the antenna. Military antennas have been developed and made to have "ultra low" sidelobes, that is, to radiate as little energy as possible outside the main beam to avoid detection and to receive as little echo energy from the sidelobes in the direction of clutter echoes and hostile jamming using planar arrays with computer controlled milling of the slot elements. Much research work has been carried out to find tapering functions to provide adaptive notches for array antennas but the processing takes time, needs large amounts of memory, and is presently unsuited to the majority of search radars that provide signals or data to create an air picture with many aircraft for those who control aircraft in flight. Development is proceeding so that radars in the future with a number of receiver channels may have this feature.

Before the era of planar arrays and large numbers of stand-off jammers producing a wide barrage of jamming signals, sidelobe cancellers were used to place notches in the antenna characteristics to reduce the jamming energy going to the receivers, one example is the Howells and Appelbaum canceler used since the 1960s on search radars.

The Howells and Appelbaum canceller has additional antennas with gains just above the gains of the sidelobes of the main antenna and correlates these signals with the jamming signals received in the main lobe in order to cancel them. A single canceler is shown in Figure 5.87.

The main radar antenna is shown as A in Figure 5.87, the signals from it are mixed down to intermediate frequency by the mixer C, have the jamming signals subtracted out E, and pass on to a conventional receiver. The signals are echo signals entering the main beam and jamming signals entering via the antenna sidelobes. (The canceler cannot cancel main beam jamming.) The signal used by E to remove the jamming must have the same amplitude and phase as the jamming signal received through the main antenna.

At first the auxiliary antennas, B, were omnidirectional antennas mounted hanging under the top of the radome or on a mast as close as possible to the main antenna with a gain to cover the sidelobes. With later radars a broadly directional antenna is mounted on the main antenna (an extra channel through the rotary joint is required) and with a gain tailored to cover the near sidelobes as well.

The signals from the auxiliary antenna are mixed down to intermediate frequency and one way goes directly to the multiplier K and the other to form a signal with the complex conjugate of the modulation for the multiplier H. The analogue circuits to form the complex conjugate depend on mixing the intermediate frequency signal to a higher carrier frequency. The modulation of the sum and difference frequencies are complex conjugates of each other and the complex conjugate is selected. When the residue of the jamming signal and the complex conjugate of the signal from the auxiliary antenna are multiplied together, correlation takes place and the correlated signal is passed to the integrator, J, (at intermediate frequency a bandpass filter). The time constant of the integrator strips off the fast modulation of the jamming signals but is able to follow slower changes in amplitude and phase in the case that the auxiliary antenna rotates with the main antenna. The signal from G representing the magnitude and phase of the jamming signals is multiplied,

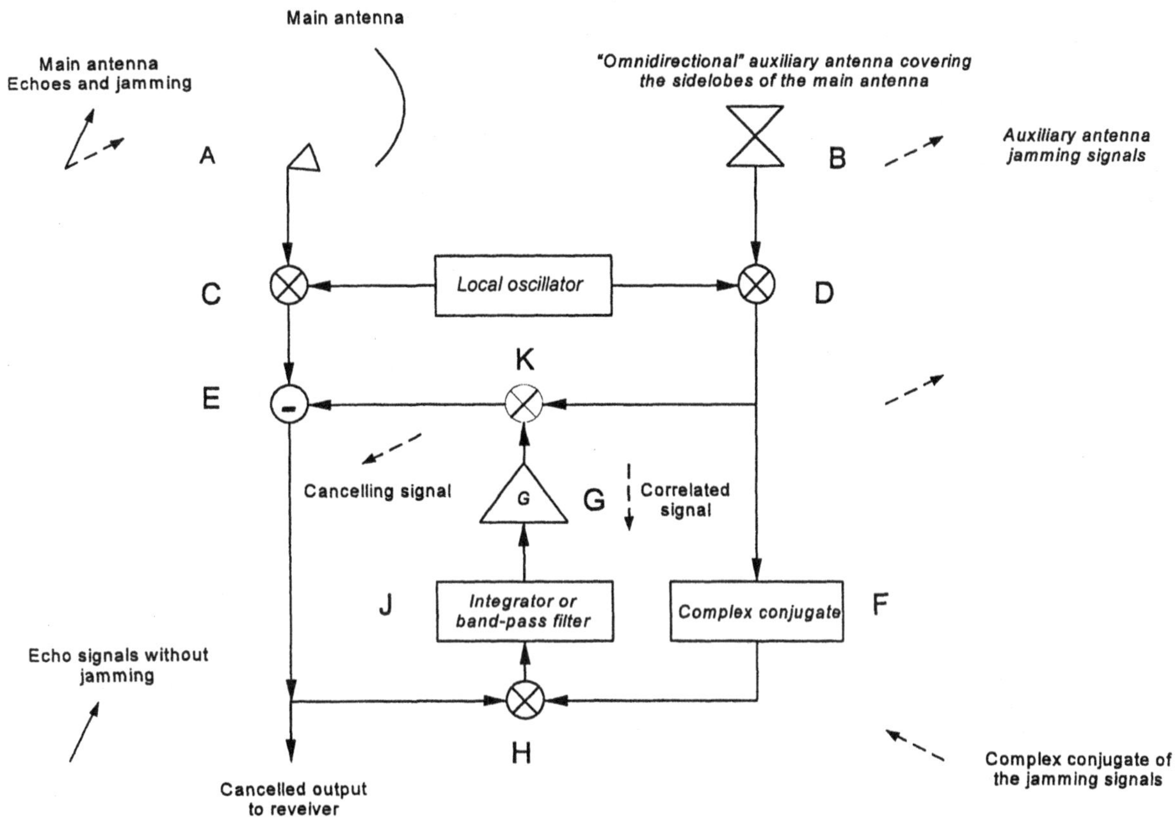

Figure 5.87 Block diagram of a Howells and Appelbaum canceller.

K, with the jamming signals directly from D to reconstruct the jamming signal received through the main antenna for subtraction or cancellation at E. The loop E, H, J, G, I takes s short time to settle. The echo and jamming signals have widths given in microseconds in contrast to the modulation of the signals caused by the rotation of the antenna or changes in jamming passed by the integrator, J, measured in tens of milliseconds. Additionally the echo signals received through the main beam are much greater than the echo signals received through the auxiliary antenna and these two characteristics stop the canceller from cancelling the wanted echo signals.

The circuitry shown in Figure 5.87 can only cancel the signals from one or the most powerful jammer and each extra jammer to be cancelled requires an extra auxiliary antenna and cancellation loop. Examples of more detailed analyses can be found in [29-31].

Other approaches use open loops with greater processing times. When this is applied in the time (range) dimension also, the technique is called space-time adaptive processing that, generally speaking, takes too long to be able to produce useful air pictures, see [32, 33].

5.13 ANTENNAS MOUNTED ON AIRCRAFT

Some of the first radar antennas mounted on aircraft were part of the airborne radars in operational use in 1940 for the detection of shipping (air to surface vessel, ASV), and other aircraft (aircraft interception, AI). The radars were manufactured to detect single targets for the aircraft to attack and their successors can be found in the radars installed in current fighter and maritime attack aircraft.

For airspace surveillance, ground surveillance radars can only see above the earth's surface so that elevated siting gives a better view out to further in range but also increases the range where ground clutter is present. Greater surveillance volumes extending to low altitudes can only be achieved by mounting the radar under a balloon or on an aircraft. There are two major types: those that are made to search the airspace above the ground that have to suppress the large amount of ground clutter and radars that map objects on the ground.

There are a number of successful radar systems that are hung under balloons or rotate within the gas envelope.

Balloons are difficult to move quickly and radars mounted on airframes are more flexible. Small radars have radomes that are carried underneath aircraft and helicopters. For longer ranges, larger antennas are necessary that are mounted inside "flying saucer" radomes to reduce aerodynamic drag and to have the same drag at all antenna positions when the antenna is rotated. Ground clutter from the vertical sidelobes is a problem and has to be reduced to an acceptable level using an airborne moving target indicator processor (Chapter 11) or a high pulse repetition pulse frequency Doppler processor and antennas with low vertical and horizontal sidelobes. Another form of antenna is to have a fixed linear array along the length of the fuselage or on top of it as a dorsal fin. Phased array scanning gives a coverage in azimuth of ± 60 degrees on each side of the aircraft with limited coverage towards the nose and tail of the aircraft using the end-fire antenna gain. Again good ground clutter suppression, using an airborne moving target indicator or pulse Doppler processing is required.

Using a fixed antenna looking sideways and allowing the movement of the aircraft to achieve the scanning was used initially for military reconnaissance to look over a front line at objects on the ground, in contrast to the signal processing in Chapter 11 devoted to suppressing ground echoes. For this use the time frame between observation and use is quite different and measured in hours, not seconds as required by air traffic control. The radar signals were recorded on film during flight, after the flight the aircraft landed, and the film was developed and evaluated. Though range resolution can be improved by using shorter pulses, the azimuth resolution depends on the beamwidth of the antenna. The length available for narrow enough beamwidths was too short so that synthetic aperture radar was developed.

5.13.1 Synthetic apertures[1]

In Figure 5.88 an aircraft is flying in a straight line and illuminating the point P with its sideways-looking radar. The pattern of the antenna mounted on the aircraft that illuminates the ground is defined in range by the vertical beamwidth. The range extent depends also on the height of the aircraft and is called the swath, a term used in mowing grass or cereals with a scythe or mower. The point P is illuminated by the radar in the aircraft within the horizontal beamwidth when the aircraft flies from A to C.

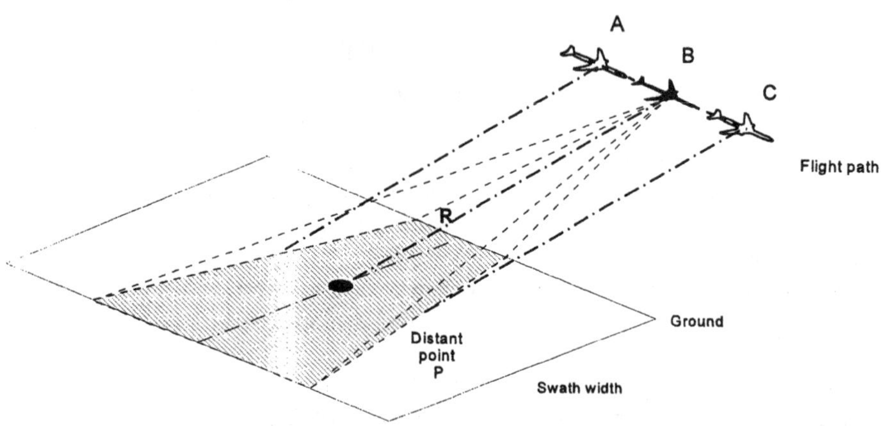

Figure 5.88 An aircraft flying by a distant point P.

Discrimination in range is achieved by using small pulse widths, if necessary using pulse compression. The width illuminated is proportional to the range from the radar as the radar flies by. The coordinates of the "picture" are expressed as Cartesian coordinates projected on the ground plane in range and cross-range (alternatively across the line of flight and along the line of flight, respectively) in contrast to the polar angle coordinates and range in conventional surveillance radars.

For a long time the offset bipolar coherent video from a ring demodulator (Section 9.2.1) was recorded on film with the range direction across the film. Optical lens systems were used to process the film image after the flight that allowed the vector summing of the many echoes from each object that the radar had seen, as shown diagrammatically in Figure 5.89.

---

[1] The general term is *synthetic aperture*. If the aperture can be considered as being made up of an array of elements, then the term *synthetic array* is also used.

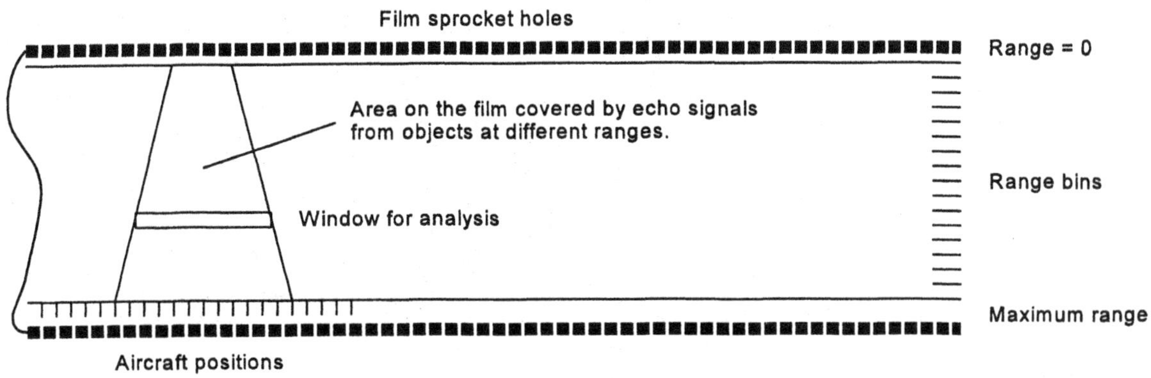

Figure 5.89 A film record of video for later analysis.

As semiconductor and magnetic memory became faster and larger, it became possible to store the radar echo vector data using them. Smaller, faster computers allow evaluation during flight and the results can be passed to the ground using radio links. Figure 5.90 shows the physical antenna and a device for storing the signal vectors as solid lines.

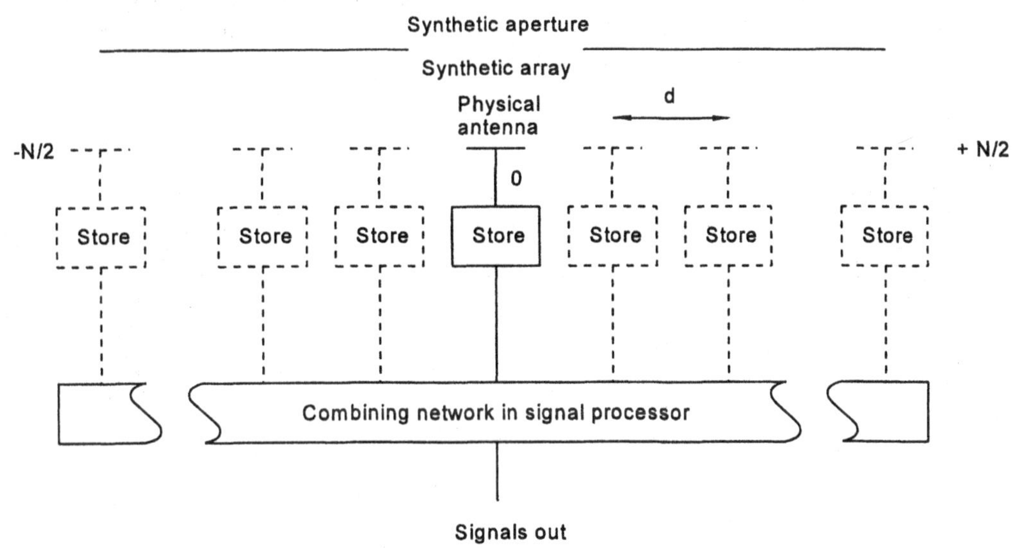

Figure 5.90 The physical antenna in N+1 positions to give a synthetic array in a synthetic aperture.

The elements shown dotted are the other positions of the antenna and storage components as the aircraft flies by the broadside position of point P.

The aircraft flies with a velocity, v m/s, and has a constant pulse repetition frequency, prf, thus the distance between the synthetic elements, d, is

$$d = \frac{v}{prf} \text{ m} \tag{5.112}$$

The velocity of the aircraft and the pulse repetition frequency must be chosen so that the positions of the synthetic elements are less than $w/2$ (w is the width of the antenna on the aircraft), much less gives oversampling.

The feed system of a physical array adds the echo signals together vectorially and if the record of the echo signals at the same range within the physical beamwidth of the antenna on the aircraft are added vectorially, the same process occurs. The left-hand part of Figure 5.91 shows an aircraft radar illuminating the point P and on the right 31 idealized echo signals at the same zero phase angle. If these signals are added together vectorially, as they would be in a feed system, the synthetic azimuth beamwidth is much narrower than the physical beamwidth.

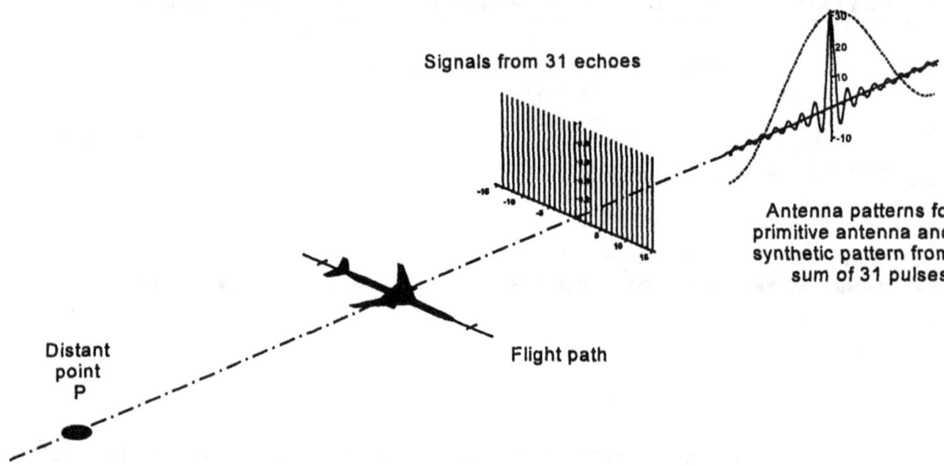

Figure 5.91 Signals taken along a line and the synthetic antenna pattern.

The phase shift from synthetic element to element is double the phase shift between elements in a physical array. With a physical array the objects of interest (P in Figure 5.92) are assumed to be illuminated by a wavefront coming nominally from the phase center of the transmitting antenna (O in Figure 5.92).

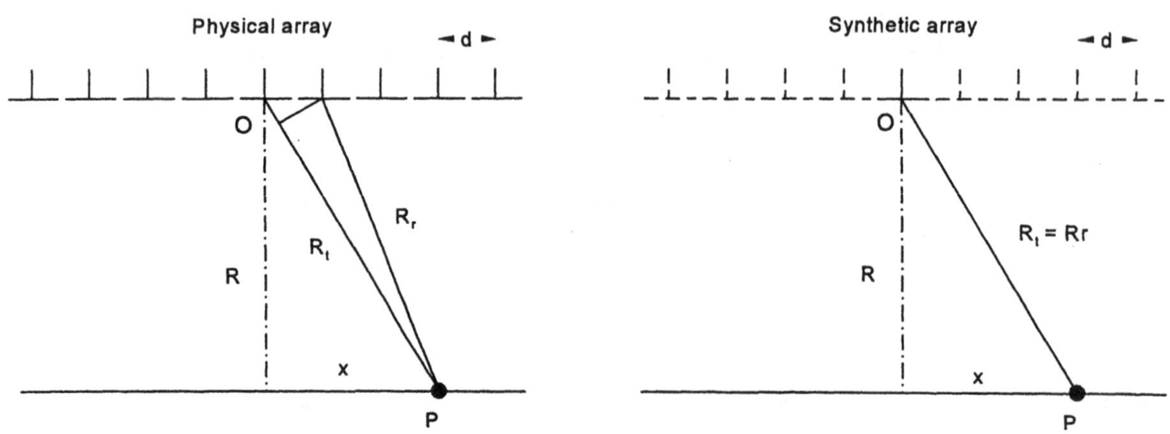

Figure 5.92 Comparison of the path lengths with physical and synthetic apertures.

The echo signals are scattered back from the point P and enter the array elements with a phase angle proportional to the two-way path length to an element, L_1, given by (Figure 5.92)

$$L_1 = \sqrt{R^2 + x^2} + \sqrt{R^2 + (x - nd)^2} \quad \text{Pythagoras} \tag{5.113}$$

where R is the distance from the center to the plane containing the point P;
x is the distance from the center-line in the same units;
d is the distance between two elements or aircraft positions;
n is the number of the element counting from the center.

The phase difference there and back is $L_1\, 2\pi/\lambda$ radians. Expanding L_1 using the first terms of the binomial series

$$L_1 = R\left(1 + \frac{x^2}{R^2} - \frac{nd}{R^2} + \frac{n^2 d^2}{2R^2} + \ldots\right) \tag{5.114}$$

With the synthetic array there is physically only one element so that the transmitting phase center moves with each pulse along the synthetic array so that the lengths out and back are the same, namely,

$$L_2 = 2\sqrt{R^2 + (x-nd)^2} \tag{5.115}$$

Expanding as the first two terms of a binomial series

$$L_2 = R\left(2 - \frac{x^2}{R^2} - \frac{2nd}{R^2} + \frac{n^2d^2}{R^2} + \ldots\right) \tag{5.116}$$

Notice the linear term for the total path length in L_2 $(2\,nd/R^2)$ is twice that of L_1 giving twice the phase change across the aperture compared to a physical array to give a beamwidth half of that of the physical aperture, width W.

$$\text{Standard beamwidth, } \Theta = \frac{\lambda}{2W} \text{ radians} \tag{5.117}$$

More simply, the echo signals reaching a physical antenna have originally come from its phase center. With the synthetic array the phase angles of the echo signals have an additional phase change caused by the fact that the antenna used for transmission has moved in contrast with the phase center of the physical antenna.

The standard beamwidth of the physical antenna of width, w, is λ/w and it covers a length on the ground of $R\lambda/w$. The resolution of the synthetic aperture is ΘR

$$\text{Resolution} = R\Theta = \frac{\lambda R}{2W} = \frac{\lambda R}{2}\frac{w}{\lambda R} = \frac{w}{2} \tag{5.118}$$

The resolution is limited by the amount of signals that are able to be collected and processed coherently, the tapering of these signals to reduce sidelobes, and irregularities of the motion of the aircraft. Part of the problem is caused by geometry in that the echoes to be integrated do not lie on a circle but on a straight line, this is called unfocused processing.

Most physical antennas are "focused at infinity" and assume that all objects appear in the far field or Fraunhofer zone. Equation (5.8) gives the range necessary to assume that far field (Fraunhofer zone) conditions exists, namely $\lambda/16$ or $\lambda/32$ distance error at the edge of the aperture, $\pi/8$ or $\pi/16$ radians phase error. Different authors give maximum errors for synthetic apertures before phase correction becomes necessary, for example, [34, p. 922, Fig. 12.1] $\lambda/8$ one way, two-way phase shift = 90°.

The length of synthetic apertures and the ranges used place the objects of interest in the near field so that the range and phase angle to the object changes during the fly past, as shown in Figure 5.93.

The distance of P from the flight path varies and is $R\sec\theta$ so that there is a phase change along the flight path, ϕ, of

$$\begin{aligned}\phi &= R\,\frac{2\pi}{\lambda}(\sec\theta - 1)\\ &= R\,\frac{2\pi}{\lambda}\left(\frac{\theta^2}{2!} + \frac{5\theta^4}{4!} + \ldots\right)\end{aligned} \tag{5.119}$$

where R is the range, λ is the wavelength measured in the same units, and $\sec\theta$ has been expanded as a series.

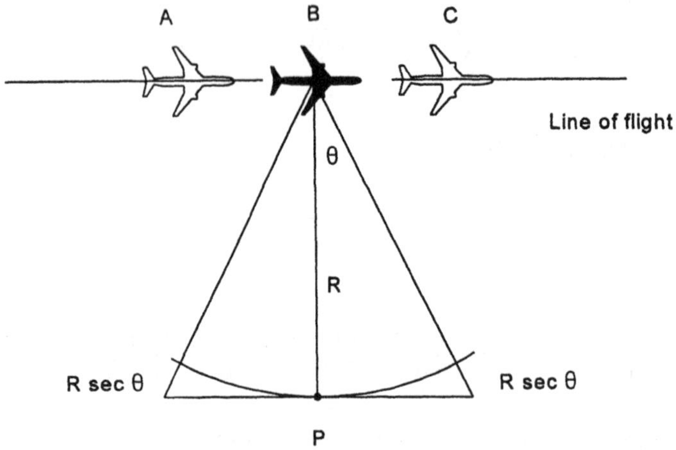

Figure 5.93 Simple geometry for a synthetic aperture.

Figure 5.94 shows the echoes returned to the physical antenna flown with a parabolic phase shift of the signals at a maximum of π/2 or 90 degrees at the ends. The same processing has been carried out as in Figure 5.91 and the phase errors in the echoes cause the complex beam synthetic pattern to warp and wilt.

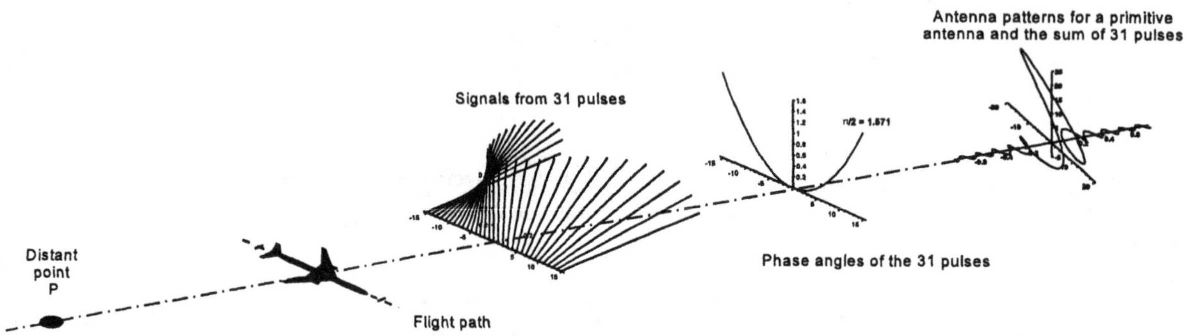

Figure 5.94 The signals and synthetic antenna pattern from a train of signals with phase shifts due to range difference.

The modulus of the beam pattern is in the common form in Figure 5.95. With π radians or 180 degrees phase shift at the ends, the synthetic pattern starts to break up. When the beam is focussed, the phase along the record of echo signals is corrected to remove the phase errors.

The resolution without phase correction and having a phase error of π/2 or 90 degrees at the start and end of the recording corresponding to a one-way distance of λ/8, is then

$$\left(R + \frac{\lambda}{8}\right)^2 = \frac{W^2}{4} + R^2 \tag{5.120}$$

giving the maximum width of the synthetic aperture of

$$W = \sqrt{R\lambda} \tag{5.121}$$

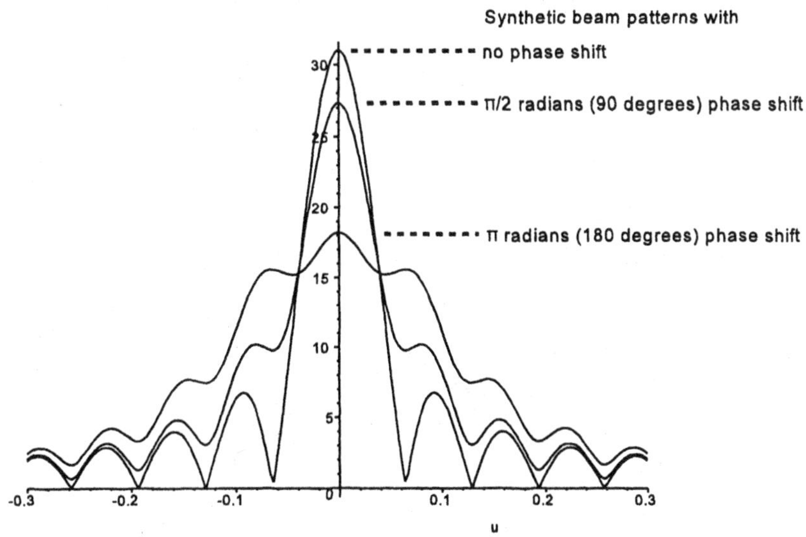

Figure 5.95 Synthetic patterns with parabolic phase shift.

with a standard beamwidth of

$$\frac{\lambda}{2W} = \frac{1}{2}\sqrt{\frac{\lambda}{R}} \tag{5.122}$$

and resolution

$$\Theta R = \frac{1}{2}\sqrt{\lambda R} \tag{5.123}$$

The unfocused resolution is independent of the size of the physical antenna and greater resolution is only obtainable by shortening the wavelength.

The three types of antenna using a physical broadside antenna 2 m wide at X-band are compared in Table 5.6.

Table 5.6
Comparison of standard beamwidths and resolutions for an antenna 2 m wide with 3 cm wavelength

	Standard beamwidth		Resolution at 20 km	
Physical antenna	λ/w	15.0 mrad	R λ/w	300.0 m
Unfocused antenna	$\sqrt{\lambda/R}$ /2	0.61 mrad	$\sqrt{\lambda R}$ /2	12.25 m
Focused antenna	λ/2W		W/2	1.0 m

5.13.1.1 Focusing

Table 5.7 shows the improvement in a typical case that can be obtained by correcting the phases of the echoes before summing. The correction is different and must be calculated for each range.

Reference [35] gives the approximate characteristics of the AN/APQ-102A radar and the characteristics are quoted as an example.

Table 5.7
Approximate characteristics of the AN/APQ-102A radar [35]

Characteristic	Value
Wavelength	3 cm
Transmitter	
Pulsewidth	1 μs
Bandwidth	15 MHz
Average power	90 W
Antenna length	1.2 m
Standard beamwidth	1.43 degrees

The number of cross-range samples collected [35] are shown in Table 5.8.

Table 5.8
Amount of data sampled by the AN/APQ-102A radar [35]

Characteristic	Values		
		Range	
	18.5 km		55 km
Processed synthetic aperture	30.8 m		92.3 m
p.r.f		1 800	
Synthetic aperture time	0.11 s		0.33 s
Number of samples collected	3 042		2 281

The performance after signal processing [35] is shown in Table 5.9.

Table 5.9
Resolution performance of the AN/APQ-102A radar [35]

Characteristic	Value		
		Range	
	18.5 km		55 km
Range resolution	10 m		10 m
Cross-range resolution	9 m		9 m
Number of samples processed	202		606

5.13.1.2 Spotlight synthetic array radar

Where flying by an area of interest does not give enough resolution, it is possible to have a clearer picture by flying an arc around the area of interest, but with greater processing effort per unit area [35, 36].

5.13.2 Mapping

Three-dimensional ground mapping may be carried out when a second receiving antenna is mounted above or below the original antenna as an interferometer. The phase differences between echoes received give an elevation angle (sometimes ambiguous) that allow contour maps to be produced for areas of the world that are normally inaccessible or have so much fog that optical measurements are not possible.

5.13.3 Radars on satellites

Synthetic aperture radars are used on satellites for, among other uses, maritime surveillance. The oceans are large and the objects of interests, ships, do not move quickly.

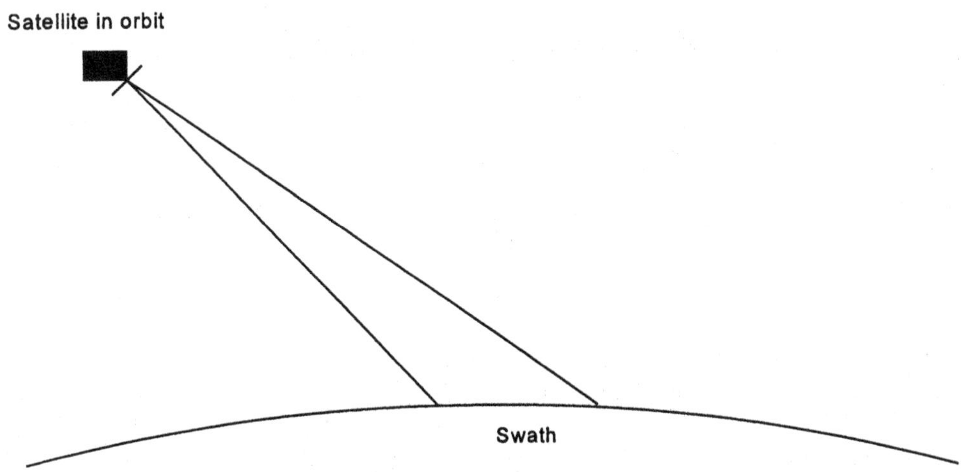

Figure 5.96 A radar in a satellite illuminating the Earth's surface.

Figure 5.96 shows a radar in a satellite illuminating the ground obliquely in order to have a range dimension. The coverage depends on the orbit and the ability to direct the antenna in angle on the satellite. A number of satellites are in orbit with heights of around 250 km or 900 km. The angle of the antenna is set to give swath widths of the order of 50 to 100 km and the resolutions have been of the order of 15 m to 100 m. The pulse repetition frequencies are chosen to cover the swath width so that there will be a number of pulses on their way to the earth and back at the same time. The radio frequencies from L- to X-band have been used.

5.13.4 Other considerations

It has been assumed that the cross-range samples (range sweeps) are collected equally spaced in time along a straight line or arc of a circle. Aircraft in flight react to buffeting in flight inversely proportional to their mass. With wavelengths of 3 cm, random movements can be an appreciable proportion of a wavelength. Signals recorded from the aircraft's navigation system may be used to be able to correct for these phase shifts. As aircraft are flexible, an inertial measurement unit (IMU) may be mounted behind the antenna to signal any short term positional changes.

In [36, p. 448] a number of methods are described for correcting a run of data after it had been collected by being able to estimate the quadratic phase errors and applying corrections.

The discussion above has been limited to sideways looking antennas. Synthetic aperture techniques have also been applied to antennas squinted off broadside.

5.14 FIGURES AFFECTING RADAR PERFORMANCE

The following figures are used in Chapter 14, Performance, to calculate radar performance.

5.14.1 Range

The main parameters affecting maximum range are listed in Table 5.10.

Table 5.10
Main parameters affecting range performance

Parameter	Value	Units
Antenna gain when transmitting the pulse	....	dB with respect to the isotropic level
Antenna gain when receiving the echoes	....	dB with respect to the isotropic level
Radome loss if not included above, two way	....	dB
Beam shape loss	....	dB
Scanning loss	....	dB
(Illumination function, blocking, spillover, and surface tolerance losses are included in the measured antenna gain)		
Noise temperature at the receiving connector or waveguide flange.	....	K

There a number of secondary parameters which are important for the reduction of unwanted echoes or clutter, electrical interference, and jamming. The secondary parameters are shown in Table 5.11.

Table 5.11
Secondary characterisics for range

Parameter	Value	Units
First azimuth sidelobe level	....	dB with respect to peak gain
First elevation sidelobe level	....	dB with respect to peak gain
Far azimuth sidelobe level	....	dB with respect to peak gain
Far elevation sidelobe level	....	dB with respect to peak gain
Integrated cancellation ratio for rain	....	dB

The first elevation sidelobe below the main beam is responsible for most of the ground clutter together with the first azimuth sidelobes, and for this reason they are often given in radar specifications. For military radars high sidelobes enable energy from jammers to enter the receiving system augmented by the gain of the sidelobe. For this reason antennas with very low near and far sidelobes have been developed for military radars.

The integrated cancellation ratio for rain is only used for calculating the maximum range in rain clutter.

5.14.2 Resolution

The beamwidth of the antenna defines the resolving of two echoes at exactly the same range in the appropriate azimuth or elevation direction. At close ranges, however, detections may be possible through the antenna sidelobes which introduce false returns into the radar picture. In pulse radars sensitivity time control is normally able to reduce the gain at short ranges enough to eliminate echoes of sizes in the range of interest. Large reflectors, for example large buildings, may still cause problems.

There is no resolution between echoes at the same range in the monopulse difference pattern.

5.14.3 Accuracy

Typical fixed, or biassing, errors are the northing error for azimuth and the leveling error for elevation. Additionally there is a random element dependent on the accuracy of angle measurement and signal-to-noise ratio which is given in Chapter 14, Performance.

The patterns from antennas which use linear feeds depend on the electrical distance, or phase, between the radiating elements. The physical distance depends on temperature and the phase depends on frequency. The displacement of the beam with frequency is called squint. Phase shifters in phased arrays may also have their characteristics varied by frequency and temperature so that corrections to the assumed beam position may be necessary.

5.14.4 Stability

If the elements of the antenna vibrate, the velocity of these elements adds to the Doppler frequency shift of the returning echoes. Antennas work outside and are often subject to strong winds and rain and have to be strong structures. In more severe climates, radomes are used to protect antennas.

The vibrations will be often less than 2 Hz. The observation times for signal samples from radars is often in the region of less than 50 ms so that antenna mechanical stability is not an appreciable quantity which limits stability.

If an outside antenna does not have a balanced sail area for winds arriving at all azimuths, the antenna rotation will accelerate and retard with azimuth. This can cause problems with azimuth change pulse generators giving extra pulses during rotation through a circle.

The radar pulse repetition frequency and the antenna turning rate are not related. Delayed radar triggers which have been divided into groups have often been used to simulate echoes traveling on a circumferential course. Where there is memory from scan to scan and clutter maps are used, the fact that the antenna does not illuminate the same clutter areas at the same aspect at each antenna revolution gives rise to fluctuating values in the clutter map. The fluctuations require higher detection thresholds and reduce sensitivity over clutter. The antenna is generally turned by induction motors whose speed depends on their load. The more modern radars which use clutter maps observe the azimuth of the antenna and control the transmitter pulses so that the groups of pulses are transmitted at the same azimuths each revolution.

REFERENCES

1. Glazier, E. V. D., and H. R. L. Lamont, *Services Textbook of Radio, Volume 5, Transmission and Propagation,* London: Her Majesty's Stationary Office, 1958, p. 428.
2. Stutzman, W. L., and G. A. Thiele, *Antenna Theory and Design*, New York: Wiley 1981.
3. Silver, S., (ed.), *Microwave Antenna Theory and Design, Volume 12 of the MIT Radiation Laboratory Series,* New York: McGraw-Hill 1949, Section 2.14.
4. Barton, D. K., and H. R. Ward, *Handbook of Radar Measurement*, Englewood Cliffs, New Jersey: Prentice-Hall, 1969.
5. Rudge, A. W., K. Milne, A. D. Olver, and P. Knight, *Antenna Design*, London: Peter Peregrinus, 1986.
6. Abramowitz, M., and I. A. Stegun, *Handbook of Mathematical Functions*, New York: Dover, 1965.
7. Dolph, C. L., "A Current Distribution for Broadside Arrays Which Optimizes the Relationships Between Beamwidth and Sidelobe Level", *Proceedings IRE* 34, 1946, pp. 335-348.
8. Skillman, W. A., *Radar Calculations Using the TI-59 Programmable Calculator*, Dedham, Massachusetts: Artech House, 1983.
9. Taylor, T. T., "One Prameter Family of Line Sources Producing Modified sin $\Pi u/\pi u$ Patterns", *1953, Report TM234,* Hughes Aircraft Co., reported in Rudge, A. W., K. Milne, A. D. Olver, and P. Knight, *Antenna Design*, London: Peter Peregrinus, 1986.
10. Gradshteyn, I. S., and I. M. Ryzhik., *Table of Integrals, Series, and Products*, New York: Academic Press, 1980.
11. Bayliss, E. T., "Design of Monopulse Antenna Difference Patterns with Low Sidelobes," *Bell System Technical Journal*, Vol. 47, May-June 1968, pp. 623-650.
12. McNamara, D. A., "Direct Synthesis of Optimum Difference Patterns for Discrete Linear Arrays Using Zolotarev Polynomials," *Proceedings of the IEE,* H Vol. 140, No. 6, December 1993.
13. McNamara, D.A., "Performance of Zolotarev and Modified-Zolotarev Difference Pattern Array Distributions," *IEE Proceedings on Microwave antennas and Propagation*, Vol. 141, 1st February 1994.
14. Weisstein, E. W., *CRC Concise Encyclopedia of Mathematics*, Boca Raton, Florida: Chapman and Hall, 1999.
15. Hansen, R. C., *Phased Array Antennas*, New York: John Wiley, 1998.
16. Cheston, T. C., and J. Frank, in M. I. Skolnik (ed), *Radar Handbook*, New York: McGraw-Hill, 1970.
17. Cheston, T. C., and J. Frank., in M. I. Skolnik (ed), *Radar Handbook*, 2nd ed., New York: McGraw-Hill, 1990.
18. Brookner, E., *Practical Phased Array Antenna Systems*, Norwood, Massachusetts, Artech House, 1991.

19. Saad, T. S., *Microwave Engineers' Handbook*, Dedham, Massachusetts: Artech House, 1971.
20. Johnson, R. C., and H. Jasic, *Antenna Engineering Handbook*, New York: McGraw-Hill, 1984.
21. Condon, E. U., and H. Odishaw, *Handbook of Physics*, New York: McGraw-Hill, 1967.
22. Barton, D. K., *Modern Radar System Analysis*, Norwood, Massachusetts: Artech House, 1988.
23. Hall, W. M., "Antenna Beam-Shape Factor in Scanning Radars," *IEEE Transactions on Aerospace and Electronic Systems, Volume AES-4*, May 1968, pp. 402-409.
24. Blake, L. V., "The Effective Number of Pulses per Beamwidth for a Scanning Radar," *Proceedings of the IRE*, Vol. 41, June 1953, pp.770-774.
25. Charton, S., in R. S. Berkowitz, *Modern Radar*, New York: Wiley, 1965.
26. Dicke, R. H., et al., "Atmospheric Absorption Measurements with a Microwave Radiometer," *Physical Review*, Vol. 70, No. 340, 1946.
27. Blake, L. V., "A Guide to Basic Pulse Radar Maximum Range Calculation," *U.S. Naval Research Laboratory Report*, 23 December 1969.
28. Blake, L. V., *Radar Range-Performance Analysis,* Norwood, Massachusetts: Artech House, 1986, pp. 152 et seq.
29. Monzingo, R. A., and T. W. Miller, *Introduction to Adaptive Arrays*, New York: Wiley, 1980.
30. Nitzberg, G., *Adaptive Signal Processing for Radar*, Norwood, Massachusetts: Artech House, 1992.
31. Nitzberg, R., *Radar Signal Processing and Adaptive Systems*, Norwood, Massachusetts: Artech House, 1999.
32. Klemm, R., *Principles of Space-Time Adaptive Processing*, Stevenage: IEE Publishing, 2002.
33. Klemm, R., (ed.), *Applications of Space-Time Adaptive Processing*, IEE Publishing, 2004.
34. Galati, G., (ed.), *Advanced Radar Techniques and Systems*, Stevenage, Herts: Peter Peregrinus, 1993.
35. Goj, W. W., *Synthetic Aperture Radar and Electronic Warfare*, Norwood, Massachusetts: Artech House, 1993.
36. Richards, M. A., *Fundamentals of Radar Signal Processing*, New York: McGraw-Hill, 2005.

5A MATHEMATICAL APPENDIX

Methods of calculation that were used to produce the illustrations are given in this appendix as a number of misprints have been found in other texts.

5A.1 Taylor distribution

The equations for the Taylor antenna patterns were given in Sections 5.1.4.1 and 5.2.2.

5A.2 Zolotarëv distribution

McNamara [12] derived expressions to proceed from x to calculate the Zolotarëv function numerically and divided abscissa range from 0 to +1 into three regions bounded by x_1 and x_3 so that (5.53) can be expressed using real values. Unfortunately there are typesetting errors in the reference and the equations are repeated in this appendix with a slightly different notation, namely: in the literature the primes of k', K', and q' are used for complementary functions that have lead to misprints, so that k_1, K_1, and q_1 are used here for clarity.

The variable x in (5.53) is complex and has three subranges with the peak occurring at x_2 and the values are given by

$$x_3 = sn(M,k), \quad x_1 = \frac{k_1 \, x_3}{dn(M,k)}, \quad \text{and} \quad x_2 = x_3 \sqrt{\frac{1 - cn(M,k)\; z(M,\,k)}{sn(M,k)\; dn(M,k)}}$$

where $z(M, k)$ is the Jacobi zeta function [6, p. 978; 14, p. 949].

In region I from 0 to x_1 the variable x is transformed to the variable t given by

$$t = x \frac{cn(M,\,k)}{\sqrt{1 - x^2}\; k_1 \; sn(M,\,k)} \tag{5A.1}$$

and

$$\phi = F(t, k) \tag{5A.2}$$

where $F(t, k)$ is the incomplete elliptic integral [6, p. 590; 14, p. 537].

The values of ϕ corresponding to the values of x are used to calculate the function

$$Z_{2n+1} = \cos\left((2n+1)\ h(M, \phi, k)\right) \tag{5A.3}$$

where $h(a, b, k) = -\dfrac{\pi}{2} - \arctan\left(\tan\dfrac{\pi a}{2K}\ \tanh\dfrac{\pi b}{2K}\right)$

$$+ 2\sum_{r=1}^{\infty}(-1)^r\ r^{-1}\ \frac{q^{2r}}{1 - q^{2r}}\ \sin\frac{r\pi a}{K}\ \sinh\frac{r\pi b}{K}$$

and q is the nome [6, p. 576; 14, p. 1238], q_1, the complementary nome given by

$$q = \exp\left(-\pi\frac{K_1}{K}\right) \quad \text{and} \quad q_1 = \exp\left(-\pi\frac{K}{K_1}\right) \tag{5A.4}$$

Having calculated the function, we pass to the region II between x_1 and x_3 with a peak at x_2. The variable x is transformed to the variable p, namely,

$$p = \frac{1}{k}\sqrt{\frac{sn^2(M, +k) - x^2}{sn^2(M, k)\ (1 - x^2)}} \tag{5A.5}$$

and then to s, given by

$$s = F(p, K) \tag{5A.6}$$

where $F(p, k)$ is the incomplete elliptic integral [6, p. 590; 14, p. 537].

The Zolotarëv function in this region is

$$Z_{2n+1} = \cos(\pi x)\ \cosh\left((n + \frac{1}{2})\ f(M, s, k)\right) \tag{5A.7}$$

where

$$f(a, b, k) = -\frac{\pi ab}{KK_1} + \ln\left(\frac{\cosh\left(\dfrac{\pi(a + b)}{2K_1}\right)}{\cosh\left(\dfrac{\pi(a - b)}{2K_1}\right)}\right) - 4\sum_{r=1}^{\infty}\frac{(-1)^r}{r}\left(\frac{q_1^{2r}}{1 - q_1^{2r}}\right)\ \sinh\left(\frac{r\pi a}{K_1}\right)\ \sinh\left(\frac{r\pi b}{K_1}\right) \tag{5A.8}$$

Finally for region III with the sidelobes x is transformed to

$$r = \frac{sn(M, k)}{cn(M, k)}\ \frac{\sqrt{1 - x^2}}{x} \tag{5A.9}$$

and

$$\phi = F(r, k_1) \tag{5A.10}$$

where F(a, b) is the complete elliptic integral of the first kind [6, p. 590; 14, p. 537].

The Zolotarëv function in region III is

$$Z_{2n+1} = \cos\left(\left(n + \frac{1}{2}\right) g(M, \phi, k)\right) \tag{5A.11}$$

where

$$g(a, b, k) = -\frac{\pi ab}{KK_1} + 2\arctan\left(\frac{\tan\frac{\pi b}{2K_1}}{\tanh\frac{\pi a}{2K_1}}\right) - 4\sum_{r=1}^{\infty} \frac{q_1^{2r}}{r(1 - q_1^{2r})} \sinh\left(\frac{r\pi a}{K_1}\right) \sin\left(\frac{r\pi b}{K_1}\right) \tag{5A.12}$$

A Zolotarëv function for $n = 4$ with a sidelobe ratio of 30 dB (voltage ratio 31.62) is shown in Figure 5A.1.

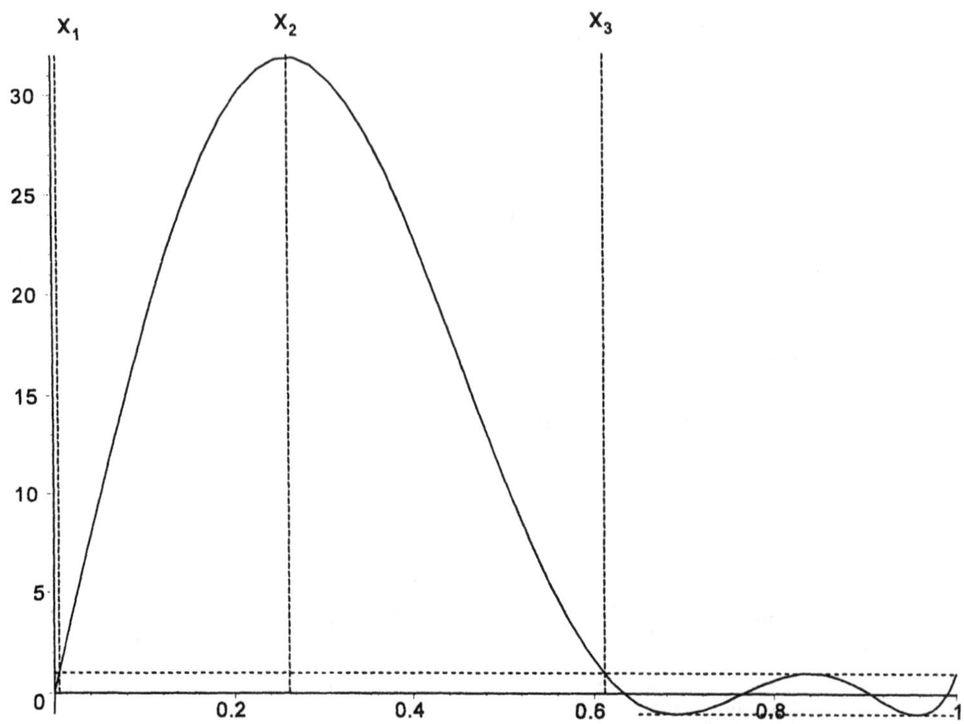

Figure 5.A.1 A Zolotarëv function for $n = 4$ (9 elements) and 30 dB sidelobe ratio.

5A.3 Bayliss distribution

The model Taylor distribution alters the position of the near-in zeroes according to a fixed formula and its derivative does not provide a satisfactory difference pattern. Bayliss used an iterative process to find polynomial approximations for the relationship between the sidelobe level (SLL) and the factor A, the positions of four near-in zeroes, u_1 to u_4, and the position of the peaks of the difference pattern given by u_0. The constants given in Table 5A.1 [15] were used to calculate the diagrams and give reducing sidelobes instead of the Taylor plateau.

Table 5A.1
Constants for calculating the values for the Bayliss antenna patterns

	C_1	$10C_2$	$10^3 C_3$	$10^5 C_4$	$10^7 C_5$
A	0.3038753	-0.5042922	-0.27989	-0.343	-0.2
u_1	0.9858302	-0.333885	0.14064	0.19	0.1
u_2	2.00337487	-0.1141548	0.4159	0.373	0.1
u_3	3.00636321	-0.0683394	0.29281	0.161	0
u_4	4.00518423	-0.0501795	0.21735	0.088	0
u_0	0.4797212	-0.1456692	-0.18739	-0.218	-0.1

To obtain A from Table 5A.1 the following is calculated:

$$A = C_1 + \frac{SLL\ C_2}{10} + \frac{SLL^2\ C_3}{10^3} + \frac{SLL^3\ C_4}{10^5} + \frac{SLL^4\ C_5}{10^7} \tag{5A.13}$$

and similarly for the zeroes, z_n, and positions of the peaks of the difference patterns.

Chapter 6

Factors outside the radar: propagation, scattering, and clutter

This chapter covers what happens to the transmitted pulse after it leaves the antenna and travels through the atmosphere until part of it is scattered back to the radar. The echo signal that enters the receiver is often not an exact copy of the transmitted pulse because the scatterer may have radial components of movement that shift the frequency of the echoes. The limited time taken to look at the objects of interest broadens the spectrum of the echo signals that are passed by the antenna to the receiver.

This chapter describes the amplitude and the phase of the returning echoes with the phenomena encountered by the signals on their way out and back. This chapter discusses the following:

- Effects of the atmosphere, refraction, and attenuation;
- Scattering at objects with fluctuation characteristics;
- Ground reflections;
- Typical scenario for a model radar;
- A number of examples of range-height paper in Appendix 6A.

6.1 AMPLITUDE AND PHASE OF THE ECHO

The pulse that leaves the antenna spreads out in space, is scattered by the objects it hits, and part of the scattered energy is directed toward the receiving antenna of the radar.

6.1.1 Amplitude of the echo

The power radiated from an isotropic radiator is spread over the inside of a sphere of radius r. The area of the surface of the sphere is $4\pi r^2$, so that the power density is inversely proportional to the square of the distance (inverse square law).

$$\textit{Power density} = \frac{1}{4\pi r^2} \tag{6.1}$$

Radars detect the echoes from objects and a dilution of energy occurs twice, once on the way out and then once on the way back, as shown in Figure 6.1. For clarity, the bistatic radar case, with separate transmitting and receiving antennas, is shown with notional square antenna beams.

If the transmitting antenna has a gain of G_t, the power intercepted by an area σ square meters at a distance R_t meters is

$$\textit{Power intercepted by the scatterer} = \frac{P_t G_t}{4\pi R_t^2}\,\sigma \quad \text{W} \tag{6.2}$$

The area, σ, for a conducting sphere that gives the same echo power as an actual object, is called the effective echoing area or radar cross-section. It is usually expressed in square meters even when other units are used in the expressions.

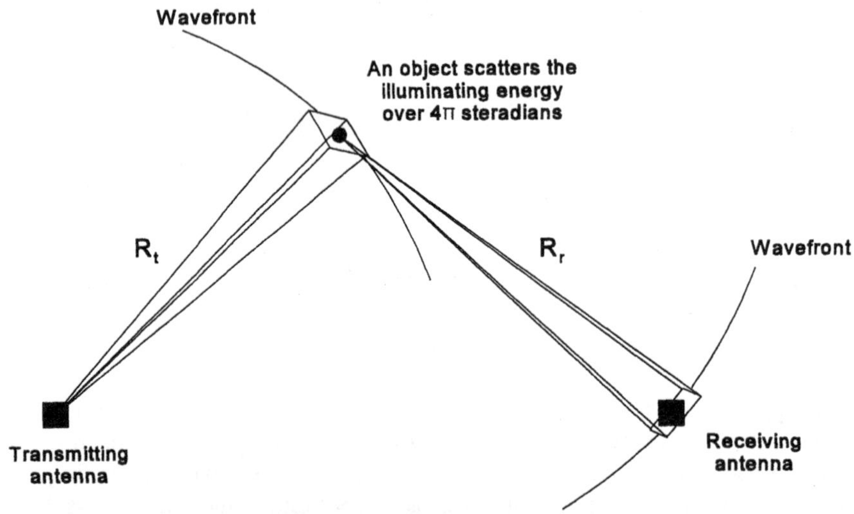

Figure 6.1 The inverse square law applied to radar.

For the equivalent sphere assumed in defining radar cross-section, the power is said to be scattered equally in all directions, over 4π steradians, in the same way as an isotropic radiator. Thus, the power intercepted by the receiving antenna of effective area A_r at a distance R_r from the scatterer is

$$\textit{Power intercepted by the receiving antenna} = \frac{P_t\, G_t}{4\pi\, R_t^2}\, \sigma\, \frac{A_r}{4\pi\, R_r^2} \quad \text{W} \tag{6.3}$$

Normally, the gain of the receiving antenna is measured and given by $G\lambda^2/4\pi$ (see Chapter 5, Antennas), so taking the monostatic radar case with common transmitting and receiving antennas with a gain G,

$$\textit{Power at monostatic receiving antenna} = \frac{P_t\, G}{4\pi\, R^2}\, \sigma\, \frac{G\, \lambda^2}{4\pi}\, \frac{1}{4\pi\, R^2} = \frac{P_t\, G^2\, \sigma^2\, \lambda^2}{(4\pi)^3\, R^4} \quad \text{W} \tag{6.4}$$

In practice the scatterer is not a perfect isotropic radiator. The Institute of Electrical and Electronics Engineers Standard Radar Definitions [1] recognizes three cases:

- Monostatic case (default);
- Forward scattering case, used for radar fences;
- General bistatic case.

6.1.2 Phases of the echoes and Doppler frequency

The phase of radio frequency energy in space completes a full circle (360 degrees or 2π radians) each wavelength. Normally, with pulse radars the scenario shows no movement during, say, a 1 μs pulse so that the phase of the echo depends on the number of wavelengths the energy travels. Search radars must be able to receive and process echo signals with any phase from 0 to 360 degrees or 2π radians.

Over longer times, 1 ms for example, a scatterer may move enough radially in range to give a measurable phase difference between radar pulses, though the time difference or range may not be perceivable. The change of phase is for the outward and return paths, double the movement of the scatterer, and is given by

$$\Delta\phi = \frac{2\, \Delta R}{\lambda} \text{ cycles} = \frac{4\pi\, \Delta R}{\lambda} \text{ radians} \tag{6.5}$$

If this effect occurs continually, then the value $\Delta\phi/\Delta T$ has the dimensions of frequency. All scatterers that have radial movement, v m/s towards the radar, cause an increase of frequency, the Doppler frequency, in their echoes.

$$\text{Doppler frequency, } f_d = \frac{2\,v}{\lambda} \text{ Hz} \tag{6.6}$$

Useful values to remember for the Doppler frequencies for particular radio frequencies in the popular primary radar bands are listed in Table 6.1.

Table 6.1

Doppler frequency constants for frequencies in the common radar bands

Frequency band			Typical frequency, MHz	Wavelength, m	Doppler frequency constant, Hz/m/s
British	United States	Electronic counter-measures			
UHF	UHF	B	450	0.667	3
UHF	—	C	600	0.5	4
L	L	D	1 350	0.222	9
S	S	E	3 000	0.1	20
C	C	G	5 400	0.056	36
—	—	G	6 000	0.05	40
X	X	I	9 000	0.033	60
—	—	J	15 000	0.02	100
J	Ku	J	16 500	0.018	110
Q	Ka	K	30 000	0.01	200

For secondary radar there are different outgoing and receiving frequencies, so the Doppler frequency shift is not multiplied by the factor two in (6.6). The Doppler frequency shifts for the common secondary radar frequencies are given in Table 6.2.

Table 6.2

Doppler frequency constants for secondary radar

	Frequency, MHz	Wavelength, m	Doppler frequency constant, Hz/m/s
Interrogator to aircraft	1 030	0.291	3.433
Aircraft transponder	1 090	0.275	3.633

International Civil Aviation Organization (ICAO) Annex 10 [2] allows the tolerances in Table 6.3.

Table 6.3

Frequency tolerances for secondary radar (ICAO)

	Frequency tolerance	
	Interrogator on the ground, 1 030 MHz	Transponder in the aircraft, 1 090 MHz
Modes A to D	200 kHz	3.0 MHz
Mode S, aircraft below 15 000 feet (4 572 m)	10 kHz	3.0 MHz
Mode S, others	10 kHz	1.0 MHz

It can be seen that equipment tolerances are much greater than any Doppler frequency change.

6.2 EFFECTS OF THE ATMOSPHERE

There are two effects when the radar waves travel through the atmosphere:

- Bending the rays in the elevation plane from and to the radar [3, p. 188] which also causes lens loss or the splaying of the rays in the elevation plane. The two paths of the rays from and to the radar are reciprocal, so a single one-way calculation is sufficient.
- Absorbing energy from the waves, or attenuating the waves.

6.2.1 Exponential atmosphere models

The refraction subroutines use a common exponential atmosphere algorithm with different inputs and outputs. First the model of the earth's atmosphere must be chosen. All exponential atmosphere models define the refractivity as

$$N = N_s \exp(-c_e h) \tag{6.7}$$

where N_s is the refractivity at the earth's surface in N units;
c_e is a constant;
h is the height above the earth's surface.

Since the refractive index of air, n, is near unity, an easier number to use is the refractivity, N, which is measured in N units. The relation is

$$n = 1 + N \cdot 10^{-6} \tag{6.8}$$

The various models define the constant c_e differently. Examples of exponential atmosphere models are as follows:
- The Central Radio Propagation Laboratory (CRPL) [4] used in the United States of America defines N as:

$$\begin{aligned} N &= N_s \exp(-0.14386\ h) \quad \text{where } h \text{ is altitude over mean sea level in km} \\ N_s &= \text{default value 313, a mean value for the United States at 700m altitude} \end{aligned} \tag{6.9}$$

- The North Atlantic Treaty Organisation (NATO) model representing northern Europe [5] gives N as:

$$\begin{aligned} N &= 320 \exp(-0.03709\ h) \quad \text{where } h \text{ is the height in thousands of feet} \\ &= 320 \exp(-0.1217\ h) \quad \text{where } h \text{ is the height in km} \end{aligned} \tag{6.10}$$

- The Comité Consultif des Radiocommunications (CCIR) published maps for N_s for the world [6] and used the model:

$$\begin{aligned} N &= N_s \exp(-0.136\ h) \quad \text{where } h \text{ is the height above the surface of the Earth in km} \\ N_s &= \text{surface reflectivity taken from maps, default 289} \end{aligned} \tag{6.11}$$

The ray from the radar is traced from the radar to the point of interest, as in Figure 6.2, which may be to:
- Trace the ray to calculate the height of the point of interest;
- Trace a height line on range-height paper;
- Trace the surface of the earth on range-height paper;
- Find the attenuation along the ray.

Figure 6.2 represents a ray leaving the radar on a pedestal or flying at a height h_r at an elevation angle θ_1. Having chosen an increment ΔR, the height to the layer boundary is calculated using the cosine rule:

$$R_2^2 = R_1^2 + \Delta R^2 + 2\ R_1\ \Delta R \sin(\theta_1) \tag{6.12}$$

Equation (6.12) gives the new height, h_2, as

$$h_2 = R_2 - a \tag{6.13}$$

where a is the earth's radius.

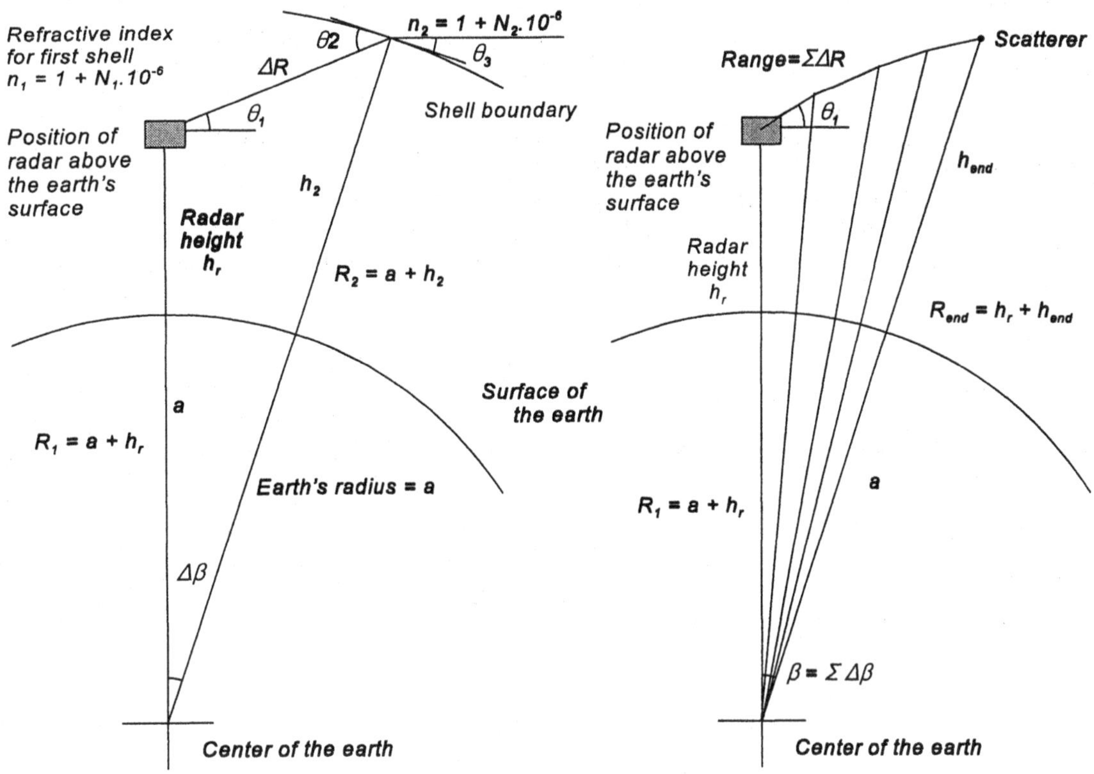

Figure 6.2 Diagram of a single segment and a complete ray trace.

The angle subtended at the center of the earth, Δβ, by the segment is given by the sine rule:

$$\frac{\sin(\Delta\beta)}{\Delta R} = \frac{\cos(\theta_1)}{R_2}$$

or

$$\Delta\beta = \sin^{-1}\left(\frac{\Delta R\ \cos(\theta_1)}{R_2}\right) \tag{6.14}$$

The angles β and θ are updated to allow the angle of incidence to be calculated:

$$\begin{aligned} \theta_2 &= \theta_1 + \Delta\beta \\ \beta_2 &= \beta_1 + \Delta\beta \end{aligned} \tag{6.15}$$

At the end of ΔR, the ray crosses a shell border and is refracted according to Snell's law, namely,

$$\frac{\cos(\theta_3)}{\cos(\theta_2)} = \frac{n_1}{n_2} = \frac{1+N_1.10^{-6}}{1+N_2.10^{-6}}$$

$$\theta_3 = \cos^{-1}\left(\frac{n_1}{n_2}\cos(\theta_2)\right) \tag{6.16}$$

Finally, R is incremented by ΔR and the values are updated to trace the ray through the next shell.

The mean refractive index of the atmosphere in Europe is said to be 1.000320 [5], and in the United States of America is 1.000313, which is at the limit of seven figure accuracy. Double precision must be used in many programming languages or the equivalent in mathematical programs.

The computation is used to find the height of a scatterer at a given range with height-finder, tracking, or three-dimensional radars or, to plot range-height paper, for the range of a scatterer at a given height. The final segment is often interpolated, depending on whether range or height has been the end criterion.

The length of the segments is chosen to give a height accuracy within a chosen tolerance. Since the refraction depends on the change of the refractive index, segment lengths inversely proportional to the refractive index may reduce calculation time.

6.2.2 Constant *k* atmosphere model

Until digital computers became common, it was usual to approximate the range curve to the scatterer in Figure 6.3 to a straight line leading to B.

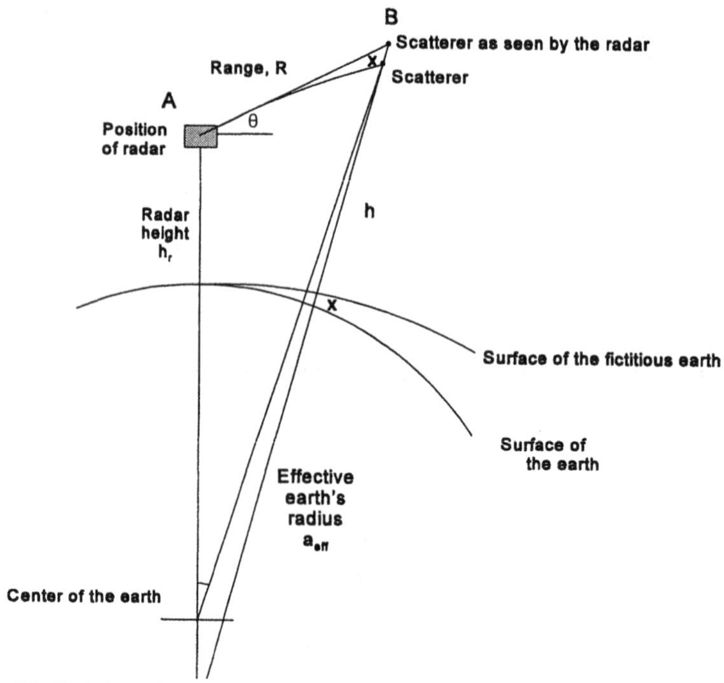

Figure 6.3 Height calculation using the cosine rule.

The scatterer at B is assumed by the radar to be at A at an elevation angle of θ. The height error is x which may be reduced by using a radius of the earth k times the real value for the earth's curvature correction. Using the cosine rule,

$$
\begin{aligned}
(h + a_{eff})^2 &= (h_r + a_{eff})^2 + R^2 + 2R(h_r + a_{eff})\sin\theta \\
h &= \sqrt{(h_r + a_{eff})^2 + R^2 + 2R(h_r + a_{eff})\sin\theta} - a_{eff}
\end{aligned}
\tag{6.17}
$$

The value of k depends on the gradient of the refractive index of the atmosphere. The refractive index, n, is given by Smith and Weintraub in [7] as

$$
(n - 1)10^6 = N = \frac{77.6p}{T} + \frac{3.73\ 10^5\ e}{T^2} \tag{6.18}
$$

where p is the atmospheric pressure, in hPa (millibars);
e is the partial pressure of the water vapor, in hPa (millibars);
T is the temperature of the atmosphere, in K.

The value k is given by [8, p. 20-12]

$$k = \frac{1}{1 + a \dfrac{dn}{dh}} \tag{6.19}$$

The gradient dn/dh is normally negative. When $dn/dh = -1/a$, then $k = \infty$. The earth in this model becomes flat and there is anomalous propagation (see Section 6.2.5). The critical gradient may be found from

$$a \frac{dn}{dh} = a \frac{dN}{10^6 dh} \tag{6.20}$$

which gives

$$\text{for ducting,} \quad \frac{dN}{dh} = -\frac{10^6}{6367} = -157 \quad \text{N units/km} \tag{6.21}$$

Figure 6.4 assumes that the earth in Figure 6.3 is parabolic, and the classical height finder formula is [8, p. 22-10]

$$h = R \sin \theta + \frac{R^2}{2a_{eff}} \quad \text{or } pR \sin \theta + qR^2 \tag{6.22}$$

Table 6.4 gives the values of the constant q for the k factor of 4/3.

Table 6.4
Constants for (6.22), the height-finder equation

Value for p	Value for q	Units for h	Units for R
6 076	0.6624	Feet	Nautical miles
1 000	0.0589	Meters	Kilometers

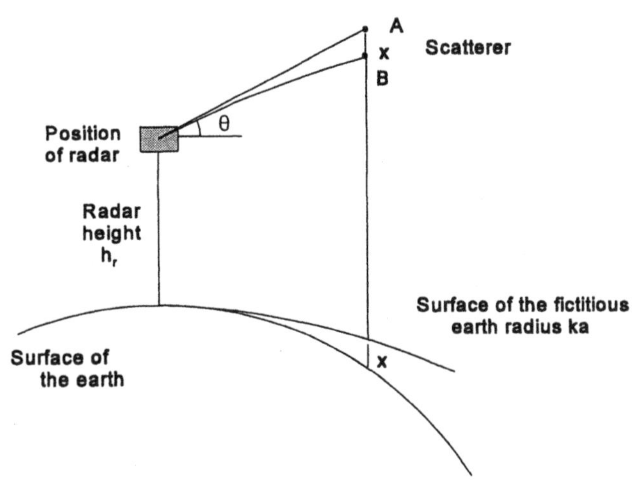

Figure 6.4 Simplified height calculation over a parabolic earth.

The k factor varies with the weather from day to day. Climatological values may be used for the time of year, or it may be forecast with the weather. The same is true for the d value, the difference between the height of an aircraft measured by a particular radar and the barometric height shown on its altimeter.

6.2.3 Range-height paper

Range-height paper is used to illustrate the vertical coverage of a radar. If equal range and height scales are chosen, the constant slant range lines are circles and the angle scale is true, but this confines the coverage diagram to a small corner. If range-height paper with a coverage of 300 nautical miles (555.6 km) by 100 000 feet (30.48 km) had equal horizontal and vertical scales, then the paper would have a width to height ratio of 18:1. For Deutches Institut für Normung (DIN) sized paper the ratio is 1.414:1. It is thus normal to expand the vertical scale to fill the page. If the vertical scale expansion is q, then the angle on the paper becomes

$$Elevation\ angle\ on\ paper = \tan^{-1}(q\ \tan(\theta)) \tag{6.23}$$

Slant range circles on equally scaled paper become ellipses and the elevation angle lines bunch near the vertical. The height lines may be drawn for a real atmosphere or a model atmosphere be it exponential or constant k.

Appendix 6A shows examples of range-height paper for three different types of ground radar in Table 6.5.

Table 6.5
The following types of range-height paper given in Appendix 6A

	Nominal range	Height coverage
Airport radar	0..60 nmi (111 km)	0..10 kft (3.05 km)
Air traffic control *en route* or military gap-filler radar	0..150 nmi (278 km)	0..60 kft (18.3 km)
Military surveillance	0..250 nmi (407 km)	0..120 kft (36.6 km)

These types of paper are drawn for the three model atmospheres:

- Conventional $k = 4/3$ atmosphere;
- North European atmosphere [5] with refractivity given by $N = 320\ \exp(-0.03709\ h)$ with h in thousands of feet or $N = 320\ \exp(-0.1217\ h)$ with h in kilometers which is an average for the colder climate there;
- Central Radio Propagation Laboratory (CRPL) model [4] with a surface refractivity of 313.

Shorter range radars either neglect refraction or use current weather data to calculate more exact positions. Figure 6.5 shows the height lines of three types of range-height paper plotted against ground range on a Cartesian grid. On the scale shown there is little difference visible between the lines plotted for refractivities of 131 and 320.

The constant $k = 4/3$ earth's curvature model is a simple model and, in spite of the availability of digital computers, is still often used. For accurate height calculation, the surface refractivity and the refractive index profile of the atmosphere must be known, so often, the constant k model is accurate enough.

6.2.4 Lens loss

Appreciable lens loss occurs at longer ranges, more than 100 nautical miles (185.2 km). It is caused by the changing refractivity of the atmosphere, which is greatest near the earth's surface. The lower parts of the beam are refracted more, so the wavefront has a greater vertical extent, as is shown in Figure 6.6.

There is no differential refraction in the azimuth dimension, so the loss is the ratio B/A. A graphs of this loss is shown in Figure 6.7 and [3, p. 192; 8; 9, pp. 51-54].

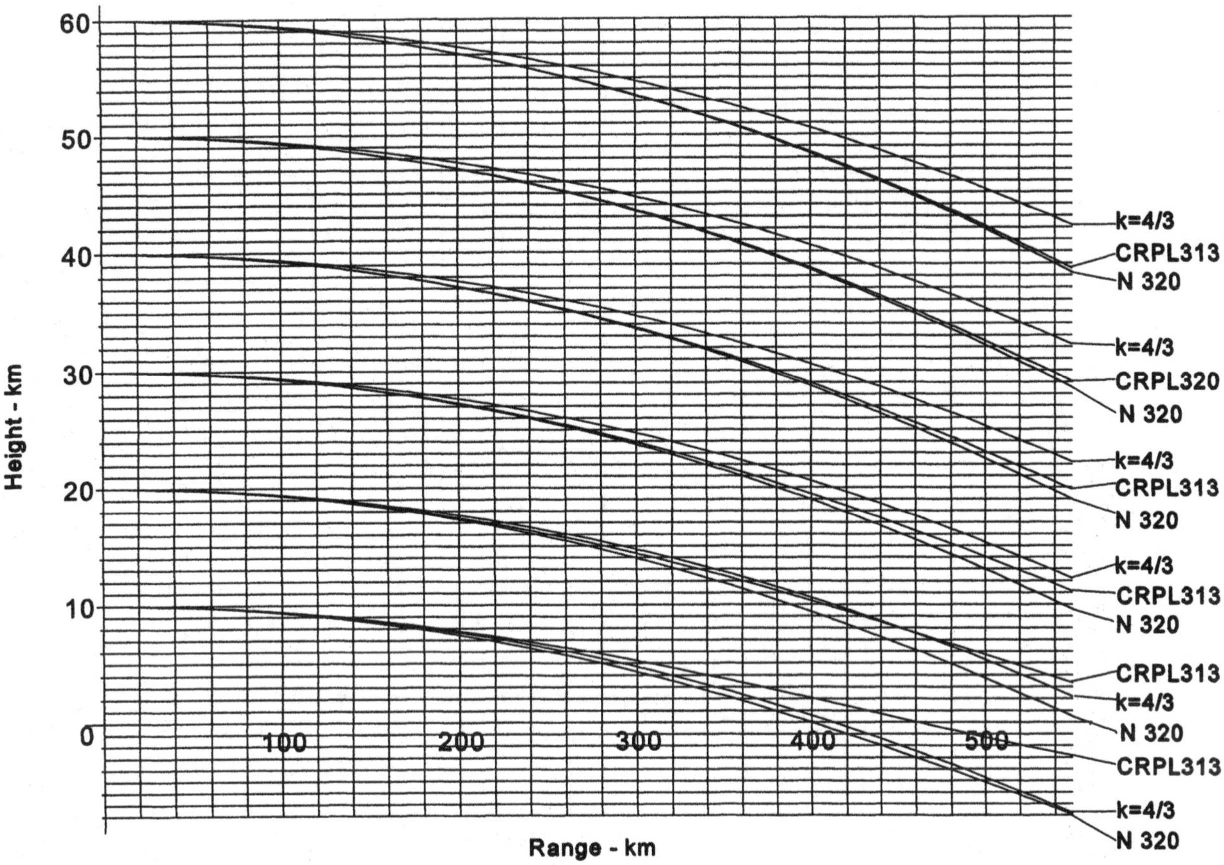

Figure 6.5 Height lines for constant $k = 4/3$, NATO, and CRPL range-height papers.

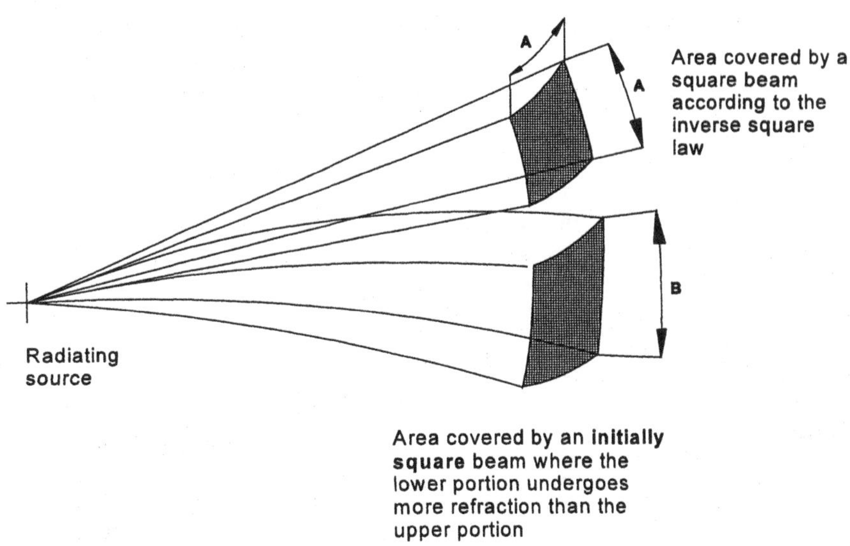

Figure 6.6 An exaggerated view of the lens loss phenomenon.

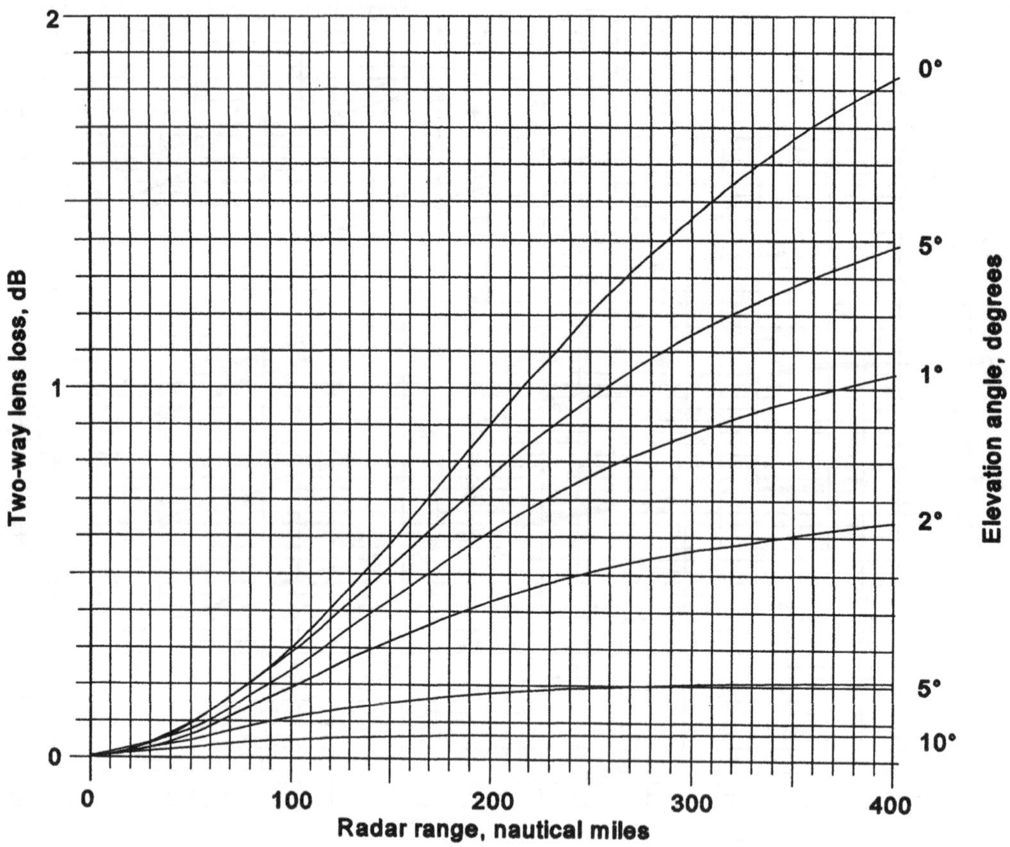

Figure 6.7 Lens loss. [From Weil, T. A., "Atmospheric Lens Effect", *IEEE Trans, AES-9*, January 1973, pp. 51-54.]

6.2.5 Anomalous propagation and superrefraction

Occasionally, there is such an abrupt change in the atmosphere between "layers" that the radar is like a fish, under water, in a pond. The fish can see straight up and down to an angle where the emerging refracted ray is along the surface of the pond. Rays at angles below this angle are reflected back into the pond. This situation is illustrated in Figure 6.8.

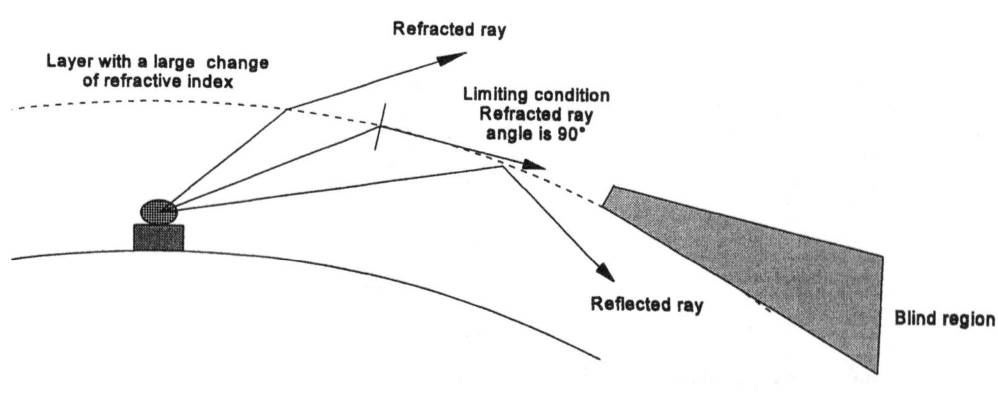

Figure 6.8 Refraction and reflection at an atmospheric layer.

This condition brings extra coverage along the surface of the earth and for low-flying aircraft and extra clutter, and a blind volume below the surface of the 90 degree refracted ray.

Total reflection causes the ray to be trapped between the earth's surface and the layer and is called trapping or ducting. These effects may be caused by temperature inversion, that is, an increase of temperature with height, such as warm dry air over a cool sea. The water vapor content decreases much more with altitude than with a standard atmosphere which gives a low duct along the surface. Temperature inversion may occur at higher altitudes, giving higher level ducts. It is usually caused by atmospheric subsidence, or the sinking and spreading of an air mass over stable air that causes a local temperature inversion. This effect is usually associated with areas of high pressure.

Normally, the refractivity decreases by 41.9 N units for the first kilometer above 700 m for the CRPL atmosphere (6.9) [3, p. 225] and 36.7 N units for the European atmosphere (6.10). As shown in Section 6.2.2, trapping occurs when the refractivity gradient is more than 157 N units per kilometer.

6.2.6 Attenuation in the atmosphere

In contrast to refraction and attenuation at optical wavelengths, the frequencies used for radar are sensitive to the water vapor content. At higher radar frequencies there are frequency bands where there is higher absorption. Those below 100 GHz are:

- Oxygen: many lines around 60 GHz;
- Water vapor: 22.235 GHz.

The attenuation or absorption losses for a two-way path have been calculated by Blake [3, pp. 210 et seq., Figs. 5.14-5.22; 10, p. 138, Figs. 22A-22L; 11] for the CRPL exponential atmosphere with N_s = 313. The temperature and pressure profiles are based on the ICAO standard atmosphere with a water vapor content of 7.5 g/m³ at the surface. The graphs are shown here as Figures 6.9 - 6.15.

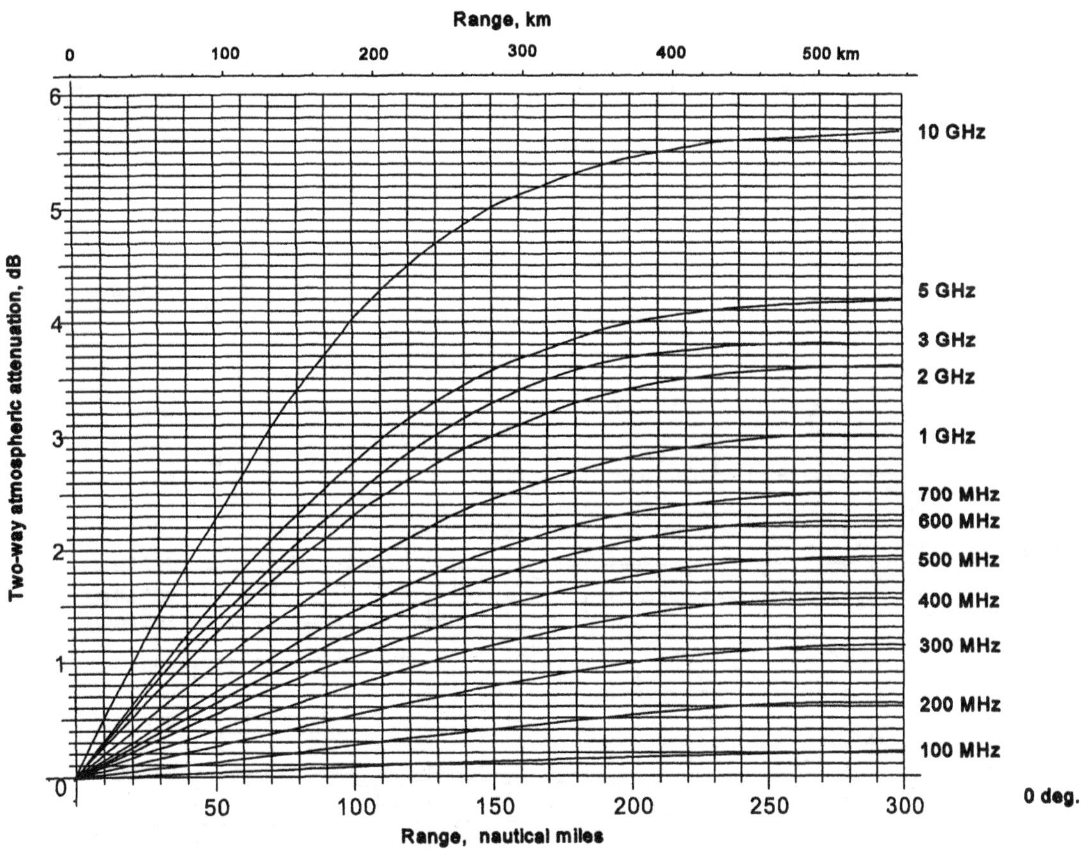

Figure 6.9 The two-way attenuation for an elevation angle of 0 degrees for the atmosphere described in Section 6.2.6. [Source: Blake, L.V., *A Guide to Basic Pulse-Radar Maximum-Range Calculation*, NRL 1970.]

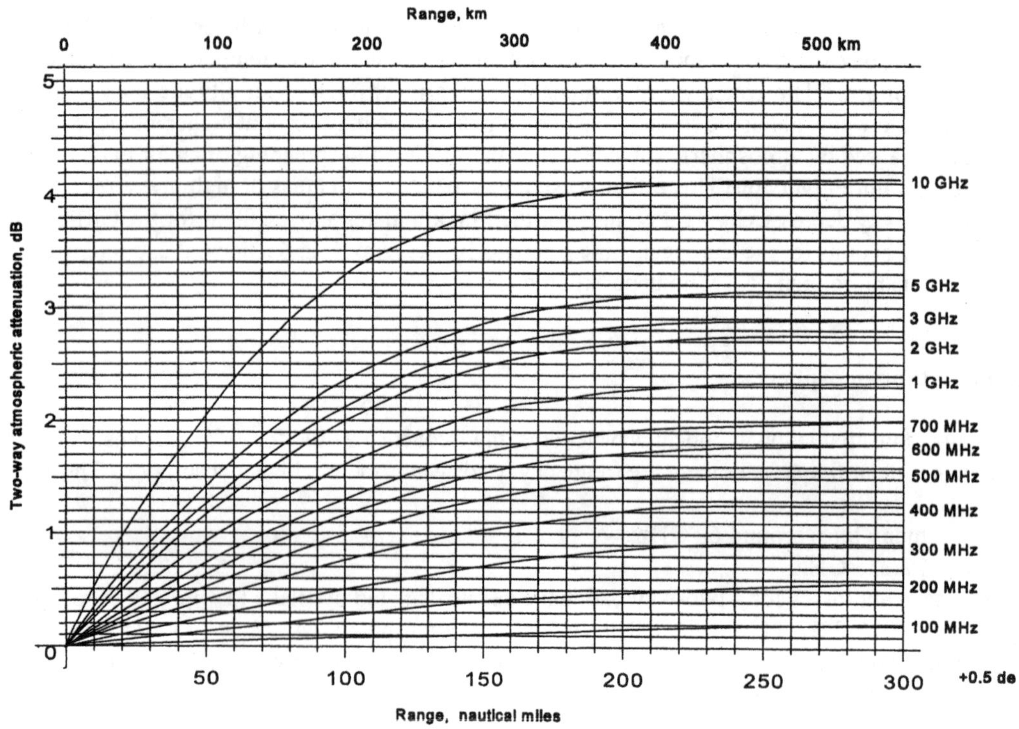

Figure 6.10 The two-way attenuation for an elevation angle of 0.5 degrees for the atmosphere described in Section 6.2.6. [Source: Blake, L.V., *A Guide to Basic Pulse-Radar Maximum-Range Calculation*, NRL 1970.]

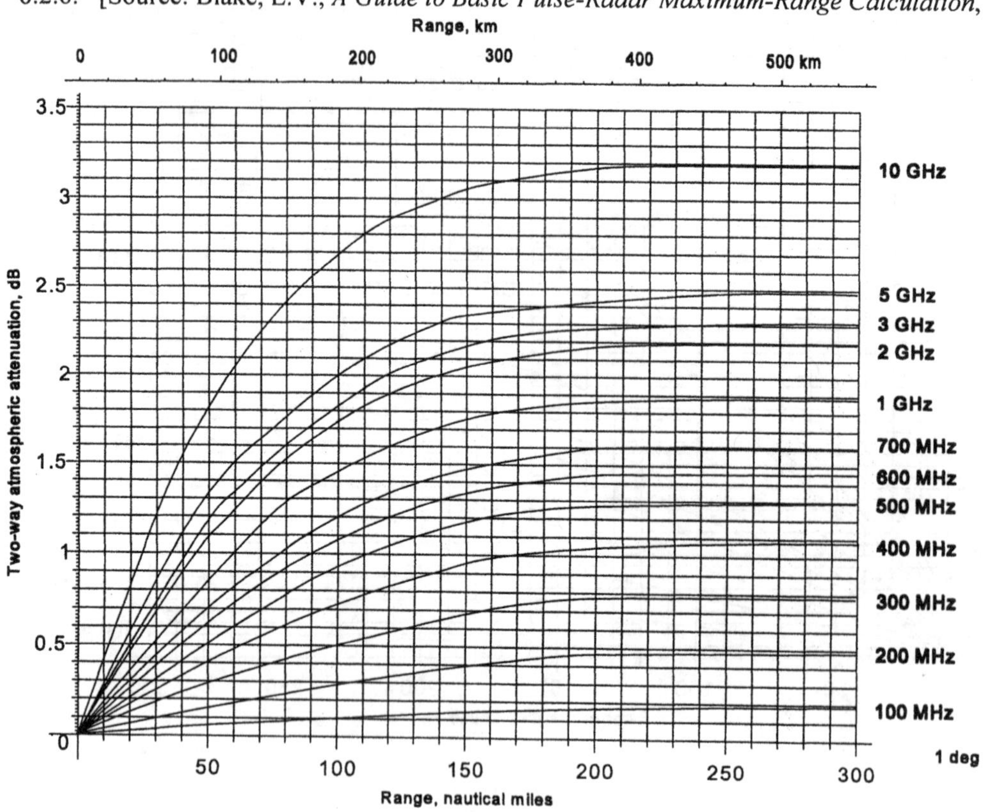

Figure 6.11 The two-way attenuation for an elevation angle of 1 degree for the atmosphere described in Section 6.2.6. [Source: Blake, L.V., *A Guide to Basic Pulse-Radar Maximum-Range Calculation*, NRL 1970.]

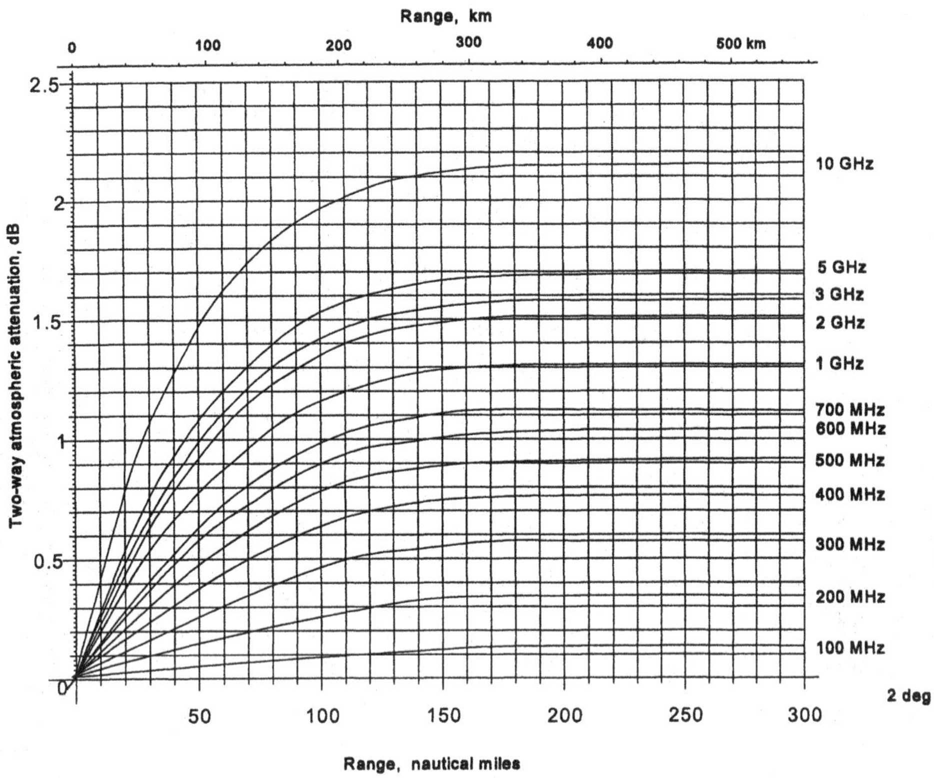

Figure 6.12 The two-way attenuation for an elevation angle of 2 degrees for the atmosphere described in Section 6.2.6. [Source: Blake, L.V., *A Guide to Basic Pulse-Radar Maximum-Range Calculation*, NRL 1970.]

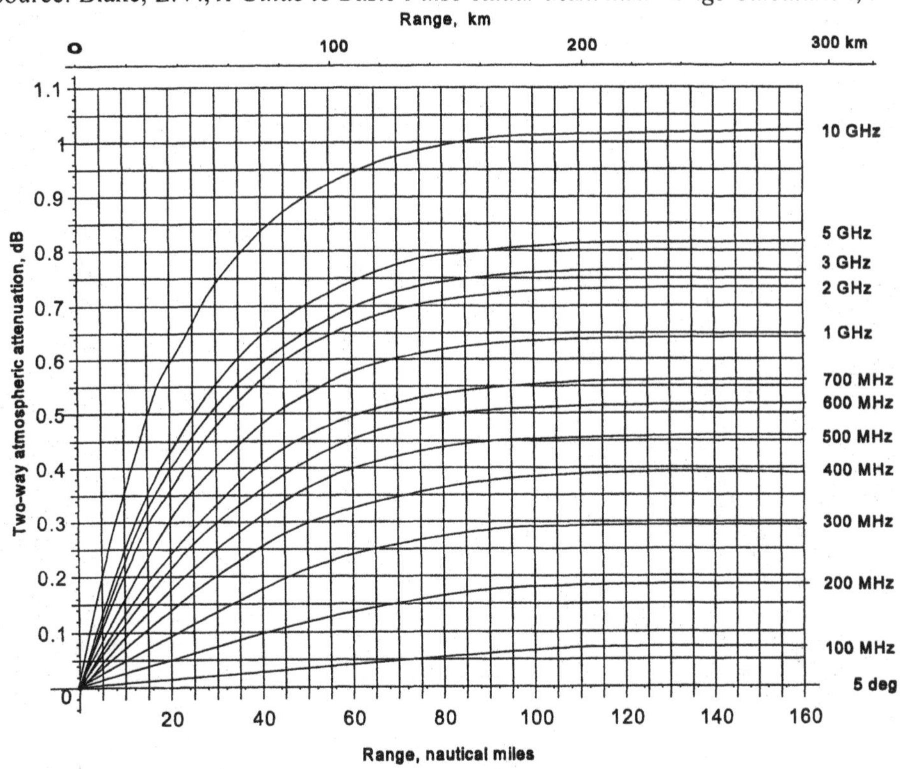

Figure 6.13 The two-way attenuation for an elevation angle of 5 degrees for the atmosphere described in Section 6.2.6. [Source: Blake, L.V., *A Guide to Basic Pulse-Radar Maximum-Range Calculation*, NRL 1970.]

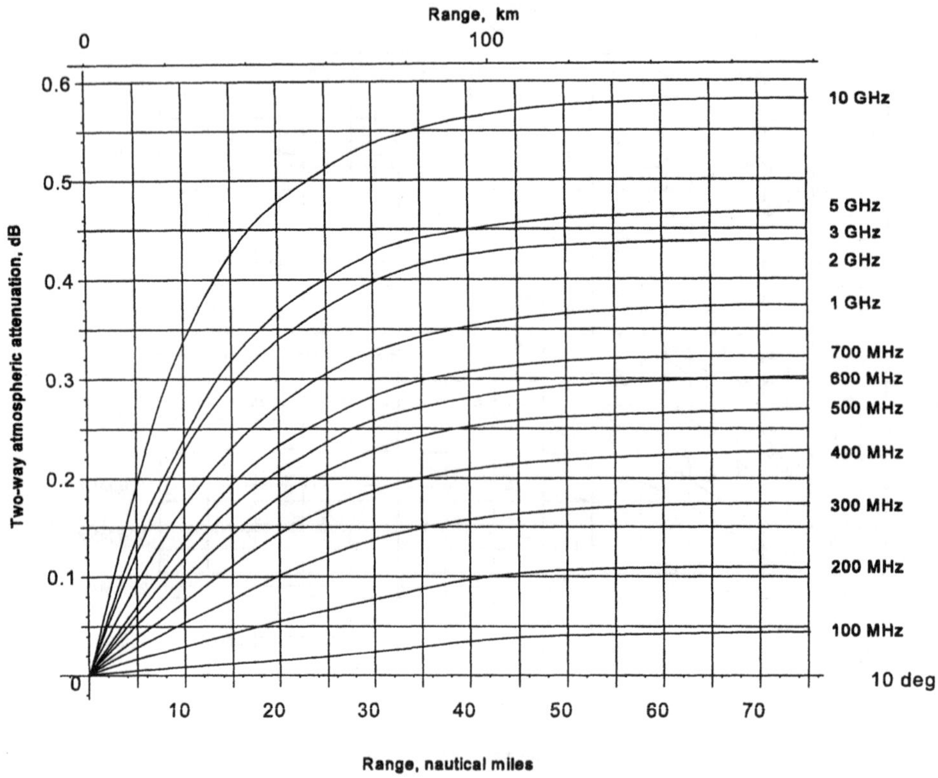

Figure 6.14 The two-way attenuation for an elevation angle of 10 degrees for the atmosphere described in Section 6.2.6. [Source: Blake, L.V., *A Guide to Basic Pulse-Radar Maximum-Range Calculation*, NRL 1970.]

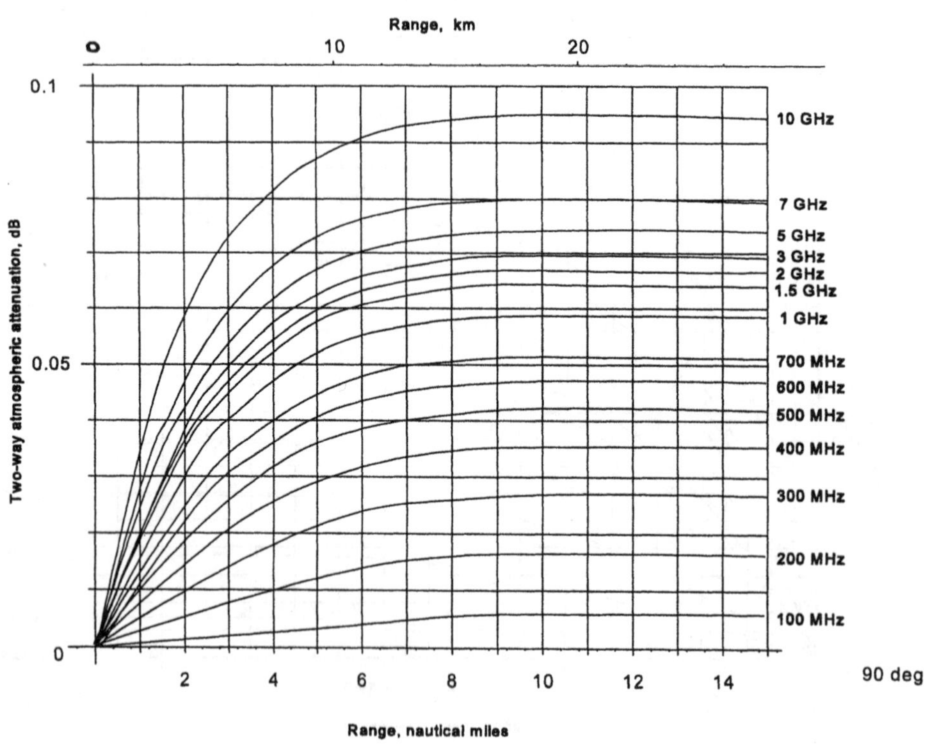

Figure 6.15 The two-way attenuation for an elevation angle of 90 degrees for the atmosphere described in Section 6.2.6. [Source: Blake, L.V., *A Guide to Basic Pulse-Radar Maximum-Range Calculation*, NRL 1970.]

6.2.6.1 Fog

A relatively simple formula has been given in [12], which, in terms of frequency, is [13]

$$Fog\ attenuation = 4.87.10^{-4}\ M\ f^2 \quad \text{dB/km} \tag{6.24}$$

where M is the water content of the fog in g/m³;
f is the frequency in GHz.

Three fog densities are plotted in Figure 6.16. These figures are for 18°C and depend on the temperature of the condensed water. The coefficient must be multiplied by 2 for temperatures near 2°C or by 0.7 for temperatures near 30°C.

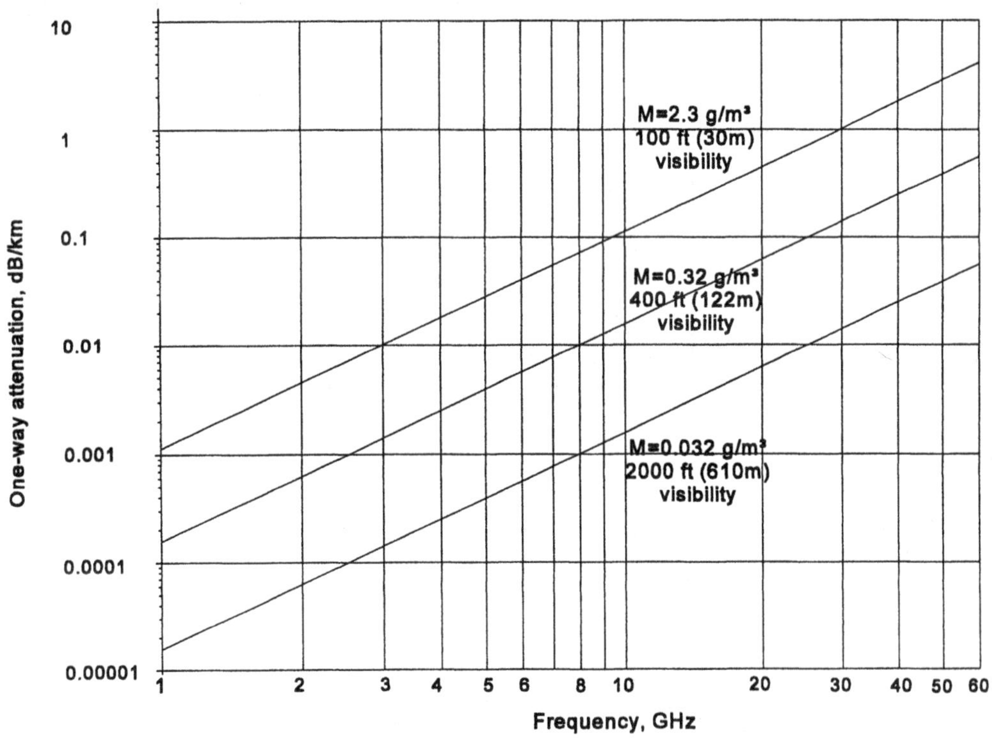

Figure 6.16 Attenuation caused by fog and clouds.

6.2.6.2 Rain

Attenuation and scattering are proportional to the drop sizes, which have been related theoretically to the rain rate. This can be measured on the ground, and the commonly used rain rates are shown in Table 6.6.

Table 6.6
Commonly used rain rates

Name of rain rate	Rain rate, mm/hour
Drizzle	0.25
Light rain	1.0
Moderate rain	4.0
Heavy rain	16.0
Tropical rain	40.0

Formulas to calculate rain attenuation are given in [14]. The results for different rain rates and frequencies at normal temperatures are shown in Figure 6.17. It should be noted that the values are for 18°C and depend on the temperature of the condensed water. The coefficient must be multiplied by 2 for temperatures near 2°C or by 0.7 for temperatures near 30°C.

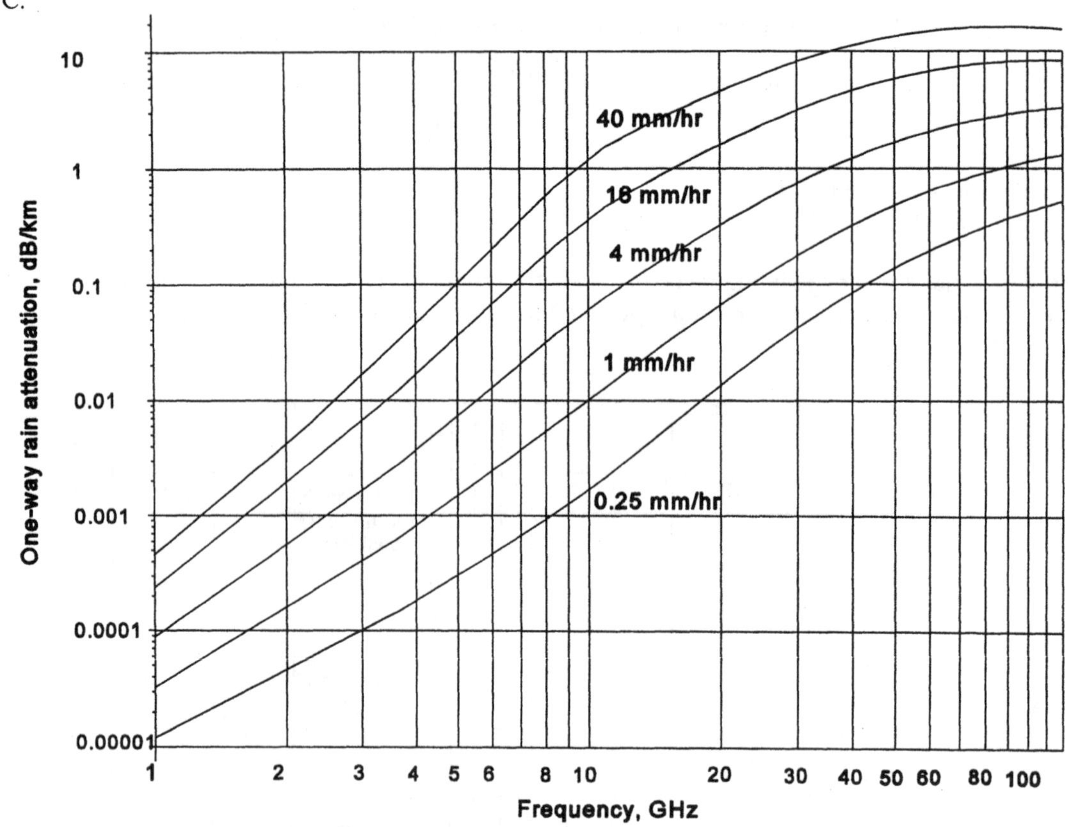

Figure 6.17 One-way rain attenuation for different rain rates.

Rain can occur at up to 2 km to 3.5 km in height. There are occurrences when rain at higher altitudes evaporates as it falls and none is perceived on the surface.

6.2.6.3 Snow

It is assumed that ice crystals do not attenuate. All the attenuation comes from the water content of wet snow. Gunn and East in [3, p. 221] give the formula

$$\textit{Wet snow attenuation} = \frac{0.00349\ r^{1.6}}{\lambda^4} + \frac{0.0022\ r}{\lambda} \quad \text{dB/km} \tag{6.25}$$

where λ is the wavelength in cm;
r is the water content of the snow in mm/hour.

Other estimates by Bell [3, p. 221] state that wet snow has an even greater attenuation than rain.

6.3 SCATTERING

The radar cross-section, σ m², first mentioned in Section 6.1.1, depends on the shape of the scatterer and how it moves during the period when it is under observation by the radar.

Scatterers are normally much smaller than the size of the beam from the antenna. The amplitude of the echo depends on the radar cross-section of the scatterer. Figure 6.18 shows that the width of an echo signal displayed on a plan position indicator (PPI) also depends on its power. A weak echo crosses the tip of the beam and is narrow, stronger echoes are wider, and the strongest echoes will be painted by the sidelobes. Echoes in the sidelobes give false alarms.

Radars normally have antennas that have much smaller sidelobes and use sensitivity time control (STC or swept gain) or sidelobe suppression (cancellation or blanking) to reduce or eliminate these effects. When looking for faults in a radar, it is often advisable to switch off these measures and observe a distant reflector in an area free of clutter. If the sidelobes are not symmetric or are too high, there is a fault in the antenna.

The energy scattered by an object depends on the material, its shape, its orientation, and the radar wavelength and polarization. The energy returned to a radar is composed of elemental reflections from the parts of the object. The waves may add or destructively interfere with each other, giving rise to fading.

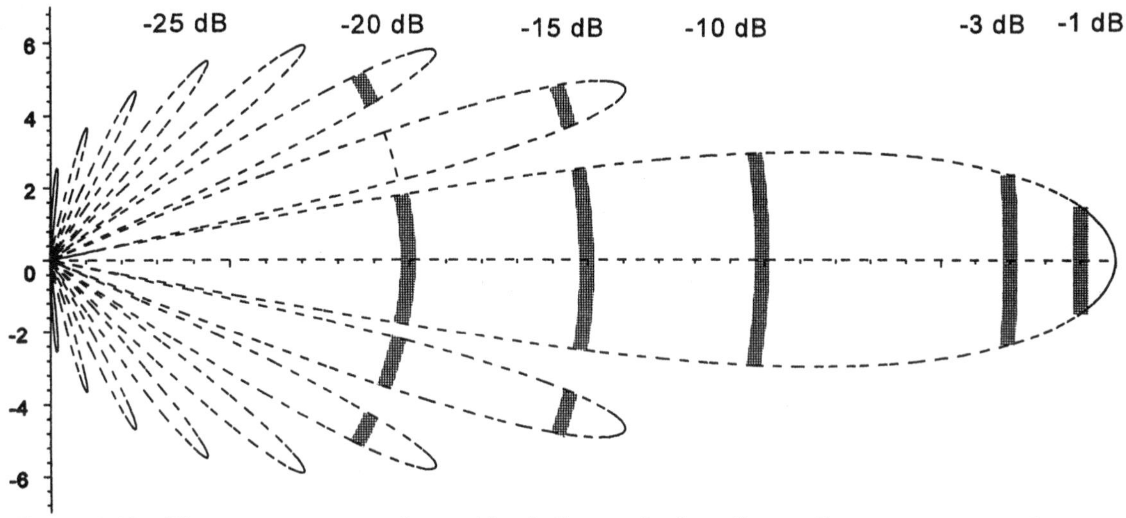

Figure 6.18 The appearance on a plan position indicator of echoes from reflectors strong enough to remain detectable in sidelobes of different gains, showing the extent of sidelobe responses.

6.3.1 Scattering without fading

Scattering without fading occurs with symmetrical shapes and direct reflectors whose radar cross-section may be calculated directly. Typically these are spheres and corner reflectors.

6.3.1.1 Polarization effects

According to optical theory, in order to stop an incident wave at a reflector, it is necessary to create a vector equal and opposite to the incident vector, and the result is zero. This created vector forms the reflected ray.

In the case of circularly polarized waves, the incident vector rotates on the surface of the reflector. A reaction vector is created that rotates with the incident vector but propagates in the opposite direction. Looking in the direction of propagation, the reaction vector in Figure 6.19 rotates in the opposite sense to give left-hand polarization. Thus, simple and odd numbers of reflections change the sense of circular polarization. Even numbers of reflections do not.

Flat and smoothly curved spherical reflectors reverse the sense of polarization. Reflectors with sharp edges and small cylinders reflect only the component parallel to themselves, and the reflected wave is the sum of two circularly polarized waves with opposite senses, which gives a loss. Reflections in corners (for example, wing roots) do not have their sense reversed. The generally accepted figure for the reduction of radar cross-section for aircraft when circular polarization is used is 50%, or 3 dB. For this reason circular polarization is used only when necessary.

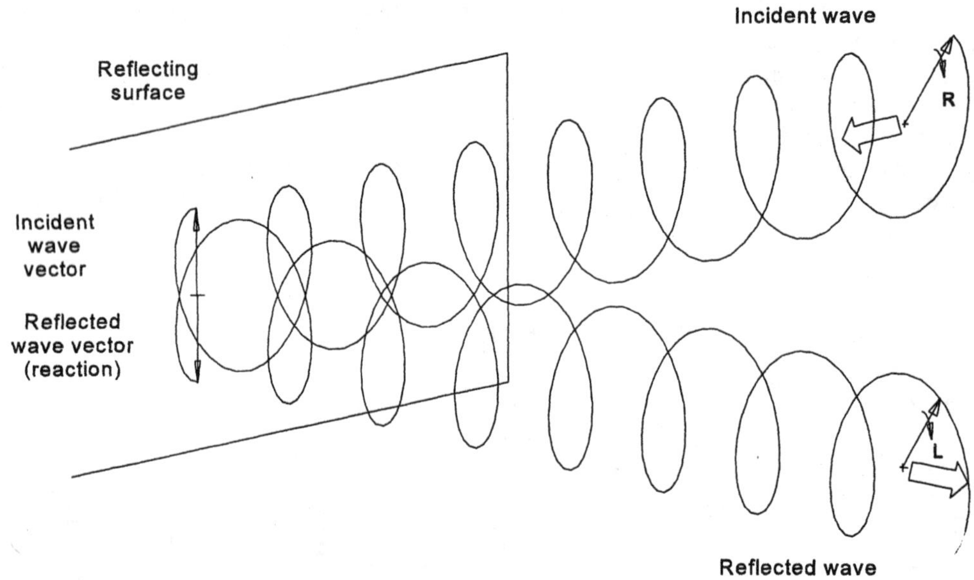

Figure 6.19 The reversal of the sense of reflected circularly polarized waves.

6.3.1.2 Sphere

The echo returned by a sphere can be obtained from [15]. The cross-section of a sphere radius, ρ, normalized to its projected area, $\pi\rho^2$, is given by

$$\text{Cross-section} = \frac{1}{\rho^2}\left(\left|\sum_{n=1}^{20}(-1)^n(2n+1)\Big(a(n,\rho)+b(n,\rho)\Big)\right|\right)^2 \tag{6.26}$$

where $a(n,\rho) = \dfrac{j_n(\rho)}{j_n(\rho) - y_n(\rho)}$ and $b(n,\rho) = \dfrac{-\dfrac{d}{d\rho}\,\rho\, j_n(\rho)}{\dfrac{d}{d\rho}\Big(\rho j_n(\rho) - j\rho y_n(\rho)\Big)}$

$j = \sqrt{-1}$;

$j_n(z)$ is a spherical Bessel function of the first kind order n and argument z [16];

$y_n(z)$ is a spherical Bessel function of the second kind order n and argument z.

For radii below 5λ, the radar cross-section oscillates in the Mie region and then tends to zero for radii below λ/10 in the Rayleigh region.

The radar cross-section for large spheres used for radar range verification is

$$\text{Optical region, } a > 4\lambda \text{ and } \sigma = \pi a^2 \tag{6.27}$$

where a is the radius of the sphere.

Table 6.7 shows a representative selection of spheres of different radar cross-sections or projected areas. There will be no problem with radars working at C-band (more than 4 GHz) and above. At lower frequencies, a sphere size must be chosen so that the radius is more than four times the wavelength and where this is not possible, the cross-section may be found from Figure 6.20.

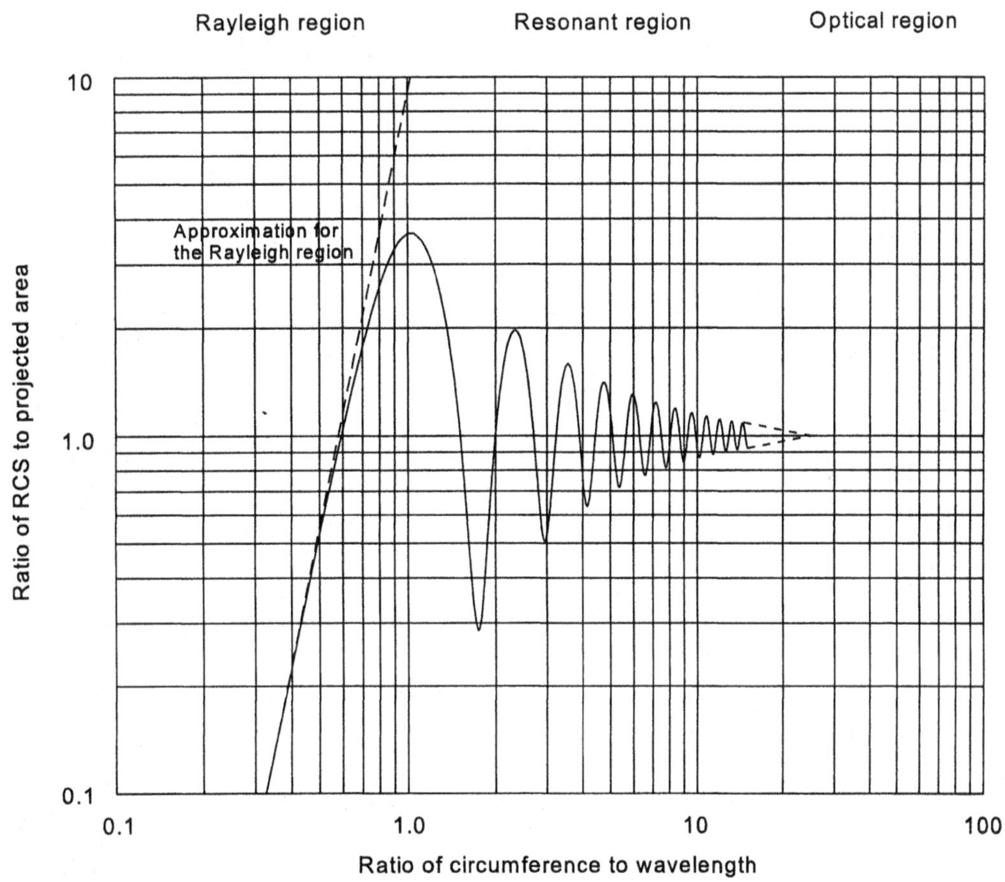

Figure 6.20 Cross-section of a sphere showing the Rayleigh, resonant, and optical regions.

Table 6.7
The limiting wavelengths and frequencies for a number of sizes of spheres

Projected area, m^2	Radius of sphere, m	Diameter of sphere, m	Limiting wavelength, m	Limiting frequency, GHz
0.5	0.3989	0.7978	0.0798	3.76
1	0.5642	1.1204	0.1128	2.66
2	0.7979	1.5958	0.1596	1.88
3	0.9772	1.9544	0.1954	1.53
5	1.2616	2.5232	0.2523	1.19
10	1.7841	3.5682	0.3568	0.84

A sphere provides a scatterer with a known, steady cross-section when viewed from any angle. The single reflection reverses the direction of circular polarization.

The radar cross-section of small spheres, such as rain in the Rayleigh region, is approximated by [3, p. 101]

$$\textit{Rayleigh region,} \quad a < \lambda/10, \quad \sigma = 9\left(\frac{2\pi a}{\lambda}\right)^4 \pi a^2 \tag{6.28}$$

where a is the radius of the sphere.

Typical values for perfectly conducting spheres of raindrop size are shown in Figure 6.21 for the popular surveillance radar bands. Small spheres also reflect circularly polarized waves in the opposite sense.

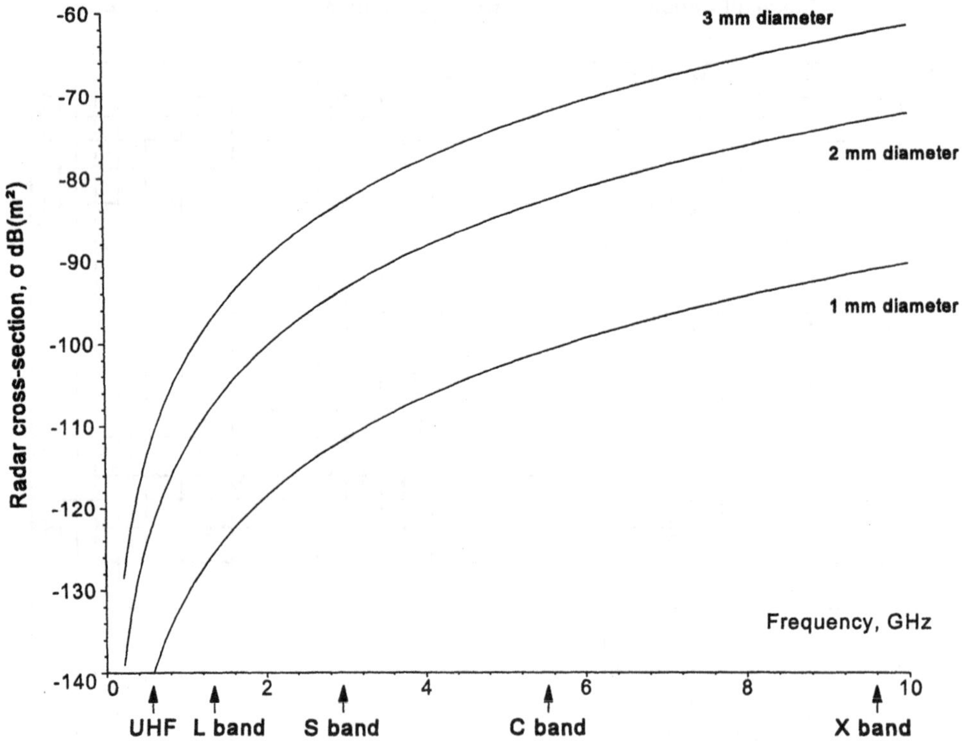

Figure 6.21 The radar cross-section of a small sphere with Rayleigh scattering.

6.3.1.3 Corner reflectors

Corner reflectors are used commonly to give reflecting points to show the navigation channel for barges under bridges. The common triangular form is shown in Figure 6.22. Each reflector is made of three right-angled triangles at 90 degrees to each other. Seven other reflectors, shown dotted, are added to give coverage over a sphere. The radar cross-section is the projected area as seen by the radar and is a maximum on the axis of symmetry. At a small angle δ off the angle of symmetry, the area is [3, p. 112]

$$\begin{aligned} \textit{Corner reflector projected area, } A &= \frac{h^2}{\sqrt{3}} (1 - 0.00076\ \delta^2) \\ \textit{Radar cross-section, } \sigma &= 4\pi \frac{A^2}{\lambda^2} \end{aligned} \tag{6.29}$$

where h is the height of one facet;
δ degrees is the small angle off the axis of symmetry.

The corner reflectors' main mode is triple reflection, which changes the sense of circular polarization.

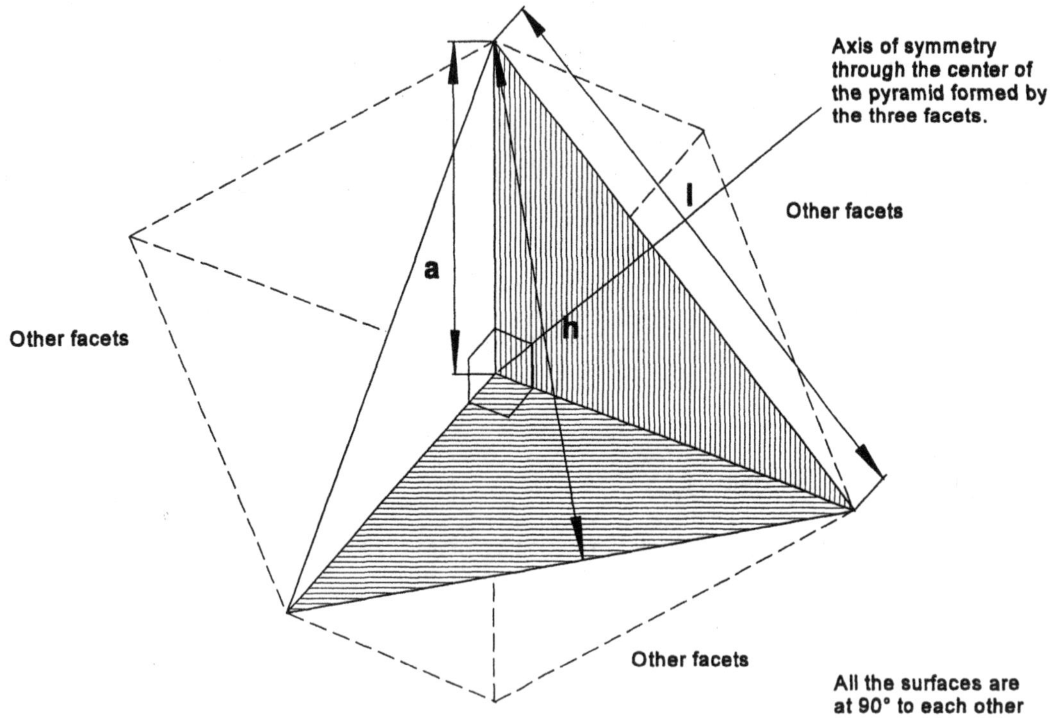

Figure 6.22 One of eight corner reflector facets.

6.3.2 Scattering with fluctuation

Most objects of interest, such as aircraft, ships, and many other types of irregular shapes, consist of groups of scattering facets that interfere with each other, and only statistical estimations are valid. These vary from spherical objects, which give a constant reflection, to an irregular cluster of equally sized reflectors. The reflections from such clusters are constant during the short radar pulse length. If the radar frequency changes, there are different interference effects that depend on the probability distribution of the scatterer. This is the same distribution as if the object were rotated when measured at a constant frequency.

Steady echoes are the easiest to detect and measure. Fluctuating echoes must be observed during a number of fading cycles to obtain a reasonable mean value for (detection and) measurement.

A number of probability distributions have been postulated (Table 6.8) with decreasing degrees of dispersion.

Table 6.8
A number of probability distributions used to describe scatterers

Distribution	Used for
Log-normal	Ground clutter, ships
Weinstock	
Swerling I and II models	Aircraft
Swerling III and IV models	Missiles
Rice	
Uniform for steady or Marcum model	Towed spheres

The radar equation uses the mean radar cross-section to calculate the average echoed signal. Except for the log-normal case, the probability distributions are related to gamma distributions (see Chapter 15, Statistics).

To illustrate how the typical shapes of the cross-section characteristics vary with angle, diagrams have been drawn using 200 random points taken from the chosen distributions. These random numbers give rise to spiked "daisies", with the white center corresponding to the steady portion of the echo. The log-normal daisies have been drawn with logarithmic scales; the others with linear scales.

In measured characteristics, larger and smaller values tend to be clustered around certain angles. Moving the spikes around to simulate measured characteristics does not change the statistical distribution *over all angles*.

6.3.2.1 Log-normal models

Where the mean-to-median ratio is small, the log-normal distribution describes a large reflector with a number of small reflecting facets, such as a spherical satellite. Large mean-to-median ratios are used to describe ground clutter where the mean value is determined by very few high valued points. The log-normal distribution is given by

$$\text{Log-normal} \quad p(x) = \frac{1}{x\sigma\sqrt{2\pi}} \exp\left(-\frac{1}{2}\left(\frac{\ln(x)-x_{median}}{\sigma}\right)^2\right) \tag{6.30}$$

where σ is the standard deviation;
the mean-to-median ratio is $\exp(\sigma^2/2)$.

If a logarithmic abscissa is used, the log-normal distribution becomes Gaussian in shape and more manageable, and commonly decibel scales are used. The standard deviation of this form is

$$\sigma_{db} = \sqrt{\frac{20}{\ln(10)}\,(\text{mean-to-median ratio dB})} \tag{6.31}$$

Equation (6.31) gives the standard deviations for a small satellite with a mean-to-median ratio of 10 [17, p. 176] as 9.31 dB and for Barton model clutter with a mean-to-median ratio of 100 as 13.18 dB.

An idea of how log-normal scatterers appear in practice is shown in Figure 6.23. Actual scatterers may have their spikes arranged differently but this does not change the distribution.

The examples shown represent a small spherical satellite with protrusions seen at various angles and Barton model clutter, if it were possible to rotate its components. Note that *the scales are in decibels.*

The use of logarithmic scales clearly shows the Gaussian distributions on the logarithmic abscissae centered on the median. Remember that the true values are 10 to the power $x/10$, where x is the value in Figure 6.23. The calculation of the mean is dominated by the few spikes that exceed the mean. The other "daisies" are shown with linear scales.

6.3.2.2 Derivatives of the gamma distribution, the χ^2 fluctuation model

These distributions are gamma distributions and are given by [18, p. 122]

$$\text{Gamma distribution} \quad p(x) = \frac{\lambda^{\eta}}{\Gamma(\eta)} x^{\eta-1} \exp(-\lambda x) \tag{6.32}$$

where η is the shape parameter;
λ is the scale parameter.

The mean and the standard deviation are given by

$$\begin{aligned} \text{Gamma distribution mean} &= \frac{\eta}{\lambda} \\ \text{Standard deviation} &= \frac{\sqrt{\eta}}{\lambda} \end{aligned} \tag{6.33}$$

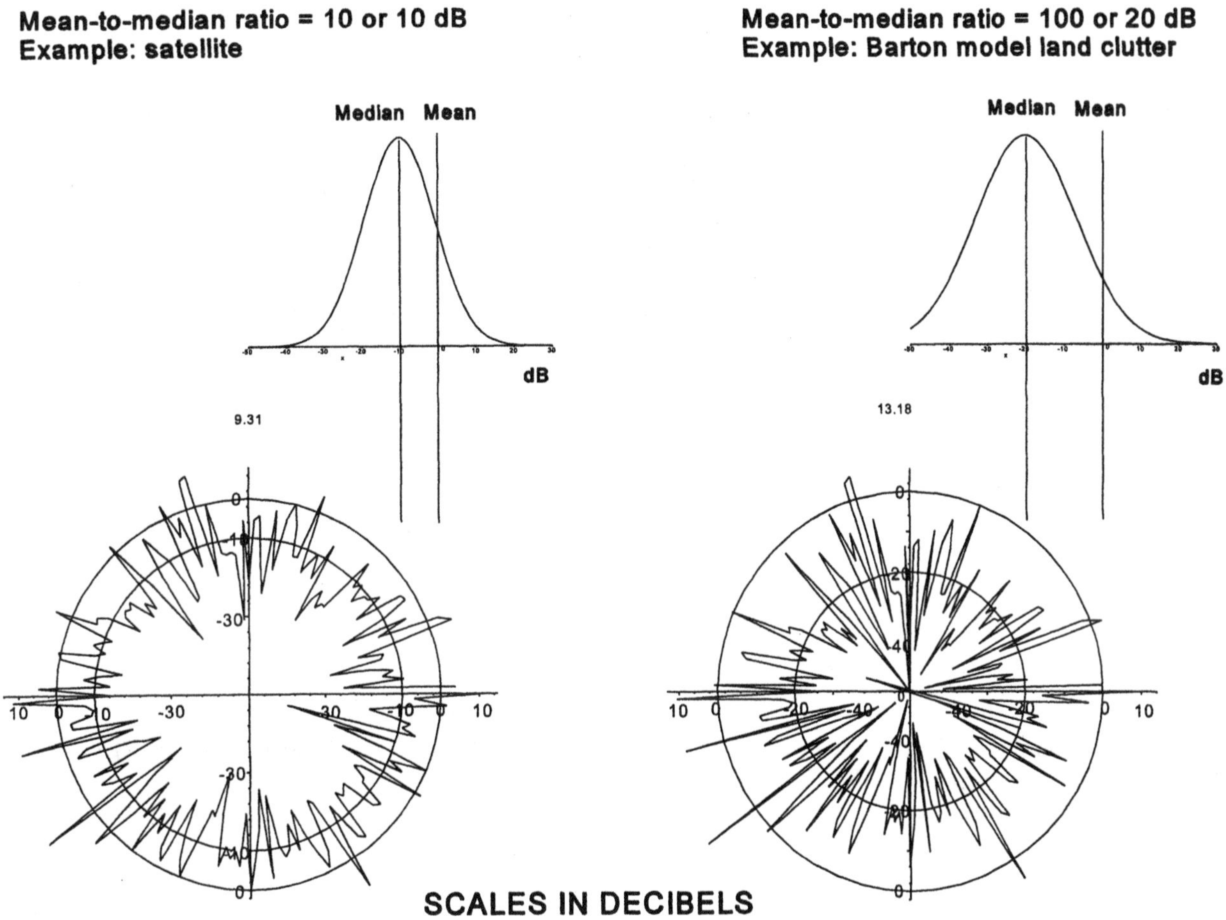

Figure 6.23 Probability distributions and simulated scattering for two examples with log-normal distributions.

The χ^2 (chi-squared) distribution is a special case of the gamma distribution where the shape parameter η is replaced by $\nu/2$, where ν is the number of degrees of freedom. The scale parameter is fixed as $\lambda = 1/2$.

$$\chi^2 \textit{ distribution} \quad p(x) = \frac{\left(\frac{1}{2}\right)^{\frac{\nu}{2}}}{\Gamma\left(\frac{\nu}{2}\right)} x^{\frac{\nu}{2}-1} e^{-\frac{x}{2}} \qquad (6.34)$$

where ν is a whole number of degrees of freedom.

The mean and the standard deviation are given by

$$\begin{aligned} \chi^2 \textit{ distribution mean} &= \nu \\ \textit{Standard deviation} &= \sqrt{2\nu} \end{aligned} \qquad (6.35)$$

In the past, the choice of statistical distribution was driven by which computed tables were available, for example, the χ^2 distribution. The term "duodegrees" of freedom was used for even values of ν. If modern mathematical software is used, it accepts only integer values of ν. The gamma distribution, from which it is derived, does not have this limitation. To maintain the integral from zero to infinity and the mean at unity, set $\eta = \lambda$.

$$\textit{Modified gamma distribution} \quad p(x) = \frac{\eta^{\eta}}{\Gamma(\eta)} x^{\eta - 1} e^{-\eta x} \tag{6.36}$$

Representative distributions are shown in Figure 6.24.

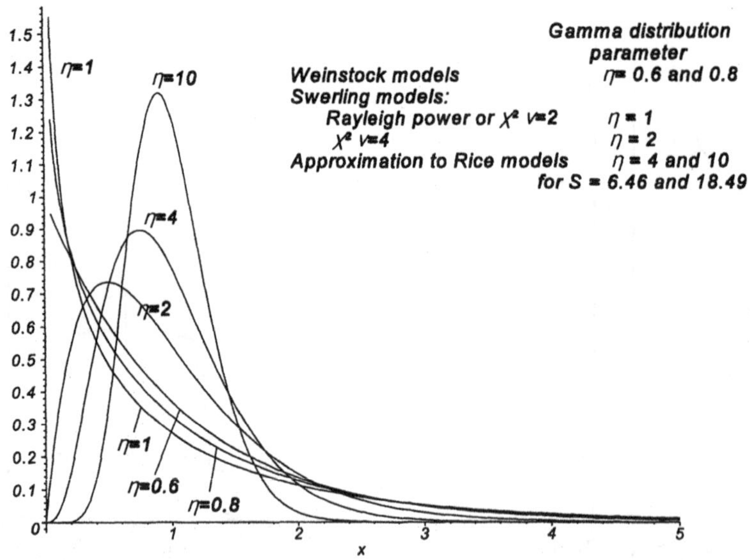

Figure 6.24 Examples of the modified gamma distribution for Weinstock, Swerling, and χ^2 models with a common mean of unity.

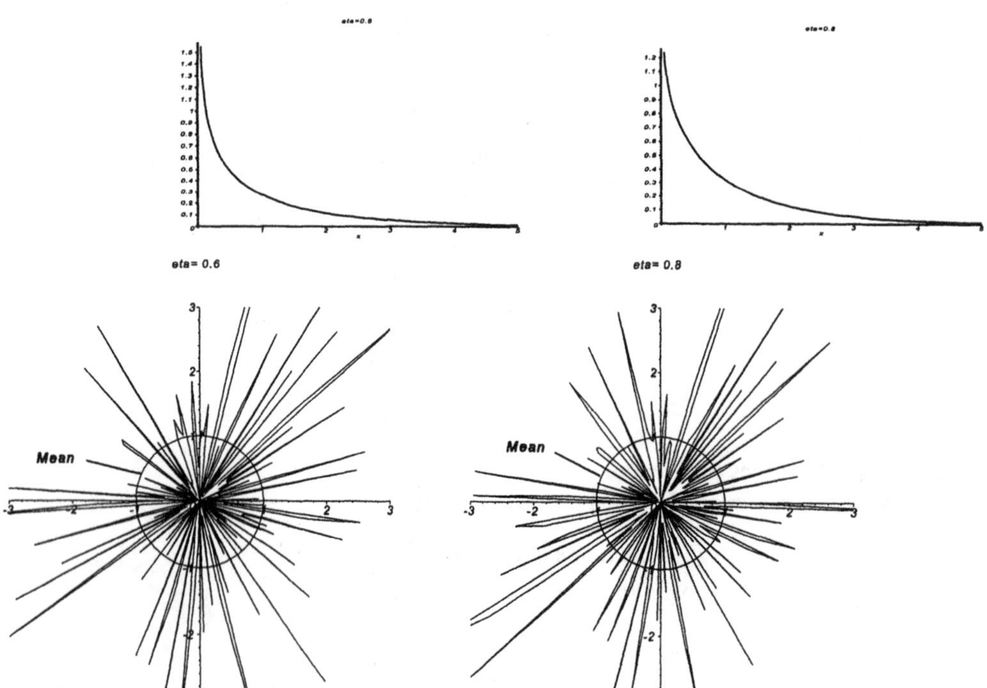

Figure 6.25 Probability distributions and simulated scattering characteristics for two Weinstock models, $\eta = 0.6$ and $\eta = 0.8$.

6.3.2.3 Weinstock models

The Weinstock model uses the modified gamma distribution with η between 0.6 and 4 (see Figure 6.24). A simulated scatter polar diagram is shown in Figure 6.25.

6.3.2.4 Swerling models, cases I and II

When a radar beam illuminates a volume containing a large number of scatterers, the number illuminated depends on either the solid angle of the beam (examples: clouds or rain) or the size of the group of scatterers (an aircraft). Generally the range extent of the transmitted pulse is a large number of wavelengths, that is, the phases of the individual echoes of a similar size cover all possible angles and are uniformly distributed. This is the classic form of the random walk described by Lord Rayleigh [19, p. 56] who first discussed the problem of "a drunkard who took one pace and then fell down, stood up, took another pace in a different direction, and fell down once again."

An example of such an excursion with nine 10 inch steps is shown in Figure 6.26.

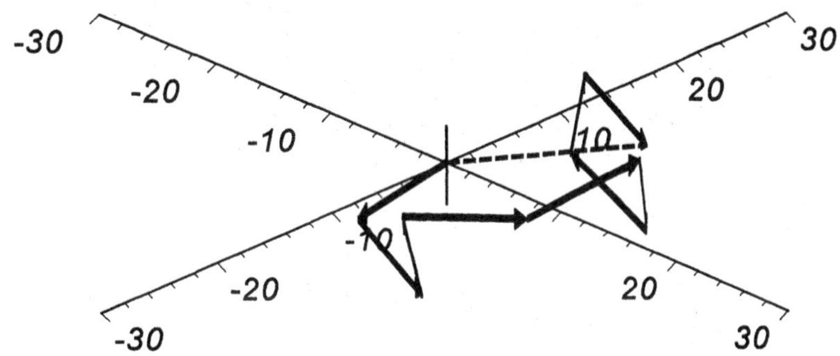

Figure 6.26 The result of summing the random vectors or random walk.

The random walk is considered, classically [20], in two orthogonal components each along a line from east to west and from south to north. There is an equal chance (p = ½) that the first and following steps will be either to the east or to the west and the range of the end of the walk of N equal steps is between $-N$ and $+N$. The probability of a particular sequence of N steps is $(½)^N$. To arrive at m, on an average $(N + m)/2$ steps must have been taken to the east and $(N - m)/2$ to the west, where N and m may be either both even or both odd so that the number of possibilities (combinations) of arriving at a point m after N steps is

$$\text{Number of combinations} = \frac{N!}{\frac{(N+m)}{2}!\ \frac{(N-m)}{2}!} \tag{6.37}$$

so that the probability of arriving at m is

$$\mathrm{p}(m) = \frac{N!}{\frac{(N+m)}{2}!\ \frac{(N-m)}{2}!}\left(\frac{1}{2}\right)^N \tag{6.38}$$

This binomial or Bernoulli distribution has a mean of zero and a variance of N. Using Stirling's formula for $N!$, for large N the binomial distribution may be represented by the Gaussian distribution

$$\mathrm{p}(m) = \sqrt{\frac{2}{\pi N}}\exp\left(-\frac{m^2}{2N}\right) \tag{6.39}$$

Extending this to two dimensions, x and y, the joint probability is

$$p(x,y) = \frac{2}{\pi N} \exp\left(-\frac{x^2+y^2}{2N} \right) \tag{6.40}$$

This is a Rayleigh distribution with the orthogonal component standard deviation of $\sqrt{N/2}$ giving a radial standard deviation of $\sqrt{N}$ steps giving the mean distance from the origin after N steps.

As an illustration, if the steps have an average length of 10 inches with a Gaussian standard deviation of unity, the distribution after the first step is shown in Figure 6.27.

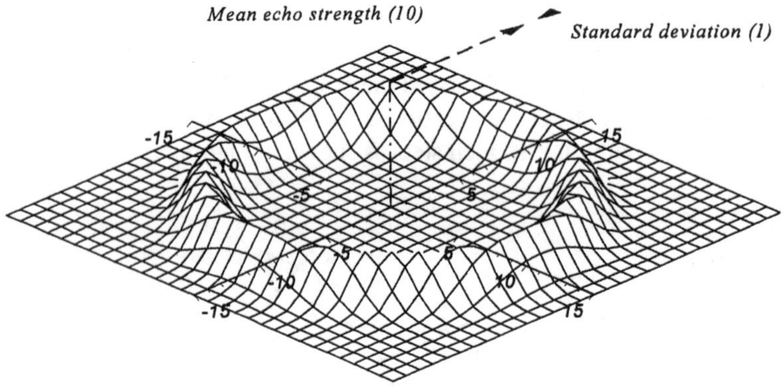

Figure 6.27 The distribution of the position after the first step.

Lord Rayleigh's drunkard will stand up again in Figure 6.27 and his second step will start at the rim in Figure 6.27 and where he lands is distributed as shown in Figure 6.28.

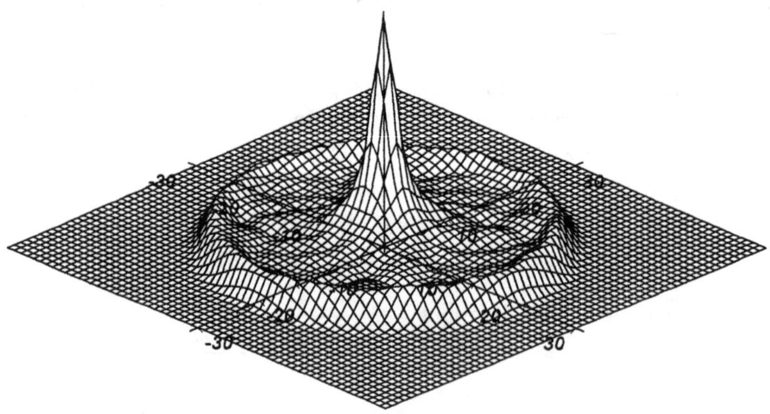

Figure 6.28 The distribution of the second step of a random walk.

For his second step he has a one-third chance of going within the circle of radius 10 inches, and two-thirds chance of stepping outside it. Stepping inside the circle builds up the center and stepping outside it builds up a ring of radius 20 inches. The features of distribution of the landing points after the third step will reinforce, rather diffusely, the ring at 10 inches both from the center and from the 20 inch ring. A ring of smaller amplitude will appear at 30 inches.

After an infinite number of steps, the accumulated probabilities will have a bivariate Gaussian distribution. The *law of large numbers* states that the sum of an infinite number of samples, irrespective of how they are distributed individually, will have a Gaussian distribution so the sums of the vectors with the same amplitude but random direction will have the same distribution. Figure 6.29 shows the distribution of where Lord Rayleigh's drunk lands after the nine steps in Figure 6.26.

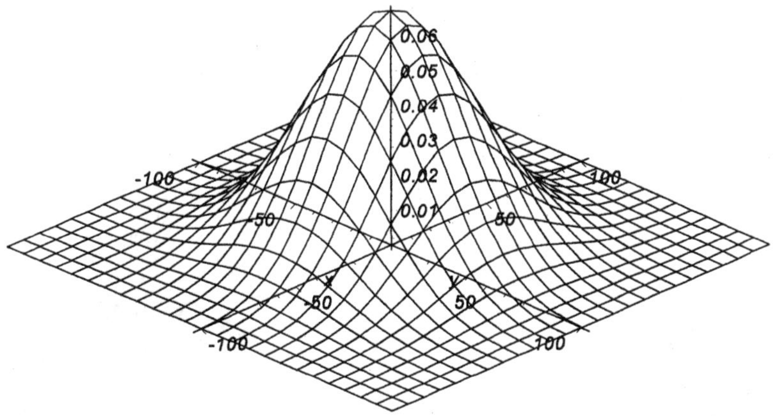

Figure 6.29 The distribution after nine steps.

The Swerling models, cases I and II, assume that the voltages of the echoes from the scatterer are parts of a Rayleigh distribution where $x^2 + y^2$ in (6.40) is replaced by R^2.

$$\text{Swerling I and II, Rayleigh voltage distribution,} \quad p(R) = \frac{R}{\sigma^2}\, e^{-\frac{R^2}{2\sigma^2}} \tag{6.41}$$

This distribution is the same as a χ^2 distribution with 2 degrees of freedom.

If $\frac{x}{\bar{x}} = \frac{R^2}{2\sigma^2}$, then the distribution function for the power received from the echoes is exponential:

$$\text{Swerling I and II, Rayleigh power, or exponential, distribution,} \quad p(x) = \frac{x}{\bar{x}}\, e^{-\frac{x}{\bar{x}}} \tag{6.42}$$

This distribution is the Rayleigh distribution in Figure 6.24. Reference [17, p. 169] states that experimental data from large aircraft fit this model.

The cross-section in square meters seen by the radar is a random distribution in three dimensions with examples shown in Figure 6.30. Such diagrams look like similar sea urchins, so that in the discussion that follows a slice through the center is taken to give a better idea of what the distribution looks like.

6.3.2.5 Swerling models, cases III and IV

The Swerling models III and IV, are defined as a χ^2 distribution with four degrees of freedom. The power distribution is

$$\text{Swerling III and IV,} \quad \chi^2_4 \text{ power distribution,} \quad p(x) = \frac{4x}{\bar{x}^2}\, e^{-\frac{2x}{\bar{x}}} \tag{6.43}$$

This distribution, too, is shown in Figure 6.24 where $\eta = 2$. Simulated scattering characteristics are shown in Figure 6.31.

6.3.2.6 Rice models

The echoes from one large scatterer accompanied by several much smaller ones, such as a spherical satellite with antennas, are represented by the Rice power distribution [3, p. 75], namely,

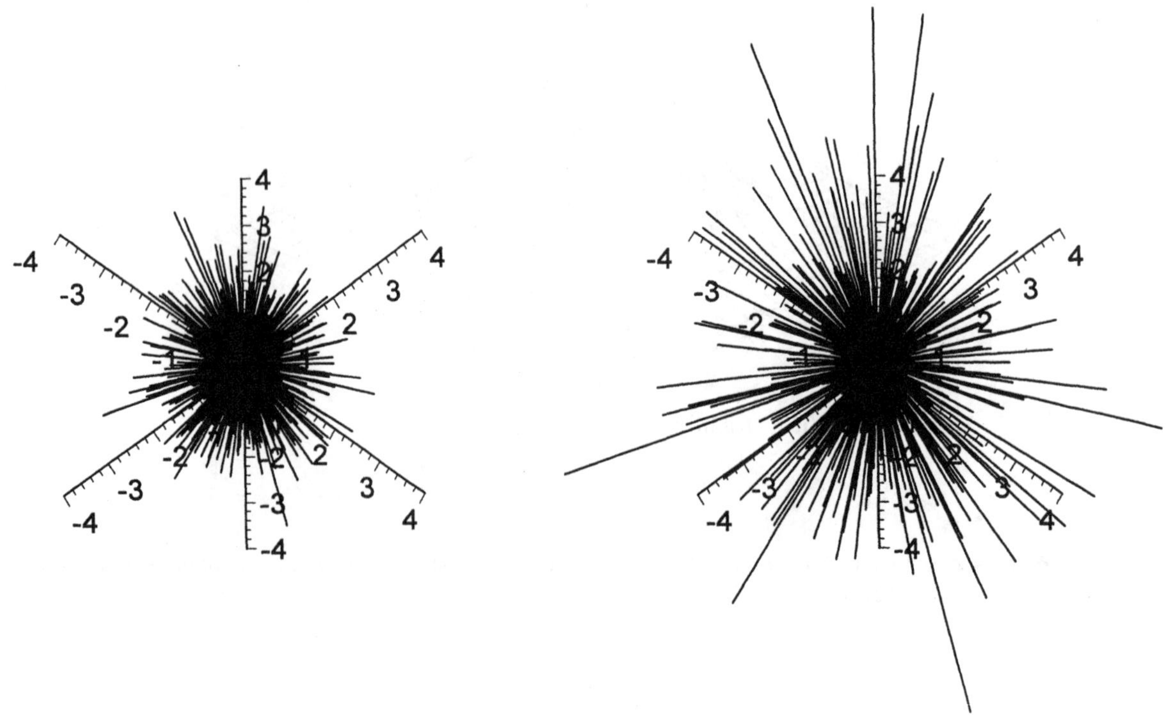

(a) Rayleigh distributed scattering

(b) Exponentially distributed scattering

Figure 6.30 Rayleigh and exponential scattering in three dimensions.

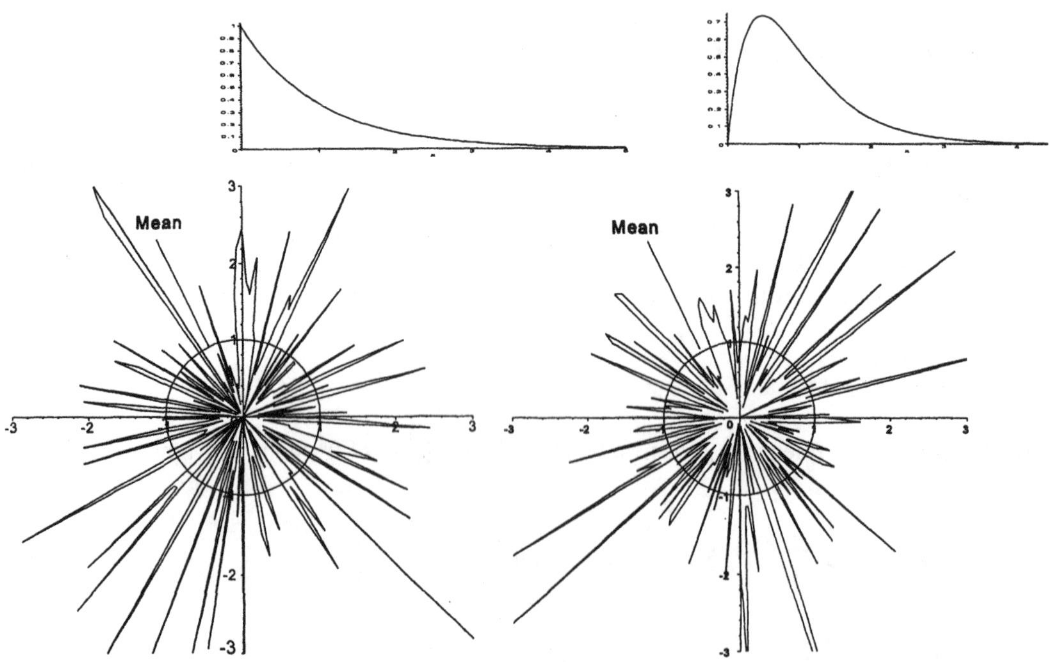

Swerling cases I and II, modified gamma or Rayleigh power distribution η=1

Swerling cases III and IV, modified gamma (η=2) or χ² (ν=4) distribution

Figure 6.31 Probability distributions and simulated scattering characteristics for the two types of Swerling models I and II $\eta = 1$ and III and IV $\eta = 2$.

$$p(x) = (1 + S)\exp(-S - (1 + S)x)\, I_0(2\sqrt{S(1 + S)x}) \tag{6.44}$$

where S is the ratio of the power of the steady component to the total power in the random components;
I_0 is the modified Bessel function of zero order.

The Rice distribution may be approximated to the modified gamma distribution by equating the means and variances of the two distributions [3, p. 76], giving

$$\eta = 1 + \frac{S^2}{1 + 2S} \tag{6.45}$$

An example of the approximation (6.45) is shown in Figure 6.32.

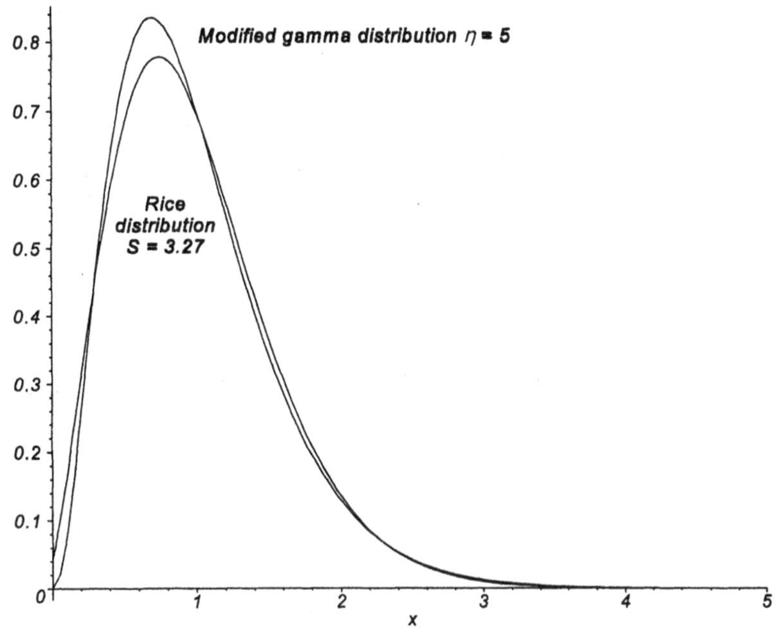

Figure 6.32 The Rice distribution with $S = 3.27$ compared to the gamma distribution approximation.

The "daisy" diagram is shown for $S = 6.46$ and $S = 18.49$ in Figure 6.33.

6.3.2.7 Slow and fast fluctuation for detection

Scanning radars look at each part of their environment normally at set times. Neglecting the effects of the antenna pattern, the amplitudes of the echoes are samples from the probability distributions of the radar cross-sections of each scatterer. Slow and fast fluctuation are defined in Table 6.9.

If the diagram in Figure 6.30 is turned, it will twinkle like a star, called scintillation. With slow scintillation, the radar cross-section does not change during the time the radar is looking at the scatterer of interest, and the echo signals have the same amplitude. Slow scintillation happens when the radar uses a constant radio frequency. If the radar is frequency agile, then each frequency used gives rise to a new random distribution of the same shape so over a number of pulses the sum approaches the average. This is fast scintillation.

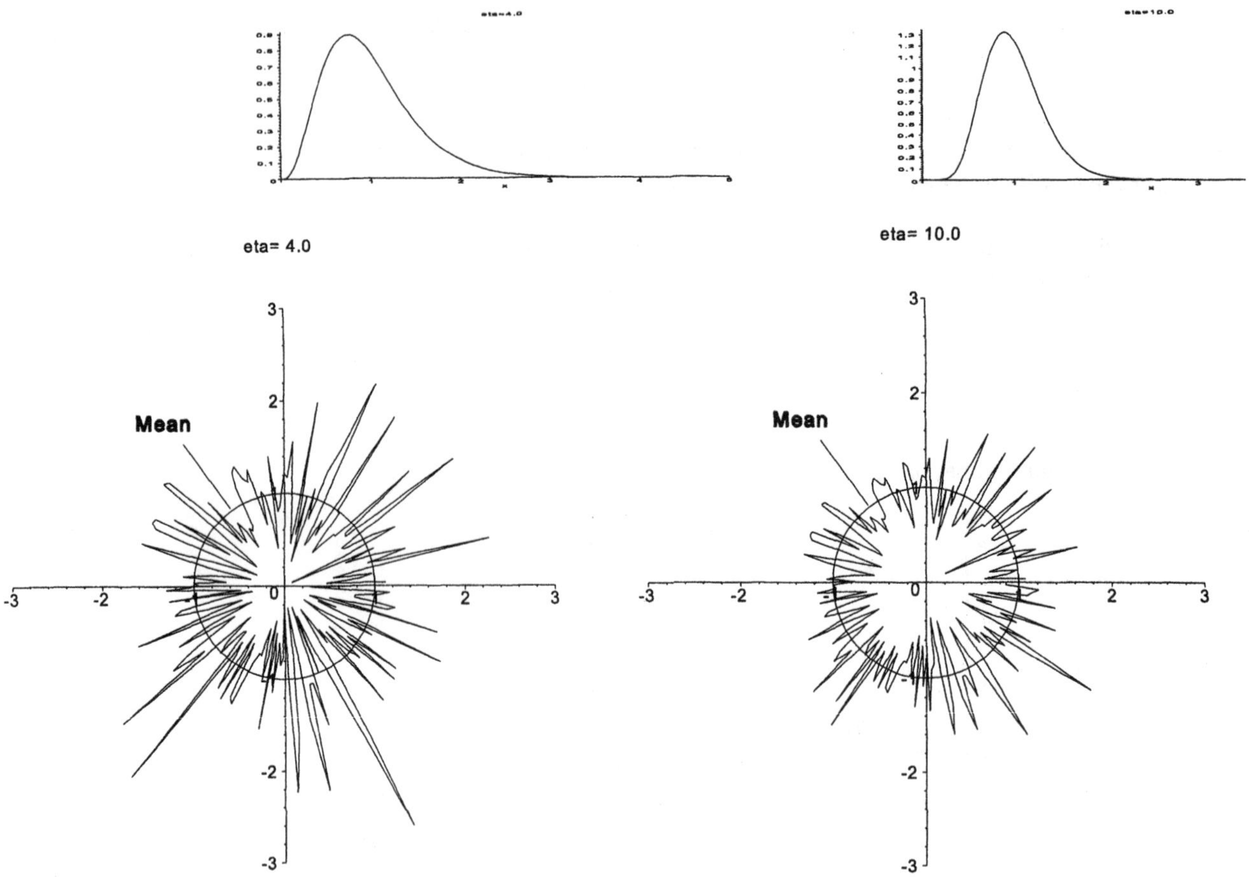

Figure 6.33 Probability distributions and simulated scattering characteristics for approximated Rice models $\eta = 4$ (Rice $S = 6.46$) and $\eta = 10$ (Rice $S = 18.49$).

Table 6.9
Slow and fast fluctuation

	Description
Slow fluctuation	The echoes have the same amplitude and are correlated during a look, that is, the same sample from the distribution occurs each pulse. If the scatterer moves partial decorrelation can occur as in Sections 6.3.2.8 to 6.3.2.10. The echo amplitudes between looks are decorrelated and are normally different.
Fast fluctuation	The echoes have different amplitudes during a look. That is, a number of independent samples from the distribution taken during a look is equal to the number of pulses per look.

A particular case of slow correlation is in moving target detector radars which use pulse repetition frequency diversity to cover clutter alias speeds (or blind speeds in moving target indicators). Neglecting the antenna pattern effects and number of pulses integrated, the amplitudes of the echoes and the probability of detection are the same for each burst, if both are near the maximum of a Doppler frequency characteristic. The centroid of the echoes during one look thus gives a good estimate of the scatterer's position.

Decorrelation of the returning echoes occurs with a change in the wavelength of the illuminating radar pulse, its polarization, or the aspect of the scatterer. Figure 6.34 shows the scattering pattern from a line scatterer.

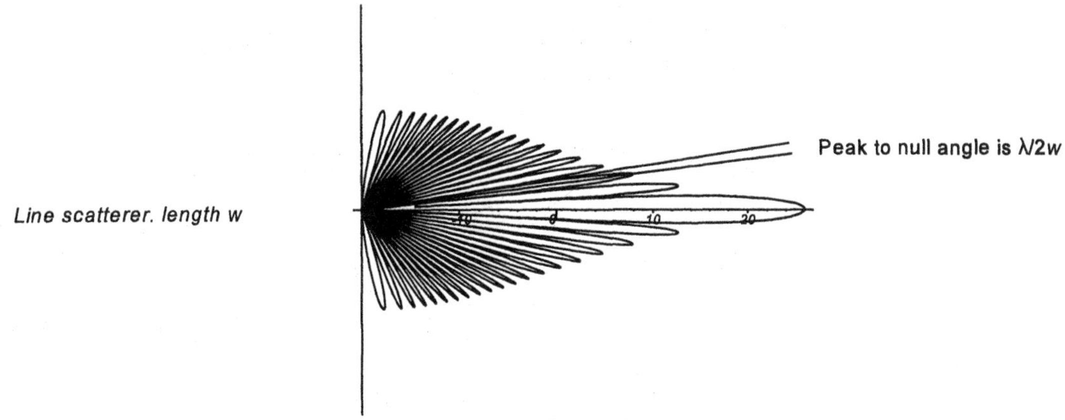

Figure 6.34 The pattern from a line scatterer.

For a uniform line scatterer, length w, the angle between a sidelobe peak and its neighboring nulls is $\lambda/2w$ for small angles. The wavelength or frequency for decorrelation may be given by

$$\begin{aligned} &\textit{Wavelength change for decorrelation} > \lambda_c = \frac{\lambda}{2\,w} \quad \text{m} \\ &\textit{Frequency change for decorrelation} > f_c = \frac{c}{2\,w} \quad \text{Hz} \end{aligned} \tag{6.46}$$

where c is the velocity of light.

Other authors use the range extent, L_r, instead of the width, w, to cause a different mix of interfering waves and resulting echoes [21, p. 86].

As an example for an aircraft covering a distance of 10m, this frequency is 15 MHz. For the jumbo jet in Figure 6.29, which is much larger, the frequency for decorrelation is less.

6.3.2.8 Fluctuation during position measurement

Whereas most surveillance radars determine the position of a scatterer in one pass, a number of radars use many independent primitive looks, and turning (Figures 6.34 and 6.35) changes the mix of interference effects. In order that the echoes remain the same during the primitive looks, the object of interest may not turn appreciably during this time, the correlation time, t_c. The decorrelation time for rigid objects is [21, p. 86]

$$\textit{Time for decorrelation, } t_c = \frac{\lambda}{2\omega_\alpha\, w} \quad \text{seconds} \tag{6.47}$$

where ω_α is the rate of rotation (radians/s) of the scatterer about the line of sight;
w is the extent of the scatterer normal to the line of sight and axis of rotation (m).

For an aircraft, extent 10m, turning through 180 degrees or π radians in one minute, the time for decorrelation for a radar in S-band ($\lambda = 0.1$ m) is 95 ms.

When the scatterer turns the different interference effects cause a change of position of the center of the scatterer, called glint. Quoting the IEEE standard [1]:

Glint: The inherent component of error in measurement of position and Doppler frequency of a complex target due to interference of the reflections from different elements of the target. Glint may have peak values beyond the target extent in the measured coordinate.

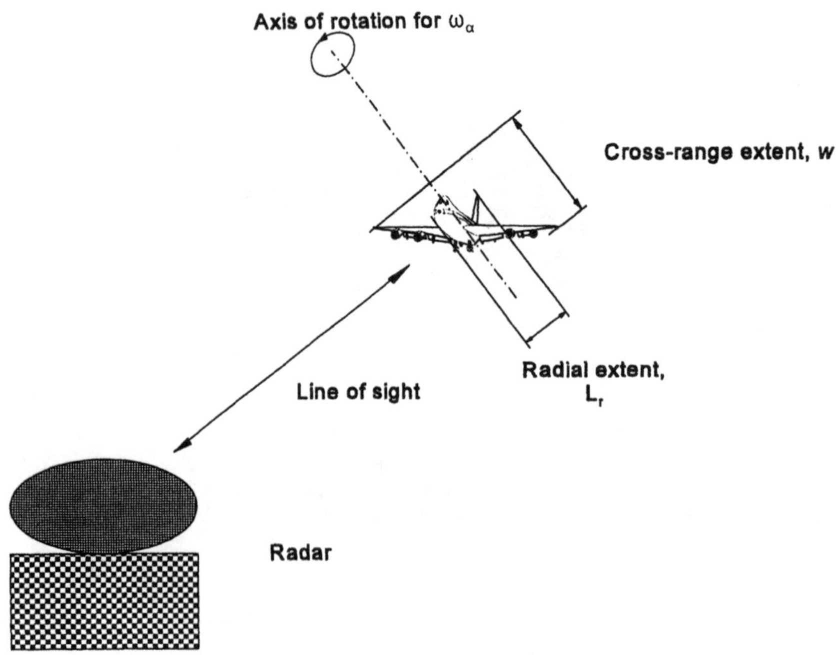

Figure 6.35 The radial and cross-range extents of a scattering aircraft.

6.3.2.9 Fluctuation during incoherent integration

Integration is used to add a number of echo video signals to give a better mean and improve the signal-to-noise ratio. The echoes remain the same during the correlation time, t_c, but the noise does not. Thus the number of independent samples for integration during the integration time, t_i, is

$$\text{Number of independent samples for integration, } n_e = 1 + \frac{t_0}{t_c} \tag{6.48}$$

6.3.2.10 Fluctuation during tracking or scintillation

After the aircraft or other object of interest has been detected, its position is monitored either continuously (tracking radars) or at each scan of the antenna (track-while-scan radars). In both cases, the complex object presents a changing aspect that affects the mix of primitive echoes that interfere with each other. The combination gives a different angle on the "daisy" diagrams and radar cross-section, an effect called scintillation.

With radars that track automatically, there may be short periods with bad signal-to-noise ratio. The automatic tracking algorithm must contain means to smooth the data during these times, called coasting. The signals in track-while-scan radars must exceed the detection threshold each time to be able to update the position information. Again the tracking algorithm must estimate the missing position data. The IEEE standard definitions are [1]:

- Scintillation: Variations in the signal received from a complex target due to changes in aspect angle or other causes;
- Scintillation error: Error in radar-derived target position or Doppler frequency caused by interaction of the scintillation spectrum with frequencies used in sequential measurement techniques;

6.4 SECOND-TIME-AROUND EFFECT

Low pulse repetition frequency pulse radars are designed so that the echoes from objects at the nominal maximum range have returned before the next transmitter pulse is sent. The time between the last echo and the next transmitter pulse is called the dead time and is often used for test or calibration waveforms for either automatic or manual procedures.

Plains surrounded by mountains may have clutter out to 100 nautical miles (185 km); ducting (see Section 6.2.5) may extend the clutter range to 200 or more nautical miles. If the radar has a smaller maximum range, echoes from the previous transmitter pulse may arrive to compete with shorter range echoes, which may cause false alarms.

With medium and high pulse repetition frequency waveforms, used to increase the number of pulses used for frequency filtering, multiple-time-around effects are the rule.

6.5 GROUND REFLECTIONS AND MULTIPLE PATHS

For an antenna on a tower less than 300 m, the flat earth model can be used. Generally, the land surrounding a ground radar is not flat but lightly rolling and must be surveyed to calculate a more exact diagram. The following gives an introduction to the effects of ground reflections.

6.5.1 Flat or nearly flat earth

Figure 6.36 shows the direct ray, R_d, and the ray, R_l, reflected by a flat earth fictionally from the image of the antenna, developed from [3, p. 328] and [7, p. 128].

The signal fluctuation at the antenna is caused by a difference of phase between the echo that returns directly and that which has been reflected. The path difference is given by

$$\delta = R_1 + R_2 - R_d \tag{6.49}$$

If the scatterer is at a large distance, that is $h >> h_{radar}$ and $R_d >> h_{radar}$, then the direct and reflected rays may be considered to be parallel, or $\theta_d = \theta_r$:

$$\delta = 2\, h_{radar} \sin\theta \tag{6.50}$$

The phase shift α is composed of the sum of the excess path length and the phase change caused by reflection, ϕ:

$$\gamma = \frac{2\pi\delta}{\lambda} + \phi \tag{6.51}$$

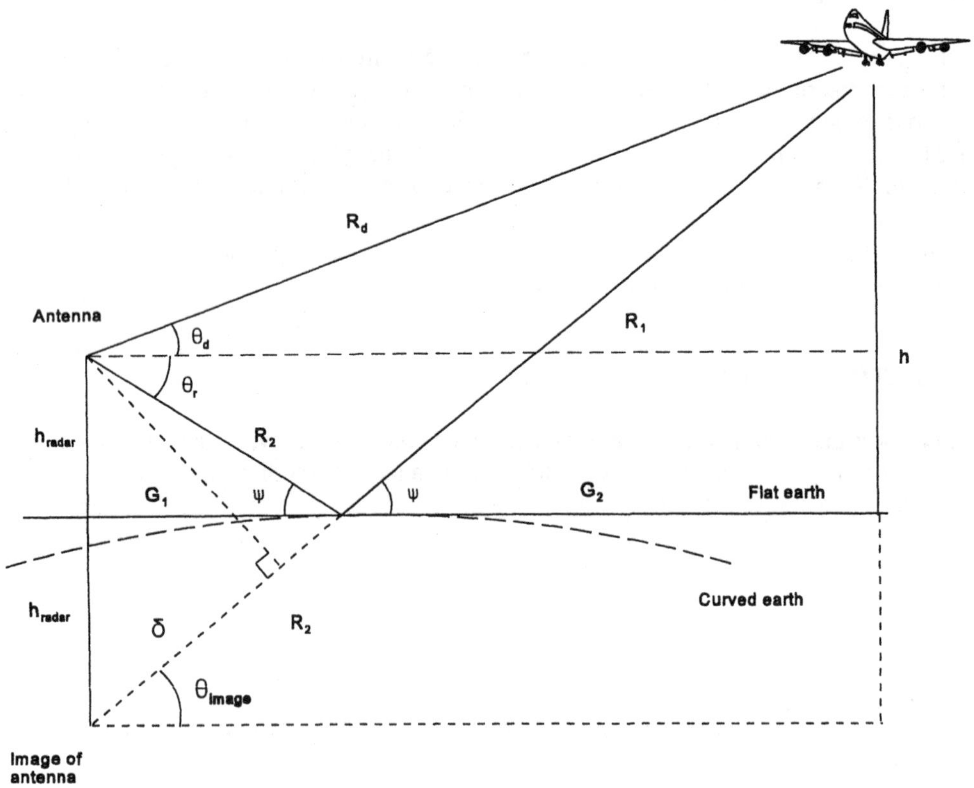

Figure 6.36 Direct and reflected rays over a flat earth.

The vector sum of the direct and reflected voltages, the propagation factor, is shown in Figure 6.37.

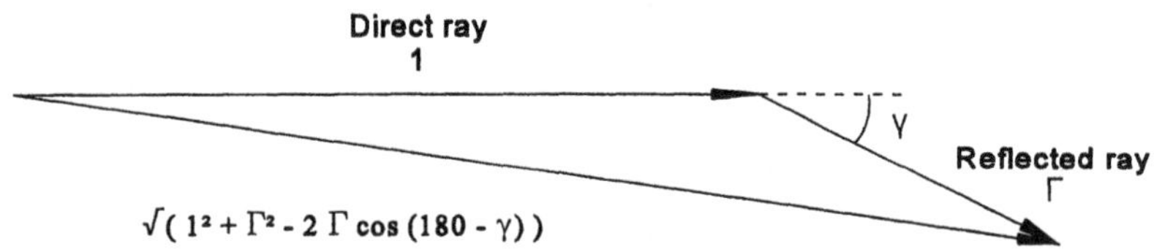

Figure 6.37 The sum of the direct and reflected rays at the antenna.

$$F = \sqrt{1^2 + \Gamma^2 - 2\ \Gamma\cos(180°-\gamma)} = \sqrt{1^2 + \Gamma^2 + 2\ \Gamma\cos(\gamma)} \tag{6.52}$$

where Γ is the complex reflection coefficient given, for horizontal polarization, in (6.60) and for vertical polarization in (6.61).

The vector sum F is the voltage gain of the antenna used to transmit and to receive. The power gain, one way, is F^2, and the two-way factor for the radar range equation is F^4. This is the same relationship as between signal strength and range, thus range is proportional to F.

6.5.1.1 Special case, sea water, $\Gamma = 1\angle 180$ degrees

Ships roll and pitch, so that an antenna must have a beam that encompasses higher elevations for aircraft, if necessary, and extends below zero degrees of elevation if it is not stabilized, as shown in Figure 6.38.

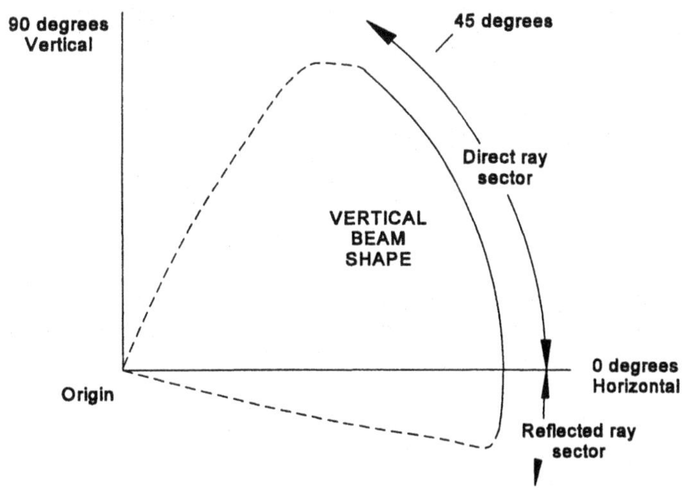

Figure 6.38 Vertical (coverage) diagram of a fan beam antenna.

If the amplitudes of the direct and reflected rays are equal and the phase of the reflection coefficient ϕ is 180 degrees, or π radians (γ becomes $-\gamma$), for horizontal polarization over sea water the vector sum of the direct and reflected signals is

$$F_1 = \sqrt{1^2 + 1^2 - 2\cos\gamma} = 2\left|\sin\frac{\gamma}{2}\right| \tag{6.53}$$

Substituting (6.51) into (6.53)

$$F_1 = 2\left|\sin\left(\frac{\pi}{\lambda}(2h_{radar}\sin\theta_d)\right)\right| \tag{6.54}$$

The peaks occur when F_1 is 2 and the nulls when F_l is zero. That is,

$$\textit{Maxima occur at}\ \arcsin\left(\frac{(2n-1)\lambda}{4h_{radar}}\right) \qquad n = 1, 2, 3, \ldots \tag{6.55}$$

$$\textit{Minima occur at}\ \arcsin\left(\frac{(n-1)\lambda}{2h_{radar}}\right) \qquad n = 1, 2, 3, \ldots \tag{6.56}$$

The number of lobes or fingers depends on the wavelength, λ m, and the antenna height, h_{radar}, and is independent of polarization. Taking the horizontal polarization over sea water for simplicity, from (6.55)

$$\frac{2n-1}{\lambda} = 4h_{radar} \tag{6.57}$$

Then

$$\textit{Number of lobes},\ n = \frac{1}{2}\left(\frac{4h_{radar}}{\lambda} + 1\right) \approx \frac{2h_{radar}}{\lambda} \quad \text{for large } n \tag{6.58}$$

For example, an L-band radar with an antenna 6 m above the ground has 52 vertical lobes.

6.5.1.2 General case

The complex index of reflection, n, is given by [7, p. 91]

$$n^2 = \epsilon_r - j\,60\sigma_s\lambda \tag{6.59}$$

where ϵ_r is the relative permittivity of the reflecting surface;
σ_s is the conductivity of the surface layer Siemens per meter (S/m, formerly mho/m);
λ is the wavelength in m.

Values for relative permittivity and conductance are listed in Table 6.10 [3, p. 260, Eq. 6.70 and 6.71; 22, p. 84].

Table 6.10
Relative permittivity and conductivity

	Relative permittivity	Conductivity S/m
Sea water	70 (for microwaves)	3 to 5
Fresh water	81	10^{-2} to 10^{-3}
Wet earth	5 to 30	10^{-1} to 10^{-3}
Dry earth	2 to 5	10^{-4} to 10^{-5}

The reflection coefficients for horizontal and vertical polarizations are different and are given by

$$\Gamma_h = \frac{\sin\psi - \sqrt{n^2 - \cos^2\psi}}{\sin\psi + \sqrt{n^2 - \cos^2\psi}} \tag{6.60}$$

where ψ is the grazing angle.

For vertical polarization,

$$\Gamma_v = \frac{n^2\sin\psi - \sqrt{n^2 - \cos^2\psi}}{n^2\sin\psi + \sqrt{n^2 - \cos^2\psi}} \tag{6.61}$$

This allows the calculation of the complex reflection factor. The phase angle of the reflection adds to the lag in the path difference term. The propagation factor F may be calculated using (6.52). These are calculated for an antenna over different surfaces in Table 6.11 [7, pp. 92 et seq.]:

Table 6.11
Square of the index of reflection, n^2, for a number of surfaces

Surface	Square of the index of reflection, n^2	Surface	Square of the index of reflection, n^2
Sea water	81 - j278.52	Dry land	15 - j0.67
Marsh land	30 - j6.67	Desert	3 - j0.667
Average land	15 - j1.67		

Figures 6.40 to 6.49 show the reflection factors with grazing angle for the following representative frequencies:

• Band	VHF	UHF	SSR	L	S	C	X
• Frequency, MHz	100	600	1045	1300	3000	5500	9000
• Wavelength, m	3	0.5	0.287	0.231	0.1	0.0545	0.0333

The normal radar range calculations assume that the propagation factor, F, is unity. When F is greater than unity, there is a range bonus. Conversely, when F is less than unity, there is less coverage. Only the amplitude of F is used, though it is complex. If the reflection coefficient is expressed in polar form as $\Gamma \angle \gamma$, then

$$\text{Total phase shift of reflected ray} = \frac{2\pi\delta}{\lambda} + \gamma \tag{6.62}$$

In the past, reflection effects were used with ground radars in that search or surveillance radars were horizontally polarized to allow increased "range" in the outer or early warning region, where a relatively low individual probability of detection was allowable. The coverage holes caused by lobing were small enough to be ignored in the inner region. Conventionally, height-finder and three-dimensional radars use vertical polarization, which has lesser ground reflections. The reflections are most pronounced over smooth sea water. The vertical coverage diagram calculated for a fan beam antenna 20 feet (6.096 m) above smooth sea water is shown in Figure 6.39. The frequency of 100 MHz has been chosen to give a small number of lobes.

Figure 6.39 shows the vertical range coverage for horizontal and vertical polarization with the plot of the propagation factor, *F*, between them. For vertical polarization, there is a kink in the propagation factor and diagram at the pseudo-Brewster angle. In optics, a ray entering an optically denser medium is refracted and reflected. At the Brewster angle (measured from the normal to the surface) the (electrically) horizontally polarized component of the ray is reflected and the vertically polarized part is refracted. At the Brewster angle (measured as a grazing angle) there is little reflection of the vertically polarized component and the propagation factor, *F*, is nearly unity. Range is neither gained nor lost. For rays above the pseudo-Brewster angle, the horizontal lobes occur between the vertical lobes, and this is one of the reasons that some radars have used polarization diversity.

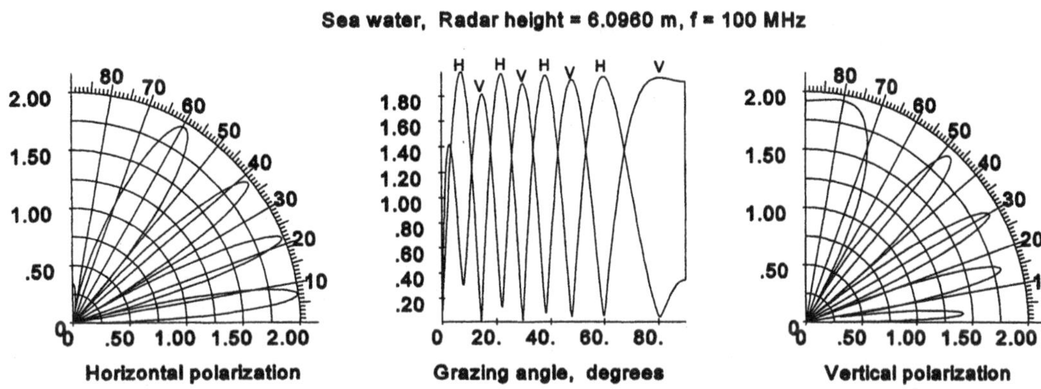

Figure 6.39 The vertical coverage diagrams for horizontally and vertically polarized radiation at 100 MHz from an antenna 20 feet (6.096 m) over smooth sea water.

The upper parts of Figures 6.40 to 6.44 show the shapes of the reflection factors, Γ, in amplitude (left) and phase (right) for the listed frequencies. Labels indicate the ends of the frequency lists for horizontal (H) and vertical (V) polarizations. The lower parts show the propagation factors, *F*, for horizontal polarization (left), vertical polarization (right), and the interleaving of the horizontally and vertically polarized lobes (center). In the diagrams the abbreviation H stands for horizontal, V for vertical, F for forward, and R for reverse.

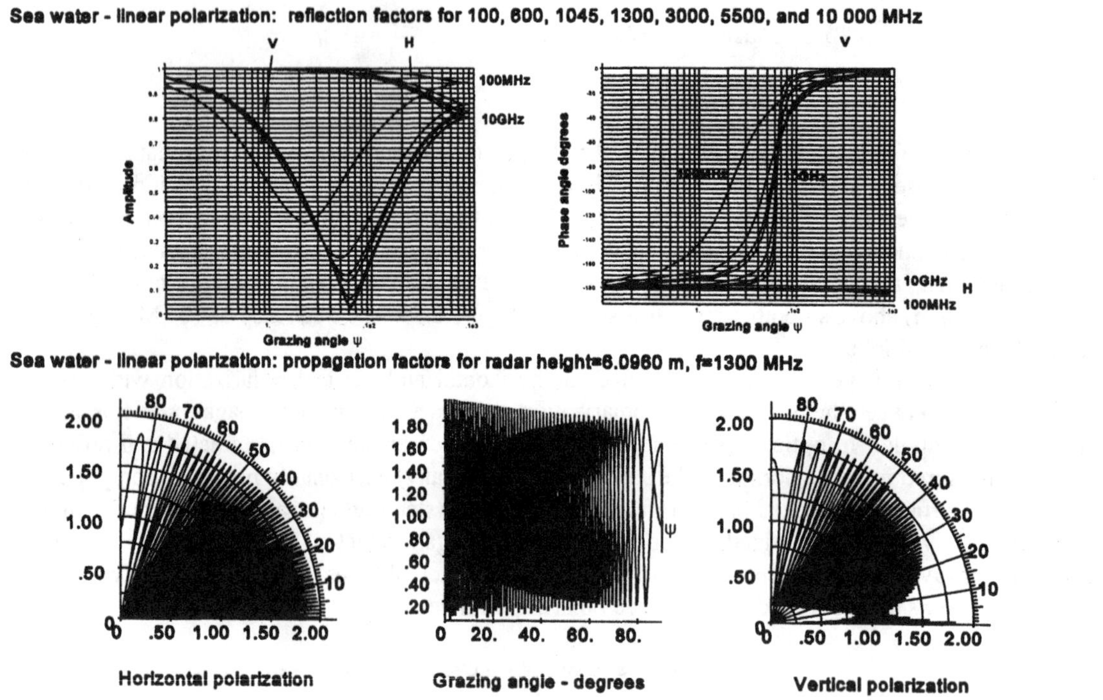

Figure 6.40 The reflection factors and the relative vertical coverage diagrams for horizontally and vertically polarized radiation over smooth sea water.

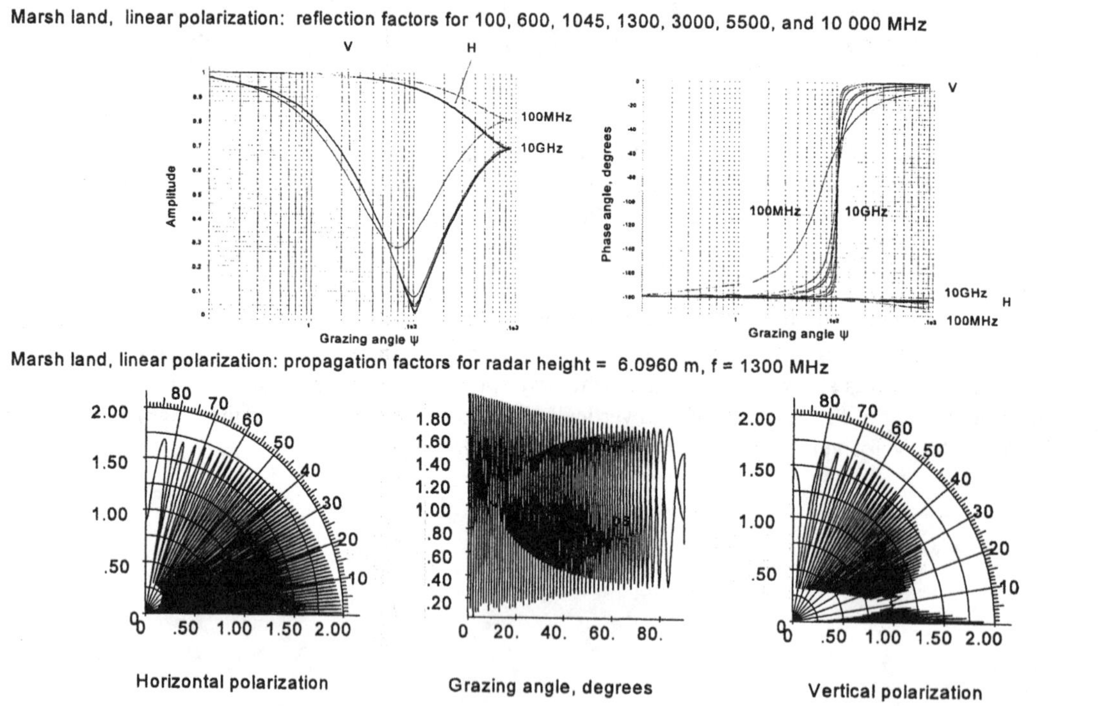

Figure 6.41 The reflection factors and the relative vertical coverage diagrams for horizontally and vertically polarized radiation over marsh land.

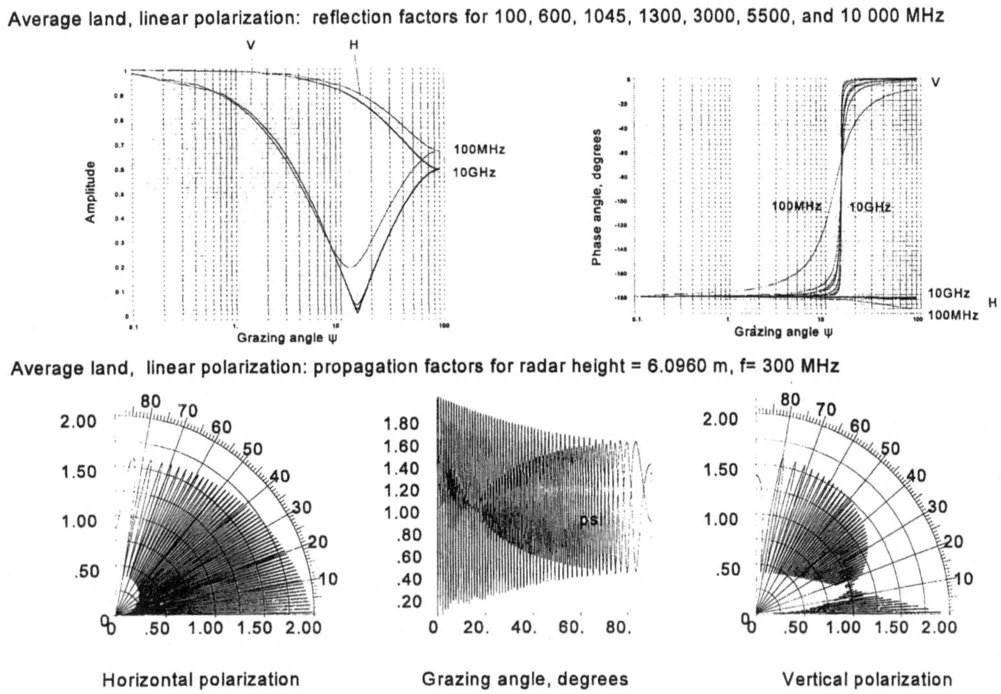

Figure 6.42 The reflection factors and the relative vertical coverage diagrams for horizontally and vertically polarized radiation over average land.

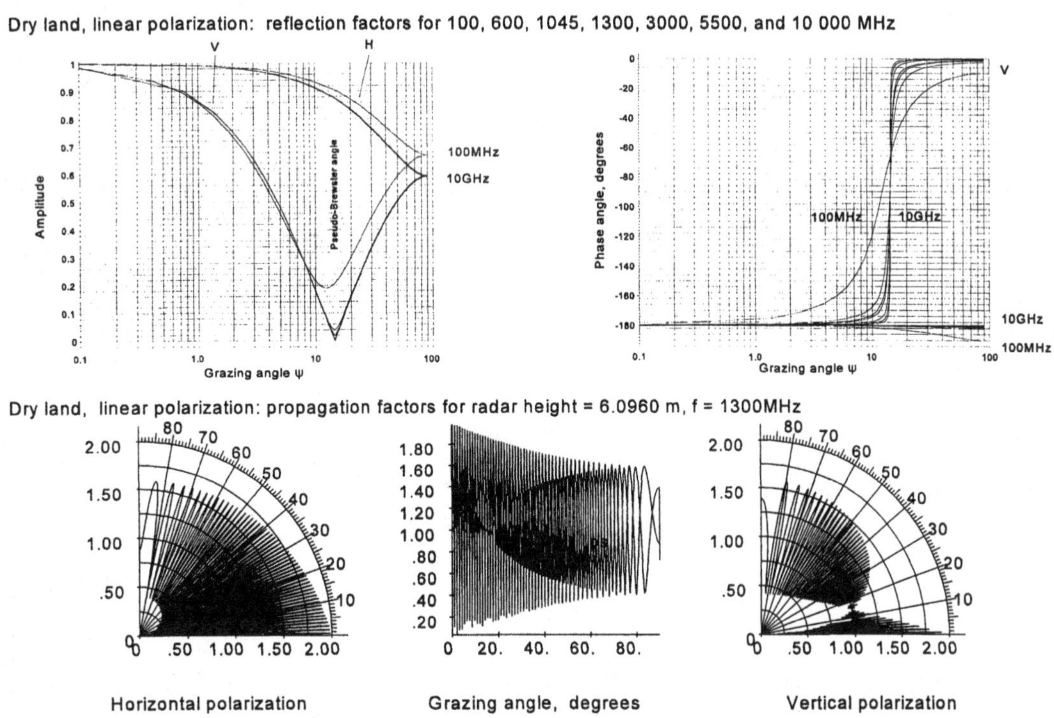

Figure 6.43 The reflection factors and the relative vertical coverage diagrams for horizontally and vertically polarized radiation over average land.

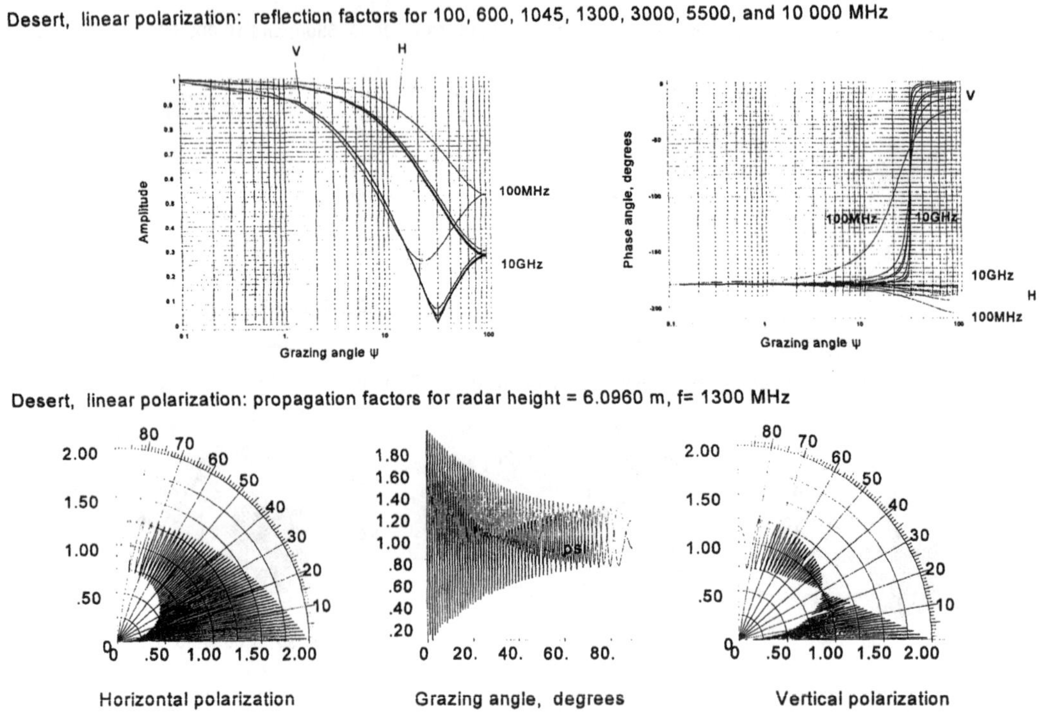

Figure 6.44 The reflection factors and the relative vertical coverage diagrams for horizontally and vertically polarized radiation over a desert.

6.5.1.3 Circular polarization

The values for the reflection factor, Γ, for circular polarization are calculated from the vertical and horizontal components, each of which has a different reflection factor. There are two reflection factors for circularly polarized reflections:

- The forward or the component with the same polarization, given by [3, p. 264]

$$\Gamma_f = \frac{\Gamma_H + \Gamma_V}{2} \tag{6.63}$$

- Unlike horizontal and vertical polarizations, where the reflected ray has the same polarization, the reflected ray is elliptically polarized. This has a forward circular component (6.63) and a reverse circular component. The reverse component is given by

$$\Gamma_r = \frac{\Gamma_H - \Gamma_V}{2} \tag{6.64}$$

The reverse component is normally rejected by the antenna and plays no further part for the radar. This component is plotted in Figures 6.45 to 6.49 for information only.

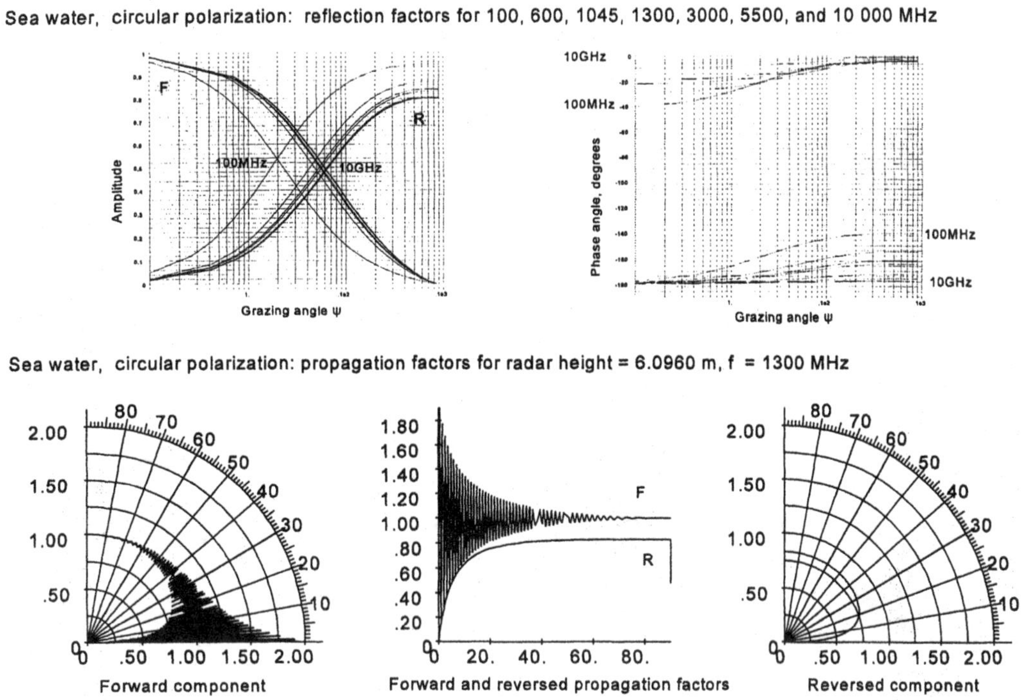

Figure 6.45 The reflection and vertical coverage diagrams for circularly polarized radiation over smooth sea water.

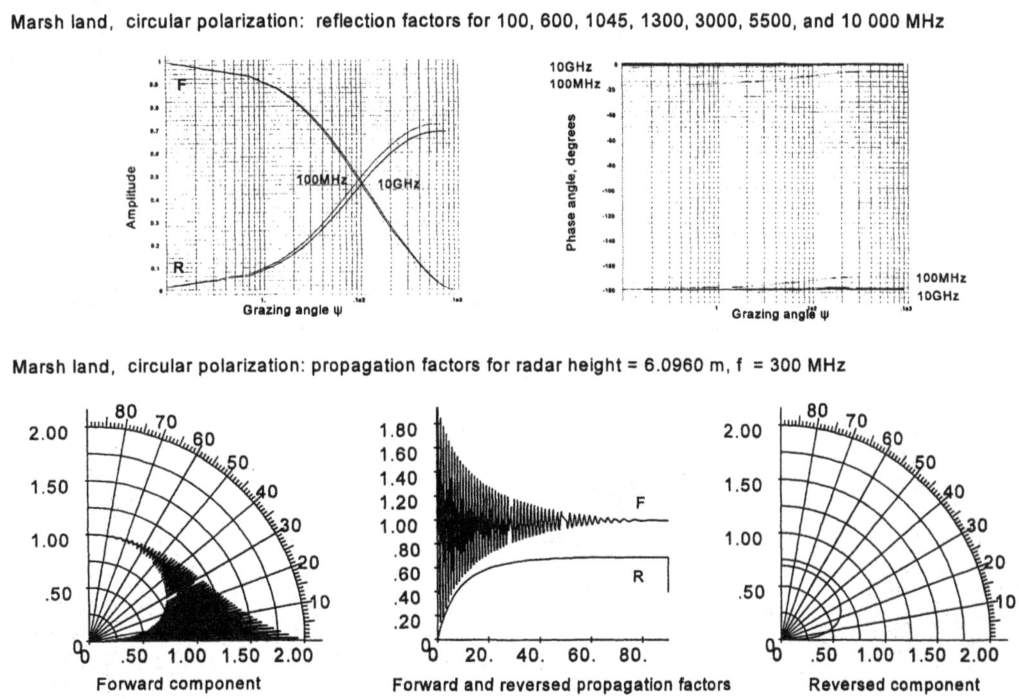

Figure 6.46 The reflection and vertical coverage diagrams for circularly polarized radiation over marsh land.

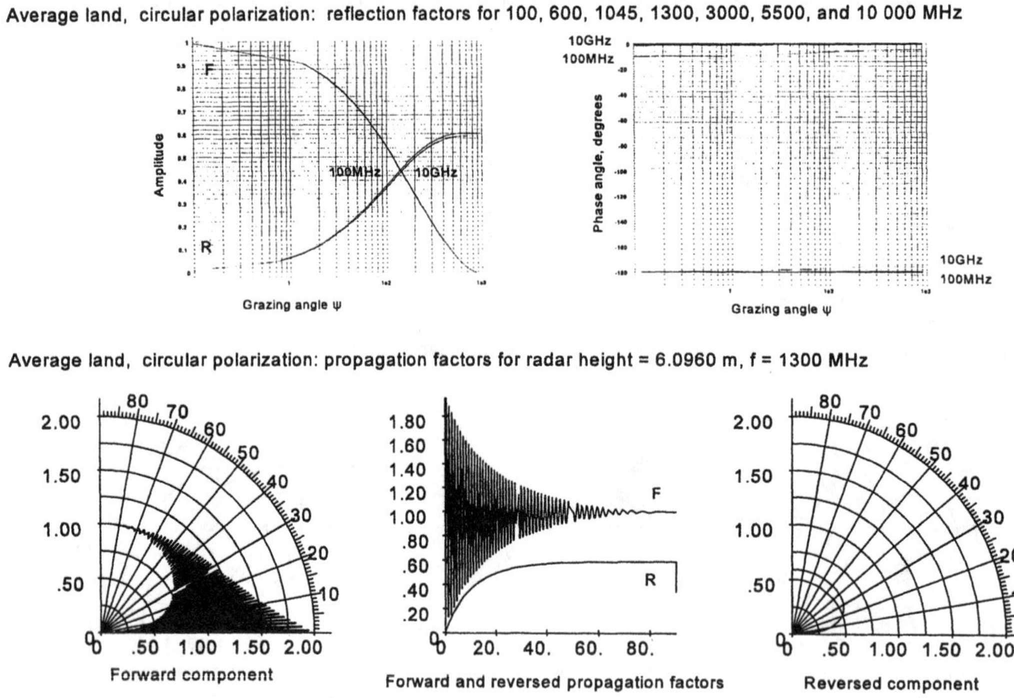

Figure 6.47 The reflection and vertical coverage diagrams for circularly polarized radiation over average land.

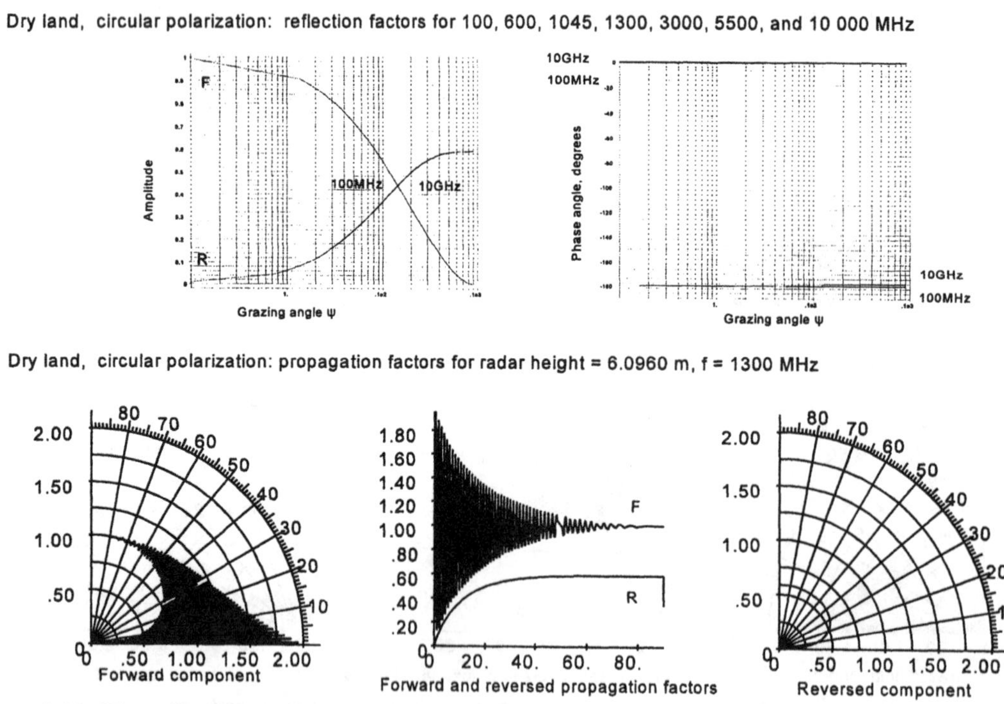

Figure 6.48 The reflection and vertical coverage diagrams for circularly polarized radiation over dry land.

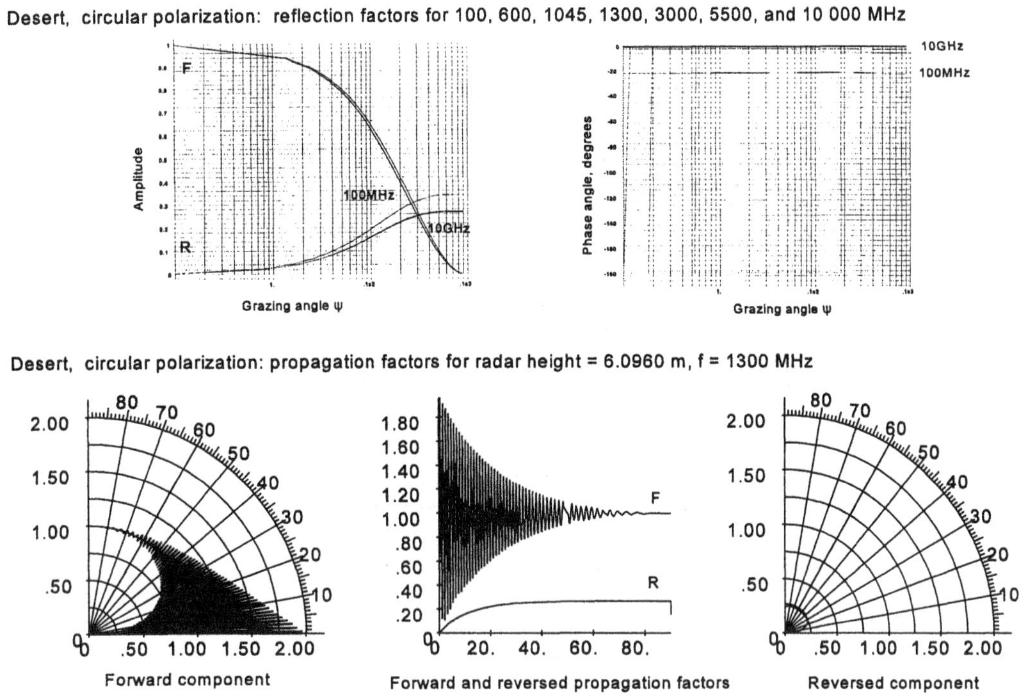

Figure 6.49 The reflection and vertical coverage diagrams for circularly polarized radiation over desert.

6.5.2 Factors that reduce reflections and lobing

Up to this point the antenna has been assumed to have a beam that has good coverage in elevation below the radar horizon. Modern antennas have shaped beams, for example, cosecant squared for aircraft surveillance radars, and the underside of the beam is shaped to reduce reflections and clutter.

The earth is flat only over short distances and rarely in Europe does a radar stand in predicably flat countryside so the interference diagrams are an indication, not an exact measure. The divergence factor is the factor that calculates the reduced power density of a reflection by a spherical surface. The reflections that cause predictable interference are specular reflections. These are reduced if the reflecting surface is rough.

6.5.2.1 Reduced gain below the horizon

If the gain of the antenna is reduced at angles where the reflections enter it, the effect of reflections is reduced. Such a vertical antenna diagram is shown in Figure 6.50.

Horizontal polarization over sea water produces the most pronounced lobes. If the radar is on land looking out over the sea, that is, fixed in elevation, the expressions for the peaks and troughs of the lobes become

$$\begin{aligned} &F\ peak\ (voltage) && \sqrt{1 + \Gamma(\Gamma + 2)} \\ &F\ trough\ (voltage) && \sqrt{1 + \Gamma(\Gamma - 2)} \end{aligned} \tag{6.65}$$

The effect on radar range is illustrated in Figure 6.51.

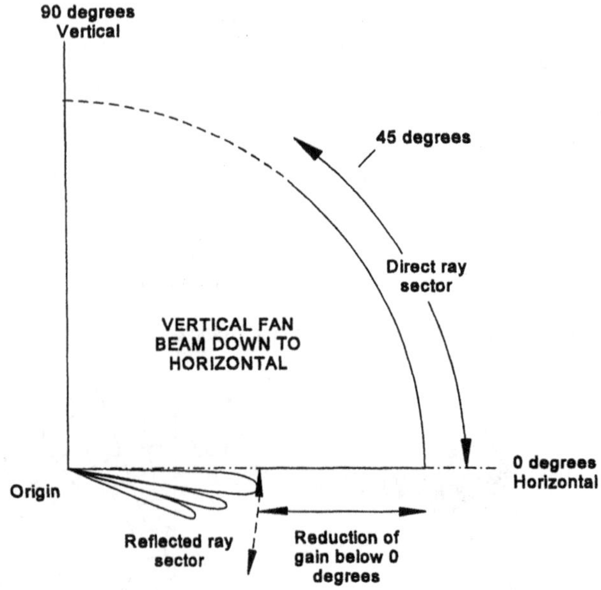

Figure 6.50 A vertical fan antenna pattern with reduced gain in the reflected ray sector.

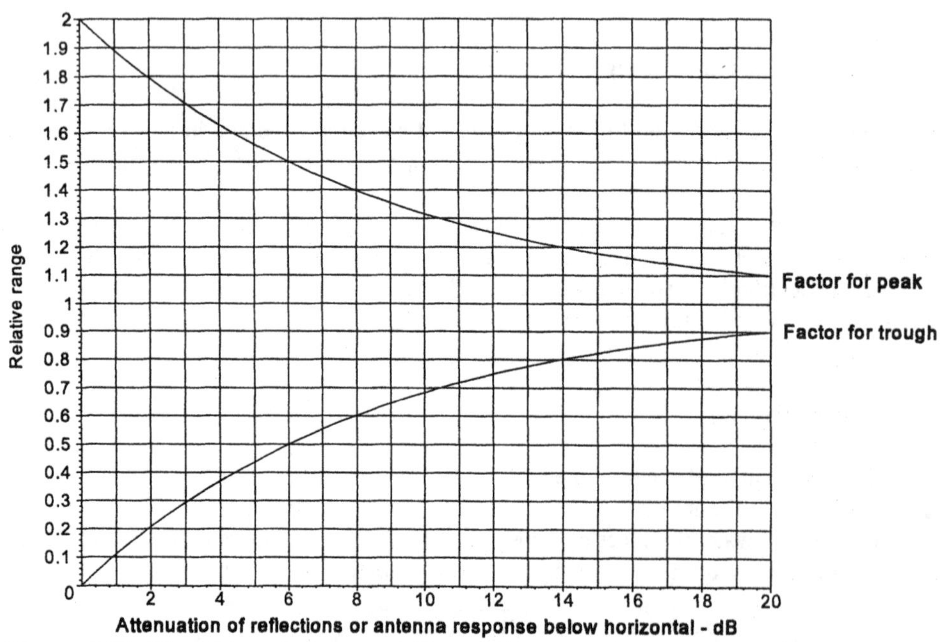

Figure 6.51 The effect of antenna response below zero degrees elevation on lobing over sea water with horizontal polarization.

6.5.2.2 Divergence factor

The earth is not flat but acts as a convex mirror. The exact formula [3, p. 270] is complex so the approximate Fishback expression which is in terms of elevation angle θ_d has been used here:

$$\text{Divergence factor, } D = \sqrt{\frac{1}{3}\left(1+\frac{2\zeta}{\sqrt{\zeta^2+3}}\right)} \tag{6.66}$$

where ζ is $\sqrt{\dfrac{a_{eff}}{2\,h_{radar}}\tan\theta_d}$;

a_{eff} is the effective earth's radius;
h_{radar} is the radar height above the ground;
θ_d is the elevation angle.

The divergence factor in (6.66) is plotted in Figure 6.52. The equations for divergence factor are only valid for elevation angles down to the first maximum [3, p. 271]. This divergence factor has been included in the foregoing coverage diagrams.

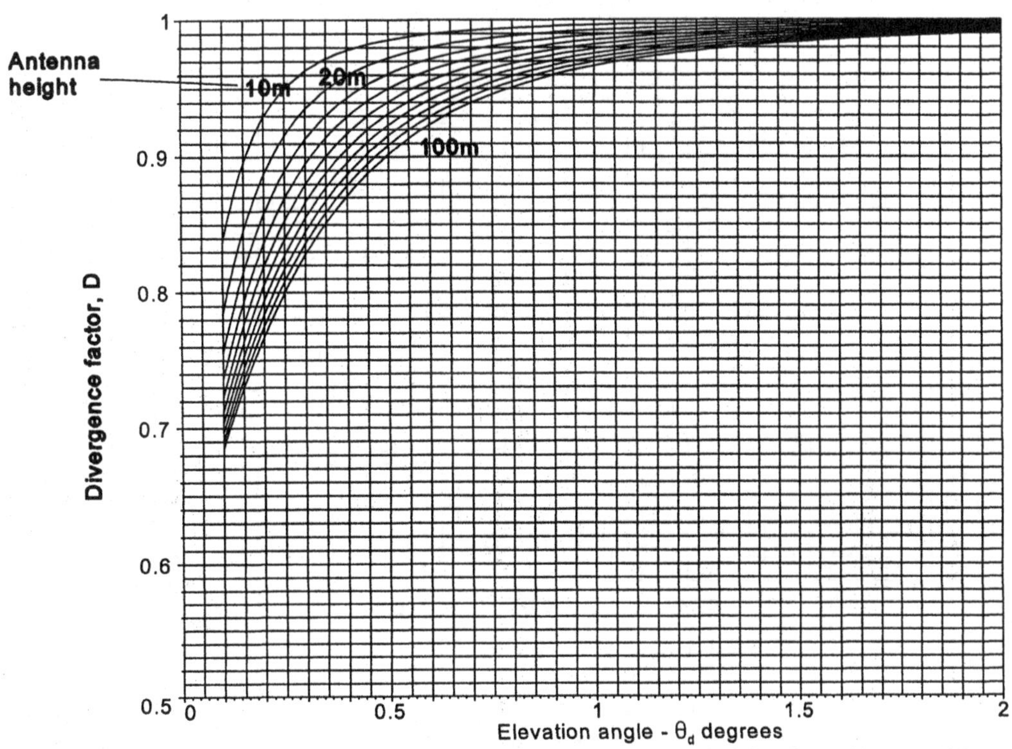

Figure 6.52 The approximate divergence factor, *D*, for antenna heights from 10 m to 100 m.

6.5.2.3 Roughness of the reflecting surface

Normal land surfaces are rough and even lakes have small waves to disturb specular reflection at higher frequencies. The amount of specular reflection from rough surfaces has been studied. For a surface that has Gaussian distributed roughness of standard deviation σ_h illuminated at a grazing angle ψ, the specular component may be given by Ament 1953 in [3, p. 266] as

$$\rho_s = \exp(-z) \tag{6.67}$$

where $z = 2\left(\frac{2\pi\sigma_h \sin(\psi)}{\lambda}\right)^2$.

The above estimation does not include the effects of shadowing at small grazing angles. Another expression for a "Gaussian collection of sinusoidal surface waves" has been given by Brown and Miller in 1974 [3, p. 266]

$$\rho_s = e^{-z} I_0(z) \tag{6.68}$$

where z is $2\left(\frac{2\pi\sigma_h \sin(\psi)}{\lambda}\right)^2$;

I_0 is a Bessel function of order zero with imaginary argument.

These are shown in Figure 6.53.

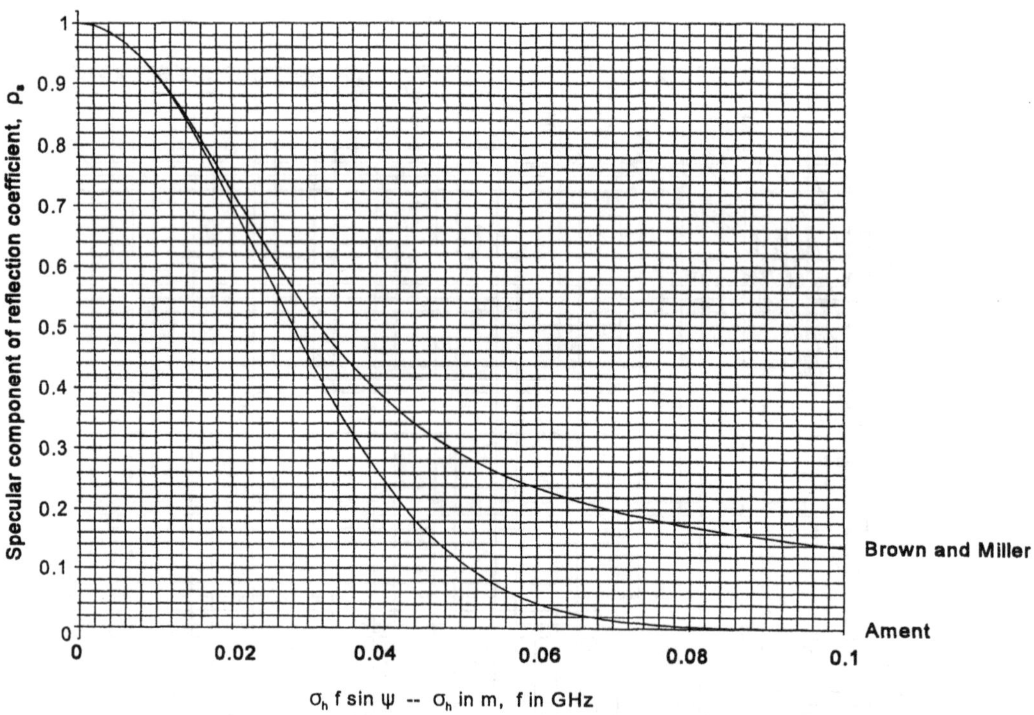

Figure 6.53 The specular component of the reflection coefficient for rough surfaces. [From Blake, L. V., *Radar Range-Performance Analysis,* Norwood, Massachusetts: Artech House, 1986. p. 267]

6.6 SCENARIO TO SIMULATE A "TYPICAL" RADAR ENVIRONMENT

Each radar is situated on a different site and each environment is different. The only way to discuss the reception and processing of signals is to use an artificial scenario. Clutter from ground echoes varies from radar to radar and is impossible to quantify, and the Barton model is used. Figure 6.54 illustrates the signals and Table 6.12 lists the signals that enter the antenna.

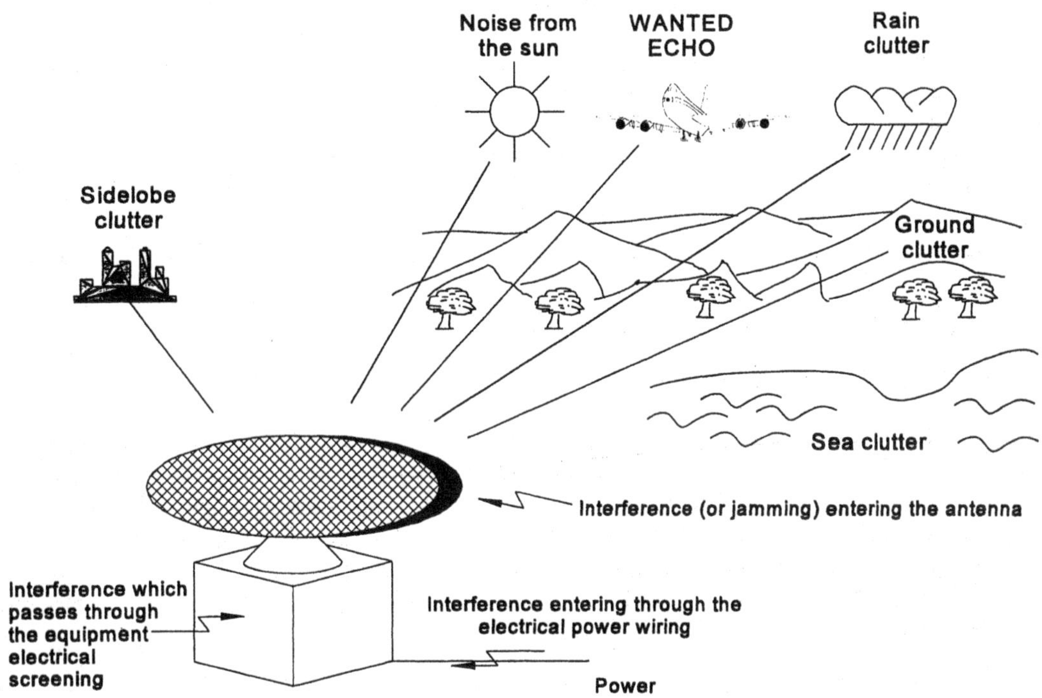

Figure 6.54 Sources of the signals at the output of the receiver.

Table 6.12
Signals entering the antenna

	Type of signal	Additionally for military radars
Wanted echoes	Echoes from objects of interest	
Competing echo signals	Echoes from around the object of interest	
	Ground clutter	
	Sea clutter	
	Rain and snow clutter	Chaff or window (military)
	Second time around clutter	
	Ground reflections interfering with returning echoes	
Interfering signals		Jamming signals (military)
	Entering the main beam of the antenna (sun)	Self protection or self screening jamming
	Entering the near sidelobes of the antenna	Escort jammers
	Entering the far sidelobes of the antenna	Barrage jammers
	Nearby interference entering the antenna (unsuppressed electric drills)	Fresnel zone jamming
	Interference from nearby radars or other radiating equipment	
	Energy entering into badly screened cabinets	
	Interference entering via the power cables	

In early radars, the signals were all displayed and radar operators had to sort them out. Echoes were distinguishable by shape and size. As this type of working is tiring, signal processors became available to help the operators sort out the radar picture. Current radars use automatic extraction and need much signal processing in order to report wanted scatterers only.

The first filter is the antenna which rejects signals away from the main beam. The gain of the sidelobes may be high enough to accept strong echoes giving rise to sidelobe clutter. This adds to the main clutter.

Jamming signals are rejected in military radars using expensive planar arrays which have low or ultra-low sidelobe response.

In this chapter the relative sizes and shapes of signals are simulated for a radar with the following characteristics:

- Radio frequency: 3 GHz (S-band);
- Pulse width: 1 μs;
- Azimuth beamwidth (3 dB) points: 1.5 degrees;
- Antenna rotation rate: 12.5 revolutions per minute;
- Signal-to-noise ratio at 100 km: 10 dB.

The aim is to show the shapes of the phenomena. For other radars with other characteristics the parameters must be scaled.

6.6.1 Echoes of interest, the reference echo

Echoes of interest are aircraft, ships, or other objects depending on the use of the radar. The shape of the returning echoes is the same as the transmitted pulse, if necessary, after a pulse-compression filter. Since the range may be any real number of wavelengths and the phase shifts in the receivers and filters are *a priori* unknown, the signals for processing may be at all possible phase angles. It is assumed that the scatterer does not accelerate measurably during the time that it is illuminated by the pulse, so that the phase of the echo is constant during this time.

The amplitude is defined by the critical signal-to-noise ratio, A, at the specified maximum range (see Chapter 12, Threshold). The signal-to-noise ratio at other ranges is

$$Signal\text{-}to\text{-}noise\ ratio = A\left(\frac{maximum\ range}{range}\right)^4 \tag{6.69}$$

For illustration, a simulated scenario has been generated using mathematical software (MAPLE) to show the shapes of the echo signals in range and the Doppler frequency. For convenience, it is assumed that a 10 dB signal-to-noise ratio is necessary to detect a 1 m² echo of interest at a given detection probability and false alarm rate at 100 km. Competing echo powers are calculated using this reference. Figure 6.55 shows the signal-to-noise ratios in decibels at different ranges.

Figure 6.55 has signals with a dynamic range of 130 dB. A further 40 dB represents the radar cross-section in the range of 1 to 10 000 m². A radar cathode ray tube is able to give a brightness range of only 10 dB. In the past, a combination of sensitivity time control and limiting was used to reduce the glare of ground clutter. Currently, the dynamic range is chosen to lie within the range of the analogue-to-digital converters, 66 dB for 12 bits.

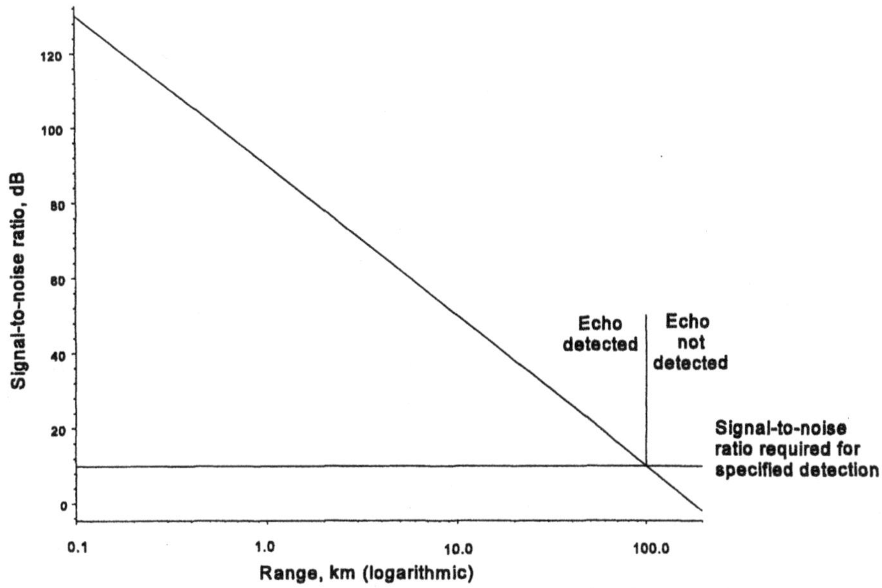

Figure 6.55 Signal amplitudes and signal-to-noise ratios for the reference echo. (In this example it has 10 dB signal-to-noise ratio at 100 km.)

6.6.1.1 Spectrum

The spectrum of the echo is determined by the movement of the scattering object and changes in the illumination. Aircraft echoes, for example, are dominated by the fuselage which moves at a near constant radial velocity v m/s. The basic one-way Doppler frequency equation is

$$f = f_0 \frac{c \pm v_{observer}}{c \mp v_{source}} \tag{6.70}$$

where c is the velocity of light, approximately $3 \cdot 10^8$ m/s;
v are the velocities of the source and observer;
f_0 is the original frequency radiated.

For a stationary radar, the scatterer is the observer of the radar pulse, and, on the way back, the radar is the observer of the returning scattered energy which gives, on expansion

$$f \approx f_0 + 2f_0 \left(\frac{v}{c} + \frac{v^2}{c^2} + \frac{v^3}{c^3} + \ldots \right) \tag{6.71}$$

where f_0 is the radar frequency;
v is the velocity of the scatterer.

The difference between transmitted and received frequencies ($f - f_0$) is the Doppler frequency. When v is much less than c, the Doppler frequency can be approximated by

$$\begin{aligned} f_{Doppler} &= 2f_0 \frac{v}{c} \text{ Hz} \\ &= 2\frac{v}{\lambda} \text{ Hz} \end{aligned} \tag{6.72}$$

An example is

- Radar frequency: 3 GHz, S-band;
  - giving a wavelength of: 0.1 m;
  - giving a Doppler frequency shift factor of : 20 Hz/m/s;
- Aircraft speed: 300 m/s (just less than the speed of sound);
  - giving a Doppler frequency shift of : 6000 Hz.

Other common values of Doppler frequencies to remember are given in Section 6.1.2.

Additional movement, such as propeller or jet engine turbine blade motion, adds to the spectrum of the echoes. The operators of radars used for ground observation use headphones to identify the types of echoes according to the rhythm of the sound of the Doppler frequency.

The act of the radar beam passing over a point scatterer while illuminating it produces additional spectral components. If shape of the tip of the antenna beam is assumed to be Gaussian, then the amplitude of the echoes will be modulated in time by this Gaussian characteristic.

In the example:

- Antenna rotation speed is: 12.5 revolutions per minute (rpm);
- Antenna (3 dB) beamwidth is: 1.5 degrees;
- Time between the 3 dB points is: 20 ms.

If the antenna has a Gaussian pattern and takes T_{tot} seconds (time on target) to scan from one (one-way) half power point to the other, the point scatterer will have its illumination power modulated by the expression (see Chapter 2, Usual and unusual concepts)

$$P(t) = \exp\left(-4\ \ln(2)\left(\frac{t}{T_{tot}}\right)^2\right) = \exp\left(-2.77259\left(\frac{t}{T_{tot}}\right)^2\right) \tag{6.73}$$

where t is the time variable.

The one-way power characteristic is the same as the two-way voltage characteristic, so that the voltage of the echoes in the receiver will be modulated additionally by the receiving pattern to give

$$E(t) = \exp\left(-4\ \ln(2)\left(\frac{t}{T_{tot}}\right)^2\right) \tag{6.74}$$

The voltage spectrum is given by (see Figure 6.56 for the radar in the example)

$$\sqrt{\frac{\pi}{2\ \ln(2)}}\ \exp\left(-\frac{\pi^2 f^2 T_{tot}^2}{2\ \ln(2)}\right) \tag{6.75}$$

This has a standard deviation of $\dfrac{\sqrt{\ln(2)}}{\pi T_{tot}} = \dfrac{0.26501}{T_{tot}}$ Hz.

The complete spectrum is centered on the Doppler frequency shift caused by the main body of the scatterer with extra spread caused by secondary movements and scanning. The frequency of the returning echoes can often exceed the pulse repetition frequency and Table 6.13 shows these for the example radar.

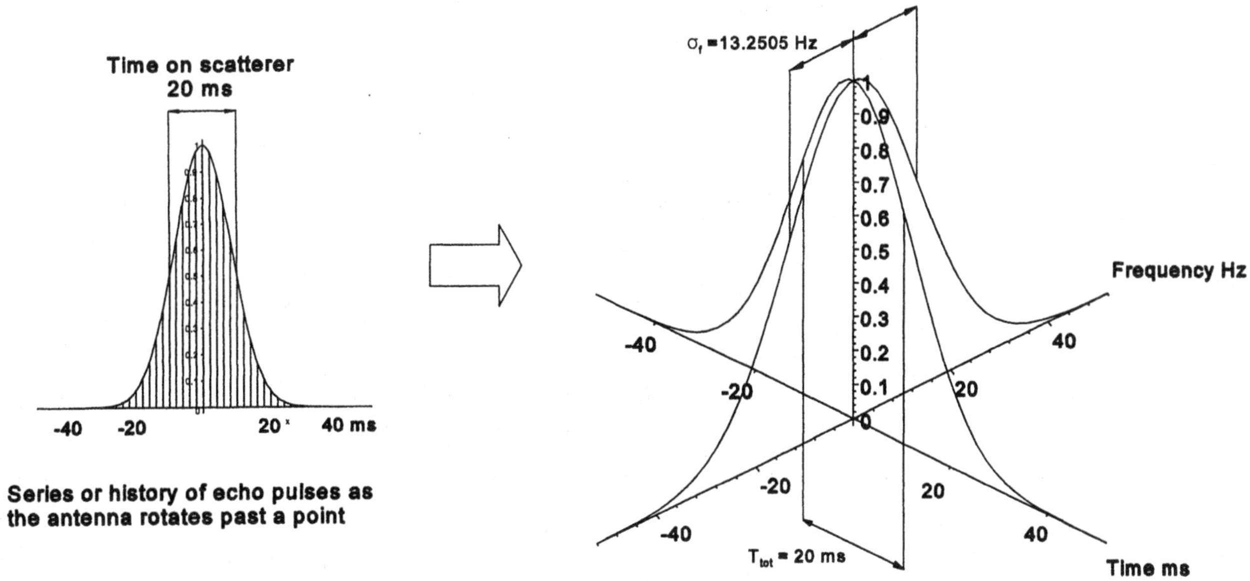

Figure 6.56 Echo pulse history as an antenna with a Gaussian pattern scans an object and its voltage spectrum.

Table 6.13
Ranges of speeds to be expected and the Doppler frequencies for an S-band radar

	Range of speeds		Range of Doppler frequencies
Airfield radar	0..Mach 1	0..330 m/s	0.. 6 600 Hz
Military ground control radar	0..Mach 3	0..990 m/s	0..19 800 Hz

The upper Doppler frequencies (Hz) for other frequencies are shown in Table 6.14.

Table 6.14
Upper Doppler frequencies for other radar bands

Radar frequency, GHz	0.6	1.2	3	6	10	15	30	90
0..330 m/s	1320	2640	6600	13200	22000	33000	66000	198000
0..990 m/s	3960	7920	19800	39600	66000	99000	198000	594000

6.6.2 A land clutter model

The echo scattered back from a land surface depends on the amount of surface that is illuminated, A, and the grazing angle, ψ. The cross-section that the radar sees is $A\sin\psi$ which leads to the constant γ model in Figure 6.57. The reflectivity, $\bar{\sigma}$, is

$$\bar{\sigma} = \gamma \sin \psi \quad \mathrm{m^2/m^2} \tag{6.76}$$

As the grazing angle, ψ, tends to zero at the radar horizon for ground-based radars, so does the average reflectivity, $\bar{\sigma}$. Reference [17, p. 323, Fig. 7.23] shows the reflectivity for rural land clutter, which agrees with an older model [23] for "rolling countryside" at low grazing angles out to 30 km.

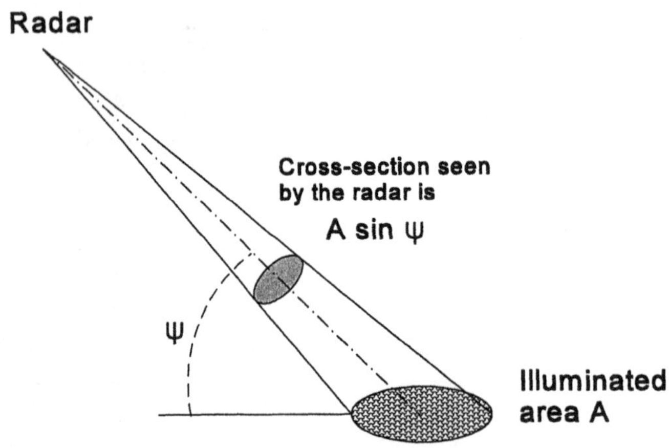

Figure 6.57 Clutter cross-sectional area as seen by a radar.

$$\textit{Average reflectivity},\ \bar{\sigma} = \frac{0.00032}{\lambda}\ \mathrm{m^2/m^2}\ \text{for } 0..30\ \mathrm{km} \tag{6.77}$$

The beam of a ground-based search radar is assumed to pass over the ground to give an amount of ground clutter proportional to the resolution cell area. It is assumed to be half the effective physical length of the transmitted pulse in the radial direction by the half-power antenna beamwidth.

$$\begin{aligned}&\textit{Average cross-section area}\\ &\quad = \textit{Resolution cell length}\ (\mathrm{m}) \times \textit{Resolution cell width}\ (\mathrm{m}) \times \textit{reflectivity}\ (\mathrm{m^2/m^2})\ \ \mathrm{m^2}\end{aligned} \tag{6.78}$$

where *Resolution cell length* = *Effective pulsewidth* (μs) × $c/2$ (m/s) m;
Resolution cell width = *Effective antenna beamwidth* (radian) × *range* (m) m.

Beyond the 30 km horizon, the transmitted pulses and the returning ground echoes are diffracted. The value at the 30 km horizon is used scaled inversely by the fourth power of the range. The effective clutter cross-sections for a radar with a 1 μs pulse and 1.5 degree antenna beamwidth are shown for the common radar frequency bands in Figure 6.58.

Clutter comes from the part of the antenna pattern that extends below the horizon, conventionally at 0 degrees elevation or horizon. With older antennas, the tilt was adjusted so that the lower 3 dB elevation point of the elevation pattern lay on the horizon. Modern antennas have a much steeper pattern below 0 degrees elevation, which reduces clutter. Additionally, a number of two-dimensional surveillance radars use an elevated reception beam at short ranges to reduce clutter further.

The amount of clutter is compared with the reference echo signal for the above radar in S-band (3 GHz) in Figure 6.59. If the power of the clutter is log-normally distributed with a 20 dB mean-to-median ratio, the probability distribution is

$$p(x) = \frac{1}{x\sqrt{2\pi}\,\sigma}\exp\left(\frac{\ln(x)-\mu}{2\sigma^2}\right),\quad x > 0 \tag{6.79}$$

where μ is the median and σ is the standard deviation.

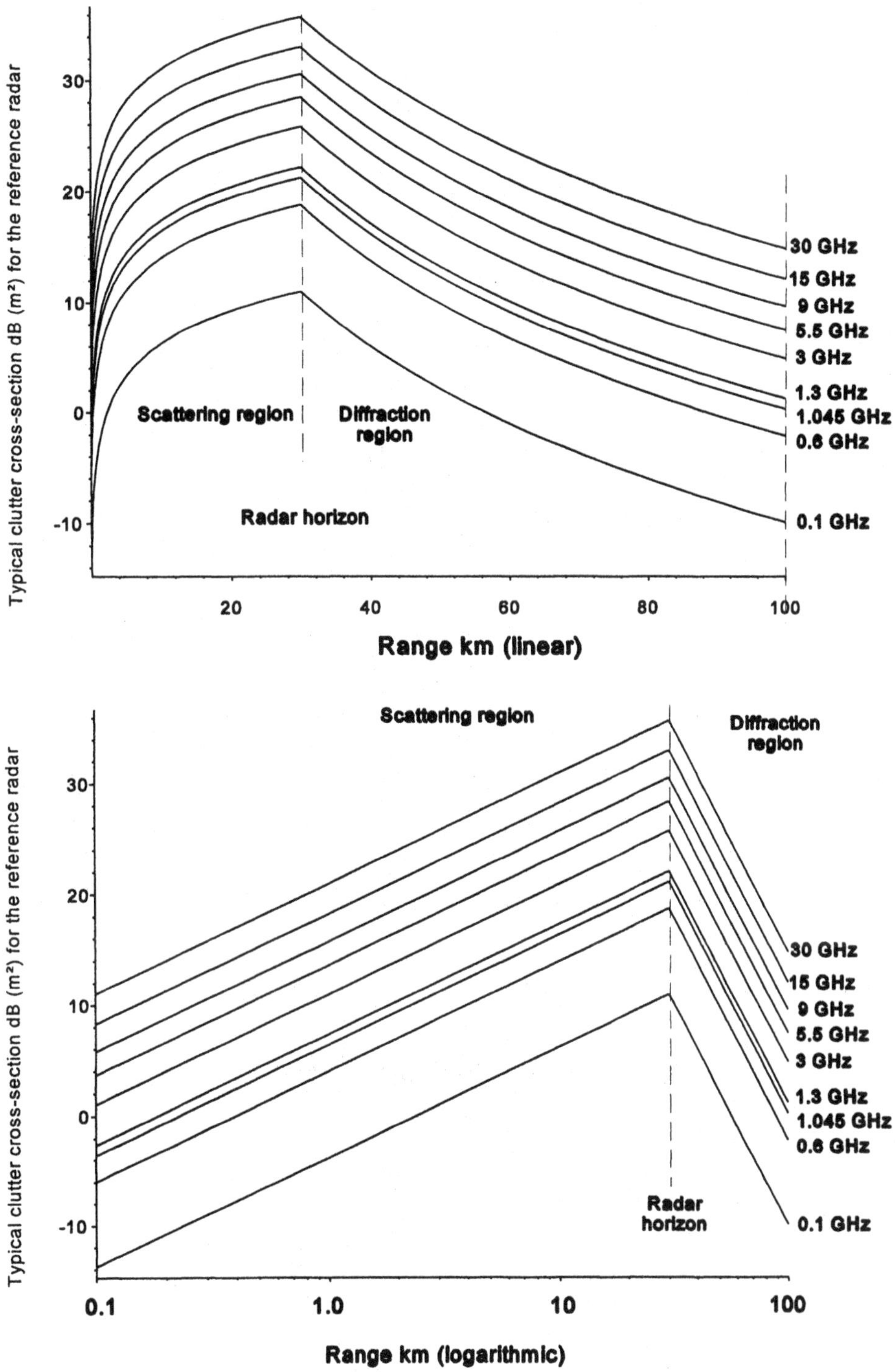

Figure 6.58 Mean clutter cross-section in decibels with respect to 1 m² plotted against range for the simple clutter model using linear and logarithmic range scales.

The cross-section of clutter is compared with the reference echo signal for the above radar in S-band (3 GHz) in Figure 6.59.

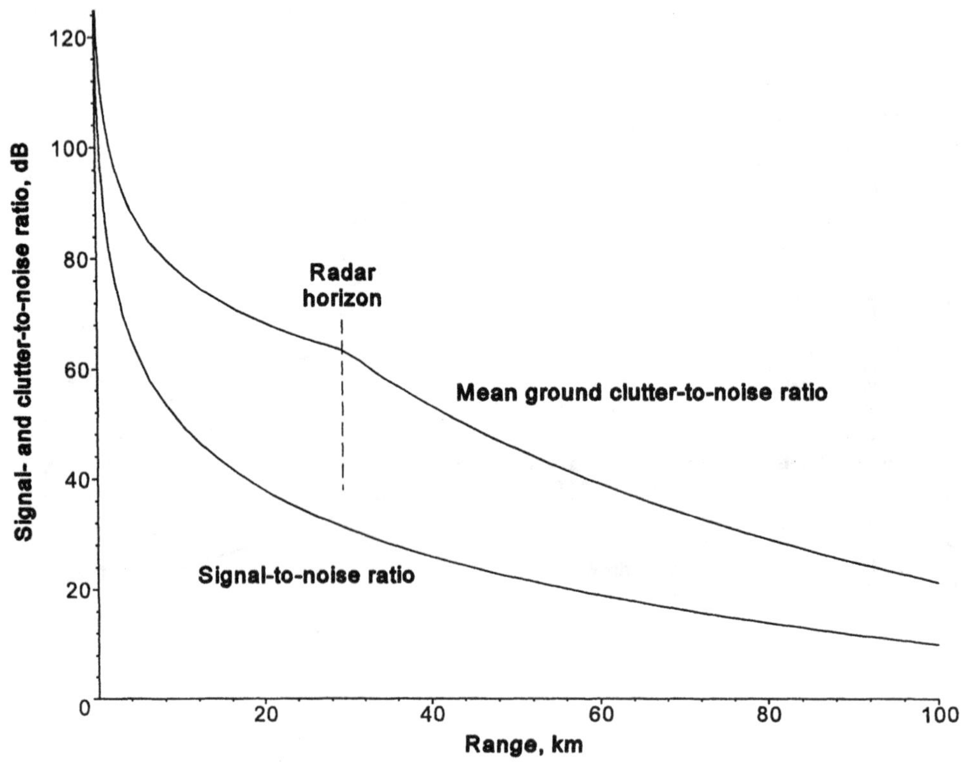

Figure 6.59 Signal- and clutter-to-noise ratios.

In the case of log-normal clutter with a mean-to-median ratio, *m*, of 100 the standard deviation in decibels is given by

$$\text{Mean-to-median ratio} = \exp\left(\frac{\sigma^2}{2}\right) \tag{6.80}$$

Converting to decibels in order to be able to plot the function on a decibel abscissa (see Figure 6.60),

$$10\ \log_{10}(100) = \frac{\sigma^2}{2}\ 10\ \log_{10}e = 20\ \text{dB} \tag{6.81}$$

Solving for σ,

$$\sigma = \sqrt{\frac{2 \times 20}{10 \times \log_{10}e}}\ \text{exponential units} \tag{6.82}$$

Converting σ to decibels,

$$\begin{aligned} \sigma &= 3.03485\ \text{exponential units} \\ &= 3.03485 \times 10\ \log_{10}e\ \text{dB} = 13.1802\ \text{dB} \end{aligned} \tag{6.83}$$

The median-to-mode ratio is given by

$$\text{Median-to-mode ratio} = \exp(\mu - \sigma^2) \tag{6.84}$$

Changing to decibels,

$$\textit{Median-to-mode ratio, dB} = (\mu - \sigma^2)\ 10\ \log_{10}e$$
$$= -3.03485^2\ .\ 4.3429\ \text{dB} = -40\ \text{dB} \tag{6.85}$$

The log-normal distribution has a peak (the mode), a median (the 50% point), and a mean, which are widely separated. It is plotted for clutter with 100 m² mean value in three parts about the mode, the median, and the mean in Figure 6.60.

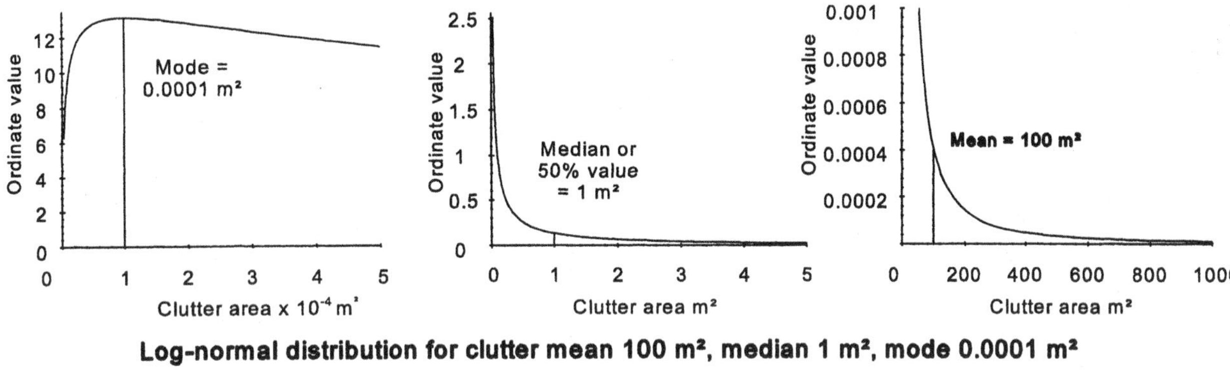

Figure 6.60 Log-normal distribution for clutter with a mean of 100 m².

The probability and cumulative distributions are plotted in decibels in Figure 6.61.

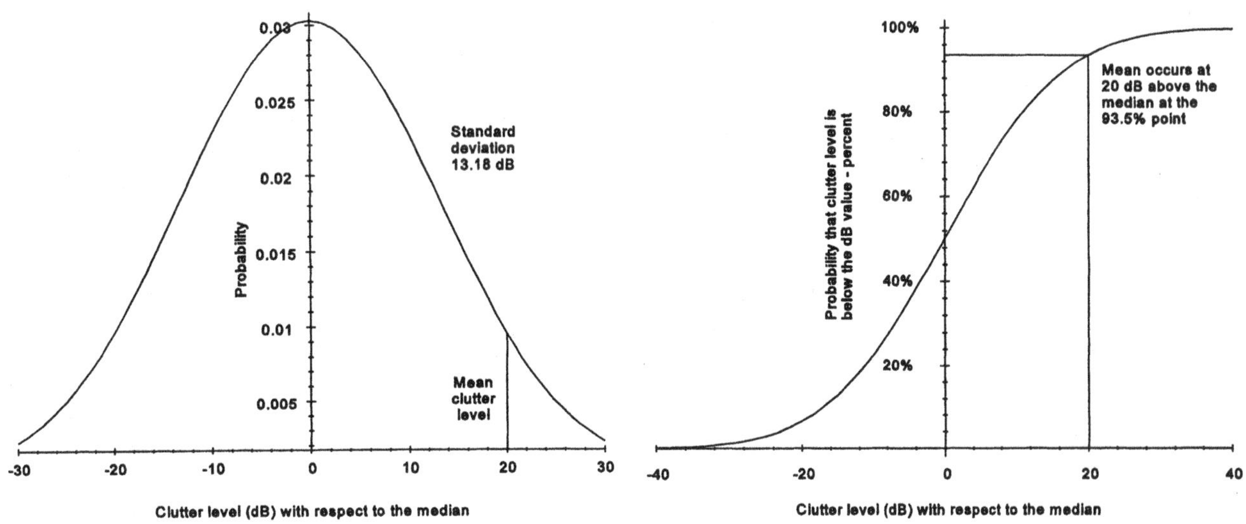

Figure 6.61 The probability and cumulative distributions of log-normal clutter with a standard deviation of 13.18 dB.

Figure 6.62 shows the probability distribution of the log-normal land clutter plotted with the clutter-reflecting area and range from (6.77). The mean clutter level is determined by a few very strong reflectors. Most of the clutter occurs around the median, or 50%, level. The reference echo level is also shown in the diagram. For example, the reference echo is stronger than the median clutter (50% of the range cells) at ranges up to 8 km and beyond 41 km.

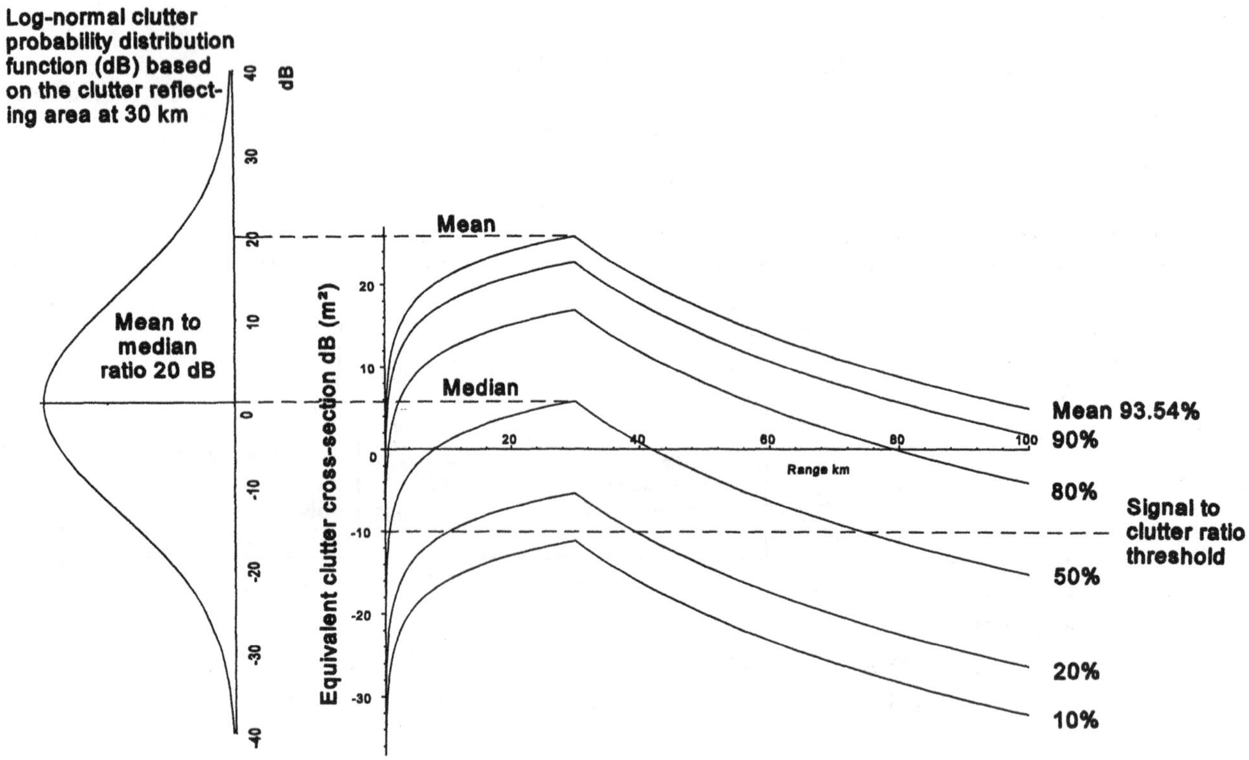

Figure 6.62 Plots of log-normal clutter distribution for the S-band radar in the example.

Table 6.15 gives percentiles with the level in decibels about the median.

Table 6.15
Percentiles referred to decibels about the mean

Percentile →	dB	dB →	Percentile	Percentile →	dB	dB →	Percentile
5	-21.68	-18	8.60	50	0.00	0	50.00
10	-16.89	-16	11.24	55	1.66	2	56.03
15	-13.66	-14	14.41	60	3.34	4	61.92
20	-11.09	-12	18.13	65	5.08	6	67.55
25	-8.89	-10	22.40	70	6.91	8	72.81
30	-6.91	-8	27.19	75	8.89	10	77.60
35	- 5.08	-6	32.45	80	11.09	12	81.87
40	-3.34	-4	38.08	85	13.66	14	85.59
45	-1.66	-2	43.97	90	16.89	16	88.76
50	0.00	0	50.00	95	21.68	18	91.40

The signal-to-noise ratio for detection depends on the nature of the land clutter echoes. They have the narrowest statistical distribution, that is, the clutter echoes are the most stable when the radar uses a fixed frequency and fixed azimuths for transmission. Then the clutter echo signals vary from scan to scan according to clutter movement and propagation changes. A threshold may be chosen empirically above the mean of the echo signals to give the required false alarm rate. This uses visibility above the clutter, or super-clutter visibility, and is the basis of area moving target indicators. In Figure 6.62, 10 dB signal-to-clutter ratio for detection has been chosen, which gives detection beyond 75 km.

If no measures are taken to make the clutter signals repeatable or frequency agility is used, the clutter echoes vary more from scan to scan, in which case a higher threshold must be used.

6.6.3 The Weibull clutter model

The log-normal model has a very long upper tail and is difficult to manipulate mathematically. Among other distributions, the Weibull distribution has been proposed [22].

The Weibull distribution has two parameters, η for shape and σ for scale, and is given by [18, p. 122]

$$\text{Weibull probability distribution function,}\quad p(x) = \frac{\eta}{\sigma}\left(\frac{x}{\sigma}\right)^{\eta-1}\exp\left(-\left(\frac{x}{\sigma}\right)^{\eta}\right) \tag{6.86}$$

Examples of the distribution function are given in Figure 6.63.

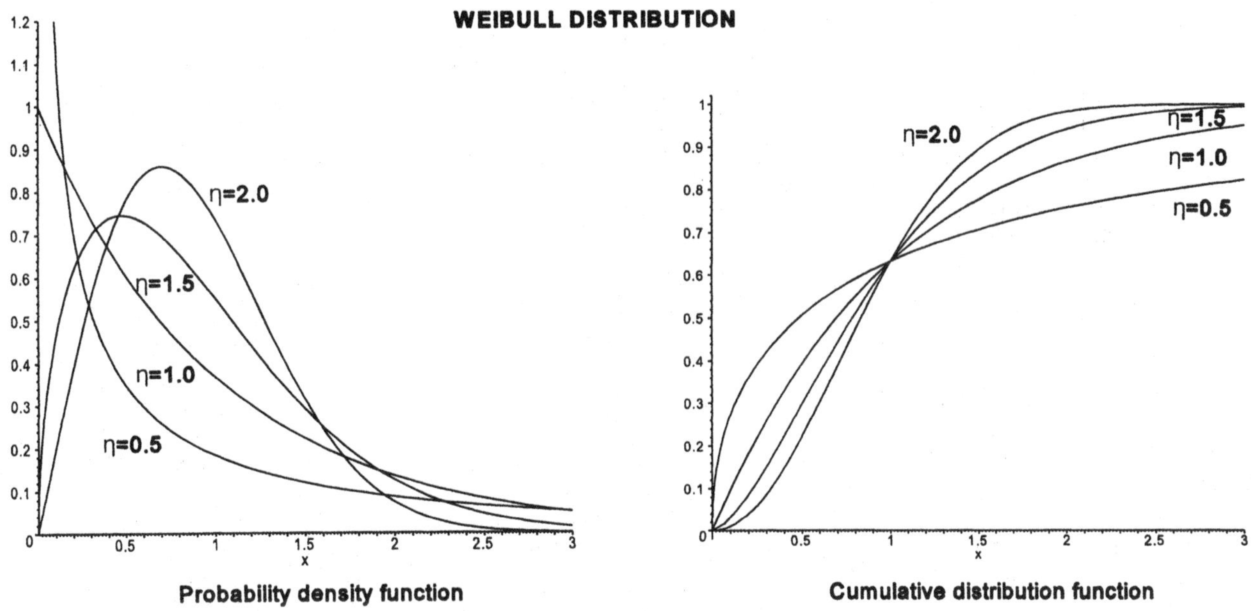

Figure 6.63 Plots of the Weibull distribution $\sigma = 1$.

The Weibull distribution has the following properties:

$$\begin{aligned} \textit{Mean} &= \sigma\,\Gamma\left(\frac{1}{\eta}+1\right) \\ \textit{Standard deviation} &= \sigma^2\left[\Gamma\left(\frac{2}{\eta}+1\right)-\left(\Gamma\left(\frac{1}{\eta}+1\right)\right)^2\right] \\ \textit{Median} &= (\ln(2))^{\frac{1}{\eta}}\,\sigma \end{aligned} \tag{6.87}$$

The cumulative distribution is

$$P(x) = 1 - \exp\left(-\left(\frac{x}{\sigma}\right)^{\eta}\right) \tag{6.88}$$

When $\eta = 2$ and σ is replaced by $\sqrt{2}\sigma$, this is the Rayleigh distribution. If the mean power echoed by clutter is $\overline{\sigma_0}$, that is, $\overline{\sigma_0} = \sigma^{\eta}$, then the probability distribution of Weibull clutter is [23, p. 5]

$$p(\sigma_0) = \frac{1}{\overline{\sigma_0}} \eta \, \sigma_0^{\eta-1} \exp\left(-\frac{\sigma_0^{\eta}}{\overline{\sigma_o}}\right) \tag{6.89}$$

When $\eta = 1$ this gives the Rayleigh power distribution. The cumulative distribution is

$$P(\sigma_0) = \int_0^{\sigma_0} p(\sigma_0) d\sigma_0 = 1 - \exp\left(-\frac{\sigma_0^{\eta}}{\overline{\sigma_0}}\right) \tag{6.90}$$

The median is

$$Median, \; \sigma_{med} = \left[\overline{\sigma_0} \ln(2)\right]^{\frac{1}{\eta}} \tag{6.91}$$

The values of η used for ground clutter are as follows [23]:

- Rolling hills: $\eta = 0.626$ in L-band;
- Rocky mountains: $\eta = 0.512$ in S-band;
- Forest: $\eta = 0.25$ to 1 in X-band.

6.6.4 Land clutter spectrum

Clutter has a spectrum, often defined by trees waving in the wind, with a standard deviation taken to be 0.3 m/s. At S-band, this gives an additional standard deviation of 0.3 . 20 = 6 Hz.

$$\begin{aligned} \textit{Clutter spectrum standard deviation} &= \frac{2}{\lambda} \textit{clutter velocity standard deviation} \text{ Hz} \\ &= 6 \text{ Hz}, \quad \text{in S-band here} \end{aligned} \tag{6.92}$$

For a scanning radar, the standard deviation of the clutter spectrum must be added to the standard deviation caused by antenna scanning to give the total standard deviation. The sum of the standard deviations is $\sqrt{13.25^2 + 6^2} = 14.55$ Hz. In order to be able to illustrate the spectrum throughout the 66 dB dynamic range in a radar using a 12 bit analogue-to-digital converter, it is plotted in decibels in Figure 6.64.

The spectral density of the clutter in square meters/hertz seen by the antenna is plotted in Figure 6.65. When the spectral density is summed over all Doppler frequencies, the total is recalled (in decibels in Figure 6.62). Land clutter at 30 km, in this example, has a cross-section of 163.36 m² or 22.13 dB (m²). If the spectrum is evenly spread over 40 Hz, the spectral density is 4 m²/Hz, which explains the low value of spectral density in Figure 6.65.

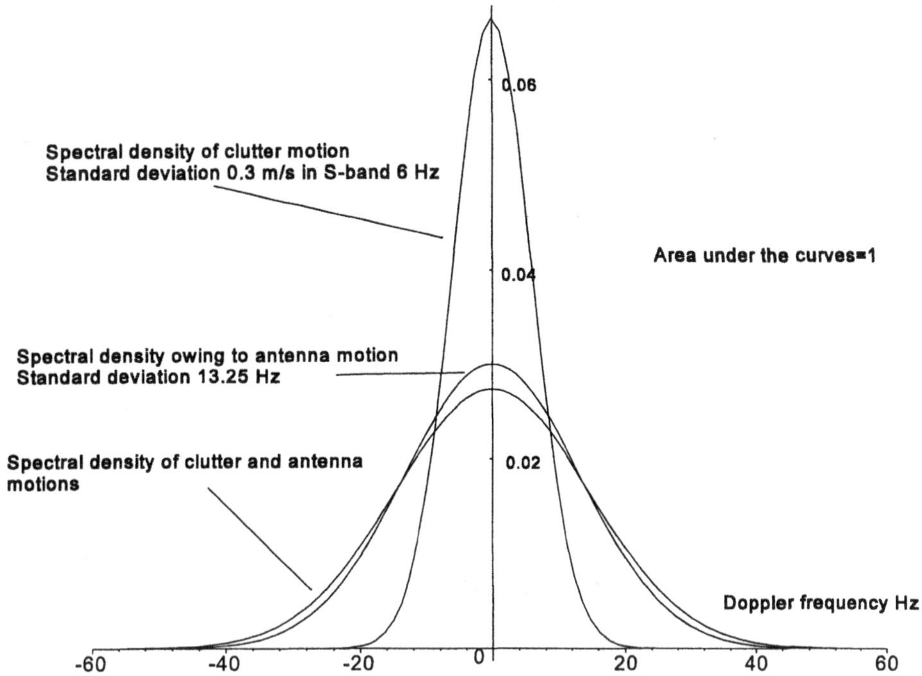

Figure 6.64 The spectral density of the clutter motion, motion caused by antenna scanning, and their sum.

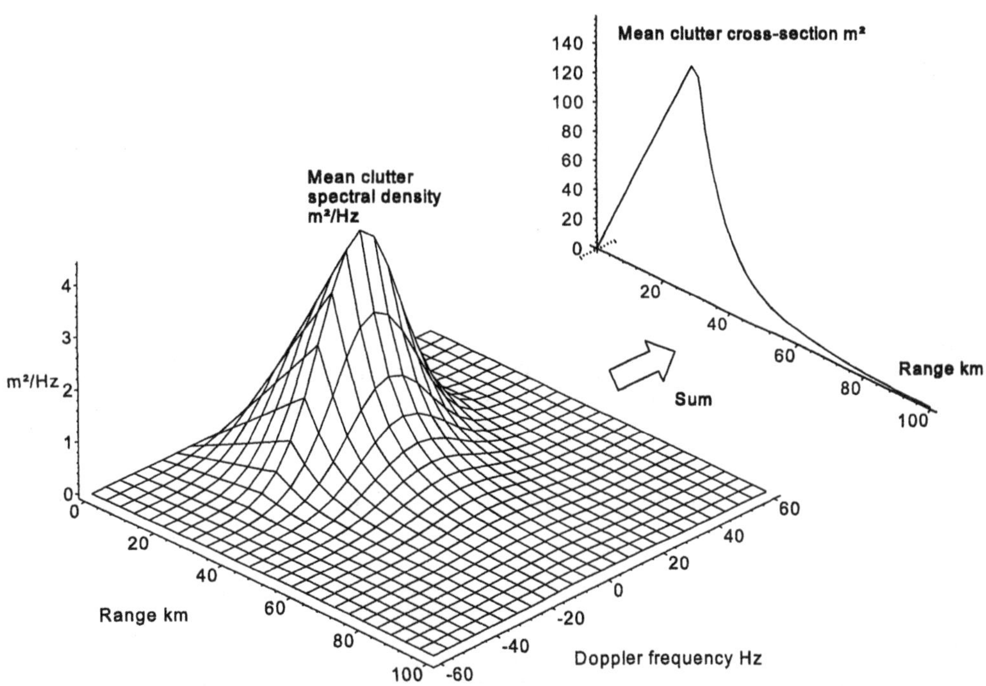

Figure 6.65 The spectral density of the mean clutter area in the example.

6.6.5 Sea clutter

The roughness of the sea depends on the wind speed which is normally measured on the Beaufort scale. Reference [21] gives the normal reflectivity, γ, as

$$10 \log \gamma = 6K_B - 10 \log \lambda - 64 \quad \text{dB/m}^2 \tag{6.93}$$

In this case, K_B is the wind speed on the Beaufort scale which is shown in Table 6.16.

Table 6.16
The Beaufort wind scale

Beaufort scale	Wind speed, knots from	Wind speed, knots to	Description	Wind speed, km/h from	Wind speed, km/h to
0		<1	Calm		<1.9
1	1	3	Light air	1.9	5.6
2	4	6	Light breeze	7.4	11.1
3	7	10	Gentle breeze	13.	18.5
4	11	16	Moderate breeze	20.4	29.6
5	17	21	Fresh breeze	31.5	38.9
6	22	27	Strong breeze	40.8	50
7	28	33	Near gale	51.9	61.1
8	34	40	Gale	63.	74.1
9	41	47	Strong gale	76.	87.1
10	47	55	Storm	87.1	101.9
11	56	63	Violent storm	103.8	116.7
12	>64		Hurricane	>118.6	

A ship's radar looks down onto the sea and the beam hits the sea at the grazing angle, ψ. The reflectivity of the sea seen by the radar is

$$\sigma_0 = \gamma \sin \psi \quad \text{m}^2/\text{m}^2 \tag{6.94}$$

where ψ is the grazing angle.

The geometry is seen in Figure 6.66. The radar antenna is situated h_{radar} meters above the sea surface. The factor $k = 4/3$ is used to modify the value of the earth's radius to take account of atmospheric refraction at radar frequencies. Using geometry,

$$\textit{Horizon slant range, } R = \sqrt{(ka + h_{radar})^2 - (ka)^2} \quad \text{m} \tag{6.95}$$

Using the cosine rule,

$$\psi = \arccos\left(\frac{(ka)^2 + R^2 - (ka + h_{radar})^2}{2\, ka\, R} \right) - \frac{\pi}{2} \quad \text{radians} \tag{6.96}$$

In the example, the antenna is 30 m above the sea surface.

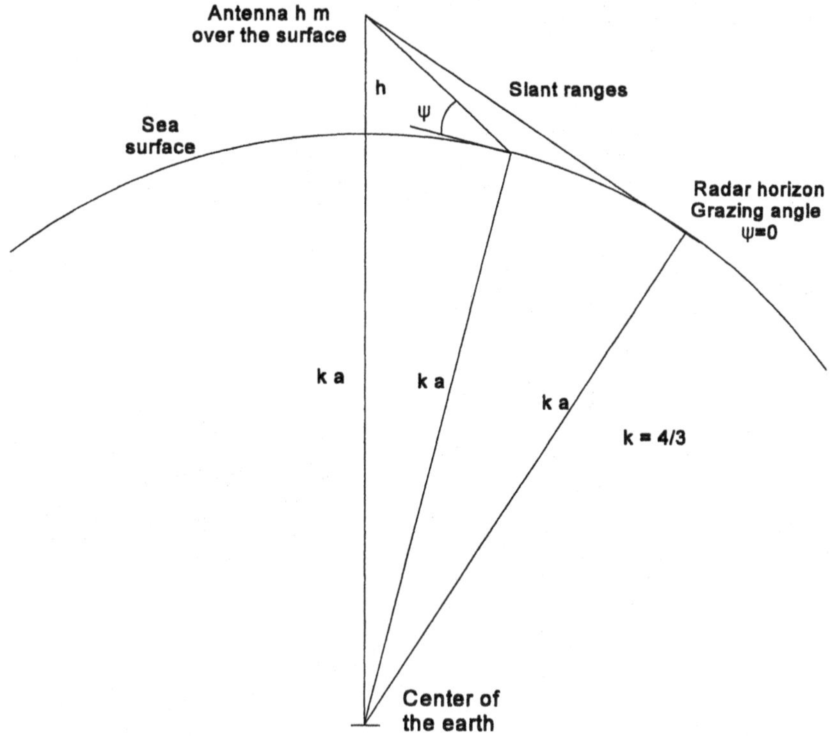

Figure 6.66 The radar's view of the sea.

The geometry of Figure 6.66 allows the reflectivity to be plotted against range, as shown in Figure 6.67.

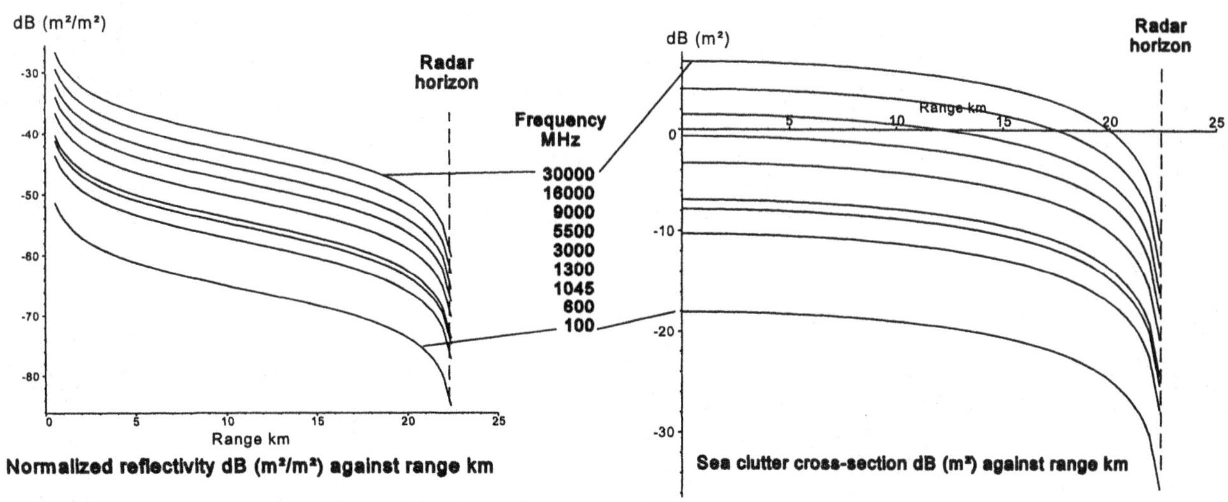

Figure 6.67 Sea clutter: normalized reflectivity, dB(m²), and clutter area seen by the antenna in the example.

The sea echo amplitudes depend on the antenna beamwidth, 1.5 degrees here, and the radar pulse width, 1 μs here. In Figure 6.67, the sea echo is plotted with the reference echo against range. The statistical distribution varies between Rayleigh, where many waves fall within the pulse footprint, to log-normal, when the waves are resolved giving clutter islands.

The clutter spectrum is assumed to be Gaussian and centered on the mean radial speed of the waves which is said [21, p. 27] to vary between an eighth and a quarter of the wind speed. The standard deviation is assumed to be one eighth of the wind speed plus the effects of antenna scanning.

For an S-band radar in a radial wind of force 5 on the Beaufort scale (say, 36 kmph) blowing towards the radar the waves are blown at 6 kmph, or 1.67 m/s. The sea clutter spectral density out to the radar horizon centered on the Doppler frequency of 33.3 Hz is shown in Figure 6.68.

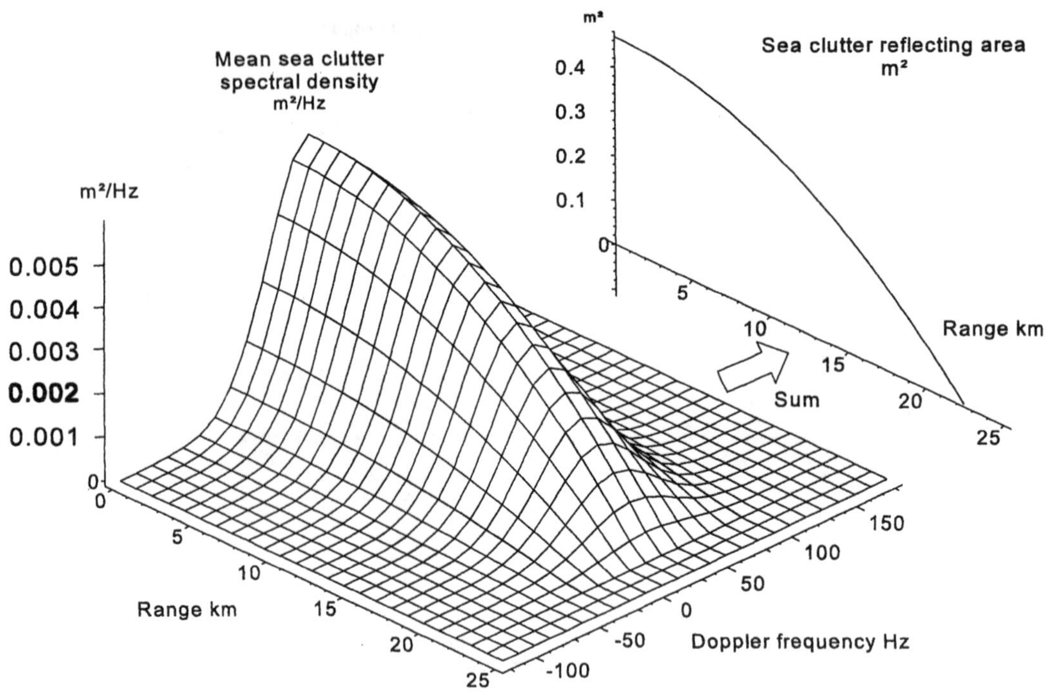

Figure 6.68 The spectral density of sea clutter in a Beaufort scale force 5 wind.

The probability distribution depends on whether the radar can resolve the individual waves in range. For low resolution radars the probability distribution is a Rayleigh distribution (mean is the 54.4% point on the cumulative distribution) and for high resolution radars it approaches log-normal. If 10 dB signal-to-noise ratio is required to detect an echo, then signal processing will be required to reduce the clutter level to -10 dB for a 1 m² reflector. Then, 54.4% of the Rayleigh sea clutter will be below this level. To the sea clutter surface in Figure 6.68 must be added the contributions from the backlobe and sidelobes. The backlobe adds an attenuated surface, which is the Doppler frequency mirror image of the main lobe clutter, and the sidelobe clutter fills the spaces in between.

When the Weibull distribution is used to represent sea clutter, the η values given in [22, p. 22] vary between 0.67 for a rough sea and 1.59 for a smooth sea and a low resolution radar.

6.6.6 Volume clutter, rain or snow clutter, and chaff or window

The clutter cross-section is calculated from the volume of the resolution cell. In addition to the two-dimensional surface area for land clutter is added the elevation beamwidth. There are two cases to be considered:

- Two-dimensional search radar with high elevation coverage;
- Pencil beam radars with a beam narrow in azimuth and elevation.

In the case of a two-dimensional search radar there is no elevation discrimination. The rain echo exists from the radar horizon up to the maximum cloud height, say, 6 000 feet or 2 000 m. The radial velocity of the rain echoes depends on the local wind speed. The local wind speed varies with altitude and this change, or wind shear, is assumed to be 4 m/s for each 1 km height difference [17, p. 207], which is an average over 360 degrees. There is a spread of the wind velocity with a standard deviation of about 1 m/s.

For rain, the reflectivity may be given by

$$\eta = \frac{\pi^5}{\lambda^4} |K^2| Z \quad \text{m}^2/\text{m}^3 \tag{6.97}$$

where $K^2 = 0.93$, the complex refractive index squared for rain;
$Z = 200\, r^{1.6}$ mm³;
r is the rainfall rate in millimeters per hour.

Reference [24] shows values for Z found by a number of authors. The values vary between $280\, r^{1.45}$ and $780\, r^{0.95}$. For the above conventional model,

$$\eta_{rain} = \frac{5.7\ 10^{-14}\ r^{1.6}}{\lambda^4} \quad \text{m}^2/\text{m}^3 \tag{6.98}$$

English descriptions for the rainfall rates, r, are given in Table 6.17.

Table 6.17
English descriptions and rainfall rates

Description	Rainfall rate mm/hour
Drizzle	0.25
Light rain	1
Moderate rain	4
Heavy rain	16
Excessive rain	40

For snow, [24] states that the reflectivity is

$$\eta = \frac{\pi^5}{\lambda^4} |K^2|\ Z \tag{6.99}$$

where $K^2 = 0.2$ the complex refractive index for snow;
$Z = 2\ 000\, r^2$ mm³;
r is the equivalent rainfall rate in mm/hour.

Thus,

$$\eta_{snow} = \frac{1.224\ 10^{-13}\ r^2}{\lambda^4} \quad \text{m}^2/\text{m}^3 \tag{6.100}$$

With narrow beam radars, the beam is filled, and the vertical extent is limited by the elevation beamwidth. The volume is multiplied by η to give the average reflecting area in square meters. Rain clutter signals are assumed to have a Rayleigh distribution.

The powers of the rain echoes received for the model S-band radar are shown in Figure 6.69 for a number of rainfall rates. If 10 dB signal-to-interference ratio is required to detect echoes of interest, the radar will detect them at all ranges in rain up to 1 mm/hour rainfall rate. There will be detection up to 2 km in 4 mm/hour rain but no detection in 16 mm/hour rain.

The average rain clutter that may be expected can be calculated. In practice, rain storms cover relatively small areas in the search volume in inverse proportion to the rainfall rate. One model [17, p. 221] is given as

$$\textit{Rain storm diameter} = 41.595 - 23.608 \log_{10}(r) \quad \text{km} \tag{6.101}$$

The CCIR reference model estimates are about one third of those in (6.101).

Chaff, or window, is sown by aircraft at up to a height of 15 km. It is specified by the reflectivity in square meters per cubic meter. It can take up to 8 hours [21] to fall from this height. The average cross-section has been given as [17, pp. 263-265]

$$\sigma_{chaff} = \frac{6600\ W}{f} = 22000\lambda \quad \text{m}^2 \tag{6.102}$$

where W is the weight of chaff in kg;
f is the frequency in GHz;
λ is the wavelength in m.

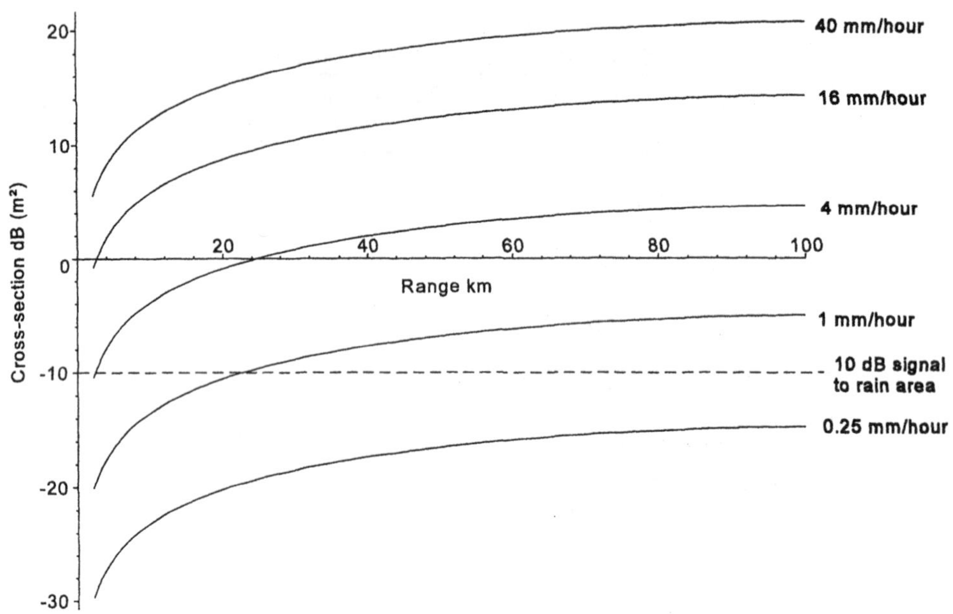

Figure 6.69 Signal-to-noise ratio of rain echoes compared with that of the reference echo without sensitivity time control for the S-band radar in the example.

6.6.7 Rain and chaff spectra

In falling, rain and chaff assume the horizontal wind speed which is not constant but normally increases with height. Values commonly used are:

- Wind shear: up to 4 m/s for each km in height, average [17, p. 242];
- Turbulence: 1 m/s standard deviation [17, p. 245];
- Falling rate variation: 1 m/s [17, p. 246].

The example shows the effects of 4 mm/hour rain with a surface wind velocity 36 km/hour (10 m/s), which has a Doppler frequency of 200 Hz in S-band. If the rain extends up to 2 000 m, then the spread in wind velocity is 8 m/s, or 160 Hz in Doppler frequency and the spectral density of the rain echoes will be as in Figure 6.70. The spectral density is also wedge shaped in the Doppler frequency dimension. This shape is caused by the earth's curvature reducing the height difference seen by the radar at longer ranges. The shapes at other rainfall rates are shown in Figure 6.71 using a logarithmic, decibel scale.

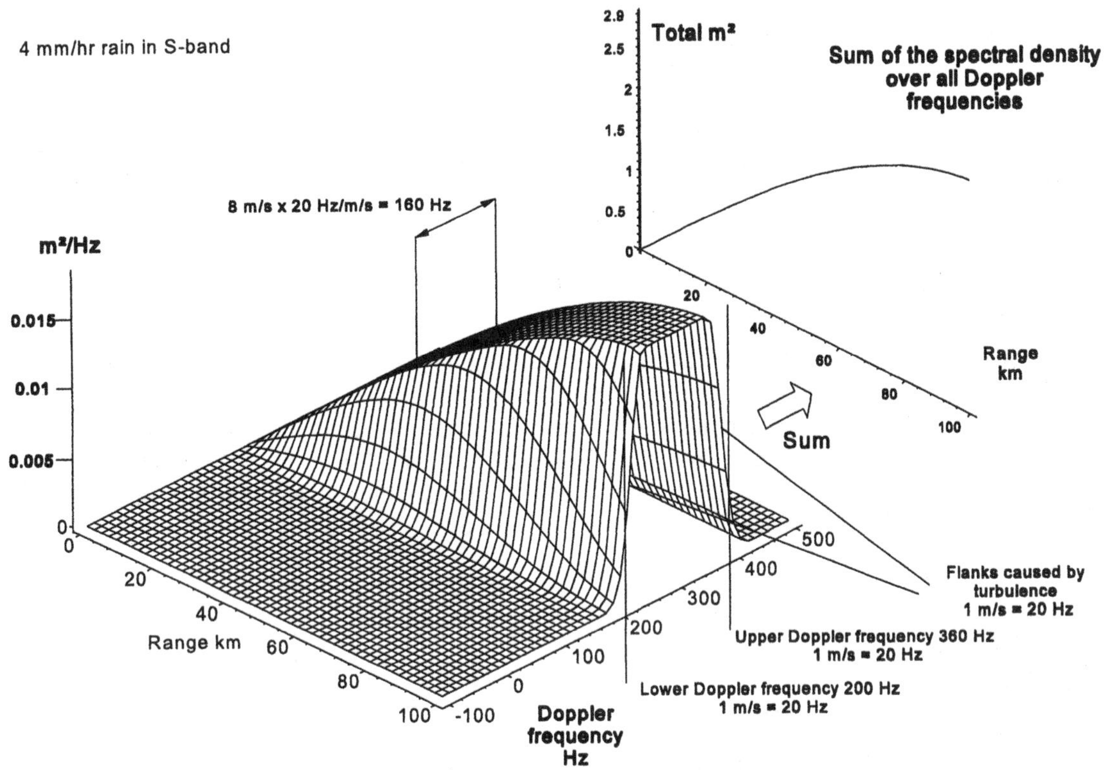

Figure 6.70 An example of the spectral density of rain clutter existing up to 2 000 m altitude for a wind surface speed of 10 kmph for the S-band radar in the example.

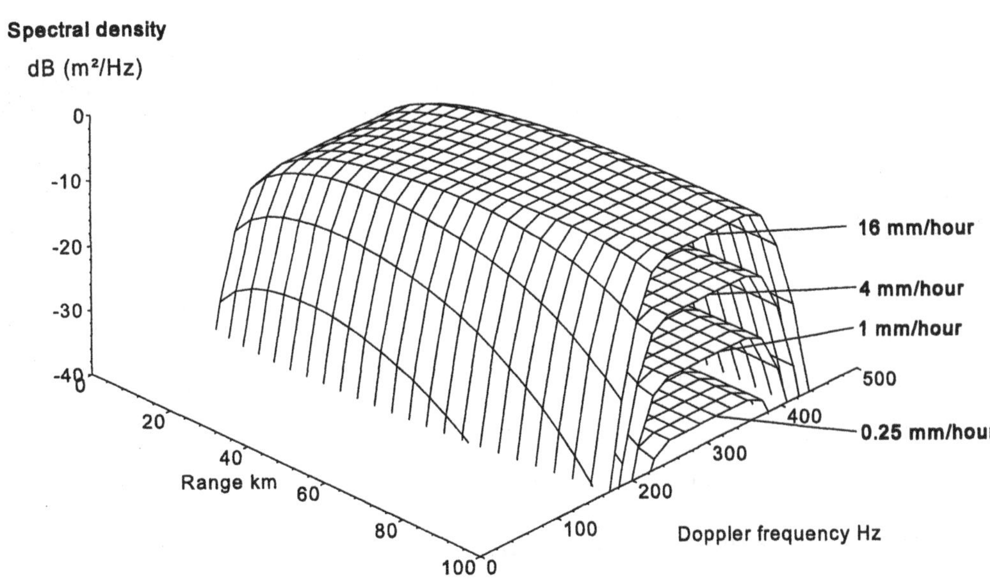

Figure 6.71 Spectral densities in dB (m²/Hz) for various rainfall rates for rain with a ground wind speed of 10 km/hour for the S-band radar in the example.

6.6.8 Total signal at the input of the receiver

The echoes that reach the receiver are:

- Wanted (aircraft) echoes;
- Weather clutter (rain and snow) echoes;
- Ground and sea clutter echoes.

All the echoes are vector voltages that add to give the resultant signal. There are, in general, two cases:

- Single or constant frequency radars;
- Frequency agile radar.

In the case of constant frequency radars, the echo from each resolution cell on the ground has a constant phase from pulse to pulse with a slight phase variation caused by, for example, wind in the trees. To this are added slowly moving rain clutter and then the wanted echo which often moves much faster. Sea clutter moves slowly. These situations are shown in Figure 6.72. Coherent signal processing is often able to separate and select the wanted signals.

Frequency agile radars change their radio frequency from pulse to pulse. The echoes have random phases from pulse to pulse and can be separated only on the basis of duration and amplitude. The dynamic range involved is dominated by the ground clutter signal, which was shown in Figure 6.59.

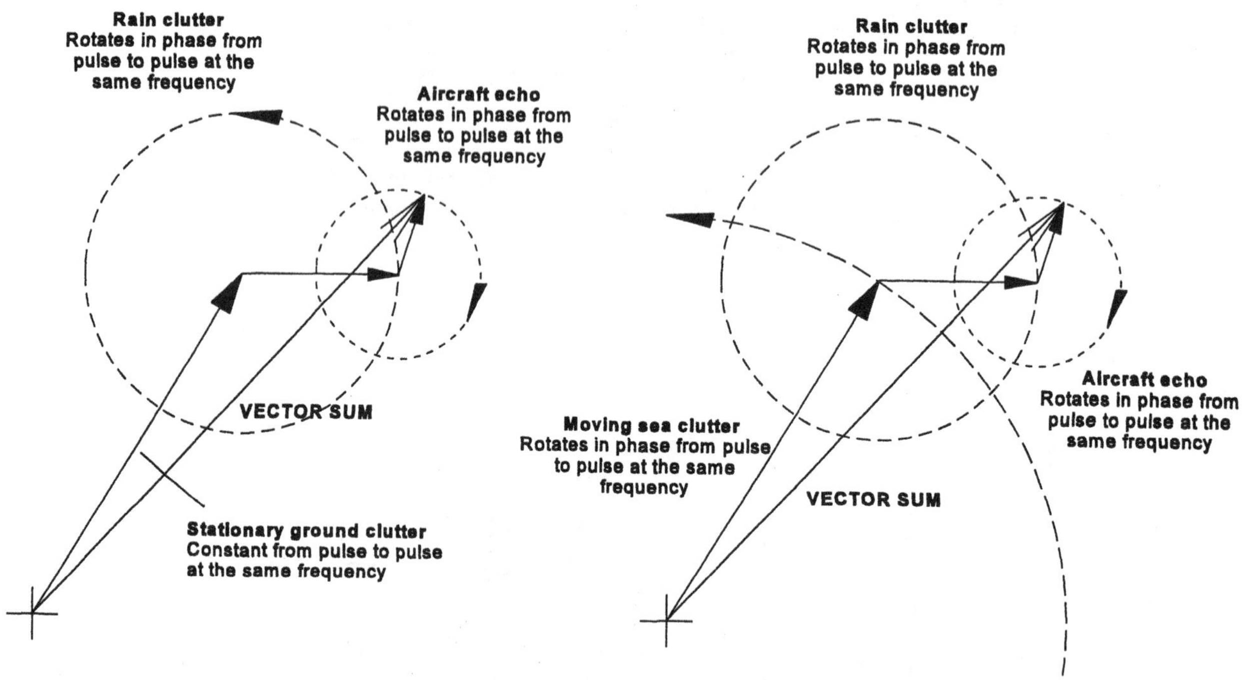

Figure 6.72 The signal for the receiver is the sum of the aircraft echo and clutter vectors.

6.7 FIGURES AFFECTING PERFORMANCE

The following figures are used in Chapter 14 to calculate radar performance, in terms of range, accuracy, and stability.

6.7.1 Range

The parameters affecting range performance are shown in Table 6.18.

Table 6.18

Parameters affecting the maximum range calculation

Parameter	Value	Units
Atmospheric attenuation from the antenna to the scatterer when necessary to include rain and snow	....	Decibels
Atmospheric attenuation from the scatterer to the antenna when necessary to include rain and snow	....	Decibels
Radar cross-section of the scatterer	....	Square meters
Echo loss when circular polarization is used, normally 3 dB	....	Decibels

6.7.2 Accuracy

Azimuth and elevation angle measurements are the angles measured at the antenna. For very accurate measurements, the refraction profile of the atmosphere must be known. Examples are:

- Azimuth calibration using the sun is not carried out when weather fronts are in the area.
- A *D*-factor is used to convert height measured by a radar to flight levels used by the controllers. This *D*-factor is often forecast.
- The *k*-factor used in the constant *k* model may be forecast.
- Radars that locate artillery use the atmospheric profiles used to aim the guns.

The effects of ground reflections are reduced by using an antenna with low sidelobe levels and by mounting the radar antennas high above the ground on towers, which increases the number and blurs the reflection lobes.

6.7.3 Stability

It is assumed that the atmosphere and the scatterer are stable during an individual pass of the antenna. For longer observation times with staring beams, for example with battlefield observation radars, moving scatterers, such as men or deer, may be recognized by the Doppler frequencies generated by movement.

REFERENCES

1. *IEEE Standard Radar Definitions, IEEE Standard 686-1990,* New York: The Institute of Electrical and Electronic Engineers, 1993.
2. *Aeronautical Telecommunications Annex 10 to the Convention on International Civil Aviation,* Vol. 1, Montreal: International Civil Aviation Organization, April 1985.
3. Blake, L. V., *Radar Range-Performance Analysis,* Norwood, Massachusetts: Artech House, 1986.
4. *CRPL Exponential Reference Atmosphere, National Bureau of Standards Monograph 4,* Washington, DC: GPO, October 1969.
5. Bijvoet, J. A., *Standard Methods for Predicting and Specifying Performance of Air Surveillance Systems,* The Hague: Supreme Headquarters Allied Powers Europe (SHAPE) Technical Centre Technical Report TR-50.
6. Comité Consultif des Radiocommunications, XVth Plenary Assembly, Vol. V, Report 563-2, Geneva: Comité Consultif des Radiocommunications, 1982.
7. Reed, H. R., and C. M. Russel, *Ultra High Frequency Propagation,* Boston: Boston Technical Publishers, 1964, and London: Science Paperbacks, 1965.
8. Skolnik, M. I., *Radar Handbook*, New York: McGraw-Hill, 1970.
9. Weil, T. A., "Atmospheric Lens Effect", *IEEE Transactions, AES-9*, January 1973.
10. Blake, L. V., *A Guide to Basic Pulse-Radar Maximum-Range Calculation*, Washington DC: U.S. Naval Research Laboratory, December 1969.
11. Blake, L. V., *Radar/Radio Tropospheric Absorption and Noise Temperature*, Washington, DC: U.S. Naval Research Laboratory Report 7461, October 1972.
12. Goldstein, H., "Attenuation by Condensed Water", in D. E. Kerr, (ed.), *Propagation of Short Radio Waves*, Volume 13 in the MIT Radiation Laboratory Series, New York: McGraw-Hill, 1951.

13. Goldstein, H., in L. V. Blake, *Radar Range-Performance Analysis,* Norwood, Massachusetts: Artech House, 1986.
14. Rivers, W. K., in L. V. Blake, *Radar Range-Performance Analysis*, Norwood, Massachusetts: Artech House, 1986, p. 220.
15. Kerr, D. E., *Propagation of Short Radio Waves*, Volume 13 of the MIT Radiation Laboratories Series, New York: McGraw-Hill, 1951, pp. 451-451.
16. Abramowitz, M., and I. A. Stegun, *Handbook of Mathematical Functions*, New York: Dover, 1965.
17. Nathanson, F. E., *Radar Design Principles*, New York: McGraw-Hill, 1991, p. 176.
18. Hahn, G. J., and S. S. Shapiro, *Statistical Models in Engineering*, New York: John Wiley, 1967, p. 122.
19. Bracewell, R. N., *The Fourier Transform and Its Applications*, 2nd ed., New York: McGraw-Hill, 1978.
20. Chandrasekhar, S., "Stochastic Problems in Physics and Astronomy", *Rev. Modern Phys.* Vol. 15, No. 1, 1943, pp. 1-89,. Reprinted in Wax, N., (ed.), *Noise and Stochastic Processes* New York: Dover, 1954, pp. 3-91.
21. Barton, D. K., *Modern Radar System Analysis*, Norwood, Massachusetts: Artech House, 1988, p. 86.
22. Barton, D. K., "Land Clutter Models for Radar Design and Analysis", *Proceedings of the IEEE,* Vol. 23, February 1985.
23. Sekine, M., and Y. H. Mao, *Weibull Radar Clutter*, Stevenage, Herts: Peter Peregrinus, 1990.
24. Morchin, W., *Radar Engineer's Sourcebook*, Norwood, Massachusetts: Artech House, 1993.

APPENDIX 6A RANGE-HEIGHT PAPER

This appendix gives examples of range-height paper that are drawn for three types of radar, for three atmosphere models, and for nautical miles and kilometer scales. The coverages for the radars are listed in Table 6A.1. Military gap filler radars cover volumes below the radar horizons of the neighboring surveillance radars and have a similar coverage to that of air traffic control *en route* radars.

Table 6A.1
Coverages for three common types of radar

	Nautical miles		Metric	
	Range	Height	Range	Height
Airport radar	0..70 nmi	0..30 kft	0..130 km	0..10 km
Air traffic control, *en route*	0..160 nmi	0..80 kft	0..300 km	0..25 km
Military gap filler				
Large military surveillance radar	0..300 nmi	0..200 kft	0..550 km	0..80 km

There are range-height papers for three model atmospheres in this appendix:

Constant k = 4/3 paper

	Nautical miles	Metric
Airport radar	Figure 6A.1	Figure 6A.4
Air traffic control *en route*	Figure 6A.2	Figure 6A.5
Military gap filler		
Large military surveillance	Figure 6A.3	Figure 6A.6

European exponential atmosphere N = 320 exp(-0.03709 h)

	Nautical miles	Metric
Airport radar	Figure 6A.7	Figure 6A.10
Air traffic control *en route*	Figure 6A.8	Figure 6A.11
Military gap filler		
Large military surveillance	Figure 6A.9	Figure 6A.12

CRPL exponential atmosphere N_s = 313

	Nautical miles	Metric
Airport radar	Figure 6A.13	Figure 6A.16
Air traffic control *en route*	Figure 6A.14	Figure 6A.17
Military gap filler		
Large military surveillance	Figure 6A.15	Figure 6A.18

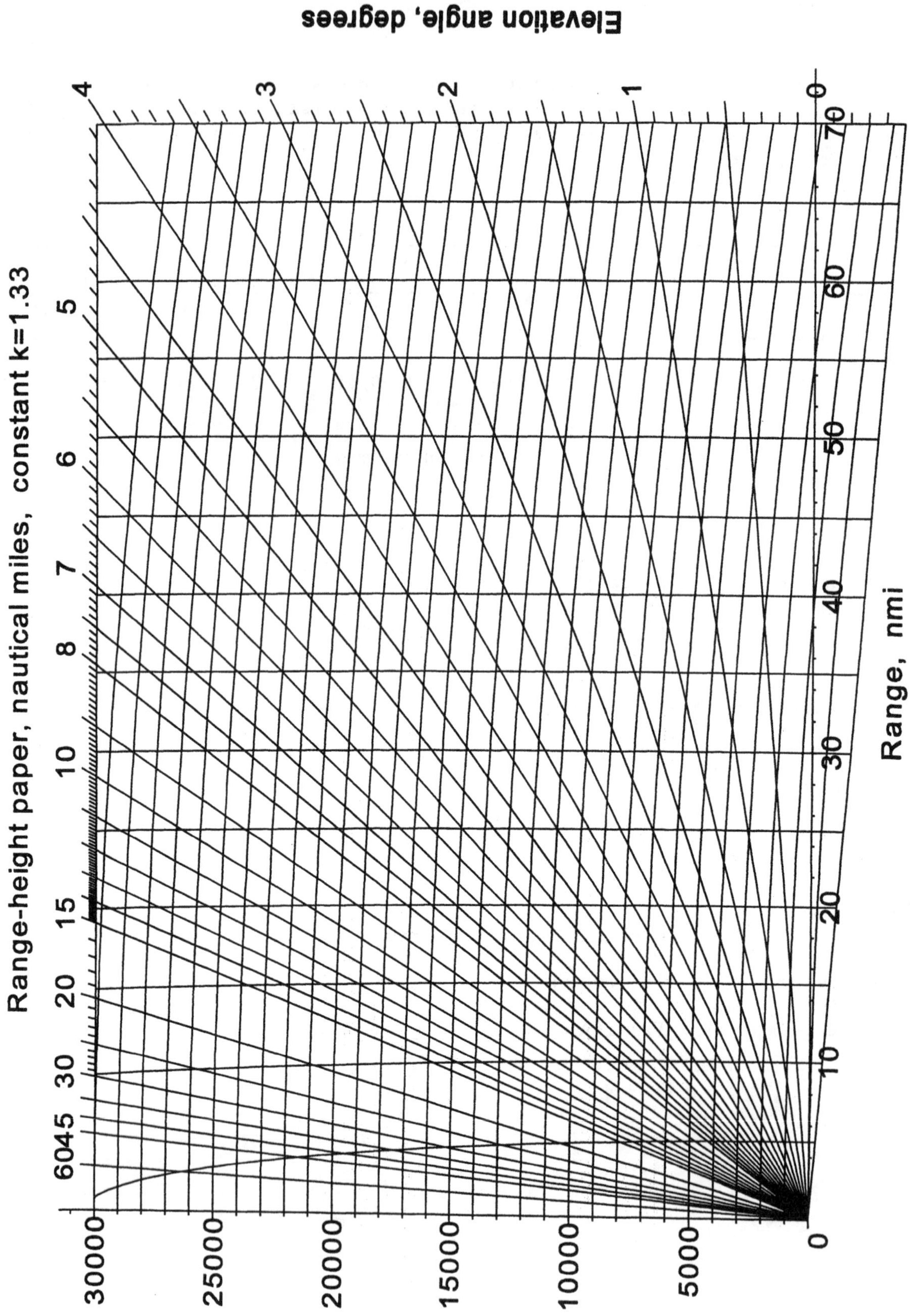

Figure 6A.1 Constant $k = 4/3$ range-height paper, range 70 nautical miles.

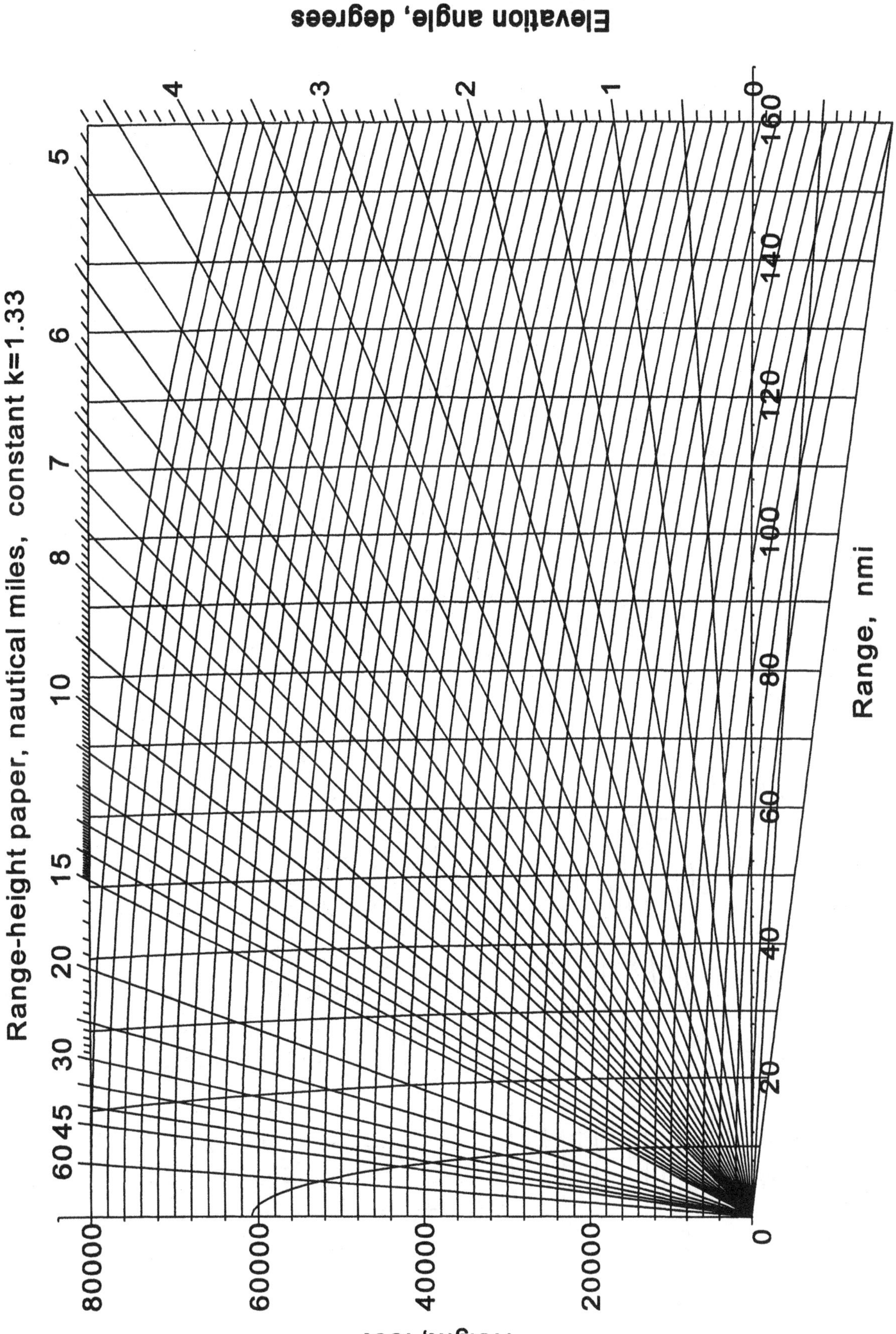

Figure 6A.2 Constant $k = 4/3$ range-height paper, range 160 nautical miles.

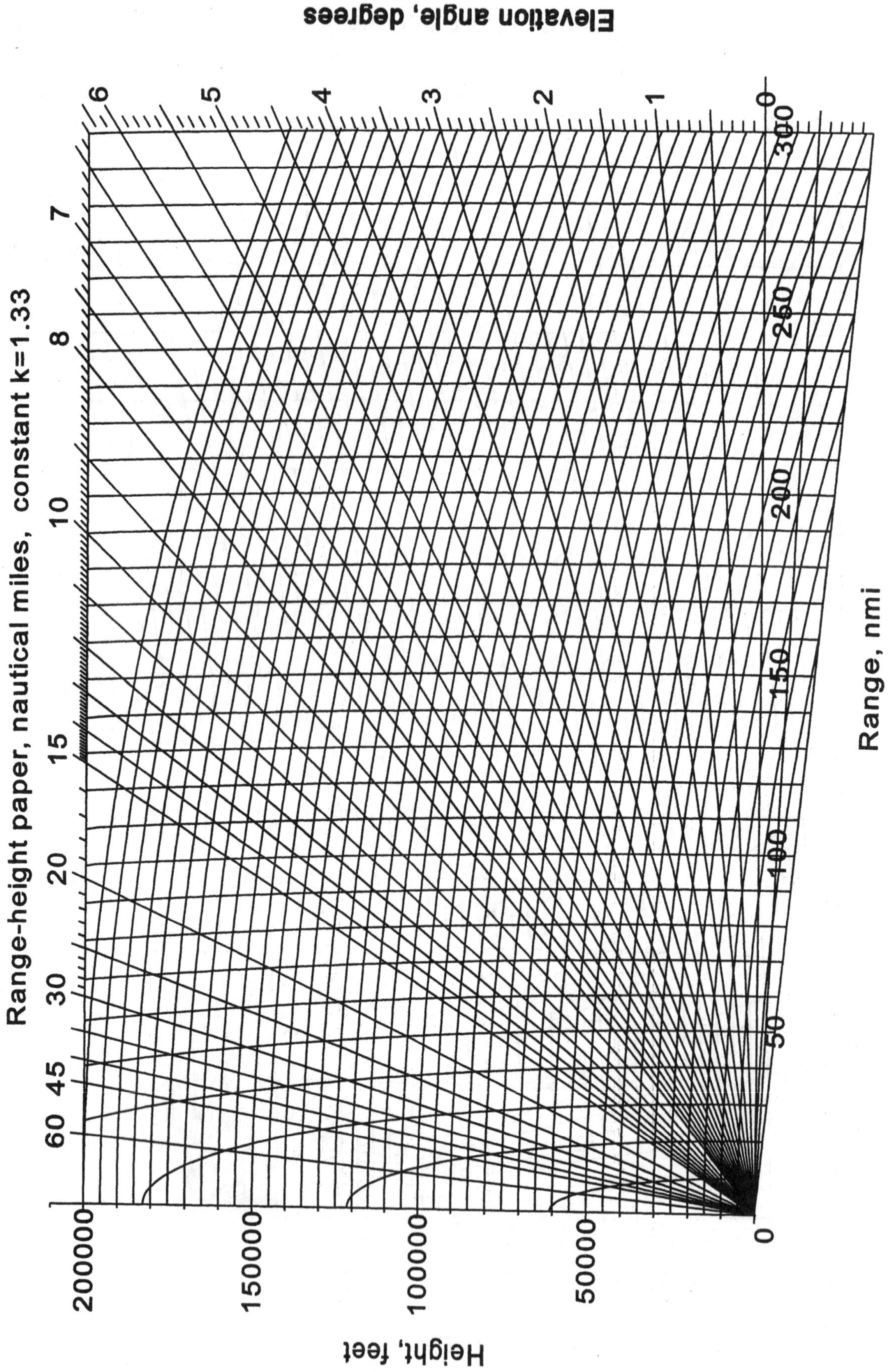

Figure 6A.3 Constant $k = 4/3$ range-height paper, range 300 nautical miles.

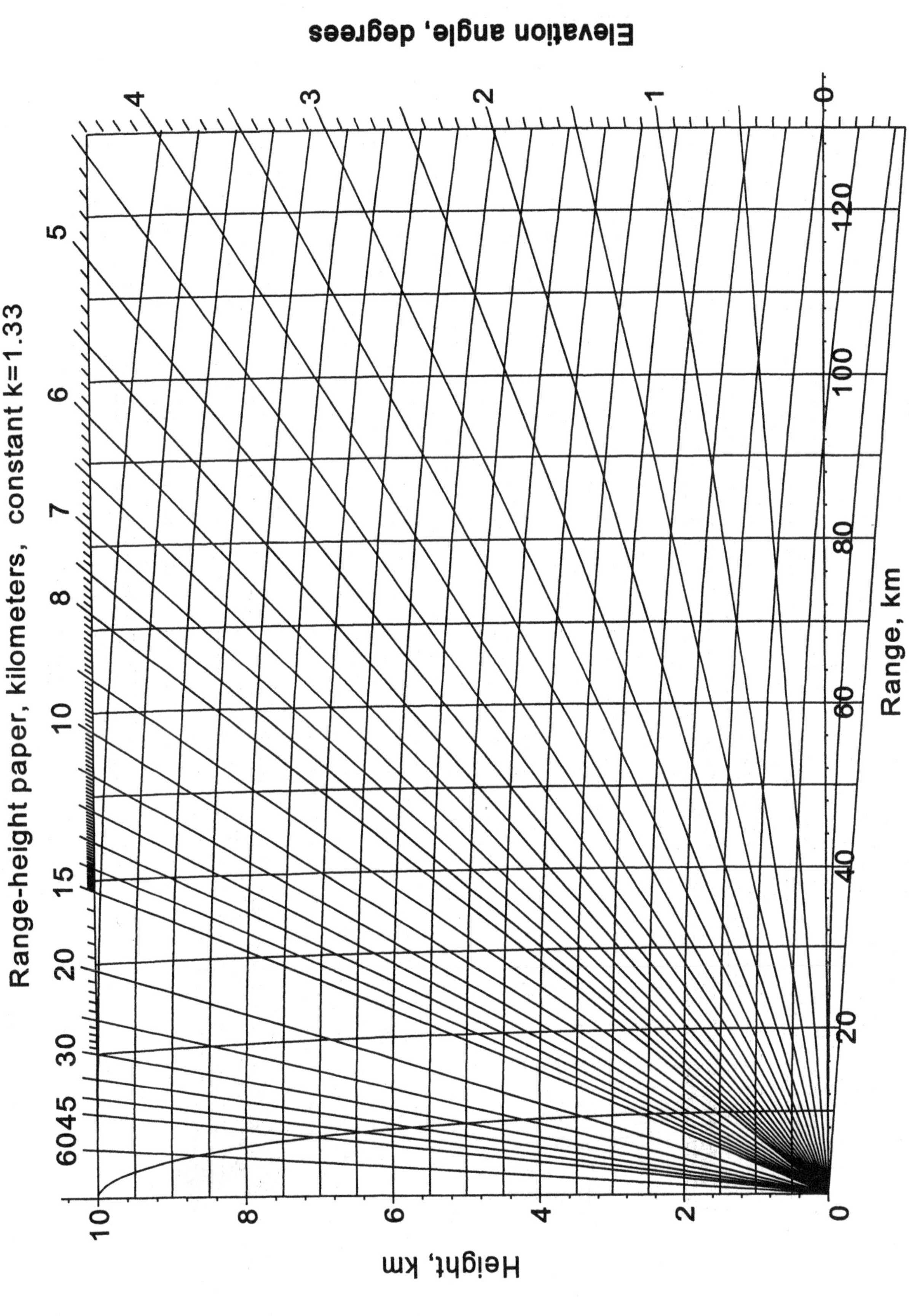

Figure 6A.4 Constant $k = 4/3$ range-height paper, range 130 km.

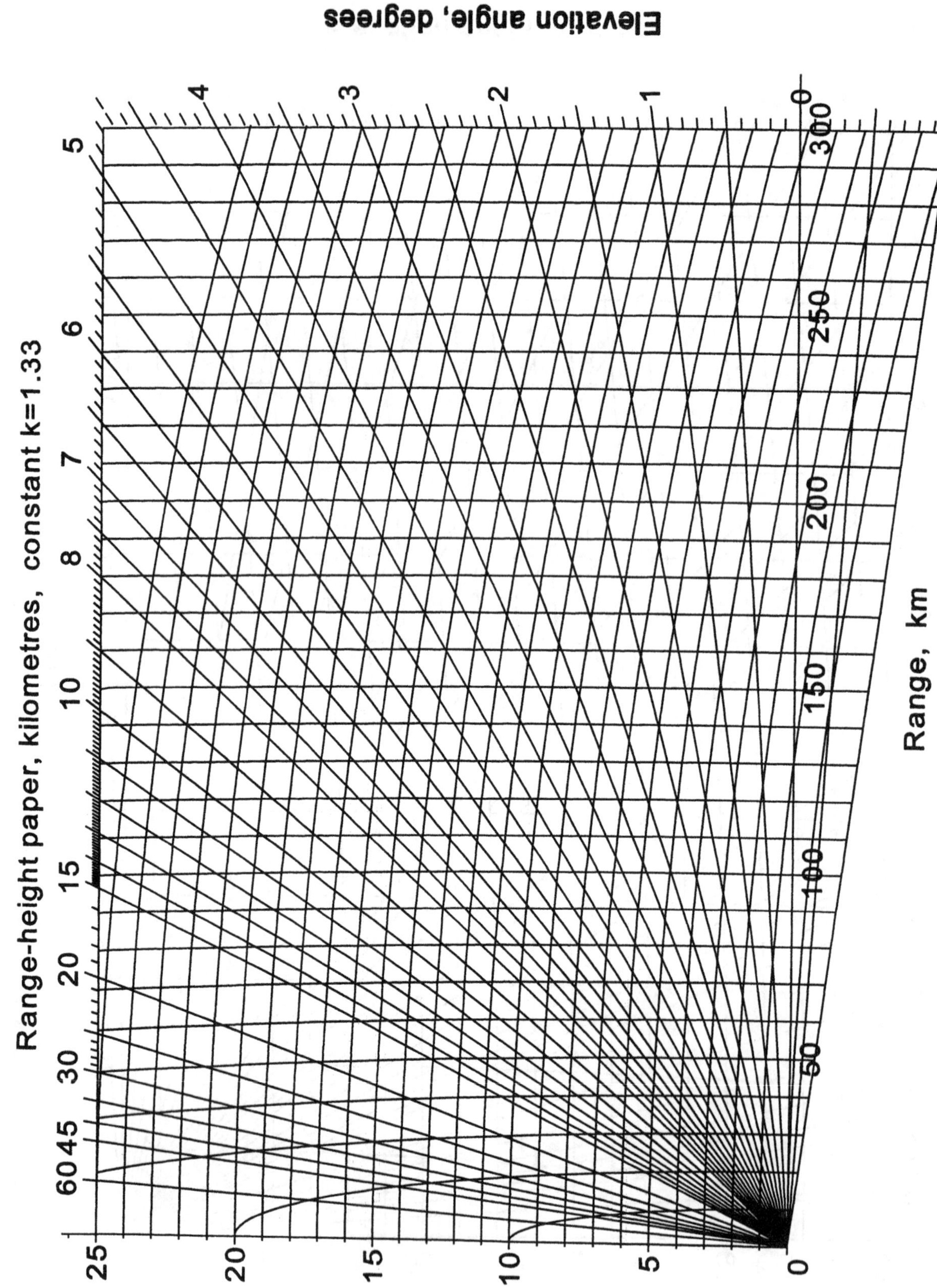

Figure 6A.5 Constant $k = 4/3$ range-height paper, range 300 km.

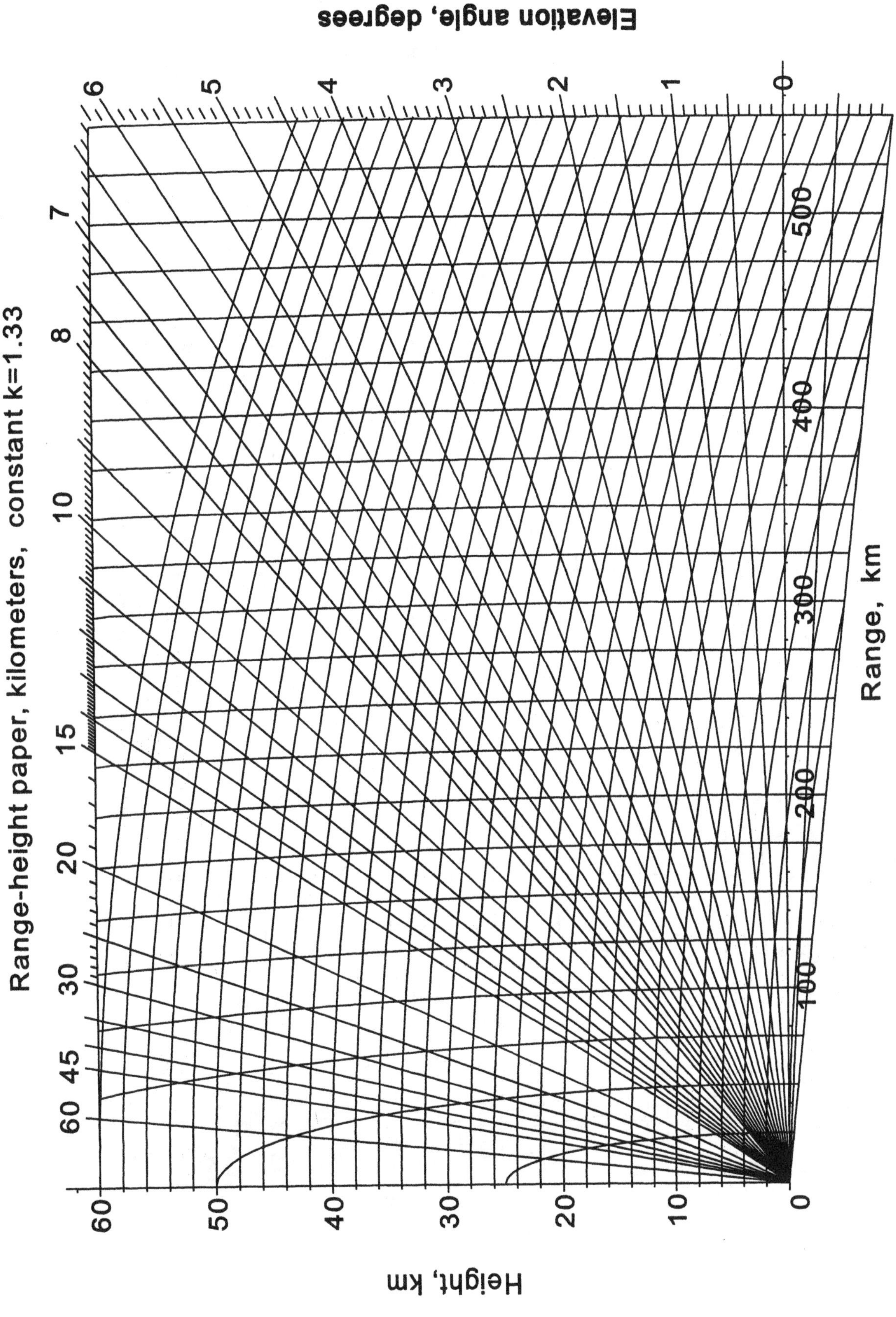

Figure 6A.6 Constant $k = 4/3$ range-height paper, range 550 km.

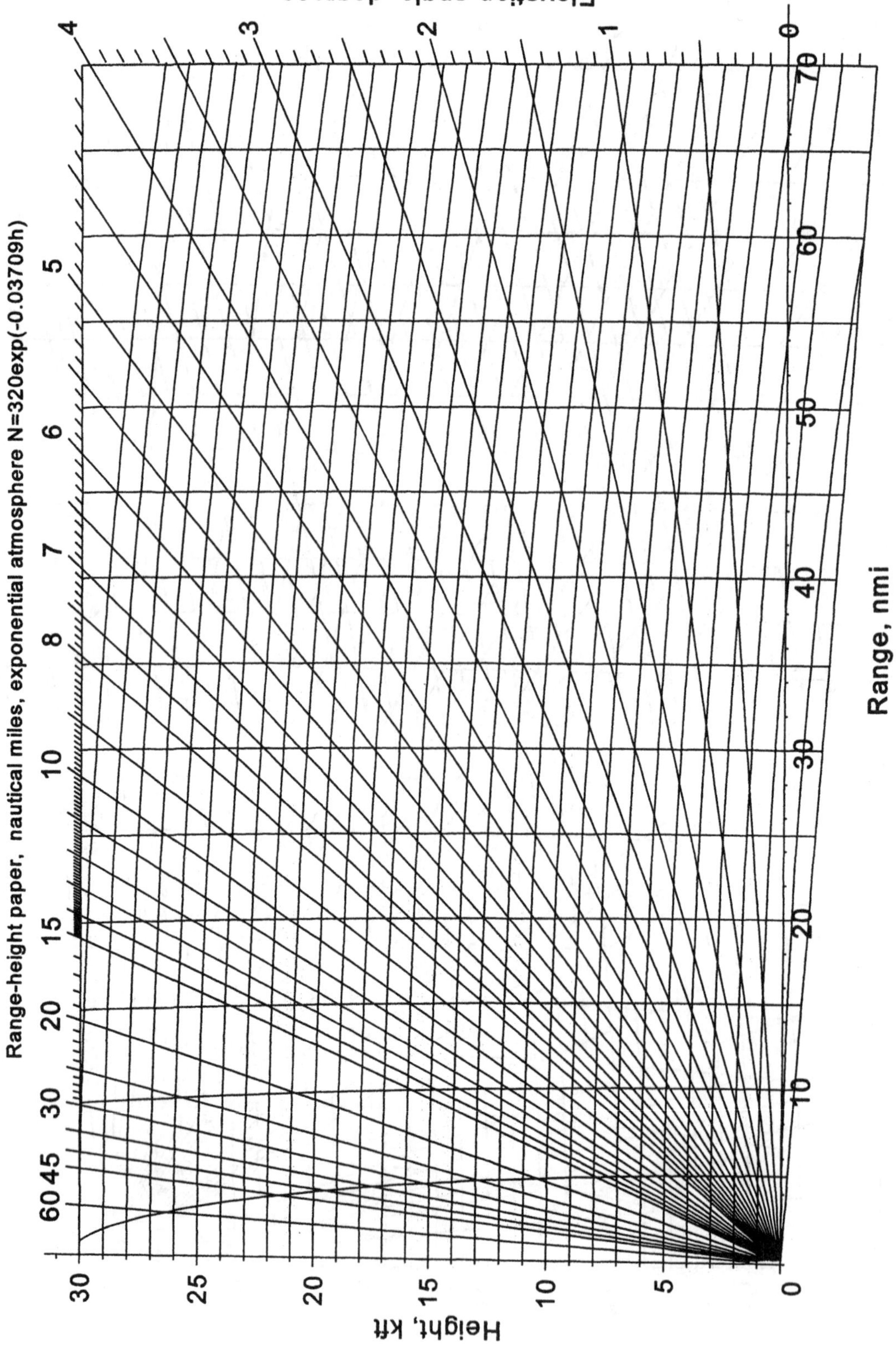

Figure 6A.7 European exponential atmosphere out to 70 nautical miles.

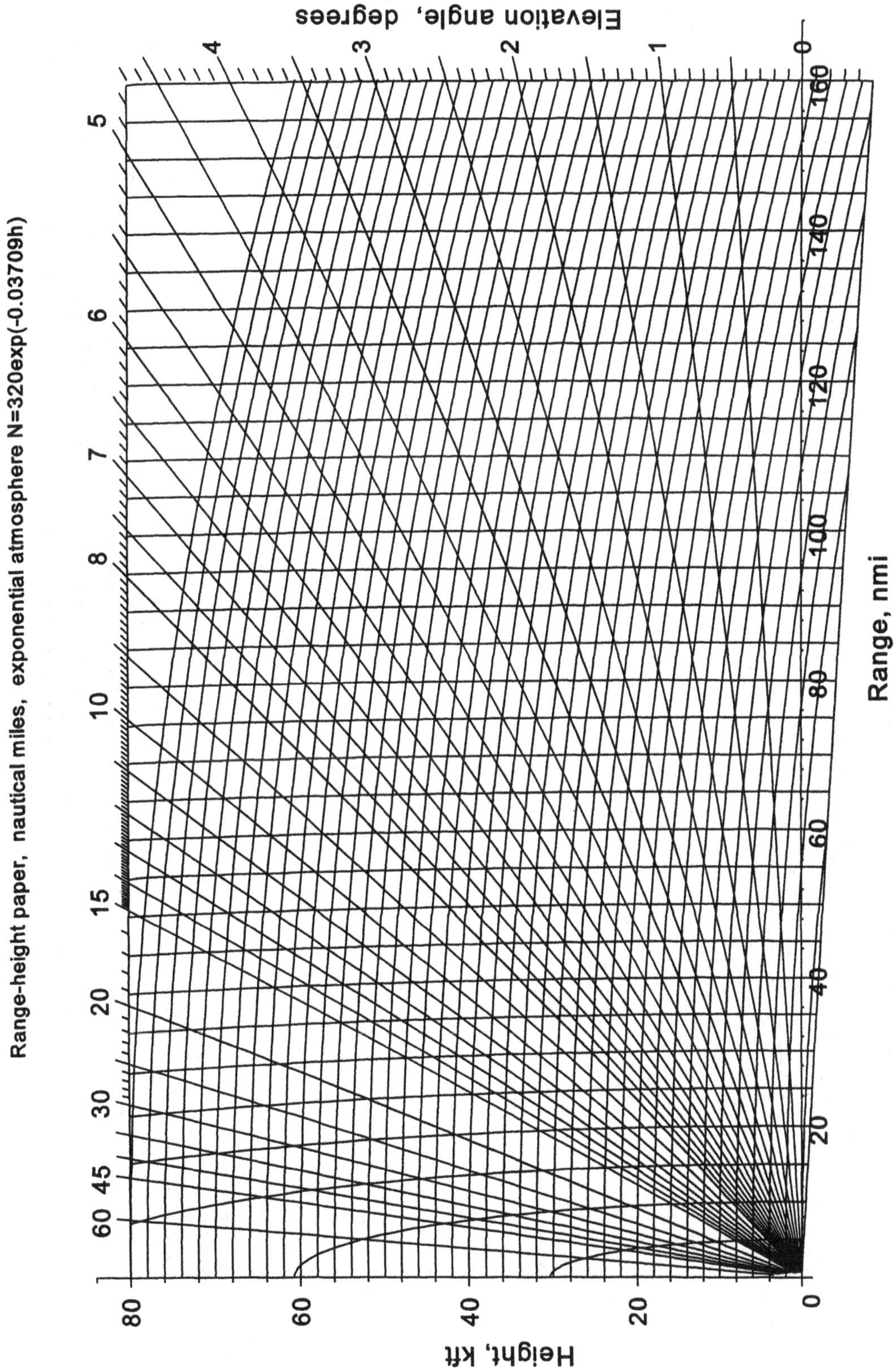

Figure 6A.8 European exponential atmosphere out to 160 nautical miles.

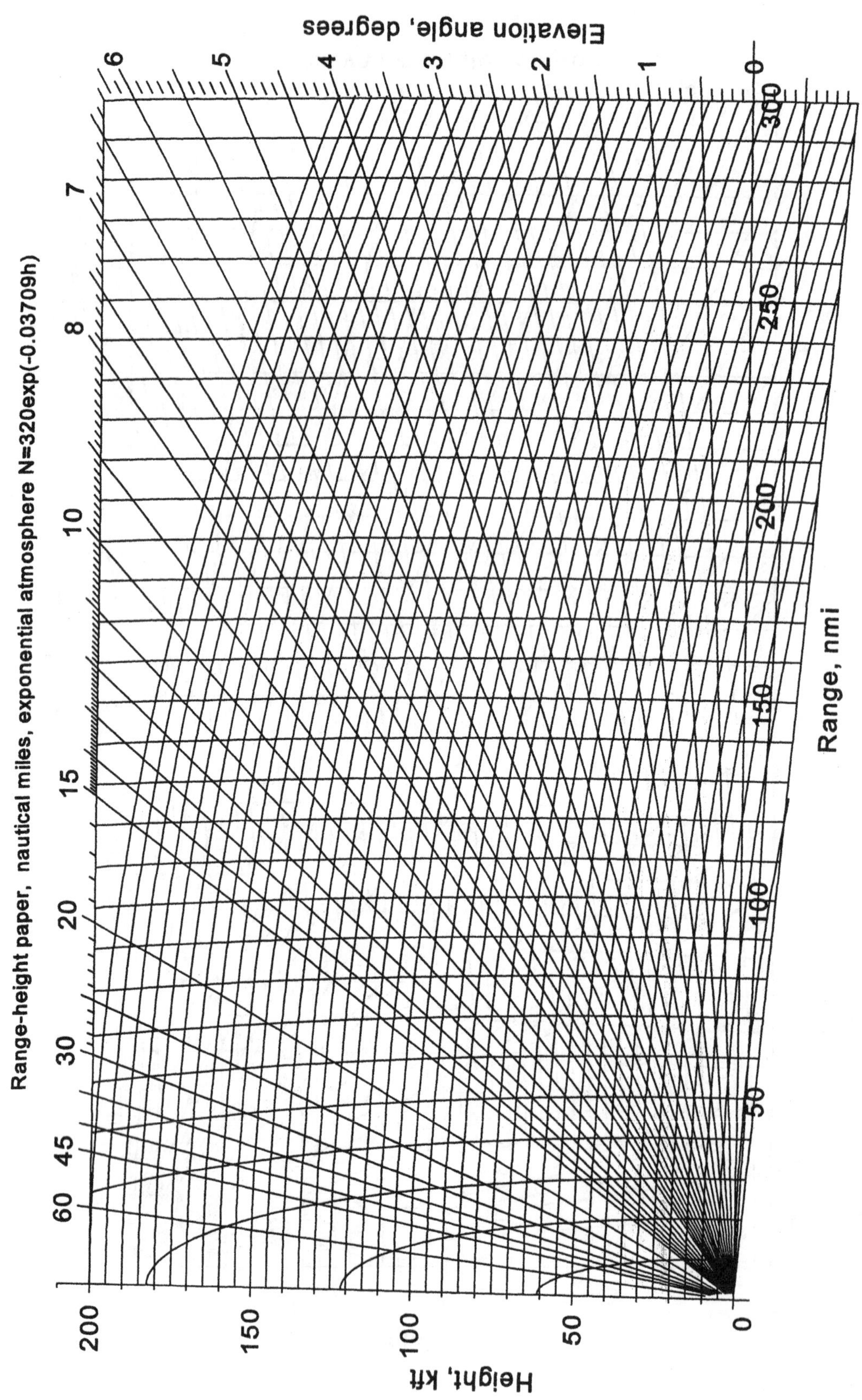

Figure 6A.9 European exponential atmosphere out to 300 nautical miles.

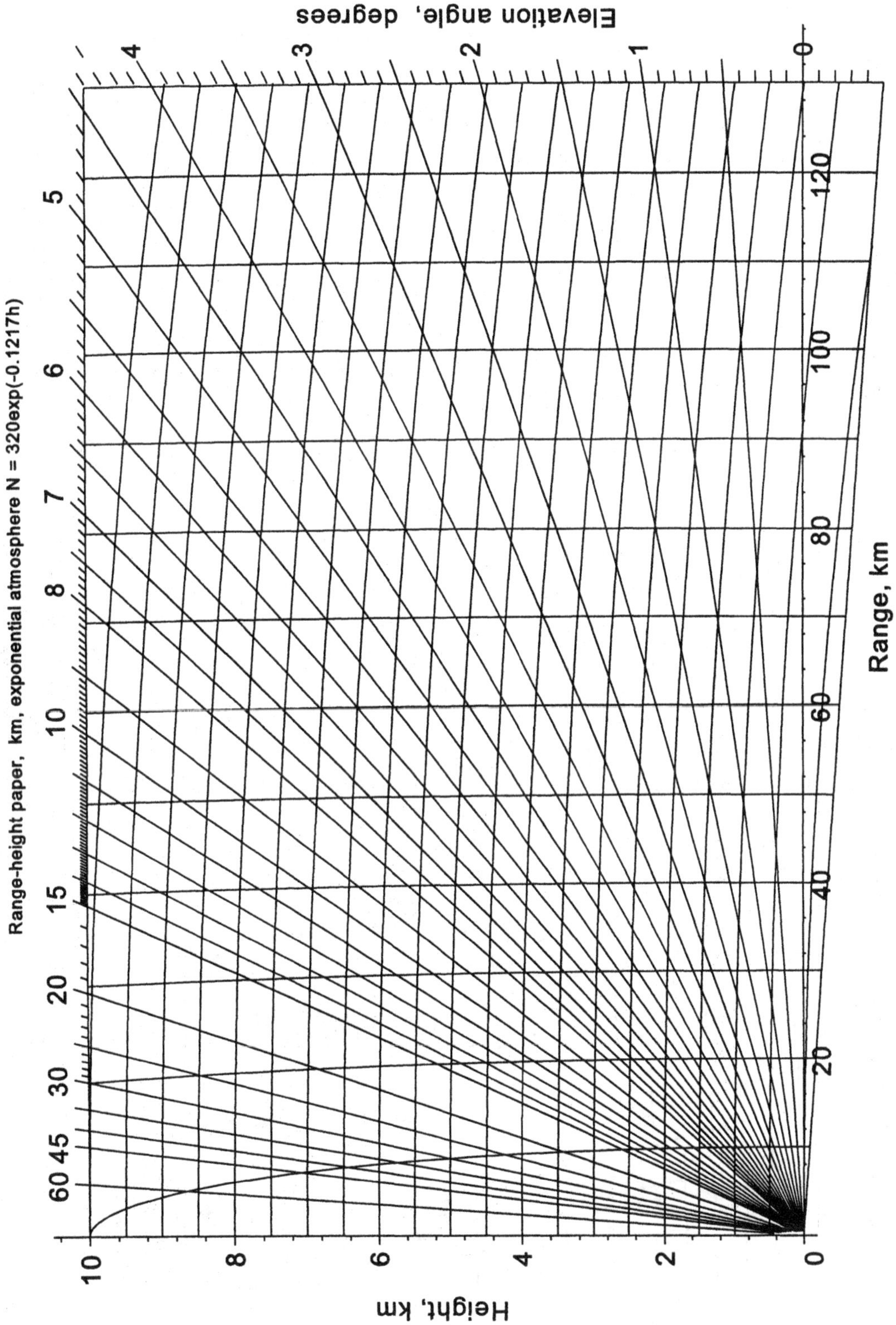

Figure 6A.10 European exponential atmosphere out to 130 km.

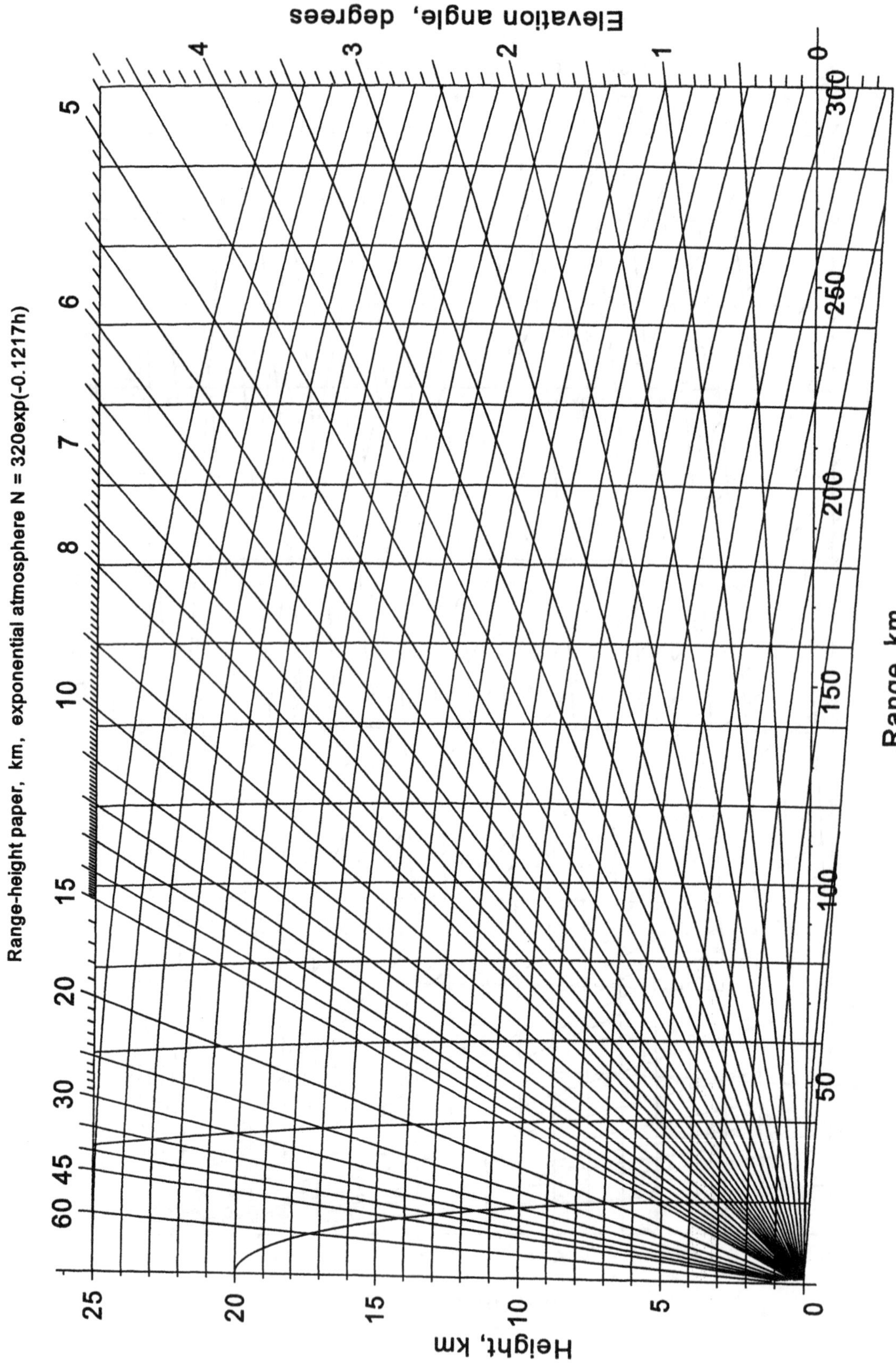

Figure 6A.11 European exponential atmosphere out to 300 km.

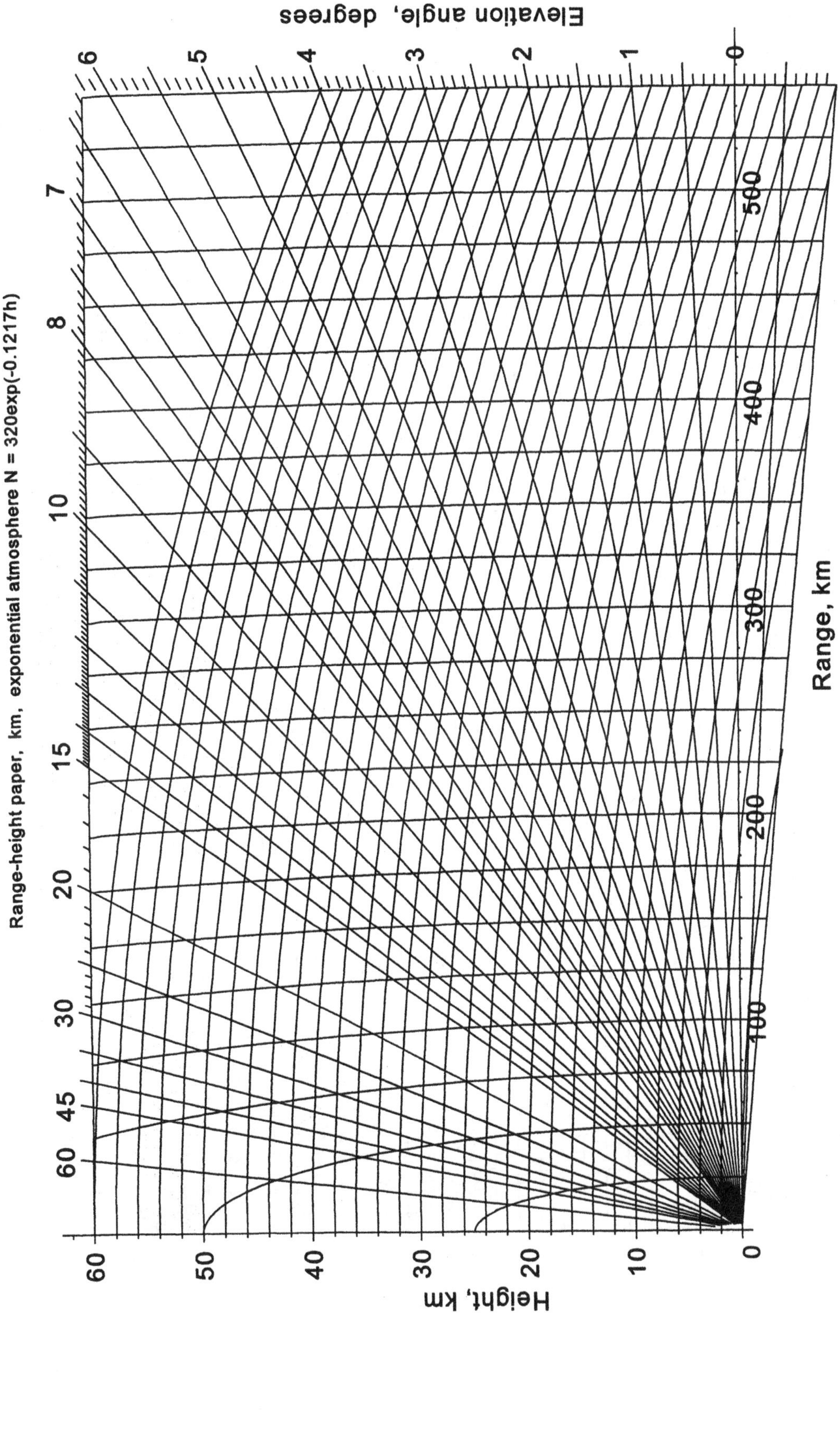

Figure 6A.12 European exponential atmosphere out to 550 km.

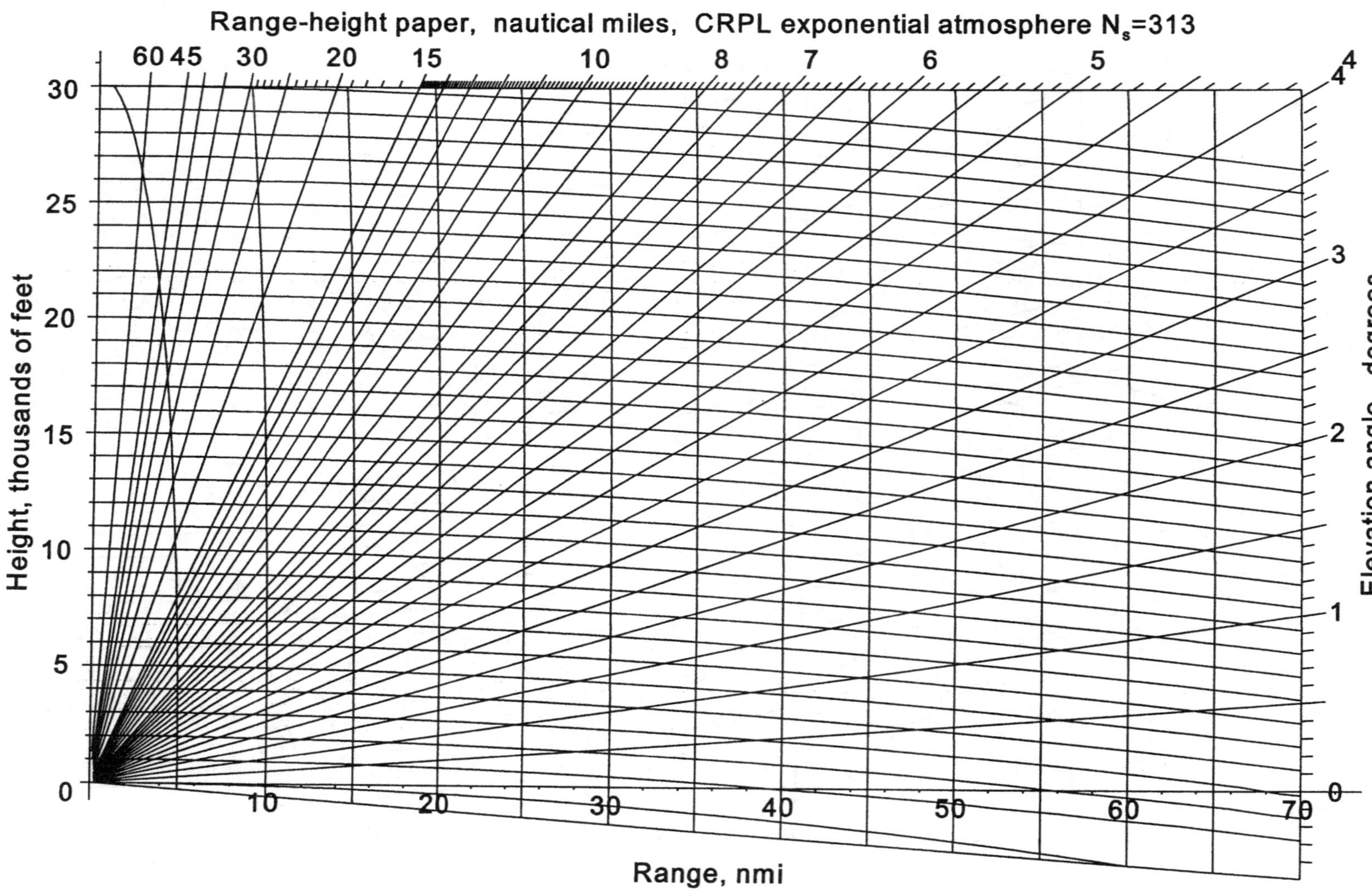

Figure 6A.13 CRPL exponential atmosphere N_s = 313 out to 70 nautical miles.

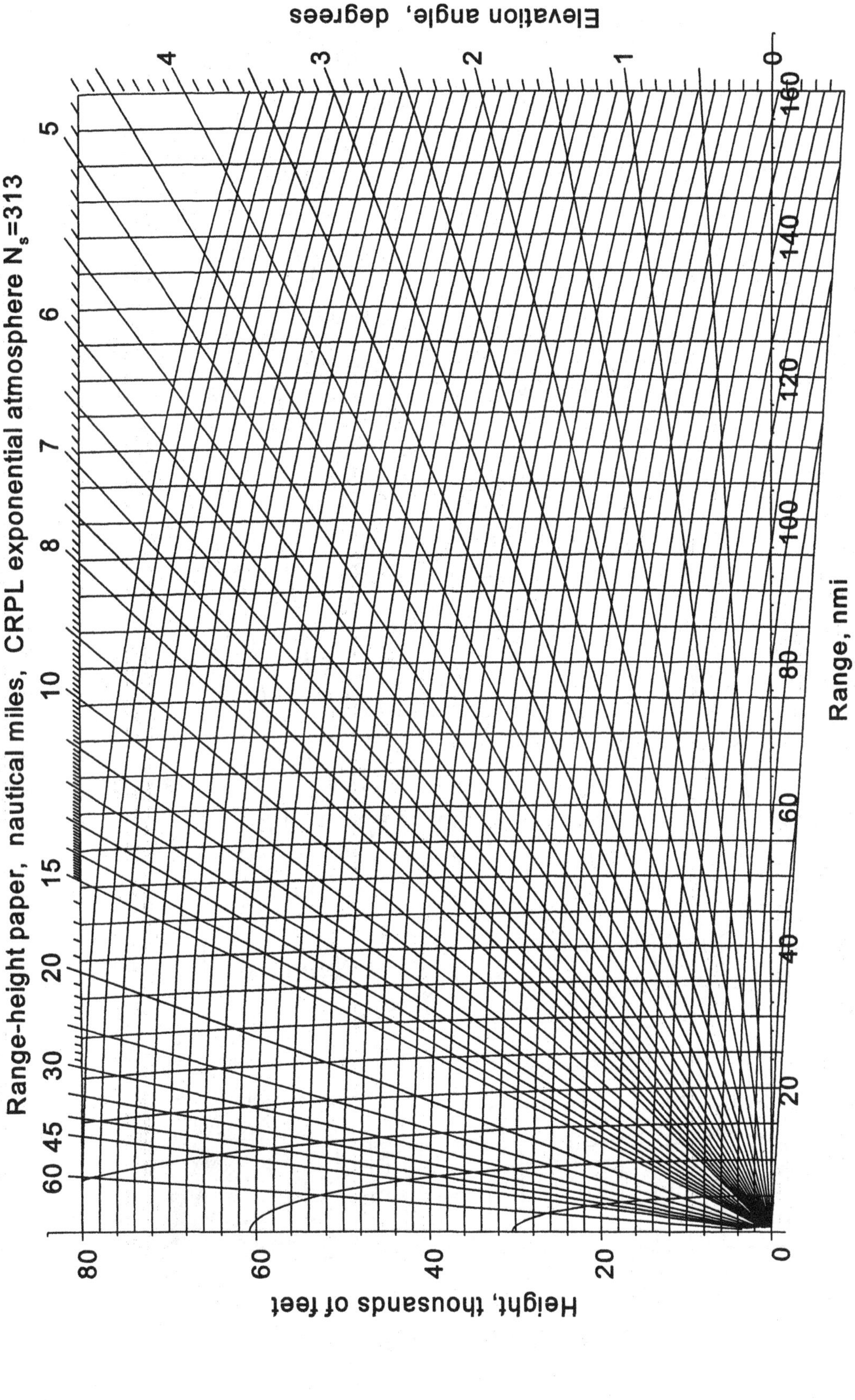

Figure 6A.14 CRPL exponential atmosphere $N_s = 313$ out to 160 nautical miles.

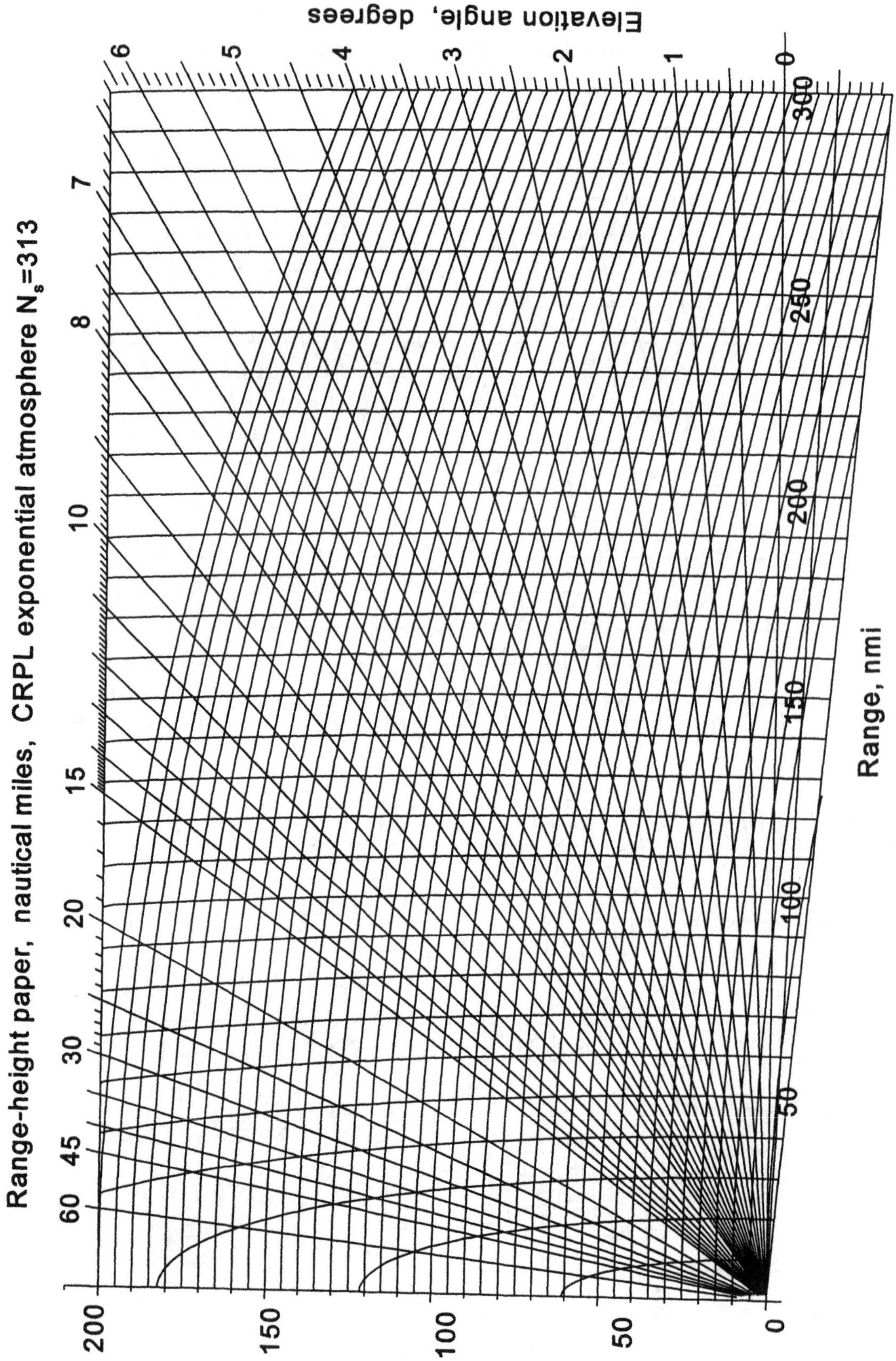

Figure 6A.15 CRPL exponential atmosphere N_s = 313 out to 300 nautical miles.

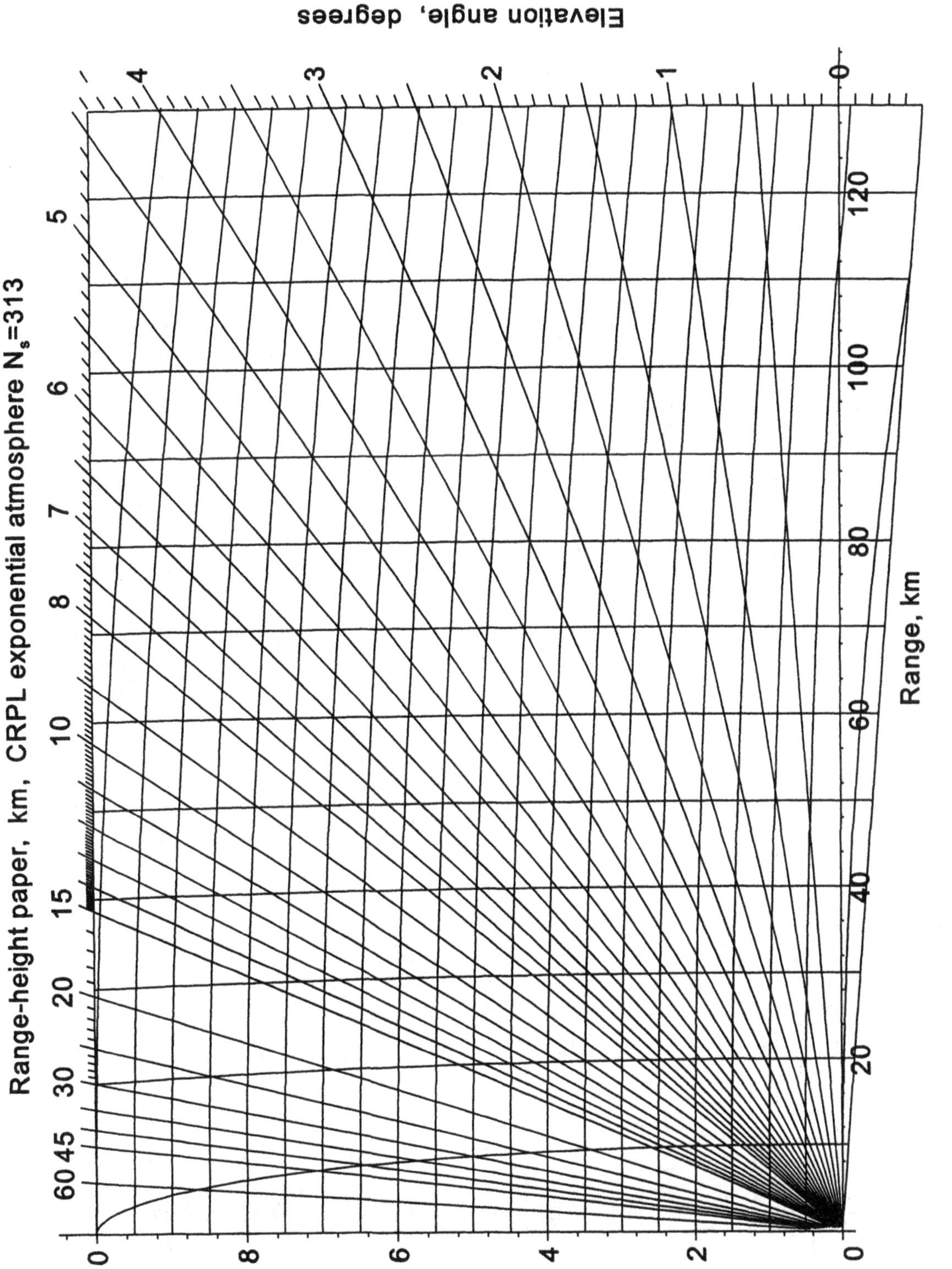

Figure 6A.16 CRPL exponential atmosphere N_s = 313 out to 130 km.

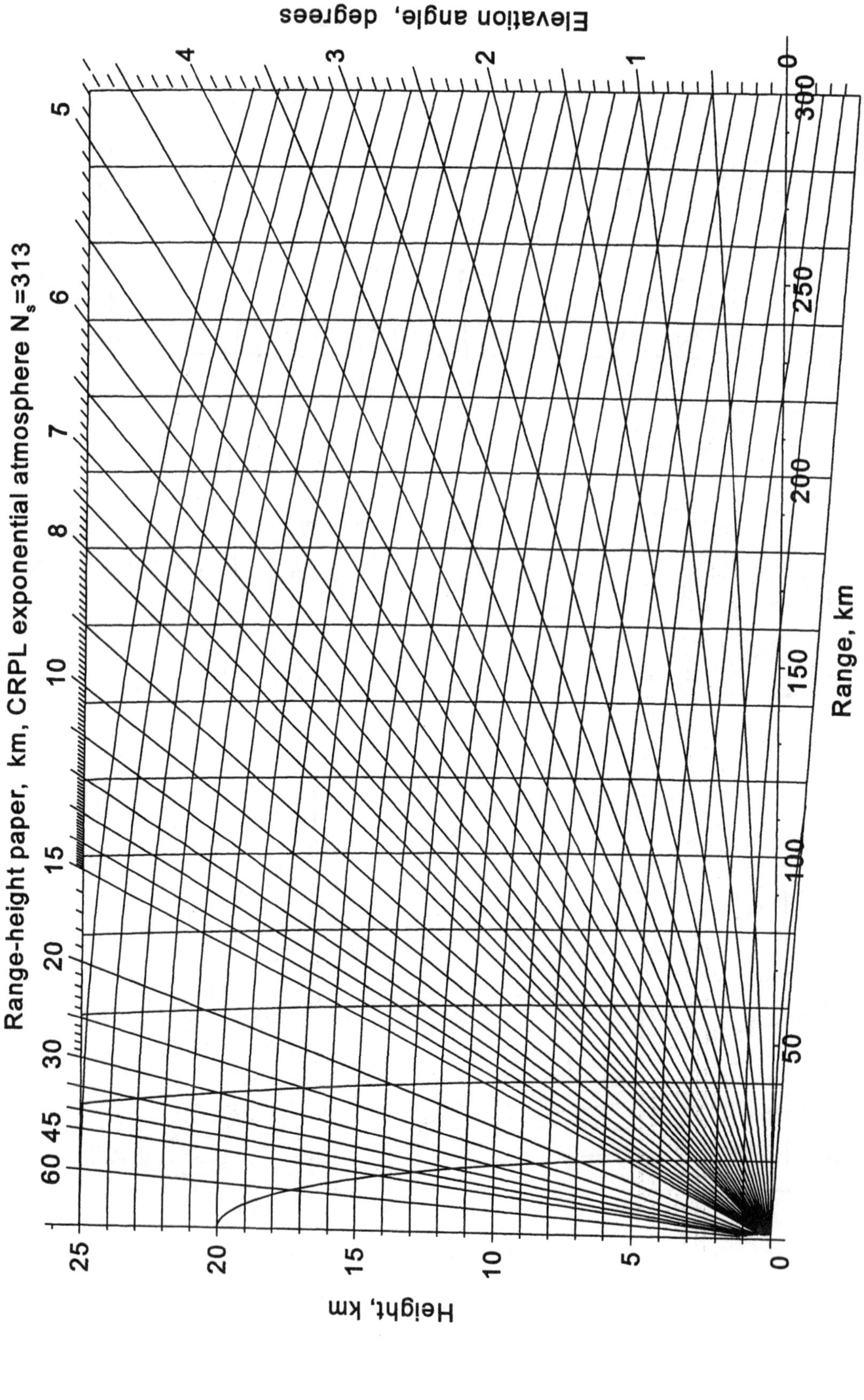

Figure 6A.17 CRPL exponential atmosphere $N_s = 313$ out to 300 km.

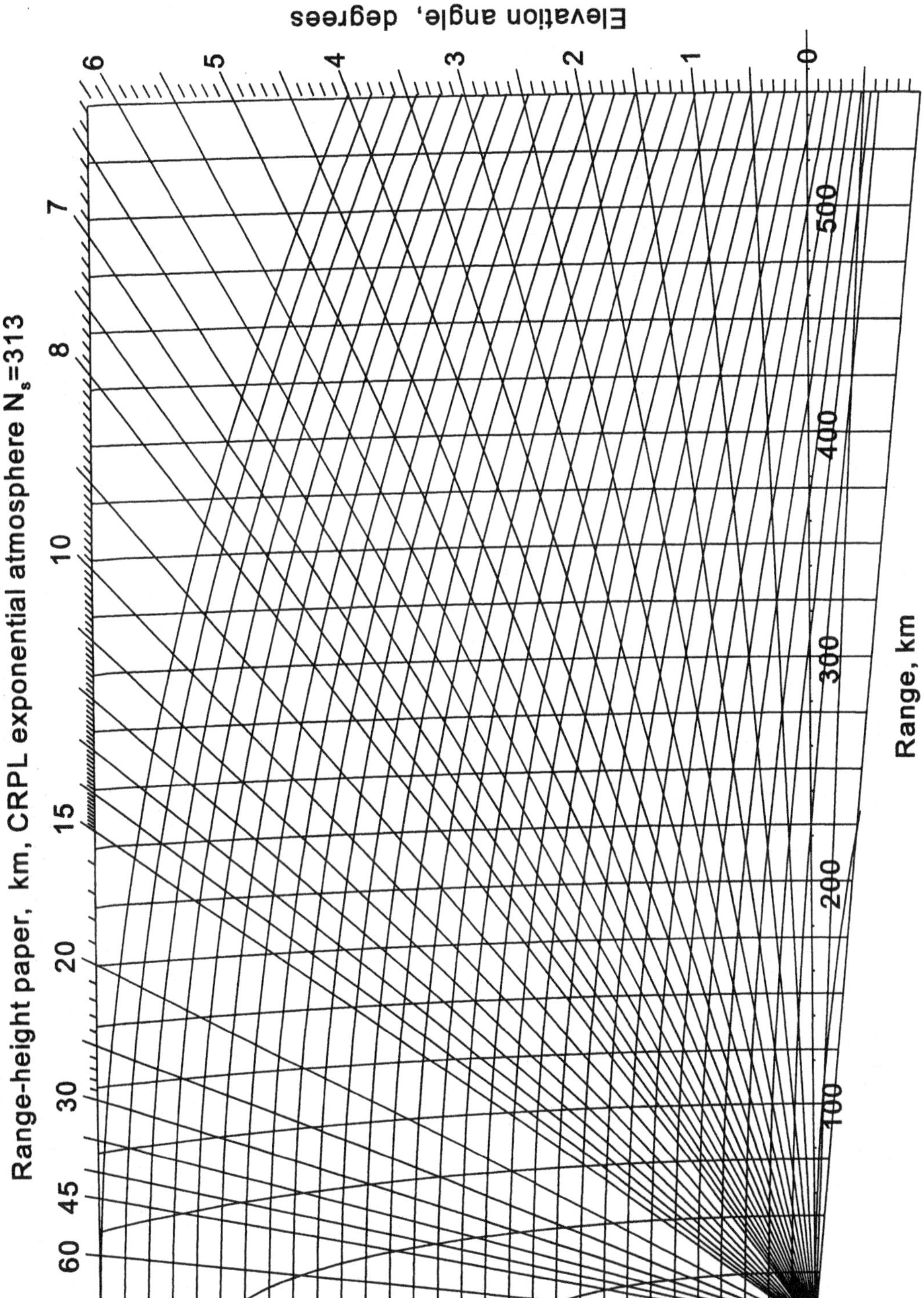

Figure 6A.18 CRPL exponential atmosphere N_s = 313 out to 550 km.

Chapter 7

Receivers

The input to the receiver consists of cosmic noise, which enters the antenna from outer space; echoes, which are returning copies of the transmitted pulse; and maybe interference from the sun (noise) or nearby unsuppressed electric devices. The returning copies of the transmitted pulse vary in amplitude, depending on the size and the range of the object causing the echo. The phases of the returning echoes depend on their range within domains of half a wavelength ($\lambda/2$). The purpose of the receiver is to amplify the echo signals, filter out all other signals, and pass the echo signals to the matched filter section.

There are simple and more complex receivers, depending on the type and cost of the radar. Modern signal processors depend on their ability to find the correct difference of large numbers over a number of pulses, so more modern receivers have a wide dynamic range. Older receivers use limiting or logarithmic amplifiers to compress the dynamic range. The blocks described in this section are shown in Figure 7.1.

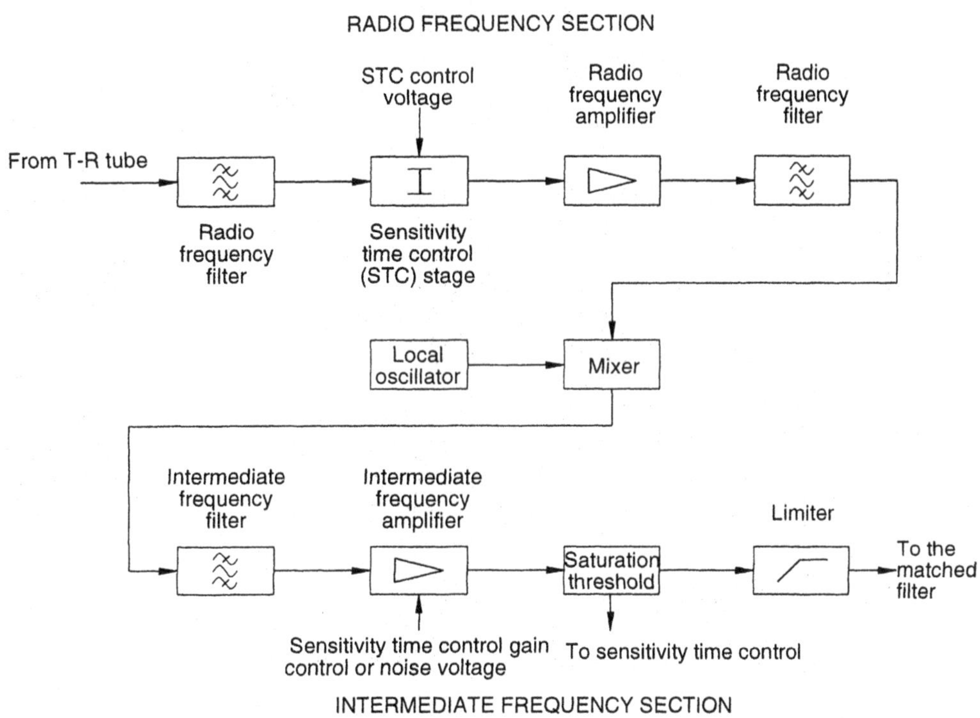

Figure 7.1 Block diagram of the receiver showing alternate positions for the radio frequency filter. The intermediate frequency amplifier may be replaced by a logarithmic amplifier.

The number of receivers depends on the type of radar:

- Single beam: one receiver;
- Dual beam on receive: one receiver switched between beams;
- Multiple or stacked beam three-dimensional radar: one receiver for each beam;
- Stationary staring arrays: one receiver for each beam.

Monopulse radars, which measure the azimuth and possibly the elevation angle with a single pulse, need at least two receivers for each dimension. They may be sum, Σ, and difference, Δ, channels from the sum and difference outputs from the antenna or the outputs from adjacent antenna beams.

The main aims for the design of radar receivers are as follows:

- It must have maximum sensitivity, that is, the lowest noise figure or temperature;
- It must be robust enough to survive the leakage of transmitter power from the local transmitters into the receiver system;
- A very large dynamic range is required for search radars. Receivers in radars which use phase detectors must not introduce phase distortion caused by widely differing amplitudes.

7.1 DYNAMIC RANGE, CONTROL OF GAIN, AND SENSITIVITY TIME CONTROL

Surveillance radars must be able to handle all echoes of interest and clutter from their minimum range to their maximum instrumented range. The dynamic range required depends on the ratio of maximum range to minimum range to the power of 4, or 12 dB per octave, 40 dB per decade. For secondary radars the inverse square law applies (constant transponder power) so the dynamic ranges in decibels are halved. Typical dynamic ranges for a number of range ratios are given in Table 7.1.

Table 7.1
Dynamic ranges

Range ratio	Dynamic ranges	
	Primary radar	Secondary radar
50:1	68 dB	34 dB
100:1	80 dB	40 dB
250:1	96 dB	48 dB

To this must be added the ratio of the cross-sections of the assumed greatest echo to the minimum scatterer cross-section. If these are 10 000 m^2 and 1 m^2, respectively, a further 40 dB of dynamic range is required. The dynamic range for amplifiers is usually defined as the ratio of the signal level, where the amplifier has 1 dB less amplification than for small signals near the noise level to the noise level. No thermionic or semiconductor components can handle such a range.

The calculated signal- and clutter-to-noise ratios for an S-band radar, azimuth beamwidth 1.5 degrees, pulse length 1 μs that has a signal-to-noise ratio of 10 dB at 100 km are shown in Figure 7.2.

Formerly, radars provided video signals for cathode ray tube displays. Intensity modulated displays (the plan position indicator (PPI) is the most common) can produce a contrast ratio on the screen of about 10:1, or 10 dB, 3.16:1 in voltage. With the advent of stable digital signal processing to sort the wanted from the unwanted, or cluttering, echoes, greater dynamic ranges are required in the receiver. The dynamic range depends on the number of bits that the analogue-to-digital converter provides by converting voltage (2:1 in voltage = 6 dB) and are shown in Table 7.2.

Table 7.2
Numbers of bits and dynamic range

Bits	Dynamic range	
	Power range, dB	Output values, voltage
7 + sign = 8	42	−128..+127
11 + sign = 12	66	−2 048..+2 047
15 + sign =16	90	−32 728..+32 727

The most common analogue-to-digital converters convert to 12 bits or 66 dB dynamic range.

The ratio of largest to the smallest echo is in the region of 40 dB and constant. The remaining amount of dynamic range depends on range and is predictable. Thus, the receiver gain may be reduced at short ranges, and this process is called sensitivity time control (STC), swept gain, or gated time control (GTC) in secondary radars. Sensitivity time control was first used to reduce the (clutter) video amplitude at close ranges, which limited the clutter glare on plan

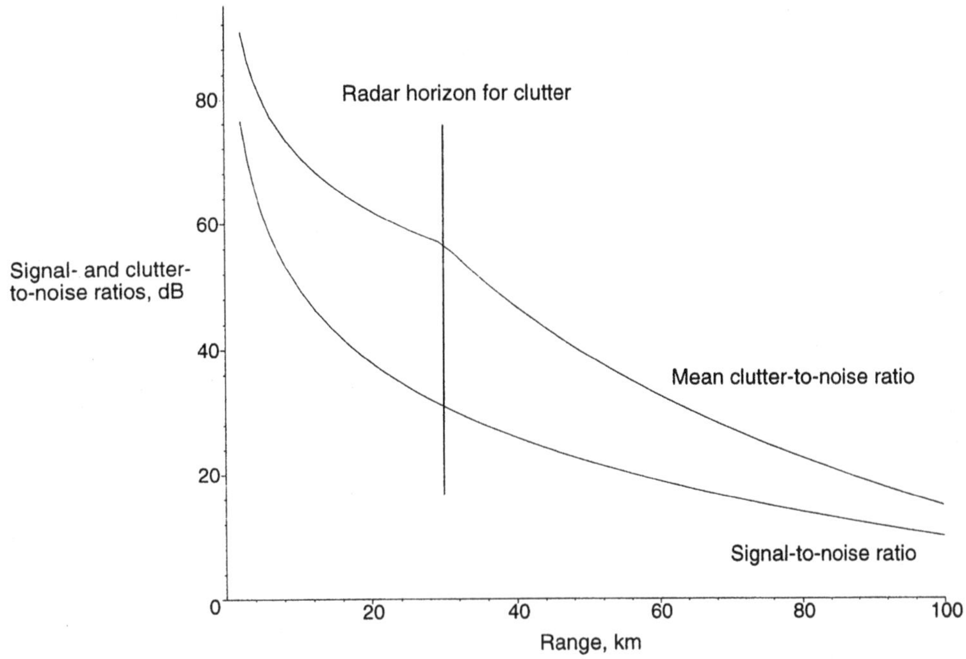

Figure 7.2 Signal- and clutter-to-noise ratios for the model radar.

position indicators. Better results were obtained when the gain of the first intermediate frequency stages was reduced. When semiconductor attenuators (PIN diodes) became available for radio frequency attenuators, a controlled attenuator was placed in the receiver input to reduce the dynamic range at the input of the first amplifier in the receiving system.

Sensitivity time control was first applied equally to a fixed range in a circle around the radar. The controlling waveform was designed to reduce close echoes to within the dynamic range of the amplifiers and to give full gain at half the instrumented range and beyond. The form of the waveform was often either $1/R^3$ (to give constant clutter strength) or $1/R^4$ (to give constant small echo strength), as is shown in Figure 7.3.

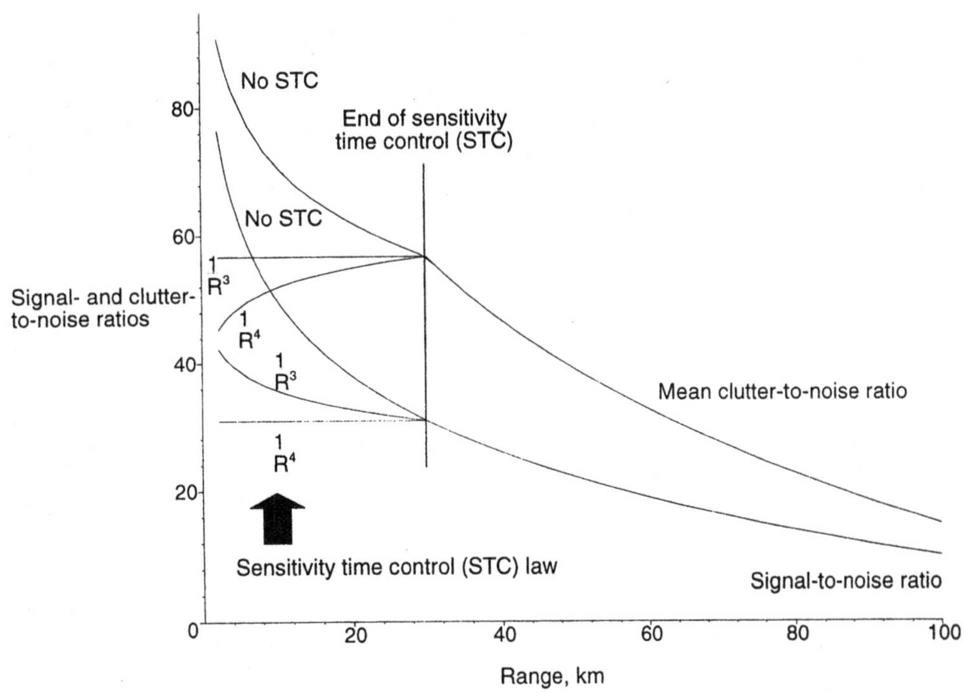

Figure 7.3 Signal- and clutter-to-noise ratios for the model radar with sensitivity time control.

The spectral densities of the signals in Figure 7.3 are shown in Figure 7.4. Theoretically, the clutter spectral density extends down to minus infinity. To emphasize the practical width of the clutter spectrum with clutter strength, the clutter spectral densities have been cut off at −10 dB.

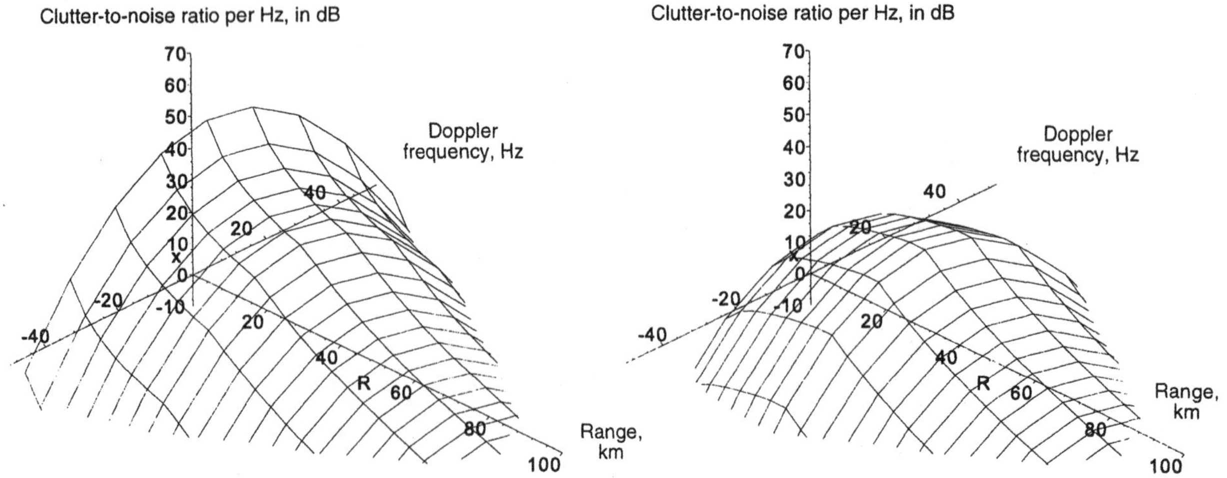

The spectral density of clutter in the model	The spectral density of clutter in the model with sensitivity time control

Figure 7.4 Spectral densities in the model without and with sensitivity time control (STC). The base is trimmed to -10dB/Hz signal-to-noise ratio.

The choice of the sensitivity time control end range and the law is governed by the following:

- The smallest wanted close range echoes have an adequate signal-to-noise ratio not only for detection but also for accurate measurement. If that is 30 dB for 1 m^2, then the largest wanted echo in the dynamic range is 36 dB larger (for a dynamic range of 66 dB), or 4 000 m^2.
- The clutter "amplitude" must fall in the linear region. In the example, the maximum value of the mean clutter lies near 56 dB, so there is 10 dB to spare when a 12-bit A/D converter is used, which covers 98.9% of the expected clutter values (see Chapter 6, Factors outside the radar).

As more memory became available, the amount of sensitivity time control could be changed from sector to sector. Currently, maps are used for control in azimuth and range. Maps have the advantage of being able to be modified for changing propagation conditions.

The sensitivity time control map in Figure 7.5 is the result of increments caused by the saturation detector and overflow pulses on a day with no abnormal clutter (anomalous propagation). For radars that are not mobile, such maps may be stored in nonvolatile memory, for example, electrically erasable programmable read only memory (EEPROM), so that the maps are available immediately after switching on the radar. Changes in propagation conditions (extra clutter) may be taken into account by using an additional dynamic map in random access memory (RAM). The fixed map may be coarse (for example, 64 resolution cells) and the dynamic map more detailed (for example, two resolution cells), to limit the amount of nonvolatile memory required. A resolution cell is one cell in range by the antenna beamwidth or a coherent processing interval in azimuth. The values of attenuation in the dynamic map are decremented at regular intervals and will be corrected by new saturation or overload pulses.

The clutter clumps are normally many antenna beamwidths wide and many transmitter pulsewidths long. A typical "fixed" map may contain sensitivity time control values for cells two pulsewidths in range by eight beamwidths in azimuth. This map is complemented by a dynamic map one pulsewidth in range and one beamwidth wide in azimuth. Saturation in the intermediate frequency amplifiers or overflow in the analogue-to-digital converters both cause an increment in the attenuation values. After several minutes, the dynamic part of the map is decremented.

Sensitivity time control has other beneficial effects:

- Echoes below a given cross-section are reduced, such as insects and birds to near or below the noise level. Sensitivity time control may be switched off to observe bird migrations.
- Close-in echoes received through the antenna sidelobes are reduced. This is specially true for replies received by secondary radar where the symmetry of the antenna sidelobes may be checked by switching off the sensitivity time control. With primary radar it is much more difficult to discern the sidelobe echoes.
- Second time around returns are reduced in the sensitivity time control region.
- Noise and false alarms in the range covered by sensitivity time control are reduced.

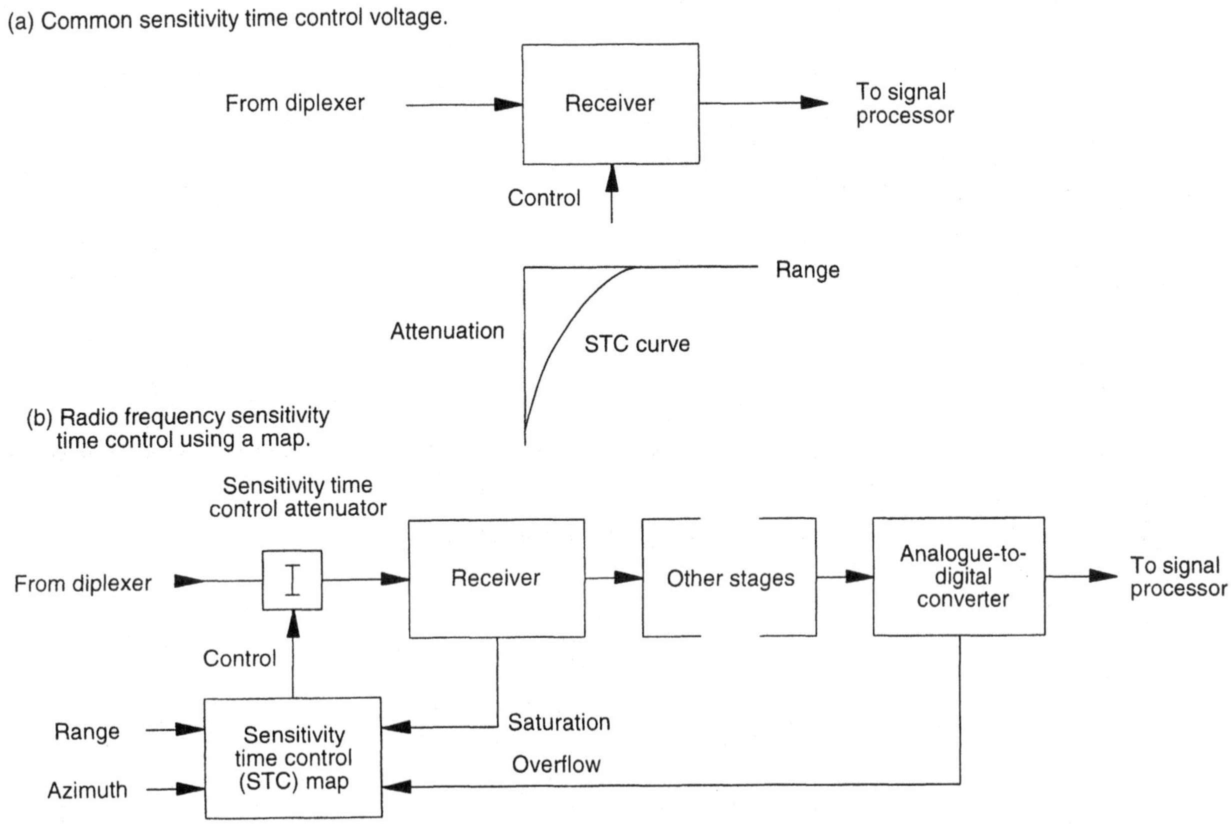

Figure 7.5 Sensitivity time control using (a) a common gain control voltage and (b) a radio frequency sensitivity time control (STC) attenuator with a map.

The effects of sensitivity time control should be the same over complete extended transmitter pulse. Because of problems with continuous wave (CW), medium and high pulse repetition frequency pulse Doppler radars, and long pulse radars, sensitivity time control is not used with these types of radars. They are unable to apply sensitivity time control at specific ranges. Long pulses for long ranges are only used for ranges beyond half the instrumented range without sensitivity time control.

Where a single object is being observed, the gain for that object can be adjusted automatically, a process called automatic gain control (AGC). General automatic gain control may be used in search radars by keeping the noise level at long ranges or beyond maximum range constant.

7.2 RADIO FREQUENCY SECTION

The radio frequency section includes input filtering, sensitivity time control attenuation, initial amplification, and frequency conversion down to intermediate frequency.

7.2.1 Radio frequency amplifier

The radio frequency amplifier is the first amplifier in the receiver and normally determines the level of noise in the system. All amplifiers add noise to the signals that they amplify which limits the sensitivity of the radar.

The noise at the input of the amplifier is represented by the token resistor representing the resistance at the input at a temperature T Kelvin, which is given by

$$Noise\ power\ at\ input = k\ T\ B \quad \text{W} \tag{7.1}$$

where k is Boltzmann's constant 1.38 . 10^{-23} J/K;
T is the absolute temperature of the resistor (K);
B is the noise bandwidth (Hz).

The voltage across an imaginary input resistor, R Ω, is

$$Noise\ voltage\ at\ input = \sqrt{k\ T\ B\ R} \quad \text{Volt} \tag{7.2}$$

Amplifiers, gain A, are not perfect and add noise of their own and give an output of $\overline{NF}\ A$ times the sum of the input power and internal amplifier noise power, the latter being expressed in terms of an equivalent input temperature component, T_e, at the input termination. The value $\overline{NF}$ is called the noise figure.

$$\begin{aligned} Output\ noise\ power &= Amplified\ input\ noise \\ &= A\ k\ T_{in}\ B \end{aligned} \tag{7.3}$$

The noise figure, or factor, is defined as [1]

$$Noise\ figure,\ \overline{NF} = \left. \frac{Noise\ power\ out}{Gain\ .\ Noise\ power\ in} \right|_{\text{when the input is terminated at 290 K}} \tag{7.4}$$

Referring all quantities to the input and rewriting (7.3) using the noise factor on the left hand side and the sum of the noise components on the right hand side

$$\overline{NF} k T_0 BA = kT_0 BA + kT_e BA \tag{7.5}$$

where T_0 is the temperature of the input terminator (by definition the standard reference temperature, 290 K [2]);
T_e is the effective system temperature referred to the input.

Rearranging (7.5), the amplifier has the equivalent temperature (referred to the standard input):

$$T_e = T_0\ (\overline{NF} - 1) \tag{7.6}$$

The system noise temperature, T_{sys}, referred to the receiver input, includes all sources of noise and receiver noise, as shown in Figure 7.6, that is,

$$T_{sys} = \frac{T_a}{L_r} + T_r + T_e \tag{7.7}$$

where T_a is the noise temperature at the output of the antenna (see Chapter 5, Antennas);
T_r is the noise temperature of the transmission line between the antenna and the receiver, including the sensitivity time control attenuator when fitted.

The noise temperature of the transmission line is $T_0 \left(1 - \frac{1}{L_r}\right)$ where L_r is the loss in the line between the antenna and the receiver (see Chapter 4).

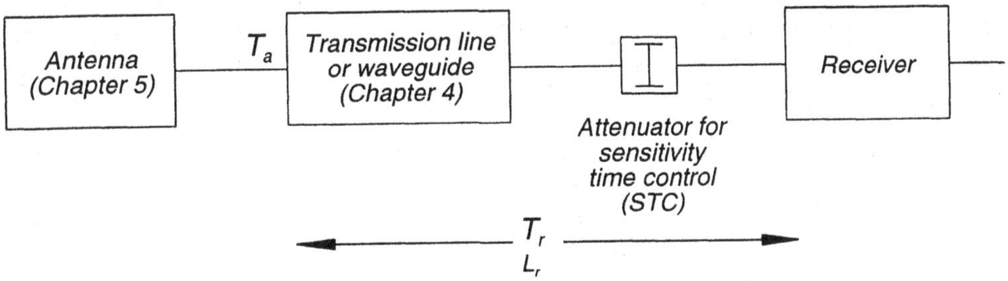

Figure 7.6 The antenna and transmission system components.

Improving technology has led to a reduction of receiver noise figures. In the past, there were no effective microwave amplifiers except parametric amplifiers which needed careful maintenance. Typical values, which include the losses for input limiters for robust radar receivers, are listed in Table 7.3.

Table 7.3
Typical noise figures for radar receivers

Type of first stage	Typical noise figure
Crystal diode frequency mixer	10 dB in L- and S-bands
Valve radio frequency stage for protection	10 dB in L-band
Traveling wave tube amplifiers	5 dB in S-band
Parametric amplifiers	4 dB in S-band
Transistors	3 dB or less

A change of noise figure from 10 dB to 3 dB is a factor 5. Under "typical" conditions with an antenna noise temperature 100 K, antenna-to-receiver line loss of 1 dB, the system noise temperatures from (7.7) are

$\dfrac{T_a}{L_r}$	$T_r = T_0\left(1 - \dfrac{1}{L_r}\right)$	$T_e = T_0(\overline{NF} - 1)$		System noise temperature
79.43 K +	59.65 K +	2 610 K =	2 749 K	for a 10 dB noise figure
79.43 K +	59.65 K +	288.63 K =	427.70 K	for a 3 dB noise figure

The ratio of the systems noise temperatures, 6.43 or 8.08 dB, is the ratio of sensitivities which gives an improvement in range of 2.02 dB, 1.59, or 59%. Increasing transmitter power by a factor of 6.43 requires a much greater transmitter with greater running costs.

The radio frequency amplifier is the first of a number of amplifiers, each of which has its own noise figure. Using (7.5), cascading, and dividing by A_1, the power amplification factor of the first stage to refer the temperature at the output of the second stage to the input of the first stage,

$$T_{out} = T_0 + T_0\,(\overline{NF_1} - 1) + \frac{T_0(\overline{NF_2} - 1)}{A_1} \tag{7.8}$$

As long as the amplification factor of the first stage is sufficiently large, the noise contribution of the second and following stages may be neglected, which is the normal case.

Note that a radio frequency attenuator in front of the receiver attenuates only the incoming noise. Noise generated in the first stage is not attenuated and continues to be the constant, dominant noise throughout the range. Sensitivity time control applied to intermediate frequency stages reduces the receiver noise level at short ranges.

7.2.2 Radio frequency filter

Very often, the radio frequency bandwidth is wide enough to allow noise at radio frequencies above and below the local oscillator frequency to be mixed down to intermediate frequency, thus increasing interference and cosmic noise. The radio frequency filter selects the band where the echoes occur. If the filter is placed in front of the first stage, it rejects interference but its losses must be added to L_r. Placing the filter after the first amplifying stage restores the receiver sensitivity, but high power interference at any frequency within the antenna passband that enters the radio frequency amplifier can cause intermodulation or signal suppression. In a frequency agile radar, the radio frequency filter, unless it is able to be tuned, must pass all frequencies used, and, unless the intermediate frequency is greater than half the agile bandwidth, the image frequency may pass through the filter. In this case, the single sideband mixer, described next, may be necessary.

7.2.3 Mixer

The mixer uses a local oscillator signal to convert the radio frequency echoes to intermediate frequency for the main amplification and filtering. A block diagram with filters is shown in Figure 7.7.

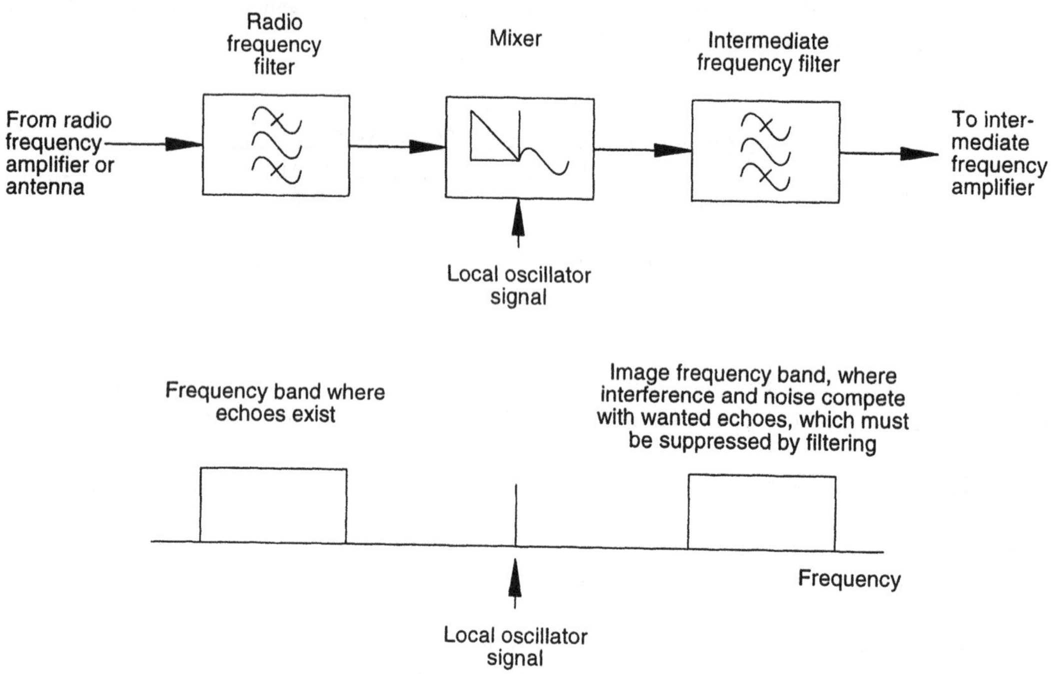

Figure 7.7 Conversion to intermediate frequency using filters for selection of the intermediate frequency signal.

The local oscillator signal itself contains noise. Commercial diode ring modulators or mixers are available in which noise is canceled at the intermediate frequency output.

If the image response cannot be eliminated by the radio frequency filter, a mixer derived from a single-sideband modulator is used, as is shown in Figure 7.8.

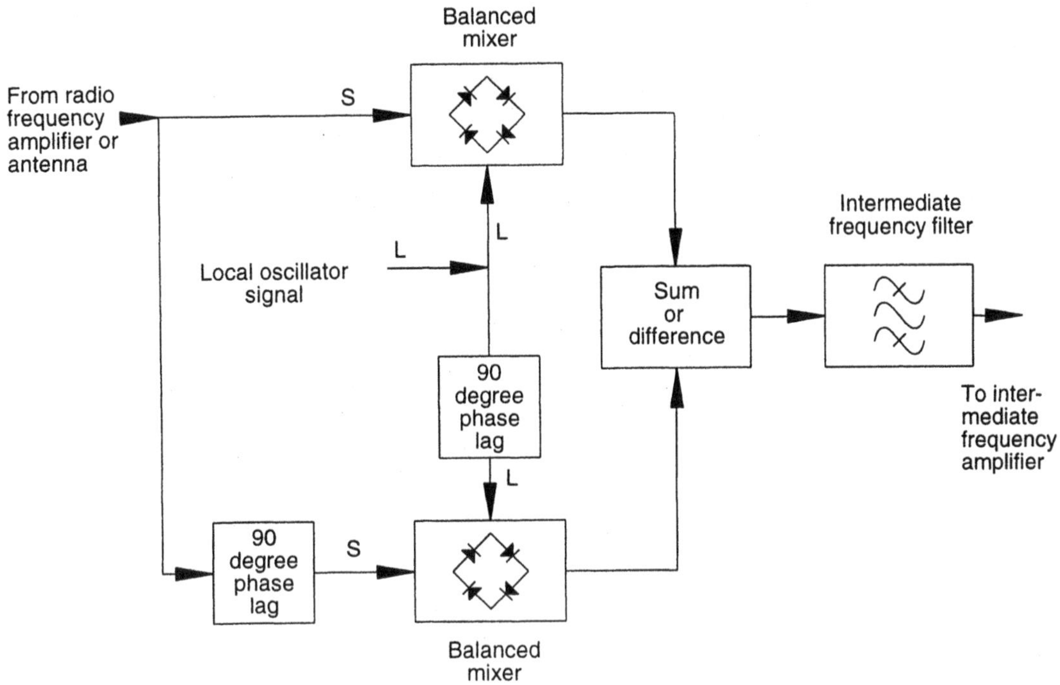

Figure 7.8 Quadrature type of single sideband mixer.

If the signal is represented by $S\cos(\omega_S t + \phi)$ and the local oscillator by $L\cos(\omega_L t)$, then the signals with a 90 degree phase lag are given by $S\sin(\omega_S t + \phi)$ and $L\sin(\omega_L t)$, the outputs of the two balanced mixers are

$$
\begin{aligned}
&\textit{Upper balanced mixer} \\
&S\cos\omega_S t\, L\cos\omega_L t = \frac{SL}{2}\left(\cos(\omega_S + \omega_L)t + \cos(\omega_S - \omega_L)t\right) \\
&\textit{Lower balanced mixer} \\
&S\sin\omega_S t\, L\sin\omega_L t = \frac{SL}{2}\left(\cos(\omega_S + \omega_L)t - \cos(\omega_S - \omega_L)t\right) \\
&\textit{sum} = SL\cos(\omega_S + \omega_L)t \\
&\textit{difference} = SL\cos(\omega_S - \omega_L)t
\end{aligned}
\tag{7.9}
$$

If the outputs from the balanced mixers are summed, the sum of the signal frequency and the local oscillator frequency is selected and the difference gives the difference frequency.

7.3 INTERMEDIATE FREQUENCY AMPLIFIER AND FILTER

The intermediate frequency is chosen large enough to avoid image frequency interference and small enough to allow adequate amplification with the bandwidth required. For example, if the radio frequency stages have a bandwidth of 200 MHz, then a 30 MHz intermediate frequency amplifier will receive echoes at the transmitter frequency and interference and noise from a frequency at 60 MHz from the transmitter frequency. Tuning at radio frequency is necessary to suppress this image frequency. If a first intermediate frequency of 250 MHz is chosen, then the image frequency interference may be filtered out by the radio frequency filter, because it will be 500 MHz away from the transmitter frequency.

After filtering, a mixer can convert the signals to a lower second intermediate frequency, and in some cases further mixing to a third or even a fourth intermediate frequency may be used for further amplification. The output level is designed to be sufficient for the matched filter or the second detector.

Traditionally, intermediate frequency amplifiers have a flat bandpass with relatively steep sides which often served as the "matched filter" in the radar. Pulse interference "charges" such narrow filters, and the pulse energy takes time to decay and may affect many range cells that follow. This is called ringing time. Amplifiers starting with a wide bandwidth, equivalent to a short ringing time, with good design limit the ringing time to one following pulsewidth. The following filter suppresses the pulse if it falls outside its bandwidth which avoids the ringing in one range cell overflowing into succeeding range cells, as in Figure 7.9.

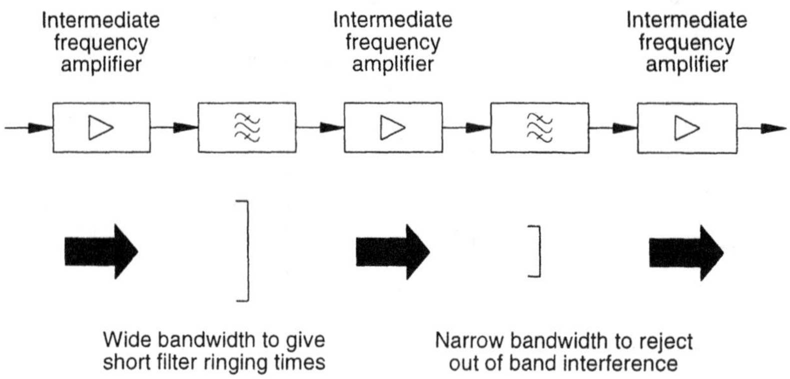

Figure 7.9 A common form of intermediate frequency amplifier with stages with decreasing bandwidth.

In older radars, a sensitivity time control waveform controlled the gain of the intermediate frequency amplifiers. The noise from the radio frequency section was also reduced. If the intermediate frequency amplifier was followed by a limiter and narrowband filter, the noise level remained more or less the same.

Where adaptive sensitivity time control is used, an amplitude detector with a threshold (saturation detector) is set to sense excessive signal peaks that are near the limits of the dynamic range. This signal is used to increase the amount of attenuation.

7.4 LIMITERS

Because the dynamic range of the stages is limited, all signals above a level necessary for signal processing or display overload the following amplifiers and thus are limited. Limiters were used to limit unnecessary glare on brightness modulated displays (the plan position indicator (PPI) is an example) and clutter residues in moving target indicator (MTI) systems.

Limiters are not perfect. The differential change of phase with different amplitudes is called phamp [3, pp. 5-31 and 5-37; 4, pp. 3-31 to 3-32], which must be measured where phase angle distortion is critical. Limiters exhibit the frequency modulation (FM) capture effect, in which a larger signal reduces a smaller signal. The smaller signal component does not itself cross the limiting threshold as often as the larger one [5, pp. 87-91; Appendix B].

One form of military receiver that used limiting is the Dicke-fix receiver which is shown in Figure 7.10 and discussed

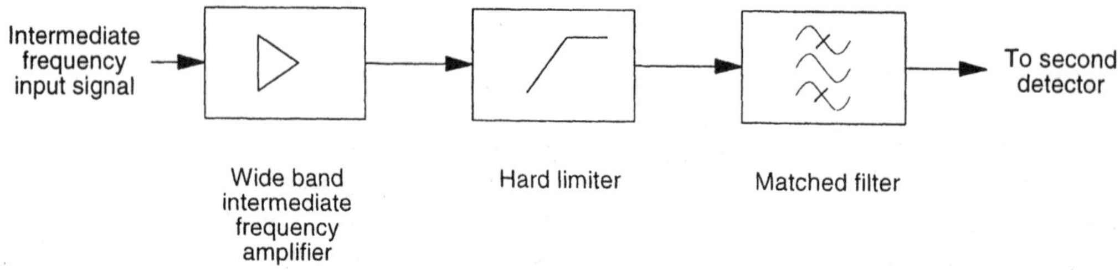

Figure 7.10 Block diagram of a Dicke-fix type of receiver.

more fully in Section 8.1.4.

The limiter cuts off the vector sum of the signals from the wideband amplifier to a fixed voltage. The output signal is at a maximum when an echo from the transmitter pulse that matches the matched filter is received. With normal wideband noise, the noise output is constant, more or less independent of the input noise. The matched filter selects a proportion of the noise that is the ratio of the narrow and wide bandwidths. An echo captures the limiter and is passed with full amplitude by a fully matched filter; thus, the ratio of the bandwidths gives the dynamic range of the output which gives a constant false alarm rate (CFAR) for a fixed threshold.

Interference (or in military radars deliberate jamming) that falls outside the matched filter bandwidth captures the limiter and creates a hole in the coverage. A wideband amplifier has a short ringing time during which interference or jamming pulses can affect the following range cells. This is often specified as less than one or two transmitter pulse widths. The holes in the coverage give the military concept of the "black scope", where the aim is to suppress jamming and only show valid echoes. One hopes for a better situation for the next look. A logarithmic receiver with a wide dynamic range (see Chapter 9, Detectors) is used to observe the jamming.

As mentioned above, search radars require a constant false alarm rate (CFAR) characteristic, while tracking radars need a constant sum signal. Automatic gain control (AGC) is applied to the intermediate frequency stages to achieve this. The automatic gain control voltage is derived either from noise samples near or beyond maximum instrumented range or from the sum signal.

7.4.1 Effects on amplitude

The signal in an intermediate frequency amplifier is the vector sum of the noise and all the received components which may be clutter, interference, and wanted echoes. The nonlinear amplification gives rise to harmonics that are filtered out by the intervening filters and mixing products of the signals and the intermediate frequency "carrier" and the signals. With hard limiting, the output is phase modulated only.

7.4.2 Effects on spectrum

In [5, pp. 87-91], the power spectrum for limited signals is given as a series of Gaussian distributions caused by mixing with the intermediate frequency "carrier" and being aliased down into the intermediate frequency bandwidth [6]. The expression for the spectrum [7, p. 137] is expanded [8, p. 51, Eq. 1.641], and is given by

$$S(\omega) = \sum_{p=0}^{\infty} C_p \, e^{-\frac{\omega^2}{(2p+1)\sigma_c^2}} \tag{7.10}$$

where
$$C_p = \frac{a_p \, b_p}{(1 + \alpha)^{2p+1}}$$

$$a_p = \frac{(2p)!}{2^{2p}\,(p!)^2\,(2p+1)}$$

$$b_p = \frac{(2p+1)!}{p!(p+1)!}$$

$$\alpha = \frac{2}{\pi}\,\frac{\textit{clutter power saturation point}}{\textit{clutter power}}$$

The spreading of a signal with a Gaussian spectrum of unit bandwidth is plotted in Figure 7.11. As the value p in (7.10) increases during summing, the width of the Gaussian curve is determined by its standard deviation $\sqrt{2p+1}$ and its amplitude by C_p. The wider skirts are better seen when plotted in decibels, as in Figure 7.12.

The effects of limiting the clutter on the clutter spectral density with sensitivity time control is shown in Figure 7.13. The original spectral density in Figure 7.4 is shown with two veils illustrating the spectra with hard limiting and limiting 10 dB below the clutter peaks. For a fixed limiting level and variable clutter signal, the spreading effects depend on the clutter amplitude, which varies with range.

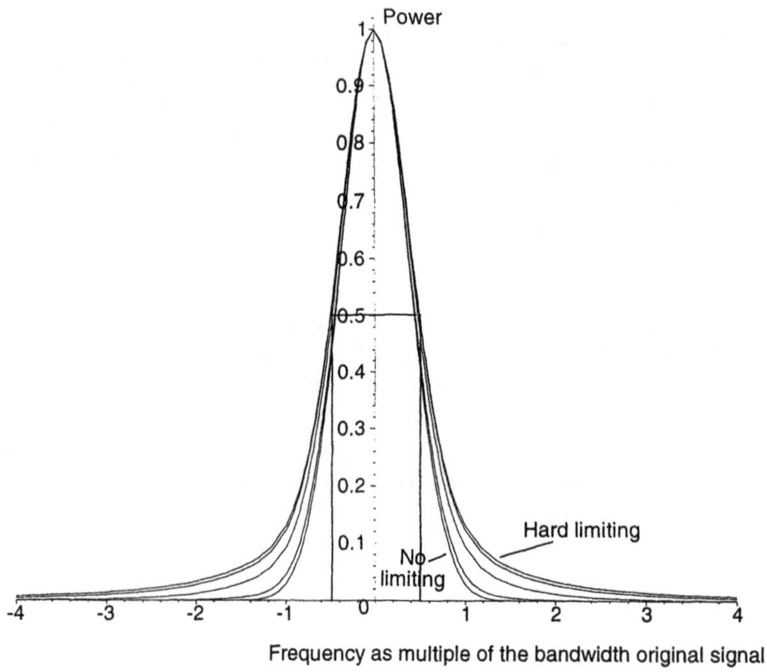

Figure 7.11 Spectra of signals originally with a Gaussian spectrum after limiting.

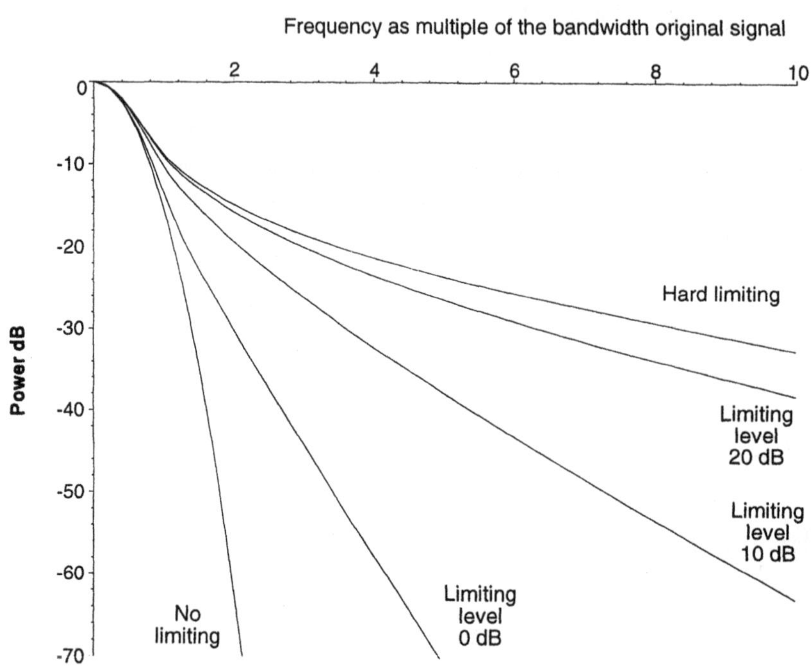

Figure 7.12 Spectra of signals originally with a Gaussian spectrum after limiting plotted in decibels.

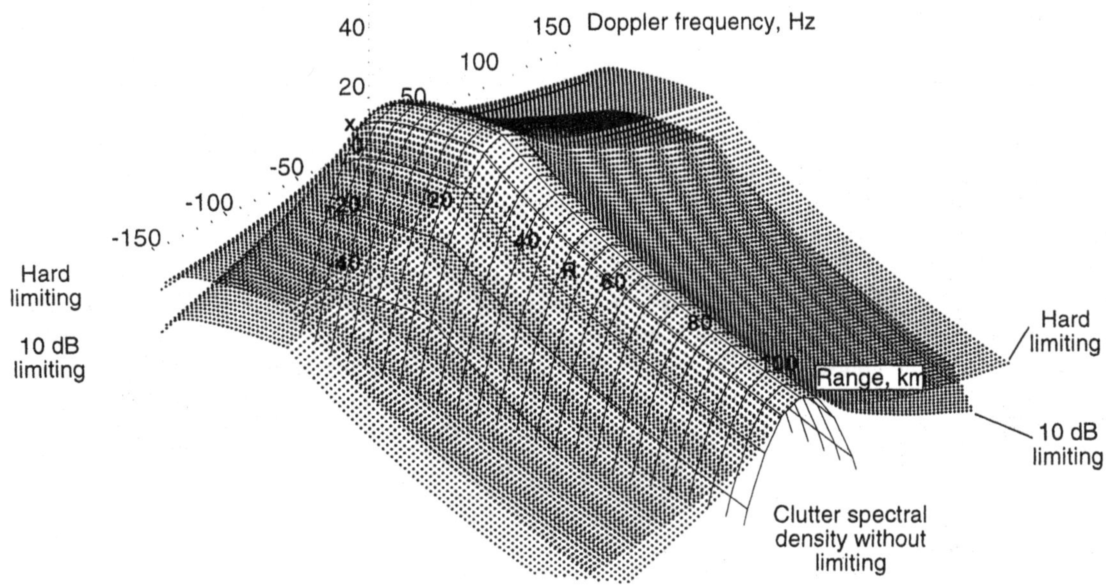

Figure 7.13 The spectral density of clutter in the model radar with sensitivity time control (STC) and limiting.

7.5 RECEIVER CHARACTERISTICS

The performance of a receiver is determined by its:

- Sensitivity;
- Gain;
- Bandwidth, filtering, ringing, and the ability to reject interference;
- Dynamic range before and after sensitivity time control (STC);
- Ability to withstand the transmitter pulse spikes;
- Constant signal or noise output.

7.5.1 Minimum range

During the transmitter pulse, the receiver has its input heavily attenuated by the (gas-filled) transmit-receive cell and the radio frequency sensitivity time control attenuator, when fitted. These prevent the destruction of the first receiver stage, which is driven into saturation. After the transmitter pulse has left the antenna, these protection circuits and the receiver itself take time to recover. The range at which a standard scatterer can be detected is one measure of minimum range.

7.5.2 Gain

Gain must be high enough to drive the matched filter or detector stages that follow.

7.5.3 Bandwidth, filtering, ringing, and the ability to reject interference

The receiver bandwidth must be wide enough to accept echo signals within the tuning range of the transmitter. Bandstop filters may be used to reduce the effects of known, close sources of interference, for example, the other transmitter-receiver channel in frequency diversity radars. It must be narrow enough to reject out-of-band interference. Also, the bandwidth must be graded to avoid overloading the matched filter and causing it to ring.

7.5.4 Dynamic range before and after sensitivity time control (STC)

The dynamic range of the echo signals leaving the antenna depends on the size of the scatterers and inversely to the fourth power of their range compared to the maximum range of the radar. Figure 7.2 shows the dynamic range for a small aircraft (1 m^2) in an airport radar. Scatterers of up to 10 000 m^2 may be present. The greater the maximum range of the radar, the greater is the dynamic range required.

The measure of dynamic range for the receiver extends from the noise level to the point where there is a one decibel compression of the gain characteristic (generally 80 dB or less). The stages that follow the receiver generally have a smaller dynamic range and digital processors currently use 12 bit quantities, which give a dynamic range of 66 dB.

Radio frequency sensitivity time control in front of the first amplifier attenuates close-in signals to reduce the dynamic range at its output to the maximum that can be handled later.

Modern sensitivity time control units are controlled from digital STC maps, and the switching time from one attenuation level to another must be held within close limits, for example, one range cell. The size of the area of one cell in the STC map must match the local clutter. The time constant of additional dynamic STC must be long enough for stability and short enough to respond to local weather conditions.

7.5.5 Ability to withstand transmitter pulse spikes

The radar receiver follows the transmit-receive tube, which shorts the input during the transmitter pulse. Before the tube fires, it allows a spike of energy to pass. Some tubes require a "keep-alive voltage" in order to be able to fire at lower power levels to protect the receiver. Special PIN diode circuits in the sensitivity time control unit allow the limiting of incoming radio frequency energy, which gives additional protection to the receiver. With no control current, the diode circuits must rectify the incoming radio frequency energy to provide the attenuation current, and this is especially the case when other radars are operating nearby. This is the reason why radar receivers do not have the sensitivity of those used for radio astronomy, for example.

7.5.6 Constant signal or noise output

The receiver provides the control element (generally for automatic gain control (AGC)) to give a constant noise output for surveillance radars or a constant sum signal for tracking radars. The automatic gain control reduces the gain in the difference channels in tracking radars by the same amount. The performance figures are as follows:

- Surveillance radar automatic gain control (AGC):
  - Working range in decibels in which the noise output is constant or the constant false alarm rate is maintained;
  - Reaction time for the radar automatic gain control.
- Tracking radar automatic gain control (AGC):
  - Working range decibels in the sum and difference channels for a constant sum channel signal;
  - Reaction time for the radar automatic gain control.

7.6 FIGURES AFFECTING RADAR PERFORMANCE

The following figures are used in Chapter 14, Performance, to calculate radar performance. Additional important parameters are the dynamic range of the receiver with and without sensitivity time control and the gain control range for manual or automatic gain control. Automatic gain control is used to keep the noise level constant for the threshold stage (Chapter 12, Threshold).

7.6.1 Range budget

The parameters affecting range performance are shown in Table 7.4.

Table 7.4

Receiver parameters affecting radar performance

Parameter	Units
Receiver noise figure	dB
(The bandwidth and the mismatch are covered in Chapter 8)	Hz

The receiver noise figure is measured where the coaxial line or waveguide is connected to the receiver. There are losses incurred when the receiving system is working outside its dynamic range which may happen with strong clutter returns or with jamming.

7.6.2 Resolution

The ability of the radar to resolve echoes may be limited by the receiver. Strong signals may cause the tuned circuits to ring, with the ringing signal occupying the following range cells. Limiting causes the spreading of the spectra of the signal components (see Section 7.4) and reduces resolution in the Doppler frequency domain.

7.6.3 Accuracy budget

The sensitivity time control law must be chosen so that the signal-to-noise ratio is high enough to allow accurate range, azimuth, elevation, and Doppler frequency measurement. The amplitude and phase changes with amplitude must be held as small as possible.

7.6.4 Stability budget

A number of effects detract from the stability budget for signal processing and extraction. Stability is required by signal processing systems, which compare the echoes from a number of transmitter pulses. If subtraction is used to remove unwanted echoes, then any differences in consecutive echoes will give a residue that limits the cancellation or selection performance. The receiver may have instabilities in amplitude and phase. Amplitude and phase instabilities may be caused by the sensitivity time control (STC) and by the local oscillator.

7.6.4.1 Sensitivity time control jitter

The trigger for starting the sensitivity time control waveform or clock must be synchronized with the master trigger, so that the same values of attenuation are used for neighboring sweeps.

When a dynamic sensitivity time control map is used, all changes must be made after the look or when the antenna has passed the scatterer. This process avoids two or more values of the scatterer amplitude and satisfies the time that data processing systems require to make a decision.

7.6.4.2 Local oscillator phase noise and stability

Coherent signal processing depends on the phases of the echoes from the same piece of clutter being the same from pulse to pulse. Any change of amplitude or phase represents movement, which will reduce the clutter cancellation. Figure 7.14 shows a coherent transmitter and receiver. To generate the transmitter pulses, the signal from the coherent oscillator (COHO) is mixed with the signal from the stable local oscillator (STALO). The extra term ϕ occurs because the COHO and STALO oscillators normally run independently and are asynchronous.

$$\begin{aligned}\textit{Mixed signal for the transmitter} &= S\cos(\omega_{STALO}t + \phi)\ C\cos\omega_{COHO}t \\ &= \frac{SC}{2}[\cos(\omega_{STALO} + \omega_{COHO}t + \phi) + \cos(\omega_{STALO}t - \omega_{COHO} + \phi)]\end{aligned} \tag{7.11}$$

where ϕ is the phase of the stable local oscillator (STALO) at the time of the transmitter pulse.

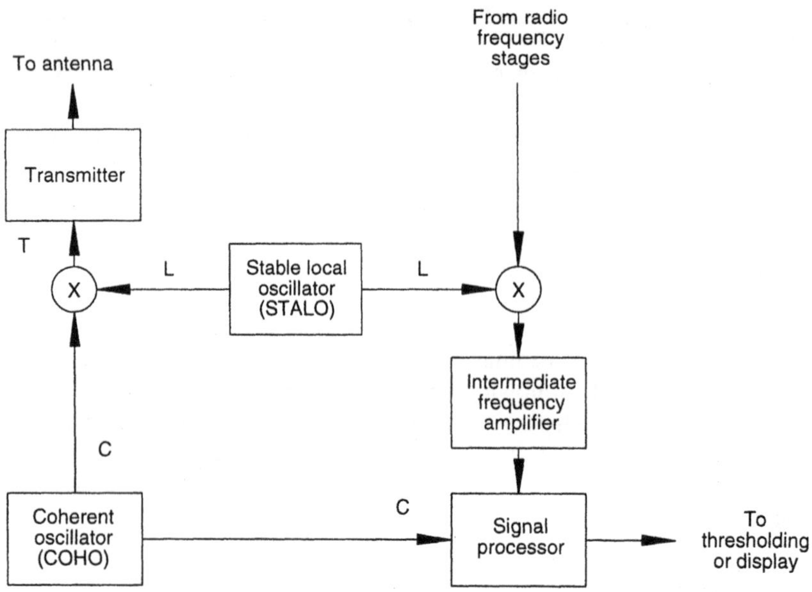

Figure 7.14 A coherent radar transmitter and receiver.

Either the sum signal or the difference signal is selected to drive the transmitter.

$$\text{Transmitter drive signal is one of} \quad \begin{aligned} &\frac{SC}{2}\left[\cos(\omega_{STALO} + \omega_{COHO}t + \phi)\right] \\ &\frac{SC}{2}\left[\cos(\omega_{STALO} - \omega_{COHO}t + \phi)\right] \end{aligned} \tag{7.12}$$

The transmitter pulse is radiated, is reflected by an object, and a very small part of it, amplitude A, is echoed to the antenna and the receiver. If the object moves towards the radar, the reflections are at a higher (Doppler) frequency. Because the reflecting objects are not placed at multiples of one half a wavelength, there is an extra phase shift of θ,

$$\text{Echo signal is one of} \quad \begin{aligned} &A\left[\cos(\omega_{STALO} + \omega_{COHO}t + \phi + \omega_{Doppler}t + \theta)\right] \\ &A\left[\cos(\omega_{STALO} - \omega_{COHO}t + \phi + \omega_{Doppler}t + \theta)\right] \end{aligned} \tag{7.13}$$

where θ is the phase term caused by the range not being a multiple of $\lambda/2$.

The same stable local oscillator (STALO) signal is used to mix down the returning echoes to the coherent oscillator (COHO) frequency, and the starting phase term, ϕ, disappears.

$$\begin{aligned} \textit{Intermediate frequency signal} &= A\left[\cos(\omega_{STALO} \pm \omega_{COHO}t + \phi + \omega_{Doppler}t + \theta)\right]\cos(\omega_{STALO}t + \phi) \\ &= \frac{A}{2}\left[\cos(\omega_{COHO}t + \omega_{Doppler}t + \theta)\right] \\ &+ \frac{A}{2}\left[\cos(2\omega_{STALO} + \omega_{COHO}t + 2\phi + \omega_{Doppler}t + \theta)\right] \quad \text{- eliminated} \end{aligned} \tag{7.14}$$

The mixer products in (7.14) are at the coherent oscillator frequency (difference) and at the sum of the coherent and the local oscillator frequencies which are eliminated by filtering.

Taking the coherent oscillator to be an exact time source, the amplitude and the phase of the echo are

$$\begin{aligned} Amplitude &= B \\ Phase &= \omega_{Doppler} t + \theta \end{aligned} \tag{7.15}$$

The coherent oscillator (COHO) signal in the radar is used as the time reference. If the stable local oscillator (STALO) frequency drifts with respect to the COHO, then it will be at two different frequencies during the time when the transmitter pulse is formed and when the echoes are received, and the STALO phase angle ϕ (7.14) is not completely eliminated. If the frequency error in the STALO oscillator is Δf, the phase error at the time t when the echo returns is

$$Phase\ error,\ \delta\phi = 2\pi\ \Delta f\ t \tag{7.16}$$

The residue when two sine waves with a phase error are subtracted is

$$\begin{aligned} A \sin 2\pi ft - A\sin(2\pi ft + \phi) &= 2A \cos\left(2\pi ft + \frac{\phi}{2}\right)\sin\left(\frac{\phi}{2}\right) \\ \text{(for small phase angles)} &\approx A\phi \cos\left(2\pi ft + \frac{\phi}{2}\right) \end{aligned} \tag{7.17}$$

Thus, the proportion of residue is ϕ radians (radians have no dimensions). The residue passes through the moving target indicator or the moving target detector systems and thus limits the input clutter to clutter residue ratio or improvement factor,

$$\begin{aligned} I_{STALO} &= 20 \log_{10}(phase\ error) \\ &= 20 \log_{10}\left(\frac{1}{2\pi\ \Delta f_{STALO}\ t}\right) \end{aligned} \tag{7.18}$$

where t is the delay time of the echo.

A number of older radars, mainly magnetron radars, use local oscillators where the frequency is controlled by an automatic frequency control (AFC) loop to follow the frequency wanderings of the magnetron. As the magnetron warms up, the dimensions of the cavities expand and thus the frequency changes. Any power supply ripple will add to the phase modulation in the automatic frequency control system and the effects will depend on range. Reference [9, p. 206, Eq. 7.31] gives the effective phase error for sinusoidal phase modulation as

$$\Delta\theta = 4A_p \left| \sin\left(\pi \frac{f_m}{f_{prf}}\right) \sin(\pi f_m t_{echo}) \right| \tag{7.19}$$

where A_p is the peak phase deviation;
f_m is the modulation frequency of the phase modulation;
f_{prf} is the pulse repetition frequency;
t_{echo} is the echo time delay.

This relation is plotted in Figure 7.15 for quarter maximum range (R/4), half (R/2), and maximum range (R).

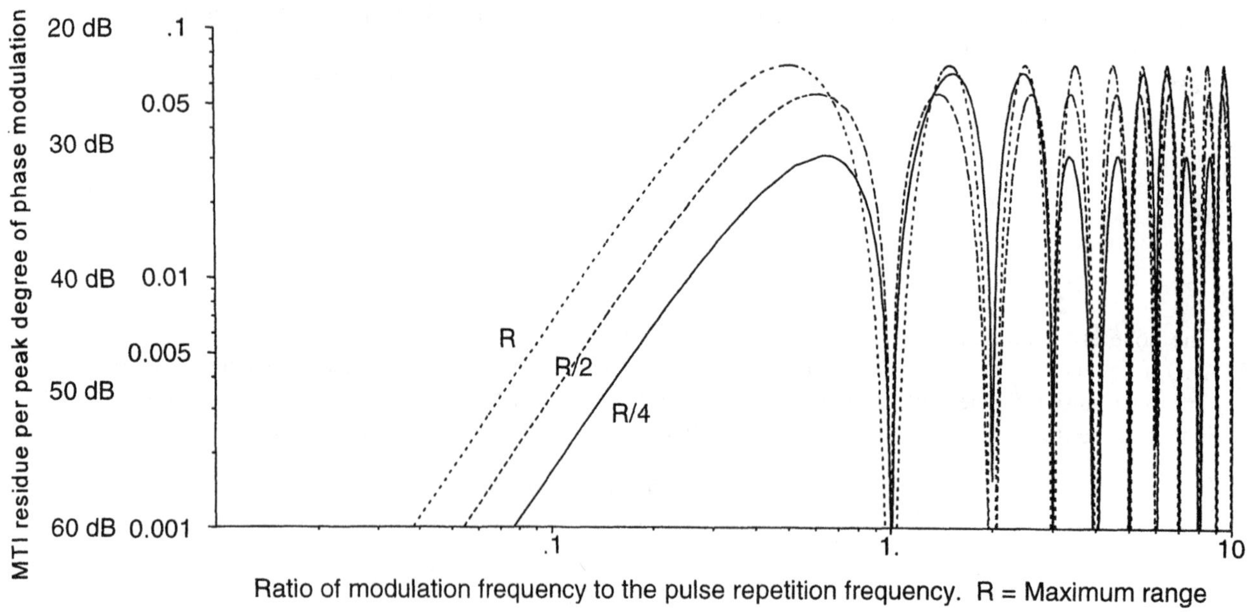

Figure 7.15 The effect of sinusoidal phase modulation of the local oscillator.

Notice that there is no degradation with modulation frequencies that are multiples of the pulse repetition frequency. Many American radars have pulse repetition frequencies that are multiples of 60 Hz and European radars have pulse repetition frequencies that are multiples of 50 Hz that are the common electrical power frequencies.

REFERENCES

1. *IEEE Standard 100*, New York: The Institute of Electrical and Electronic Engineers, 1993, Noise figure (factor).
2. *IEEE Standard Radar Definitions, IEEE Standard 686-1990,* New York: The Institute of Electrical and Electronic Engineers, 1993, noise temperature, (2)(standard).
3. Taylor, J. W. and J. Mattern, in M. I. Skolnik, (ed.), *Radar Handbook*, New York: McGraw-Hill, 1970.
4. Taylor, J. W., in M .I. Skolnik, (ed.), *Radar Handbook*, 2nd ed., New York: McGraw-Hill, 1990.
5. Vaccaro, D. D., *Electronic Warfare Receiving Systems*, Norwood, Massachusetts: Artech House, 1993, Appendix B.
6. Ward, H. R., "The effect of bandpass filtering on noise with a Gaussian spectrum", *Proc IEEE* 57, No. 11, November 1969, pp. 2089-2090.
7. Schleher, D. C., *MTI and Pulsed Doppler Radar*, Norwood, Massachusetts: Artech House, 1991.
8. Gradshteyn, I. S., and I. M. Ryzhik, *Table of Integrals, Series, and Products*, New York: Academic Press, 1980.
9. Barton, D. K., *Radar System Analysis*, Englewood Cliffs, New Jersey: Prentice-Hall, 1964.

Chapter 8

Matched and matching filters

Radar echo signals are converted down to intermediate or video frequency and amplified. Early British radars used the design of the 45 MHz radio frequency stages of pre-1939 television receivers as intermediate frequency amplifiers. The filtering was performed by stagger tuning to give a flat frequency response over the required bandwidth. Currently, commercial intermediate frequency amplifier blocks with a flat response are used with separate shaped bandpass filters.

The optimum peak echo signal-to-mean-noise power is obtained using a matched filter. In the case of echoes from moving scatterers, the filter must be matched for the Doppler frequency shifts of the echo signals. For a rectangular pulse with a sin x/x spectrum, the matched filter must have a sin x/x characteristic. If the filter is fully matched with a sin x/x bandpass characteristic, the output of the filter has a $(\sin x/x)^2$ spectrum which is the spectrum of the delayed triangular pulse shown in Figure 8.1.

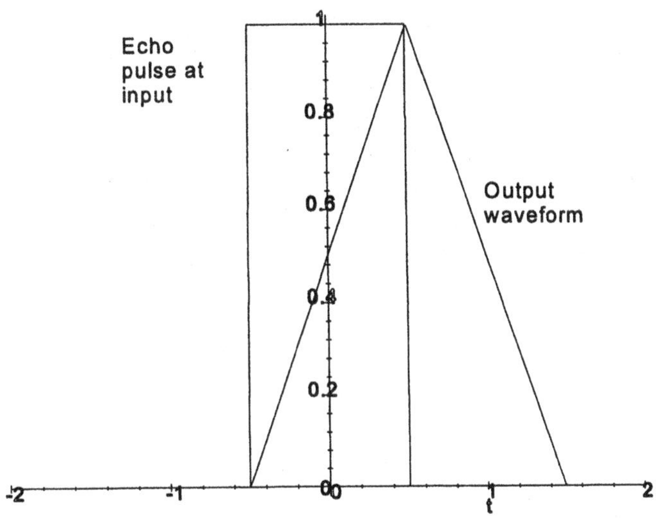

Figure 8.1 The input waveform of a filter fully matched to a rectangular pulse compared to the output waveform.

The output waveform too has the same width at the 50% points on the voltage plot but no overshoots. Commonly, a filter with a simple rectangular or Gaussian shape is used. Such a filter is called a *matching* filter.

Mostly the filters are designed to match the initial "ideal" pulse generated in the transmitter before amplification and it must be remembered that this pulse is influenced on its way from initial generation to its arrival at the filter. The influences are:

- Droop (amplitude) and phase distortion during amplification in the transmitter;
- Generally speaking the atmosphere and the objects that scatter the transmitted pulse are not frequency dispersive during a radar pulse;
- Restricted bandwidths in the receiver circuits before filtering;
- Distortion caused by limiting, restricted dynamic range, or errors caused by analogue to digital conversion;
- With long pulses the effects of sensitivity time control on the amplitude of the pulse during the pulse.

The tuned intermediate frequency amplifiers can be made to ring by strong pulses at or outside the intermediate frequency. Modern radars use a number of amplifiers and filters to narrow the bandwidth in stages to avoid ringing. The final filter, which optimizes the ratio of signal to wideband noise, is often placed before the detector. Alternatively, it may be a lowpass filter placed after a detector.

This final filter is a discrete component and is used to recognize the echo of the transmitter pulse again. The transmitter pulses are of two types:

- Simple or uncompressed pulses;
- Frequency or phase modulated transmitter pulses used in pulse compression radars.

The Maple mathematical program has been used as a calculating engine for the waveforms and spectra. For convenience, the pulses have been centered on zero time to give symmetrical spectra. In real life, there is no "negative time" and the spectra are not balanced about zero frequency (see Chapter 3, Transmitters).

8.1 UNCOMPRESSED PULSES

This section considers two classical cases:

- Rectangular echo pulse passing through a rectangular filter (Figure 8.2);
- Rectangular echo pulse passing through a Gaussian shaped filter (Figure 8.6).

8.1.1 Rectangular transmitter pulse and rectangular filter

A rectangular transmitter pulse centered on zero has a real sin x/x spectrum. The filter for this pulse is of the same form that would be used in the line modulator in the transmitter, and it needs a wider bandwidth intermediate amplifier than that used in the past. The following shows what happens when a rectangular pulse passes through a rectangular filter and how the bandwidth of this filter may be optimized to have the maximum signal-to-noise ratio at the output.

This filtering process is illustrated in Figure 8.2 with the axes labeled in the upper right hand corner. Figure 8.2(a) shows the rectangular echo pulse and its sin x/x spectrum. The receiver noise through a wide bandpass amplifier is shown with its rectangular spectrum in Figure 8.2(b). The left hand sequence is for a notional filter without any time delay. Real filters delay the signals, have a helical phase characteristic, and the form of filtering is shown in the right hand sequence. The optimum rectangular matching filter bandpass is shown in Figure 8.2(c) and the result in Figure 8.2(d).

The input pulse of width τ seconds has a spectrum given by its Fourier transform (see Figure 8.2(a)).

$$S(f) = \int_{-\tau/2}^{\tau/2} \exp(-j\,2\pi f t)\,dt = \frac{\sin(\pi f \tau)}{\pi f \tau} \tag{8.1}$$

The wideband input noise has a spectrum limited by the intermediate amplifier bandwidth with a density of $N/2$ W/Hz which is shown in Figure 8.2(b). The rectangular filter in Figure 8.2(c) cuts into both the signal and the noise spectra. The integrated signal voltage at the output of the filter is defined by its bandwidth, B Hz, given by the limits of integration.

$$Filt_{out} = \int_{-B/2}^{B/2} \frac{\sin(\pi f \tau)}{\pi f \tau}\,df = \frac{4\,\mathrm{Si}\left(\frac{B\tau}{4}\right)}{\tau} \tag{8.2}$$

where Si is the sine integral $\mathrm{Si}(x) = \int_0^x \frac{\sin t}{t}\,dt$

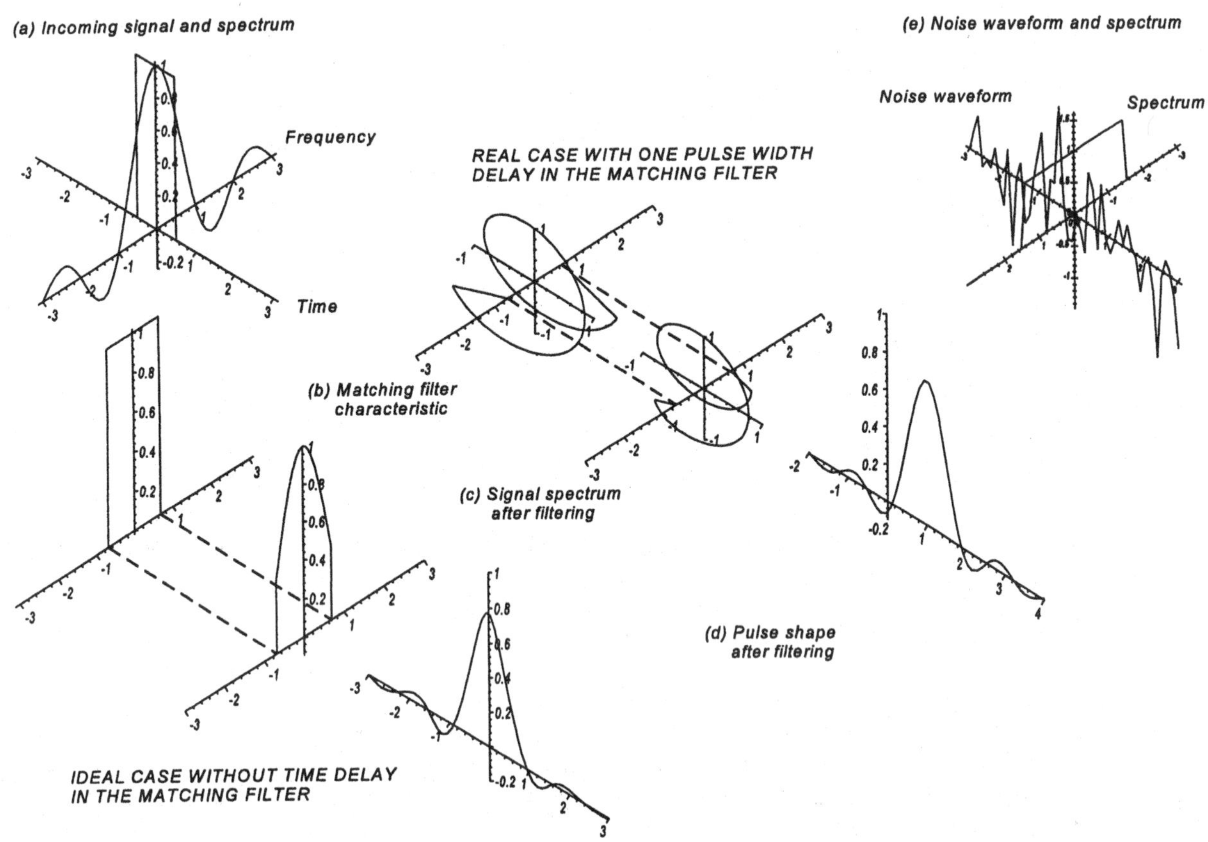

Figure 8.2 The filtering process with a rectangular pulse and a rectangular bandpass filter.

The output signal power is then

$$\text{Power out} = \frac{\left[\dfrac{4\,\mathrm{Si}\left(\dfrac{B\tau}{4}\right)}{\tau}\right]^2}{2} \tag{8.3}$$

The noise is incoherent, and thus output noise power is the sum of the squares of the individual voltage components:

$$\text{Noise power} = \int_{-B/2}^{B/2} n^2 df \tag{8.4}$$

The output signal-to-noise ratio relative to a matched filter is given by

$$\text{Signal-to-noise ratio} = \frac{\left[\displaystyle\int_{-B/2}^{B/2} \frac{\sin(\pi f\tau)}{\pi f\tau} df\right]^2}{\displaystyle\int_{-B/2}^{B/2} n^2 df} \tag{8.5}$$

The output signal-to-noise ratio is plotted against the bandwidth B in Figure 8.3 for a one second pulse. The abscissa thus represents the time-bandwidth product $B\tau$. The upper limit has been extended to $8\ B\tau$ to show the effects of the sine integral shape.

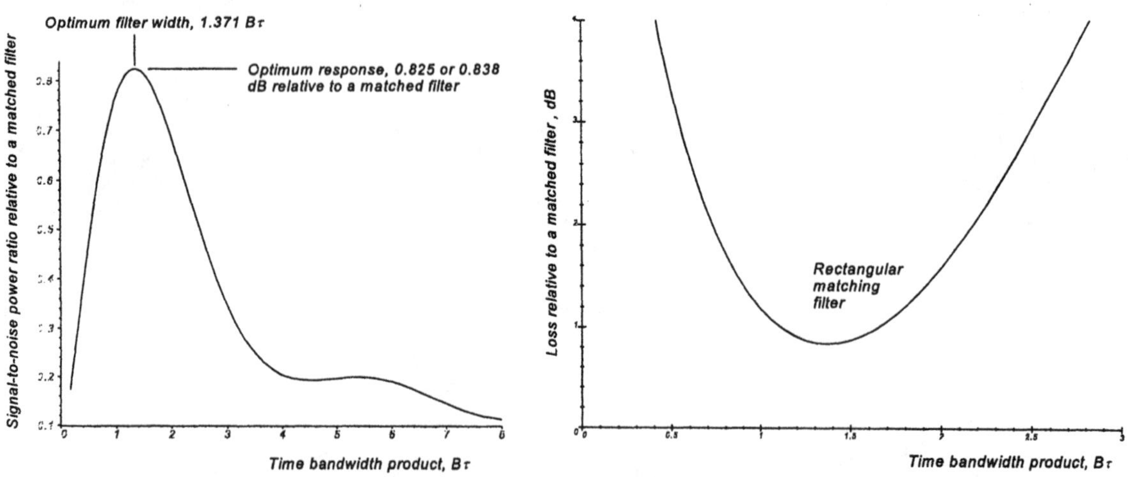

Figure 8.3 The reduction of the signal-to-noise ratio from a rectangular pulse at the output of the rectangular filter.

The results of the above are:

- Optimum filter width: $1.371/\tau$;
- Output signal-to-noise ratio at this width relative to a matched filter: 0.825;
- Loss at the optimum filter width relative to a matched filter: 0.838 dB.

This is the basis for the rule-of-thumb 0.8 dB to 1.0 dB matched filter loss [1, p. 5-25; 2, p.3.21; 3, pp. 144-148]. Figure 8.4 shows the input pulse spectrum with the optimum filter passband biting into it, which narrows the spectrum, thus stretching the output waveform.

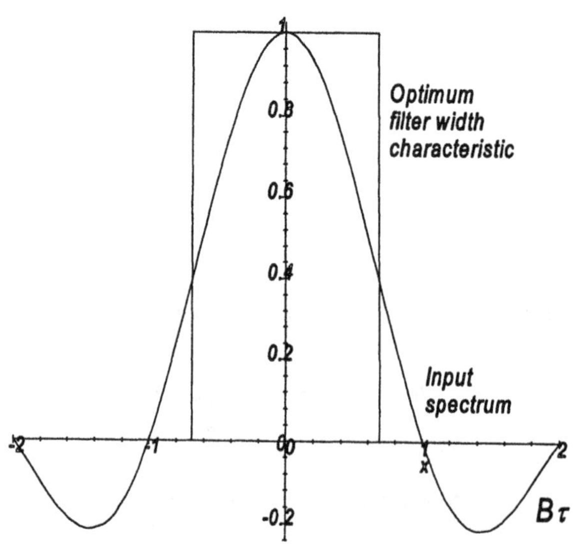

Figure 8.4 The spectrum of a rectangular pulse of width unity with the optimum rectangular filter drawn over it.

The input and output waveforms are shown in Figure 8.5. This is the classical matching filter approximation. The rectangular pulse has been blurred into a waveform with overshoots. More modern radars use a Gaussian filter, which is a better approximation without noticeable overshoot.

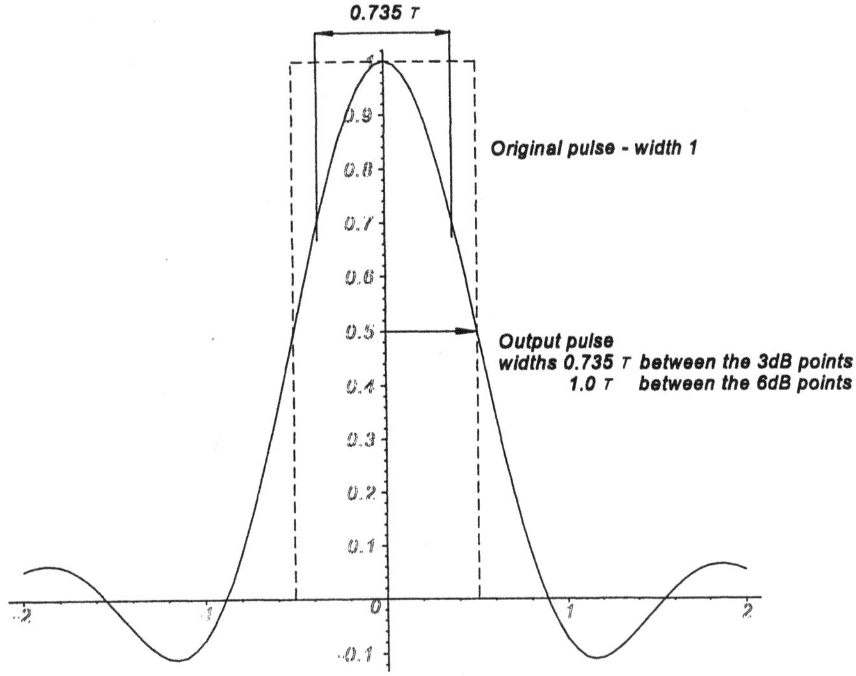

Figure 8.5 The input waveform compared to the output waveform without time delay.

In the late 1940s radar operators performed the detection decisions from their cathode ray tube displays. The critical parameter was found to be the average observed signal power. The minimum loss for this was found experimentally to be 2.3 dB at $B\tau = 1.2$ [4, p. 172]. Haeff found an approximation for the extra loss factor at time-bandwidth products other than $B\tau = 1.2$ which is given by [5, p. 367]

$$C_b = \frac{B\tau}{4\alpha}\left(1 + \frac{\alpha}{B\tau}\right)^2 \text{ dB} \tag{8.6}$$

where $\alpha = 1.2$.

8.1.2 Rectangular transmitter pulse and Gaussian filter

A Gaussian filter characteristic provides a better match to the sin *x/x* spectrum of the returning echoes. The filter bandwidth for optimum signal causes less of the echo signal to be rejected, giving lower losses. As before, this process is illustrated in Figure 8.6, with the axes labeled in the upper right corner of Figure 8.6. Figure 8.6(a) and (b) are the same as in Figure 8.2. Figure 8.6 (c) shows the input spectrum with the optimum Gaussian characteristic without and with a real time delay in the matching filter. The resulting spectrum and waveforms are shown in Figure 8.6(d). The spectrum of the input pulse is from (8.1):

$$\begin{aligned} S(f) &= \int_{-\tau/2}^{\tau/2} e^{-j2\pi f t}\,dt \\ &= \frac{\sin(\pi f\tau)}{\pi f\tau} \end{aligned} \tag{8.7}$$

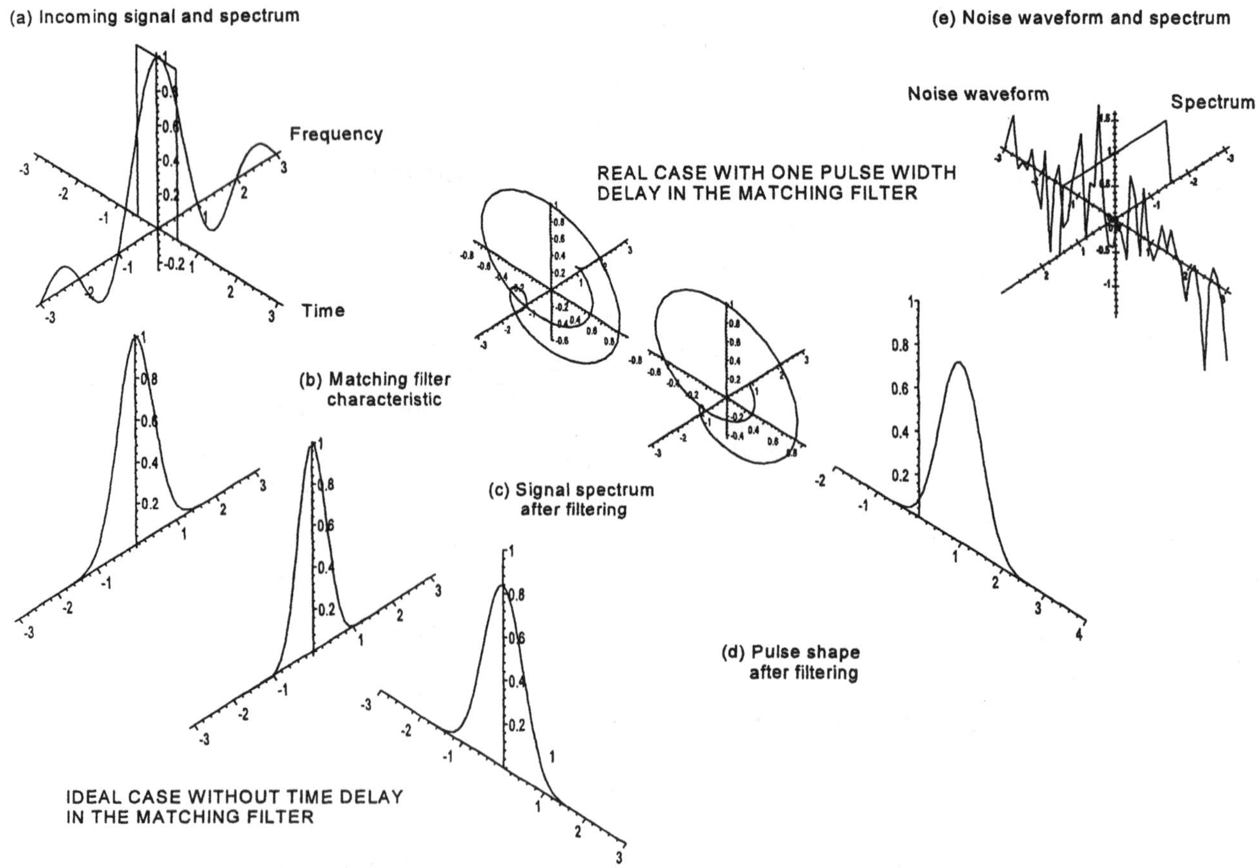

Figure 8.6 The filtering process with a rectangular pulse and a Gaussian bandpass filter.

The filter characteristic with a 3 dB width of B is given by

$$\textit{Filter characteristic}\ (f) = \exp\left(-\frac{f^2\tau^2\ 2\ln(2)}{B^2}\right) \tag{8.8}$$

Thus, the signal spectrum leaving the filter is

$$\textit{Spectrum out of filter}\ (f) = \frac{\sin(\pi f\tau)}{\pi f\tau}\exp\left(-\frac{f^2\tau^2\ 2\ln(2)}{B^2}\right) \tag{8.9}$$

Because the noise spectrum is the same as the filter characteristic, the output signal-to-noise ratio relative to a matched filter is given by

$$\textit{Signal-to-noise ratio} = \frac{\left[\int_{-\infty}^{+\infty}\frac{\sin(\pi f\tau)}{\pi f\tau}\exp\left(-\frac{f^2\tau^2\ \ln(2)}{B^2}\right)df\right]^2}{\int_{-\infty}^{+\infty}\left[n\exp\left(-\frac{f^2\tau^2\ \ln(2)}{B^2}\right)\right]^2 df} \tag{8.10}$$

As in (8.4), the noise power is the incoherent sum of the squares of the individual components.

The optimum filter width is found by varying B and finding the maximum resulting signal-to-noise ratio at the output. This is illustrated in Figure 8.7 and the shapes may be contrasted with Figure 8.3.

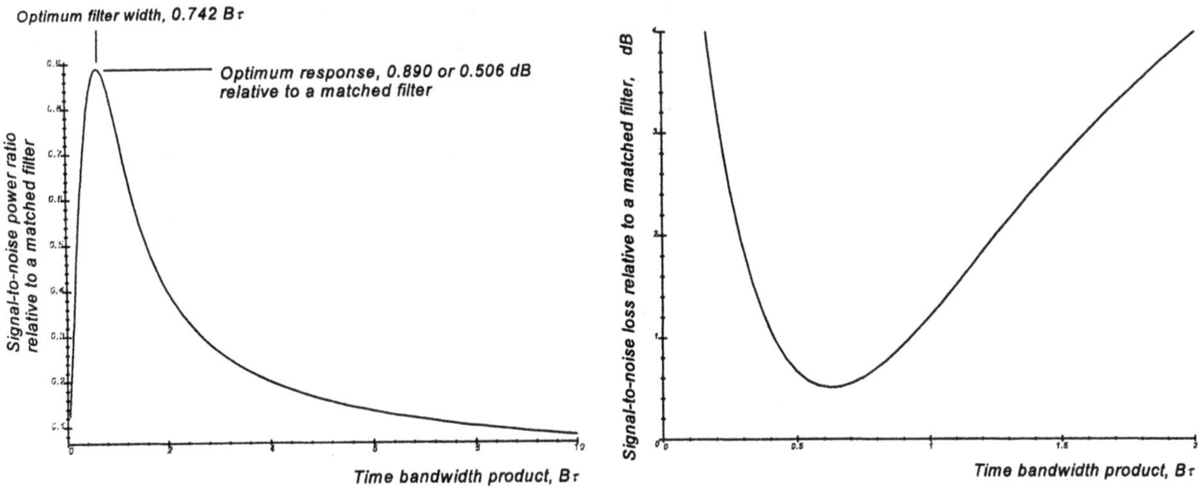

Figure 8.7 The reduction of the signal-to-noise ratio at the output of the Gaussian filter

The results of the square pulse and the Gaussian filter are:

- Optimum 3dB filter width: 0.742 /τ;
- Output signal-to-noise ratio at this width: 0.890;
- Loss at this width: 0.506 dB.

Figure 8.8 shows the input and output spectra and the filter characteristic. Here it can be seen that the Gaussian filter provides a much better match to the sin *x*/*x* shaped spectrum. The filter characteristic bites into the input spectrum to give an output spectrum that is narrower and almost Gaussian. If it is assumed to be Gaussian, then the output is a Gaussian pulse with a 3 dB width 1.743 times the width of the input pulse, as shown in Figure 8.9.

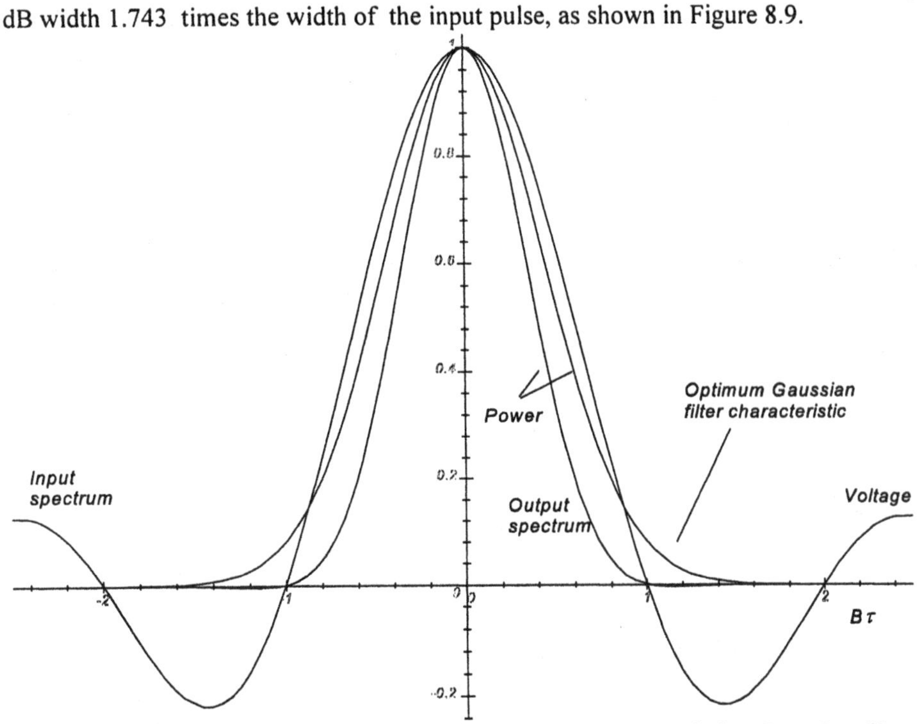

Figure 8.8 The spectra of the input and output pulses and the Gaussian filter characteristic.

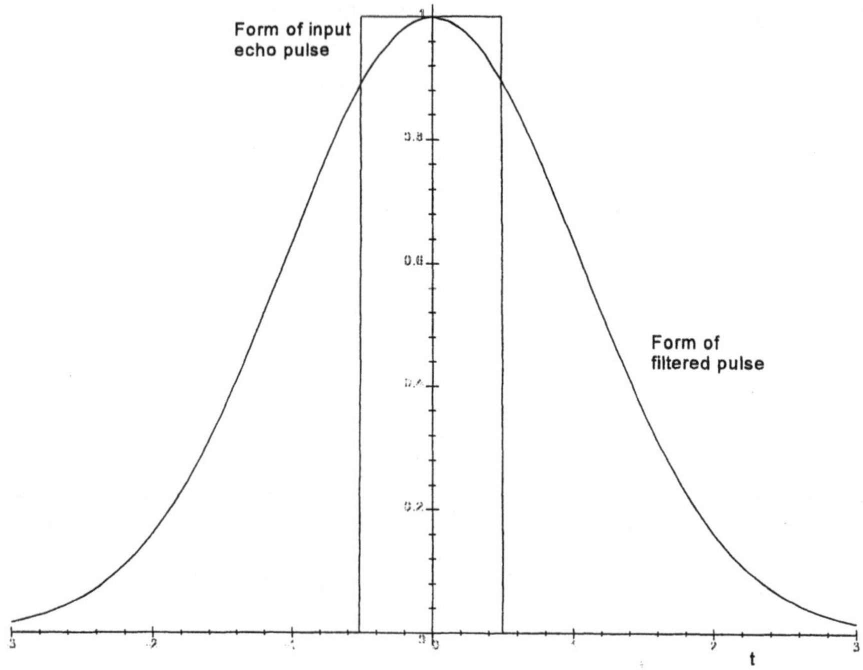

Figure 8.9 Comparison of rectangular input echo pulse with the filtered output pulse.

8.1.3 Doppler frequency shift, detuning

The matched filter has a definite bandwidth. A check must be made to see if Doppler frequency shifts in possible echoes may fall outside the filter bandwidth, though normally it seldom happens. For example, in S-band at 3 GHz, the Doppler shift is 20 Hz/m/s. The edges of an optimum rectangular filter occur at $\pm$ 0.685 $B\tau$. For a 1 μs pulse, it corresponds to $\pm$ 685 kHz, or 34.275 km/s, more than Mach 100. When pulse compression is used, this is not always the case.

8.1.4 Filtering after limiting, Dicke-fix receiver

During the Second World War radar military operators were trained to adjust their receivers in order to see through jamming. This was a matter of great skill to adjust the receiver gain control and only applicable to a single echo, principally for fire control radars. The Dicke-fix circuit, first proposed as the Lamb noise canceler [6, p. 11], is one way of working during jamming with the effective receiver gain being adjusted automatically and its block diagram is shown in Figure 8.10. This form of gain control is suitable for surveillance radars.

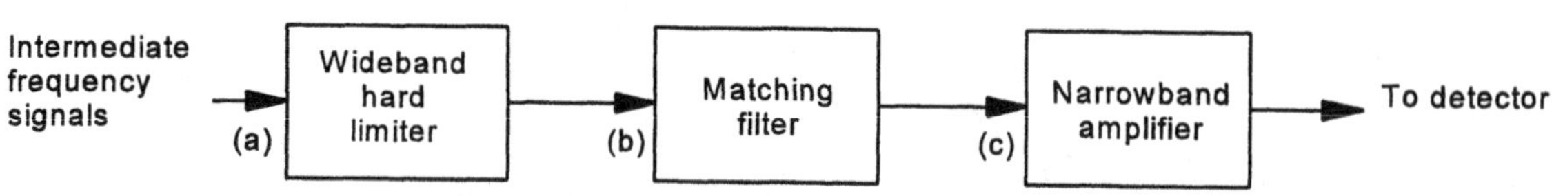

Figure 8.10 Block diagram of the Dicke-fix receiver.

The radar receiver is constructed to have a wide bandwidth and the signals are passed through a hard limiter. The voltage modulation of the wideband signals on a (notional) carrier wave is shown for an echo pulse for noise alone in Figure 8.11. The sum of echo signal and noise is shown in Figure 8.12(a).

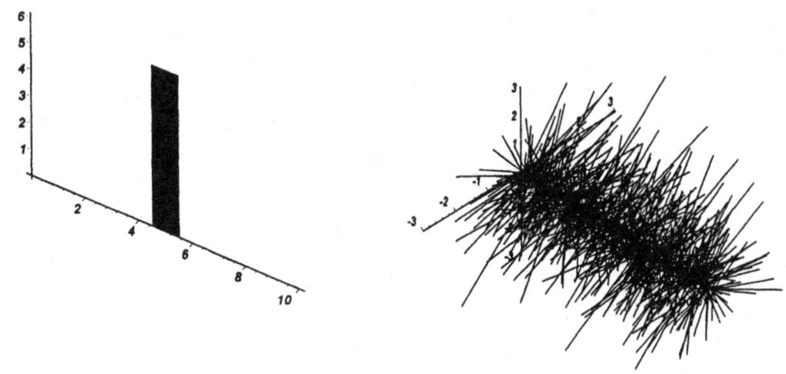

(a) Signal pulse *(b) Bottle-brush noise*

Figure 8.11 Echo signal pulse and bottle-brush (Gaussian) noise modulation.

In Figure 8.12(a) there are 900 samples of noise and, between 450 and 550, 100 samples of signal plus noise. After limiting (Figure 8.12(b)) all the noise tips of the noise and signals touch the inside of a cylinder and the signal plus noise region is only recognizable by the similarity in phase (note: Figure 8.12(b) has been rotated for clarity, also observe the change of scale). The effects of a matched filter that performs a running average on 100 samples is shown in Figure 8.12(c). In the noise region the resulting sum snakes randomly around the time or sample number axis and the triangular signal from the matched filter has been deformed in amplitude and phase by the noise.

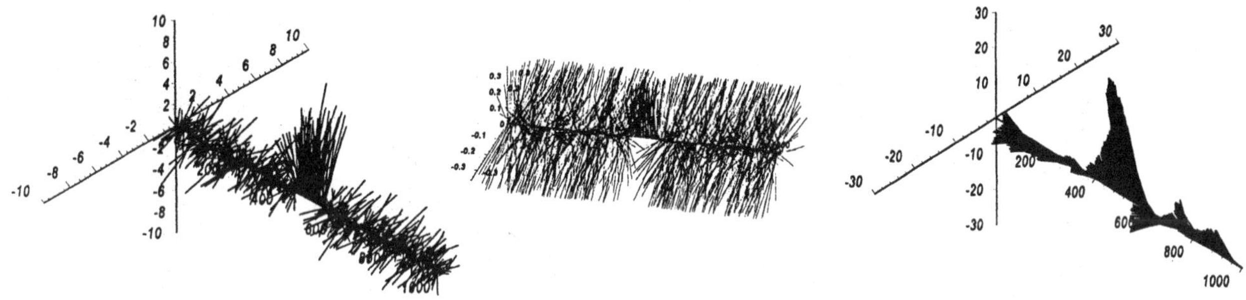

(a) Signal + noise *(b) Signal + noise after limiting* *(c) Signal + noise after limiting and filtering*

Figure 8.12 Samples of echo signals and noise at the input (a), after limiting (b), and after limiting and filtering (c).

The signal is shown in the more usual form as the modulus in Figure 8.13.

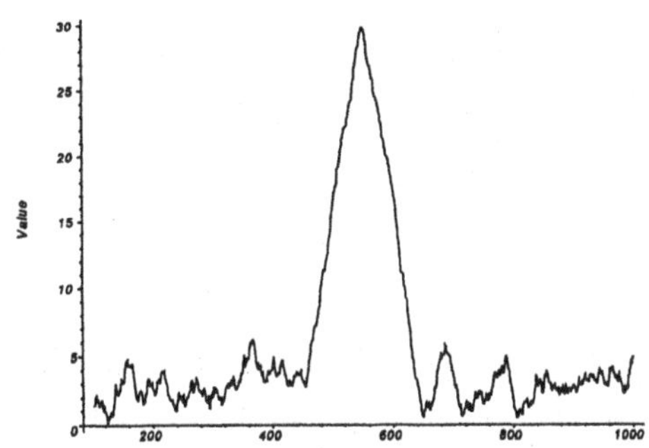

Figure 8.13 Modulus of the detected signal voltages (signal width = 100).

Interference or jamming with an amplitude greater than noise that occurs outside the bandwidth of the matching filter produces a helical form at the output of the limiter. The filter output is the running average, for a helix, zero. A strong signal outside the bandwidth can thus suppress a weak signal inside the bandwidth but no false alarm is generated.

The Dicke-fix receiver limits the dynamic range to the ratio of the bandwidths of the wideband to narrowband filters. In the example this ratio, say N, is the number of samples averaged in the matched filter. Consider N Gaussian noise voltage samples: their sum is $\sqrt{N}$ and consider the same number of echo signal samples with infinite signal-to-noise ratio, that is with the same phase to give a sum of N. The resulting signal-to-noise ration is $\sqrt{N}$ in voltage or N in terms of power. In the example, the signal and noise samples are limited to 0.3 V so that the sum of 100 samples of signals gives 30 V and the sum of the noise is $\sqrt{100} \times 0.3 = 3$ V. The dynamic range is 10 in voltage or 100 (20 dB) in terms of power, the number of samples averaged.

The noise coming from the limiting stage in Figure 8.12(b) and which is then filtered fulfills the characteristics required for a random walk so that the noise voltage has a Rayleigh distribution with a standard deviation only dependent on the limiting level and independent of the input level. Changing the variable to power, the probability distribution becomes a negative exponential distribution (see Chapter 15, Statistics).

The main use of the Dicke-fix receiver is to present a constant noise level to an operator or an extractor in spite of changes in gain or noise level in the preceding stages or in the presence of interference and jamming. If the limiting level is not well below the input noise level, there is a finite chance that a wideband noise sample has an amplitude below the limiting level as shown in Figure 8.14. An increase in the input noise will fill in the gaps so that the output noise level and false alarm rate will rise slightly.

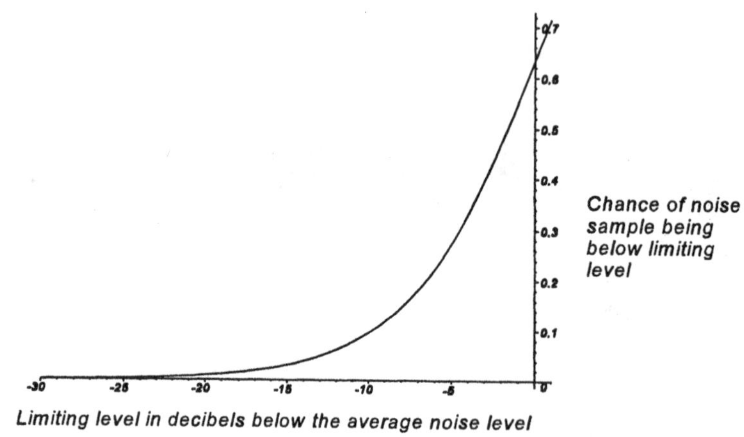

Figure 8.14 Chance of no sample being below the limiting level.

When the limiting level is well below the noise level, this effect is much less and the false alarm rate more constant so that (12.12) for the probability of false alarm for one filtered noise pulse becomes

$$P_{fa} \leq \exp(-Y_b) \tag{8.11}$$

where Y_b is the threshold level.

Gaussian noise has a Gaussian spectrum [7, Chapter 7] so that the spectra of limited signal plus noise is also of bottlebrush form in which the signal cannot be readily identified. Thus there is no treatment here in the spectral domain.

Sensitivity time control (STC) may be effected by injecting extra wide band noise at point (a) in Figure 8.10. As before the echo signals, receiver noise, and additional noise add to increase the phase excursions in Figures 8.12(a) and (b) and noise sensitivity time control avoids having to adjust and hold bias levels for the intermediate frequency amplifiers. Note that modern radars use an attenuator before the first radio frequency amplifier.

Radars can give a constant false alarm rate characteristic to linear or other videos by using a Dicke-fix receiver to gate the videos as shown in Figure 8.15 before automatic detection and display.

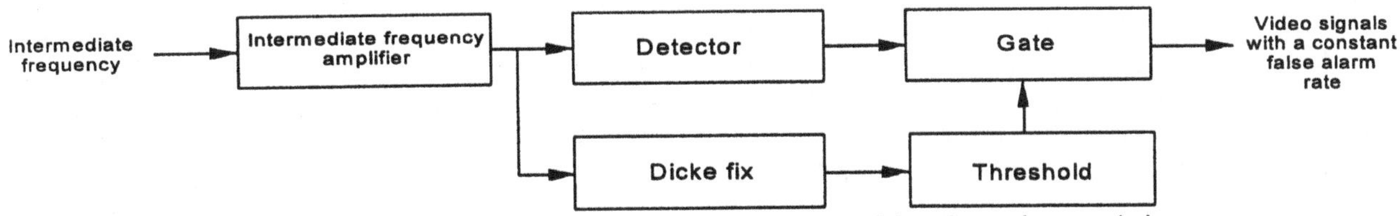

Figure 8.15 The use of Dicke fix to give a conventional video constant false alarm characteristics.

A loss of approximately π/4 or 1 dB is associated with this process.

8.2 PULSE COMPRESSION USING FREQUENCY MODULATION

The range measuring capacity and resolution are normally provided by pulse modulating the transmitter signal: the shorter the pulse, the better the range resolution. The range of the radar depends on the energy in the pulses, so that higher peak powers are required but there are technical limitations for the maximum peak power, such as maximum high voltage or power from the output stage, or waveguide breakdown.

Further resolution may be obtained by frequency or phase modulation during the transmitter pulse. In early pulse compression systems, a short pulse is stretched to form a modulating signal, which modulates the transmitter pulse, and on reception, the echoes are compressed to near the original short pulse length as shown in Figure 8.16. Taken to the extreme there are frequency modulated continuous wave (FMCW) radars where the range resolution is governed by the bandwidth as with frequency modulated pulse compression.

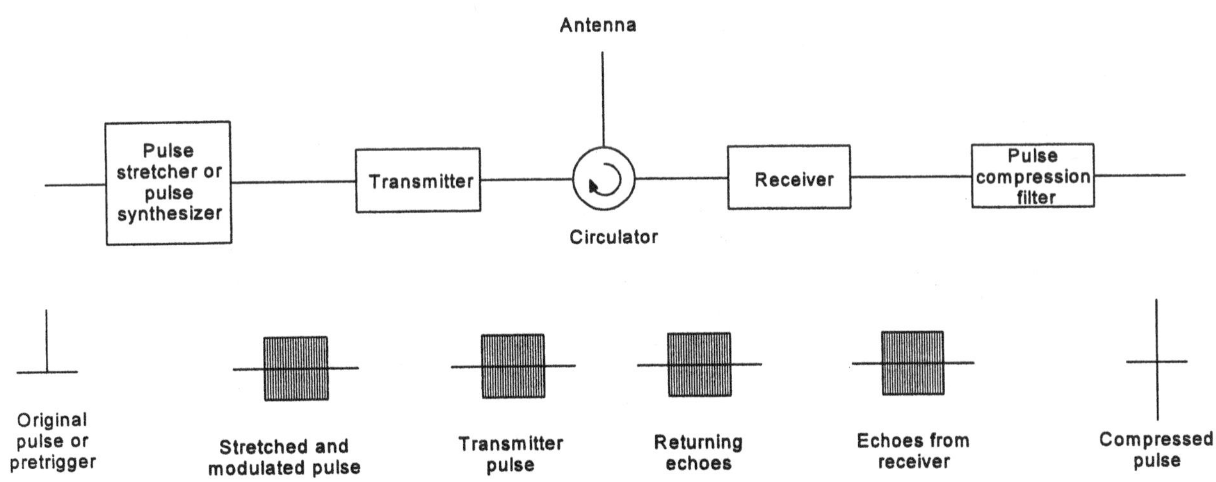

Figure 8.16 Block diagram showing pulse compression components.

8.2.1 Linear frequency modulation

On the transmitting side (see Chapter 3, Transmitters) the short video (trigger) pulse is either stretched by a filter, often a surface acoustic wave (SAW) device, or formed in a synthesizer and then amplified by the transmitter. The returning echo signals are converted to intermediate frequency and amplified. The compression filter, which implements the matched filter, is the complement of the pulse stretcher in the transmitter. This filter may be placed at one of a number of points when linear processing is used: as an intermediate frequency filter, a filter at baseband following a coherent vector detector (see Section 9.2 and Section 10.3.4), or following coherent signal processing. Baseband processing is mostly performed with digital filters. The analogue hardware may be active frequency or phase modulators, dispersive filter circuits, or surface acoustic wave (SAW) or other dispersive delay lines described in standard textbooks [2, p.

10.10-12; 8, pp. 225-239]. For pulse compression after detection, equivalent digital methods are used.

The losses for the complementary filters are determined by the hardware and must be measured. The wider spectrum is often nearly rectangular, and thus the compressed signal pulse has a sin *x/x* form. That is, there is a signal output at a time before the compression filter fully reassembles the echo and residues after it has been assembled. Additional pulse shaping (tapering or sidelobe suppression) widens the compressed pulse and reduces the time sidelobes at the expense of extra losses (see Chapter 16, Transforms).

Alternatively the echo pulses may be correlated with the conjugate of a copy of the transmitted pulse. The components for this are shown in Figure 8.17.

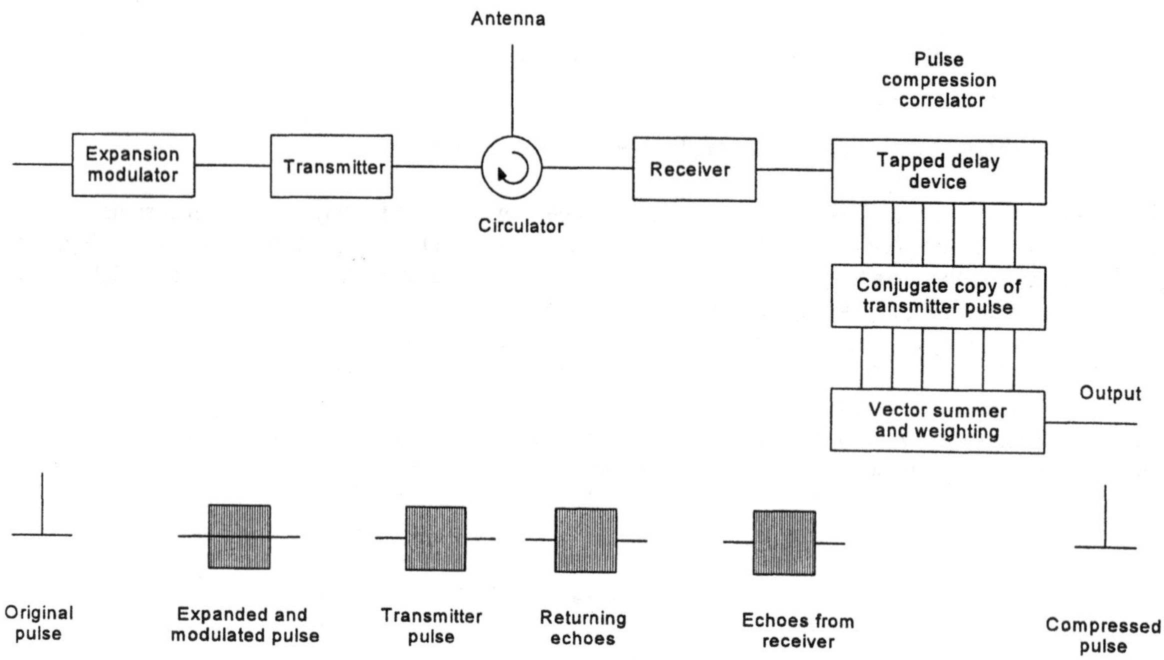

Figure 8.17 Block diagram of the components of a pulse compression system using correlation.

With the first linear frequency modulation systems the dispersive pulse stretcher took a narrow (wide bandwidth) pulse, and the different frequency components took different times to pass through the block to give a frequency modulation along the pulse. The frequency modulation is equivalent to the constant change of phase with time of an alternating current plus an accelerating change of phase with time of the modulation.

$$\textit{Expanded pulse} = \sin(2\pi f_c t \pm k t^2) \tag{8.12}$$

where f_c Hz is the carrier frequency and the modulation coefficient is $k/2$.

A positive sign represents a rising frequency and a negative sign a falling frequency. The accelerating phase component, kt^2 (radians), may be represented in polar form as points around a circle. If time is taken as the third dimension, this representation becomes a helix. The accelerating helix may be represented by

$$\textit{Modulation of the expanded pulse} = \exp\left(j \ \pi \ \frac{k}{2} \ t^2 \right) \tag{8.13}$$

The echo signals have the same form as the transmitted pulse and the helical form of the modulation is shown as the input in Figure 8.18. In order to show the classical forms, a compression ratio of 100 is used as an example for the diagrams. The recovery of the original short pulse may be carried out either by filtering or by correlation.

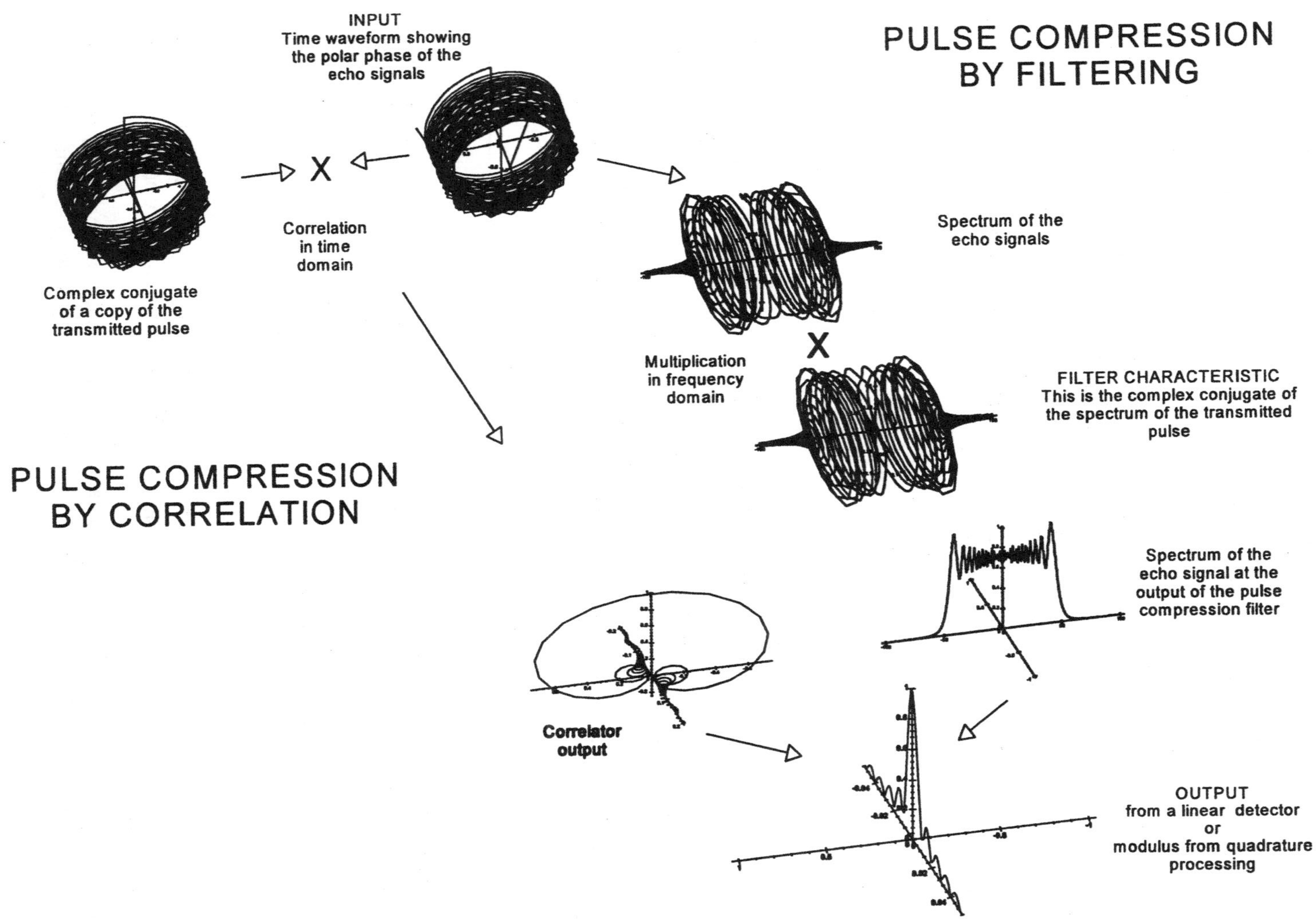

Figure 8.18 Pulse compression principle using either correlation or filtering. Waveforms are centered on an artificial zero time.

The spectrum of the echo signals is the same as the transmitted pulse (see Chapter 3, Transmitters) shown to the right of Figure 8.18. Since the echo signals are complex, so is their spectrum. The modulus of this spectrum is shown in Chapter 3, Transmitters. The filter characteristic required to extract the compressed pulse has the opposite time characteristic to the expansion filter and is the conjugate of the spectrum of the received echoes. In the filtering process the components at each frequency are multiplied to give the output spectrum in Figure 8.18, namely,

$$\text{for each frequency} \quad Output(f) = Signal\ spectrum(f) \times Compression\ filter\ characteristic(f) \tag{8.14}$$

The spectrum is nearly rectangular if the overshoots are neglected. This rectangular spectrum gives a sin x/x time waveform at the output. The output spectrum of the receiving filter may be tapered (see Chapter 16, Transforms) to reduce the time sidelobes. The tapering affects the wanted signals and noise equally, but the signal-to-noise ratio at the output is reduced by the processing loss and the width is stretched. The tapering loss is the matching loss and examples of tapering are given in Appendix B.

Looking at the left hand side of Figure 8.18, it can be seen that the same result may be obtained with correlation. In Figure 8.17, the correlation process is

$$Correlator\ output = \int_{-\tau/2}^{\tau/2} signal(t) \cdot copy(t - x)dt \tag{8.15}$$

where *copy* is the complex conjugate of the transmitted pulse.

Equation (8.15) represents the incoming echo signal passing over the conjugate copy, and the output is the vector sum of the multiplications. In practice, a limited number of taps are available, so that the integral in (8.15) becomes a sum. The phase of the correlator output changes as one time helix passes the other. The time waveform is the modulus of the output helix which is the same as the waveform from a pulse compression filter.

The pulse coming from the compression stage in Figure 8.18 has a sin x/x shape with 13.26 dB principal time sidelobes occurring before and after the main pulse. The causes of the sidelobes are shown in Table 8.1.

Table 8.1
Equivalence of filtering and correlation

Filter	Correlation
The modulus of the spectrum of the expanded pulse is nearly rectangular, and the modulus of the filter has the same shape, however, the phase sequence is in the opposite sense. The modulus of the Fourier transform of the filtered spectrum thus has a sin x/x shape.	The polyphase signal is cross-correlated (vector multiplied and summed over the length of the signal) with a waveform of the opposite sense. Near coincidence, the output has a sin x/x shape with time as the coils move "in and out of phase". Away from coincidence, the differences in the pitches of the coils give near zero output.

The time sidelobes limit the visibility of wanted echoes between clutter patches and can be reduced by tapering the filter characteristic or correlation waveform (see Appendix B). The processing loss may be reduced by predistorting the spectrum of the transmitter as in Section 8.2.2.

For the purposes of illustration, the frequency modulation on a carrier pulse may be considered to be a pulse at a single frequency that has an accelerating phase modulation. The example that follows uses the following expression, which is illustrated in Figure 8.19,

$$h(t) = \exp(-j\ 40\ t^2) \tag{8.16}$$

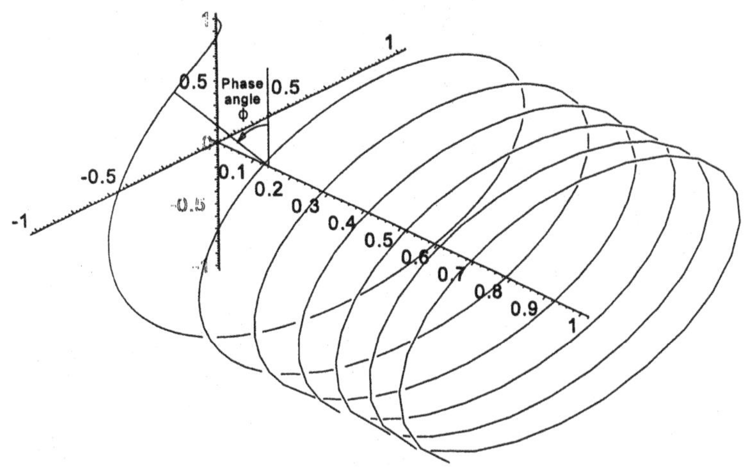

Figure 8.19 Helix showing the acceleration of phase during frequency modulation.

The autocorrelation function is the result of the waveform multiplied by its complex conjugate as one waveform is dragged by the other as in Figure 8.20. In radar, the signal waveform would be the returning echoes, and the reference is the complex conjugate of the transmitter pulse modulation (the helix with the reverse sense):

$$\int_0^{\tau} h(t)\; h^*(\tau - t)\; dt \tag{8.17}$$

where h^* is the complex conjugate, or, in this case, the reverse helix.

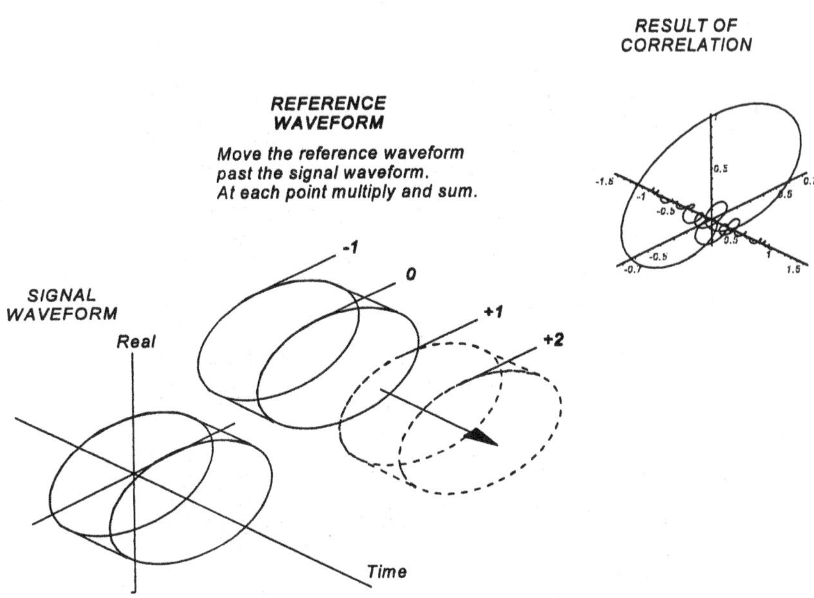

Figure 8.20 The positions of the signal and reference signal during cross-correlation.

The complex result of echo signal correlation with the reference is shown in Figure 8.21(a). The amplitude, in decibels, is shown in Figure 8.21(b). It consists of a narrow peak and has neighboring sidelobes as shown in the decibel representation. Cook in [9, p. 221] gives the closed form for the correlated time function as

$$g(t) = \frac{\sin\left(\dfrac{\omega_d + kt}{2}(\tau - |t|)\right)}{\omega_d + kt} \tag{8.18}$$

The sine function has a variable period, especially wide at $t = \tau/2$.

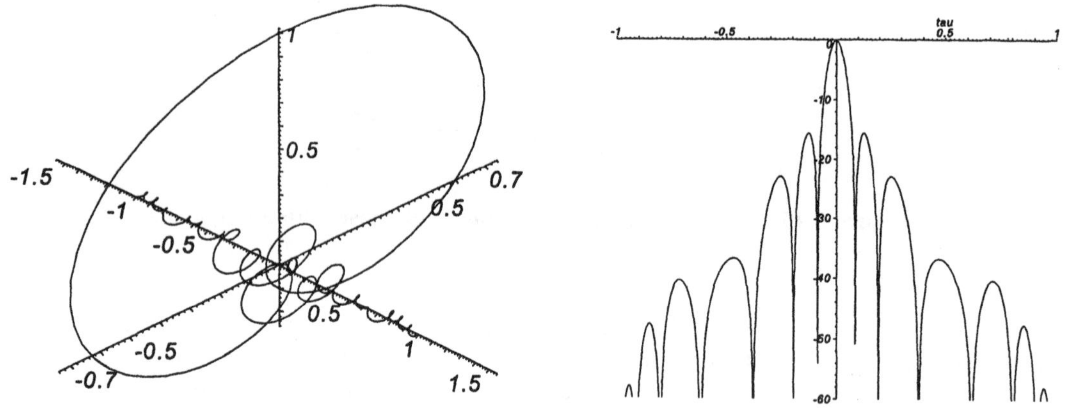

(a) Correlation function (b) Modulus of the correlation function, dB

Figure 8.21 The result of correlation in vector and absolute (dB) forms.

If the echoes are clutter signals, then the time sidelobes extend the range extent of the clutter and reduce the probability of detection of wanted echo signals between clutter. Tapering the products of the two waveforms reduces the time sidelobes, as with filtering and antenna theory. If cosine squared tapering is used as an example,

$$h(t) = \cos^2\left(\pi\frac{t}{\tau}\right) \tag{8.19}$$

If the reference waveform is thus tapered, the appearance is as shown in Figure 8.22.

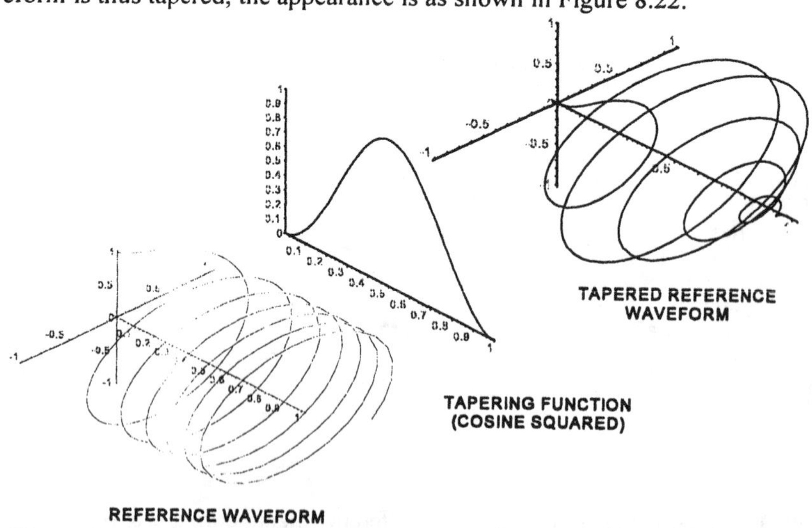

Figure 8.22 The reference waveform, a cosine squared tapering function, and the tapered reference.

Tapering gives lower time sidelobes with a widened main signal lobe, as in Figure 8.23. Within a dynamic range

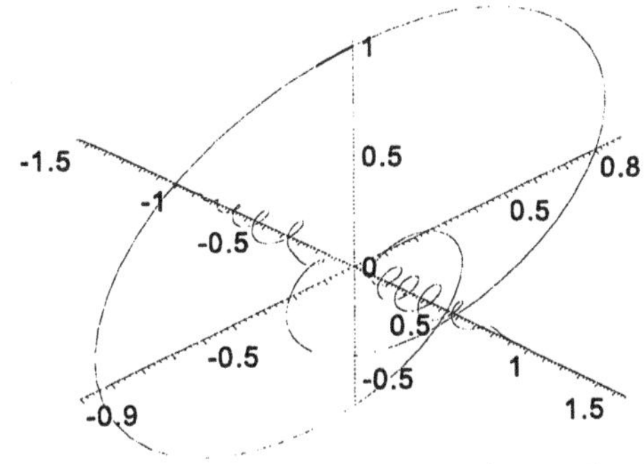

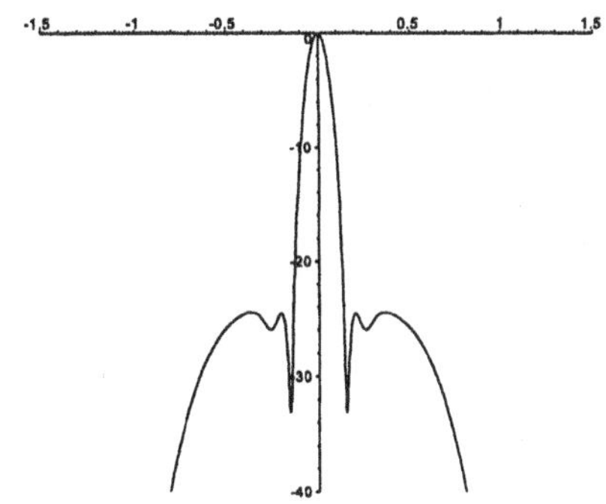

Figure 8.23 The result of correlation with tapering.

of 25 dB, better resolution is obtained. Clutter has a much wider dynamic range, so this tapering with pulse compression does not provide the full resolution over the full dynamic range. Generally, low sidelobe tapering functions, for example, Taylor, are used. Tapering also reduces the signal-to-noise ratio (processing loss). The characteristics with and without cosine squared tapering are shown in Figure 8.24.

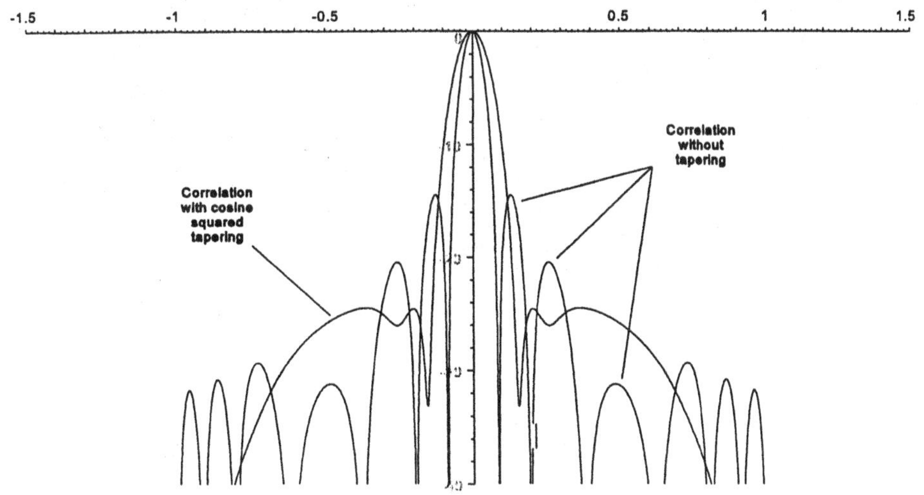

Figure 8.24 Comparison of the absolute values (dB) of the correlation functions with and without tapering.

The negative sequence components of the signal (caused by analogue-to-digital converter errors) are also correlated with the reference waveform. A complex result is shown in Figure 8.25. The negative sequence components add to the range sidelobes.

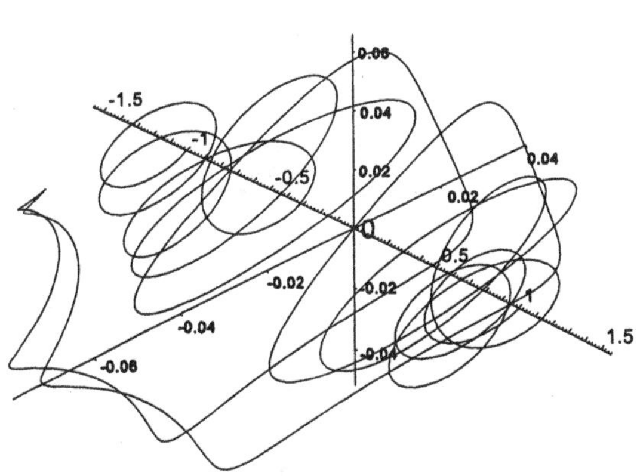

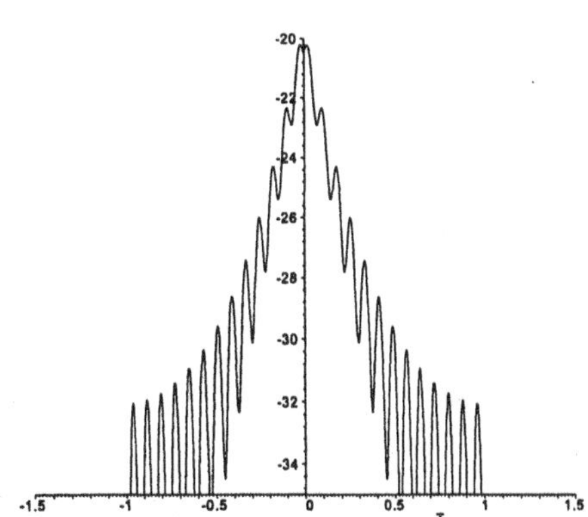

Figure 8.25 The correlation of the negative phase sequence components showing the vector characteristic and its modulus.

Up to now echoes without Doppler frequency shift have been considered. When the echoes are shifted in frequency, they no longer have a matching spectrum for the filter or match the stored copy of the transmitter pulse. The shape of (8.18) extended into the Doppler frequency region is shown in Figure 8.26. The diagram has a roof-shaped central ridge with time and Doppler frequency sidelobes spreading from it. Cross-sections through Figure 8.26 are shown in Figure 8.27 (note that this is the same as the time curve in Figure 8.21).

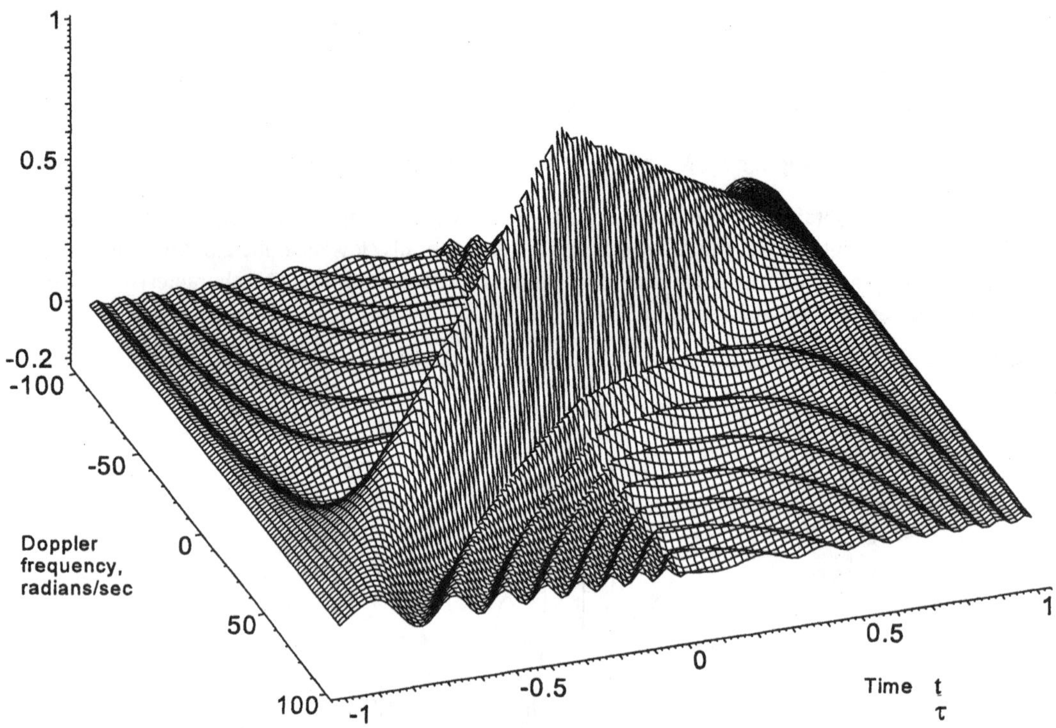

Figure 8.26 Time and Doppler frequency ambiguity diagram for a time-bandwidth product ($B\tau$) of 100. Note that the 25 by 25 grid does not show the detail of the central spike.

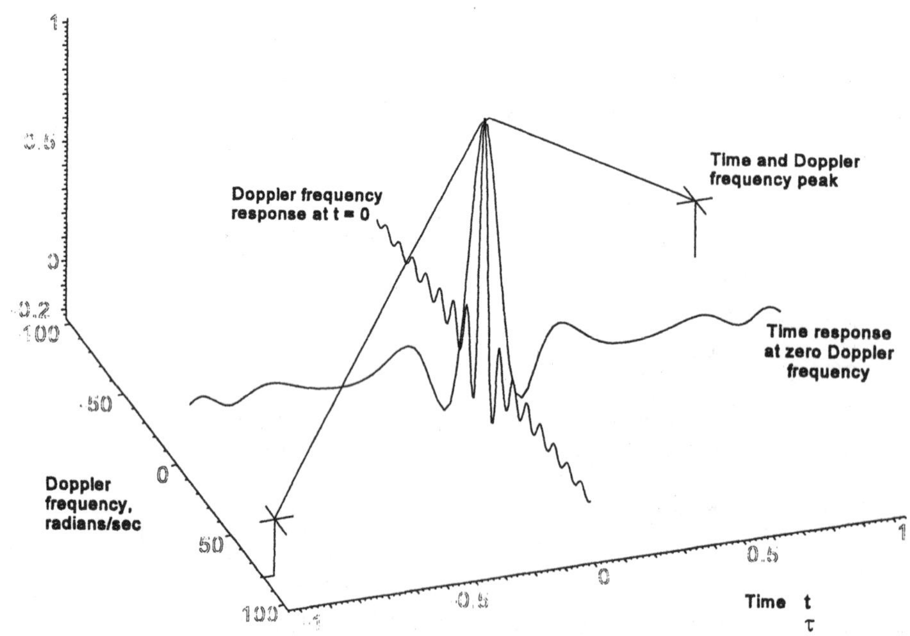

Figure 8.27 The curves in the principal planes in Figure 8.20 with the roof shaped peak of the curve.

Figure 8.27 and Figure 8.28 show that the center peak is slanted so that the Doppler frequency affects the timing of the pulse. This is called range Doppler frequency coupling.

If the pulse is frequency modulated through Δf over a time τ, then an echo with a Doppler frequency of f_d will experience at the filter output an extra time delay given by [10, p. 246]

$$\textit{Extra time delay,}\ \Delta t = -\tau\frac{f_d}{\Delta f} \tag{8.20}$$

Such a system correctly indicates the true range delay at a time shifted by Δt from the actual time of measurement, where $\Delta t = \tau\Delta f/f_0$, gives the measurement offset time for any echo, independent of its Doppler shift and dependent on the carrier frequency f_0.

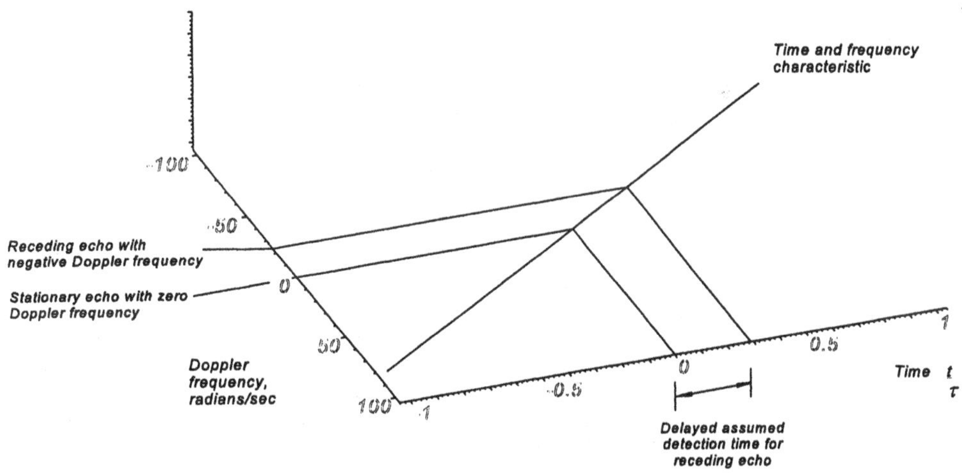

Figure 8.28 The central spike of the time and Doppler frequency ambiguity diagram.

As an example, for a 100 μs transmitter pulse at 30 GHz swept through 1 MHz, an aircraft approaching at 100 m/s (360 km/hour) gives a Doppler frequency of 20 000 Hz. This gives a time error of -2 μs or -300 m in range. The other way of making the correction is to take the average ranges for up and down swept frequency modulated or chirped pulses.

The peak of the ambiguity diagram occurs at zero time and Doppler frequency. Echoes with a Doppler frequency shift pass through the pulse compression filter with lesser amplitude representing a loss. Frequency modulated waveforms have a much greater tolerance to Doppler frequency shifts than the waveforms described in the next section.

8.2.2 Nonlinear frequency modulation

The compressed pulse from linear frequency modulation is of sin x/x form with a first time sidelobe 13.46 dB below the power of the wanted compressed pulse. Reference [11, quoting 12] gives an empirical formula for the frequency modulation for the frequency

$$\textit{Modulation frequency} \equiv \frac{t}{T}\left(\beta_L + \beta_C\frac{1}{\sqrt{1 - 4\frac{t^2}{T^2}}}\right) \tag{8.21}$$

where t is within $\pm\tau/2$;
β_L is the slope of the linear component of the modulation;
β_C is the slope of the Chebyshëv component of the modulation;
T is the length of the expanded pulse.

An example with $\beta_L = 50$ and $\beta_C = 20$ is shown in Figure 8.29.

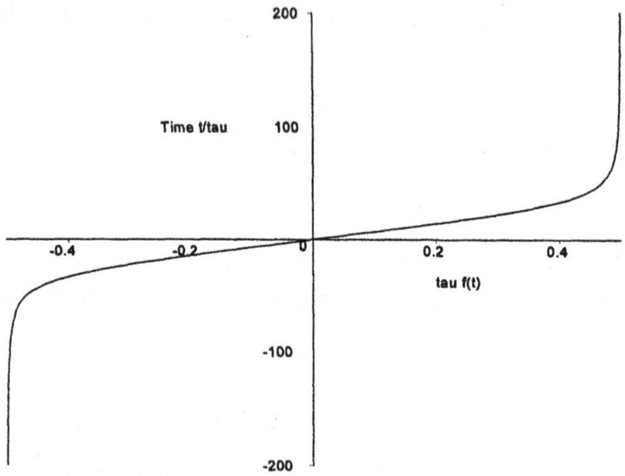

Figure 8.29 The curve of frequency against time for a nonlinear frequency modulated signal.

When (8.21) is integrated and multiplied by 2π, to change from cycles to radians, the phase of the modulation is

$$\theta(t) = \frac{\pi\beta_L}{T} t^2 + \frac{\beta_C T}{2} \left(1 - \pi\sqrt{1 - \frac{4t^2}{T^2}} \right) \tag{8.22}$$

The progression of phase angles from the center of the pulse and the form of the spectrum are shown in Figure 8.30.

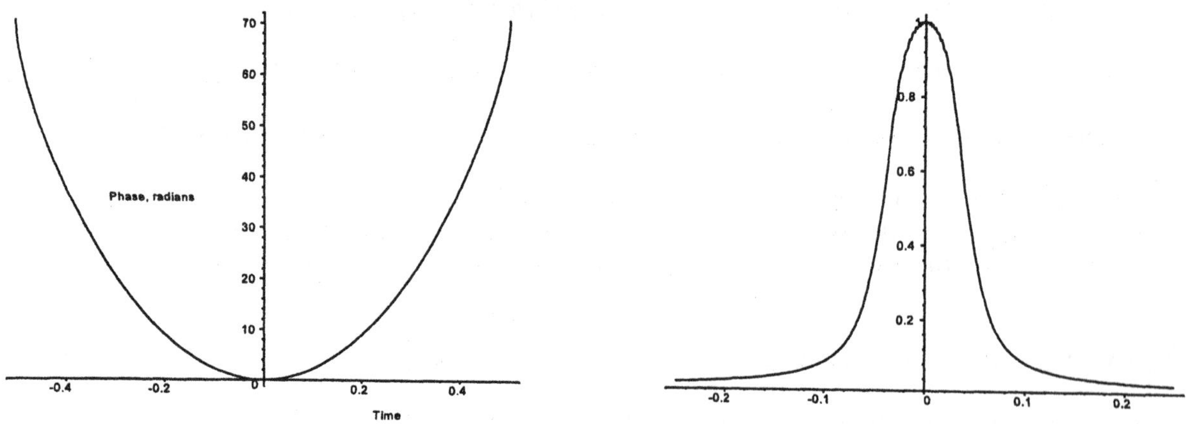

(a) The phase of the modulating waveform (b) The spectrum of the modulating waveform

Figure 8.30 The phase and spectrum of the nonlinear frequency modulation in the example.

The complex modulation function is shown in Figure 8.31(a) with a part of the complex spectrum in Figure 8.31(b).

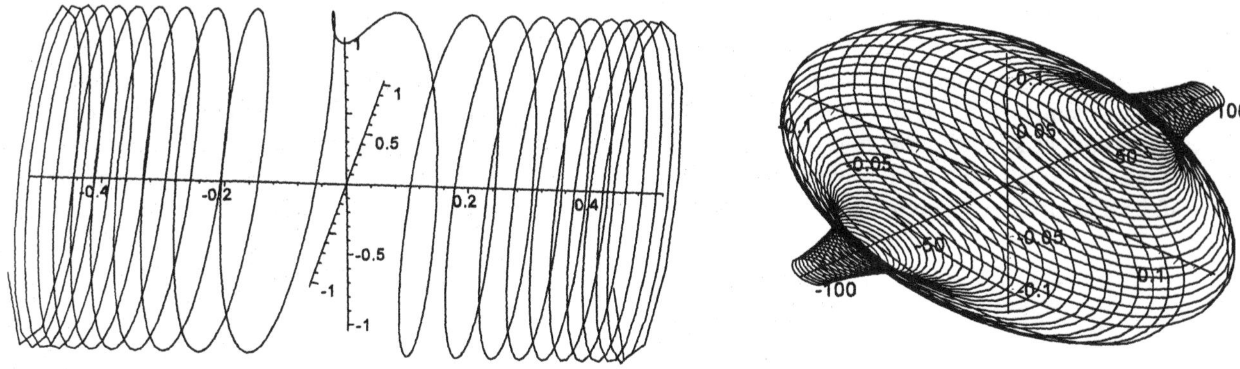

(a) The complex modulation function in time (b) The complex spectrum of (a)

Figure 8.31 The complex modulation function and spectrum for a nonlinear frequency modulated pulse.

The autocorrelation function of the waveform in (8.22) is shown in Figure 8.32.

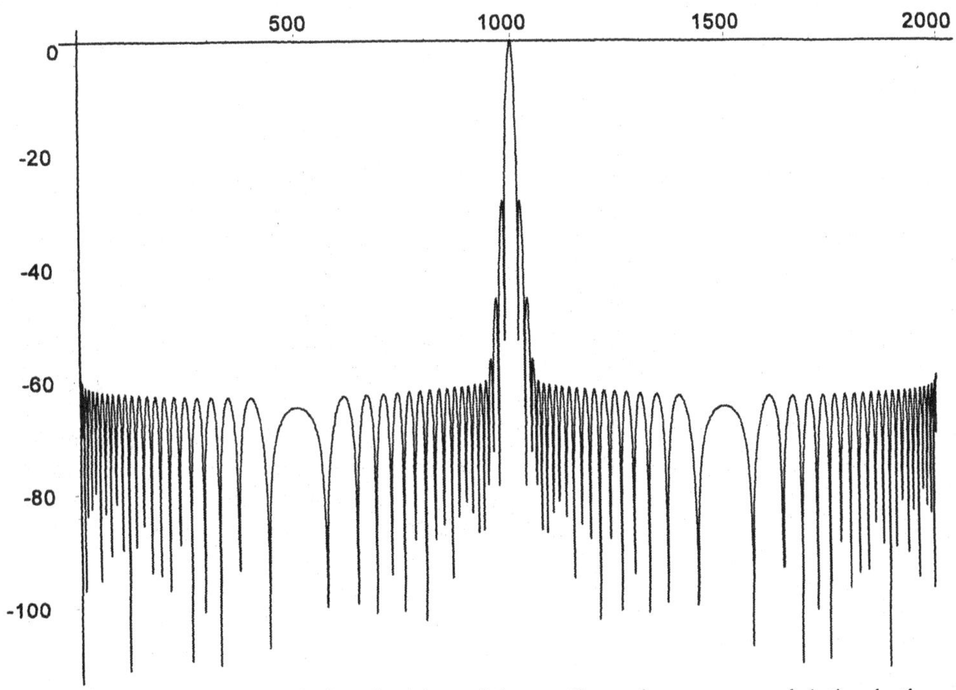

Figure 8.32 The autocorrelation function of the nonlinear frequency modulation in the example.

The price for lower sidelobes is a broader peak, given by the bandwidth of the main linear part, and the larger bandwidth required in the transmitting and receiving stages.

Other techniques for designing nonlinear frequency modulated waveforms may be found in [13].

8.2.3 The effects of limiting before the pulse compression filter

The effects of limiting before the linear frequency modulated pulse compression filter are the same as the Dicke fix receiver in Section 8.1.4 but are difficult to illustrate. The line of limited signals of the same phase is replaced with signals of accelerating phase, impossible to recognize, and to distinguish from noise, in an illustration.

8.2.4 General correlator

A general correlation block diagram is shown in Figure 8.33. The echo signals pass through the receiver, are converted to a digital data stream, and the signals for one listening time (sweep) are converted to their combined spectrum using a Fourier transform block: fast Fourier transform blocks have long been used in pulse Doppler radars. The transmitted pulse signal passes through a similar receiver and its complex conjugate is formed by reversing the sign in the quadrature (Q) part of the data stream. The references in brackets are the same in Figures 8.33 and 8.34.

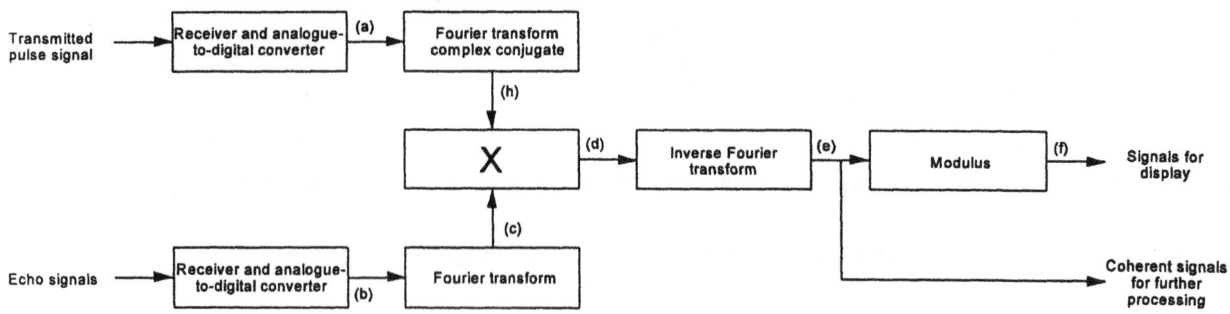

Figure 8.33 General correlator.

An example using a coded pulse one time unit long with five cycles of phase modulation in the form of a conically shaped helix with a negative phase sequence is shown in Figure 8.34(a). The pulse is amplified and echo signals return occurring between 5 and 6 time units and between 9 and 10 time units, illustrated in Figure 8.34(b), and the Fourier transform for the sweep is calculated, see Figure 8.34(c). The spectrum is around -1 000 reflecting the range and modulation of the echo signals and shows the effect of the beating of the two echo signals. In parallel the Fourier transform of the transmitter signal is found (Figure 8.34(g)) and the complex conjugate is found (Figure 8.34(h)). Correlation is carried out by multiplying the spectrum of the echo signals with the complex conjugate of the transmitter spectrum to give the spectrum in Figure 8.34(d) that only contains range and phase information with the transmitter pulse coding removed (or when applicable compressed). Inverting the spectrum gives the correlation function in Figure 8.34(e) with the echo signals occurring in the same positions as in Figure 8.34(b) that provides a coherent data stream for further signal processing and the absolute value of this data stream provides signals for display.

This type of correlation could be used to provide a coherent signal data stream for a radar having an unruly transmitter or a the receiving site of a bistatic radar.

8.3 DISCRETE PHASE SHIFT MODULATED PULSE COMPRESSION

Modulation in waveforms that occur at discrete times may be processed digitally as long as the analogue to digital converter is fast enough to have at least one conversion for each element: better two to avoid straddling losses (see Section 10.4.1.1). Each sample is independent of its neighbors so that there is no ringing from range cell to range cell as in intermediate frequency amplifiers when overloaded. Long delays are available using memory with no losses and the hardware is made up from usually available, standard components. There are waveforms available, discussed later, that allow low time sidelobes with no tapering giving losses. For radars with different switchable pulsewidths, changing the clock frequency and, if necessary, changing the number of delays enables the same hardware to be used for different compression ratios. The digital data stream at the output is of the same form as that needed for later signal processing.

With discrete phase modulation the transmitter pulse is divided into a number of sections, or chips, with a phase modulator to change the phase of each chip. The phase increment may be 180 degrees for binary phase modulation or a lesser fraction of a circle. Discrete phase shifts may be achieved simply by using a limited number of phase shifters and switches.

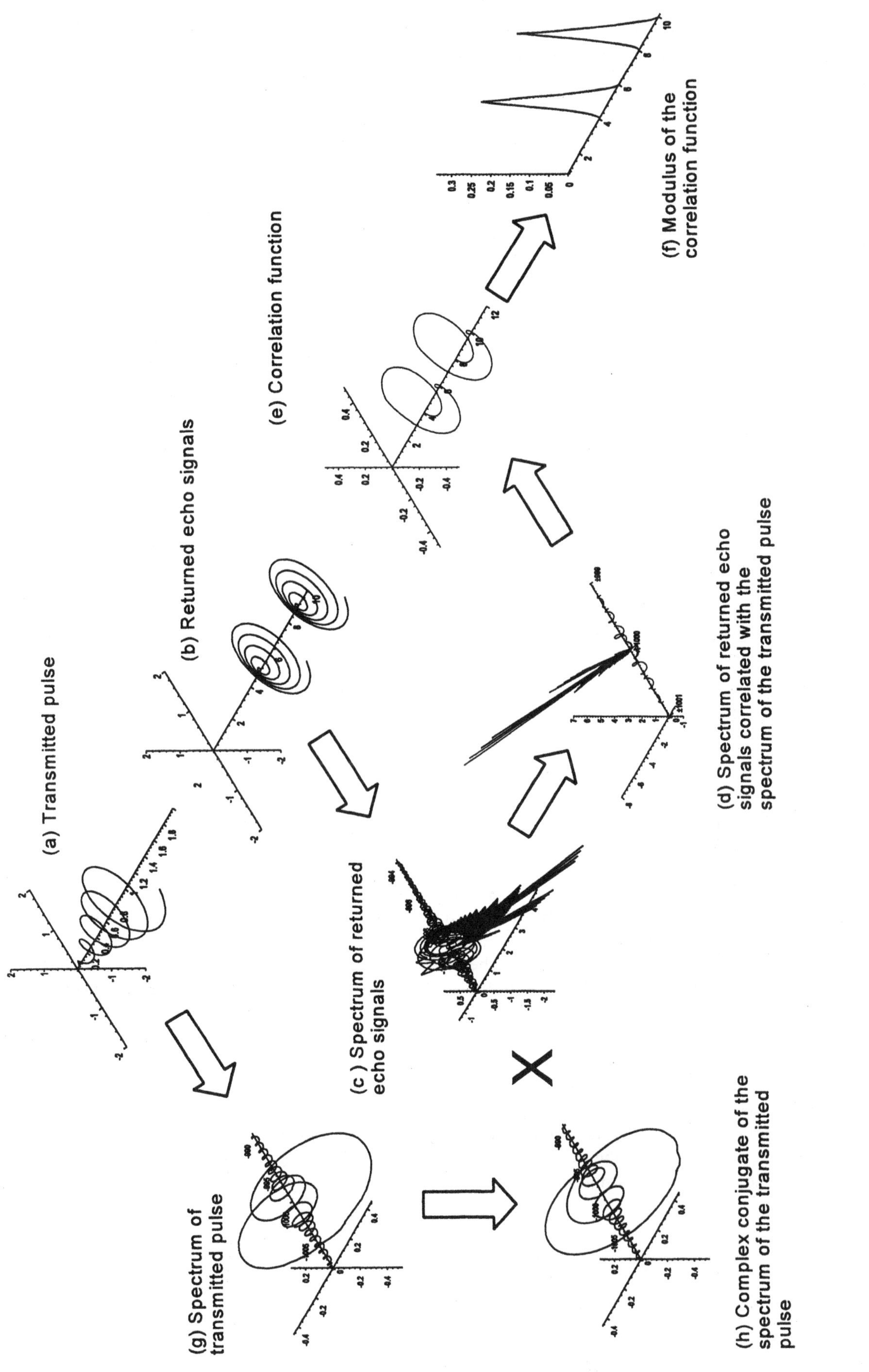

Figure 8.34 The processing of signals with a correlator using Fourier transforms.

On the receiving side, the returning echoes pass through a tapped delay device, shown in Figure 8.35. The outputs from the taps are samples of the echo signal at intervals of the subpulse spacing. The samples pass through phase shifters of opposite sign to those in the transmitting section. A vector summer sums the phase shifted samples to give the output.

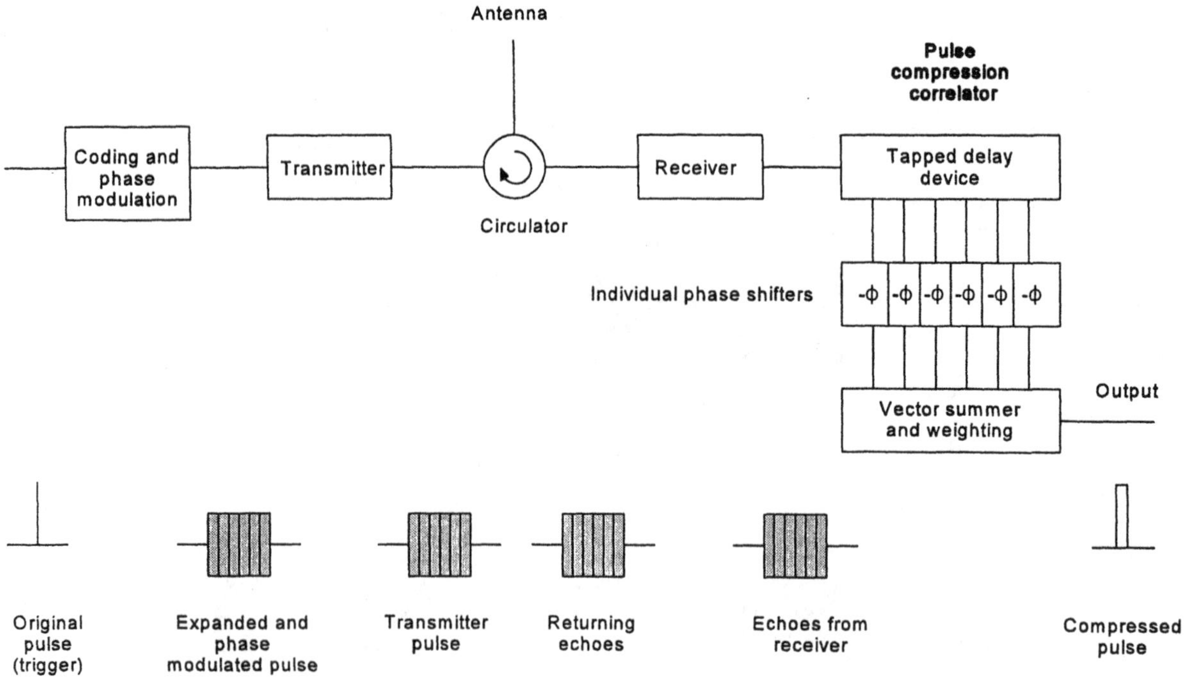

Figure 8.35 Block diagram showing phase modulated pulse compression.

As digital components can be used to generate the modulation for the transmitter and correlate the returned echoes, so there has been much research into suitable modulation codes. The criteria for the codes are:

- There is a central, narrow ambiguity spike of amplitude nearly equal to the compression ratio, ρ.
- The response around the spike, the time sidelobes, is low to allow the detection of wanted echoes as close as possible to clutter.
- The ambiguity function should be relatively insensitive to Doppler frequency shift.

There is an inherent contradiction between the first and third criteria for phase coded waveforms [D. Barton, personal correspondence, 2000]. If the central spike of the ambiguity function is narrow in both coordinates, the sensitivity to Doppler frequency shift is high. The basic difference between linear and nonlinear frequency modulation and phase coding alone is that the former can be made very insensitive to Doppler frequency shift (for shifts up to 10% of the signal bandwidth, Δf), while the latter is sensitive to shifts approaching 10% of the reciprocal of the pulse width, $1/\tau$. Thus, linear frequency modulation tolerates Doppler frequency shifts greater than phase coding by a factor equal to the pulse compression ratio $\Delta f \tau$.

There has been considerable research into longer maximum length sequence codes generated by shift registers with feedback. For each length, there are often a number of different codes, which give a noise like modulation on a long, low amplitude transmitter pulse. The pulses are difficult to detect by radar warning receivers that are not programmed to be sensitive to that code, which is one of the bases for low probability of intercept (LPI) radars.

The types of codes are shown in Table 8.2.

Table 8.2
Examples of pulse compression waveforms or codes

Code	Comment
Binary forms	
Barker binary phase	See Sections 3.4.3.1 and 8.3.2.1
Pseudo-random binary phase, maximum length sequence codes	See Sections 3.4.3.1 and 8.3.1.3
Random binary phase	See Sections 3.4.3.1 and 8.3.1.3
Polyphase forms	
Step frequency derived polyphase, Frank code	See Sections 3.4.3.2 and 8.3.2.1
P1 code	See Sections 3.4.3.3 and 8.3.2.1
P2 code	See Sections 3.4.3.4 and 8.3.2.3
P3	See Sections 3.4.3.5 and 8.3.2.4
P4	See Sections 3.4.3.6 and 8.3.2.5
Other discrete codes	
Costas codes, separate filters matched to chip frequency and delays, Huffman	See Sections 3.4.3.7 and 3.4.3.8
Frequency modulation	
Linear frequency modulation	See Sections 3.4.1 and 8.2.1
Nonlinear frequency modulation	See Section 3.4.1 and 8.2.2

The representation of the effects of Doppler frequency are given in terms of the phase shift over the pulse length. For one cycle phase shift the velocity is given by

$$\begin{aligned} f_d\tau &= 1 \quad \text{cycle} \\ f_d &= \frac{2v}{\lambda} \quad \text{Hz} \\ v &= \frac{\lambda}{2\tau} \quad \text{m/s} \end{aligned} \tag{8.23}$$

where f_d is the Doppler frequency, Hz;
τ is the pulse width, s;
λ is the wavelength, m;
v is the velocity of the scatterer, m/s.

8.3.1 Binary codes

Binary phase modulation with compression ratios of up to 13 are shown in Figures 8.36 and the phase modulation codes are shown in Table 8.3 [14, p. 541].

Table 8.3
The first 13 codes for binary phase modulation

Length	Code	Barker code	Length	Code	Barker code
2	10	B	8	10010111	
3	110	B	9	111010110	
4	1101 or 1110	B	10	1110011010	
5	11101	B	11	11100010010	B
6	110100		12	111101011010	
7	1110010	B	13	1111100110101	B

8.3.2.1 Barker codes

The correlation curves in Figures 8.36 show voltage sums which may be positive or negative. Codes called Barker codes (marked with a B in Table 8.3) have lengths of 2, 3, 4, 5, 7, 11, and 13, and have maximum sidelobes of amplitude 1. The others have maximum sidelobes of 2.

The sidelobe levels are defined as

$$\begin{aligned} \textit{Peak sidelobe level, PSL} &= 10\log_{10}\max\left(\frac{x_i^2}{x_0^2}\right) \text{ dB} \\ \textit{Integrated sidelobe level, ISL} &= 10\log_{10}\sum_{i=0}^{i=n}\frac{x_i^2}{x_0^2} \text{ dB} \end{aligned} \tag{8.24}$$

Where x_0 is the (voltage) level of the main lobe;
x_i is the (voltage) level of the ith sidelobe.

The sidelobe levels may be calculated as in Table 8.4.

Table 8.4
Sidelobe levels for binary codes up to length 13

Code length	Sidelobe levels		Code length	Sidelobe levels	
	Peak, dB	Integrated, dB		Peak, dB	Integrated, dB
2	6.0205999	3.0103	8	12.0411998	6.0205999
3	9.5424251	6.5321251	9	13.0642503	5.2827378
4	12.0411998	6.0205999	10	13.9794001	5.8502665
5	13.9794001	7.9588002	11	20.8278537	10.8278537
6	9.5424251	4.1017447	12	15.563025	8.573325
7	16.9019608	9.1204483	13	22.2788671	11.4870546

In an ideal system all the echo energy is compressed into the pulse leaving the matched filter, that is, N times (voltage). If the sum representing x_0 is less than N, there is a loss given by

$$\textit{Pulse compression processing loss} = 10\log_{10}\left(\frac{x_0^2}{N^2}\right) \text{ dB} \tag{8.25}$$

There is, therefore, no net "pulse compression" gain. There is increased resolution and, mostly, an increased matching loss caused by tapering which is the system cost for pulse compression.

The codes in Tables 8.3 and 8.4 have phase shifts of 180 degrees and are conveniently correlated using digital circuits. These codes are very sensitive to Doppler frequency shift in the echoes.

Compressed binary (and Barker) coded echo signals transmitter pulses are sensitive to uncompensated Doppler frequency shift. The phase shift over the length of the expanded pulse caused by a moving scatterer is given by $2\pi f_d\tau_t$ and the effects over 360 degrees of phase shift are shown in Figure 8.37. The response at the time of interest falls to zero with two displaced returns so that often a criterion of 90 degrees phase shift over the transmitted pulse length is used to keep the compressed pulse narrow and the time sidelobes low. As an example, if an S-band radar with an expanded pulse width of 13 μs is used, a 360 degree phase shift occurs at a Doppler frequency of 1/(13 μs) or 76.923 kHz. This Doppler frequency corresponds to a radial velocity of $f_d\,\lambda/2$ or 3 846 m/s so that if the 90 degree criterion is used to keep the narrow peak and low time sidelobes, the maximum allowable radial velocity is 962 m/s or approximately Mach 3. With longer pulses there is a greater degradation and with, for example, a counter battery radar used to track artillery shells, the local oscillator must be offset to bring down the expected Doppler frequency of incoming fire to near zero or in the

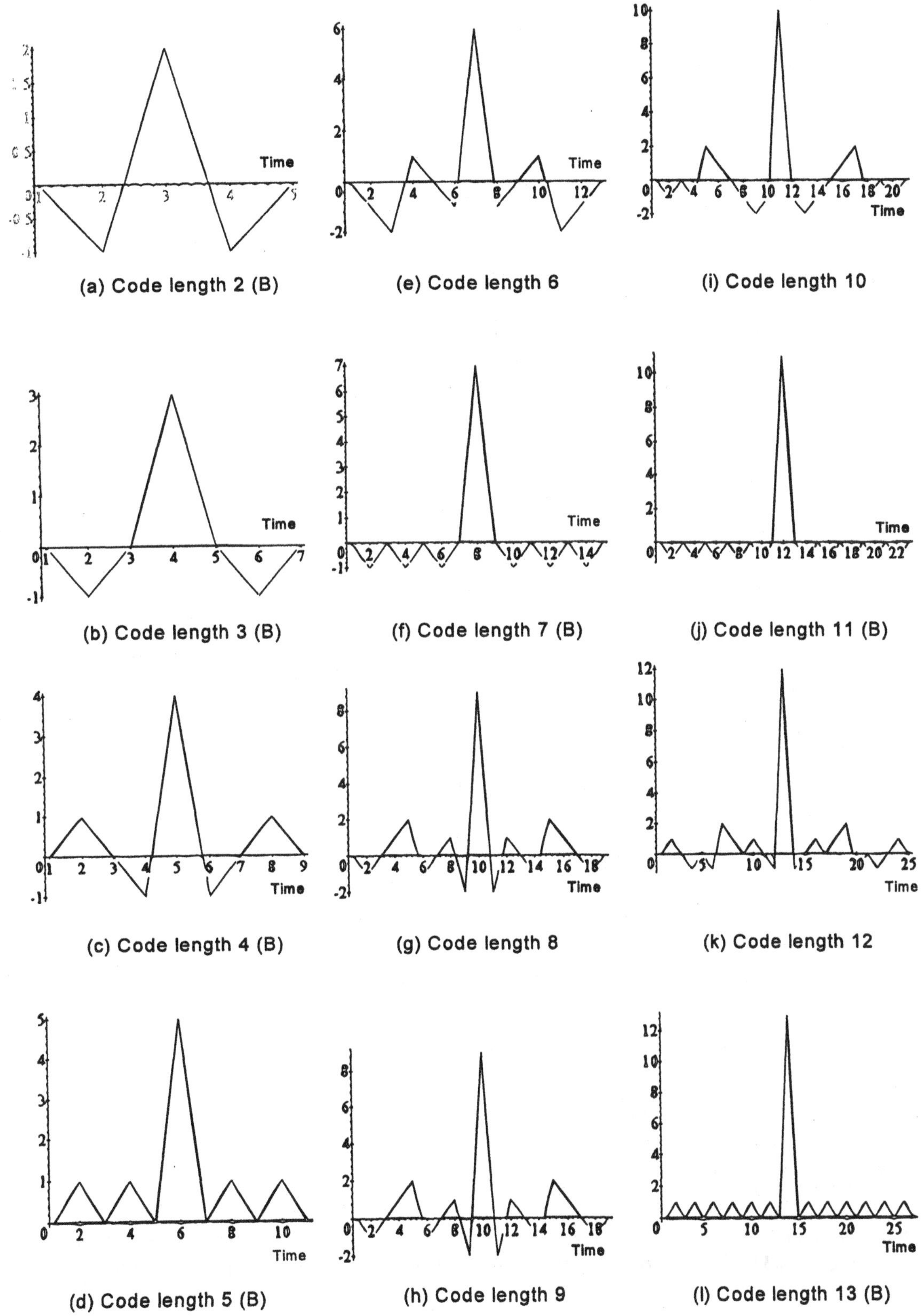

Figure 8.36 Binary codes of length 2 to 13. Barker codes are marked with (B).

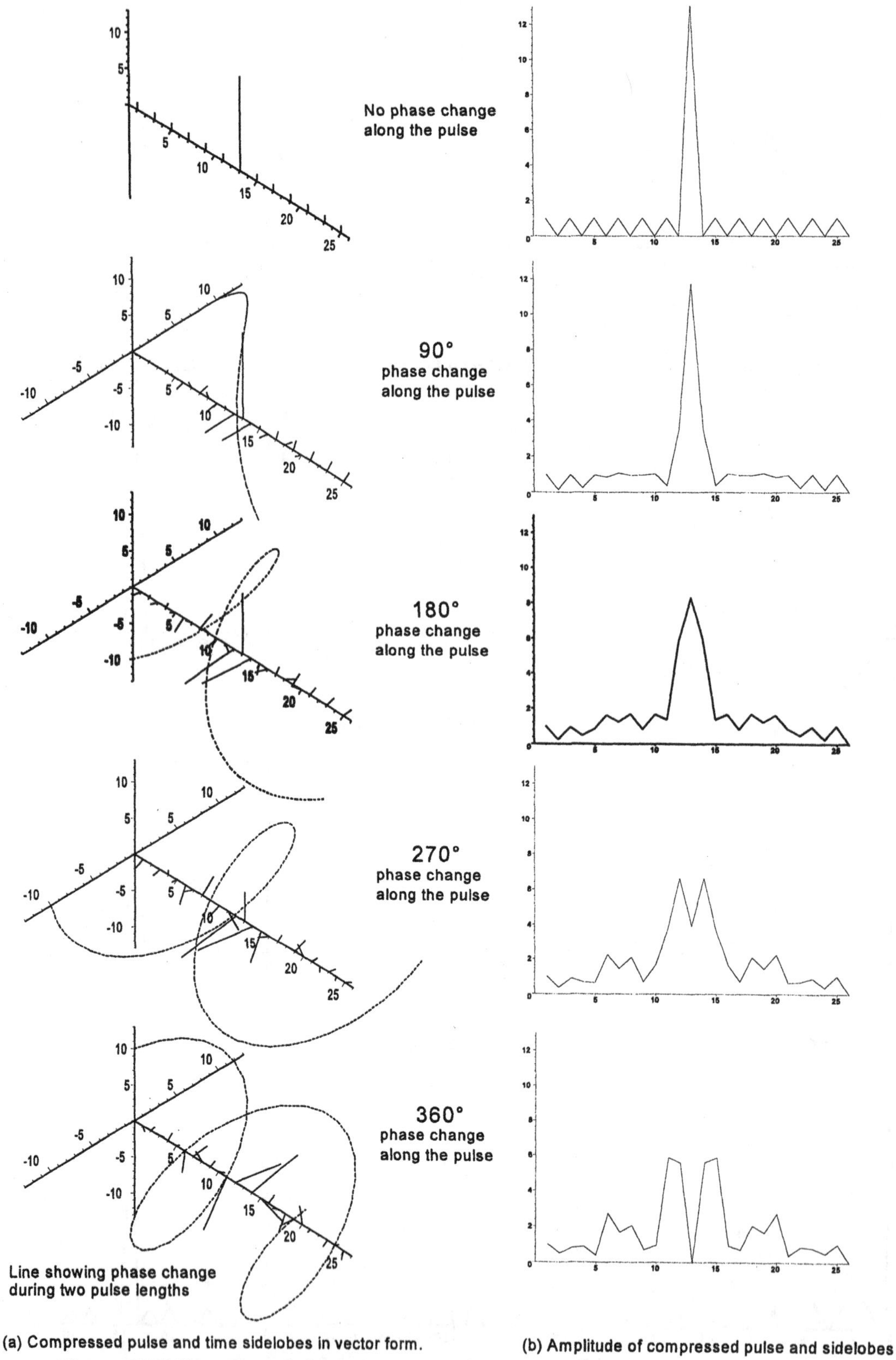

Figure 8.37 The effect of phase change of the echo signal along a binary coded pulse.

other direction for outgoing fire.

The sensitivity to Doppler frequency is shown more generally in Figure 8.38.

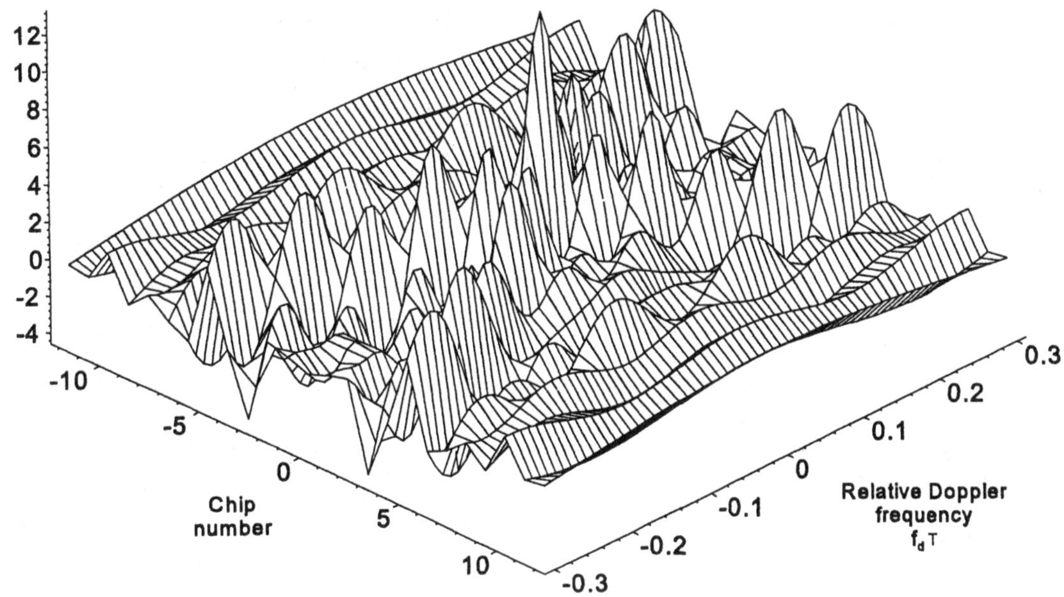

Figure 8.38 The response of a 13 bit Barker coded pulse to moving echoes with Doppler frequency shift.

There is a danger that one small echo signal can be lost in the sidelobes of a near large (clutter) signal as in the example in Figure 8.39.

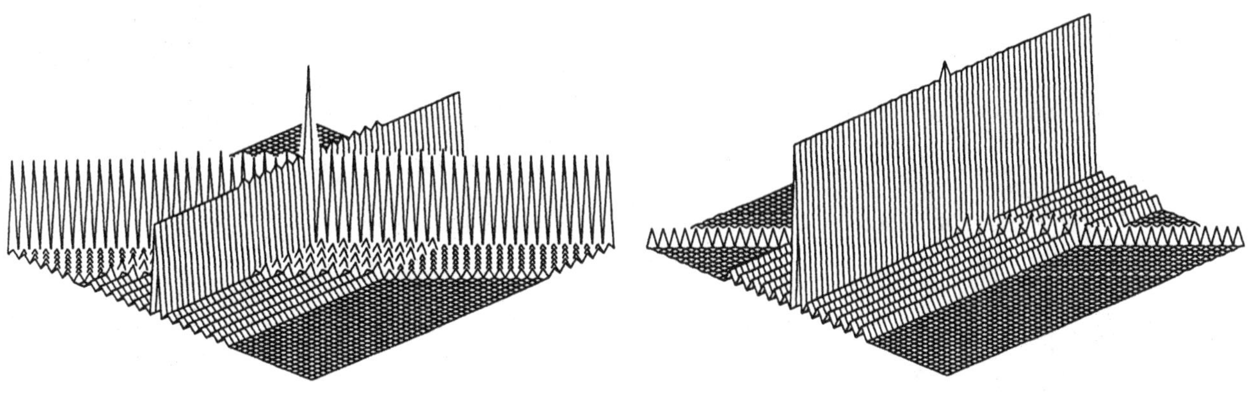

Figure 8.39. The interaction of two Barker coded signals after pulse compression.

8.3.1.2 The effects of limiting with binary pulse compression

There are a number of successful radars turning today that use limiting or Dicke-fix processing with pulse compression. One of the first used a Barker code to stretch the transmitter pulse and the echoes were hard limited before being recompressed, called by one manufacturer the coded pulse anticlutter system or CPACS shown in Figure 8.40 [2, p. 3-49].

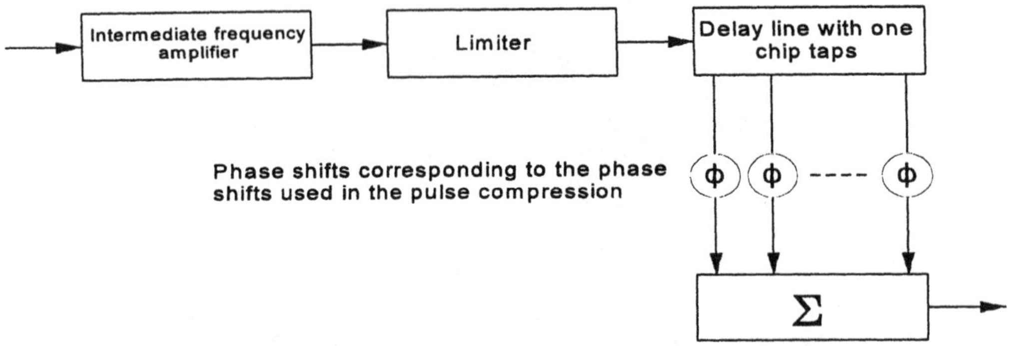

Figure 8.40 Diagram of a pulse compression system with limiting.

Taking a Barker code pulse compression system as an example, the phase shifts and the compressed signal without noise are shown in Figure 8.41.

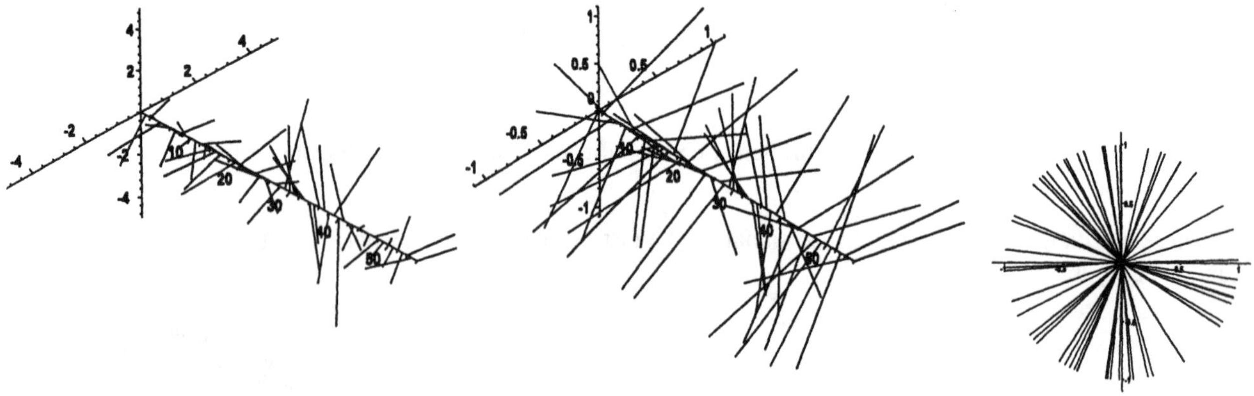

(a) Stretched Barker code with Gaussian noise

(b) Limited Barker signal with noise (notice change of scale)

(c) End on view of (b)

Figure 8.41 A Barker coded stretched signal with noise and after limiting.

After pulse compression the limited signals become those in Figure 8.42.

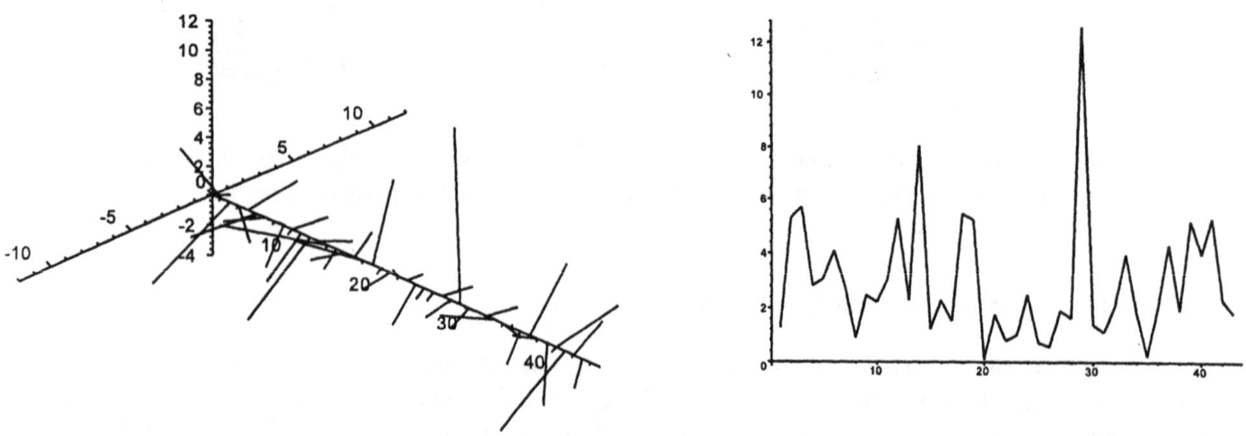

(a) Limited signal after pulse compression

(b) Conventional representation of (a)

Figure 8.42 A Barker coded signal after pulse compression.

Hard limited signals have the same property as frequency modulated radio broadcast signals that are also hard limited to suppress light pulsed interference. A weaker signal is repressed by a stronger signal and this applies to echoes that overlap in time before passing through the pulse compression filter as shown in Figure 8.42(b).

Barker codes may be nested but in each nest there is a higher sidelobe caused by the peak for the Barker code it contains.

8.3.1.3 Longer binary codes

Longer binary codes may be generated from sequences of ones and zeroes read from memory, generated at random, or as maximum length codes discussed in Section 3.4.3.1, and much effort has been expended to find codes of length greater than 13 that add to the full compression ratio, have low sidelobes, and good Doppler frequency tolerance.

The example of a random code of length 100, in Figure 8.43 does not fulfill these wishes.

For echoes from stationary objects the peak adds up to 55 and not 100, the sidelobes have a triangular envelope typical of the autocorrelation function of a rectangular pulse with the highest sidelobe being 9 dB below the main lobe.

Phase shifts over the pulse reduce the peak to sidelobe level until the signal cancels with 360 degrees, or 2π radians, over the length of the pulse.

8.3.2 Polyphase codes

There are a number of methods of modulating pulses giving simple operation and designed to have low sidelobes in their autocorrelation functions without tapering. The U.S. Naval Research Laboratory P-codes (NRL P-codes) [15] are described in this section and the generation of such codes is discussed in Section 3.4.3.2 to 3.4.3.6.

8.3.2.1 Frank code

The generation of Frank codes is illustrated in Section 3.4.3.2 and as mentioned in that section the Frank code can be expanded and recompressed using digital circuit blocks as used for the fast Fourier transform. An example for a 100 element ($N = 10$) is shown in Figure 8.44.

The phases of a 100 element Frank coded echo signal are many (see Figure 3.23), and it is not possible to follow the phase twist in the echo signals with Doppler frequency. In Figure 8.43 for zero phase shift the quadrature scale is exaggerated (-3..3) to show that the sidelobes are in quadrature.

The theoretical peak sidelobe ratio for Frank codes is [15, p. 18, (2-24)]

$$\frac{Peak^2}{Peak\ sidelobes^2} = \frac{(N^2)^2}{\left(\frac{N}{\pi}\right)^2} = \pi^2 N^2 = \pi^2 \rho \tag{8.26}$$

For $N = 10$, 100 elements, the central peak has the value 100 and the sidelobes are $100\pi^2 = 987$ or 29.9 dB and the mean square sidelobe level is $10\pi^2 N^2$, for $N = 10$ this is 39.9 dB.

In the 90 degree column the compressed peak is reduced by 0.9 dB and the skirt has moved to the right with similar sidelobe levels. At 180 degrees the peak shows signs of splitting and is 3.92 dB below the peak with no Doppler frequency, the other sidelobes remain between 30 dB and 40 dB on the zero degree scale (or 26 dB to 36 dB below the peak). At 360 degrees, grating lobes appear at each N elements giving sidelobes of 20 dB spaced 10 elements away from the main peak (see Section 5.4.4) that has moved one element to the right. The time shift is reminiscent of linear frequency modulation.

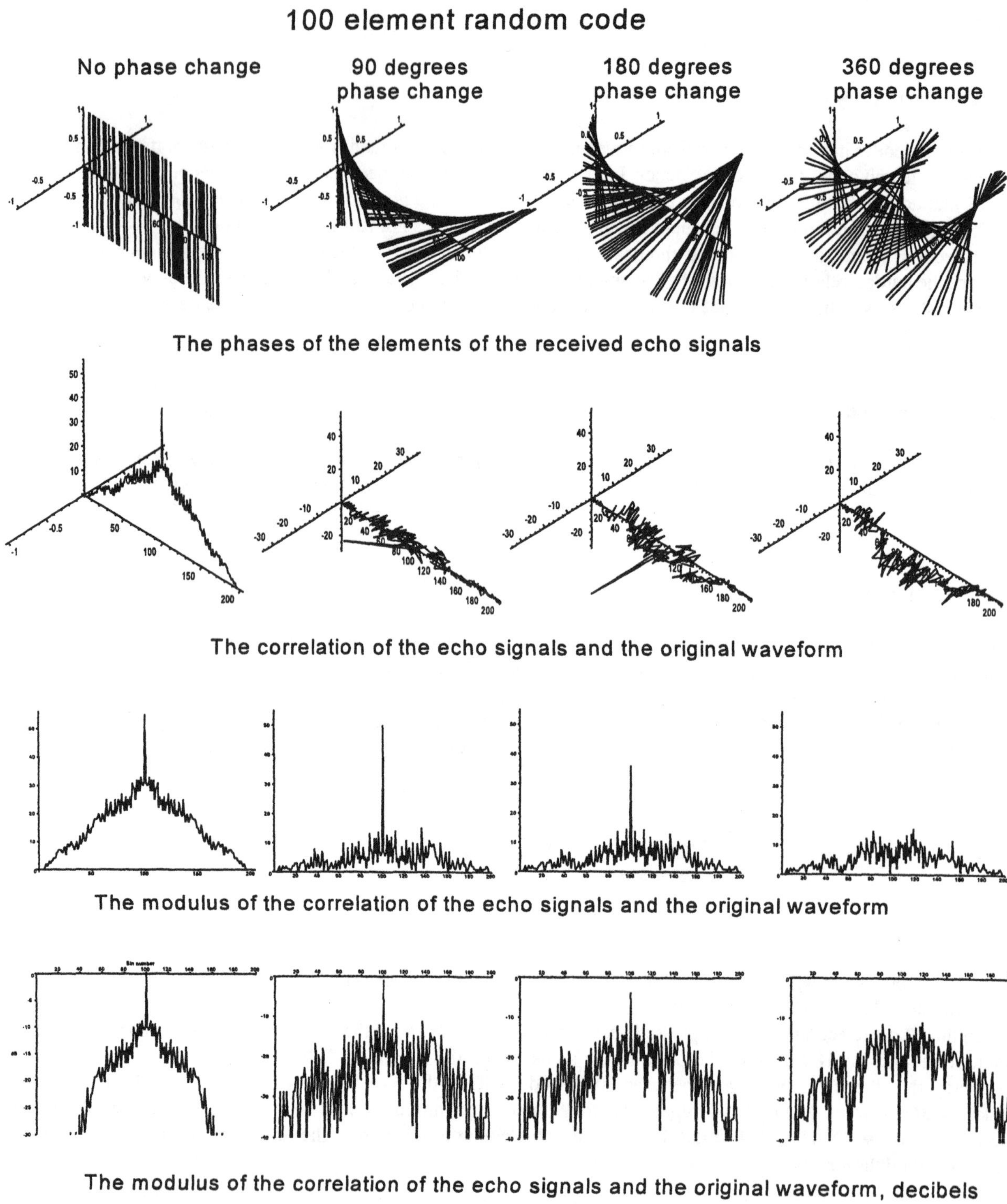

Figure 8.43 The characteristics of a 100 element random binary code.

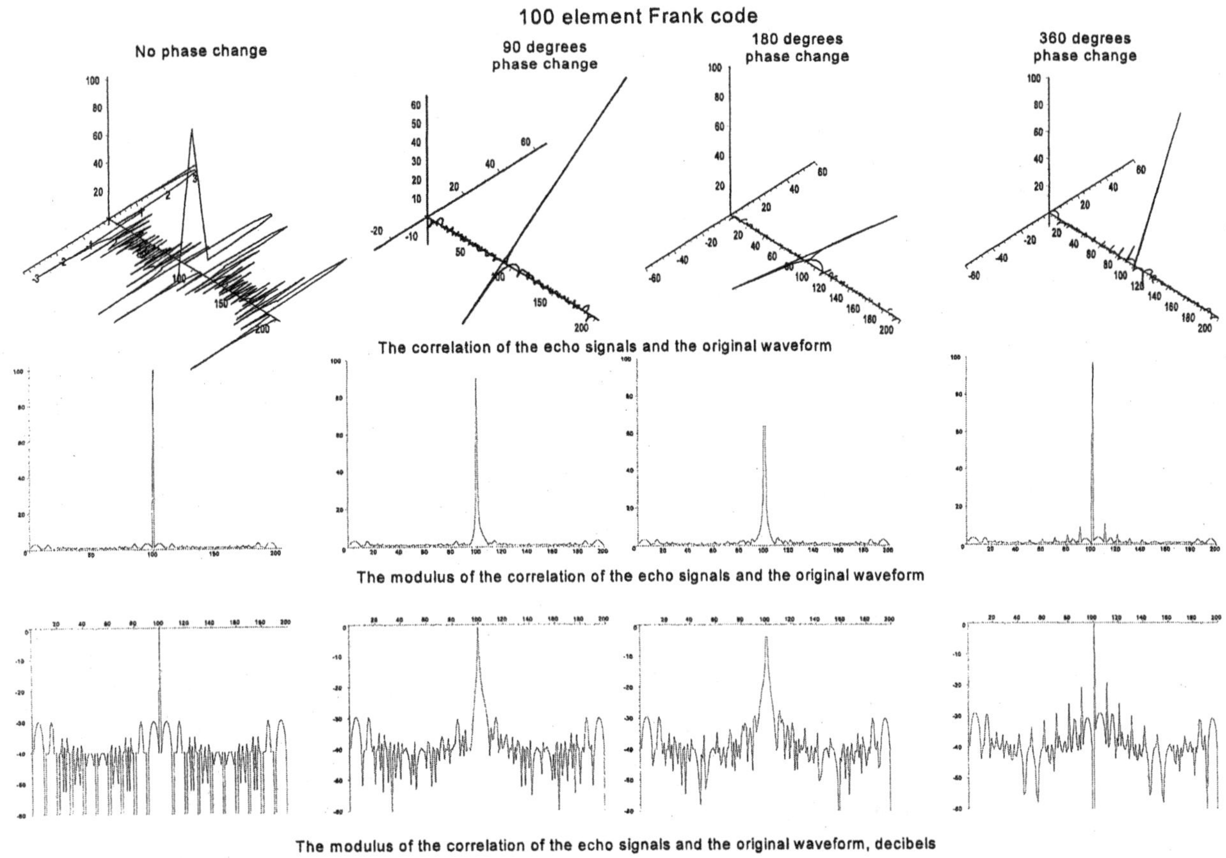

Figure 8.44 A 100 element Frank code and the sensitivity to Doppler frequency.

8.3.2.2 P1 code

The form of the P1 is shown in Section 3.4.3.3. When Frank coded echo signals arrive back at the matched filter, they are often distorted by the transmitter and receiver stages not being able to carry the full bandwidth. The P1 code has been developed to provide good time sidelobe performance in the presence of bandwidth clipping, the time sidelobes are hardly affected (see Figure 8.45).

The shapes of the characteristics are similar to the Frank code in the previous section, except that the peak has a loss of 2 dB at 360 degree phase change along the pulse. Grating lobes appear spaced at N elements from a phase change of 90 degrees along the pulse giving raised sidelobes at a level of -21 dB and rising to -16 dB for 360 degrees phase change.

8.3.2.3 P2 code

The P2 code imitates a Butler matrix for extracting a number of beams from a linear array and its generation is shown in Section 3.4.3.4. The characteristic of a P2 code with N = 10 or 100 elements is shown in Figure 8.46.

The sidelobe levels are similar to the P1 code with the grating lobes in this instance appearing later. The peak has less loss and moves one element in position for 360 degrees phase variation over the pulse.

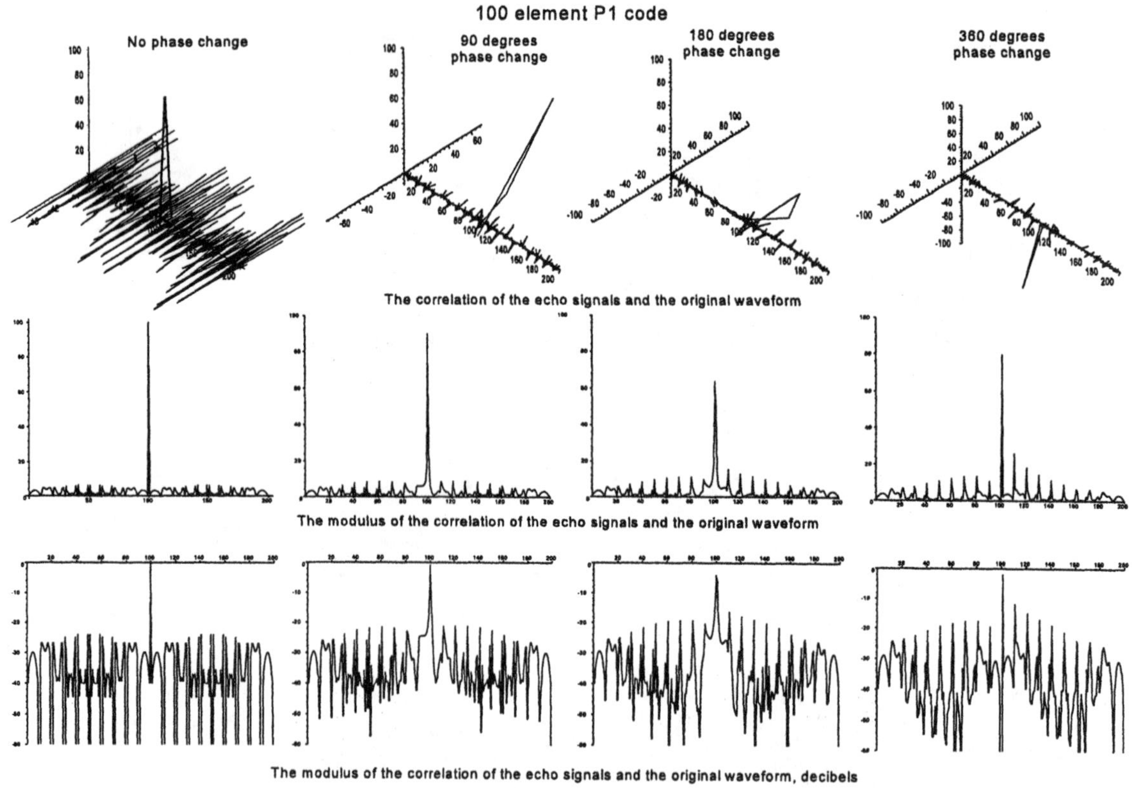

Figure 8.45 The characteristics of a 100 element ($N = 10$) P1 code.

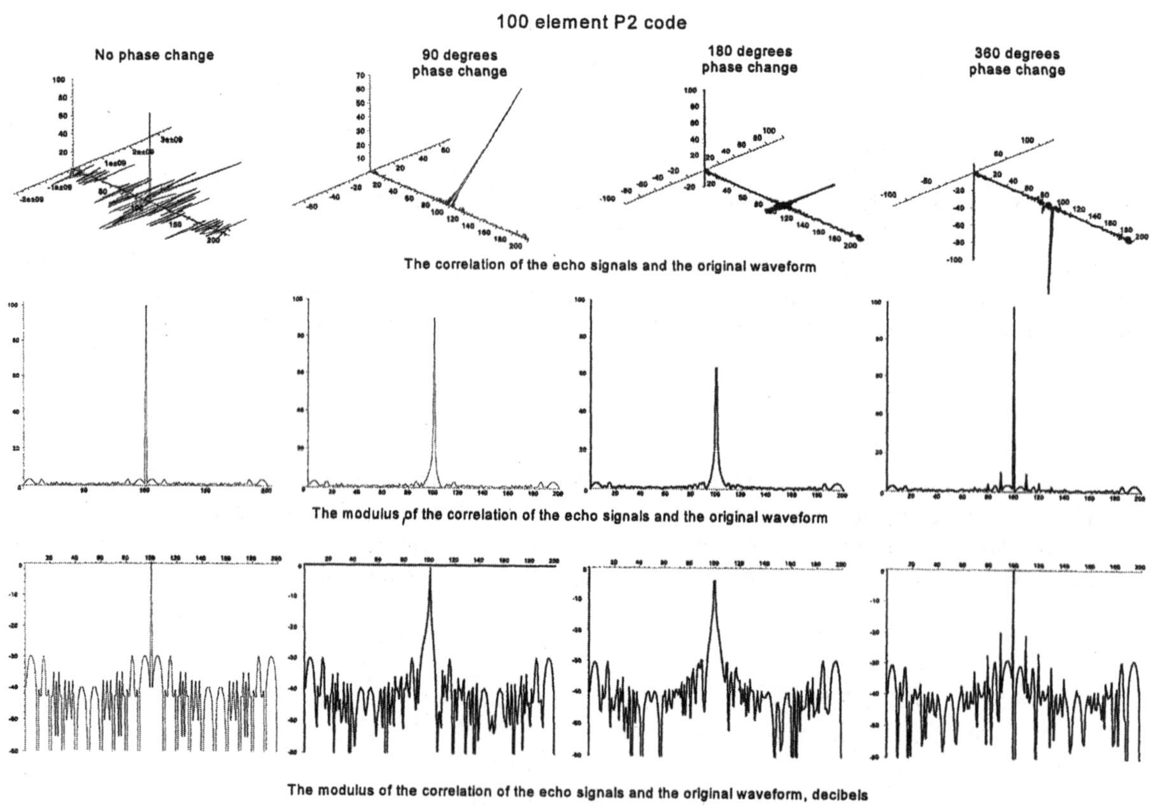

Figure 8.46 The characteristics of a 100 element ($N = 10$) P2 code.

8.3.2.4 P3 code

The P3 code was introduced in Section 3.4.3.5 as an approximation to linear frequency modulation for any code length, and the characteristics of the compressed pulses for a 100 element P3 code are shown in Figure 8.47.

The near and root mean square sidelobes are 3 dB to 3.5 dB higher than those for Frank codes without any grating lobes forming. The peak has a 4 dB droop with a phase change of 180 degrees over the pulse and moves one element with 360 phase difference.

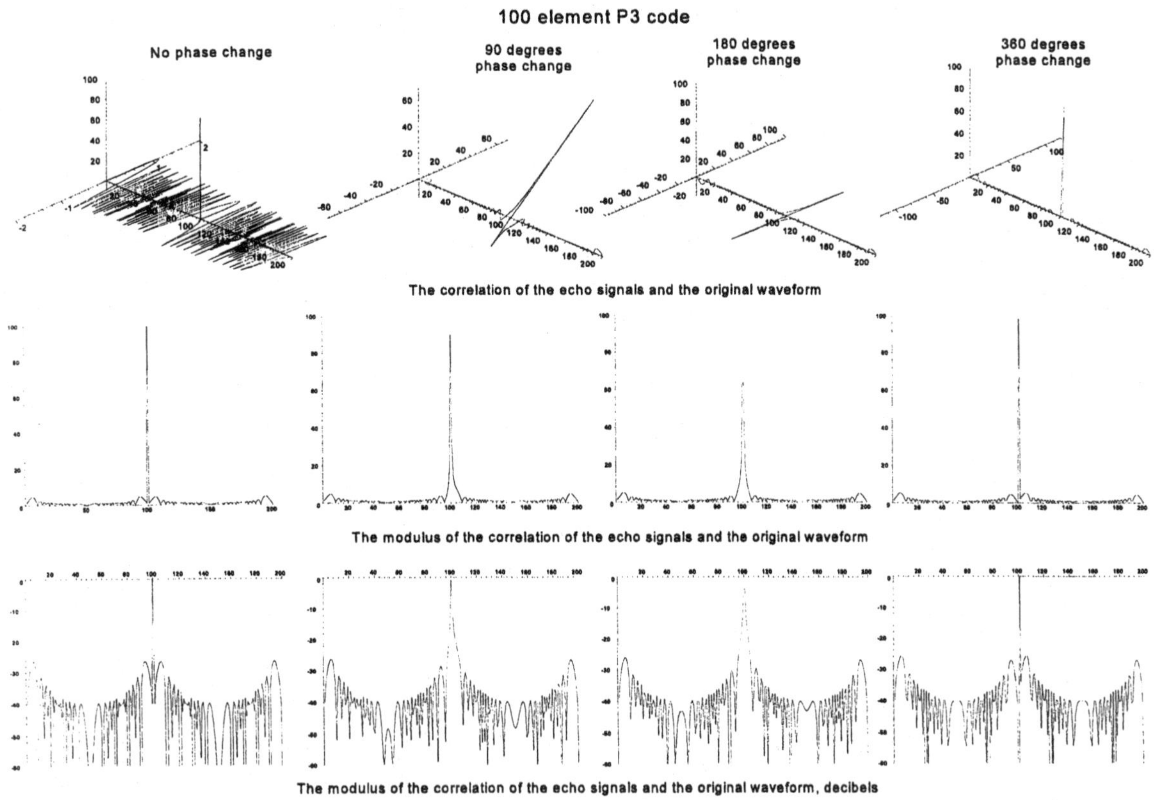

Figure 8.47 The characteristics of a 100 element P3 code.

8.3.2.5 P4 code

The P4 code gives better performance than the P3 code when bandwidth clipping occurs and was introduced in Section 3.4.3.6 as an approximation to linear frequency modulation for any code length (see Figure 8.48). Again, the near and root mean square sidelobes are between 3 dB and 3.5 dB worse than Frank codes.

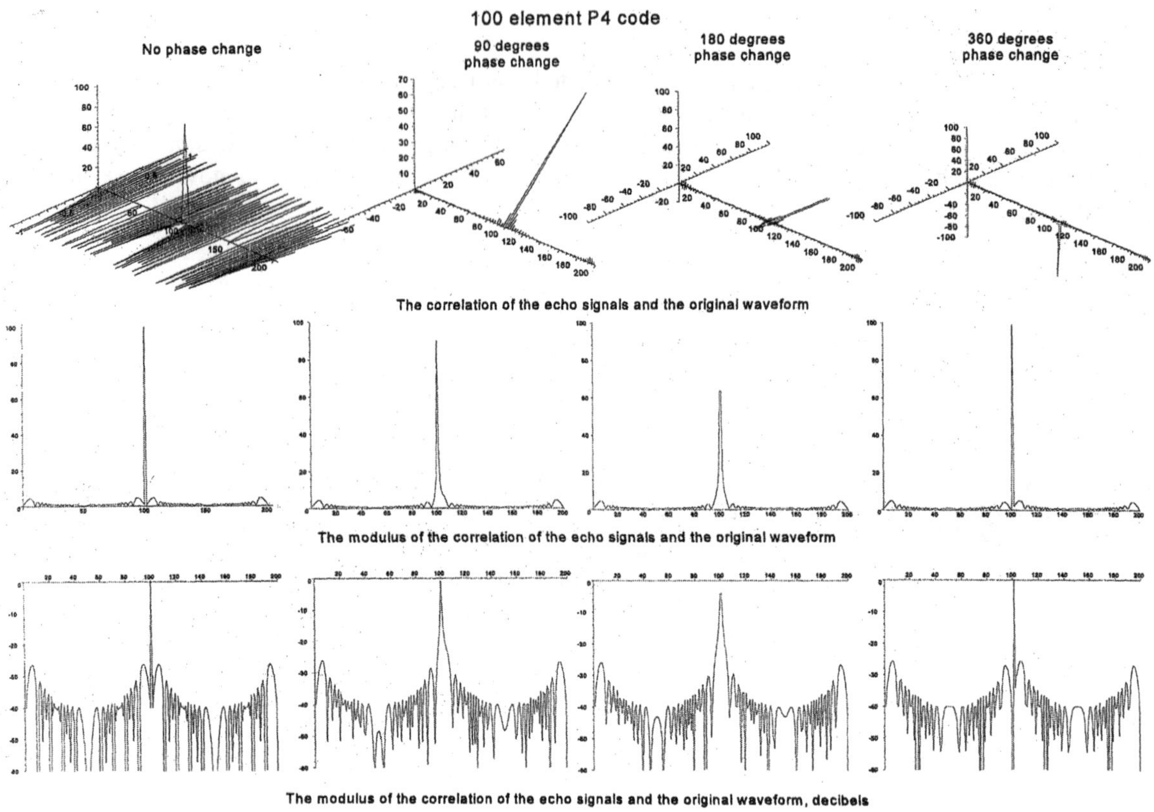

Figure 8.48 The characteristics of a 100 element P4 code.

8.4 OTHER FORMS OF MODULATION

Forms of modulation for the transmitted pulse have been proposed, for example Huffman codes [see Kretchmer and Lin in 15, p. 94], that involve amplitude and phase modulation with the promise of no near time sidelobes. Amplitude modulation within the pulse reduces the power out of the transmitter and is undesirable and these codes will not be discussed further.

Frank codes are an approximation to stepped frequency modulation whereas in Costas codes the frequency steps do not occur in linear order. Though the use of Costas codes may be useful for low probability of intercept in military radars, the consideration of Costas codes is beyond the scope of this book.

8.5 NEGATIVE PHASE SEQUENCE SIGNALS

Whenever two phase signals are used for pulse compression instead of intermediate frequency signals, the negative phase sequence component convolves in the pulse compression filter or correlator, adding to the amplitudes of the time sidelobes.

8.6 OUTPUT SIGNALS

The matching filter fishes out the best signal-to-wide-band-noise ratio, and the signals and noise have the same bandwidth. During the time corresponding to the reciprocal of the bandwidth, the range cell, neither the signal nor the noise changes much in amplitude or phase. The range cell amplitudes, or after conversion, binary values provide the signals for signal processing later.

8.7 FIGURES AFFECTING RADAR PERFORMANCE

The following characteristics are used in Chapter 14, Performance.

8.7.1 Range budget

The matching filter section, in spite of providing the best signal-to-noise ratio for the given filter shape, represents a loss for the range budget. The matching filter losses, m, are given for optimum filters in Sections 8.1.1 and 8.1.2. The loss is increased by bandwidth mismatching, C_b. The factor $m\ C_b$ may be taken from Figures 8.3 or 8.7.

With pulse compression, generally the loss in signal-to-noise ratio is measured and is composed of the loss caused by Doppler frequencies and the tapering loss (see Appendix B).

8.7.2 Accuracy and resolution budget

Instead of finite, crisp limits defined by an almost rectangular pulse, the matched filter stretches the skirts of the pulse, especially in regions occupied by signal processing residues (-30 dB to -50 dB). The width of the echo is defined by its half power (3 dB) or half voltage (6 dB) points.

The stretched pulse is often further stretched by signal processing that reduces accuracy and resolution.

8.7.3 Stability budget

Generally, there is no effect on the stability budget for analogue filters when the digital clocks are synchronized to the main system clocks in digital systems.

REFERENCES

1. Skolnik, M. I., *Radar Handbook*, New York: McGraw-Hill, 1970.
2. Skolnik, M. I., *Radar Handbook*, 2nd ed., New York: McGraw-Hill, 1990.
3. Lawson, J. L., and G. E. Uhlenbeck, *Threshold Signals*, Volume 13, M. I. T. Radiation Laboratory Series, New York: McGraw-Hill, 1951.
4. Barton, D. K., *Radar System Analysis*, Englewood Cliffs, New Jersey: Prentice-Hall, 1964.
5. Blake, L. V., *Radar Range-Performance Analysis,* Norwood, Massachusetts: Artech House, 1986.
6. Lamb, J. J., "A Noise Silencing IF Circuit for Superheterodyne Receivers," *QST*, Vol. 20, February 1936.
7. Meikle, H. D., *A New Twist to Fourier Transforms*, Weinheim: Wiley-VCH, 2004.
8. Galati, G., (ed.), *Advanced Radar Techniques and Systems*, Stevenage, Herts: Peter Peregrinus, 1993.
9. Cook, C. E., in R. S. Berkowitz, (ed.), *Modern Radar*, New York: Wiley, 1965.
10. Rihaczek, A. W., *Principles of High Resolution Radar*, New York: McGraw-Hill, 1969.
11.. Richards, M. A., *Fundamentals of Radar Signal Processing*, New York: McGraw-Hill, 2005.
12. Price, R., "Chebyshev Low Pulse Compression Sidelobes via a Nonlinear FM", *URST National Radio Science Meeting,* Seattle, Washington, 18 June 1979.
13. Levanon, N., and E. Mozeson, *Radar Signals*, New York: Wiley, 2004.
14. Nathanson, F. E., *Radar Design Principles*, New York: McGraw-Hill, 1991.
15. Lewis, B. L., F. F. Kretschmer, Jr., and W. W. Shelton, *Aspects of Radar Signal Processing*, Norwood, Massachusetts: Artech House, 1986.

Chapter 9

Detectors

The detector stage converts the intermediate frequency signals into video signals as in a classical superheterodyne receiver. The intermediate frequency signal has the following types of modulation:

- Amplitude modulation;
- Phase (or frequency) modulation.

In early radars the output of the detector was amplified and displayed, as shown in Figure 9.1. In radars that observe or track individual objects (fire control radars, battlefield observation radars, for example) as an aid to identifying each type of scatterer the range gated echoes are passed to headphones. With a staring radar the sound in the headphones allows the operator to identify, for example, propeller modulation of aircraft in chaff and the movement of animals, people, or vehicles.

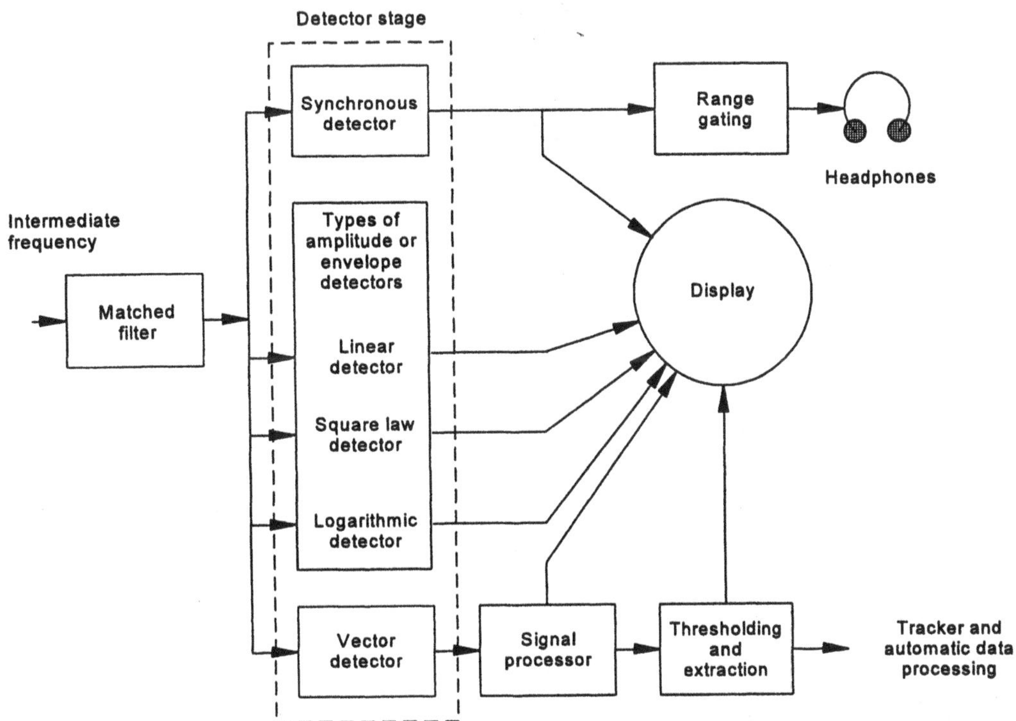

Figure 9.1 Detectors.

Between the echoes there is principally receiver noise without any perceptible carrier frequency. The noise pulses occur at random and have a bidirectional Gaussian distribution shown in Figure 2.19, called polyphase or bottlebrush noise. The echo signals are modulated on a notional carrier, the intermediate frequency representation of the transmitter pulse, and have an amplitude proportional to the scatterer size and inversely proportional to the fourth power of their range. The phase of the echo signal with respect to the coherent oscillator (COHO) depends on its fine range. The phase angles repeat themselves each half of a wavelength in range. The frequency of the carrier is the intermediate frequency plus or minus the Doppler frequency of the scatterer.

There is a problem illustrating signals mixed with noise. The form of a single echo signal coming from the matched filter is an inverted U, ∩. The inverted U contains the sum of one sample of the signal plus one sample of noise, and the two cannot be separated. If an echo 100 pulse lengths long is considered, portions of the echo will interfere with each other to give Rayleigh distributed echo signals. Here, a fictional situation is used to illustrate signals with noise with 100 point reflectors of the same cross-section spaced one range cell apart so their reflections do not interfere but close enough so there is no visible variation of signal strength with range. The echoes arrive without any phase shift which allows the illustration of a signal plus noise and is a convention used for the rest of this chapter. This is labeled as 100 adjacent point echoes in the diagrams.

A noise signal with 100 adjacent point echoes as it would appear on an oscillograph is shown in Figure 9.2. Notice that the signals are symmetrical about zero and that the noise on the echo signal shows that it is many pulse widths wide.

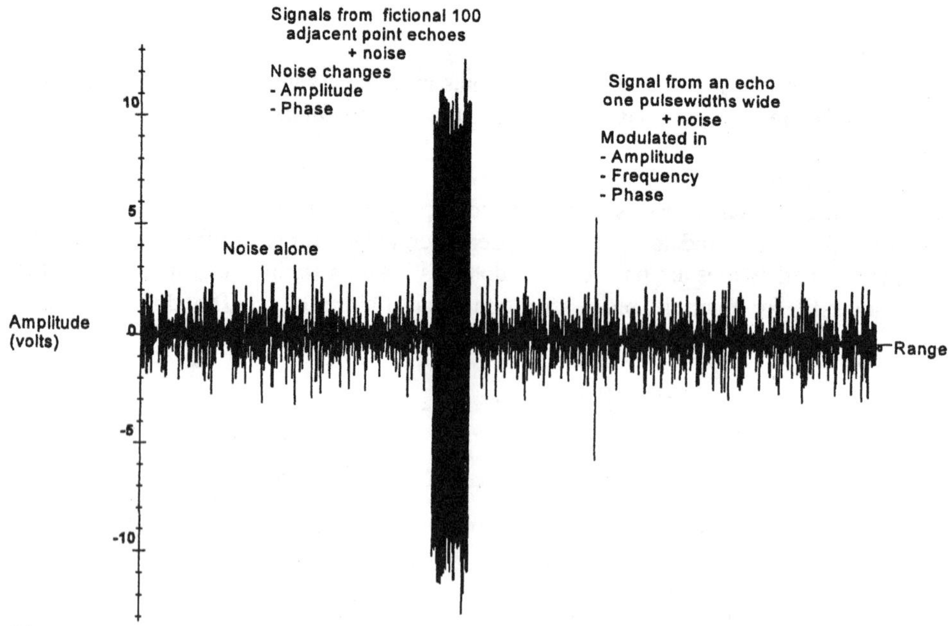

Figure 9.2 Intermediate frequency noise with the fictional 100 adjacent point echoes.

An echo signal may be represented by

$$IF\ signal = A \sin(2\pi(f_{if}+f_{Doppler})t + \phi) \tag{9.1}$$

where A is the amplitude, V;
f_{if} is the intermediate frequency, Hz;
$f_{Doppler}$ is the Doppler frequency, Hz;
ϕ is the fine range phase, radians.

The noise between the echo signals is a suppressed carrier modulated by two orthogonal Gaussian components. These components are filtered in the intermediate frequency amplifier and give a random modulation at intermediate frequency in amplitude and phase.

$$IF\ noise = e_{ni} \sin(2\pi f_{if}\, t) + e_{nq} \cos(2\pi f_{if}\, t) \tag{9.2}$$

where e_{ni} and e_{nq} are independent, identically distributed Gaussian voltages.

There are three main types of detector discussed:

- Incoherent detectors;
- Coherent detectors;
- Vector detectors.

9.1 INCOHERENT DETECTORS

Incoherent detectors give an output in relation to the amplitude of the intermediate frequency signal. The phase and Doppler frequency information is lost.

If the probability distributions of the modulating random noise in the x and y directions are plotted (standard deviation or rms value unity), Figure 9.3 is obtained which is a bivariate Gaussian distribution in x and y. It is also the probability distribution of noise on a notional intermediate frequency carrier in amplitude and phase.

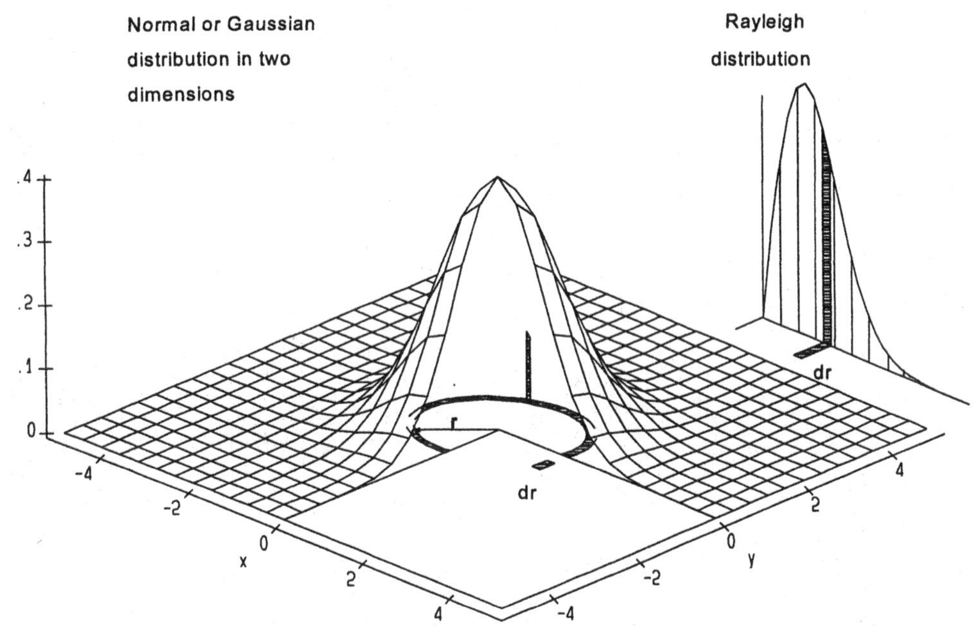

Figure 9.3 Bivariate Gaussian distribution of noise.

This probability distribution of the noise around the carrier is, in x and y,

$$p(x,\ y) = \frac{\exp\left(-\frac{x^2}{2\sigma^2}\right)}{\sqrt{2\pi\sigma^2}} \frac{\exp\left(-\frac{y^2}{2\sigma^2}\right)}{\sqrt{2\pi\sigma^2}} = \frac{\exp\left(-\frac{x^2+y^2}{2\sigma^2}\right)}{2\pi\sigma^2} \tag{9.3}$$

where σ is the rms value of the noise voltage.

Substituting for x and y and noting that $x^2 + y^2 = r^2$

$$\textit{Height of hump at } (x,y),\ h = \frac{\exp\left(-\frac{r^2}{2\sigma^2}\right)}{2\pi\sigma^2} \tag{9.4}$$

The probability at the radius r is

$$p(r) = 2\ \pi\ r\ dr\ h = \frac{r\ \exp\left(-\frac{r^2}{2\sigma^2}\right)}{\sigma^2} \tag{9.5}$$

Equation (9.5) is the Rayleigh distribution and is the probability distribution of the linearly (envelope) detected noise and is shown on the right of Figure 9.3. The echo signal from 100 adjacent reflectors and noise from such a detector are shown in Figure 9.4.

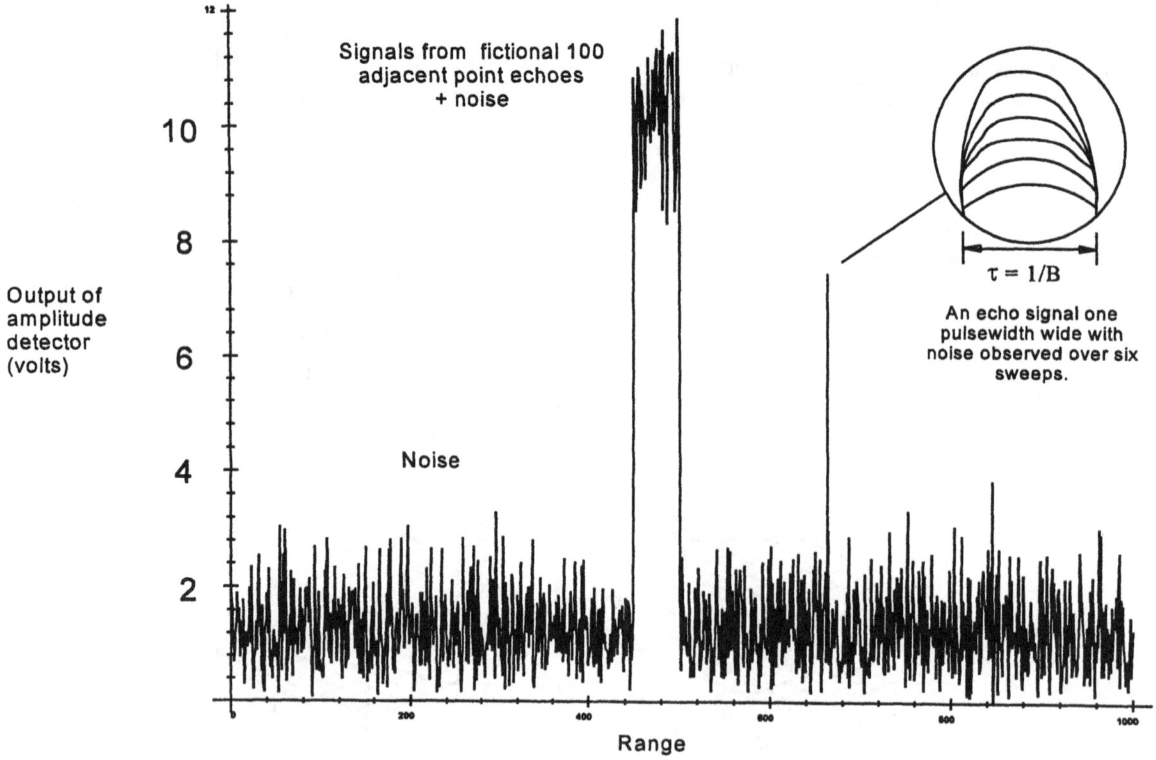

Figure 9.4 Video and Rayleigh noise linearly detected from the intermediate frequency signal in Figure 9.1.

The following relationships apply if σ is the standard deviation (or rms value) of the original Gaussian distributions:

$$
\begin{aligned}
\textit{rms value} &= \sqrt{2}\,\sigma &&= 1.4142\ \sigma \\
\textit{mean} &= \sqrt{\frac{\pi}{2}}\,\sigma &&= 1.2533\ \sigma \\
\textit{median} &= \sqrt{2\ \ln 2}\ \sigma &&= 1.1774\ \sigma \\
\textit{mode} &= \sigma
\end{aligned}
\tag{9.6}
$$

σ = standard deviation of the original Gaussian distributions

Figure 9.4 has been drawn using σ = 1 V. It can be seen that the "shoulder value" of the noise, as would be measured on an oscilloscope, is between approximately 2.5σ and 3σ.

The original detectors were simple diodes that produced an output proportional to the square of the amplitude of the input signal, called square law detectors. If the signal from Figure 9.1 is passed through a square law detector, the result is as shown in Figure 9.5. A true square law detector emphasizes the peaks of the signal and noise.

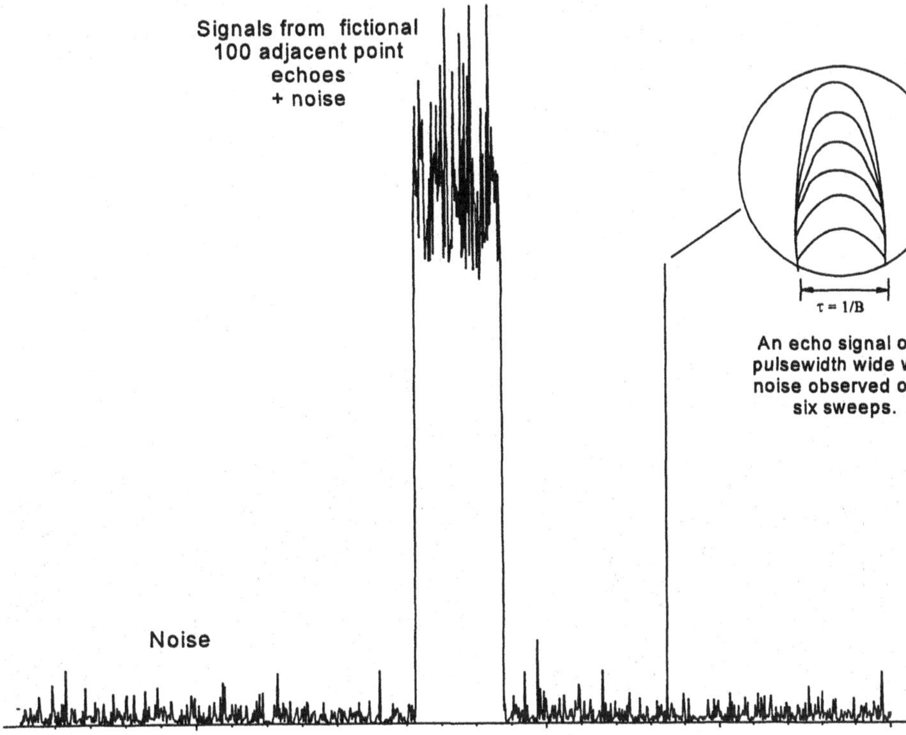

Figure 9.5 Echo signals from 100 fictional adjacent reflectors + noise and noise at the output of a square law detector.

9.1.1 Logarithmic intermediate frequency amplifiers

Intermediate frequency amplifiers can be wired with cascaded stages to produce a signal proportional to the logarithm of the intermediate frequency input signal, as in Figure 9.6.

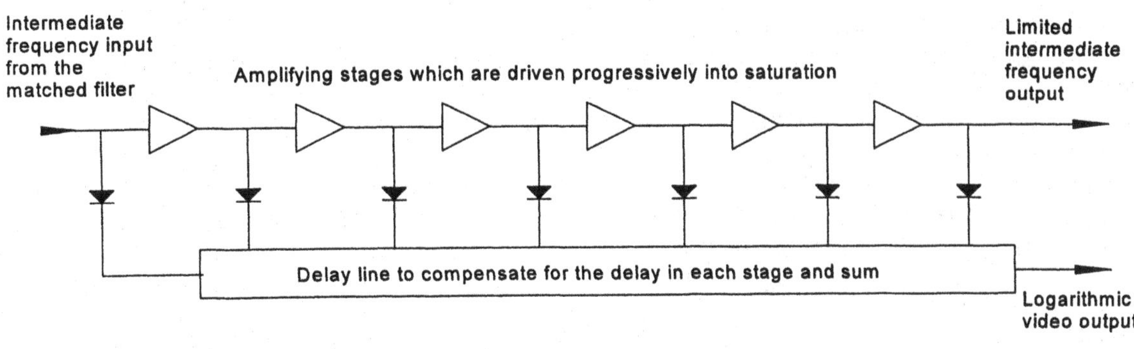

Figure 9.6 Block diagram of a logarithmic amplifier and detector.

A number of stages are cascaded. With increasing intermediate frequency signal the stages, starting at the last stage, saturate and limit. The output of the logarithmic detector is the sum of the signals present between each of the stages.

The logarithmic output curve is made up of a number of segments approximating a logarithmic characteristic. The greater number of stages, the better the approximation.

With a limited number of stages, often the noise level is small enough not to saturate the last stage. These are often called lin-log amplifiers and are good enough to observe signals in the full dynamic range of the radio frequency stages and for logarithmic approximations of large signals for three-dimensional radars. These are not good enough for signal processing which depends on logarithmically shaped noise as an input. The logarithmic output has a compressed dynamic range, as shown in Figure 9.7. Notice that the signal amplitude is compressed compared to the noise and that the noise amplitude mixed with the signal is inversely proportional to the size of the signal.

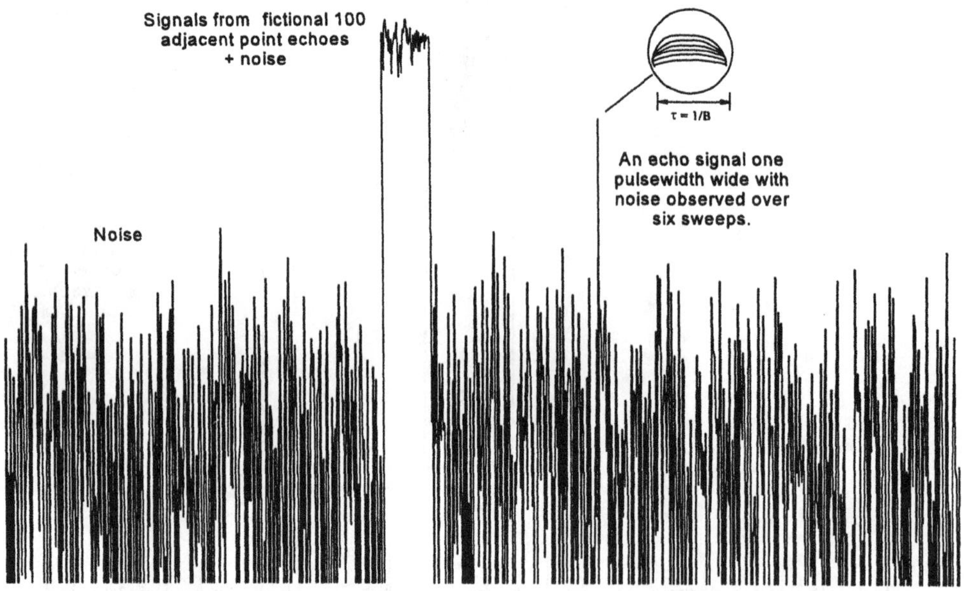

Figure 9.7 Log-video from the intermediate frequency signal in Figure 9.2.

9.2 COHERENT DETECTORS

Coherent detectors compare the echo signals with a phase reference in order to be able to use the Doppler frequency information later. The simplest coherent detector is used in continuous wave radars, where the returning echoes mix with the transmitted signal. They beat together and a simple detector gives a signal at the Doppler frequency as an output which is adequate for artillery and other proximity fuses. When the projectile is close enough to damage its target, the beat signal is large enough to trigger the fuse. Today, the major use of continuous wave radars is in vehicle speed measuring equipment used by many police forces.

Coherent pulse radars have to maintain a continuous frequency reference between the pulses they transmit. A block diagram of such a radar is shown in Figure 9.8. The coherent oscillator block (COHO) provides the reference signal and this is the phase reference for the whole radar. Master oscillator power amplifier (MOPA) transmitters mix the frequencies of the coherent and local oscillators to generate pulses at the transmitter frequency.

Many simpler radars use magnetrons as transmitting valves. These are fed with direct current pulses only, and the transmitted pulses are not referenced to any phase source and have a random phase shift with respect to each other. The solution is to stop the coherent oscillator and restart it with a sample of the transmitter pulse converted down to the coherent oscillator frequency. Thus the phase coherence between the transmitted pulse and the coherent oscillator is restored, as is shown in Figure 9.9.

Another method is to measure the phase of the locking pulse. The phase angle measured is used to correct the phase of the echoes once they have been converted to digital vector form (Chapter 10).

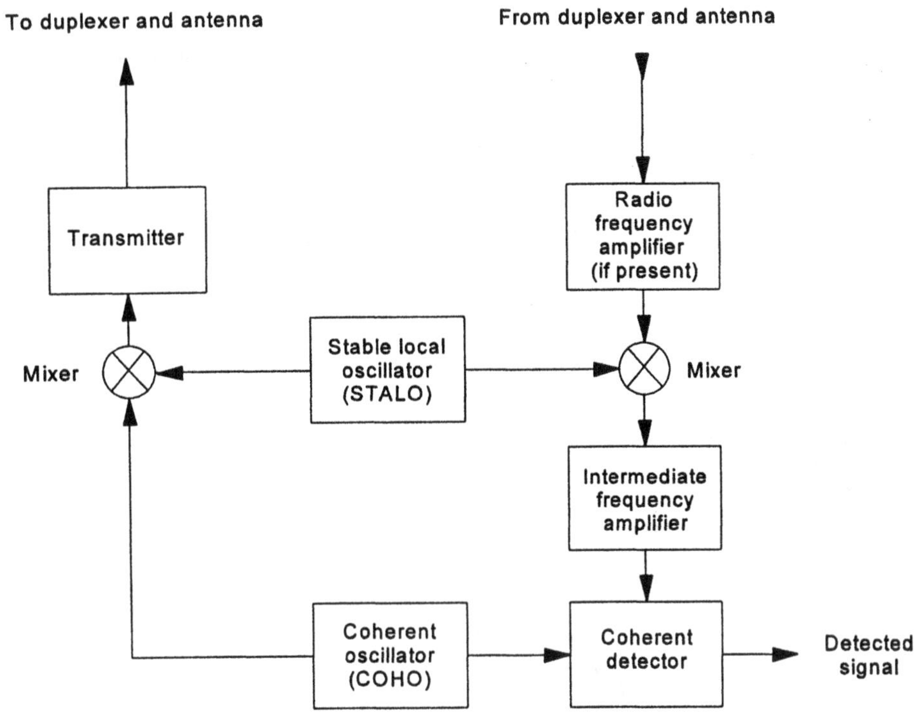

Figure 9.8 Block diagram of a radar with coherent transmission and reception.

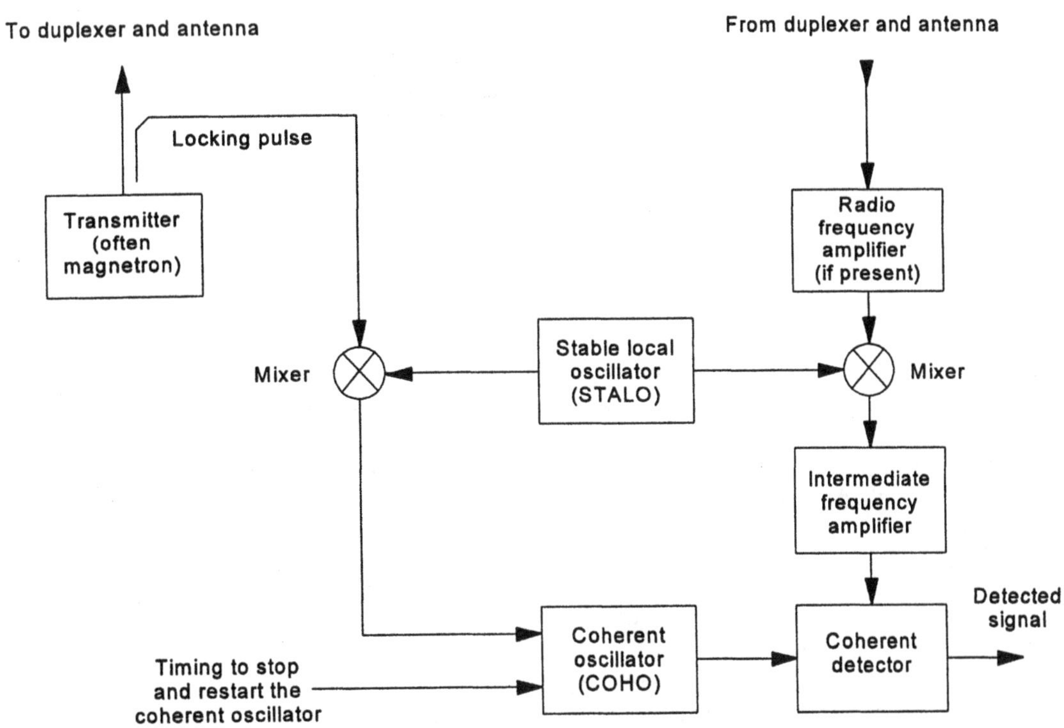

Figure 9.9 Block diagram of a radar with coherent reception only.

9.2.1 Ring modulator or demodulator

Figures 9.8 and 9.9 show the coherent oscillator giving a reference signal to the coherent detector. One such detector is the ring demodulator which is a balanced multiplier circuit and is illustrated in Figure 9.10. The signals enter via the transformer in the left and are mixed with the balanced reference waveform entering through the transformer from below. The reference waveform switches either the diodes A and B or the diodes C and D into conduction to connect either the upper or lower input transformer terminal to the output. Taking cosine signals,

$$\begin{aligned} Input &= A\cos(2\pi st+\phi) \\ Reference &= \cos(2\pi rt) \\ Output &= A\cos(2\pi st+\phi)\cos(2\pi rt) \\ &= \frac{A}{2}\left(\cos\left(2\pi(r+s)t+\phi\right) + \cos\left(2\pi(r-s)t + \phi\right)\right) \end{aligned} \tag{9.7}$$

where A is the amplitude of the signal;
s is the frequency of the signal;
ϕ is the phase of the signal;
r is the frequency of the reference waveform.

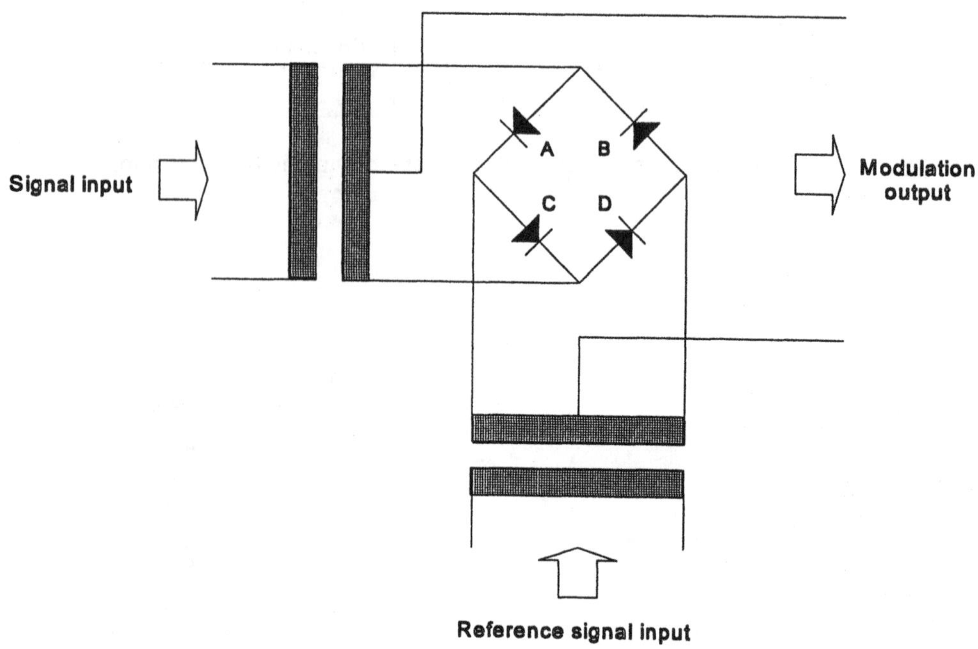

Figure 9.10 Principle of the ring demodulator.

The output consists of signals of frequencies at the sum of the signal and reference frequencies (modulation) and the difference of the frequencies (demodulation). The choice between modulation and demodulation is made by filtering out the unwanted output.

The demodulated output is the component of the signal which is synchronous with the reference signal, namely, $A/2\cos\phi$. This detector is called a synchronous detector. (See [1] which defines I and Q videos as coming from a pair of synchronous detectors.)

Alternatively the reference signal may be represented by its sine which gives

$$\begin{aligned} Input &= A\cos(2\pi st+\phi) \\ Reference &= \sin(2\pi rt) \\ Output &= A\cos(2\pi st+\phi)\sin(2\pi rt) \\ &= \frac{A}{2}\left(\sin\left(2\pi(r+s)t+\phi\right) - \sin\left(2\pi(r-s)t + \phi\right)\right) \end{aligned} \tag{9.8}$$

The demodulated output is $\frac{A}{2}$ sin ϕ.

If the input signal is limited to a constant unity, then the output is sin ϕ or, for small angles, ϕ. This has the function of a phase detector which is approximately linear over about ±0.1 radian or ±5.7 degrees. The use of the same circuit as a modulator and three types of demodulator is the cause of much confusion and the demodulators are shown in Table 9.1.

Table 9.1
Uses of the ring demodulator

Signal	Video output	Detection type or use
Signal with full dynamic range	Bipolar	Synchronous detector
Softly limited signal	Bipolar	Used in old moving target indicator (MTI) radars and detects a mixture of amplitude and phase
Hard limited signal	Phase video	Phase detector

The video output from a synchronous detector is the component of detected echo which is synchronous with the reference phase and one component of the bivariate Gaussian noise, as is shown in Figure 9.11. The noise is Gaussian around zero, leading to the term bipolar video. A signal is shown which may be either positive or negative with the accompanying noise.

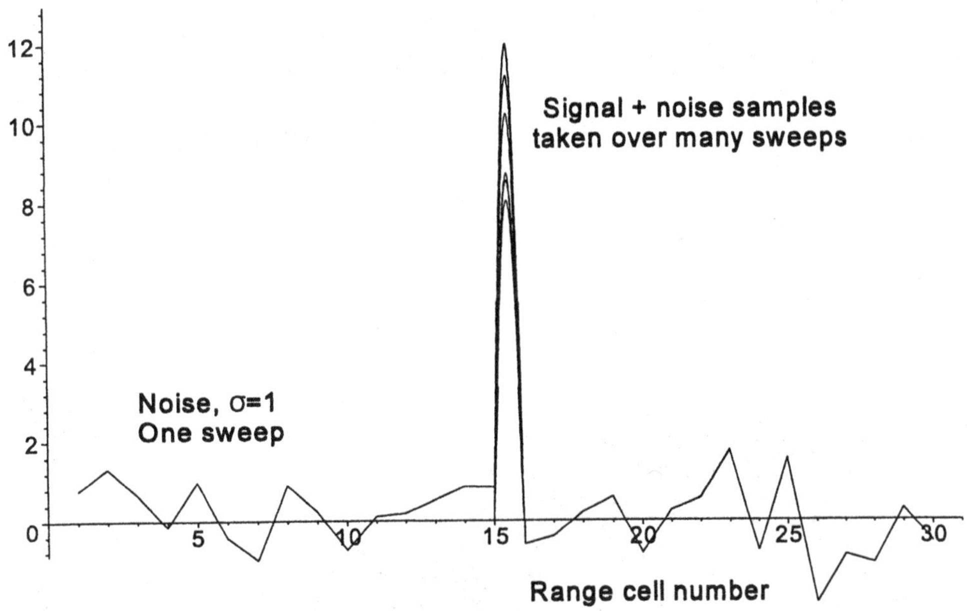

Figure 9.11 A coherently detected echo signal with no Doppler frequency shift with noise.

If the echo is moving, the phase angle, ϕ, changes from pulse to pulse, leading to the amplitude of the component changes from pulse to pulse. Such an echo is shown in Figure 9.12. The fluttering of the echo signal is a characteristic of bipolar video that has been used to distinguish between fixed and moving echoes on A-scopes and displays derived from them.

If the range cell is gated, the Doppler frequency component may be fed to headphones or a loudspeaker for the identification of the scatterer giving the echo. This is used in continuous wave tracking radars to indicate an aircraft's change of radial velocity or course, and radars that watch over terrain use this to identify the type of animal or human intruder.

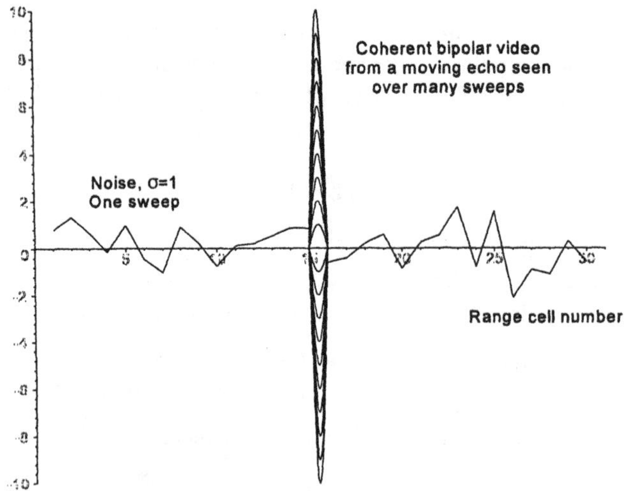

Figure 9.12 A coherently detected signal with Doppler frequency change.

9.3 VECTOR DETECTORS

Figure 9.13 shows the vectors of the modulation of signals and bottle-brush noise during one sweep.

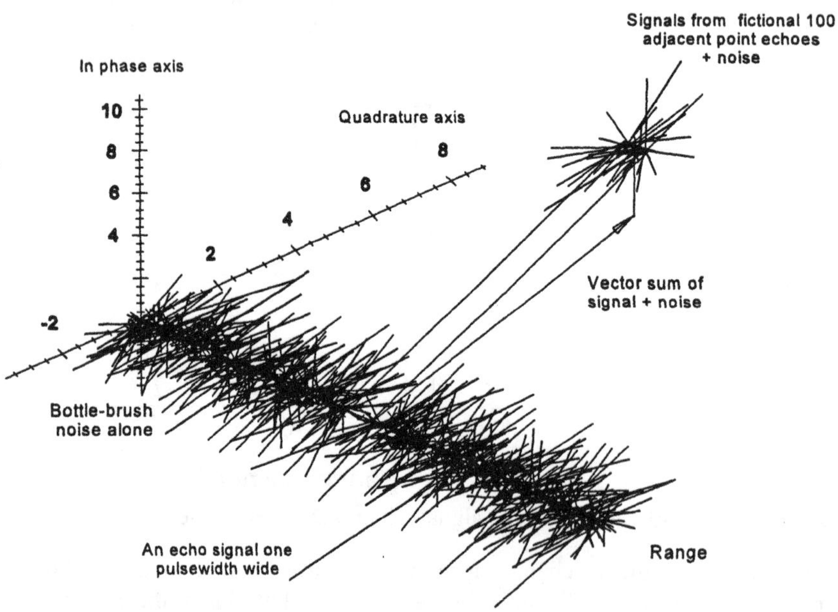

Figure 9.13 A vector representation of the modulation on the intermediate frequency.

When signals are modulated on an intermediate frequency carrier (often suppressed), they may be added, subtracted, and multiplied for processing in vector form. When the signals are processed after detection, as is the case with digital processing, the vectors must be represented by their two-dimensional components as two signals which may be either:

- Amplitude and phase from a polar detector;
- Two orthogonal components from a Cartesian detector. One component corresponds to the amplitude component in phase with the radar's coherent oscillator (COHO), and the other in phase quadrature. This has the form of a two-phase alternating current.

9.3.1 Polar detectors

The representation of a vector in polar coordinates consists of the amplitude of the signal and its phase angle as two separate signals. Typical signal and noise outputs from the arrangement in Figure 9.14 are shown in Figure 9.15. The phase of the noise is uniformly distributed over ±180 degrees or $\pm\pi$ radians. The phase is only stable when a stationary echo dominates the noise.

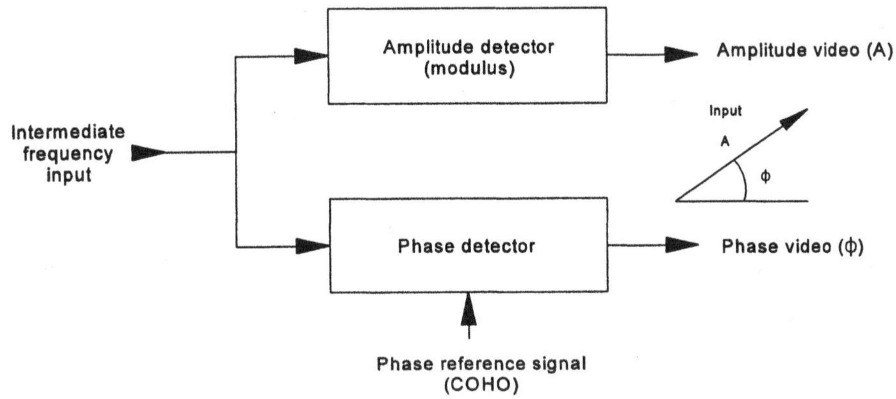

Figure 9.14 Block diagram of a polar detector.

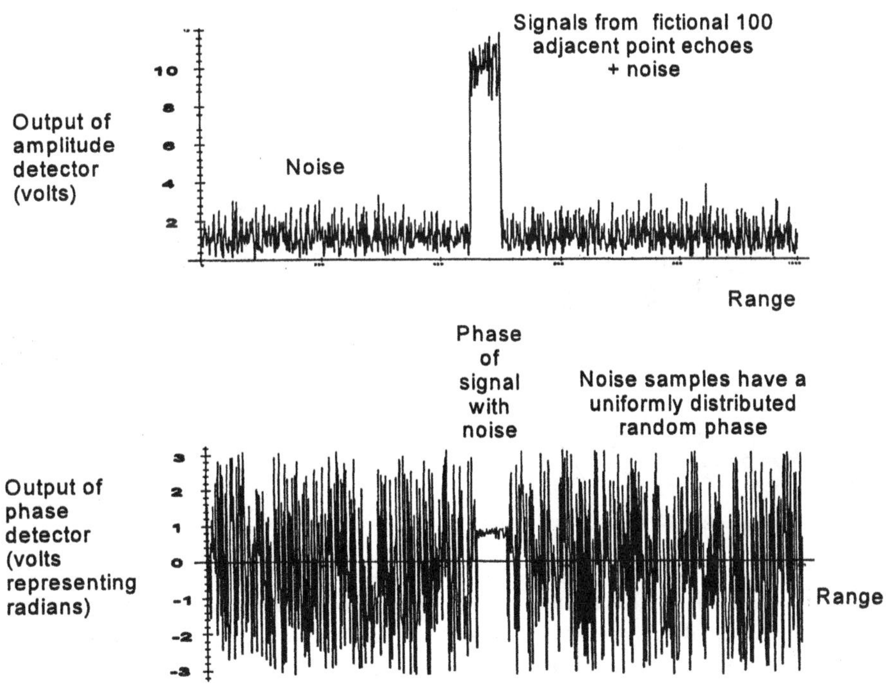

Figure 9.15 The outputs of the amplitude and phase detectors for the vectors in Figure 9.13.

9.3.2 Cartesian or two-phase detector

While it is possible to provide vector processing in amplitude and phase, the components that have been developed for digital signal processors are for the x and y (or I and Q) components of signals. This is normally performed by two separate synchronous detectors, as shown in Figure 9.16. One has the in-phase coherent oscillator (COHO) signal as reference and produces the I video and the other uses the quadrature signal as reference to produce Q video. The positive Doppler frequencies are represented by a positive phase sequence in the I and Q videos, and "negative" frequencies by a negative phase sequence. This convention is used in polyphase alternating current.

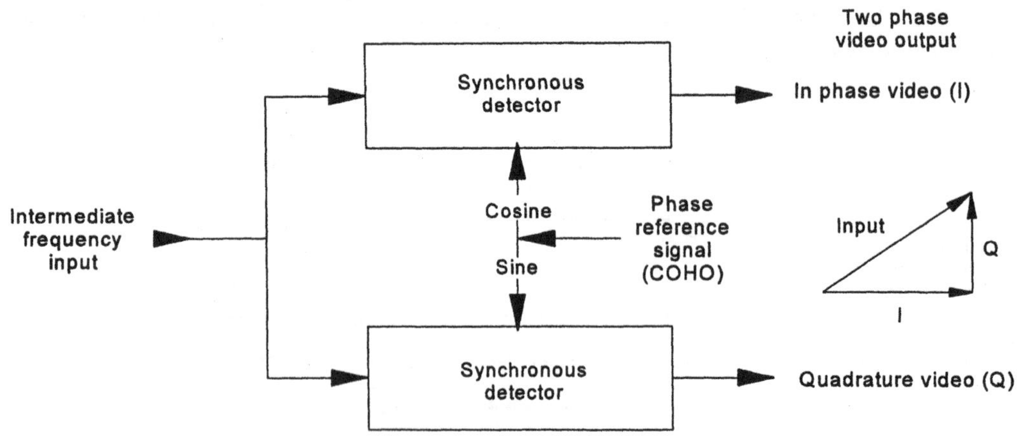

Figure 9.16 Block diagram of a Cartesian detector.

The I and Q videos derived from Figure 9.13 are shown in Figure 9.17. Synchronously detected videos are bipolar and depend on the phase of the echoes.

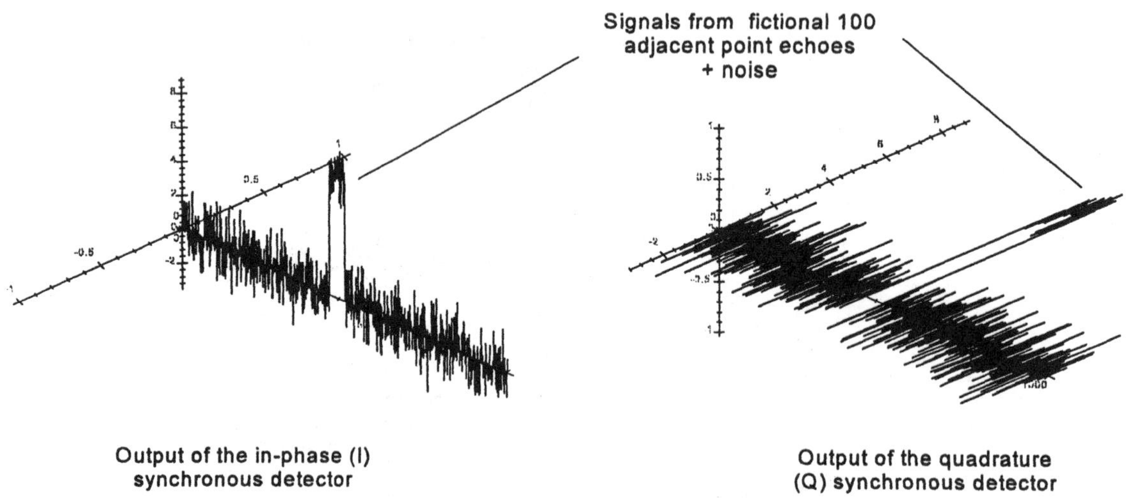

Figure 9.17 I and Q videos from the vectors in Figure 9.13.

A problem with I and Q processing is that degradation in the later processing occurs when there is a direct voltage offset, the gain in the two channels is not the same, or when the reference phases are not exactly 90 degrees apart.

Such asymmetry in polyphase power systems is represented by a positive sequence component, a negative sequence component, and a constant vector. The same occurs in this system.

- The positive sequence component represents the main signal.
- The negative sequence component represents the unbalance in amplitude and the phase error in the coherent oscillator (COHO) phases.
- The direct component represents the direct voltage zero offset errors.

The first signal is the true signal, with its Doppler frequency components to be processed later. The second component gives a signal with a "negative" Doppler frequency and adds to the range sidelobes in pulse compression systems. The third component, the offset, adds to the clutter component. In modern radars with digital signal processing, errors in the second detector are measured after conversion to digital form. This allows them to be corrected (see Chapter 10).

9.4 SAMPLING WAVEFORMS AND SPECTRA

Sampling takes place in various forms, each of which has its own spectrum. It is best illustrated by assembling the video signals from a number of sweeps as the antenna scans past an object and presenting them in three dimensions in Figure 9.18. In the range dimension, the spectrum is limited by the bandwidth at the output of the matched filter. After detection, the signal spectrum extends from zero to half the matched filter bandwidth.

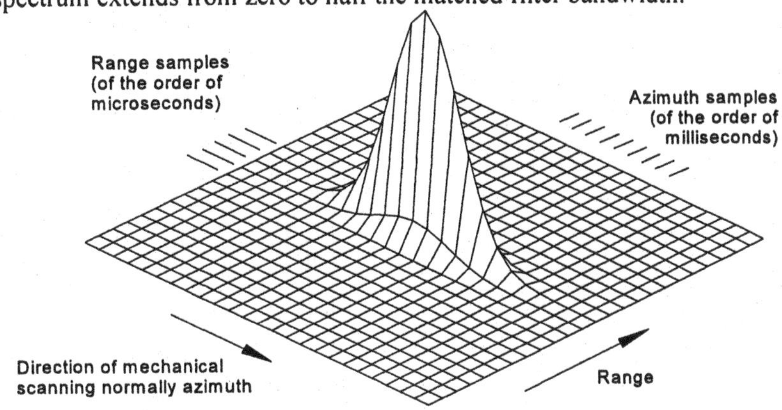

Figure 9.18 The power densities of echo signals in the range and azimuth dimensions.

The shape of the assembled samples in the direction where a mechanical antenna scans the object is much wider in time, generally tens of milliseconds. The Fourier transform of this scanning modulation gives the spectrum that is filtered during signal processing.

9.4.1 Simple or single phase sampling

In pulse radar, the echo signal is sampled at the pulse repetition frequency. In signal processing, samples spaced by a range cell or some fraction of a range cell may be taken at that frequency. Figure 9.19(a) shows sampling points with unity spacing representing the sampling frequency, which is 8 times that of a sine waveform. The straight lines show the estimate of the waveform after sampling.

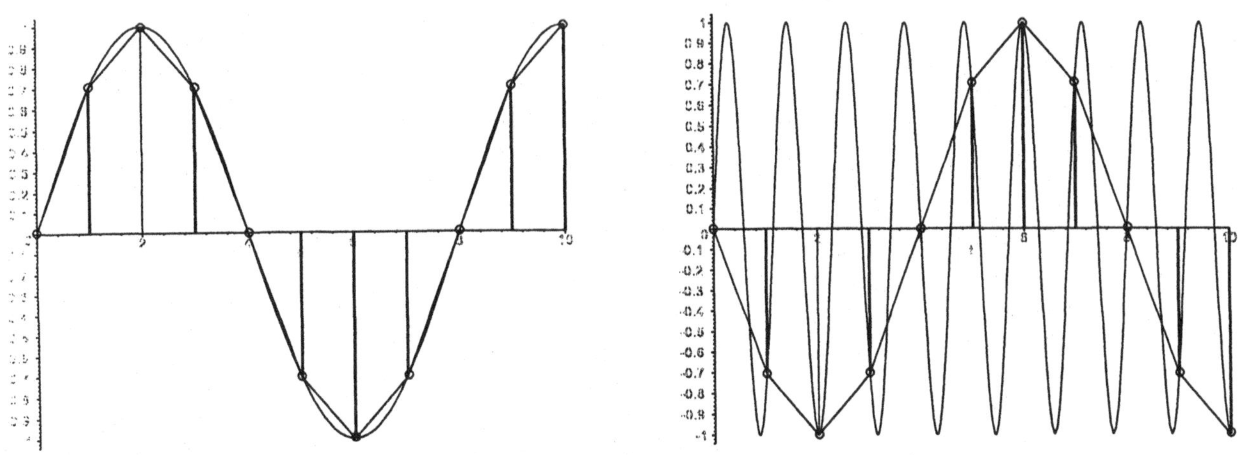

(a) Sine wave frequency=1, sampling frequency =8 (b) Sine wave frequency=7, sampling frequency =8

Figure 9.19 Sampling at 8 times the frequency of a sine wave and undersampling at 7/8ths of the sine wave frequency.

If the frequency of the sine wave is increased, the sine voltage moves much more between samples. When the waveform is sampled less than each half cycle, the samples are no longer a true representation of the original, as defined by the Nyquist criterion. In Figure 9.19(b) the sampled waveform is the same as that in Figure 9.19(a) except for a reversal of phase. This is called undersampling.

The echoes often return with Doppler frequencies above the pulse repetition (sampling) frequency. Figure 9.20(a) shows a sine wave at 1 1/8 F_s, that is above the sampling frequency. Figure 9.20(b) shows a sine wave at 1 7/8 times the sampling frequency.

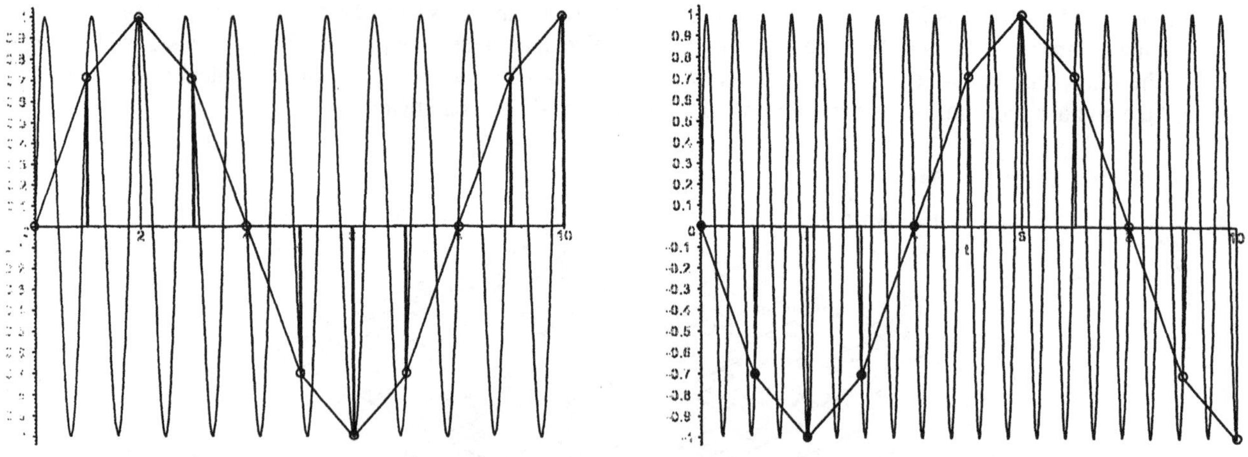

(a) Sine wave frequency=8, sampling frequency =9 (b) Sine wave frequency=8, sampling frequency =15

Figure 9.20 Examples of undersampling and aliasing.

All the above sampled waveforms at 1/8, 7/8, 1 1/8, and 1 7/8 of the sampling frequency are folded or aliased down to the same waveform, 1/8th of the sampling frequency, neglecting the phase. The only frequency range not subject to aliasing, the unambiguous range, extends from zero to half the sampling frequency. This transformation is illustrated in the aliasing diagram shown in Figure 9.21 for a sampling frequency of 1 000 Hz.

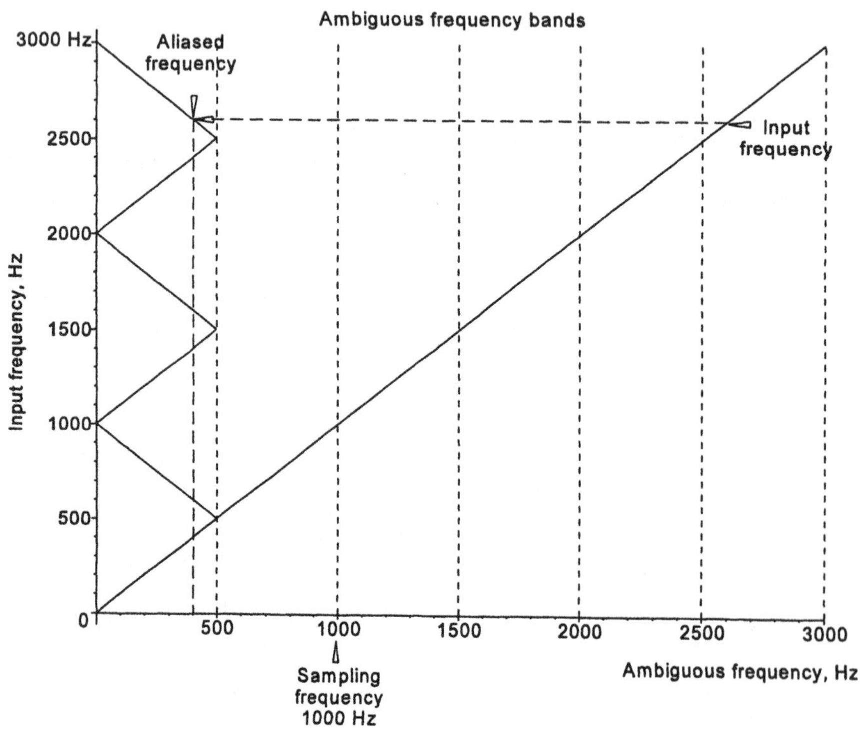

Figure 9.21 Single-phase scalar sampling showing the effects of aliasing with a sampling frequency of 1 000 Hz.

Where the input frequencies are less than half the sampling frequency, it is common to use antialiasing filters to exclude all harmonics that could be folded down by sampling and cause interference.

9.4.2 Complex or two-phase sampling

Polyphase (normally two-phase) sampling is used to represent vector voltages, normally in the form of in-phase (I) and quadrature phase (Q) video or digital values. The phases of the returns from echoes which have undergone a Doppler frequency shift change from pulse to pulse and may be represented by the spiral in Figure 9.22. This spiral is sampled to measure the orthogonal in phase and quadrature components to give the sequence of two-phase signals in Figure 9.22.

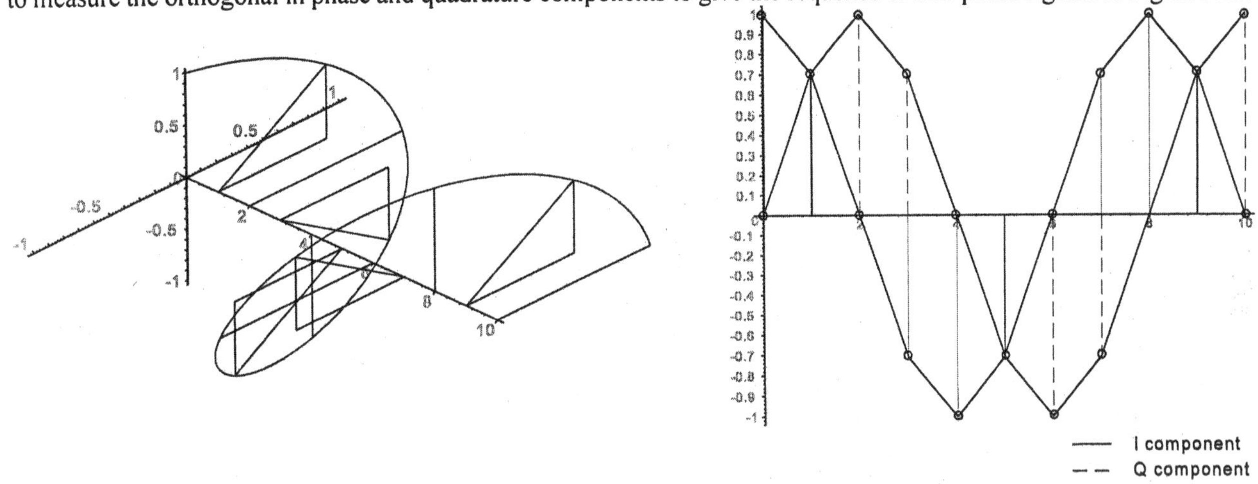

Polyphase sine wave frequency=1, sampling frequency=8

Figure 9.22 A polyphase sine waveform, its Cartesian components, and complex sampling at 8 times its frequency.

When the frequency is raised to 7/8 of the sampling frequency, the situation is shown in Figure 9.23. The sequence of the two-phase samples rotates in the opposite direction to Figure 9.22, so that when the two phases are taken together, the frequency range between zero and the sampling frequency is uniquely defined. The aliasing diagram for this is shown in Figure 9.24.

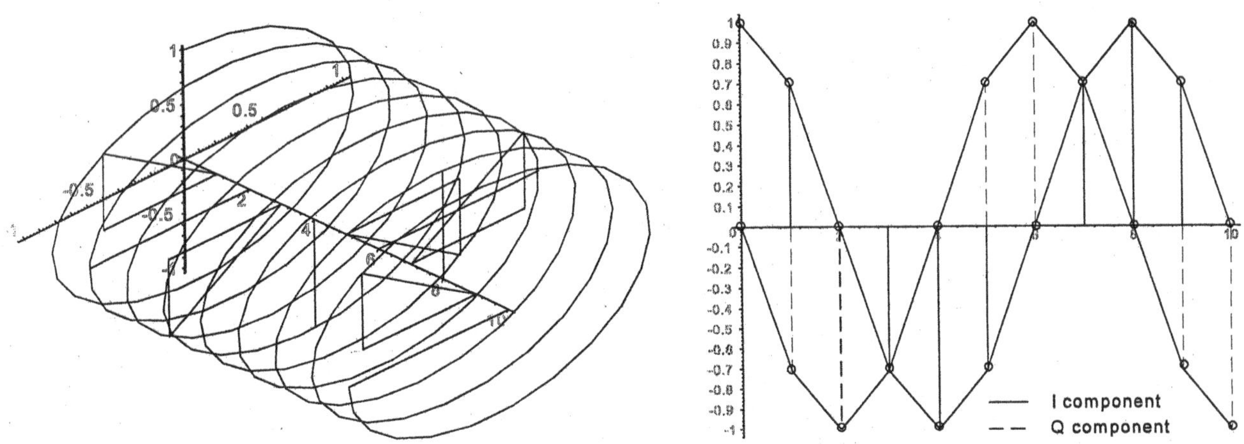

Polyphase sine wave frequency=7, sampling frequency=8

Figure 9.23 Complex or two-phase sampling of a waveform at 7/8 of the sampling frequency.

Figure 9.24 has been drawn with "negative frequencies" which represent the negative phase sequence components. An echo with a negative phase sequence Doppler frequency of 300 Hz is sampled as 700 Hz, which is in contrast with the single-phase sampling of echoes with Doppler frequencies conventionally lower than the transmitter frequency, where the sampling characteristic is symmetric.

Twice the effective number of samples thus gives a frequency range from zero to the sampling frequency, which complies with the Nyquist criterion. Since in a two-phase waveform the ordinate of one phase is proportional to the slope of the other, two-phase sampling is an example of ordinate and slope sampling [2].

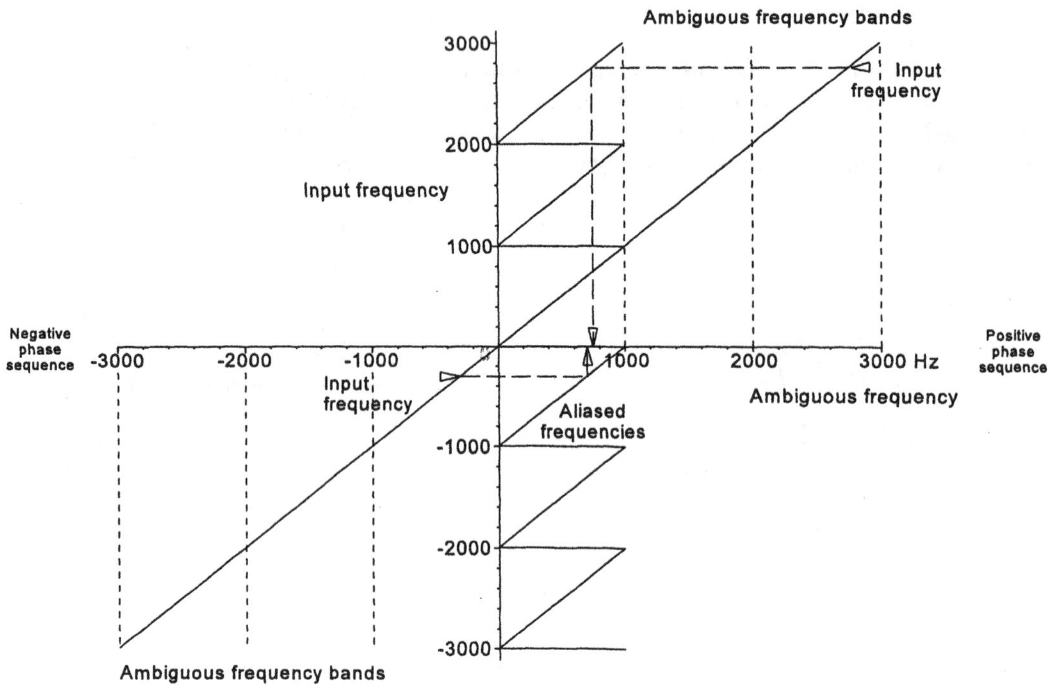

Figure 9.24 Polyphase vector sampling showing the effects of aliasing with a sampling frequency of 1 000 Hz.

Table 9.2 shows a summary for the two types of sampling with a sampling frequency F_s Hz.

Table 9.2
A summary of the properties of single- and two-phase sampling

	Single-phase scalar	Two-phase vector
Number of samples	F_s per second	2 F_s per second
Frequency range	0 Hz to $F_s/2$ Hz	0 Hz to F_s Hz
Negative sequence ("negative" frequency)		Negative phase sequence samples
components > F_s	Frequency inversion of components between $(n-1/2)$ F_s and F_s	Frequency displacement by $+n\,F_s$ No inversion

9.4.3 Sampling at intermediate frequency

Sampling may take place at (a low) intermediate frequency and some analogue-to-digital converters are able to assume the function of detector and are described in Section 10.3.4.

9.5 MEASUREMENT OF NOISE

Noise is difficult to measure on an oscilloscope. For many years the "shoulder" value has been used which has little to do with the root mean square value, which is best measured with a true rms voltmeter. Problems have occurred when connecting equipment with interfaces defined by different types of noise measurement.

There are two shortcuts with an oscilloscope to measure Gaussian and Rayleigh noise, but they should be used only when neither true root mean square voltmeter nor a power meter is available or appropriate. An example is when the noise level has to be estimated between signals. Both methods use a double trace oscilloscope, with preferably a switching frequency many times the noise bandwidth. With averaging over time, that is, over the length of the trace, sequential switching between the two beams is acceptable. The noise is applied to the Y1 and Y2 inputs, with the Y2 input switched to negative display as shown in Figure 9.25.

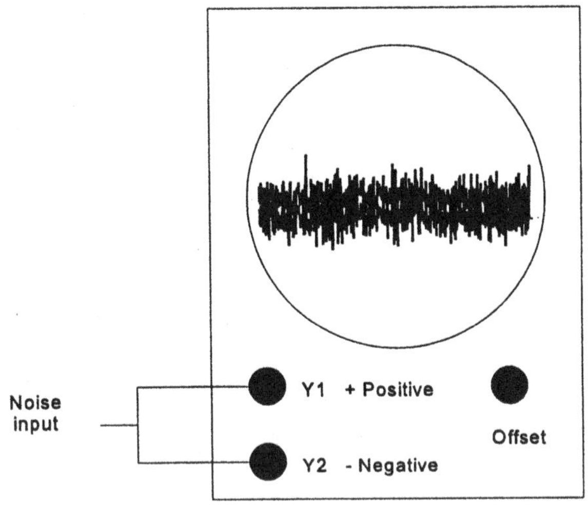

Figure 9.25 The connection of an oscilloscope to measure noise.

9.5.1 Gaussian noise

The noise signal is applied to both channels. The direct voltage offset of one channel is adjusted first so that the noise waveforms coincide. Then the offset is adjusted until the two waveforms just start to separate. Figure 9.26 shows one trace of both channels. The afterglow on an oscilloscope screen shows the average of many sweeps so that the crossovers and gaps average out.

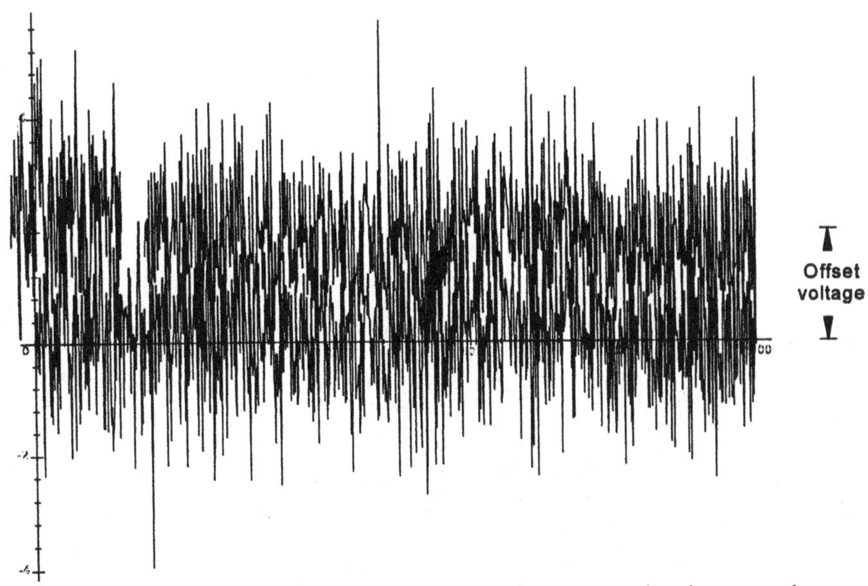

Figure 9.26 Two Gaussian noise traces offset by 2σ, or twice the rms voltage.

Figure 9.27 shows two Gaussian probability distributions separated by 2σ and $2\sigma\pm20\%$. The σ-20% curve shows the waveforms crossing, the σ+20% curve shows a definite saddle point, with the 2σ separation giving a broad bright band. Then the rms voltage is half the offset voltage.

$$Gaussian\ rms\ voltage = \frac{Offset\ voltage}{2} \tag{9.9}$$

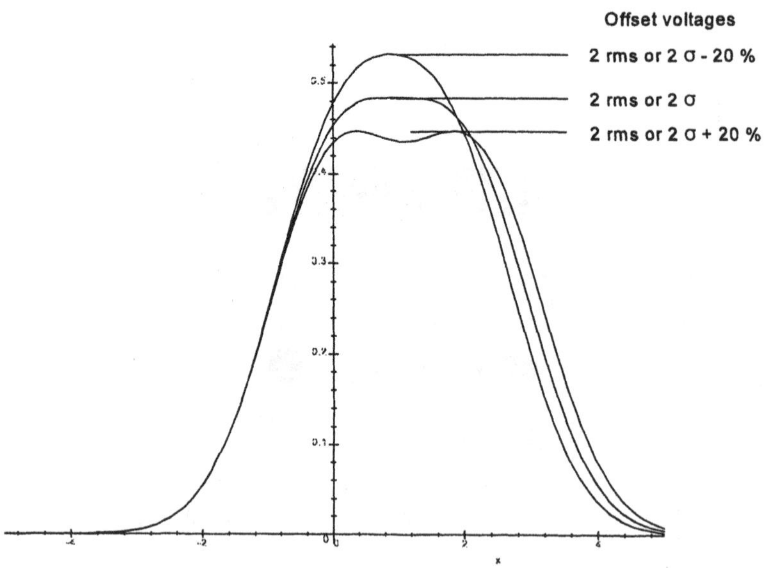

Figure 9.27 The sum of two Gaussian probability distributions, or the brightness on a cathode ray tube, with offsets of 1.8σ, 2σ, and 2.2σ.

9.5.2 Rayleigh noise

The Rayleigh distribution has separate means, medians, and modes. Only the mode, the brightest value may be guessed at, rather inaccurately, and the measurement is open to question (Figures 9.28 and 9.29).

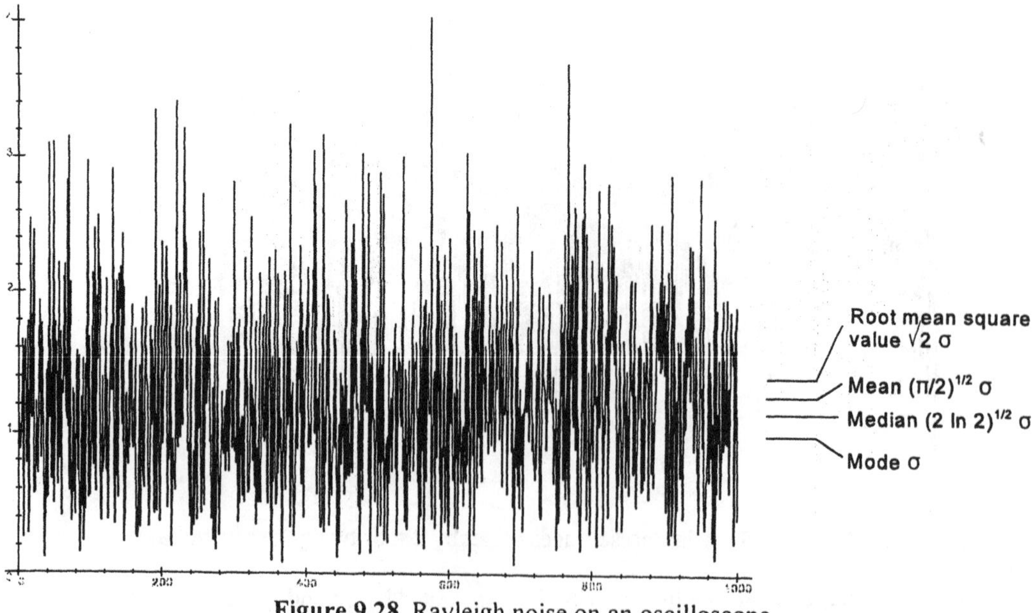

Figure 9.28 Rayleigh noise on an oscilloscope.

If the noise signal is applied to one channel and the negative value to the other (switch the polarity in one channel), the noise voltage may be estimated more accurately. One method is to alter the offset and brightness to give the narrowest and brightest line. In this case the modes are aligned and the offset is 2σ.

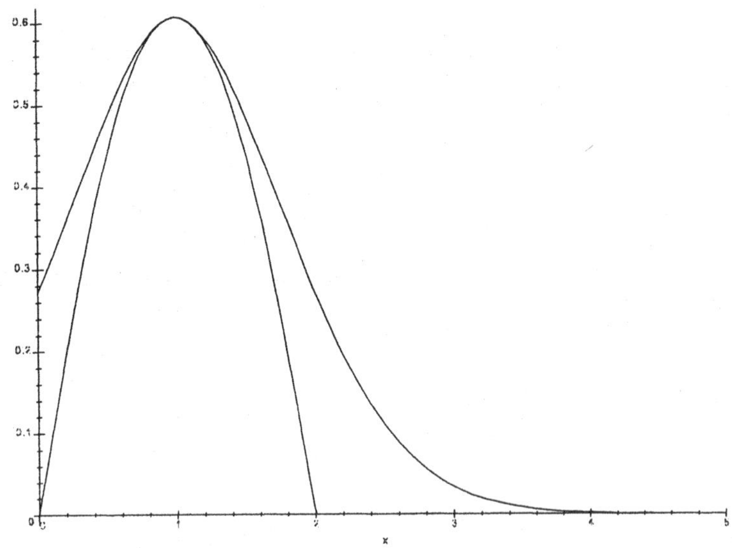

Figure 9.29 Saddle diagram of Rayleigh noise with the modes aligned.

As with Gaussian noise, the direct voltage offset of one channel is adjusted so that the noise waveforms start to separate as shown in Figure 9.30.

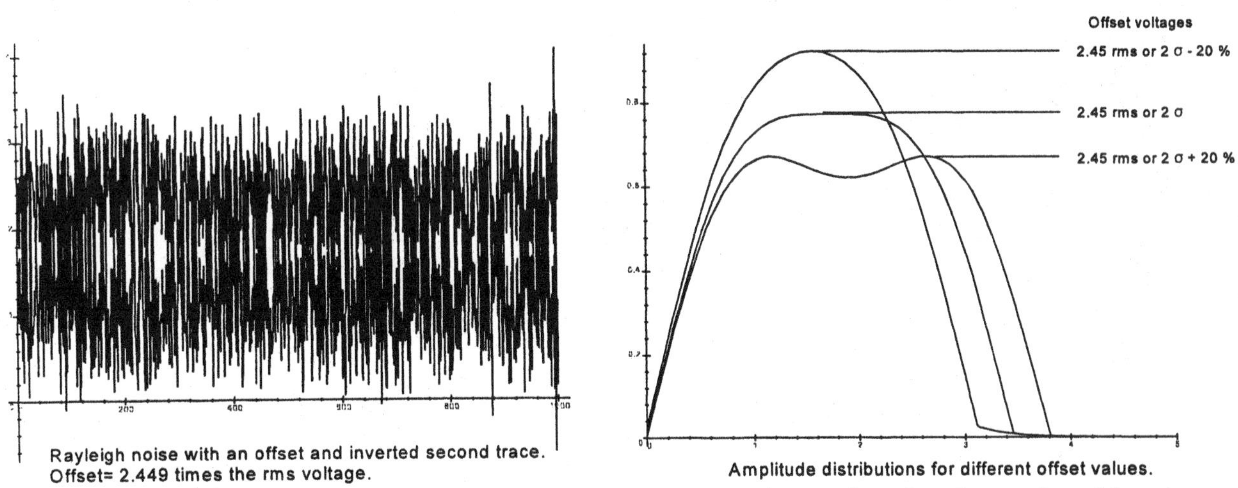

Figure 9.30 Two Rayleigh noise traces offset by 2.449 times the rms voltage at the point of separation of the two traces and the probability distributions of the sums with offsets of 2.449 the rms value and errors of ±10%.

Then the offset is adjusted until the two waveforms just start to separate and the rms voltage is

$$\textit{Rayleigh rms voltage} = \frac{\textit{Offset voltage}}{\sqrt{6}} = \frac{\textit{Offset voltage}}{2.449} \tag{9.10}$$

9.6 FIGURES AFFECTING RADAR PERFORMANCE

The following value is used in the range, accuracy, and stability budgets in Chapter 14.

9.6.1 Range budget

Generally, no loss is associated with linear detection at signal-to-noise ratios which are high enough to detect an echo. There are losses associated with logarithmic detection, but this type of detector is primarily used to observe the radar environment, especially jamming.

REFERENCES

1. *IEEE Standard Radar Definitions, IEEE Standard 686-1990,* New York: The Institute of Electrical and Electronic Engineers, 1993, Synchronous detector.
2. Bracewell, R. N., *The Fourier Transform and Its Applications*, 2nd ed., New York: McGraw-Hill, 1978.

Chapter 10

Analogue-to-digital conversion

The analogue-to-digital converter changes analogue (usually bipolar) video signals into (parallel) digital values, which can be processed by hardware similar to that in computers. Radar hardware normally provides vector processing, so two converters are necessary. Two types of conversion are possible:

- The two Cartesian components from the synchronously detected intermediate frequency signal, I and Q videos;
- Amplitude and phase from a linear detector and phase detector, respectively.

Generally signal processors use data streams formed from the Cartesian components. The matched filter determines the bandwidth of the videos from the detectors. The range clock tells the analogue-to-digital converter to take samples at fixed intervals during the listening or processing time.

10.1 PRINCIPLE

The converter changes the input voltage into a stream of measurements of the voltage expressed as binary integers. Each binary digit represents a doubling of the voltage, that is, a 6 dB power change. Analogue-to-digital converters are available as commercial integrated circuits.

The videos from the synchronous detectors are bipolar videos so that the voltage range is from a negative maximum to a positive maximum. One binary bit is used for the sign. Common analogue-to-digital converters used in radar convert 8, 12, or 16 bits. When one bit is used for the sign, the dynamic ranges are shown in Table 10.1.

Table 10.1
Conversion ranges

Number of bits	Conversion range with sign	
	Voltage	Power
8 bits	± 127	42 dB
12 bits	± 2 048	66 dB
16 bits	± 32 768	90 dB

Amplitude and phase videos are often offset to avoid negative values and the value zero, which allows an extra bit for resolution. The dynamic range available for signal processing depends on how many bits are used to convert noise alone. The voltage samples are taken at specific ranges from the master trigger. Normally, two samples are taken for every processed pulse width to reduce straddling losses irrespective of whether pulse compression precedes or follows analogue-to-digital conversion.

Converting the signals to digital form is accompanied by errors and losses which are added to the errors that occur in the second detector and the filters and amplifiers that follow for correction.

10.2 DYNAMIC RANGE

The analogue-to-digital converter receives bipolar video from the synchronous detector. Each bit in a binary converter represents a doubling of the voltage, or 6.01 dB. The returning echoes from a moving object give a sampled alternating voltage. The positive and negative maxima must fit in the dynamic range of the converter. The ratio of peak-to-peak to root mean square voltage is $2\sqrt{2}$, or 9.03 dB.

The scale of the converter is adjustable by altering the input amplifier gain. Thus, the number of least significant bits (LSB) representing the noise level is adjustable and reduces the dynamic range by 20 $\log_{10}$(*rms noise voltage/LSB voltage*). The ranges are shown in Figure 10.1.

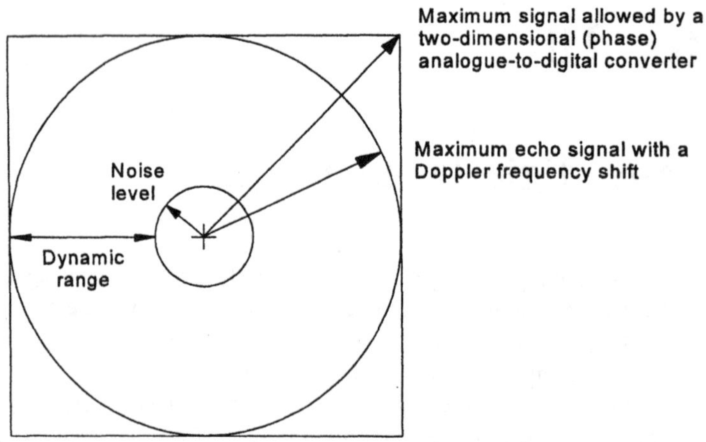

Figure 10.1 The dynamic range covered by a two-phase analogue-to-digital converter.

The dynamic range for echo signals with Doppler frequency shift is given by [1, p. 3-40]

$$\textit{Dynamic range} \text{ (Doppler frequency shifted echoes)} = 6.01N - 9.03 - 20 \log_{10}\frac{\sigma}{LSB} \quad dB \tag{10.1}$$

where N is the number of bits in the analogue-to-digital converter;
σ/LSB is the noise level in units of one LSB.

Clutter signals are fixed (direct voltage) vectors, where the rms value is half the peak to peak range extending into the corners of Figure 10.1, which gives a clutter dynamic range of

$$\textit{Dynamic range (clutter)} = 6.02(N - 1) - 20 \log_{10}\frac{\sigma}{LSB} \quad dB \tag{10.2}$$

A 3 dB greater dynamic range is available for diagonal vectors and because the echo signals may arrive with any phase, this extra range is of no use.

Pulse compression and coherent integration stages, which follow analogue-to-digital conversion, expand the dynamic range by the pulse compression ratio and the number of pulses that are integrated. For example, if the analogue-to-digital converter has an output 8 bits wide, and the pulse compression ratio is 256, the dynamic range becomes 256 × 256, which requires a data word 16 bits wide in subsequent stages.

10.3 NATURE AND TREATMENT OF ERRORS

Errors are measured by injecting a test signal at intermediate frequency and analyzing the digital data stream. Measurement of the test signals results in correction factors that can be applied before signal processing starts.

10.3.1 Types of errors

The principal errors in a single converter are:

- Symmetry errors;
- Step or quantization errors;
- Nonlinearity errors;
- Saturation errors.

10.3.1.1 Symmetry errors

The treatment of symmetry errors is similar to the symmetrical component theory in polyphase power systems. Instead of the common three phases, here there are two: I or in-phase and Q or quadrature phase. The three symmetrical components are shown in Table 10.2.

Table 10.2
Positive, negative, and zero phase sequence components

Component	Signal description
The positive sequence component	This is the useful echo signal.
The negative sequence component	This is a by-product caused by distortion that gives a component at the negative of the radial velocity of the original echo signal. It is caused by unequal gain in the I and Q channels and the phase of the component by errors in the phase (orthogonality error) of the I and Q reference waveforms to give an additional image, or "ghost" signal.
The zero sequence component	This is a direct voltage offset in the converter that gives an extra zero velocity (clutter) component.

These converters are constructed to give minimum negative sequence and direct voltage components. For greater fidelity or accuracy, these error components must be measured and corrected later.

10.3.1.2 Step or quantization errors

The voltage steps for setting each bit are not exactly equal and the ratio is not precisely 2:1, which leads to small errors that are often neglected because they are consistent from sweep to sweep for clutter.

Additionally, errors are caused by the difference between the input voltage and the representation at the output. The difference, which is shown in Figure 10.2, is triangular in form with an amplitude of one least significant bit peak to peak. The root mean square value is $LSB/\sqrt{12}$ V, or 10.79 dB below the power of a single least significant bit (LSB) [2, p. 76]. Additional noise is present when the steps are unequal or missing.

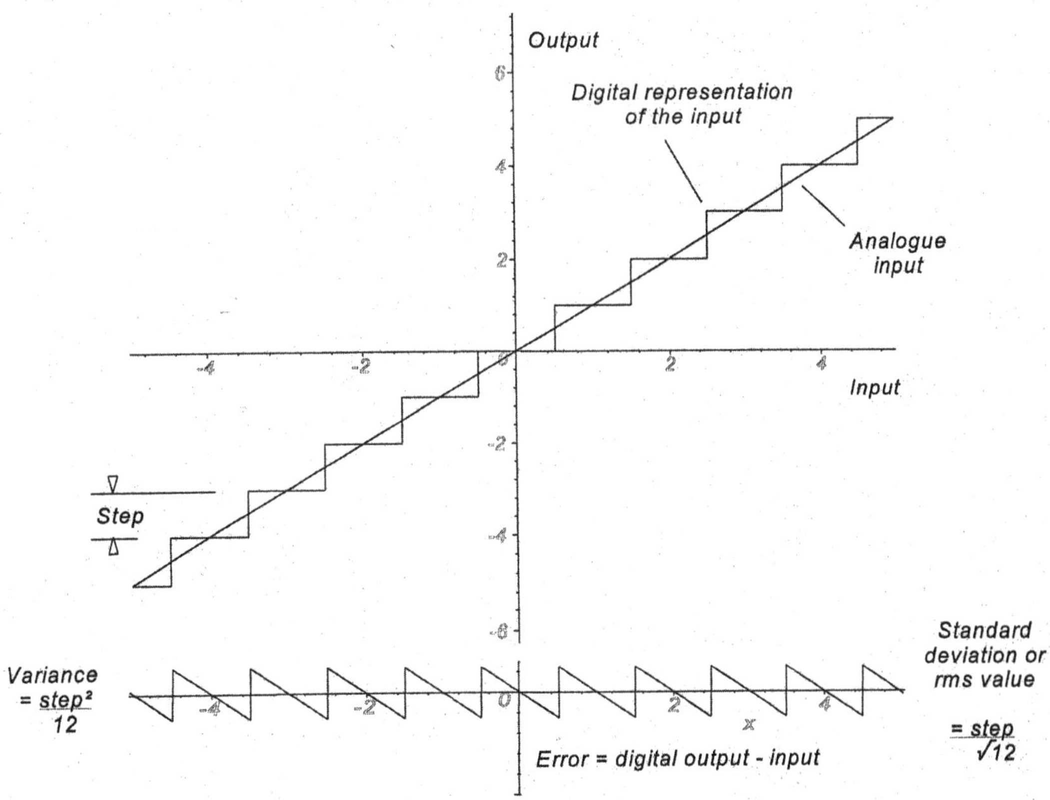

Figure 10.2 The conversion characteristic and errors.

10.3.1.3 Nonlinearity errors

The detectors and amplifiers are not always linear and the gain in the (two) channels is not always the same.

10.3.1.4 Saturation errors

Hitting the floor or the ceiling of the conversion range is the equivalent of limiting, producing harmonics so that the signal processing becomes nonlinear and clutter discrimination is reduced. The overflow bit is set on the analogue-to-digital converter. Where dynamic sensitivity time control is used, this signal is used to increase the attenuation of the range cell during the next sweep or look. The attenuation has been set correctly when overflow no longer occurs.

The saturation indication is used to set a flag on the echo detection decision to help in the second censoring, extraction, and tracking processes. A limiter in the following intermediate frequency stages limits symmetrically and avoids the more serious effects of asymmetrical saturation [3].

10.3.2 Measurement of errors

Errors are measured by injecting a test signal into the intermediate frequency input for the second detector. Two forms of signal are commonly used:

- Intermediate frequency signal with constant power and offset slightly in frequency. Demodulation gives a rotating vector, which should rotate in a circle.
- A noise signal, which, over time will contain samples at all phase angles which are able to cross a voltage threshold. The number of threshold crossings in each quadrant are counted.

A general scheme with in phase (I) and quadrature (Q) signals and their loci around an ellipse is shown in Figure 10.3. Note that the axes are skewed by the orthogonality error.

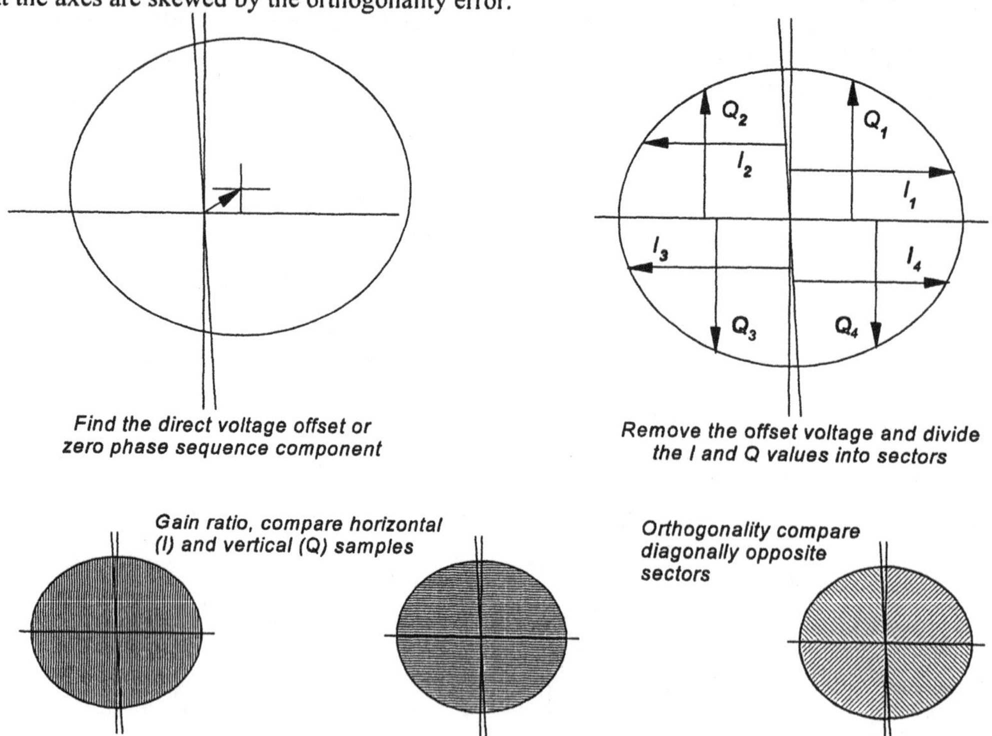

Figure 10.3 In-phase (I) and quadrature (Q) values from a test signal with gain and orthogonality errors.

Since the means of the in-phase (I) and quadrature (Q) values from the test signal samples should be zero, the averages of the I and Q values give the direct voltage offsets. These are the Cartesian components of the zero-phase sequence component.

$$\begin{aligned} \textit{In-phase offset voltage} &= \frac{\Sigma I}{n} \\ \textit{Quadrature offset voltage} &= \frac{\Sigma Q}{n} \end{aligned} \tag{10.3}$$

Once the direct voltage offset in each phase has been found, they can be subtracted from the samples to give the corrected I′ and Q′ values. For a large sample, the sum of the absolute I values should match the sum of the absolute Q values. For an ellipse with a greater horizontal diameter, the geometrical eccentricity, ϵ, is given by

$$\epsilon^2 = \frac{(\Sigma|I'|)^2 - (\Sigma|Q'|)^2}{(\Sigma|I'|)^2} \tag{10.4}$$

The ellipticity is given by

$$\textit{Ellipticity} = \frac{\Sigma|Q'|}{\Sigma|I'|} \tag{10.5}$$

The skewing of the axes gives rise to an unbalance in the area of the diagonally opposed quadrants. For a large number of samples that are uniformly distributed, the error angle, ϕ, is given by the difference in the number of samples in the diagonally opposed quadrants.

The gain or ellipticity errors and the orthogonality or axis skewing errors are not correlated, and the total voltage error is the square root of the sum of the squares of the voltage ratios [4, p. 5-32].

$$\textit{Image frequency power ratio} = \frac{\phi^2 + \epsilon^2}{4} = 10\log_{10}\left(\frac{\phi^2 + \epsilon^2}{4}\right) \text{ dB} \tag{10.6}$$

A plot of (10.6) is shown in Figure 10.4.

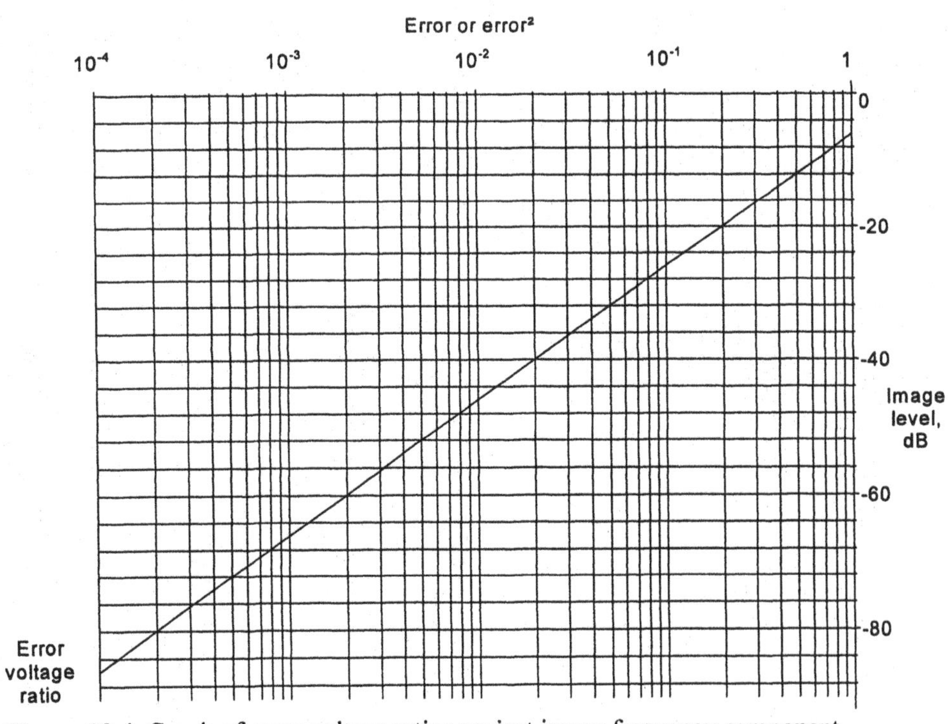

Figure 10.4 Graph of error voltage ratios against image frequency component.

10.3.3 Correction of errors

Commercially available components give reasonably good results without correction. Approximations for small angles may be used, namely,

$$\begin{aligned} \sin\phi &\text{ tends } to\ \phi \\ \cos\phi &\text{ tends } to\ 1 \\ \epsilon\sin\phi &\text{ tends } to\ 0 \end{aligned} \qquad (10.7)$$

Practical considerations limit the orthogonality of uncorrected I and Q videos [4, p. 5-38]. The corrected videos and the actual videos may be represented by:

$$\begin{aligned} \textit{Corrected I video, } S_I &= A\cos\alpha \\ \textit{Corrected Q video, } S_Q &= A\sin\alpha \end{aligned} \qquad (10.8)$$

and the actual video values are

$$\begin{aligned} S'_I &= A\cos\alpha \\ S'_Q &= A(1+\epsilon)\sin(\alpha+\phi) \\ &= S_Q\cos\phi + S_I\sin\phi + \epsilon S_Q\cos\phi + \epsilon S_I\sin\phi \end{aligned} \qquad (10.9)$$

If the phase error, ϕ, is small, then these may be rearranged to give the corrected video values

$$\begin{aligned} S_I &= S'_I \\ S_Q &= S'_Q - \epsilon S'_Q - \phi S'_I \end{aligned} \qquad (10.10)$$

10.3.3.1 Magnetron radars

Starting and stopping the coherent oscillator itself leads to instability. Where digital signal processing is used, it is better to measure the phase of the locking pulse and use this measured phase angle to correct for the phase difference with the coherent oscillator. One such arrangement with a separate locking pulse mixer is shown in Figure 10.5. In addition to the correction for analogue-to-digital conversion errors, which uses approximations for small angles, the correction for the transmitter phase, ψ, must be undertaken at angles up to a full rotation. There are no small angles, namely

$$\begin{aligned} x &= I'\cos\psi - Q'\sin\psi \\ y &= I'\sin\psi + Q'\cos\psi \end{aligned} \qquad (10.11)$$

The values $\sin\psi$ and $\cos\psi$ are given by the conversion of the locking pulse by the analogue-to-digital converter. Should polar detection be used, the correction is a simple subtraction of phase angles.

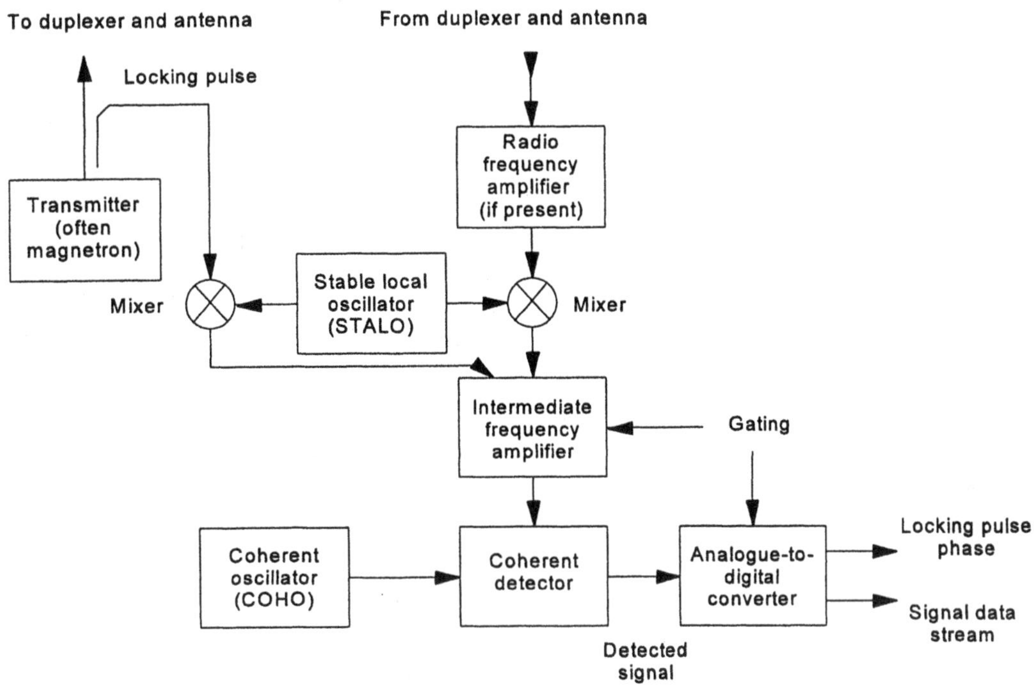

Figure 10.5 Phase compensation for a magnetron radar using digital signal processing.

10.3.4 Analogue-to-digital conversion using intermediate frequency signals

Synchronous detection followed by two separate analogue-to-digital converters allows differential errors in the two digital data streams, leading to negative phase sequence components. The gains and linearity of the two paths may not be the same and the two paths may not be orthogonal. The paths may be combined by initial sampling and digital conversion of the signal on a (notional) intermediate frequency carrier and selecting the two phases by further choosing the samples later as in [5].

The process depends on fast analogue-to-digital converters and using a low intermediate frequency. If the bandwidth of the echo signals is B Hz, the echo signals are converted down to a center frequency of B Hz and exist in the band $B/2$ to $3B/2$ around the notional carrier at B. Sampling at a frequency of $4B$ requires fast analogue-to-digital converters especially when pulse compression is used as with many current radars that use pulse compression.

Amplitude and phase modulated echoes on a notional sine wave carrier may be represented, in exponential form by

$$\cos(2\pi ft+\phi) = \frac{e^{j(2\pi ft+\phi)}}{2} + \frac{e^{-j(2\pi ft+\phi)}}{2} \tag{10.12}$$

The equation represents two frequency components at $\pm f$ Hz components representing two contra-rotating helices in time that add together to form a cosine wave (see Figure 2.16). Figure 10.6 shows the progress of the signals starting with the intermediate frequency signal (f_{if}) and its spectrum. Echoes from stationary objects are exactly at the intermediate frequency, while moving echoes deviate by the Doppler frequency and this is illustrated at the top of Figure 10.6. If the bandwidth after the matching filter is B Hz, then the intermediate frequency is mixed down to a center frequency of B Hz having sidebands existing between $B/2$ and $3B/2$. The sum frequencies, out of band noise, and direct voltage components are removed by a bandpass filter.

The analogue to digital converter converts at a rate of $4B$ samples per second and the data words representing the intermediate frequency signals are passed to a filter that removes the "negative frequency" sideband. Reference [5] shows

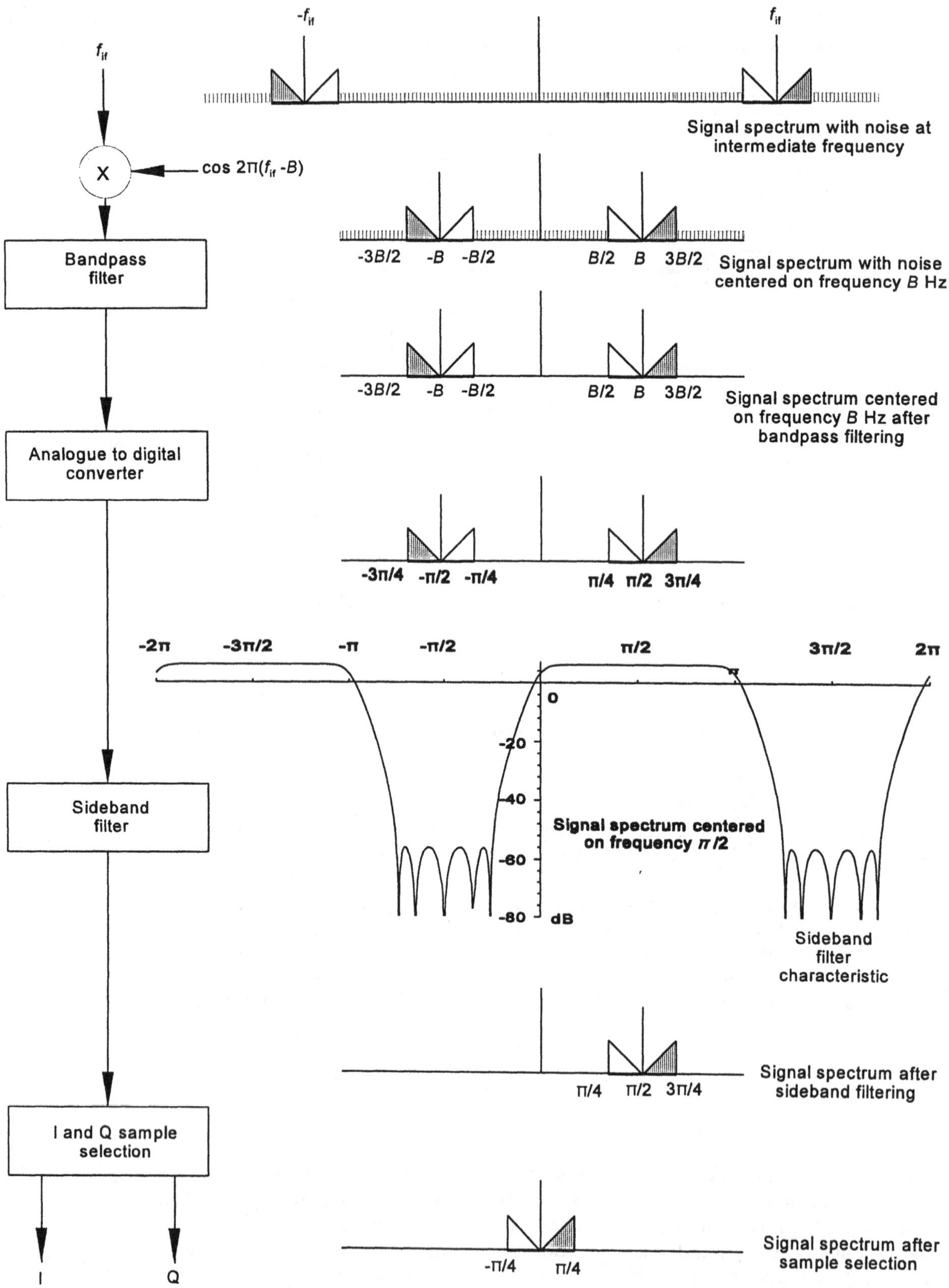

Figure 10.6 The filtering process for the intermediate frequency analogue-to-digital converter.

a digital filter that is designed to pass the frequencies from $B/2$ to $3B/2$, has a nominal voltage gain of 2, and has a stop-band response of -60 dB, that is 2/sinϵ, the error ϵ is less than 2 mrad or 0.1145 degrees. The filter is illustrated in Figure 10.7 [5].

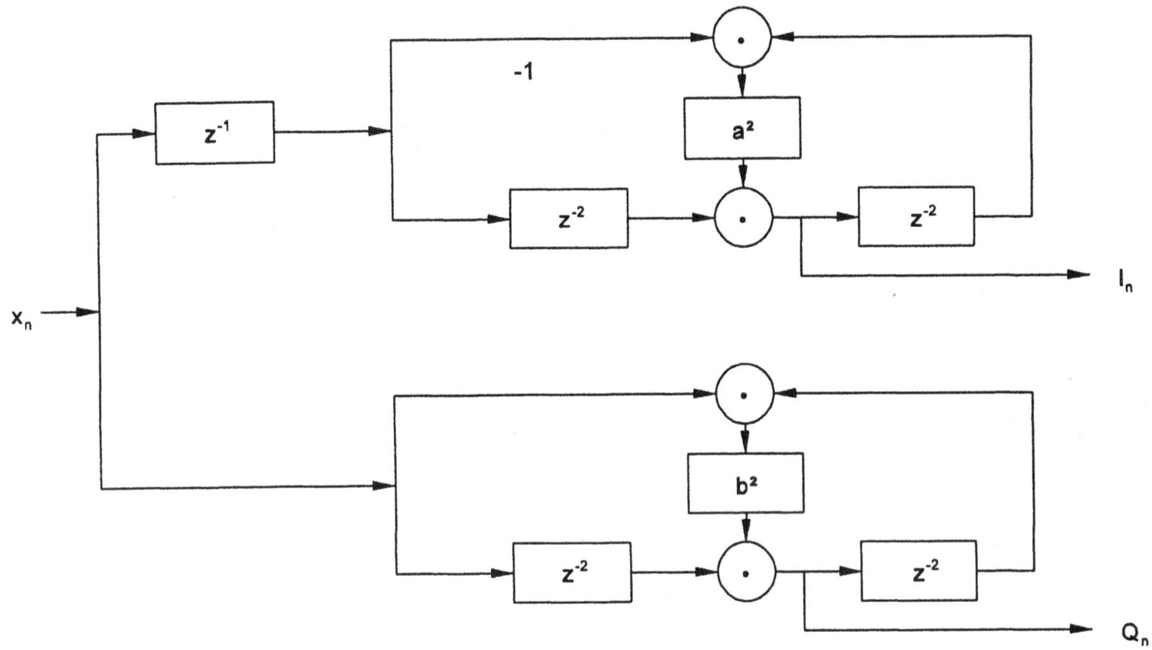

Figure 10.7 Filter to remove negative phase sequence components. [Source: Rader, C.M., "A Simple Method for Sampling In-Phase and Quadrature Components", *IEEE Transactions on Aerospace and Electronic Systems*, Vol. 20, No. 6, 1984.]

Its characteristic is given in [5] by

$$H(z) = H_1(z) + jH_2(z) \tag{10.13}$$

where

$$H_2(z) = -\frac{z^{-2} - b^2}{1 - b^2 z^{-2}} \qquad b^2 = 0.1380250 \tag{10.14}$$

and

$$H_1(z) = z^{-1}\frac{z^{-2} - a^2}{1 - a^2 z^{-2}} \quad \text{or} \quad = -\frac{H_2(z)}{z} \qquad a^2 = 0.5846832 \tag{10.15}$$

The filter is based on phase-splitting filters [6, pp. 90-92] and the frequency characteristic is shown in Figure 10.6 and in vector form in Figure 10.8.

A single spectral line gives a rotating modulation vector (see Figure 16.3) and the sampling at a frequency of $4B$ of a notional carrier is shown in Figure 10.6. More generally the echo signals are not discrete lines but are a combination signals with positive and negative phase sequence components.

Taking the odd and even samples at $4B$ selects the in-phase (I) and quadrature (Q) samples and taking every second I or Q sample effectively translates the data stream representing the signal to base-band (Figure 10.9). This data stream is passed to the signal processor in the next chapter.

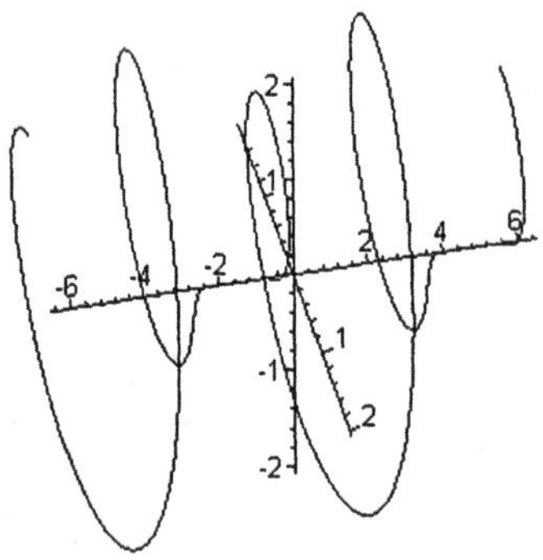

Figure 10.8 The vector form of the sideband selection filter characteristic.

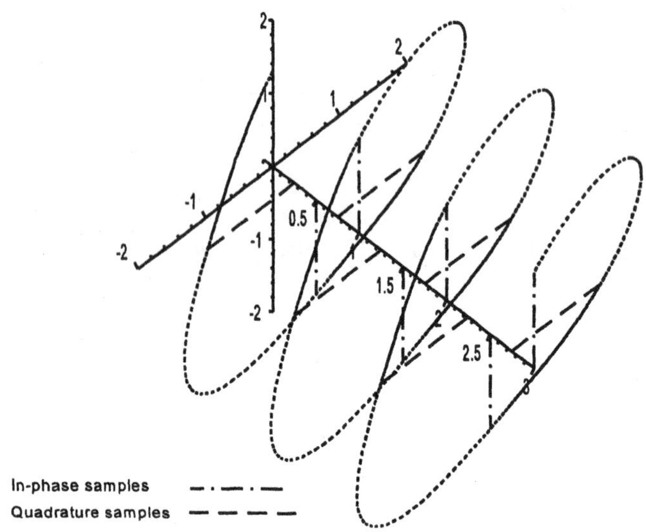

Figure 10.9 The selection of the in-phase (I) and quadrature (Q) samples.

Experiments have shown that the errors in conversion at intermediate frequency with a single analogue to digital converter are about one order smaller than with two converters [7].

10.4 FIGURES AFFECTING RADAR PERFORMANCE

The advantages of digital signal processing greatly outweigh the losses incurred in converting to digital data stream form.

10.4.1 Range budget

The following losses are not correlated and are therefore summed:

- Straddling loss;
- Range jitter loss;
- Quantization noise.

10.4.1.1 Straddling loss

Strictly speaking, the conversion window should be centered on the echo to give a measure of its amplitude. In search radar, the echoes may arrive at any time and may be divided between two fixed time windows or range gates. In that case, only part of the amplitude is converted, which represents a loss. The loss depends on how much the echo pulses are stretched in the matched filter and is greater for greater probabilities of detection. The solution is to increase the number of range strobes or overlapping range gates used for analogue-to-digital conversion so that there will be one centered near the signal peak.

The use of a wide sampling gate adds its own form of filtering [D. Barton, personal correspondence, 2000], which gives a weighting function corresponding to the convolution of the filter impulse response and the sampling window. If a wideband intermediate frequency amplifier precedes the analogue-to-digital conversion, the optimum window is one matched to the pulse width or the reciprocal of the signal bandwidth if pulse compression follows conversion. There will be a loss owing to a reduced signal-to-noise ratio applied to the threshold stage.

The straddling loss may be taken from [8, Fig. 5.4.11] repeated in Figure 10.10. Commonly, two samples are taken for every transmitter pulse width in more modern equipment with better processors to reduce straddling losses to approximately 0.2 dB [4, p. 5.37; 8, p. 269 for different values of $B\,\tau$].

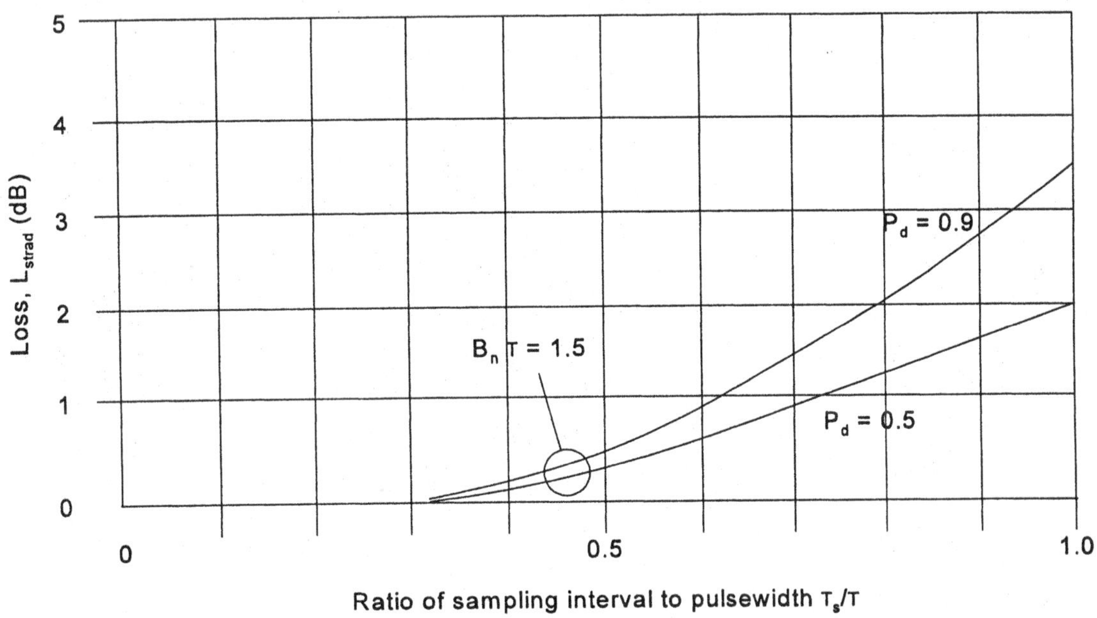

Figure 10.10 Range straddling loss for rectangular pulses. [Source: Barton D. K., *Modern Radar System Analysis*, Norwood, Massachusetts: Artech House, 1988, Figure 5.4.11.]

10.4.1.2 Range jitter loss

Range jitter loss occurs when the transmitter pulse does not occur exactly in the same position in the zeroth range gate. Echoes from consecutive transmitter pulses fall in neighboring range gates giving a similar loss to straddling loss. In most radars, this is eliminated by synchronizing the transmitter (pre-)trigger with a particular range gate. When coherent processing is used, this limits the improvement factor, so that range jitter must be held to levels much smaller that which could cause a loss.

10.4.1.3 Quantization noise loss

Quantization noise is additional noise caused by transforming a linear signal using a number of steps. The steps in this waveform cause the extra noise (power = step/12), and therefore the signal-to-noise ratio must be raised by this value to obtain the required signal at the threshold point [D. Barton, personal correspondence, 2000]. If the noise level is represented by m quanta or 2^q bits then the quantization loss is given by

$$L_{quant} = 1 + \frac{1}{Quantization\ variance} = 10\ \log_{10}\left(1 + \frac{1}{12}\ \frac{1}{2^q}\right)\ \text{dB} \tag{10.16}$$

This is shown in Figure 10.11.

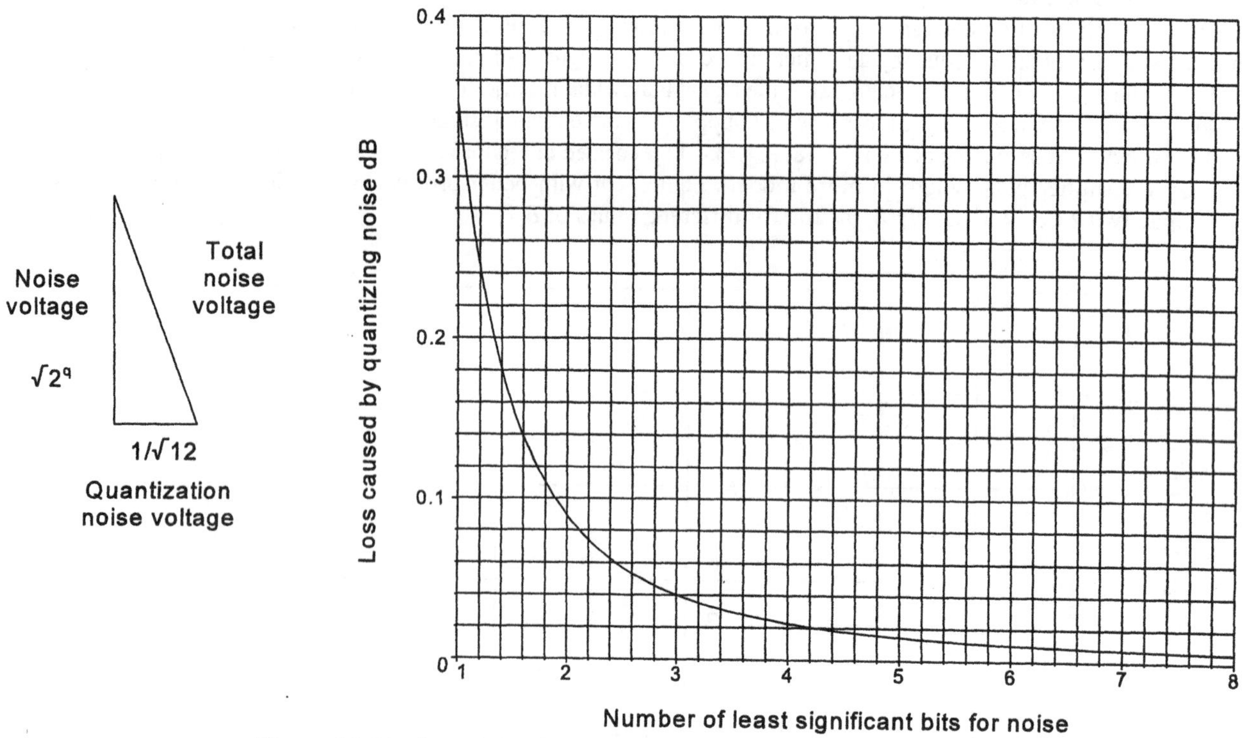

Figure 10.11 Quantizing loss against the number of bits for noise.

10.4.2 Accuracy and resolution budget

If there is no interpolation of signals in neighboring samples, quantizing in range adds $\frac{quantizing\ interval^2}{12}$ to the range variance. The standard deviation is $\frac{1}{\sqrt{12}}$ of the range quantizing interval. Normally the quantizing interval is small enough not to stretch pulses so that resolution is not affected.

10.4.3 Stability budget

The range jitter in the sampling pulses used for conversion limits the cancellation ratio, CR, just as transmitter pulse jitter. Quantization noise, $1/\sqrt{12}$ of a bit, is a constant voltage and adds to the signal. The proportion is thus inversely proportional to the signal vector. From [9, p. 415] the limit to the improvement factor is

$$CR_{quant} = 10\ \log_{10}\frac{1}{12\left(2^{q-1} - \frac{1}{2}\right)^2} \tag{10.17}$$

This is shown in Figure 10.12. It is thus important to choose a clutter sample that is near saturation when measuring signal processor performance. The maximum improvement factor limits for q bits caused by quantization is given in Table 10.3.

Table 10.3
Maximum improvement factor limits

Number of bits, q	Maximum CR_{quant}	Number of bits, q	Maximum CR_{quant}	Number of bits, q	Maximum CR_{quant}	Number of bits, q	Maximum CR_{quant}
1	4.771	5	34.598	9	58.94	13	83.038
2	14.314	6	40.758	10	64.969	14	89.059
3	21.673	7	46.847	11	70.994	15	95.08
4	28.293	8	52.902	12	77.016	16	101.101

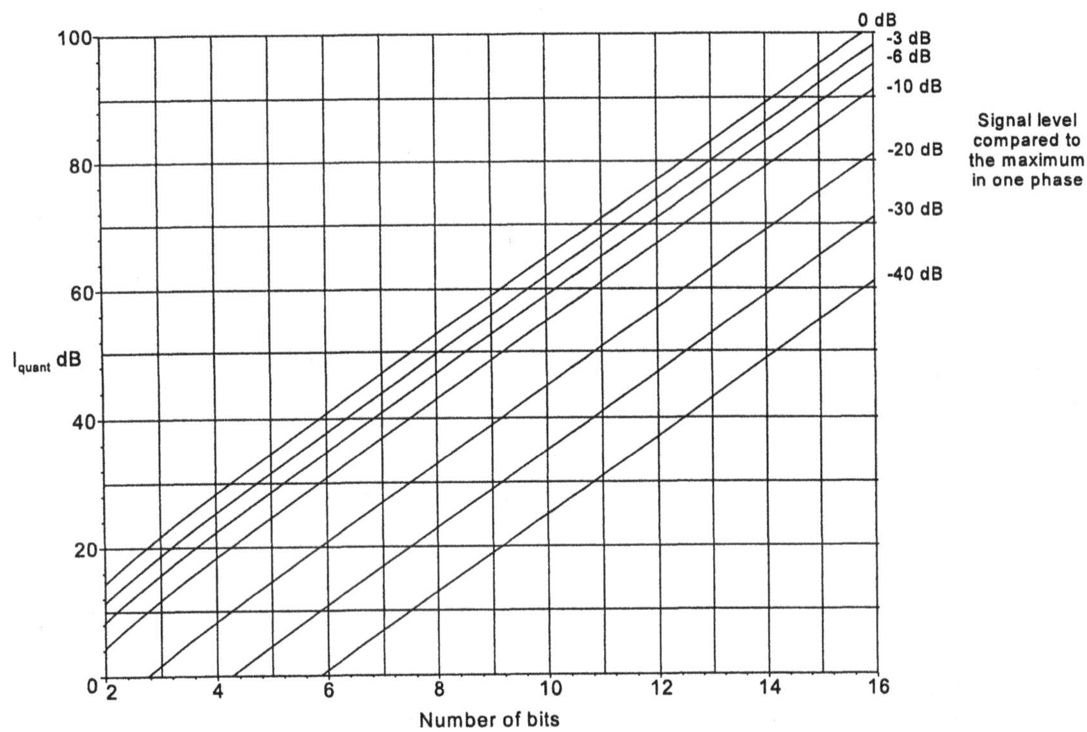

Figure 10.12 The limit to the improvement factor caused by analogue-to-digital conversion.

REFERENCES

1. Skolnik, M. I., *Radar Handbook*, 2nd ed., New York: McGraw-Hill, 1990.
2. Scheer, J. A., and J. L. Kurtz, *Coherent Radar Performance Estimation*, Norwood, Massachusetts: Artech House, 1993, p. 76.
3. Zeoli, G. W., "IF Versus Video Limiting for Two-Channel Coherent Signal Processors", *IEEE Transactions*, IT-17, No. 5, September 1971, pp. 579-586.
4. Ludloff, A., *Handbuch Radar- und Radarsignalverarbeitung*, Braunschweig, Germany: Vieweg, 1993.
5. Rader, C. M., "A Simple Method for Sampling In-Phase and Quadrature Components", *IEEE Transactions on Aerospace and Electronic Systems*, Vol. 20, No. 6, 1984.
6. Gold, B., and C. M. Rader, *Digital Processing of Signals*, Malabar, Florida: Krieger Publishing Co., 1983.
7. Skolnik, M. I., *Introduction to Radar Systems*, New York: McGraw-Hill, 2001
8. Barton, D. K., *Modern Radar System Analysis*, Norwood, Massachusetts: Artech House, 1988.
9. Nathanson, F. E., *Radar Design Principles*, 2nd ed., New York: McGraw-Hill, 1991.

Chapter 11

Signal Processing

Radar echoes are a mixture of wanted echoes (for example, aircraft) and clutter echoes (often from land, sea, weather, or birds), which are generally of no interest. Signal processing tries to filter the echoes for the user and match them to his display. Modern digital processing often causes the echoes to be displayed as synthetic shapes that show only the information for which it has been preprogrammed. In contrast, radar operators who use older equipment are used to identifying echoes by size, shape, and movement. The evaluation of these criteria takes time and greatly reduces the number of echoes that an operator may follow, but it may be critical in unusual or unforeseen situations.

This chapter describes the filtering of echo signals according to Doppler frequency and also the judging of the correct threshold when clutter is present. The statistical consideration of signals and noise with Swerling model echoes is described in Chapter 12.

The echo signals from the receiver are the sum of the following, as shown in Figure 11.1:

- Echoes from objects illuminated by the main antenna lobe (leg A out, leg A back);
- Echoes from objects illuminated by the antenna sidelobes and received through the sidelobes (leg C out, leg C back);
- Echoes from objects illuminated by reflections from the antenna mainlobe and received through a sidelobe (leg C out, leg B, leg A back) when the mainlobe points towards the town clutter;
- Echoes from objects illuminated by the antenna mainlobe and received by reflection through a sidelobe (leg A out, leg B, leg C back);
- Interference signals (in military language, jamming) entering the receiving system through the antenna main and sidelobes (X);
- Interference caused by inadequate screening of the receiving system (Y);
- Interference carried in by the main power system (Z).

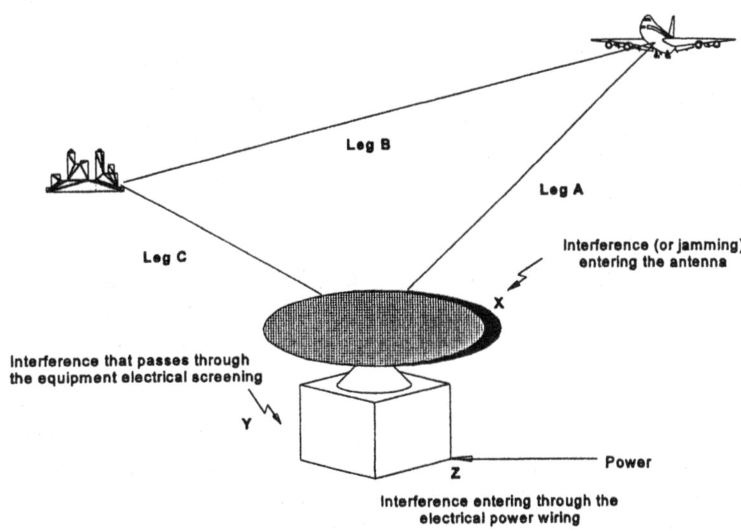

Figure 11.1 Sources of signals at the receiver output.

The several types of echo signals leaving the receiver may be resolvable in time and Doppler frequency, or they may be unresolvable. The object of signal processing is to separate the echo signals from the objects to be observed from all other signals, called clutter. The antenna is the first stage capable of selecting the returning echo signals by angle of arrival at the antenna which is why military radars have antennas with low sidelobes, almost regardless of expense in order to select the signals entering the main lobe only. The ability to reject sidelobe interference may be augmented by sidelobe cancellation (SLC) which is covered in Section 5.12 or by sidelobe blanking (SLB) in Section 11.1.8.

After the echo signals have passed through the receiver and the matched filter, it is normal to process the video before the decision process. The decision process may be a human operator watching a cathode ray tube, an electrical threshold followed by an alarm (buzzer), or a data processor.

Signal processing changes the balance of the signals that are displayed or otherwise used for detection so as to favor the echo signals of interest. The amount of signal processing hardware depends on the types of radars, which are principally:

- Radars that move their antennas to concentrate on one echo of interest (for example, fire control radars which track one target) that have processing and memory for one echo. Phased array radars are now preferred for fire control radars and can have processing and memory capacity for tens of echoes.
- Radars that are able to watch the space around them, often up to a hemisphere. The antennas may rotate to scan the space around them or they may be phased arrays where the beam is steered electrically.

The first device that was able to accept and display the radar picture from one complete scan was the phosphor used in cathode ray tubes for plan position indicators (PPI) and B-scopes. The display allowed an operator to see the complete picture and allowed him to distinguish between aircraft echoes and clutter and to track aircraft echoes between clutter patches. The brightness on a cathode ray tube has a dynamic range of about 10 dB so that the logarithm of the signal amplitude has to be used to display large dynamic ranges. Cathode ray tubes were not generally successfully used to store and replay data held as an electrostatic image on the inside of a cathode ray storage device. The video signals from sideways-looking reconnaissance radars in aircraft (later synthetic array radars) were recorded on photographic film.

The emergence of acoustic delay lines to store one sweep, 4 ms in the longer range radars, led to the development of moving target indicators (MTI). Mercury delay lines in the 1950s were used at first to delay video signals and then the more stable quartz delay lines made vector moving target indicators in the 1960s. Quartz delay lines were also used in video and coherent integrators such as the velocity indicating coherent integrator (VICI), see [1, p. 17-56].

In the past the more elaborate ways of processing, such as pulse Doppler processing, were reserved for single echo radars. Later large semi-conductor memories and synchronous digital logic have enabled large quantities of hardware to be so organized that the hardware works together and does not drift. The stability, that has been achieved, has allowed the development of the moving target detector (MTD) by the Lincoln Laboratory, which applied pulse Doppler radar methods to air traffic control radars.

In principle, there is no difference between the old analogue signal processing and the current digital signal processing. Digital signal processing has the following advantages:

- There is no degradation of signal-to-noise ratio in stages following the analogue-to-digital converter. There is a degradation caused by arithmetic rounding errors.
- Synchronous logic (retiming after each stage) eliminates differential timing and jitter problems which allows the use of much more complex parallel processing. The central clock acts as the conductor of the signal processing orchestra.
- Digital multipliers are used to eliminate gain drift problems in analogue systems.
- Digital memory is fast, cheap, and extremely flexible. There is no longer a need to process in delayed real time in step with the pulse repetition frequency of the radar. Large amounts of cheap memory make detailed clutter maps possible.
- Wider data words allow a wide dynamic range, commonly 12 bits or 66 dB plus sign.

Drifts in gain, temperature, and supply voltage stability made the maintenance of the more complex analogue systems a critical factor. The drifts are eliminated in synchronous digital systems. The wider dynamic range more or less eliminates limiting (if sensitivity time control (STC) is used before the first active stage) which allows the linear processing of signals to occur in almost any order.

Signal processing may be divided into at least three types, as shown in Figure 11.2:

- Processing during one sweep:
  - Altering the form of the intermediate frequency signal before the detector, such as reducing the dynamic range and limiting (see Chapter 7, Receivers);
  - Altering the form of the video after the detector. Examples are limiting, logarithmic video, differentiating the video, and constant false alarm rate thresholding.

- Processing during many sweeps:
  - Integrating the video.
  - Differencing the video between sweeps: coherent and noncoherent moving target indication.
  - Differencing the intermediate frequency signal between sweeps: vector moving target indication.
  - Differencing the polyphase video signal from synchronous detectors: coherent moving target indication.
  - Integration of the intermediate frequency signals or polyphase vector video signals to sort and filter the echoes of interest: pulse Doppler or moving target detection processing.
- Processing over many scans:
  - Storing the clutter picture over many scans in order to look for changes that may represent echoes of interest: area moving target indication (AMTI).
  - Building up maps with difficult detection areas for special processing, rain, blanking areas, and so on.

The processing of signals may take place as the echo signals pass through to the displays or in batches, as is mostly the case with digital pulse Doppler processors. The method used depends on the hardware and both are equivalent.

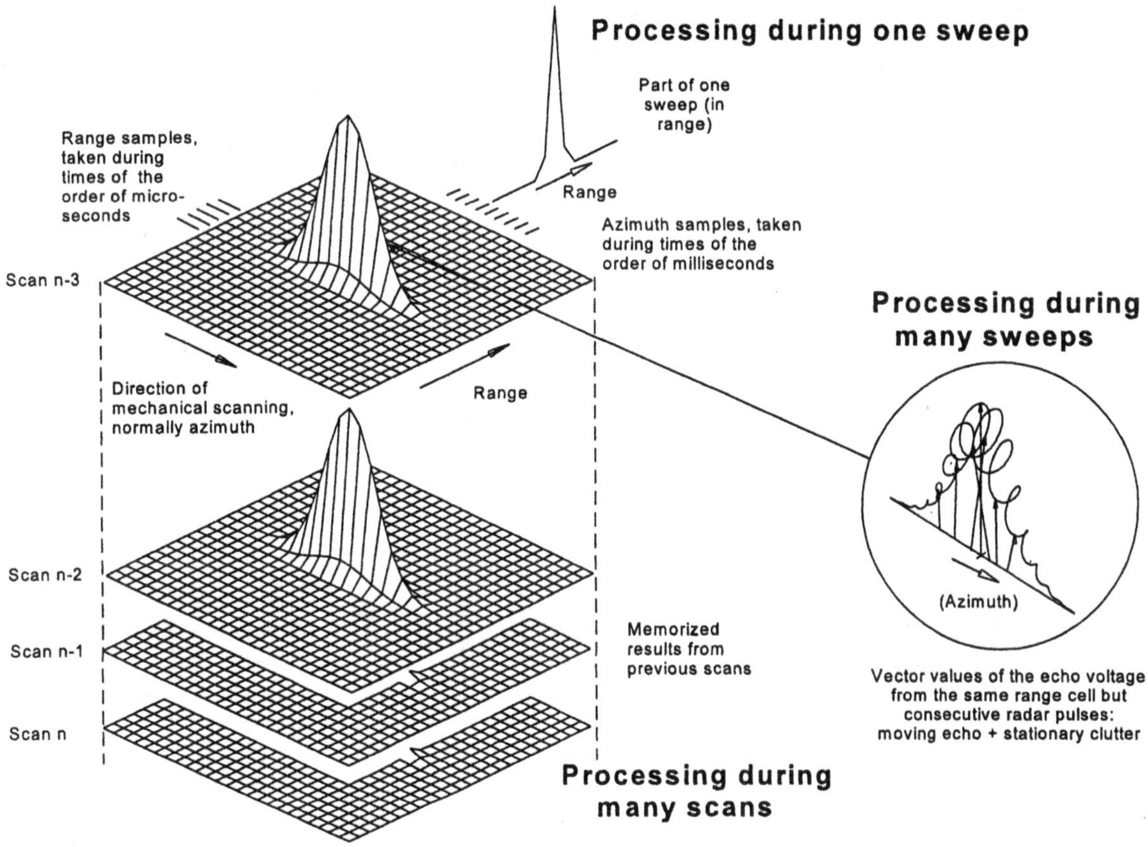

Figure 11.2 The concepts of signal processing during one sweep, many sweeps, and many scans.

The result of signal processing is hopefully a clean "picture" that shows the wanted echo signals only, which are matched to the user's needs. The descriptions that follow are more or less in historical order. First the video signal was shaped or filtered, followed by the processing of the intermediate frequency signal, until fully coherent digital processing became possible.

11.1 ALTERING THE FORM OF THE VIDEO DURING ONE SWEEP

Early radar literature is full of "fixes" to improve the display of wanted echoes for operators. Some of these forms of processing are described in this section. The changes are often permanent, and any signals lost may not be recovered later.

11.1.1 Limiting

Limiting is the oldest of all processes. Commonly the video signals used for A-scopes have a peak signal-to-noise ratio of 10:1 or 20 dB, and plan position indicator (PPI) displays have a signal-to-noise ratio of 3:1, in voltage, or about 10 dB. To avoid overdriving the cathode ray tube and producing too much glare for an operator to look at, larger signals are limited, to reduce the dynamic range to 10 dB, which gives a display less tiring to the eyes but maintains full sensitivity at longer ranges. The amplitude information above the limiting level is lost.

Intermediate frequency limiting followed by a matching filter, commonly called the Dicke-fix receiver, is described in Chapter 8.

11.1.2 Differentiation of the video: fast (or short) time constant

Video limiting results in large, solid clutter patches. One method to break up the patches is to pass the video through a fast time constant (FTC) or short time constant (also abbreviated to STC) resistor capacitor circuit (see Figure 11.4(b)) before it is limited. The time constant is chosen to pass the short wanted echoes with no degradation and to attenuate the large blocks. This helps the video circuits and the cathode ray tube from becoming saturated. The gaps in the large clutter blocks allow the detection (by an operator) of the moving echoes between them, giving limited greater interclutter visibility. The information contained in the wide blocks is lost, as is information in the shadows of the large blocks.

11.1.3 Pulse length (or width) discriminator

The pulse length (or width) discriminator (PLD) is a more elaborate circuit to break up large clutter blocks and accentuate echoes from scatterers of similar length to the transmitted pulse length (m), as shown in Figure 11.3.

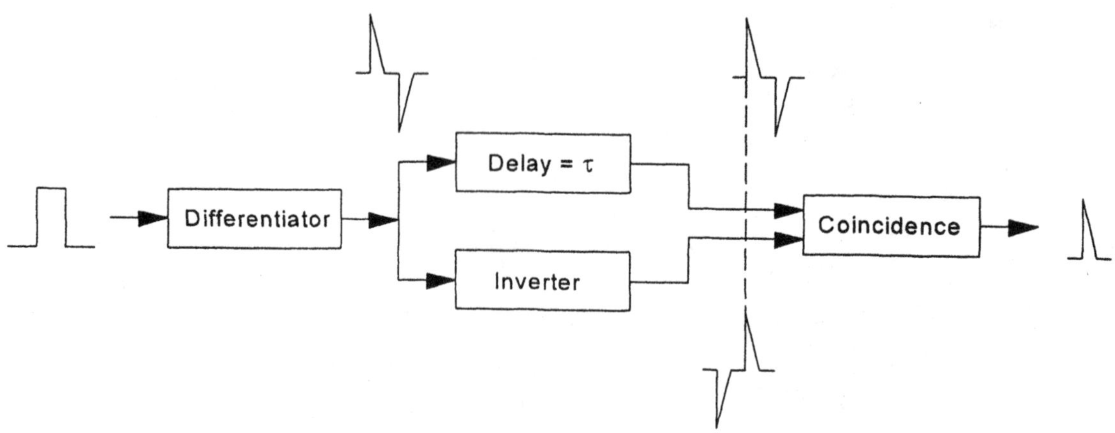

Figure 11.3 Pulse length (or width) discriminator.

11.1.4 Logarithmic video and log FTC

Detection with a logarithmic law is normally carried out using special intermediate frequency amplifiers. It allows the display of signals with a large dynamic range on a cathode ray tube (see Chapter 9, Detectors). Logarithmic detectors may be followed with a fast time constant (FTC) circuit to reduce clutter still further by breaking up the large clutter patches. Croney [2] describes theoretical and experimental work to reduce sea clutter and rain on radar displays. He was looking for an automatic gain control as an alternative to sensitivity time control to give a constant background or false alarm rate (CFAR) in the presence of sea and rain clutter. The block diagram is shown in Figure 11.4.

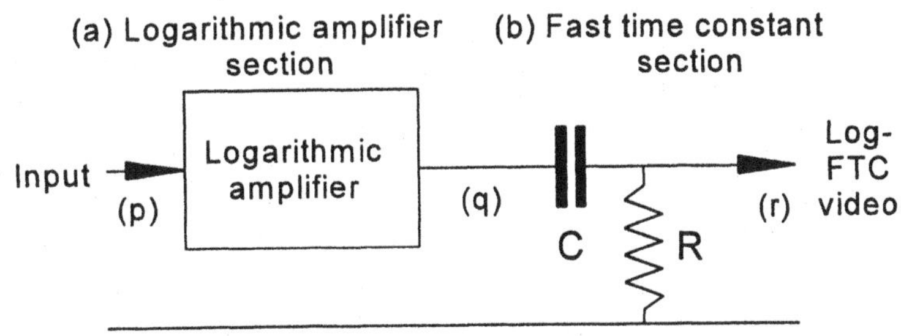

Figure 11.4 Logarithmic amplifier followed by a fast time constant stage.

The time constant, *RC* seconds, is approximately the radar pulse width or the expected range extent of wanted echoes. All clutter signals longer than the time constant will be attenuated.

The effects of log-FTC processing can be demonstrated by generating Rayleigh distributed random numbers with a standard deviation inversely proportional to the fourth power of the range, taking their logarithms (as in a logarithmic amplifier), and taking the differences between consecutive samples (the equivalent of fast time constant, FTC). Range samples beyond 900 are noise alone. This is shown in Figure 11.5.

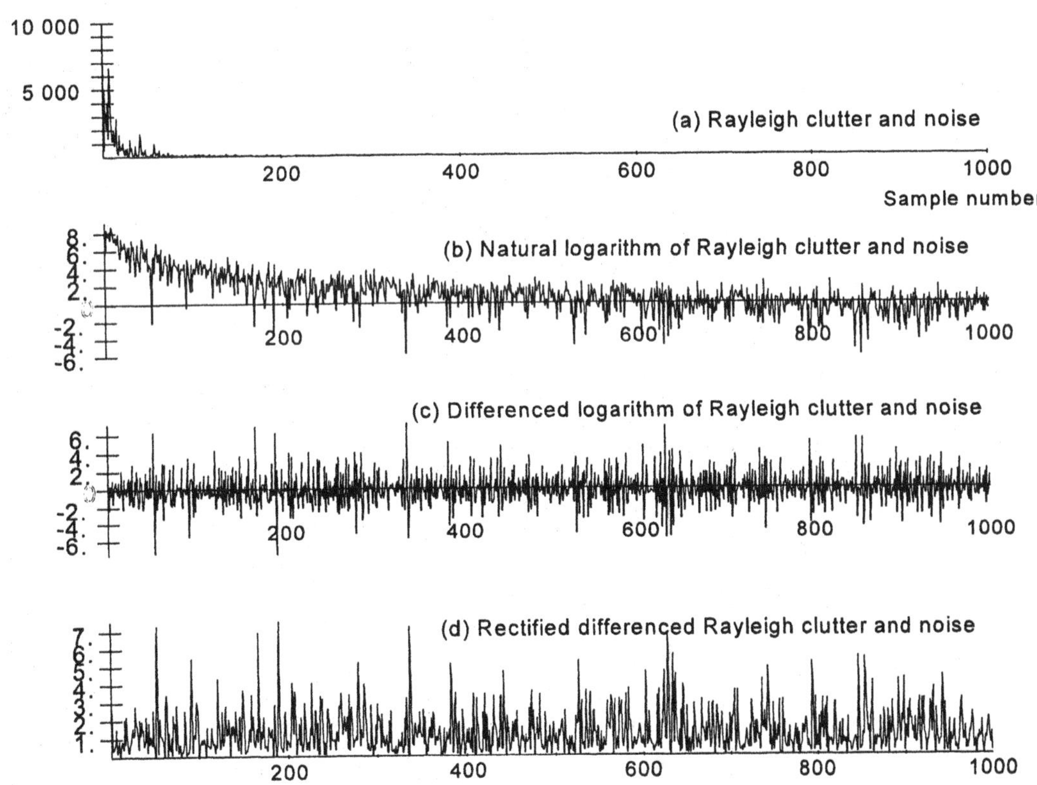

Figure 11.5 Demonstration of the effects of log-FTC video processing on Rayleigh clutter and noise.

The Rayleigh random numbers have too wide a dynamic range to be seen properly in Figure 11.5(a); note that the maximum ordinate value of the topmost trace in Figure 11.5 is 10 000. Taking logarithms enables them to be seen to the end of the sweep and the differencing (fast time constant) removes the direct voltage component. The logarithmic amplifier or detector law constant for the simulation is taken to be unity.

The voltage of Rayleigh distributed clutter has the probability distribution from (15.35) with σ as the scale factor

$$p(R,\ \sigma) = \frac{R}{\sigma^2} e^{-\frac{R^2}{2\sigma^2}}\ dR \qquad R > 0 \tag{11.1}$$

if u is the clutter to noise ratio R^2/σ^2, then in terms of power

$$p(u) = e^{-u}\ du \tag{11.2}$$

The signals pass through a logarithmic receiver with the law

$$y = a\ \ln(bx) = \frac{a}{2}(\ln\ b^2 x^2\ ) \tag{11.3}$$

changing the variable to u, the probability distribution becomes

$$y = \frac{a}{2}\ (\ln(b^2\sigma^2) + \ln u) \tag{11.4}$$

The first moment about the origin, mean, or expected value of the signal in Figure 11.5(b) is, for the integration see [3, p. 573, Eq. 4.331.1],

$$\begin{aligned} E(y) &= \frac{a}{2}\ [\ln(b^2\sigma^2) + \int_0^\infty \ln(u)\ p(u)\ du] \\ &= \frac{a}{2}\ [\ln(b^2\sigma^2) - \gamma] \end{aligned} \tag{11.5}$$

where γ is Euler's constant (0.5772156649).

The second moment about the origin is, for the integration see [3, p. 574, Eq. 4.335.1],

$$\begin{aligned} E(y^2\ ) &= \frac{a^2}{4}\ (\ln\ b^2\sigma^2 + \ln\ u)^2\ p(u)\ du \\ &= \frac{a^2}{4}\ ((\ln\ b^2\sigma^2)^2 + 2\ln b^2\sigma^2 \int_0^\infty \ln u\ p(u)\ du + \int_0^\infty (\ln u)^2\ p(u)) \\ &= \frac{a^2}{4}\ ((\ln\ b^2\sigma^2)^2 + 2\ln b^2\sigma^2\gamma + \frac{\pi^2}{6} + \gamma^2) \end{aligned} \tag{11.6}$$

The variance is the difference of the second and the first moments (where the mean is removed as in Figure 11.5(c))

$$E(y^2\ ) - (E(y))^2 = \frac{a^2}{4}\frac{\pi^2}{6} = \frac{a^2}{24} \tag{11.7}$$

The root mean square residue is $a/\sqrt{24}$ and is independent of the original clutter voltage, σ, and the logarithmic law, b (see Figure 11.5(c) and (d)). The probability distributions are shown in Figure 11.6.

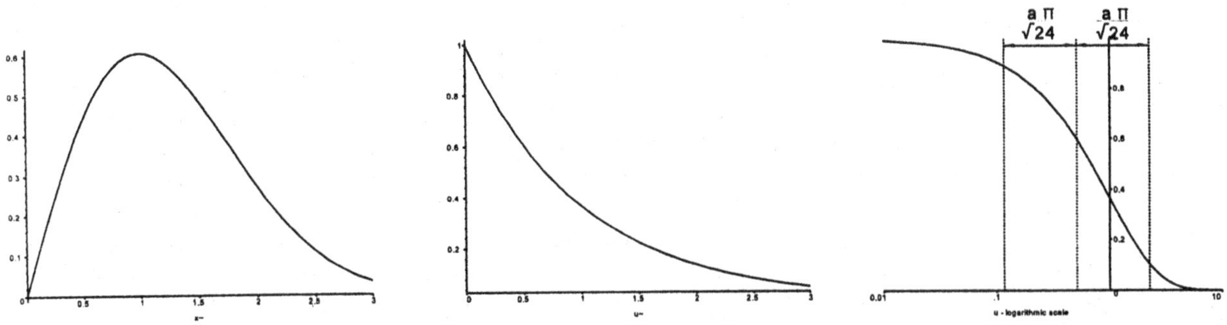

(a) Rayleigh distribution (σ = 1) (b) Rayleigh power or negative exponential distribution (σ = 1) (c) Negative exponential distribution (σ = 1) on a logarithmic scale.

Figure 11.6 The probability distributions used for Croney's log-FTC processing.

This analysis depends on the logarithmic conversion of the signals being perfect. Commonly, logarithmic detectors are linear in the noise region, and maybe in a small range above the noise level, and logarithmic above this so that they are called lin-log detectors or amplifiers. Croney recommends that the amplifier is logarithmic at least from 20 dB below the root mean square noise level.

Both effects limit the effectiveness of this type of signal processing. In addition, the correlation of sea and rain clutter in azimuth may be broken by using different frequencies for neighboring radar pulses, or frequency agility. This type of clutter reduction was proposed for marine radars searching for ships and coastlines so that the loss associated with logarithmic detection can be neglected. Another form of constant false alarm rate processing is discussed in Section 11.1.7.

As mentioned in Chapter 6 (Factors outside the radar), the Weibull clutter model may be used [3] which is an extension of the Rayleigh distribution with

$$p(x) = \frac{c}{b}\left(\frac{x}{b}\right)^{c-1} \exp\left(\left(\frac{x}{b}\right)^{c}\right) \qquad (11.8)$$

where b and c are scale and shape parameters, respectively.

If $c = 2$, the distribution becomes a Rayleigh distribution.

In contrast to log-FTC signal processing with Rayleigh distributed echoes, the value of the shape parameter, c, must be known to enable constant false alarm rate processing. Different types of clutter give different values of the shape parameter, c.

11.1.5 Bandwidth or stretching of the video signals for display

During the Second World War there were many experiments, referenced in [1, p. 2.12] to discover the optimum bandwidth of the signals for display. Modern signal processing and synthetic symbols for display make this small correction unnecessary.

Where pulse compression on search radars is used, the raw video signals may be too short in duration to excite the phosphor on the plan position displays. Pulse stretching may be necessary to achieve this.

11.1.6 Noise clipping

A fixed threshold is set just below the "peak" level of the noise. Only signals and noise exceeding the threshold are displayed or processed further. The use of variable thresholds is discussed in the next section.

11.1.7 Constant false alarm rate processing by cell averaging

Human radar operators accustom themselves very quickly to any changes in noise level, and after a number of scans the operator has memorized the clutter or clutter residue picture so that he is able to follow moving objects between clutter patches or see them above weak clutter. Visibility between fixed land clutter, with information accumulated over many scans, is described in Section 11.3.1 and this section deals with principally slowly moving weather clutter. An increase in noise fogs the display and causes automatic extractors to declare plots from noise and saturate themselves or the systems that follow. In the past, the emphasis was to reduce the load on radar operators so that a video thresholded just above the fogging level gives a clearer picture (Section 11.1.6). A common form of constant false alarm rate processing for signals in which the threshold is a multiple of the average is shown in Figure 11.7.

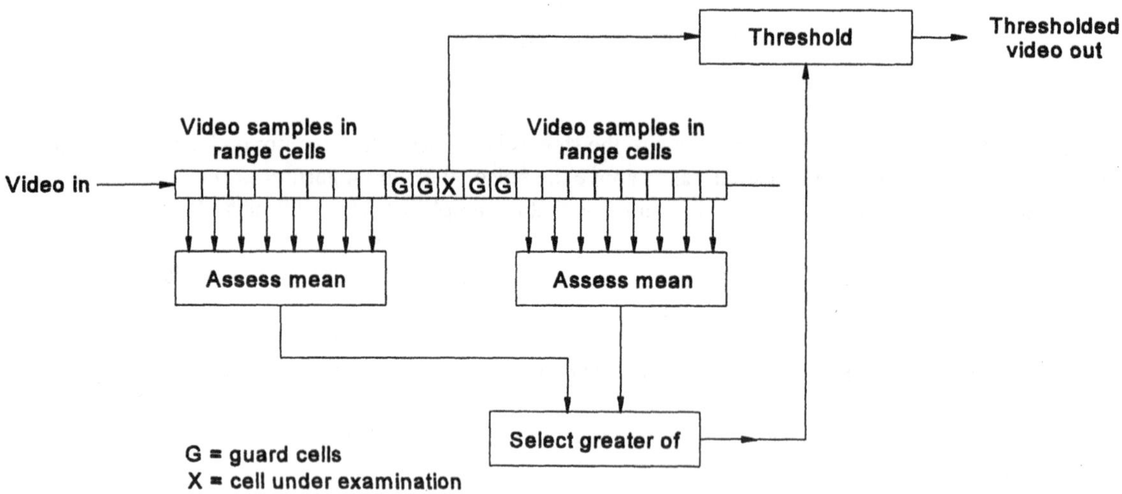

Figure 11.7 A typical constant false alarm (CFAR) stage block diagram.

The problem is to find the correct threshold for the range gate marked X. The signal in this range cell may correlate with the information in the neighboring cells, as is the case with oversampling or the time sidelobes after pulse compression. Guard cells, marked G, separate the cell in question (statistically) from those used to assess the threshold. The amplitudes from a number of cells outside the guard cells are used to assess the rain clutter + noise + interference (+ chaff + jamming) mean level. To avoid problems at the edges of clutter patches, separate means are calculated before and after the cell in question and the greater mean is chosen.

The constant false alarm rate loss depends inversely on the number of cells (or the accuracy of the mean and the safety margin required) used to assess the mean. A large number of cells give the minimum loss in constant clutter, and shorter lengths reduce the loss at the edges or where the clutter amplitude varies. The level of the threshold above the mean defines the false alarm rate (FAR) and depends on the statistical distribution of the signals. Though this principle may be extended to Weibull distributed clutter [4], the clutter is assumed to have a Rayleigh distribution, see [5, pp. 9-1 et seq.; 6, Chapter 9; 7, Chapter 7; and 8, Chapter 7].

Cell averaging constant false alarm rate thresholds are often used on the sweep information after Doppler frequency processing such as the moving target detector in Section 11.2.4.3.

The form of cell averaging constant false alarm rate (CA-CFAR) processing in Figure 11.6 uses the sum of the noise plus interference samples in order to estimate a threshold that keeps the false alarm rate constant. The distribution of Rayleigh distributed interference, from rain or chaff, is fully described by its mean (see Section 15.2.3) and the distribution of the power in the Rayleigh distributed interference, average power b, for each sample is a negative exponential distribution.

$$p(x) = \frac{1}{b}\exp\left(-\frac{x}{b}\right) \tag{11.9}$$

The false alarm probability depends on the power ratio above basic noise b and the threshold Y_b, namely

$$\text{False alarm probability} = \int_{Y_b}^{\infty} \frac{1}{b}\exp\left(-\frac{Y}{b}\right)\, dY = \exp\left(-\frac{Y_b}{b}\right) \tag{11.10}$$

The sum of statistically independent samples from an exponential distribution follows the gamma distribution so that any threshold derived from this sum also follows a gamma distribution. If the threshold multiplier is α then the derived sample threshold, Y_T is

$$Y_T = \frac{\alpha}{N} \sum_{i=1}^{N} y_i \tag{11.11}$$

Figure 15.19 shows the probability distributions of the sums of N exponentially distributed noise (power) samples. The curves have a peak or mode of N–1, an average of N, with the area under the curve unity. In Figure 11.8 the curves have been replotted with the abscissa scaled by the number of pulses, N, so that they all have a common mean of unity.

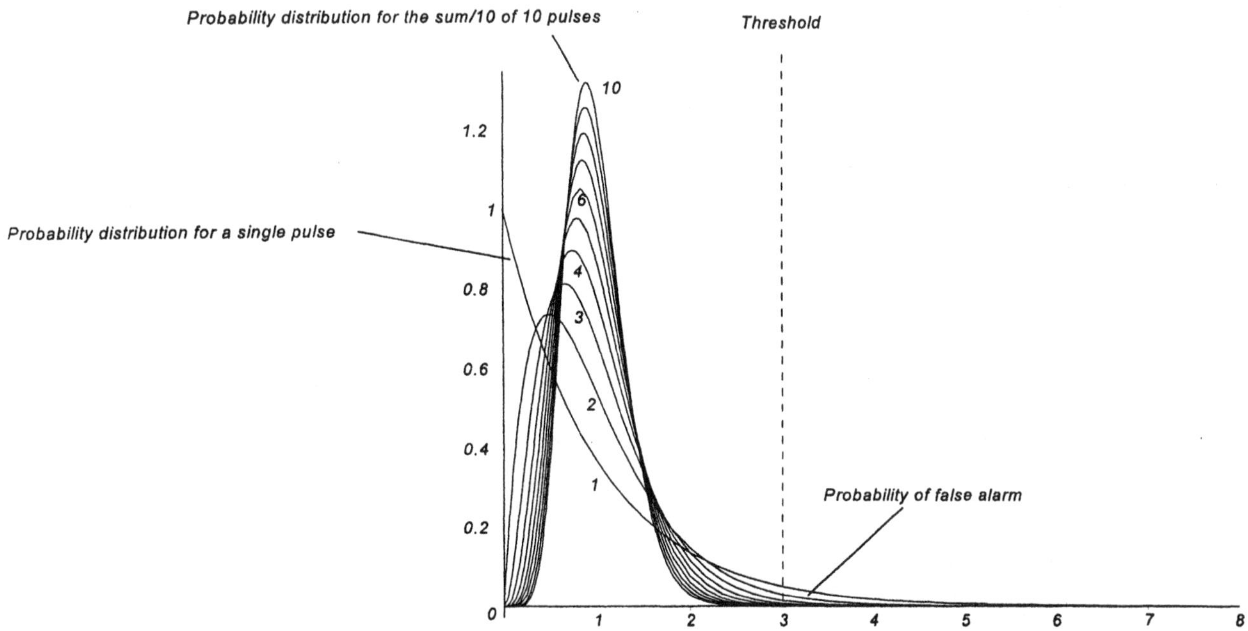

Figure 11.8. Family of gamma function curves p(NY) for the sum of one to 10 Rayleigh (power) pulses.

The modes occur at (N–1)/N to give a narrower grouping with higher values of N and a lower probability of false alarm, or area under the curve to the right of the threshold Y_T in (11.11). The result is that the greater the number of pulses summed, the better is the estimate of the average, and the threshold may be reduced to maintain the same probability of false alarm. Mathematically, the gamma distribution from (12.11) and (15.44) is

$$p(Y) = Y^{N-1} \frac{\exp(-Y/b)}{\Gamma(N)\, b^N} \tag{11.12}$$

The mean of a gamma distribution is ab and the mode $(a-1)/b$. For noise alone, b is set to unity and for N samples with a normalized mean of unity for the curves in Figure 11.8 are given by

$$p(NY) = N \frac{(NY)^{N-1} \exp(-NY)}{\Gamma(N)} \quad \text{or} \quad N \frac{(NY)^{N-1} \exp(-NY)}{(N-1)!} \tag{11.13}$$

Note that the leading factor N is $d(NY)/dY$ required to keep the areas below the curves as unity.

The expected value or mean of the false alarm probability is [8] (the expression does not exist for Y_T less than zero)

$$\overline{P_{fa}} = \int_0^{+\infty} \exp\left(-\frac{Y_T}{b}\right) p_T(Y_T)\, dY_T$$
$$= \left(\frac{N}{\alpha b}\right)^N \frac{1}{\Gamma N} \int_0^{+\infty} Y_T^{N-1} \exp\left(-\left(\frac{N}{\alpha}+1\right)\frac{Y_T}{b}\right) dY_T \tag{11.14}$$

Using the integral [3, p. 310, Eq. 3.351.3]

$$\overline{P_{fa}} = \left(1 + \frac{\alpha}{N}\right)^{-N} \tag{11.15}$$

or

$$\alpha = N\left(\overline{P_{fa}}^{\,-\frac{1}{N}} - 1\right) \tag{11.16}$$

The threshold multipliers are shown in Figure 11.9.

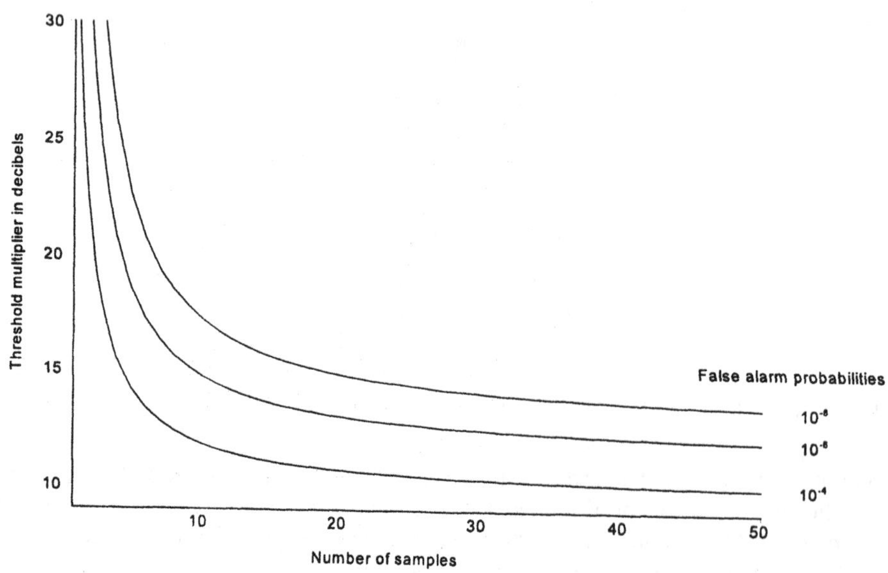

Figure 11.9 Threshold multipliers (in dB) for different numbers of samples and false alarm rates.

The probability of detection for Swerling case I with a signal-to-noise ratio, R, is

$$P_d = \exp\left(-\frac{Y_b}{1+R}\right) \tag{11.17}$$

and gives an expected or mean value of, for the integral see [3, p. 310, Eq. 3.351.3],

$$\overline{P_d} = \int_0^{+\infty} \exp\left(-\frac{Y_b}{R+1}\right) p_T(T)\, dT = \left(1 + \frac{\alpha}{N(1+R)}\right)^{-N} \tag{11.18}$$

or

$$R = \frac{\left(\dfrac{\overline{P_d}}{P_{fa}}\right)^{\frac{1}{N}} - 1}{1 - \overline{P_d}^{\frac{1}{N}}} \tag{11.19}$$

There is a loss with this type of false alarm rate regulation using a limited number of exponential clutter plus noise samples. The loss is given by the signal-to-noise ratio using the threshold from N samples divided by the signal-to-noise ratio from an infinite number of samples (in Chapter 12), namely

$$\begin{aligned} Loss &= \frac{\textit{Signal-to-noise ratio for detection with CFAR}}{\textit{Signal-to-noise ratio for detection with noise alone}} \\ &= \frac{\left(\left(\dfrac{P_d}{P_{fa}}\right)^{\frac{1}{N}} - 1\right)}{\left(1 - P_d^{\frac{1}{N}}\right)} \frac{\ln P_d}{\ln\left(\dfrac{P_{fa}}{P_d}\right)} \end{aligned} \tag{11.20}$$

This loss is plotted with logarithmic axes [6] in Figure 11.10.

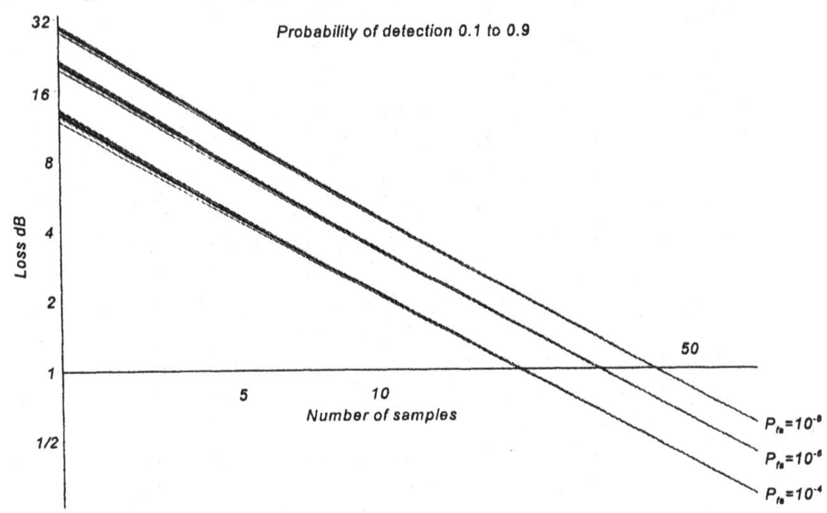

Figure 11.10 Loss for cell averaging constant false alarm rate processing.

Using the relations and approximations that neglect higher terms,

$$\left(\frac{P_d}{Pfa}\right)^{\frac{1}{N}} = \exp\left(\frac{\ln\left(\dfrac{P_d}{Pfa}\right)}{M}\right), \qquad \frac{1}{1+x} \approx 1 - x, \qquad \text{and} \quad \ln(1+x) \approx x$$

approximate loss for large numbers of samples can be given by [6]

$$Loss_{dB} \approx \frac{5}{N}\frac{\ln P_{fa}}{\ln 10} = 2.17147\,\frac{P_{fa}}{N} \tag{11.21}$$

and are shown, together with the exact values, in Figure 11.11.

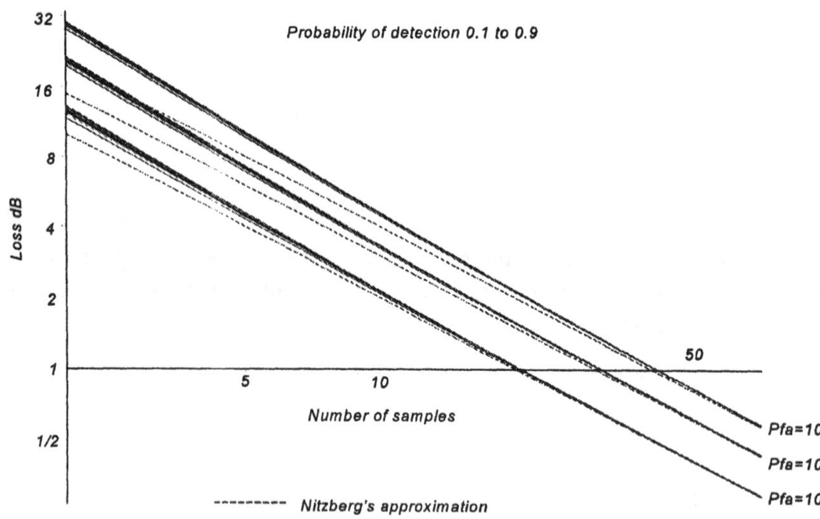

Figure 11.11 Loss and Nitzberg's approximation for cell averaging constant false alarm rate processing.

Cell averaging constant false alarm rate processing has a number of limitations and is subject to continuing research. If there are a number of strong wanted echoes in the calculation of the threshold, the threshold will be raised leading to the masking of some of them. This effect can be reduced (with greater loss) by estimating the threshold from sums of the logarithms of the interference. There is a similar effect if there are too few guard cells.

False alarms occur at the edges of clutter and if the number of samples used to calculate the threshold is reduced, the range over which false alarms occur is diminished but with a greater false alarm loss. One such logic, see Figure 11.12, is to divide the cells outside the guard cells into groups and compare the sums of the lower and upper group (A) nearest the guard cells. If the difference is small, then a further group (B) is compared, and so on with group C. When sums from the groups of A's differ by a threshold, say at the edge of clutter, then only the greater of the A's is used. If sums from the two groups of A's differ only slightly then the sums from the two groups of B's are added in and so on for the two groups of C's. Alternatively the outer groups are weighted to have less influence in order to determine the threshold from the greatest number of cells possible to have the least loss.

Other ways of determining the threshold are the nonparametric ordered statistic constant false alarm rate methods. Instead of the signals in the cells being added, they are placed in a list and the threshold level is calculated from the strength of the kth echo in the list. This statistic is not influenced by a few strong echoes.

More complicated cell processing produces the variance in addition to the mean to be used for the thresholds for Weibull and log-normal models of clutter.

The description above refers to cell averaging constant false alarm rates for the video signals soon after the detector. More modern radars use the moving target detector (Section 11.4.4) that separates the echo signals according to their Doppler frequency so that Rayleigh distributed rain echoes generally appear in separate outputs from the ground clutter (see area MTI, Section 11.3.1). Cell averaging constant false alarm rate processing is commonly used after the moving target detector.

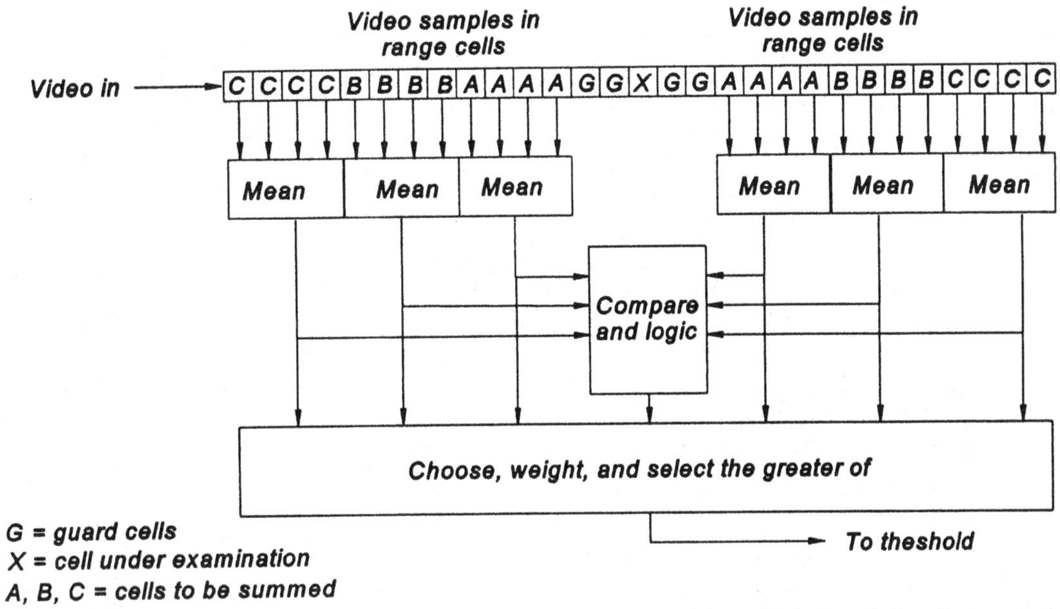

Figure 11.12 Cell averaging constant false alarm rate processor with variable length for clutter edges.

11.1.8 Gating the video

There are a number of gating circuits for confirming echo signals and suppressing others. Two common forms of gating the videos from receivers are sidelobe blanking and frequency diversity operation.

11.1.8.1 Sidelobe blanking

Sidelobe blanking (SLB) can be used to cut out signals which are presumed to have entered through the antenna sidelobes, as shown in Figure 11.13.

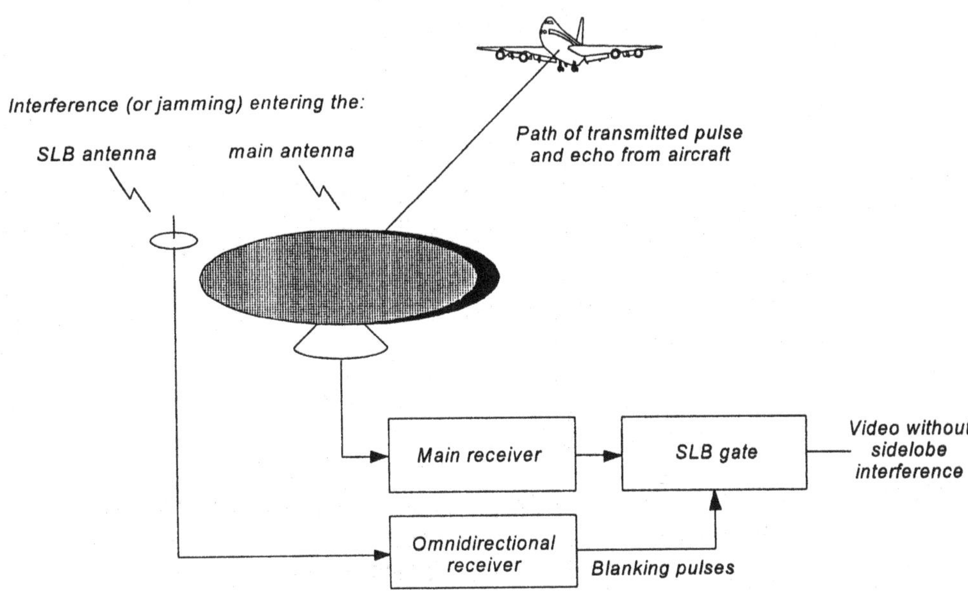

Figure 11.13 Block diagram of receivers using sidelobe blanking (SLB).

11.1.8.2 Frequency diversity operation

In frequency diversity radars, the signals from the two channels may be .AND.ed together to reject interference (or jamming) received in a single channel only, as shown in Figure 11.14.

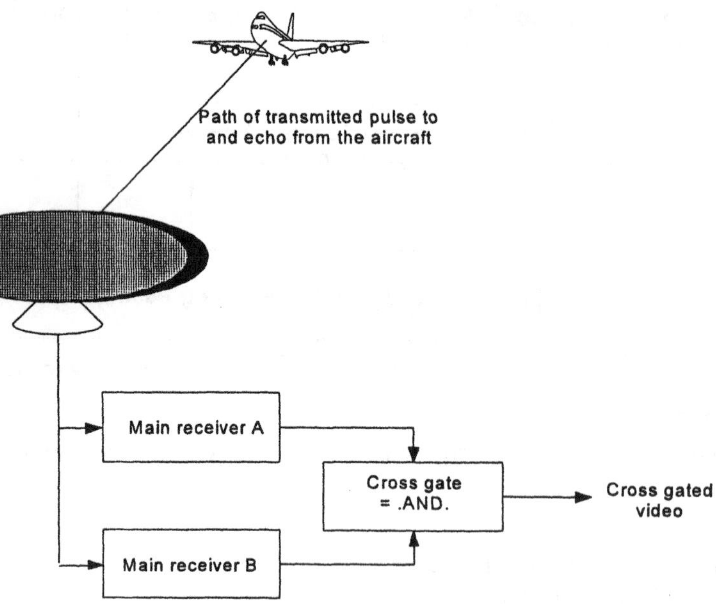

Figure 11.14 Block diagram of diversity radar receivers with cross gating.

Cross-gating is used to filter out random false alarms or interference, which occur in each channel from time to time, which alters the probability of detection (P_d) and the false alarm probability (P_{fa}).

$$\textit{Combined false alarm probability, } P_{fa} = P_{fa1} \cdot P_{fa2} \tag{11.22}$$

$$\textit{Combined probability of detection, } P_d = P_{d1} \cdot P_{d2} \tag{11.23}$$

As an example, if the false alarm probability for each channel is 10^{-6}, then the combined probability is 10^{-12}. If the probability of detection is 80%, then the combined probability of detection is 64%. Threshold adjustments for a false alarm probability of 10^{-3} in each channel lower the threshold, which raises the probability of detection. For *m* channels see [9], and for different types of cross gating, see [10].

Similarly if the cross-gate is replaced by an .OR. gate, then the false alarm probability is raised to the sum of the rates in each channel. The probability of detection becomes

$$\textit{Probability of detection with .OR. gate, } P_d = 1 - (1 - P_{d1})(1 - P_{d2}) \tag{11.24}$$

Thus, the 80% probability of detection in each channel gives 96% when either channel is able to report a detection.

Modern radars, which use frequency diversity, label the plots with quality flags such as detected by one channel and detected by two channels. Higher quality detections in both channels are often required to initiate a track and one or two channel detections to update it.

11.1.9 Combining the videos from several beams

Stacked beam surveillance radars use a number of beams in order to estimate the elevation angle of an aircraft, as shown in Figure 11.15. The displays are only two dimensional, so the videos from the beams are mixed for display and use.

Adding the video signals adds the noise attached to them. At short ranges, the echoes are strong enough to be seen above the noise and the intermediate frequency sensitivity time control has reduced the gain on echo signals and noise at short ranges. The added noise gives rise to a greater number of false alarms which requires a raising of the threshold, giving the collapsing loss. The collapsing ratio for n samples of signal + noise with an additional m samples of noise is given by Marcum [11]:

$$\text{Collapsing loss, } \rho = \frac{m + n}{n} \tag{11.25}$$

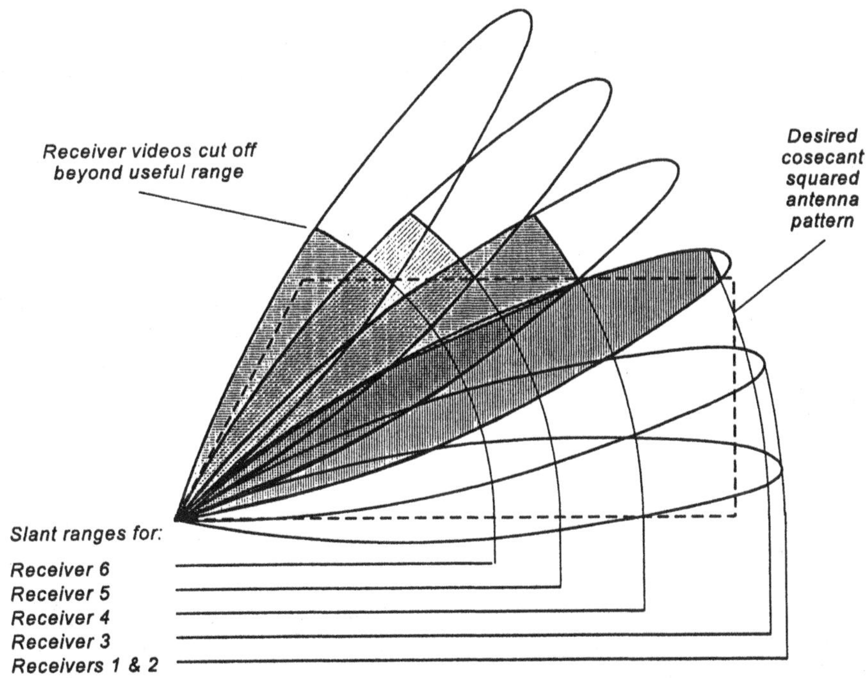

Figure 11.15 Beam gating to reduce collapsing loss.

There are two cases of collapsing ratio [12; 13, p. 77]:

- Addition of signals so that the false alarm rate is increased by the factor ρ so that the thresholds must be raised. Examples are:
  - The addition of signals from m independent radar receivers, $\rho = m$.
  - Collapsing coordinates on a display, for example, a C-scope. The collapsing ratio, ρ, is the number of range cells that are collapsed.
- The number of opportunities for false alarms is decreased. Examples are:
  - The video bandwidth, B_v, is less than half the intermediate frequency bandwidth ($B_{IF} = 1/\tau$) or the intermediate frequency bandwidth is excessive ($>1/\tau$)

$$\rho = \frac{2B_v + B_{IF}}{2B_v} \tag{11.26}$$

- The sweep speed on a cathode ray tube is too slow to resolve the signals:

$$\text{Collapsing ratio for cathode ray tubes, } \rho = \frac{d + s\tau}{s\tau} \tag{11.27}$$

where s is the sweep speed;
τ is the signal pulse width;
d is the spot diameter.

- When the range gates are too wide, τ_g:

$$\rho = \frac{\tau_g}{\tau} \tag{11.28}$$

11.1.9.1 Collapsing loss

In Figure 11.15, the video from one receiver is mixed with the noise from all m receivers, which gives a collapsing ratio of m. The false alarm rate is m times the false alarm rate of a single receiver. The threshold must be raised to bring back the false alarm rate to the design value, namely

$$\text{Collapsing loss, } L_c = \frac{\text{Signal-to-noise ratio for } \rho \text{ times false alarm rate}}{\text{Signal-to-noise ratio for original false alarm rate}} \tag{11.29}$$

The collapsing loss depends on the type of echo signal fluctuation, the type of detector, the number of pulses that are integrated, and the probability of detection. Curves for collapsing loss for steady scatterers have been given in the literature by Marcum [11] and Trunk [quoted in 14, p. 8-4]. For very large numbers of pulses (no fluctuation), the loss tends to $\sqrt{\rho}$, or 5 log ρ dB, but is less for smaller numbers.

Currently, software allows the calculation of the signal-to-noise ratio for detection at other increased false alarm rates. The new value of S_{min} is used directly.

11.2 SIGNAL PROCESSING OVER A NUMBER OF SWEEPS

The first form of signal processing over a number of sweeps takes place on the face of a cathode ray tube. The phosphor is repeatedly painted by the video over the width of the echo and the noise paints different areas on an A-scope or adds to the brightness of the video "painted" on a plan position indicator (PPI). The modern form is the video integrator or correlator.

The analysis of echo signals over a number of sweeps allows enough time for the recognition and filtering of the signals with different Doppler frequencies. If, for example, an S-band radar (3 GHz) has a pulse width of 1 μs, there will be 3 000 cycles within the pulse. With a Doppler frequency of 1 000 Hz, there will be 3 000.001 cycles within the pulse, and the Doppler frequency is impossible to recognize over so short a time. If the echoes are observed over tens of milliseconds, this becomes possible. Doppler frequency processing may take place at:

- Intermediate frequency: The amplitude and phase modulation of a notional carrier represent a voltage vector;
- Video: Polyphase signals (two or more phases) represent the video vector voltages.

Figure 11.16 shows the complex echo video signal or modulating function sampled in range as an antenna with a Gaussian beam scans past a moving object over clutter. The clutter signal has been drawn, conventionally, with a relative phase angle of zero. In reality, clutter echoes may arrive with any phase angle. The moving echo signal vector rotates around the clutter vector, giving the impression of a bent spiral spring. Since the echoes occur at discrete times, they will be discrete sampling vectors within the envelopes in Figure 11.16 (see also Figures 11.61 and 11.62).

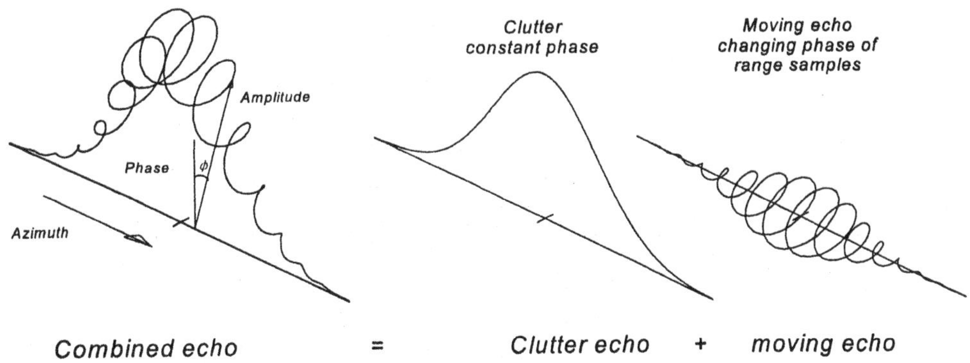

Figure 11.16 Form of the vector video or modulation of a carrier as the antenna scans a moving echo over clutter.

Signal processing is normally analyzed in the frequency domain, and the problem is to separate the spiral, which represents the moving echo, from the larger Gaussian function, which represents the clutter. The development of signal processing reflects a constant improvement in the ability to separate the two signals and detect the wanted echo signals. Typically, in the past, the moving echoes one hundredth of the clutter power (20 dB) could be detected and today less than one hundredth of the voltage (40 dB).

The effects of signal processing can be demonstrated and illustrated following the steps in Figure 11.17. This allows the illustration of the clutter spectra and their width with respect to clutter amplitude. This gives a much better insight than time or spectral considerations alone. The results for the radar used as an example are shown in Figure 11.18.

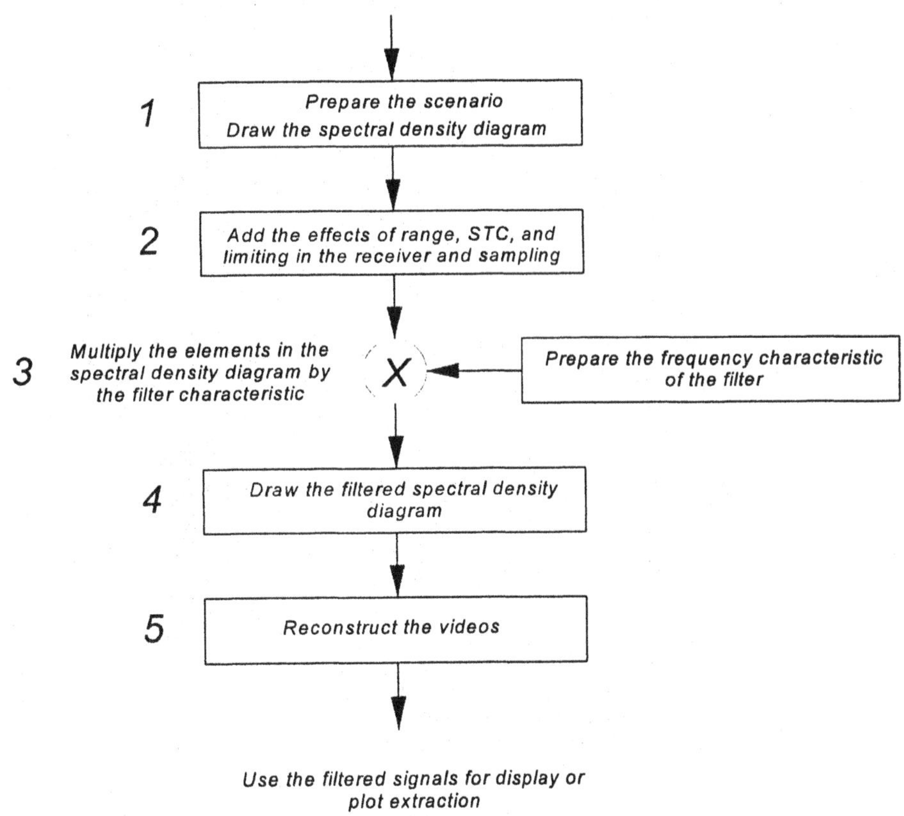

Figure 11.17 The steps used to calculate the scenario and show the effects of filtering.

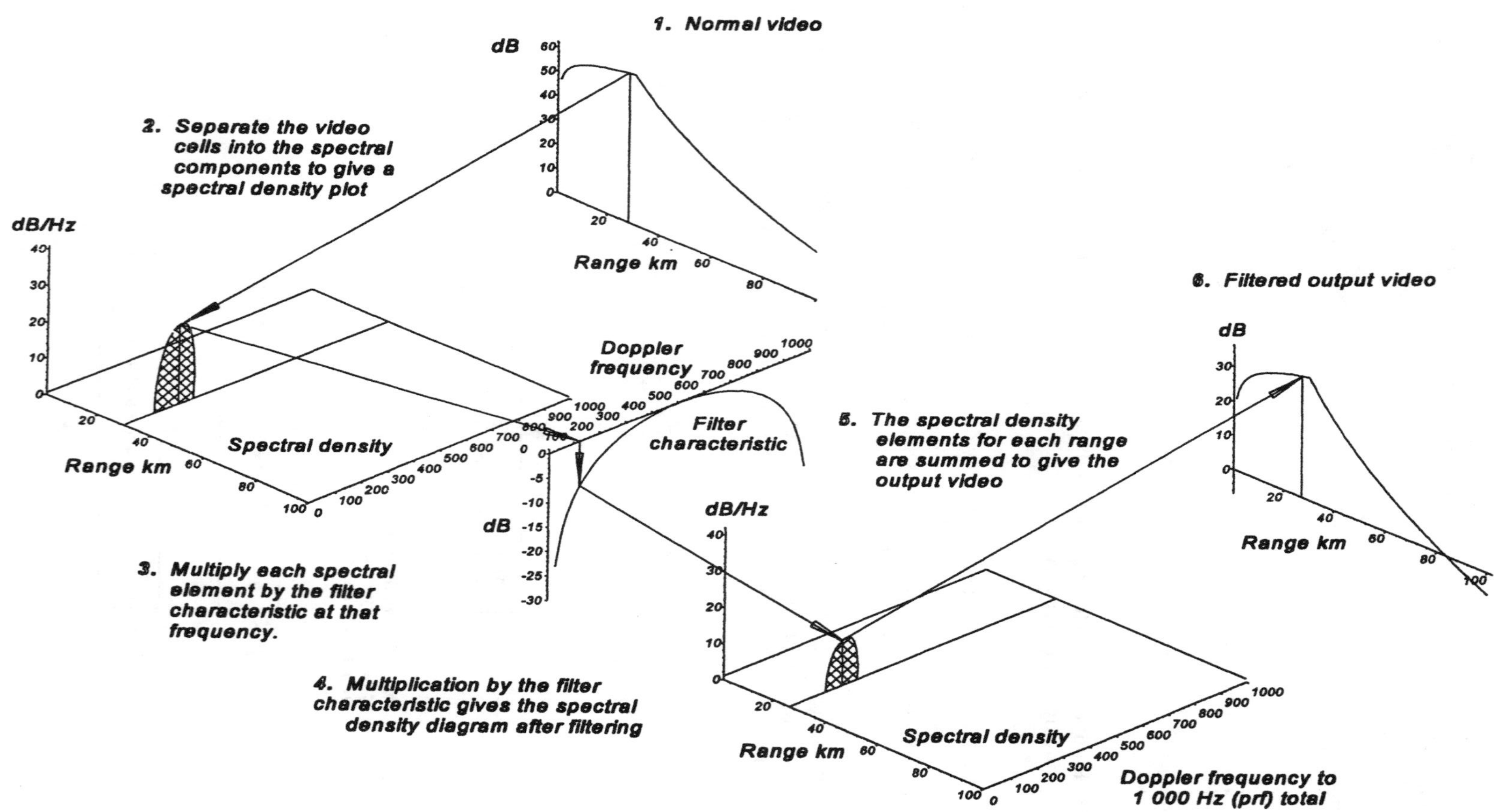

Figure 11.18 Steps in calculating the effects of filtering in a situation with wanted echoes and clutter.

The signal returns are estimated (1) and a spectral density diagram is drawn (2) (as in Chapter 6, Factors outside the radar). The characteristic of the signal processing filter is shown in (3). This characteristic is multiplied by the spectral density at each range in (4). The spectral density for each range is summed to give the output video (5).

Processing using the Doppler frequency of the echoes as a criterion is historically divided into moving target indication (MTI), or the suppression of stationary echoes, and pulse Doppler frequency processing (PD), the selection of echoes at a chosen Doppler frequency. Ground radars, which use a low pulse repetition frequency (long unambiguous range), have used MTI in the past. The more recent moving target detector (MTD) now used in search radars has enough Doppler frequency filters to process any echo Doppler frequency selectively, and thus falls into the pulse Doppler radar category.

In the past, the only way to store signals was to use delay lines, which has colored the treatment of filtering. A number of delay lines with a delay equal to the pulse repetition time, T, are shown in Figure 11.19. The output represents the sum of signals from scatterers at the same range from the present, over the direct path, and echo signals from past transmitter pulses.

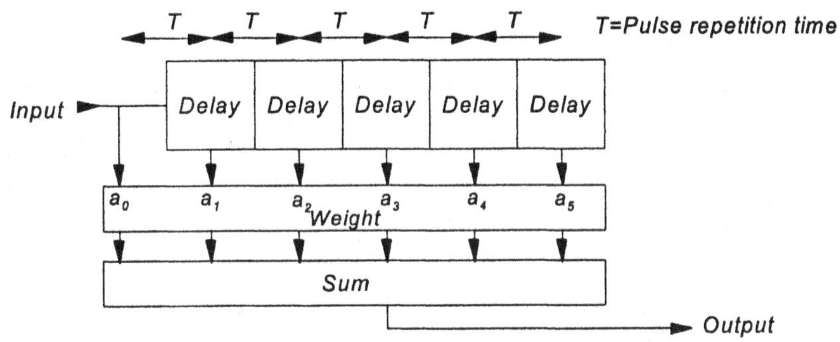

Figure 11.19 Block diagram of a finite impulse response filter.

If a single needle pulse traverses the delay lines, it gives a finite number of needle pulses at the output. This is called a finite impulse response (FIR) filter. The amplitude and the phase of the pulses from the delay lines for summing are modified by the weighting block to define the shape of the filter characteristic.

The outputs of the delay lines from a single needle pulse give the impulse response. The single pulses are never summed. If the single pulse is replaced by a train of samples then the sum is the discrete Fourier transform of the tapering function. The tapering function may be designed to give a lowpass, a highpass, a bandpass, or a bandstop response [15] (as shown in Section 16.7.1).

The notation of Laplace transforms is used, as in control theory, with $p = s + j\omega$. A delay may be applied to a signal by multiplying it by e^{-pT}. Then the output from the filter with N taps in Figure 11.19 is

$$\textit{Finite impulse response (FIR) filter output,} \quad H(p) = \sum_{k=0}^{N-1} a_k \, e^{-kpT} \tag{11.30}$$

In the steady state ($s = 0$), and (11.31) is a form of the discrete Fourier transform

$$\textit{Finite impulse response (FIR) filter output,} \quad H(n) = \sum_{k=0}^{N-1} a_k \, e^{-jk\frac{n}{N}T} \tag{11.31}$$

If e^{pT} is replaced by z, then

$$H(z) = \sum_{k=0}^{N-1} a_k \, z^{-k} \qquad k \text{ is an integer} \tag{11.32}$$

As k is an integer, this has the form of the z transform. The sum in (11.32) can be represented as a polynomial

$$H(z) = a_0 + a_1 z^{-1} + a_2 z^{-2} + \ldots + a_N z^{-N} \tag{11.33}$$

or as factors containing the roots of the polynomial:

$$H(z) = \left(\frac{1}{z} - r_1\right)\left(\frac{1}{z} - r_2\right)\left(\frac{1}{z} - r_3\right) \dots \left(\frac{1}{z} - r_N\right) \tag{11.34}$$

The other form of filtering using delay lines uses feedback, as shown in Figure 11.20. A needle pulse circulates around the filter with ever decreasing amplitude until it finally becomes imperceptible (when $b_k < 1$), hence the name infinite impulse response (IIR) filter.

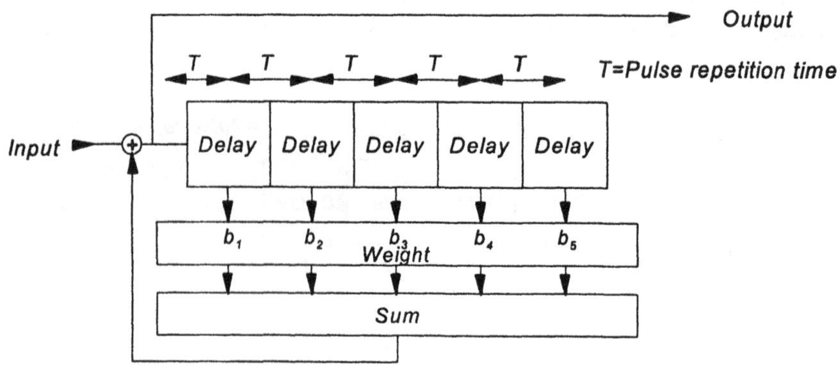

Figure 11.20 Infinite impulse response filter.

The negative feedback equation from control theory applies. Using Laplace transform notation,

$$\textit{Infinite impulse response (IIR) filter output,} \quad H(p) = \frac{1}{1 - \sum_{k=1}^{N} b_k e^{-kpT}} \tag{11.35}$$

Writing e^{pT} as z, we have

$$H(z) = \frac{1}{1 - \sum_{k=1}^{N} b_k z^{-k}} \tag{11.36}$$

The sum in the denominator may also be expressed as a polynomial or its factors. The two types of filters can be combined to give the filter shown in Figure 11.21.

The transfer function is the product of those from the two types of filters with N_1 and N_2 taps:

$$H(p) = \frac{\sum_{k=0}^{N_1} a_k e^{-kpT}}{1 - \sum_{k=1}^{N_2} b_k e^{-kpT}} \tag{11.37}$$

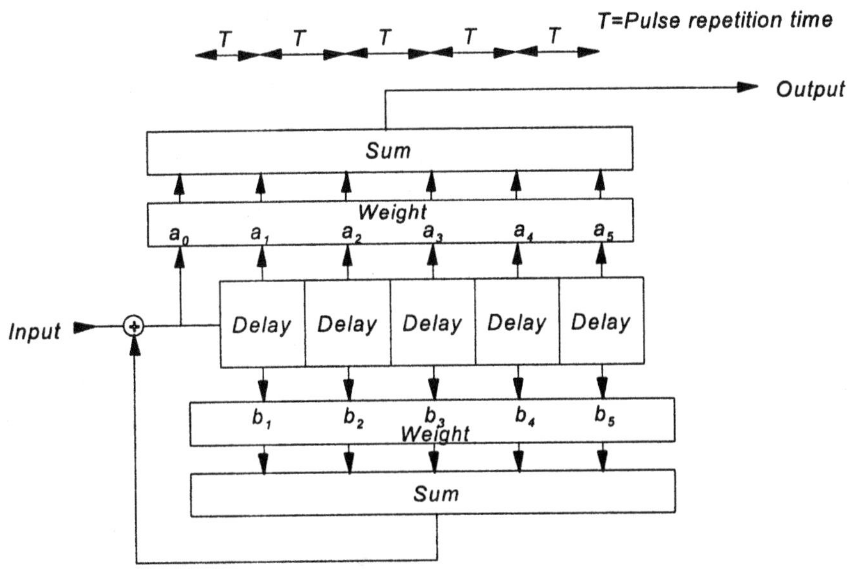

Figure 11.21 General filter, canonical form.

or in terms of z,

$$H(z) = \frac{\sum_{k=0}^{N_1} a_k z^{-k}}{1 - \sum_{k=1}^{N_2} b_k z^{-k}} \tag{11.38}$$

11.2.0.1 Continuous processing versus store and dump

In radars that use analogue receivers and signal processing, the signals are continually fed into the delay lines and they are cast out at the end of the processing interval (the time given by the number of delays) or they die away. The filter circuitry remains full of signals, and the center of a plot may be found by finding the maximum or by averaging the

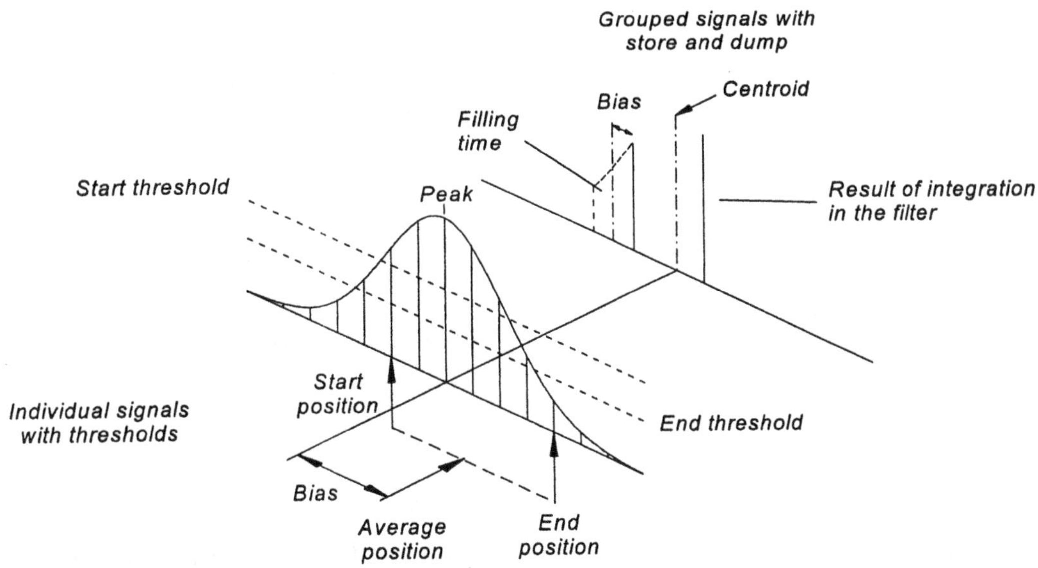

Figure 11.22. Comparison of the methods for finding the center of a group of echoes.

positions (normally measured in azimuth change pulses, ACP) in which the echo signal passes through (start) and falls below (end) a threshold. Often, the end threshold is lower than the starting threshold, to counter the effects of fading. The granularity of the estimate is the azimuth between two transmitter pulses which is shown on the left of Figure 11.22.

With store and dump, the filter is filled with signals (maybe in the form of data). After the required number of samples have been received, the sum is read out, as in the back of Figure 11.22. The store and dump process results in sparsely spaced output samples that must be interpolated, normally centroided, to obtain an accurate result. When moving target indicator filtering stages are used before integration, extra transmitter pulses are necessary to fill the delay lines, which increases the mean power, the transmitter size, and running costs without an increase in range.

11.2.0.2 Azimuth shift or bias

Both procedures in Figure 11.22 require time to collect and produce results which result in fixed azimuth errors or biases.

11.2.1 Video integration

The radar detection decision is based on the sum of the signals plus noise during one look at an echo. In older radars, the video was summed from pulse to pulse on the phosphor of the cathode ray tube. The video is also summed on A-scopes (and their relations); when the spot passes over the same point again and again, the trace brightens. On plan position indicators the brightening for each pulse adds together to give a brighter point, or banana, with a rotating antenna: the phosphor decay automatically frees the display for new information.

Later the process has been duplicated using circuitry to ease the recognition of wanted echoes amongst the surrounding noise on a cathode ray tube display, video integration. An analogue hardware implementation is shown in Figure 11.23, where the value of β represents the decay characteristic when the trace has passed by.

Automatic video extractors use video integration to sum the signal energy before a threshold decision.

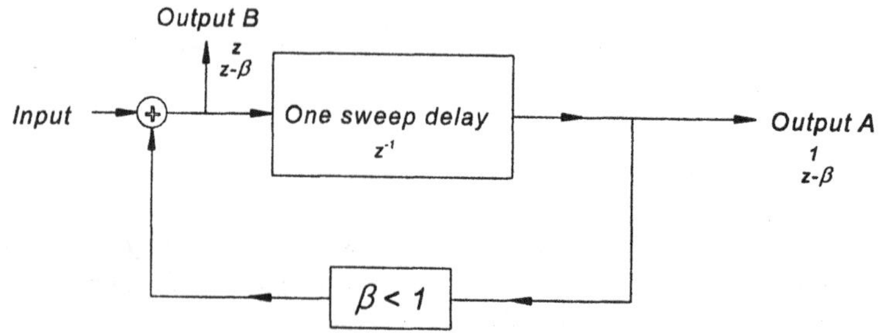

Figure 11.23 Video integration.

A finite response integrator is shown in Figure 11.23. If the addition function is replaced by combining logic, the integrator becomes a correlator (Figure 11.24) and blocks asynchronous interference (commonly called "running rabbits").

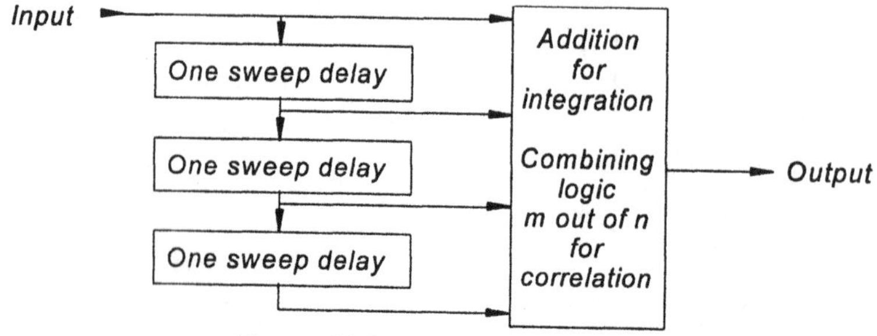

Figure 11.24 Video correlation.

The combining logic makes an m-out-of-n decision to pass on the video which is often used in secondary radar signal processors and is called a "defruiter" (from the acronym FRUIT, for "false replies unsynchronized to interrogator transmission" [16]).

11.2.2 Binary integration

Binary integration is a special case of video integration. First the video passes through a thresholding stage. The results of the threshold crossings are subjected to an m-out-of-n test to give a final result. The chances of a final threshold crossing are given, as in quality control, by the binomial distribution [12; 13, p. 74].

$$p(x) = \frac{n!}{m!\,(n-m)!}\, p^{m}\,(1-x)^{n-m} \tag{11.39}$$

The cumulative distribution is the sum of the probabilities from m to n:

$$P(x) = \sum_{j=m}^{n} \frac{n!}{j!\,(n-j)!}\, p^{j}\,(1-x)^{n-j} \tag{11.40}$$

For small values of x, the first term is dominant and (11.40) can be used for the cumulative distribution. For example, for an 8 out of 10 second threshold, a false alarm rate of 0.1 becomes:

$$\begin{aligned} 45 \times 0.1^{8} \times (0.9)^{2} &= 0.3645 \times 10^{-6} \\ 10 \times 0.1^{9} \times 0.9 &= 0.009 \times 10^{-6} \\ 0.1^{10} &= \underline{0.0001 \times 10^{-6}} \\ &= 0.3736 \times 10^{-6} \end{aligned}$$

and the approximate value is 0.45×10^{-6} which reduces a 10% false alarm rate to an acceptable value of less than one in a million.

For wanted echoes, the single pulse probability is reduced to give a loss of between 1 and 2 dB. Binary integration was primarily used as a second threshold in the hard wired extractors installed between the 1950s and 1970s. These extractors have been superseded by equipment using larger integrated circuits that are capable of performing the integration more or less exactly, which eliminates the loss.

11.2.3 Rejection of echoes that have unwanted Doppler frequencies, moving target indicator

Moving target indication (MTI) was the first effective measure against ground clutter. Figure 11.25 is a block diagram of a single canceler. The technology used was first mercury delay lines, then quartz delay lines, and currently semiconductor memories are used. The memory technology first supported amplitude video signals (mercury delay line),

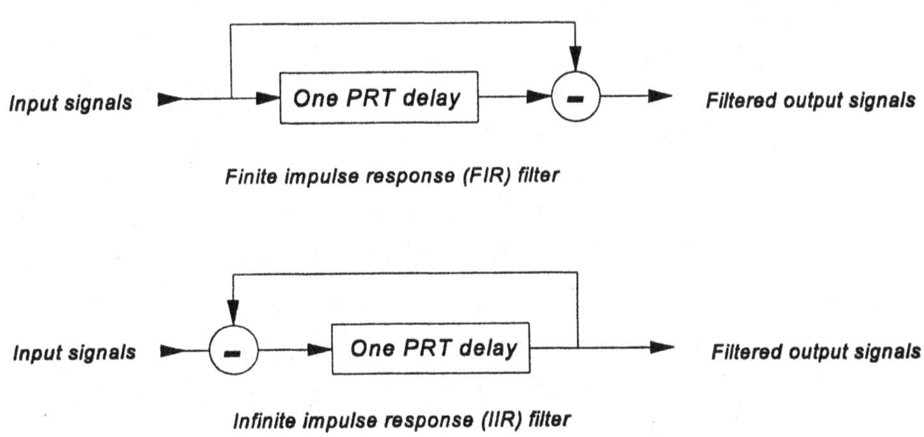

Figure 11.25 Single pulse repetition time (PRT) delay filters for moving target indicators in finite and infinite impulse response forms.

then phase detected videos (quartz and semiconductor delay elements), intermediate frequency signals (quartz delay line), and polyphase (I and Q) synchronously detected video data. Semiconductor memories also allow signal processing, which is not strictly synchronous with the pulse repetition frequency. Principally, clutter echoes with no Doppler frequency shift are subtracted out. Additionally other Doppler frequencies may be suppressed by:

- Separate stages with complex feedback;
- Offsetting the coherent oscillator (COHO) frequency which is sometimes used against rain (or chaff) outside clutter areas.

11.2.3.1 Amplitude processing or noncoherent moving target indicator (MTI)

Amplitude processing was the first of all the types of moving target indicators and is called noncoherent MTI. If the video on an A-scope is taken as a reference, the video is stored in a delay line. The old video is subtracted from the video being currently received. The result over a number of sweeps is the beating of the echo having a Doppler frequency shift against another echo in the same range cell which allows the display of the beats on an A-scope or, when rectified, on a plan position indicator (PPI). The ability to detect moving echoes in clutter of greater scattering area gives subclutter visibility.

If the clutter amplitude is C, and the moving echo amplitude is M, then the signal received is the sum $M^2 + C^2 - 2\,MC\cos\psi$. If two such signals from consecutive sweeps are detected by a square law detector and subtracted, then

$$\begin{aligned}
&\textit{Sweep1:} && C^2 + M^2 - 2CM(\cos\psi_1)\\
&\textit{Sweep2:} && \underline{C^2 + M^2 - 2CM(\cos\psi_2)}\\
&\textit{Subtract} && 2CM(\cos\psi_2 - \cos\psi_1)\\
& && = -4CM\sin\frac{\psi_1+\psi_2}{2}\sin\frac{\psi_2-\psi_1}{2}
\end{aligned} \tag{11.41}$$

where ψ_1 and ψ_2 are the phase angles for different pulse repetition intervals in Figure 11.26.

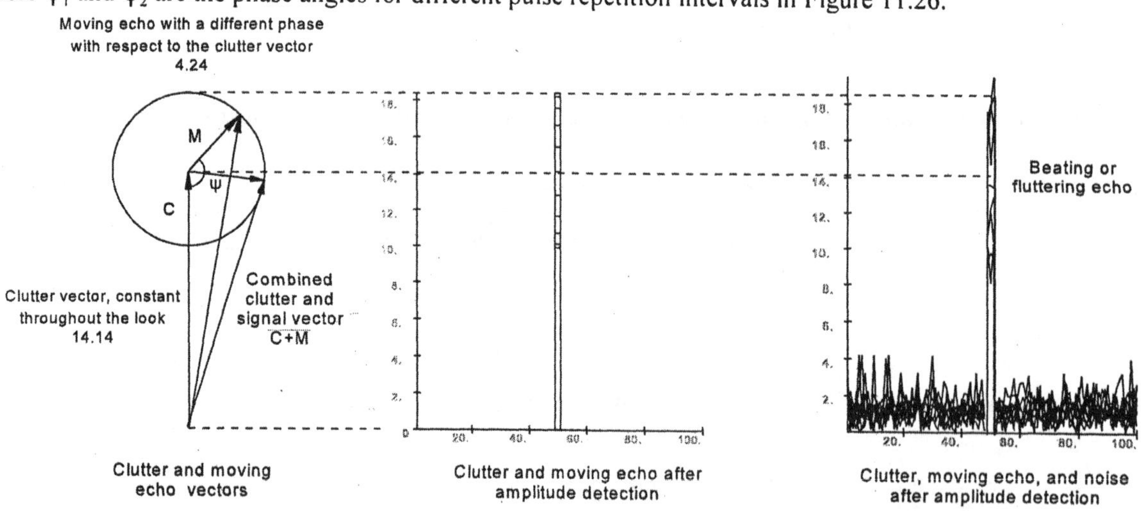

Figure 11.26 The relationships between the echo signal vectors and the detected signals in a noncoherent moving target indicator.

The description above shows the main weakness of noncoherent MTI. If a constant moving echo is not surrounded by clutter or rain, there is no beating and thus no detection. There is thus no interclutter visibility.

11.2.3.2 Coherent moving target indicator using phase processing

If a phase detector uses the signal from the coherent oscillator (COHO) as a reference phase, a phase detected video is produced. The clutter signals have the same phase from sweep to sweep, moving echo signals do not; so that simple subtraction gives a bipolar video, which is rectified for display. After limiting, the amplitudes of two echo signals from the same object in consecutive sweeps have an amplitude of unity as shown in Figure 11.27.

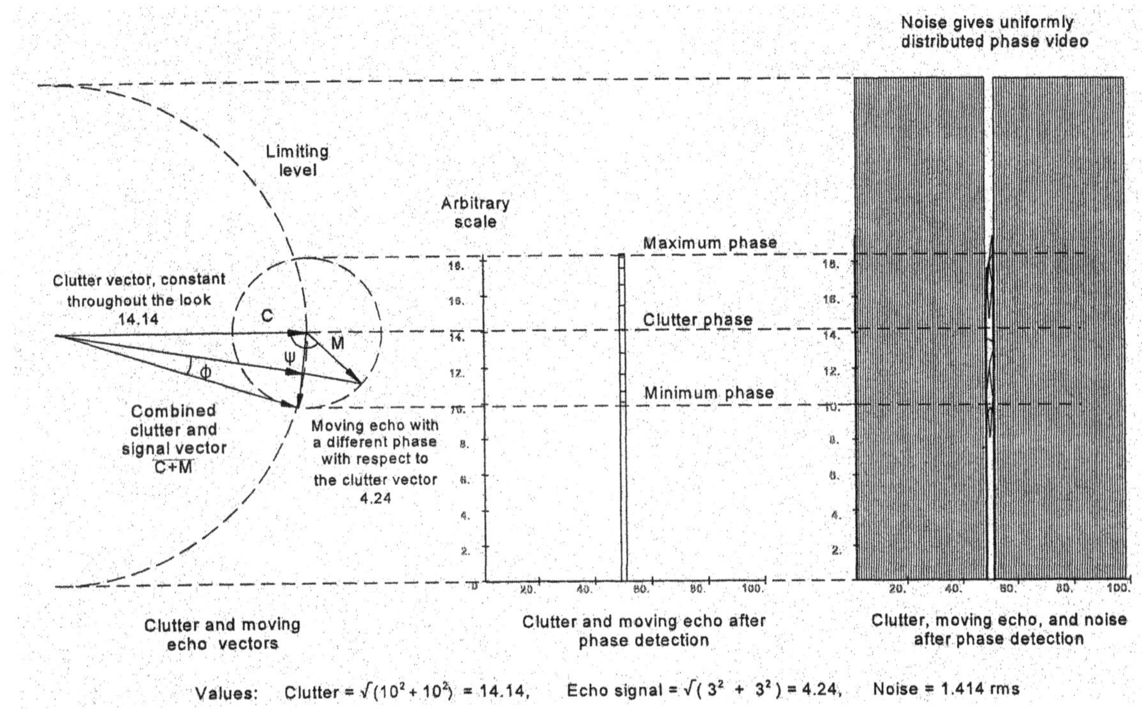

Figure 11.27 Moving target indication using pure phase detection.

In older radars, a ring demodulator was used with an amount of limiting in the preceding intermediate frequency stages. This demodulator gave synchronous detection in one phase at low levels, which reduced the noise to that of a single phase synchronous detector, and phase detection at echo strengths where limiting took place. This is shown in Figure 11.28.

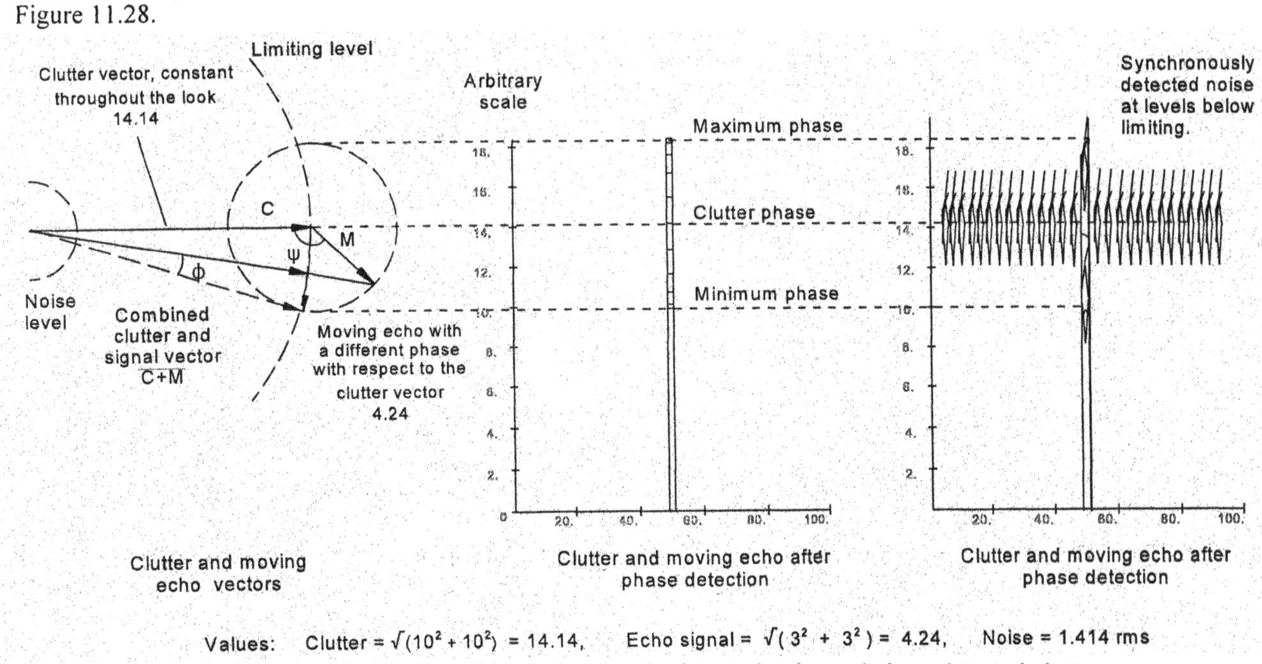

Figure 11.28 Moving target indication with synchronously detected noise and phase detected clutter.

If the signal from the first sweep is $\sin(2\pi ft+\theta)$, then the next echo from the same stationary object will be $\sin(2\pi ft+\theta)$. If the object moves between sweeps, the small movement gives an additional phase shift of ϕ which will give a residue after subtraction, namely,

$$\sin(2\pi ft + \theta) - \sin(2\pi ft + \theta + \phi) = 2\cos\left(2\pi ft + \theta + \frac{\phi}{2}\right)\sin\frac{-\phi}{2} \tag{11.42}$$

where f is the "carrier" (normally intermediate) frequency;
θ is the phase in the first sweep;
ϕ is the phase shift caused by movement.

The incremental phase difference may vary from zero to 360 degrees, or 2π radians. The modulating function in (11.42) varies between zero and 180 degrees, or π radians, and gives the sine characteristic. If the scattering object has moved half a wavelength radially, the (two-way) phase difference is 360 degrees between echoes from consecutive pulses. Thus at certain speeds, the blind speeds, there will be no subtracted signal, namely,

$$\frac{\lambda}{2} = n\, v_b\, T \tag{11.43}$$

where λ is the transmitter wavelength in meters;
T is the time between two transmitter pulses, s;
v_b is the first blind speed, m/s;
n is an integer.

A further disadvantage is that the cosine function in (11.42) may have a value near zero giving near zero subtracted signals which are called blind phases. If two-phase processing (I and Q) is used, a signal at zero in one phase is at a maximum in the other phase. The echoes may be of any phase so that single phase processing loses half the signal.

The notch given by the sine function is often not wide enough to cancel stronger clutter where the width of the Gaussian skirt is outside the notch. Matters may be improved by cascading two moving target indicator stages and the characteristic is

$$\left(2\cos\left(2\pi ft + \theta + \frac{\phi}{2}\right)\sin\frac{-\phi}{2}\right)^2 \tag{11.44}$$

This effect is valid both when the scatterer is moving alone or through clutter so that there is both interclutter and subclutter visibility.

11.2.3.3 Vector moving target indicator using amplitude and phase

The development of quartz delay lines made the storage of intermediate frequency signals possible. The stored intermediate frequency signal is subtracted from incoming signal, and the output is available for further processing at the intermediate frequency. Detection gives a relatively clutter free video signal suitable for display, further processing, or video extraction.

Quartz delay lines are expensive and analogue circuits need regular adjustment. Long chains of amplifiers and filters used in analogue processing tend to drift. Digital synchronous semiconductor logic has its gain and timing defined by design, and semiconductor memory has become cheap and robust. Current signal processing systems use the in and quadrature phase signals converted to data for vector processing.

Synchronous detection eliminates the intermediate frequency components to give the video vectors C_1, from one sweep, and C_2, from the next sweep, as shown in Figure 11.29.

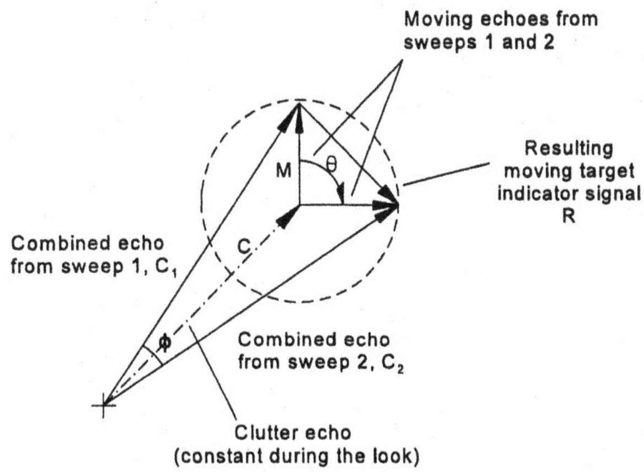

Figure 11.29 The subtraction of two video vectors for moving target indication.

The video vectors comprise the clutter echo vector, which is constant from sweep to sweep, and the vector from the moving echo, which rotates. The resulting video signal is (cosine rule)

$$\textit{Resulting video}^2,\ R^2 = C_1^2 + C_2^2 - 2\,C_1\,C_2\cos(\phi) \quad \text{volts}^2 \tag{11.45}$$

or, in terms of the I and Q components,

$$\begin{aligned} \textit{Resulting I component},\ R_I &= C_{I1} - C_{I2} \\ \textit{Resulting Q component},\ R_Q &= C_{Q1} - C_{Q2} \\ \textit{Resulting video},\ R &= \sqrt{R_I^2 + R_Q^2} \quad \text{volts} \end{aligned} \tag{11.46}$$

The resulting video depends on the subtraction of two large numbers, so that any distortion accruing between the pulses will add to the resulting video.

The phase shift of the moving component between consecutive returning echoes is

$$\begin{aligned} \textit{Phase shift between pulses},\ \theta &= \frac{2\pi\ v}{\lambda\ f_{prf}} \quad \text{radians} \\ \textit{Doppler frequency},\ f_{Dop} &= \frac{2\ v}{\lambda} \quad \text{Hz} \end{aligned} \tag{11.47}$$

The resulting video (cosine rule) is

$$\begin{aligned} \textit{Resulting video}^2,\ R^2 &= 2M^2(1 - \cos\theta) \quad \text{volts}^2 \\ \textit{Resulting video},\ R &= 2M\sin\left(\frac{\theta}{2}\right) \quad \text{volts} \end{aligned} \tag{11.48}$$

An alternate way is to use the Fourier transform. The video pulses that represent the echoes are shown as two Dirac functions +1 and –1 separated by one pulse repetition time as shown in Figure 11.30(a):

$$\textit{two\_pulse}(t) = \delta(t + 1/2) - \delta(t - 1/2) \tag{11.49}$$

The filter characteristic is found by taking the Fourier transform, shown in Figure 11.30(b). The two pulses produce a sinewave in the imaginary plane (see Section 16.2.2). The Fourier transform is given by

$$F_{two\ pulse}(f) = \exp(-j\pi f) - \exp(j\pi f) = -j2\sin(\pi f) \tag{11.50}$$

The amplitude of the characteristic is shown in Figure 11.30(c).

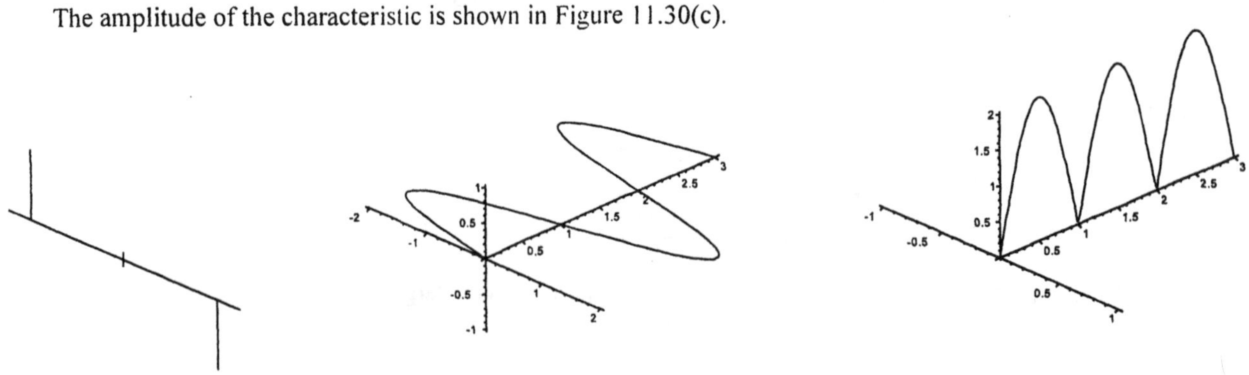

(a) Two pulses separated by one pulse repetition time.

(b) The Fourier transform of (a). The frequency scale is in multiples of the pulse repetition frequency.

(c) The frequency characteristic, the(modulus of (b). The frequency scale is in multiples of the pulse repetition frequency.

Figure 11.30 Two pulses for a canceler, their Fourier transform, and the frequency characteristic.

As an example, an S-band radar at 3 GHz ($\lambda = 0.1$ m) has a pulse repetition frequency of 1 000 Hz. The blind speed is 50 m/s, or 180 km/h, and the moving target indicator filter characteristic is shown in Figure 11.31.

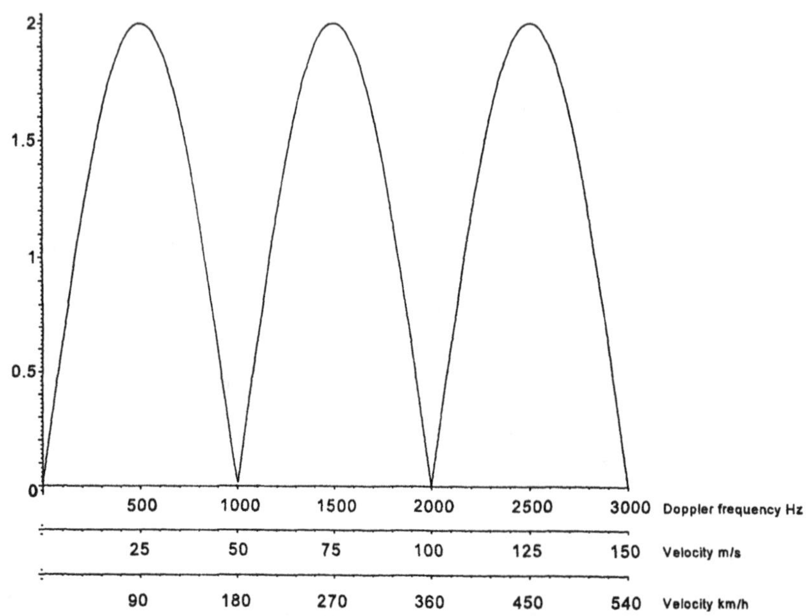

Figure 11.31 Moving target indicator characteristic for a radar frequency of 3 GHz and a pulse repetition frequency of 1 000 Hz.

The Doppler frequencies are folded down into the range from zero to the pulse repetition frequency so that the first arch in Figure 11.31 describes the moving target indicator characteristic fully. The frequency characteristic also affects wanted echoes, which will be enhanced or suppressed according to their Doppler frequencies.

The relative movement of the rotating antenna and the stationary clutter in the scenario broadens the clutter spectrum. The single canceler notch is too narrow, and there are components of clutter that are not attenuated enough. The mean

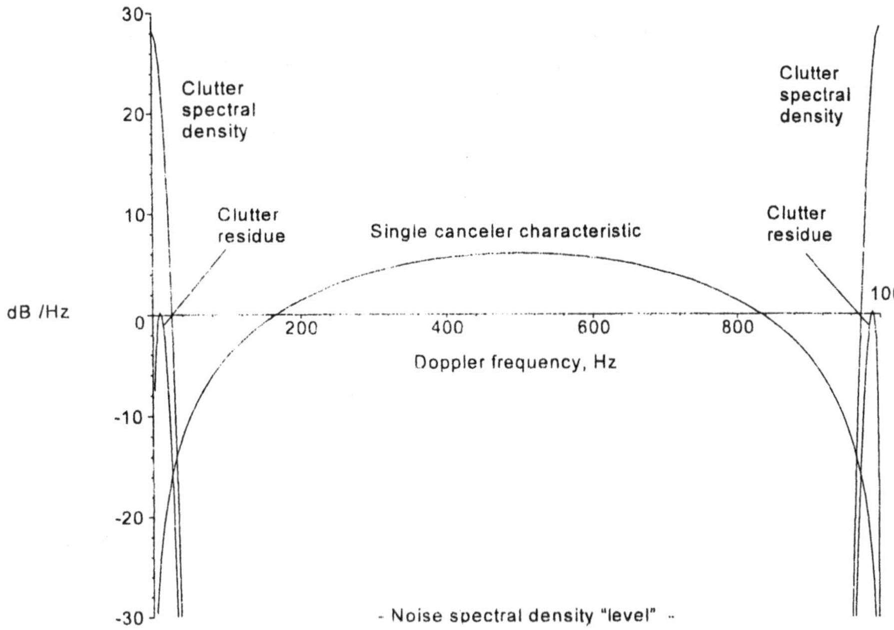

Figure 11.32 The effect of a single canceler moving target indicator filter on clutter in the model.

clutter residue at 30 km (maximum) is shown in Figure 11.32. The residue, at the 94th percentile of the clutter distribution, is more than enough to cross a detection threshold of, say, 10 dB.

The spectral density diagrams have been drawn by spreading the clutter in decibel units with respect to noise ratio over the ambiguous Doppler frequency space. Thus the noise spectral density "level" is divided, in the example here, by 1 000 Hz or 30 dB.

Double cancelers were often used in the past and they consist of two single cancelers in series. This results in the three pulses having the weights −1, 2, −1 and can be represented by the Dirac functions, see Figure 11.33(a),

$$three\_pulse(t) = -\delta(t+1) + 2\delta(t) - \delta(t-1) \tag{11.51}$$

The Fourier transform is purely real, a negative cosine wave around the value 2 as shown in Figure 11.33(b), given by

$$F_{three\ pulses}(f) = -\exp(-j2\pi f) + 2 - \exp(j\pi f) = -2\cos(2\pi f) + 2 \tag{11.52}$$

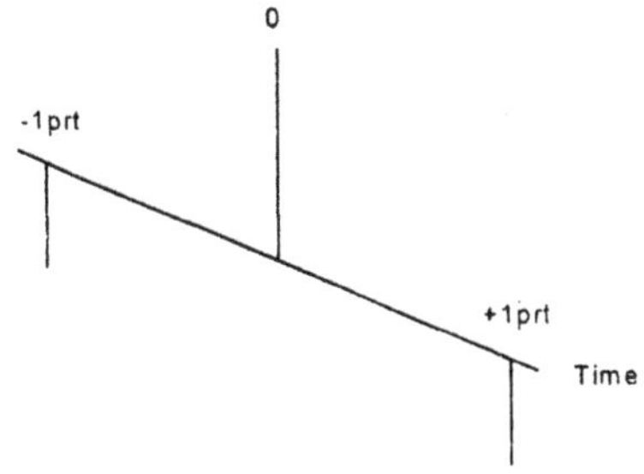

(a) Three pulses each separated by one pulse repetition time (prt).

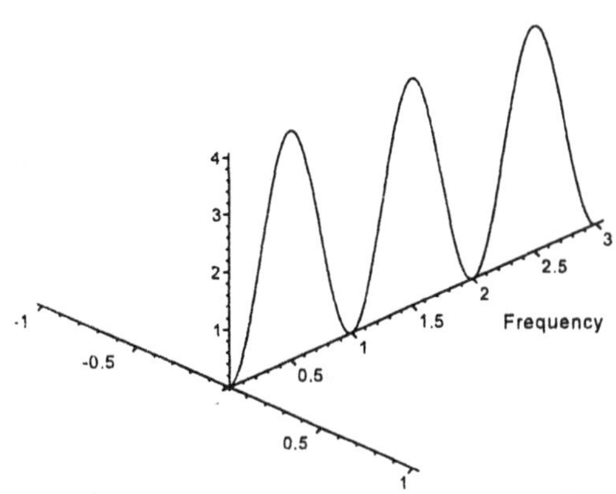

(b) The Fourier transform and frequency characteristic of (a) The frequency scale is in multiples of the pulse repetition frequency.

Figure 11.33 Three pulses for a canceler and their Fourier transform or frequency characteristic.

Double, or three pulse cancelers have a wider and deeper $\sin^2 x$ rejection notch and the effect on the model clutter is shown in Figure 11.34. Single and double cancelers are compared in Figure 11.35. Better control of the notch characteristic can be obtained by using more cancellation stages and controlling the feed-forward and feed-back paths.

The effect of moving target processing using a single or two pulse canceler with the clutter in the model radar is shown in Figure 11.35. A double canceler reduces the clutter on the spectral density so much that they are too small for illustration using the same scale.

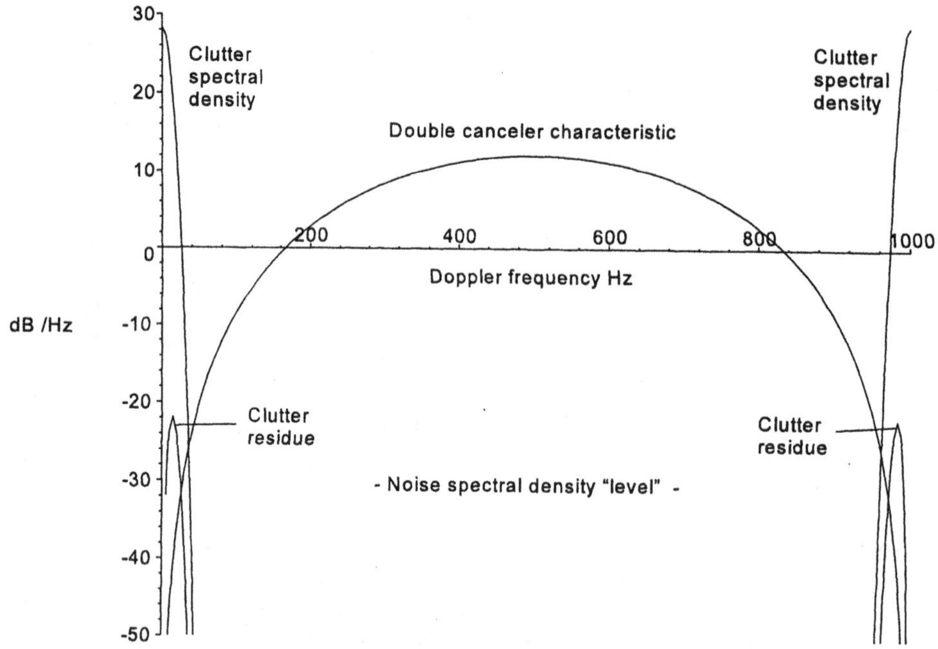

Figure 11.34 The effect of a double canceler moving target indicator filter on clutter in the model.

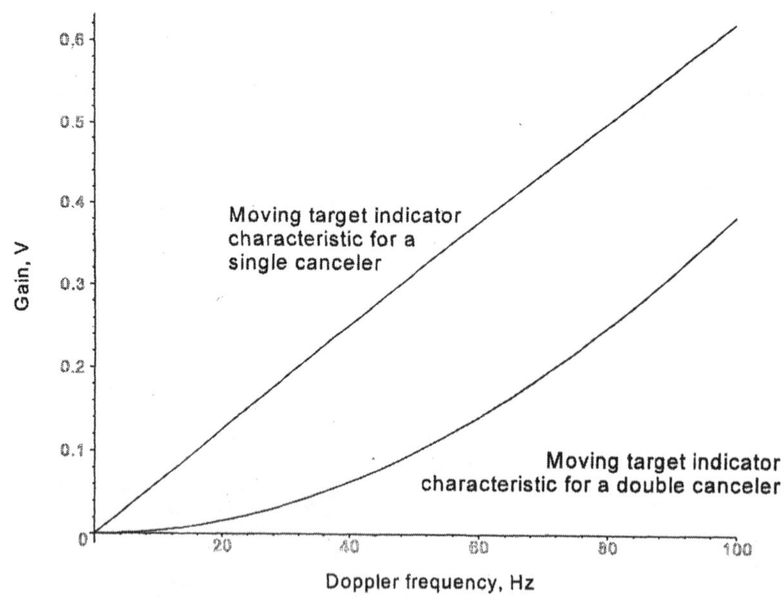

Figure 11.35 Comparison of the single and double canceler moving target indicator characteristics in the clutter region.

The effect of a single canceler on the clutter landscape is shown in Figure 11.36.

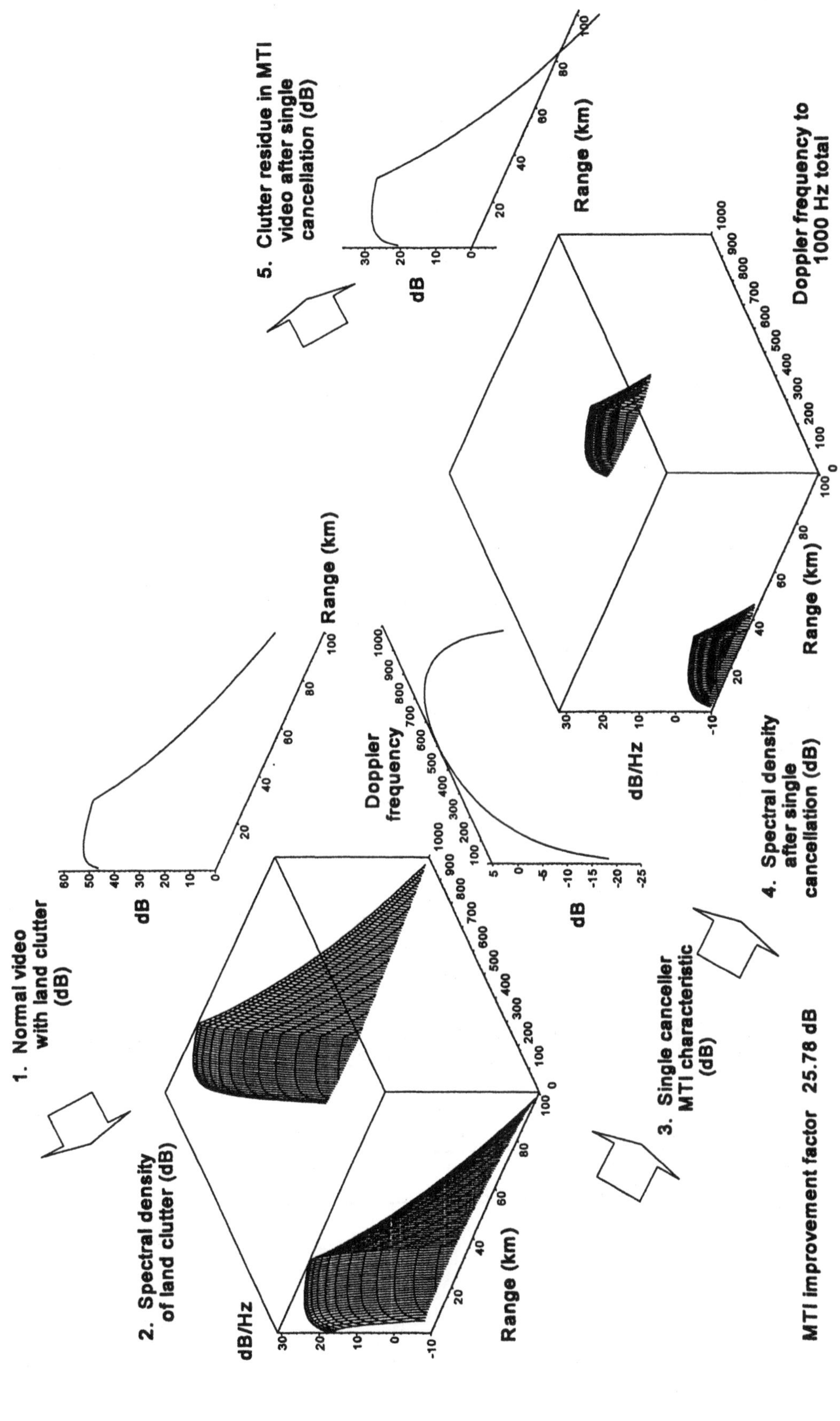

Figure 11.36 The effects of single moving target indicator (MTI) filtering on clutter in the scenario.

A more general treatment of MTI with a number of canceling stages may be made by changing the form of the filter to the general finite impulse response filter form as shown in Figure 11.37.

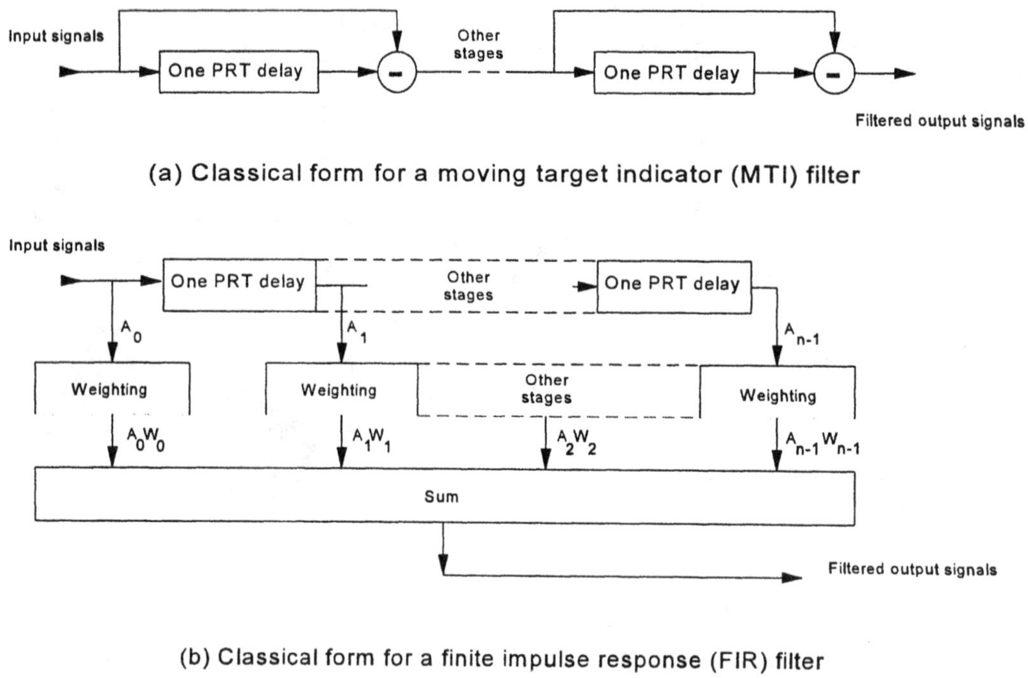

Figure 11.37 Classical forms of moving target indicator (MTI) and finite impulse response (FIR) filters.

If a is the input to the time delay and b is the output, then the output from a single canceler is $a - b$. If the cancelers are cascaded, we have

$$\begin{aligned}
a - b &= a - b \\
(a - b)^2 &= a^2 - 2ab + b^2 \\
(a - b)^3 &= a^3 - 3a^2b + 3ab^2 - b^3 \\
(a - b)^4 &= a^4 - 4a^3b + 6a^2b^2 - 4ab^3 + b^4 \\
(a - b)^5 &= a^5 - 5a^4b + 10a^3b^2 - 10a^2b^3 + 5ab^4 - b^5 \\
&\text{and so on}
\end{aligned} \tag{11.53}$$

The weighting functions of the finite impulse response filter are the binomial coefficients in (11.53) in the form of a Pascal triangle shown in Table 11.1. The polarities of the weighted pulses may be also reversed.

Table 11.1
Pascal triangle weighting constants

Number of delays	Number of pulses	Weighting
1	2	+1, −1
2	3	+1, −2, +1
3	4	+1, −3, +3, −1
4	5	+1, −4, +6, −4, +1
5	6	+1, −5, +10, −10, +5, −1

Examples for four and five pulse cancelers are given in (11.54) to (11.57). The pulses for a four pulse canceler may be represented by

$$four\_pulse(t) = \delta(t + 1.5) - 3\delta(t + 0.5) + 3\delta(t - 0.5) - \delta(t-1.5) \tag{11.54}$$

The Fourier transform is

$$\begin{aligned} F_{four\ pulses}(f) &= \exp(-j\pi f) - 3\exp(-j3\pi f) + 3\exp(j\pi f) - \exp(j3\pi f) \\ &= -j2\ \sin(3\pi f) + j6\ \sin(\pi f) \end{aligned} \tag{11.55}$$

These are shown in Figure 11.38.

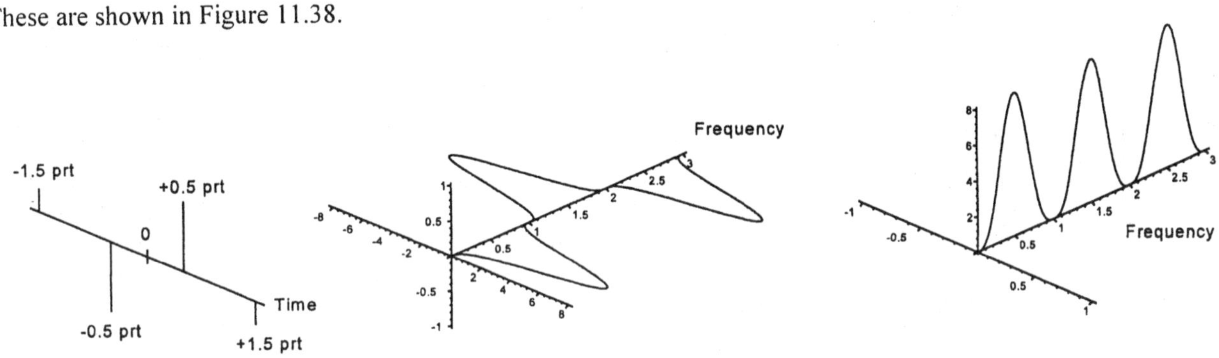

(a) Four pulses each separated by one pulse repetition time (prt).

(b) The Fourier transform of (a). The frequency scale is in multiples of the pulse repetition frequency.

(c) The frequency characteristic, the modulus of (b). The frequency scale is in multiples of the pulse repetition frequency.

Figure 11.38 A four-pulse canceler, Fourier transform, and frequency characteristic.

For a five pulse canceler the input pulses are

$$five_{pulse}(t) = \delta(t + 2) - 4\delta(t + 1) + 6\delta(t) - 4\delta(t - 1) + \delta(t - 1); \tag{11.56}$$

to give a Fourier transform

$$\begin{aligned} F_{five\ pulses}(f) &= \exp(-j4\pi f) - 4\exp(-j2\pi f) + 6 - 4\exp(j2\pi f) + \exp(j4\pi f) \\ &= 2\cos(4\pi f) + 6 - 8\cos(2\pi f) \end{aligned} \tag{11.57}$$

The waveform and the Fourier transform or frequency characteristic are shown in Figure 11.39.

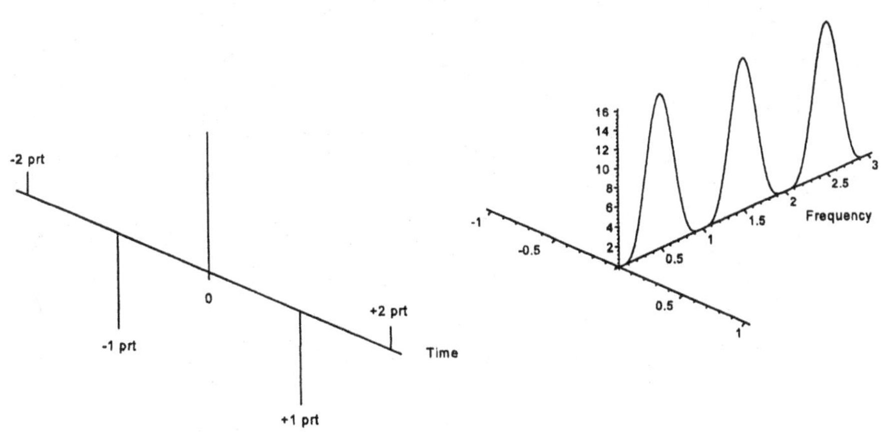

(a) Five pulses each separated by one pulse repetition time (prt).

(b) The frequency characteristic, the modulus of (b). The frequency scale is in multiples of the pulse repetition frequency.

Figure 11.39 A five-pulse canceler, Fourier transform, and frequency characteristic.

Reference [17, p. 110] gives the equation for the improvement factor for coherent moving target indicator processing using n pulses with binomial weights as

$$\text{Improvement factor, } I = \frac{\displaystyle\sum_{j=0}^{n}\left(\frac{n!}{j!(n-j)!}\right)^2}{\displaystyle\sum_{j=0}^{n}\sum_{k=0}^{n}(-1)^{k+j}\frac{n!}{j!\,(n-j)!}\,\frac{n!}{k!\,(n-k)!}\,\rho(j-k)} \tag{11.58}$$

where the correlation factor, ρ, is

$$\rho(j-k) = \exp\left(\frac{-(\,(j-k)\,2\pi\sigma_{clutter\ spectrum}T\,)^2}{2}\right) \tag{11.59}$$

where $\sigma_{clutter\ spectrum}$ is the standard deviation of the clutter spectrum;
T is the pulse repetition time.

The improvement factor is plotted for clutter spectrum relative standard deviation in Figure 11.40 and for the number of pulses between the antenna half power points in Figure 11.41. The dynamic range of all the preceding stages must exceed that of the desired improvement factor.

Simpler expressions have been used in the past [14, p. 15.14 quotes 9, pp. 210-219], namely,

$$I_1 = \frac{N^2}{1.39} \qquad I_2 = \frac{N^4}{3.84} \qquad I_3 = \frac{N^6}{16.0} \tag{11.60}$$

where I_1, I_2, and I_3 are the improvement factors with one, two, and three delays, respectively.

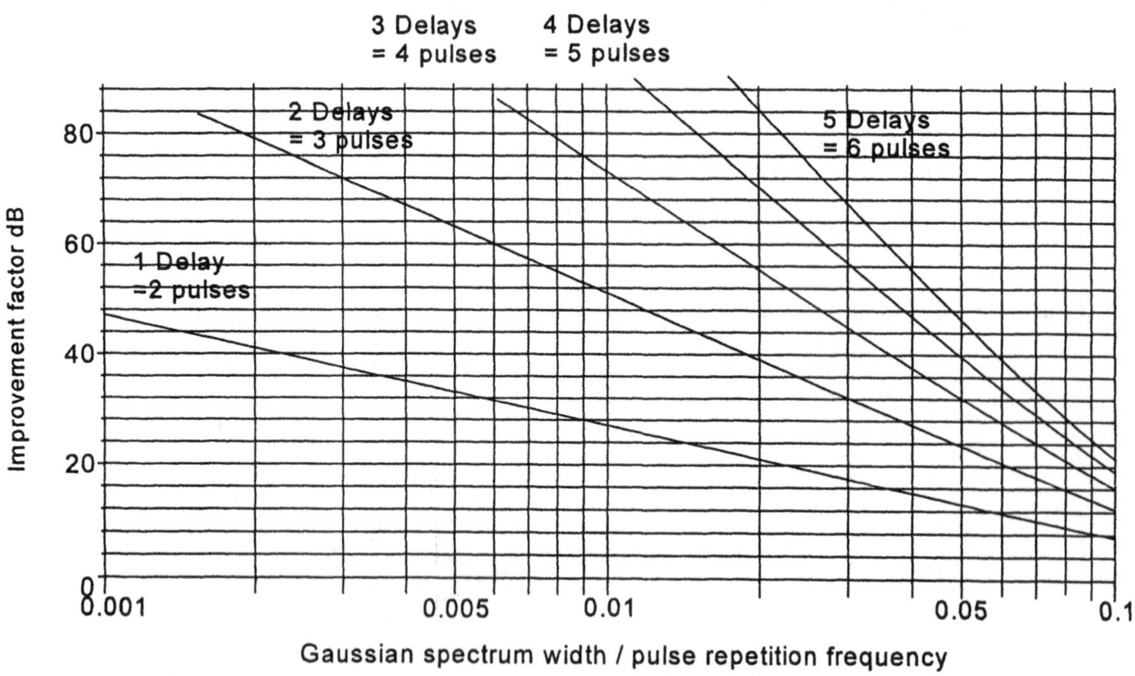

Figure 11.40 Graphs of maximum improvement factor against clutter spectrum width.

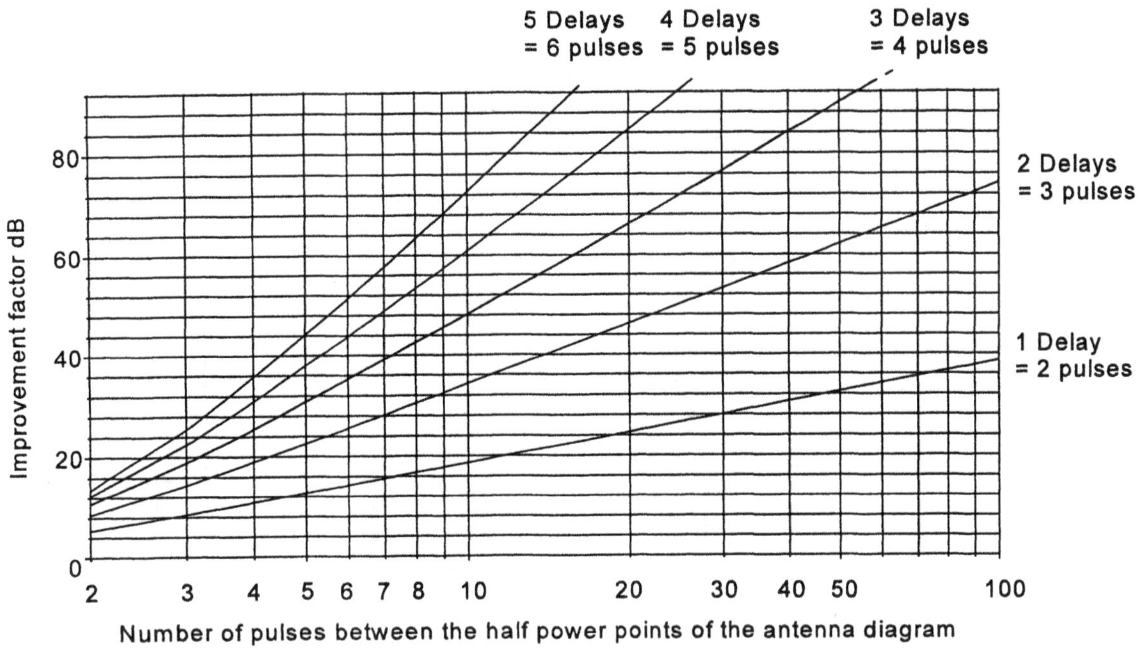

Figure 11.41 Plot of maximum improvement factor against the number of pulses between the half power points of the antenna pattern.

11.2.3.4 Correlation of neighboring pulses (in azimuth) at the output

The contents of the range cells after moving target indicator filtering are not independent from pulse to pulse. This subject is treated in [18, 19] for a radar with definite dwells, such as a phased array radar. For example, in a two-pulse canceler the subtracted output pulses consist of components of both input pulses and are no longer independent. For the purposes of summing the signal samples and noise samples, the effective number of independent samples is reduced and is given by the following equations.

For a two pulse canceler:

$$N_2 = \frac{n^2}{n + \frac{n-1}{2}} \approx \frac{2}{3} n \quad \text{for large } n \tag{11.61}$$

For a three pulse canceler:

$$N_3 = \frac{n^2}{n + \frac{8(n-1)}{9} + \frac{n-2}{18}} \approx \frac{18}{35} n \quad \text{for large } n \tag{11.62}$$

For a four pulse canceler:

$$N_4 = \frac{n^2}{n + \frac{9(n-1)}{8} + \frac{9(n-2)}{50} + \frac{n-3}{200}} \approx 0.433\, n \quad \text{for large } n \tag{11.63}$$

The relationships in (11.61 to 11.63) are shown in Figure 11.42 and the reduced number of pulses is used to calculate the probability of detection after filtering (see Chapter 12, Threshold and detection). The relationship is approximate for mechanically rotating radars.

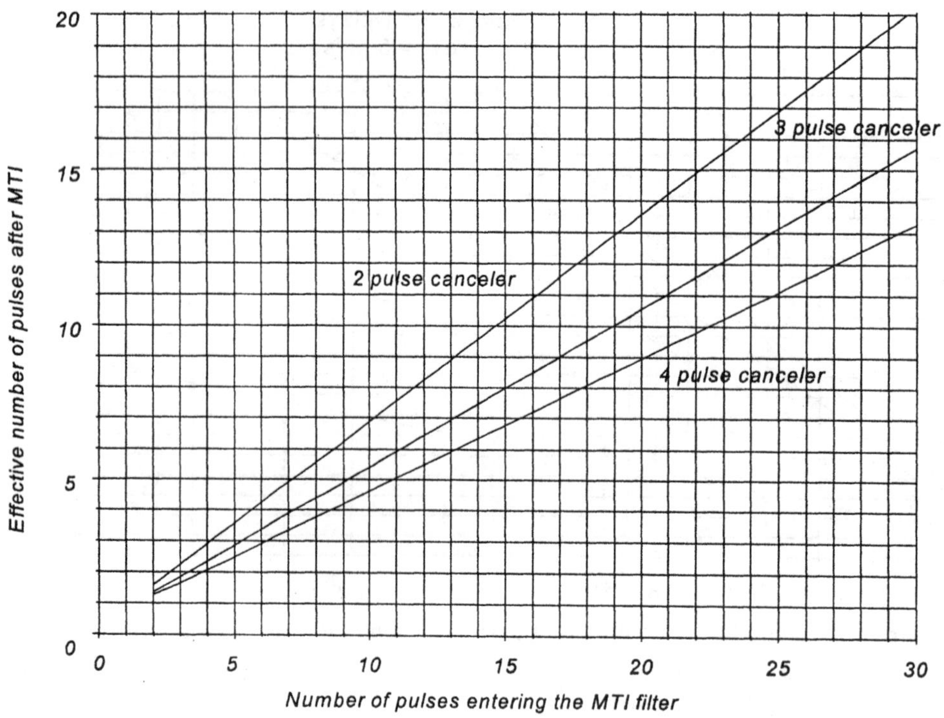

Figure 11.42 The reduction of the effective number of independent samples after moving target indicator filtering.

11.2.3.5 Staggering the pulse repetition frequencies

The notches in the moving target indicator velocity characteristic occur at the blind speeds, often speeds normally used by objects of interest. Examples are shown in Tables 11.2 and 11.3 for the frequency bands in United States (U.S.), United Kingdom (U.K.), and electronic countermeasures (ECM) nomenclatures.

Table 11.2
Examples of first blind speeds in metric units for a number of radar bands

Pulse repetition frequency, Hz				10 000	1 000	250
Maximum range, km				15.00	150.00	600.00
		Band		First blind speeds, m/s		
Frequency	U.S.	U.K.	ECM			
600 MHz	UHF	UHF	C	2.50	250.00	62.50
1 300 MHz	L	L	D	1153.85	115.38	28.85
3 000 MHz	S	S	E..F	500.00	50.00	12.50
5 500 MHz	C	C	G	272.73	27.27	6.82
10 000 MHz	X	X	I	150.00	15.00	3.75
16 000 MHz	Ku	J	J	93.75	9.38	2.34
30 000 MHz	Ka	Q	K	50.00	5.00	1.25

Table 11.3
Examples of first blind speeds in nautical miles for a number of radar bands

Pulse repetition frequency, Hz				10 000	1 000	250
Maximum range, nautical miles				8.11	81.08	324.32
		Band		First blind speeds, knots		
Frequency	U.S.	U.K.	ECM			
600 MHz	UHF	UHF	C	4875.00	487.50	121.88
1 300 MHz	L	L	D	2250.00	225.00	56.25
3 000 MHz	S	S	E..F	975.00	97.50	24.38
5 500 MHz	C	C	G	531.82	53.18	13.30
10 000 MHz	X	X	I	292.50	29.25	7.31
16 000 MHz	Ku	J	J	182.81	18.28	4.57
30 000 MHz	Ka	Q	K	97.50	9.75	2.44

If different pulse repetition frequencies are used for successive transmitter pulses, the notch at zero speed remains and the blind speeds become blurred. The results of moving target indicator processing at the different pulse repetition frequencies are integrated during display or extraction.

If a three pulse moving target indicator processor is considered (Steinberg in [20, Part VI, Chapter 2]) it will have three pulse repetition times, T, $T(1 + \epsilon)$, and $T(1 - \epsilon)$, as shown in Figure 11.43. The regular rhythm is broken and becomes the rhythm of someone staggering around, hence the name.

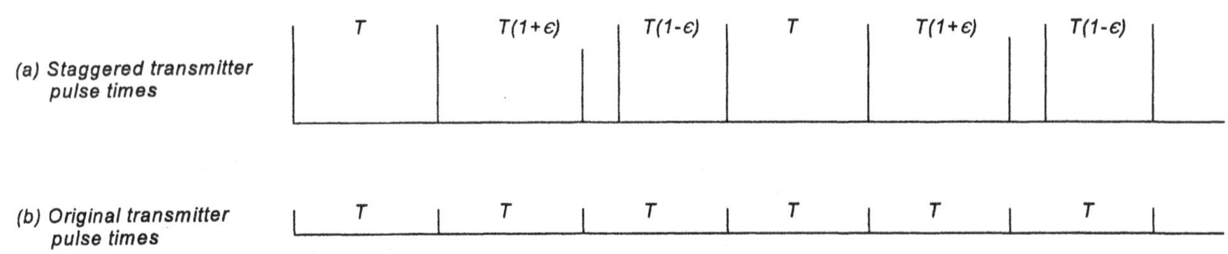

Figure 11.43 Comparison of (a) staggered and (b) normal transmitter pulse times.

The frequency characteristic can be given by the sum of the three vectors

$$\begin{aligned} h(t) &= \delta(t + T(1 - \epsilon) - 2\delta(t) + \delta(t - T(1 + \epsilon) \\ H(f) &= e^{j2\pi fT(1 - \epsilon)} - 2 + e^{-j2\pi fT(1 - \epsilon)} \\ &= -2(1 - e^{j2\pi fT\epsilon}\cos 2\pi fT) \end{aligned} \tag{11.64}$$

When ϵ=0, this is the same as the double or three-pulse canceler without staggering. For illustration, a system with 10% pulse repetition frequency stagger is chosen, and Figure 11.44 shows the effect of this timing turning into a phase shift in the frequency domain to give the characteristic a twist. Commonly, the absolute value of this twisted characteristic is used for the staggered moving target indicator characteristic.

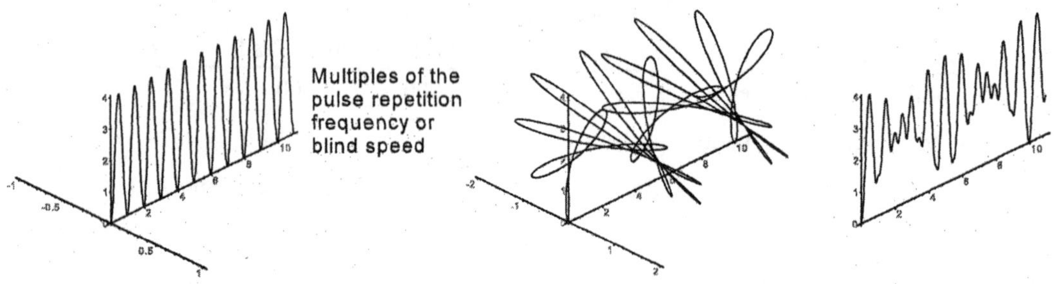

(a) No stagger (b) 10% stagger, complex (c) 10% stagger, modulus

Figure 11.44 The effects of 10% staggering on the moving target indicator characteristic.

The frequency axis in Figure 11.44 is in units of the pulse repetition frequency or the first blind speed. When the terms $e^{j\omega T\epsilon}$ and $\cos\omega T$ are both +1 or −1, the curve touches the frequency axis and a genuine blind speed occurs. In the example, this occurs for Doppler frequencies of 10 times the pulse repetition frequency or at 10 times the first unstaggered blind speed. The curves are shown in Figure 11.45, and in both cases the average value of the processing gain is 2.0.

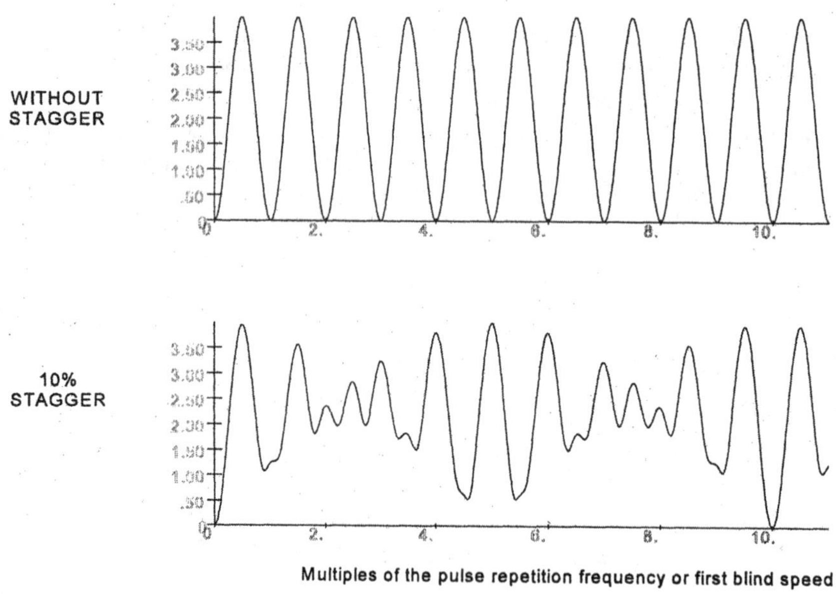

Figure 11.45 Three pulse moving target indicator without and with 10% stagger.

For small values of ϵ ($\epsilon \ll 1$), the first null amplitude is

$$\frac{H\left(\frac{2\pi}{T}\right)}{H_{\max}} = \frac{2\left|1 - e^{-j2\pi\epsilon}\right|}{4} \approx \pi\epsilon \tag{11.65}$$

In the example with 10% staggering, this is 0.4π, or 1.26, in Figure 11.46.

Staggering does have penalties: limited improvement factors and stationary echoes beyond maximum range (second time around) arrive at staggered times and are not canceled. The literature [1, p. 7-15; 14, p. 15.19; 18, p. 146] gives the limitation as

$$I_{\lim} = 20\log_{10}\frac{2.5}{\gamma - 1}\ \text{dB} \tag{11.66}$$

where $\gamma = \dfrac{\epsilon + 1}{\epsilon - 1}$ or $\epsilon = \dfrac{\gamma - 1}{\gamma + 1}$.

This is plotted in terms of γ and ϵ in Figure 11.46 and Figure 11.47, respectively.

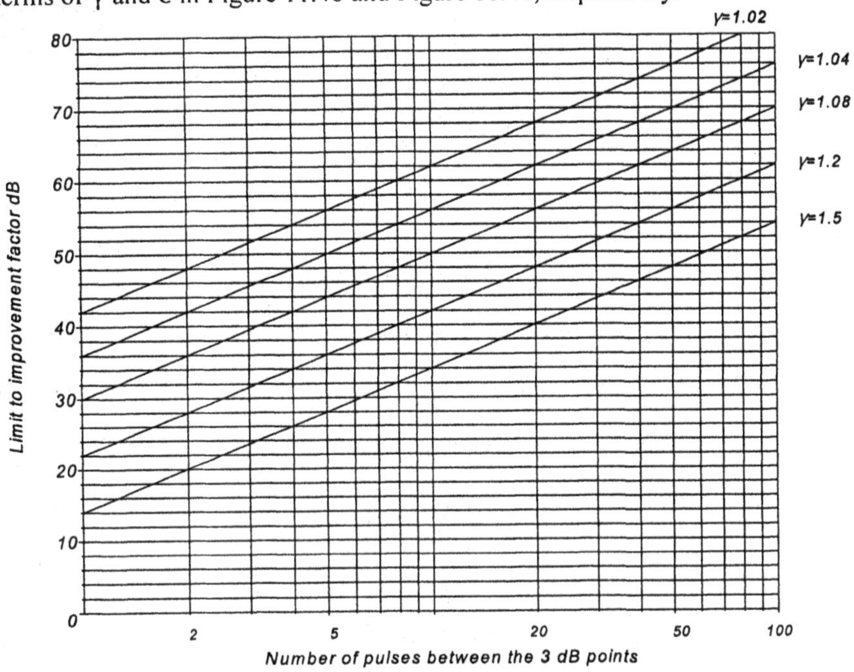

Figure 11.46 Limits to improvement factors with staggering in terms of γ.

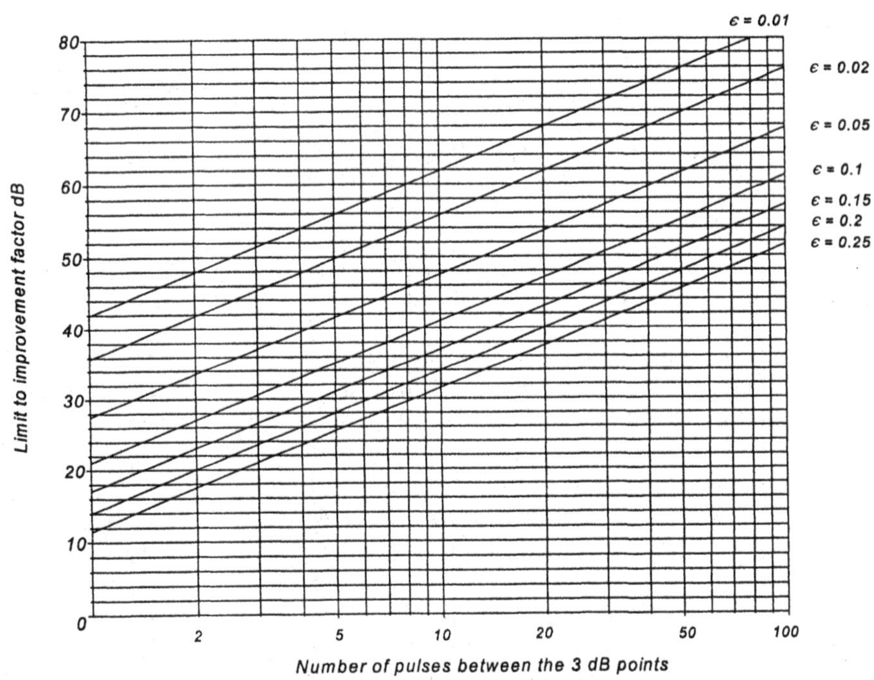

Figure 11.47 Limits to improvement factors with staggering in terms of ϵ.

Figures 11.48 and 11.49 show the form of the Doppler frequency characteristics of moving target indicators stages using 4 pulses (three delays) and five pulses (four delays). The stagger ratios are expressed mostly in whole numbers as the relative spacing between the pulses. Diagrams with even numbers of pulses remind one of the diagram for two pulse cancelers with their sine wave Fourier transform characteristic.

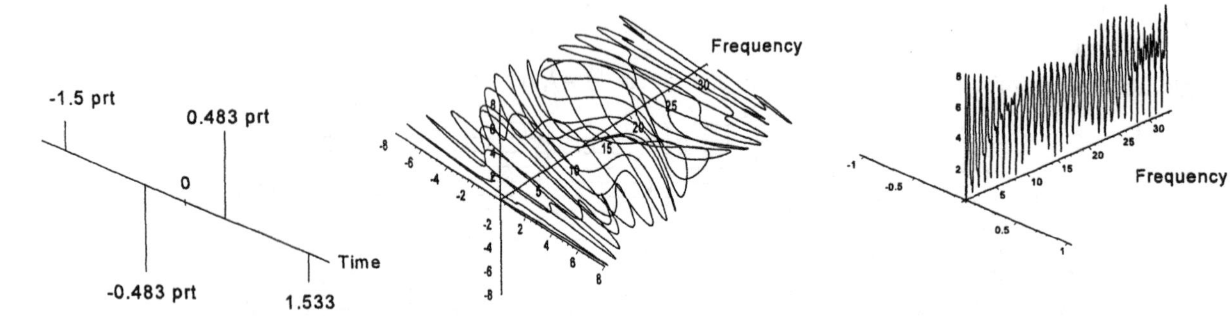

(a) Four pulses each separated by approximately one pulse repetition time (prt).

(b) The Fourier transform of (a). The frequency scale is in multiples of the pulse repetition frequency.

(c) The frequency characteristic, the modulus of (b). The frequency scale is in multiples of the pulse repetition frequency.

Figure 11.48. The weighted staggered pulses and Doppler frequency characteristics for four pulses with stagger ratios of 30:29:31.

With five pulses the diagram resembles that for three staggered pulses with a raised axis for the circular whorls.

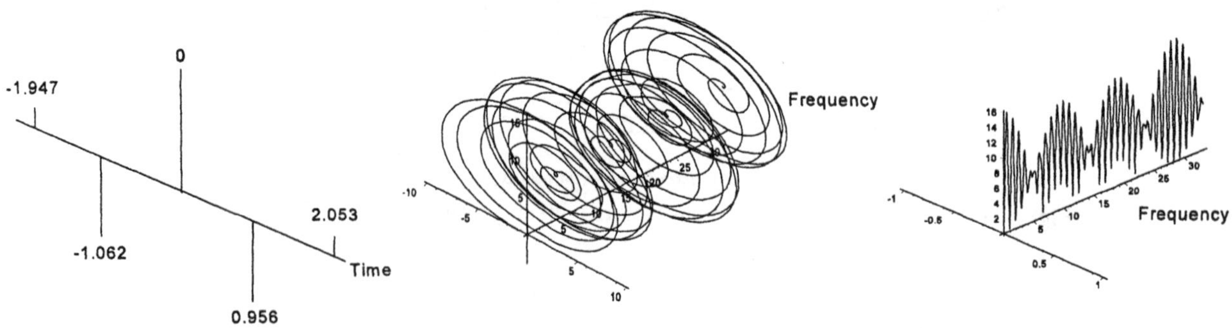

(a) Five pulses each separated by approximately one pulse repetition time (prt).

(b) The Fourier transform of (a). The frequency scale is in multiples of the pulse repetition frequency.

(c) The frequency characteristic, the modulus of (b). The frequency scale is in multiples of the pulse repetition frequency.

Figure 11.49. The weighted staggered pulses and Doppler frequency characteristics for five pulses with stagger ratios of 25:30:27:31.

11.2.3.6 The effects of limiting

The wishes of operators only to see moving objects on their displays are not completely fulfilled by moving target indicator processors. The residues of large clutter echoes exist because their amplitude is greater than the improvement factor. The clutter is not canceled to below the noise level and may be great enough to show on displays or to be extracted to give plots.

The situation for the operators may be improved if the clutter is limited in amplitude. That allows the peaks to be canceled to below the noise level. The process of limiting widens the clutter spectrum (see Chapter 7, Receivers, Figures 7.12 and 7.13) so that parts of the spectrum pass through the moving target indicator filter. This reduces the effective improvement factor.

Figures 11.50 and 11.51 show the response of single and double cancelers to clutter in the model radar at a range of 30 km. The mean clutter spectral density is 28 dB/Hz with reference to a signal-to-noise ratio of 10 for a 1 m^2 scatterer at 100 km. With the log-normal clutter model, around 6% of the clutter will be greater than this level.

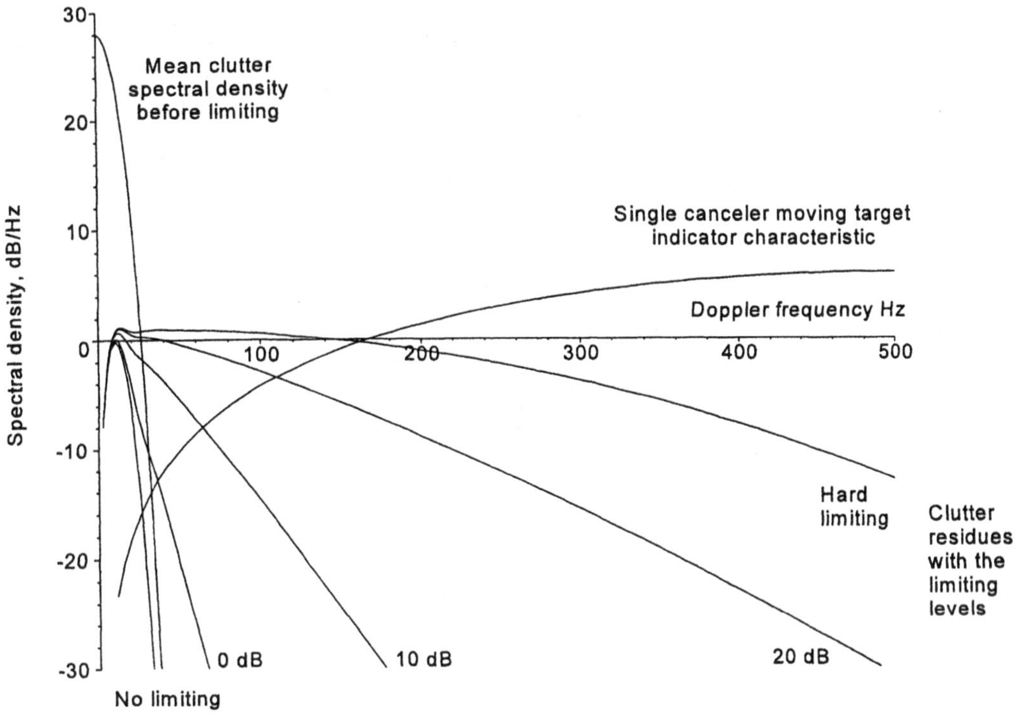

Figure 11.50 The effects of limiting on the spectral densities of clutter residues with a single canceler.

The video signal is the integral or the sum across all unambiguous Doppler frequencies. The amount of attenuation of the clutter is given by [17, pp. 138-139] for a single canceler as

$$I_1 = \frac{\sum_{p=0}^{\infty} C_p}{\sum_{p=0}^{\infty} C_p \left(1 - e^{-\frac{(2p+1)x^2}{2}}\right)} \tag{11.67}$$

where

$$C_p = \frac{a_p b_p}{(1 + \alpha)^{2p+1}}, \qquad a_p = \frac{(2p)!}{2^{2p}(p!)^2(2p + 1)}, \qquad b_p = \frac{(2p + 1)!}{p!(p + 1)!}$$

and

$$\alpha = \frac{2}{\pi} \frac{\textit{Clutter power at limiting point}}{\textit{Clutter power before limiting}}, \qquad x = 2\pi \frac{\textit{Doppler frequency}}{\textit{Pulse repetition frequency}}$$

The improvement factor has been plotted against the ratio of clutter Doppler frequency standard deviation to pulse repetition frequency in Figure 11.51. (Note that Schleher used $\sigma_v / f_r \lambda$, which gives a difference factor of 2.)

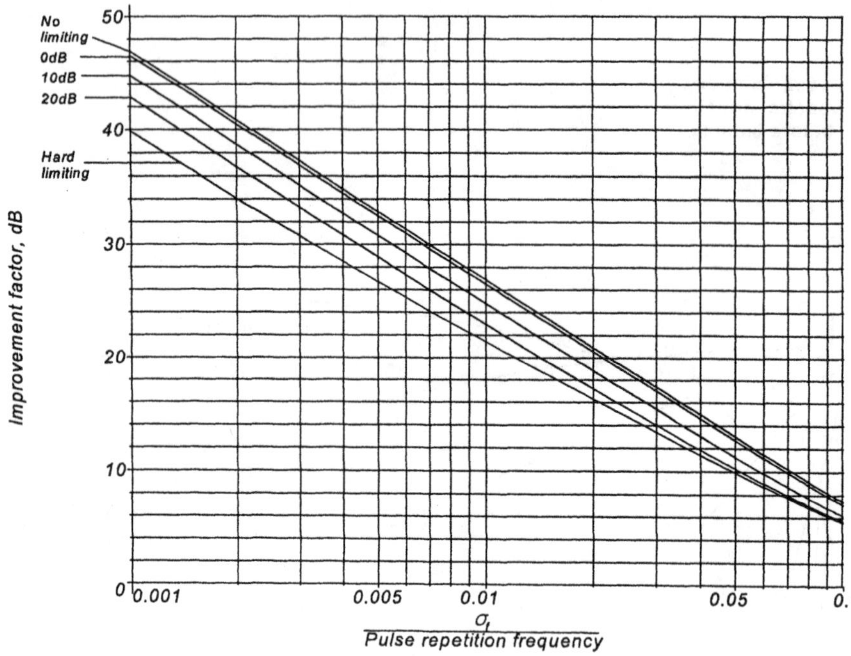

Figure 11.51 The improvement factors for a single canceler for clutter after limiting.

Double cancellation gives better clutter suppression. The effect on the limited spectral densities is shown in Figure 11.52. The improvement factors with double canceling are given by

$$I_2 = \frac{3\sum_{p=0}^{\infty} C_p}{\sum_{p=0}^{\infty} C_p \left(3 - 4e^{-\frac{(2p+1)x^2}{2}} + e^{-2(2p+1)x^2}\right)} \tag{11.68}$$

Limiting severely reduces the ability of a moving target indicator stage to reduce clutter. For example, with a single canceler, the number of hits required increases from 11 to 22 for a maximum improvement factor of 20 dB and from 36 to 82 for 30 dB. For a double canceler, the increase is from 10 to 12 hits for 35 dB or from 13 to 102 hits for 40 dB maximum improvement factor. It is unlikely that there will be enough hits from a surveillance radar to give acceptable clutter suppression in the presence of strong clutter with reasonable resolution in azimuth. That is why it is imperative that there is linear amplification of the signals starting from the input to the receiver.

11.2.3.7 Gating moving target indicator videos

Though moving target indicator processing brings advantages, it has also many losses associated with it. It alters the amplitudes of the wanted echoes and suppresses them near the blind speeds. Luckily, clutter exists mostly at short ranges, where the wanted echoes have good signal-to-noise ratios so the losses are tolerable. For this reason, the video that is displayed or used for an extractor is moving target indicator processed (MTI video) within the clutter region and linear (normal) video beyond and is called MTI-normal video. Later, clutter maps were introduced to remember the positions of the clutter areas and to switch the video for display and extraction.

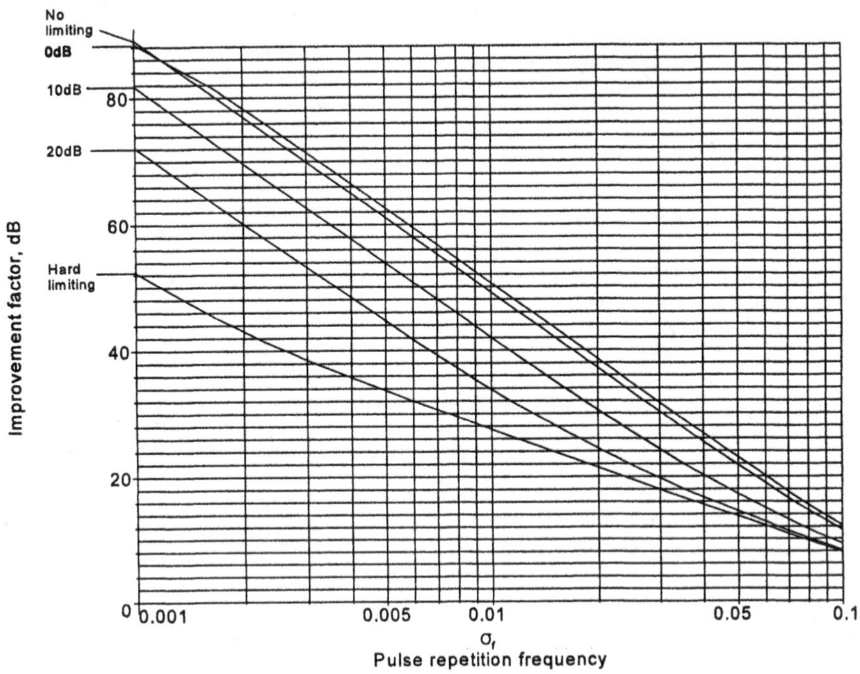

Figure 11.52 Improvement factors for a double canceler for clutter after limiting.

11.2.3.8 Moving target indicator systems which use three or more pulses or two delays

The processing of more than three pulses brings more degrees of freedom when designing the filter. This allows the following:

- Increasing the number of pulses;
- Changing the weights;
- Notches;
- Feedback;
- Better staggering.

Increasing the number of pulses
When the number of pulses in the moving target indicator filter is increased, the filling time for the filter is greater and the passband of the filter is narrowed, as shown in Figure 11.53. The use of such narrow filters is discussed in Section 11.2.4.

Changing the weights
Changing the weights from binomial to optimum brings a slight improvement [17, p. 113].

Notches
Notches in the characteristic can be used to suppress moving clutter, such as rain. Pairs of notches may be generated by changing the value of k_1 in Figure 11.41, where $_1k = -2$ is the normal case. Normally, these are too narrow to accommodate the spread of Doppler frequencies caused by wind shear.

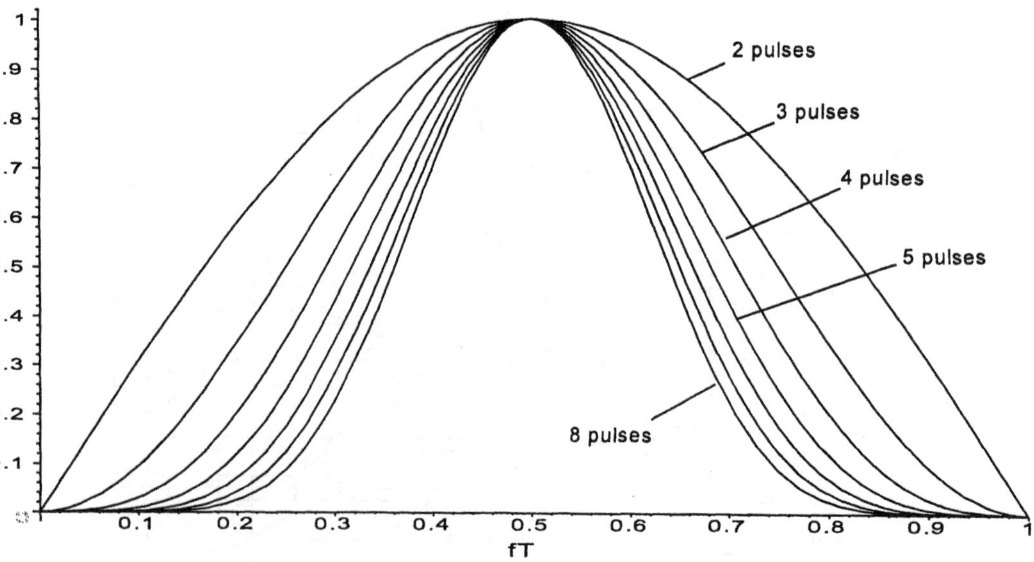

Figure 11.53 The change from a moving target indicator notch to a pulse Doppler binomial filter with an increasing number of pulses.

The notches could be moved to cover the worst of the rain echoes, but their use is limited by the width of the notch and losses to wanted echoes.

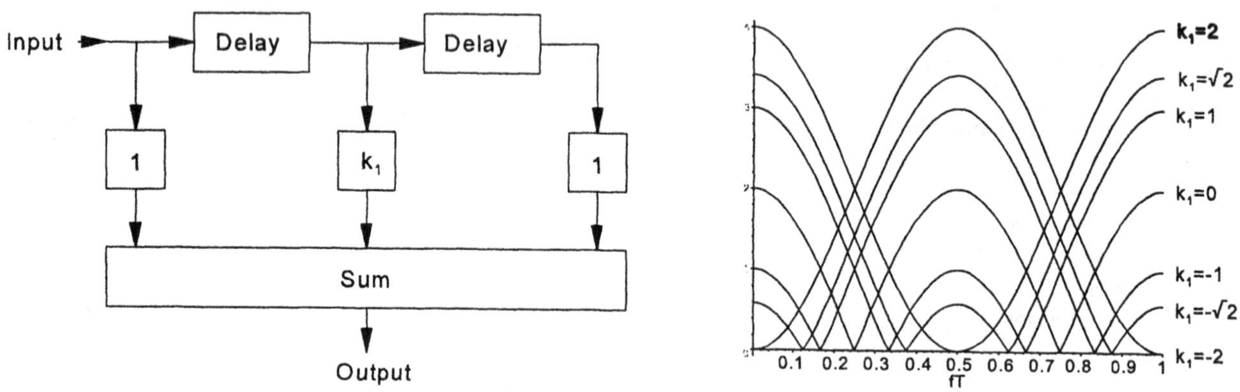

Figure 11.54 The effect of varying the value of k_1 to produce notches in the double delay moving target indicator characteristic.

Feedback

The use of feedback allows the better shaping of the passband (see Steinberg in [20]), to give a steeper rise outside the clutter band and a flatter response [17, p. 628]. The frequency characteristic is given by

$$Characteristic = \frac{(e^{-j2\pi fT} - k_{01})(e^{-j2\pi fT} - k_{02})}{(e^{-j2\pi fT} - k_{p1})(e^{-j2\pi fT} - k_{p2})} \tag{11.69}$$

Figure 11.54 shows the effect of a double canceler with $k_{01} = k_{02} = 1$ and additionally feedback with $k_{p1} = \sqrt{2}+j\sqrt{2}$ and $k_{p2} = \sqrt{2}-j\sqrt{2}$ compared with the simple double canceler curve scaled to 1/8. The characteristic is flatter and has much less gain.

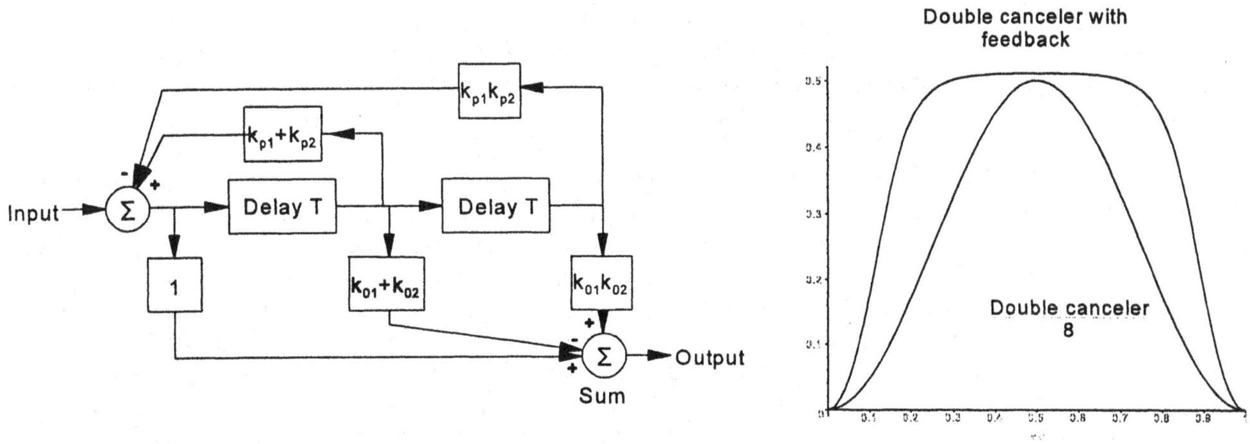

Figure 11.55 The use and effects of feedback on the moving target indicator characteristic.

Better staggering
More staggered pulse repetition frequencies give a flatter response up to and extend the first real blind speed.

There is a limit to the number of pulses that are processed. First, the filter must fill in order to work, which gives an azimuth shift. Filters without feedback give an increasingly narrow response so as to be selective filters for a small frequency band. Such filters are used for pulse Doppler processing.

11.2.4 Doppler frequency processing which selects the Doppler frequencies of interest

In contrast to moving target indicator processors which eliminate echoes with (normally) zero Doppler frequency, Doppler processors choose signals at a particular Doppler frequency. Where pulsed transmissions are used to measure range, the radar is called a pulse-Doppler radar. Three common forms of pulse Doppler, classified by the pulse repetition frequencies, are shown in Table 11.4.

Table 11.4
Types of pulse Doppler radar

	Ambiguities	Use
High	Range	Fighter aircraft radars, Airborne surveillance (AWACS) radars
Medium	Range and Doppler frequency	Fighter aircraft radars
Low	Doppler frequency	Ground surveillance (MTD) radars

The high and medium pulse repetition frequency radars use multiple pulse repetition frequencies to resolve range ambiguities. The medium and low pulse repetition frequency radars use a number of pulse repetition frequencies to resolve the Doppler frequency. The amount of clutter received at low elevations by ground search radars often saturates the dynamic range available today, so a low pulse repetition frequency and range gating must be used to separate clutter unambiguously.

When clutter maps are used to control the thresholds after pulse-Doppler processing in a low pulse repetition frequency radar, the term moving target detector (MTD) is used.

11.2.4.1 Echoes with single known Doppler frequencies or recognizable waveforms

The simplest Doppler radar is a continuous wave radar (CW) when the echo signal is integrated in a filter tuned to the Doppler frequency. If the Doppler frequency is known, then the filter may be pretuned. Otherwise, a tunable Doppler frequency filter may search for the echo. Examples are ground-to-air missile radars which search from high to low Doppler frequencies since it is most likely that their targets will scatter back the highest Doppler frequency to the radar. Speed measuring radars normally operate in the continuous wave mode and count the Doppler frequency as a measure of speed.

Radars that are interested in single echoes gate the signals at the range of the single echo. A filter tuned to the Doppler frequency of the echo or the Fourier transform of a train of signals is found. A simple example is a continuous wave missile seeker head that is pointed at its target. Knowing the frequency of the illuminating radar, it searches from the maximum expected Doppler frequency downwards until the target echo is found and acquired. The missile can then be fired.

Even simpler radars use an oscillator to generate their radio frequency energy, for example, a Gunn oscillator. Signals from moving objects beat with the outgoing energy to provide a (video) signal. When the video signal exceeds a threshold, an alarm is given. This principle is used in burglar alarms to indicate intruders in their surveillance volume. With variable time (VT) for (anti-aircraft) artillery shell fuses, simple oscillators in the noses of shells radiate their energy through an antenna. When the energy is reflected by the ground or a nearby aircraft, it has a Doppler frequency that beats with the outgoing signal. When the beats are strong enough, a signal is given to explode the shell.

If the echoes in pulse radars are gated, then the gated echo, after synchronous detection, is modulated by the Doppler frequency. Some tracking radars are able to lock onto and track the Doppler frequency of the echo after a search.

If the object being observed moves at a constant radial speed with no acceleration, the gated echoes will be heard in a loudspeaker as a single tone. The other movements of the object (for example, with aircraft, compressor turbine blades, or propellers) will give subsidiary tones. Propeller modulation was used by the German operators wearing headphones in the Second World War to track aircraft in chaff.

In radars used to watch for intruders, a moving object may be identified by a trained operator using headphones. Vehicles, people, and many types of animals may be distinguished by their characteristic tones and rhythms.

11.2.4.2 Many echoes with unknown Doppler frequencies

In the case of surveillance radars that receive many echoes with any Doppler frequency, all Doppler frequencies of interest must be handled. One of the first applications, the coherent memory filter [1, p. 17-57], used a delay line to give a form of discrete Fourier transform, as is shown in Figure 11.56.

Coherent memory filter

The intermediate frequency input (normally 30 MHz) contains both the amplitude and the phase of the echo. The signal is delayed by one pulse repetition time and then mixed with the shift frequency, f_s (ca. 500 kHz, normally upwards), to reenter the delay line at the new frequency. Without the bandpass filter, this would continue indefinitely. The passband is set to pass the number of pulses that are to be integrated.

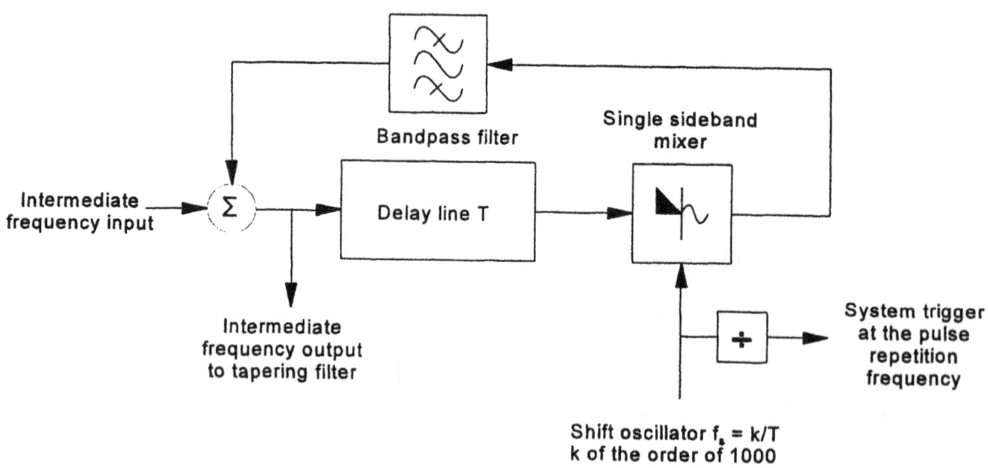

Figure 11.56 Coherent memory filter.

At the output there is a comb of spectral lines which add to give narrow pulses. The shift oscillator has a frequency that is a multiple of the pulse repetition frequency and is high enough to separate the individual sweeps in the delay line. The spectrum may be compared with the individual channels in a frequency division multiplex telephone transmission system. A definite number of cycles at intermediate frequency fit in the delay line in the pulse repetition interval so that stationary echoes at the intermediate frequency enter and leave the delay line with the same phase which also applies to the stationary echo when shifted by multiples of the shift frequency. The sum is a narrow pulse with a width of the reciprocal of the shift frequency divided by the number of pulses. If the echo has a range extent greater than the reciprocal of the shift frequency, it is represented by a number of such narrow pulses.

Echoes with a Doppler frequency shift leave the delay line with a phase change proportional to the Doppler frequency. The complex spectrum at the summing point consists of spectral lines forming a spiral, as shown in Figure 11.57. In the time domain, the rate of the spiral determines a time displacement, so that fixed and moving echoes at the same range occur at different times at the output. The spectrum, neglecting antenna modulation, is a rectangular comb so that the time function is a sin *x/x* function with time sidelobes. The sidelobes may be reduced with a tapering filter to alter the shape of the spectrum.

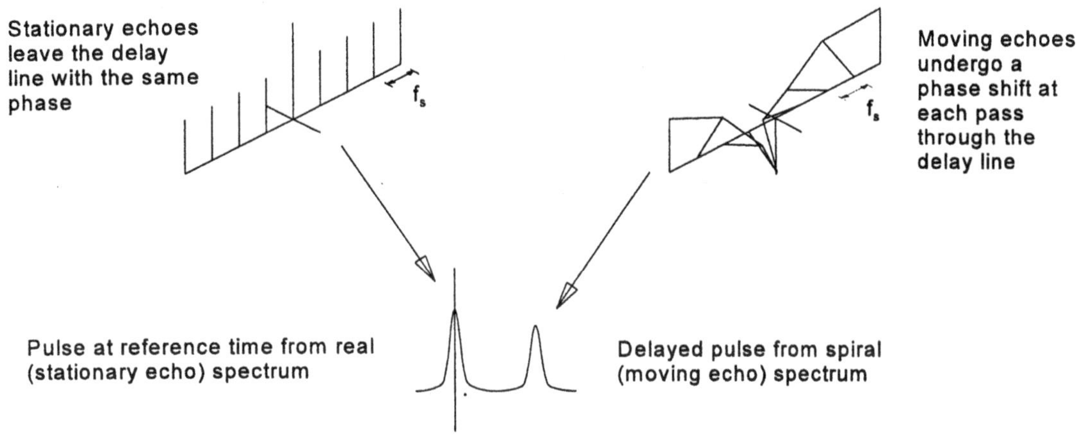

Figure 11.57 Spectra of stationary and moving echoes at the output of a coherent memory filter.

For a shift frequency of 500 kHz, the pulses for moving echoes may occur at any time in a range of 2 μs, which adds $2^2/12$ μs² to the range variance (standard deviation $2/\sqrt{12}=0.577$ μs or 85.6 m). Commonly an automatic frequency control circuit controls the shift frequency so that a test pulse at the intermediate frequency has no phase change as it passes through the delay line. The shift frequency is divided down to generate the pulse repetition frequency.

The signal voltage vectors from each pulse add whatever their Doppler frequency is. For N pulses the sum is N times the input. In contrast, the noise samples have random phases so that their sum is $\sqrt{N}$ times the noise voltage input. This gives a power gain of $(N/\sqrt{N})^2 = N$; alternatively the system may be considered to use a single pulse with N times the transmitter power.

The intermediate frequency echoes are detected with a linear detector after the tapering filter as in the block diagram in Figure 11.58. The autogate is a form of pulse length discriminator and is shown in the lower part of Figure 11.58.

The video signal passes through the delay lines each T/k seconds delay or the time between two video pulses at the same Doppler frequency. The m out of n logic is activated by long echoes, such as clutter, and blocks the gate long enough to stop the narrow clutter pulses. Wanted echoes, at other Doppler frequencies, occur at different times, are short enough in range not to activate the blocking logic, and are passed.

Owing to limitations in the tapering filter, the integrator is normally preceded by a moving target indicator stage. The coherent memory filter gave greater sensitivity and the autogate was effective in eliminating clutter residues and rain. These analogue circuits needed careful adjustment and maintenance.

Synchronous digital circuitry, which replaced the memories in moving target indicators, presented the opportunity of having a large amount of circuitry working together in parallel. The ability to keep large amounts of hardware stable was the basis for the development of the moving target detector (MTD).

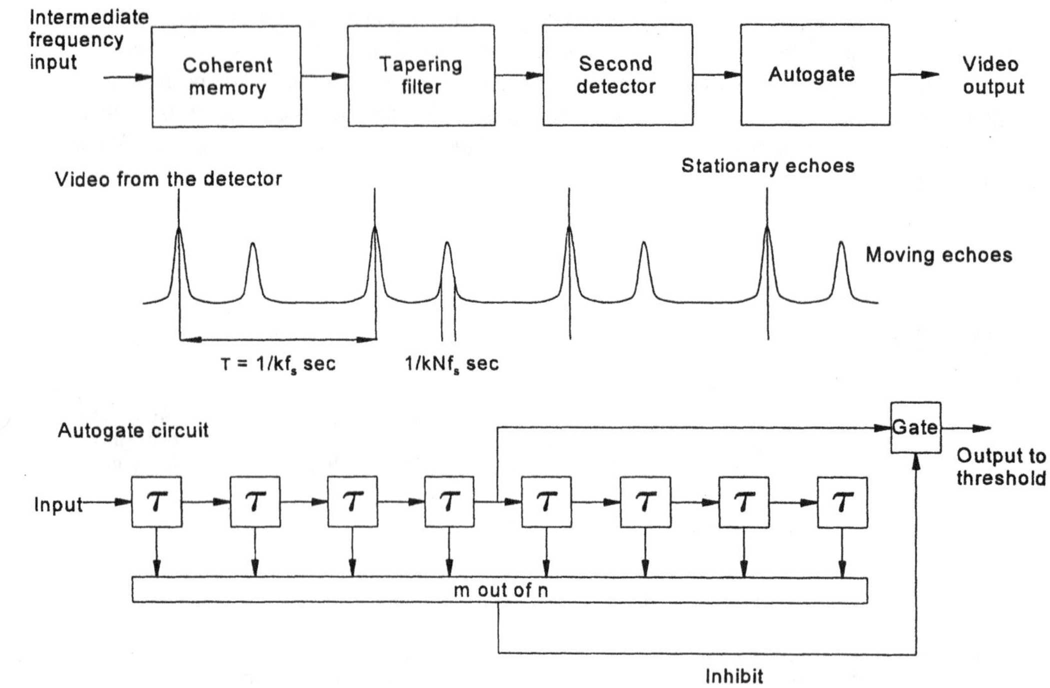

Figure 11.58 The autogate circuit.

11.2.4.3 Moving target detector (MTD)

Digital systems eliminated the need for continuous maintenance and readjustment. Synchronous logic solved the timing instabilities which plagued the coherent memory filter. Cheap digital memory has solved the storage problems, and large scale integration the space problem. Commercial digital integrated circuits are commonly manufactured for vector processing as x and y (or I and Q) components.

For pulse radars, the analogue filters must be separate for each range gate. This type of processing requires much hardware, which is difficult to adjust and maintain, and was used in fighter aircraft. These radars used a high pulse repetition frequency (ambiguous range) with few range gates. These filters were replaced by fast Fourier transform blocks and led to the development of the first moving target detectors for ground surveillance radars.

The first step is to store each vector video component of a range cell. Commonly the Cartesian I and Q components are stored as data values. This is shown in Figure 11.59.

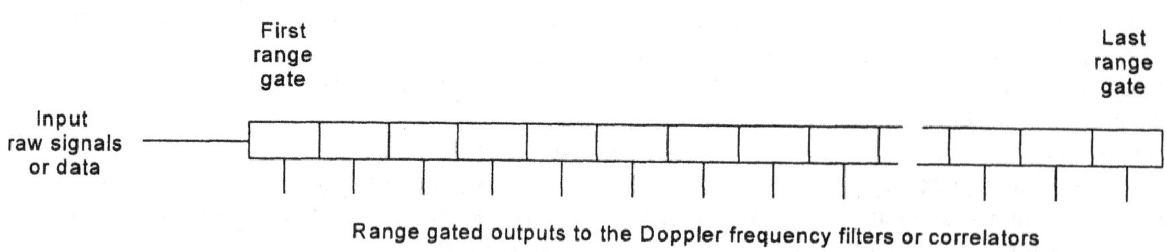

Figure 11.59 The allocation of video signals or data to range gates.

The aim is to sort the clutter into a minimum number of filter outputs and leave the others free of clutter. When the clutter spectral width is small enough, the clutter leaves one filter only. With a wider clutter spectrum the neighboring filters pass clutter signals also, namely three. If the filter center frequencies are offset, then the clutter signal outputs may be confined to two filters.

In the example used, the pulse repetition frequency is 1 000 Hz and eight pulses are integrated during the coherent processing interval (CPI). The sum of the clutter and moving echo vectors during one look is shown in Figure 11.60. There will be a 180 degree phase shift each pulse repetition time (here 1 ms) at 500 Hz Doppler frequency. Thus for 250 Hz and eight pulses, there will be two cycles in the integration time or coherent processing interval. An expanded view of the eight vectors is shown in Figure 11.61.

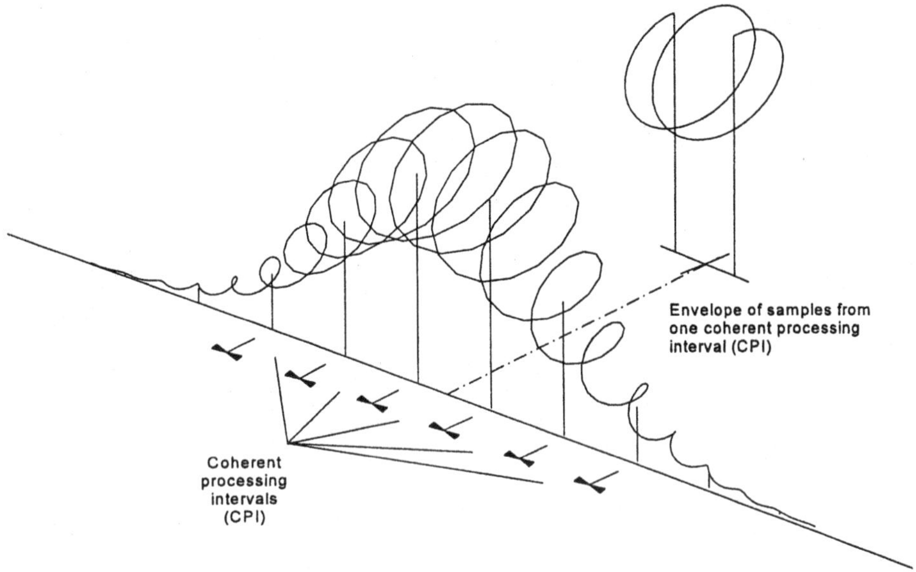

Figure 11.60 The division of the echo vectors during a scan past a moving scatterer over clutter into coherent processing intervals (CPI).

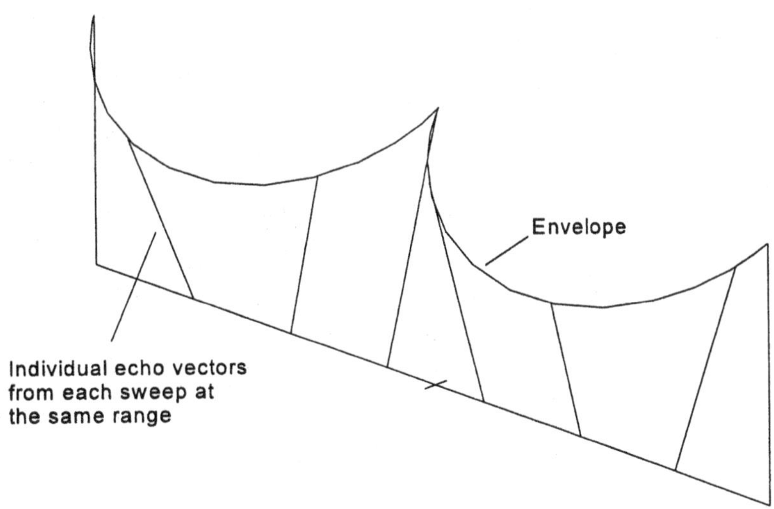

Figure 11.61 The vectors of a compound echo during the chosen coherent processing interval in Figure 11.60.

The simple sum of the vectors averages out the moving component and gives the mean clutter level during the coherent processing interval, as shown in Figure 11.62. The polyphase convention has been used, that is, the reference vector is vertical. If only stationary clutter is present, the vectors in Figure 11.62 will be in a straight line.

If each vector is given a progressive phase shift of $-2\pi fT$, the vectors in Figure 11.62 have their directions changed by this angle and tend to form a circle at each frequency, f, except when f is the Doppler frequency of the echo. Figure 11.63 shows the summing of the vectors at four values of f near the Doppler frequency of the moving echo. The result of the summed vectors is shown by a circle. The peak is not exactly on the real axis owing to the influence of the clutter signal. The diagrams have been drawn with a stationary-to-moving-signal voltage ratio of 3:1. For larger ratios, the moving signals will be the result of subtraction of large numbers with all the care that the subtraction of large numbers implies. The vector sum of the echoes that have been subjected to a phase shift of $-2\pi fT$ radians each pulse repetition time T for a chosen frequency is given by

$$\text{Vector sum} = \sum_{k=0}^{N-1} V_{echo}[k+1]\, e^{-j2\pi fkT} \tag{11.70}$$

where $V_{echo}[k+1]$ is the voltage of the k+1 th echo;
$e^{-j\,2\pi fkT}$ is the unit vector for the phase shift;
T is the pulse repetition interval.

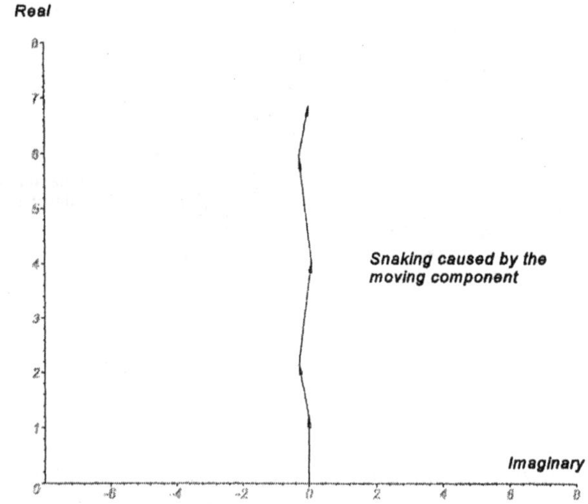

Figure 11.62 The sum of the eight vectors in Figure 11.60.

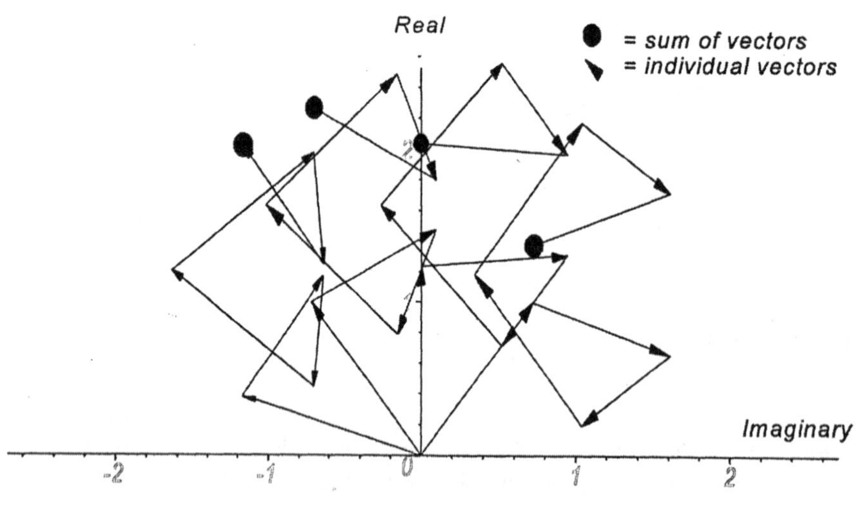

Figure 11.63 The sum of stationary and moving echo vectors with phase shifts near the Doppler frequency.

Equation (11.70) is very similar to the discrete Fourier transform of V_{echo}. If $f = 0$ this is the simple sum of the vectors and the moving component is averaged out leaving the clutter vector. If f is the Doppler frequency, the "clock is turned back" on the moving component, so that its vector components add in a straight line. The clutter components ideally form a circle and do not contribute.

The calculation of Fourier transforms takes up much processing power. Early moving target detectors used the fast

Fourier transform (FFT) [21, 22]. The fast Fourier transform is a form of discrete Fourier transform normally implemented in hardware and receives 2^N complex input samples. The parallel hardware transforms the samples into their discrete Fourier transforms with 2^N complex spectrum points. The discrete Fourier transform is only true when the number of time points is the same as the number of spectrum points. An extreme example is shown in Figure 11.64 (see [23, Fig. 4.11]) where the discrete Fourier is found for a number of pulses and the inverse transform with a continuous transform drawn between the points. The relationships between time and frequency are shown in Figure 11.65.

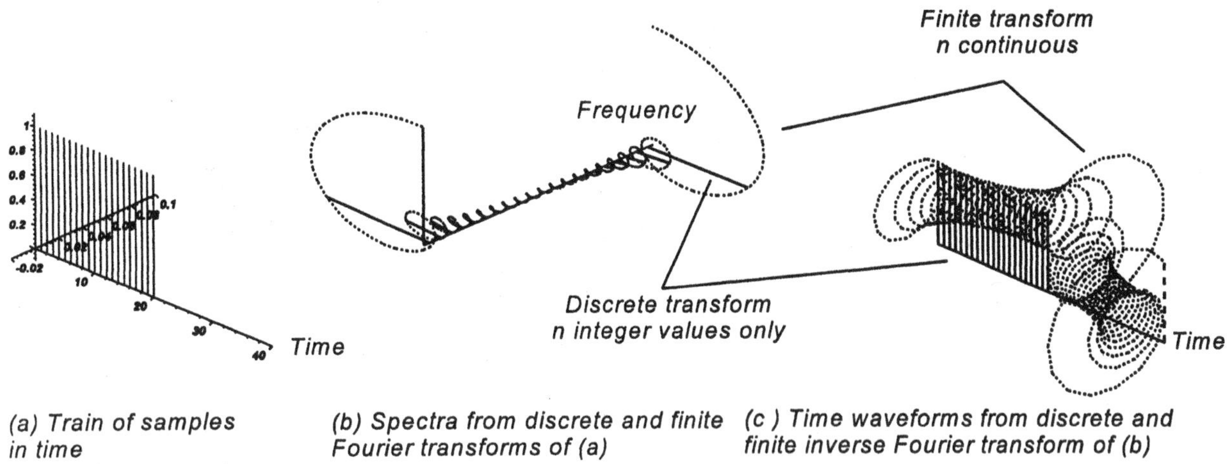

Figure 11.64 A discrete Fourier transform of a train of pulses and its inverse with the continuous transform of the pulses. [Source: Meikle, H.D., *A New Twist to Fourier Transforms*, Weinheim: Wiley-VCH, 2004.]

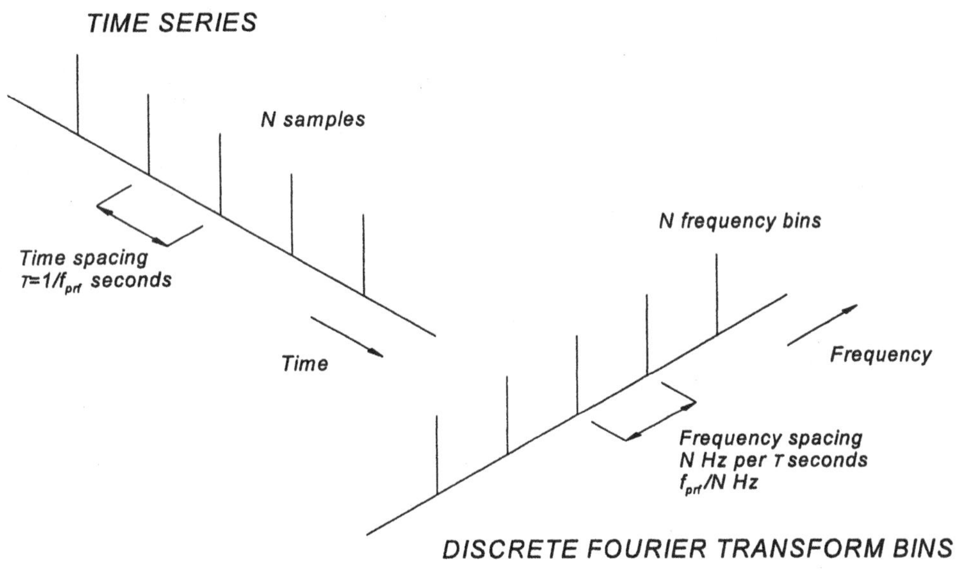

Figure 11.65 The relationship between the number of samples and the number of frequency bins.

The lower values of 2^N are 1, 2, 4, 8, 16, 32, and so on. Typical values for the number of hits between the half power points for airport radars lie around 20, and it is customary to divide the train of hits into two coherent processing intervals. The discrete Fourier transform is much more flexible and multiplier-accumulator (MAC) chips have been developed to perform the calculations for the finite impulse response filter in Figure 11.66.

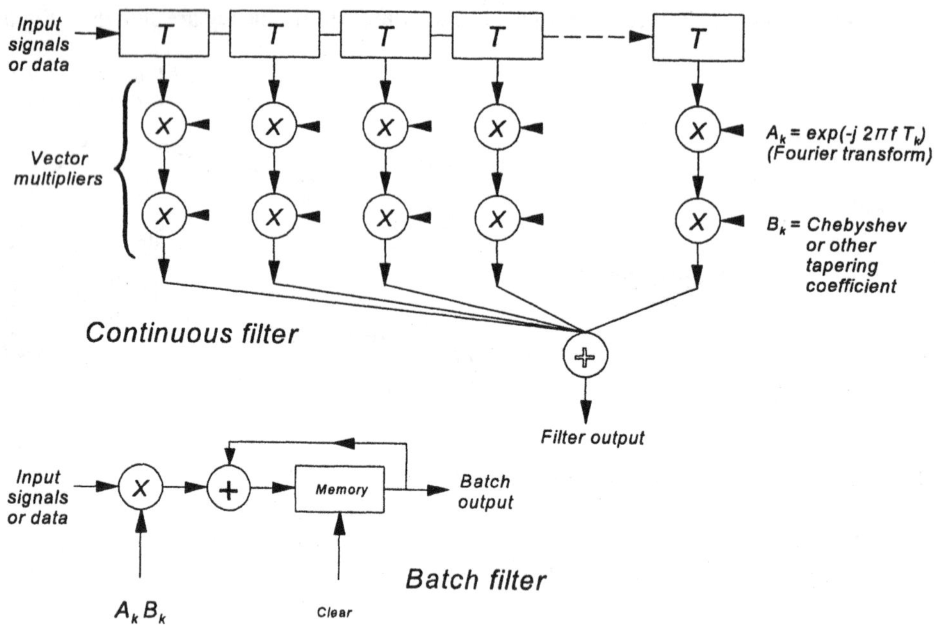

Figure 11.66 Continuous and batch forms of the finite impulse response filters using vector processing.

Figure 11.67 shows the vector multiplication of the envelope of signal samples during one cycle of the Doppler frequency gathered during part of one scan of the antenna. At the epoch zero, both waveforms have a phase angle of zero (vertical). When this signal is multiplied by a signal of the same frequency but opposite sequence, all products are +1. If there is a different phase relation, the result will show the phase relation.

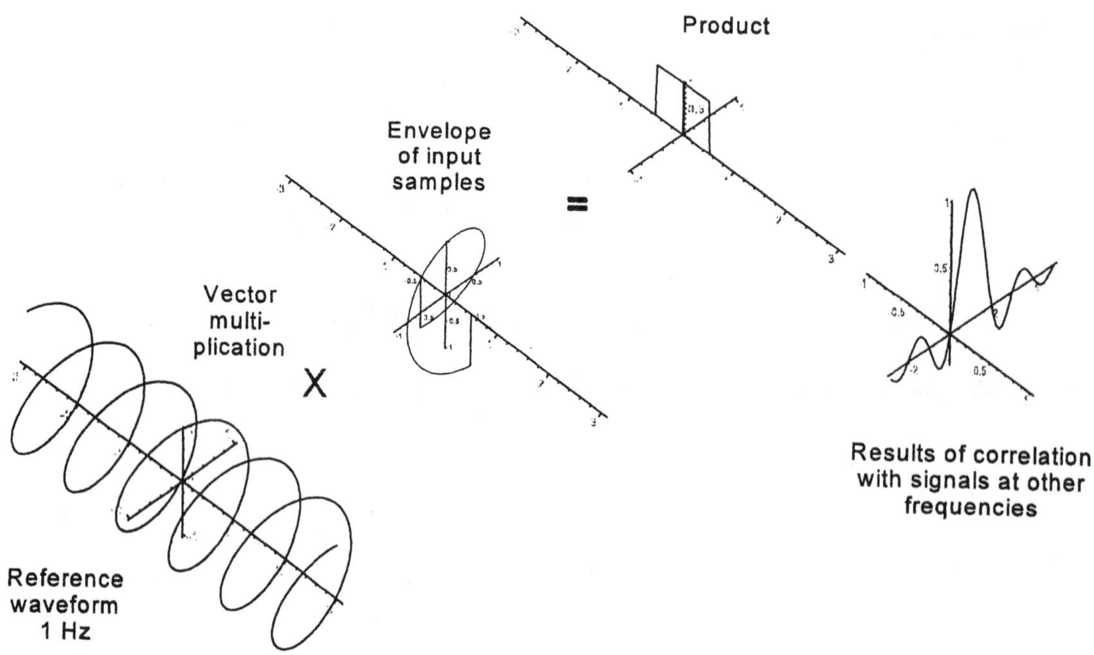

Figure 11.67 The vector multiplication of a signal and a reference waveform.

If the reference is not at the same frequency, then the result will not be +1. In the example both waveforms cross +1 at zero time. If the signal has frequency f Hz and a phase angle ϕ radians at zero time, then the result of multiplying with a reference signal at the same frequency but opposite sequence (polyphase) is

$$\begin{aligned} Signal &= Ae^{-j(2\pi ft+\phi)} \quad \text{for } |t| < \frac{1}{2}, \text{ otherwise } 0 \\ Reference &= Be^{j2\pi ft} \\ Product &= ABe^{-\phi} \quad \text{for } |t| < \frac{1}{2}, \text{ otherwise } 0 \end{aligned} \tag{11.71}$$

In practice, the output is the modulus of the product, so the phase angle (representing fine range in the domain $\lambda/2$) is lost. If the signal does not have the same frequency, then the product is

$$\begin{aligned} Signal &= Ae^{-j(2\pi f_{signal}t+\phi)} \quad \text{for } |t| < \frac{1}{2}, \text{ otherwise } 0 \\ Reference &= Be^{j2\pi f_{reference}t} \\ Product &= ABe^{j(2\pi(f_{reference}-f_{signal})t-\phi)} \quad \text{for } |t| < \frac{1}{2}, \text{ otherwise } 0 \end{aligned} \tag{11.72}$$

If the product is integrated over its length, correlation, it gives a form of the finite Fourier transform:

$$\left| AB\int_{-T/2}^{T/2} e^{-j(2\pi(f_{signal})t+\phi)}\, e^{j2\pi f_{reference}t}\, dt \right| = \frac{\sin\left(\pi(f - (f_{reference} - f_{signal}))T\right)}{\pi\left(f - (f_{reference} - f_{signal})T\right)} \tag{11.73}$$

which is shown on the right of Figure 11.67.

The echo signals at a particular range are in the form of discrete samples. The discrete Fourier transform is given by [24, 25]

$$F(n) = \frac{\sum_{n=0}^{N-1} f(n)e^{-j2\pi\frac{n}{N}T}}{N} \tag{11.74}$$

where T is the sampling interval and the value n/N is in units of cycles per sampling interval.

Without tapering, (11.74) has the form of a geometric progression. Using the relationship $\sum_{n=0}^{N-1} x^n = \frac{1 - x^N}{1 - x}$ and changing from cycles per sampling interval to Hz, (11.74) becomes

$$F(n) = e^{j\pi\frac{n}{N}T\left(1-\frac{1}{N}\right)} \frac{\sin \pi nT}{N \sin \pi\frac{n}{N}T} \tag{11.75}$$

The exponential phase term in (11.75) has no effect on the modulus and is often omitted.

Discrete Fourier transforms are cyclic and are unique over the range $-N/2$ to $N/2$ representing frequencies from $-1/2T$ to $+1/2T$ Hz with n/N representing cycles per sampling interval, as above. The modulus of (11.76) becomes

$$H(f) = e^{j\pi fT\left(1-\frac{1}{N}\right)} \frac{\sin \pi NfT}{N \sin \pi fT} \tag{11.76}$$

Figure 11.68 shows the shapes of the central peaks for various numbers of integrated pulses in the normalized frequency range 1. The greater the number of pulses, the narrower the peak.

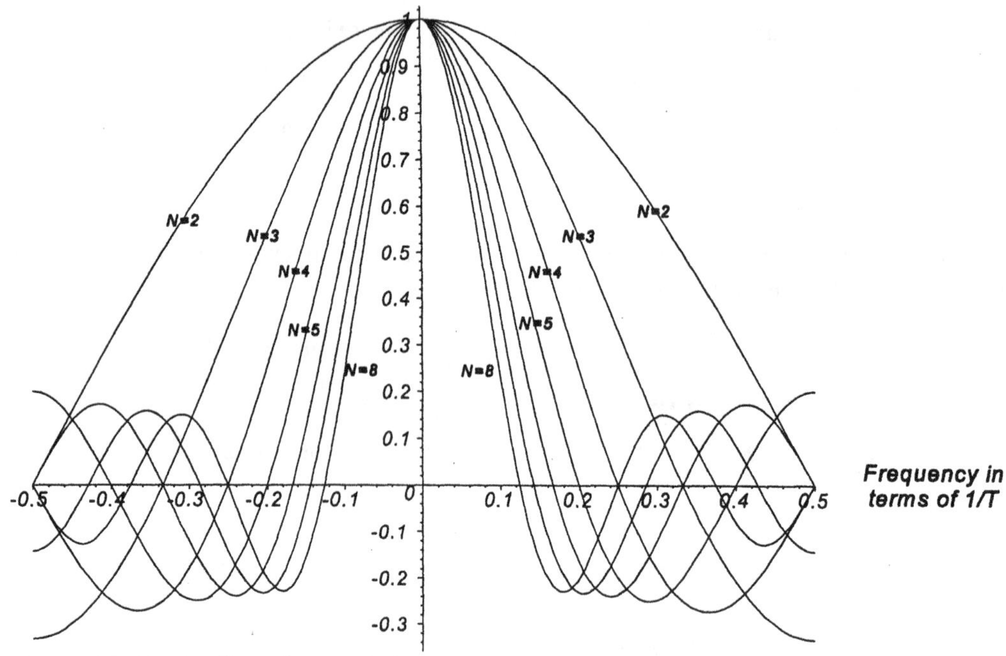

Figure 11.68 The filter characteristic with an increasing number of samples.

For $N = 2$ the characteristic is $\cos \pi x$ and for higher values of N the curves tend to $\frac{\sin \pi NfT}{\pi NfT}$. Figure 11.69 shows the half power filter width and the first sidelobe levels for different values of N.

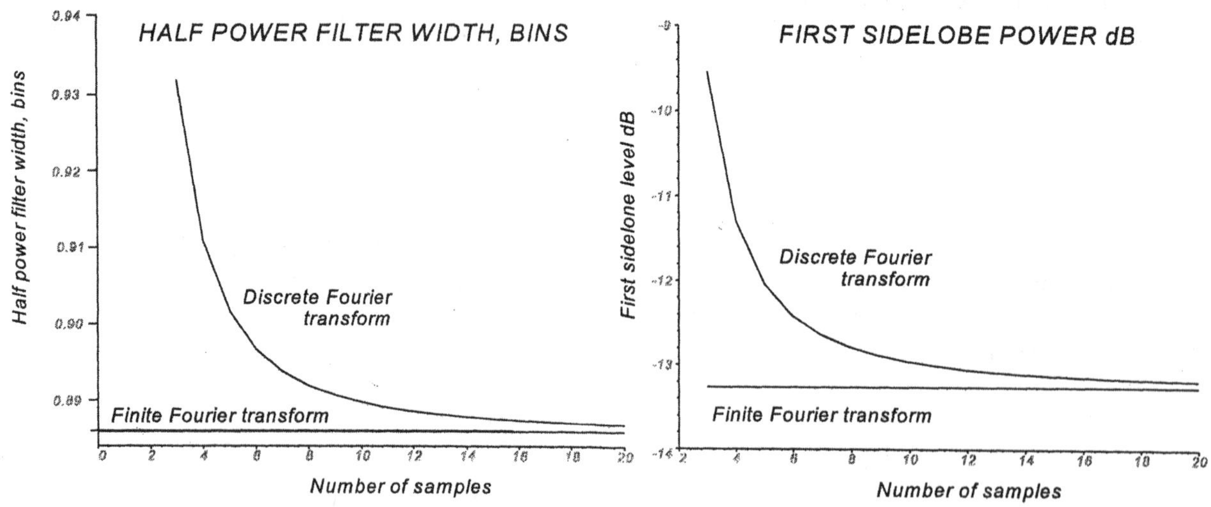

Figure 11.69 Half power filter width and first sidelobe power with an increasing number of samples.

Notice that for a two-pulse moving target indicator canceler and a two-pulse coherent integrator we have

$$\begin{aligned} \textit{Canceler gain} &= e^{j\pi fT} - e^{-j\pi fT} = 2\sin \pi fT \\ \textit{Integrator gain} &= e^{j\pi fT} + e^{-j\pi fT} = 2\cos \pi fT \end{aligned} \tag{11.77}$$

The sine function is the two-pulse moving target indicator characteristic, and the cosine function is the two-pulse integrator characteristic.

Continuous processing requires the storage of the last N echo signals for that range gate. The filter outputs change little from pulse to pulse as they are highly correlated. The sum of N signals may be accumulated in a single memory cell, as shown in the batch filter in Figure 11.66. With more than one coherent processing interval for each scatterer, two or more outputs are available whose centroid gives the accurate azimuth.

Unlike heavy power components (three phase motors, and so on) there are no commercial polyphase semiconductor components, so that two single phase processing lines are necessary, as shown in Figure 11.70.

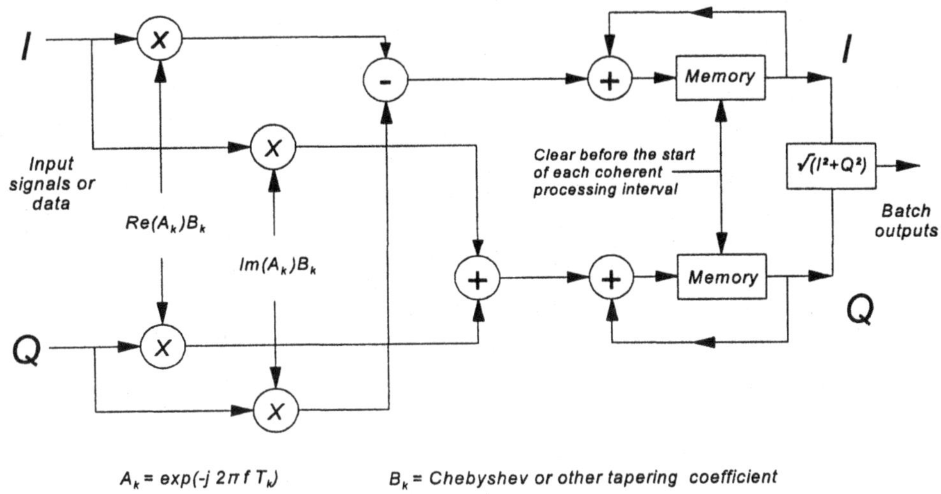

Figure 11.70 Two phase batch processing for vectors.

Each of the I and Q phases is processed separately and the vector sum or modulus is formed at the output. The multipliers or tapers A_k B_k are often stored in memory as premultiplied fixed values.

As with the fast Fourier transform, it is usual to have N correlators or filters for N input samples, so that N correlators in parallel are required.

In the radar in Figure 11.71, the eight filters for 1 000 Hz pulse repetition frequency are equally spaced 125 Hz apart.

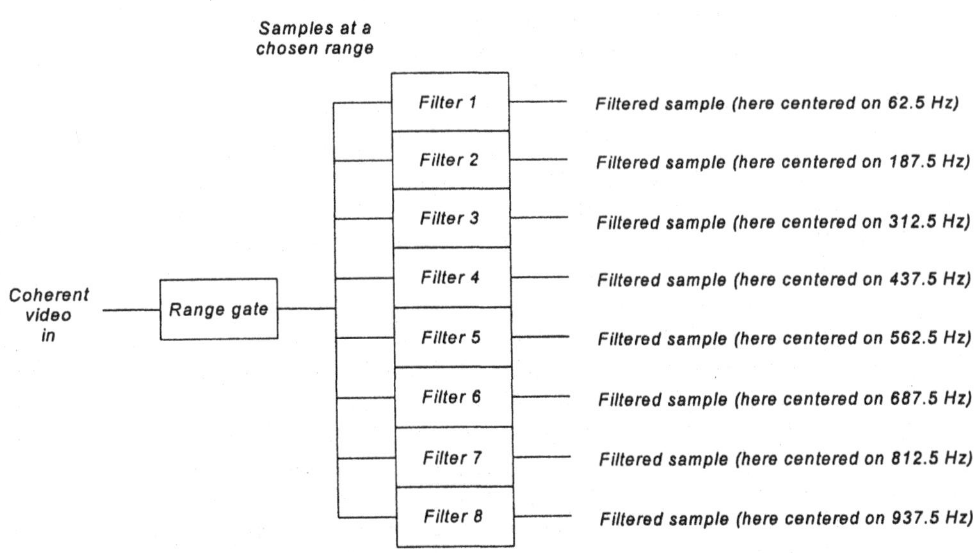

Figure 11.71 A range gate followed by a bank of Doppler frequency filters.

If one of the filters is centered on zero, three filters will pass clutter echoes. If all the filters are offset by half the separation, only two filters will be affected. Remember, a 12 bit analogue-to-digital converter has a 66 dB dynamic range. The filter spacings for both cases are shown in Figure 11.72.

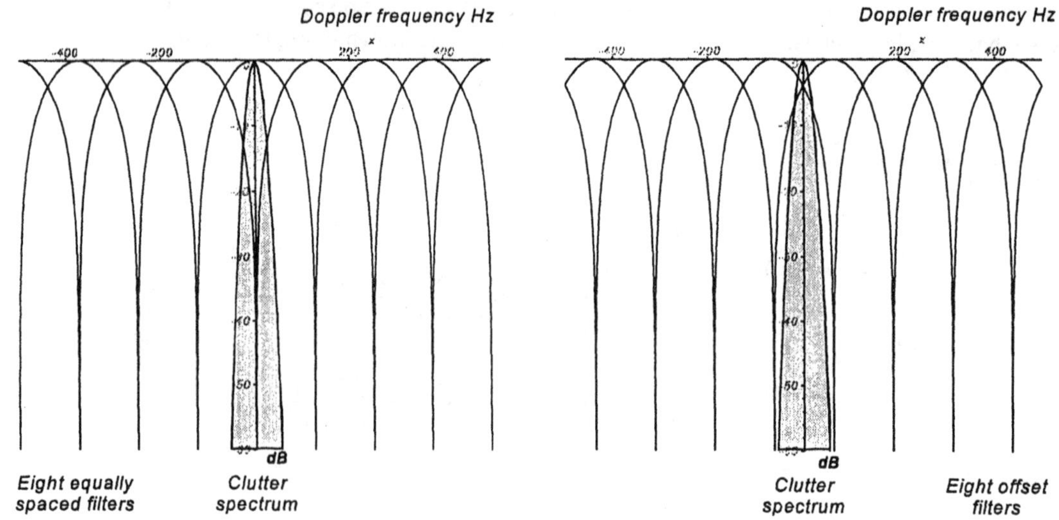

Figure 11.72 The characteristics of eight filters centered on zero and offset. These are ideal without sidelobes.

In practice, the samples are of nearly the same amplitudes and a phase that reflects the movement of the echo, or part of it. The rate of change of phase rarely corresponds to a frequency of $\frac{\nu}{N} f_{prf}$, so that the response to other frequencies is a $\frac{\sin(N\pi f)}{N\pi f}$ characteristic. The response for three neighboring frequency bins is shown in Figure 11.75 (see also Figure B.1).

As has been already mentioned, after the coherent memory filters (military), the fast Fourier transform was used to sort the echoes with their different Doppler frequencies into bins. Then came the discrete Fourier transform filters which allowed the use of any number of pulses for the coherent processing intervals. In the past, all three types of processing needed a single canceler or two pulse canceler moving target indicator stage to remove large clutter blocks. The short history is shown in Figure 11.73. As with normal moving target indicator stages, this is gated in range and often in azimuth.

The two pulse canceler must first be filled before it can supply the integrator or filter stages with containing reduced clutter. The moving target indicator filter requires two extra pulses for each coherent processing interval, that is, to integrate eight pulses, ten transmitter pulses are necessary.

Developments in the design of filters using tapering have allowed better improvement factors without a moving target indicator stage. These modern moving target detector processors have the following improvements:

- They do not need a preceding moving target indicator stage and save the two or more extra transmitter pulses in each coherent processing interval. For eight integrated pulses, there is a reduction of mean transmitter power of 10:8, or 20%, which means lower purchase and running costs. No moving target indicator processing means no blind speeds. The moving target indicator hardware needs to be neither purchased nor maintained.
- Two or more pulse repetition frequencies are used to give pulse repetition frequency diversity. Echo signals with Doppler frequencies that are multiples of the pulse repetition frequency are folded down into the clutter spectrum, making them difficult to detect in clutter. A second chance at another pulse repetition frequency allows the aliasing down into frequency bins that are free of clutter to allow detection. Detections at two pulse repetition frequencies usually allow the true Doppler frequency and radial velocity to be determined.
- In order to keep the angle swept by the antenna during a coherent processing interval the same at each pulse repetition frequency, the number of pulses may be made proportional to the pulse repetition frequency. This requires a reorganization of the number of filters for each type of coherent processing interval.

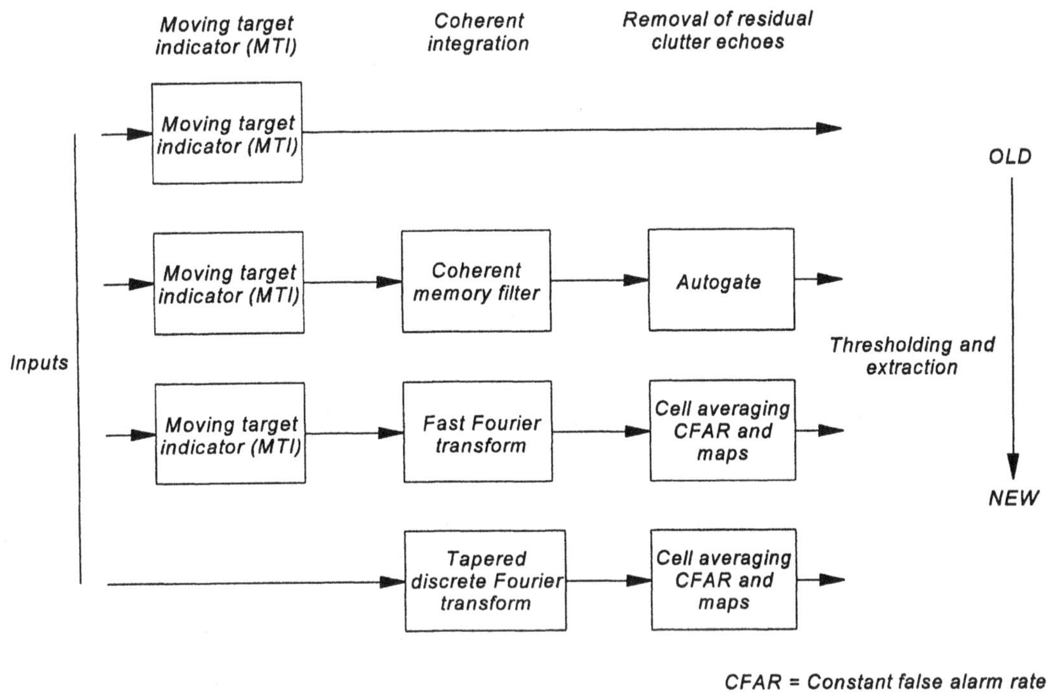

Figure 11.73 The history of the development of moving target detector processors.

- The use of the discrete Fourier transform allows the moving of the center frequencies. When the clutter amplitude is high, the skirts of the clutter spectrum may touch more than two filters, as shown in Figure 11.74. In the case of large amounts of clutter, the clutter filters may be widened (see shaped antenna patterns) and the center frequencies of others moved away to keep them free of clutter. The minimum width of the filter depends on the number of pulses integrated and does not change, and there is a greater overlap. The wider filters for clutter often have high sidelobes that stretch into the bins for moving echo signals. If moving echo signals contaminate large clutter signals, the difference will not be noticeable.

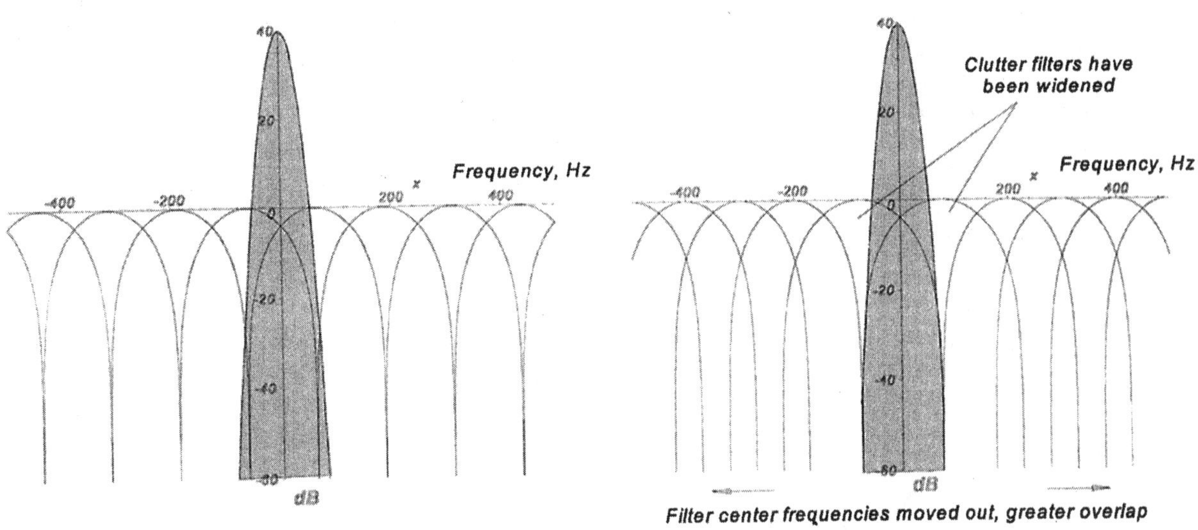

Figure 11.74 The effect of moving the center frequencies to keep filters free of clutter.

In practice, all filters have sidelobes that have not been shown in the preceding diagrams. Figure 11.75 shows the $\frac{\sin(N\pi f)}{N\pi f}$ shape of the frequency characteristics for three of the eight identical integrated pulses.

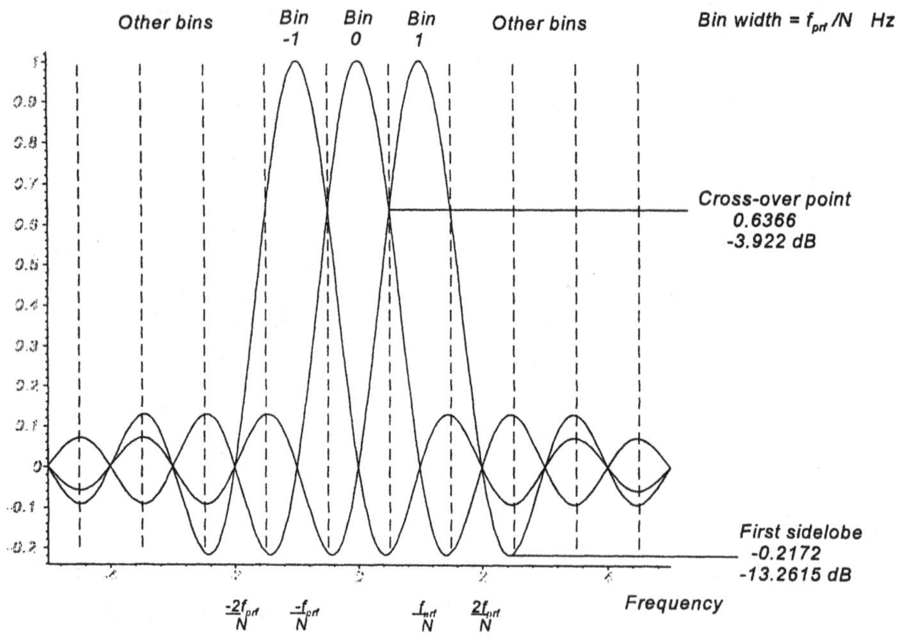

Figure 11.75 The filter characteristics of three neighboring frequency bins without tapering.

The contents of a bin consist of the energy in that bin plus the energy in the sidelobes. The highest lobe is outside the bin and has an attenuation of 13.26 dB, and would be the improvement factor when clutter appears in this lobe. All the sidelobes may be reduced by using tapering or weighting functions.

11.2.4.4 Tapering functions

Unlike the coherent memory integrator, in which a standard version was often wide enough to integrate twice the number of pulses between the one-way half power points of the scanning antenna, moving target detector integrators commonly use half the number of pulses between the half power points. Thus, the pulses within a coherent processing interval are of a similar amplitude, forming a quasi-rectangular envelope. Their Fourier transform or filtering characteristic tends to a sin *x*/*x* function. As with antennas, tapering functions are used to tailor the inputs to the summing function which softens the envelope of the pulses and reduces the frequency sidelobes. The effect of tapering on the envelope of samples

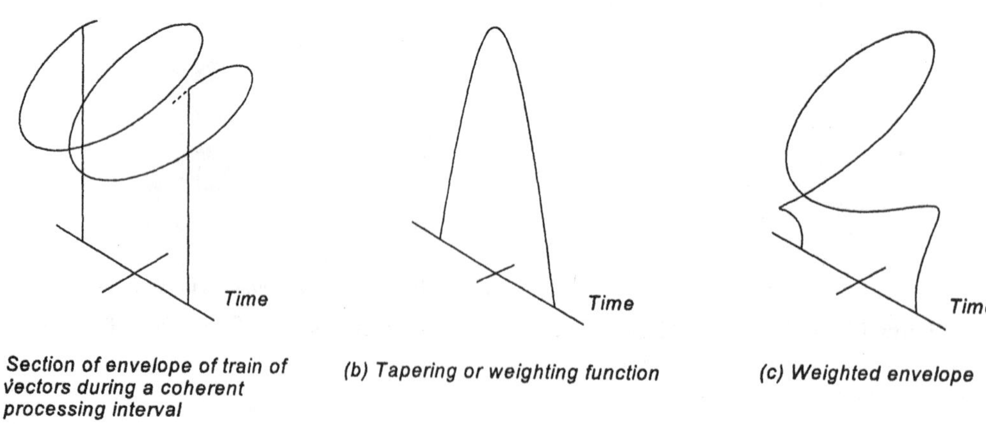

Figure 11.76 The envelope of the *N* samples used for integration and their tapering before summing.

(illustrated in Figure 11.60 and Figure 11.61) is shown in Figure 11.76.

The improvement in the frequency sidelobes comes at a cost. The ends of the train of samples are reduced in amplitude, the integrated sum is reduced, and the signal-to-noise ratio for thresholding is reduced. In order to compare tapering functions, there are a number of figures of merit [26]. As with antennas, the basic reference function is that without tapering. The coherent voltage gain is N, the number of samples integrated. If each of the samples is multiplied by a weighting factor $w[n]$, the coherent (voltage) gain is reduced to

$$\text{Voltage, or coherent, gain} = \sum_{n=0}^{N-1} w[n] \tag{11.78}$$

The relative coherent gain is

$$\text{Relative coherent (voltage) gain} = \frac{\sum_{n=0}^{N-1} w[n]}{N} \tag{11.79}$$

Tapering also reduces the noise through the filter. The sum of the squares of the tapered noise voltages represents the noise power through the filter.

$$\text{Power gain for noise} = \sum_{n=0}^{N-1} \left(w[n]\right)^2 \tag{11.80}$$

If the signal-to-noise ratio is S_{in}/N_{in} at the input to the filter, then the signal-to-noise ratio at the output is

$$\text{Signal-to-noise ratio at output} = \frac{S_{in}}{N_{in}} \frac{\text{Signal power gain}}{\text{Noise power gain}} = \frac{S_{in}}{N_{in}} \frac{\left(\sum_{n=0}^{N-1} w[n]\right)^2}{\sum_{n=0}^{N-1} \left(w[n]\right)^2} \tag{11.81}$$

Without tapering the output signal-to-noise ratio is N and the ratio to (11.81) is less than N and thus gives a loss, the processing loss.

$$\text{Processing loss} = 10 \log_{10} \frac{\left(\sum_{n=0}^{N-1} w[n]\right)^2}{\sum_{n=0}^{N-1} \left(w[n]\right)^2} \text{ dB} \tag{11.82}$$

Without tapering, the spectrum for a (constant) rectangular input T seconds long is $\sin(\pi fT)/(\pi fT)$ which is a function of the frequency-time product. Units of the frequency-time product are commonly called bins. This loss is less than the ratio of the coherent gain to the number of samples, N, as the noise also is reduced by the tapering. As with antennas the width of a rectangular area that has the same surface area as the main frequency lobe is called the noise bandwidth and is shown in Figure 11.77. The characteristics of adjacent filters are shown in Figure 11.78.

The equivalent noise bandwidth varies as the reciprocal of the processing loss and is given by

$$\textit{Equivalent noise bandwidth} = \frac{\sum_{n=0}^{N-1} \left(w[n]\right)^2}{\left(\sum_{n=0}^{N-1} w[n]\right)^2} \text{ bins} \tag{11.83}$$

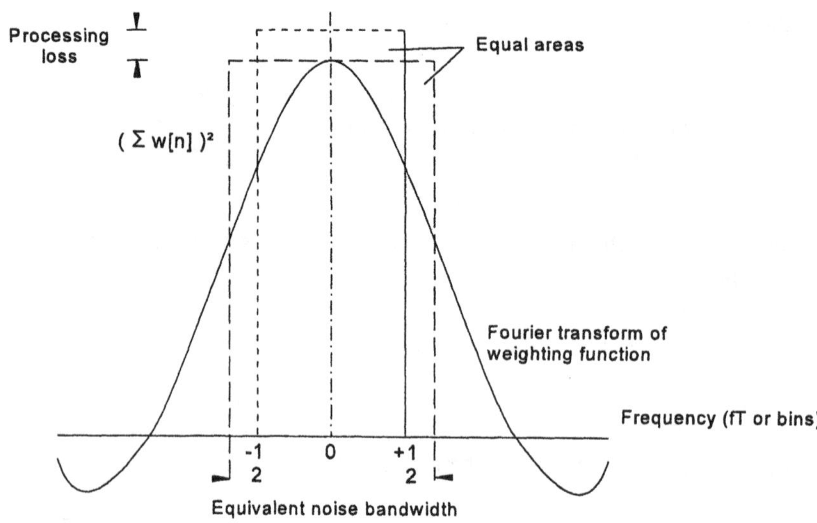

Figure 11.77 The equivalent noise bandwidth of a frequency passband.

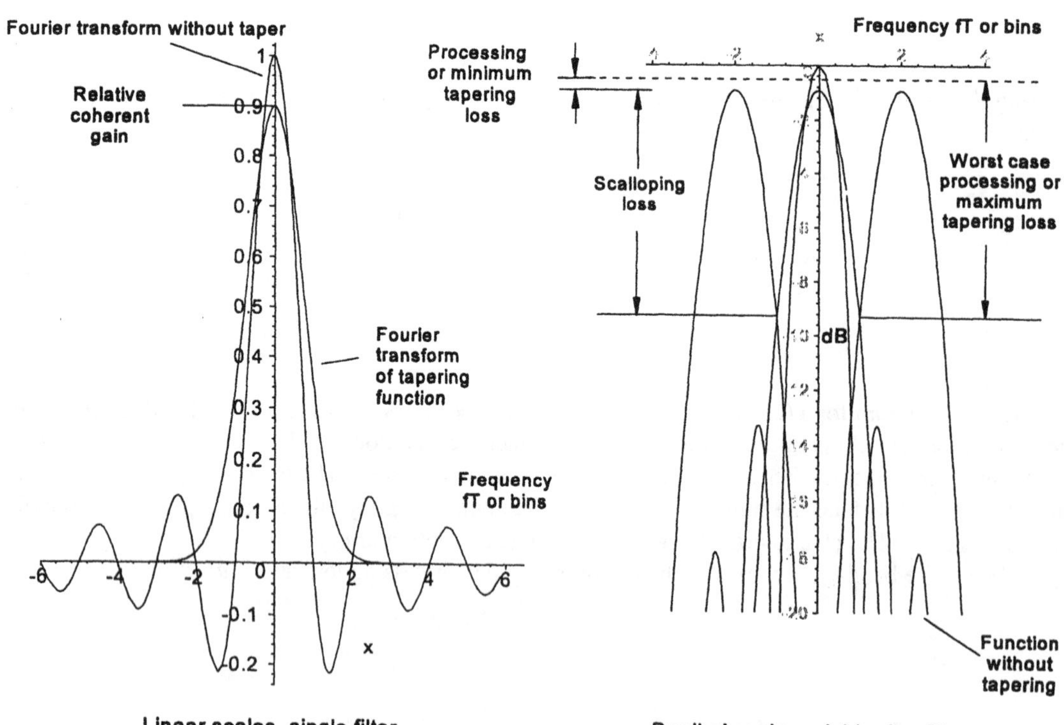

Figure 11.78 Gain and loss definitions with tapering.

With most tapering functions, the characteristic outside the main passband has a number of lobes, the sidelobes. A filter near a filter that passes clutter echoes will have these same clutter echoes in its sidelobes. The amplitude of the first sidelobe gives a good indication of the quality of the clutter suppression quality.

The use of discrete processing methods brings an additional disadvantage. There are extra sidelobes or spectral leakage when a frequency component falls between two bins. Figure 11.79 shows this for two signals in a 100 sample filter. The wanted signal is in bin 16, which is 40 dB smaller than the interfering or clutter signal, which is moved between bins 10 and 11. When the major signal has its frequency corresponding to a bin center, the two signals may be separated. When the major signal is moved in frequency between bins, the minor signal is swamped. A number of tapering functions are in the literature and are given in Appendix B [26].

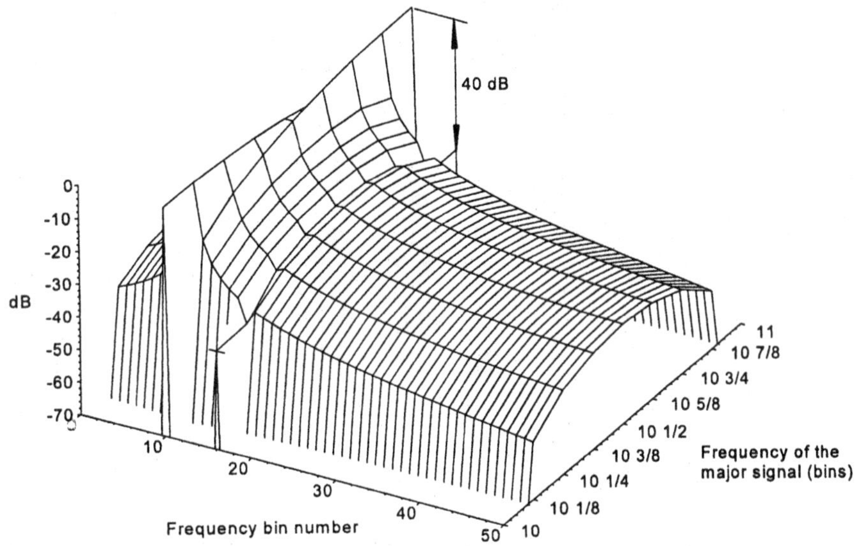

Figure 11.79 Masking of a small signal when the frequency of a larger signal falls between bins.

Example. In this example, a bank of equally spaced Chebyshëv filters is used. The stopband attenuation is the major characteristic for the clutter improvement factor in the clear filters, here 50 dB. The design calculations proceed as follows:

$$\begin{aligned} \textit{Stopband attenuation}, R &= 50 \text{ dB} \\ r &= 316.228 \quad \text{voltage ratio} \end{aligned} \tag{11.84}$$

The factor α is for an eight pulse filter, $N = 8$:

$$\begin{aligned} \alpha &= \tanh^2\left(\frac{\ln\left(r + \sqrt{r^2 - 1}\right)}{N - 1}\right) \\ &= 0.527 \end{aligned} \tag{11.85}$$

The weights for the tapering are given by [27]

$$W_k = \frac{N - 1}{N - K} \sum_{S=0}^{K-2} \frac{(K - 2)!\,(N - K)!\,\alpha^{S+1}}{S!\,(K - 2 - S)!\,(S + 1)!\,(N - K - S - 1)!} \tag{11.86}$$

which gives the normalized tapering coefficients for eight points as *weight*[1] := 0.0945 *weight*[2] := 0.3493 *weight*[3] := 0.7182 *weight*[4] := 1.0

weight[5] := 1.0 *weight*[6] := 0.7182 *weight*[7] := 0.3493 *weight*[8] := 0.0945

A train of eight pulses with these weights is shown graphically in Figure 11.80.

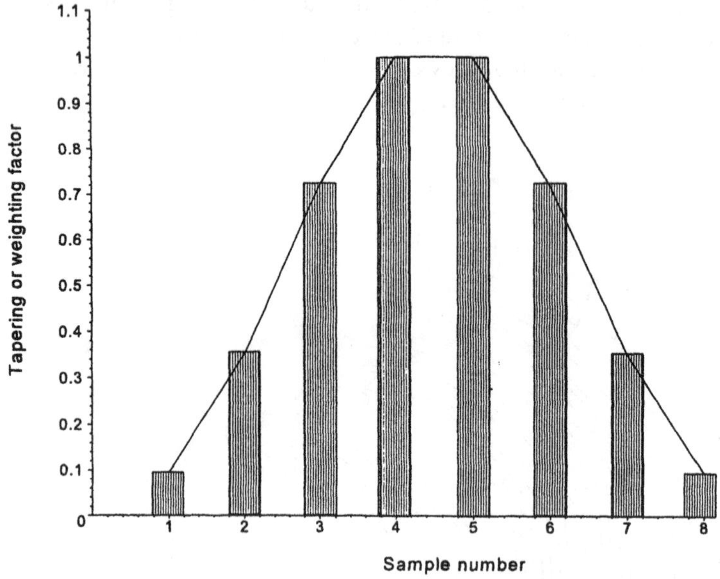

Figure 11.80 Tapering weights for a train of eight pulses for a 50 dB Chebyshëv filter.

A typical bank of Chebyshëv filters with 50 dB of attenuation outside the passband would have the characteristics shown in Figure 11.81. These have been offset from zero by half the filter width.

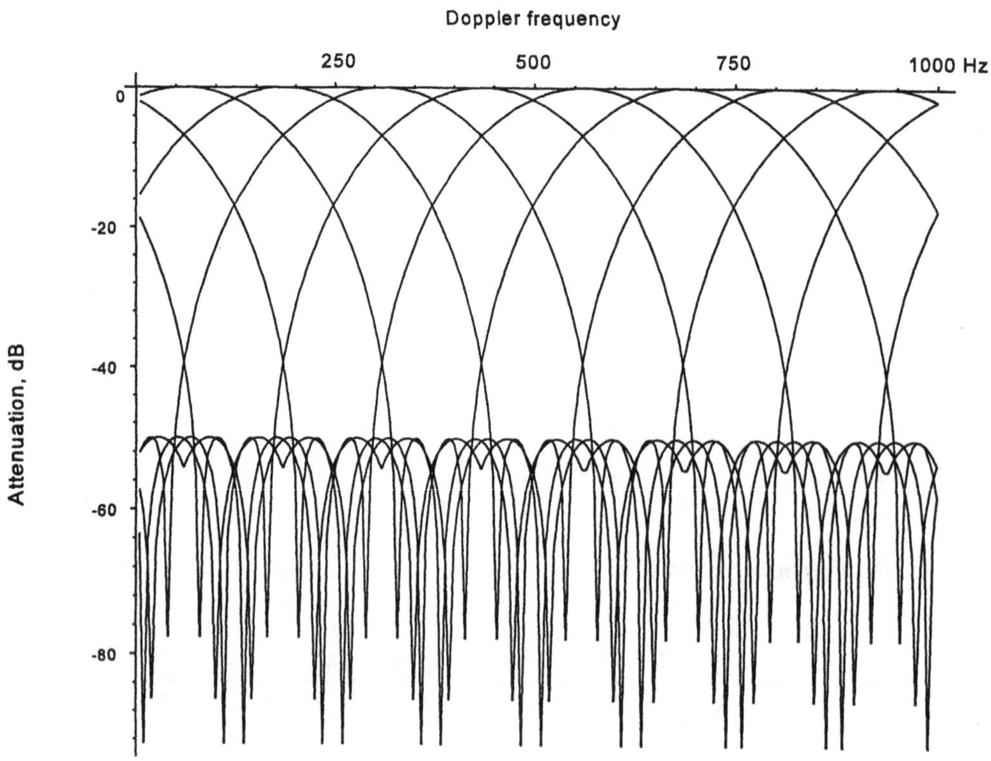

Figure 11.81 The frequency characteristics of a bank of eight Chebyshëv filters with 50 dB sidelobes in a band of 1 000 Hz.

Signal components at near zero frequency will fall in the main characteristic of filters 1, 2, 7, and 8. Clutter falls in the sidelobes of filters 3, 4, 5, and 6. For 60 dB clutter and 50 dB filter attenuation, there will be 10 dB clutter residue. The threshold for the detection must be raised for the signal outputs of the Doppler frequency filters containing clutter at the affected ranges using a stored map, as in area moving target indication.

The signals selected by the individual filters were calculated by taking the dot product (MAPLE dotprod function) of the spectral density (vector) at that range and the filter characteristic (as a vector). The dot product is the sum of the products of the elements of two vectors.

$$dotprod([a_1, a_2, a_3, .., a_n], [b_1, b_2, b_3, .., b_n]) = a_1b_1 + a_2b_2 + a_3b_3 + .., + a_nb_n \tag{11.87}$$

Figure 11.82 shows a cut through Figure 11.81 (4), where the filter output has been divided by the mean clutter power. The maximum improvement factor in filters 3, 4, 5, and 6 is greater than the peak Chebyshëv sidelobe and is derived from the average of the frequency sidelobes. The "improvement factor" in the clutter filters (1 and 8) shows that the center frequencies are not zero.

The improvement factor of the center filters reflects the average sidelobe level of the Chebyshëv filter characteristic. The filters next to the clutter filters attenuate it by 16.6 dB and the clutter filters by 1.7 dB.

The sorting of clutter echoes in the range and Doppler frequency dimensions is shown in Figure 11.83 in the same form as that for the moving target indicator. The mean clutter amplitude over range is shown in (1) and its spectral density over the 1 000 Hz pulse repetition frequency, in the example radar, in (2). The characteristics of the filters in Figure 11.81 are repeated in (3).

For a standard object that gives a 10 dB signal-to-noise ratio at 100 km, the signal-to-noise ratio at 30 km (no sensitivity time control) is 30.9 dB, which gives good detection for signals leaving filters 3, 4, 5, and 6. The 16.6 dB improvement for the near clutter filters is not enough to give a wanted echo-to-clutter ratio of 10 dB. Moving the center frequency of these filters to avoid the clutter spectrum would give a better improvement factor.

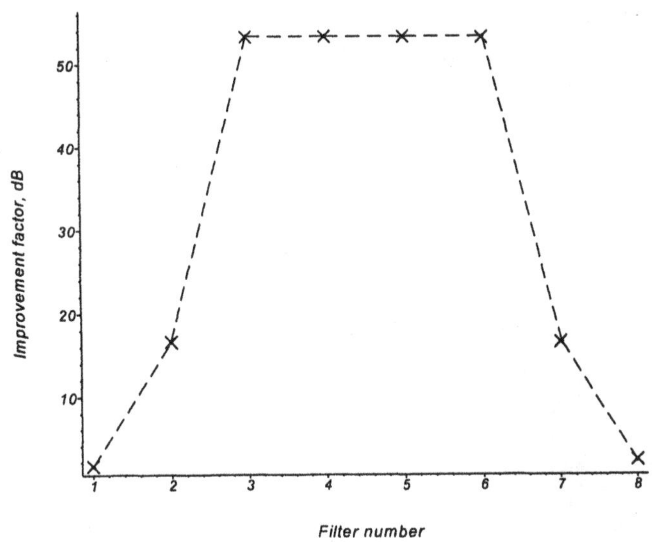

Figure 11.82 The improvement factor in the example moving target detector.

For moving objects whose echo signals fall in the bands of filters 1 and 8, these signals will be among ground clutter. Luckily, ground clutter is patchy, and wanted echoes may be seen between the patches. Clutter maps (described in Section 11.3.1) can be used to set the thresholds above the clutter level to give superclutter visibility. Between clutter patches, the clutter maps are set to almost zero, giving good interclutter visibility. Beyond ground clutter, these signals with near zero Doppler frequency will be in noise as for the other filters but will enjoy the full benefits of coherent integration.

Unlike moving target indicator processing, which suppresses these echo signals, it is thus disadvantageous to switch off the moving target detector signals beyond clutter, and it also applies to moving objects—such as aircraft—flying around the radar on a circumferential course (with zero radial velocity).

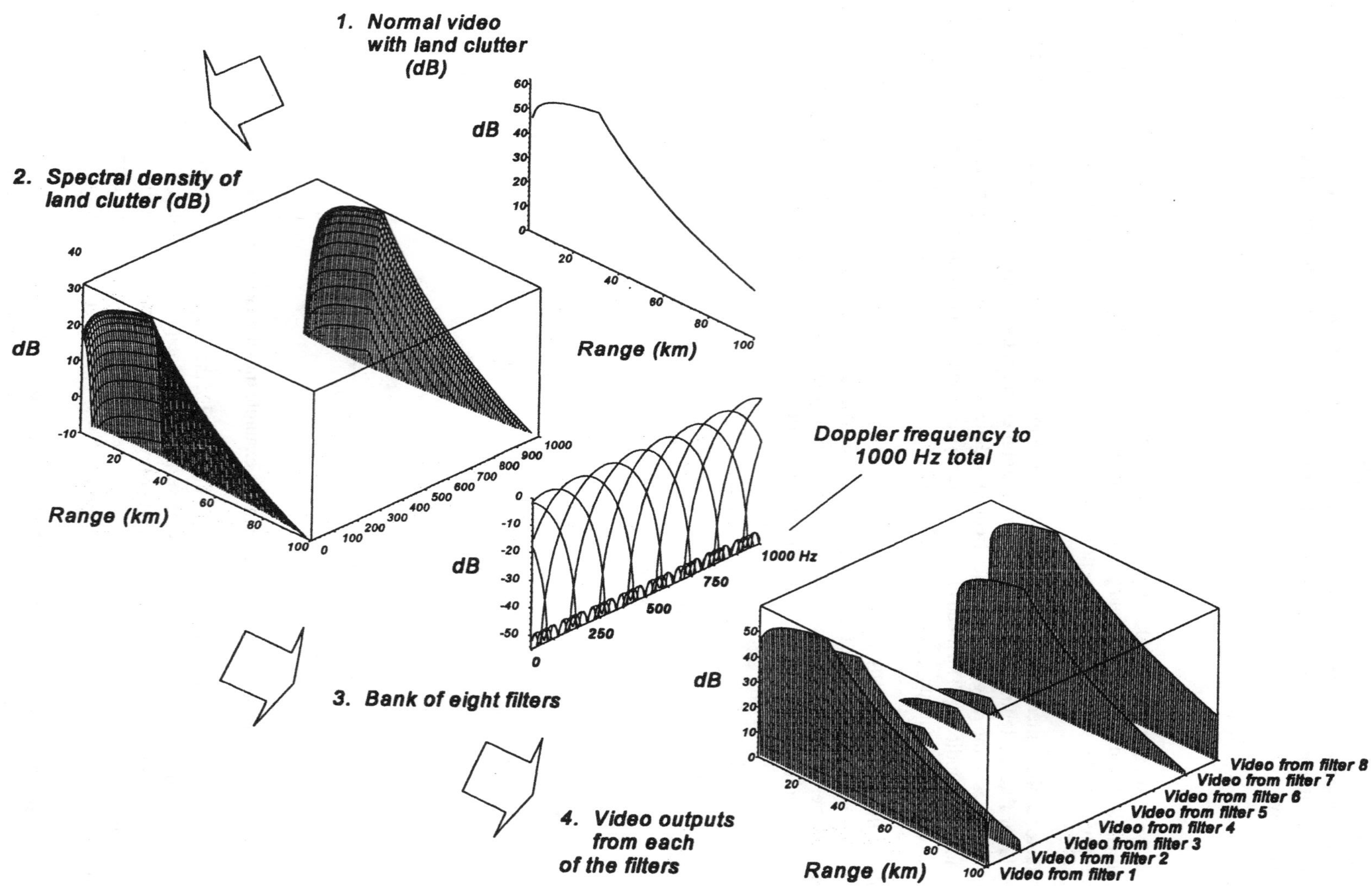

Figure 11.83 Clutter, clutter spectral density, and filtering in a moving target (MTD) detector processor. Note that the clutter level shown is average clutter level at the 94% point in the log-normal distribution. The median is 20 dB lower.

The choice of tapering function affects the influence of clutter signals on smaller signals with discrete processing. In Figure 11.79, the smaller of two signals in a 100 pulse (bin) filter is suppressed most of the time by the discrete processing as the frequency of the larger signal is moved from bin 10 to bin 11. Figure 11.84 shows this effect for the same filter with Chebyshev 50 dB tapering. The main responses are wider, but the minor signal is visible all of the time as the major signal frequency moves between bins 10 and 11. The development of such filters rendered a preceding moving target indicator stage unnecessary.

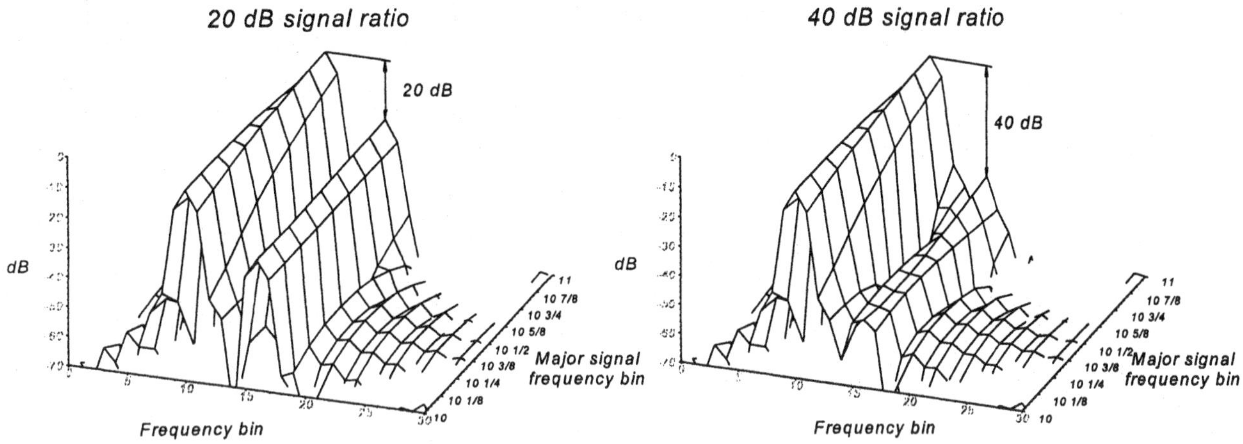

Figure 11.84 The two-tone characteristics with Chebyshev tapering.

The sums in the batch filter in Figure 11.66 accumulate during a coherent processing interval and the result is available at the end of it. Two memory blocks may be used: one to accumulate the data and the other to supply data for analysis. The memory blocks are shown in Figure 11.85.

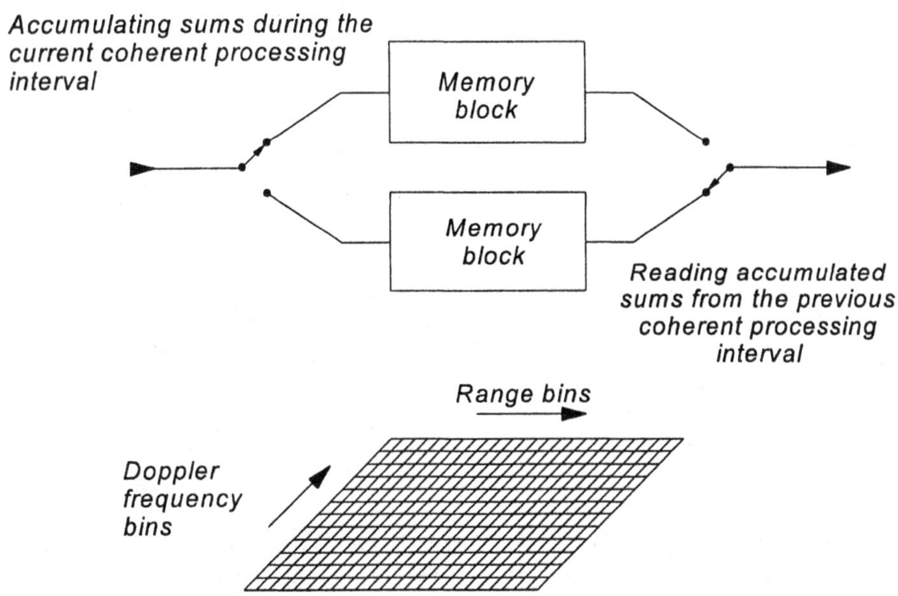

Figure 11.85 The organization of memory for the accumulation and analysis of signals during the coherent processing intervals.

The memories are arranged with separate areas for the information in each range and Doppler frequency bin or cell. Data from the previous coherent processing interval are read out and compared with a threshold chosen from

- Noise alone;
- Stationary clutter;
- Typically a cell averaging constant false alarm rate threshold.

The contents of the Doppler frequency bins at one range are shown graphically in Figure 11.86. Bins 1 and 8 contain clutter and the threshold is set to a level greater than the the value stored in clutter maps for the individual bins. Mass memory allows the storage of clutter levels in each resolution cell in range, azimuth, and Doppler frequency. Because clutter changes slowly, the thresholds referenced to the clutter map values are lower than those for noise alone. Bins 2 and 7 contain clutter residues, and again the threshold is determined by the clutter maps. The Doppler frequency bins 3, 4, 5, and 6 contain no clutter, so that the noise threshold is used. As the threshold-to-noise ratio is greater than the threshold-to-clutter ratio, the choice may be made automatically by choosing the greater.

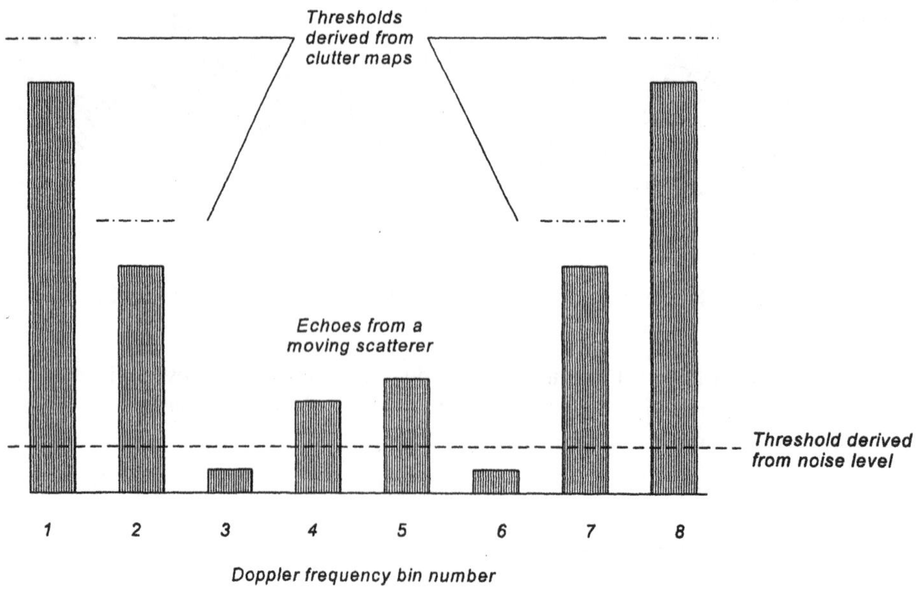

Figure 11.86 The contents of the Doppler frequency bins in one range cell in clutter.

The choice of threshold depends on the amount of clutter that is expected in the bins, and this is shown in Table 11.5.

Table 11.5
Choice of threshold in a moving target detector radar

Signal strength in the clutter filters	Thresholds in Clutter filters	Thresholds in Adjacent filters	Thresholds in Remaining filters
High	Clutter map	Clutter map	Clutter map
Medium	Clutter map	Clutter map	Noise threshold
Low	Clutter map	Noise threshold	Noise threshold
Noise	Noise threshold	Noise threshold	Noise threshold

The information stored in each Doppler frequency bin during the previous coherent processing interval is subjected to single sweep video processing. Rain echoes form clusters that may be detected as areas that exceed the noise threshold (by 3 dB, say) in a rain map. The radial velocity of the rain is large enough, normally, to be able to affect all Doppler frequencies. This map is used to switch in cell averaging constant false alarm rate threshold.

Rain is blown by the wind, the velocity of which increases with altitude. A value of 4 m/s/km is often used in radar specifications for wind shear. With no wind shear, the rain accumulates in one Doppler frequency cell, leaving the others free. Otherwise, the wind shear spreads the rain echo energy over a number of Doppler frequency bins, decreasing the rain echo amplitude in the individual bins. Both improve the detection of moving echoes compared to the moving target indicator processor with its (single) narrow notch.

Rain storms may lead to high amplitudes of rain clutter. Care must be taken that the negative sequence components from the detectors and analogue-to-digital converters do not give an appreciable signal. The negative sequence signal appears as a "negative" frequency out of the filters and can double the number of Doppler frequency bins affected by rain.

In addition to processing the echoes in the range domain, it is also possible to find a threshold in the Doppler frequency domain, which is derived from the average level of the Doppler frequency bins that are free of clutter. This type of threshold gives protection against short noise bursts that occupy the whole Doppler frequency domain.

Moving scatterers rarely have radial velocities that correspond to the frequency of one Doppler frequency filter alone. Figure 11.86 shows a detection at a Doppler frequency between filters 4 and 5. The ambiguous Doppler frequency (from zero to the pulse repetition frequency) can be estimated from the center of gravity or centroid.

11.2.4.5 Unambiguous Doppler frequency or radial velocity

The use of two or more pulse repetition frequencies generally allows the determination of the true Doppler frequency and thus radial velocity. This is best illustrated by an example.

A radar with 1.5 degree beam width has 360/1.5 = 240 distinct beam positions in a circle. If the rotation speed is 12 revolutions in a minute, 1.5 degrees will take 20.8 ms. If the maximum range of the radar is 100 km, the maximum pulse repetition frequency is $(3 \times 10^8)/(2 \times 100\,000) = 1\,500$ Hz. Allowing 20% for dead time, this calculation gives a maximum pulse repetition frequency of 1 200 Hz. A number of solutions are possible that satisfy the equation:

$$\frac{N_1}{prf_1} + \frac{N_2}{prf_2} = Dwell\ time, \text{ here } \frac{20.833}{1000} \text{ seconds} \tag{11.88}$$

For example, 11 pulses at 1 000 Hz followed by 9 pulses at 915.25 Hz is a possible solution.

For some solutions for the ambiguous Doppler frequency, an integer ratio of the pulse repetition frequencies is necessary. The frequencies of 1 000 Hz and 925 Hz are in the ratio of 40:37.

Since separate detections are made at the two pulse repetition frequencies, the combined detection is the result of pulse repetition frequency diversity. Estimates of the ambiguous Doppler frequencies at the two or more pulse repetition frequencies can be used to calculate the true Doppler frequency and thus radial velocity. The true Doppler frequency is

$$f_{Doppler} = m\ prf_1 + f_{filter1} = n\ prf_2 + f_{filter2} \tag{11.89}$$

where $f_{Doppler}$ is the true Doppler frequency;
prf_1 and prf_2 are the two pulse repetition frequencies;
$f_{filter1}$ and $f_{filter2}$ are the interpolated estimates of ambiguous Doppler frequency at the two pulse repetition frequencies;
m and n are integers.

For an aircraft flying at 200 knots (370.4 km/h, or 102.89 m/s) radially towards an S-band radar operating at 3 GHz, the Doppler frequency is 102.89 m/s . 20 Hz/m/s = 2 057.78 Hz. The ambiguous frequencies at the two pulse repetition frequencies are:

2057.78 mod 1000 = 57.78 Hz;
and 2057.78 mod 915.25 = 227.278 Hz.

At a pulse repetition frequency of 1 000 Hz, the first Doppler frequency folds down into the clutter spectrum. Finding the unambiguous Doppler frequency is possible normally only outside or between clutter patches. At the second pulse repetition frequency, the Doppler frequency folds down into a part of the spectrum free of clutter.

Two common algorithms are used:

- Increasing the integers until coincidence;
- The Chinese remainder theorem.

Coincidence

If burst-to-burst frequency agility is used, then the ambiguous Doppler frequency depends on both the radar radio frequency and the pulse repetition frequency. Only the echo velocity is constant from burst to burst.

$$Velocity = \frac{f_d\ c}{2\ f_{rf}} \quad \text{m/s} \tag{11.90}$$

where f_d is the Doppler frequency, Hz;
f_{rf} is the radar frequency, Hz;
c is the velocity of light, m/s.

The estimated Doppler frequencies are ambiguous so that trial velocities must be calculated using integer values for n and m for each burst until

$$Velocity = \frac{(f_{ambd1} + n\ prf_1)\ c}{2\ f_{rf1}} = \frac{(f_{ambd2} + m\ prf_2)\ c}{2\ f_{rf2}} \quad \text{m/s} \tag{11.91}$$

where f_{ambd1} is the ambiguous Doppler frequency, Hz;
f_{rf1} is the radar frequency, Hz;
prf_1 is the pulse repetition frequency of the first burst, Hz (values for the second burst have the subscript 2);
c is the velocity of light, m/s.

In this example, the aircraft radial speed is –200 knots (–370.4 km/h, or –102.89 m/s) and the Doppler frequency is –2 057.78 Hz. Again, the pulse repetition frequencies are 1 000 Hz and 915.25 Hz for a radar working on a fixed frequency of 3 Ghz, giving ambiguous Doppler frequencies of -57.778 Hz and -227.28 Hz. The n times the pulse repetition frequency is added to the ambiguous Doppler frequencies until there is coincidence. In this example, for n from –4 to +4 we have the following values for each pulse repetition frequency (prf):

```
prf 1 trials   -4057.78, -3057.78, -2057.78, -1057.78,   -57.78, 942.22, 1942.22, 2942.22, 3942.22 Hz
prf 2 trials   -3888.28, -2973.03, -2057.78, -1142.53, -227.28, 687.97, 1603.22, 2518.47, 3433.72 Hz
                                   |======|
                                   Coincidence
```

Thus, the true Doppler frequency is –2 057.78 Hz. The results are the same if the complimentary Doppler frequencies (942.22 Hz and 687.97 Hz) are used. When burst-to-burst frequency agility is used, the trial velocities must be compared with a small tolerance.

Chinese remainder theorem

Commonly, civil radars are allocated fixed operating frequencies for each radar channel by the regulatory agencies so that the same radio frequency is used for all the coherent processing intervals. If the pulse repetition frequencies are related by integers (by design) then the Chinese remainder theorem, which depends on integers, may be used.

The Chinese remainder theorem [28] states that for positive integers r and s that are relatively prime and for a and b which are any other two integers, then there will be a number N, such that

$$a \bmod r = N = b \bmod s \tag{11.92}$$

For each r and s, there will be two numbers where $b_1\ r \bmod s = 1$ and $b_2\ s \bmod r = 1$.

The two pulse repetition frequencies are derived from a common frequency divided by r and s. In the preceding example, the nearest integer ratio is 37:40, giving pulse repetition frequencies of 1 000 Hz and 925 Hz. The common frequency is 37 × 1 000 = 40 × 925 = 37 000 Hz, which is the range of the unambiguous Doppler frequencies. Integers are used that give 37 × 40 = 1 480 steps of 25 Hz.

For an aircraft flying on a radial course at 200 knots (370.4 km/h, or 102.89 m/s) the Doppler frequency is 2

057.78 Hz. This will be aliased down by the pulse repetition frequencies (1 000 Hz and 925 Hz) to 57.78 Hz and 227.28 Hz. The calculations to find the unambiguous Doppler frequency are made in the following steps [27]. MAPLE list notation is used instead of subscripts, that is x_1 is $x[1]$.

Constants

Pulse repetition frequencies, Hz	[1 000, 925]
Ratios *r, s*	[37, 40]
Chinese remainder constants, *a* and *b* (11.64)	[13, 25]

Example

	"Positive" Doppler frequency	"Negative" Doppler frequency
Doppler frequencies	[57.78, 207.78]	[–57.78, –207.78]
Fractional ambiguous Doppler frequencies,		

$$x = \frac{Doppler\ frequency}{Pulse\ repetition\ frequency}$$

	[0.0578, 0.2246]	[–0.0578, –0.2246]
Intermediate calculations	a = [trunc($x[1]b[2]$), trunc($x[2]b[1]$)]	
	[2, 8]	[–2, –8]

The number of frequency steps is given by

$$Frequency\ steps = \left(r\ b[1]\ a[1] + s\ b[2]\ a[2]\right) \bmod (r\ s) \qquad (11.93)$$

	82 step widths	1 398 step widths
Estimation of true Doppler frequency, steps times step width		
	2 050 Hz	34 950 Hz
		34 950–37 000 = –2 050 Hz

"Negative" Doppler frequencies are to be found in the range from half the common frequency upwards.

Generally, only speed measuring radars display the velocity. Unambiguous velocity is useful as an extra sorting factor for tracking and classification of the echoes of interest.

In contrast with pulse repetition frequency staggering in moving target indicators where the processed signal from a number of pulse repetition frequencies is presented to the threshold, the results of each integration at each pulse repetition frequency are presented separately to the threshold for detection. This is pulse repetition frequency diversity. The difference is illustrated in Figure 11.87.

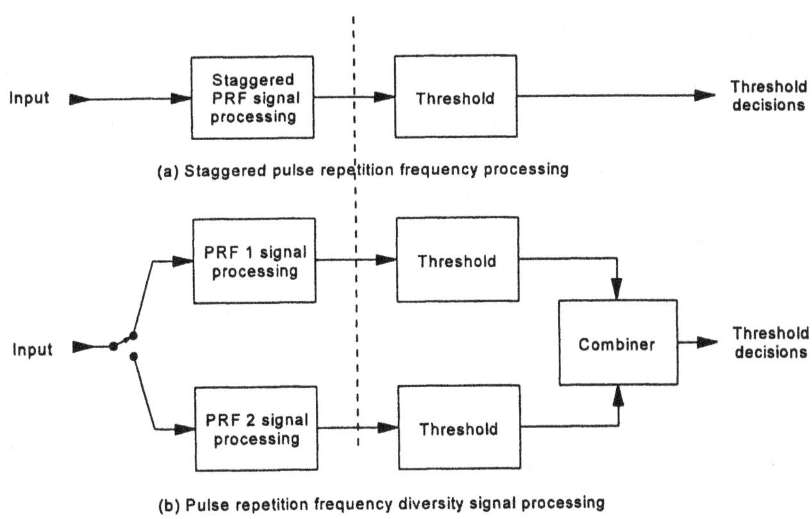

Figure 11.87 (a) Staggered pulse repetition frequencies and (b) pulse repetition frequency diversity.

11.2.5 Comparison of moving target indicators and detectors

Moving target indicator processing has served radar operators for many tens of years and is simpler and cheaper than moving target detection. Moving target indicator and detection (without an intervening moving target indicator stage) processing may be compared in Table 11.6.

Table 11.6
A comparison of moving target indicator and moving target detector signal processing

	Moving target indicator (MTI)	Moving target detector (MTD)
Sensitivity	Less, the MTI processing must be restricted to where necessary.	Greater, is used at all ranges and gating is not desirable.
Blind speeds	The effects are reduced by using a number of staggered pulse repetition frequencies.	Where the Doppler frequencies of wanted echoes are folded down into the clutter spectrum, a second pulse repetition may be used.
True Doppler frequency		Possible using coincidence or Chinese remainder theorem.
Superclutter visibility	Clutter and wanted echoes at the same ambiguous Doppler frequency are suppressed.	Possible.
Interclutter visibility	Possible only with detailed maps to switch in linear video.	Possible.
Tangential course scatterers	Not seen in MTI region.	Seen in regions free of clutter.
Rain and snow	Poor suppression owing to single narrow notch.	Suppressed by following cell averaging thresholds.

11.3 PROCESSING OVER MANY SCANS AND MAPS

A number of attempts were made in the past to clean up moving target indicator residues and switch the videos using maps. Owing to the numbers of resolution cells, typically for surveillance radars between 100 000 and 1 000 000, cathode ray tube storage had to be used and many were more trouble than they were worth. As computers developed, so did memory technology. The experimental drum memories were replaced by stable, reliable semi-conductor memories as they became available.

Currently maps are used also where, in the past, simple range switching was used. Examples are:

- Transmitter power control, azimuth only;
- Antenna beam switching and steering;
- Sensitivity time control for the receivers (see Chapter 7, Receivers);
- Moving target indicator video switching;
  - Areas for rain gates;
- Moving target detector;
  - Clutter maps for each Doppler frequency bin;
  - Filter bandwidth maps;
  - Rain maps;
- Detection control maps and second censoring;
  - Blanking maps;
  - CFAR control;
  - False alarm control;
  - Quality selection;
- Weather or rain intensity maps for air traffic control.

The resolution of the maps has to be great enough to represent the environment around the radar, typically the antenna (noise) beamwidth in azimuth by the effective pulsewidth in range.

For functions where less precision is required, such as rain maps, the amount of memory may be reduced by increasing the area represented by each memory cell.

The choice of the type of memory for each type of map depends on the following:

- Whether the data should be retained after the radar has been switched off. Electrically erasable programmable read only memories (EEPROM) are used where the data are rarely changed. Static random access memory, in which storage is maintained using batteries, is used for information that is changed often and requires faster access.
- Dynamic random access memory (DRAM) is used when large amounts of cheap memory are required, as in most personal computers. The disadvantages are that the contents must be refreshed (normally internally) each 20 ms and that the contents must be replaceable if lost. This type of memory may be used for clutter maps where the contents are built up during the first few antenna rotations after the radar is switched on.

11.3.1 Area moving target indication (AMTI) or clutter map

In the discussion of clutter models, it was shown that in many cases the signal-to-clutter ratio is high enough to allow detection. The detection of wanted echoes in a background of clutter echoes is only possible when the operator remembers the clutter picture and is able to see the change when an echo of interest is present.

Area moving target indicators remember the radar picture gathered previously and report the changes only. Large amounts of memory are necessary. For example, the model radar has 800 range increments and 360/1.5 = 240 antenna beam positions in a circle giving at least 192 000 cells to store in memory or double this when the cells are commonly half a beamwidth wide.

A typical early area moving target indicator for a search radar used 6 bits to store the logarithmic clutter level within the dynamic range of the receiver, namely, up to 64 dB. The updating of the levels was carried out nonparametrically, namely, for a video pulse of amplitude greater than that stored in memory reported a hit and caused the value stored to be raised by 4 dB. The stored values were reduced by 1 dB after N scans. If $N = 5$ then the threshold needs 20 scans to return to the former value, which reduces the alarms to 5% of the previous value. To this threshold a noise threshold is added which, in practice, gives a “black” display with interclutter visibility, that is visibility between the clutter patches (see Figures 1.12 to 1.15). The equivalent threshold was so high that there was limited superclutter visibility. Developments in memory technology have allowed the information for smaller cells to be stored and have made parametric processing possible.

Commonly, a running average algorithm is used to maintain the clutter picture. Typically [14, p. 15.65],

$$U_n = (1 - \beta)\, U_{n-1} + \beta\, U_{in} \tag{11.94}$$

where U_n is the estimate at the nth scan;
U_{n-1} is the estimate at the $n-1$th scan;
U_{in} is the current measurement;
β is the smoothing factor.

If α is 1/8, then the information in the clutter map is refreshed during eight scans. When the radar is switched on, α may be set to 1 for the first scan and halved on subsequent scans in order to accelerate the establishment of the clutter map.

Since the stored signals represent strong smoothed clutter with a very small proportion of noise, lower thresholds, often in the range of 7 dB, may be used to detect wanted echoes. The clutter in each cell has the same amplitude from scan to scan with a fixed frequency radar and has Gaussian noise added to it, a Ricean distribution (see Sections 6.3.2.6 and 15.2.4). The distribution is given by

$$p_{Rice}(y) = \exp(-(y + R))\, I_0(\sqrt{4yR})\, dy \tag{11.95}$$

where $I_0(z) = \dfrac{1}{\pi}\displaystyle\int_0^{\pi} \exp(\pm z\cos\theta)\, d\theta.$

The distribution exceeds a threshold Y for a proportion of time P_{Rice} given by

$$P_{\text{Rice}}(y) = 1 - \int_0^Y \exp(-(y + R))\, I_0(\sqrt{4yR})\, dy \tag{11.96}$$

and depends the clutter-to-noise ratio, R. Thresholds for different false alarm rates and clutter to noise ratios are given in Figure 11.88.

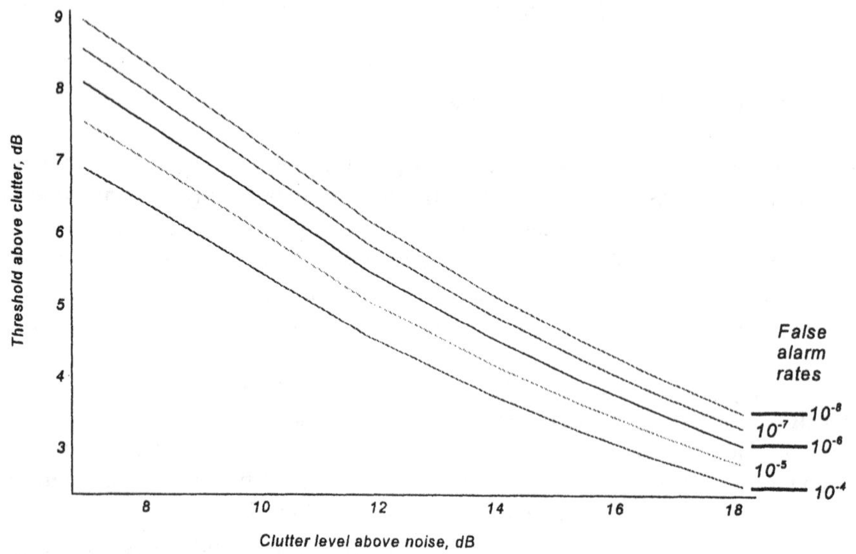

Figure 11.88 Thresholds above clutter for area moving target indicators.

Figures 1.12 and 1.13 show the signal to clutter relationships for a model radar with log-normal model clutter. In practical clutter the peaks may be grouped together but this does not alter the statistical properties. The area moving target indicator would reduce the clutter to below the black level on a display allowing the echoes of interest to be displayed.

Lower thresholds require that the clutter signals are almost exactly the same from scan to scan. Although radar range is accurate and stable, in older radars the triggers for the transmitter were not adjusted for the antenna rotation, so that clutter patches were illuminated from a different part of the beam during each scan, which caused a variation in the clutter signal to be stored. The variation of the clutter amplitudes stored may be minimized by dividing the 360° into equal sectors, each being a coherent processing interval. Most scanning antennas are driven by induction motors, so the pulse repetition frequency of each burst must be made variable within small limits. This mechanism allows the same patch of clutter to be struck at the same angle each scan.

References [5 to 7] extend the treatment to the variations of the clutter level to include variations in the running average and give the probability of detection as

$$P_d = \prod_{m=0}^{\infty} \frac{1}{1 + \dfrac{\alpha}{1 + R}\beta(1 - \beta)^m} \tag{11.97}$$

and the probability of false alarm as

$$\overline{P_{fa}} = \prod_{m=0}^{\infty} \frac{1}{1 + \alpha\beta(1 - \beta)^m} \tag{11.98}$$

Reference [6] gives an approximation for the loss as

$$Loss \approx -5.5 \log_{10} \frac{\overline{P_{fa}}}{1 + \frac{2}{\beta}} \tag{11.99}$$

Equation (11.100) is shown graphically in Figure 11.89.

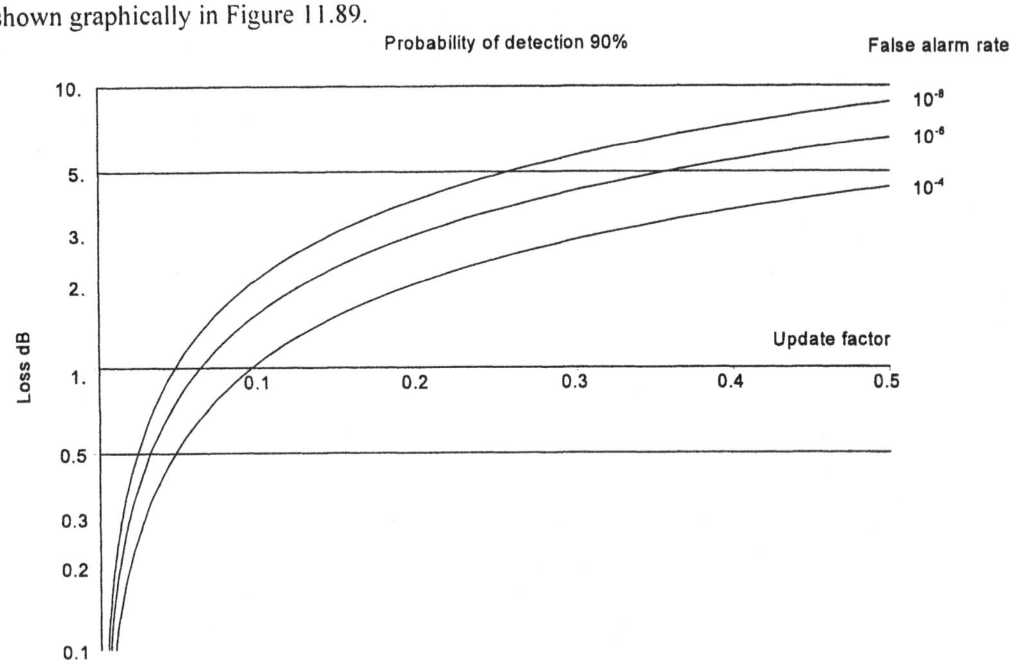

Figure 11.89 The loss factor for area moving target indicator. [After: 6, 7.]

A common value for the update factor is 1/8 giving a 1.94 dB loss for a 10^{-6} false alarm rate.

In a moving target detector the information is dumped after each coherent processing interval and new information is gathered during the next. The store and dump process is in contrast with "traditional" processing, where echo information is integrated over the whole beamwidth. Thus the maps for the clutter filters are often dimensioned as one pulse width by one coherent processing interval, which gives the maximum resolution to enhance interclutter visibility.

With a 12 bit analogue-to-digital converter and up to 16 data samples (echoes) added together, the data width has to be increased to 16 bits to avoid overflow of the registers.

Some older radars use a clutter map based on threshold crossings in map cells. If the echo signals cross the current threshold for the cell, a hit is declared, and the threshold is raised by, say 4 dB, and after a number of scans the clutter map is established. An aircraft must move from cell to cell in order to be seen and the threshold levels are reduced by 1 dB after *n* scans in order to keep the thresholds just above the clutter.

11.3.2 Maps for a moving target detector

The moving target detector has maps in order to optimize its operation and present a "clean" picture.

11.3.2.1 Clutter maps

Clutter maps clear up the picture after moving target indicators or detector processing, and they are another term for an area moving target indicator described in Section 11.3.1.

11.3.2.2 Maps to control filter bandwidths

The amplitude gives an estimate of the width of the clutter spectrum. Often the clutter filter bandwidth is widened to accommodate all the ground clutter and the center frequencies of the others moved to keep them free of clutter.

The aim of a moving target detector is to sort all clutter signals into the clutter filters, so the remaining filters contain only useful information. Unfortunately, the width of a Gaussian spectrum depends on its amplitude, so that for strong clutter the clutter filters are widened (for example, using the Woodward-Levinson method in Chapter 5, Antennas) and the center frequencies of the remaining filters are moved away from the clutter bands.

A map notes the clutter amplitude bands in blocks of resolution cells from a previous scan, which are used to set the filters in the following scan.

For moving target detector filtering, Figure 11.83 shows the sorting of signals into Doppler frequency bands. Even in bands thought to be free of clutter, there may well be clutter residues. The whole of the "picture" for each Doppler frequency band in the scan can be stored to provide a base for setting modified thresholds. Where the clutter amplitude is high, the residue is probably fixed clutter; where the fixed clutter is low, any area returns are probably rain (chaff in military situations) or snow.

11.3.2.3 Rain maps

Air traffic control radars often double as weather radars and report areas of difficult weather so aircraft may be directed to fly around them. A rain map is displayed as a background for the air traffic controller. Rain maps, typically, are of the following types:

- Rain areas to switch in the rain cell averaging constant false alarm rate thresholds;
- Rain maps showing rain echo contours for the aircraft controllers;
- Automatic selection of circular polarization when it rains. Circular polarization reduces the sensitivity of the radar, typically 3 dB. An aircraft that is just detectable at 100 km will be seen only out to 84 km when circular polarization is used.

11.3.2.4 Detection control maps and second censoring

In addition to the maps used for frequency filtering, a number of maps are used for the censoring of the detections. Examples are:

- Blanking maps;
- False alarm control maps;
- Cell averaging constant false alarm rate maps.

Blanking maps

Blanking maps define areas where detections are not to be reported to the extractor. There are a number of echo signals coming from the environment that are not wanted, as, for example:

- Extremely strong fixed echoes which in spite of sensitivity time control and clutter maps still give detections.
- Ventilator fans for buildings. The fan blades rotate quickly giving a moving echo. In this case, and in that above, the echoes occupy few resolution cells. Blanking these cells hardly affects the radar coverage but reduces the load on the extractor, the tracker (which will probably reject a stationary object), and the operators.
- Traffic on roads and motorways gives moving echoes along thin lines which are often broken by shadows [5, pp. 9-25]. Speed limits have been introduced near airports to keep road traffic below the radar's moving target indicator speed threshold, but many drivers are more than equal to this challenge. One way to register the positions of roads is to record the positions of the echoes during two periods. The periods are chosen so that the road traffic is the same during these periods but the wanted echoes follow different paths. The recordings of the two periods are compared and all cells showing detections during both periods are marked for blanking.

False alarm control

When a radar operator sees a cloud of detections he automatically makes the best guess when seeking wanted echoes on his display. Data processing systems are quickly overloaded. To avoid this, the threshold reports are counted in discrete areas, and when their number exceeds the number set, the threshold voltage is increased (up to 6 dB, say) for a set time.

The volume where the sensitivity has been reduced should be shown to the operators as a map for warning.

Cell averaging constant false alarm rate control
Cell averaging thresholds are chosen in the presence of rain or other extended echoes in the rain map. The sample chosen for cell averaging ideally is long, has equal echo amplitudes, and does not contain an edge. The length used is shortened so as not to include edges. When edges occur, the "greater than" function reduces false alarms but decreases the sensitivity. When necessary the choice of length may be restricted in particular areas using maps.

11.3.3 Quality selection

Modern radars often add binary flags to label their plot reports as a further indication of quality that is used in the tracking algorithms. Examples of these flags are shown in Table 11.7.

Table 11.7
Examples of quality flags

	Indication	Possible cause	Possible action
1	The signal caused an overload in the receiver.	Probably strong clutter or interference.	Reject the echo.
2	The signal caused an overflow in the analogue-to-digital converters.	Probably strong clutter or interference.	Reject the echo.
3	The sequence of echoes in a CPI shows more amplitude variation than the antenna pattern allows.	Asynchronous interference.	Use the plot for updating only.
4	The noise integrated over all ranges exceeds a threshold.	Interference (military — jamming) is coming from the azimuth sector. This can be caused by the sun.	Use the plot for updating only.
5	Detection in ground clutter filters.		If there is a detection in another filter, then use the plot normally; otherwise, use for update only.
6	Detection in rain (or chaff).		Has the detection occurred at both repetition frequencies? If yes, good.
7	Detection during one coherent processing interval only.		Mark for update only.

Quality labeling allows otherwise dubious threshold crossings to be used to complete the information from more reliable threshold crossings.

11.4 AIRBORNE MOVING TARGET SIGNAL PROCESSING

When a radar is mounted on a moving vehicle, the velocity of the vehicle adds vectorially to the radial velocity of the vehicle, mostly aircraft or ships. Some older ground radars allowed a gate to be placed over a patch of rain (or in military radars, window or chaff) and the coherent oscillator signal used for detection was offset by the Doppler frequency of the rain. The rain echo was canceled by the moving target indicator but not the ground echo nor scatterers (aircraft) moving at other radial velocities. The azimuth and range gate that defined the area and the Doppler frequency offset were set by hand. This section provides an introduction and references for some types of processing to detect moving scatters from moving platforms; examples are the TACCAR initially developed by the Lincoln Laboratory for the E2C aircraft for the United States Navy, displaced phase center antenna, and moving target indicators for sideways looking radars. Space and time adaptive processing (STAP) [29, 30] is not covered here.

The main problem is adjusting the reference frequency for coherent detection automatically so that land or sea clutter is canceled. The Doppler frequency of the land or sea clutter depends on the velocity of the aircraft, and the angle in bearing with respect to the aircraft is as in Figure 11.90. If the velocity of the aircraft is v, then the radial velocity of the clutter patch is $v \cos \theta$. The radial velocity component is weakly dependent on the aircraft height as shown in [14, p. 16.3]. The antenna rotates slowly at 5 revolutions per minute and the problem is to estimate the average Doppler frequency for each sweep and use this frequency to center the clutter Doppler frequency on zero hertz for conventional processing later.

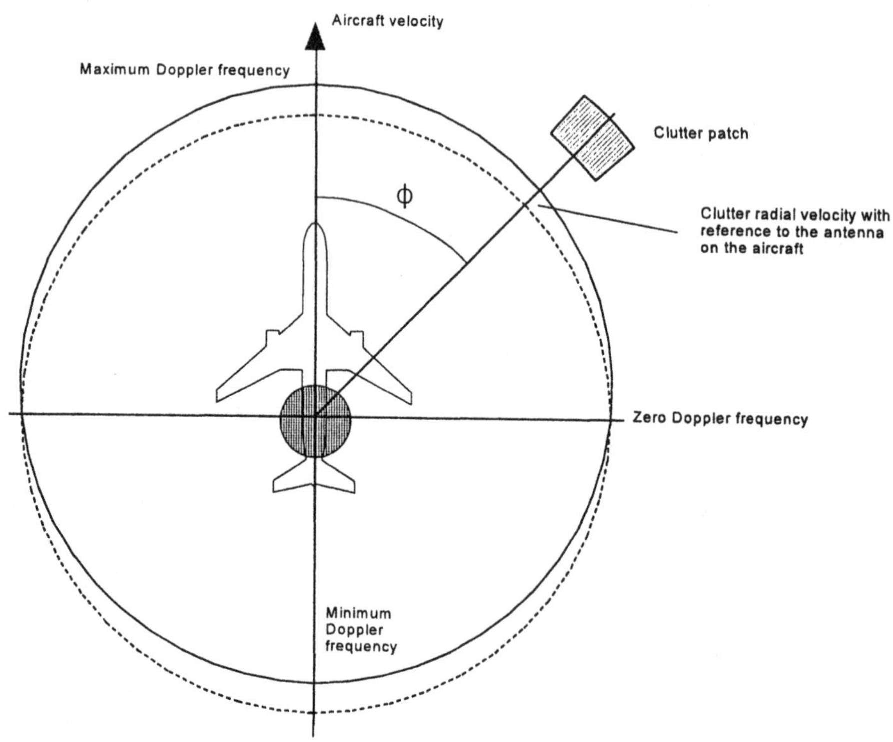

Figure 11.90. The relative radial velocities and Doppler frequencies of ground clutter that varies with the antenna bearing.

11.4.1 The TACCAR moving target indicator

The acronym TACCAR stands for time averaged clutter coherent airborne radar. It uses one clutter signal either from the ground, the sea, or weather to correct for the varying radial velocities as the antenna rotates. Older systems for magnetron radars offset the oscillator frequency and phase for synchronous or phase detection, and this was used as the reference frequency for the detectors [1, p. 17.32]. Other modern systems offset the transmitter frequency and use the fixed coherent oscillator frequency for synchronous detection [14, p. 16.5].

In Figure 11.91 the intermediate frequency signals with large clutter content are coherently detected (formerly a phase detector, later a synchronous demodulator) and pass through a delay line, length one pulse repetition time. The difference in the phase of the clutter signals between the input and output of the delay stage is found by the phase detector and passed to the range gate. The range gate may be set by hand to ranges that contain clutter, or an automatic clutter detector can open the gate to pass the phase differences to a lowpass filter that removes all frequencies above the rhythm of the rotating antenna.

The voltage controlled oscillator (VCO) is part of a phase locked loop using the coherent oscillator (COHO) as a reference. The error signal from the frequency discriminator (B) in Figure 11.91 also passes through a lowpass filter to remove fast changes to a summing circuit. The summing circuit has an input (A), the phase error signal and, in later systems, an estimate of the Doppler frequency from the aircraft's navigation system (C).

The sum of the offset voltages change the frequency of the voltage controlled oscillator to match the Doppler frequency of the clutter so that:

- For magnetron radars — dashed lines in Figure 11.91:
  The frequency phase of the voltage from the lock pulse block allows the clutter to be detected with zero Doppler frequency for further processing.
- For modern master oscillator power amplifier (MOPA) transmitters:
  The voltage from the voltage controlled oscillator (here coherent master oscillator – COMO) is mixed with the stable local oscillator (STALO) signal to form radio frequency pulses for the transmitter (dot dot dash lines). The echoes from the clutter patch then have the same frequency as the coherent oscillator signal that is used in the coherent (synchronous) detector.

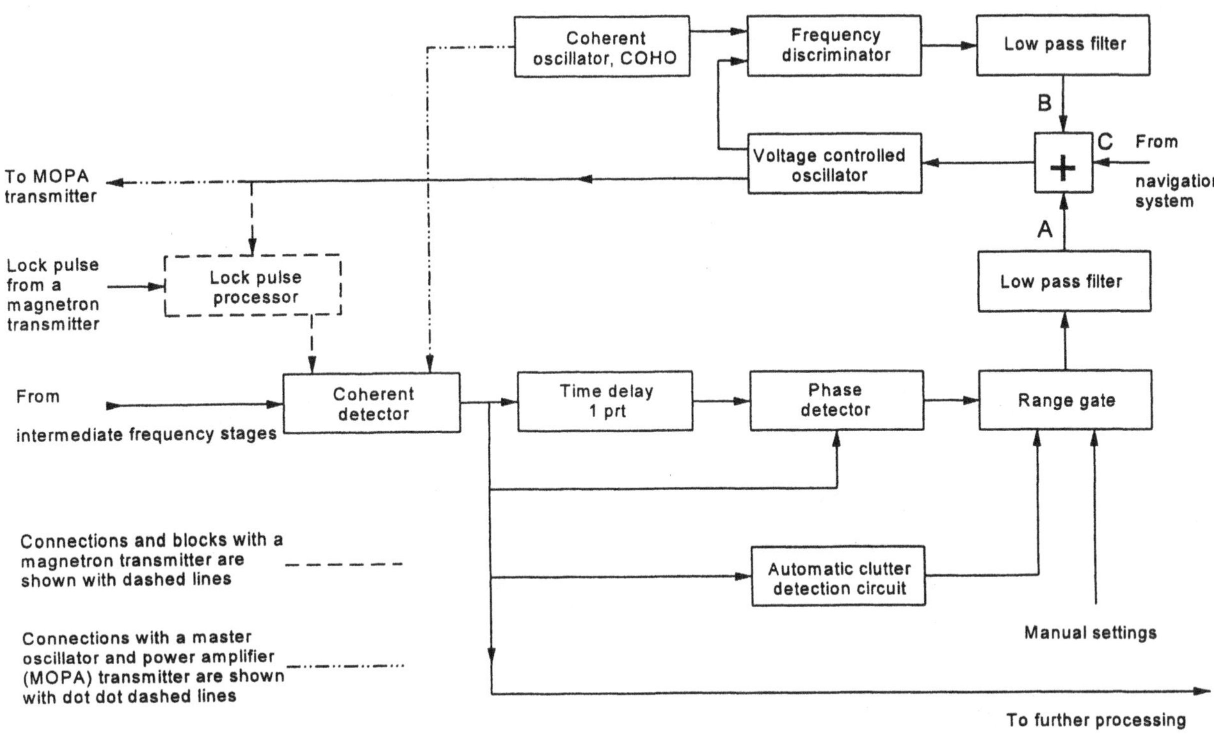

Figure 11.91 The TACCAR processor for clutter locked signal processing.

11.4.2 Displaced phase center antenna

In sideways-looking radars mounted on aircraft, the effect of differing Doppler frequencies depending on the position of the scatterer in the beam may be reduced by having effectively many antennas mounted along the length of the aircraft. One or a combination of the antennas are chosen during the look so that the phase center of the antenna remains stationary while the aircraft moves; see Figure 11.92. This technique is called the displaced phase center antenna (DPCA).

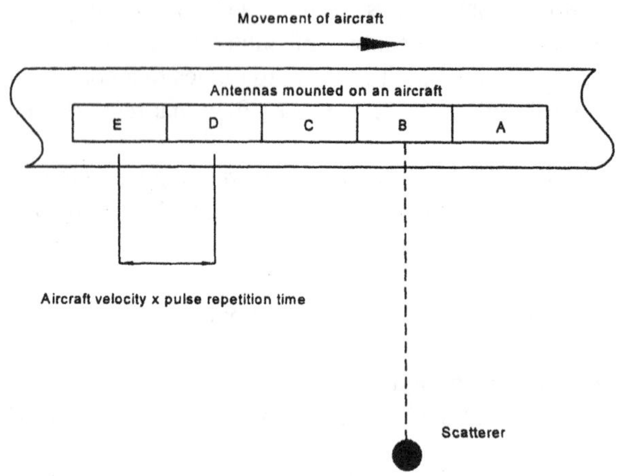

Figure 11.92 Side-looking radar antennas mounted on an aircraft.

The first pulse is transmitted using antenna A in Figure 11.92 and the echoes are received. By the time for the next pulse, antenna B is in the physical position of antenna A and B is used for this pulse, and so on. The velocity component in the direction of the aircraft becomes zero. In practice a small number of antenna sections are used.

The echo signals in radars mounted in aircraft with rotating antennas also experience differing Doppler frequencies according to their position in the beam. The beam rotation can be effectively "stopped" using a difference output from the antenna [14, 31]. The general scheme is shown in Figure 11.93.

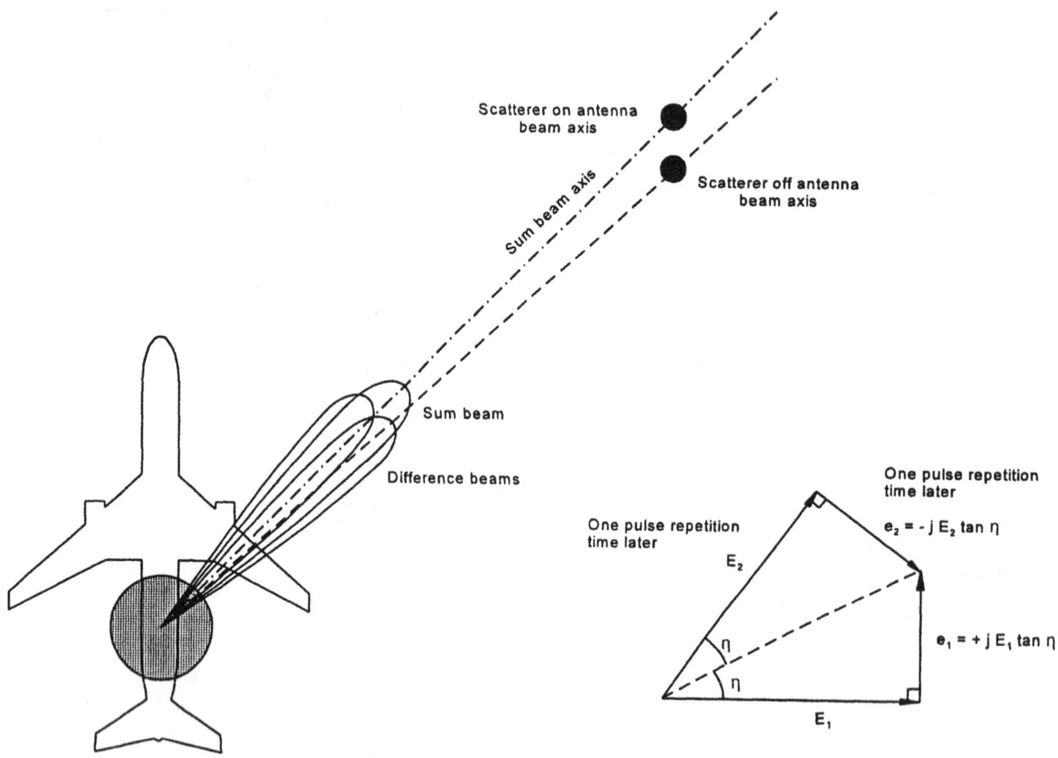

Figure 11.93 The geometry for a rotating search antenna mounted on an aircraft.

The component of velocity along the axis of the beam for a stationary scatterer on the axis is

$$\textit{Velocity along axis of the beam} = V_g \cos(\alpha) \tag{11.100}$$

where V_g is the velocity of the aircraft along the ground and α is the bearing from the roll axis of the aircraft. Other scatterers within the beam also return echoes with a Doppler frequency depending on their position in the beam, θ, namely,

$$Velocity\ seen\ by\ the\ radar = V_g \cos(\alpha + \theta) \tag{11.101}$$

The radar thinks that the bearing is given by the angle of the axis so that there is a spread of the clutter spectrum depending on the angles of the scatterers off the antenna axis. For coherent processing (moving target indicator, moving target detector, Doppler frequency processor, and so on) this effect changes the phases of a train of echoes for processing. The phase change from echo pulse to echo pulse is

$$\delta\phi_c = 2\eta = 2\pi\ \delta f_c\ T \tag{11.102}$$

where δf_c is the frequency change from pulse to pulse each spaced T seconds apart.

The change of frequency from pulse to pulse is given by (11.101)

$$\delta f_c = \frac{2V}{\lambda}\cos(\alpha - \theta) - \frac{2V}{\lambda}\cos\alpha \approx \frac{2\theta}{\lambda} V \sin\alpha, \text{ assuming that } \theta \text{ is small} \tag{11.103}$$

Using (11.104) the phase variation 2η, depending on α and θ, is

$$2\eta = 4\pi T \frac{\theta}{\lambda} V \sin\alpha \tag{11.104}$$

An example of a UHF radar with a wavelength of 50 cm flying at 200 m/s, the spread of the spectrum is shown in Figure 11.94.

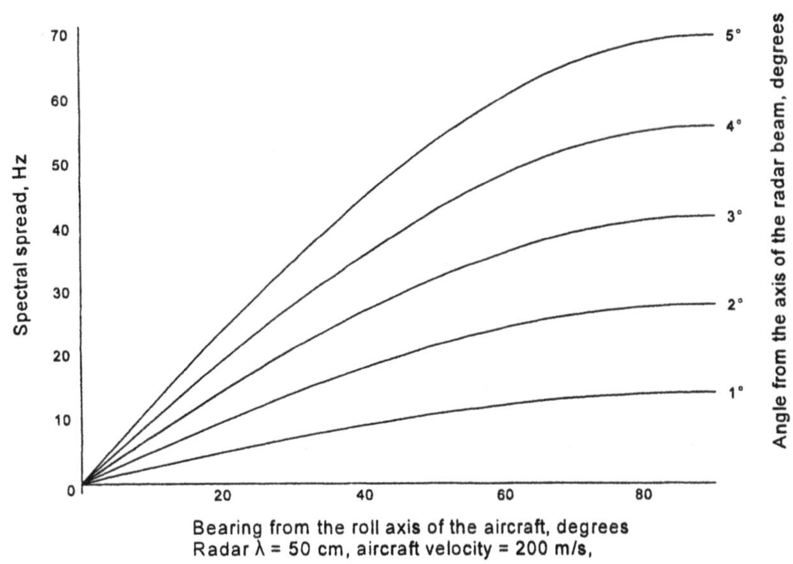

Figure 11.94 The spread of the clutter spectrum with bearing for a UHF radar.

For a UHF antenna in a rotating dome with 7 degrees beamwidth the spread is ±49 Hz. For a pulse repetition frequency of 360 Hz the clutter spread abeam occupies 27% of the Doppler frequency domain and this is unacceptable. Thus a way has to be found to add correction signals, e_1 and e_2, to the echo signals E_1 and E_2 compensate for this change.

The pulses are transmitted using the sum (or Σ) channel on the antenna and the echo signals are received using both the sum and difference (Δ) channels. The two-way characteristic, subscripted with r, may be represented as

$$\begin{aligned} \Sigma_r(\theta) &= k_1\ \Sigma^2(\theta) \\ \Delta_r(\theta) &= k_2\ \Sigma(\theta)\ \Delta(\theta) \end{aligned} \tag{11.105}$$

with k_1 and k_2 as constants to account for the radar equation. The amplitudes of the two echo signals are assumed to be equal (as in Marcum or Swerling case I or III), namely, $|k_1\ \Sigma^2(\theta)|$.

For linear antennas, width W, that are uniformly excited, the sum and difference patterns are

$$\Sigma(\theta) = \frac{\sin\left(\frac{\pi W}{\lambda}\sin\theta\right)}{\frac{\pi W}{\lambda}\sin\theta} \qquad \Delta(\theta) = \frac{1 - \cos\left(\frac{\pi W}{\lambda}\right)\sin\theta}{\left(\frac{\pi W}{\lambda}\right)\sin\theta} \tag{11.106}$$

The correction signals in the vector diagram in Figure 11.93 and using (11.104) and (11.106) are given by

$$e_1 = E_1\tan\eta = j\ k_1 \frac{\sin^2 \frac{\pi W}{\lambda}\sin\theta}{\left(\frac{\pi W}{\lambda}\sin\theta\right)^2} \tan\left(\frac{2\pi T\theta}{\lambda}\ V\sin\alpha\right) \approx j\ k_1 \frac{2\pi\ T\ \theta}{\lambda}\ V\sin\alpha \tag{11.107}$$

where k_1 is a constant for the sum pattern. For the second pulse $e_2 = -e_1$.

For small offset, θ, the two-way signal received in the difference channel (11.105) using (11.106) is approximated by

$$\Delta_r(\theta) \approx k_2 \frac{\pi W}{2\lambda}\ \theta \tag{11.108}$$

Substituting (11.108) into (11.107),

$$e_1 = j\frac{k_1}{k_2}\frac{4\ V\sin\alpha}{W}\ T\ \Delta_r(\alpha) = j\frac{k_1}{k_2}\ V\sin\alpha\ \Delta_r(\theta) \quad \text{and} \quad e_2 = -e_1 \tag{11.109}$$

Adding the correction pulses e_2 and e_1 to the sum signals for cancellation gives

$$\begin{aligned} \textit{Corrected pulse No 1} &= \Sigma_r(\theta) + j\ k\ V\sin\alpha\ \Delta_r(\theta) \\ \textit{Corrected pulse No 2} &= \Sigma_r(\theta) - j\ k\ V\sin\alpha\ \Delta_r(\theta) \end{aligned} \tag{11.110}$$

The process described above may be implemented with equipment in Figure 11.95.

11.4.3 Pulse Doppler radars

Low pulse repetition frequency pulse Doppler signal processing was described in Section 11.2.4 as a moving target detector (MTD) radar and in Section 11.2.4.2 as the coherent memory filter, with unambiguous range but ambiguous Doppler frequency. The true Doppler frequency has to be estimated using two or more pulse repetition frequencies.

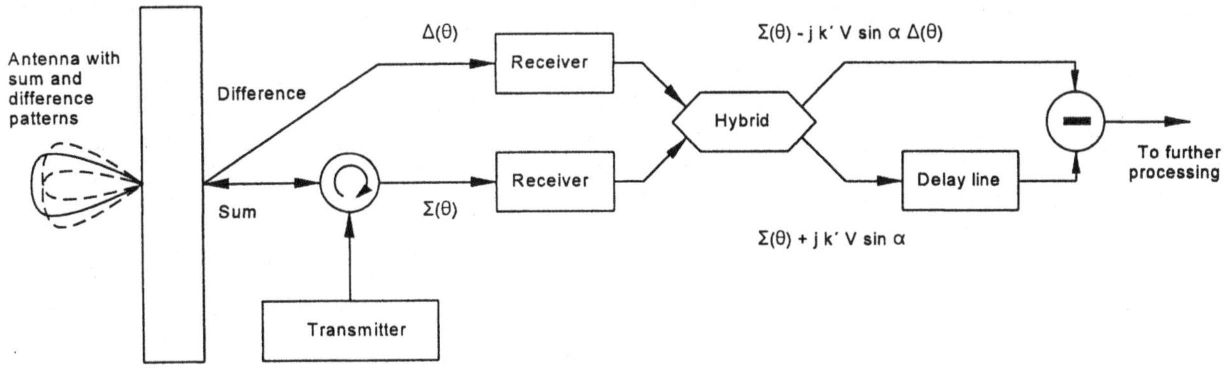

Figure 11.95 A hardware implementation for a displaced phase center correction for a rotating antenna.

Medium and high pulse repetition frequency pulse Doppler radars are generally used in aircraft and used as sights for weapons with one or few targets. With medium pulse repetition frequency radars both range and Doppler frequency are ambiguous and frequent changes of pulse repetition frequency are necessary to place the targets out of the eclipsing interval, to separate them on the ambiguous scale from clutter, and to gather enough information for the algorithms for true range and Doppler frequency. Further discussion is beyond the scope of this book.

The radar used by the AWACS surveillance aircraft to observe the airspace covered by a large ground surveillance radar runs, when flying over land, with a high pulse repetition frequency. Advances in digital signal processing allow the construction of Doppler frequency filter banks with a large number of filters for each range gate in the ambiguous range. A number of different pulse repetition frequencies and unique Doppler frequencies allow the true ranges of the aircraft observed to be found and eclipsing intervals to be bridged over. The critical component is the ultra-low sidelobe antenna that discriminates in angle to reduce ground clutter received in the sidelobes.

11.4.4 Sideways-looking radars

Synthetic aperture radars (see Section 5.13.1) produce a picture of an area of the ground after the radar has travelled over the area. Signal processing for the signals from synthetic apertures is a specialized subject not covered here. In general echos from scatterers moving in the range direction beat with the ground echo signals so that noncoherent moving target indicator processing may be used. In the cross-range direction, the echos from moving scatterers return with a different Doppler frequency and are reconstructed with a shift in position. Such echoes may be recognized by using interferometric processing.

11.5 FIGURES AFFECTING RADAR PERFORMANCE

The signal processor is the stage that filters the wanted echoes from other echoes that clutter the display. Generally, useful filtering involves a loss in signal-to-noise ratio, which must be accounted for in maximum range and accuracy calculations.

11.5.1 Range budget

Losses depend on the type of signal processing used. The most simple is the display of the video signal on a cathode ray tube without any intervening processing. During a radar look, the echoes from a number of pulses reinforce each other (or are integrated) on the cathode ray tube phosphor. The signal-to-noise ratio for a particular probability of detection and false alarm rate is given for the Marcum and Swerling cases (with and without echo signal fluctuation) in Chapter 12.

Moving target indicator (MTI) processing involves removing the clutter signals by subtraction and, as a sideeffect, the wanted echo signals if their Doppler frequencies represent the blind speeds. The subtraction process increases the correlation of the noise and reduces the number of effective hits (see Figure 11.43) to be used for the signal-to-noise ratio for detection. The difference in the signal-to-noise ratios with and without moving target indicator processing represents a loss.

Coherent integration involves the adding of the wanted signals so that, ideally, the unwanted signals add vectorially to zero. Conceptually, all the signals and noise from the *N* pulses are combined into one. In order to reduce the responses to signals at nearby frequencies (compare with antenna sidelobes in Chapter 5), tapering of signals in the train is used. That is, we choose not to use all the returned echo signal energy for processing, which represents a loss.

Often the signals after moving target indicator and detector processing are subjected to higher cell averaging constant false alarm rate thresholds, which too represents a loss.

Known losses for receiver "fixes", for example, "Dicke-fix" in Chapter 8, must be included when used.

11.5.2 Accuracy and resolution budget

The signal-to-noise ratio used in Chapter 13 for accuracy must include the losses incurred in signal processing.

Signal processing takes time. For analogue signal outputs, there is a delay of often tens of microseconds, so that the zero range trigger for the displays must be delayed also. Processing over a number of sweeps gives an additional delay or shift in azimuth equivalent to the antenna rotation for half the number of processed hits.

11.5.3 Stability budget

Improvement factors for moving target indicator processing may be taken from Figures 11.41 and 11.42. When limiting has taken place, Figures 11.51 and 11.52 should be used. Instabilities in the moving target indicator system caused by pulse repetition staggering can be found from Figures 11.46 and 11.47.

The improvement factor for Chapter 14 becomes

$$\frac{1}{I_{Sp}} = \frac{1}{I_{MTI}} + \frac{1}{I_{Stagger}} \tag{11.111}$$

The improvement factor in moving target detector systems depends on how the integrating filters are designed and how far in frequency they are from the filter containing clutter.

Fourier transforms of amplitude measurements of a clutter return taken over, say, 128 antenna scans are calculated to give the spectrum of the clutter signal. The power ratio of the peak representing the clutter signal to the background gives the final improvement factor, including all the instabilities.

REFERENCES

1. Skolnik, M. I., *Radar Handbook*, New York: McGraw-Hill, 1970.
2. Croney, J., "Clutter on Radar Displays", *Wireless Engineer,* April 1956.
3. Gradshteyn, I. S., and I. M. Ryzhik., *Table of Integrals, Series, and Products*, New York: Academic Press, 1980.
4. Sekine, M., and Y. H. Mao, *Weibull Radar Clutter*, London: Peter Peregrinus, 1990.
5. Ludloff, A., *Handbuch Radar- und Radarsignalverarbeitung*, Braunschweig, Germany: Vieweg, 1993.
6. Nitzberg, R., *Adaptive Signal Processing for Radar*, Norwood, Massachusetts: Artech House, 1992.
7. Nitzberg, R., *Radar Signal Processing and Adaptive Systems*, Norwood, Massachusetts: Artech House, 1999.
8. Richards, M. A., *Fundamentals of Radar Signal Processing,* New York: McGraw-Hill, 2005.
9. Poelman, A. J., *Supplement 2 to Technical Report TR-50-U (Standard methods for predicting and specifying performance of air surveillance radar systems)*, SHAPE Technical Centre, The Hague, April 1970.
10. Poelman, A. J., and E. J. Bennée, *Supplement 3 to Technical Report TR-50-U (Standard methods for predicting and specifying performance of air surveillance radar systems)*, SHAPE Technical Centre, The Hague, May 1973.
11. Marcum, J. I., *A Statistical Theory of Target Detection by Pulsed Radar*, Rand Corporation, Mathematical Appendix to RM-754, April 1952, p. 213.
12. Barton, D. K., *Radar System Analysis*, Englewood Cliffs, New Jersey: Prentice-Hall, 1964, pp. 128-132.

13. Barton, D. K., *Modern Radar System Analysis*, Norwood, Massachusetts: Artech House, 1988.
14. Skolnik, M. I., *Radar Handbook*, 2nd ed., New York: McGraw-Hill, 1990.
15. Bozic, S. M., *Digital and Kalman Filtering*, London: Arnold, 1979.
16. Stevens, M. C., *Secondary Surveillance Radar*, Norwood, Massachusetts: Artech House, 1988.
17. Schleher, D. C., *MTI and Pulsed Doppler Radar*, Norwood, Massachusetts: Artech House, 1991.
18. Blake, L. V., *Radar Range-Performance Analysis,* Norwood, Massachusetts: Artech House, 1986, p. 345.
19. Hall, W. M., and H. R. Ward, "Signal-to-Noise Loss in Moving Target Indicator", *Proceedings IEEE* 56, No. 2, February 1968, pp. 233-234.
20. Berkowitz, R. S., (ed.), *Modern Radar, Analysis, Evaluation, and System Design*, New York: Wiley, 1965.
21. Cooley, J. W., and J. W. Tukey, "An Algorithm for the Machine Calculation of Complex Fourier Series", *Mathematics of Computation*, Vol. 19, No. 90, 1965, pp. 297-305.
22. Bergland, G. D., "A Guided Tour of the Fast Fourier Transform", *IEEE Spectrum*, Vol. 6, July 1969, pp. 41-52.
23. Meikle, H. D., *A New Twist to Fourier Transforms*, Weinheim: Wiley-VCH, 2004
24. Bracewell, R. N., *The Fourier Transform and Its Applications*, 2nd ed., New York: McGraw-Hill, 1965, p. 358.
25. Elliott, D. E., and R. K. Rao, *Fast Transforms*, Orlando, Florida: Academic Press, 1982, p. 182.
26. Harris, F. J., "On the Use of Windows for Harmonic Analysis with the Discrete Fourier Transform", *Proc. IEEE,* Vol. 86, No. 1, January 1978.
27. Skillman, W. A., *Radar Calculations Using the TI-59 Programmable Calculator*, Dedham, Massachusetts: Artech House, 1983.
28. Weisstein, E. W., (ed.), *CRC Concise Encyclopedia of Mathematics*, Boca Raton, Florida: CRC Press, 1999.
29. Klemm, R., *Principles of Space-Time Adaptive Processing*, Stevenage: IEE Publishing, 2002.
30. Klemm, R., (ed.), *Applications of Space-Time Adaptive Processing*, Stevenage: IEE Publishing, 2004.
31. Skolnik, M. I., *Introduction to Radar Systems*, 3rd ed., New York: McGraw-Hill, 2001.

APPENDIX 11A AN APPROXIMATION TO SOLVE FOR THRESHOLDS ABOVE CLUTTER

A simple linear approximation used to estimate the limits for solving the equation for thresholds, Y_0 above clutter is

$$Y_0 = 4.46 + 3.737p + (1.236 + 0.061p)\,x \tag{11A.1}$$

where $p = -\log_{10} P_{fa}$;
x is the clutter to noise ratio.

The threshold above clutter is given by the factor Y_C.

$$Y_C = \frac{Y_0}{x} \tag{11A.2}$$

Figure 11A.1 shows Y_C in decibels plotted against the clutter-to-noise ratio in decibels with the approximation (11A.3) shown as dashed lines.

$$\begin{aligned} \textit{Approximate threshold} = \; & (-0.0000432\,p^2 + 0.000669\,p + 0.0118)x^2 \\ & + (0.00319\,p^2 - 0.0656\,p - 0.531)x \\ & - 0.0592\,p^2 + 1.41\,p + 6.703 \end{aligned} \tag{11A.3}$$

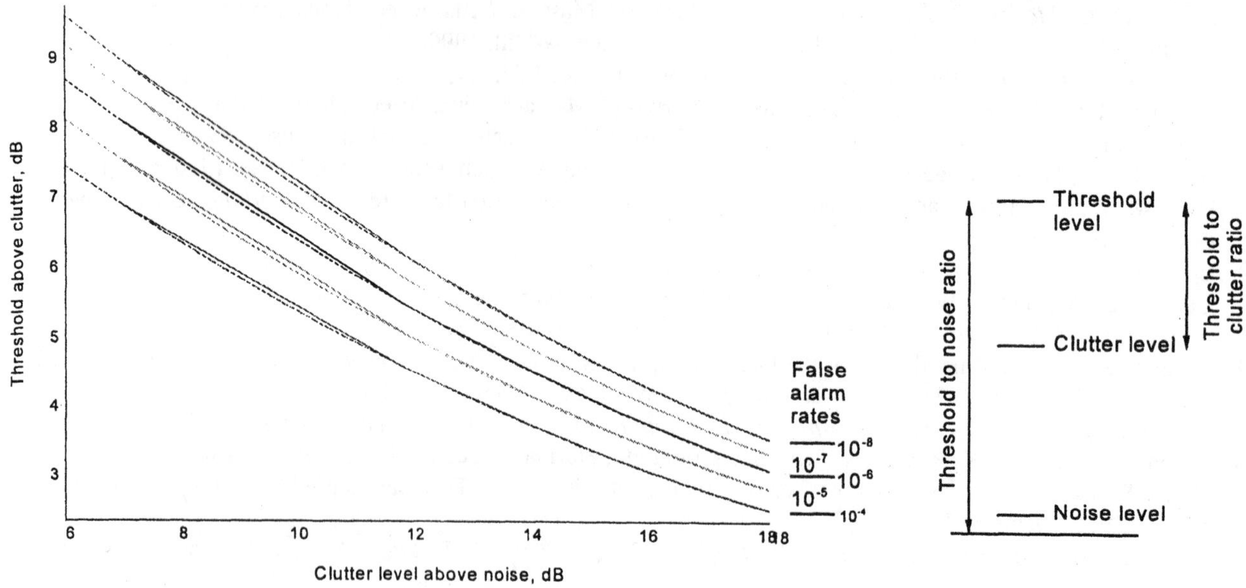

Figure 11A.1 The real and approximate threshold levels from (11A.3).

Chapter 12

Threshold and detection

After the signals have been filtered to exclude out-of-band noise, the pulse has been compressed (when necessary), and the clutter and interference have been reduced to the noise level, a decision is made as to whether the echo signal represents an object of interest. The decision may be made automatically or by an operator.

For radars that observe ground areas for moving objects, a gate may be set over the echo of interest and the operator is able to listen to the signal using earphones. Each type of animal, human, and vehicle gives its characteristic "music", which a trained operator is able to identify.

Most scatterers are irregular, and their fluctuation or fading characteristics have been described in Chapter 6 (Factors outside the radar). This chapter describes the probabilities of detection of five of the classical cases.

This chapter deals with the statistics of signals and noise giving the probabilities of:

- Random noise impulses adding together and by chance having the characteristics of a valid echo signal, the probability of false alarm, P_{fa};
- A wanted echo being recognized as such, or the probability of detection.

Basic gamma distributions are used in this chapter in place of Pearson's form of the incomplete gamma distribution and the chi-square distribution for which there are tables as used by older texts. Modern computers are able to calculate Bessel and gamma functions and their integrals directly.

Fourier transforms, described in Chapters 15 and 16, are used to calculate the probability density functions of signal plus noise. To be consistent with the rest of the text and the radar literature in general, the $-f$ convention is used for Fourier transforms.

12.1 DWELL TIME AND THE NUMBER OF ECHOES

Unlike the human eye, which observes the whole field of view simultaneously, radars, generally, either view one object of interest (tracking radars) or scan their surroundings and view many objects one after another (surveillance radars). The amount of time devoted to looking in the direction of an object is called the dwell time.

During the dwell time, a number of echoes are received and the energy summed or integrated. In the past, this process took place by building up brightness and the afterglow on a cathode ray tube. In the armed forces, a bright spot that looked like a star in the surrounding noise or a deflection above the noise on an A-scope was deemed to be a target since, originally, radars were used to aim weapons. Operators are easily influenced and may become tired. Automatic detectors do not suffer from such problems and carry out what they have been designed to do, but they are completely inflexible when confronted by the unforeseen.

An operator sees the sum of the signals during the afterglow time on his display. The amount of deflection or afterglow depends on the number of echoes received during the time that the antenna is pointing towards the object of interest. In detection theory, the number is the number of pulses integrated (or summed). In a scanning radar, the number of pulses is taken to be the number of pulses during the time that the radar looks in the direction of the echoing object between the 3 dB points of the antenna pattern.

$$Dwell\ time = \frac{Beamwidth\ \text{(degrees)}}{360}\ \frac{60}{N\ \text{(rpm)}} \quad \text{seconds} \tag{12.1}$$

If the dwell time is multiplied by the pulse repetition frequency (prf), the number of pulses is

$$\begin{aligned} \textit{Number of pulses} &= \textit{Dwell time} \times \textit{prf} \\ &= \frac{\textit{prf} \times \textit{Beamwidth}\ \text{(degrees)}}{6\ \ N\ \text{(rpm)}} \end{aligned} \tag{12.2}$$

The dwell time, or the time for one look, is conventionally defined between the 3 dB points on the antenna pattern. The noise, principally receiver noise, is constant during this time. The echo signal strength depends on the instantaneous antenna gain giving rise to the pattern loss (see Chapter 5, Antennas).

For example, if the radar has a pulse repetition frequency of 1 000 Hz, its antenna has a half-power beamwidth of 1.5 degrees and rotates at 12.5 revolutions per minute, then the number of pulses between the half-power points is

$$\textit{Number of pulses} = \frac{1\ 000 \times 1.5}{6 \times 12.5} = 20$$

During this time, conventionally, N samples of the echo signals and noise are added together. Strictly speaking, the noise beamwidth should be used, which is the beamwidth of the sector with maximum gain as its radius with the same area as the antenna pattern. The noise beamwidth is seldom measured on a test range. Outside these beamwidths at longer ranges the echo signals of interest are small and may be neglected.

12.2 FALSE ALARM PROBABILITIES, TIMES, AND THRESHOLDS

In the spaces between echoes, noise samples are added together. There is a small statistical chance that the energies in fortuitous noise spikes add up to a size that would represent an echo of interest (for the gunners: a target). This gives a false alarm. There are a number of ways to describe the number of false alarms [1]. If the noise samples occur at a rate of B_n, then the false alarm times and probabilities can be defined.

12.2.1 False alarm time

The false alarm time is the average time between false alarms. In the older literature, notably Marcum, it was the time for which the probability of one or more false alarms was one half, $t_{fa\text{-}old}$. This use is deprecated according to [1].

$$\text{Old (Marcum):}\quad t_{fa} = \frac{\ln(2)}{B_n\ P_{fa}} \tag{12.3}$$

Currently the false alarm time is defined as the average times between false alarms [1]:

$$\text{Current:}\quad t_{fa} = \frac{1}{B_n\ P_{fa}} \tag{12.4}$$

12.2.2 False alarm number

The false alarm number is the number of possible independent detection decisions during the false alarm time, which leads to the Marcum definition for his false alarm time,

$$\text{Old (Marcum):}\ \textit{False alarm number} = \frac{\ln(2)}{P_{fa}} \tag{12.5}$$

The current definition is [1]

$$\text{Current: } \textit{False alarm number} = \frac{1}{P_{fa}} \tag{12.6}$$

12.2.3 False alarm probability

The false alarm probability is the probability that noise or other interfering signals will erroneously cause a detection decision at each detection trial. The signal and noise inputs to thresholding stage are those from the second detector (see Chapter 9, Detection) as modified by signal processing (see Chapter 11, Signal processing). Rayleigh distributed noise after a square law detector may be simulated by adding the squares of two separate Gaussian distributed samples. The distribution of the sums of squares from a Gaussian distribution form a chi-squared, or more generally, a gamma distribution.

A radar operator has to have confidence in his equipment: an indication of an echo signal must represent an object that has been seen and not an unfortunate sum of noise spikes. The probability distribution of noise power is a negative exponential distribution (see Section 15.3.4 [2, Chapters 6 and 7]) of mean power unity is a gamma distribution with $N = 1$, namely,

$$p(Y) = \exp(-Y) \tag{12.7}$$

as is shown in Figure 12.1(a). It has the characteristic function from (15.45) $\lambda = 1$

$$C(\xi) = \frac{1}{1 + j2\pi\xi} \tag{12.8}$$

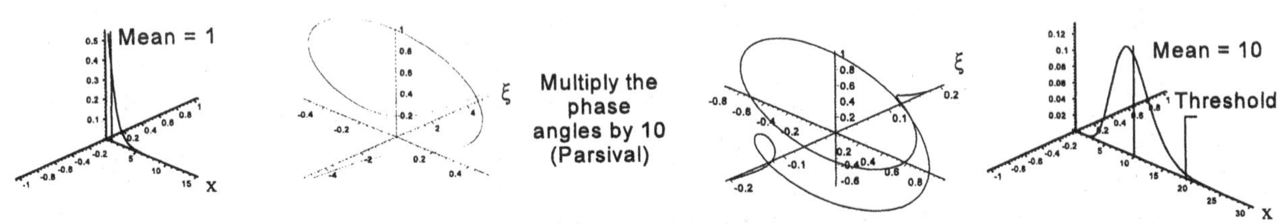

(a) Probability distribution for one sample of Rayleigh (power) noise

(b) Characteristic function for (a)

(c) Characteristic function for the sum of 10 samples of noise

(d) Anticharacteristic function or probability distribution for the sum of 10 samples of noise.

Figure 12.1 The probability distributions of one and the sum of 10 samples of noise. [After: [3]. Meikle, H. D., "Another way of representing echo signals, their spectra, and their statistics", *Proceedings of the German Radar Symposium GRS 2002*, Bonn, Germany: DGON, 5th to 9th of September 2002, pp. 509–513.]

If, for example, N noise samples are added together, the characteristic function of the sum is the characteristic function to the power N, $C(\xi)^N$, as shown in Figure 12.1(c) [2, Chapter 7]. Note that the range of the ξ axis is one tenth of that in Figure 12.1(b).

$$C(\xi)^N = \frac{1}{(1 + j2\pi\xi)^N} \tag{12.9}$$

The anticharacteristic function is a gamma distribution (Section 15.3) with $\eta = N$ that has a mode at $N - 1$ and is shown in Figure 12.1(d).

$$p(Y) = \frac{Y^{N-1}\exp(-Y)}{\Gamma(N)} \qquad (12.10)$$

Note: $\Gamma(n) = (n-1)!$

This distribution for a small number of pulses is shown in Figure 12.2.

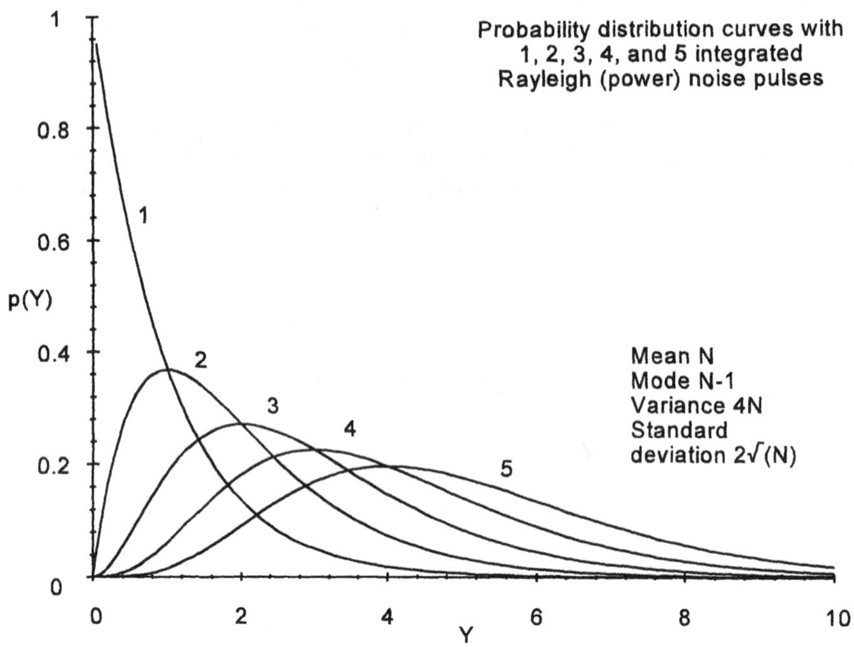

Figure 12.2 Probability distributions (gamma) of the sum of 1, 2, 3, 4, and 5 noise (power) samples.

The main characteristics are as follows:

- Mean: The number of pulses summed or integrated;
- Mode: The peak of the curve occurs at N - 1. A longer right hand tail is necessary to balance the curve around the mean, N;
- Variance: $4N$;
- Standard deviation: $2\sqrt{N}$.

The area under the curve in Figure 12.1(d) to the right of an abscissa value representing a threshold gives the false alarm probability, 10^{-6}. If Y_b is the threshold level, then the false alarm probability is

$$\text{P(False alarm) or } P_{fa} = \int_{Y_b}^{\infty} \frac{Y^{N-1}\exp(-Y)}{\Gamma(N)}\, dY \qquad (12.11)$$

The threshold for a false alarm probability of 10^{-6} for the sum of 10 noise pulses in Figure 12.1(d) is 32.71.

Changing the limits of integration, the area under the curve less than the threshold may be found to give

$$P_{fa} = 1 - \int_0^{Y_b} \frac{Y^{N-1}\exp(-Y)}{(N-1)!}\, dY \tag{12.12}$$

Equations (12.11) and (12.12) are normally solved numerically. The limits for the solution can be found using Johnson's approximation [4, p. 98, Eq. 3-1.6] which in itself is reasonably accurate, namely,

$$Y_b \approx N - \sqrt{N} + 2.3\sqrt{-\log_{10}P_{fa}}\left(\sqrt{-\log_{10}P_{fa}} + \sqrt{N} - 1\right) \tag{12.13}$$

Having found the threshold for a given false alarm probability, the amount of signal plus noise, which exceeds the threshold for an amount of time given by the probability of detection (P_d, percent), can be calculated. These calculations are available from graphs or tables [5] for a limited number of probabilities of detection. Today, further processing follows the threshold, which may well require calculations using values of probabilities of detection and false alarm not covered by the tables.

In the examples that follow, the distribution for the sums of 10 samples or hits is used. The threshold that is crossed in one case in one million, false alarm probability 10^{-6}, is given by

$$10^{-6} = \int_{Y_b}^{\infty} \frac{Y^{10-1}\exp(-Y)}{(10-1)!}\, dY \tag{12.14}$$

Numerical solution gives $Y_b = 32.710$, or 15.147 dB. The starting point, given by Johnson's approximation, was 32.810, or 15.161 dB.

Other results are shown graphically in Figure 12.3.

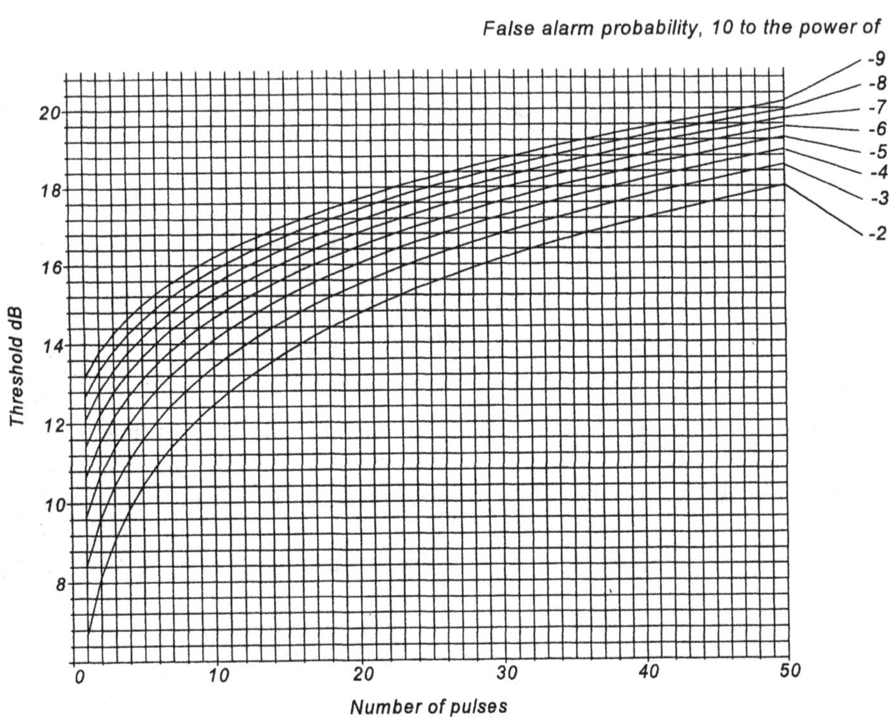

Figure 12.3 Threshold (dB) above the noise level for false alarm probabilities, from 10^{-2} to 10^{-9} against the number of noise pulses integrated.

12.2.4 Changing the threshold levels

Normally, the threshold level is set so that the false alarm probability is low enough for the stages that follow. Locally, the constant false alarm rate stages may not function correctly and there may be clusters of false alarms that will saturate the following data processing systems. Operators of radars with simple videos on displays react automatically to this change. To reduce the clusters, the threshold in the area (or volume in three-dimensional radars) may be raised which lowers the probability of detection of the wanted echoes in these areas.

12.3 PROBABILITY OF DETECTION

Having discussed the probability of a false alarm caused by noise, the size of a signal plus noise sample required for detection is estimated. The probability distribution of the sum of N noise samples is a gamma distribution, as shown in Figure 12.4. The threshold is set to give a specified false alarm probability. The distribution of the sum of N signal plus noise samples is shown in the lower right of Figure 12.4. Part of the distribution does not cross the threshold, and the area represents the probability of the echo NOT being detected. Thus, the area to the right of the threshold is the probability of detection.

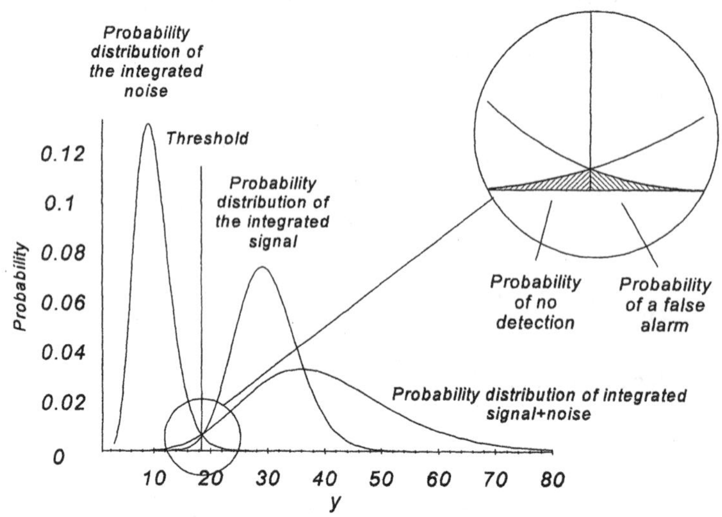

Figure 12.4 Probability distributions involved in detection with a threshold.

The echo signals may be constant, or they may fluctuate. Five classical cases are considered in Table 12.1 (see Chapter 6, Factors outside the radar, Section 6.3.2).

Table 12.1
The five classic cases of fluctuation

Case	Description
0	No fluctuation, the Marcum case [6].
1	The echoes are part of a Rayleigh power distribution (equivalent to a gamma distribution with one degree of freedom or a chi-square distribution with two degrees of freedom), and do not fluctuate during a look but fluctuate between looks. This case was investigated by Swerling [7] and is known as Swerling case I.
2	As case 1, but the echoes fluctuate from pulse to pulse during each look: known as Swerling case II [7].
3	The echoes are part of a chi-square distribution with four degrees of freedom but do not fluctuate during a look but fluctuate between looks: known as Swerling case III [7].
4	As case 3, but the echoes fluctuate from pulse to pulse during each look: known as Swerling case IV [7].

There are two ways of adding the energies of the received echoes in order to calculate the probabilities of detection: coherent and incoherent. For the Marcum, Swerling I, and Swerling III cases, the echoes received during one dwell are constant in amplitude and the phase changes only when the object moves in range. The predictable phase angle allows coherent integration, which may be treated as if the energy of the N pulses were concentrated into one single pulse. In this case, the variance of the sum of the Gaussian noise pulses before square law detection is divided by the number of noise samples (in statistics, the standard error of the mean). In the cases where the signal is coherent during the dwell time (Marcum case and Swerling cases I and III), the echo signals may be added (or integrated) coherently in amplitude and phase (see Chapter 11, Signal Processing). The sum of N unit voltage samples is N instead of $\sqrt{N}$ for noise or incoherent addition, which is the same as when all the energy contained in the many pulses is concentrated in one pulse. The signal-to-noise ratio for a given probability of detection is thus that for one pulse divided by the number of pulses integrated coherently.

The curves for the five classical cases give the signal-to-noise ratios for a given probability of detection and the numbers of pulses which are incoherently integrated are shown for each case. All have the form of Figure 12.5, and separate curves are given for false alarm probabilities of 10^{-4}, 10^{-6}, and 10^{-8}. Other values must be estimated by interpolation.

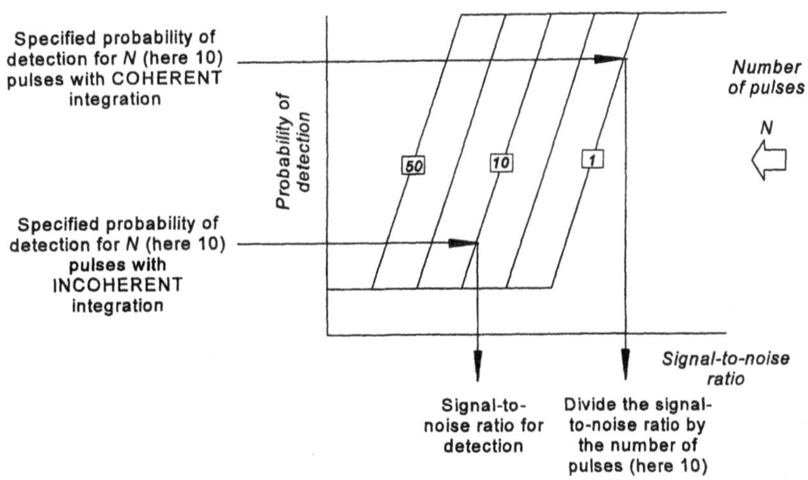

Figure 12.5 Conventions used in the probability of detection diagrams.

Coherent integration means that the signals and noise are added together or integrated before detection. The signal voltages add and grow in contrast to the bottlebrush Gaussian noise voltages that tend to cancel each other so that the combined power for N hits may be considered to be concentrated into one hit alone. Remember that the signal-to-noise ratios are given in decibels, which means for coherent integration the decibel value of the number of hits $10 \log_{10} N$ must be subtracted from the signal-to-noise ratio (dB) for one hit.

12.3.1 Marcum case: no fluctuation

The nonfluctuating, or Marcum case, sometimes called Swerling case 0 or 5, is for echoes from symmetrical objects, such as spheres, which have the same reflecting area in all directions. The signal is constant, so that its statistical distribution is a line (Dirac delta function). A constant signal and ten samples added of noise are shown in Figure 12.6.

The Ricean probability distribution, in terms of power (15.30), for a single sample may be given by

$$p_{Rice}(Y) = \exp(-(Y + R))\ I_0(\sqrt{4YR}) \tag{12.15}$$

where $I_n(z)$ is the modified Bessel function, given by $I_n(z) = \int_0^{\pi} \exp(z\cos\theta)\cos(n\theta)\, d\theta$.

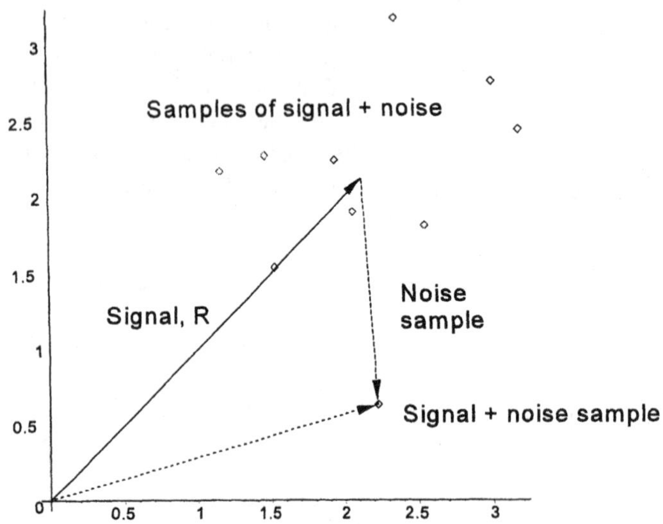

Figure 12.6 A constant signal and 10 noise samples added to it.

The characteristic function is

$$C_{\text{Rice}}(\xi) = \frac{j \exp\left(-\dfrac{2\pi R\xi}{2\pi\xi - j}\right)}{j - 2\pi\xi} \tag{12.16}$$

and for N samples, the characteristic function for the Rice distribution is from Chapter 15 and [8, Eq. 10.4–12b]

$$C_{\text{Rice sum of N samples}}(\xi) = (C_{\text{Rice}}(\xi))^N = \left(\frac{j \exp\left(-\dfrac{2\pi R\xi}{2\pi\xi - j}\right)}{j - 2\pi\xi}\right)^N \tag{12.17}$$

where ξ is the variable, here using the $-f$ notation;
R is the signal-to-noise ratio;
N is the number of pulses added;
j is the square root of minus one.

An example of the characteristic function of the Ricean distribution of the example for one and ten samples of signal plus noise are shown in Figure 12.7.

For N independent samples the characteristic function is $(C_{\text{Rice}}(\xi))^N$ and has the anticharacteristic function from Campbell and Foster's pair 650.0 to give the probability distribution function

$$p(Y) = \left(\frac{Y}{RN}\right)^{\frac{N-1}{2}} \exp(-Y - NR)\, I_{N-1}\left(\sqrt{4NRY}\right) \tag{12.18}$$

where Y is the variable;
R is the signal to noise ratio;
N is the number of pulses;
$I_n(z)$ is a modified Bessel function of the first kind, order n, with imaginary argument [9, Section 9.6.3, p. 375].

For comparison with the Swerling cases later, a threshold of 32.71 from Figure 12.2(d) is required for a false alarm probability of 10^{-6} with 10 integrated pulses and a signal-to-noise ratio of 3 to give a probability of detection of 80.5%. An alternate form is shown in Figure 12.8 with the conventional representation in Figure 12.9.

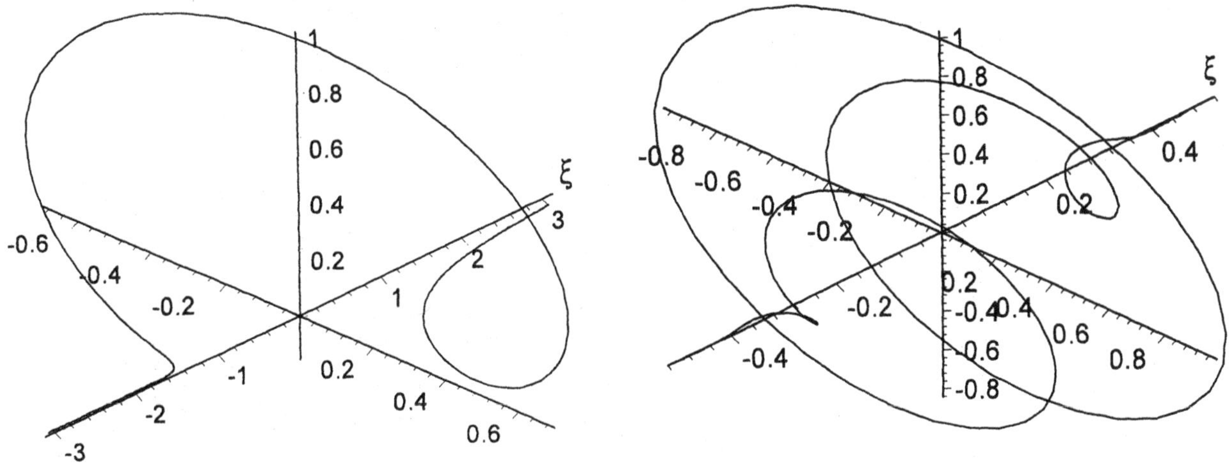

(a) Characteristic function of one sample of signal + noise (b) Characteristic function of 10 samples of signal + noise

Figure 12.7 Characteristic functions for one and 10 samples of signal plus noise from the Ricean distribution.

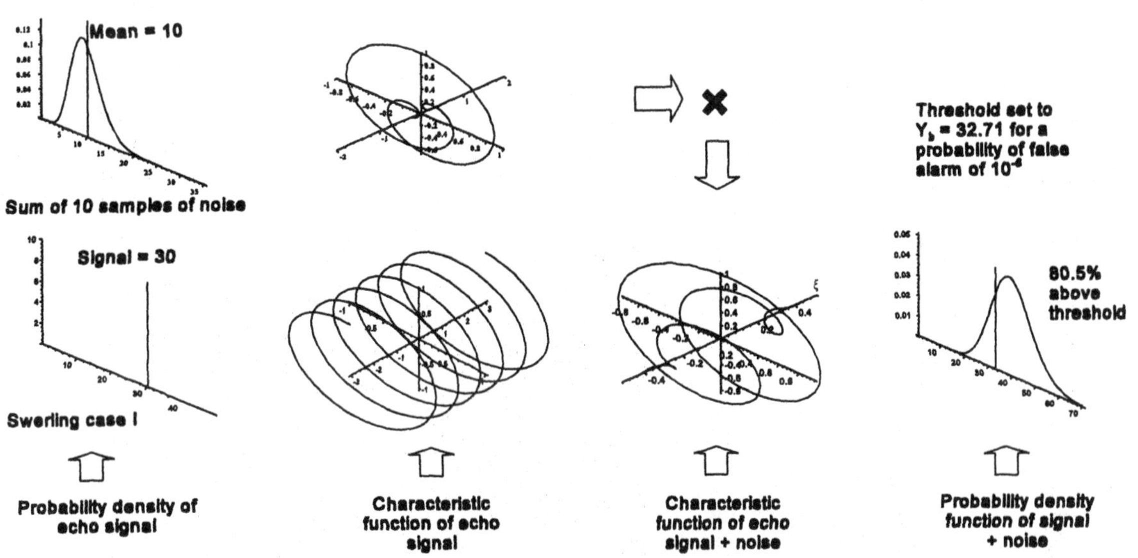

Figure 12.8 The probability distributions and characteristic functions for the steady signal or Marcum case. [From: Meikle, H. D., *A New Twist to Fourier Transforms*, Weinheim: Wiley-VCH, 2004, Figure 7.14.]

The probability of detection, or the probability that the sum of signal plus noise exceeds the threshold, is the integral of (12.19) to the right of the threshold [9, Eq. 10.4-28]:

$$
\begin{aligned}
P_d &= \int_{Y_b}^{\infty} \left(\frac{Y}{NR}\right)^{\frac{N-1}{2}} \exp(-Y - NR)\; I_{N-1}\left(\sqrt{4NRY}\right) dY \\
&= 1 - \int_{0}^{Y_b} \left(\frac{Y}{NR}\right)^{\frac{N-1}{2}} \exp(-Y - NR)\; I_{N-1}\left(\sqrt{4NRY}\right) dY
\end{aligned}
\qquad (12.19)
$$

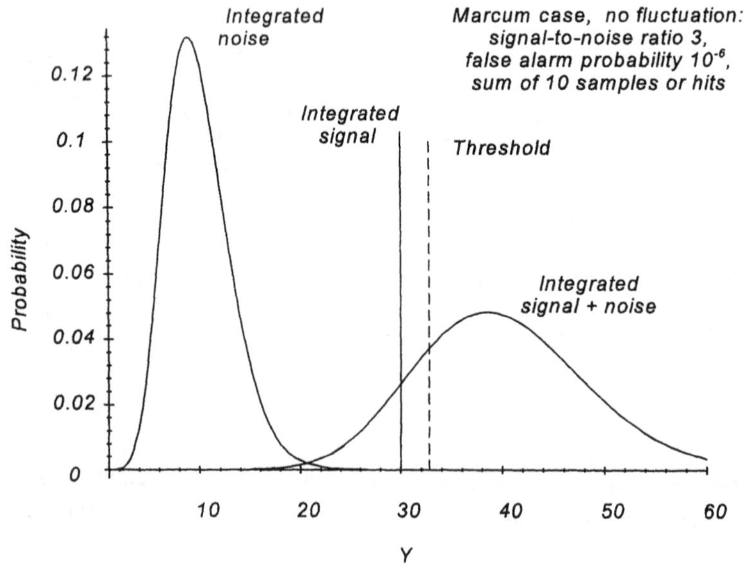

Figure 12.9 The Marcum case, no echo fluctuation: probability distributions for 10 integrated pulses.

The radar range equations use the signal-to-noise ratio for each pulse, and values for signal-to-noise ratio may be taken for three false alarm probabilities from Figure 12.10, Figure 12.11, and Figure 12.12. The numbers of pulses are 1, 2, 3, 4, 5, 6, 8, 10, 12, 15, 20, 25, 30, 40, and 50.

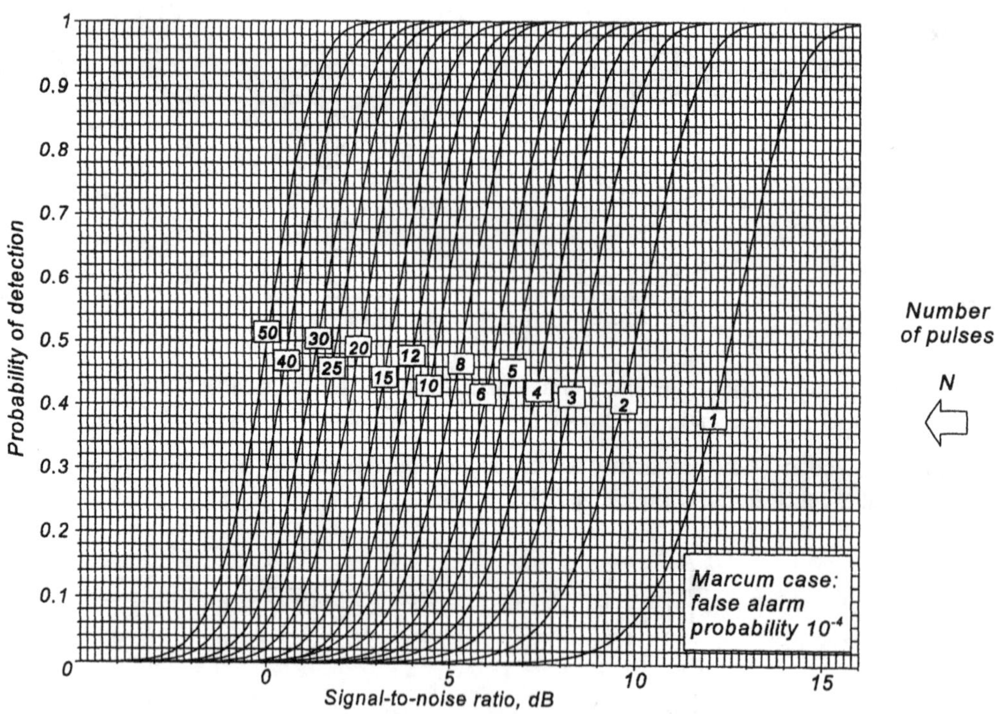

Figure 12.10 Marcum case, no echo fluctuation: probability of detection against signal-to-noise ratio (dB) for N integrated pulses and a false alarm probability of 10^{-4}.

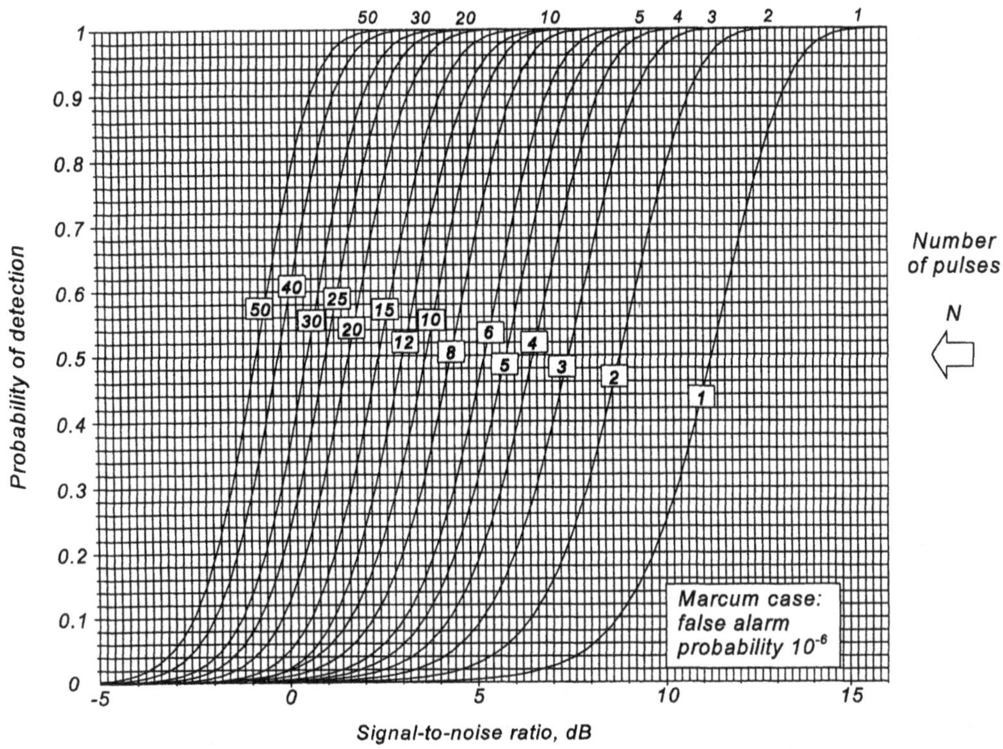

Figure 12.11 Marcum case, no echo fluctuation: probability of detection against signal-to-noise ratio (dB) for N integrated pulses and a false alarm probability of 10^{-6}.

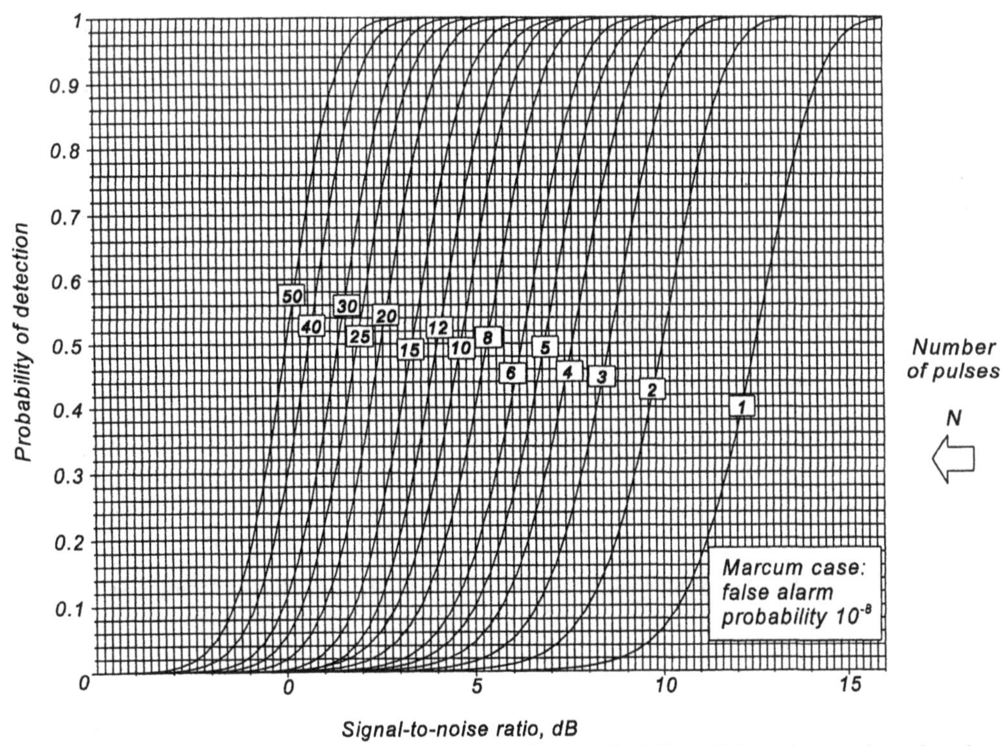

Figure 12.12 Marcum case, no echo fluctuation: probability of detection against signal-to-noise ratio (dB) for N integrated pulses and a false alarm probability of 10^{-8}.

Approximate values for large numbers of hits may be found from [8, Eq. 10.7-17]

$$P_d \approx 1 - \int_0^{Yb} \frac{1}{\sqrt{2\pi N(1 + 2R)}} \exp\left(- \frac{[Y - N(1 + R)]^2}{2N(1 + 2R)} \right) dY \tag{12.20}$$

where R is the power signal-to-noise ratio;
N is the number of hits.

Coherent integration
In the case of coherent integration, the voltages of the signal and noise samples are added together, in contrast to adding the powers in incoherent integration. The voltage sum is compared to the threshold. The signal-to-noise ratios for detection using coherent integration are shown in Figures 12.39 to 12.42.

12.3.2 Swerling case I: slow fluctuation

Swerling case I is the case for most rotating, fixed frequency, single channel air surveillance radars. Aircraft are illuminated by a number of pulses for a short time during each antenna revolution or scan. The illumination time is taken to be too short for the aircraft to change its aspect during this time, so that the returning echoes have the same amplitude. The amplitude of the echoes is taken from a Rayleigh voltage distribution or negative exponential power distribution corresponding to objects that have a large number of equal small reflectors (see Chapter 6, Factors outside the radar).

Slow fluctuation means that the signal sample is chosen from the gamma distribution with one degree of freedom, for power a negative exponential distribution, and remains at the same for the complete train of echoes. The probability distribution of the sum of the independent noise samples is shown in the upper part of Figure 12.13. The run of N signals is N times the power of the single signal sample and the probability distribution keeps its negative exponential form in the lower left of Figure 12.13. Its characteristic function is shown to the right of it and is multiplied by the characteristic function of the sum of the noise samples and the probability distribution of sum of signals plus noise is shown on the right of the diagram.

The probability distributions for the signal fluctuations in the Swerling cases may be derived from the gamma distribution in Chapter 15

$$p(Y, \eta, r) = \frac{1}{r\,\Gamma(\eta)} \left(\frac{Y}{r}\right)^{\eta - 1} \exp\left(-\frac{Y}{r}\right) \tag{12.21}$$

where Y is the power variable;
η is the shape factor;
r represents the echo signal-to-noise ratio = $1/\lambda$ in Chapter 15.

For Swerling case I, $\eta = 1$ giving the negative exponential distribution for a single pulse, compare with Table 12.1,

$$p_{\text{Swerling I}}(Y, 1, R) = \frac{1}{R} \exp\left(-\frac{Y}{R}\right) \quad \text{For a single pulse} \tag{12.22}$$

The slow fluctuation from scan to scan means that the run of N individual echoes has the same power giving an average power for the sum of NR.

$$p_{\text{Swerling I}}(Y, 1, NR) = \frac{1}{NR} \exp\left(-\frac{Y}{NR}\right) \quad \text{For } N \text{ pulses} \tag{12.23}$$

In the example in Figure 12.13, $N = 10$ and $R = 3$.

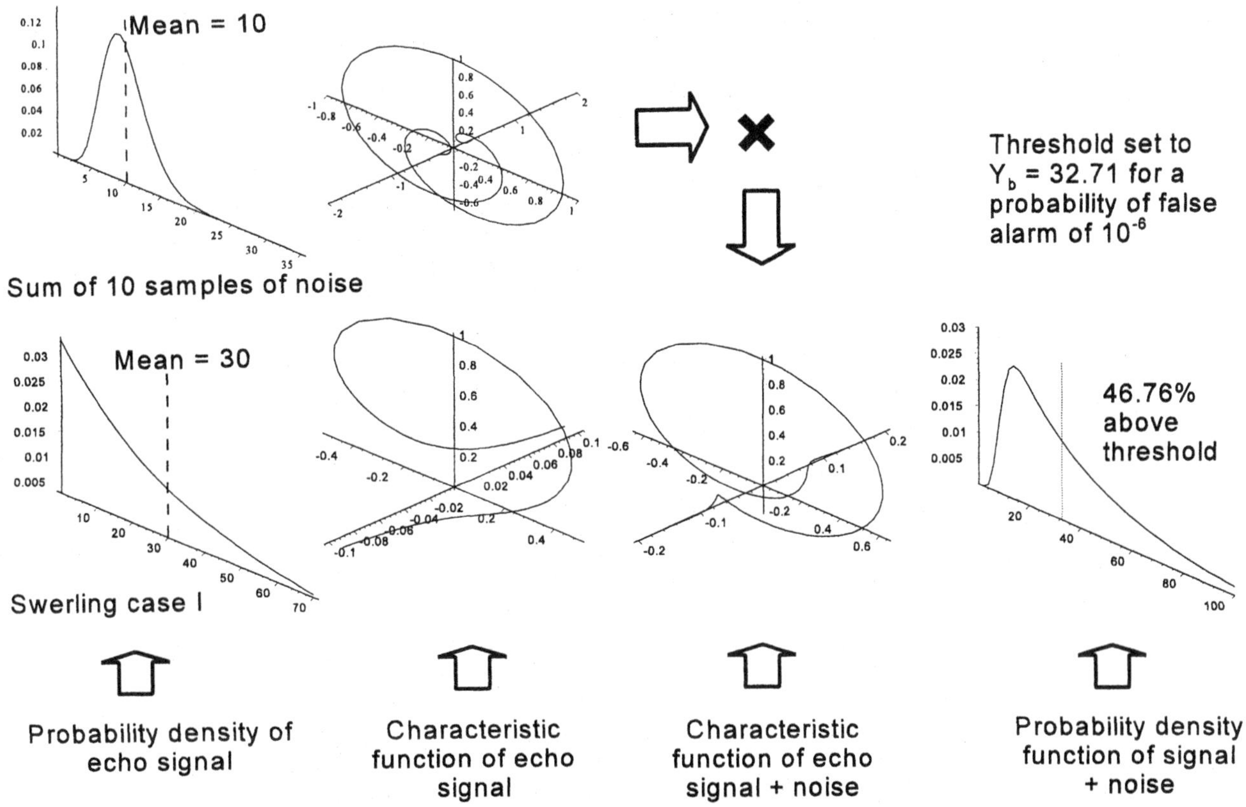

Figure 12.13 The probability distributions and characteristic functions for Swerling's case I. [After: [2]. Meikle, H. D., *A New Twist to Fourier Transforms*, Weinheim: Wiley-VCH, 2004, Figure 7.14.]

In order to find the probability function of signal plus noise, the characteristic function must be found (see Chapter 15):

$$C_{\text{Swerling I}}(\xi) = \frac{1}{1 + j2\pi\xi NR} \tag{12.24}$$

The characteristic function of N samples of signal plus noise is the product of the characteristic functions of signal (12.24) and noise (12.9), given by

$$C_{\text{signal+noise Swerlng I}} = \frac{1}{(1 + j2\pi\xi)^N \, (1 + j2\pi\xi NR)} \tag{12.25}$$

The probability distribution function for signal plus noise is found from Campbell and Foster's pair 581.7 [10] and given by [8, Eq. 11.2-32]

$$\begin{aligned} p(Y) &= \frac{Y^{N-2}}{(N-2)!} - (1 + NR)\frac{dp(Y)}{dY} \\ &= \frac{1}{NR}\left(1 + \frac{1}{NR}\right)^{N-2} \exp\left(-\frac{Y}{1 + NR}\right) \gamma\left(N - 1, \frac{Y}{1 + \dfrac{1}{NR}}\right) \end{aligned} \tag{12.26}$$

where $\gamma(a, x)$, the incomplete γ function $= \dfrac{1}{\Gamma(a)} \int_0^x e^{-t}\, t^{a-1}\, dt$ [9, p. 260, Eq. 6.5.1].

The distributions of the sum of 10 noise samples, signal samples, and signal plus noise samples are shown in Figure 12.14.

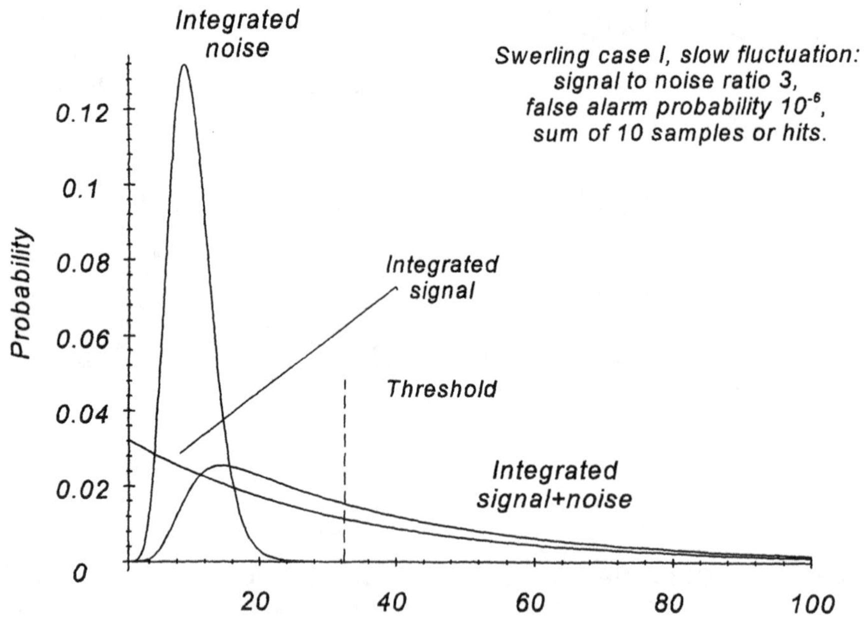

Figure 12.14 Swerling case I, slow Rayleigh (power) echo fluctuation, probability distributions for 10 integrated pulses.

The echo has a power three times the noise so the number of degrees of freedom for signal and noise may not be simply added to obtain the distribution of signal plus noise (see Chapter 15, Statistics). With slow fluctuation, the echo samples are all the same as the first to return, so that the distribution for the sum is that for a single pulse. For single pulse detection, the threshold is

$$\begin{aligned} P_{fa} &= \exp(-Y_b) \\ Y_b &= -\ln(P_{fa}) \end{aligned} \tag{12.27}$$

The amount of signal plus noise, $1 + R$, that exceeds the threshold P_d (percent) of the time is given by

$$\begin{aligned} P_{dI} &= \exp\left(-\frac{Y_b}{1 + R}\right) \\ \ln(P_{dI}) &= \frac{-Y_b}{1 + R} \end{aligned} \tag{12.28}$$

Rearranging gives an exact solution for the signal-to-noise ratio required for a single pulse detection, and is shown in Figures 12.15 and 12.16:

$$P_d(1) = \exp\left(\frac{-\ln P_{fa}}{1 + R}\right) \tag{12.29}$$

and

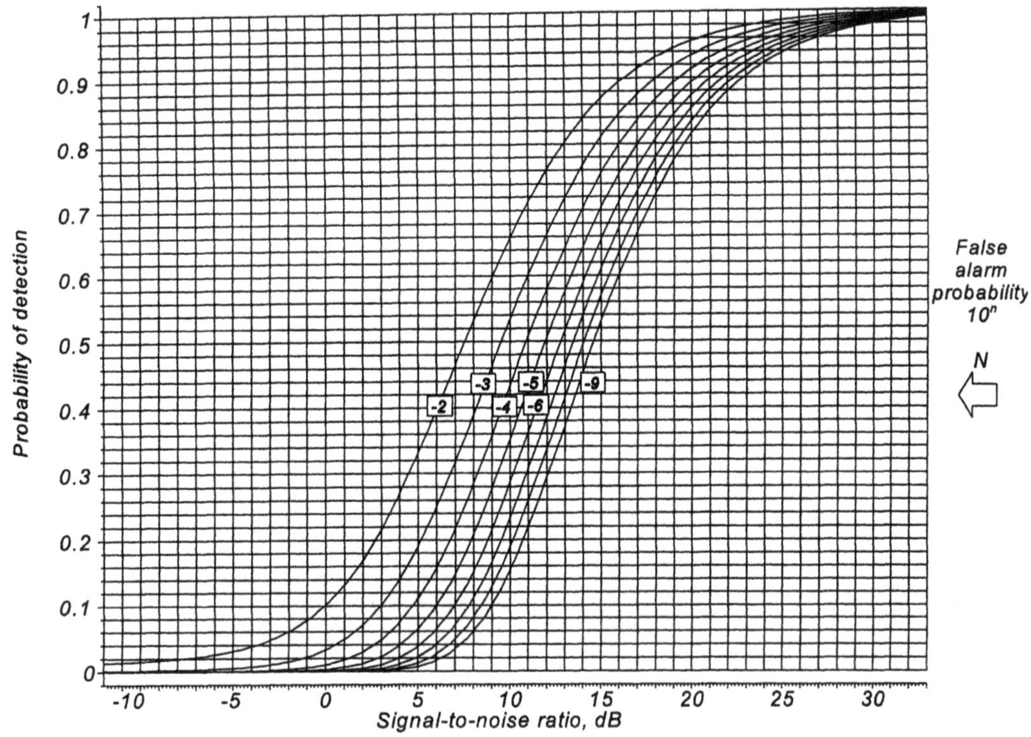

Figure 12.15 Probability of detection versus signal-to-noise ratio, with 10^{-n} probability of false alarm, for the detection of a single pulse, Swerling case I.

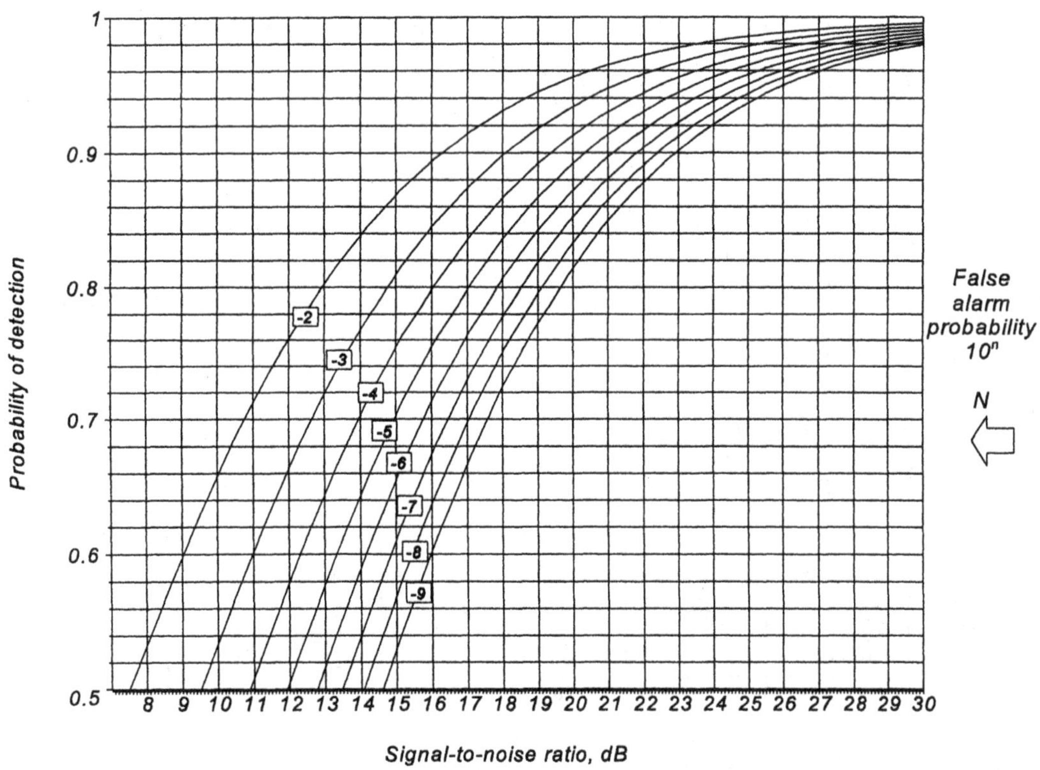

Figure 12.16 Probability of detection versus signal-to-noise ratio, with 10^{-n} probability of false alarm, for the detection of a single pulse, Swerling case I. (Upper part of Figure 12.15.)

$$R = \frac{\ln P_{fa}}{\ln P_d} - 1 \qquad (12.30)$$

For more than one pulse, the signal plus noise distribution must be found to find the probability of detection [4, p. 100, Eq. 3-1.15; 8, Eq. 11.2-38]

$$P_d = 1 - \gamma(N-1, Y_b) + \left(1 + \frac{1}{NR}\right)^{(N-1)} \exp\left(-\frac{Y_b}{1+NR}\right) \gamma\left(N-1, \frac{Y_b}{1+\frac{1}{NR}}\right) \qquad (12.31)$$

where $\gamma(a, x) = \int_0^x \exp(-t)\, t^{a-1}\, dt$, the incomplete gamma function.

Curves for signal-to-noise ratios for a given probability of detection are shown for three false alarm probabilities in Figures 12.17, 12.18, and 12.19. The numbers of pulses are 1, 2, 3, 4, 5, 6, 7, 8, 10, 12, 15, 20, 25, 30, 40, and 50.

12.3.2.1 Coherent integration

In the case of coherent integration, the voltages of the signal and noise samples are added together, in contrast to adding the powers in incoherent integration. The voltage sum is compared to the threshold. The signal-to-noise ratios for detection using coherent integration are shown in Figures 12.39 to 12.42.

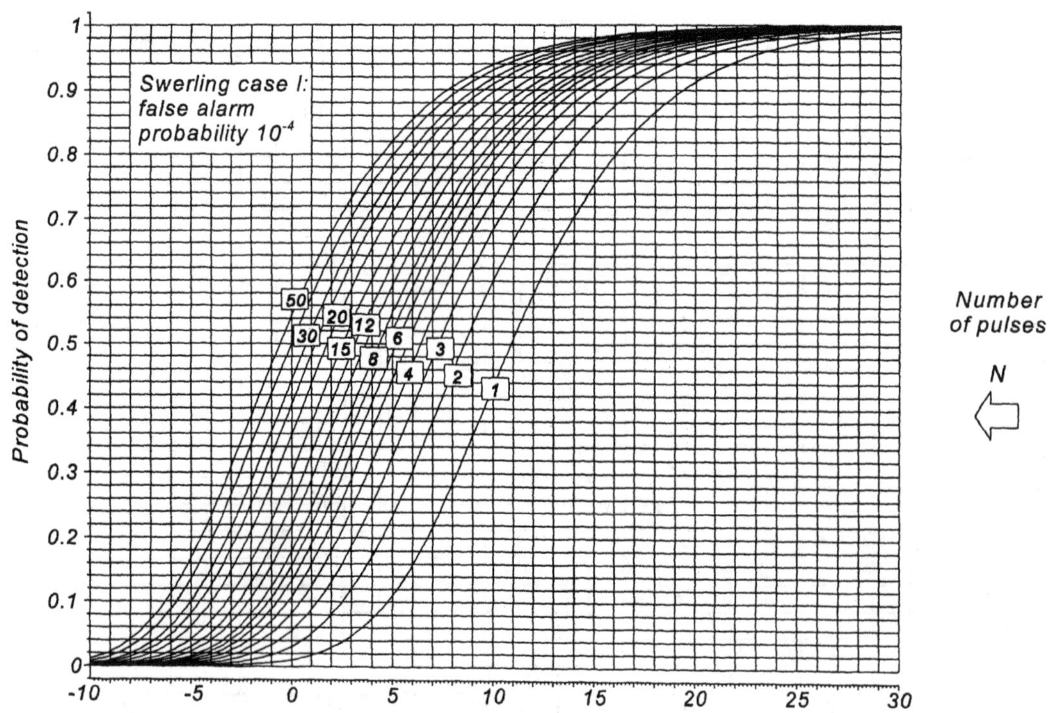

Figure 12.17 Swerling case I, slow Rayleigh (power) echo fluctuation: probability of detection against signal-to-noise ratio (dB) for a false alarm probability of 10^{-4}.

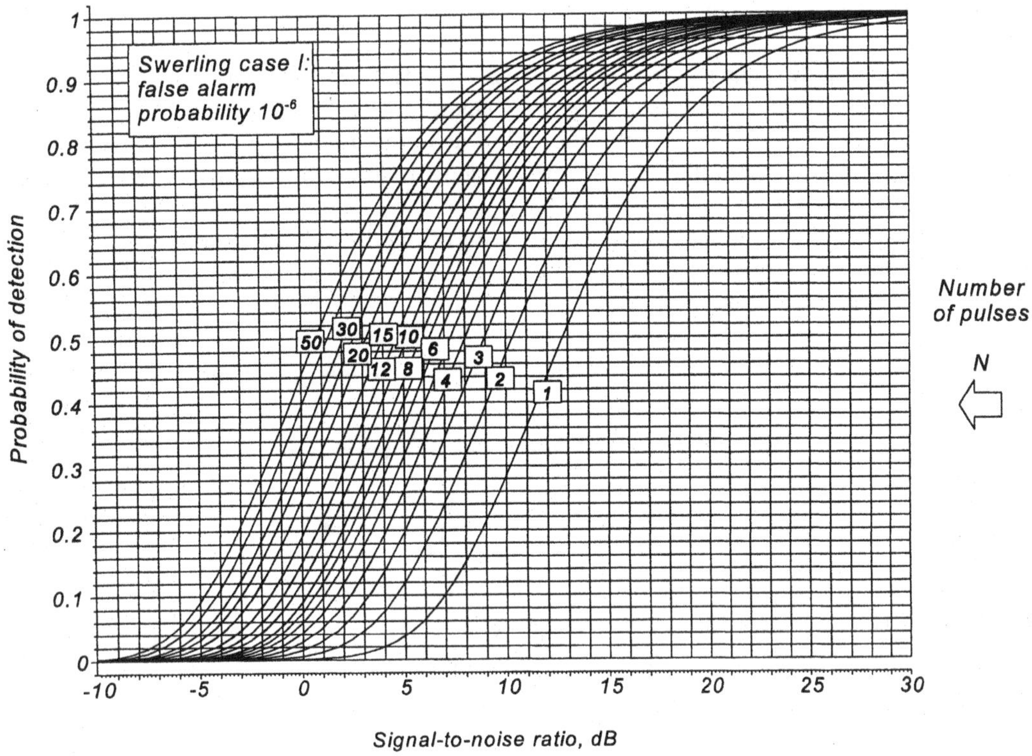

Figure 12.18 Swerling case I, slow Rayleigh (power) echo fluctuation: probability of detection against signal-to-noise ratio (dB) for a false alarm probability of 10^{-6}.

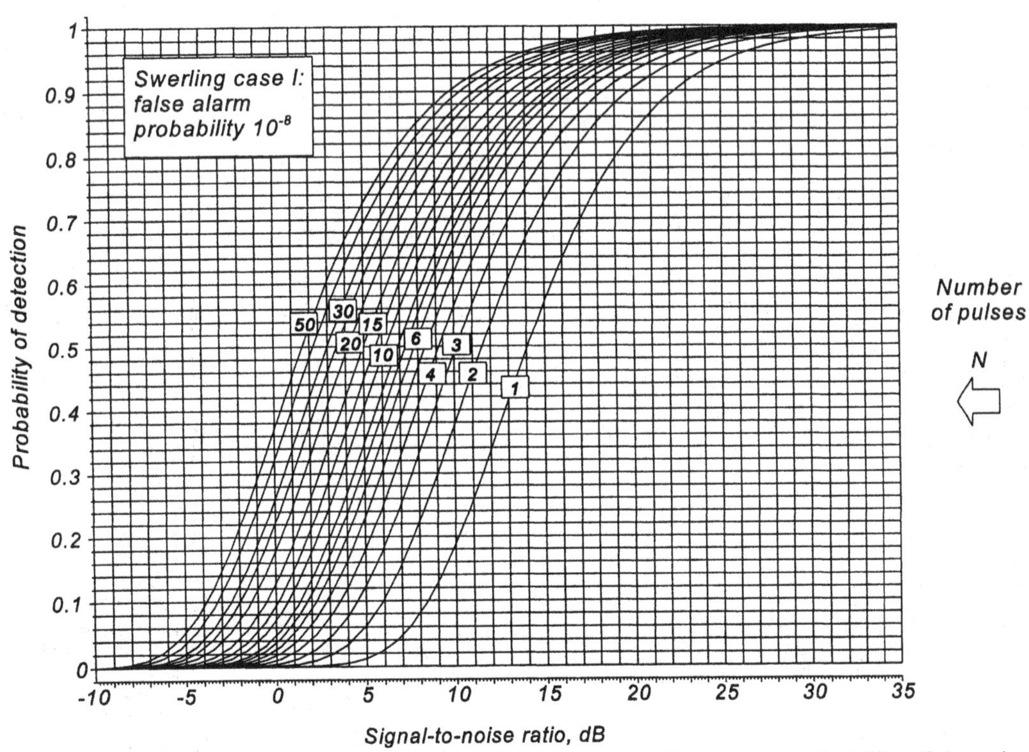

Figure 12.19 Swerling case I, slow Rayleigh (power) echo fluctuation: probability of detection against signal-to-noise ratio (dB) for a false alarm probability of 10^{-8}.

12.3.3 Swerling case II: fast fluctuation

Swerling case II covers the case when the returning echoes are independent samples from a Rayleigh distribution, which happens with radars which change their transmitter frequency significantly from pulse to pulse (also called frequency agile radars). Each returning echo is the sum of the echoes from each small reflector, as in Swerling case I. The vector sum depends on the distance between the reflectors and the wavelength. If the wavelength changes each pulse, so does the echo amplitude.

A change of frequency gives changes in phase for the echoes from the individual reflectors, depending on their range. If the range of the near side of the object is R and the radial extent is d, then the phases of the echoes from the center of the near and far halves are at wavelength λ_1 [11]

$$\begin{aligned} \textit{Phase from the center of the near half} &= 2\pi\frac{2(R + d/2)}{\lambda_1} \quad \text{radians} \\ \textit{Phase from the center of the far half} &= 2\pi\frac{2(R + 3d/2)}{\lambda_1} \quad \text{radians} \\ \textit{Phase difference} &= 2\pi\frac{d}{\lambda_1} \quad \text{radians} \end{aligned} \tag{12.32}$$

The amount of frequency change to give 180 degree phase difference, or complete decorrelation, of the echo from the near and far halves of the object at the two frequencies (f_1, f_2) is given by

$$\begin{aligned} \pi &= 2\pi d\left(\frac{1}{\lambda_1} - \frac{1}{\lambda_2}\right) \\ 1 &= 2d\left(\frac{f_1}{c} - \frac{f_2}{c}\right) \\ \Delta f &= \frac{c}{2d} \end{aligned} \tag{12.33}$$

The probability distribution of the sum of the echo signals is that for the sum of N independent samples or a gamma distribution, as for noise. The signal echoes are generally many times larger than the noise echoes and have the same probability distribution function as noise (compare with (12.21) and (12.22)).

$$\mathrm{p}(Y,\ 1,\ R) = \frac{1}{R}\exp\left(-\frac{Y}{R}\right) \qquad \text{For a single pulse} \tag{12.34}$$

where Y is the echo signal power variable;
R is the signal-to-noise ratio ($R = 1$ for noise).

Their characteristic function from (15.36) for the sum of N signal echoes is (see Figure 12.20)

$$C_{\text{Swerling II}} = \frac{1}{(1 + j2\pi\xi R)^N} \tag{12.35}$$

The characteristic function of the sum of N samples of signal plus noise is given by the product of the characteristic functions for signals and noise (12.9)

$$C_{\text{signal + noise Swerling II}} = \frac{1}{(1 + j2\pi\xi)^N}\,\frac{1}{(1 + j2\pi\xi R)^N} \tag{12.36}$$

The anticharacteristic function is found from Campbell and Foster's pair 431 [10] and is the probability distribution function for the gamma distributed signal sum plus gamma distributed noise sum and depends on the signal-to- noise ratio. It is given by [8, p. 379, Eq. 11.3-18]

$$p(Y) = \frac{1}{(1 + R)^N} \frac{1}{(N - 1)!} Y^{N-1} \exp\left(-\frac{Y}{1 + R}\right) \tag{12.37}$$

The distributions of the sums of 10 samples of noise, echo signal, and signal plus noise are shown in Figures 12.20 and 12.21. The false alarm probability gives the threshold Y_b. The probability of detection is given by [8, Eq. 11.3-21]

$$P_d = 1 - \gamma\left(N, \frac{Y_b}{1 + R}\right) \tag{12.38}$$

where $\gamma(a, x) = \int_0^x \exp(-t)\, t^{a-1}\, dt$.

The shape factor, N, corresponds to the number of independent pulses integrated in the Swerling models and are given in Table 12.1. In the example there are 10 pulses with a signal-to-noise ratio of 3.

Curves for signal-to-noise ratios for a given probability of detection are shown for three false alarm probabilities in Figures 12.22, 12.23, and 12.24. The numbers of pulses are 1, 2, 3, 4, 5, 6, 7, 8, 10, 12, 15, 20, 25, 30, 40, and 50.

The signal-to-noise ratios for detection are shown in Figures 12.39 to 12.42.

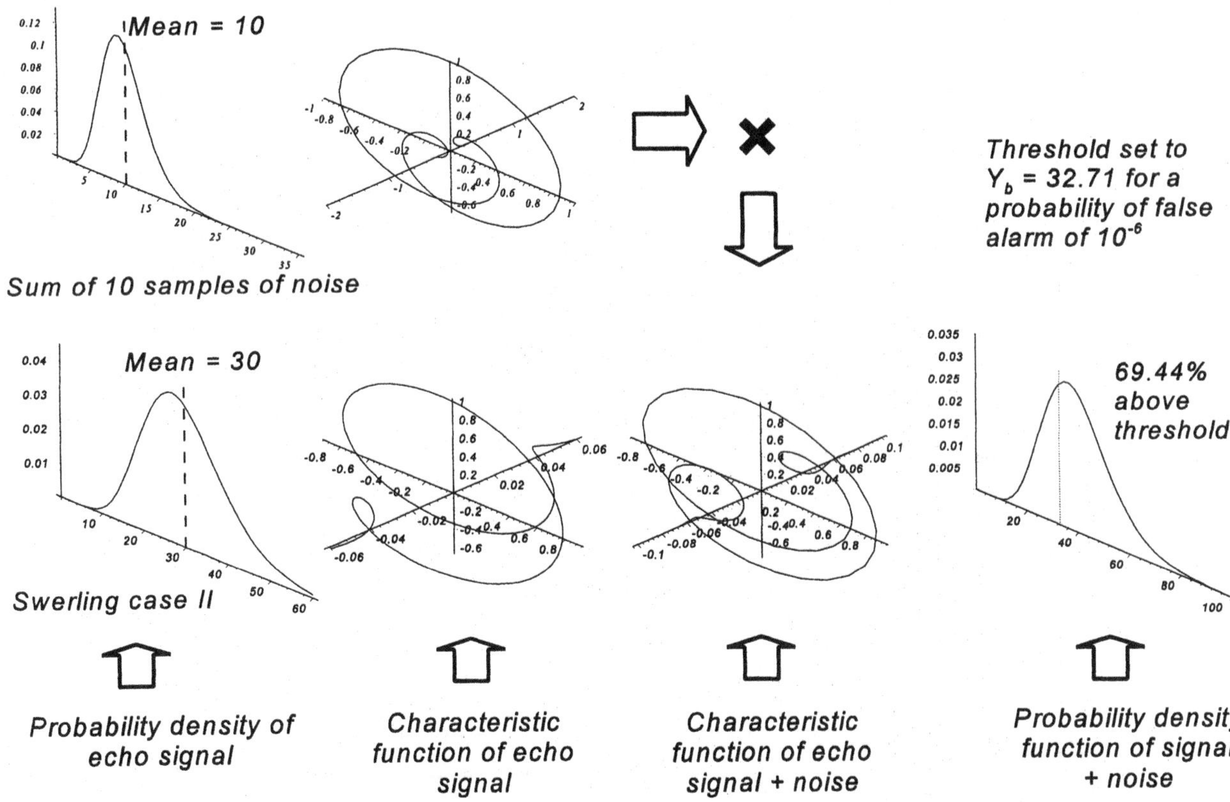

Figure 12.20 The probability distribution and characteristic functions for Swerling's case II. [After: [2]. Meikle, H. D., *A New Twist to Fourier Transforms*, Weinheim: Wiley-VCH, 2004, Figure 7.14.]

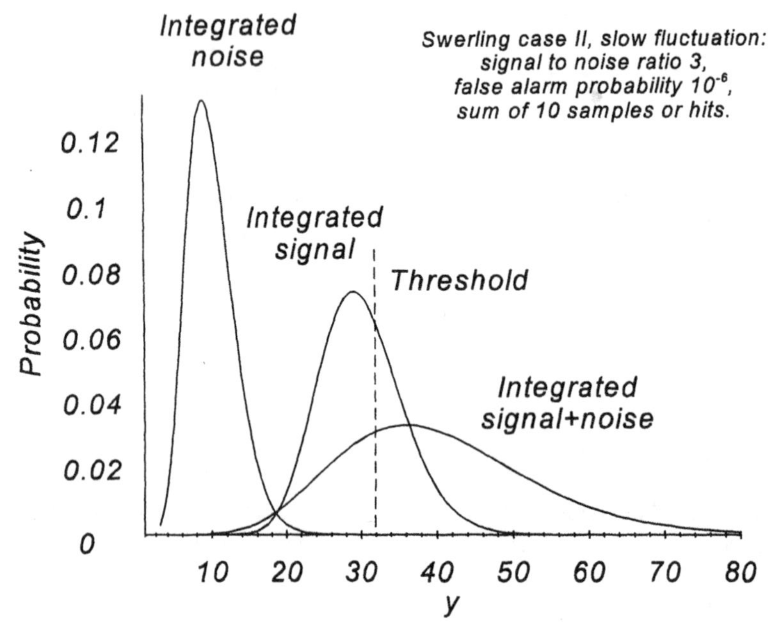

Figure 12.21 Swerling case II, fast Rayleigh echo fluctuation: probability distributions for 10 integrated pulses.

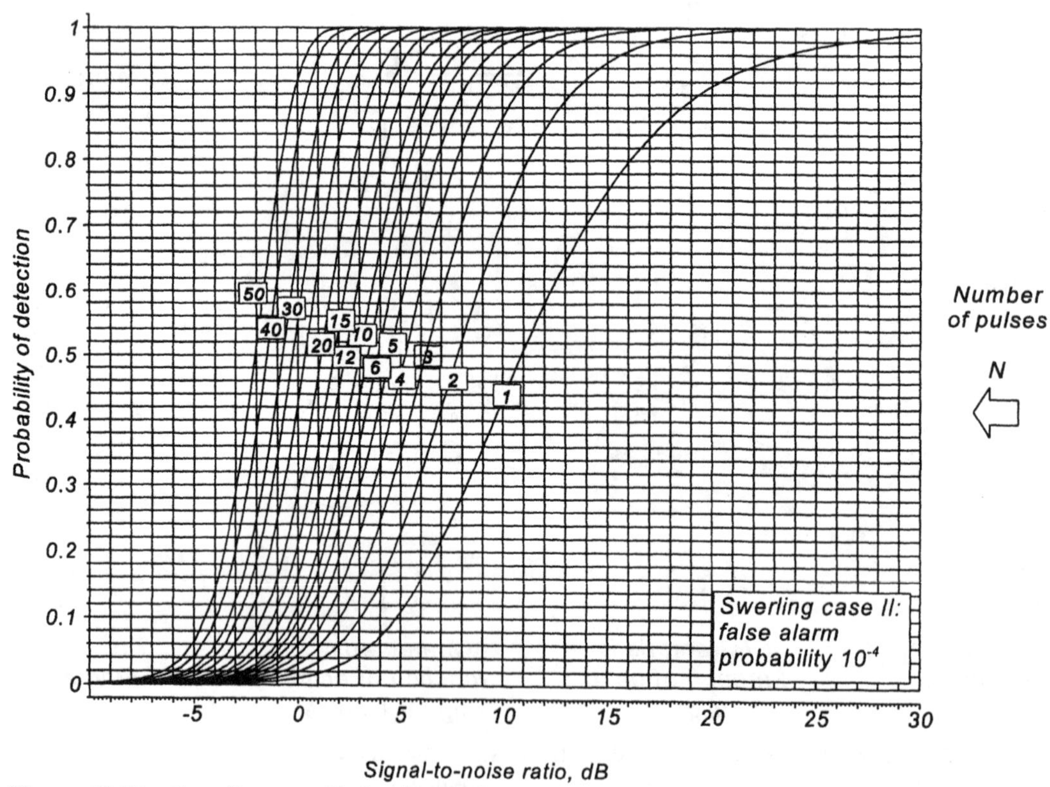

Figure 12.22 Swerling case II, fast Rayleigh (power) echo fluctuation: probability of detection against signal-to-noise ratio (dB) for a false alarm probability of 10^{-4}.

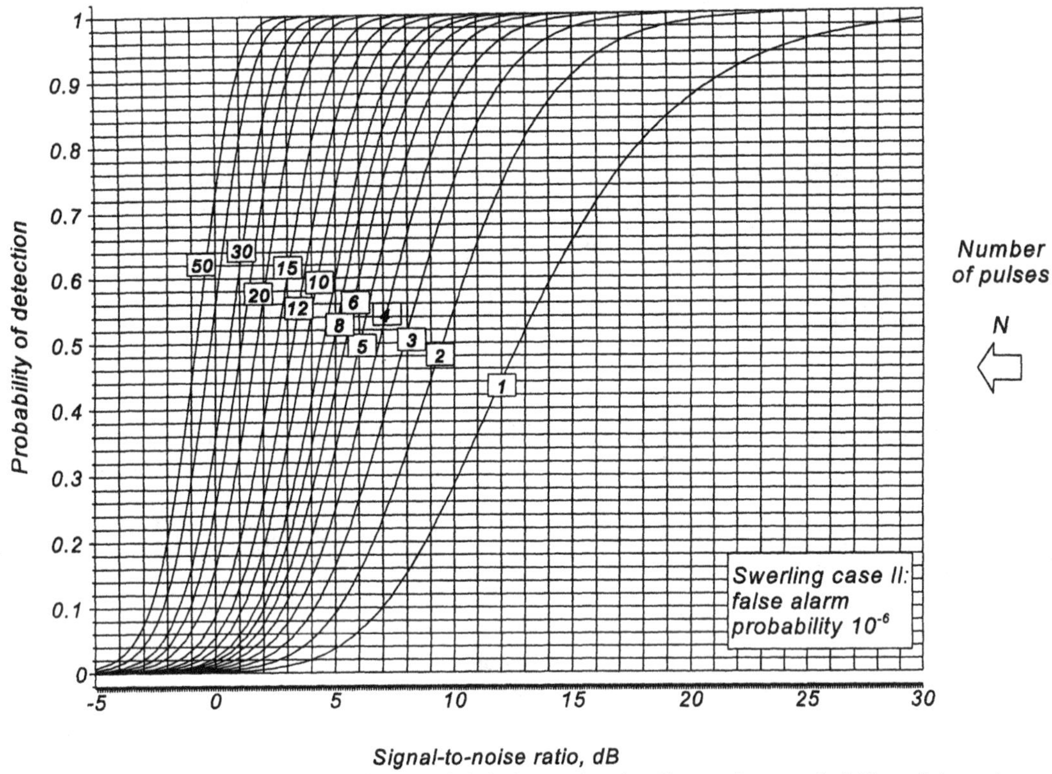

Figure 12.23 Swerling case II, fast Rayleigh (power) echo fluctuation: probability of detection against signal-to-noise ratio (dB) for a false alarm probability of 10^{-6}.

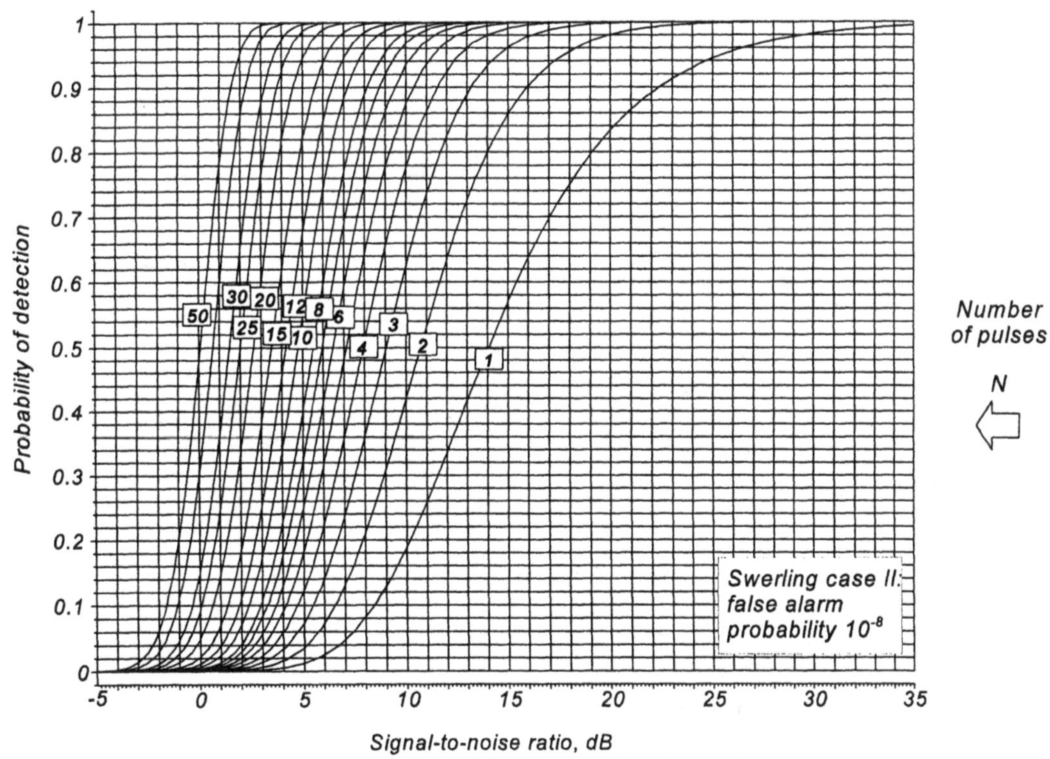

Figure 12.24 Swerling case II, fast Rayleigh (power) echo fluctuation: probability of detection against signal-to-noise ratio (dB) for a false alarm probability of 10^{-8}.

Owing to the decorrelation of the echoes from each of the radar pulses, no coherent addition is possible.

12.3.4 Swerling case III: slow chi-squared fluctuation

Swerling cases III and IV are similar to cases I and II, but the object of interest has one major reflector and many minor ones. The echoes have a chi-squared probability distribution function with four degrees of freedom or a gamma distribution with a shape factor, $\eta = 2$ as in (12.21), namely,

$$p_{\text{Swerling III}}(Y,\ 2,\ R/2) \quad = \quad \frac{4Y}{R^2}\exp\left(-\frac{2Y}{R}\right) \qquad \text{For a single pulse} \tag{12.39}$$

The slow fluctuation from scan to scan means that the run of N individual echoes has the same power giving an average power for the sum of NR (compare with (12.23).

$$p_{\text{Swerling III}}(Y,\ 2,\ NR/2) \quad = \quad \frac{4Y}{(NR)^2}\exp\left(-\frac{2Y}{NR}\right) \qquad \text{For } N \text{ pulses} \tag{12.40}$$

In the example used for illustration there are 10 pulses with a signal-to-noise ratio of 3 and this is shown in Figure 12.25.

The characteristic function of the probability distribution of the signals is

$$C_{\text{Swerling III}}(\xi) \quad = \quad \frac{1}{(1 + j\pi\xi R)^2} \tag{12.41}$$

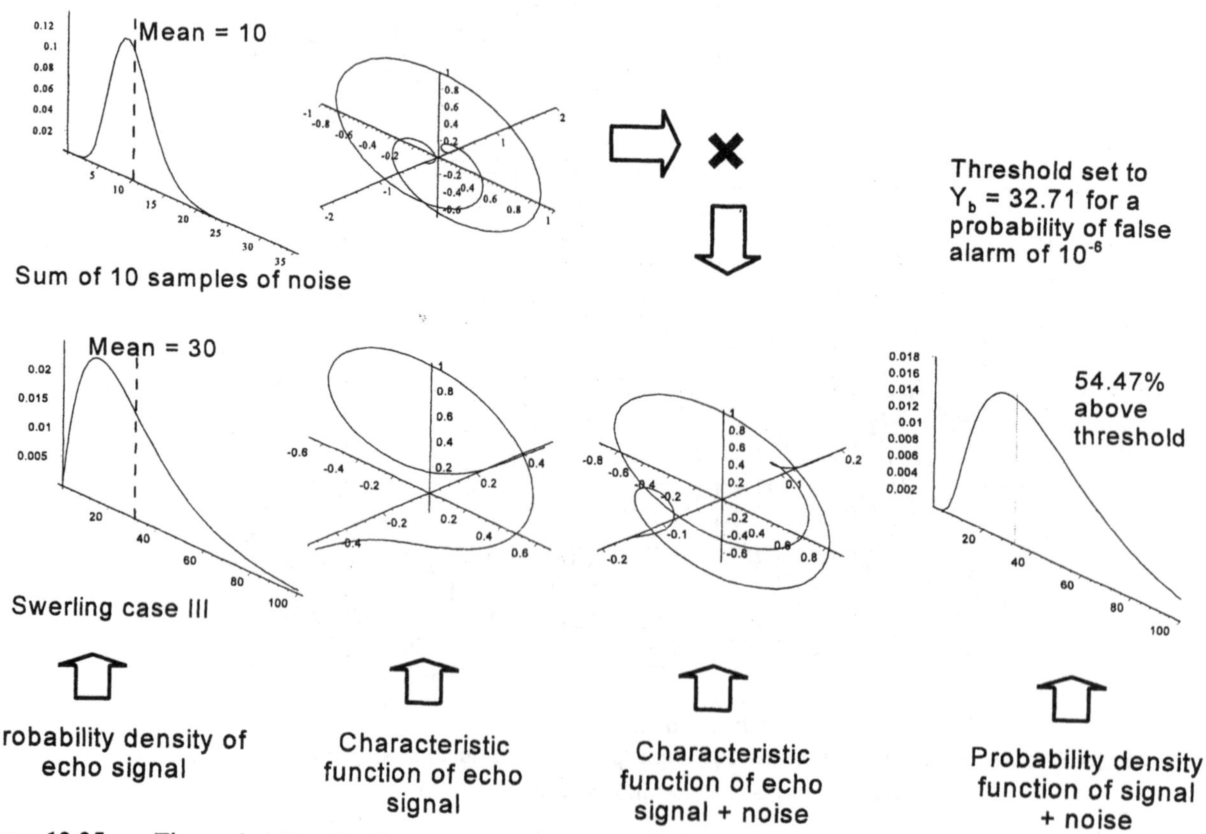

Figure 12.25 The probability distributions and characteristic functions for Swerling case III. [After: [2]. Meikle, H. D., *A New Twist to Fourier Transforms*, Weinheim: Wiley-VCH, 2004, Figure 7.14.]

The characteristic function of signal plus noise is the product of the characteristic functions for signals and noise (12.9), namely,

$$C_{\text{signal + noise Swerling III}} = \frac{1}{(1+j\pi\xi R)^2}\,\frac{1}{(1+j2\pi\xi)^N} \tag{12.42}$$

The probability distribution function for signal plus noise is the anticharacteristic function and is found from Campbell and Foster's pair 581.1 [10] and is given by [8, p. 410, Eq. 11.4-18]

$$p(Y) = \frac{Y^{N-1}\exp(-Y)}{\left(1+\dfrac{NR}{2}\right)(N-1)!}\cdot {}_1F_1\left(2,\ N;\ \frac{Y}{1+\dfrac{2}{NR}}\right) \tag{12.43}$$

where ${}_1F_1$ is the confluent hypergeometric function given by [9, Eq. 13.2.1; 9]

$${}_1F_1(a;\ b;\ z) = \frac{\Gamma(b)}{\Gamma(b-a)\,\Gamma(a)}\int_0^1 \exp(zt)\, t^{a-1}\,(1-t)^{b-a-1}\, dt \tag{12.44}$$

The probability distributions for noise, echo signal, and echo plus noise are shown in Figure 12.26.

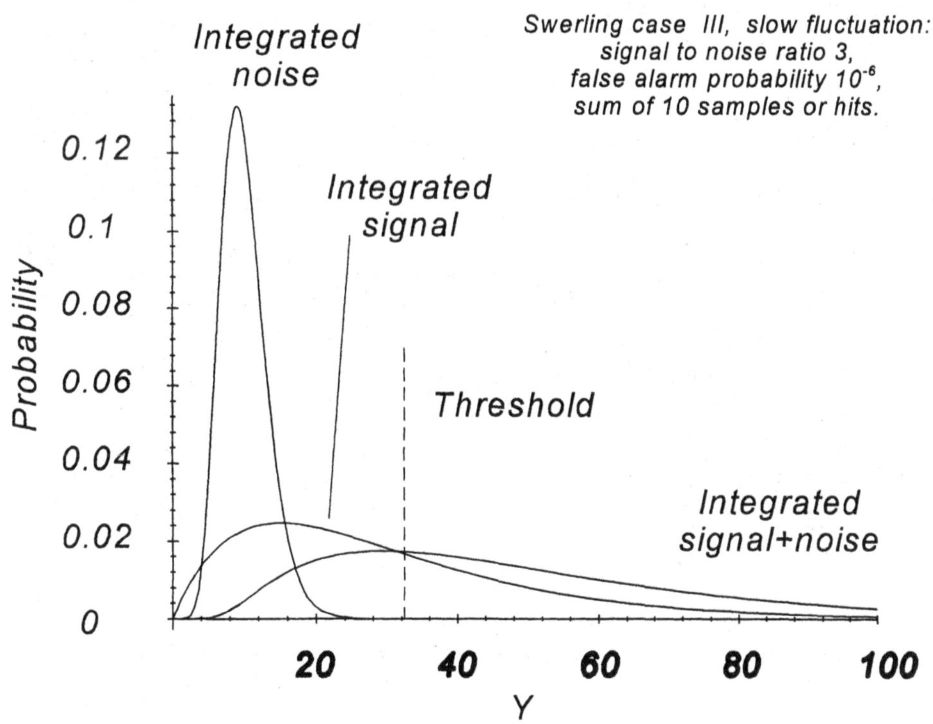

Figure 12.26 Swerling case III, slow chi-squared echo fluctuation: probability distributions for 10 integrated pulses.

The probability of detection is given by [8, p. 421, Eq. 11.4-24]

$$P_d = \left(1 + \frac{2}{NR}\right)^{N-2}\left(1 + \frac{Y_b}{1 + \frac{NR}{2}} - 2\frac{N-2}{NR}\right)\exp\left(-\frac{Y_b}{1 + \frac{NR}{2}}\right) \tag{12.45}$$

Equation (12.45) is exact for one or two pulses and can be solved numerically to obtain the signal-to-noise ratios for different numbers of pulses integrated. Curves for signal-to-noise ratios for a given probability of detection are shown for three false alarm probabilities in Figures 12.27, 12.28, and 12.29. The numbers of pulses are 1, 2, 3, 4, 5, 6, 7, 8, 10, 12, 15, 20, 25, 30, 40, and 50.

12.3.4.1 Coherent integration

In the case of coherent integration, the voltages of the signal and noise samples are added together, in contrast to adding the powers in incoherent integration. The voltage sum is compared to the threshold. The signal-to-noise ratios for detection using coherent integration are shown in Figures 12.39 to 12.42.

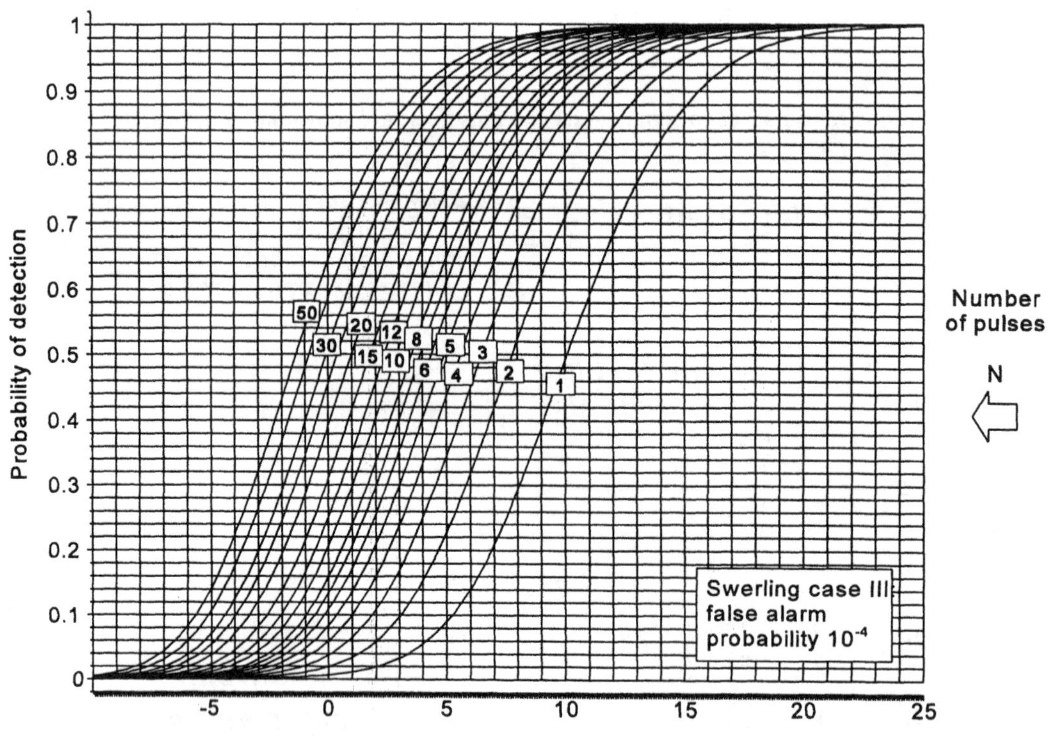

Figure 12.27 Swerling case III, slow chi-squared echo fluctuation: probability of detection against signal-to-noise ratio (dB) for a false alarm probability of 10^{-4}.

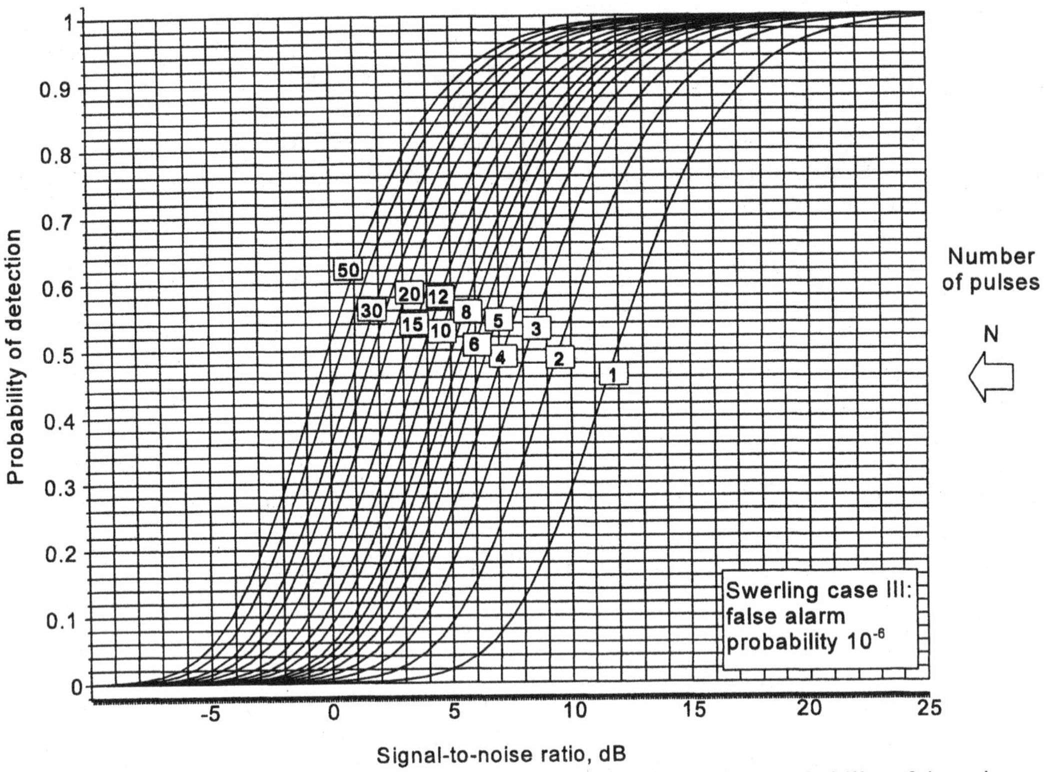

Figure 12.28 Swerling case III, slow chi-squared echo fluctuation: probability of detection against signal-to-noise ratio (dB) for a false alarm probability of 10^{-6}.

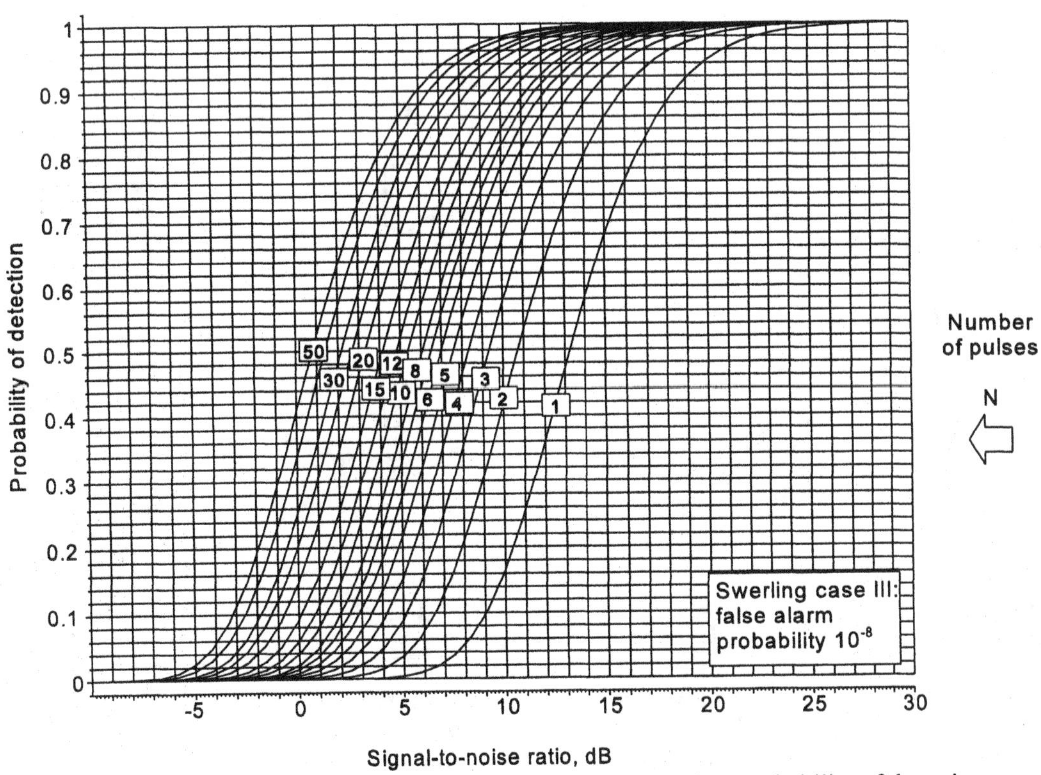

Figure 12.29 Swerling case III, slow chi-squared echo fluctuation: probability of detection against signal-to-noise ratio (dB) for a false alarm probability of 10^{-8}.

12.3.5 Swerling case IV: fast chi-squared fluctuation

Swerling case IV is similar to case II, but the object of interest has one major reflector and many minor ones. The cross-section has a chi-squared probability distribution function with four degrees of freedom and varies from echo to echo. It is the case with frequency agile radars, where the frequency steps are larger than the minimum frequency for decorrelation, as discussed in Section 12.3.3 for Swerling case II. As for Swerling case III the probability distribution for a single echo pulse is from (12.39)

$$p_{\text{Swerling IV}}(Y,\ 2,\ R/2) \;=\; \frac{4Y}{R^2}\exp\left(-\frac{2Y}{R}\right) \qquad \text{For a single pulse} \tag{12.46}$$

where Y is the echo signal power variable;
R is the signal-to-noise ratio.

The probability distribution function for the sum of N independent echo signals is found from the characteristic function

$$C_{\text{Swerling III or IV}} \;=\; \frac{1}{(1 + j\pi\xi R)} \qquad \text{For a single pulse} \tag{12.47}$$

and for N independent pulses the characteristic function is

$$C_{\text{Swerling IV}} \;=\; \frac{1}{(1 + j\pi\xi R)^{2N}} \qquad \text{For } N \text{ independent pulses} \tag{12.48}$$

and is shown for $N = 10$ and $R = 3$ in Figure 12.30.

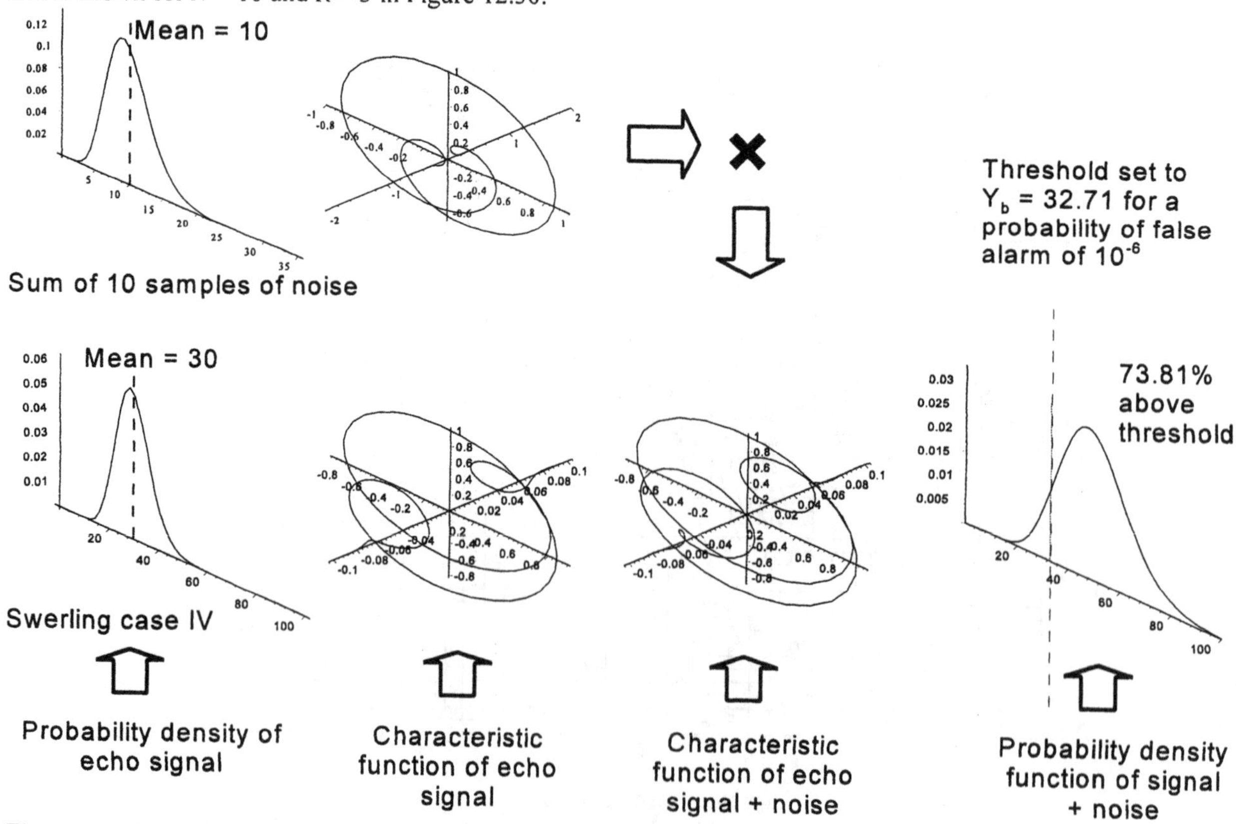

Figure 12.30 The probability distributions and characteristic functions for Swerling case IV. [After: [2]. Meikle, H. D., *A New Twist to Fourier Transforms*, Weinheim: Wiley-VCH, 2004, Figure 7.14.]

The characteristic function of N samples of signal plus noise is the product of the individual characteristic functions (12.48) and (12.49), given by

$$C_{\text{signal + noise Swerling IV}} = \frac{1}{(1+j2\pi\xi)^N} \frac{1}{(1+j\pi\xi R)^{2N}} \quad (12.49)$$

The anticharacteristic function gives the probability distribution function for signal plus noise using Campbell and Foster's pair 581.1 [10] and is given by [8, p. 427, Eq. 11.5-14]

$$p(Y) = \frac{Y^{N-1} \exp\left(-\dfrac{Y}{1+\dfrac{R}{2}}\right)}{(1+R)^{2N} (N-1)!} \, {}_1F_1\left(-N, N; -\frac{R}{1+R} Y\right) \quad (12.50)$$

where ${}_1F_1$ is the confluent hypergeometric function given by [9, 12]

$${}_1F_1(a; b; z) = \frac{\Gamma(b)}{\Gamma(b-a)\,\Gamma(a)} \int_0^1 e^{zt} t^{a-1} (1-t)^{b-a-1} \, dt \quad (12.51)$$

The probability distributions of the summed echo signal, noise, and signal plus noise are shown in Figure 12.31.

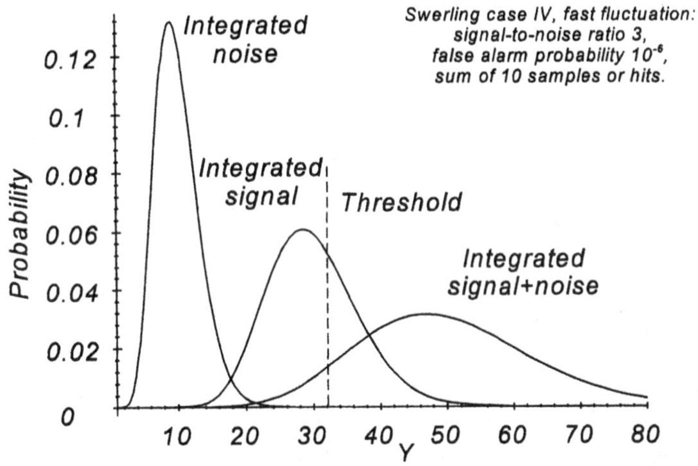

Figure 12.31 Swerling case IV, fast chi-squared echo fluctuation: probability distributions for 10 integrated pulses.

The probability of detection is given by [8, p. 427, Eq. 11.5-19]

$$P_d = 1 - \frac{N!}{\left(1+\dfrac{2}{R}\right)^N} \sum_{k=0}^{\infty} \left(\frac{R}{2}\right)^k \gamma\left(N+k, \frac{Y_b}{1+\dfrac{R}{2}}\right) \quad (12.52)$$

where $\gamma(a, x) = \int_0^x e^{-t} t^{a-1} \, dt.$

Curves for signal-to-noise ratios for a given probability of detection are shown for three false alarm probabilities in Figures 12.32, 12.33, and 12.34. The numbers of pulses are 1, 2, 3, 4, 5, 6, 7, 8, 10, 12, 15, 20, 25, 30, 40, and 50.

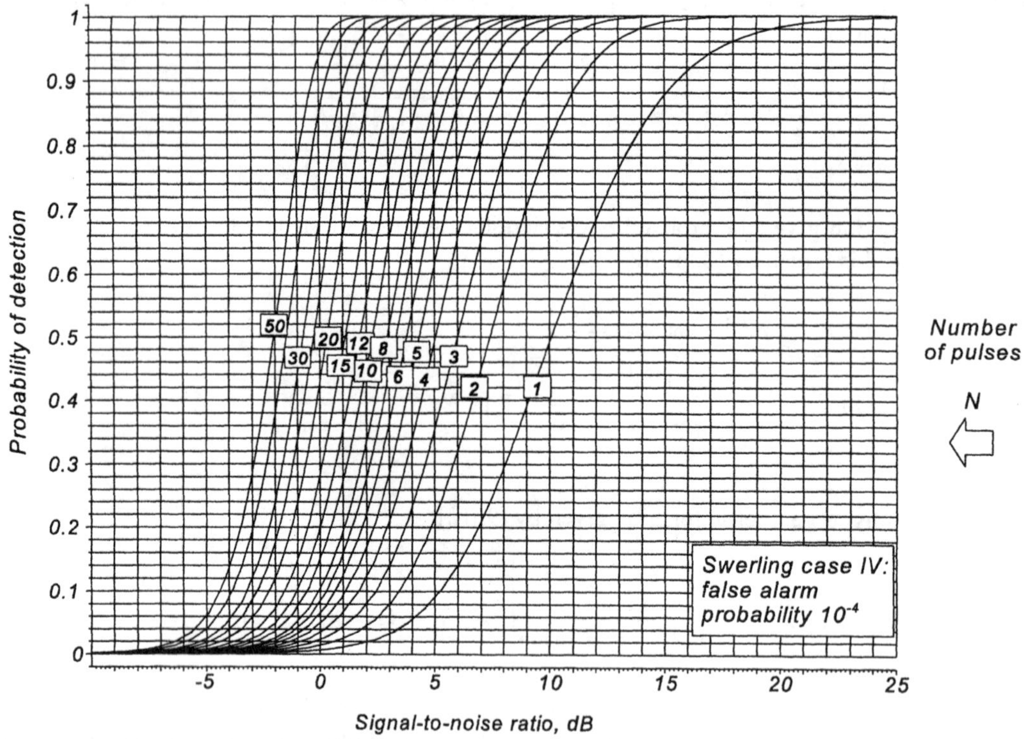

Figure 12.32 Swerling case IV, fast chi-squared echo fluctuation: probability of detection against signal-to-noise ratio (dB) for a false alarm probability of 10^{-4}.

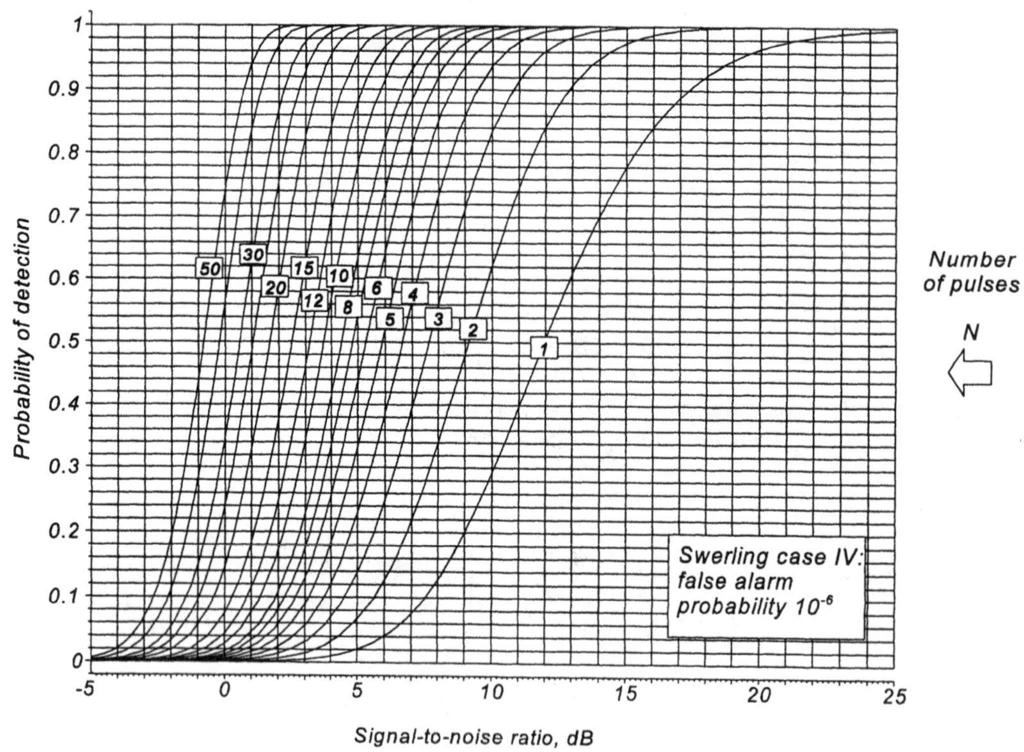

Figure 12.33 Swerling case IV, fast chi-squared echo fluctuation: probability of detection against signal-to-noise ratio (dB) for a false alarm probability of 10^{-6}.

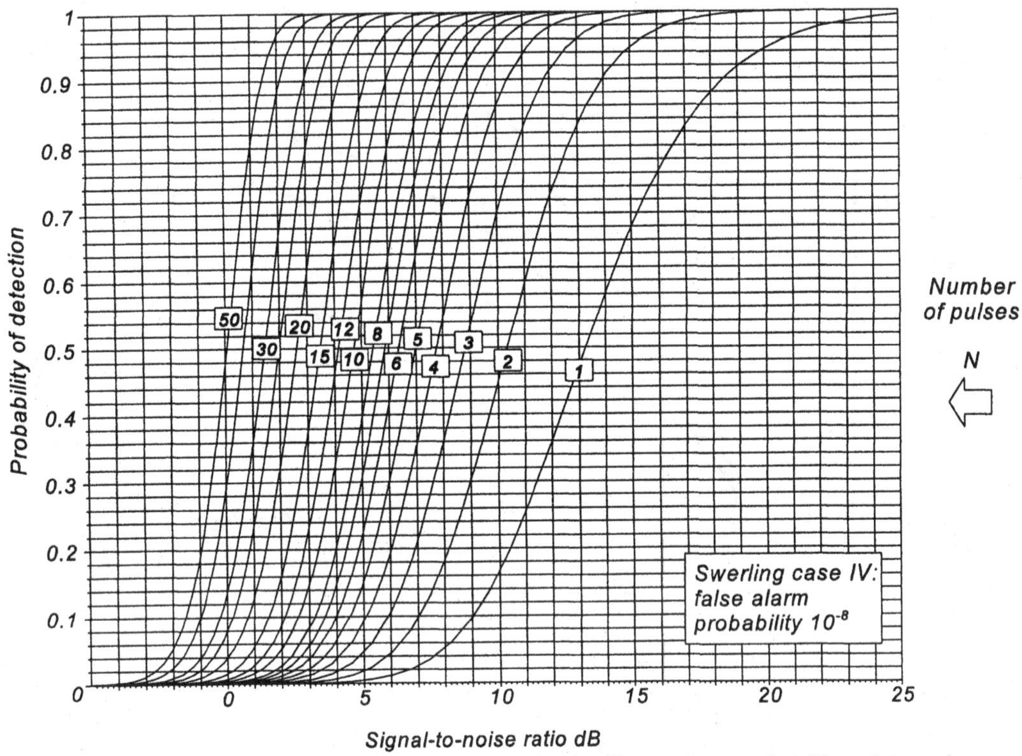

Figure 12.34 Swerling case IV, fast chi-squared echo fluctuation: probability of detection against signal-to-noise ratio (dB) for a false alarm probability of 10^{-8}.

The signal-to-noise ratios for detection are shown in Figures 12.39 to 12.42.

Because of the decorrelation of the echoes from each of the radar pulses, no coherent addition is possible.

12.4 COMPARISON OF PROBABILITY OF DETECTION CASES

The choice of detection case is often forced on the radar designer. Often fixed frequencies are allocated by the regulating authorities, and the presence of clutter requires moving target indication or detection signal processing, which limits the detection cases to:

- Marcum: Steady echoes;
- Swerling case I: Rayleigh (power) distributed echoes;
- Swerling case III: Chi-squared with four degrees of freedom.

Frequency agile radars may take advantage of the fast echo signal fluctuation during the dwell to give lower required signal-to-noise ratios with:

- Swerling case II: Rayleigh fading echoes;
- Swerling case IV: Chi-squared with four degrees of freedom.

Most aircraft are either jet aircraft, which have Rayleigh distributed echoes, or propeller driven aircraft, which may have Rayleigh or chi-squared distributed echoes.

A comparison of the shapes of the Marcum case and the four Swerling cases with incoherent integration (or summing) is shown in Figure 12.35. Echoes with no fluctuation require the lowest signal-to-noise ratios for probabilities of detection of greater than 50%. The Marcum curve has the greatest slope showing that the signal plus noise and noise probability distribution functions are well separated, as shown in Figure 12.9. The opposite occurs in Swerling case I, in which the probability density functions run into each other, as shown in Figure 12.14. In Swerling cases II (Figure 12.21) and IV (Figure 12.31), the signal plus noise and noise probability distributions are well separated. In the case of coherent integration, for example, a moving target detector (MTD), the single pulse curve is used but with all the energy of the N pulses concentrated in one pulse. That is, the signal-to noise-ratio required is the single pulse value (N = 1) divided by the number of pulses.

Very often, the amplitudes of the echoes that are summed are weighted to improve signal processing, which gives a loss.

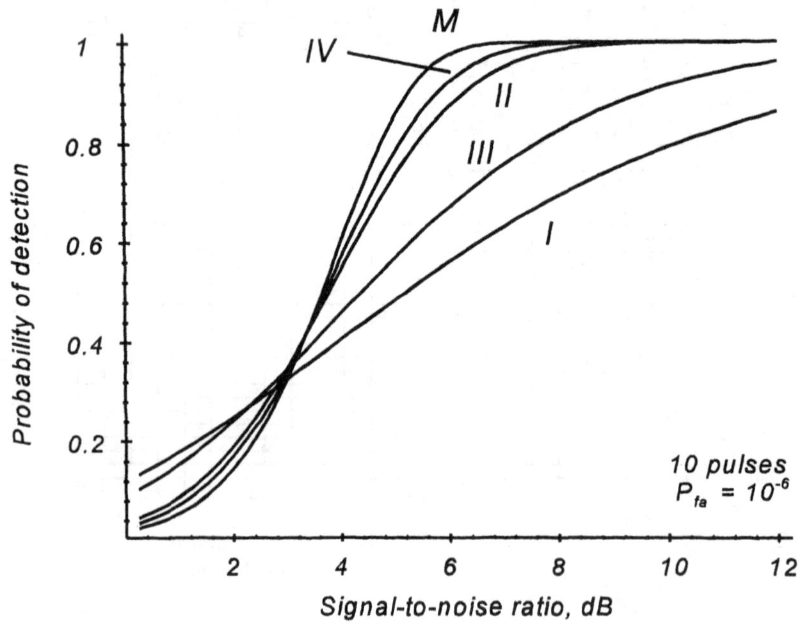

Figure 12.35 A comparison of probabilities of detection against signal-to-noise ratios (dB) for incoherently 10 integrated pulses for:
M Marcum case, no echo fluctuation;
I Swerling case I, slow Rayleigh fluctuation; III Swerling case III, slow χ^2 fluctuation;
II Swerling case II, fast Rayleigh fluctuation; IV Swerling IV, fast χ^2 fluctuation.

12.4.1 Coherent integration: slow fluctuation

With coherent integration, the voltages from N echo pulses are added together into one pulse. The noise voltages contained in the N echo pulses add incoherently, so that the signal-to-noise ratio (power) is given by

$$\textit{Coherently integrated signal-to-noise ratio (power)} = \frac{N^2}{N} = N \tag{12.53}$$

The signal-to-noise ratio for each pulse required for detection is thus that for one pulse divided by the number of pulses. The curves in Figure 12.36 have a slope of -3 dB/octave or -10 dB/decade.

Fluctuation losses are the difference in signal-to-noise ratio (in decibels) for detection between that with no fluctuation and the fluctuation model considered. Those with coherent integration are constant and real curves are shown in Figures 12.39 to 12.42.

12.4.2 Noncoherent integration: slow fluctuation

The shape of these curves is shown in Figure 12.37. After a faster drop after a single hit, the curves settle to the noncoherent integration slope of -1.5 dB/octave or 5 dB/decade.

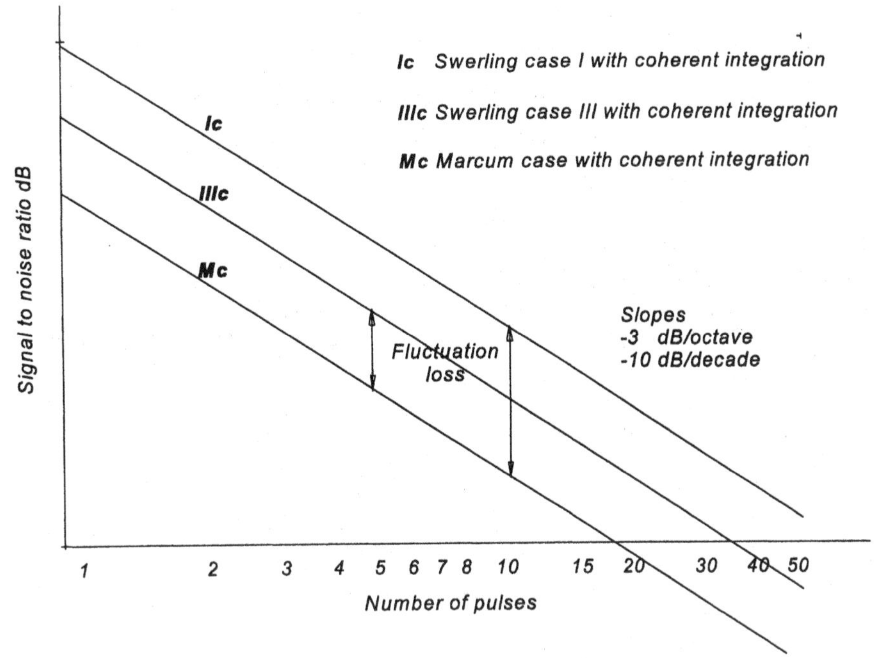

Figure 12.36 The shape of detection curves with coherent integration.

Fluctuation losses vary and are smaller with greater numbers of hits. Real curves are shown in Figures 12.39 to 12.42.

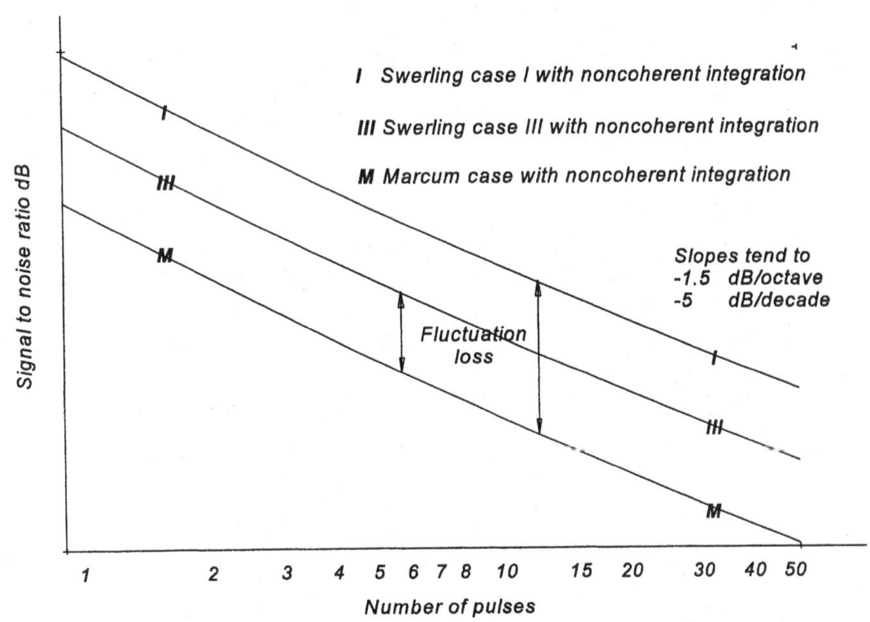

Figure 12.37 The shape of detection curves with noncoherent integration and slow fluctuation.

12.4.3 Noncoherent integration: fast fluctuation

Here the fluctuation of the echo signals during the dwell is dominant. The greater the number of signals that form the sample, the more the sample resembles a Gaussian or normal distribution. The mean is the same as for a steady Marcum [6] train of echoes. The shapes with the Marcum (steady) case reference are shown in Figure 12.38.

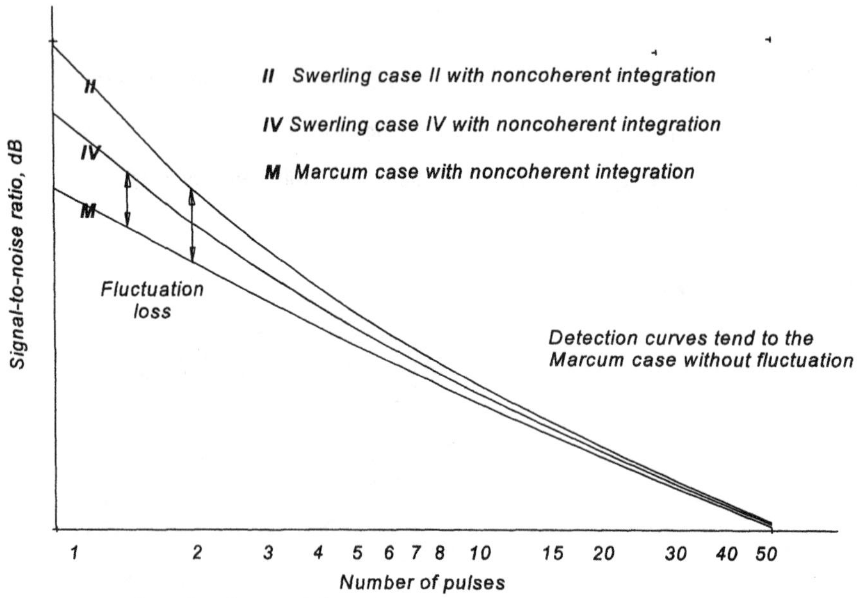

Figure 12.38 The shape of detection curves with noncoherent integration and fast fluctuation.

The fluctuation losses decrease rapidly with higher probabilities of detection and greater numbers of hits. A number of curves are shown in Figures 12.39 to 12.42.

12.5 JOINT PROBABILITIES OF DETECTION

Very often, a radar is constructed so that a simple threshold crossing does not automatically lead to a detection. Two common cases are binary integration and detection with frequency diversity radars in Section 12.5.2.

12.5.1 Other forms of integration after the threshold

In the past, the video signals used for display were also used to feed automatic extractors. The "banana" form of the signal on a plan position indicator is assessed in order to find the center of the "banana" or the azimuth (or elevation on a height finder radar) of the echo. Commonly an m-out-of-n threshold was used to declare the start of a run of hits during a scan in azimuth (or elevation). Because echo signals may be missing during the run of echoes, the end is declared after $m/2$ misses (see Section 11.2.2).

If p is the false alarm probability, then the probability of m noise pulses giving a detection is p^m. For m-out-of-n pulses, this becomes

$$\textit{Integrated false alarm rate} = \frac{n!}{(n-m)!m!}\, p^{m} \tag{12.54}$$

Example. If the threshold for Rayleigh noise is set at 3 dB above the root mean square value, the false alarm probability (at 2σ) is 13.53%. For an 8 out of 10 criterion the false alarm probability is reduced to

$$\frac{10!}{(10-8)!8!}\, 0.1353^{8} = 45 \times 1.125 \times 10^{-7} = 5.064 \times 10^{-6}$$

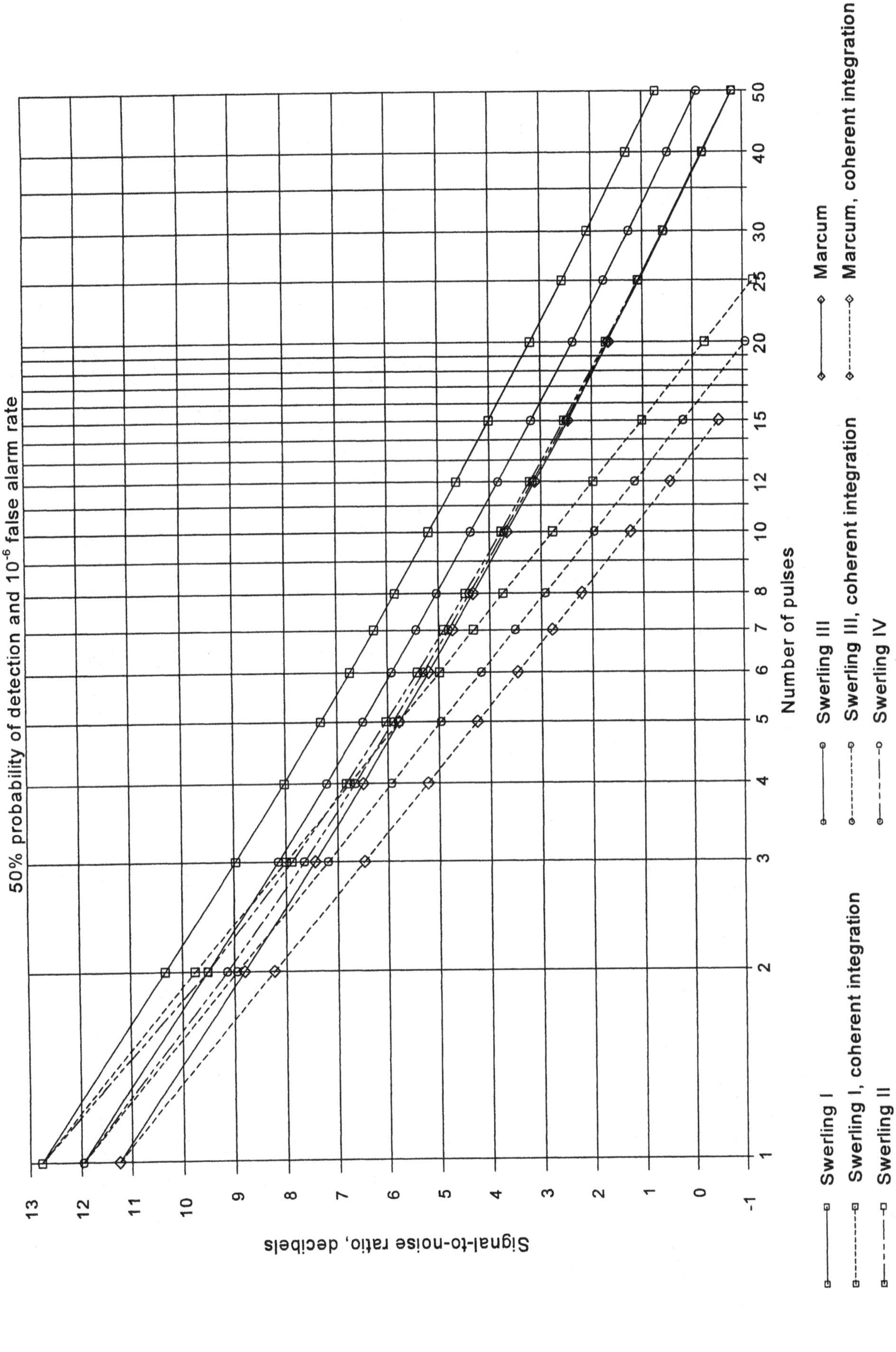

Figure 12.39 Detectability curves for 50% probability of detection with a false alarm probability of 10^{-6} for the classical cases.

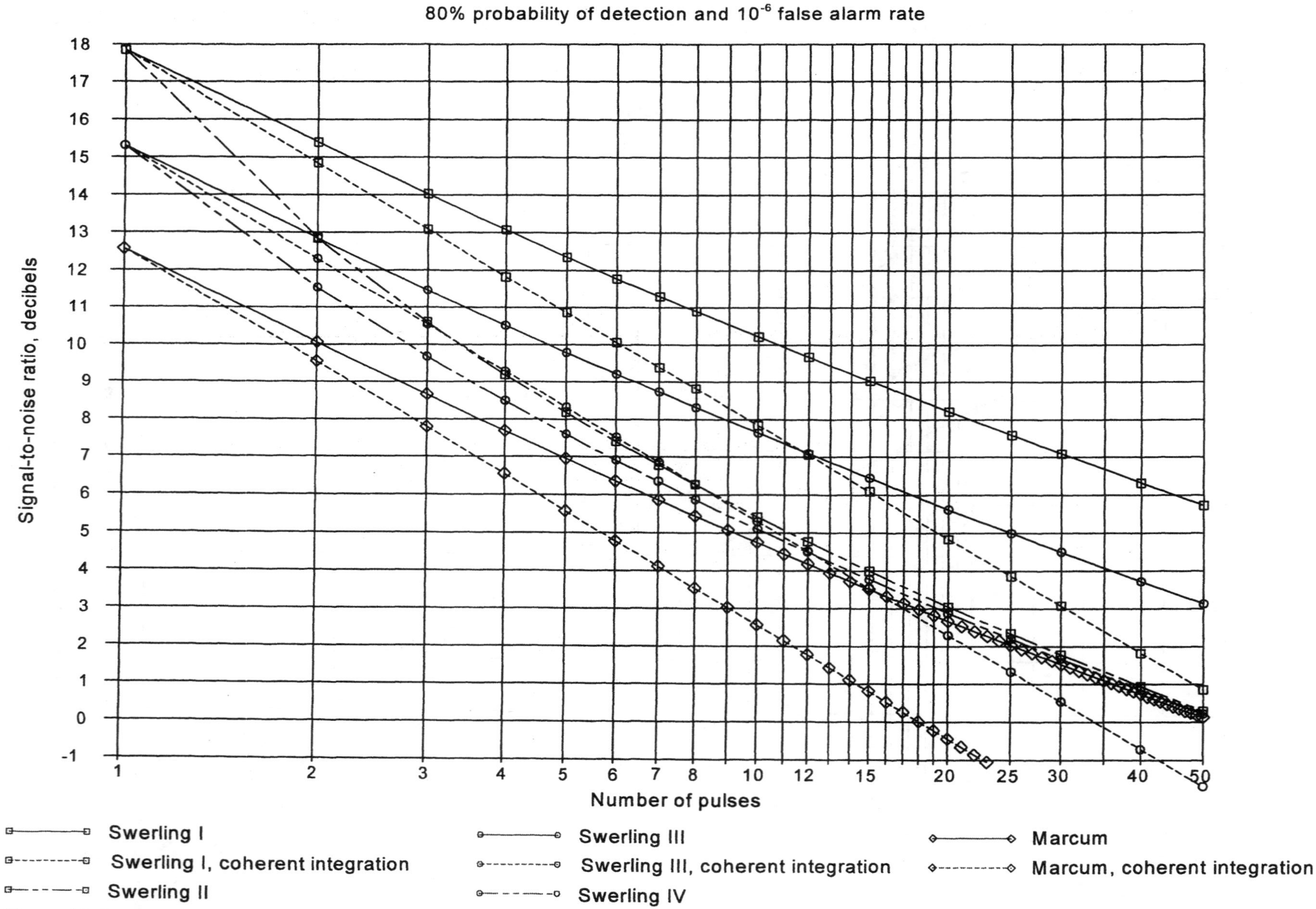

Figure 12.40 Detectability curves for 80% probability of detection with a false alarm probability of 10^{-6} for the classical cases.

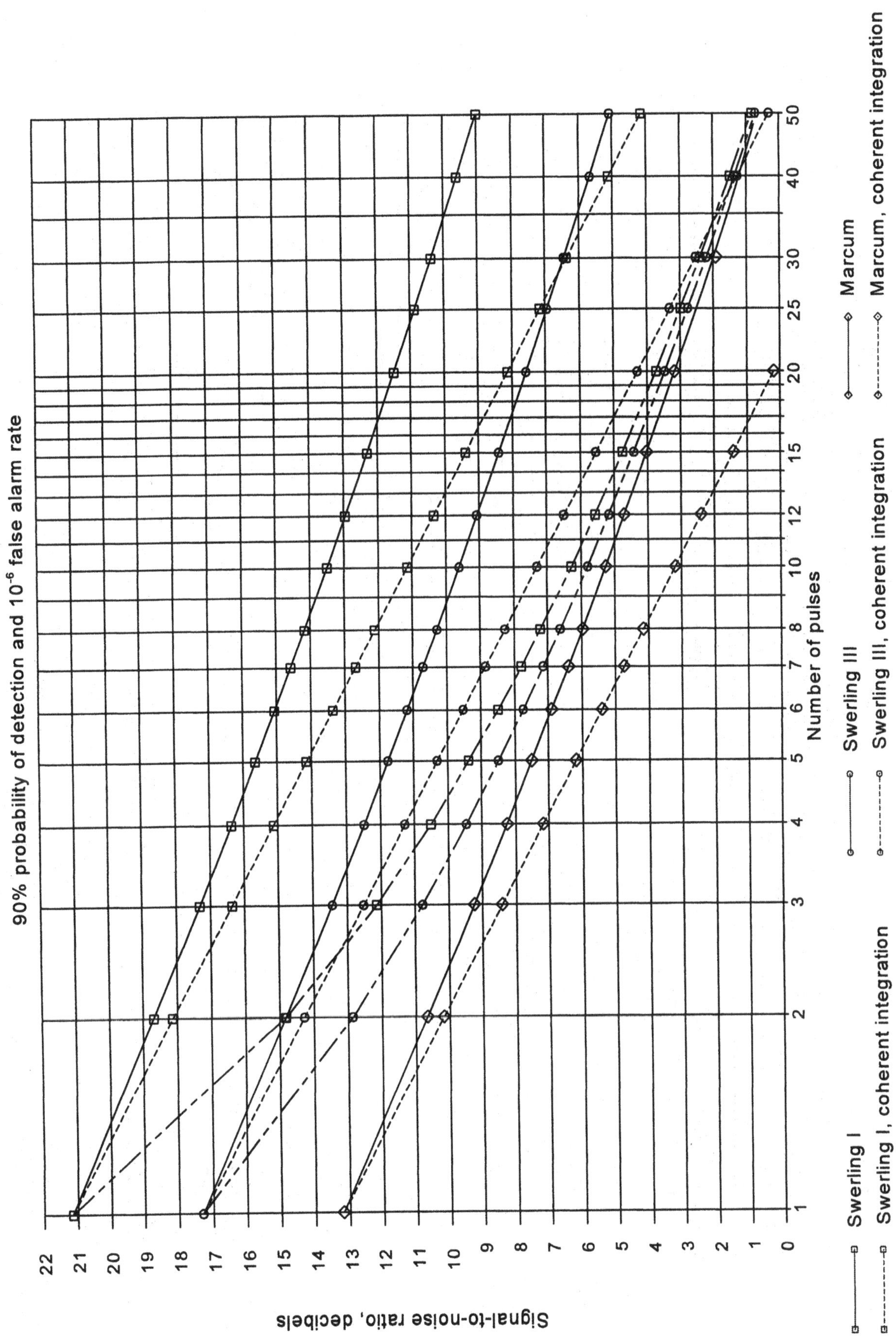

Figure 12.41 Detectability curves for 90% probability of detection with a false alarm probability of 10^{-6} for the classical cases.

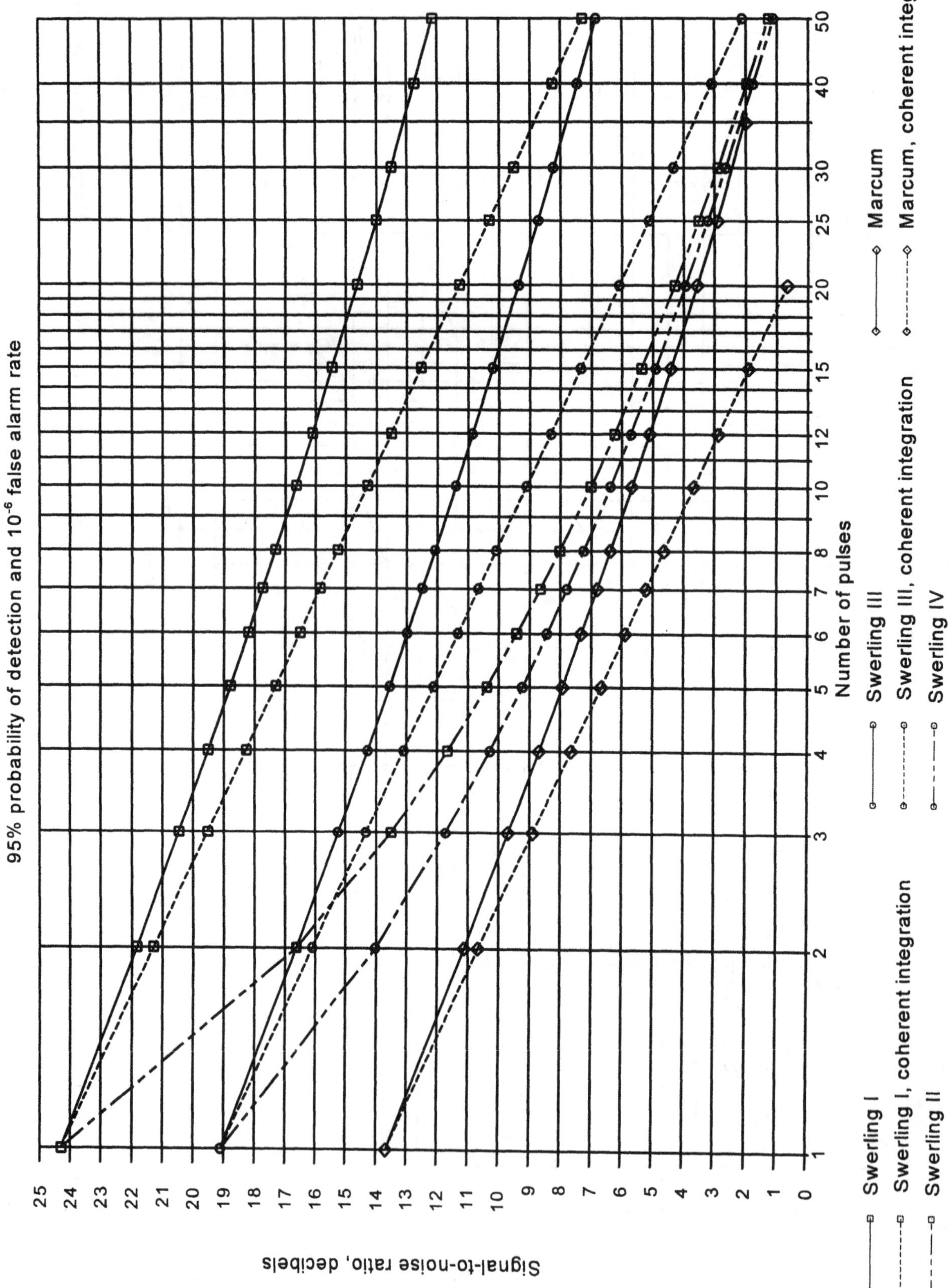

Figure 12.42 Detectability curves for 95% probability of detection with a false alarm probability of 10^{-6} for the classical cases.

12.5.2 Frequency diversity radars

Frequency diversity radars are those with a number of independent transmitters and receivers with a common antenna. Each transmitter and receiver combination uses a separate radio frequency, and the radar frequencies must be far enough apart so that the fading of the echo signals in each channel is completely uncorrelated. (The frequency difference is discussed in Section 12.3.3, which describes Swerling case II.)

Dual frequency diversity is often used today for highly reliable installations, at airports for example, so that if one channel fails, operation continues with the remaining channel. In the past up to three diversity channels were used to spread the frequencies used as an antijamming measure. Frequency agile radar systems can cover better a wider bandwidth without the losses and complexity incurred by the diplexer (see Figures 4.2 and 4.3).

The signals from the channels may be combined by adding the videos, as shown in Figure 12.43 or logically as shown in Figure 12.44. The addition of the videos from fixed frequency channels changes the probability distribution of the sum from the two Swerling case I distributions to the form of a Swerling case III distribution.

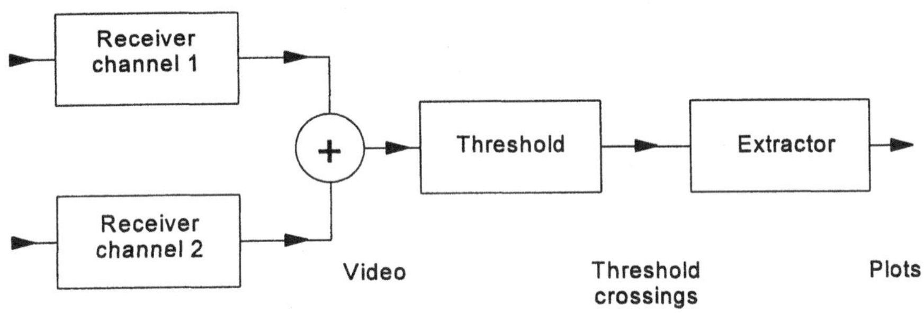

Figure 12.43 Block diagram of two receivers with summing before the threshold.

The probability distribution function for each of the signals coming from the receivers is (12.23)

$$p(Y) = \frac{1}{NR} \exp\left(-\frac{Y}{NR}\right) \tag{12.55}$$

The Fourier transform or characteristic function for N pulses in each channel is (12.24)

$$C(\xi) = \frac{1}{1 + j2\pi NR\xi} \tag{12.56}$$

For the sum of N pulses in each of two channels the characteristic function is

$$C(\xi) = \frac{1}{(1 + j2\pi R\xi)^2} \tag{12.57}$$

If the signal is divided equally between the two channels, R becomes $R/2$ the and the characteristic function becomes

$$C(\xi) = \frac{1}{(1 + j\pi R\xi)^2} \tag{12.58}$$

The anticharacteristic function, or the probability distribution is that of Swerling case III, is

$$p(Y) = \frac{4Y}{(NR)^2}\exp\left(-\frac{2Y}{NR}\right) \tag{12.59}$$

The values of signal-to-noise ratio for 50% and 80% probability of detection for probabilities of 10^{-6} and 10^{-10} are given by Poelman [13] for up to four channels in frequency diversity in terms of the total numbers of echo signals in all channels. Thus for a dual diversity radar with 10 hits per channel, noncoherent integration, and a probability of detection of 80% in Figure 12.40, the signal-to-noise ratio necessary for a detection in one channel alone is 10.2 dB (Swerling I) and for the sum of videos from two frequency diversity channels is 5.7 dB (Swerling III, 20 hits).

For the case of frequency agility in both channels, the Swerling case II curve is used with the total number of hits from both channels, in the case above with 10 hits in each of both channels, the total number of hits is 20 hits, giving 2.9 dB.

Poelman defines the diversity gain as the change in the required signal-to-noise ratio compared to a single channel radar neglecting the increase in input (transmitter) power. Diversity efficiency is defined by Poelman as "the gain in required signal-to-noise ratio of the actual system compared with the case in which additional power was used to increase the single channel pulse power".

Signal-to-noise ratio in Figure 12.40	10.2	dB
Signal-to-noise ratio with dual diversity	5.7	dB
Diversity gain	4.5	dB
Number of channels (dB)	3.0	dB
Diversity efficiency	1.5	dB

The losses in the diplexer have not been taken into account.

Currently the majority of civil radars using diversity have two independent channels and the plots from the extractor are combined and labeled to show if they came from a single channel or both. A combined plot from both channels is considered to be confirmed and not a false alarm. The channels can be considered as two separate radars from the point of view of adjustment and maintenance. The false alarm rate from the combined signals is twice that of the single channel alone.

The performance of the radar is mostly defined on a single channel basis.

Logical combination from a number of receivers give rise to conditional probabilities of false alarm and detection, depending on what logic is used to combine them, as shown in Figure 12.44. With .AND. logic and individual probabilities of detection of p, q, and r, the probability of NOT detecting the signal is

$$\textit{Probability of NOT detecting the echo} = (1-p)(1-q)(1-r) \tag{12.60}$$

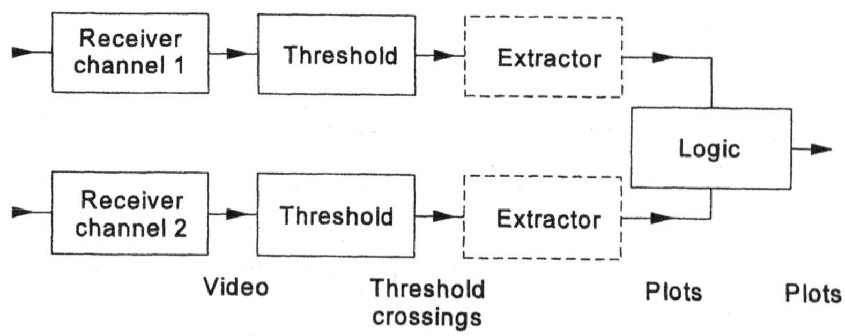

Figure 12.44 Logical echo report combination.

Thus, the probability of detection in at least one channel is

$$\textit{Probability of detection in one or more channels,}\ P_{d,\,OR} = 1-(1-p)(1-q)(1-r) \tag{12.61}$$

The probability of detection in all three channels simultaneously is

$$\text{Probability of detection in all channels, } P_{d,\,AND} = p\, q\, r \tag{12.62}$$

For false alarms with .OR. logic, the false alarms from all receivers are summed, which gives the collapsing loss. With .AND. logic, the false alarm probabilities are multiplied, giving a reduction. With more than two receivers, probabilities between pure .AND. and .OR. are possible with m-out-of-n logic. Frequency diversity improves the quality of detection. Reliability is increased compared to a single channel radar, because if one channel fails or is being repaired, another is still operating.

Example. The coverage for a two channel search radar is defined as the range in which the probability of detection is 80% in one channel. It is assumed that the two channels have radio frequencies that are separated widely enough so that the echo signals in the two channels are completely uncorrelated.

- The probability that an echo is NOT seen in both channels is

  $$(1 - 0.8)(1 - 0.8) = 0.04$$

  Thus, the probability of detection at the range in question in either or both channels is increased to 96% (12.61). Some tracking algorithms look for a detection in both channels to start a track. In this case, the probability of detection in both channels simultaneously at this range is reduced to 0.8 × 0.8 or 64%. To summarize:

Probability of detection in both channels:	0.8 × 0.8	64%;
Probability of detection in channel A alone:	0.8 × 0.2	16%;
Probability of detection in channel B alone:	0.8 × 0.2	16%;
Probability of detection in either or both channels:		96%.

- If the false alarm probability is 10^{-6} in one channel, then the coherently integrated signal-to-noise ratio for detection is 17.847 dB. To have 80% probability of detection with .AND. logic, the probability of detection in each channel must be $\sqrt{0.8} = 0.8963$. The false alarm probability in each channel may be increased to 10^{-3} to give a combined false alarm probability of 10^{-6}. The signal-to-noise ratio required in each channel is again 17.847 dB.
- If .OR. logic is used, the false alarm probability in each channel must be halved to 5×10^{-7}. The probability of detection in each channel, p, to give 80% as the combined value is the solution of

  $$1 - (1 - p)^2 = 0.8,$$

  which gives $p = 55.278\%$. The required signal-to-noise ratio is 13.706 dB. Thus, the use of .OR. logic gives a diversity gain of 4.141 dB.

Poelman and Bennée [14] give curves for further logic combinations for three and more channels used in military radars to spread the bandwidth used by the radar system though the gains expressed may be reduced by the triplexers..

12.6 USEFUL APPROXIMATIONS

The calculations for probability of detection are difficult and in the past published tabulated values of distributions (for example, Toronto functions chi-square distributions, and Pearson's form of the incomplete gamma function [8] and a number of series) were used to find the probabilities of detection. In this book the fundamental Bessel and gamma functions have been used for calculation often requiring the numerical solution of integrals. For numerical solution the Maple program requires the bounds for the solution in order to find it more quickly. Albersheim's and Snidman's approximations are a great help.

12.6.1 Albersheim's approximation for the Marcum case

For the case of a steady signal plus noise, or Marcum case, with a linear detector and noncoherent integration, Albersheim, quoted in [15, p. 329] gives the following approximation for the signal-to-noise ratio, D_{Marcum}:

$$A = \ln\left(\frac{0.62}{P_{fa}}\right) \qquad B = \ln\left(\frac{P_d}{1 - P_d}\right) \tag{12.63}$$

$$D_{Marcum} = -5\log_{10}N + \left(6.2 + \left(\frac{4.54}{\sqrt{N+0.44}}\right)\right)\log_{10}(A + 0.12AB + 1.7B) \text{ dB} \tag{12.64}$$

The error in the estimate of D_{Marcum} is said to be of the order of 0.2 dB for P_{fa} between 10^{-7} and 10^{-3}, P_d between 0.1 and 0.9, and the number of pulses N between 1 and 8 096. Figures 12.39 to 12.42 were plotted using exact values, but the differences using Albertsheim's approximation would have been as shown in Figure 12.45.

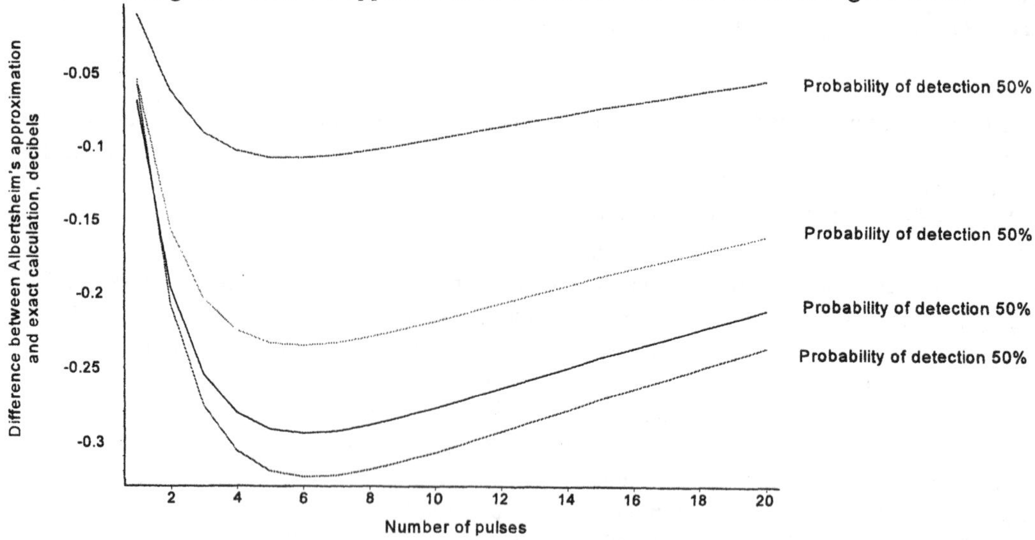

Figure 12.45 Differences in Figures 12.39 to 12.42 between exact calculation of the signal-to-noise ratios and Albertsheim's approximation.

12.6.2 Snidman's approximation for the Marcum and Swerling cases

Snidman's approximation [15, 16] is for a square law detector and the Swerling case is determined by the choice of the variable K.

$K = \infty$ for the Marcum, or nonfluctuating, case;
1 for Swerling case I;
N for Swerling case II, where N is the number of pulses integrated;
2 for Swerling case III;
$2N$ for Swerling case IV, where N is the number of pulses integrated;
$\alpha = 0$ when $N < 40$ and 0.25 when $N \geq 40$.

$$\eta = \sqrt{-0.8\ln(4P_{fa}(1-P_{fa}))} + \text{signum}(P_d-0.5)\sqrt{-0.8\ln(4P_d(1-P_d))} \tag{12.65}$$

$$X_\infty = \eta\left(\eta+2\sqrt{\frac{N}{2}+\left(\alpha-\frac{1}{4}\right)}\right) \tag{12.66}$$

$$C_1 = \frac{((17.7006P_d-18.4496)P_d+14.5339)P_d - 3.525}{K} \tag{12.67}$$

$$C_2 = \frac{1}{K}\left(\exp(27.31P_d - 25.14) + (P_d - 0.8))\left(0.7\ln\left(\frac{10^{-5}}{P_{fa}}\right) + \frac{2N-20}{80}\right)\right) \tag{12.68}$$

$$\begin{aligned} C_{dB} &= C_1 && \text{for } 0.1 \le P_d \le 0.872 \\ &= C_1 + C_2 && \text{for } 0.872 < P_d \le 0.99 \end{aligned} \tag{12.69}$$

$$C = 10^{C_{db}/10} \tag{12.70}$$

$$D_{Swerling} = \frac{C\,X_\infty}{N} \quad \text{and} \quad D_{Swerling\ dB} = 10\log_{10} D_{Swerling} \ \text{dB} \tag{12.71}$$

The errors that would have occurred with Snidman's approximation in Figures 12.39 to 12.42 are shown in Figure 12.46.

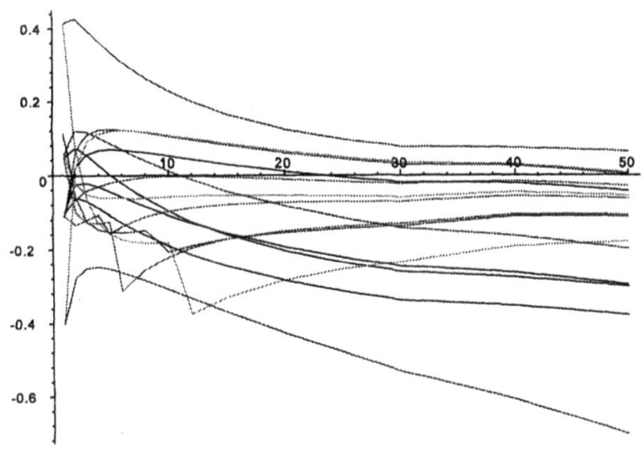

Figure 12.46 Differences between Snidman's approximation and exact values for signal-to-noise ratio.

12.7 FIGURES AFFECTING RADAR PERFORMANCE

The critical signal-to-noise ratio to achieve a given probability of detection for a given false alarm ratio in this chapter, $D_{,}$, is a critical value in the radar equation in Chapter 14 and determines whether the echo of interest is seen. The actual signal-to-noise ratio is used in Chapter 14 to determine the accuracies.

REFERENCES

1. *IEEE Standard Radar Definitions, IEEE Standard 686-1990,* New York: The Institute of Electrical and Electronic Engineers, 1993.
2. Meikle, H. D., *A New Twist to Fourier Transforms,* Weinheim, Germany: Wiley-VCH, 2004.
3. Meikle, H. D., "Another Way of Representing Echo Signals, Their Spectra, and Their Statistics", *Proceedings of the German Radar Symposium GRS 2002*, Bonn, Germany: DGON, 5th - 9th of September 2002, pp. 509–513.
4. Skillman, W. A., *Radar Calculations Using the TI-59 Programmable Calculator*, Dedham, Massachusetts: Artech House, 1983.
5. Meyer, D. P., and H. A. Mayer, *Radar Target Detection*, London: Academic Press, 1973.
6. Marcum, J. I., *A Statistical Theory of Target Detection by Pulsed Radar*, Rand Corporation, Mathematical Appendix

to RM-754, April 1952.

7. Swerling, P., *Probability of Detection for Fluctuating Targets*, Rand Corporation RW-1217, March 1954.
8. Di Franco, J. V. and W. L. Rubin, *Radar Detection*, Englewood Cliffs, New Jersey: Prentice-Hall, 1968, reprinted by Artech House, 1980.
9. Abramowitz, M., and I. A. Stegun, *Handbook of Mathematical Functions*, New York: Dover, 1965.
10. Campbell, G. A., and R. M. Foster, *Fourier Integrals for Practical Applications*, Princeton, New Jersey: D. van Nostrand Company, , 1948. This book replaces the paper "The Practical Application of the Fourier Integral", *Bell System Technical Journal*, October 1928, pp. 639–707.
11. Barton, D. K., *Modern Radar System Analysis*, Norwood, Massachusetts: Artech House, 1988, p. 86.
12. Weisstein, E. W., (ed.), *CRC Concise Encyclopedia of Mathematics*, Boca Raton, Florida: CRC Press, 1999, p. 297.
13. Poelman, A. J., *Supplement 2 to Technical Report TR-50-U (Standard methods for predicting and specifying performance of air surveillance radar systems)*, SHAPE Technical Centre, The Hague, April 1970. Reprinted in Barton, D. K.,ed. *Radars, Vol. 6, Frequency Diversity and Agility*, Dedham, Massachusetts: Artech House, 1977.
14. Poelman, A. J., and E. J. Bennée, *Supplement 3 to Technical Report TR-50-U (Standard methods for predicting and specifying performance of air surveillance radar systems)*, SHAPE Technical Centre, The Hague, May 1973.
15. Richards, M. A., *Fundamentals of Radar Signal Processing*, New York: McGraw-Hill, 2005.
16. Snidman, D. A., "Determination of the Required SNR Values", *IEEE Transactions on Aerospace and Electronic Systems,* Vol. AES-38, No. 3, July 2002, pp. 1059-1064.

Chapter 13

Determination of position

Radar is an acronym for radio detection and ranging. Once the echo signals have been detected, the position of the object that caused them must be estimated. The echo signals are combined with noise and the measurement accuracy depends on the signal-to-noise ratio. Traditionally, radars are divided into a number of types and examples of these types are:

- Radars that turn their antennas towards the single "targets" that interest them, typically fire control radars and radars that track weather balloons or other single flying objects;
- Radars that measure many target positions sequentially, typically modern fire control radars and artillery locating radars;
- Radars that scan the volume around them continually, typically surveillance radars.

This chapter describes the measurement of position with a general discussion of accuracy.

13.1 FIRE CONTROL RADARS

Fire control radars began to be developed before the start of the Second World War, to measure the positions of attacking aircraft, which were the targets for anti-aircraft artillery. First the antenna was moved by hand until the maximum signal was visible on an A-scope, which allowed the range of the target to be measured, but the angles were better measured by an optical director. Switching the antenna lobe to fixed positions on either side of the target direction allowed the two signals to be compared, as in Figure 13.1. When the two signals were equal, the antenna was pointing towards the target, which gave accurate angle measurements. Such radars required operators for azimuth, elevation, and range.

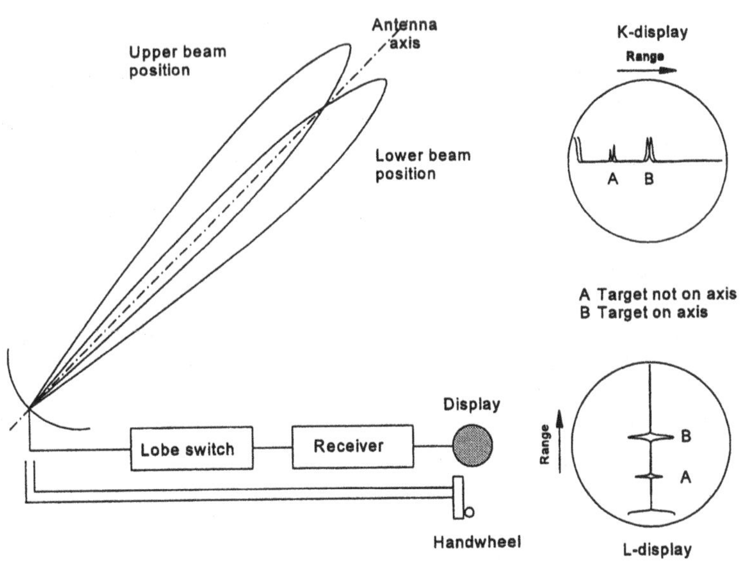

Figure 13.1 Angle measurement using lobing.

13.1.1 Conical scanning

A rotating, slightly offset antenna feed provides a cone of illumination and reception around the target, as shown in Figure 13.2. This is called conical scanning, and the switch behind the antenna separates the echo signals into those received to the left, to the right, above, and below the target. First, a range gate has to be set to choose the echo signal of interest and to reject all others. In order to handle the large dynamic range, automatic gain control (AGC) is used to keep the echo signal to a constant level. The differences in the signals received from the left and from the right of the target are used to drive the azimuth motors, to reduce the difference to zero: similarly in elevation. Typically the feed would be rotated at 1 500 revolutions per minute for a radar with a pulse repetition frequency of 1 500 Hz, so that one error signal sample contains the information from 60 pulses.

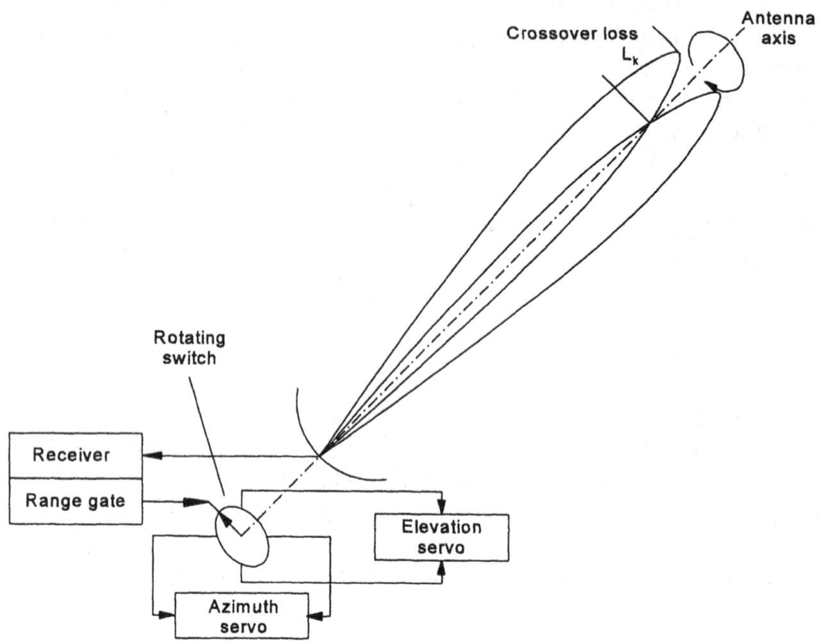

Figure 13.2 A conical scan tracking system.

An amplitude modulated jammer is able to modify the echo signals so that a false difference signal is generated that throws the antenna off track. This led to the development of a radar receiving system that is capable of providing error signals from each single echo. A second output from the antenna and receiving chain is necessary, and the angle of the incoming echoes can be determined from the amplitudes or the phases of their signals.

13.1.2 Amplitude monopulse receivers

The echo signals return and have amplitudes over a large dynamic range. The off-center signals are normalized by division which has led to two common forms of amplitude monopulse systems, shown in Figures 13.3 and 13.7.

The form used by fire control radars has a transmitter which feeds the antenna normally, and for reception it uses separate receivers for the sum or Σ channel, and the extra difference or Δ channel. The antenna illumination functions and patterns of monopulse antennas are described in Section 5.3. The difference pattern in Figure 5.24 is the difference of the two components in Figure 13.3. The off-axis signal is found by dividing the difference signal, Δ, by the sum signal, Σ,

$$\textit{Off-axis signal or error} = p\,\frac{\Delta}{\Sigma} \tag{13.1}$$

where p is a constant.

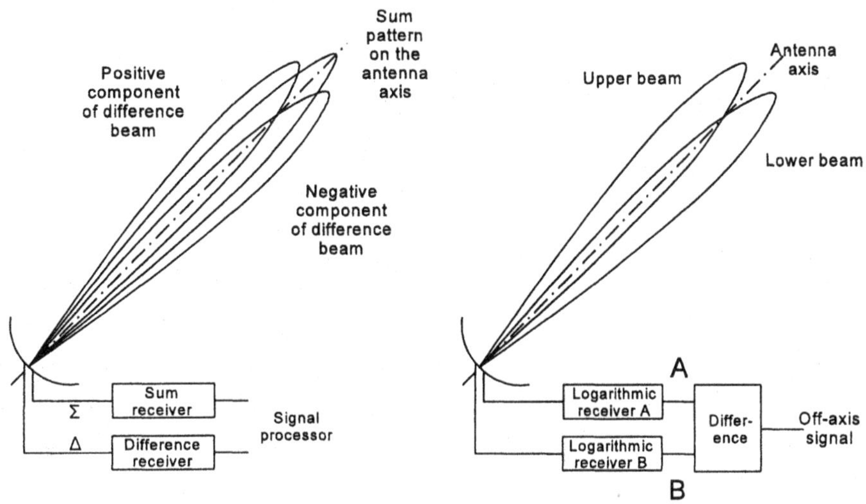

Figure 13.3 Types of amplitude monopulse receiving systems.

Often the phase of the signals is lost in the receivers so that in (13.1) the output signal is ambiguous in sign. One form of signal processor is shown in Figure 13.4. A separate phase detector extracts the sign of the off-axis signal, which is used to drive servomotors to turn the antenna or a signal to give a finer angle measurement.

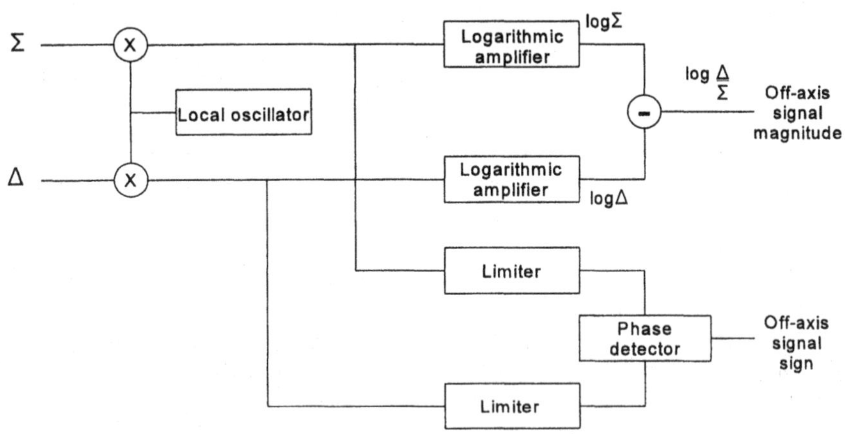

Figure 13.4 Logarithmic sum and difference processor.

Alternatively, a phase detector may be used to perform the division, as illustrated in Figure 13.5. The off-axis signal is proportional to 2 $\tan^{-1} \Delta/\Sigma$, which is linear enough to control motors that drive the antenna so that its axis points towards a target. The difference signal is often not linear enough to refine angle measurements, so that a table is used to provide linearly proportional output values (as with stacked beam radars and the azimuth angle measurement with mode-S secondary radars).

On the right hand side of Figure 13.3, there are two separate beams and receivers A and B. To simplify the dynamic range and division problems, logarithmic receivers are used. The off-axis signal is given by

$$\textit{Off-axis signal or error} = q\,\frac{A}{B} \quad \text{or} \quad r\,(\log A - \log B) \tag{13.2}$$

where q and r are constants.

For Gaussian beams, log A - log B is a linear function; for others it is only approximately linear. The Gaussian beam is given by (from Section 2.9.2)

$$F(x) \;=\; \exp\left(-4\ \ln(2)\left(\frac{x}{\theta_3}\right)^2\right) \tag{13.3}$$

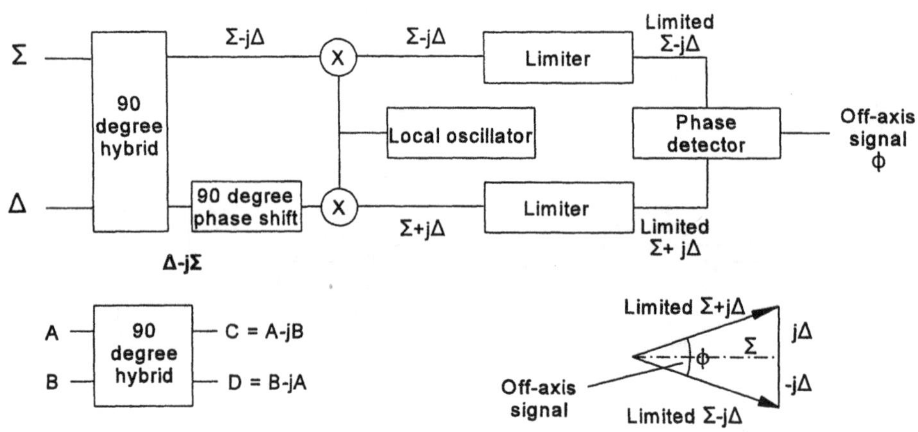

Figure 13.5 Amplitude monopulse receiver using a phase detector.

If the two beams are offset by p beamwidths from the axis, the off-axis signal is given by (using natural logarithms)

$$\begin{aligned} \mathit{Off\text{-}axis\ signal} &= \ln\left(\exp\left(-4\ln(2)(x+p)^2\right)\right) - \ln\left(\exp\left(-4\ln(2)(x-p)^2\right)\right) \\ &= -4\ln(2)(x+p)^2 + 4\ln(2)(x-p)^2 \\ &= -16\ln(2)xp \end{aligned} \tag{13.4}$$

where p is the offset of the two beams from the axis in beamwidths.

The antenna patterns, difference patterns, and the line of the off-axis signal are shown in Figure 13.6. The difference pattern is similar to that of the center fed monopulse antenna shown in Figure 5.24.

The noise on the difference signal (divided by the sum signal) represents an uncertainty or inaccuracy in angle when reflected onto the angle axis, as in the magnified circle in Figure 13.6. A critical parameter is the slope of the difference curve, which depends on the illumination function (see Appendix B) and the spacing of the two separate beams.

13.1.3 Phase monopulse receivers

Phase monopulse systems use interferometer antennas and measure the phase difference of the arriving wavefront at the two feed points. Such an antenna is shown in Figure 13.7.

The off-axis signal is the phase angle, ϕ, from the phase detector. If the phase difference at the two receiving elements is ψ, then

$$\phi \;=\; \sin^{-1}\frac{\psi}{2\pi d} \tag{13.5}$$

Again, the signal is linear only over small angles. For larger angles, a table must be used.

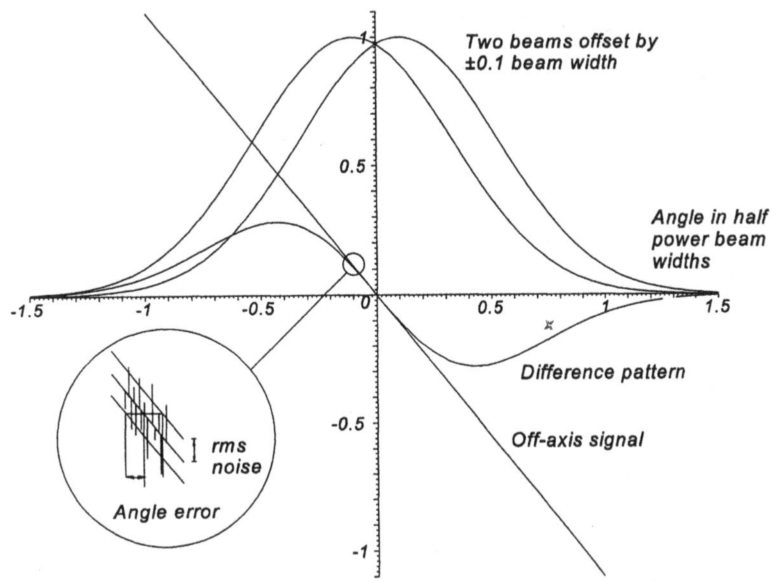

Figure 13.6 Gaussian beams offset from the axis by ±0.1 beam widths with difference pattern and off-axis signal.

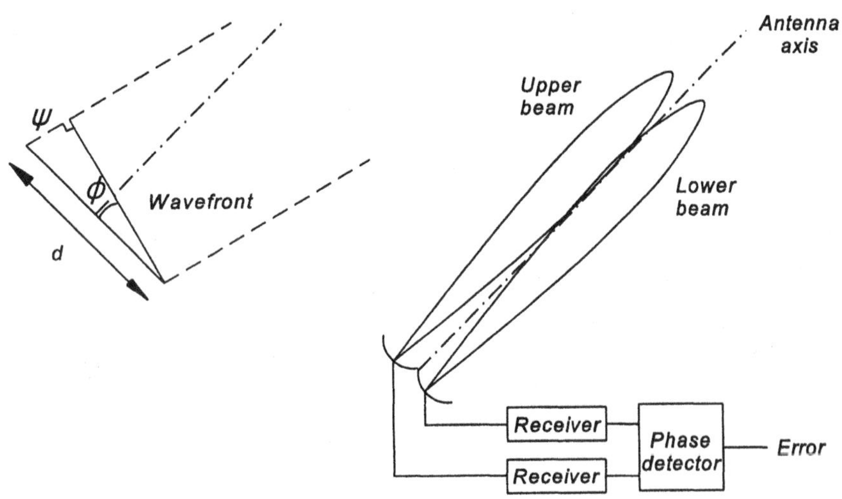

Figure 13.7 A phase monopulse receiving system.

13.1.4 Measurement of range

Normally, aircraft are much shorter than the length of the effective transmitted pulse length in space. The echo duration in time is the effective pulse width, so that range gates of fixed duration are used to select the target of interest. Automatic tracking in range is then possible.

The signal in the range gate is split into two, and both halves are integrated and compared. When the integrated voltages are not equal, the range servo moves the gates so that the voltages in the two halves are equal once more. The setting of the range servo shows the range.

Reflections from the ground lengthen and interfere with the echo pulse, which gives a false center and thus an incorrect range. Jammers on aircraft, which are being targeted by a fire control radar, can emit a pulse which augments the normal echo. The pulse is then moved so that the radar range becomes greater and greater and the radar tracks the artificial echo and indicates an incorrect range: the artificial echo "steals" the range. At a safe range interval, the jammer is switched off, the echo is no longer present in the range gate, and the radar has lost the target.

Leading edge tracking was evolved to defeat the effects of ground reflections and range gate stealers. The leading edge of the echo arrives first over the shortest path. If the edge is differentiated, the positive part of the pulse may be used for early and late gates. A much larger pulse is then required to steal the gate.

13.1.5 Extracting the Doppler frequency

When a tunable Doppler frequency filter is used, it too must follow the Doppler frequency. This frequency or radial velocity should have had the ambiguities removed (see Section 11.2.4.5) and is part of the plot message.

13.2 SECTOR SCAN RADARS

Precision approach radars are still used at some airfields as an aid to landing in bad visibility for aircraft without functioning landing aids. Two radars are used: one has an antenna that scans a small azimuth sector, and the other in elevation. An approach controller watches a B-scope for azimuth and a range-height indicator for elevation and is able to guide the pilot along the correct glide path for landing.

The other use for such radars is the nodding height finder radar used to estimate the height of aircraft above sea level (ASL). The height finding function has mostly passed to three-dimensional surveillance radars in military radar systems. Radar height finding is not necessary for civil aircraft, because they report their altimeter height at each secondary radar mode C interrogation.

The most common form of height finder was a nodding antenna that scanned up and down. The antenna pattern is shaped like a beaver's tail, narrow in elevation and broader in azimuth. Height requests came from the air controller who sent the azimuth and range information to the height finder radar. The azimuth information was used to turn the antenna towards the aircraft and the range information to indicate the range to the height finder operator on his range-height display. The height finder operator had to find the aircraft echo on his display, move the height line over it, and transmit the height information by pressing a button. A number of circuits were developed for driving the vertical deflection of the display to include elevation angle, the effects of the earth's curvature, and atmospheric refraction. Commonly, the constant k atmosphere (see Section 6.2.2) was used and good results were obtained if the k factor was calculated from the atmospheric measurements or forecast. Some range-height indicators used an exponential atmosphere model. Nodding height finders are slow and only 20 or more height measurements per minute are possible. Automatic height extractors reduced the load on height operators until they were superseded by three-dimensional search radars, which measure the height of almost all aircraft each scan.

The accuracy of the height measured by a radar depends also on the atmospheric model used. Air traffic control uses standard pressure altimeter height for heights above a transition level. A table provided by meteorologists must be used to convert between the two types of height measurements.

13.3 FAST SCANNING RADARS

Increasing requirements for fire control radars to be able to set weapons onto a number of targets quasi-simultaneously led to the development of radars with electrically steered arrays for fire control. The beam may be steered in azimuth and elevation using phase shifters, although more economical radars have used frequency scanning in one direction. For many applications a mechanically rotating antenna in azimuth with phased array or frequency scanning in elevation is adequate.

These radars measure the positions of all the targets in turn fast enough to keep the position information current and to allow accurate targeting of the weapons that they control. The approximate positions of objects seen by the radar in a surveillance mode or from an external source are held in a memory. When the antenna beam is due to pass by the target, monopulse reception is used to give accurate angle information. Once one object has been measured, the antenna beam is moved to the next. These are relatively short range radars, so that the amount of time spent on tracking is small which allows an interleaved surveillance mode.

The combined surveillance and tracking radar is now preferred for multifunction radars and radars used for artillery and mortar location.

13.4 SURVEILLANCE RADARS

For a long time, operators watching plan position indicators (PPI) marked the positions of the echoes with grease pencils and wrote the flight information, such as identity and height, beside the track. As operators were required to handle more aircraft, the manual detection and tracking work had to be mechanized, which led to the development of extractors. Extractors extract the positions of the clumps of echo signals from the surrounding empty space and present them as plot data to a tracking function. There are many types of plot messages. An example of the information content is given in Table 13.1.

Table 13.1

Examples of the information in a plot message

Range	Amplitude
Azimuth angle	Quality
Elevation angle or height	Detection time
Doppler frequency or radial velocity	

13.4.1 Binary detection decisions

A run of echoes from a rotating surveillance radar generally occupies the same range cell over a small sector of azimuth angles. Such a run of echoes is shown in Figure 13.8 along a range ring. The echoes that exceed the threshold are shown outside. The azimuth angles between the threshold decisions are measured in azimuth change pulses, and there is mostly no relationship between the pulse repetition frequency and that of the azimuth change pulses. Additionally, staggered pulse repetition frequencies may be used.

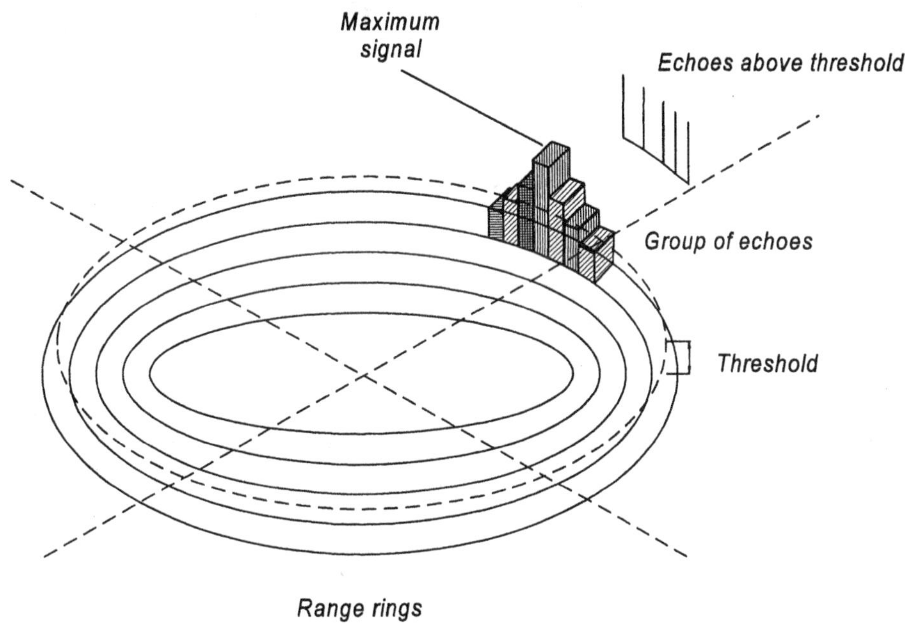

Figure 13.8 A group of echoes at the same range.

The number of echoes in a run depends on the echo strength and the antenna pattern and is variable. A clump of echoes is recognized by an m- out-of-*n* threshold, often 8 out of 10. As an example, if the threshold is set to give a 13.6% false alarm rate (3 dB above Rayleigh rms noise), the probability of eight consecutive pulses above the threshold is $(0.136)^8 = 1.170 \times 10^{-7}$. With 8 out of 10, there are $10!/(8! \times 2!) = 45$ chances, so that the final false plot rate is 5.267×10^{6}. When adjusting such second thresholds by hand there must be a comparatively large number of pulses at the input.

Once the threshold has been crossed, the starting azimuth is noted. When the threshold crossings cease, this point is the ending azimuth. Near the end of the maximum range of the radar a number of echoes in the run will not exceed the threshold and cause an early end of run. For this reason, the end criterion is often half the start criterion so that these "holes" are covered. The average of the starting and ending azimuths is taken to be the center azimuth, which is the same as an operator judging the center of an echo on his display, called beam splitting. Note that this gives an additional azimuth shift.

There is a loss between 1.5 dB and 1.8 dB compared to video integration associated with this form of detection [1, p. 75]. The accuracy using this form of beam splitting is said to be 20% worse than the optimum given in Section 13.5 [2] and approaches the rms error caused by the angle between the transmitter pulses $\Delta\theta$, namely, $\Delta\theta/\sqrt{12}$.

13.4.2 Maximum signal

With scanning radars using narrow beams, which have a small number of hits (for example, three), the central pulse is noticeably larger than its neighboring pulses. The highest central pulse determines the azimuth. The angle error is $\Delta\theta/\sqrt{12}$.

13.4.3 Center of gravity or centroid

Moving target detector radars commonly provide groups of two, maybe three, threshold crossings with spacings of half a beamwidth. The only way to estimate the true azimuth is to find the position of the center of gravity, or centroid, of the two or three pulses. This is shown in Figure 13.9.

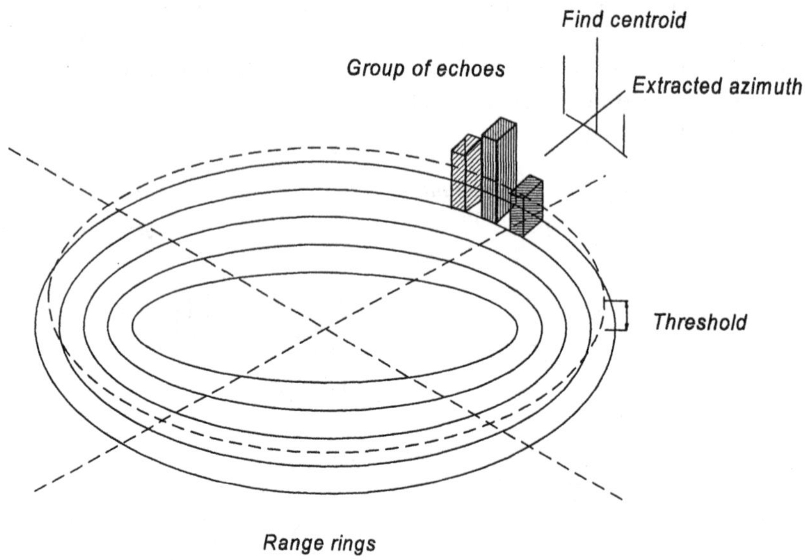

Figure 13.9 Centroiding of threshold crossings to estimate azimuth.

The expression to find the center of gravity, or centroid, is

$$\mathit{Centroided\ azimuth} = \frac{\sum_{1}^{n} \mathit{Azimuth}_n \times \mathit{Amplitude}_n}{\sum_{1}^{n} \mathit{Amplitude}_n} \tag{13.6}$$

In modern radars, which often use oversampling, centroiding is also performed in the range dimension to increase the range accuracy. In moving target detector radars, the Doppler frequency may be reported as the amplitudes of the signals present in the filters. Centroiding estimates the Doppler frequency, which is often ambiguous. Methods of resolving these ambiguities are given in Section 11.2.4.5.

13.4.4 Monopulse angle measurement with search radars

Three-dimensional stacked-beam radars use monopulse measurement for the elevation angle, as shown in Figure 13.10. Commonly, logarithmic signal processing is used with a lookup table to extract the elevation angle. The lookup table is created using data from measurements on an antenna test range.

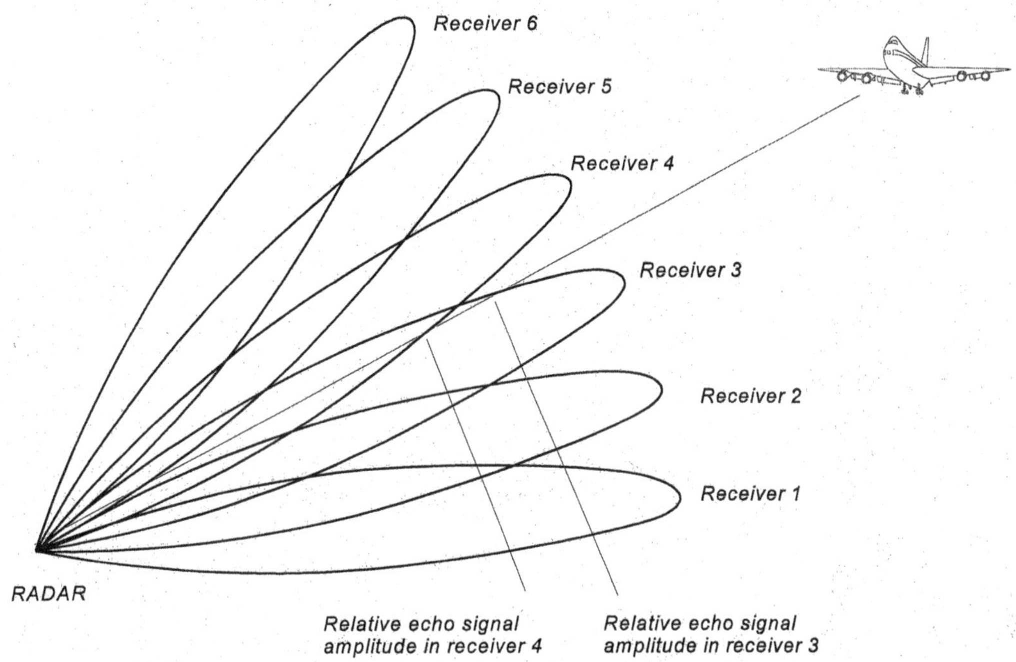

Figure 13.10 Monopulse elevation angle measurement with a stacked-beam radar.

A number of primary three-dimensional radars use mechanical scanning in azimuth and phased array scanning in elevation. They use low pulse repetition frequencies for longer ranges and higher pulse repetition for shorter ranges, where a moving target indicator processor is used. There are fewer long range pulses and often only a single echo returned, which increases the azimuth error, and is the reason that some radars use monopulse reception to increase azimuth accuracy.

The problem of missing replies is relatively common in secondary radar systems. The transponder in an aircraft has a dead time of up to 120 μs [3] after it has replied with a maximum reply frequency of 1 200 per second. In dense radar environments, the aircraft will be interrogated by a large number of secondary radars, and the runs of replies will contain a large number of holes. Holes in the runs of echoes or replies will give azimuth errors when the procedure in Section 13.2.1 is used. Monopulse reception allows the exact azimuth of each reply to be estimated instead of the group, and the individual azimuth measurements may then be averaged. Monopulse azimuth angle measurement has solved the azimuth accuracy problem, allows much lower interrogation rates, and is part of the Mode S recommendations in [3].

13.5 ACCURACY

The errors in the measurement of range were first treated by Woodward [4] who gave the relationship between the time (or range) standard deviation, σ_τ, and the second moment of bandwidth, β, in terms of the numerical energy ratio R.

$$\sigma_\tau = \frac{1}{\beta\sqrt{R}} \tag{13.7}$$

It must be noted that (13.7) is of the same form as the expression for the standard error of the mean in statistics. This expression is limited to the larger values of R generally necessary for detection.

The second moment, or the root mean square bandwidth of the signal spectrum $A(f)$ is given by

$$\beta^2 = \frac{\int_{-\infty}^{+\infty} (2\pi f)^2 \, |A(f)|^2 \, df}{\int_{-\infty}^{+\infty} |A(f)|^2 \, df} \tag{13.8}$$

Skolnik [5] extended the ideas behind (13.7) and (13.8) to angle errors and Barton and Ward [6] to errors generally. Table 13.2 shows the extension to the second moment, or root mean square, antenna width and time duration. Tapering functions used for antennas and signal processing worsen the accuracies, and the factors are given in Appendix B.

Table 13.2
Root mean square (rms) apertures, bandwidths, and durations without tapering

Component	Equation for standard deviation	Parameter	
Angle, θ radians	$\sigma_\theta = \dfrac{\lambda}{\mathcal{L}\sqrt{R}}$	$\dfrac{\mathcal{L}}{\lambda}$	Number of wavelengths for the rms aperture
Delay time, t_d seconds	$\sigma_t = \dfrac{1}{\beta\sqrt{R}}$	β	rms bandwidth
Frequency, f Hz	$\sigma_f = \dfrac{1}{\alpha\sqrt{R}}$	α	rms duration of measurement

For uniform illumination or no tapering, the slope of the difference curve is

$$K_0 = \frac{\mathcal{L}_0}{\lambda} = \frac{\pi}{\sqrt{3}} \frac{w}{\lambda} = 1.8138 \frac{w}{\lambda} \tag{13.9}$$

Figure 13.11 shows a Gaussian shape of standard deviation unity, which represents the tip of a pulse after a matching filter or the tip of an antenna beam pattern. If, instead of early and late gates, two narrow gates are used to sample the amplitudes to find the center, then as the interval between the gates approaches zero, the difference signal tends to the derivative of the Gaussian curve.

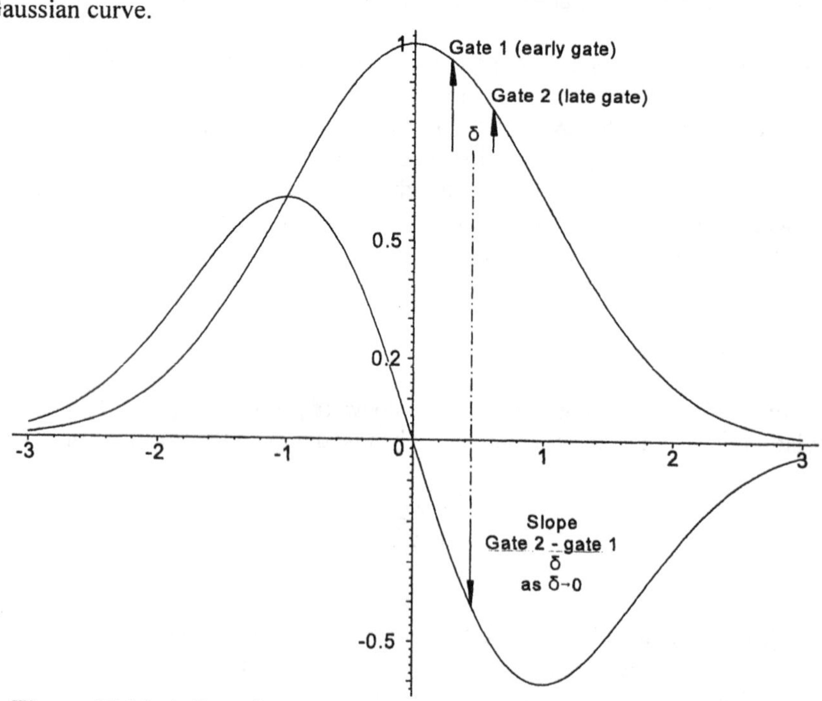

Figure 13.11 A Gaussian curve with sampling gates and its derivative.

Figure 13.12 shows the case of a Gaussian antenna pattern with returning echo signals, namely, the bandwidth of the samples is much greater than the bandwidth of the Gaussian curve. Figure 13.13 shows the case where the noise bandwidth is very similar to the width of the Gaussian curve. In both cases, the noise on the "difference" signal reflected onto the abscissa determines the standard deviation of the random error. The slope of the "difference" signal (the second differential of the pattern or waveform at the center) determines the sensitivity to noise.

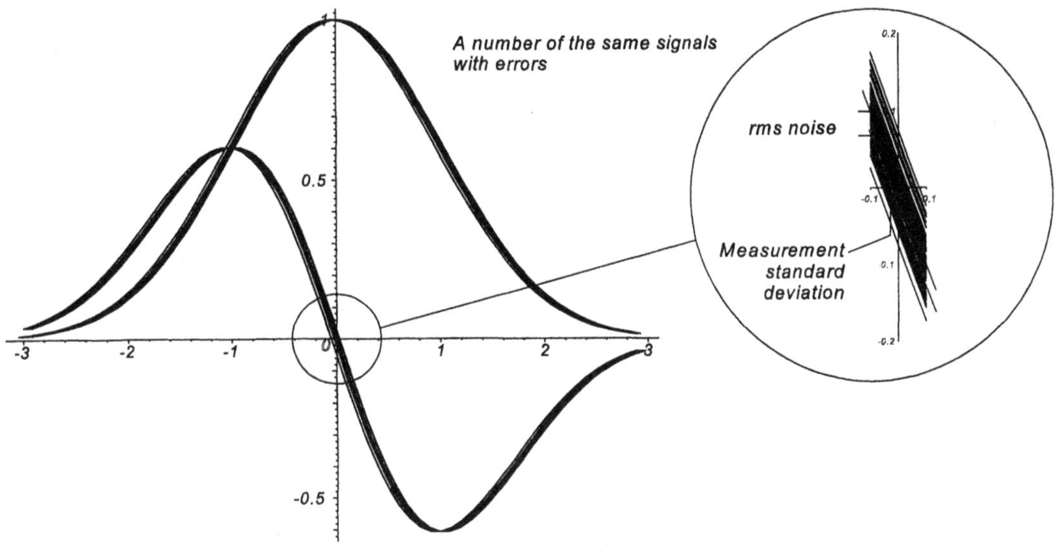

Figure 13.12 A Gaussian curve with noise of the same bandwidth with the difference curve.

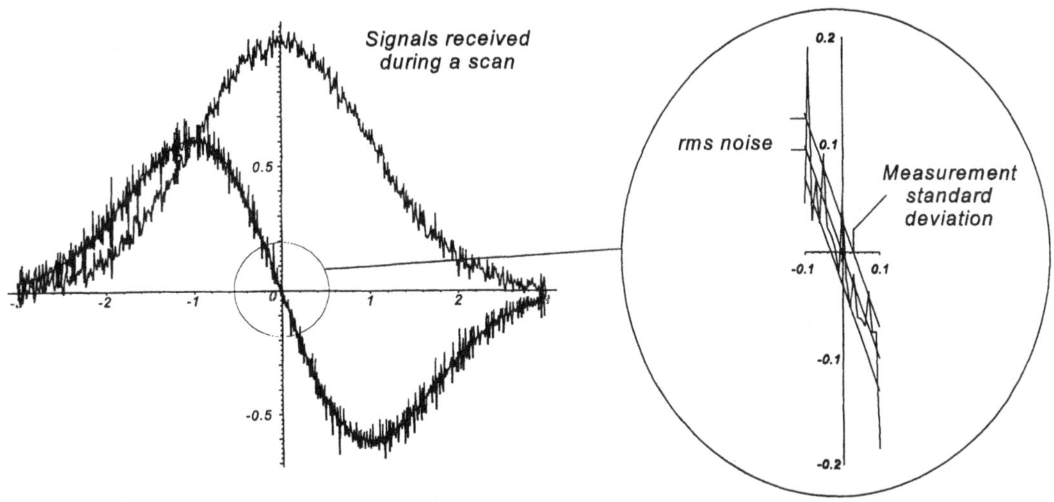

Figure 13.13 A Gaussian curve with wideband added noise and the difference curve.

The energy ratio R is the sum of the energy received during the duration of the measurement divided by the noise density. For the matched filter case,

$$R = \frac{2E}{N_0} = 2n\,\frac{S}{N} \tag{13.10}$$

Appendix B gives the relative slopes, K_r, defined as the true slope $K = K_0 K_r$.

13.5.1 Angular accuracies and root mean square aperture

The general principle for accuracy can be applied to common forms of radar. Search or surveillance radars, monopulse radars, and conical scan tracking radars have different types of losses.

13.5.1.1 Search radar

There are differences in the losses for primary and secondary surveillance radars. A secondary surveillance radar receives replies from an omnidirectional antenna in an aircraft. The main losses are the one-way beam shape or pattern loss, L_{p1}, which reduces the energy ratio from that on the axis R_m.

$$\sigma_\theta = \frac{\lambda}{\mathcal{L}_s \sqrt{R}} \tag{13.11}$$

where σ_θ is the standard deviation of the azimuth error;
$\mathcal{L}_s$ is the rms aperture (from Appendix B);
R is the energy ratio, given by

$$R = \frac{R_m}{L_{p1}} = \frac{2n\left(\frac{S}{N}\right)_m}{L_{p1}} \tag{13.12}$$

The two-way beamwidth is reduced, and the beam shape losses are increased to L_{p2} when, as with primary radar, the antenna illuminates the object.

$$\sigma_\theta = \frac{\lambda}{\mathcal{L}_{s2} \sqrt{R}} \tag{13.13}$$

where R is

$$R = \frac{R_m}{L_{p2}} = \frac{2n\left(\frac{S}{N}\right)_m}{L_{p2}} \tag{13.14}$$

Generally, if $n(S/N)$ is greater than 25 (14 dB) the error will be less than $\theta_3/10$.

When frequency agility is used there is a fluctuation in the amplitude of each of the echoes, which gives an additional scintillation error, σ_s, given by [6, p. 180, Eq. 6.14]

$$\frac{\sigma_s}{\theta_3} = \frac{0.215}{\sqrt{n}} \tag{13.15}$$

13.5.1.2 Monopulse radar

The accuracy of monopulse angle estimation is given by [6, p. 2.27, Eq. 2.27]

$$\sigma_\theta = \frac{\lambda}{K_r \mathcal{L}_0 \sqrt{R_0}} \tag{13.16}$$

where K_r is the relative slope with tapering (see Appendix B);
$\mathcal{L}$ is the rms aperture with no tapering;
R_0 is the energy ratio of the sum pattern.

Monopulse radars estimate the angle for each pulse, so that they are not subject to scintillation error. Variations of angle caused by glint, or the movement of the centroid of the radar cross-section with angle, remain.

13.5.1.3 Conical scan radar

The conical scan radar scans around the object it observes, never illuminating it using the peak of the antenna pattern as with the monopulse sum pattern. The same applies to sequential lobing.

The scan pattern has the parameters beam width and offset angle, so that there is no simple expression for the tapered relative slope. On the axis of the cone of the scan, the signal voltage is reduced by a factor of $\sqrt{L_k}$. The azimuth and elevation error signals are produced cyclically, each occupying half the rotation, which halves the energy ratio for each channel increasing the error by $\sqrt{2}$. For a measured slope k_s, the error is

$$\sigma_\theta = \frac{1}{k_s}\sqrt{\frac{2}{R}} \tag{13.17}$$

Section 6.3.2 describes echo fluctuations during measurement, which add to the measurement error when the correlation time is shorter than that required to receive the signals from all the lobes concerned. The additional scintillation error with a number of independent samples, $n > 2$, is, as for frequency agility in the search radar case [1, p. 396],

$$\frac{\theta_s}{\theta_3} = \frac{0.215}{\sqrt{n}} \tag{13.18}$$

Equation (13.18) also applies to frequency scan radars with decorrelated returns (see Section 6.2.3.7).

13.5.2 Time accuracies and root mean square bandwidth

The expression for range accuracy was given in (13.7) and is repeated for completeness:

$$\sigma_t = \frac{1}{\beta\sqrt{2\frac{E}{N_0}}} \tag{13.19}$$

The values for β/B or $\beta\tau$ are the same as $\mathcal{L}_s/w$ given for the different signal spectrum shapes (antenna patterns) in Appendix B.

13.5.3 Frequency accuracy and root mean square signal duration

The accuracy of the estimate of frequency depends on the measurement duration, τ, and the weighting (or tapering) of the signal samples during this time.

$$\sigma_f = \frac{1}{\alpha \sqrt{2 \frac{E}{N_0}}} \tag{13.20}$$

where α is the root mean square time duration.

The normalized second moments of area or rms time duration, α/τ, with tapering are the same as those for $\mathcal{L}_s/w$ in Appendix B.

13.6 THE DISPLAY OF POSITION

Historically, before the time of digital computers, the position of the objects of interest that were often the targets of weapons systems were displayed on the screens of cathode ray tubes. The types of displays, as with frequency bands, were given code letters and may be divided into those for surveillance, those for approach, and those for the aiming of weapons. Live displays may be more difficult and take more time to interpret than synthetic symbols, but they avoid an airliner being mistaken for a fighter-bomber. In the era of digital processing, graphic displays of analogue signals can be very useful for understanding what a radar sees, its electromagnetic environment, any unusual happening that may be misinterpreted by automatic processing, and for diagnostics.

13.6.1 Displays used to measure range

When a pulse radar was first used to measure range, an oscillograph was used. This was the first type of display built into equipment and, being the first, was given the code name A-scope. Later, when a longer range display was necessary, a circular display was used. These types of displays are shown in Figure 13.14.

When the oscilloscope screens were too small, a way of increasing the length of the trace was to display range around the circumference of a circle, the J-scope. Both the A-scope and the J-scope needed a scale in order to be able to read off range with the left hand starting mark on the breakthrough signal from the transmitter pulse on the left of the display. A more accurate way to measure range, without parallax, was to brighten the spot over a small distance, often called a strobe, move the spot over the object of interest, and read off the range from a calibrated potentiometer. Phosphors in the cathode ray tubes at that time were not very bright so that a better way was to move a notch or a step that could be seen more easily to the echo signal of the object and then read off the range from scale on the potentiometer. Later the timebase around the strobe was expanded to provide a magnified view, as on modern oscilloscopes, and calibration signals (cal. pips) in yards, meters, or nautical miles were generated using a quartz crystals.

13.6.2 Displays for surveillance

Surveillance radars often rotate continuously, but others, such as fixed phased arrays, only scan over a sector. The plan position indicator (PPI) gives an exact geographical representation of the airspace around the radar and is shown in Figure 13.15(a) with ground clutter at the center and one aircraft echo near north. Range is measured from the center to the echo. The effect of the signal bandwidth on range performance has been discussed in Section 8.1.1.

Sector scan radars in azimuth often use the B-scope (Figure 13.15(b)) that magnifies the azimuth scale for nearer echoes. For heightfinder radars that nod in elevation the equivalent is the E-scope although the range-height indicator (RHI) is more common.

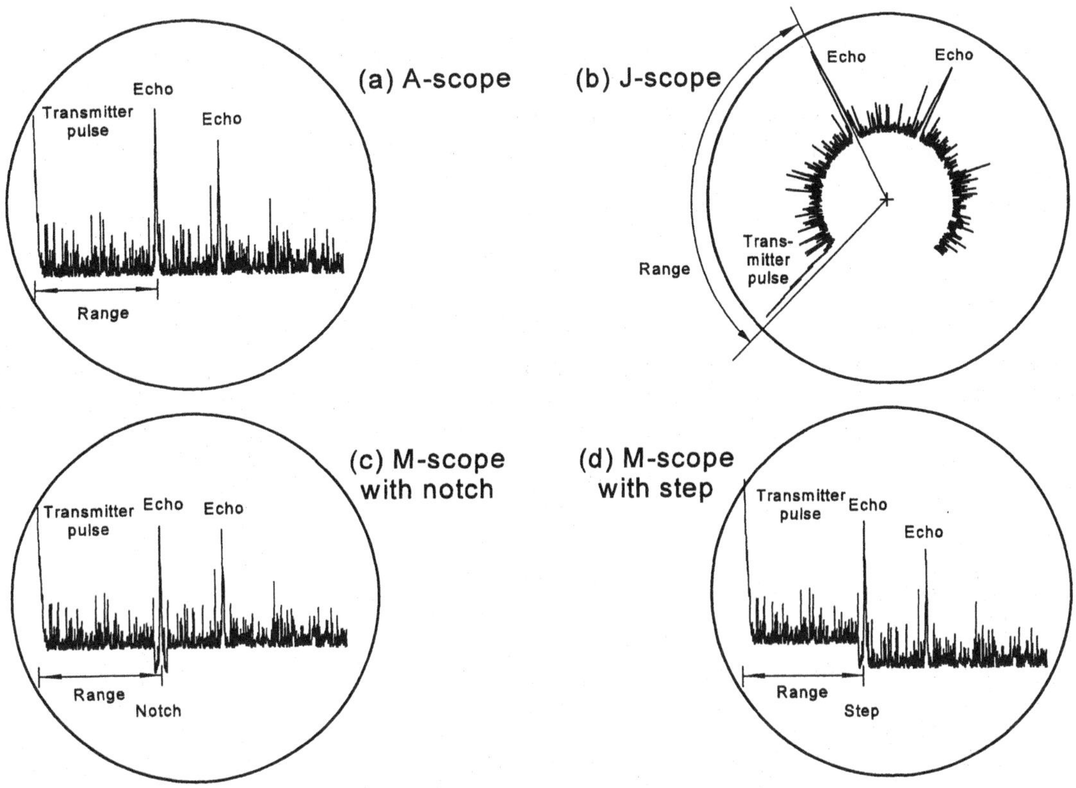

Figure 13.14 Displays for measuring range.

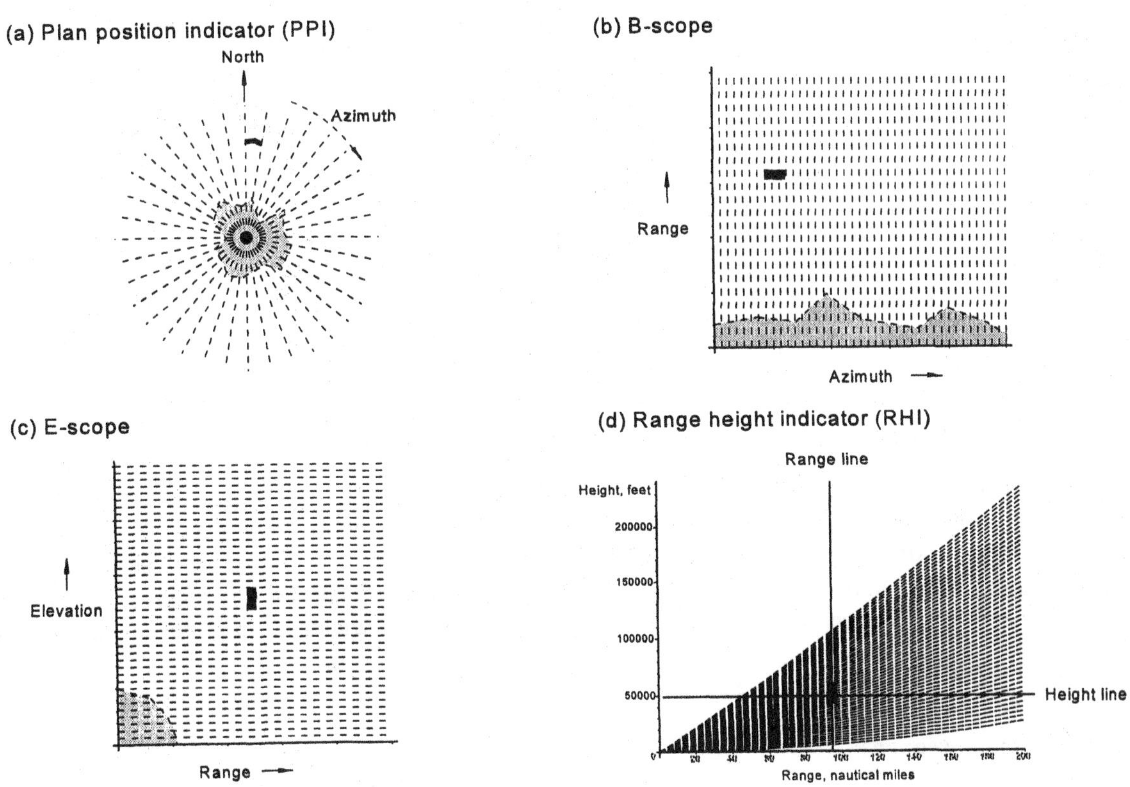

Figure 13.15 Displays used for surveillance.

Surveillance personnel use the aircraft's height as the third dimension, and this is not readily available from the E-scope in Figure 13.15(c). With the range-height indicator in Figure 13.15(d) the aircraft for height measurement is pointed to with the range line and a height line, controlled by a potentiometer with a height scale, is moved over the echo to measure its height. The sweeps in Figure 13.15(d) bend upwards to compensate for refraction and the curvature of the Earth, see Section 6.2.

Usually the displays may be expanded to allow detailed examination of a smaller area.

13.6.3 Airborne displays

Displays in aircraft were first used for airborne intercept (AI) radars and, as radars improved, surveillance radars were used. The primary need was to indicate the azimuth and elevation angle of the aircraft being sought or chased, range was not displayed as in the C-scope and F-scope in Figure 13.16 or was secondary as in the other types of display.

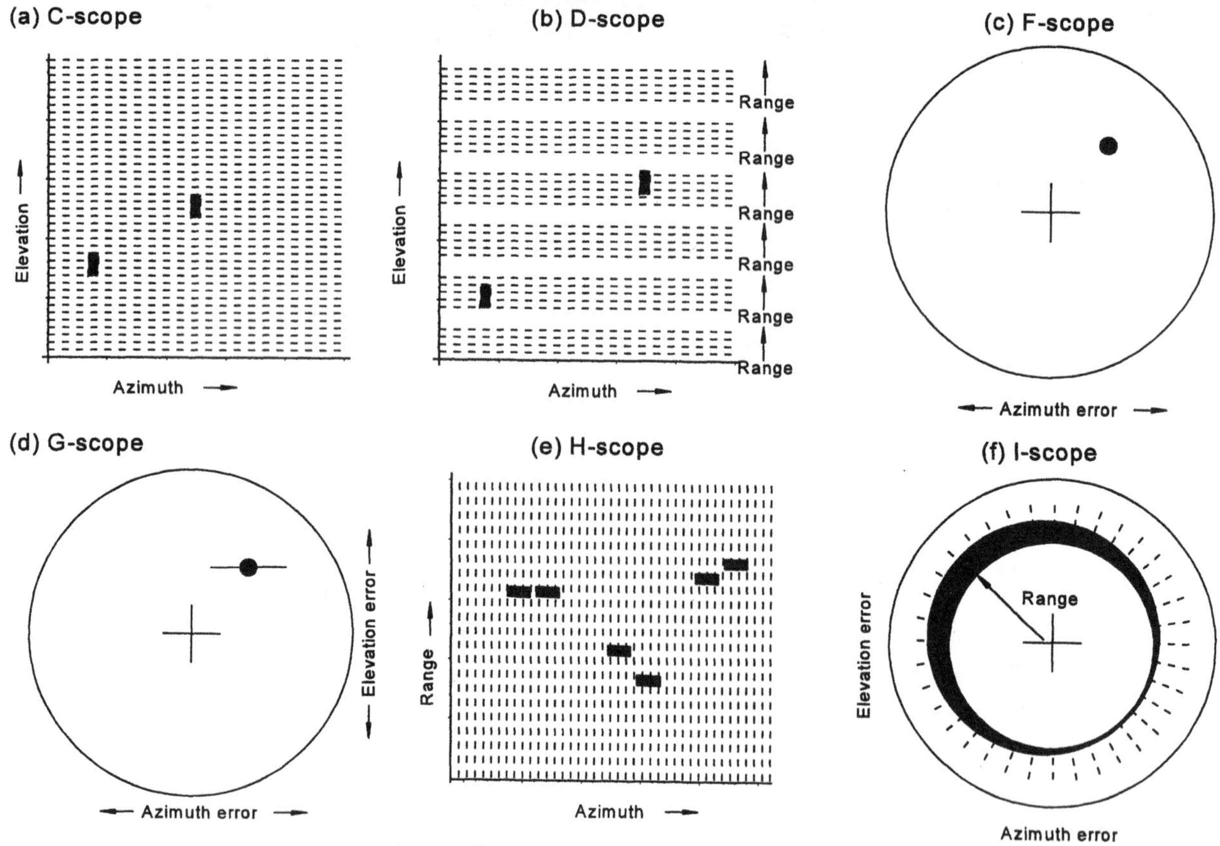

Figure 13.16 A number of displays used by airborne intercept radars.

Range was of secondary importance and first the C-scope display was divided into bands with the vertical position within the band giving coarse range. The G-scope was upgraded from the F-scope to show range by giving the representation of the echo wings of span inversely proportional to range, as in real life. On the H-scope each aircraft is represented by two dots: the left dot at the position of azimuth and range with the right dot showing relative elevation. Where conical scan is used, the I-scope shows the range as a ring with the thickest part giving the azimuth and elevation error.

13.6.4 Displays for aiming weapons

Until the coming of radar the aiming of anti-aircraft guns had the problem that, although optical directors could measure the angles accurately, is was hardly possible to measure range accurately enough. Bad weather or clouds obscured the target aircraft and made sighting for the guns impossible.

Early microwave radars, for example, the Radar AA No.3 Mk2, had separate operators to measure and follow in range, azimuth, and elevation. The lobing switch or conical scanner (see Figure 13.1) commutated between sweeps in real range and sweeps in delayed range so that the azimuth operator could turn the antenna until the two echo signals were equal and announce "on target", see Figure 13.17. Later, manually operated fire control radars used N-scopes where the target was selected by the range operator and it was obvious to the azimuth and elevation operators which echo had to be tracked. The elevation operator had his own separate display.

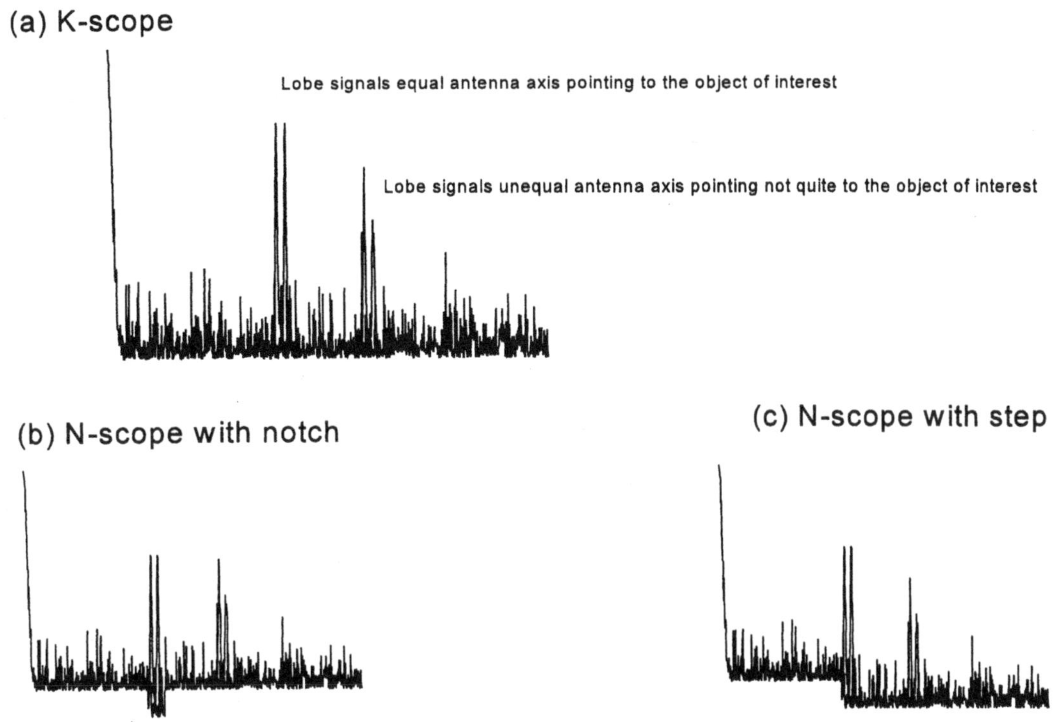

Figure 13.17 Displays for fire control radars.

A different display was the L-scope with the lobe signals from the echoes displayed on different sides of the base line, see Figure 13.18. It will be noticed that the noise on both sides occurs at different times and is not correlated.

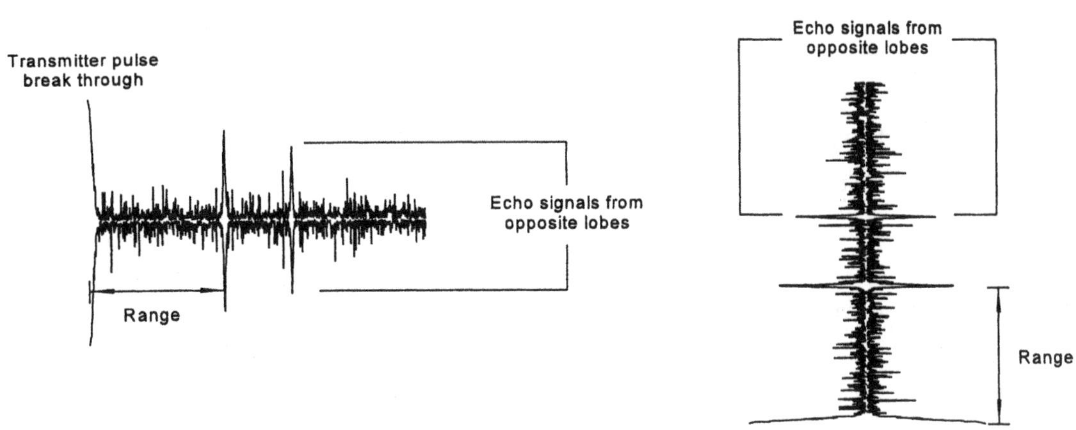

Figure 13.18 Displays for fire control radars (2).

13.6.5 Displays to indicate interference or jamming

Military radars need to be able to locate jammers and estimate the effects of active jamming by an adversary. The strobe display is an artificial line from the center pointing to the azimuth of the jammer and the strobes from a number of radars may be used to locate the jammer, see Figure 13.19(a). The amplitude of the jamming is indicated on the azimuth versus amplitude display (AVA), which also shows jamming in the sidelobes, or the effective range for a reference aircraft by a ring decreasing inwards in Figures 13.19(b) and (c), respectively.

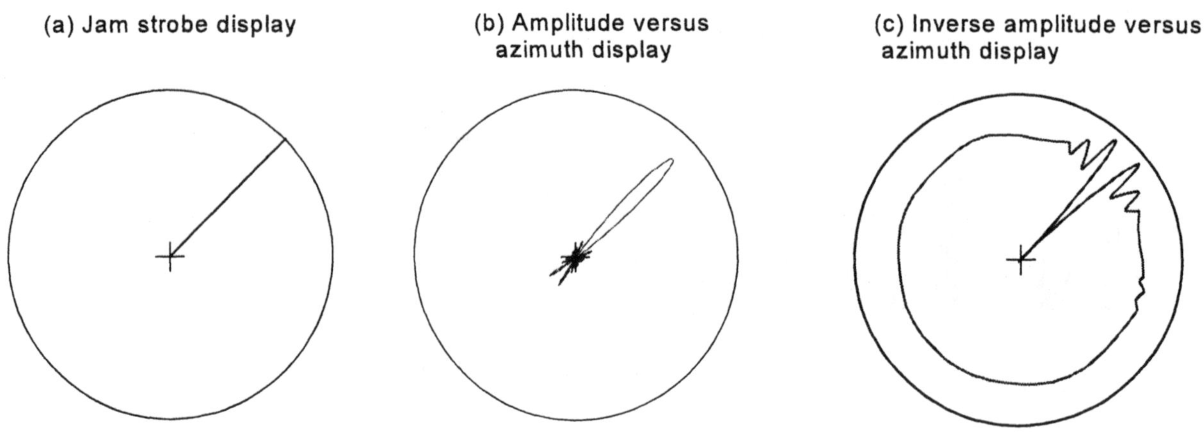

Figure 13.19 Displays used to display jamming signals.

13.6.6 Signals used for displays

The signals to drive the displays are listed in Table 13.3.

Table 13.3
Signals used for displays

Display	X deflection	Y deflection	Brightness signal
A-scope	Range sawtooth	Video signal	Strobe, when used
B-scope	Azimuth signal	Range sawtooth	Video signal
C-scope	Azimuth signal	Elevation signal	(Gated) video signal
D-scope	Azimuth signal	Elevation + range signal	Video signal
E-scope	Range sawtooth	Elevation signal	Video signal
F-scope	Azimuth signal	Elevation signal	(Gated) video signal
G-scope	Azimuth signal	Elevation signal	(Gated) video signal + wings
H-scope	Azimuth signal	Range sawtooth	(Gated) video signal + elevation spots
I-scope	Cosine conical scan	Sine conical scan	(Gated) video signal
J-scope	Cosine range voltage + video	Sine range voltage + video	(Strobe)
K-scope	First lobe range sawtooth Displaced second lobe range sawtooth	First lobe video Second lobe video	(Strobe)

Table 13.3 (continued)
Signals used for displays

Display	X deflection	Y deflection	Brightness signal
L-scope	Range sawtooth	First lobe Inverted second lobe	(Strobe)
M-scope	Range sawtooth	Video signal + notch or step	(Strobe)
N-scope	First lobe range sawtooth Displaced second lobe range sawtooth	First lobe video Second lobe video + notch or step	(Strobe)
Plan position indicator	Sine azimuth	Cosine azimuth	Video
Range height indicator	Range sawtooth	See Table 6.4	Video
Strobe	Sawtooth sine azimuth	Sawtooth cosine azimuth	
Amplitude versus azimuth and inverse	Sawtooth sine azimuth	Sawtooth cosine azimuth	Spot delayed by amplitude signal

13.7 FIGURES AFFECTING PERFORMANCE

Radar measurement errors may be divided into constant setting or bias errors and random errors. Adjustment can eliminate bias errors, and are typically those in Table 13.4.

Table 13.4
Bias errors

Range	Signal processing delay	Delay the zero range trigger
Azimuth	North setting error Signal processing delay	Realign radar to north Correct plot azimuths
Elevation	Antenna tilt error	Correct
Doppler frequency	Offset errors	Correct

Radar random errors or accuracies are commonly measured at higher signal-to-noise ratios (for example, 30 dB) in the middle of their range than for detection (10 dB) at maximum range. The greater signal-to-noise ratios improve the standard deviations of the accuracies by a factor of 10.

The figures for the standard deviations of the accuracies must include the uncompensated bias errors and the transmission errors. Analogue transmission systems (synchros, for example) have a transmission error. The variance for this error must be added to the measurement variance. Digital systems normally have an error rate too small to calculate. The main error is 1/12 of the least significant bit of additional variance on conversion to digital form.

$$\sigma_{after\ quant} = \sqrt{1 + \frac{x^2}{12}} \tag{13.21}$$

where x is the size of the least significant bit in terms of the error standard deviation.

Equation (13.21) is evaluated for steps from zero to two standard deviations in Figure 13.20. The total error for Chapter 14 is given by

$$\sigma_{Total} = \sqrt{\sigma^2_{random} + \sigma^2_{bias} + \sigma^2_{transmission}} \tag{13.22}$$

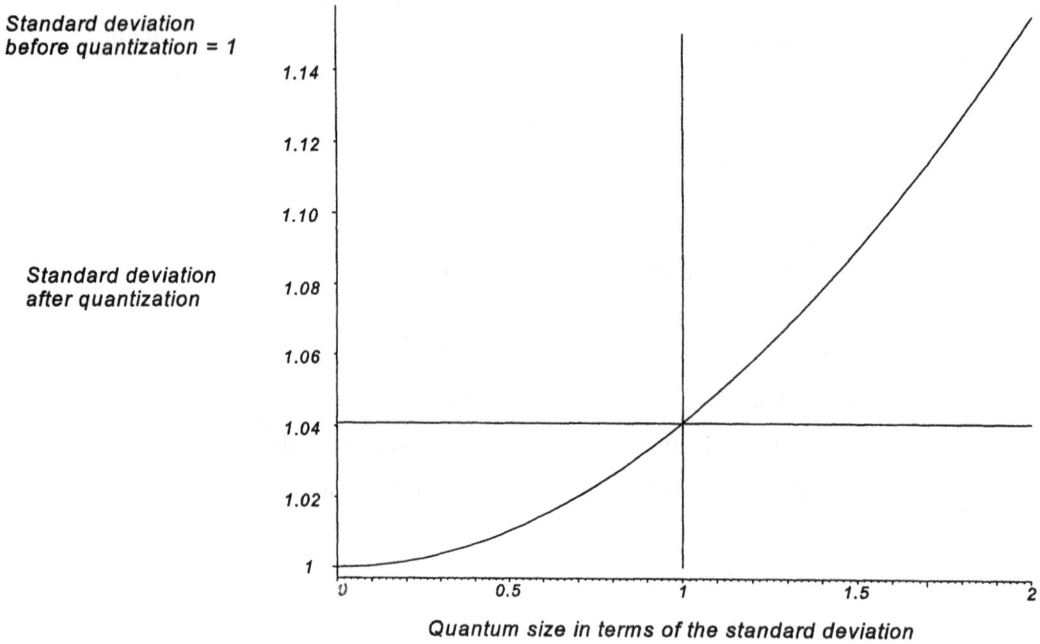

Figure 13.20 The increase in standard deviation caused by quantization.

REFERENCES

1. Barton, D. K., *Modern Radar System Analysis*, Norwood, Massachusetts: Artech House, 1988.
2. Trunk, G. V., in M. I. Skolnik, (ed.), *Radar Handbook*, 2nd ed., New York: McGraw-Hill, 1990.
3. *Aeronautical Telecommunications Annex 10 to the Convention on International Civil Aviation,* Vol. 1, Montreal: International Civil Aviation Organization, April 1985.
4. Woodward, P. M., *Probability and Information Theory with Applications to Radar*, 2nd ed., Oxford: Pergamon Press, 1964.
5. Skolnik, M. I., *Introduction to Radar Systems*, New York: McGraw-Hill, 1962.
6. Barton, D. K., and H. R. Ward, *Handbook of Radar Measurement*, Englewood Cliffs, New Jersey: Prentice-Hall, 1969.

Chapter 14

Performance

The performance of a radar describes how far it can see (range), how accurately it can measure the position of an object (accuracy), how well it can resolve objects near to each other (resolution), and how well it can see objects against a background (clutter improvement). Managing the performance of a radar during development is a matter of punctilious control of budgets to achieve the specified end values. In this chapter, performance is divided into:

- Range;
- Accuracy;
- Resolution;
- Stability.

Each of these characteristics is the final result of a budget, that is best controlled by using a spreadsheet on a computer.

14.1 RANGE

To cross the detection threshold in Chapter 12, the critical signal to background ratio, D_s, must be exceeded. The background in clear conditions is always thermal noise, kT W/Hz, plus

- Over ground clutter or in weather clutter (and chaff) the clutter signal or its residue (see Section 6.6);
- Interference or jamming signals present in the receiver (see Section 14.6).

Range in clear conditions neglecting siting is determined by the range equation

$$Range,\ R = \left(\frac{P_t\ \tau_t\ G_t\ G_r\ \sigma}{(4\ \pi)^3\ k\ T_n\ D_s\ L} \frac{c^2}{f^2} \right)^{\frac{1}{4}} \text{ m} \tag{14.1}$$

where the variables and their units are given in Table 14.1.

Table 14.1

Symbols and units used in (14.1)

Symbol	Parameter	Units	From Chapter
P_t	Transmitter peak power	W	3
τ_t	Transmitter pulse width, s	s	3
G_t	Gain of the antenna for transmission		5
G_r	Gain of the antenna for reception		5
σ	Scatterer cross section	m²	6
F	Propagation factor		6
c	Velocity of light $2.9979.10^8$ [1]	m/s	
f	Radar frequency	Hz	1
k	Boltzmann's constant, $1.380622.10^{-23}$ [1]	J/K	
T_n	Effective temperature at the receiver	K	7
D_s	Signal-to-noise ratio to detect the echo		12
L	Represents the combined losses		All

When siting is taken into account, the square of the propagation factor, F^2, must be inserted as a factor in the numerator inside the fourth root in (14.1).

Common forms of (14.1) take the constant factors out of the fourth root, and more practical units are used, namely,

$$Range,\ R = K_{range} \left(\frac{P_t(\text{kW})\ \tau_t(\mu\text{s})\ G_t\ G_r\ \sigma}{f^2(\text{MHz})\ k\ T_n\ D_s\ L} \right)^{\frac{1}{4}} \tag{14.2}$$

The constant K_{range} 239322. m;
239.322 km;
129.223 nautical miles.

For most radars the antenna gain on transmission and reception are the same, so that the factor G^2 may be used.

Maximum range is a fundamental aim in a radar design. During the development of a radar, some components may not achieve their specified characteristics, while others may exceed them. The same may also happen in operational systems caused by maintenance problems, which leads to the concept of a range budget. The losses are treated at the block level except for those between the transmitter and the antenna, shown in Figure 14.1.

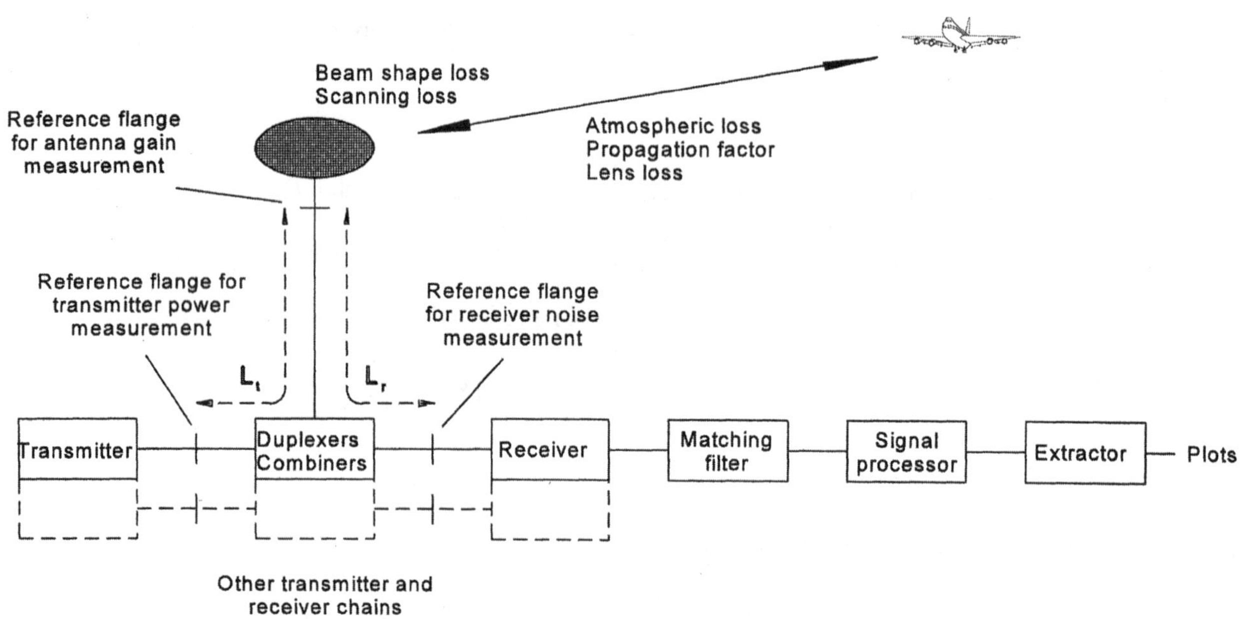

Figure 14.1 Quantities in the radar equation.

The only quantity remaining to be explained is the effective temperature at the receiver, T_n, which is given by [2, p. 25] and illustrated in Figure 14.2.

$$T_n = \frac{T_{output} - T_0}{L_r} + \overline{NF}\ T_0 \tag{14.3}$$

where T_{output} is the temperature at the output flange or connector of the antenna, K;
T_0 is the standard temperature, 290 K;
L_r is the loss between the antenna and the receiver;
$\overline{NF}$ is the noise factor of the receiver (Chapter 7).

One of the first generally accepted forms for the calculation of radar range to manage the range budget was published by Blake [3]. It had columns for the parameters and their values in decibels. Adding the decibels gave the calculation

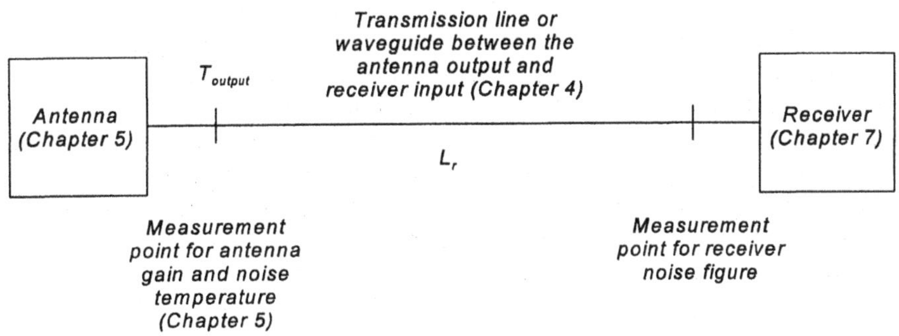

Figure 14.2 Components for the system noise temperature.

inside the fourth root in (14.2), dividing by 4 gave the fourth root, and adding the constant in decibels gave the range in decibels. Currently, computer spreadsheets constructed to resemble the Blake charts give the most practical way of managing the range budget. In operational systems, the calculated range budget may be backed up by quality control statistics, such as the range at which a known small aircraft is first observed.

The losses, L, often reduce radar range considerably and are shown in (14.1). A brief list is given in Table 14.2.

Table 14.2
Losses in a radar

Loss	Chapter or Section
Transmitter to antenna	3, 4
Antenna losses	5.7
Beam shape loss	5.8
Scanning loss	5.9
Atmospheric losses	6 and 6.2
Propagation factor	6.5
Lens loss	6.2.4
Dry atmosphere	6.2.6
Rain	6.2.6.2
Snow	6.2.6.3
Antenna to receiver	4, 7
Receiver noise factor	7
Matching filter	8
m	8.1
C_b	8.1.1
Tapering	Appendix B
Detection	9
Analogue-to-digital converter	10
Quantizing noise	10.4.1
Signal processing	11
- Constant false alarm rate losses	11.1.7
- Moving target indicator	
- Reduced number of hits	11.2.3.4
- Moving target detector	
- Tapering loss	Appendix B
Extraction loss	

14.2 ACCURACY

The fundamental errors or standard deviations of measurement have been discussed in Chapter 13. The distributions are assumed to be Gaussian, and the standard deviations are increased by other factors to give the final standard deviation.

There are two forms of error: bias error, which is constant, and random error, which depends heavily on signal-to-noise ratio. Bias errors can be compensated by offsets during calibration.

14.2.1 Bias errors

Normally radars operate at distances in the Fraunhofer region of their antennas, so that test reflectors to return signals for the measurement of angles must themselves be in the Fraunhofer region. When antennas are tested, often an optical sight is calibrated to be parallel with the axis of the beam. For two-dimensional surveillance radars, the sight may be in azimuth only with a separate scale to set the tilt angle and, if precautions are not taken, the mounting point for the sight may be used as a step and displaced. The only regular source of radiation for larger antennas is the sun, which may be used near sunrise and sunset for calculating azimuth. The movement of the sun is not regular and navigational or, better, surveyors' astronomical tables are needed to calculate the position of the sun. Above 15 degrees of elevation, the expression for the correction of atmospheric refraction at an elevation angle ϕ is more usable [4, 5].

$$\Delta\phi = N_s \cot \phi \, . \, 10^{-6} \quad \text{radians} \tag{14.4}$$

where ϕ is the elevation angle and N_s is the surface refractivity. Refractivity is given by Smith and Weintraub [5, p. 70] as

$$N = \frac{77.6\, p}{T} + \frac{3.73 \, . \, 10^5\, e}{T^2} \tag{14.5}$$

where p is the atmospheric pressure, hPa (mb);
T is the atmospheric temperature, K;
e is the partial pressure of water vapor, hPa (mb).

The course of the sun in azimuth and, when required, elevation is plotted together with the reports from the radar. There must be no weather fronts in the paths of the sun's rays. The azimuth correction may be made directly, and the elevation correction after correcting for atmospheric refraction.

A helicopter flying above a known position above the ground (possibly in the shadow of clutter) can be used to measure bias and random errors in range and azimuth. Random errors may be found from the statistical distributions of the plots around a regression line for fixed wing aircraft flying on straight courses.

14.2.2 Random errors

The fundamental random errors of signal measurement have been discussed in Chapter 13. The quantities are expressed as voltages, shaft positions, or digital words and are subject to additional distortion and quantization errors. After conversion to a digital form, it is assumed that no further errors occur.

The root mean square sum of the errors is

$$\sigma_{total} = \sqrt{\sigma_1^2 + \sigma_2^2 + \sigma_3^2 \, .. \, + \, .. \, \sigma_n^2} \tag{14.6}$$

Two types of errors common to all equipment are quantization error, which is $1/\sqrt{12}$ of the least significant bit representing the quantity (compare with quantization noise in Chapter 10), and round-off errors, which occur as a result of calculations.

14.2.2.1 Range errors

The additional components of range error are shown in Figure 14.3. The components are the jitter introduced by the distribution of the master trigger, the effects of amplification and modulation on the timing of the transmitter pulse (Section 3.6.4), any variations of the atmospheric refractive index from the assumed value (usually small), variations in the delay of the receiver (usually small), and the error of range estimation by the range gates or extractor (Chapter 13). Where the range gates follow the echo signal, there will be a servo lag, or range error, depending on the rate of change

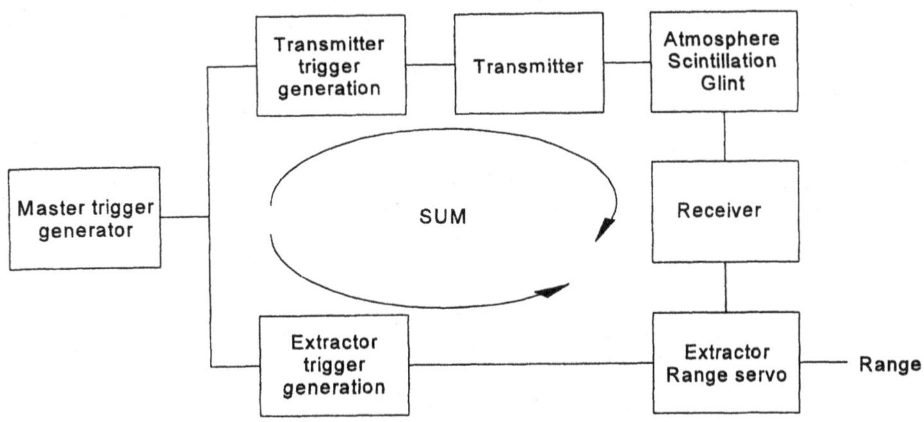

Figure 14.3 Additional components of range error.

of range.

Some radars use a pulse derived from the transmitter pulse as the range measurement trigger, which cuts out the error in the trigger distribution system.

Where the effective length of the transmitter pulse compares to the object being observed, the position of the center of reflection may move between measurements to give a scintillation error, with a change of radar frequency, or glint, with a change of aspect angle.

14.2.2.2 Angle errors

Whereas radars measure range in a similar manner, angle measurement is different for the following types of radar:

- Tracking radars with mechanically pointed antennas;
- Search radars that also use monopulse angle measurement to increase accuracy:
  - Primary radars with few returning echo pulses;
  - Secondary radar with mode S azimuth measurement improvement;
  - Stacked beam three-dimensional radars;
- Normal search radars;
- Phased array radars.

Radar antennas may be installed in aircraft, on vehicles and ships, or may be mounted on the tops of towers or masts. Aircraft and ships maintain a vertical reference, which is subject to error. Masts, especially those to raise mobile antennas above trees, are subject to twist and sway when the antenna is accelerated to point it in a different direction and with wind.

Tracking radars with mechanically pointed antennas

Errors in radar on the left of Figure 14.4(a) must be accounted for separately in both angle dimensions, normally azimuth and elevation. In addition to the errors caused by the signal-to-noise ratio (Chapter 13), there are errors caused by servo lag and the angle transmitter.

Search radars that use an auxiliary monopulse angle measurement

This type is shown in Figure 14.4(b), and describes a radar that sends too few pulses, or the echoes (or secondary radar replies) fluctuate too much to estimate the angle of a scatterer (or transponder) accurately. The source of coarse angle is the antenna, and the angle is read using mechanical angle transmitter, normally a digital encoder, which has bias and quantization errors. Examples of such radars are surveillance radars using frequency agility (scintillation) and secondary radars where a number of replies may be missing.

Stacked beam three-dimensional radars rotate in azimuth and may use conventional azimuth angle measurement.

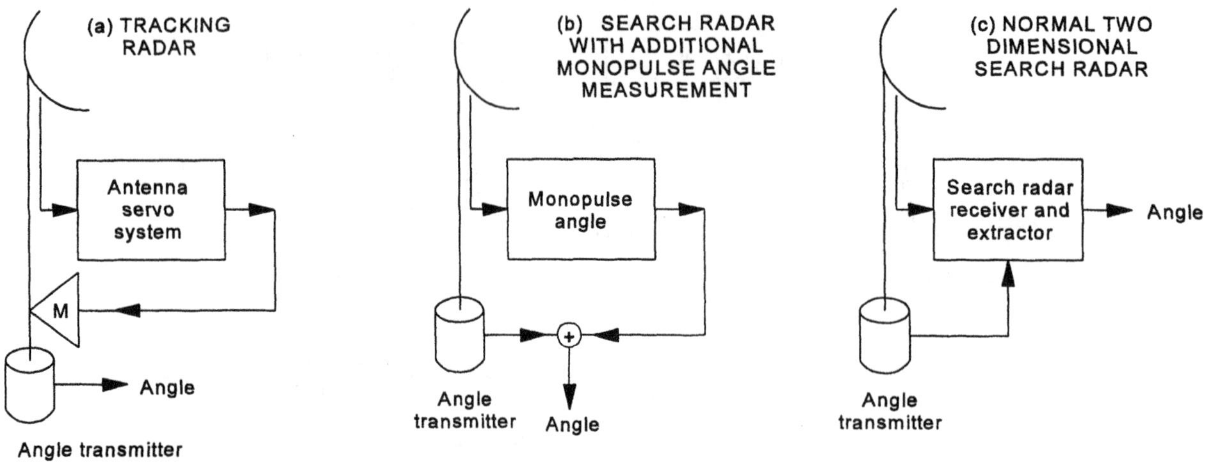

Figure 14.4 Error sources for angle measurements with mechanically scanned antennas.

The radar uses a number of reception beams and receivers to detect the echoes and provide coarse elevation angles. Monopulse angle estimation is used for the fine elevation angle. In addition to the signal-to-noise ratio errors (Chapter 13), there are errors in the positions of the beams in elevation and the interpolation curves normally stored in tables.

Normal search radars

A typical arrangement of a two-dimensional surveillance radar is shown in Figure 14.4(c). The error of the angle transmitter, usually a digital encoder, must be added to the signal-to-noise errors in Chapter 13.

Phased array radars

Phased array radars with digitally controlled beam steering components are able to point their beams only at certain fixed azimuth and elevation angles that are close enough to cover the volume of interest. Where fixed frequencies are used (no scintillation fluctuation), centroiding may be used. Otherwise the beam steering network must provide a difference output to allow monopulse fine angle estimation as shown in Figure 14.5.

A variation of this are three-dimensional radars that rotate in azimuth and use beam steering to scan in elevation.

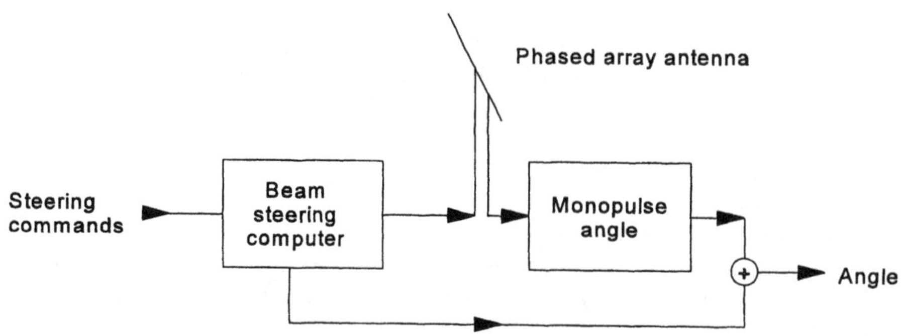

Figure 14.5 Angle error sources for a phased array radar.

14.2.2.3 Doppler frequency errors

The measurement of Doppler frequency was first the domain of tracking radars, then continuous wave radars, later pulse Doppler radars, and currently includes moving target detector (MTD) radars.

Continuous wave radars are used mostly for speed measurements of single objects, such as by the police for cars. The returning echo signals are detected by beating them with the outgoing signal, which gives a continuous wave at the

Doppler frequency. Counters can be used to measure the Doppler frequency. For example, for a 10 GHz radar the Doppler frequency for an object on a radial course is $2v/\lambda$, or 66.7 Hz/m/s. For a car traveling at 30 mph, or 13.83 m/s, the Doppler frequency is 892.2 Hz. One count error in an observation time of 0.1 seconds represents 0.6 m/s, 1.34 mph, or 2.16 km/h. When a large number of measurements are made, the standard deviation of the counting error is $1.34/\sqrt{12} = 0.386$ mph in addition to the signal-to-noise errors (Chapter 13). Detection using a Cartesian detector (Section 9.3.2) allows the sequence of the echoes ("positive" or "negative" Doppler frequency) to distinguish between positive and negative radial velocities.

Pulse Doppler radars are often used as radars in aircraft and, for short ranges, may use a pulse repetition frequency that is higher than the highest expected Doppler frequency. A tunable filter may be used to track and extract the Doppler frequency. The tuning of these filters is subject to all the errors in servo-mechanisms.

When the Doppler frequency is above the pulse repetition frequency, ambiguities occur. Such ambiguities may be resolved at extra cost with extra looks at different pulse repetition frequencies (pulse repetition frequency diversity), as discussed in Section 11.2.4.5; otherwise, gross errors can occur.

Moving target detector radars do not seek to provide accurate radial velocity measurement. The errors are the interpolation errors between two Doppler frequency filters.

14.3 RESOLUTION

Resolution has two meanings:

- For correlation it means:

  The necessary distance in any dimension between two objects seen by a radar, so that the contents of neighboring resolution cells are uncorrelated or statistically independent of each other.

- For resolution of two plots it means:

  The distance in any dimension between two objects so that they are reported as independent plots. An amount of empty space or a saddle point must be recognized by the extractor between the two objects.

Resolution can be defined for all four radar dimensions: range, azimuth angle, elevation angle, and Doppler frequency. Few radars offer resolution in all radar dimensions. Two-dimensional surveillance radars, for example, offer no resolution in elevation angle and often none in Doppler frequency.

Pulse radars are able to resolve two or more echoes in range, antenna characteristics allow resolution in angle, and Doppler radars may be designed to resolve objects by their Doppler frequencies. Noise on the echo signals makes the discrimination between one or two echoes subject to statistics so that specifications often state that "when two objects are x units apart they will be resolved y% of the time". The separation must be in at least one dimension.

Figure 14.6 shows the sum of two Gaussian voltage curves of half power width unity—$\exp(-2\ln 2\, x^2)$, see Section 2.9.2—with varying separations. Formerly there had to be at least one resolution cell that was completely empty before two objects were resolved. Better radar operators and more sophisticated extractors are able to detect saddle points as two signals separate from each other and declare two objects on the basis of saddle points. This is illustrated on the right of Figure 14.6 and shows a 3 dB saddle point from a separation of $\sqrt{3}$ pulse widths.

For other separations, y in units of the Gaussian curve width, the height of the saddle point (voltage) is given by

$$\textit{Ordinate of saddle point} = 2\exp\left(-\frac{1}{2}\ln(2)\, y^2\right) \tag{14.7}$$

The Gaussian curves start to separate when the spacing is $1/\sqrt{\ln 2}$ width units and a deeper and deeper saddle is formed with increased separation. This may be compared with the measurement of Gaussian noise on an oscilloscope in Section 9.5.1 and is illustrated in Figure 14.7.

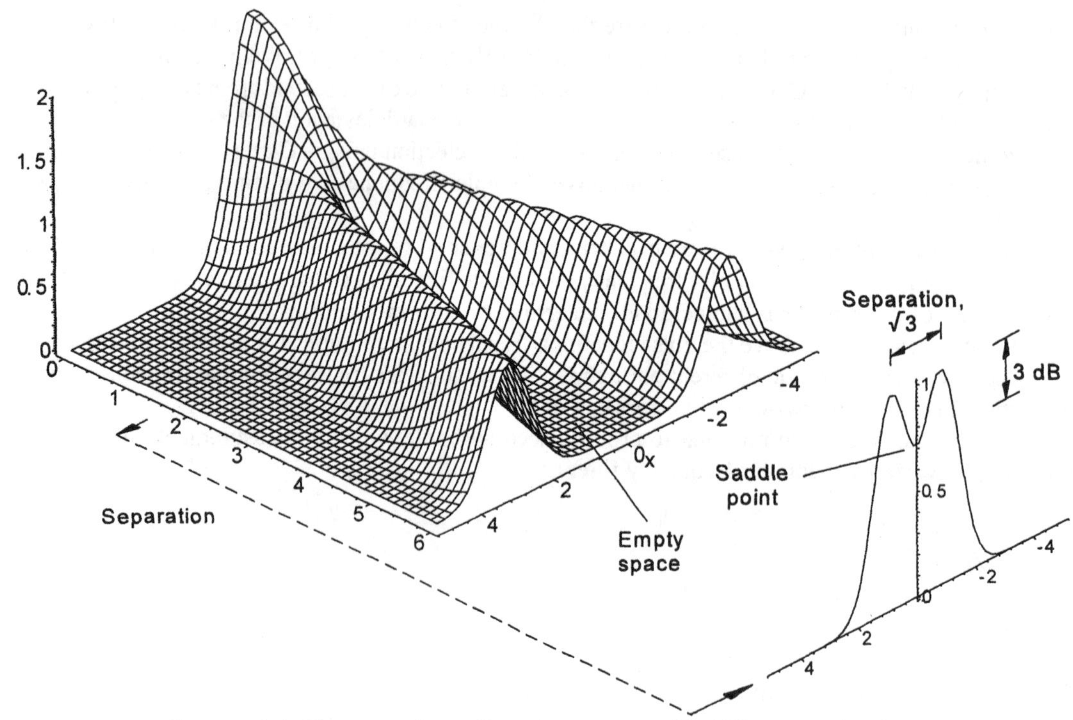

Figure 14.6 The sum of two Gaussian curves with different separations.

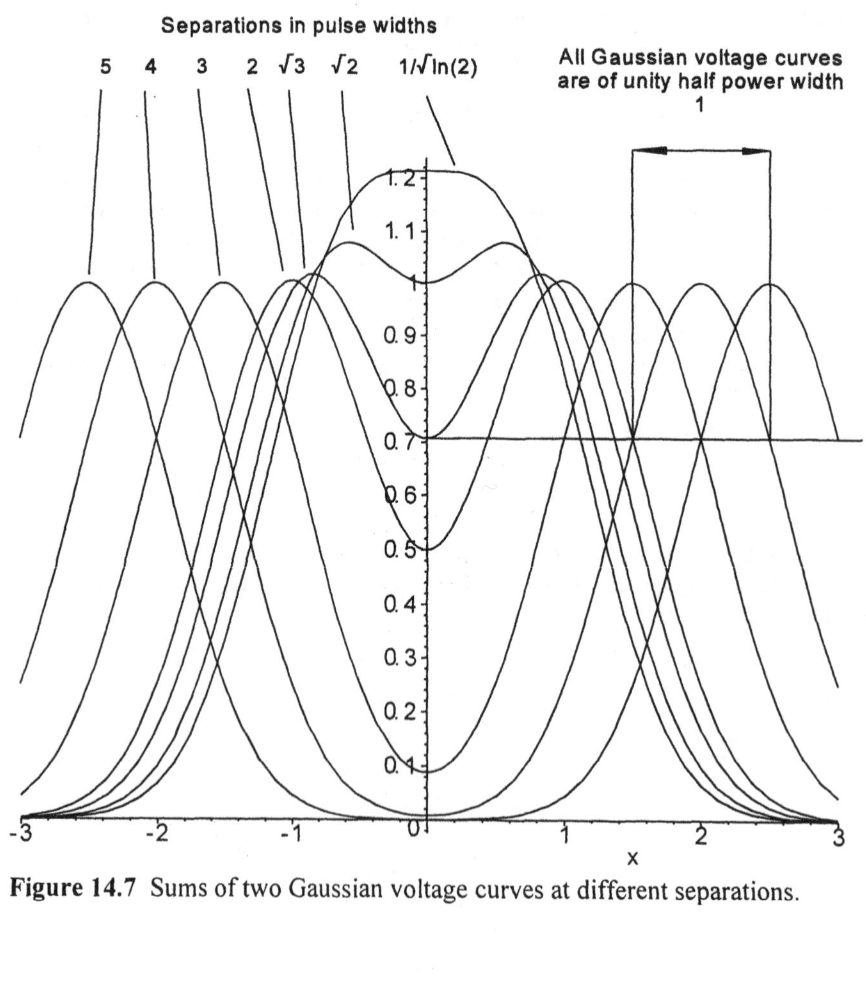

Figure 14.7 Sums of two Gaussian voltage curves at different separations.

Some representative values are given in Table 14.3.

Table 14.3

Separation distances and saddle points for two Gaussian curves

Separation	Saddle point ordinate	Peak to saddle point ratio, dB	Separation	Saddle point ordinate	Peak to saddle point ratio, dB
$1/\sqrt{\ln(2)}$	1.213	0.0	2	½	-6.021
$\sqrt{2}$	1.000	-0.628	3	0.08839	-21.072
$\sqrt{3}$	$1/\sqrt{2}$	-3.010	4	0.007812	-42.144
2	½	-6.021	5	0.000345	-69.237

The Gaussian shape may be assumed, with little error, between the half power points ($1/\sqrt{2}$ in voltage or 3 dB). Beyond the half power points, the actual shapes of the antenna patterns, spectra, or time weighting must be used.

This discussion has concerned two signals of equal strength, but signals of unequal strength are more common and require greater separation.

14.3.1 Resolution in range

Resolution in range depends on recognizing the gap or saddle point between two echoes. In radars that use pulse compression, the range sidelobes must be taken into account, as shown in Figure 14.8. Figure 14.8 shows two compressed echo signals of width unity (6 dB) separated by two units with a power ratio of 10. There may be problems in detecting the smaller pulse if it is less than or of a similar size to the sidelobes of the larger pulse. In receivers where limiting occurs, the frequency modulation capture effect may cause the smaller signal to be swamped, leaving only the uncovered portion for partial pulse compression.

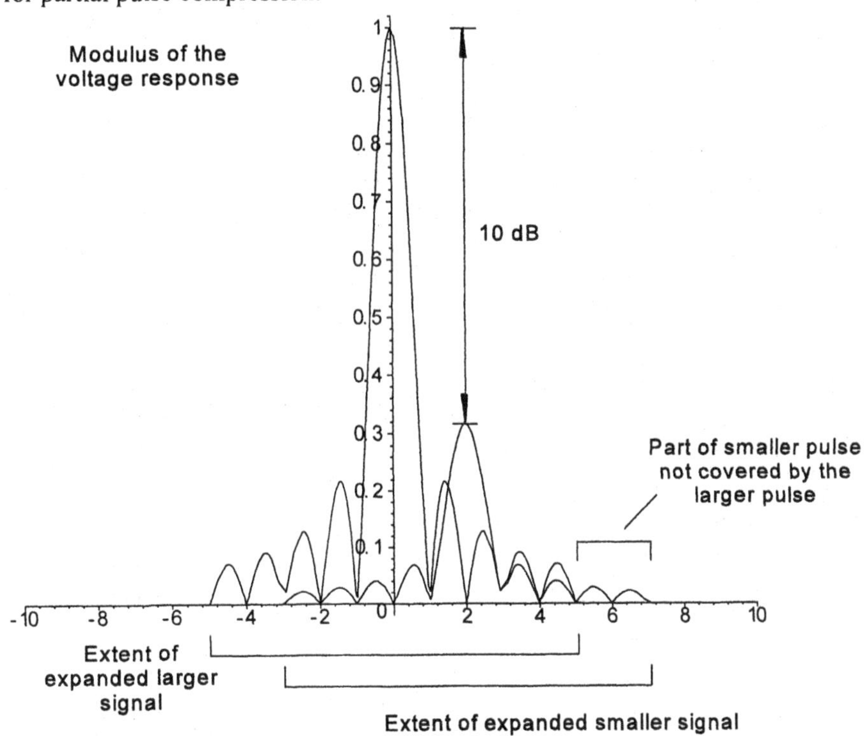

Figure 14.8 Two nearby echoes with 10 dB amplitude difference.

14.3.2 Resolution in azimuth angle

As with range, azimuth resolution depends primarily on the azimuth beamwidth and, for echoes with widely different amplitudes, on the azimuth sidelobe levels.

14.3.3 Resolution in elevation angle

Resolution in elevation angle depends on elevation beamwidth and sidelobe levels. Resolution in the elevation plane is further complicated by ground reflections.

14.3.4 Resolution in Doppler frequency

With a number of continuous wave speed measuring radars, the Doppler frequency is counted directly. Some use two-phase detection, which allows the direction of the radial movement to be found. The count is the average Doppler frequency (no resolution) of the signals received, which is why the operator must ensure that only one car, for example, is in the coverage of the radar.

Tracking radars can use a Doppler frequency filter to select an echo to follow. They have no resolution in Doppler frequency.

Many pulse-Doppler radars use Fourier transform filters which allow them to be able to resolve signals containing a number of Doppler frequencies. It was shown in Section 11.2.4.3 that the filter width is the pulse repetition frequency divided by the number of pulses processed, which is equivalent to the α term in (13.20). These radars use the amplitudes from adjacent filters to estimate the Doppler frequency so that the resolution has to be a number of these filters. There is a further problem with echoes having Doppler frequencies not at the center of the filters when the discrete Fourier transform is used, which was discussed in Section 11.2.4.4. (Remember that the fast Fourier transform is also a discrete Fourier transform.)

A number of radars that use moving target detector processing use Doppler frequency resolution only to reject clutter but not to resolve two other echoes in the same range cell. The ability to partition out the clutter depends on the cancellation ratio limitation brought about by the general stability of the radar.

14.4 STABILITY, THE CANCELLATION RATIO

The effectiveness of systems to select or reject echoes of fixed Doppler frequencies is determined by the stability of the whole radar. This stability may be measured by taking a large number of amplitude measurements of a stable scatterer, such as a large building, that occupies the full dynamic range of the receiver (most of the bits of the analogue-to-digital converter) and taking the discrete Fourier transform. The stability or cancellation ratio, *CR*, limitation is the power ratio of the clutter peak to the background of the spectrum. The factors are commonly quoted in decibels and are shown in Table 14.4.

Table 14.4

Stability factors for the clutter cancellation ratio

Stability	Value, dB	Reference, last section of
Trigger pulse jitter between the components		Chapter 1
Transmitter stability		Chapter 3
Waveguide stability		Chapter 4
Antenna stability		Chapter 5
Local oscillator stability		Chapter 7
Quantizing limit		Chapter 10

The stability or cancellation ratio limit for the entire radar is

$$\frac{1}{CR_{Total}} = \frac{1}{CR_{Transmitter}} + \frac{1}{CR_{Antenna}} + \frac{1}{CR_{Waveguide}} + \frac{1}{CR_{STALO}} + \frac{1}{CR_{quant}} \tag{14.8}$$

Very often, $\dfrac{1}{CR_{Antenna}} = \dfrac{1}{CR_{Waveguide}} = 0$

where all the cancellation ratio components must be converted from decibel to linear values.

14.5 INTERFERING OR JAMMING SIGNALS

Radar allows an air force commander to "see" the aircraft in the air space in his charge and to deploy his forces with greater effectiveness. Anti-aircraft weapons often use radar for sighting or guidance. Radars used by the army are used for battlefield surveillance at night or to look through fog or smoke and to locate mortars, artillery pieces, and rocket launchers by tracking their projectiles. Primitive radars in the noses of shells or rockets explode the warheads when they are close enough to cause damage (VT or variable time fuses). Once, in the author's experience, an old, unsuppressed electric drill jammed a radar until it was found and switched off.

Those being observed by a radar thus have an interest in making the data from it as unusable as possible and resort to passive jamming (chaff or window) or active jamming.

14.5.1 Chaff or window

Chaff or window is a cloud of half-wave dipoles cut to the wavelength of the radar to reflect the transmitted pulses. Chaff may be thrown out of an aircraft to provide a larger radar cross-section so that the radar and its operator are deceived. Clouds extending over tens of kilometers may be formed to increase clutter levels so much that aircraft flying through the cloud may not be detected.

In specifications, the chaff density is given in square meters per cubic meter, m^2/m^3, so that the radar cross-section and spectrum may be found as for rain (see Sections 6.6.6 and 6.6.7). When the chaff echoes are greater than the thermal noise the wanted echo must be greater by the factor D_s (see Chapter 12).

14.5.2 Active jamming

Active jamming consists of generating and transmitting electromagnetic waves to impede detection or to deceive a radar. In clear conditions, the thresholding stage sees the echo signal, thermal noise, and, if present, clutter residues.

When they arrive at the threshold stage, active jamming signals the noise background, requiring the threshold to be raised. Only stronger signals will be detected, thus reducing the effective range of the radar. Active jamming signals may come from a number of sources, such as an aircraft being tracked that does not wish to be a target for the weapons that the radar is controlling (called self-protection jamming), from an aircraft in the same formation (escort jamming), and from aircraft flying out of range of weapons providing barrage jamming.

The radar range equation (14.1) can be rewritten placing the background noise term at the end:

$$Range,\ R \quad = \quad \left(\frac{P_t\ \tau_t\ G_t\ G_r\ \sigma}{(4\ \pi)^3\ D_s\ L} \ \frac{c^2}{f^2} \ \frac{1}{k\ T_n} \right)^{\frac{1}{4}} \quad \text{m} \qquad (14.9)$$

The power received from a single jammer by the radar is

$$Jammer\ power\ density\ at\ radar\ receiver \quad = \quad \frac{E_j\ G_{rj}}{(4\ \pi)^2\ R_j^2\ L_r\ L_{aj}} \ \frac{c^2}{f^2} \quad \text{W/Hz} \qquad (14.10)$$

where E_j is the effective radiated power density of the jammer, W/Hz;
G_{rj} is the gain of the receiving antenna in the direction of the jammer;
R_j is the range of the jammer, m;
L_{aj} is the atmospheric loss between the jammer and the radar.

In a military situation, there will be a number of jammers, n, and the jamming power in the receiver must be much greater than the noise to be effective. The noise term in (14.9) is replaced by the sum of the jamming from (14.10):

$$\text{Range in active jamming, } R_x = \left(\frac{P_t\ \tau_t\ G_t\ G_r\ \sigma}{(4\ \pi)^3\ D_s\ L} \frac{c^2}{f^2} \sum_1^n \frac{(4\pi)^2\ R_{j,n}^2\ L_{aj,n}}{E_{rp,n}\ G_{rj,n}} \frac{f^2}{c^2} \right)^{\frac{1}{4}} \text{ m}$$

$$= \left(\frac{P_t\ \tau_t\ G_t\ G_r\ \sigma}{4\ \pi\ D_s\ L} \sum_1^n \frac{R_{j,n}^2\ L_{aj,n}}{E_{rp,n}\ G_{rj,n}} \right)^{\frac{1}{4}} \text{ m} \tag{14.11}$$

In more practical units, (14.11) becomes [2, pp. 34 et seq.]

$$\text{Range in active jamming, } R_x = K_j \left(\frac{P_t(\text{kW})\ \tau_t(\mu\text{s})\ G_t\ G_r\ \sigma}{4\ \pi\ D_s\ L} \sum_1^n \frac{R_{j,n}^2\ L_{aj,n}}{E_{rp,n}(\text{W/MHz})\ G_{rj,n}} \right)^{\frac{1}{4}} \text{ m} \tag{14.12}$$

The values for K_j for different range units are 94.449 m
0.094449 km
0.069403 nautical miles.

There are three common forms of jamming calculations: self-protection, escort, and barrage. Self-protection jamming is used by an aircraft to mask its return and deny range information that is important for anti-aircraft guns that fire shells with timed fuses. There is the danger for missile systems that can home on the jamming, that the visibility of the direction of the aircraft may be improved. The jammer range is the same as the aircraft so that (14.12) becomes

$$\text{Range for self-protection jamming, } R_{sp} = \left(\frac{P_t\ \tau_t\ G_t\ \sigma}{4\ \pi\ D_s\ L_j} \frac{1}{E_{rp}} \right)^{\frac{1}{2}} \text{ m} \tag{14.13}$$

In practical units,

$$\text{Range for self-protection jamming, } R_{sp} = K_{sp} \left(\frac{P_t(\text{kW})\ \tau_t(\mu\text{s})\ G_t\ \sigma}{D_s\ L_j} \frac{1}{E_{rp}(\text{W/MHz})} \right)^{\frac{1}{2}} \text{ m} \tag{14.14}$$

The values for K_{sp} for different range units are 8.9206 m;
0.0089206 km;
0.0048167 nautical miles.

Specialized aircraft carrying jammers may fly in formations of aircraft to deny range information to radars. If the formation is small enough to be completely in the main beam, the self-protection case applies. For larger formations, when the escort jammer is in the first sidelobe, the mutual screening case applies. G_{rj} is the gain of the first sidelobe and $n = 1$ in (14.11). The jamming from such jammers allows their direction to be determined.

Large jammers outside the range of weapons systems provide barrage jamming that enters the antenna via the distant sidelobes. In that case, there are a larger number of jammers and the antenna gain for jammers is the average gain for the distant sidelobes.

Active jamming may also illuminate chaff clouds to augment their effectiveness as a radar countermeasure.

14.5.3 Deception jamming

Deception jamming is more economical with power and, with some forms, ideally does not alert the radar operator. One form is a drogue towed behind an aircraft carrying a jammer so that a weapons system shoots at the drogue. Other forms are carried in the parent vehicle. Jammers that confuse surveillance radars produce extra signals that resemble echoes to distract operators from the real danger. Aircraft may produce signals imitating a whole formation so as to give the operators the task of finding which is the real echo. These types of deception are called spoofing.

Deception signals sent to weapons radars depend on the type of radar. It is possible to detect the conical scan of older weapons radars and signals may be generated to interfere with the azimuth and elevation tracking circuits so that the radar loses track. Monopulse radars are not susceptible to this form of deception, but they are often sensitive to signals with orthogonal polarization. Another form of deception is to simulate a stronger echo and have it steal the range gate or Doppler frequency gate so that the radar tracks the simulated echo. In all these cases the weapons radar data no longer represents the position of its target.

14.6 TABLES

These tables are not all inclusive but an example because there is always a problem in remembering to present all the data in a uniform way to perform standard calculations. For radars which have a number of modes of operation, these tables must be extended to cover each mode. Military requirements are not included here.

14.6.1 Basic radar requirements

A radar has to fulfill a purpose for which the main characteristics are to be found in Table 14.5.

Table 14.5

Basic characteristics

Characteristic	Units	Value	From section
Maximum range	Length	....	1.7
for a radar cross section of σ	m^2	....	1.7
Minimum range	Length	....	1.7 and 4.2.5
Azimuth coverage	Degrees	....	1.7
Elevation coverage	Degrees	....	1.7
or height coverage	m or ft	....	1.7
Doppler frequency or radial velocity range	Hz, m/s	....	1.7
Probability of false alarm, P_{fa}		....	1.7
Probability of detection, P_d		....	1.7
Dynamic range required	dB	....	1.7
Clutter cancellation ratio	dB	....	1.7
System stability	dB	....	1.7
Resolution in range	Length	....	1.7
Resolution in azimuth	Degrees	....	1.7
Resolution in elevation	Degrees	....	1.7
Accuracy in range, standard deviation	Length	....	1.7
Accuracy in azimuth, standard deviation	Degrees	....	1.7
Accuracy in elevation, standard deviation	Degrees	....	1.7
Accuracy in Doppler frequency, standard deviation	Hz	....	1.7
Refreshment or update rate	1/sec	....	1.7

Notes:
1. Ranges may be quoted in many units, from meters to nautical miles. "Length" has been used above for length units.
2. Elevation coverage is often defined as coverage within a number of defined points, as shown in Figure 14.9. The points may be defined in ground or slant range and height or elevation angle. Generally, a maximum elevation angle is given. The volume above the maximum elevation angle is the cone of silence.

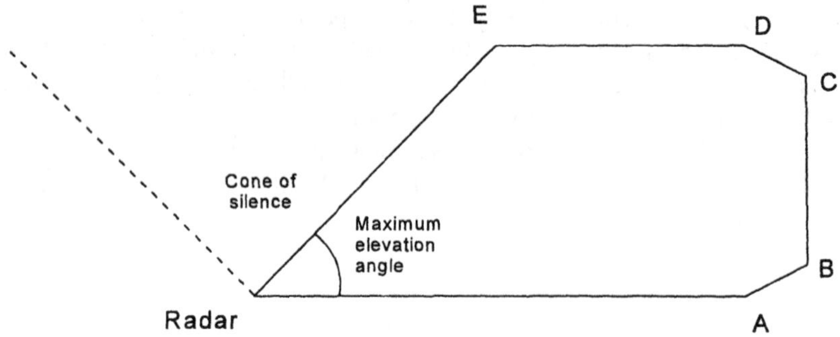

Figure 14.9 Elevation coverage.

Phased array radars have a coverage that is a part of a sphere, normally over an azimuth sector up to, or as near as possible to, the zenith. Figure 14.10 shows such a coverage over a spherical triangle.

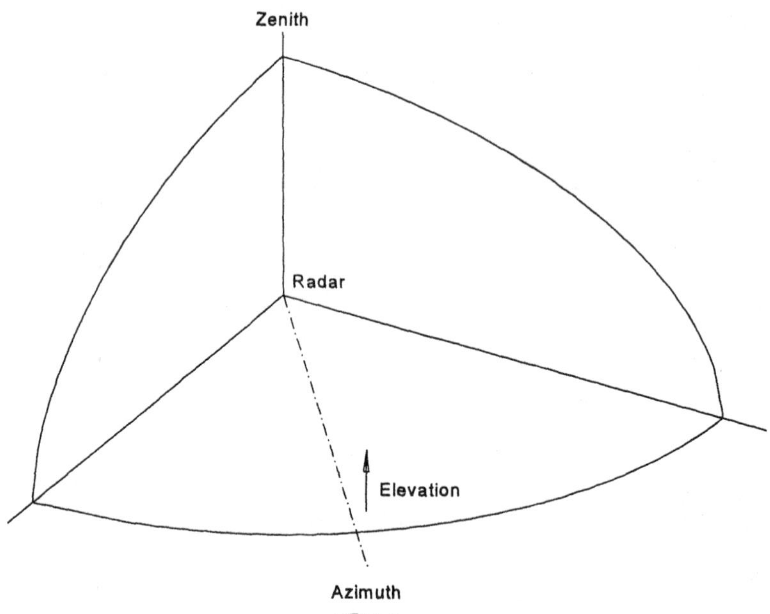

Figure 14.10 Coverage of a phased array radar.

The Fourier transform describing the antenna pattern is in terms of the sines of the azimuth and elevation angles, so an alternative way of defining the angles is to use the sines of the angles, or sine space. The transformations are shown in Table 14.6.

Table 14.6
Normal to sine space transformations

Angle		transforms to	Sine space
Degrees	Radians		
-90 to +90	-π/2 to π/2		-1 to +1
0 to 30	0 to π/6		0 to ½

14.6.2 Derived characteristics

The following values are derived directly from the values in the Table 14.7.

Table 14.7

Derived characteristics

Characteristic	Units	Value	Reference, chapter
Antenna scanning rate		....	1
Pulse repetition frequencies	Hz	....	1
Blind Doppler frequencies or speeds	Hz, m/s	....	11

14.6.3 Factors for calculating range

The main factors for the calculation of range are listed in Table 14.8.

Table 14.8

Factors to calculate range

Characteristic	Units	Value	Reference, Section
Eclipsing loss, L_{ec}	dB	....	1.7.1
Transmitter peak power, P_t	W	....	3.6.1
Transmitter pulse width, τ_t	s	....	3.6.1
Losses during transmitting, L_t	dB	....	4.41
Losses during reception, L_r	dB	....	4.41
Antenna gain during transmission, G_t	dB	....	5.14.1
Antenna gain during reception, G_r	dB	....	5.14.1
Radome loss, two way, L_{radome}	dB	....	5.14.1
Loss caused by circular polarization, L_{pol}	dB	....	5.14.1
Beam shape loss, L_{beam}	dB	....	5.14.1
Scanning loss, L_{scan}	dB	....	5.14.1
Noise temperature at antenna, T_{output}	K	....	5.14.1
Scatterer cross section, σ	m²	....	6.7.1
Echo loss with circular polarization	dB	....	6.7.1
Two-way attenuation in clear weather, L_{clear}	dB	....	6.7.1
Two-way attenuation in x mm/hour rain, L_{rain}	dB	....	6.7.1
Two-way attenuation in x mm/hour snow, L_{snow}	dB	....	6.7.1
Receiver noise figure, F	dB	....	7.6.1
Effective noise temperature at the receiver, T_n	K	....	14.1
Limiting level	dB	....	7.6.1
Matching filter loss, m	dB	....	8.7.1
Bandwidth loss, C_b	dB	....	8.7.1
Pulse compression filter loss, $L_{pulse\ comp}$	dB	....	8.7.1
Straddling loss, L_{strad}	dB	....	10.4.1
Range jitter loss, L_{jitter}	dB	....	10.4.1
Quantization loss, L_{quant}	dB	....	10.4.1
Signal processing losses, clear, $L_{sp\ clear}$	dB	....	11.5.1
Signal processing losses, clutter, $L_{sp\ clut}$	dB	....	11.5.1
CFAR losses, L_{CFAR}	dB	....	11.5.1
Signal-to-noise ratio for detection, D_x	dB	....	12.3
Signal-to-noise ratio with joint probabilities of detection	dB	....	12.5

Once the maximum range has been calculated without atmospheric loss, the loss for that range is included in a new calculation (see Section 6.2.6). Atmospheric loss gives a reduced range for which a new, lesser atmospheric loss is obtained. The iterative process is repeated until the range figure is inside the required tolerance.

Table 14.9
Losses for a clear atmosphere, rain, and snow

Characteristic	Units	Value	Reference, Section
Two-way attenuation in clear weather, L_{clear}	dB	....	6.7.1
Two-way attenuation in x mm/hour rain, L_{rain}	dB	....	6.7.1
Two-way attenuation in x mm/hour snow, L_{snow}	dB	....	6.7.1

14.6.4 Resolution

The main characteristics for resolution are shown in Table 14.10.

Table 14.10
Factors for radar resolution

Characteristic	Units	Value	Reference, Section
Azimuth beamwidth	Degrees	....	5.14.2
Elevation beamwidth	Degrees	....	5.14.2
Filtered pulse width	s	....	8.7.2
Dwell time	s	....	

14.6.5 Accuracy

The main figures to determine accuracy are given in Table 14.11.

Table 14.11
Figures for the determination of accuracy

Characteristic	Units	Value	Reference, Section
Nominal signal-to-noise ratio	dB	....	6.7.2
rms pulse width	s	....	13.5
rms azimuth beamwidth	Degrees	....	13.5
rms elevation beamwidth	Degrees	....	13.5
rms Doppler spectrum width	Hz	....	13.5

14.6.6 Stability

Stability limits the cancellation ratios for moving target indicator and detector radars. The components are shown in Table 14.12.

Table 14.12
Factors for radar stability

Characteristic	Units	Value	Reference, Section
Transmitter stability	dB	....	3.6.4
Waveguide stability	dB	....	4.2.2
Effects of wind shear and clutter spectrum width	Hz	....	6.7.1
Effect of limiting on the rms clutter spectrum width	Hz	....	7.4.2
Cancellation ratio limit, STALO, I_{STALO}	dB	....	7.6.3.2
Cancellation ratio limit caused by quantization, I_{quant}	dB	....	10.4.3

REFERENCES

1. Jordan, E. C., *Reference Data for Engineers*, 7th ed., Indianapolis: Howard Sams, 1986, p. 3-14.
2. Bijvoet, J. A., *Standard Methods for Predicting and Specifying Performance of Air Surveillance Systems,* The Hague: Supreme Headquarters Allied Powers Europe (SHAPE) Technical Centre Technical Report TR-50.
3. Blake, L. V., *A Guide to Basic Pulse-Radar Maximum-Range Calculation*, Washington, DC: U.S. Naval Research Laboratory, December 1969, p. 94.
4. Bean, P. R., and E. J. Dutton, *Radio Meteorology*, National Bureau of Standards Monograph No. 92, Washington: GPO 1966. Reprinted by Dover Publications, New York, p. 53.
5. Campen, C. F., and A. E. Cole, *Tropospheric Variations of Refractive Index at Microwave Frequencies*, Bedford, Massachusetts: Air Force Surveys in Geophysics, No. 79, October 1955.

Chapter 15

Statistics

In Chapter 12, Threshold and detection, a number of statistical distributions are used, some of which are related. In the past, before the personal calculator or the computer age, distributions had to be chosen for which there were tables of the values. This text is not a mathematical treatment but an introduction to the families of distributions and their relationships. References [1-3] furnish the theory behind this chapter.

There are many more statistical distributions than those described in this chapter, and only those mentioned in other chapters are discussed. Discrete distributions, for example, the Poisson distribution, which do not have any use in antenna theory or signal processing, are omitted.

This chapter also introduces the Fourier transforms, or characteristic functions, of statistical probability distribution functions. The discussion is limited to the requirements of this book and the reader is referred to statistical texts for a more fundamental treatment. In contrast to statistical texts that use the $+\omega$ convention for their Fourier transforms, the $-f$ convention is used here to be compatible with other radar tests [4]. The sense of the loops representing characteristic functions is the reverse of those in [5, Chapter 6] and there is a factor of 2π in the scaling of the ξ axis.

Fourier transforms for for spectra and antenna diagrams are discussed in Chapter 16.

15.1 TERMS

A number of terms are used to describe the sizes and shapes of the distributions illustrated in Figure 15.1. A difference must be made between sample statistics (for which Roman letters are used) and the statistics for a whole population (for which Greek letters are used). The statistics from a small sample, that is the data that one has to hand, are an estimate of the population statistics, and there is always a small difference or error between the two.

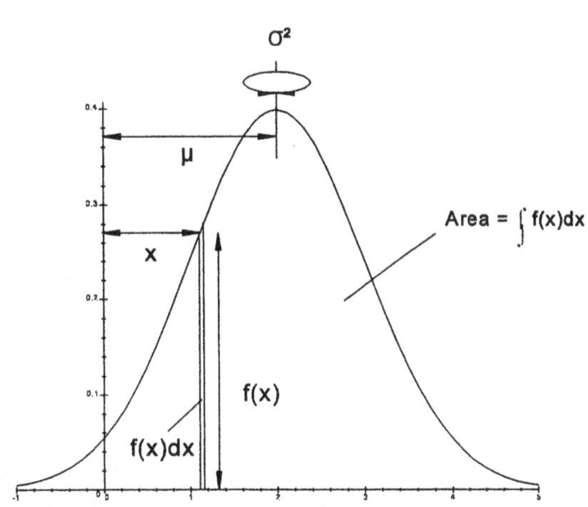

Figure 15.1 First (mean) and second (standard deviation) moments of the population.

15.1.1 Mean and expected values

The first value that occurs is the mean, or the average. This is simply the sum of the samples, x_i, divided by the number of samples, n.

$$\text{Mean, } m = \sum \frac{x_i}{n} \tag{15.1}$$

The mean is extended as an integral for the population. It can be seen that (15.2) is also the formula for the center of gravity or the direct voltage component in a waveform. Here, x represents the distance from the origin and $f(x)$ the ordinate, as shown in Figure 15.1.

$$\text{Population mean, } \mu = \frac{\int \text{First moments of area}}{\text{Area}} = \frac{\int_{-\infty}^{\infty} x f(x)\, dx}{\int_{-\infty}^{\infty} f(x)\, dx} \tag{15.2}$$

In the theoretical models, the denominator in (15.2) evaluates to unity. Thus, the expected (mean) value for the population is

$$\text{Expected value, } E\{X\} = \int_{-\infty}^{\infty} x f(x)\, dx \tag{15.3}$$

15.1.2 Variance

Many probability distributions are distributed about the mean. In order to find the differences around the mean the differences are first squared so that they are all positive and can be added. The average square of the differences is called the variance. The sample variance is

$$\text{Sample variance, } s^2 = \frac{\Sigma(x - m)^2}{n - 1} \tag{15.4}$$

The population variance is given by (see Figure 15.1)

$$\text{Population variance, } \sigma^2 = \frac{\int \text{Second moments of area around the mean}}{\text{Area}} = \frac{\int_{-\infty}^{\infty} (x - \mu)^2 f(x)\, dx}{\int_{-\infty}^{\infty} f(x)\, dx} \tag{15.5}$$

Note that the denominator is equal to the number of members of the population, so that the sample variance is greater than the population variance. That is because the sample mean has an error, whereas the population mean is exact. The variance represents the moment of inertia of the distribution about the mean or the power represented by an alternating voltage. In mechanics, the moment of inertia is at a minimum when the axis passes through the center of gravity, μ, and increases as the axis is displaced, to m.

15.1.3 Standard deviation

The standard deviation is the positive square root of the variance. The symbols used here are:

- Sample standard deviation, s;
- Population standard deviation, σ.

Mechanically the standard deviation represents the radius of gyration or, in electricity, the root mean square value of an alternating voltage or current.

15.1.4 Histogram and probability distribution

When, for example, a cross is made in a bin when a number of the same parts from a production run are measured, the result is a histogram similar to that shown in Figure 15.2.

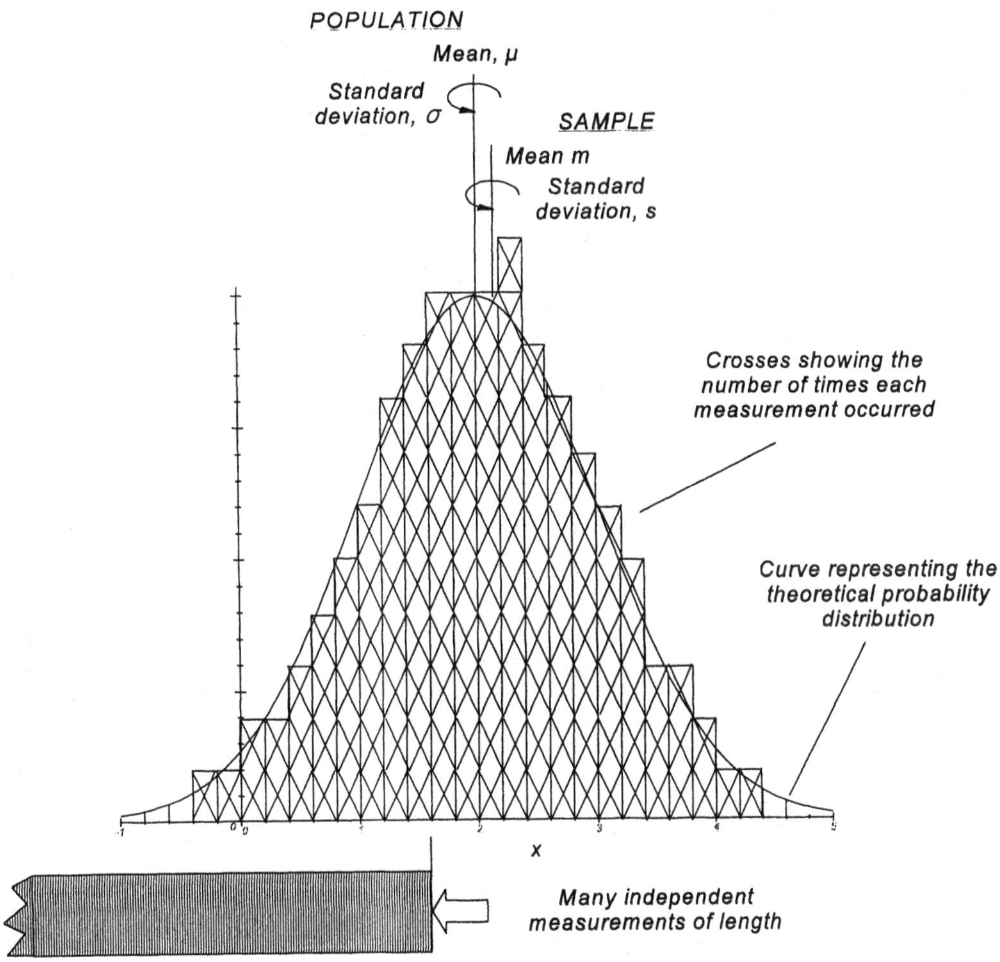

Figure 15.2 An example of a histogram.

The sample mean and the standard deviation may be calculated using (15.1) and (15.4). The position of the peak of the distribution is called the mode, and the theoretical probability distribution function (pdf) may be drawn on the histogram. The probability distribution function for the population has an area of unity.

15.1.5 Cumulative distribution function

If crosses are drawn in a row in the bins where the measurement is less than the abscissa value, the cumulative distribution diagram is obtained. Again, a curve representing the theoretical cumulative distribution function (cdf) can be drawn. An example is shown in Figure 15.3. The cumulative distribution function is the integral of the probability distribution function, and its ordinate has a range from zero to unity. It represents the proportion of the population up to the chosen abscissa point.

15.1.6 Percentiles and quartiles

The sample or the population can be divided into percentiles or quartiles, normally first (25%) or third (75%). The median of a distribution is the value that represents the 50% point for the population.

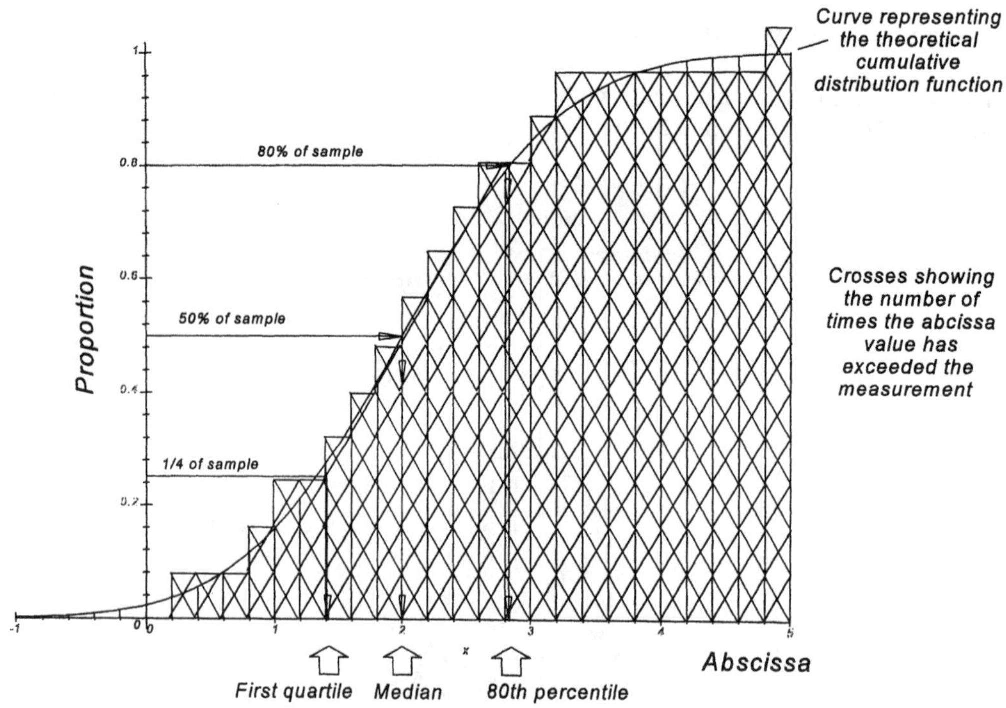

Figure 15.3 An example of a cumulative distribution.

It is sometimes easier to define antenna characteristics in terms of percentiles. An example is the definition of far sidelobes. The more stringent requirement is the root mean square value, where one larger lobe can fail the antenna. Alternatively if 95% of the sidelobes must be below a certain value, the remaining 5% of the lobes may be of any amplitude.

15.1.7 Moment generating functions

This section and Section 15.1.8 are much shortened introduction for use as an aid to understanding the mathematics in Chapter 12. A fuller treatment is to be found in [2, 3].

The statistics of probability distributions resulting from a number of completely independent statistical processes are found by adding the moments of the individual probability distributions. One way of considering this is using the *moment generating function* of p(x) over the range of x and over a range of values of t given by the expected value

$$M_0(t) = \int \exp(tx)\, p(x)\, dx \quad \text{over the range of } x \tag{15.6}$$

Expanding the exponential and integrating the terms gives

$$M_0(t) = 1 + \mu'_1 t + \mu'_2 \frac{t^2}{2!} + \mu'_3 \frac{t^3}{3!} + \ldots \tag{15.7}$$

where for the rth term

$$\mu'_r = \int x^r \, p(x) \, dx \tag{15.8}$$

The rth term is the moment of order r about the origin, where $x = 0$, and the sum plus unity is the moment generating function about the origin. The moment generating function about a value $x = a$ is the expected value

$$\begin{aligned} M_a(t) &= \int \exp(t(x - a)) \, p(x) \, dx \\ &= \int \exp(-at) \exp(tx) \, p(x) \, dx \end{aligned} \tag{15.9}$$

The moment generating function of the sum of two independent statistical variates is the product of their moment generating functions. Taking the expected values denoted by E{ },

$$E\{\exp(t(x + y))\} = E\{\exp(tx) \exp(ty)\} = E\{\exp(ty)\} E\{\exp(ty)\} \tag{15.10}$$

The origin may be chosen at will and the theorem holds for moment generating functions about any specified value.

If μ_r and μ'_r are the moments of x about the mean and the origin and m and m' are those of y, the moment generating function of $x+y$ is

$$M_\mu(t) = \left(1 + \mu'_1 t + \mu'_2 \frac{t^2}{2!} + \mu'_3 \frac{t^3}{3!} + ..\right)\left(1 + m'_1 t + m'_2 \frac{t^2}{2!} + m'_3 \frac{t^3}{3!} + ...\right) \tag{15.11}$$

Multiplying and collecting the first two terms,

$$M_{m\mu}(t) = (m'_1 + \mu'_1)t + (m'_2 + 2m'_1 \mu'_1 + \mu'_2)\frac{t^2}{2!} \tag{15.12}$$

The first moment about the origin is $m_1 + \mu_1$. The second moment about the mean is

$$\begin{aligned} m_2 \ or \ \sigma^2 &= m'_2 + 2m'_1 \mu'_1 + \mu'^2_1 - (m'_1 + \mu'_1)^2 \\ &= (\mu'_2 - \mu'^2_1) + (m'_2 - m'^2_1) \\ &= \mu_2 + m_2 \end{aligned} \tag{15.13}$$

Thus the variance, σ^2, of the sum (or difference) of two independent variates is equal to the sum of the individual variances.

This can be extended to any number of uncorrelated variates [2, p. 366, Section 8.4] so that the total variance is the sum of the individual variances and the standard deviation is the square root of the sum of the variances.

It may be observed that the expression for the moment generating function is similar to that of the Fourier transform described in the next section.

15.1.8 Fourier transform

Though moment generating functions are often used in the mathematics of statistics, they are ambiguous: that is, a specific moment generating function does not have a unique probability distribution function. This problem is solved using the Fourier transform, or characteristic function, given by

$$\Xi(\xi) = \int_{-\infty}^{+\infty} p(x) \exp(ix\xi) \, dx \tag{15.14}$$

where ξ is the variable in the characteristic function $\Xi(\xi)$;
x is the variable in the probability density function p(x);
i is $\sqrt{-1}$.

The use of $\Xi(\xi)$ and i here to show that the $+\omega$ convention is used.

To be consistent with the remaining parts of this book and the radar literature in general the $-f$ convention is used, namely,

$$C(\xi) = \int_{-\infty}^{+\infty} p(x) \exp(-j2\pi x\xi)\, dx \tag{15.15}$$

where ξ is the variable in the characteristic function C(ξ);
x is the variable in the probability density function p(x) and is integrated over the valid range of x;
j is $\sqrt{-1}$.

As with moment generating functions, the characteristic function of the sums of variates is the product of their individual characteristic functions giving, as the result, the Fourier transform of the sum. The inverse, called the anticharacteristic function, is unique, and is the final probability distribution function thus allowing the calculation of percentiles. This process is similar to convolution.

Probability distribution functions (always real) and their characteristic functions may be illustrated in the same manner as with signals and spectra, though disparities in vertical scale may involve splitting the diagrams. An example using the Gaussian distribution is shown in Figure 15.4. This form of illustration is extensively used in Chapter 12.

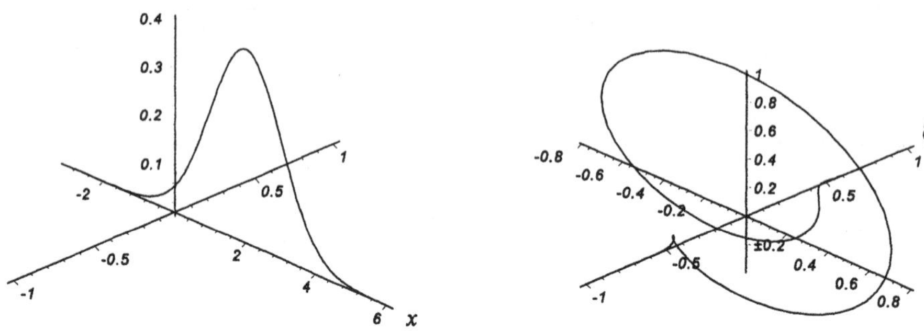

(a) Probability density function (b) Characteristic function

Figure 15.4 The Gaussian probability density function and the characteristic function.

Differentials of the characteristic function may be used to find the moments of a probability distribution. If the characteristic function is

$$C(\xi) = \int p(x) \exp(-2\pi\xi x)\, dx \tag{15.16}$$

Differentiating with respect to ξ

$$\begin{aligned} C'(\xi) &= \int -j2\pi x\, p(x) \exp(-j2\pi\xi x)\, dx \\ &= -j2\pi \int x\, p(x) \exp(-j2\pi\xi x)\, dx \end{aligned} \tag{15.17}$$

Setting $\xi = 0$

$$C'(0) = -j2\pi \int x\, p(x)\, dx \tag{15.18}$$

giving the first moment μ'_1 as

$$\mu'_1 = \frac{C'(0)}{-j2\pi} \tag{15.19}$$

Extending this to the rth moment, we have

$$C^r(\xi) = \int x^r\, p(x)\, dx = \frac{C^r(0)}{(-j2\pi)^r} \tag{15.20}$$

The example that follows shows this, using a uniform distribution of width unity centered on unity, see Figure 15.5.

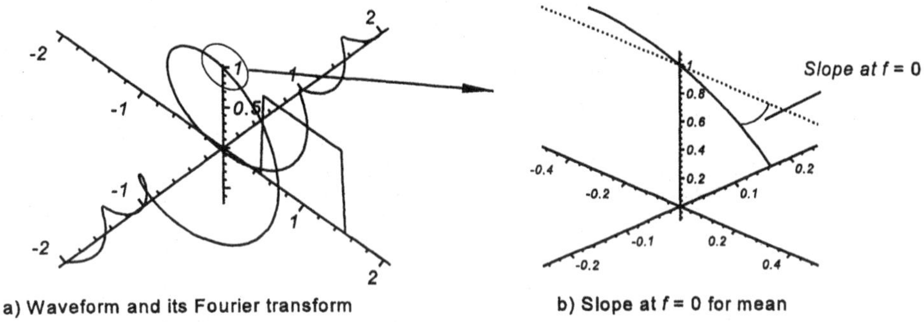

Figure 15.5 An example with a uniform distribution width 1 centered on +1. [From Meikle, H. D., *A New Twist to Fourier Transforms,* Weinheim, Germany: Wiley-VCH, 2004.]

The Fourier transform, or characteristic function, is

$$\begin{aligned} C(\xi) &= \int_{1/2}^{3/2} 1 \exp(-j2\pi\xi x)\, dx \\ &= \exp(-j2\pi\xi)\, \frac{\sin(\pi\xi)}{\pi\xi} \end{aligned} \tag{15.21}$$

The characteristic function consists of two parts: the exponential function that expresses the mean and the sin x / x function that describes the shape. The first derivative is

$$C'(\xi) = \left(\frac{\cos(\pi\xi)}{\xi} - \frac{\sin(\pi\xi)}{\pi^2\xi^2} - j\frac{2\sin(\pi\xi)}{\xi} \right) \exp(-j2\pi\xi) \tag{15.22}$$

and this is shown in Figure 15.6.

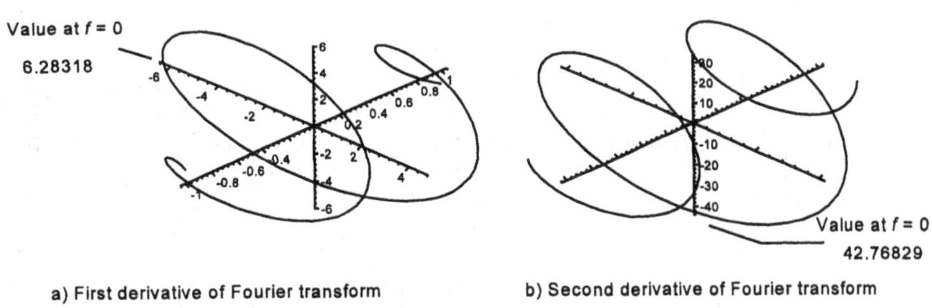

Figure 15.6 The first and second derivatives of a uniform distribution. [From Meikle, H. D., *A New Twist to Fourier Transforms,* Weinheim, Germany: Wiley-VCH, 2004.]

The first moment or mean from Figure 15.6 is found from where the curve cuts the $\xi = 0$ plane, namely at $-j2\pi$

$$\mu'_1 = \frac{C'(0)}{-j2\pi} = \frac{-j2\pi}{-j2\pi} = 1 \tag{15.23}$$

Differentiating once more

$$\begin{aligned} C''(\xi) &= \int (-j2\pi)^2\, x^2\, p(x)\, \exp(-j2\pi\xi x)\, dx \\ &= -4\pi^2 \int x^2\, p\, \exp(-j2\pi\xi x)\, dx \end{aligned} \tag{15.24}$$

Setting $\xi = 0$, the second moment about the origin is

$$\mu'_2 = -\frac{C''(0)}{4\pi^2} \tag{15.25}$$

giving the second moment about the mean

$$\mu_2 = \frac{\mu'_2}{\mu_0} - \left(\frac{\mu'_1}{\mu_0}\right)^2 \tag{15.26}$$

In the example

$$\begin{aligned} \mu_2 &= \frac{-42.768}{-4\pi^2} - 1 \\ &= \frac{1}{12} \end{aligned} \tag{15.27}$$

The variance of a uniform distribution width unity is 1/12.

15.2 FAMILIES OF DISTRIBUTIONS

Only continuous distributions are considered here. For convenience they are divided into:

- Two sided from $-\infty$ to $+\infty$;
- Bounded on one side from 0 to ∞ .

A plan showing the relationships between the distributions mentioned in this chapter is shown in Figure 15.7.

15.2.1 Gaussian or normal distribution: two sided from $-\infty$ to $+\infty$

The normal or Gaussian distribution is the most used distribution and commonly represents manual measurement errors. The central limit theorem states that the distribution of the average, and therefore the sum, of independent observations from distributions with finite mean and variance, tends to a Gaussian distribution as the number of observations becomes large. That is, if the processes leading to the observations are complex enough, the result tends to a Gaussian distribution.

Normal (Gaussian)

$$p(x, \mu, \sigma) = \frac{1}{\sigma\sqrt{2\pi}} e^{-\frac{(x-\mu)^2}{2\sigma^2}} \qquad -\infty < x < \infty$$

$x \rightarrow \ln(x)$ **Log-normal**

$$p(x, \mu, \sigma) = \frac{1}{\sigma x\sqrt{2\pi}} e^{-\frac{(\ln(x)-\mu)^2}{2\sigma^2}} \qquad x > 0, \; -\infty < \mu < \infty, \; \sigma > 0$$

$= 0$ elsewhere

Normal (Gaussian) in two dimensions

$$p(x, y, \sigma) = \frac{1}{\sigma\sqrt{2\pi}} e^{-\frac{x^2+y^2}{2\sigma^2}} \qquad -\infty < x, y < \infty$$

$x^2 + y^2 \rightarrow R^2$ ➔

Weibull

$\eta = 2$ ➔

$$p(t, \eta, \sigma) = \frac{\eta}{\sigma}\left(\frac{t}{\sigma}\right)^{\eta-1} e^{-\left(\frac{t}{\sigma}\right)^{\eta}} \qquad t > 0, \; \sigma > 0, \; \eta > 0$$

0 elsewhere

$\eta = 1$ ➔

Rayleigh

$$p(R, \sigma) = \frac{R}{\sigma^2} e^{-\frac{R^2}{2\sigma^2}}$$

$R^2/2\sigma^2 \rightarrow X/X_0$ $\lambda = 1, \; x = X/X_0$

Rayleigh power

$$p(X, X_0) = e^{-\frac{X}{X_0}}$$

Gamma family

Basic distribution

$$p(x, \eta, \lambda) = \frac{\lambda^{\eta}}{\Gamma(\eta)} x^{\eta-1} e^{-\lambda x}, \quad x > 0, \; \lambda > 0, \; \eta > 0$$

0 elsewhere

$\lambda = 1/2, \; \nu = 2\eta$ **Chi- squared**

$$p(x, \nu) = x^{\frac{\nu-2}{2}} \frac{e^{-\frac{x}{2}}}{2^{\frac{\nu}{2}}\Gamma\left(\frac{\nu}{2}\right)}, \quad x > 0, \quad \nu > 0$$

$y = \sqrt{\frac{x}{\nu}}$ **Chi**

$$p(y) = \frac{2^{1-\frac{\nu}{2}} y^{\nu-1} e^{-\frac{y^2}{2}}}{\Gamma\left(\frac{\nu}{2}\right)}$$

$\nu = 1$ 🡐

Gamma distribution with $\eta = 1$ 🡐

Negative exponential

$$p(x, \lambda) = \lambda e^{-\lambda x}$$

Gamma distribution with η integer and $b = 1/\lambda$

Erlang

$$p(x, \eta, b) = \frac{x^{\eta-1} e^{\frac{x}{b}}}{b^{\eta}(\eta - 1)!}$$

Figure 15.7 The relationships between the normal (Gaussian), Weibull, Rayleigh, and gamma distribution families (neglecting constants).

The Gaussian probability distribution (pdf) is given by

$$f(x,\ \mu,\ \sigma) \ = \ \frac{1}{\sigma\sqrt{2\pi}} e^{\frac{(x-\mu)^2}{2\sigma^2}} \qquad -\infty < x < \infty \tag{15.28}$$

where μ is the mean;
σ is the standard deviation.

This is shown in Figure 15.8.

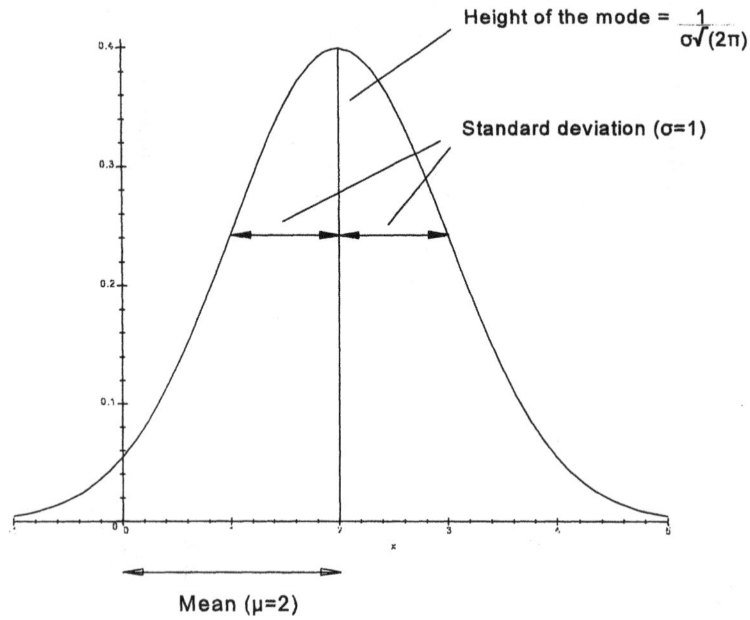

Figure 15.8 The Gaussian or normal probability density function.

The cumulative distribution function is given by

$$F(x,\ \mu,\ \sigma) \ = \ \frac{1}{\sigma\sqrt{2\pi}} \int_0^q e^{\frac{(x-\mu)^2}{2\sigma^2 dx}} \qquad -\infty < x < \infty \tag{15.29}$$

The mean, the median, and the mode occur at the same abscissa value. The median, or 50% point, is shown in Figure 15.9.

Narrowband thermal noise (modulated on a notional carrier) as seen on an oscilloscope has a Gaussian distribution. The Gaussian, or normal, distribution has two close relations: the log-normal and the Rayleigh distributions.

If independent variates x_i $(i = 1..n)$ have a Gaussian distribution about a common mean, a, variance, σ^2, this mean has also a Gaussian distribution about a with a variance of σ^2/n, standard deviation $\sigma/\sqrt{n}$. This standard deviation is called the standard error of the mean.

The characteristic function for a Gaussian distribution is given by

$$C(\xi) \ = \ \exp\left(j2\pi\xi(-\mu + j2\pi\xi\sigma^2)\right) \tag{15.30}$$

A characteristic function with $\mu = 1$ and $\sigma = 1$ is shown in Figure 15.4(b).

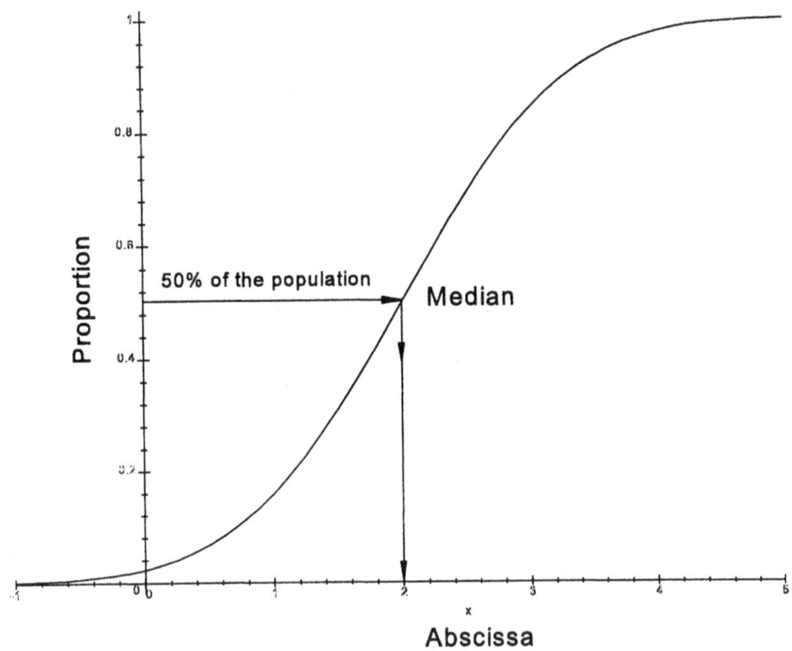

Figure 15.9 The cumulative density function (cdf) for the Gaussian or normal distribution.

15.2.2 Log-normal distribution

If the amplitude distribution of a signal plotted on a logarithmic scale (nepers) has a normal distribution, then the signal has a log-normal distribution when mapped onto a linear scale. It has two parameters: the parameter μ is the scale, and σ is the shape parameter. The distribution is given by

$$f(x,\ \mu,\ \sigma) = \frac{1}{\sigma x\sqrt{2\pi}}e^{-\frac{(\ln(x)-\mu)^2}{2\sigma^2}} \quad x > 0,\ -\infty < \mu < \infty,\ \sigma > 0 \qquad (15.31)$$

$$= 0 \quad \text{elsewhere}$$

The log-normal distribution and its parent Gaussian, or normal, distribution are shown in Figure 15.10.

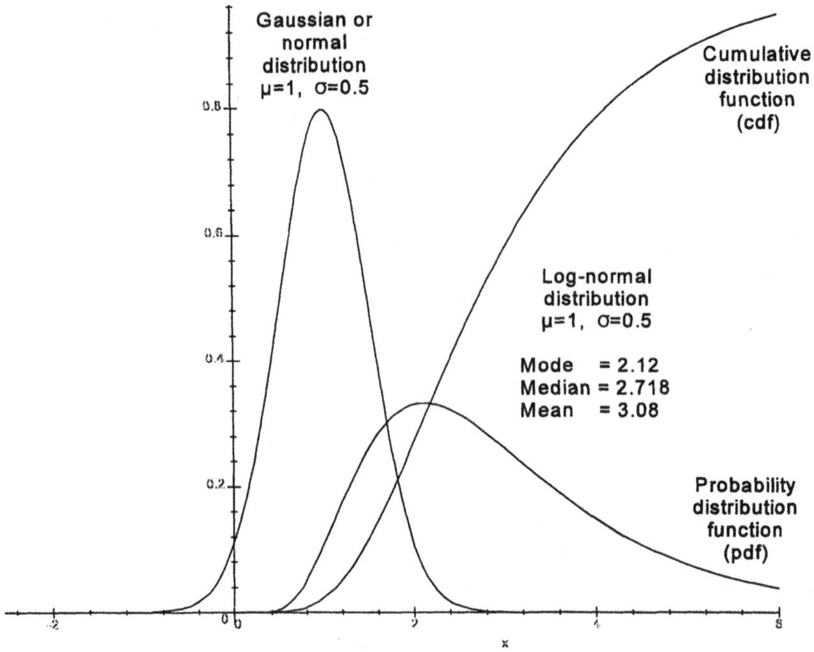

Figure 15.10 The log-normal distribution and its parent Gaussian or normal distribution.

The mode, the median, and the mean are given by [6]

$$\text{Mode} = e^{\mu - \sigma^2}$$

$$\text{Median} = e^{\mu}$$

$$\text{Mean} = e^{\mu + \frac{\sigma^2}{2}}$$

The mean-to-median ratio is a parameter which is often given, which is related to the shape parameter by

$$\frac{Mean}{Median} = e^{\frac{\sigma^2}{2}} \tag{15.32}$$

The log-normal distribution represents the distribution of the product of small errors, and is similar to the sums of the logarithms that are Gaussian distributed. In radar, some target models (ships) or some clutter models are said to be log-normally distributed when the probability distribution is very spiky.

15.2.3 Rayleigh distribution

When two independent processes have Gaussian distributions, the joint distribution about the mean looks like a hump, as shown in Figure 15.8. If the two distributions have the same standard deviation, σ, then the hump is symmetrical. When the distributions in the x and y directions are Gaussian with the same standard deviation, σ, then the distribution along the radius is a Rayleigh distribution. The original Gaussian or normal distributions are expressed in terms of x and y:

$$f(x,\ 0,\ \sigma) = \frac{1}{\sigma\sqrt{2\pi}}e^{-\frac{x^2}{2\sigma^2}} \qquad -\infty < x < \infty$$
$$f(y,\ 0,\ \sigma) = \frac{1}{\sigma\sqrt{2\pi}}e^{-\frac{y^2}{2\sigma^2}} \qquad -\infty < y < \infty \tag{15.33}$$

The joint distribution is, in x and y,

$$f(x,\ y,\ \sigma) = \frac{1}{2\pi\sigma^2}e^{-\frac{x^2+y^2}{2\sigma^2}} \tag{15.34}$$

In Figure 15.11, the volume above the circular stripe width dR, length $2\pi R$, and height $\frac{e^{-\frac{R^2}{2\sigma^2}}}{2\pi\sigma^2}$ is

$$f(R,\ \sigma) = 2\pi R dR\frac{1}{2\pi\sigma^2}e^{-\frac{R^2}{2\sigma^2}} = \frac{R}{\sigma^2}e^{-\frac{R^2}{2\sigma^2}}dR \qquad R > 0 \tag{15.35}$$

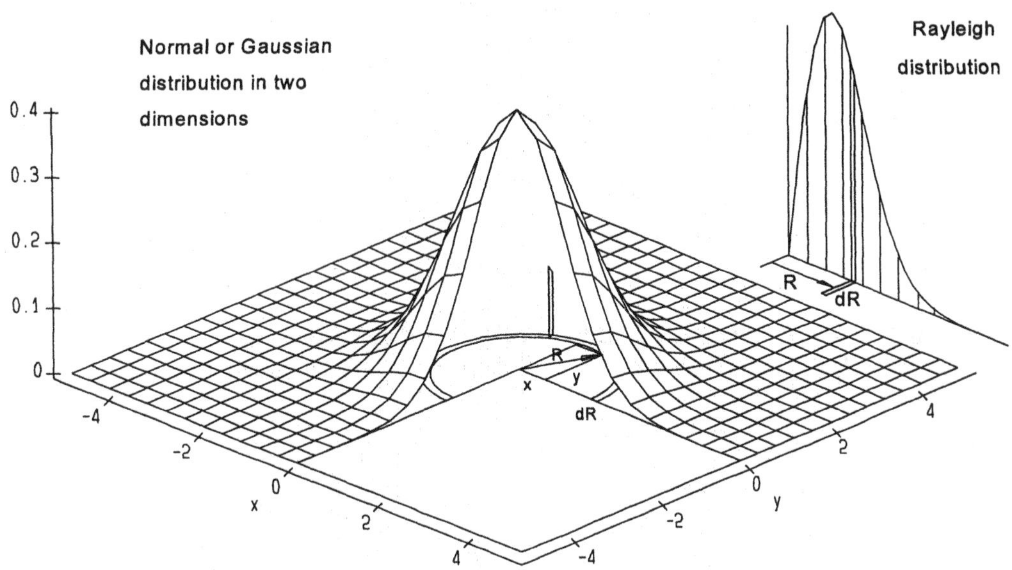

Figure 15.11 Derivation of the Rayleigh distribution from a two-dimensional Gaussian or normal distribution.

The Rayleigh distribution can be normalized using the variable $q = R/\sigma$. That is,

$$\sigma f(q) = qe^{-\frac{q^2}{2}} \tag{15.36}$$

The normalized probability distribution function is shown in Figure 15.12.

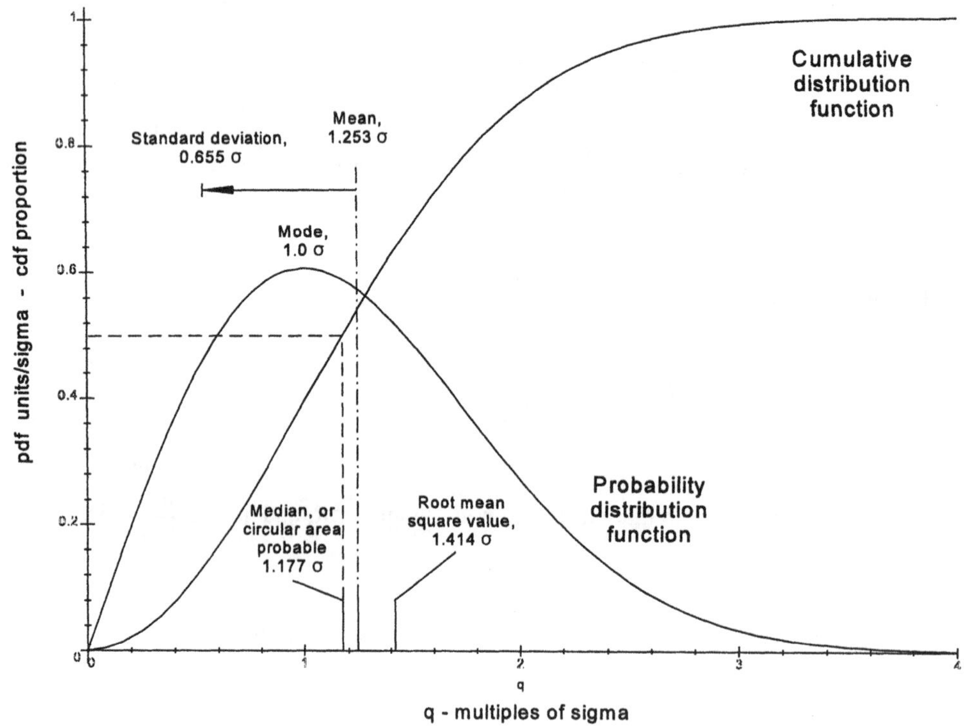

Figure 15.12 The Rayleigh distribution with normalized parameters.

The main parameters of the Rayleigh distribution are

Mean (mean radial error) $= \sqrt{\left(\frac{\pi}{2}\right)}\ \sigma \quad = 1.253\ \sigma$

Variance $= (2 - \pi/2)\sigma^2$

Standard deviation $= \sqrt{(2 - \pi/2)}\ \sigma \quad = 0.655\ \sigma$

Mode $= \sigma$

It must be noted that the value for σ is that from the original component Gaussian distributions, and the values for the Rayleigh distribution are scaled from this value of σ. The variance is the variance about the mean. The relationships are

Root mean square (rms) value $= \sqrt{2}\ \sigma \quad = 1.414\ \sigma$

Median (circular error probable, cep) $= \sqrt{2\ \ln(2)}\ \sigma \quad = 1.177\ \sigma$

The root mean square value is the total noise voltage from both the x and the y components. The median is more often used than the mean in ballistics to measure weapon accuracy. The median is the radius of the circle inside which 50% of the shots fall or 50% of plots for an echo are situated. The cumulative distribution function (cdf) is

$$F(x,\ \sigma) = \int_0^x \frac{y}{\sigma^2} e^{-\frac{y^2}{2\sigma^2}}\, dy = 1 - e^{-\frac{x^2}{2\sigma^2}} \tag{15.37}$$

The normalized cumulative distribution function is also shown in Figure 15.12.

When X represents the power $\left(= \frac{x^2}{2}\right)$ and X_0 the average power, the Rayleigh distribution becomes

$$\text{Probability distribution function (power),}\quad p(R)dR = e^{-\frac{X}{X_0}} d\frac{X}{X_0} \tag{15.38}$$

Cumulative distribution function (power), $F(X) \quad = 1 - e^{-\frac{X}{X_0}}$ (15.39)

The Rayleigh power distribution represents the process when a carrier, which has been modulated with in-phase and quadrature noise, is detected. The phase information is discarded, and the output signal distribution is represented by the distribution along the radius. The Rayleigh distribution is used to represent video noise, sea clutter, and sometimes land clutter.

The characteristic function is given by

$$C(\xi) \;=\; 1 + j\pi^{\frac{3}{2}}\xi\sigma\,\sqrt{2}\exp(-2\pi^2\xi^2\sigma^2)\left(\mathrm{erf}(j\pi\xi\sqrt{2}\sigma\,) - 1\right) \qquad (15.40)$$

where $\mathrm{erf}(x) \;=\; \frac{2}{\sqrt{\pi}}\int_0^x \exp(-t^2)\;dt$ and the curve with $\sigma = 1$ is shown in Figure 15.13.

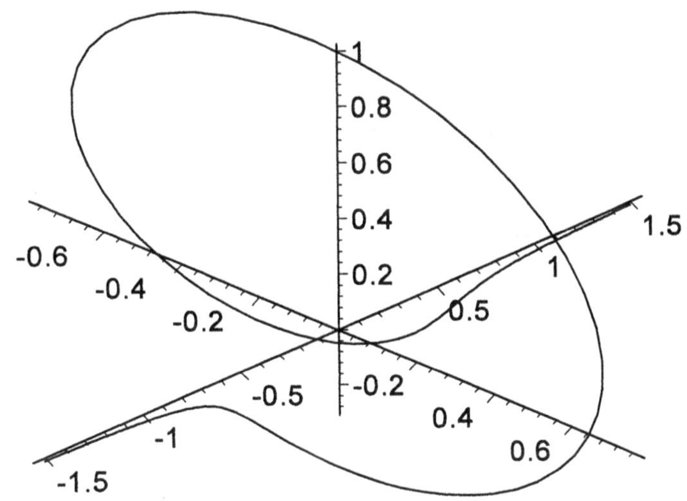

Figure 15.13 The characteristic function of a Rayleigh distribution with $\sigma = 1$.

15.2.4 Ricean distribution

If a steady signal is added to noise, the center of hump in Figure 15.11 is displaced onto the end of the signal vector as shown in Figure 15.13, an offset Rayleigh distribution [5, Section 6.3.3.2] ans is shown in Figure 15.14. If the voltage vector has the length S then the Ricean distribution may be given by

$$p_{\mathrm{Rice}}(r) \;=\; \frac{r}{\sigma^2}\exp\left(-\frac{r^2 + S^2}{2\sigma}\right)\,I_0\left(\frac{rS}{\sigma^2}\right)\,dr \qquad (15.41)$$

where $I_0(z) \;=\; \frac{1}{\pi}\int_0^{\pi}\exp(\pm z\cos\theta)\;d\theta$.

Examples of this form of the Ricean distribution are plotted in Figure 15.15.

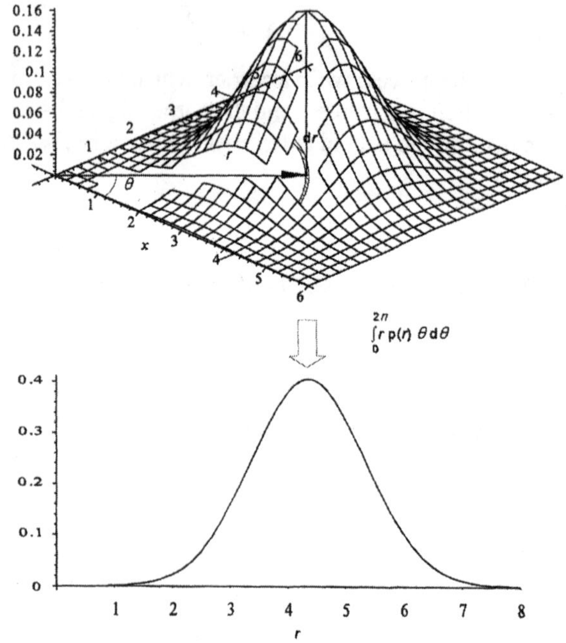

Figure 15.14 The the Ricean distribution.[From Meikle, H. D., *A New Twist to Fourier Transforms,* Weinheim, Germany: Wiley-VCH, 2004.]

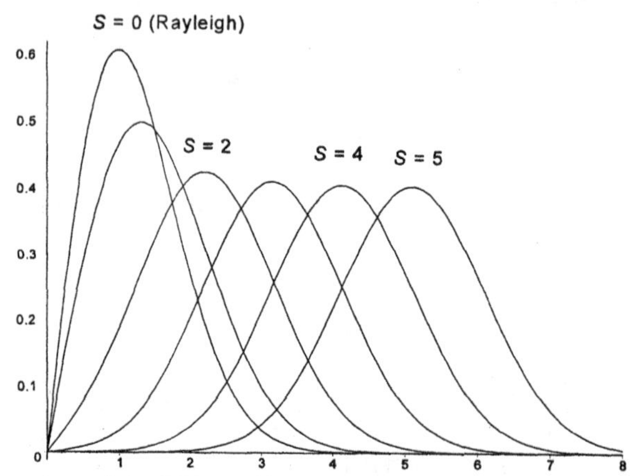

Figure 15.15 Examples of the Ricean distribution with σ=1. [From Meikle, H. D., *A New Twist to Fourier Transforms,* Weinheim, Germany: Wiley-VCH, 2004, Figure 6.14.]

In radar detection calculations the Ricean distribution is used in terms of power: y replaces $r^2/2\sigma^2$, R is substituted for $S^2/2\sigma^2$, and the Jacobean is r/σ^2, giving

$$p_{Rice}(y) = \exp(-(y + R))\, I_0(\sqrt{4yR})\, dy \tag{15.42}$$

Curves are shown in Figure 15.16. The characteristic function is given by

$$C_{\text{Rice power}}(\xi) = j\,\frac{\exp\left(-\dfrac{2\pi R\xi}{2\pi\xi - j}\right)}{j - 2\pi\xi} \tag{15.43}$$

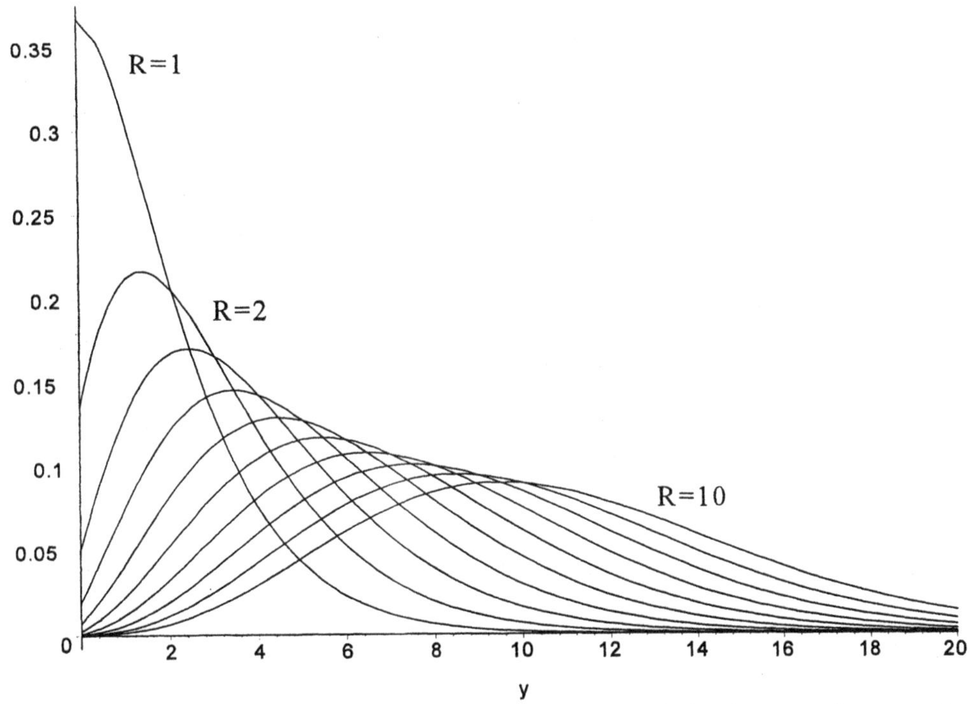

Figure 15.16 Curves of the Ricean distribution with the abcissa in terms of power ratio.

and shown for $R = 3$ in Figure 15.17.

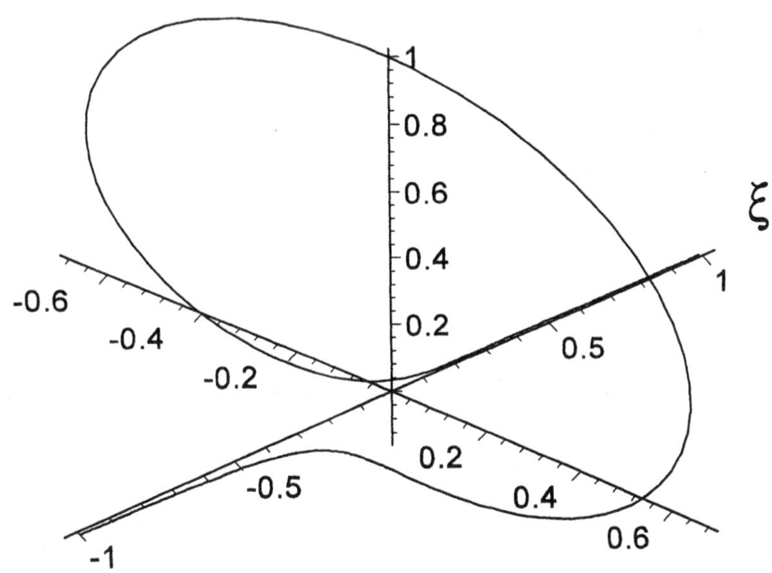

Figure 15.17 An example of the characteristic function of a Ricean distribution in terms of power, $R = 3$.

15.3 GAMMA DISTRIBUTION FAMILY: BOUNDED ON ONE SIDE FROM 0 TO ∞

The gamma family contains a number of distributions used to describe summed noise samples, echo fluctuations, and clutter in radar. The probability distribution has two parameters [1, p. 83]: η is the shape parameter, and λ is the scale parameter.

$$f(x, \eta, \lambda) = \frac{\lambda^{\eta}}{\Gamma(\eta)} x^{\eta - 1} e^{-\lambda x} \quad x > 0, \quad \lambda > 0, \quad \eta > 0 \qquad (15.44)$$
$$= 0 \quad \text{elsewhere}$$

where $\Gamma(\eta) = \int_0^{\infty} x^{\eta-1} e^{-x} \, dx$

$= (\eta - 1)!$ for integer values of η.

The main parameters of the gamma distribution are

$$\text{Mean} = \frac{\eta}{\lambda}$$

$$\text{Variance} = \frac{\eta}{\lambda^2}$$

$$\text{Standard deviation} = \frac{\sqrt{\eta}}{\lambda}$$

$$\text{Mode} = \frac{\eta - 1}{\lambda}$$

Examples of gamma distributions are shown in Figures 15.18 and 15.19.

The cumulative distribution function is

$$F(x, \eta, \lambda) = \frac{\lambda^{\eta}}{\Gamma(\eta)} \int_0^{x} t^{\eta-1} e^{-\lambda t} \, dt \quad \text{for } x > 0 \qquad (15.45)$$

This is the simple incomplete gamma function, $\lambda = 1$ [8, p. 260]:

$$\textit{Incomplete gamma function, } \gamma(a, \eta) = \int_0^{a} \frac{t^{\eta-1} e^{-t}}{\Gamma(\eta)} \, dt \qquad (15.46)$$

Most of the radar literature uses Pearson's form of the incomplete gamma function, which has been tabulated [7, p. 262, Eq. 6.5.6].

$$\textit{Pearson's form of the incomplete } \gamma \textit{ function, } I(u, p) = \int_0^{u\sqrt{p+1}} \frac{t^{p} e^{-t}}{\Gamma(p+1)} \, dt \qquad (15.47)$$

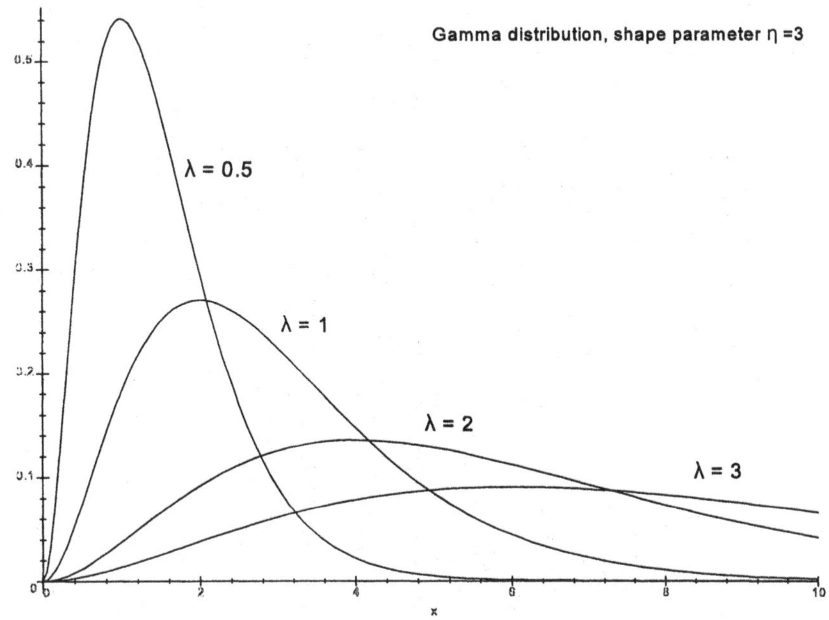

Figure 15.18 Gamma probability distribution with various scale factors (0.5..3) and constant shape (3).

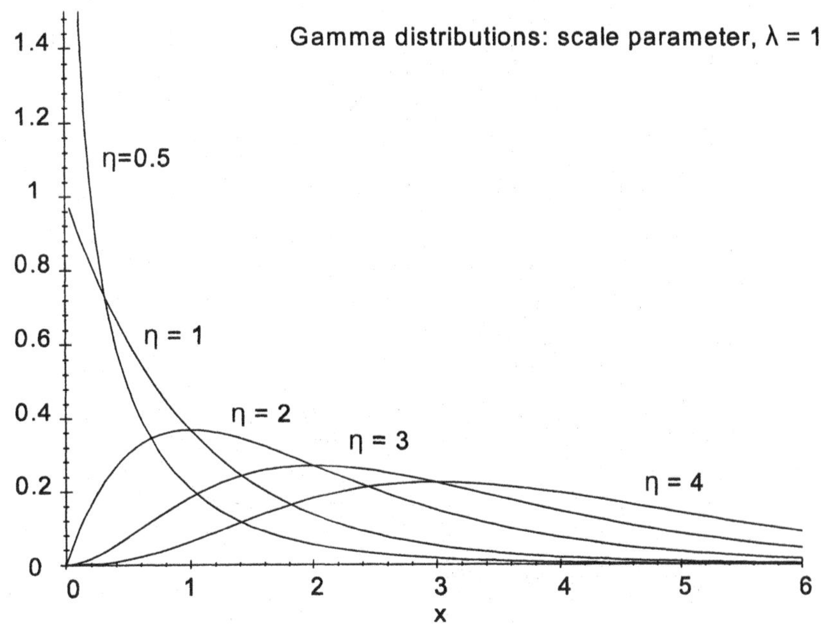

Figure 15.19 Gamma probability distribution with constant scale factor (1) and various shapes (0.5..4).

There are programs for personal computers and pocket calculators that give values for the incomplete gamma function directly. The gamma distribution has a number of properties:

- If x has a Gaussian or normal distribution with a mean μ and standard deviation σ, then $\dfrac{(x - \mu)^2}{2\sigma^2}$ is a gamma variate with $\eta = 1/2$ [3, p. 150].
- The sum of two independent gamma variates, with shape factors m and n, is a gamma variate with a shape factor

$m + n$ [3, p. 151].

The gamma distribution is the "father" of many other distributions. Special cases of the gamma distribution are shown in Table 15.1.

Table 15.1
Distributions derived from the gamma distribution

Distribution	Used for
Erlangian distribution	Queuing, telephone exchanges
Chi-squared distribution	Noise sums, χ^2 test
Negative exponential distribution	Rayleigh power distribution, Rayleigh noise power

The characteristic function of the gamma distribution is given by

$$C(\xi) = \left(1 + \frac{j2\pi\xi}{\lambda}\right)^{-\eta} \tag{15.48}$$

and an example is shown in Figure 15.20.

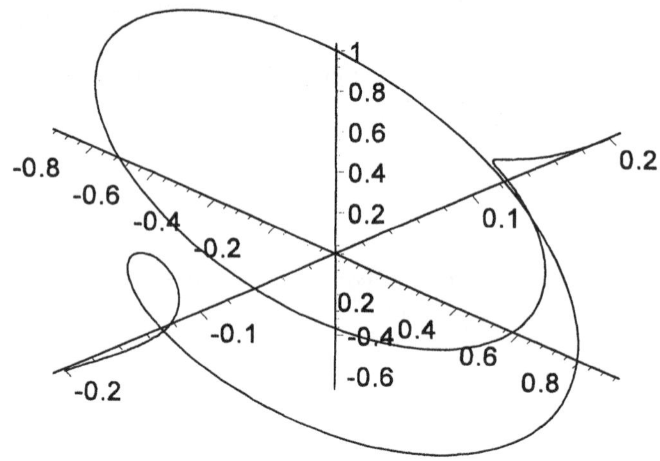

Figure 15.20 The characteristic function of a gamma distribution $\eta = 10$, $\lambda = 1$.

15.3.1 Erlangian distribution

When η in the gamma distribution is limited to integers, the resulting distribution is Erlangian. It is used for telephone exchanges with an average service time of $1/\lambda = b$ and the coefficient of variation of service time is $\sqrt{\eta}$ [8]. The probability density function is given by

$$p(x) = \frac{x^{\eta-1}\, e^{\frac{x}{b}}}{b^{\eta}\,(\eta - 1)!} \tag{15.49}$$

and the cumulative distribution function by [9]

$$P(x) = 1 - \sum_{r=1}^{\eta-1} \frac{\left(\frac{x}{b}\right)^{r} e^{-\frac{x}{b}}}{r!} \tag{15.50}$$

The Erlangian distribution is used in queuing and for the traffic capacity of telephone exchanges.

15.3.2 Chi-squared distribution

The chi-squared distribution is a special case of the gamma distribution with the following limitations:

- $\lambda = ½$;
- η is multiple of ½;
- $\nu = 2\eta$, the number of degrees of freedom.

The chi-squared distribution is given by [10]

$$p(x,\ \nu) = x^{\frac{\nu-2}{2}} \frac{e^{-\frac{x}{2}}}{2^{\frac{\nu}{2}}\Gamma\left(\frac{\nu}{2}\right)}, \quad x > 0, \quad \nu > 0 \tag{15.51}$$

The chi-squared distribution represents the distribution of sums of squares. In radar a sample of bottle-brush noise power is the sum of the squares from two Gaussian, or normally, distributed samples. For example, the sum of 10 noise samples has a chi-squared distribution with 20 degrees of freedom. Examples are shown in Figure 15.21.

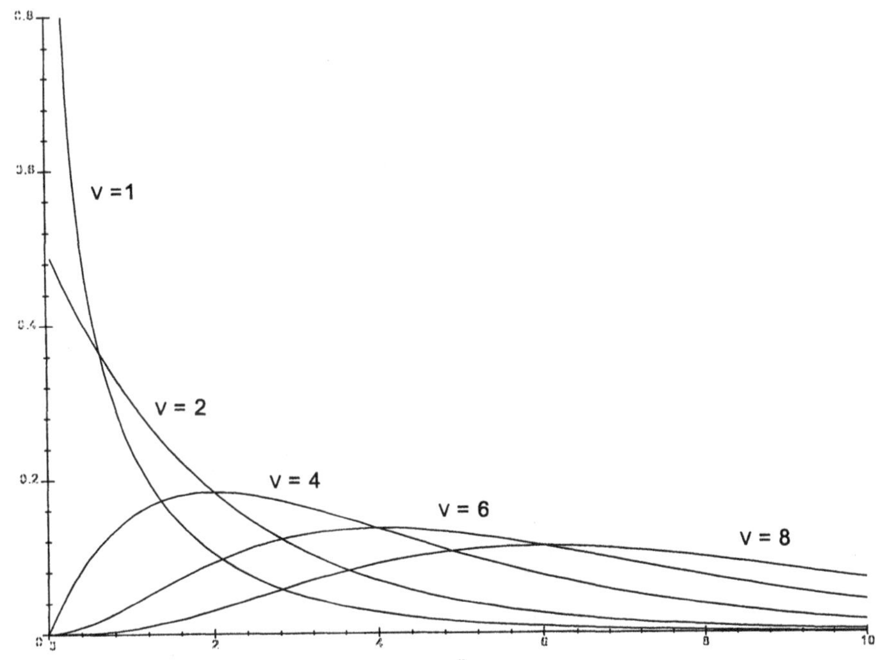

Figure 15.21 The chi-squared probability function for 1, 2, 4, 6, and 8 degrees of freedom.

The statistical parameters for the chi-squared distribution are:

Mean $= \frac{\nu}{2}$

Variance $= \frac{\nu}{2}$

Standard deviation $= \sqrt{\frac{\nu}{2}}$

Mode $= \nu - 2$

The cumulative distribution is given by [3, p. 240]

$$P(\chi^2) = \int_0^{\chi^2} \frac{t^{\frac{\nu}{2}-1} e^{-\frac{t}{2}}}{2^{\frac{\nu}{2}} \Gamma\left(\frac{\nu}{2}\right)} = \frac{\gamma\left(\frac{\nu}{2}, \frac{\chi^2}{2}\right)}{\Gamma\left(\frac{\nu}{2}\right)} \tag{15.52}$$

The chi-squared distribution represents the probability distribution of the squares of the differences of Gaussian distributed data about the mean.

15.3.3 Chi distribution

The chi distribution is formed from the chi-squared distribution by replacing the x in the distribution function by $\sqrt{x/n}$ [2, p. 181].

The probability distribution function is [10]

$$p(x) = \frac{2^{1-\frac{\nu}{2}} x^{\nu-1} e^{-\frac{x^2}{2}}}{\Gamma\left(\frac{\nu}{2}\right)} \tag{15.53}$$

The probability distribution function for different numbers of degrees of freedom, ν, is shown in Figure 15.22.

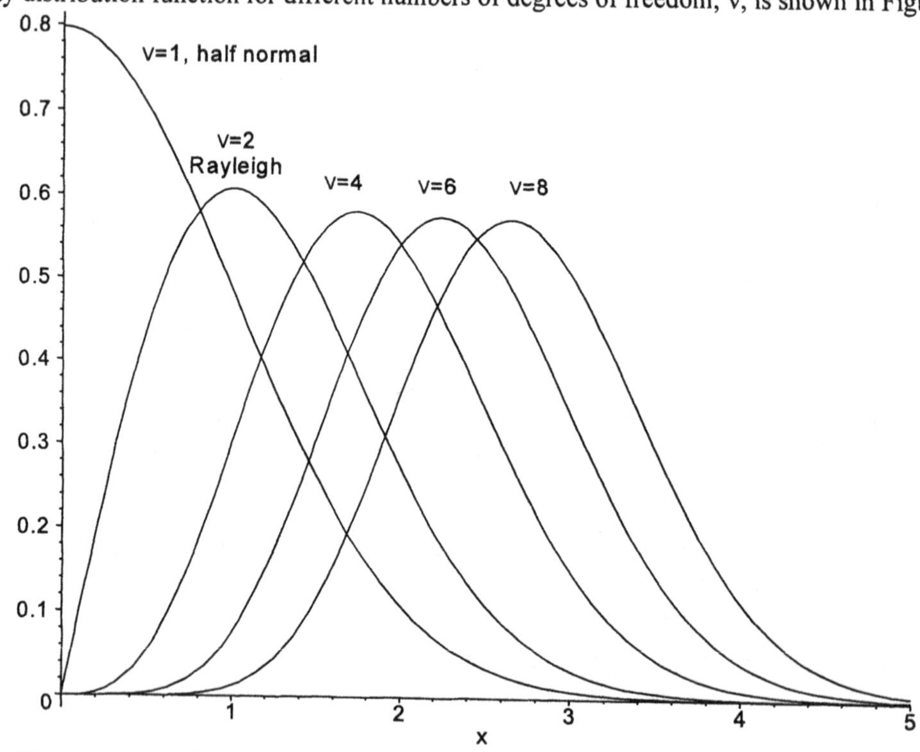

Figure 15.22 The chi probability distribution.

The main values for this distribution are [10]

$$\text{Mean} = \frac{\sqrt{2}\,\Gamma\left(\frac{1}{2}(\nu+1)\right)}{\Gamma\left(\frac{\nu}{2}\right)}$$

$$\text{Variance} = \frac{2\left(\Gamma\left(\frac{\nu}{2}\right)\Gamma\left(1+\frac{\nu}{2}\right)-\Gamma^2\left(\frac{\nu+1}{2}\right)\right)}{\Gamma^2\left(\frac{\nu}{2}\right)}$$

Two special cases are when:

- ν=1 Half-normal distribution;
- ν=2 Rayleigh distribution.

The cumulative distribution function is [10]

$$P(x) = 1 - \frac{\gamma\left(\frac{\nu}{2}, \frac{x^2}{2}\right)}{\Gamma(\nu)} \tag{15.54}$$

where $\gamma(a, \eta) = \int_0^a \frac{t^{\eta-1}e^{-t}}{\Gamma(\eta)}\, dt$.

15.3.4 Negative exponential distribution

The negative exponential distribution is a gamma distribution when $\eta = 1$:

$$f(x, \lambda) = \lambda e^{-\lambda x} \tag{15.55}$$

This has a cumulative distribution function

$$F(x, \lambda) = 1 - e^{-\lambda x} \tag{15.56}$$

A number of negative exponential distributions are shown in Figure 15.23.

The negative exponential distribution represents the power distribution for a Rayleigh voltage distribution. A typical example is the chance that a single noise pulse exceeds a threshold.

The characteristic function is

$$C(\xi) = \frac{1}{1 + j2\pi\frac{\xi}{\lambda}} \tag{15.57}$$

An example with $\lambda = 1$ is shown in Figure 15.24.

15.4 OTHER DISTRIBUTIONS BOUNDED ON ONE SIDE

Other common distributions bounded on one side are the Weibull distribution and a special case, the Rayleigh distribution.

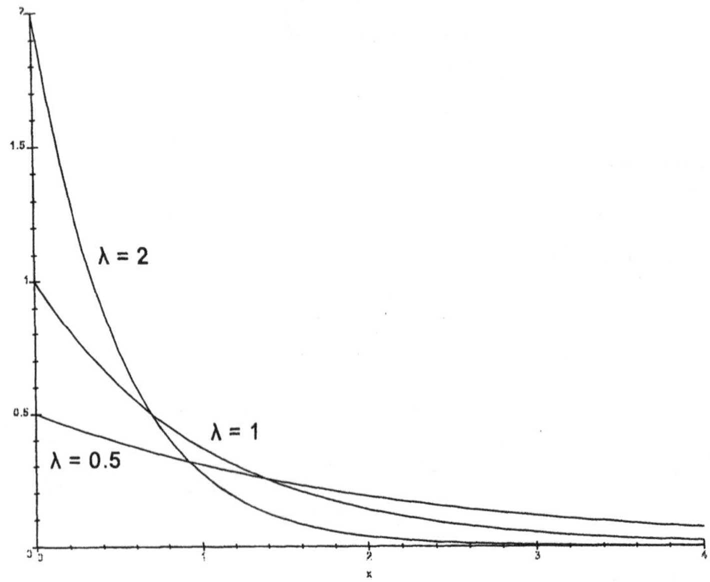

Figure 15.23 Negative exponential distribution for $\lambda = 0.5$, 1, and 2.

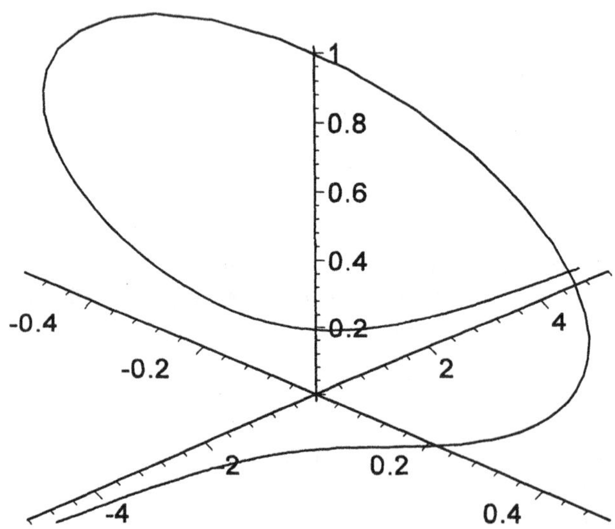

Figure 15.24 An example of the characteristic function of the negative exponential distribution, $\lambda = 1$.

15.4.1 Weibull distribution

The Weibull distribution is used in reliability calculations and to describe Weibull clutter [11]. The Weibull probability distribution function is given by

$$f(t, \eta, \sigma) = \frac{\eta}{\sigma}\left(\frac{t}{\sigma}\right)^{\eta-1} e^{-\left(\frac{t}{\sigma}\right)^{\eta}} \quad t > 0,\ \sigma > 0,\ \eta > 0 \tag{15.58}$$

$$= 0 \quad \text{elsewhere}$$

The parameter, σ, is the scale parameter and η is the shape parameter. The cumulative distribution is

$$F(t, \eta, \sigma) = 1 - e^{-\left(\frac{t}{\sigma}\right)^{\eta}} \tag{15.59}$$

Typical Weibull distributions are shown in Figure 15.25.

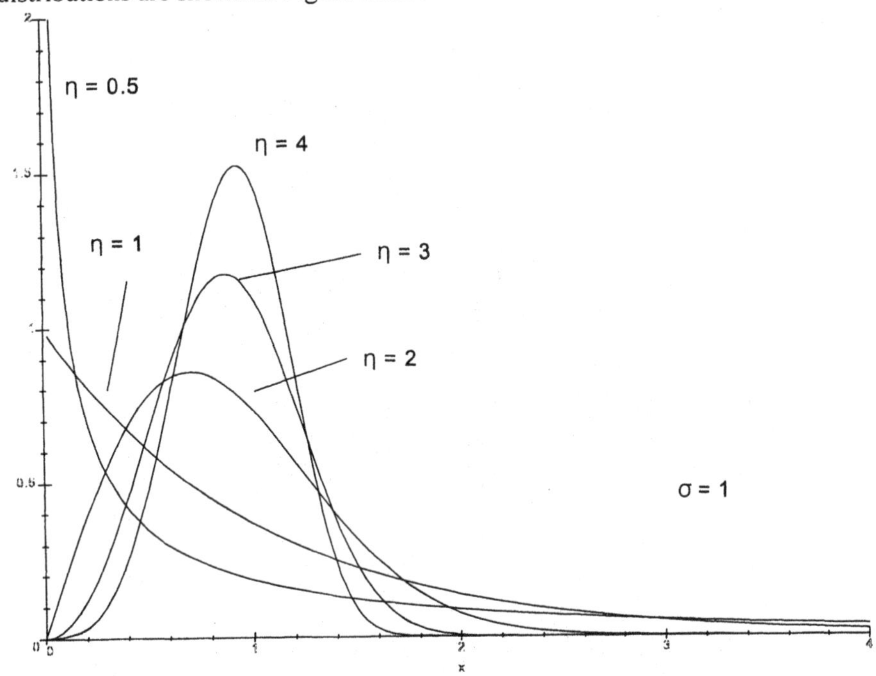

Figure 15.25 Weibull distributions with $\eta = 0.5$, 1, 2, 3, and 4 with $\sigma = 1$.

When $\eta = 2$ the Weibull distribution becomes the Rayleigh distribution.

$$f(t, \sigma) = \frac{2}{\sigma^2}\, t e^{-\left(\frac{t}{\sigma}\right)^{\eta}} \quad t > 0,\ \sigma > 0,\ \eta > 0 \tag{15.60}$$

$$= 0 \quad \text{elsewhere}$$

15.5 DISCRETE DISTRIBUTION: BINOMIAL DISTRIBUTION

Unlike the other distributions, the binomial distribution exists only for integer values of the ordinate. It describes the chance of x successes in n trials if the probability of success in each trial is p [1, p. 138]. The probability distribution is given by

$$p(x, n, p) = \frac{n!}{x!(n - x)!}\, p^{x} (1 - p)^{n-x} \tag{15.61}$$

This is shown for $n = 10$ and $p = 50\%$ in Figure 15.26(a).

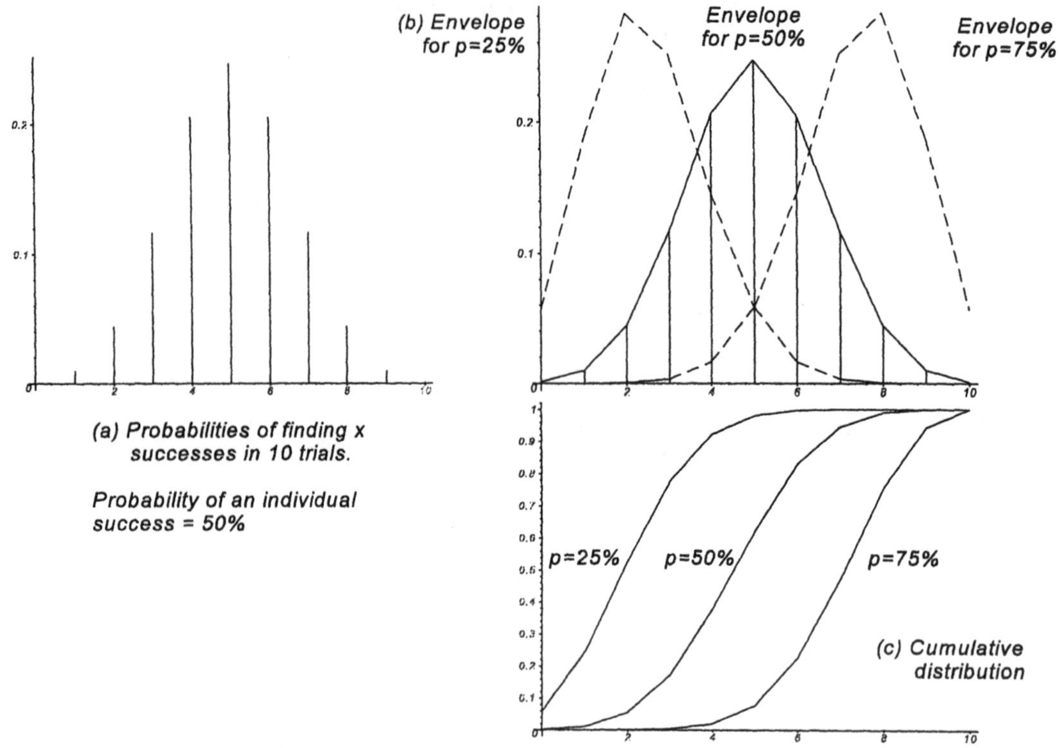

Figure 15.26 The binomial probability distributions and cumulative distributions.

The envelope of the probability distribution is drawn connecting the tops of the individual probabilities. These are shown for $p = 25\%$, 50%, and 75% in Figure 15.26(b).

The cumulative distribution is given by the area to the left of x, that is

$$P(q,\ n,\ p) = \sum_{x=0}^{q} \frac{n!}{x!(n-x)!} p^x (1-p)^{n-x} \tag{15.62}$$

These are shown in Figure 15.16(c). For the binomial distribution:

- Mean $= np$;
- Standard deviation $= \sqrt{np(1-p)}$.

This approximates to the normal or Gaussian distribution for larger values of n.

15.6 RANDOM NUMBERS

In a number of cases random numbers are used in simulations, for example, clutter. This uses the property that the area under all probability distribution functions is unity. The ordinate of the cumulative distribution function therefore is limited to the range between zero and unity. If ordinate values are given the values from a uniform distribution, the values for these on the abscissa are distributed according to the desired distribution [1, p. 242; 5, p. 950]. An example is shown in Figure 15.27.

This method has been used to generate the random numbers that simulate quantities for:

- Noise voltages in Chapter 2, Usual and unusual concepts;
- Echo characteristics in Chapter 6, Factors outside the radar, Section 6.3, Scattering;
- Noise voltages in Chapter 9, Detection;
- Noise voltages in Chapter 11, Signal processing.

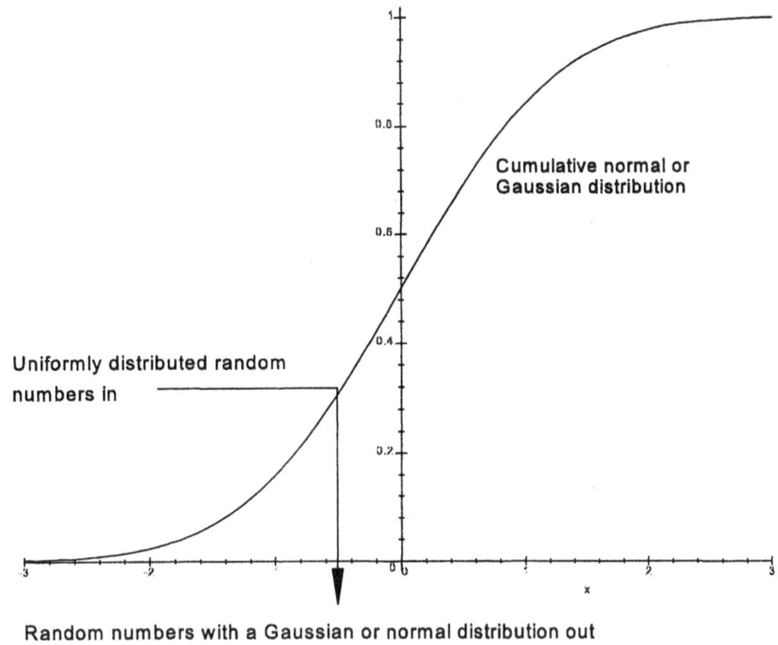

Figure 15.27 A method of generating random numbers using the inverse method.

REFERENCES

1. Hahn, G. J., and S. S. Shapiro, *Statistical Models in Engineering*, New York: John Wiley, 1967.
2. Parzen, E., *Modern Probability Theory and Its Applications*, New York: Wiley, 1960.
3. Weatherburn, C. E., *A First Course in Mathematical Statistics*, Cambridge: Cambridge University Press, 1968.
4. Di Franco, J. V., and W. L. Rubin, *Radar Detection*, Englewood Cliffs, New Jersey: Prentice-Hall, 1968, reprinted by Artech House, 1980.
5. Meikle, H. D., *A New Twist to Fourier Transforms,* Weinheim, Germany: Wiley-VCH, 2004.
6. Rhoda, G., *Tropospheric Scatter Links*, Norwood, Massachusetts: Artech House, 1988.
7. Abramowitz, M., and I. A. Stegun, *Handbook of Mathematical Functions*, New York: Dover, 1965.
8. Cox, D. R., and W. L. Smith, *Queues*, London: Chapman and Hall, 1961.
9. Cox, D. R., *Renewal Theory*, London: Methuen, London, 1961, p. 20.
10. Weisstein, E. W., (ed.), *CRC Concise Encyclopedia of Mathematics*, Boca Raton, Florida: CRC Press, 1999.
11. Sekine, M., and Y. H. Mao, *Weibull Radar Clutter*, Stevenage, Herts: Peter Peregrinus, 1990.

Chapter 16

Transforms

Transforms have been used in a number of chapters:

- Chapter 2, Usual and unusual concepts, introduced the method of illustrating complex functions and their Fourier transforms used in this book.
- In Chapter 3, Transmitters, Fourier transforms were used to calculate transmitter spectra with and without pulse compression.
- Chapter 5, Antennas, used Fourier and Hankel transforms to calculate antenna patterns, and inverse Fourier transforms to calculate the illumination function for a specified pattern.
- Chapter 8, Matched and matching filters, showed the use of inverse Fourier transforms to calculate the shape of filtered pulses, with and without pulse compression.
- Chapter 11, Signal processing, showed the use of Fourier and *z* transforms for calculating the effects of signal processing in the frequency and time domains.
- Chapter 12, Threshold and detection, used Fourier transforms to add the probability distributions of signals and noise.
- Chapter 15, Statistics, employed Fourier transforms to find the statistical sums of probability distributions.

Common aspects of the use of transforms in this book are discussed in this chapter, and full mathematical descriptions are to be found in textbooks, examples are [1-3]. The notation used in this book is shown in Table 16.1.

Table 16.1
Common notation for transforms

Basic function		Transformed function	
Symbol	Function	Symbol	Function
$h(t)$	Time function	$H(p)$	Laplace transform
		$H(\omega)$	Fourier transform, ω convention
		$H(f)$	Fourier transform, f convention
		$H(z)$	z transform
$h(x)$	Distance function	$H(u)$	Fourier transform, $u' = w/\lambda \sin\theta$
$p(x)$	Probability distribution	$C(\xi)$	Characteristic function with $-f$ notation
$h(r)$	Radius function	$H(u)$	Hankel transform, $u' = r/\lambda \sin\theta$

Tapering functions are illustrated in Appendix B.

16.1 CONVENTIONS FOR THE FOURIER TRANSFORM

An electrical signal is normally represented as a change of voltage with time which uniquely defines the signal. This signal extends over a range of frequencies, called the spectrum. The complex spectrum also defines the signal uniquely. The link between the two is the Fourier transform.

There are a number of conventions for the Fourier transform. Two, in terms of frequency (f) and angular frequency (ω), are listed in [1-5]. In this book the following definition, the $-f$ notation, is used:

$$H(f) = \int_{-\infty}^{\infty} h(t)\, e^{-j\,2\pi f t}\, dt \tag{16.1}$$

where j is $\sqrt{-1}$;
f is the frequency in Hz;
t is the time in seconds.

The inverse transform is

$$h(t) = \int_{-\infty}^{\infty} H(f)\, e^{j\,2\pi f t}\, df \tag{16.2}$$

Note that statisticians, mathematicians, and some physicists define the Fourier transform in terms of time in seconds, ω radians/s ($\omega = 2\pi f$ radians/s), and positive exponent called the +ω notation in this book

$$H(\omega) = \int_{-\infty}^{\infty} h(t)\, e^{i\omega t} dt \tag{16.3}$$

The inverse transform is

$$h(t) = \frac{1}{2\pi}\int_{-\infty}^{\infty} H(\omega)\, e^{-i\omega t}\, d\omega \tag{16.4}$$

where i is $\sqrt{-1}$ and $\frac{d\omega}{dt} = 2\pi$.

There is also a convention using +ω and the factor $1/\sqrt{2\pi}$ for the Fourier transform and its inverse [5].

Normally closed functions are sought to plot the function $H(f)$ or its inverse $h(t)$. Mathematical programs are capable of finding them or performing a large number of numerical integrations for various values of $h(t)$ or $H(f)$ to give a good estimate of the spectrum or its inverse.

16.2 SOME POLYPHASE AND SINGLE PHASE FOURIER TRANSFORMS

As can be seen the Fourier transform is a complex transformation which extends over all time and all frequencies, positive and negative. Starting with a polyphase waveform, frequency f_0 Hz

$$h(t) = e^{j\,2\pi f_0 t} \qquad (volt) \tag{16.5}$$

its Fourier transform is

$$H(f) = \int_{-\infty}^{\infty} e^{j\,2\pi f_0 t}\, e^{-j\,2\pi f t}\, dt = \int_{-\infty}^{\infty} e^{j\,2\pi (f_0 - f)\, t}\, dt \tag{16.6}$$

The right hand part of (16.6), the “test” waveform, is shown in Figure 16.1 for the frequency f of 1 Hz.

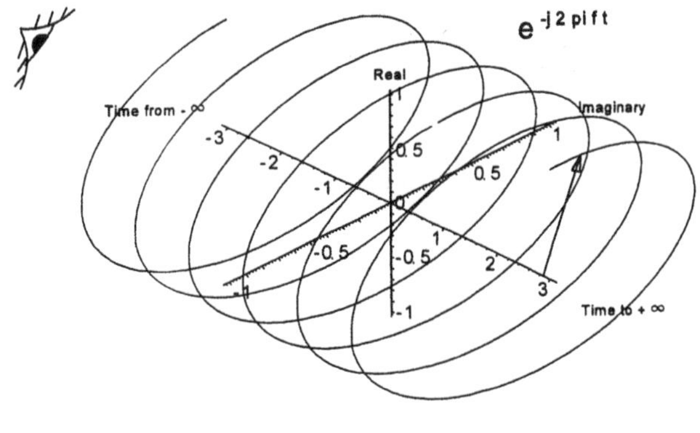

Figure 16.1 The "test" waveform at the frequency f (here 1 Hz).

To obtain the Fourier transform, the helix in Figure 16.1 is multiplied by the input waveform $h(t)$ and summed for each value of f. If in the sum or integral in (16.6) the frequency f is not equal to f_0, then the result is a helix at the difference frequency (see Figure 16.2). The integral is the sum of the complex values times the ordinate between minus and plus infinity, here zero.

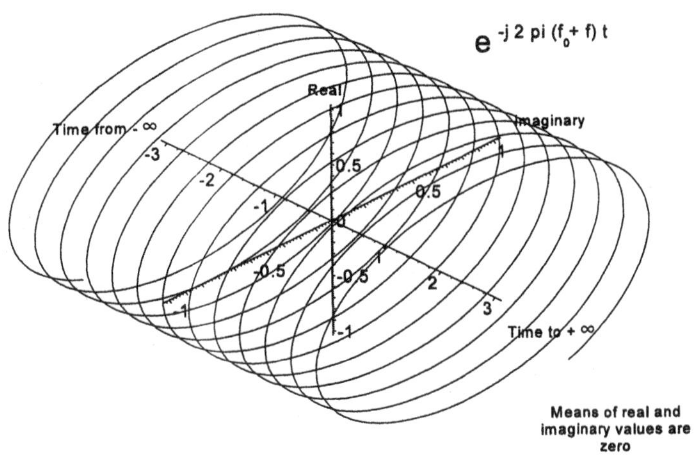

Figure 16.2 The helix that results when the "test" frequency, f, does not match any component of the input frequency, f_0.

The eye in Figure 16.1 looking from minus infinity sees a vector rotating clockwise at an angular velocity of $-2\pi f$ radians/s as it moves towards more positive time. Only when the frequency f is exactly equal to f_0 in the integral in (16.6) is the polyphase waveform helix "untwisted" and the result is a straight line (see Figure 16.3). The sum is e^0 or unity times the ordinate. If the values of the integral are plotted on the spectrum diagram in Figure 16.3, the resulting spectrum is a Dirac spike, here +1 Hz. Dirac functions have an undefined amplitude, width tending to zero, and area unity [1, p. 69].

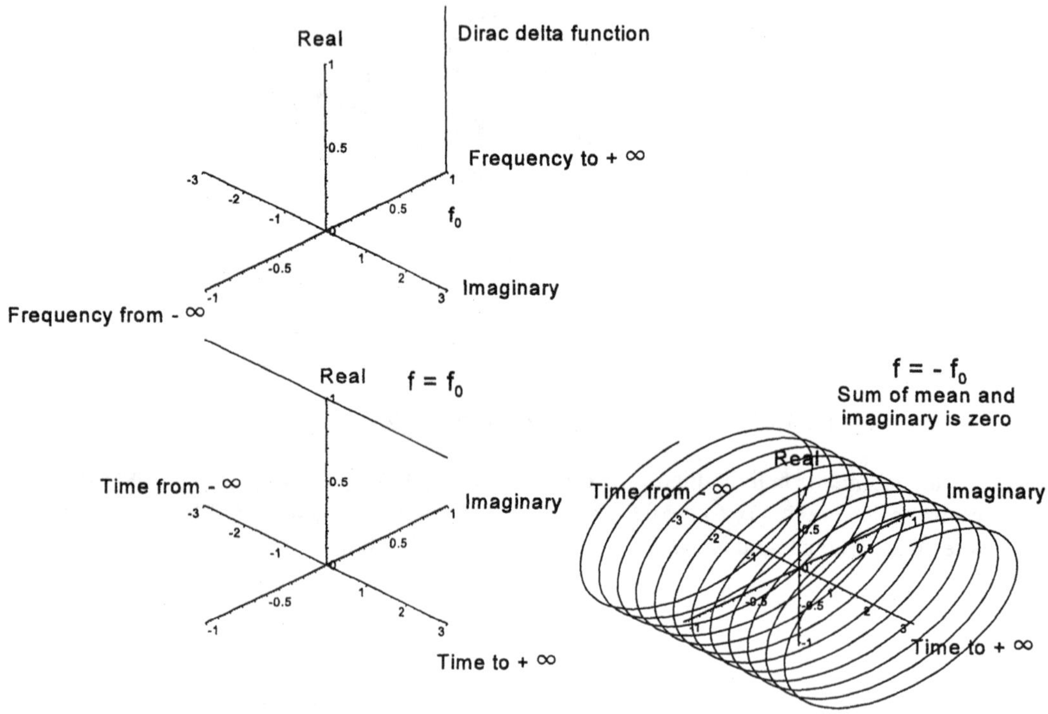

Figure 16.3. The function for integration for a polyphase signal when the test frequency $f = f_0$ and $f = -f_0$.

If the polyphase waveform has a negative sequence, that is $h(t) = e^{-j\,2\pi f_0 t}$, then it is equivalent to Figure 16.3 with a "negative" frequency f_0. The Fourier integral then gives a Dirac spike at $-f_0$. It may be shown that the direction of the Dirac spike, as a vector, corresponds to the phase difference between the input waveform, *h(t)*, and the test waveform when it passes through f_0.

In most cases of signal processing (radar is an exception), alternating voltage signals are treated, namely sine and cosine waveforms. These are discussed in the next sections.

16.2.1 Single phase cosine wave

The Fourier transform of a cosine wave is

$$H(f) = \int_{-\infty}^{\infty} \cos 2\pi f_0 t \; e^{-j\,2\pi f t}\, dt \tag{16.7}$$

A cosine wave is the sum of two helices $\cos\, 2\pi f t = \left(e^{j2\pi f t} + e^{-j2\pi f t}\right)/2$. As the test frequency is swept from minus infinity to plus infinity, the expression inside the integral is two helices of frequency $\pm|f_0 - f|$. The mean values of the coils of the helix are zero. Only in the two cases when $|f_0 - f|$ is zero does the helix has a mean value of +1/2 as shown in Figure 16.4. This then has a spectrum with two real spectral lines.

Converting the expression inside the integral to exponential form when $f_0 = |f|$

$$\begin{aligned} \cos 2\pi f_0 t \;\; e^{-j\,2\pi f t} &= \frac{e^{+j\,2\pi f_0 t} + e^{-j\,2\pi f_0 t}}{2}\; e^{-j\,2\pi f t} \\ &= \frac{1}{2} + \frac{e^{-j\,4\pi f_0 t}}{2} \qquad \textit{when}\;\; |f| = f_0 \end{aligned} \tag{16.8}$$

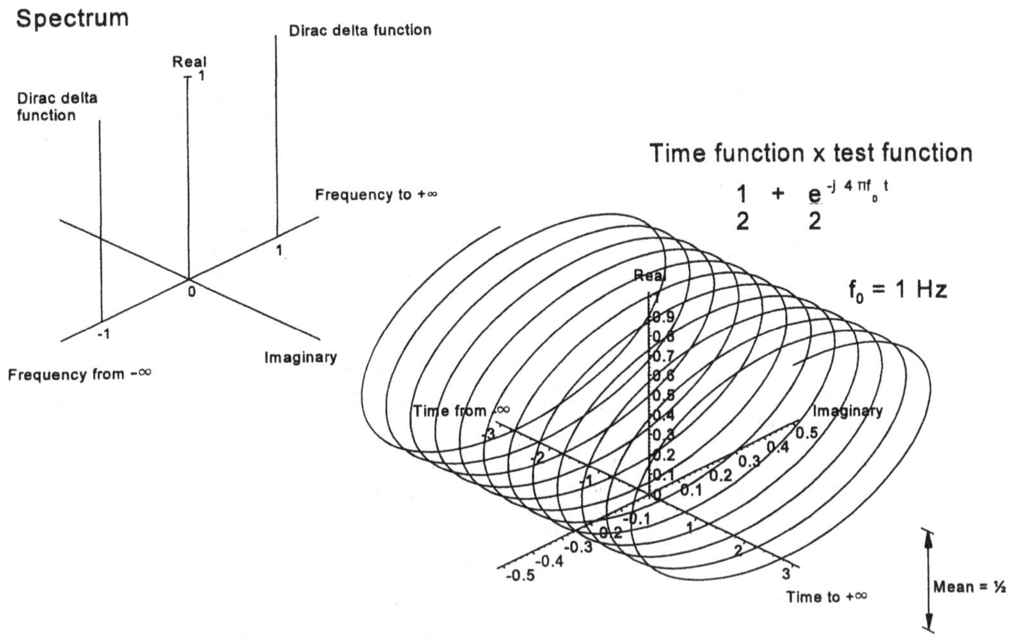

Figure 16.4 The function inside the integral for the Fourier transform of a cosine wave and its spectrum.

The means, or the position of the centers of the helices, are +½. The cosine function is an even function (cos x = cos–x) so the helices are the same whether f is positive or "negative". The energy in the spectrum is divided two "half" delta spikes at frequencies of $\pm f_0$ Hz (here $\pm$ 1 Hz).

The projection in Figure 16.4 has been chosen so that the real axis appears to be the abscissa when looking into the diagram with positive direction of the independent variables in time and frequency to the right.

16.2.2 Single phase sine wave

A sine wave is the sum of two helices $\sin 2\pi ft = \left(e^{j2\pi ft} - e^{-j2\pi ft}\right)/2j$. The divisor j indicates that Figure 16.2 is viewed from the top. In contrast to the cosine function, the sine function is odd, that is, sin–x = –sin x. The Fourier transform of a sine wave is

$$H(f) = \int_{-\infty}^{\infty} \sin 2\pi f_0 t \; e^{-j 2\pi f t}\, dt \tag{16.9}$$

As the test frequency, f, is varied between minus infinity and plus infinity, the expression inside the integral is again a helix of frequency $\pm|f_0 - f|$. Again this difference frequency has a mean value of zero and does not contribute to the spectrum. When the test frequency f passes through $-f_0$ and $+f_0$, the axis helix moves to –1/2 and +1/2 away from the time axis.

When $f = +f_0$, the function inside the transform is

$$\begin{aligned} \sin 2\pi f_0 t \; e^{-j 2\pi f t} &= \frac{e^{+j 2\pi f_0 t} - e^{-j 2\pi f_0 t}}{j\,2}\, e^{-j 2\pi f t} \\ &= \frac{1}{j2} - \frac{e^{-j 4\pi f_0 t}}{j2} \qquad when\ f = +f_0 \\ &= -j\frac{1}{2} - j\frac{e^{-j 4\pi f_0 t}}{2} \end{aligned} \tag{16.10}$$

If the test helix has a negative phase sequence (loosely negative frequency), then the function inside the transform simplifies to

$$\begin{aligned} \sin(2\pi f_0 t)\; e^{-j\,2\pi\,(-f)\,t} &= \frac{e^{+j\,2\pi\,f_0\,t} - e^{-j\,2\pi\,f_0\,t}}{j\,2}\, e^{j\,2\pi\,f\,t} \\ &= \frac{e^{+j\,4\pi\,f_0\,t}}{j2} - \frac{1}{j2} \qquad when\ f = -f_0 \\ &= -j\frac{e^{+j\,4\pi\,f_0\,t}}{2} + j\frac{1}{2} \end{aligned} \tag{16.11}$$

These cases are illustrated in Figure 16.5. The means of the helixes are $-j/2$ and $+j/2$ showing the division of the spectral energy into two lines in the imaginary plane. The expressions for the Fourier transforms for the cosine and sine waveforms give the following rule:

- Even functions transform to Real, even Fourier transform;
- Odd functions transform to Imaginary, odd Fourier transform.

Real life is not so simple as the waveforms are often neither fully even or odd and often displaced in time (phase shift).

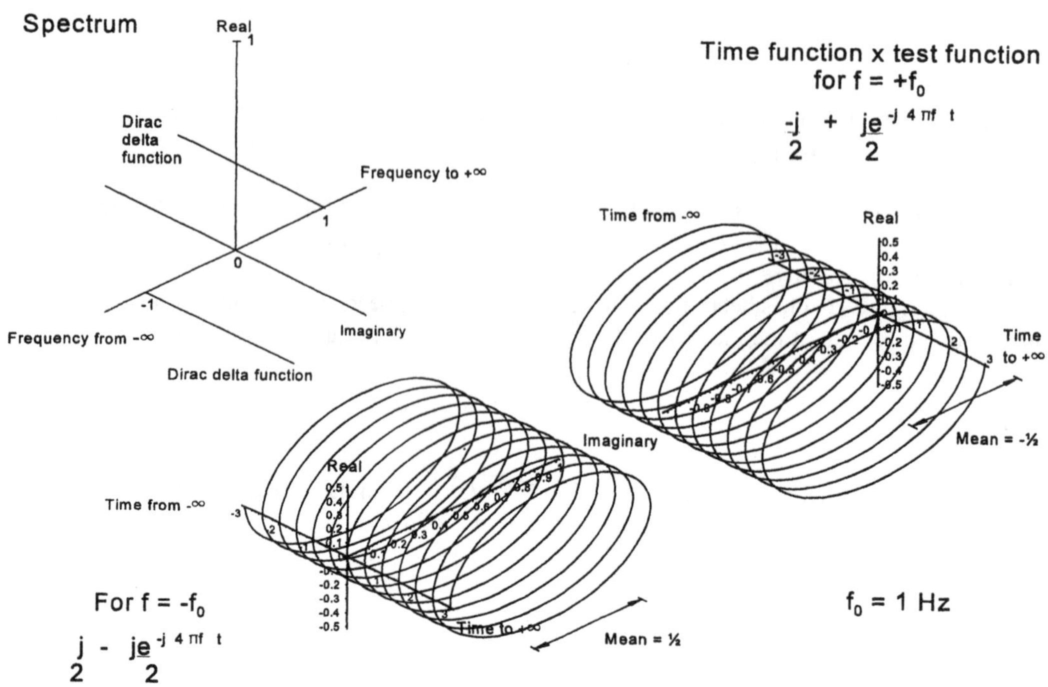

Figure 16.5 The function inside the integral for the Fourier transform of a sine wave and its spectrum.

16.2.3 Rectangular pulse

A rectangular pulse of width T seconds, amplitude A volt centered on zero time exists only between the times $-T/2$ to $+T/2$. This gives the integration limits for a voltage A to give the Fourier transform

$$
\begin{aligned}
H(f) &= A\int_{-\frac{T}{2}}^{+\frac{T}{2}} 1e^{-j2\pi ft}\,dt \\
&= A\,\frac{e^{-j2\pi f\frac{T}{2}} - e^{+j2\pi f\frac{T}{2}}}{-j2\pi f\frac{T}{2}} \\
&= A\,\frac{\sin\,\pi fT}{\pi fT}
\end{aligned}
\tag{16.12}
$$

The rectangular pulse, its spectrum, and a combined representation are shown in Figure 16.6. Note that when times shorter than from the creation up to the end of the world are considered, the components of the spectrum are no longer infinitely narrow lines and become definable shapes.

The projection for the combined time and spectrum diagrams has been chosen so that the real components of both functions are dominant when looking onto the page.

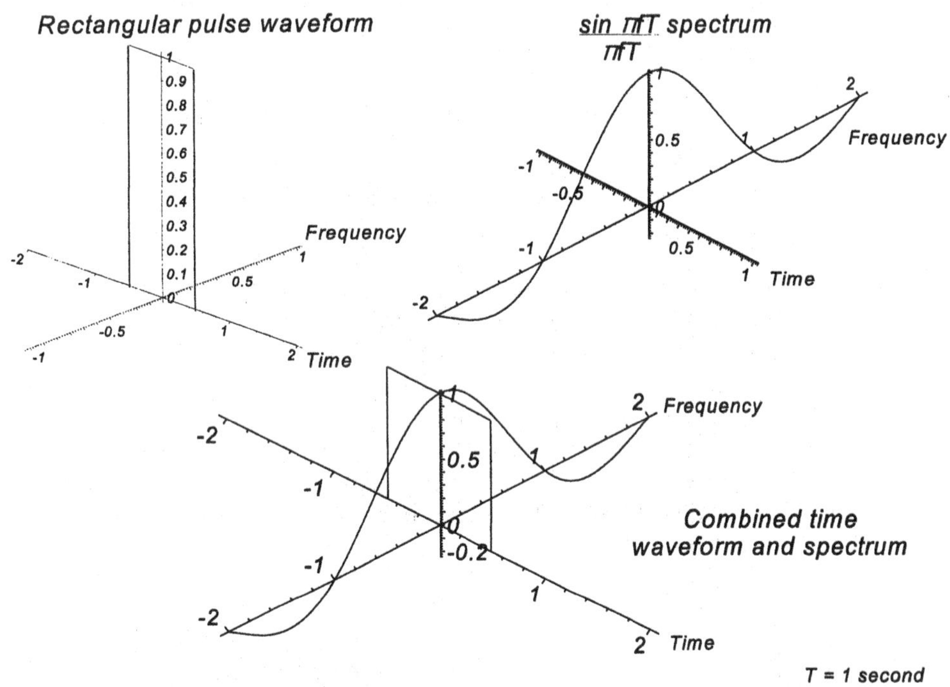

Figure 16.6 A one second rectangular pulse, its spectrum, and a combined representation.

16.2.4 Time shift

In normal situations there is no time less than zero. If the pulse centered on zero with a Fourier transform H(f) is moved into positive time by x seconds, then

$$
\begin{aligned}
\textit{Fourier transform (time shift x)} &= \int_{-\infty}^{+\infty} h(t-x)e^{-j2\pi ft}\,dt \\
&= \int_{-\infty}^{+\infty} h(t-x)e^{-j2\pi f(t-x)}\,d(t-x) \qquad (16.13)\\
&= H(f)\; e^{-j2\pi fx}
\end{aligned}
$$

The second term in (16.13) represents a helix with a frequency proportional to the time shift. This may be illustrated by moving the rectangular pulse in Figure 16.6 away from being centered on zero time. The results are shown in Figure 16.7.

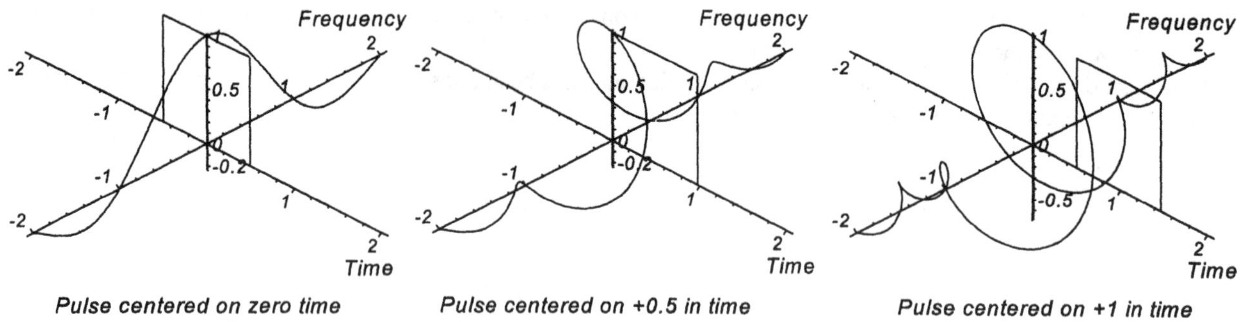

Figure 16.7 The effects on the spectrum of moving a rectangular pulse in time.

As the pulse is moved, the spectrum becomes a shaped helix with more turns/Hz proportional to the time shift. If the pulse is moved into negative time, the direction of the helix changes. Thus the complex spectrum is able to describe the times of occurrence and shapes of waveforms over all time. The modulus of the spectrum remains the same irrespective of time shift.

An interesting application of this occurs with pulse compression (see Chapter 3, Transmitters and Chapter 8, Matched and matching filters). The modulation waveform for a frequency modulated pulse has the shaped helical spectrum, depending on timing, shown in Figure 16.8. If the modulating waveform is centered on zero time, the accelerating helical (phase) away from zero shows the time shifts of the components. These components may be reassembled about time zero as the compressed pulse.

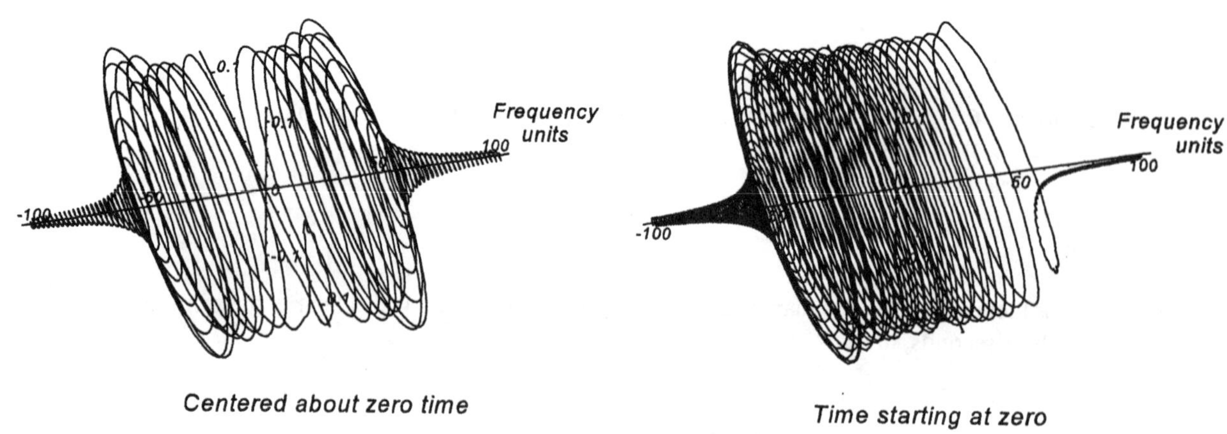

Figure 16.8 The shaped helical spectra for the modulation on a frequency modulated pulse.

In the real world there is no negative time. The pulse must be delayed by at least half a pulse width (center of Figure

16.7) in order to start at zero time. The spectrum on the left of Figure 16.7 is multiplied by the helix representing the time shift (16.13). Thus the spectrum of the modulation of a frequency modulated pulse may be moved to positive time only as illustrated on the right of Figure 16.8. Here the components are assembled at the end of the pulse on compression. The modulus of the modulation spectrum is the same in both cases.

16.2.5 Phase shift

The Fourier transform also expresses the phase of the waveform, or the angle at which the waveform crosses the plane formed by the frequency axis and the real axis.

$$\begin{aligned} \textit{Fourier transform (phase shift } \phi) &= \int_{-\infty}^{+\infty} h(t)\, e^{-j\phi}\, e^{-j2\pi ft}\, dt \\ &= H(f)\, e^{-j\phi} \end{aligned} \tag{16.14}$$

When $f = 0$, the Fourier transform of a Gaussian pulse of phase ϕ is twisted by the angle ϕ as shown in Figure 16.9.

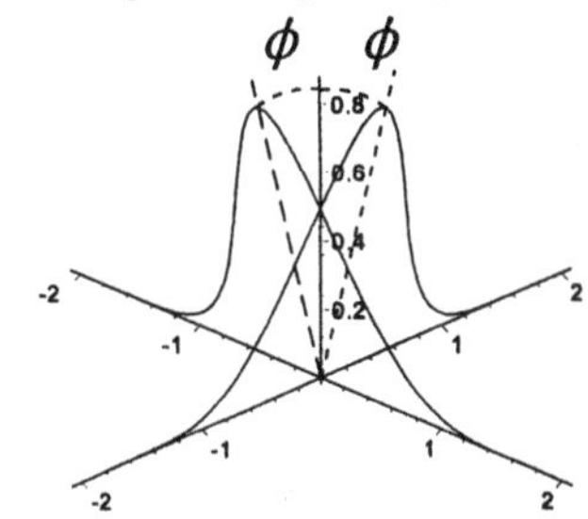

Figure 16.9 The effect of phase shift on a Gaussian pulse and spectrum. [Source: Meikle, H. D., *A New Twist to Fourier Transforms*, Weinheim: Wiley-VCH, 2004.]

16.2.6 Examples of Fourier transforms in the complex plane

Displaying Fourier transforms of the data collected from sensors in the complex plane has advantages when analyzing real data, as in Figure 16.10. The software used expands the real and imaginary scales so that the data fills the picture. The imaginary scales show interference and noise, whereas the real scale chosen compresses the noise so that it is not visible.

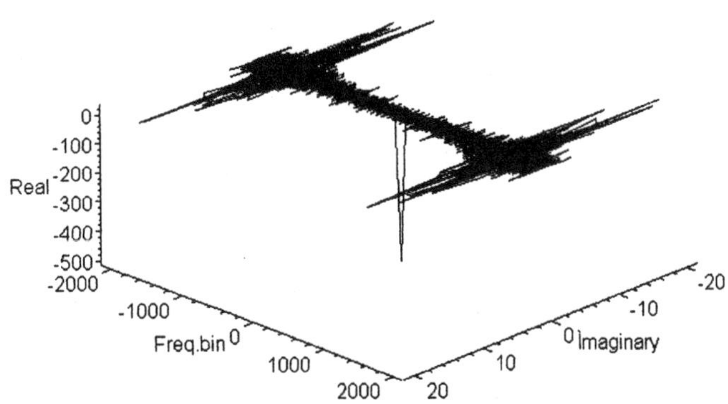

Figure 16.10 An example of a discrete Fourier transform in the complex plane.

An expanded center spike is shown in Figure 16.11.

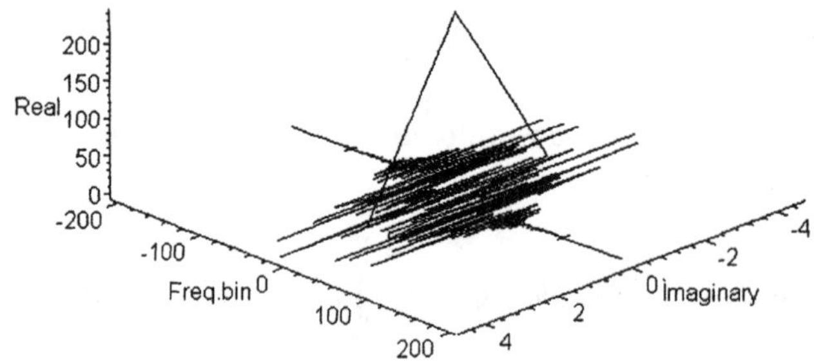

Figure 16.11 An example of a discrete Fourier transform in the complex plane.

16.2.7 Addition and subtraction

The spectrum of a complex waveform is the sum of the spectra of its components, namely,

$$\begin{aligned} H_1(f) + H_2(f) &= \int_{-\infty}^{+\infty} h_1(t)\, e^{-j2\pi ft}\, dt + \int_{-\infty}^{+\infty} h_2(t)\, e^{-j2\pi ft}\, dt \\ &= \int_{-\infty}^{+\infty} \left(h_1(t) + h_2(t)\right) e^{-j2\pi ft}\, dt \end{aligned} \tag{16.15}$$

With simple amplitude modulation a cosine wave is multiplied by a modulating function A exp(–j $2\pi f_m t$), where f_m is the modulation frequency.

$$\begin{aligned} \textit{Modulated waveform} &= A \cos(2\pi ft)\, e^{-j2\pi f_m t} \\ &= A\, e^{+j2\pi ft}\, e^{-j2\pi f_m t} + A\, e^{-j2\pi ft}\, e^{-j2\pi f_m t} \\ &= A\, e^{+j2\pi(f - f_m)\, t} + A\, e^{-j2\pi(f + f_m)\, t} \end{aligned} \tag{16.16}$$

The spectral lines occur at plus and minus the modulating frequency but rotate in different directions, a Hermite form as one component is the complex conjugate of the other.

16.2.8 Differentiation

The differentials of Fourier transforms are used for the difference patterns of monopulse antennas and for differentiating filters [1, p. 117].

$$\int_{-\infty}^{+\infty} f'(t) = \int_{-\infty}^{+\infty} limit \frac{h(t+\Delta t)-h(t)}{\Delta t} e^{-j2\pi ft}\, dt$$
$$= \frac{1}{\Delta t}\left(\int_{-\infty}^{+\infty} limit\ h(t+\Delta t)e^{-j2\pi ft}\, dt - \int_{-\infty}^{+\infty} limit\ h(t)e^{-j2\pi ft}\, dt\right) \quad (16.17)$$
$$= limit \frac{e^{j2\pi f\Delta t}H(f) - H(f)}{\Delta t}$$
$$= j\ 2\pi f\ H(f)$$

Thus the differential of a Fourier transform is itself times $j\,2\pi f$.

16.2.9 Convolution, the multiplication of Fourier transforms

Convolution is the mathematical process involving loss of detail by which a pulse is stretched, a point is optically defocussed on its way through a system, the reduction of sound quality caused by misalignment of the heads on tape recorders, or the spread of characteristics when an item is made up of a number of components. The loss of detail may be also used to smooth, or remove, the spikes from measured data. In this book convolution is used to add the sum of two probability distributions in Chapter 12 and implicitly for finding the sum of the accuracy budget in Chapter 14.

When summing two statistical distributions it must be assumed that points with high probability will add to points of low probability. As an example, two negative exponentials are used of the form

$$\begin{aligned} \mathrm{f}(t) &= a\ \exp(-\alpha t) \quad t > 0 \\ &= 0 \qquad\qquad\quad \text{otherwise} \end{aligned} \quad (16.18)$$

The curve $f(t) = \exp(-t)$ is shown in Figure 16.12 and additionally the curve $g(t) = 0.5\exp(-2t)$ in Figure 16.13(a). If they are probability distributions, what will be the probability distribution of the sum? The probability distribution of their sum is defined as

$$\mathrm{h}(t) = \int_{-\infty}^{+\infty} f(t)\ g(t-u)\ du \quad \text{or} \quad \int_{-\infty}^{+\infty} f(t-u)\ g(t)\ du \quad (16.19)$$

Figure 16.12 shows $f(t)$ having $g(t)$ dragged across it backwards (the German term is *Faltung* or folding) and integrated over u for each value of t for which both curves exist so that

$$\begin{aligned} a\exp(-\alpha t) \otimes b\exp(-\beta t) &= ab\int_{-\infty}^{+\infty} \exp(\alpha u)\ \exp\beta t - \beta u)\ du \quad \text{(a)} \\ &= ab\ \exp(\beta t) \int_0^t \exp(\alpha u - \beta u)\ du \quad \text{(b)} \\ &= ab\ \exp(\beta t)\ \frac{\exp(\alpha t - \beta t) - 1}{\beta - \alpha} \quad \text{(c)} \\ &= ab\ \frac{\exp(-\alpha t) - \exp(-\beta t)}{\beta - \alpha} \quad \text{(d)} \end{aligned} \quad (16.20)$$

where the symbol, $\otimes$, means convolved with.

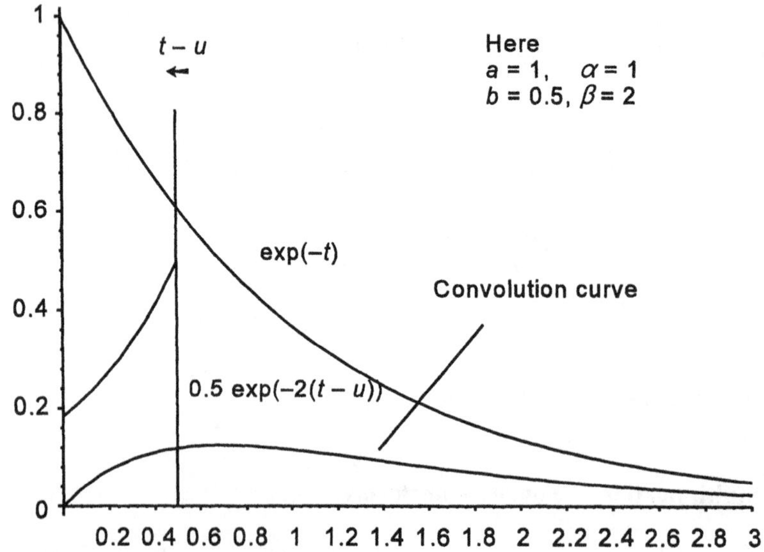

Figure 16.12 The convolution of two negative exponential curves in the example (16.20).
[Source: Meikle, H. D., *A New Twist to Fourier Transforms*, Weinheim: Wiley-VCH, 2004.]

In terms of Fourier transforms, the exponential functions have Fourier transforms of the form

$$H(f) = \frac{a}{\alpha - j2\pi f} \tag{16.21}$$

The product of the transforms is

$$\frac{a}{\alpha - j2\pi f}\,\frac{b}{\beta - j2\pi f} = \frac{ab}{\alpha\beta + 4\pi^2 f^2 - j2\pi f(\alpha - \beta)} \tag{16.22}$$

The convolution curve (the inverse Fourier transform) may be found using Campbell's and Foster's pair 448 [6]

$$\text{Convolution function} = ab\frac{\exp(\alpha t) - \exp(-\beta t)}{\beta - \alpha} \tag{16.23}$$

and is the same as the function found directly. In general the Fourier transform method is easier and the example in (16.20) is illustrated in Figure 16.13.

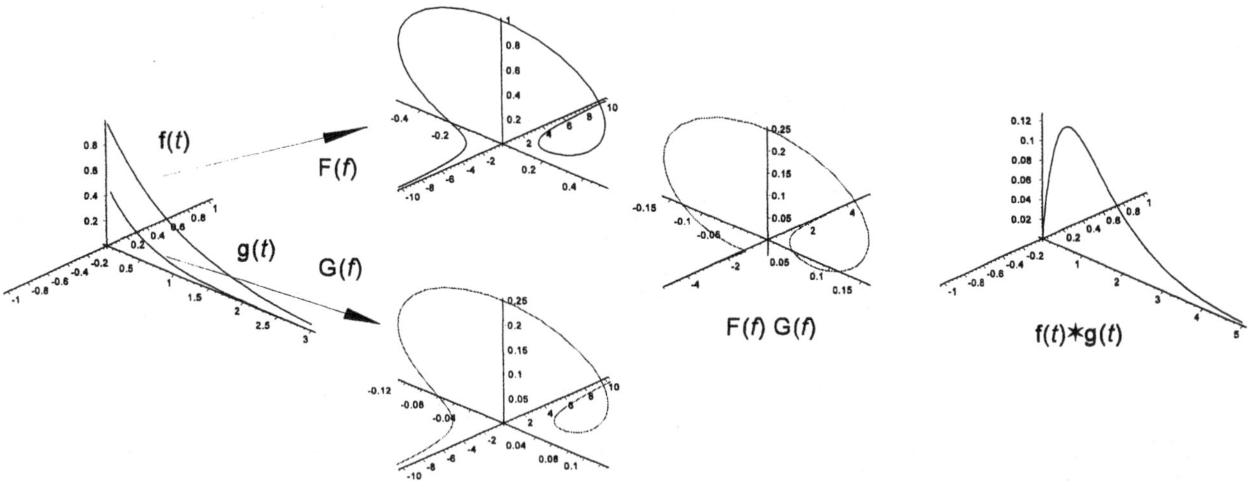

(a) Two exponential functions (b) Fourier transforms of the two functions (c) Product of the Fourier transforms (d) Convolution function

Figure 16.13 The convolution of two exponential functions. [Source: Meikle, H. D., *A New Twist to Fourier Transforms*, Weinheim: Wiley-VCH, 2004.]

The great majority of the statistical distributions for accuracy are Gaussian and the aim is to find the sum of biases, standard deviations, higher moments, and the shape of the resulting distribution. This treatment is taken from [3].

Taking two Gaussian distributions

$$h_1(t) = \frac{1}{\sqrt{2\pi\sigma}} \exp\left(-\left(\frac{t - \mu_1}{2\sigma_1} \right)^2 \right) \quad \text{and} \quad h_2(t) = \frac{1}{\sqrt{2\pi\sigma}} \exp\left(-\left(\frac{t - \mu_2}{2\sigma_2} \right)^2 \right) \tag{16.24}$$

Two such pulses with means of 3 and 6 are shown in Figure 16.14(a).

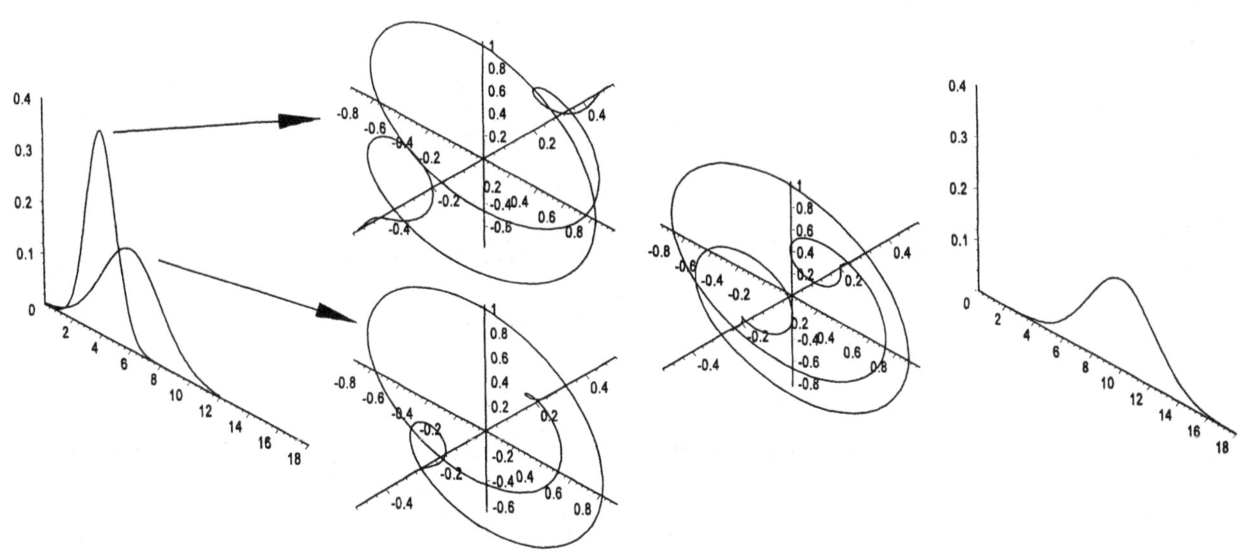

(a) Two Gaussian pulses (b) Fourier transforms (c) Fourier transform of covolution (d) Convolution function

Figure 16.14 The convolution of two Gaussian distributions. [Source: Meikle, H. D., *A New Twist to Fourier Transforms*, Weinheim: Wiley-VCH, 2004.]

The Fourier transform of the Gaussian distribution using the $-f$ notation is (see Figure 16.14(b))

$$H(f)) = \int_{-\infty}^{+\infty} \frac{1}{\sqrt{2\pi}\sigma} \exp\left(-\left(\frac{t-\mu}{2\sigma}\right)^2\right) \exp(-j2\pi f\, t)\, dx \tag{16.25}$$

$$= \exp(-j2\pi\mu f)\, \exp(-j2\pi^2\sigma^2 f^2)$$

and the product of the Fourier transforms is (Figure 16.14(c))

$$H_1(f)\, H_2(f) = \exp(-j2\pi\, \overline{\mu_1 + \mu_2}\, f)\, \exp(-j2\pi^2\, \overline{\sigma_1^2 + \sigma_2^2}\, f^2) \tag{16.26}$$

The rules of complex multiplication in polar coordinates are followed in that the phase angles are added so that the first term of the product of the Fourier transforms increases the number of turns per hertz and the second term narrows the shaped helix. As with moment generating functions, the moments μ and σ^2 add in the Fourier transform. The resulting distribution is the inverse of the Fourier transform (Figure 16.14(d)), namely,

$$\textit{Combined distribution} = \frac{1}{\sqrt{2\pi\, \overline{\sigma_1^2 + \sigma_2^2}}} \exp\left(-\frac{(t - \overline{\mu_1 + \mu_2})^2}{2\, \overline{\sigma_1^2 + \sigma_2^2}}\right) \tag{16.27}$$

The resulting distribution is Gaussian, the mean $(\mu_1 + \mu_2)$ is the sum of the means, and the variance (σ^2) is the sum of the variances $(\sigma_1^2 + \sigma_2^2)$ that results in a displaced, flattened distribution.

16.2.10 Cross-correlation, multiplication with complex conjugates

Cross-correlation is used in Chapter 8 to reassemble compressed pulses and in Chapter 11 in finite impulse response filters and is a measure of how two waveforms match. The cross-correlation function is given as

$$\mathrm{h}(t) = f(t) \bigstar g(t) = \int_{-\infty}^{+\infty} f^*(t)\, g(u - t)\, du \quad \text{or} \quad \int_{-\infty}^{+\infty} f(u - t)\, g^*(t)\, du \tag{16.28}$$

where the five-pointed star means "correlated with", $f^*(t)$ is the complex conjugate of $f(t)$ and $g^*(t-u)$ is the complex conjugate of $g(t-u)$.

Notice that the variable u is positive and one curve is dragged over the other without time reversal but there is a reversal in phase angle.

In terms of Fourier transforms, the transform of the cross-correlation function is given by

$$H(f) = F^*(f)\, G(f) = F(f)\, G^*(f)$$

where $F^*(f)$ is the complex conjugate of $F(f)$ and $G^*(f)$ is the complex conjugate of $G(f)$.

Using the curves in (16.24) and the Fourier transform in (16.25), the product of the Fourier transforms with the complex conjugate is

$$H_1(f) \bigstar H_2(f) = \exp(-j2\pi\, \overline{\mu_1 - \mu_2}\, f)\, \exp(j2\pi^2\, \overline{\sigma_1^2 + \sigma_2^2}\, f^2) \tag{16.29}$$

The sign of the exponent in the second Fourier transform has been reversed to reverse its direction of rotation and this can be seen in Figure 16.15(b). The result of the multiplication gives less turns per hertz to give a curve centered on the difference of the means but the same narrowing of the shaped helix (Figure 16.15(c)). The resulting inverse of the transform in Figure 16.15(d) shows the difference of the means or lag but retains the width of the convolution function, namely

$$\textit{Cross-correlated distribution} \;=\; \frac{1}{\sqrt{2\pi\,\overline{\sigma_1^2+\sigma_2^2}}}\exp\left(-\frac{(t-\overline{\mu_1-\mu_2})^2}{2\,\overline{\sigma_1^2+\sigma_2^2}}\right) \tag{16.30}$$

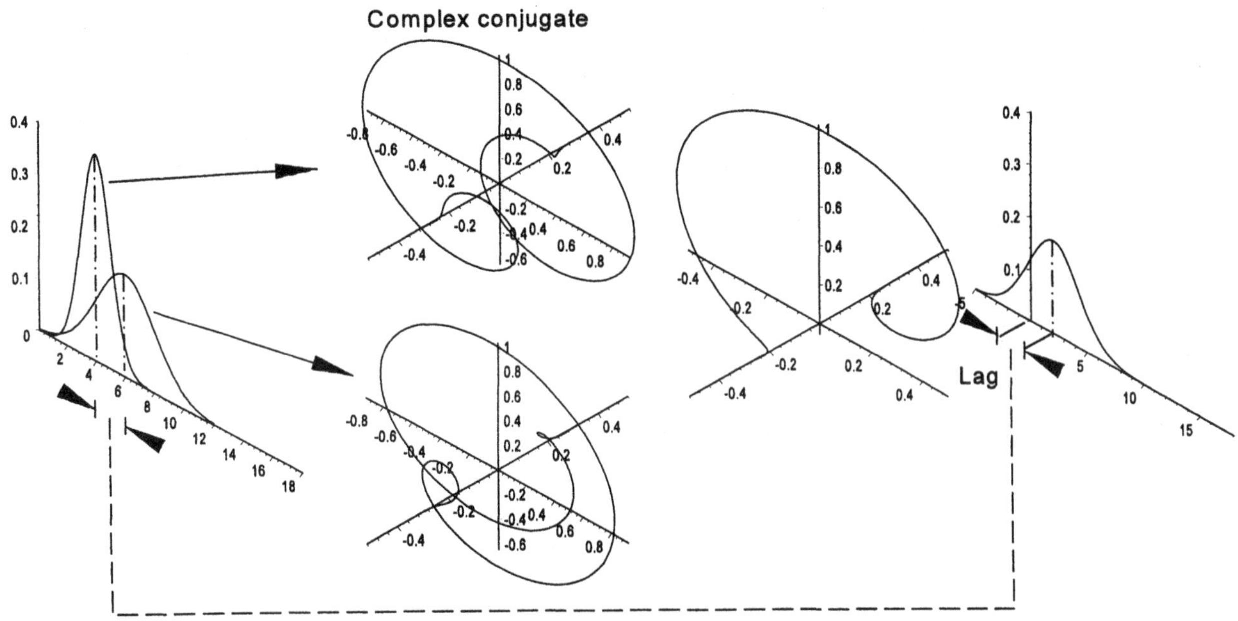

Figure 16.15 The cross-correlation of two Gaussian waveforms. [Source: Meikle, H. D., *A New Twist to Fourier Transforms*, Weinheim: Wiley-VCH, 2004.]

Conventional cross-correlation is unsatisfactory in that with the product with a complex conjugate, the rules of complex multiplication are applied to the amplitude and division to the phase angle. A true measure of correlation may be obtained when the Fourier transforms are divided so that the rules of complex division are applied to amplitude and phase. Division of the Fourier transforms results in infinity for identical waveforms, and a curve of width showing the difference of moments, principally standard deviation, of the different waveforms. There is no simple analogue circuit for division and division takes time in digital circuitry, and the curves for Fourier transforms often pass through zero leading to division by zero so that ordinary cross-correlation is used. A simple treatment is shown in the appendix to this chapter.

16.2.11 Autocorrelation, multiplication with its own complex conjugate

Autocorrelation follows the definition of power in alternating current theory where the voltage is multiplied by the complex conjugate of the current (2.8). The current is the same as the voltage across one ohm so that the power is VV*, where V* is the complex conjugate of V. The autocorrelation function is given by

$$f(t)\star f(t) \;=\; \int_{-\infty}^{+\infty} f^*(t)\, f(u-t)\, du \tag{16.31}$$

where ★ is used to indicate correlation.

The Fourier transforms are the mirror image of each other (Figure 16.16(b)), so that for a Gaussian function from (16.29) the transform is

$$\begin{aligned} H_1(f)\ H_2(f) &= \exp(-j2\pi\ \mu\ f)\ \exp(-j2\pi^2\ \sigma^2\ f^2) \\ &= \exp(-4\pi^2 f^2 \sigma^2) \end{aligned} \tag{16.32}$$

The product of the Fourier transforms that have equal and opposite phase angles and are always real is shown in Figure 16.16(c) and its inverse is shown in Figure 16.16(d). The peak of the autocorrelation function is the energy in the waveform.

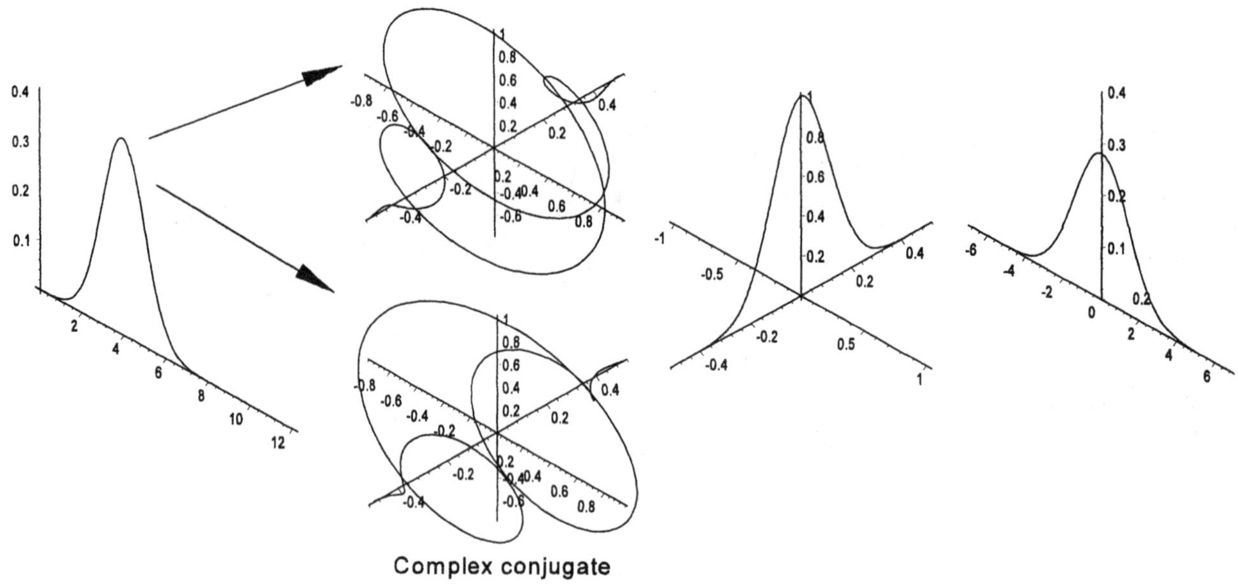

(a) Gaussian pulse (b) Fourier transforms (c) Fourier transform of autocorrelation (d) Autocorrelation function

Figure 16.16 The autocorrelation function of a Gaussian waveform. [Source: Meikle, H. D., *A New Twist to Fourier Transforms*, Weinheim: Wiley-VCH, 2004.]

Autocorrelation functions may have odd shapes so that the width is conventionally the width of a rectangle of the same height and area.

16.2.12 Energy and power

The energy in a waveform is equal to the energy in its spectrum, commonly called Parseval's theorem [7, p. 1317]. Where, starting with the time function for the energy in a pulse, the value is equated to the energy in the spectrum, namely,

$$\int_{-\infty}^{+\infty} h(t)\ f^*(t)\ dt = \int_{-\infty}^{+\infty} |h(t)|^2\ dt = \int_{-\infty}^{+\infty} |H(f)|^2\ df = \int_{-\infty}^{+\infty} H(f)\ H^*(f)\ df \tag{16.33}$$

16.3 DISCRETE FOURIER TRANSFORM

The input for signal processing is often a list of data, say $h[]$ comprising of ordered time and voltage samples.

$$h[] = [\ [time_0,\ voltage_0],\ [time_1,\ voltage_1],\ [time_2,\ voltage_2],\ \ldots\ [time_{N-1},\ voltage_{N-1}]\ ]$$

The Fourier transform may be computed from the sum

$$H(f) \quad = \quad \sum_{k=0}^{N-1} h[k,\ 2]\ e^{-j2\pi f h[k,\ 1]} \tag{16.34}$$

This gives the discrete Fourier transform for the simplest waveform that is described by the N points The Nyquist criterion states that T must be smaller than $1/2f$ where f is the highest frequency component present. Whatever happens between the points is not taken into account.

Usually in digital systems the analogue-to-digital converter output delivers a stream of data words representing the signal at fixed intervals which makes the time information redundant. The list may be reduced to

$$h[] = [\ voltage_0,\ voltage_1,\ voltage_2,\ \ldots\ \ voltage_{N-1}]]$$

The discrete transform assumes that the waveform repeats itself beyond the ends of the list. This may be represented by the cyclic representation on the left in Figure 16.17(a) [1, p. 363]. Then the discrete Fourier transform is

$$H(n) \quad = \quad \frac{1}{N} \sum_{k=-\frac{N}{2}}^{\frac{N-1}{2}} h[k]\ e^{-j\,2\pi\,\frac{n}{N}\,k} \tag{16.35}$$

and its inverse

$$H(k) \quad = \quad \sum_{n=-\frac{N}{2}}^{\frac{N-1}{2}} H[n]\ e^{-j\,2\pi\,\frac{n}{N}\,k} \tag{16.36}$$

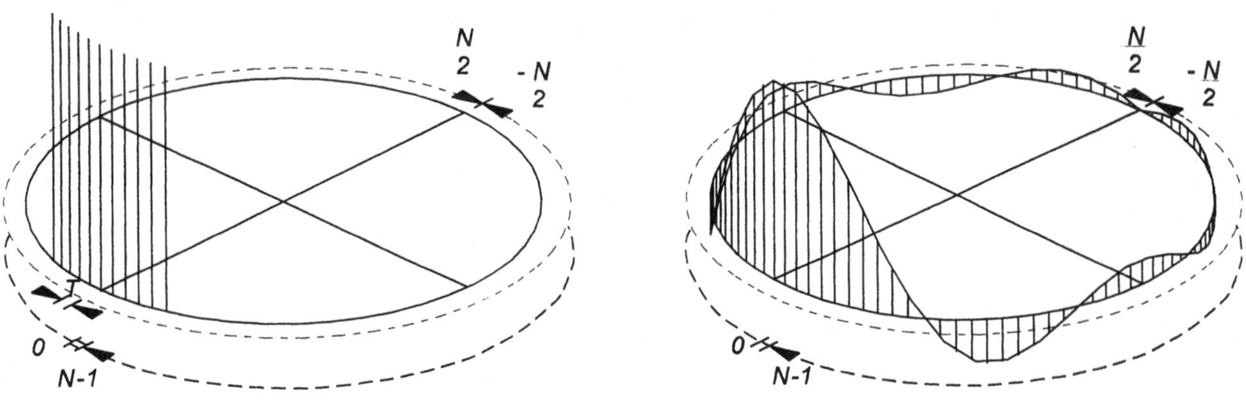

Figure 16.17 The circular nature of the discrete Fourier transform. [After Bracewell, R. N., *The Fourier Transform and Iits Applications*, 2nd ed., New York: McGraw-Hill, 1978.]

Commonly lists are not numbered symmetrically about zero but start with the index zero. This gives the definition for the discrete Fourier transform

$$H(n) \quad = \quad \frac{1}{N} \sum_{k=0}^{N-1} h[k]\ e^{-j\,2\pi\,\frac{n}{N}\,k} \tag{16.37}$$

and its inverse

$$H(k) = \sum_{n=0}^{N-1} H[n]\, e^{-j\, 2\pi \frac{n}{N} k} \tag{16.38}$$

The transform in (16.37) represents a time shift of half a cycle compared to that in (16.35). In order that an even function represents a real transform, the samples are numbered from the center of the pulse in (16.38) as shown in Figure 16.18.

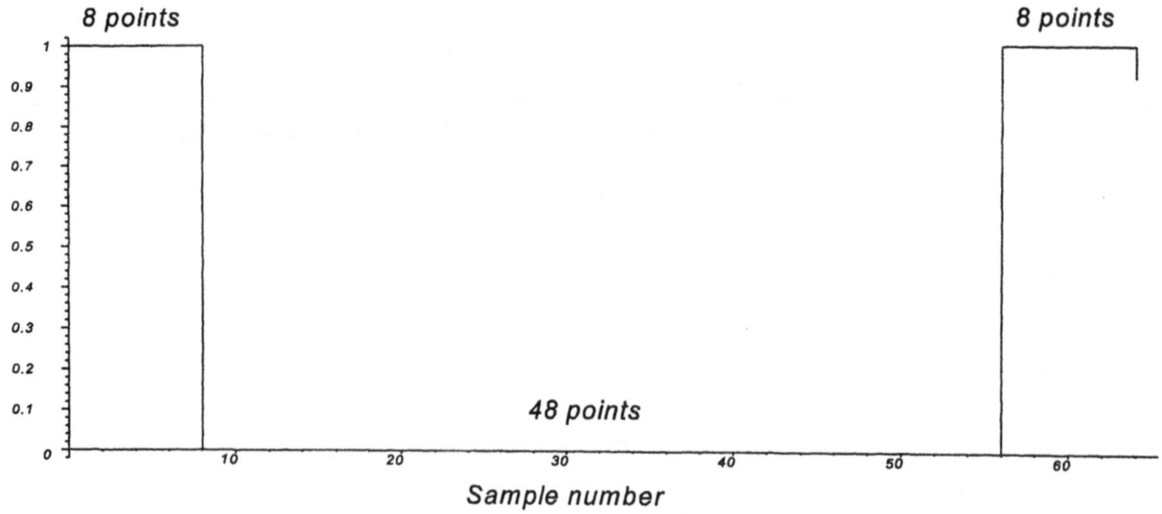

Figure 16.18 Arrangement of data from a rectangular pulse to give a real discrete Fourier transform.

The following points from Figure 16.17 are shown in Table 16.2.

Table 16.2
Steps and cycles in Figure 16.11

In Figure 16.17(a)	In Figure 16.17(b)
The points at $\pm N/2$ are the same	
One cycle represents $N\tau$ seconds	One cycle represents $\frac{1}{\tau}$ Hz
One step represents τ seconds	One step represents $\frac{1}{N\tau}$ Hz

This waveform and its discrete Fourier transform is shown in Figure 16.19. With an even number of points for the rectangular wave, there is still an asymmetry of half a time increment. This gives an imaginary sine wave component of $1/N$ [5, pp. 51-63]. The time waveform and spectrum are shown in Figure 16.19 and Figure 16.20.

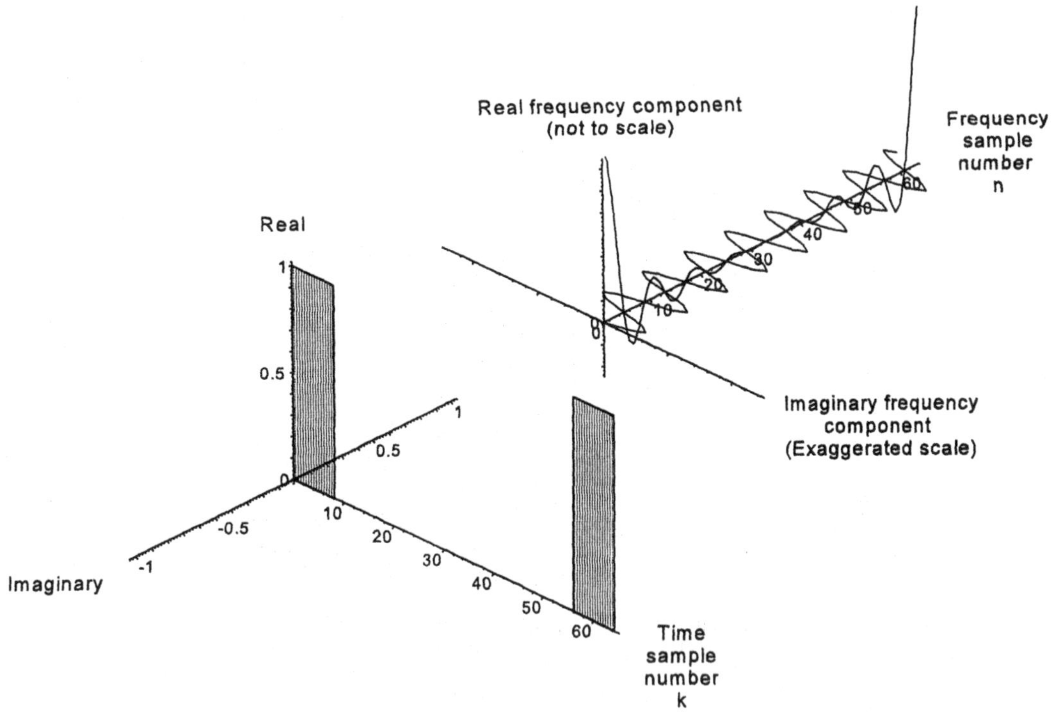

Figure 16.19 Time samples from the waveform and its spectrum components for an even number of points.

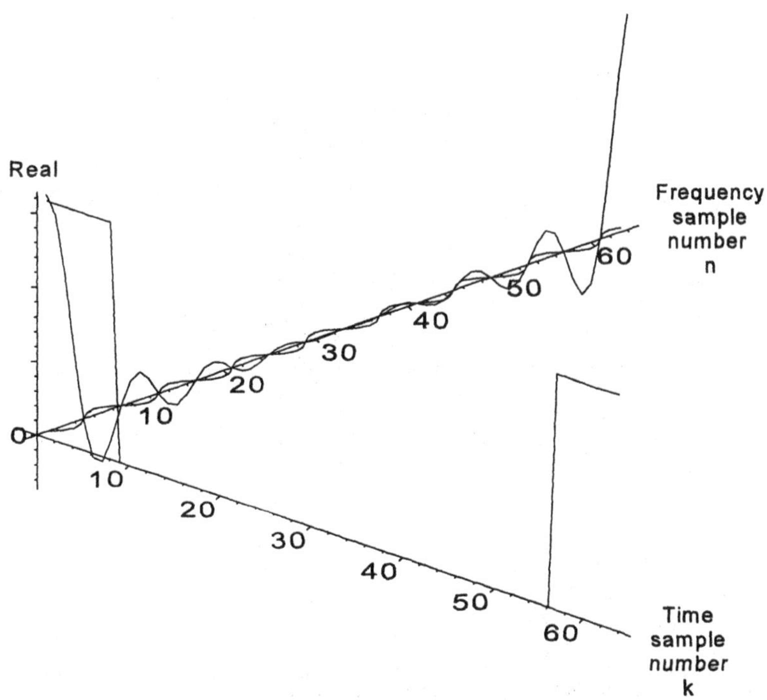

Figure 16.20 Combined sampled waveform and spectrum diagram for a rectangular pulse.

If an odd number of points are considered (namely, 9 ones, 48 zeroes, 8 ones, 65 points total compared to Figure 16.18), the transform becomes balanced and the imaginary component disappears an in Figure 16.21.

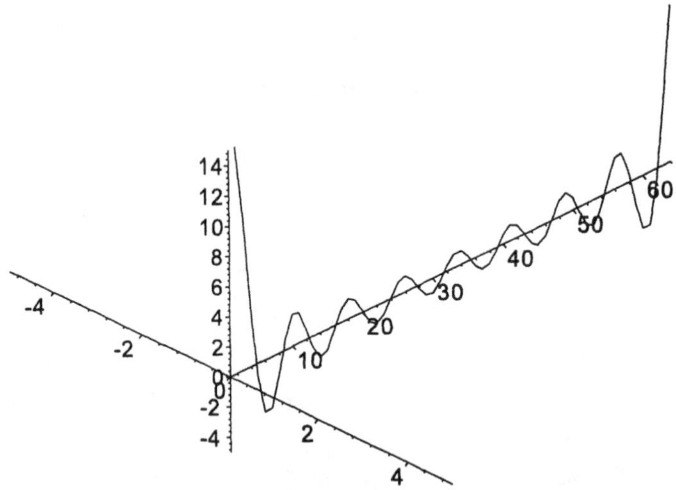

Figure 16.21 Sampled spectrum diagram for a rectangular pulse with an odd number of sampling points.

The convention for drawing a discrete Fourier transform spectrum with samples numbered from 0 to N with the direct voltage component shown at zero has been used to describe sampling and signal processing (see Chapter 9 Detectors, and Chapter 11, Signal processing).

Just as the signal is only defined at the measurement points, the discrete Fourier transform is only defined for the same number of points. Figure 16.22 shows what happens with a Fourier transform and its inverse when this is not followed [3].

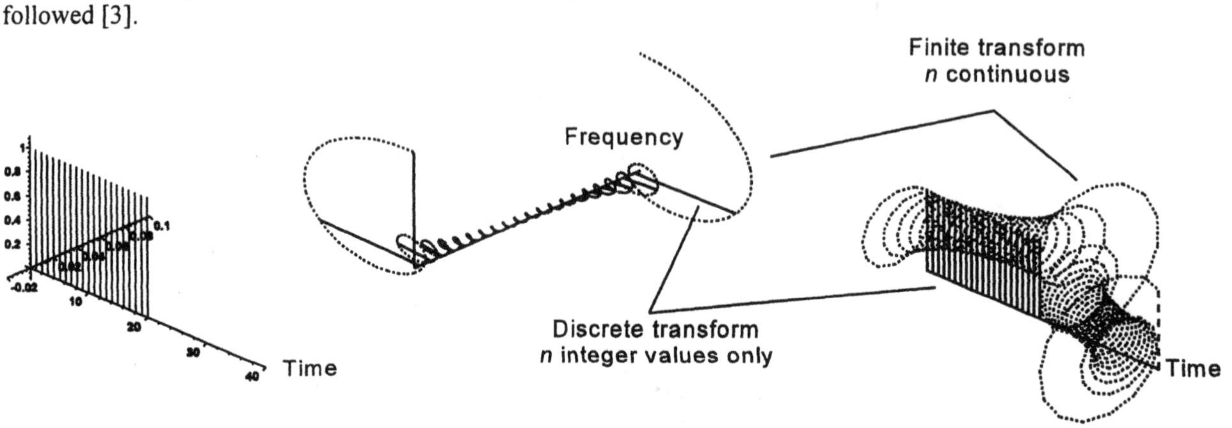

(a) Train of samples in time (b) Spectra from discrete and finite Fourier transforms of (a) (c) Time waveforms from discrete and finite inverse Fourier transform of (b)

Figure 16.22 Time samples (a) with their discrete and finite Fourier transforms (b) and the inverse back into the time domain (c). [Source: Meikle, H. D., *A New Twist to Fourier Transforms*, Weinheim: Wiley-VCH, 2004.]

The Fourier transform has a very important place in signal processing. The discrete Fourier transform requires N^2 complex multiplications and N^2 additions so that much effort was expended to reduce this number. Cooley and Tukey [8] developed an algorithm which reduced the proportionality to $N \log_2 N$ complex multiplications and additions. This is restricted to discrete Fourier transforms with lengths of 2 to the power m [9].

16.3.1 Differences between continuous and discrete transform functions

As with any integral, continuous and discrete transforms start to show differences when the number of samples for the curve is low. In the literature there is a difference between integrals taken using continuous functions and sums which are formed from a train of delta functions representing their samples. The continuous, or nonperiodic [2, p. 183], form is

$$H(f) = \int_{-\tau/2}^{\tau/2} e^{-j\,2\,\pi\,f\,t}\,dt = \frac{\sin(\pi\,f\,\tau)}{\pi\,f\,\tau} \qquad \textit{continuous or nonperiodic form} \tag{16.39}$$

The sum of the needle functions of unity amplitude which represent the samples is

$$\begin{aligned} H(n) &= \frac{1}{N}\sum_{k=0}^{N-1} e^{-j\,2\,\pi\,\frac{n}{N}\,k} && \textit{discrete or periodic form} \\ &= \frac{1}{N}\,\frac{1-e^{-j\,2\,\pi\,n}}{1-e^{-j\,2\,\pi\,\frac{n}{N}}} && \textit{sum of a geometric progression} \\ &= \frac{1}{N}\,\frac{\sin(\pi\,n)}{\sin\left(\pi\,\frac{n}{N}\right)}\,e^{-j\,\pi\,n\left(1-\frac{1}{N}\right)} && \textit{amplitude and phase} \end{aligned} \tag{16.40}$$

where n is $f\tau$, in the case of signal processing or $\frac{2\,\pi\,d}{\lambda}\,\sin(\theta+\alpha)$ for antennas.

Taking out the exponential phase term caused by the asymmetric summing interval, this periodic function is

$$|H(n)| = \frac{1}{N}\,\frac{\sin(\pi\,n)}{\sin\left(\pi\,\frac{n}{N}\right)} \tag{16.41}$$

As N becomes large, the denominator $N\,\sin\left(\pi\,\frac{n}{N}\right)$ tends to $N\,\pi\,\frac{n}{N} = \pi\,n$ as in the continuous or nonperiodic form in (16.39).

These differences are shown for numbers of samples N between 2 and 8 for the continuous and discrete transforms of a rectangular pulse. These are shown in Figure 16.23. The upper left shows both types of curves plotted together, lower left their differences (note the change in the scale of the ordinate), and lower right in decibels (voltage) with reference to unity.

Note the following:

- The continuous transform curve is narrower than the discrete curve;
- The sidelobe amplitudes are greater with small numbers of samples for the discrete transform.

Since the shape of the curve depends on the number N, the transform curves shown later are for continuous transforms. Discrete transforms have an additional disadvantage in their two-tone characteristics, see Section 16.5.3.

Note for $N = 2$ the discrete form is

$$|H(f)| = \frac{1}{2}\,\frac{\sin(\pi\,f\,\tau)}{\sin\left(\pi\,f\,\frac{\tau}{2}\right)} = \cos\left(\frac{\pi\,f\,\tau}{2}\right) \tag{16.42}$$

A plot of (16.42) has a similar shape as for a two pulse moving target indicator characteristic offset in frequency by half the pulse repetition frequency.

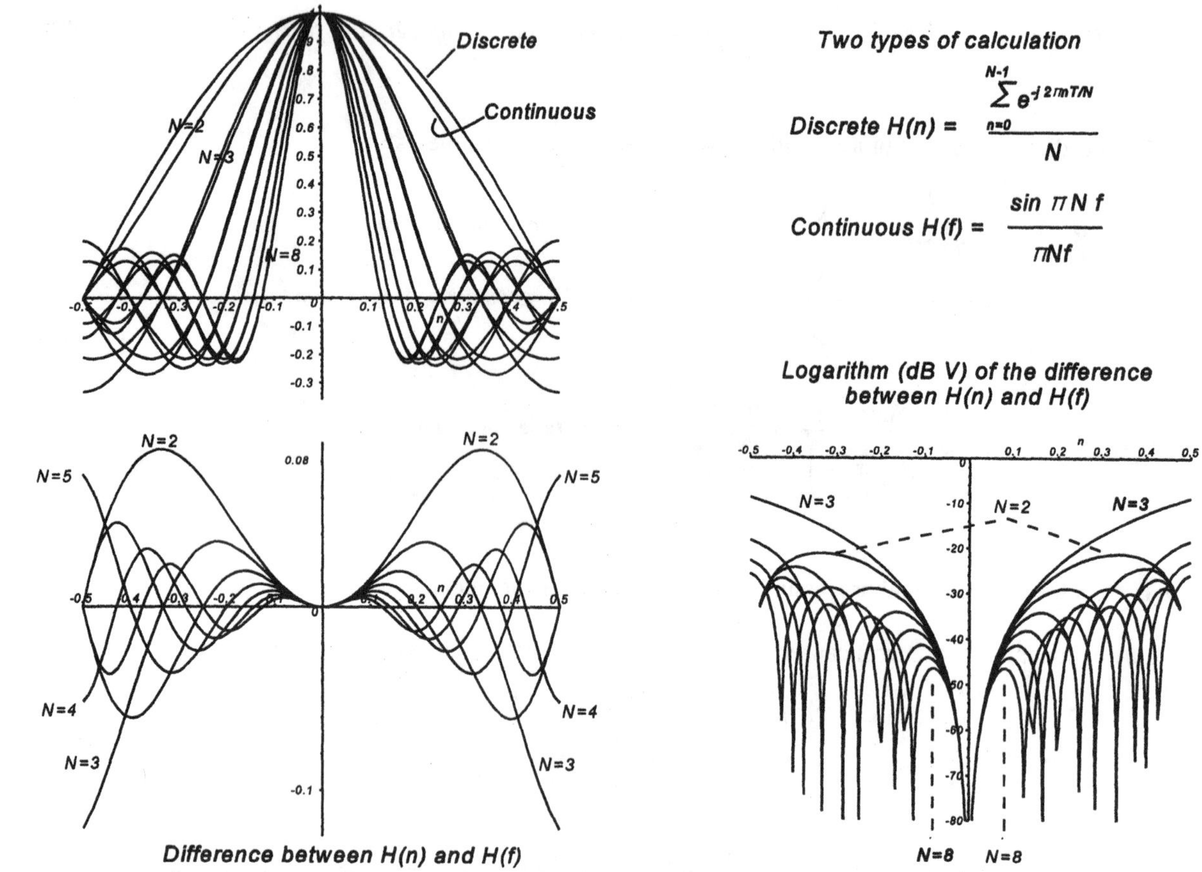

Figure 16.23 Differences between continuous and discrete rectangular transform functions.

16.3.2 Fast transforms

The calculation of discrete Fourier transforms required much manipulation of numbers so that there was considerable interest in reducing this. This lead to the Cooley and Tukey algorithm for discrete Fourier transforms or the fast Fourier transform [6, 7] which reduces the N^2 complex operations (multiplication followed by an addition) to $N/2\ \log_2 N$. The cost of this is the constraint that the number of complex samples is limited to 2^m.

16.4 SUMMARY OF PROPERTIES OF THE FOURIER TRANSFORM

There are a number of properties that are useful when dealing with Fourier transforms and a number of them are shown in Table 16.3 [1, p. 183]. For convenience a time function is transformed into a frequency function. The inverse transform changes the frequency function into a time function.

Table 16.3
Some properties of the Fourier transform

Property	Time function	Transform
Scaling	$h(at)$	$\dfrac{H\left(\dfrac{f}{a}\right)}{\lvert a \rvert}$
Addition	$h_1(t) + h_2(t)$	$H_1(f) + H_2(f)$
Shift	$h(t-x)$	$H(f)\ e^{-j2\pi fx}$
Phase shift, ϕ	$h(t)\ e^{-j\phi}$	$H(f)\ e^{-j\phi}$
Modulation	$h(t)\cos(2\pi f_1 t)$	$\frac{1}{2} H(f-f_m) + \frac{1}{2} H(f+f_m)$
Convolution	$h_1(t) \otimes h_2(t)$	$H_1(f)\, H_2(f)$
Cross-correlation	$h_1(t) \bigstar h_2(t)$	$H_1(f)\, H_2^*(f)$
Autocorrelation	$h(t) \bigstar h(-t)$	$H(f)\, H^*(f)$
Differentiation	$h'(t)$	$j2\pi f\, H(f)$
Integration	$\int h(t)$	$1/j2\pi f\, H(f)$
Rayleigh, Parseval	$\int_{-\infty}^{+\infty} (h(t))^2\, dt$	$\int_{-\infty}^{+\infty} H(f)\ H^*(f)\ df$

Notes:
$\otimes$ means convolved with
$\bigstar$ means correlated with

16.5 TAPERING

François Rabelais (*c.*1494– *c.*1553) wrote *Natura vacuum abhorret* (Nature abhors a vacuum) [10]: corners also cause trouble. Tapering smooths the corners and is usually treated for each stage of the radar separately. There is a fundamental division between those processes that add up the vectors to trade efficiency for lower sidelobe levels (tapering reduces the signal) and those which trade signal-to-noise ratio for lower sidelobes (tapering reduces signal and noise). A number of applications are shown in Figure 16.24.

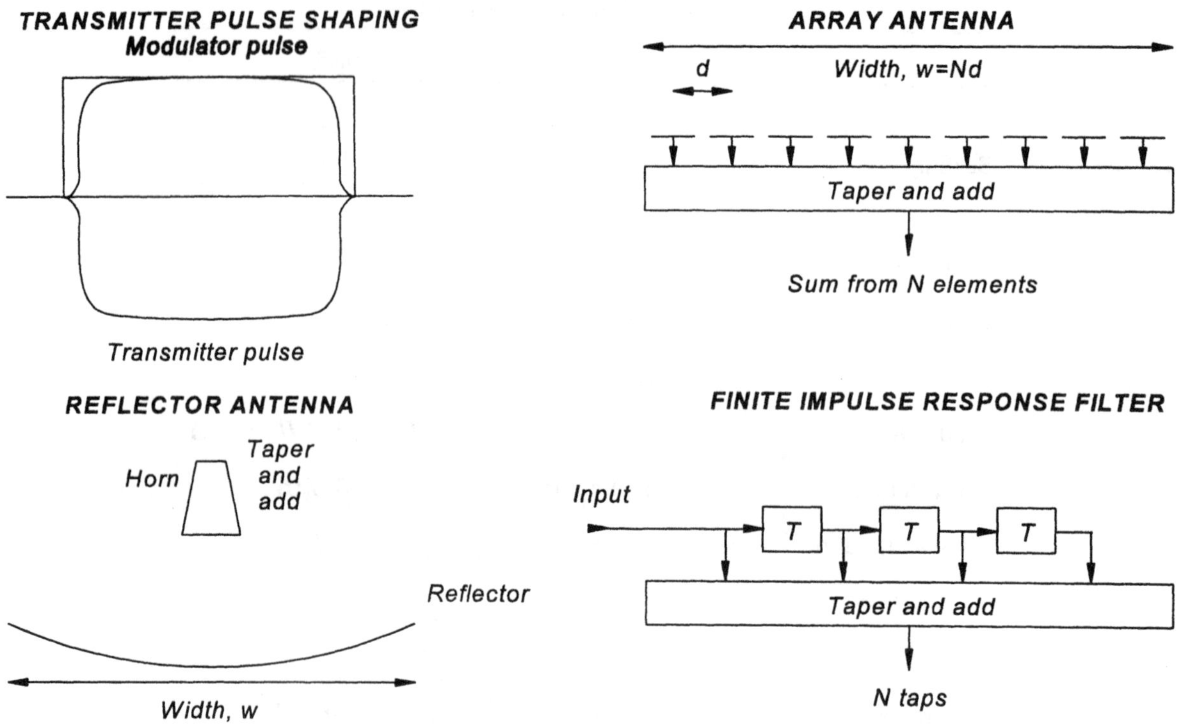

Figure 16.24 The uses of tapering for transmitter pulses, antennas, and filters.

The quantities of interest are shown in Table 16.4.

Table 16.4
Comparison of the uses of tapering

	Reduced efficiency for lower sidelobe levels	Reduced signal-to-noise ratio for lower sidelobe levels
Applications	Transmitters (pulse shaping) Antennas (beam forming)	Pulse compression Signal processing
Criterion	Greatest integral of the vectors for maximum allowable sidelobes.	Least loss in signal-to-noise ratio for the greatest allowable sidelobes
Costs	Extra modulator power is required that brings greater size and running costs. Reduction of antenna gain and a wider beam. The loss in coherent gain may be compensated for by using a larger surface	Reduction of signal-to-noise ratio Wider compressed pulse Less resolution for the signal processing filters

All the tapering functions are even and linear (symmetrical about zero) except for difference patterns for monopulse antennas (and differentiating filters which are not covered here) and are shown in Figure 16.25. Additionally a limited number of tapering functions are given for circular antennas.

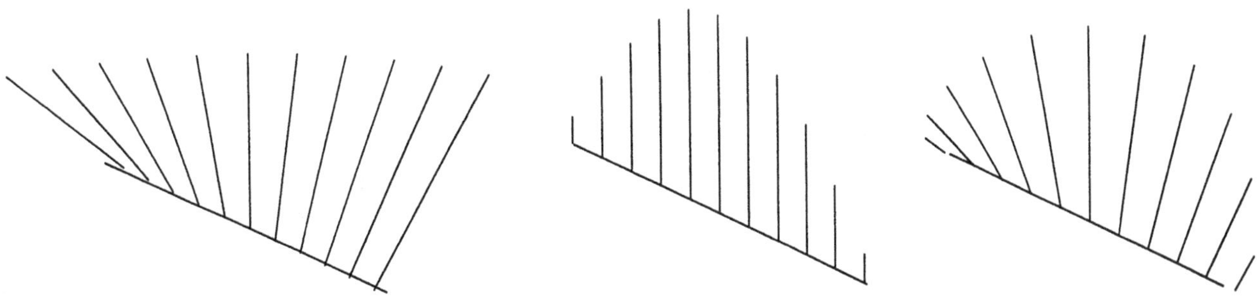

Figure 16.25 The effects of tapering on vectors.

The effects of tapering on signals with a noise background is shown in Figure 16.26. Both the signals and noise are affected so that the signal-to-noise ratio at the output is less than the losses when the signals from the elements of an antenna are assembled.

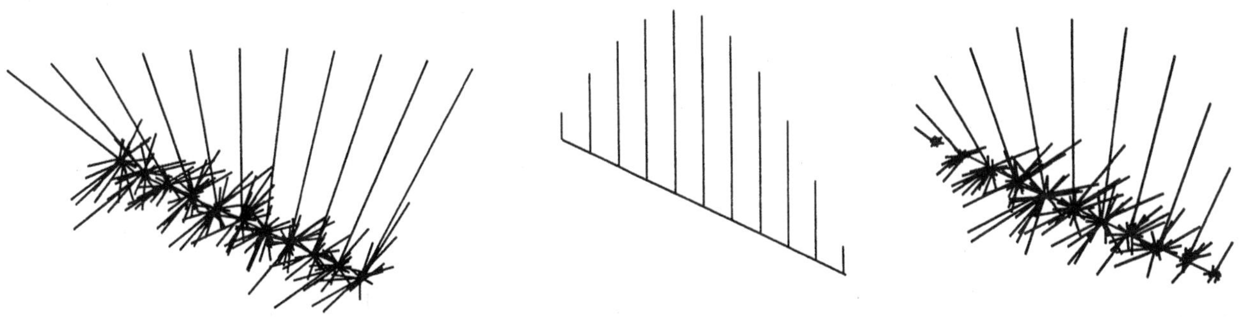

Figure 16.26 The effects of tapering on signal and noise.

A number of tapering functions for discrete Fourier transforms are in the literature [11]. In contrast with most antennas, which have one main beam and are often illuminated with a continuous function, filters in moving target detection systems work in banks of neighboring filters, which allow the sorting of the incoming echoes according to their Doppler frequencies. Historically, the two groups of applications have led to the development of separate tapering waveforms for each purpose.

16.5.1 Gains and losses

The gains and losses are shown in Figure 16.27. All use the Fourier transform of a uniform distribution as a normalizing reference ($\sin\pi Nx/\pi Nx$ shape). Processing or coherent gain here is, as with antennas, the peak of the frequency characteristic.

The coherent voltage gain without tapering is N, the number of taps, or w/λ in the case of antennas. The peak gain with tapering is

$$\textit{Maximum gain, } G_{\max} = H(0) = w\int_{-\frac{1}{2}}^{+\frac{1}{2}} h(x)\,dx \quad \text{or} \quad \sum_{N} h(n) \tag{16.43}$$

The relative coherent voltage gain compared with no tapering, N, or the sum of the vectors in Figure 16.25(a), is illustrated in Figure 16.25(c), and is

$$Relative\ coherent\ gain = \frac{\sum_N h(n)}{N} \quad \text{voltage} \tag{16.44}$$

The relative coherent gain is less than unity and represents a loss, which would be the gain from a smaller antenna or a filter with a smaller number of taps. The effective length of a linear antenna or number of taps is

$$A = \int_{-w/2}^{+w/2} h(x)\ dx = \sum_N h(n) \tag{16.45}$$

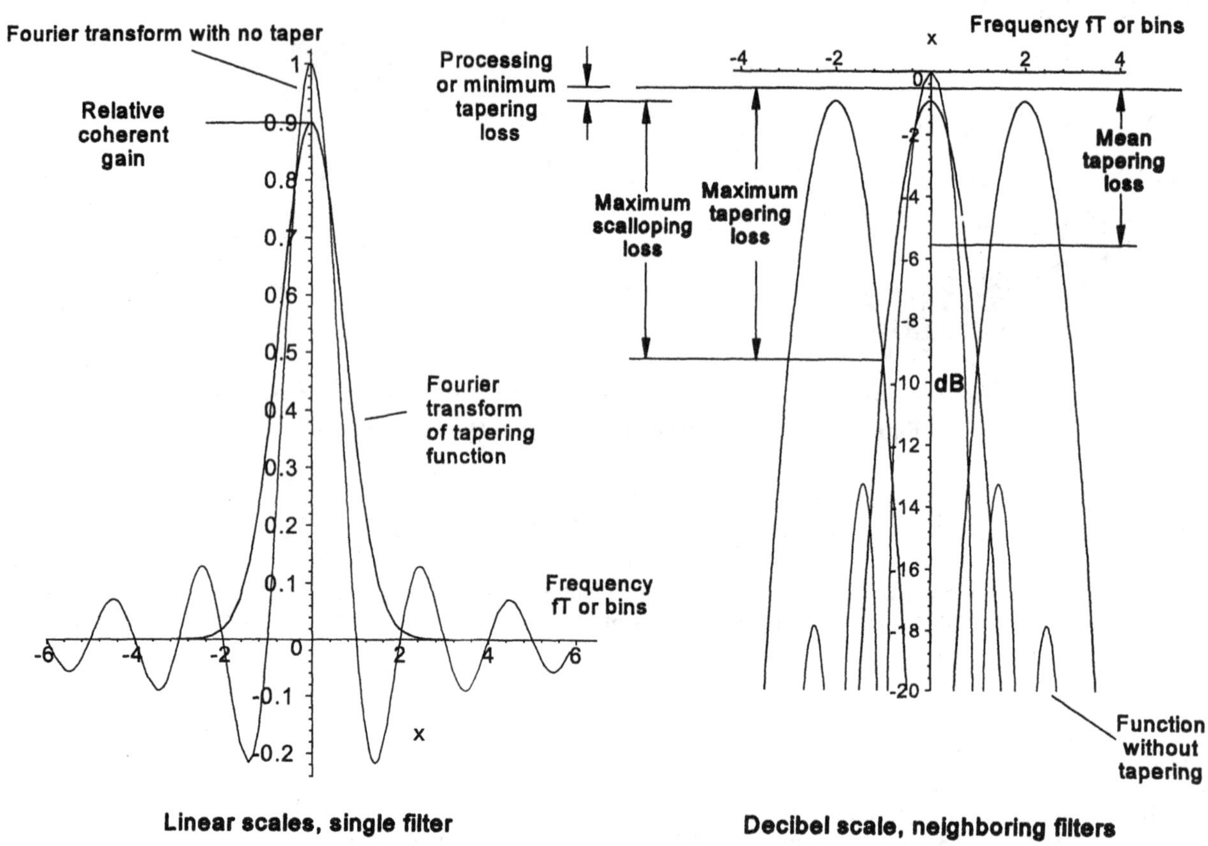

Figure 16.27 Gain and loss definitions.

The elements of the tapering function may be vectors, so that the total power is represented by the squares of the moduli of the voltages:

$$\begin{aligned} C &= \int_{-\infty}^{+\infty} |H(u)|^2\ du = \int_{-\frac{w}{2}}^{+\frac{w}{2}} |H(x)|^2\ dx = w \int_{-\frac{1}{2}}^{+\frac{1}{2}} |H(x')|^2\ dx \\ &= \sum_N |h(n)|^2 \end{aligned} \tag{16.46}$$

Antennas are normally in two dimensions, so that for a rectangular antenna the effects are the product of those in each of the two dimensions. For circular symmetry, Hankel transforms are used [1, p. 248; 12]. The Hankel transform is given by

$$H(q) = 2\pi \int_0^\infty h(r)\ J_0(2\pi q r) r\ dr \tag{16.47}$$

The inverse is

$$H(r) = 2\pi \int_0^\infty H(q)\ J_0(2\pi q r) q\ dq \tag{16.48}$$

where $J_0(x)$ is the Bessel function of the first kind, argument x [7, p. 119].

The idea of noise beamwidth of an antenna or the bandwidth of a filter is shown in Figure 16.28. If the area under the power characteristic is redrawn as a rectangle with the same height as the tapered gain, its width is the noise beam or bandwidth.

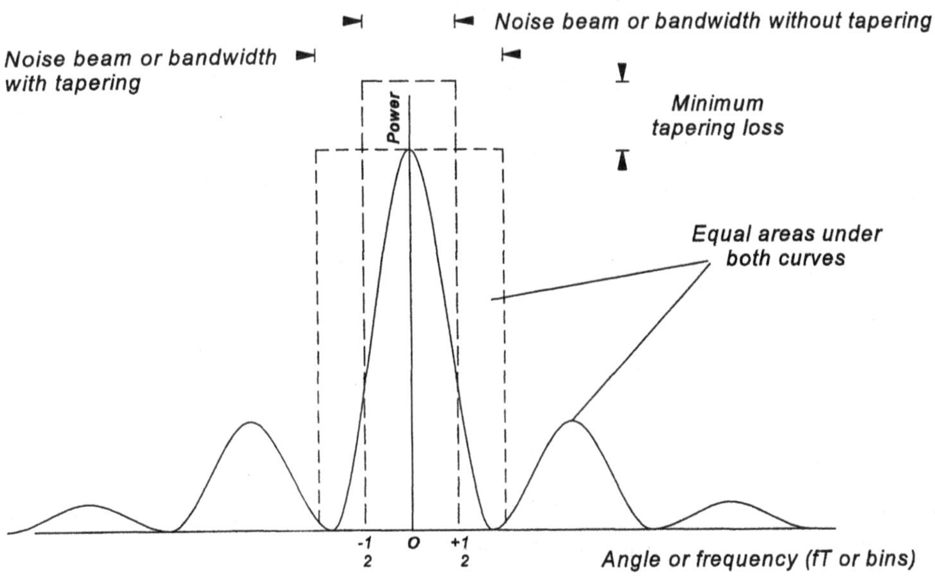

Figure 16.28 The concept of noise beamwidth or bandwidth.

If equally distributed noise is applied around an antenna or to a filter; the noise power passed depends on the beamwidth or bandwidth, which can be expressed as

$$\text{Noise beamwidth or bandwidth} = \frac{\text{Sum of tapered power}}{|\text{Peak tapered voltage}|^2} \tag{16.49}$$

For antennas, the noise beamwidth is

$$\theta_n = \frac{\lambda}{w} \frac{\int |H(u)|^2\ du}{|H_m|^2} = \frac{\lambda}{\eta\ w} \text{ radians} \tag{16.50}$$

and, for filters, the noise bandwidth is

$$\textit{Equivalent noise bandwidth, } ENBW = \frac{\sum_N \left(h(n)\right)^2}{\left(\sum_N h(n)\right)^2} \text{ bins} \tag{16.51}$$

As with tapering losses, the size of an untapered antenna with the noise beamwidth gives the noise aperture width

$$w_n = \frac{w \int_{-\frac{1}{2}}^{+\frac{1}{2}} |h(x)|^2 \, dx}{|h_m|^2} \tag{16.52}$$

The change of signal-to-noise ratio for a signal passing through a filter is [11, p. 56]

$$\begin{aligned} \frac{\textit{Signal out}}{\textit{Noise out}} &= \frac{\textit{Signal in}}{\textit{Noise in}} \frac{|\textit{Coherent voltage sum in filter}|^2}{\textit{Sum of incoherent noise power}} \\ &= \frac{\textit{Signal in}}{\textit{Noise in}} \frac{\left(\sum_N h(n)\right)^2}{\sum_N \left(h(n)\right)^2} \end{aligned} \tag{16.53}$$

Equation (16.53) leads to the processing loss

$$\textit{Processing or minimum tapering loss} = 10 \log_{10} \frac{\left(\sum_n h(nT)\right)^2}{\sum_n \left(h(nT)\right)^2} \text{ dB} \tag{16.54}$$

The loss in (16.54) represents the proportion of signals passing through the filter. Similarly, the proportion of signals passing to the output of an antenna is the antenna efficiency, which is

$$\begin{aligned} \textit{Efficiency, } \eta &= \frac{|\textit{Coherent gain}|^2}{\textit{Power radiated}} \\ &= \frac{A^2}{C} = \frac{|H_m|^2}{\int_{-\infty}^{+\infty} |H(u)|^2 \, du} = \frac{\left| \int_{-\frac{1}{2}}^{+\frac{1}{2}} h(x) \, dx \right|^2}{\int_{-\frac{1}{2}}^{+\frac{1}{2}} |h(x)|^2 \, dx} \end{aligned} \tag{16.55}$$

The efficiency is the reciprocal of the noise beamwidth or bandwidth.

Antenna gains and beamwidths are measured empirically on a test range, and the results are less than "what might have been" had the antenna been 100% efficient.

The processing or minimum tapering loss applies to signals at the center frequencies of the filters. The scalloping loss is the difference in gain between the peak of the filter and the crossover point to the next filter in the bank.

$$\text{Scalloping loss} = \frac{\text{Voltage at crossover point}}{\text{Voltage at peak}} = \frac{\left|\sum_N h(n)\, e^{-j\frac{\pi n}{2.N}}\right|}{\sum_N h(n)} \tag{16.56}$$

The mean scalloping loss is the average taken over one bin and is sometimes called the Doppler filter straddling loss. When the scalloping loss is added to the processing or minimum tapering loss, it becomes the maximum tapering loss or worst case processing loss. The mean tapering loss is taken to be the minimum tapering loss plus the mean scalloping loss.

Transmitter pulse shaping is a special case. With magnetron radars, the shapes of the leading and trailing edges of the modulator pulse must be controlled, so that the magnetron starts to oscillate in the chosen mode reliably at the beginning of each pulse. Radars with power amplifiers generally use a fixed modulator pulse form but shape the radio frequency driving pulse to achieve a much better control of the spectrum of the transmitted pulse. The power in the part of the modulator pulse outside the drive pulse is wasted (see Figure 16.24).

The similarity of definitions led to universal curves for the shapes of the tapering functions. The definitions are illustrated in Figure 16.29. The units used are shown in Table 16.5.

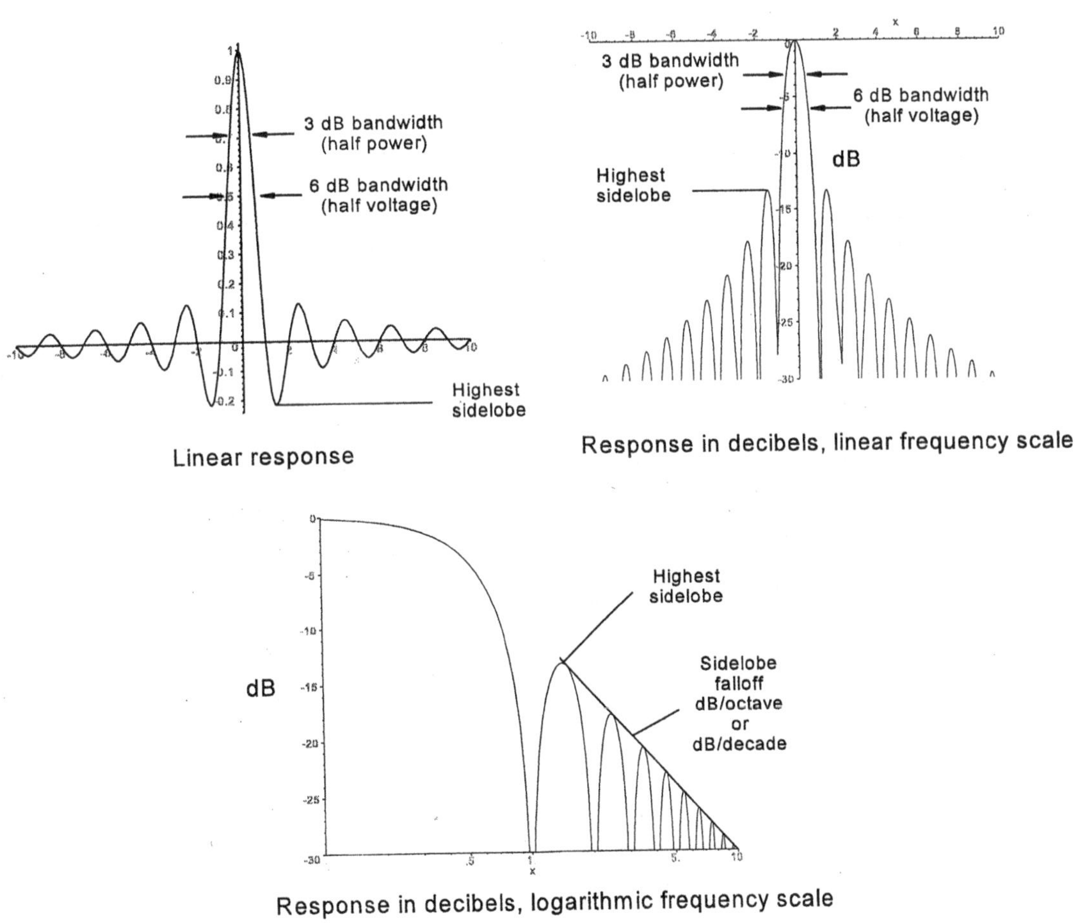

Figure 16.29 Various representations of the Fourier transforms of the tapering functions, in this example, the uniform tapering function.

Table 16.5
Units for curves

Quantity	Application	Units
Bandwidth	Filters	Bins, units of frequency multiplied by time.
Beamwidth	Antennas	Sine space angle from −1 to +1 This must be converted to degrees (or radians) in real space

The following common widths which may be calculated or measured are used:

- 3 dB width (between half power points);
- 6 dB width (between half voltage points);
- Noise width.

Low sidelobes levels are desirable, starting with the highest sidelobe level with a rapid falling off of sidelobe levels.

16.5.2 Spectral leakage or two-tone characteristics

Spectral leakage is the "leakage" of signals that should fall in one filter which fall into the output of another. It is caused by the summing of the signals at other frequencies in the sidelobes of the filter.

In addition to slightly increased sidelobes with discrete Fourier transforms, there is an additional disadvantage when the signal frequency lies between two filters. In search radars, the echoes may return with any Doppler frequency within the radar's bandwidth, and clutter signals often have ten thousand times (40 dB) the power of the wanted signals. Examples are given for each tapering characteristic for a bank of filters that use 100 samples filled with two signals.

- Clutter between 10 and 11 bins;
- A wanted signal in bin number 16.

Figure 16.30 shows the outputs of the individual filters joined by straight lines as the major signal frequency is moved between 10 and 11 bins in steps of 1/8th of a bin.

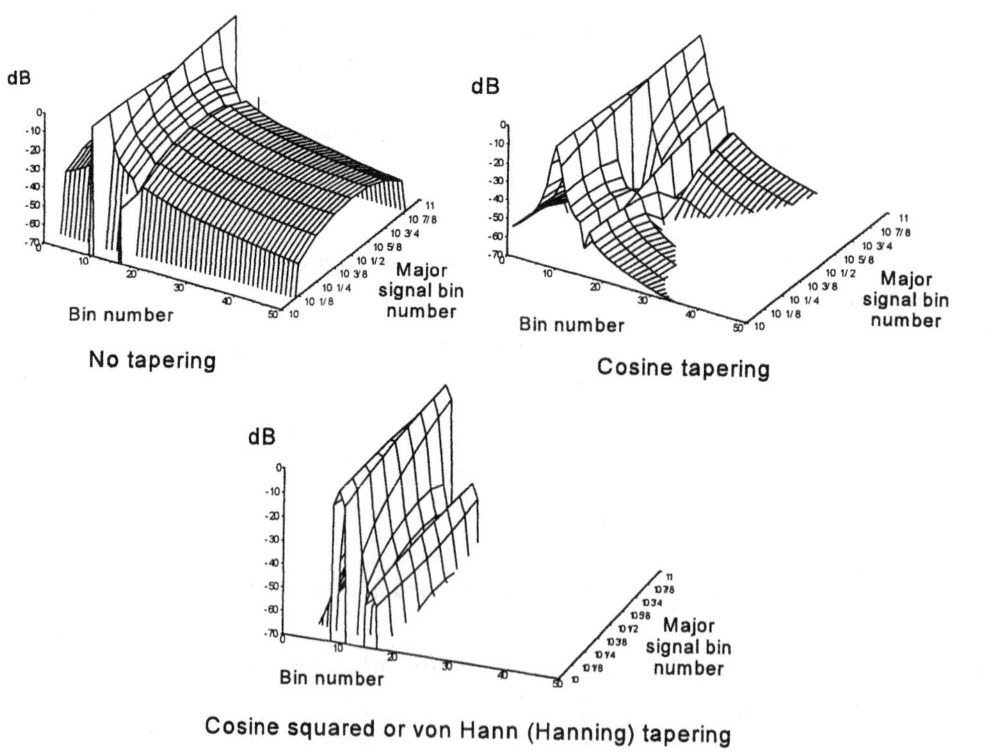

Figure 16.30 The effect of a major signal moving between 10 and 11 bins on a minor signal 40 dB smaller.

Tapering increases the width of the main lobe, and notice how crisp the main lobe is without tapering. Increasing amounts of tapering spread the main lobe over a small number of bins, as shown in Figure 16.30. With rectangular or no tapering, both major and minor signals are resolved when the major signal is at 10 bins. As the major signal is divided between 10 and 11 bins (moving to the back of the diagram), the amplitude is reduced and the sidelobes increase to swamp the minor signal. With cosine tapering, the minor signal appears when the major signal is at 10.5 bins. Cosine squared tapering divides the two signals over the whole range.

It must be remembered that generally the first threshold crossing and sorting for extraction takes place in the range direction. A deep valley between the clutter and the wanted signals in the Doppler frequency domain is required for velocity resolution.

16.5.3 Resolution

When the two characteristics are separated, they form a common hump (6 dB in [11, p. 57]). The hump starts to have a saddle point when the characteristics cross at near their 6 dB (half voltage points), as shown in Figure 16.31.

In order to decide whether there are one or two echoes, there must be a recognizable saddle point with peaks and a valley. The separations for peak-to-valley ratios of 3 dB and 6 dB are shown in Figure 16.31(b). The separation to produce a saddle point with Gaussian curves is discussed in Chapter 9, Detectors.

Resolution in Doppler frequency is rarely required in search radars other than to separate out the clutter. In the Doppler frequency domain, surveillance radars often choose only the largest echo in a range bin for the extraction threshold.

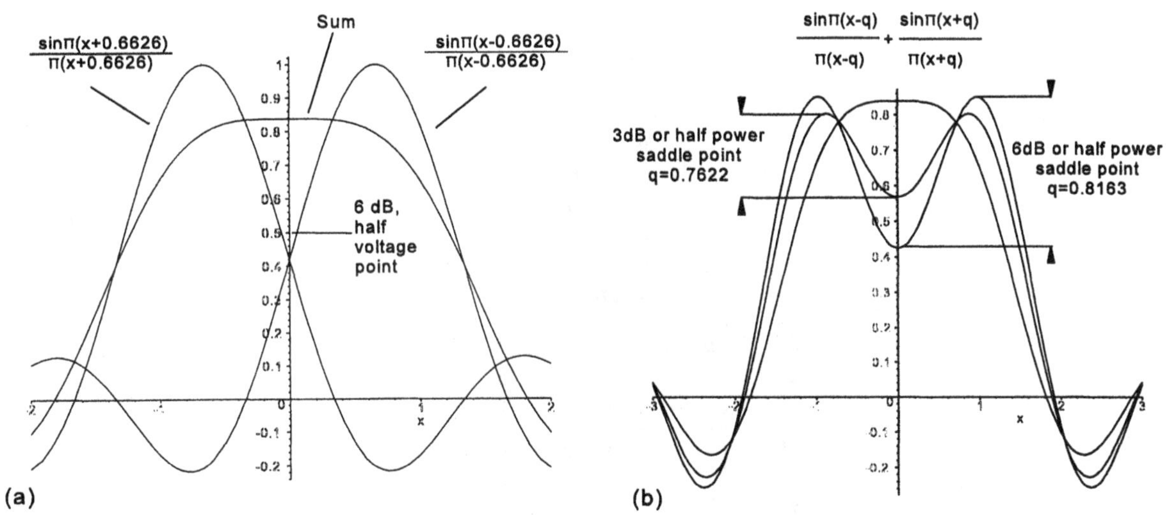

Figure 16.31 Resolution: the formation of saddle points when two sin $\pi x/\pi x$ curves are separated.

16.5.4 Example: von Hann and Hamming tapering functions

Julius von Hann suggested one of the first tapering functions, which bears his name. It is the cosine squared function:

$$\begin{aligned} h(t) &= \cos^2\left(\pi\frac{t}{\tau}\right) \qquad t \text{ between } -\frac{\tau}{2} \text{ to } \frac{\tau}{2} \\ &= \frac{1}{2} + \frac{1}{2}\cos\left(2\pi\frac{t}{\tau}\right) \end{aligned} \tag{16.57}$$

In discrete form

$$h(n) = \cos^2\left(\pi\frac{n}{N}\right) \qquad -\frac{N}{2} < n < \frac{N}{2}$$
$$= \frac{1}{2} + \frac{1}{2}\cos\left(2\pi\frac{n}{N}\right) \tag{16.58}$$

This is illustrated in Figure 16.32 and can be generalized to

$$h(n) = p + q\cos\left(2\pi\frac{n}{N}\right) \tag{16.59}$$

The Fourier transform may be derived by the sums of the transforms of the elements. The first term is constant and the modulation property is used to find the Fourier transform of the second term. With τ (seconds) and f (Hz) this is

$$H(f) = p\,\frac{\sin(\pi f\tau)}{\pi f} + \frac{q}{2}\left(\frac{\sin(\pi(f\tau - 1))}{\pi\left(f - \frac{1}{\tau}\right)} + \frac{\sin(\pi(f\tau + 1))}{\pi\left(f + \frac{1}{\tau}\right)}\right) \tag{16.60}$$

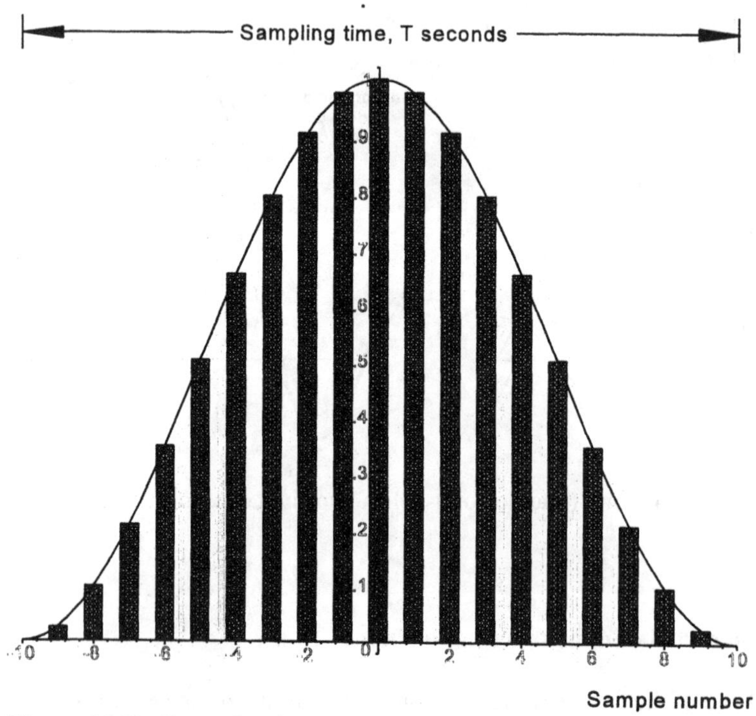

Figure 16.32 Example of 21 point von Hann (Hanning) tapering.

In the discrete form τ is replaced by N and f is expressed in cycles per sampling time. This is shown together with the spectrum of the original rectangular pulse, width 1 second, in Figure 16.33.

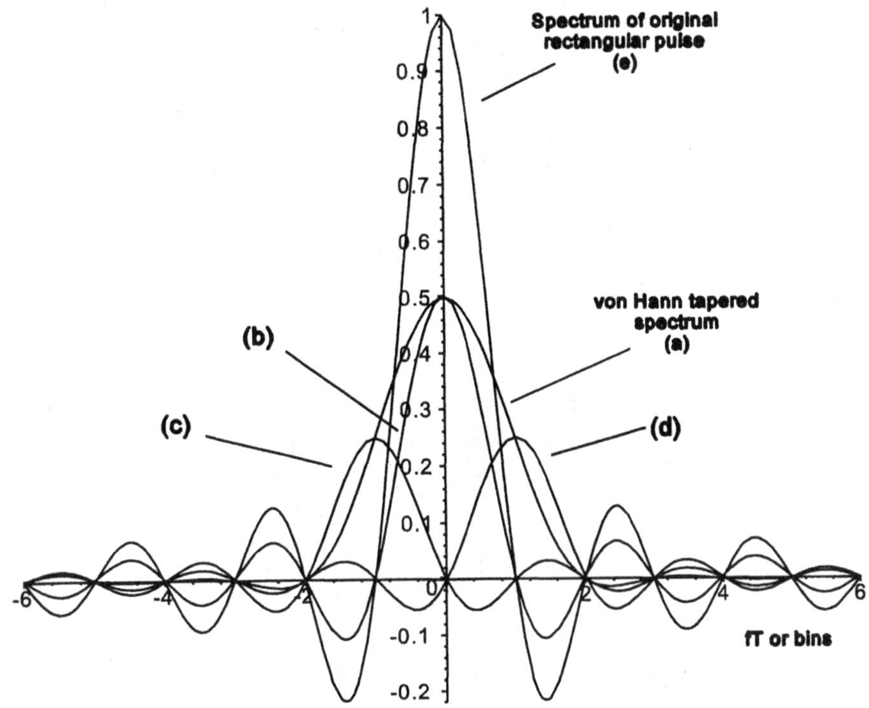

Figure 16.33 The spectrum of a rectangular pulse weighted by a von Hann (Hanning) tapering function.
The components are:

(a) $$W(f) = \frac{\sin \pi fT}{2\pi fT} + \frac{\sin \pi(fT - 1)}{4\pi(f - 1/T)} + \frac{\sin \pi(fT + 1)}{4\pi(f + 1/T)}$$

(b) $$\frac{\sin 2\pi fT}{2\pi fT}$$

(c) $$\frac{\sin \pi(fT + 1)}{4\pi(f + 1/T)}$$

(d) $$\frac{\sin \pi(fT - 1)}{4\pi(f - 1/T)}$$

(e) $$\frac{\sin \pi fT}{\pi fT}$$

The areas under all tapered spectra are all unity. The tapered spectrum is wider, so the peak value is no longer unity. The sidelobes of the original function are partially canceled by the displaced functions. For still smaller sidelobes with other tapering functions, the result depends on the accurate subtraction of large numbers.

Often the requirement is to reduce the first sidelobe. Richard Hamming (Bell Telephone Laboratory) found that by optimizing the values of p and q a characteristic with a smaller sidelobe could be created, namely,

$$h(t) = \frac{25}{46} + \frac{21}{46}\cos\left(\pi \frac{t}{\tau}\right) \quad \text{Exact Hamming} \qquad (16.61)$$

and

$$h(t) = 0.54 + 0.46\cos\left(\pi \frac{t}{\tau}\right) \quad \text{Hamming} \qquad (16.62)$$

The spectra of the tapering functions are

$$H(f) = \frac{25}{46}\frac{\sin(\pi f\tau)}{\pi f\tau} + \frac{21}{46}\left(\frac{\sin(\pi(f\tau - 1))}{2\pi\left(f - \frac{1}{\tau}\right)} + \frac{\sin(\pi(f\tau + 1))}{2\pi\left(f + \frac{1}{\tau}\right)}\right) \quad \text{Exact Hamming} \tag{16.63}$$

and

$$H(f) = 0.54\frac{\sin(\pi f\tau)}{\pi f\tau} + 0.46\left(\frac{\sin(\pi(f\tau - 1))}{2\pi\left(f - \frac{1}{\tau}\right)} + \frac{\sin(\pi(f\tau + 1))}{2\pi\left(f + \frac{1}{\tau}\right)}\right) \quad \text{Normal Hamming} \tag{16.64}$$

The normal Hamming function (not to be confused with the von Hann or Hanning function in the literature in English) gives almost as good a spectrum with simple tapering factors as the exact function. The four spectra from the original rectangular pulse, von Hann tapering, and the two forms of Hamming tapering are shown in Figure 16.34.

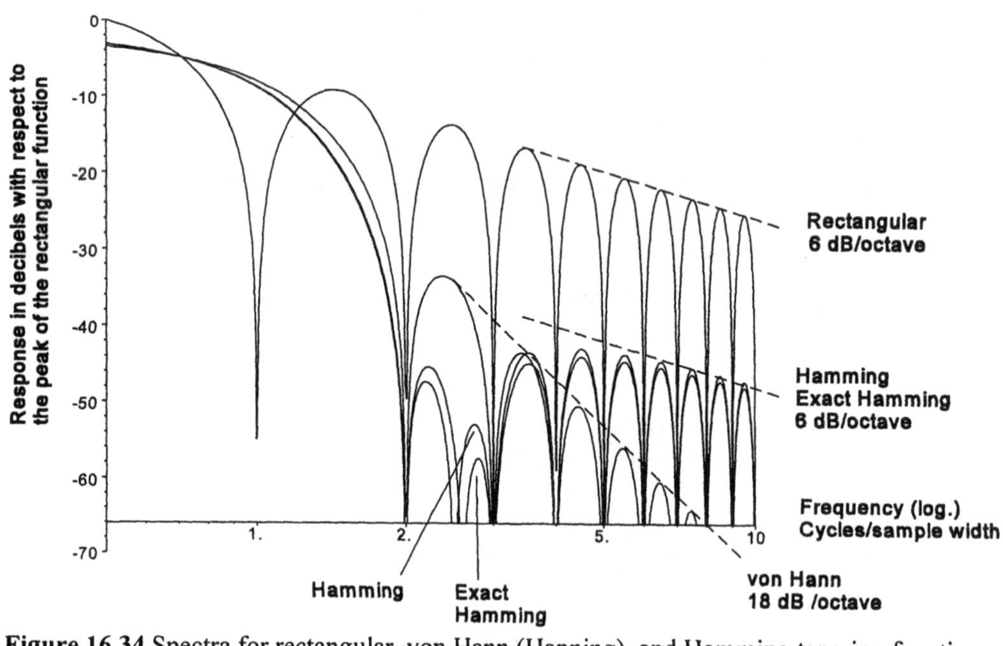

Figure 16.34 Spectra for rectangular, von Hann (Hanning), and Hamming tapering functions.

The choice of a tapering function is often a compromise. In Figure 16.34, there is a balance between the width of the peak, the first sidelobe level, and the falloff of the distant sidelobes. (The widths of the main peak and the sidelobe levels are given in later sections.)

The two-tone characteristics of the von Hann (Hanning) tapering functions are shown in Figures 16.35 and 16.36 for minor to major signal voltage ratios of 0.1 (20 dB) and 0.01 (40 dB). The three-dimensional diagrams show what happens when the frequency of the major signal is changes in steps of 1/8 from bin 10 to bin 11. The cross-sections where the highest sidelobes occur are shown in Figures 16.35(b) and 16.36(b).

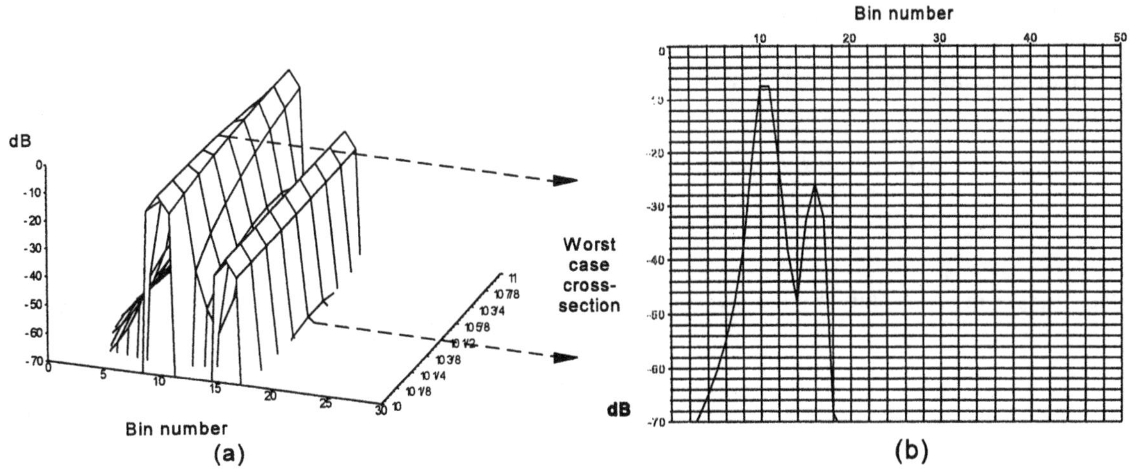

Figure 16.35 The two-tone characteristics of von Hann (Hanning) tapering with a signal ratio of 20 dB.

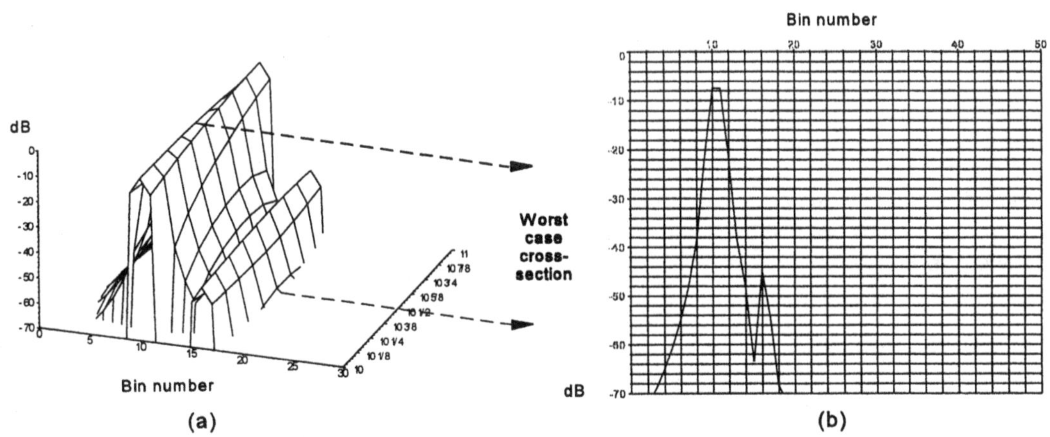

Figure 16.36 The two-tone characteristics of von Hann (Hanning) tapering with a signal ratio of 40 dB.

An example of Hamming tapering is shown in Figure 16.37. In contrast with Figure 16.36, the presence of the minor component is shown by an interfering kink in the worst case spectrum, which may not be extracted automatically.

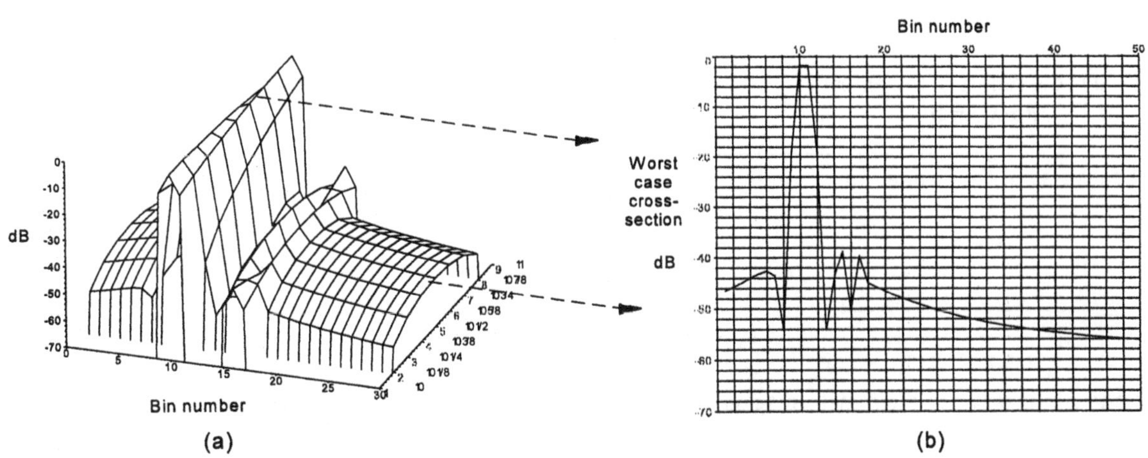

Figure 16.37 The two-tone characteristics of Hamming tapering with a signal ratio of 40 dB.

16.6 RELATIONSHIPS TO OTHER TRANSFORMS

The father of all the transforms used here is the double sided Laplace transform, which is given by

$$\textit{Double sided Laplace transform, } H(p) = \int_{-\infty}^{+\infty} h(t)\, e^{-pt}\, dt \tag{16.65}$$

The Laplace transform is normally used to investigate systems in which the time starts at zero to give the normal form

$$H(p) = \int_{0}^{+\infty} h(t)\, e^{-pt}\, dt \tag{16.66}$$

The variable p has real and imaginary parts, where $p = s + j\omega$. If the time variable has only discrete values $n\tau$, where n is an integer, then when $z = e^{p}$ [1, p. 257; 13]

$$H(z) = \sum_{n=0}^{\infty} h(n)\, z^{-n} \tag{16.67}$$

The Fourier transform can be seen to be the steady state part of the double sided Laplace transform.

In rectangular planar antenna arrays, to a first order, the patterns may be resolved in the two dimensions. The distance on a circular array is defined by its radius, and its beam is the Hankel transform of the illumination function. The Hankel transform is given by

$$F(q) = 2\pi \int_{0}^{\infty} f(r)\, J_0(2\pi q r) r\, dr \tag{16.68}$$

In the case of the normalized circular antennas pattern in sine space, $g(0) = 1$, is given by [12, p. 293]

$$g(u') = 2\pi a^2 \int_{0}^{1} J_0(u'r) r\, dr \tag{16.69}$$

where a is the radius of the antenna;
$u' = 2\pi a/\lambda \sin\theta$.

The inverse of the Hankel transform is

$$f(r) = 2\pi \int_{0}^{\infty} F(q)\, J_0(2\pi q r) q\, dq \tag{16.70}$$

The relationships between the transforms mentioned in this book are shown in Figure 16.38.

16.6.1 The *z* transform

The *z* transform was defined formally in the 1950s. Because it is relatively new, there is no mention of it in the older literature, such as that describing Schelkunov's polynomial for linear arrays [12]. As intimated, it allows discrete Fourier transform functions to be treated as polynomials, which expand the mathematical processes available for the design and analysis of antennas with discretely spaced elements and signal processing using data representing signals from analogue-to-digital converters with regular clocks.

The z transform is related to the Laplace and Fourier transforms when the exponential e^{pt} is replaced by z, so that [1, p. 259]

$$H(z) = \sum_{k=0}^{\infty} h[k]\, z^{-k} \tag{16.71}$$

The inverse is [13, p. 232]

$$h(n) = \frac{1}{j2\pi} \int_{\Gamma} \frac{1}{T} H\left(\frac{1}{T} \ln z\right) z^{n-1}\, dz \tag{16.72}$$

where Γ is the unit circle.

Double sided Laplace transform

$$\mathcal{L}(p) = \int_{-\infty}^{+\infty} h(t)\, e^{-pt} dt$$

where $p = s + j\omega$

s represents the transient part and $j\omega$ represents the steady state part

s ⇓ $j\omega$ ⇓

Laplace transform
Starts at zero time

$$\mathcal{L}(s) = \int_{0}^{+\infty} h(t)\, e^{-st} dt$$

Fourier transform
Steady state

$$\mathcal{F}(f) = \int_{-\infty}^{+\infty} h(t)\, e^{-j\,2\pi f t}\, dt$$

Hankel transform
Two dimensions with circular symmetry

$$H(u) = 2\pi \int_{0}^{\infty} h_0(r)\, J_0(2ur) r\, dr$$

***z* transform**
Sampling at equally spaced times

$$z = e^{pt}$$

$$\mathcal{F}(z) = \sum_{n=0}^{\infty} h(t)\, z^{n}$$

Discrete Fourier transform
Sampled values

$$\mathcal{F}(n) = \sum_{n=0}^{\infty} h(t)\, e^{j\,2\pi n t}$$

Fast Fourier transform
N is limited to powers of 2

Figure 16.38 The relationships between the transforms mentioned in this book.

The z transform has most of the properties of the Fourier transform [13, pp. 4 et seq.] and the most important are given in Table 16.6.

Table 16.6
Some properties of the z transform

	Time function	Transform
Addition	$h_1(t) + h_2(t)$	$H_1(z) + H_2(z)$
Scaling	$a^n\, h(t)$	$H\left(\dfrac{z}{a}\right)$
Shift	$h(t-a)$	$z^{-a}\, H(z)$

A number of commonly used transforms are shown in Table 16.7.

Table 16.7
Selected z transforms

	Time function	Transform
Delta function	$\delta(k)$	1
Heaviside step	1	$\dfrac{1}{1 - z^{-1}}$
Exponential waveform	c^k	$\dfrac{1}{1 - cz^{-1}}$
Ramp	k	$\dfrac{z}{\left(1 - z^{-1}\right)^2}$
Cosine wave	$a^k \cos(bkT)$	$\dfrac{z\left(z - a \cos bT\, z^{-1}\right)}{1 - 2a \cos bT\, z^{-1} + a^2 z^{-2}}$
Sine wave	$a^k \sin(bkT)$	$\dfrac{z\left(a \sin bT\, z^{-1}\right)}{1 - 2a \cos bT\, z^{-1} + a^2 z^{-2}}$

16.7 THE USE OF FOURIER TRANSFORMS FOR FINITE IMPULSE RESPONSE FILTERS

Finite impulse response (FIR) filters derive their name from the fact that a single needle pulse at the input gives a single series of needles spaced in time by T at the output. Such a filter is shown in Figure 16.39. The amplitudes and phases of the needles are defined by the shaping coefficients which is the impulse response $h(n)$. Normally, the filter is filled and the output is the sum of the delayed elements. As this sum has often undesirably high responses outside the pass band, they may be reduced by multiplying the individual shaped vector values by an additional tapering coefficient.

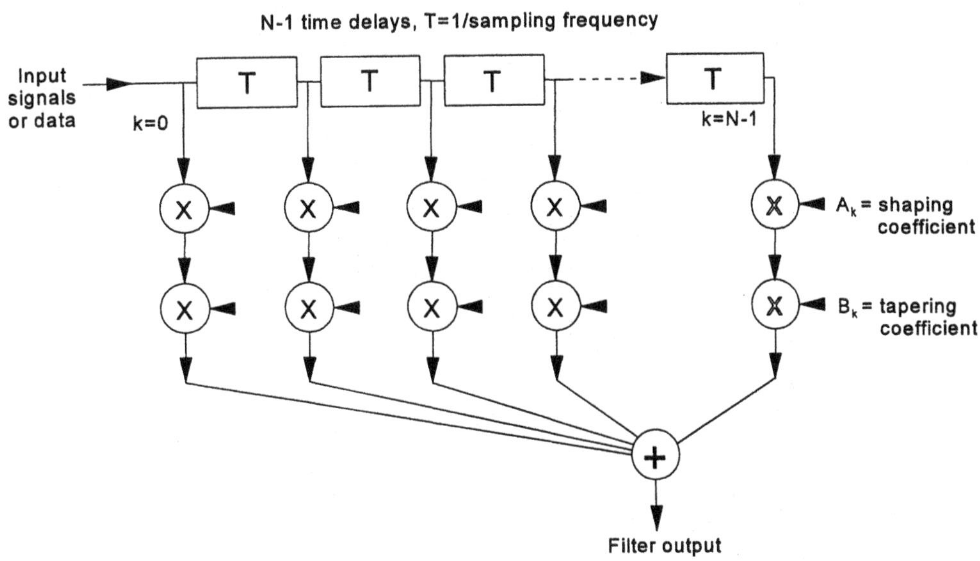

Figure 16.39 Block diagram of a finite impulse response (FIR) filter.

16.7.1 Single-phase signal filtering

The characteristic of an ideal lowpass filter with a cutoff frequency ω_c for the waveform sampled at a frequency ω_s is shown in Figure 16.40. In a sampled environment, all other frequencies are aliased into the band zero to $\omega_s/2$.

Single-phase alternating currents can be represented by the sum of two vectors that rotate in opposite directions. The number of rotations in each second represents the frequency, and both directions of rotation must be taken into account.

$$\textit{Alternating voltage} = \frac{e^{j\omega t} + e^{-j\omega t}}{2} = \cos \omega t \tag{16.73}$$

16.7.1.1 Lowpass filter

The perfect lowpass filter characteristic in Figure 16.40 passes frequencies between zero and ω_c. The "negative" frequency characteristic is shown dotted. To satisfy (16.73), the integration ranges must contain both the "negative" and the positive frequency ranges. Note that dashed lines are used for "negative" frequencies, as they are notional only in single-phase systems.

The lowpass filter has a finite time delay represented by λ, which gives a helix to the characteristic, $-\lambda\omega T$ radian/radian, and thus a phase characteristic to the response of the lowpass filter. The constant, λ, is chosen in the examples so that $(n - \lambda)$ gives a symmetrical range of integer values around zero. An odd number of samples are used in the examples.

The inverse Fourier transform to calculate the time shaping function is given in (16.74). The e^{jx} term represents a vector only representable by a polyphase voltage. As above, alternating vectors can be represented by two contrarotating vectors (16.73), hence the negative limit $-\omega_c$.

$$
\begin{aligned}
h_{LP}(n) &= \frac{1}{\omega_s}\int_{-\omega_c}^{\omega_c} e^{j(n-\lambda)\omega T}d\omega \\
&= j\,\frac{e^{-j\omega_c(n-\lambda)T} - e^{j\omega_c(n-\lambda)T}}{\omega_s(n-\lambda)T} \\
&= \frac{2\sin\left(\omega_c(n-\lambda)T\right)}{\omega_s(n-\lambda)T}
\end{aligned}
\tag{16.74}
$$

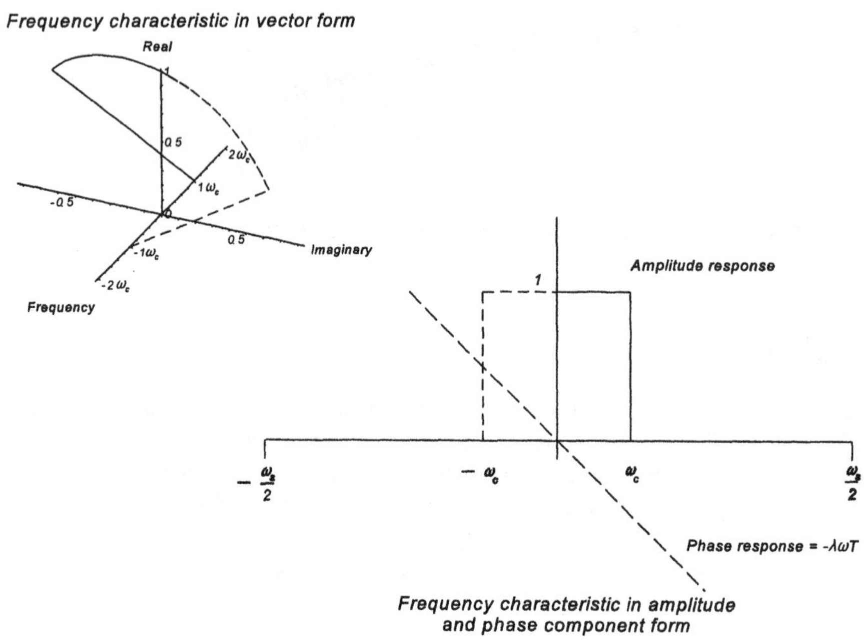

Figure 16.40 Frequency response of a lowpass filter.

Substituting $\omega_s = 2\pi/T$ and multiplying the numerator and the denominator by ω_c, the following is obtained:

$$
\begin{aligned}
h_{LP}(n) &= \frac{2\sin\left(\omega_c(n-\lambda)T\right)}{\omega_s(n-\lambda)T} \\
&= \frac{\omega_c T}{\pi}\,\frac{\sin\left(\omega_c(n-\lambda)T\right)}{\omega_c(n-\lambda)T}
\end{aligned}
\tag{16.75}
$$

The frequency response of the filter with $N-1$ delays or N pulses is

$$
H(\omega) = \sum_{n=0}^{N-1} h_{LP}(n)\, e^{j(n-\lambda)\omega T} \tag{16.76}
$$

If λ is chosen so that at the center point $(n-\lambda) = 0$ for an odd number of N points, then

$$
\begin{aligned}
H(\omega) \;=\; & h_{LP}(0)e^{-j\frac{N-1}{2}\omega} + h_{LP}(N)e^{j\frac{N-1}{2}\omega} \\
& + h_{LP}(1)e^{-j\left(\frac{N-1}{2}-1\right)\omega} + h_{LP}(N-1)e^{j\left(\frac{N-1}{2}-1\right)\omega} \\
& + \ldots \\
& + h_{LP}\left(\frac{N}{2}\right)e^{j0\omega}
\end{aligned}
\tag{16.77}
$$

The exponential terms for a symmetrical impulse response reduce to

$$
\begin{aligned}
H(\omega) &= h_{LP}\left(\frac{N-1}{2}\right) + 2\sum_{n=0}^{\frac{N-1}{2}} h_{LP}(n)\cos\left[\left(\frac{N-1}{2} - n\right)\omega T\right] \\
&= \sum_{n=0}^{N-1} h_{LP}(n)\cos\left[\left(\frac{N-1}{2} - n\right)\omega T\right]
\end{aligned}
\tag{16.78}
$$

and for an asymmetrical impulse response

$$
H(\omega) = \sum_{n=0}^{N-1} h_{LP}(n)\,\sin\left[\left(\frac{N-1}{2} - n\right)\omega T\right] \tag{16.79}
$$

Example: If the cutoff frequency ω_c is $\omega_s/8$ and there are $N = 21$ taps, then

$$
h_{LP}(n) = \frac{1}{4}\,\frac{\sin(n\,\pi/4)}{n\,\pi/4} \tag{16.80}
$$

The shaping weights are illustrated in Figure 16.41.

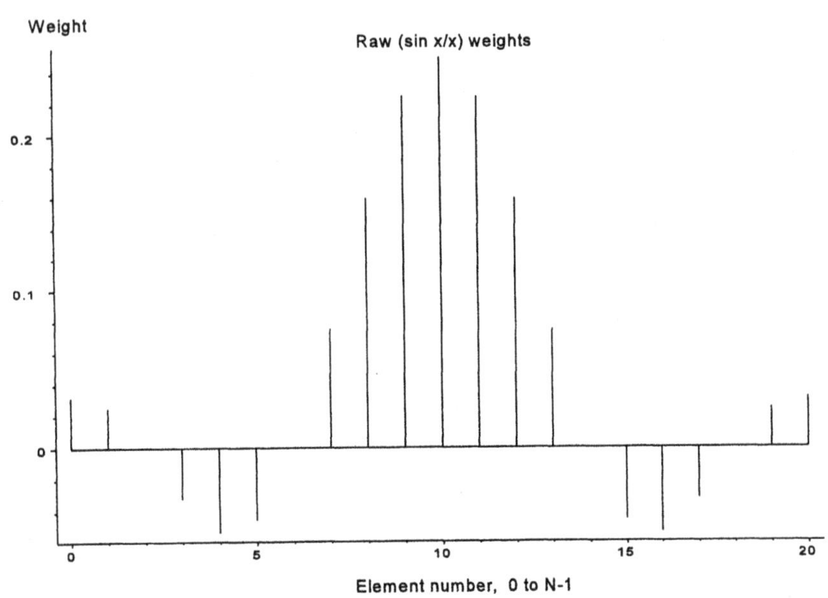

Figure 16.41 The shaping weights for a lowpass filter with 21 points and a cutoff frequency of $\omega_s/8$.

The attenuation at different frequencies for the lowpass filter is shown in Figure 16.42 and for other filters in Figure 16.43.

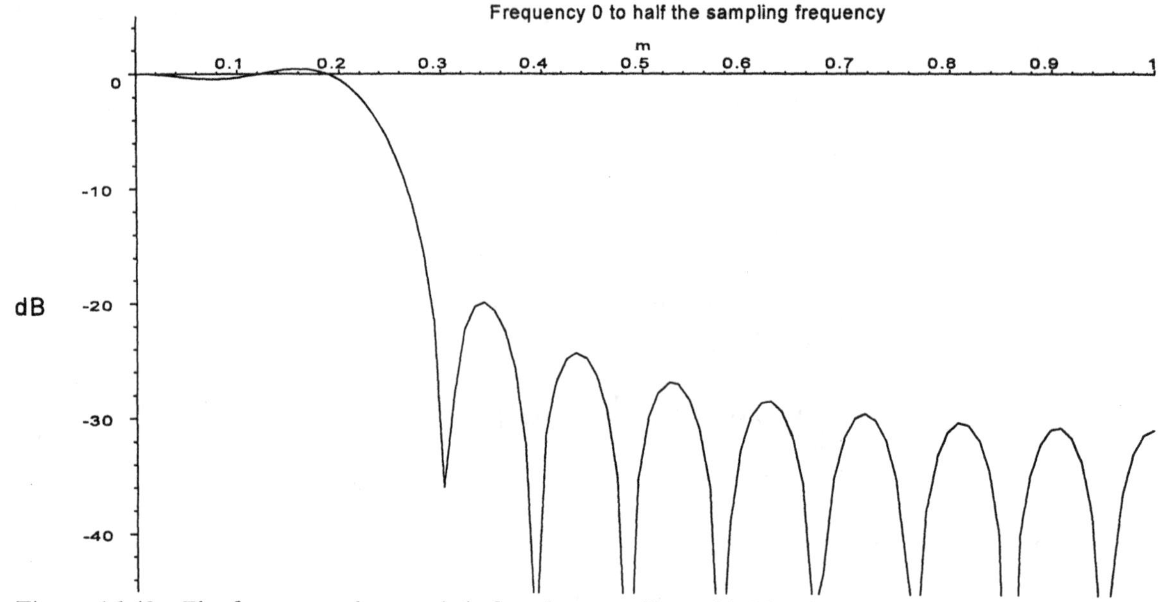

Figure 16.42 The frequency characteristic for a lowpass filter with 21 points and a cutoff frequency of $\omega_s/8$.

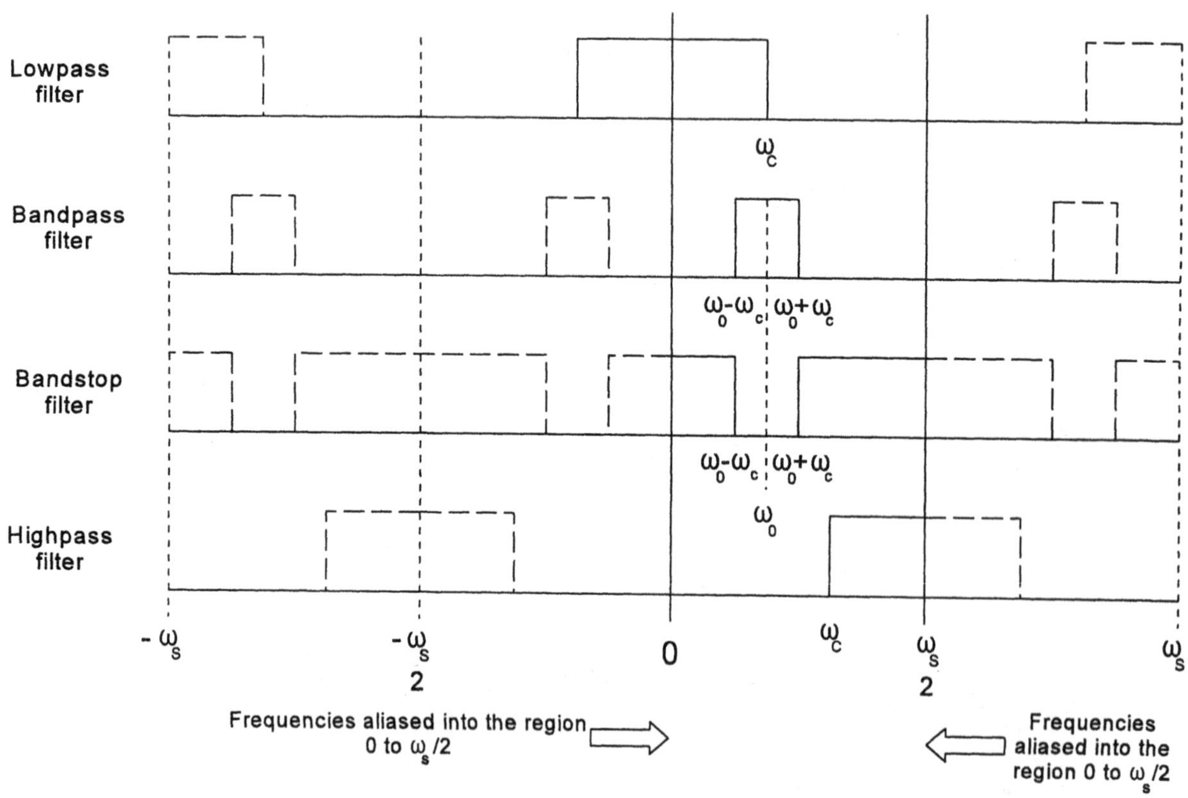

Figure 16.43 Bandpass, bandstop, and highpass filters derived from the lowpass filter.

16.7.1.2 Highpass filter

When ω_0 is $-\omega_s/2$, a highpass filter is formed with a cutoff frequency ω_c.

$$\begin{aligned} h_{HP}(n) &= h_{LP}(n)\cos\left(\omega_0\,(n-\lambda)\,T\right) \\ &= h_{LP}\cos((n-\lambda)\,\pi) \\ &= h_{LP}\,(-1)^{n-\lambda} \end{aligned} \tag{16.81}$$

Example: If the cutoff frequency ω_c is $\omega_s/8$ and there are $N = 21$ taps, then the weights are those in Figure 16.44 and its frequency characteristic is that in Figure 16.45.

16.7.1.3 Bandpass filter

In the case of a bandpass filter centered on ω_0 and that as a half-width of ω_c, the limits of integration for the contra-rotating vectors are from $-\omega_0-\omega_c$ to $-\omega_0+\omega_c$ and from $+\omega_0-\omega_c$ to $+\omega_0+\omega_c$. The bandpass characteristic with a center frequency of ω_0 and width $2\omega_c$ is obtained when

$$h(n)_{BP} = 2\cos(n\omega_0 T)\;h(n)_{LP} \tag{16.82}$$

Example: If the cutoff frequency ω_c is $\omega_s/8$ and there are $N = 21$ taps, the weights and frequency characteristics are those in Figure 16.46 and Figure 16.47, respectively.

16.7.1.4 Bandstop filter

Similarly, for a bandstop filter centered on ω_0 and width $2\omega_c$ the characteristic is derived from the lowpass and bandpass filters.

$$h(n)_{BP} = 2\cos(n\omega_0 T)\;h(n)_{LP} \tag{16.83}$$

$$\begin{aligned} h_{BS}(0) &= 1 - h_{BP}(0) \\ h_{BS}(n) &= -h_{BP}(n) \end{aligned} \tag{16.84}$$

Example: If the cutoff frequency ω_c is $\omega_s/8$ and there are $N = 21$ taps, then the weights are those in Figure 16.48 and its frequency characteristic is that in Figure 16.49.

16.7.2 Video integration in a finite impulse response filter

Video integration is lowpass filtering over many sweeps. The range gated video pulses from the sweeps are added together either to give a clearer display or for automatic detection. The pulses are tapered, if necessary, and added as in the center part of Figure 16.41.

16.7.3 Noncoherent moving target indicator finite impulse response filter

For video integration, the neighboring pulses are added; for moving target indication, they are subtracted to give a high-pass filter, as in the center part of Figure 16.44.

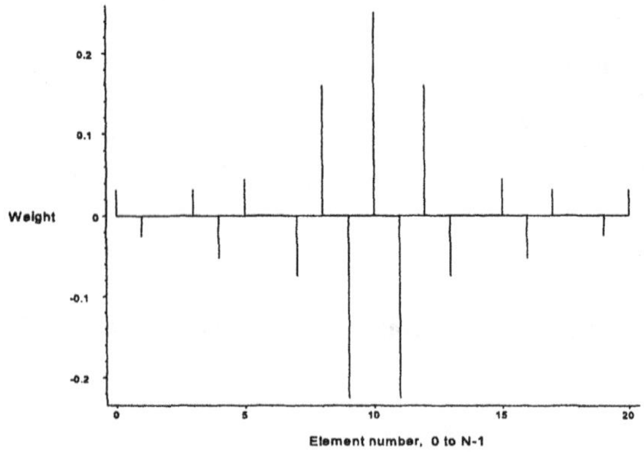

Figure 16.44 The weights for a highpass filter with 21 points and a cutoff frequency of $\omega_s/8$.

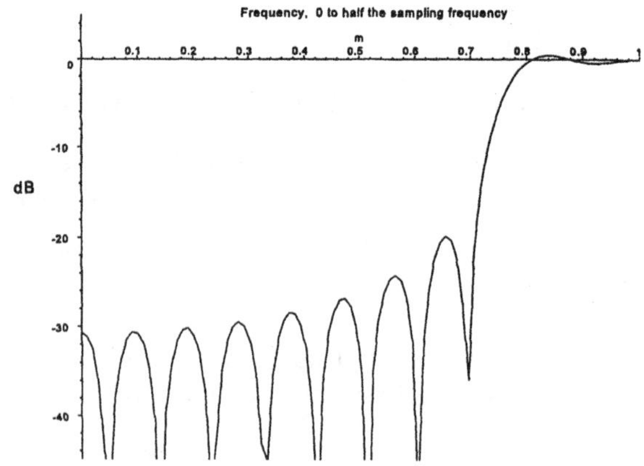

Figure 16.45 The frequency characteristic for a highpass filter with 21 points and a cutoff frequency of $\omega_s/8$.

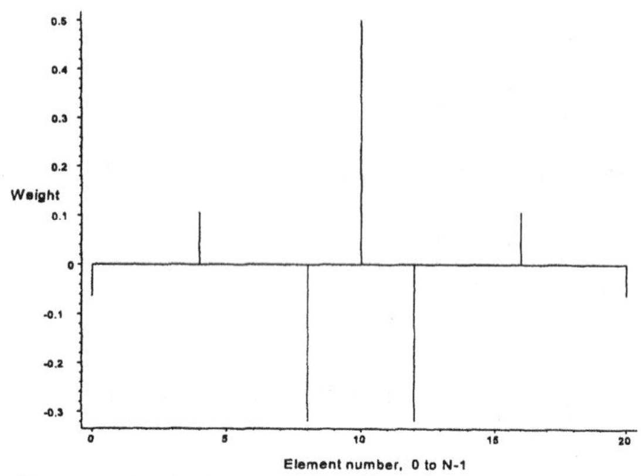

Figure 16.46 The weights for a bandpass filter with 21 points, center frequency $\omega_s/4$, and a cutoff frequency of $\omega_s/8$.

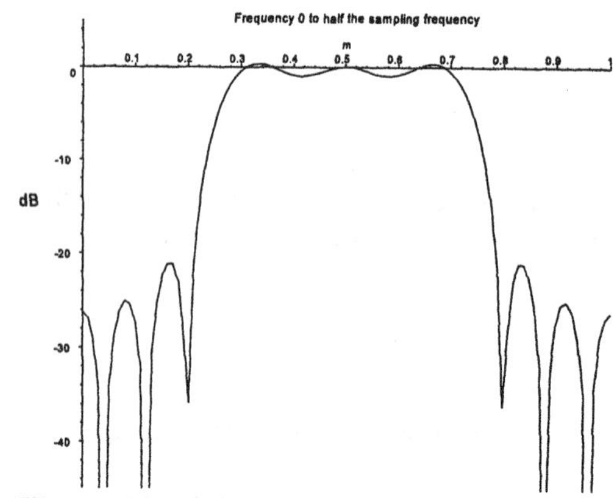

Figure 16.47 The frequency characteristic for a bandpass filter with 21 points, center frequency $\omega_s/4$, and a cut off frequency of $\omega_s/8$.

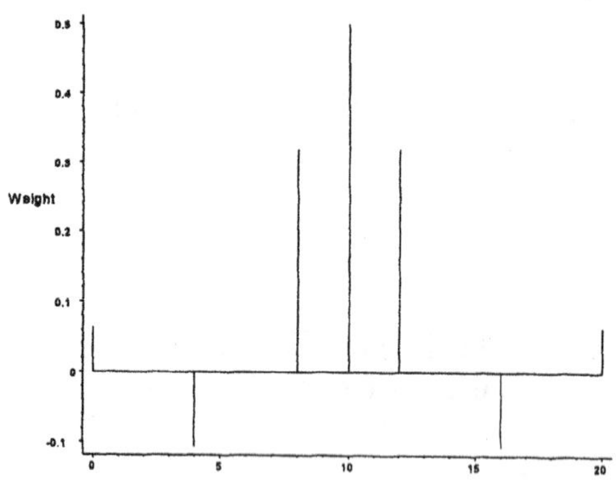

Figure 16.48 The weights for a bandstop filter with 21 points, center frequency $\omega_s/4$, and cutoff frequency $\omega_s/8$.

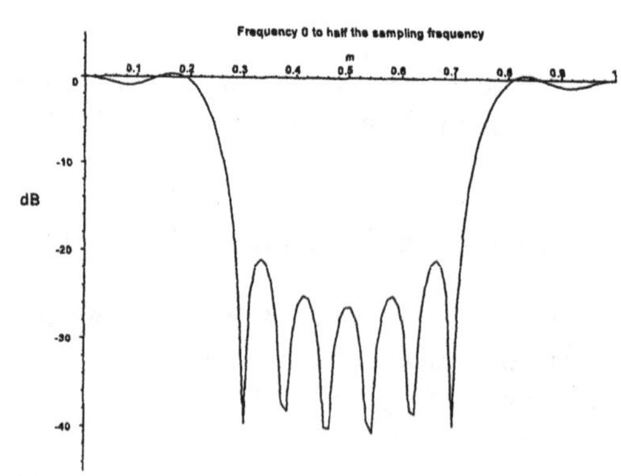

Figure 16.49 The frequency characteristic for a bandstop filter with 21 points, center frequency $\omega_s/4$, and a cutoff frequency of $\omega_s/8$.

16.7.4 Polyphase signal filtering

Polyphase signals have effectively double the number of samples and are unique in the frequency range that can be defined as zero to ω_s or $-\omega_s/2$ to $+\omega_s/2$. Vectors may rotate either to the right or to the left. Separate filters are usually used for each of two phases. A highpass filter may be used for moving target indication, for example.

In radar moving target detection filters select particular frequencies by multiplying the incoming shaped helical waveform, frequency f_i, by a reference helical waveform, f_r, or correlation. When the frequencies of the helices are the same but their sequences are opposite, a maximum is obtained over time. For N discrete samples,

$$Output\ voltage = \sum_{1}^{N} e^{+j\,2\pi f_i t}\, e^{-j\,2\pi f_r t} = \sum_{1}^{N} e^{j\,2\pi (f_i - f_r)\,t} \tag{16.85}$$

If the samples are equally spaced, the output (16.85) then becomes

$$\begin{aligned} Output\ voltage &= \frac{1}{N}\sum_{k=0}^{N-1} e^{-j\,2\pi \frac{n}{N} k} && \text{discrete or periodic form} \\ &= \frac{1}{N}\,\frac{1 - e^{-j\,2\pi n}}{1 - e^{-j\,2\pi \frac{n}{N}}} && \text{sum of a geometric progression} \\ &= \frac{1}{N}\,\frac{\sin(\pi n)}{\sin\left(\pi \frac{n}{N}\right)}\, e^{-j\pi n\left(1-\frac{1}{N}\right)} && \text{amplitude and phase} \end{aligned} \tag{16.86}$$

The bandwidths are totally dependent on the number of samples N. The curves are in terms of $f_s(f_i - f_r)/N$, where f_s is the sampling frequency and is shown in Figure 16.50. The half-power (3 dB) and the half-voltage (6 dB) beamwidths with increasing numbers of samples are shown in Figure 16.51.

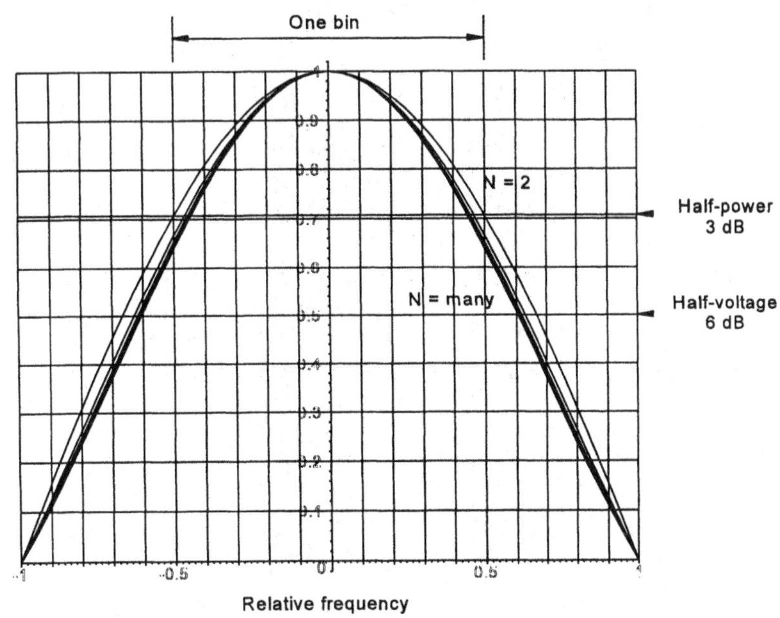

Figure 16.50 The response of a polyphase filter against relative frequency $\frac{f_i - f_r}{N} f_s$, where f_i is the signal frequency, f_r is the reference frequency, and f_s is the sampling frequency.

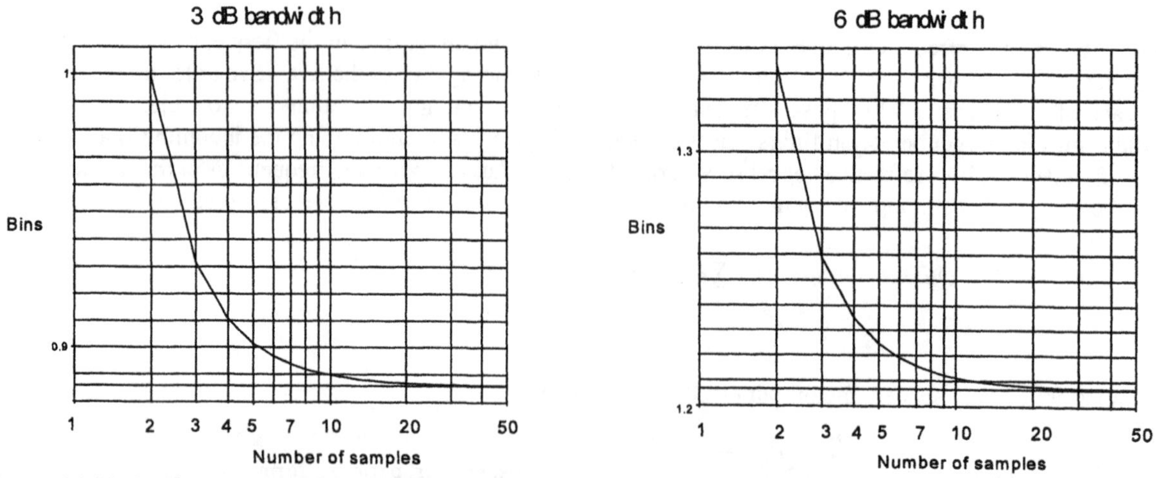

Figure 16.51 Half-power (3 dB) and half-voltage (6 dB) bandwidths for sampled filters.

Special cases of polyphase filtering are the coherent moving target indicator and detector. In both cases the amount of shaping of the response is limited by the small number of pulses that are processed. This shaping may be for:

- Tapering to reduce sidelobes;
- Widening to collect large amplitude clutter into one filter. Woodward-Levinson methods, for example, may be used to change the tapering to give a wider filter bandwidth. Clutter is the major contributor in this case; other signals collected by the higher sidelobes will have little or no effect.

REFERENCES

1. Bracewell, R. N., *The Fourier Transform and Iits Applications*, 2nd ed., New York: McGraw-Hill, 1978.
2. Elliott, D. E., and R. K. Rao, *Fast Transforms*, Orlando, Florida: Academic Press, 1982.
3. Meikle, H. D., *A New Twist to Fourier Transforms*, Weinheim: Wiley-VCH, 2004.
4. Press, W. H., B. P. Flannery, S. A. Teukolsky, and W. T. Vetterling, *Numerical Recipes*, Cambridge: Cambridge University Press, 1986, p. 381.
5. Bronshtein, I. N., and K. A. Semendyayev, *Handbook of Mathematics*, New York: van Nostrand Reinhold and Thun and Frankfurt am Main: Verlag Harri Deutsch, 1985.
6. Campbell, G. A., and R. M. Foster, *Fourier Integrals for Practical Applications*, Princeton, New Jersey: D. van Norstrand, 1948.
7. Weisstein, E. W., (ed.), *CRC Concise Encyclopedia of Mathematics*, Boca Raton, Florida: CRC Press, 1999.
8. Cooley, J. W., and J. W. Tukey, "An Algorithm for Machine Calculation of Complex Fourier Series", in L. R. Rabiner and C. M. Rader, *Digital Signal Processing*, New York: IEEE Press, 1972.
9. Cooley, J. W., and J. W. Tukey, "An Algorithm for the Machine Calculation of Complex Fourier Series", *Mathematics of Computation*, Vol. 19, No. 90, 1965, pp. 297-305.
10. *Oxford Dictionary of Quotations*, 4th ed., Oxford: Oxford University Press, 1995.
11. Harris, F. J., "On the Use of Windows for Harmonic Analysis with the Discrete Fourier Transform", *Proceedings of the IEEE*, Vol. 86, No. 1, January 1978.
12. Silver, S., ed., *Microwave Antenna Theory and Design, Volume 12 of the MIT Radiation Laboratory Series,* New York: McGraw-Hill, 1949.
13. Jury, E. I., *Theory and Application of the z-Transform Method*, New York: Wiley, 1964.

APPENDIX 16A COMPLETE CORRELATION

Conventional cross-correlation is unsatisfactory in that with the product with a complex conjugate, the rules of complex multiplication are applied to the amplitude and division to the phase angle. A true measure of correlation may be obtained when the Fourier transforms are divided so that the rules of complex division are applied to amplitude and phase. Division of the Fourier transforms results in infinity for identical waveforms, and a curve of width showing the difference of moments, principally standard deviation, of the different waveforms. There is no simple analogue circuit for division and division takes time in digital circuitry. The curves for Fourier transforms often pass through zero leading to division by zero so that ordinary cross-correlation is used.

Gaussian waveforms never have a value of zero so that they may be used as an example in (16A.1).

$$p_1(x) = \frac{1}{\sqrt{2\pi\sigma}}\exp\left(-\left(\frac{x-\mu_1}{2\sigma_1}\right)^2\right) \quad \text{and} \quad p_2(x) = \frac{1}{\sqrt{2\pi\sigma}}\exp\left(-\left(\frac{x-\mu_2}{2\sigma_2}\right)^2\right) \tag{16A.1}$$

Their Fourier transforms are of the form

$$\begin{aligned} C(\xi) &= \int_{-\infty}^{+\infty}\frac{1}{\sqrt{2\pi\sigma}}\exp\left(-\left(\frac{x-\mu}{2\sigma}\right)^2\right)\exp(-j2\pi x\xi)\,dx \\ &= \exp(-j2\pi\mu\xi)\exp(-j2\pi^2\sigma^2\xi^2) \end{aligned} \tag{16A.2}$$

shown in Figures 16A.1(a) and 16A.1(b) for two Gaussian curves:

- Curve 1: Mean = 4, standard deviation = 1;
- Curve 2: Mean = 6, standard deviation = 2.

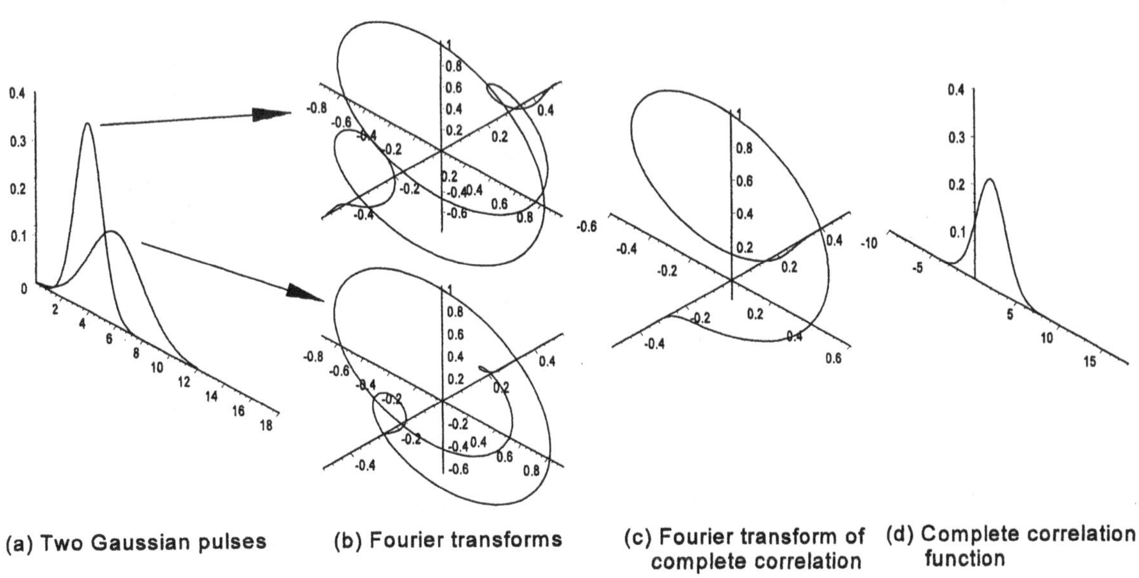

(a) Two Gaussian pulses (b) Fourier transforms (c) Fourier transform of complete correlation (d) Complete correlation function

Figure 16A.1 Complete correlation.

The division of Fourier transforms in Figure 16A.1(c) is

$$\frac{C_1(\xi)}{C_2(\xi)} = \exp(-j2\pi\,\overline{\mu_1-\mu_2}\,\xi)\exp(-j2\pi^2\,\overline{\sigma_1^2-\sigma_2^2}\,\xi^2) \tag{16A.3}$$

Invert (16A.3) to obtain the combined distribution

$$\textit{Combined distribution} = \frac{1}{\sqrt{2\pi\;\overline{\sigma_1^2-\sigma_2^2}}}\exp\left(-\frac{(x-\overline{\mu_1-\mu_2}\,)^2}{2\;\overline{\sigma_1^2-\sigma_2^2}}\right) \tag{16A.4}$$

The combined distribution has the statistics shown in Table 16A.1.

Table 16A.1
The statistics of the correlation curve in Figure 16A.1(d)

Statistic	Expression	Value in Figure 16A.1
Mean	$\mu_2 - \mu_1$	2
Variance	$\lvert\sigma_1^2 - \sigma_2^2\rvert$	3
Standard deviation	$\sqrt{\lvert\sigma_1^2 - \sigma_2^2\rvert}$	$\sqrt{3}$

Appendix A

Language and glossary

Radar grew up in military surroundings of the Second World War (1939-1945) which has colored the choice of words. Today words are used often in full ignorance of their origins, especially by those who do not normally use English.

Hopefully times have changed, and, unless it would cause misunderstanding, normal civilian English has been used in this text with the exception of *range, moving target indicator,* and *moving target detector.*

The letter and the spirit of IEEE standards [1, 2] and their substandards have been followed, with the exception that the radar vocabulary has been disarmed.

The oblique stroke is used here for division only. Uses other than that lead to woolly and incomplete thinking and often have local meanings, such as: U/S for unserviceable, M/T for empty, N/A not applicable, N/K not known, and so on.

A.1 UNIFIED TERMINOLOGY

The terminology used in this book has been demilitarized. The main changes are shown in this section.

A.1.1 Military words that have been avoided

Target — A target is something at which one aims or shoots. It is no longer the purpose of civil radar to aim weapons at the objects seen in order to destroy them. The words *scatterer, echo,* or *echo signal* are used here. The terms *moving target indication* and *detection* are retained.

Boresight — A boresight is a sight, normally cross-wires, that fits into the bore of an artillery piece. This is the basic reference used to align the optical sight or the fire control radar used as a sight, as is the normal case with anti-aircraft artillery. The term *electrical axis* is used here.

I have seen literature which referred to the enhancement of the accuracy of the measurement of azimuth angle using monopulse techniques in a secondary radar at a civil airport as "off-boresight error"! There was no bore, no sight, and no error.

Jamming — There seems to be confusion between unintentional interference components in the received signals and intentional jamming by an enemy, saboteurs, or terrorists, which seems to be more dramatic. Civil radars, unless they have a military capability, should therefore detect interference. Such sources of interference are unsuppressed electrical equipment and the daily wandering of the sun through the coverage of the radar.

A.1.2 Unfortunate words

Execute — Execution is the punishment reserved for convicted murderers, traitors, and spies condemned to death by a court. It has no place in the civil air industry (not even for hostages). Executing maintenance (people) could have unfortunate meanings. The term *execute* is used here only for carrying out a single computer instruction or program.

Plumbing — This derogatory slang was used in the past by those who did not understand waveguides. The term *waveguide system* is used here.

A.1.3 Technical words

In the past, a relatively small core of radar engineers defined the language. Experts in other branches of knowledge, for example, mathematicians, used the same vocabulary. The advent of digital processing brought in software specialists. Unfortunately, they do not have an education in electrical engineering and still use the terms for early, primitive moving target indicator components for other, more modern ways of processing. Examples are shown here.

Phase detector — This term is used only for a detector that gives a voltage proportional to the phase difference between the input signal and the reference phase; thus, the detector is, by definition, insensitive to amplitude.

Synchronous detector — This term is used for a linear amplitude detector that gives a voltage proportional to the component of signal amplitude at the reference phase. In order to obtain all the information in a signal of unknown phase, which may be in quadrature to the reference phase and give zero output, two synchronous detectors are required: the detector for the reference phase (in phase) and a detector using as a reference a signal in quadrature to the reference phase. The resulting signals are two-phase signals labeled I and Q that correspond to the waveforms found in polyphase power distribution systems [1].

Stagger — The word *stagger* is used when the parameters in a group of processes are set to provide a broadening of a characteristic in this group of processes. Examples are:

- The individual stages of intermediate frequency amplifiers are tuned to different chosen frequencies within a range in order to broaden the bandwidth of the complete amplifier. This is a stagger tuned amplifier [2].
- A group of chosen pulse repetition frequencies is used in moving target indicator radars in order to provide detection at the first and higher blind speeds. The detection decision is based on the sum of the outputs at the different pulse repetition frequencies [1]. Pulse repetition frequency staggering contrasts with pulse repetition frequency diversity, in which there are individual detection decisions at each of the individual pulse repetition frequencies. This is also the case with (radio) frequency diversity.

Gaussian distribution — This is the other term for the "normal" distribution, which is used in English speaking countries. Since, to misquote George Orwell, all distributions are normal but some are more normal than others, the term *Gaussian distribution* is used here.

A.1.4 Words retained

Broadside — The primary meaning of this word is for a row of cannons firing together out of the side of an older warship for consistent aim. The term has been retained for antenna elements radiating (figuratively, firing) in phase to aim a beam of energy.

Defruiter — A defruiter removes asynchronous interference, or fruit ("False Replies Unsynchronized to Interrogator Transmission") [3].

MTI, MTD — Moving target indicator and moving target detector. The use of these expressions is so common that it would be foolish to try to replace them.

Range — Range has many meanings in English not limited to military ballistics. It is extremely ingrained as one of the radar coordinates and has therefore been retained.

A.1.5 New words

Owing to the misuse of the term *phase detector* to indicate a detector that has a two-phase amplitude output, two other terms are used for the two forms of vector detectors.

Polar detector — At the input: Amplitude and phase modulated signals on a suppressed carrier at normally the coherent oscillator frequency.

At the output: An amplitude signal representing the instantaneous magnitude of the input signal, and a second signal representing the instantaneous phase of the input signal with respect to a phase reference, normally the coherent oscillator (COHO). The signal may be offset to avoid zero.

This is the representation of the signal vector in polar coordinates. Doppler frequencies above and below the input carrier frequency are represented by the increasing or decreasing phase angle video values with time or from pulse to pulse.

Cartesian detector — At the input: Amplitude and phase modulated signals on a suppressed carrier at normally the coherent oscillator frequency.

At the output: An amplitude signal corresponding to the input signal times the cosine of the phase reference signal: a lowpass filter selects the video component, called the in-phase, or I, video.

The second signal is an amplitude signal corresponding to the input signal times the sine of the phase reference signal: a lowpass filter selects the video component, called the quadrature, or Q, video.

This is the representation of the signal vector in Cartesian coordinates in a two-phase circuit. Doppler frequencies above and below the input carrier frequency are represented by the positive and negative phase sequence of the signals with time or from pulse to pulse.

Other new words are:

Bottle brush noise — Bottle-brush noise is an idea of Gaussian noise modulated on a notional carrier, see Figure 2.19.

Rotating voltage — A voltage that may be represented by a rotating vector as in three-phase power systems. In signal processing each dimension of the vector has to be represented by a separate voltage on two phases, normally labeled I and Q.

A.1.6 Russian names

The names P.L. Chebychëv (Пафнутий Львович Чебышёв) and E.I. Zolotarëv (Егор Иванович Золотарёв) both end in -ëv (-ёв). There are a number of transliterations in Latin script depending on the European language being used and whether the transliteration is phonetic, for example Chebyshev or the phonetic transliteration Chebyshov in English. The problem is with the Russian ë (pronounced yo or o after an consonant) and as an aid to the correct pronunciation for those with an elementary reading knowledge of Russian, the ë is retained.

A.2 GLOSSARY

This describes the terms used here and their abbreviations.

A-display — See A-display, Section A.3.4.

ACP — Azimuth change pulse.

Acquisition — The process of establishing a track for a tracking device. This may be a radar that follows one sole object of interest, for example, a weather balloon, or a number of objects using track while scanning.

AEW	Airborne early warning (military).
AGC	Automatic gain control.
AGL	Above ground level. Normally the height, often in feet, above the local ground surface.
AI	Airborne intercept (military).
Air route surveillance radar	A radar used to control aircraft between airports and their airport surveillance radars (ASR). Typically they have a range of 150 to 200 nautical miles.
Airport surface detection equipment	A radar that covers the airfield surface used to observe taxiing aircraft and vehicles.
Airport surveillance radar	A radar which is used to control aircraft leaving, landing, and flying near an airport. Typically they have a range of 60 nautical miles.
ARSR	See air route surveillance radar.
ASDE	See airport surface detection equipment.
A-scope	See A-display.
ASR	See airport surveillance radar.
Autocorrelation	The correlation of a waveform with itself, often used as a measure of power.
Azimuth angle	The angle clockwise from north in the horizontal plane.
Azimuth change pulse	A pulse representing an azimuth angle increment. The number of azimuth change pulses in a circle is normally a power of 2. Typical values are 4 096, 8 192, and 16 384.
Azimuth resolution	The minimum distance by which two aircraft flying at the same elevation and range have to be separated in order to be distinguished as two separate objects by the radar *P*% of the time.
B-display	See B-display in Section A.3.4.
B-scope	See B-display in Section A.3.4.
Bandwidth	There are a number of measures of bandwidth: Half-power (3 dB): This is the bandwidth measured between the half power points of the frequency response. Half-voltage (6 dB): This is the bandwidth measured between the half voltage points of the frequency response. Noise: The first moment of the response about the mean

$$\frac{\int_{-\infty}^{+\infty} | f \quad f_0 | \, |H(f)| \, df}{\int_{-\infty}^{+\infty} |H(f)| \, df} \text{ Hz}$$

Root mean square (rms) The second moment of the response about the mean

$$\sqrt{\frac{\int_{-\infty}^{+\infty} | f \quad f_0 |^2 \, |H(f)|^2 \, df}{\int_{-\infty}^{+\infty} |H(f)|^2 \, df}} \text{ Hz}$$

Beacon	A secondary radar transponder.
Blanking	The controlled interruption of a signal during the blanking interval.
Blind phase	If a sine wave is sampled, there is an unlucky situation when it can be sampled as it crosses zero, which gives zero output or the blind phase.
Blind range	In a radar where the pulse repetition time is shorter than that for maximum range, there will be gaps in the listening time for echoes from the the transmitter pulses. During these times the radar is blind, giving blind ranges.
Blind speed	Moving target indicator (MTI) radars suppress ground clutter by highpass frequency filtering. Sampling causes the notch to be repeated at each multiple of the pulse repetition frequency. Objects moving radially to give Doppler frequencies at multiples of the pulse repetition frequency will not be detected. These speeds are the blind speeds.
Burst	A number of transmitter pulses similar to a burst from a machine gun.
C-display or C-scope	See C-display in Section A.3.4.
Cancellation ratio	In MTI, the ratio of canceler voltage amplification for fixed-target echoes received with a fixed antenna to the gain for a single pulse passing through the unprocessed channel of the canceler [1].
CAPACS	Coded pulse anticlutter system.
Capture effect	The tendency of limiting receivers to suppress the weaker of two signals.
CFAR	See constant false alarm rate.
Characteristic function	The Fourier transform of a probability density function.
Collapsing loss	The increase of signal-to-noise ratio required to compensate for the extra noise caused by collapsing or when video signals are added. See collapsing ratio.

Collapsing ratio
When video signals are added together, the collapsing ratio is the ratio of the number of noise samples from all inputs divided by the number of noise samples from the inputs containing signals.

Constant false alarm rate
A process to limit the number of false alarms to an almost constant number in the presence of weather, angel, and clutter echoes. In military situations, jamming is added.

Convolution
A process to ascertain the spreading of one shape by another.

Correlation
A measure to ascertain the amount of match between two waveforms.

Cross-correlation
The correlation of two waveforms for providing a measure of match.

Cross-polarization
The polarization direction orthogonal to that used by the radar. In military jamming, this is used to disrupt some monopulse tracking radars.

Crossover loss
Where signals are selected by two separate beams or filters, the minimum response, or maximum loss, is where the two characteristics cross.

D-display or D-scope
See Section A.3.4.

Data rate
Rate of renewal of positional information, normally in numbers per minute.

Defruiter
A correlator that removes asynchronous replies, called fruit, from secondary surveillance radars.

Detection
The decision process to determine whether an echo signal is from an object of interest.

Display
See Section A.3.4.

Dicke-fix receiver
A form of intermediate frequency processing involving limiting and filtering (Section 8.1.4).

Diplexer
In secondary radar, a device that couples a transmitter to an antenna using one frequency and to a receiver on another frequency. In primary radar, it couples two separate radar channels working on different frequencies to a single antenna.

Diversity
Originally, the use of two or more separate channels for communication subject to fading. With radar the term is used for echoes which are subject to fading or fluctuation. Examples are:
- Frequency diversity with two or more transmitter-receiver systems;
- Polarization diversity with orthogonally polarized transmission and reception from the antenna;
- Pulse repetition frequency diversity in which the detection reports using two or more separate pulse repetition frequencies are evaluated.

DPCA
Displaced phase center antenna.

Duty cycle, factor, or ratio
This is the ratio of the on time to the pulse repetition time for a train of pulses.

E-display or E-scope
See Section A3.4.

ECCM
Electronic countercountermeasures (military).

Eclipsing	Medium and high pulse repetition frequency radars blank the receiver around the times for the transmitter pulses. Echoes returning during the blanking time are also blanked or eclipsed.
Eclipsing loss	The increase of signal-to-noise ratio required for detection when a signal is subject to eclipsing.
ECM	Electronic countermeasures (military).
Elevation angle	The angle between the object of interest and the horizontal plane at the observer.
ESM	Electronic warfare support measures.
F-display or F-scope	See Section A.3.4.
False alarm	A false detection is made in the absence of a suitable echo signal.
False alarm number	The average number of detection decisions possible between two false alarms.
False alarm probability	The probability that a detection will be declared caused by noise or other interfering signals.
False alarm time	Current: The average time between false alarms. Marcum (obsolete): The time for which the probability that a false alarm will occur is 50%.
Far sidelobe	Sidelobe occurring at an angle differing from the main lobe direction by more than ten times the width of the main lobe between -3 dB points [4].
Fast time constant	A (differentiator) circuit which passes short pulses with less attenuation than longer pulses.
Fluctuation	The change of amplitude of echoes during one look (fast) or from scan to scan (slow).
Frank code	A type of waveform for pulse compression (Sections 3.4.3.2 and 8.3.2.1).
Frequency agility	The ability of a radar to change its transmitter and receiver radio frequency from pulse to pulse.
Fruit	The acronym for "false replies unsynchronized to the interrogator transmission" or asynchronous secondary surveillance radar interference.
FTC	See fast time constant.
G-display or G-scope	See displays, Section A.3.4.
GCA	See ground controlled approach.
GCI	See ground controlled intercept (military).
GDOP	See geometric dilution of precision.
Geometric dilution of precision	A reduction of precision caused by defining a point by lines crossing at small acute angles. It occurs mainly in range only, range difference (hyperbolic), or angle only position measurement systems [2].

Ghost echo	An echo position indicated by misinterpretation of the echo signals received. Examples are wrong resolution of range and Doppler frequency ambiguities, false angle estimates caused by antenna sidelobes or reflections, and multiple crossings of jam strobe lines.
Glint	The error of position or Doppler frequency caused by reflections from different parts of a complex scatterer. These reflected signals interfere with each other.
Ground controlled approach radar	Radars used for the management of aircraft movements in the air around an airfield. See also precision approach radar.
Ground controlled intercept	Radars used by controllers on the ground to deploy fighter aircraft to intercept intruding aircraft.
H-display or H-scope	See displays, Section A.3.4.
Height resolution	The minimum distance by which two aircraft flying at the same azimuth and range have to be separated in order to be distinguished as two separate objects by the radar *P*% of the time.
Helix, shaped	The true helix may be drawn around a cylinder. The shaped helix is any other helical shape.
Hybrid coupler	A four-port coupler that divides the input between two output ports with no output from the fourth port.
ICAO	International Civil Aviation Organisation.
I-display or I-scope	See displays, Section A.3.4.
Identification friend or foe	The original military form of secondary radar in which equipment (transponder) in the aircraft transmits a coded reply when interrogated.
IFF	See identification friend or foe.
Integration	The addition of radar signals to improve detectability and accuracy.
Integration loss	The difference in signal-to-noise ratio for the same probability of detection when signals are added or integrated coherently and incoherently.
Interclutter visibility	The ability of a radar (operator) to recognize an echo of interest in the clear areas (or volumes) between clutter patches.
Interference	Signals in the receiver caused by radiation from other sources, normally accidentally.
Interpulse	Between pulses.
Interrogator	The transmitter and receiver combination used by secondary radars on the ground. It sends interrogation pulses (normally 1 030 MHz) to the transponders in the aircraft, which reply (normally with pulses at 1 090 MHz).
Intrapulse	Inside a pulse.
J-display or J-scope	See displays, Section A.3.4.
Jamming	The intentional (military) interference of a radar to reduce the probability of detection of the objects of interest.

Jitter	The random timing errors between pulses in a train of pulses.
K-display or K-scope	See displays, Section A.3.4.
L-display	See displays, Section A.3.4.
Look	One observation by a scanning radar.
M-display	See displays, Section A.3.4.
Magic tee or T	A waveguide hybrid coupler
MDS	See minimum detectable signal.
Minimum detectable signal	Commonly a radio frequency pulse is increased in amplitude until it is just detectable on a display. The level is often measured in decibels with reference to one milliwatt, dBm. The results using this method depend on the operator. For more accurate results, the threshold is set by counting the false alarm rate and the signal is increased to achieve a given probability of detection.
Monopulse	A means of determining the angle of reception of an echo using two receiving channels for each dimension. The antenna for the two channels may have two separate overlapping beams (A and B) or sum (Σ) and difference (Δ) channels. The comparison may be in amplitude, A/B or Δ/Σ, or the phase angle between the echo signals.
MOPA	Master oscillator power amplifier.
Moving target indicator	A method of reducing clutter by canceling ground echoes. This is a military term that has been retained.
Moving target detector	A method of detecting echo signals by separating them into a number of Doppler frequency bands. The signals in each band are processed separately to give the best probability of detection. This is a military description that has been retained.
MSL	(Above) mean sea level.
MTD	See moving target detector.
MTI improvement factor	The signal-to-clutter ratio at the output of the clutter filter divided by the signal-to-clutter ratio at the input of the clutter filter, averaged uniformly over all target radial velocities of interest [1].
MTI	See moving target indicator.
N-display or N-scope	See displays, Section A.3.4.
O-display or O-scope	See displays, Section A.3.4.
Near sidelobe	Sidelobe occurring at an angle differing from the direction of the main lobe by less than ten times the width of the main lobe between the -3 dB points [4].
PAR	See precision approach radar.
Permanent echo	An echo that is permanently present, usually a small piece of point clutter. Observation of the echo amplitude gives an indication of the radar's sensitivity, and its position indicates the radar's alignment.

Plan position indicator	See displays, Section A.3.4.
Plot	A report containing the position of a wanted scatterer. Depending on the radar this contains: range, azimuth angle, elevation angle, radial velocity, and time of detection.
Plumbing	A derogatory term for waveguide systems.
PPI display	Plan position indicator, see displays in Section A.3.4.
PR	See primary radar.
Precision approach radar	A radar with two antennas that measures an aircraft's position as it comes in to land.
PRI	See pulse repetition interval.
Primary radar	A radar that depends on echoes from its transmitter pulses.
Probability of detection	The probability that a sample of signal plus noise will cross the threshold and give a detection.
Probability of false alarm	The probability that a sample of noise alone will cross the threshold and give a detection.
Pulse repetition interval, period, or time	The time between two pulses in a train.
Pulse repetition frequency	The frequency of the pulses in a train of pulses.
Pulse repetition interval, period, or time	The time between two pulses in a train.
Range resolution	The minimum distance by which two aircraft flying at the same azimuth and elevation have to be separated in order to be distinguished as two objects by the radar.
Rat race	A form of (waveguide) hybrid coupler.
RHI	See range-height indicator in Section A.3.4.
Ring time	The time taken by resonating signal to decay.
SAR	Synthetic array radar.
Scan	The complete exploration of the search volume by the radar.
Scintillation	The error of position or Doppler frequency caused by reflections from different parts of a complex scatterer as it changes its aspect angle. These reflected signals interfere with each other.
Scope	Another term used for display.

Secondary surveillance radar — The combination of an interrogator (transmitter, normally on 1 030 MHz), antenna, receiver (normally 1 090 MHz), signal processor, and possibly an extractor to give a secondary radar picture. The civil modes are A (shared with military mode 3), B, and D for identification and mode C, which returns the altimeter height. The mode-S system gives major improvements and data transfer possibilities.

Secondary radar — A radar system that relies on a transponder (receiver-transmitter) to return signals that show the position of an aircraft, or a buoy, and so on.

Selective identification friend — A military secondary radar system using the frequencies 1 030 MHz ground to air and 1 090 MHz air to ground. Commonly, the interrogation modes are 1, 2, and 3. Mode 3 is shared with mode A in the civil system. The reply is up to 12 pulses between two bracket pulses.

Sensitivity time control — The reduction of receiver gain at short ranges to reduce the amplitude of short range echoes so as to reduce the dynamic range required in the receiver.

Shaped helix — See helix, shaped.

Sidelobe — Unwanted responses from an antenna or filter outside the main lobe.

Sidelobe blanker — A circuit which blanks the receiver when signals would be via the antenna sidelobes.

Sidelobe canceler — A circuit that subtracts the signal that would have been received via the sidelobes from the main lobe receiver signal.

SIF — Selective identification friend (military secondary radar).

Spoofing — According to the *Concise Oxford Dictionary*, the word was invented by A. Roberts, an English comedian, for a parody, hoax, or swindle. In radar, spoofing is a form of deception jamming in which signals representing objects of interest are deliberately fed to an enemy radar to cause confusion, as, for example, with a swarm of fish.

Squint — The movement of the main beam of an antenna when the radio frequency is changed.

SR — See surveillance radar.

SSR — See secondary surveillance radar.

Stable local oscillator — A local oscillator stable enough for moving target indicator, moving target detector, or pulse Doppler signal processing.

Staggering —

- Staggered amplifier tuning. The center frequencies of the high frequency transformers between the stages are spread or staggered to give greater bandwidth.
- Staggered pulse repetition frequencies. The times between the transmitter pulses are changed cyclically in order to suppress blind speeds in moving target indicator processors for the range of Doppler frequencies of interest.

STALO — Stable local oscillator.

STANAG — Standard NATO agreement.

Stretched pulse — The long pulse transmitted by the transmitter when pulse compression is used.

Subclutter visibility	The ratio by which the target echo power may be weaker than the coincident clutter echo power and still be detected with specified detection and false alarm probabilities. Target and clutter powers are measured on a single pulse return and all target radial velocities are assumed equally likely. From [1].
Superclutter visibility	The ability of a radar (operator) to recognize an echo of interest that is stronger than the clutter echoes, for example, a ship in sea clutter.
Surveillance radar	A radar used to observe objects within its search volume.
Sweep	The movement of the spot from one part of the display to its end position during the time between two (or a number of) transmitter pulses.
Swept gain	The gain that is increased with range, as with sensitivity time control.
Synchro	A device to transmit angle information electrically to a remote device. Examples are Selsyns, Magslips, and similar devices.
Synthetic aperture	An aperture formed by moving an antenna along a line and integrating the signals later.
Synthetic array	An alternate form of synthetic aperture where the positions of the physical antenna are treated as an array.
TACCAR	Time averaged clutter coherent airborne radar.
TR switch	See transmit-receive switch.
Track	The knowledge of the position and course (speed and direction) of an object.
Transmit-receive switch	An automatic switch to direct the transmitter pulse to a common antenna and the echoes from the antenna to the receiver in a pulse radar.
Transponder	Secondary radar equipment that receives an interrogating pulse and transmits a (coded) reply.
Triplexer	It couples three separate radar channels working on different frequencies to a single antenna.
TWS	Track while scan.
VCO	Voltage controlled oscillator.
Zenith angle	The angle, z, from the vertical (zenith) down to the object of interest. It is 90 degrees or $\pi/2$ radians minus the elevation angle.

A.3 DISPLAYS

The display (sometimes called a scope) is the principal means of showing what a radar "sees" to an operator. Watching the radar image is only part of the operator's working function, so that displays have been created for each special task. As radar developed for various military purposes to detect, identify, and aim weapons at possible targets, various specialized forms of display evolved. The development can be divided between surveillance radars, anti-aircraft artillery radars, and airborne intercept radars.

A.3.1 Surveillance radars

After the A-display, methods were developed to show information in two-dimensional map form, which resulted in the B-display for sector scanning radars, and the plan position indicator (PPI) for continuously rotating radars. The first displays were used militarily to detect and deploy resources against intruding aircraft.

A.3.2 Anti-aircraft (AA) artillery radars

Anti-aircraft radars are used to find the position in azimuth angle, elevation angle, and range of targets for artillery. The displays are two dimensional, which allow the range operator to measure range (A-display, J-display, M-display), and the bearing and elevation operators (K-display, L-display, N-display) to move the antenna to point at the target. Later radars were able to lock onto and follow the targets automatically.

Most of these radars have a search function using a B-display or plan position indicator for the case when the associated surveillance radar is not available.

A.3.3 Airborne intercept (AI) radars

Airborne intercept radars are used in fighter aircraft to find and engage targets beyond where the pilots can see, at night, for example. Some indicate a third dimension (3D) to the pilot or operator. Common search displays were B, C, D (3D), and H (3D). Having found and identified an echo as a target, F, G, and I displays were used to aim the weapons.

After the Second World War, radars started to be used for air traffic control, meteorology, and many other applications. Digital equipment and computers allowed the extracted radar information to be presented in fully synthetic form to ease the work load of the operator.

A.3.4 Display types

Short descriptions of displays are given here [1, pp. 12-13].

A-display — The first type of cathode ray tube display used for radar that is similar to an oscilloscope. The time base (horizontal) represents range and the vertical deflection represents signal amplitude.

B-display — An area display for scanning radars with azimuth displayed in the x direction, range in the y direction, and echo amplitude as brightness. The B-display is used for radars which watch a sector.

C-display — The display for scanning radars where azimuth angle is displayed in the x direction and elevation angle in the y direction. Signals are shown by brightening the spot. The C-display was used in fighter aircraft. Crude range is available in the D-display.

D-display — The D-display is an extension of the C-display. The elevation angle (vertical) is divided into a number of horizontal stripes. In each stripe range is shown in the vertical direction.

E-display — Similar to the B-display for scanning radars, but range is shown in the x direction and elevation angle in the y direction.

F-display — This shows the azimuth (horizontal) error and elevation error (vertical) of a (single target) radar using a continuous blob. F-displays were used in fighter aircraft.

G-display — The G-display shows the azimuth (horizontal) error and elevation error (vertical) of a (single target) radar. It is similar to an F-display but with range indicated by wings on the blob, and was used in fighter aircraft.

H-display	A B-display in which the echoes are shown as two bright dots. An imaginary line drawn between the dots shows the elevation angle.
I-display	Displays the signal from a conical scan radar. The echo is displayed as an arc of a circle showing azimuth and elevation errors.
J-display	This is similar to an A-display, but the trace is (part of) a circle. Range is indicated along the circumference and intensity by a radial deflection, so that greater range resolution is available on small cathode ray tubes.
K-display	A modified A-display, that was used in lobing (and conical scan) radars. Two traces are used for the two lobes, which may be offset in range and allow an operator to turn the antenna to reduce the aiming error. Separate displays are needed for azimuth and elevation.
L-display	In this display, range is shown vertically from the bottom. Signals from the two lobes deflect the trace to the left and the right. When the two signals are equal, the antenna is correctly aimed.
M-display	Similar to an A-display, but has a movable pedestal or step which is calibrated in range. This is used to measure range.
Magic carpet	The magic carpet display gives a three-dimensional representation used to estimate the number of individual objects in an echo or group of echoes. Commonly, the azimuth is shown across the display and range on a slant from lower left to upper right. The amplitude of the signals is shown by a vertical deflection.
N-display	Similar to an M-display but with two traces for a lobing radar (see K-display).
O-display	The O-display is similar to an A-display but has a movable notch that is calibrated in range.
P-display or plan position indicator (PPI) display	The PPI is a two-dimensional map display that shows the range of the echoes along the radius, their azimuth angle as an angle from north, and their amplitude as brightness.
Range height indicator (RHI)	The RHI displays a vertical slice. The sweep starts at the lower left hand corner, and the length along the sweep represents range. The vertical deflection represents height. The echoes are shown by brightening the trace, and a horizontal height line is used to measure the height of the objects giving the echoes.

A.3.5 Displays for showing the effects of jamming (military)

Very often, operators using normal displays switch to a less sensitive treated video signal when jamming is encountered. The aim is to have a "black scope"; a display that shows echo signals only and not jamming signals. A number of displays have been developed to help operators in the presence of jamming.

AVA	Azimuth versus amplitude display occurs in two forms: • A display which shows jamming signal amplitude as range; • A display which shows the effective maximum range where a scatterer (maybe a target) of a certain size is still visible.
Jam strobe	The display of a line from zero to maximum range is shown in the direction of a jammer. If the displays from a number of radars are combined, the points where the lines cross show the possible positions of jammers. Crossing points not corresponding to jammer positions are

ghost jammers.

Panoramic display — Panoramic displays are used to show frequency across the display and the amplitudes of signals occurring at those frequencies vertically, which allows the jamming density to be observed and the least jammed frequencies to be chosen.

Waterfall — The waterfall display is a panoramic display with frequency shown horizontally. The vertical sweep is time from top to bottom with a sweep rate of seconds or tens of seconds. The amplitudes of the signals are shown by the brightness of the display.

A.4 SYMBOLS

The symbols used in some diagrams are shown in Figure A.1.

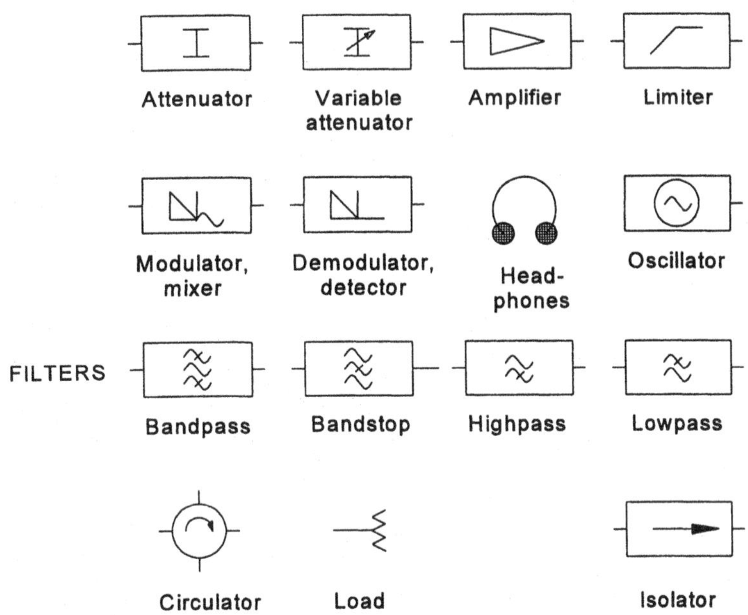

Figure A.1 Symbols used in diagrams.

REFERENCES

1. *IEEE Standard Radar Definitions, IEEE Standard 686-1990,* New York: The Institute of Electrical and Electronic Engineers, 1993.
2. *IEEE Standard 100*, New York: The Institute of Electrical and Electronic Engineers, 1993.
3. Stevens, M. C., *Secondary Surveillance Radar*, Norwood, Massachusetts: Artech House, 1988.
4. Bijvoet, J. A., *Standard Methods for Predicting and Specifying Performance of Air Surveillance Systems,* The Hague: Supreme Headquarters Allied Powers Europe (SHAPE) Technical Centre Technical Report TR-50, 1969.

Appendix B

Tapering functions

This appendix treats tapering functions and their Fourier transforms in a common way irrespective of their usual use for antenna patterns or signal processing. The tapering functions in this chapter are grouped into:

- Even functions: Transmitter pulse shaping, antenna patterns, and filtering;
- Circular functions: Circular antennas;
- Odd functions: Monopulse antennas. Odd functions are also used for differentiating filters which are not covered in this book.

The notation used follows [1].

B.1 CONVENTIONS AND NORMALIZATION

This appendix shows functions limited in distance, time, or bandwidth and their Fourier transforms. In order to be able to show these for all the tapering functions for all circumstances, they must be normalized to unit width. In order to present the parameters and results in a unified form, the variables p' and q' are used. Thus:

$h(p)$ transforms to $H(q)$ and

$h(p')$ transforms to $H(q')$, with normalized variables.

The conventions for describing limited distance, time, and bandwidth functions are shown in Table B.1.

Table B.1

Conventions and symbols used in Appendix B

	ANTENNA		WAVEFORM	SPECTRUM
	Linear	Circular		
Width	w, m	D, m	τ, s	B, Hz
Width limits	$-\frac{w}{2}\,..\,+\frac{w}{2}$	$0..\frac{D}{2}$	$-\frac{\tau}{2}\,..\,+\frac{\tau}{2}$	$-\frac{B}{2}\,..\,+\frac{B}{2}$
p	x, m	r, m	t, s	f, Hz
p'	$p' = x' = \frac{x}{w}$	$p'=r'=\frac{r}{D}$	$p'=t'=\frac{t}{\tau}$	$p'=f'=\frac{f}{B}$
Functions	$g(x)$, $g(x')$	$g(r)$, $g(r')$	$h(t)$, $h(t')$	$H(f)$, $H(f')$
	...transform to ...			
q	u, m^{-1}	u, m^{-1}	f, Hz	t, s
q'	$q' = u' = wu$	$q' = u' = wu$	$q' = f' = f\tau$	$q' = t' = tB$
	$q' = u' = \frac{w}{\lambda}\sin\theta$	$q' = u' = \frac{D}{\lambda}\sin\theta$		
Functions	$F(u)$, $F(u')$	$F(u)$, $F(u')$	$H(f)$, $H(f')$	$h(t)$, $h(t')$

A number of tapering functions are a recalculation of those in [1, pp. 268-333] which are extended by examples from [2]. The sections for each type include a formula for the tapering function; a table of calculated values as in reference [1]; graphs showing the tapering function and its Fourier transform; an example of the two-tone characteristic with discrete elements; and closed forms (where they exist). The tabulated values cover a number of parameters, which are included for completeness and which are explained in [1]. A summary is given in Table B2.

Table B.2

Expressions used for the parameters in Appendix B

Parameter		Linear antennas, time waveforms, and spectra	Circular antennas (where different)
A, the effective aperture length, processing time or bandwidth *Coherent voltage gain*		$A = \int_{-\frac{1}{2}}^{+\frac{1}{2}} h(p') \, dp'$	$A = 8 \int_{-\frac{1}{2}}^{+\frac{1}{2}} h(r') \, dr'$
C, the effective power in the aperture length, time, or bandwidth *Incoherent power gain*		$C = \int_{-\frac{1}{2}}^{+\frac{1}{2}} (h(p'))^2 \, dp'$	$C = 8 \int_{-\frac{1}{2}}^{+\frac{1}{2}} (h(r'))^2 \, dr'$
D		$D = \int_{-\frac{1}{2}}^{+\frac{1}{2}} \left(\frac{dh(p')}{dp'} \right)^2 dp'$	
F		$F = 2\pi \int_{-\frac{1}{2}}^{+\frac{1}{2}} p' h(p') \, dp'$	
G		$G = \int_{-\frac{1}{2}}^{+\frac{1}{2}} (p' h(p'))^2 \, dp'$	
H		$H = \int_{-\frac{1}{2}}^{+\frac{1}{2}} (p')^2 \, \lvert h(p') \rvert \, dp'$	
K_r, the voltage slope at the center of the beam relative to linear odd illumination		$K_r = \frac{\sqrt{12} \, F}{2\pi \sqrt{C}}$	
Efficiency, η *Processing gain*		$\eta = \frac{A^2}{C}$	
Noise width in terms of	$\frac{w}{\lambda}$, τ, B, bins	$\theta_n, t_n, B_n = \frac{C}{A^2}$	
Half-power width in terms of	$\frac{w}{\lambda}$, τ, B, bins	θ_3, t_3, B_3	
RMS width in terms of	$\frac{w}{\lambda}$, τ, B, bins	$\Theta, \tau_{rms}, B_{rms} = \sqrt{\frac{D}{C}}$	
RMS aperture width in terms of	$1/w$, $1/\tau$, $1/B$	$\mathcal{L} = 2\pi \sqrt{\frac{G}{C}}$	

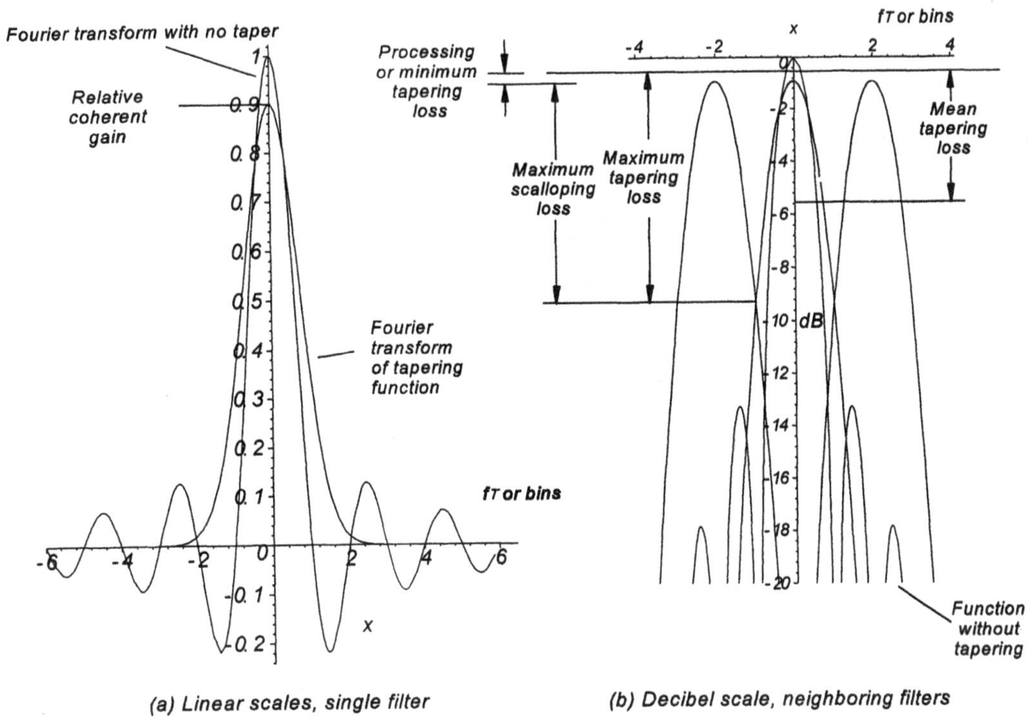

Figure B.1 Gains and losses with tapering functions.

B.1.1 Scalloping loss and worst-case processing loss

Scallopping loss and worst-case processing loss [2, pp. 51-63] are the losses that occur when a signal has a frequency between the center frequencies of two filters, as shown in Figure B.1. The tables in Sections B.2 to B.5 show the maximum scalloping loss. When the scalloping loss is added to the processing loss then this gives the worst-case processing loss.

B.1.2 Spectral leakage or two-tone characteristics

Examples are given of the effect when a discrete Fourier transform is used where the components of the time waveform do not fall exactly on the bin frequencies. The examples show one case of two signals in a 100 sample or bin filter. The major signal has its frequency varied between the 10th and 11th bin and the minor signal has a frequency corresponding to the 16th bin. Two amplitude ratios are chosen, 20 dB and 40 dB.

B.1.3 Other names used for the tapering functions

Tapering functions are used in many branches and a number are called by a number of different names, as in Table B.3 [3].

Table B.3

Names for the tapering functions in Appendix B

Name	Name used here	Section	Name	Name used here	Section
Bartlet	Triangular	B.2.1	Hamming		B.2.5.1
Bayliss		B.5.6	von Hann	Cos²	B.2.3
Blackman		B.2.5.2	Hanning	Cos²	B.2.3
Blackman-Harris		B.2.5.3	Parzen	Parabolic	B.2.2
Bochner	Parabolic	B.2.2	Raised cosine	Cosine on pedestal	B.2.4
Chebyshëv	Dolph-Chebyshëv	B.3.1	Riesz	Parabolic	B.2.2
Dirichlet	Rectangular	B.2.1	Weierstrass	Gaussian	B.2.5
Frejer	Triangular	B.2.1	Zolotarëv		B.7.7

B.2 TAPERING FUNCTIONS

B.2.1 Trapezoidal tapering

The parameter k is the flat portion of the tapering over half the antenna length. Thus $k = 0$ is for triangular tapering and $k = 0.5$ for uniform tapering and the characteristics are shown in Figures B.2 and B.3 and Table B.1. The Fourier transforms are in Figures B.4 and B.5.

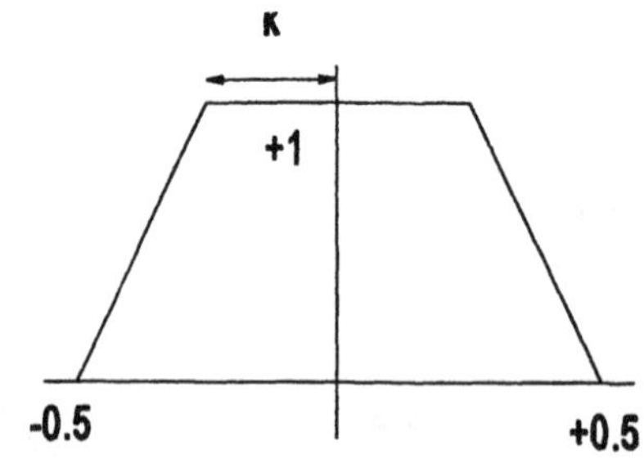

Figure B.2 Trapezoidal function.

Table B.4
Table of values for trapezoidal tapering

	Values of k					
	0.0 Triangular	0.1	0.2	0.3	0.4	0.5 Uniform
A, effective length, *coherent gain*	0.5000	0.6000	0.7000	0.8000	0.9000	1.0000
A, dB	-6.0206	-4.4370	-3.0980	-1.9382	-0.9151	0.0000
C, effective power, *power gain*	0.3333	0.4667	0.6000	0.7333	0.8667	1.0000
C, dB	-4.7712	-3.3099	-2.2185	-1.3470	-0.6215	0.0000
D (see Section B.1)	4.0000	5.0000	6.6667	10.0000	20.0000	0.0000
G (see Section B.1)	0.0083	0.0129	0.0211	0.0345	0.0547	0.0833
H (see Section B.1)	0.0208	0.0260	0.0338	0.0453	0.0615	0.0833
Efficiency η, *processing gain*	0.7500	0.7714	0.8167	0.8727	0.9346	1.0000
Efficiency η, *processing gain*, dB	-1.2494	-1.1270	-0.8796	-0.5912	-0.2937	0.0000
Noise beamwidth	1.3333	1.2963	1.2245	1.1458	1.0700	1.0000
Half-power beamwidth	1.2757	1.2474	1.1740	1.0784	0.9791	0.8859
RMS beamwidth	3.4641	3.2733	3.3333	3.6927	4.8038	0.0000
RMS aperture	0.9935	1.0460	1.1792	1.3635	1.5790	1.8138
First sidelobe, dB	-26.5229	-28.7572	-18.9939	-15.1281	-13.6238	-13.2615
Falloff, dB/octave	-12.00					-6.00
Scalloping loss	0.8106	0.8030	0.7807	0.7444	0.6958	0.6366
Scalloping loss, dB	-1.8242	-1.9055	-2.1506	-2.5634	-3.1506	-3.9224
Worst-case loss	0.7020	0.7053	0.7055	0.6955	0.6726	0.6366
Worst-case loss, dB	-3.0736	-3.0325	-3.0302	-3.1546	-3.4443	-3.9224

The closed form for the Fourier transform is

$$H(q') = \frac{\sin[\pi\, q'\,(0.5 + k)]}{\pi\, q'\,(0.5 + k)}\;\frac{\sin[\pi\, q'\,(0.5 - k)]}{\pi\, q'\,(0.5 - k)}$$

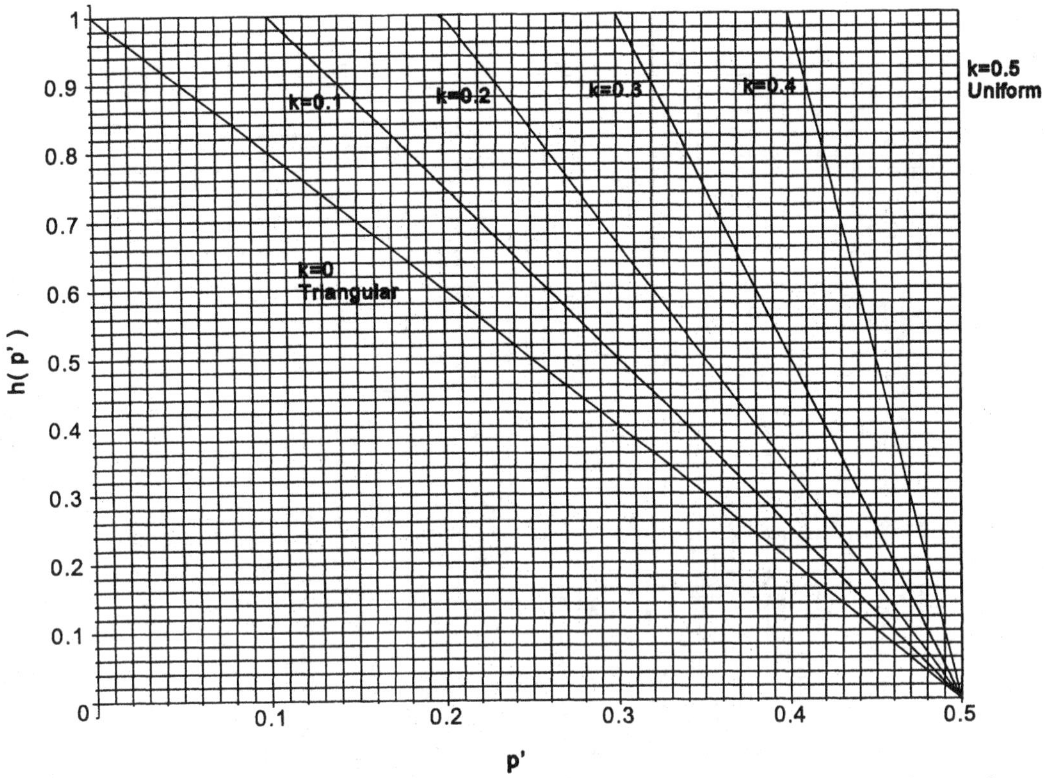

Figure B.3 The trapeziodal tapering functions for $k = 0$ for triangular tapering to $k = 0.5$ for uniform tapering.

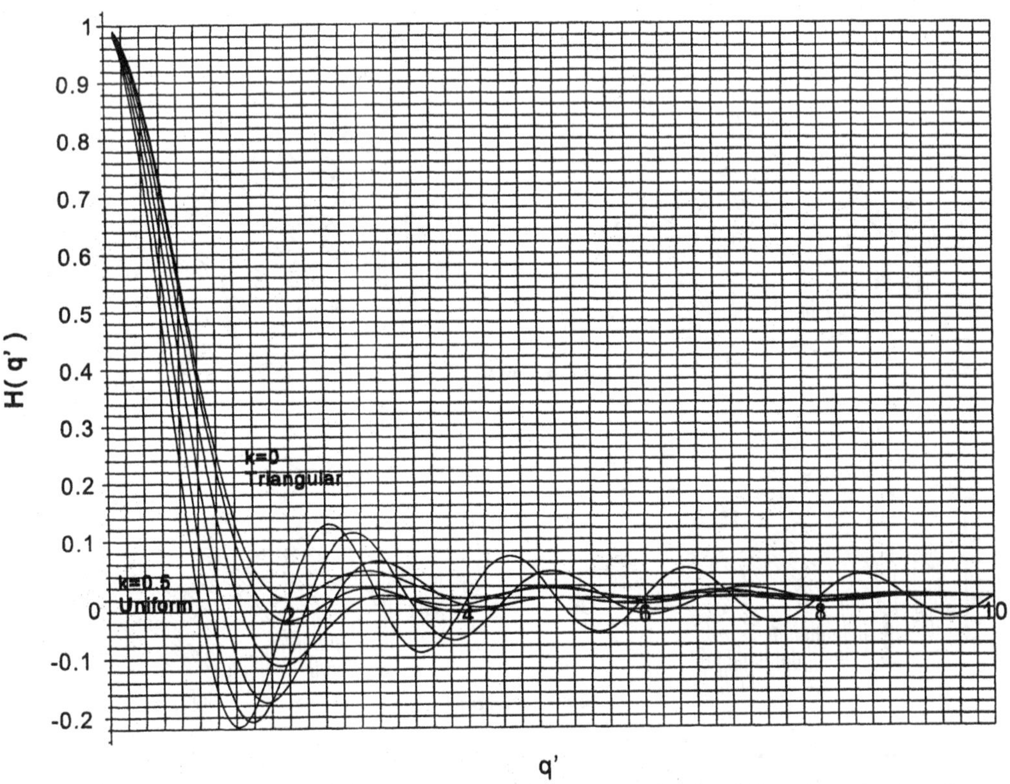

Figure B.4 The Fourier transforms for trapezoidal tapering for $k = 0$ for triangular tapering, $k = 0.1, 0.2, 0.3, 0.4$, and $k = 0.5$ for uniform tapering.

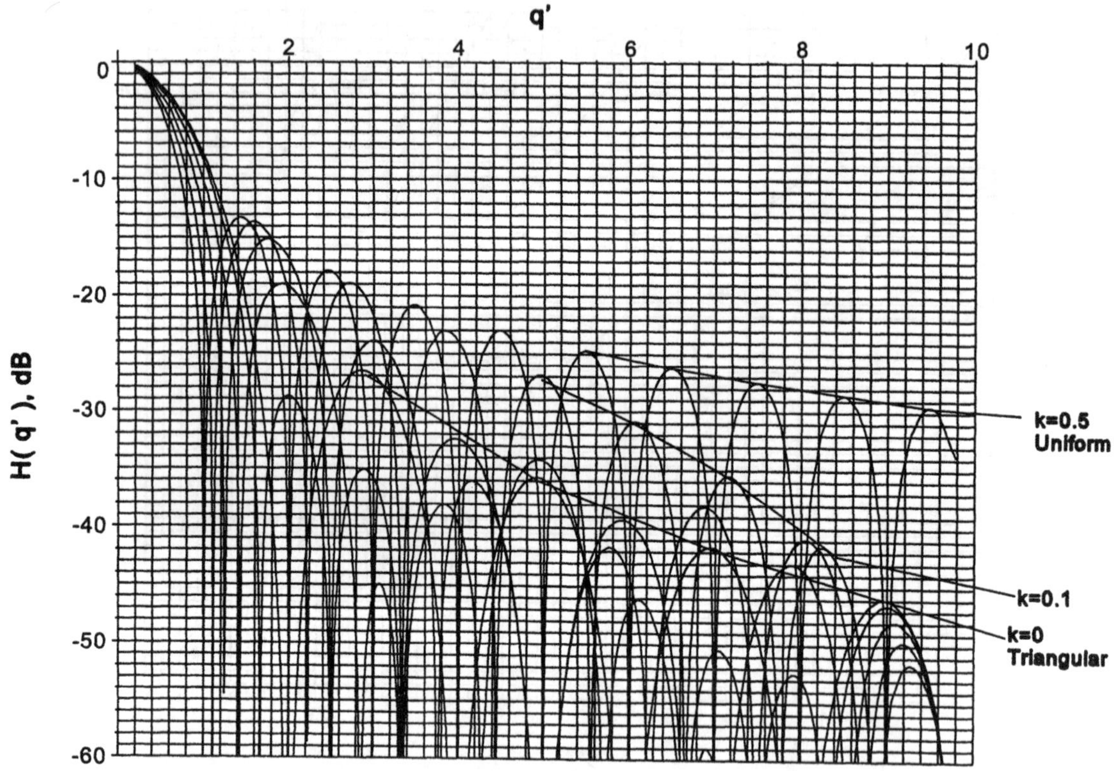

Figure B.5 The Fourier transforms in decibels for trapezoidal tapering for $k = 0$ for triangular tapering, $k = 0.1, 0.2, 0.3, 0.4$, and $k = 0.5$ for uniform tapering.

Figure B.6 shows an example of the effects of sampling on spectral leakage or two-tone characteristics when the discrete Fourier transform is used for 100 samples. The minor signal has a frequency corresponding to the 16th bin, and the major signal has its frequency varied between the 10th and 11th bins.

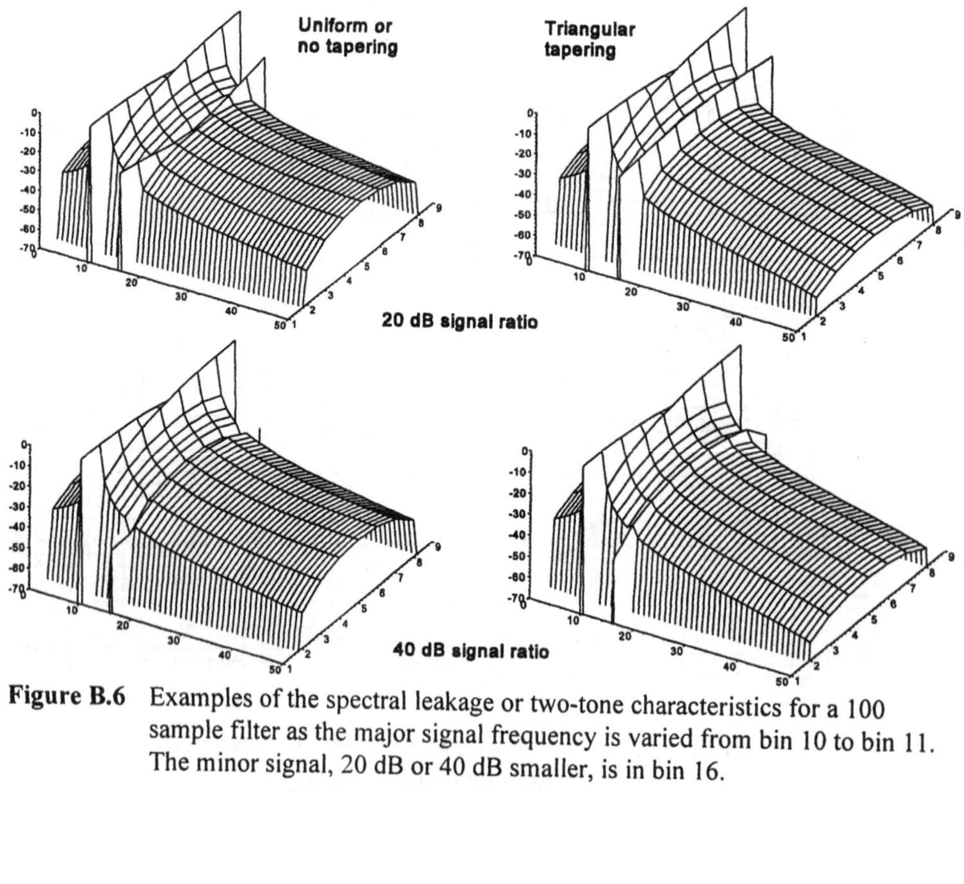

Figure B.6 Examples of the spectral leakage or two-tone characteristics for a 100 sample filter as the major signal frequency is varied from bin 10 to bin 11. The minor signal, 20 dB or 40 dB smaller, is in bin 16.

The worst case occurs when the major signal has a frequency corresponding to 10.5 bins, in Figure B.6. This case is illustrated in Figure B.7.

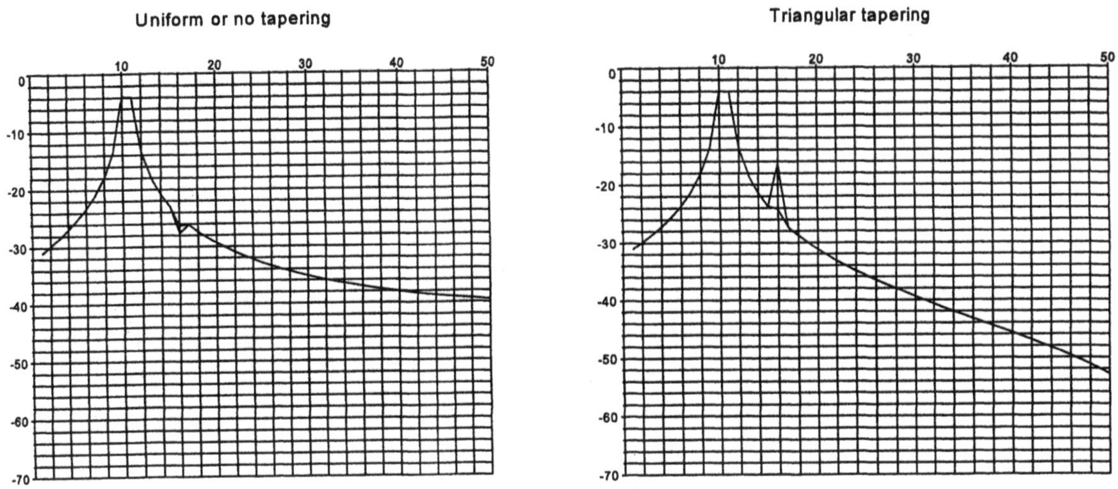

Figure B.7 Worst-case masking of the minor signal with uniform or no tapering and triangular tapering.

B.2.2 $(1-4p'^2)^n$ tapering.

The tapering function is given by $h(p') = (1 - 4p'^2)^n$ and the characterisics are shown in Table B.5.

Table B.5
Table of values for $(1-4p'^2)^n$ tapering

	Power of n					
	0 Uniform	1 Parabolic	2	3	4	5
A, effective length, *coherent gain*	1.0000	0.6667	0.5333	0.4571	0.4063	0.3694
A, dB	0.0000	-3.5218	-5.4600	-6.7990	-7.8220	-8.6499
C, effective power, *incoherent power gain*	1.0000	0.5333	0.4063	0.3410	0.2995	0.2703
C, dB	0.0000	-2.7300	-3.9110	-4.6726	-5.2355	-5.6822
D (see Section B.1)	0.0000	5.3333	4.8762	5.3195	5.8196	6.3061
G (see Section B.1)	0.0833	0.0190	0.0092	0.0057	0.0039	0.0029
H (see Section B.1)	0.0833	0.0333	0.0190	0.0127	0.0092	0.0071
Efficiency η, *processing gain*	1.0000	0.8333	0.7000	0.6129	0.5512	0.5049
Efficiency η, *processing gain,* dB	0.0000	-0.7918	-1.5490	-2.1264	-2.5865	-2.9677
Noise beamwidth	1.0000	1.2000	1.4286	1.6317	1.8141	1.9805
Half-power beamwidth	0.8859	1.1554	1.3748	1.5646	1.7341	1.8886
RMS beamwidth	0.0000	3.1623	3.4641	3.9497	4.4078	4.8305
RMS aperture	1.8138	1.1874	0.9472	0.8112	0.7207	0.6551
First sidelobe, dB	-13.2615	-21.2937	-27.7229	-33.3464	-38.4766	-43.2727
Falloff, dB/octave	-6.00	-12.00	-18.00	-24.00	-30.00	-36.00
Scalloping loss	0.6366	0.7740	0.8354	0.8704	0.8930	0.9089
Scalloping loss, dB	-3.9224	-2.2248	-1.5622	-1.2059	-0.9827	-0.8298
Worst-case loss	0.6366	0.7066	0.6989	0.6814	0.6630	0.6458
Worst-case loss, dB	-3.9224	-3.0166	-3.1112	-3.3323	-3.5693	-3.7975

Closed forms for the Fourier transform are shown in Table B.6.

Table B.6
Closed forms for the Fourier transforms for the first three values of n for $(1-4p'^2)^n$ tapering

Tapering function	Fourier transform
$(1-4p'^2)^0$	$\dfrac{\sin(\pi q')}{\pi q'}$
$(1-4p'^2)^1$	$-2\dfrac{\cos(\pi q')\pi q'-\sin(\pi q')}{\pi^3 q'^3}$
$(1-4p'^2)^2$	$-8\dfrac{\sin(\pi q')\pi^2 q'^2+3\cos(\pi q')\pi q'-3\sin(\pi q')}{\pi^5 q'^5}$

The curve is shown in Figure B.8 and the Fourier transforms are shown in Figures B.9 and B.10.

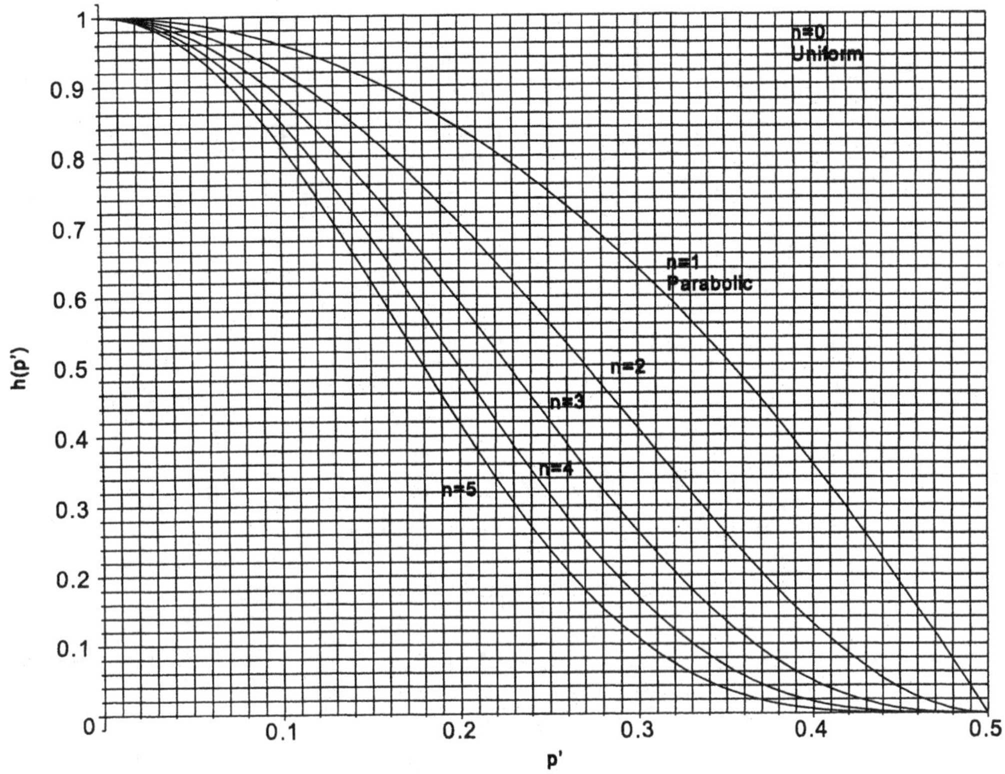

Figure B.8 $(1 - 4p'^2)^n$ tapering functions for n from 0 to 5.

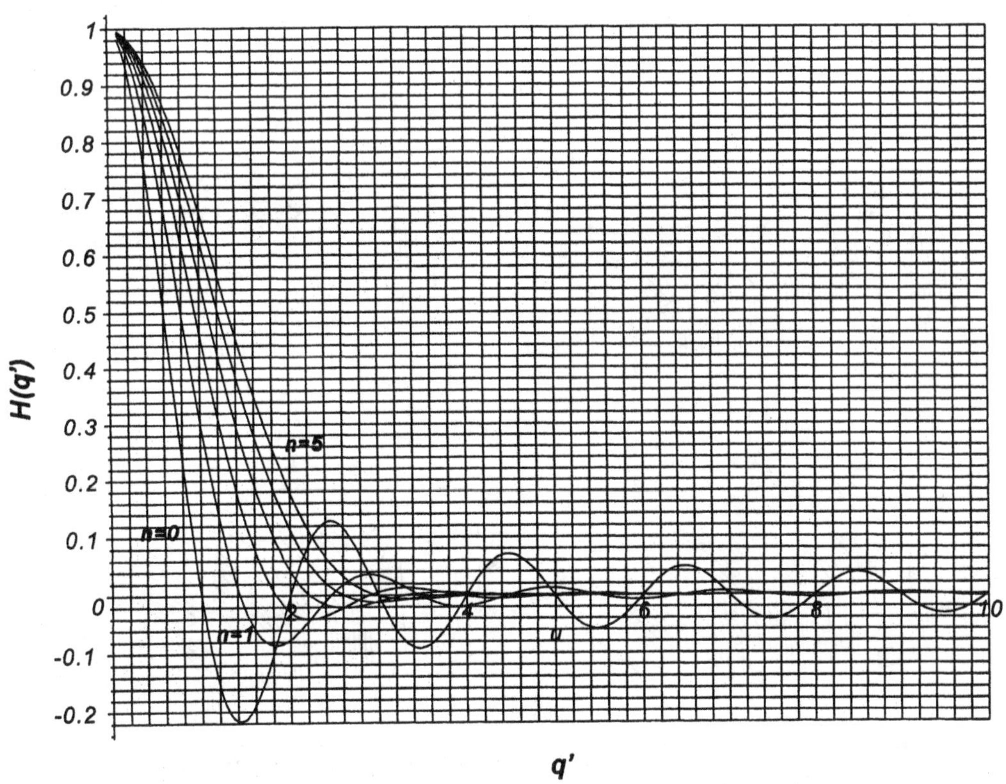

Figure B.9 Fourier transforms for $(1 - 4p'^2)^n$ tapering for n from 0 to 5.

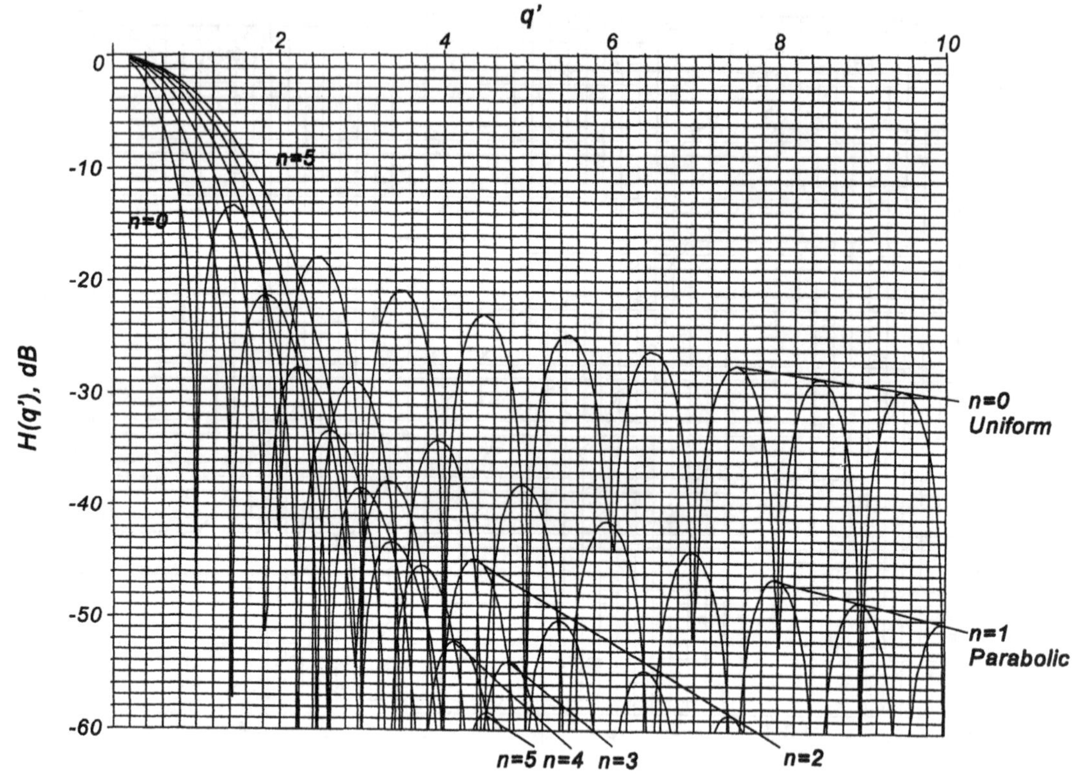

Figure B.10 Fourier transforms in decibels for $(1 - 4p'^2)^n$ tapering for n from 0 to 5.

Figure B.11 shows an example of the effects of sampling on spectral leakage, or two-tone characteristics, when the discrete Fourier transform is used for 100 samples. The minor signal has a frequency corresponding to the 16th bin, and the major signal has its frequency varied between the 10th and 11th bins.

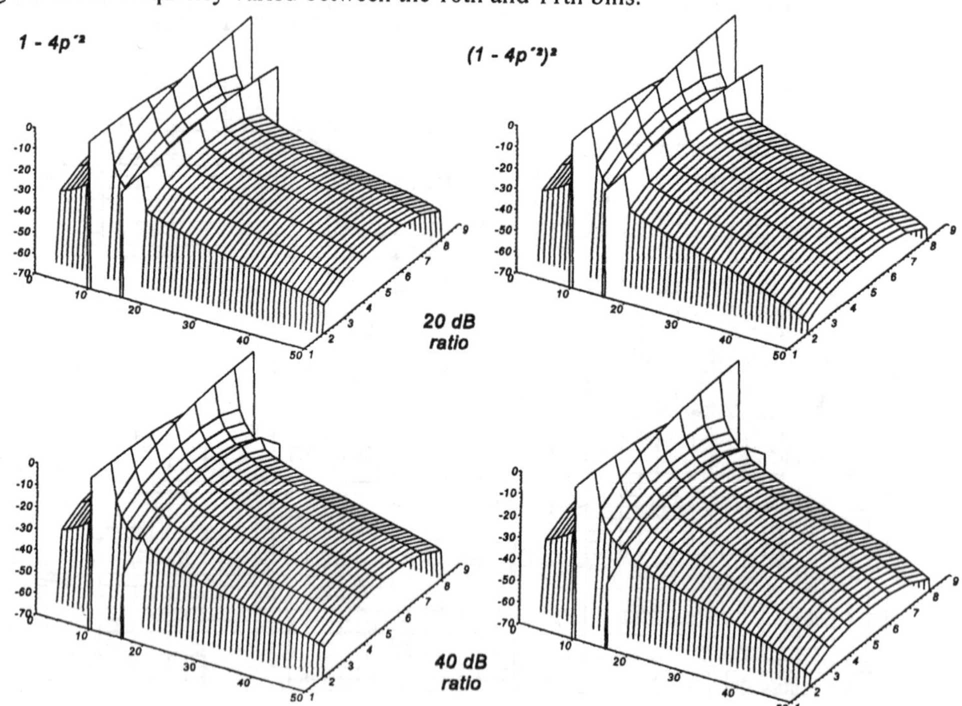

Figure B.11 Examples of the spectral leakage or two-tone characteristics for a 100 sample filter as the major signal frequency is varied from bin 10 to bin 11. The minor signal, 20 dB or 40 dB smaller, is in bin 16.

Figure B.12 shows worst case cuts through Figure B.11.

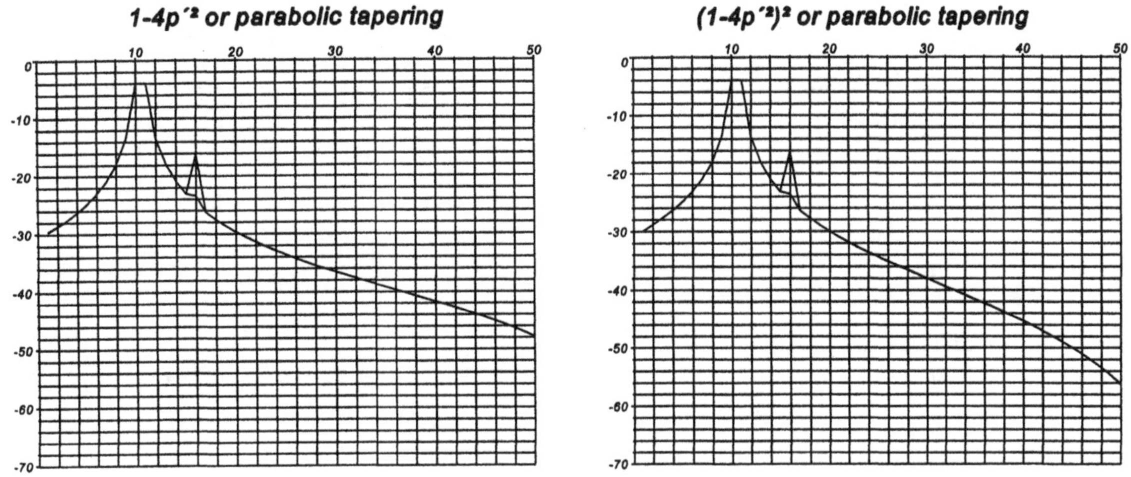

Figure B.12 Worst case masking of the minor signal with parabolic and $(1-4p'^2)^2$ tapering.

B.2.3 Cosine to the power *n* tapering

The tapering function is $\cos^n \pi p'$ and it is plotted in Figure B.13, and calculated characteristics are shown in Table B.7.

Table B.7
Table of values for cosine to the power *n* tapering

	Power of *n*					
	0	1	2	3	4	5
	Uniform	Cosine	Cosine², von Hann, Hanning	Cosine³	Cosine⁴	Cosine⁵
A, effective length, *coherent gain*	1.0000	0.6366	0.5000	0.4244	0.3750	0.3395
A, dB	0.0000	-3.9224	-6.0206	-7.4442	-8.5194	-9.3824
C, effective power, *incoherent power gain*	1.0000	0.5000	0.3750	0.3125	0.2734	0.2461
C, dB	0.0000	-3.0103	-4.2597	-5.0515	-5.6314	-6.0890
D (see Section B.1)	0.0000	4.9348	4.9348	5.5517	6.1685	6.7468
G (see Section B.1)	0.0833	0.0163	0.0075	0.0045	0.0031	0.0023
H (see Section B.1)	0.0833	0.0301	0.0163	0.0105	0.0075	0.0057
Efficiency η, *processing gain*	1.0000	0.8106	0.6667	0.5764	0.5143	0.4684
Efficiency η, *processing gain*, dB	0.0000	-0.9121	-1.7609	-2.3927	-2.8880	-3.2934
Noise beamwidth	1.0000	1.2337	1.5000	1.7349	1.9444	2.1347
Half-power beamwidth	0.8859	1.1890	1.4406	1.6585	1.8527	2.0292
RMS beamwidth	0.0000	3.1416	3.6276	4.2149	4.7496	5.2360
RMS aperture	1.8138	1.1357	0.8887	0.7534	0.6653	0.6022
First sidelobe, dB	-13.2615	-22.9988	-31.4675	-39.2956	-46.7417	-53.9352
Falloff dB/octave	-6	-12	-18	-24	-30	-36
Scalloping loss	0.6366	0.7854	0.8488	0.8836	0.9054	0.9204
Scalloping loss dB	-3.9224	-2.0982	-1.4236	-1.0752	-0.8630	-0.7206
Worst case loss	0.6366	0.7071	0.6931	0.6708	0.6493	0.6299
Worst case loss, dB	-3.9224	-3.0103	-3.1845	-3.4679	-3.7510	-4.0140

Closed forms for the Fourier transform are given in Table B.8 and are plotted in Figures B.14 and B.15.

Table B.8

Fourier transforms for cosine to the power n tapering normalized to unity at $q' = 0$

Tapering function	Fourier transform	Tapering function	Fourier transform
$\cos^0 \pi p'$	$\dfrac{\sin(\pi q')}{\pi q'}$	$\cos^3 \pi p'$	$\dfrac{9\cos(\pi q')}{(4q'^2 - 1)(4q'^2 - 9)}$
$\cos^1 \pi p'$	$\dfrac{\cos(\pi q')}{1 - 4(q')^2}$	$\cos^4 \pi p'$	$\dfrac{\sin(\pi q')}{\pi q'}\,\dfrac{4}{(q'^2 - 4)(q'^2 - 1)}$
$\cos^2 \pi p'$	$\dfrac{\sin(\pi q')}{\pi q'}\,\dfrac{1}{q'^2 - 1}$	$\cos^5 \pi p'$	$\dfrac{-225\cos(\pi q')}{(4q'^2 - 9)(4q'^2 - 25)(4q'^2 - 1)}$

The Fourier transforms are plotted in Figures B.14 and B.15.

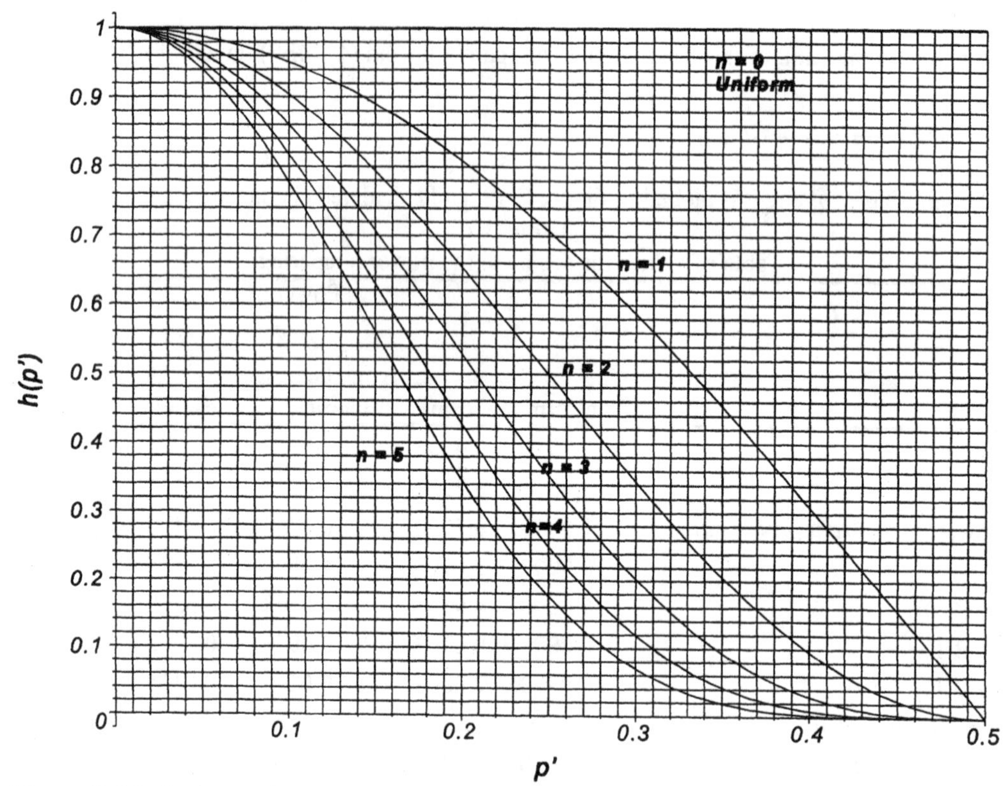

Figure B.13 Cosine to the power n tapering functions for $n = 0, 1, 2, 3, 4$, and 5.

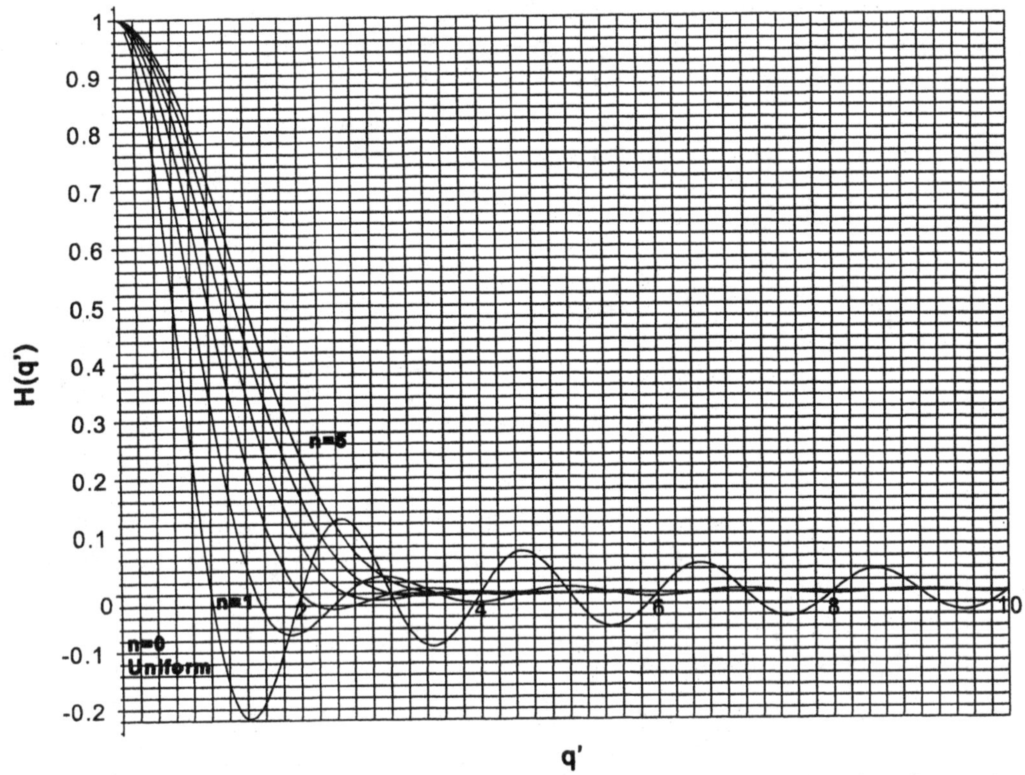

Figure B.14 Fourier transforms of the cosine to the power n tapering functions for $n = 0$, 1, 2, 3, 4, and 5.

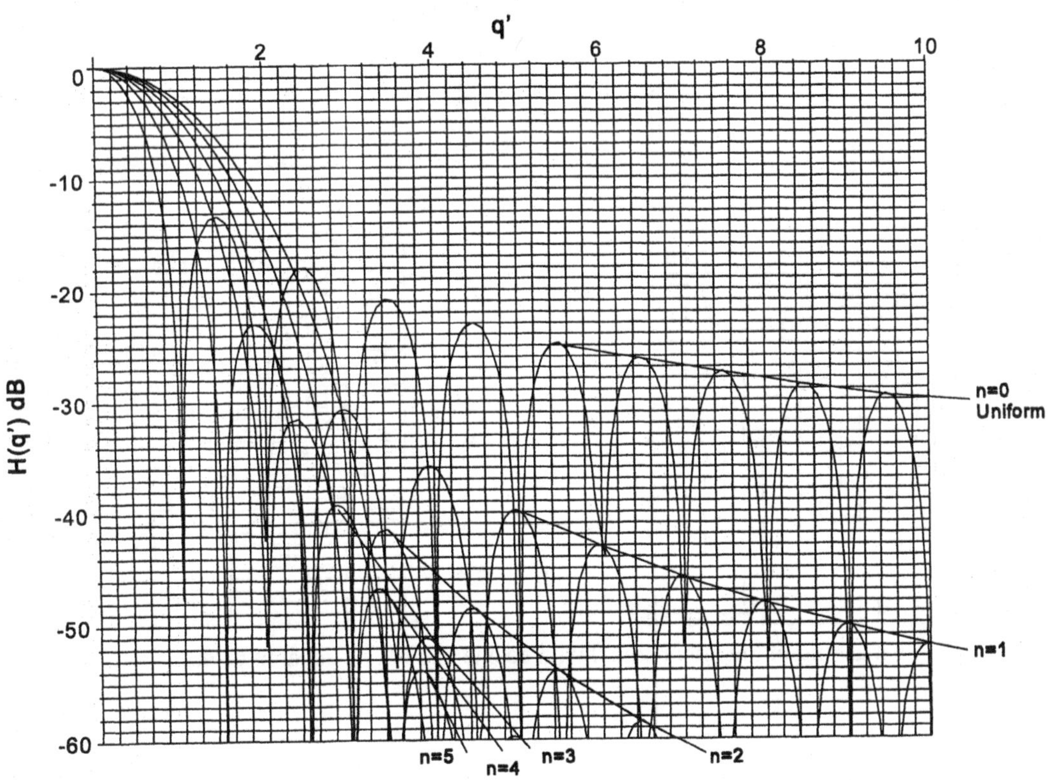

Figure B.15 Fourier transform in decibels for cosine to the power n tapering for $n = 0$, 1, 2, 3, 4, and 5.

Figure B.16 shows an example of the effects of sampling on spectral leakage, or two-tone characteristics, when the discrete Fourier transform is used for 100 samples. The minor signal has a frequency corresponding to the 16th bin, and the major signal has its frequency varied between the 10th and 11th bins.

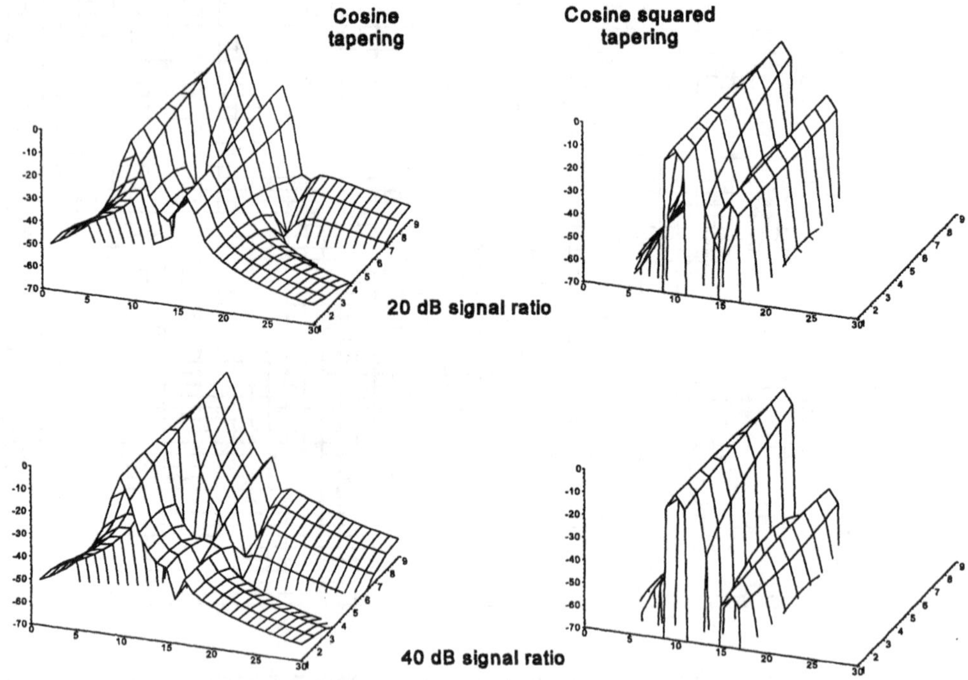

Figure B.16 Examples of the spectral leakage or two-tone characteristics for a 100 sample filter as the major signal frequency is varied from bin 10 to bin 11. The minor signal, 20 dB or 40 dB smaller, is in bin 16.

The worst cases occur with the major signal at the bin frequency, and the worst cases in Figure B.16 are shown in Figure B.17.

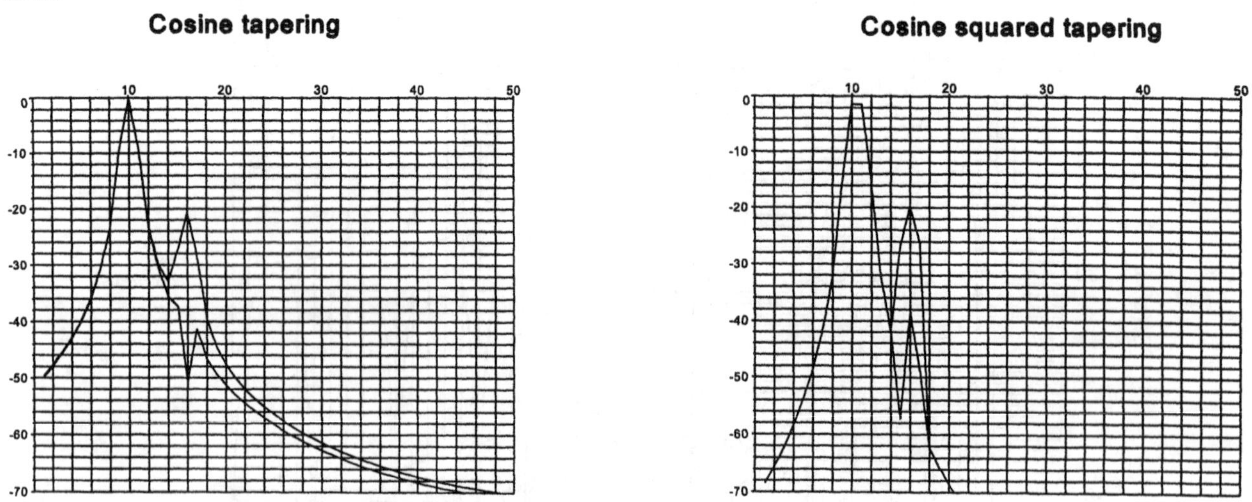

Figure B.17 Worst case masking of the minor signal with cosine and cosine squared tapering.

B.2.4 Cosine on a pedestal tapering

The tapering function is given by the edge value k is $k + (1 - k)\cos\pi p'$, is plotted in Figure B.18, and calculated values are shown in Table B.9.

Table B.9
Table of values for cosine on a pedestal tapering

Edge value	0.5	0.3	0.2	0.1	0.05	0.03
Edge value, dB	-6	-10.5	-14	-20	-26	-30.5
A, effective length, *coherent gain*	0.8183	0.7456	0.7093	0.6730	0.6548	0.6475
A, dB	-1.7416	-2.5495	-2.9835	-3.4402	-3.6780	-3.7749
C, effective power, *incoherent power gain*	0.6933	0.6024	0.5637	0.5296	0.5142	0.5084
C, dB	-1.5907	-2.2013	-2.4894	-2.7606	-2.8884	-2.9379
D (see Section B.1)	1.2337	2.4181	3.1583	3.9972	4.4537	4.6432
G (see Section B.1)	0.0400	0.0282	0.0234	0.0195	0.0178	0.0172
H (see Section B.1)	0.0567	0.0461	0.0408	0.0355	0.0328	0.0317
Efficiency η, *processing gain*	0.9658	0.9230	0.8925	0.8551	0.8338	0.8247
Efficiency η, *processing gain*, dB	-0.1509	-0.3482	-0.4941	-0.6797	-0.7895	-0.8370
Noise beamwidth	1.0354	1.0835	1.1205	1.1694	1.1994	1.2125
Half-power beamwidth	0.9766	1.0372	1.0771	1.1265	1.1558	1.1686
RMS beamwidth	1.3340	2.0035	2.3670	2.7473	2.9429	3.0221
RMS aperture	1.5090	1.3587	1.2811	1.2054	1.1695	1.1557
First sidelobe, dB	-17.6517	-20.2859	-21.6520	-22.7175	-22.9873	-23.0253
Falloff, dB/octave		-10				-12
Scalloping loss	0.6945	0.7255	0.7434	0.7633	0.7740	0.7785
Scalloping loss, dB	-3.1667	-2.7868	-2.5750	-2.3462	-2.2248	-2.1748
Worst case loss	0.6825	0.6970	0.7023	0.7058	0.7068	0.7070
Worst case loss, dB	-3.3176	-3.1350	-3.0691	-3.0259	-3.0143	-3.0118

The closed form for the Fourier transform is

$$H(q') = \frac{4kq'^2 \sin(\pi q') - k\sin(\pi q') + 2kq'\cos(\pi q') - 2q'\cos(\pi q')}{\pi q'(4q'^2 - 1)\left(k + \frac{2}{\pi} - \frac{2k}{\pi}\right)}$$

The Fourier transforms are shown in Figures B.19 and B.20.

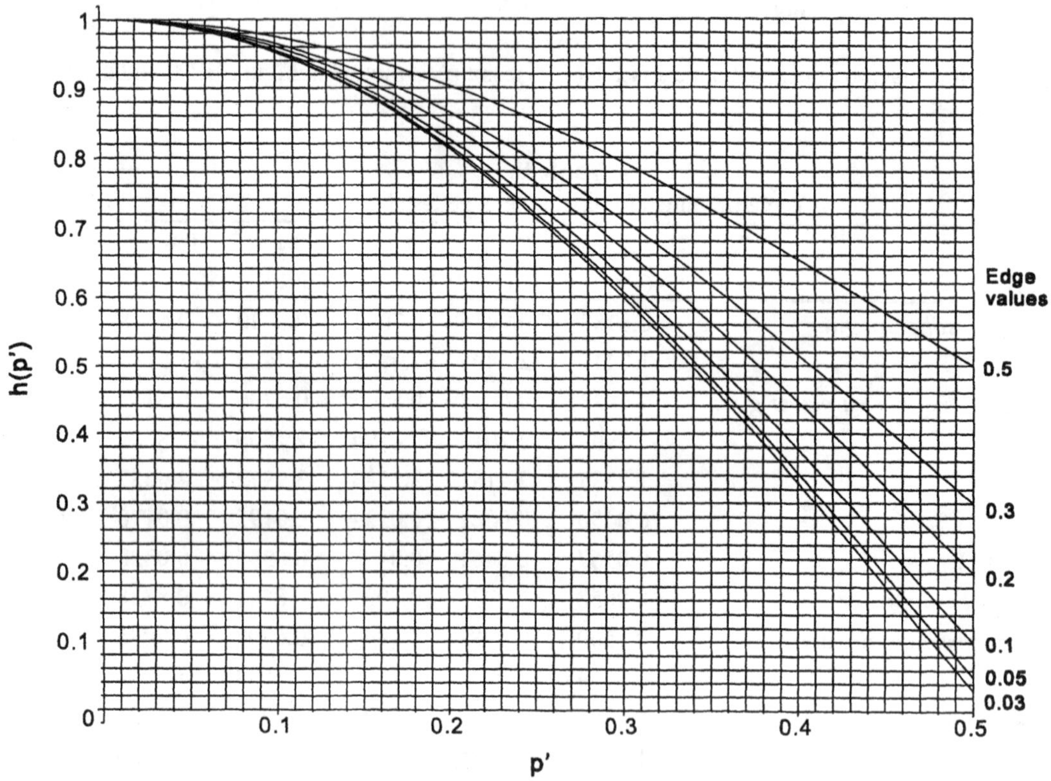

Figure B.18 Cosine on pedestal tapering with edge values *k* from 0.5 to 0.03.

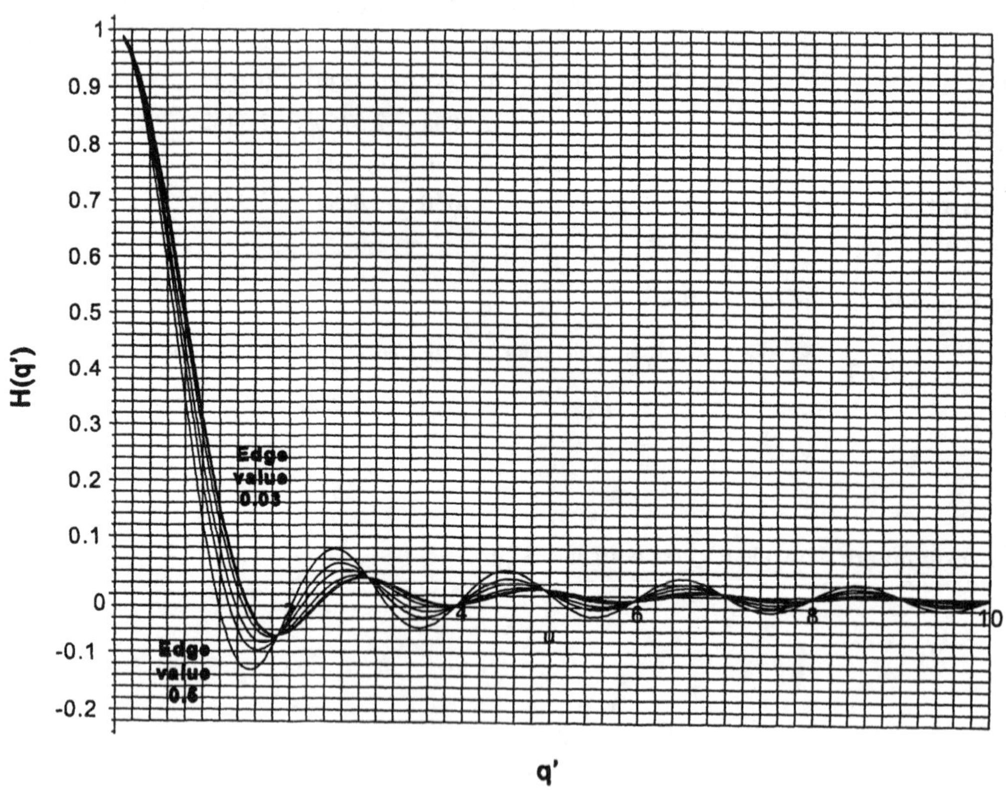

Figure B.19 Fourier transforms for cosine on pedestal illumination with edge values *k* from 0.5 to 0.03.

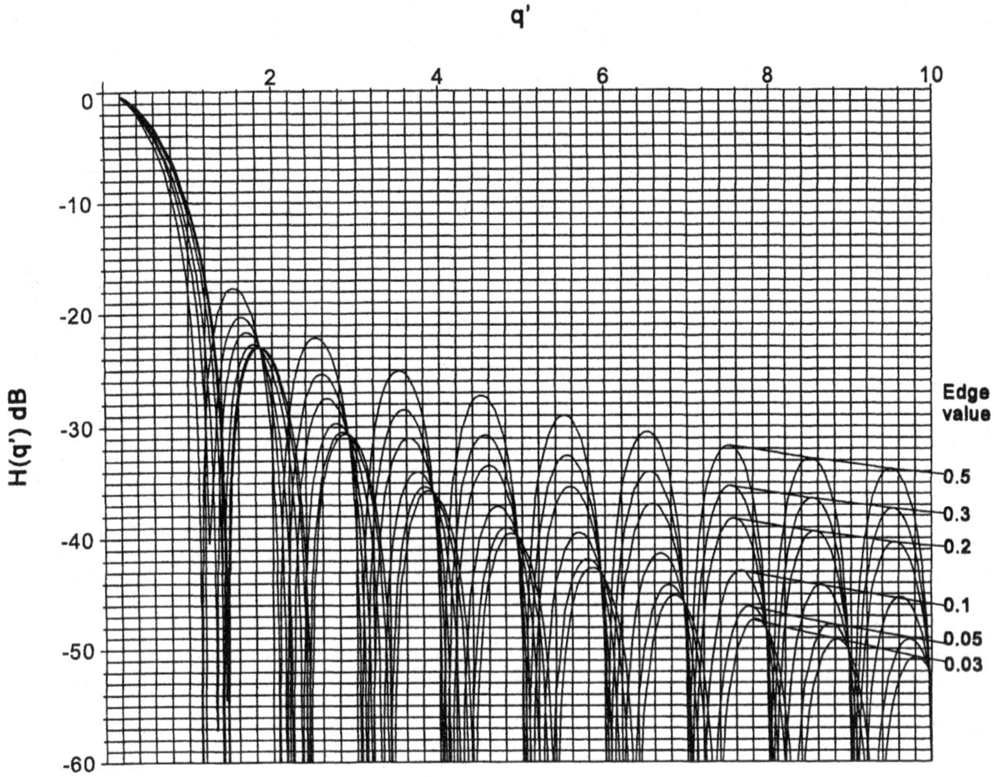

Figure B.20 Fourier transforms in decibels for cosine on pedestal tapering with edge values k from 0.5 to 0.03.

Figure B.21 shows an example of the effects of sampling on spectral leakage, or two-tone characteristics, when the discrete Fourier transform is used for 100 samples. The minor signal has a frequency corresponding to the 16th bin, and the major signal has its frequency varied between the 10th and 11th bins.

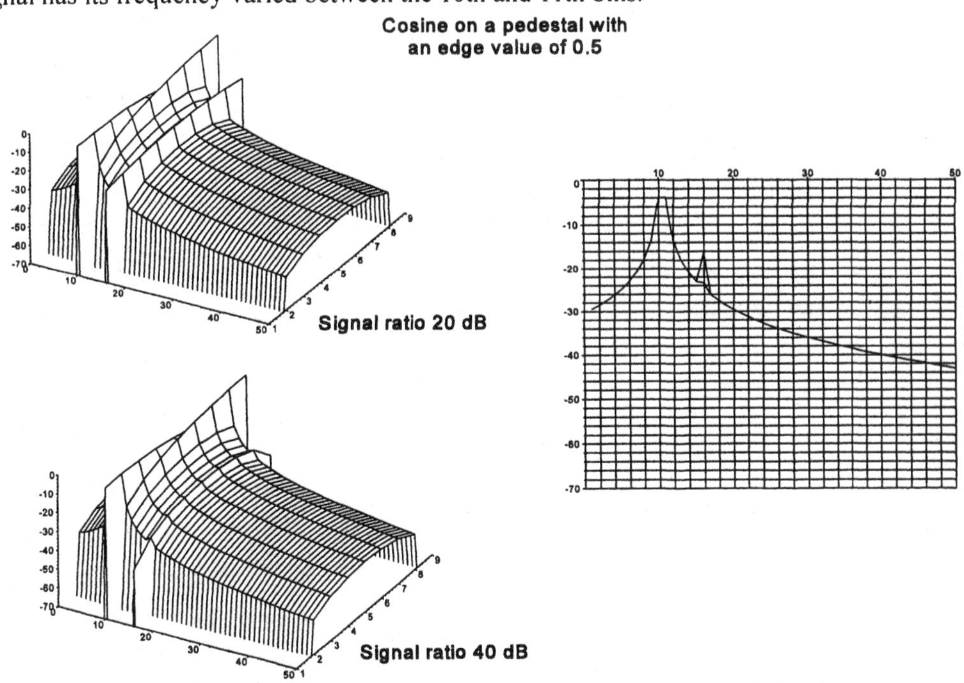

Figure B.21 Examples of the spectral leakage or two-tone characteristics for a 100 sample filter as the major signal frequency is varied from bin 10 to bin 11. The minor signal, 20 dB or 40 dB smaller, is in bin 16.

B.2.5 Hamming, Blackman, and Blackman-Harris tapering

The aperture functions for examples of the Hamming, Blackman and Blackman-Harris tapering functions are given in Table B.10 [2, 3].

Table B.10
The aperture functions for Hamming, Blackman, and Blackman-Harris tapering

Name	Aperture function
Hamming	$h(p') = 0.54 + 0.46 \cos(2\pi p')$
Exact Hamming	$h(p') = \frac{25}{46} + \frac{21}{46}\cos(2\pi p')$
Blackman	$h(p') = 0.42 + 0.5\cos(2\pi p') + 0.88\cos(4\pi p')$
Exact Blackman	$h(p') = \frac{3969}{9304} + \frac{1155}{2326}\cos(2\pi p') + \frac{715}{9304}\cos(4\pi p')$
Blackman-Harris 2 term	$h(p') = 0.53856 + 0.46144\cos(2\pi p')$
Blackman-Harris 3 term 67 dB	$h(p') = 0.42323 + 0.49755\cos(2\pi p') + 0.07922\cos(4\pi p')$
Blackman-Harris 3 term 61 dB	$h(p') = 0.44959 + 0.49364\cos(2\pi p') + 0.05677\cos(4\pi p')$
Blackman-Harris 4 term 92 dB	$h(p') = 0.35875 + 0.48829\cos(2\pi p') + 0.14128\cos(4\pi p') + 0.01168\cos(6\pi p')$
Blackman-Harris 4 term 74 dB	$h(p') = 0.40243 + 0.49804\cos(2\pi p') + 0.09831\cos(4\pi p') + 0.00122\cos(6\pi p')$

Values calculated from the aperture functions in Table B.10 are shown in Table B.11.

Table B.11

Table of values for Hamming, Blackman, and Blackman-Harris tapering

	Hamming		Blackman		Blackman-Harris				
					2 term	3 term		4 term	
		Exact		Exact		67 dB	61 dB	92 dB	74 dB
A, effective length, *coherent gain*	0.5400	0.5435	0.4200	0.4266	0.5386	0.4232	0.4496	0.3588	0.4024
A, dB	-5.3521	-5.2964	-7.5350	-7.3998	-5.3753	-7.4685	-6.9437	-8.9042	-7.9062
C, effective power gain, *incoherent gain*	0.3974	0.3996	0.3046	0.3082	0.3965	0.3060	0.3256	0.2580	0.2908
C, dB	-4.0077	-3.9840	-5.1627	-5.1114	-4.0175	-5.1422	-4.8734	-5.8844	-5.3640
D (see Section B.1)	4.1768	4.1139	5.4401	5.3334	4.2030	5.3821	5.0645	6.3066	5.6596
G (see Section B.1)	0.0093	0.0095	0.0043	0.0045	0.0092	0.0044	0.0053	0.0026	0.0038
H (see Section B.1)	0.0217	0.0222	0.0107	0.0114	0.0215	0.0111	0.0132	0.0069	0.0095
Efficiency, η, *processing gain*	0.7338	0.7392	0.5791	0.5904	0.7315	0.5853	0.6208	0.4989	0.5569
Efficiency, η, *processing gain,* dB	-1.3444	-1.3123	-2.3723	-2.2884	-1.3579	-2.3262	-2.0703	-3.0197	-2.5422
Noise beamwidth	1.3628	1.3528	1.7268	1.6937	1.3671	1.7085	1.6108	2.0044	1.7956
Half-power beamwidth	1.3030	1.2933	1.6437	1.6087	1.3071	1.6235	1.5321	1.8994	1.7055
RMS beamwidth	3.2420	3.2087	4.2261	4.1598	3.2558	4.1936	3.9440	4.9444	4.4115
RMS aperture	0.9606	0.9677	0.7464	0.7571	0.9577	0.7513	0.7988	0.6364	0.7141
First sidelobe, dB	-44.0359	-46.0095	-58.1088	-69.4129	-43.2904	-70.8261	-74.5237	-92.0097	-66.4205
Maximum sidelobe, dB	-42.6750	-41.6899	-58.1088	-68.2360	-43.1220	-70.8261	-62.0465	-92.0097	-66.4205
Falloff, dB/octave	-6	-6	-18	-6	-6	-6	-6	-6	-6
Scalloping loss	0.8174	0.8149	0.8812	0.8760	0.8184	0.8781	0.8643	0.9093	0.8889
Scalloping loss, dB	-1.7514	-1.7782	-1.0989	-1.1500	-1.7403	-1.1287	-1.2671	-0.8256	-1.0226
Worst-case loss	0.7002	0.7006	0.6706	0.6731	0.7000	0.6718	0.6810	0.6423	0.6634
Worst-case loss, dB	-3.0958	-3.0905	-3.4712	-3.4384	-3.0981	-3.4549	-3.3374	-3.8453	-3.5649

B.2.5.1 Hamming, Exact Hamming, and two term Blackman-Harris tapering

The Hamming, Exact Hamming, and two term Blackman-Harris tapering characteristic is plotted in Figure B.22 and the Fourier transforms in Figures B.23 and B.24.

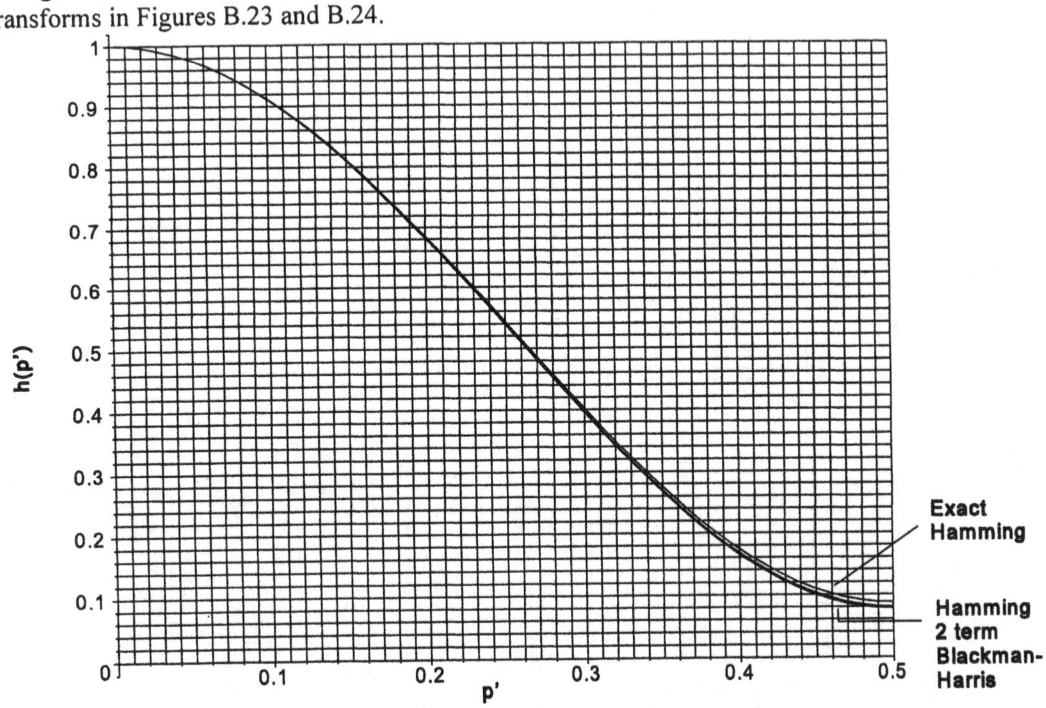

Figure B.22 Hamming, exact Hamming, and two term Blackman-Harris tapering function.

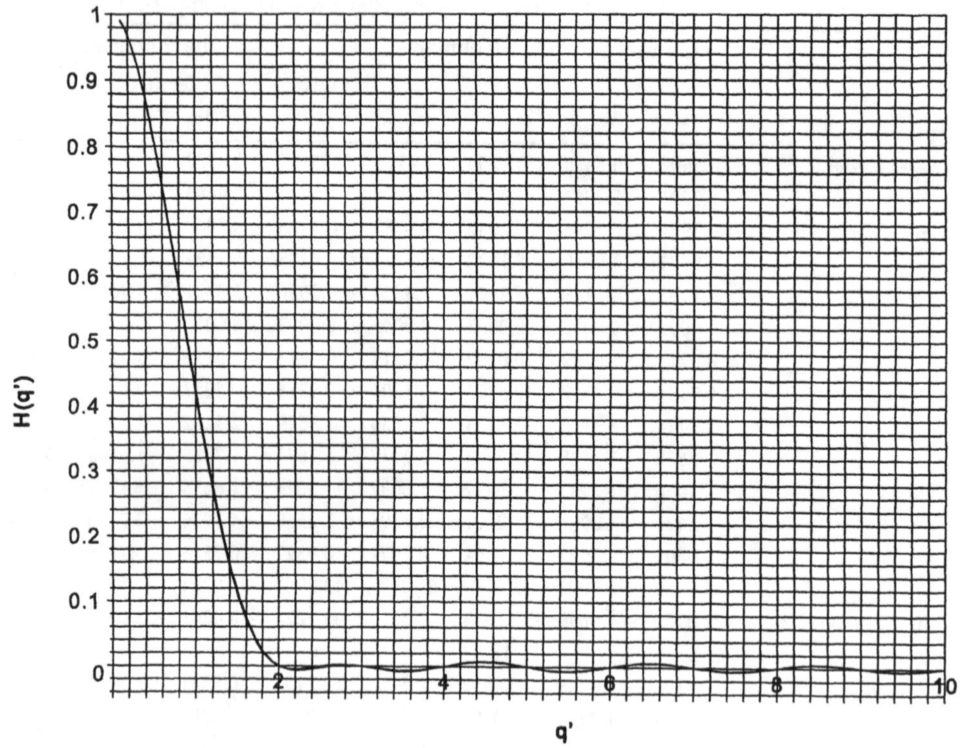

Figure B.23 Hamming, exact Hamming, and two term Blackman-Harris Fourier transforms. The curves are too close to separate.

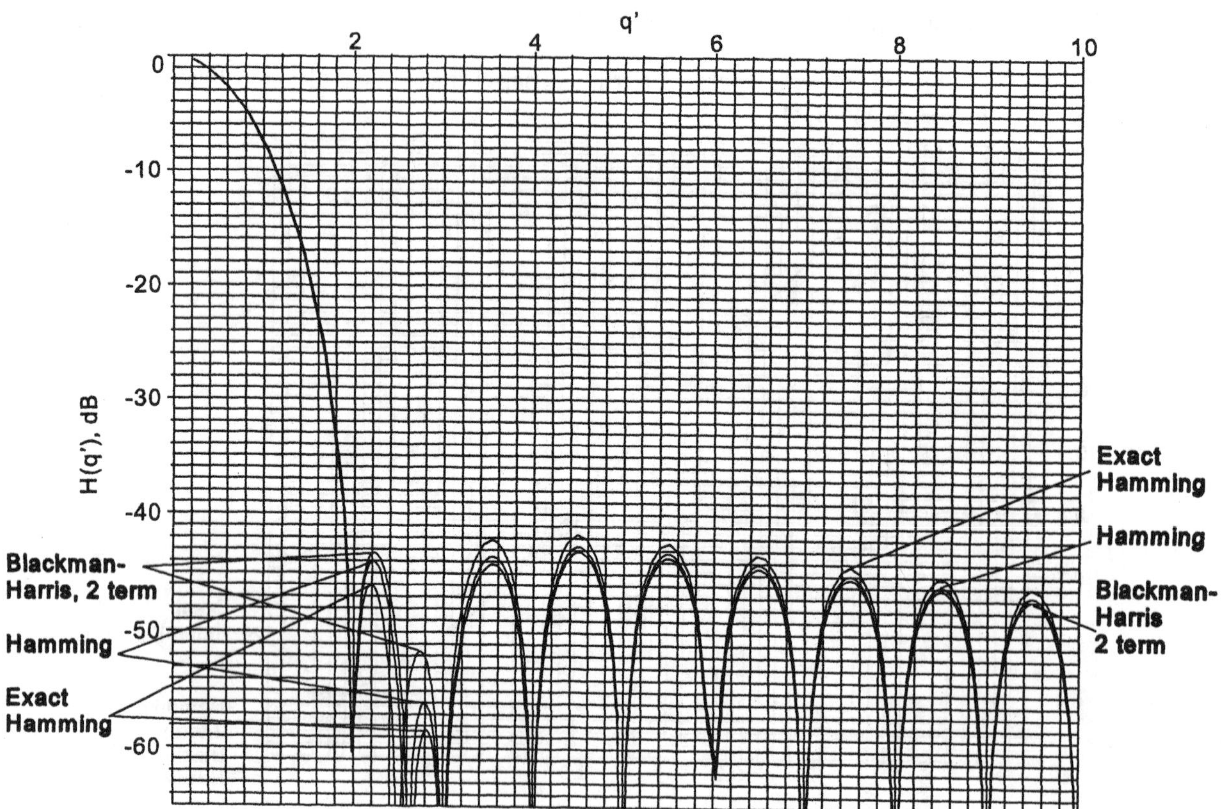

Figure B.24 Hamming, exact Hamming, and two term Blackman-Harris Fourier transforms in decibels.

The closed forms for the Fourier transforms are given in Table B.12.

Table B.12
Fourier transforms for the Hamming, exact Hamming, and two term Blackman-Harris aperture functions

Aperture function	Fourier transform
Hamming	$H(q') = \dfrac{-2(0.01273239544q'^2-0.08594366925)\sin(\pi q')}{q'(q'^2-1)}$
Exact Hamming	$H(q') = \dfrac{-2\left(-\dfrac{q'^2}{23}+\dfrac{25}{92}\right)\sin(\pi q')}{\pi q'(q'^2-1)}$
Two term Blackman-Harris	$H(q') = \dfrac{-2(-0.01227402921q'^2+0.08571448614)\sin(\pi q')}{q'(q'^2-1)}$

Figure B.25 shows an example of the effects of sampling on spectral leakage, or two-tone characteristics, when the discrete Fourier transform is used for 100 samples. The minor signal has a frequency corresponding to the 16th bin, and the major signal has its frequency varied between the 10th and 11th bins.

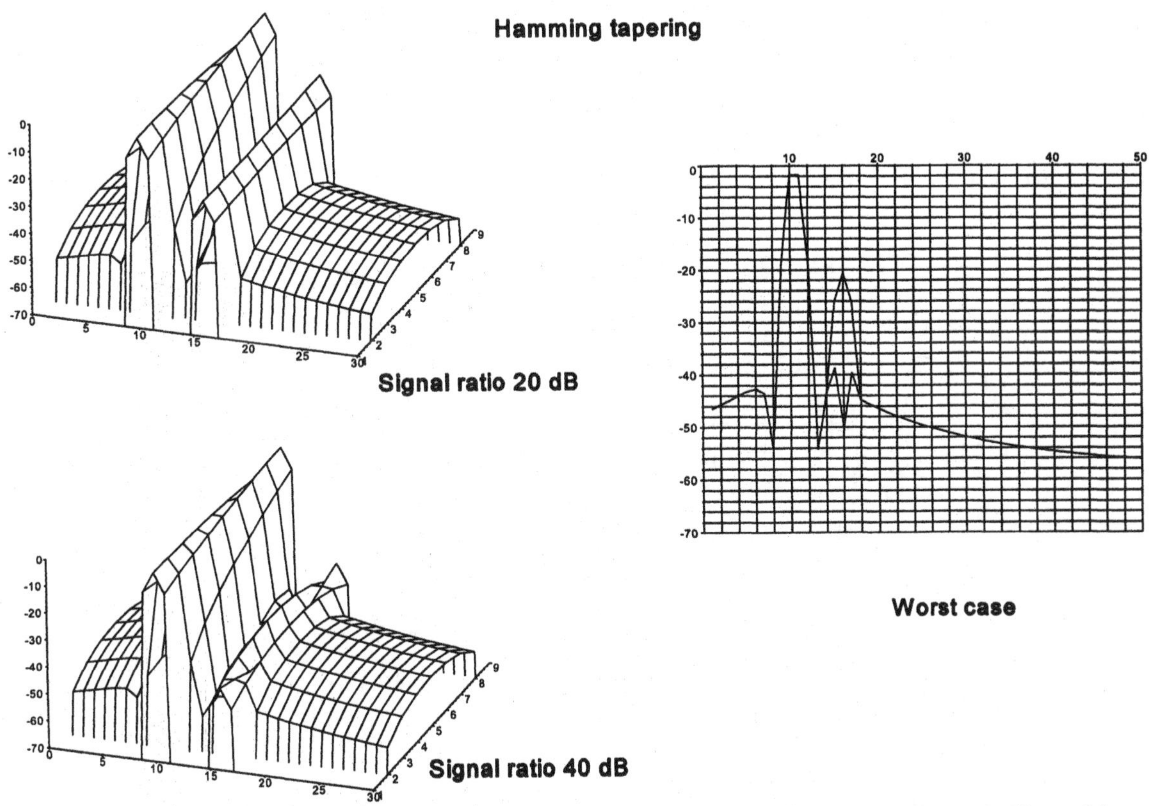

Figure B.25 Examples of the spectral leakage, or two-tone characteristics, for a 100 sample filter with Hamming tapering as the major signal frequency is varied from bin 10 to bin 11. The minor signal, 20 dB or 40 dB smaller, is in bin 16.

B.2.5.2 Blackman and exact Blackman tapering

The Blackman and exact Blackman functions are plotted in Figure B.26 and their Fourier transforms in Figures B.27 and B.28.

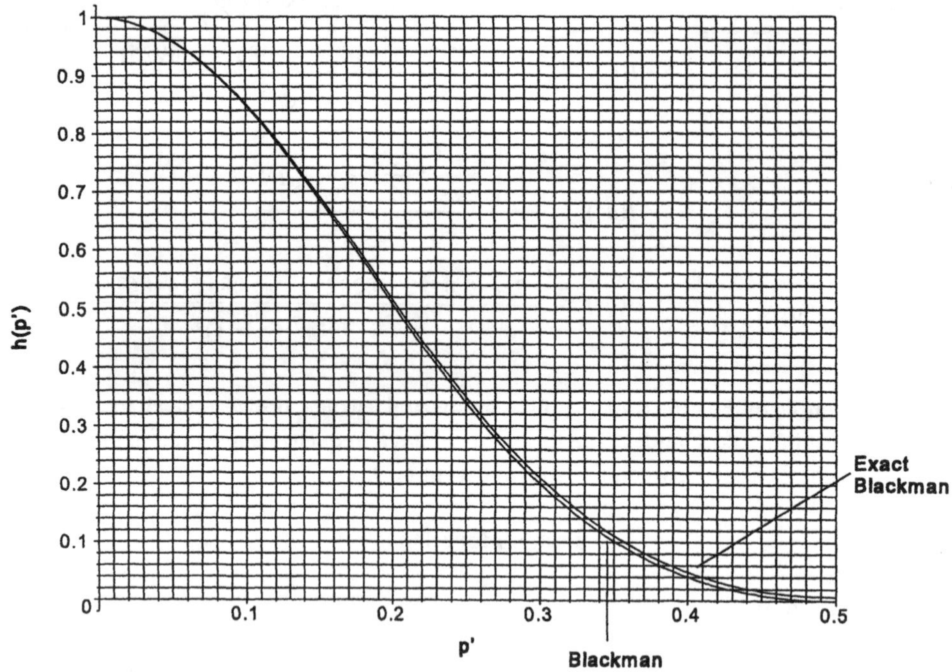

Figure B.26 Blackman and exact Blackman tapering functions.

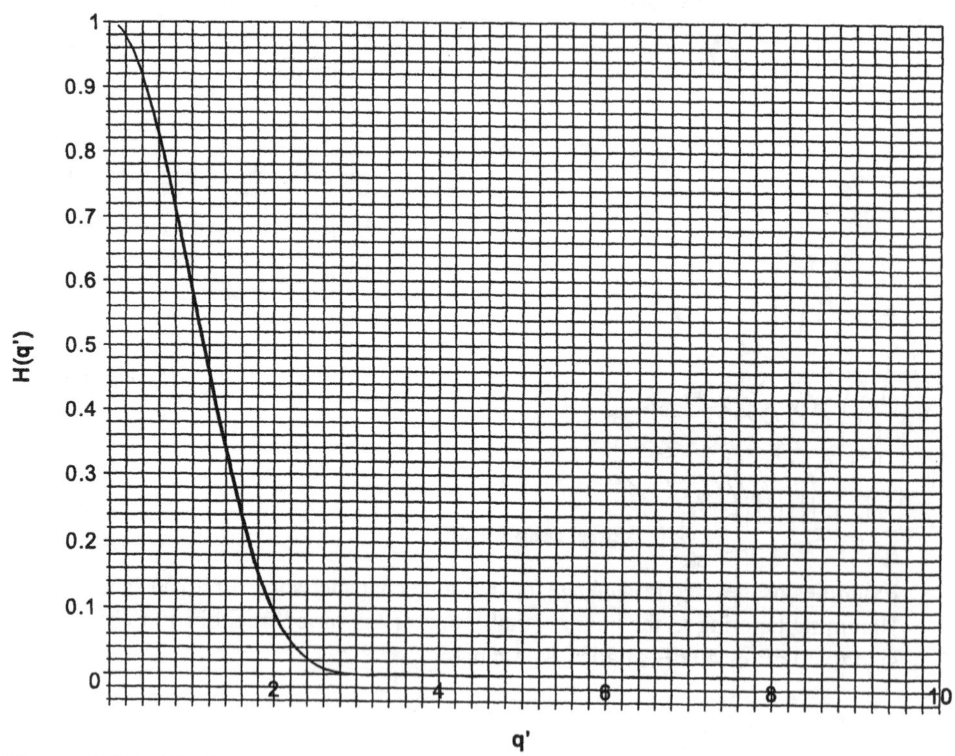

Figure B.27 Blackman and exact Blackman Fourier transforms. The curves are too close to separate.

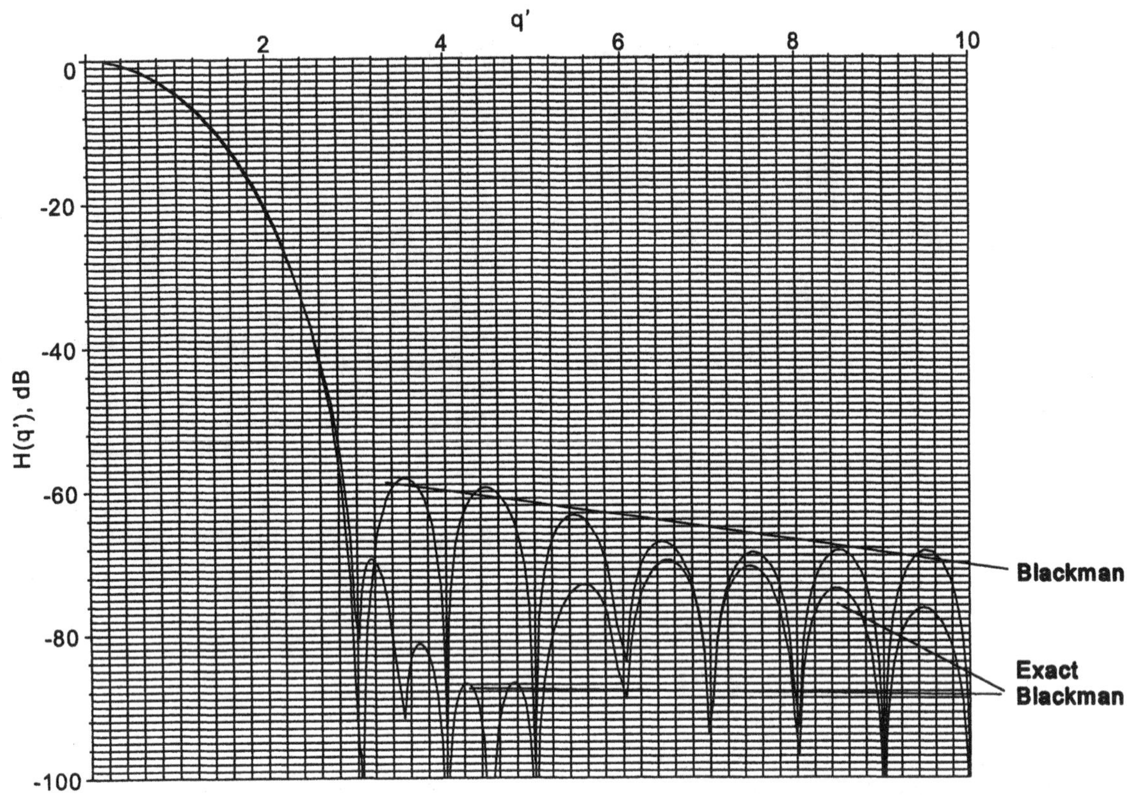

Figure B.28 Blackman and exact Blackman Fourier transforms in decibels.

Figures B.29 and B.30 each show an example of the effects of sampling on spectral leakage, or two-tone characteristics, when the discrete Fourier transform is used for 100 samples. The minor signal has a frequency corresponding to the 16th bin, and the major signal has its frequency varied between the 10th and 11th bins.

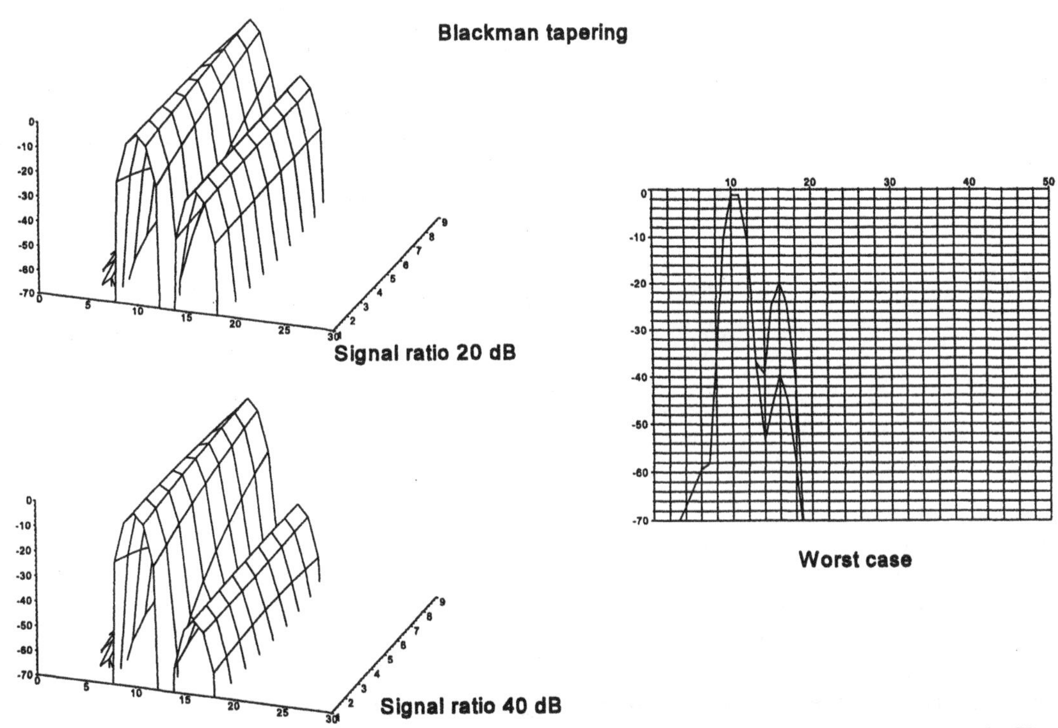

Figure B.29 Examples of the spectral leakage, or two-tone characteristics, for a 100 sample filter with Blackman tapering as the major signal frequency is varied from bin 10 to bin 11. The minor signal, 20 dB or 40 dB smaller, is in bin 16.

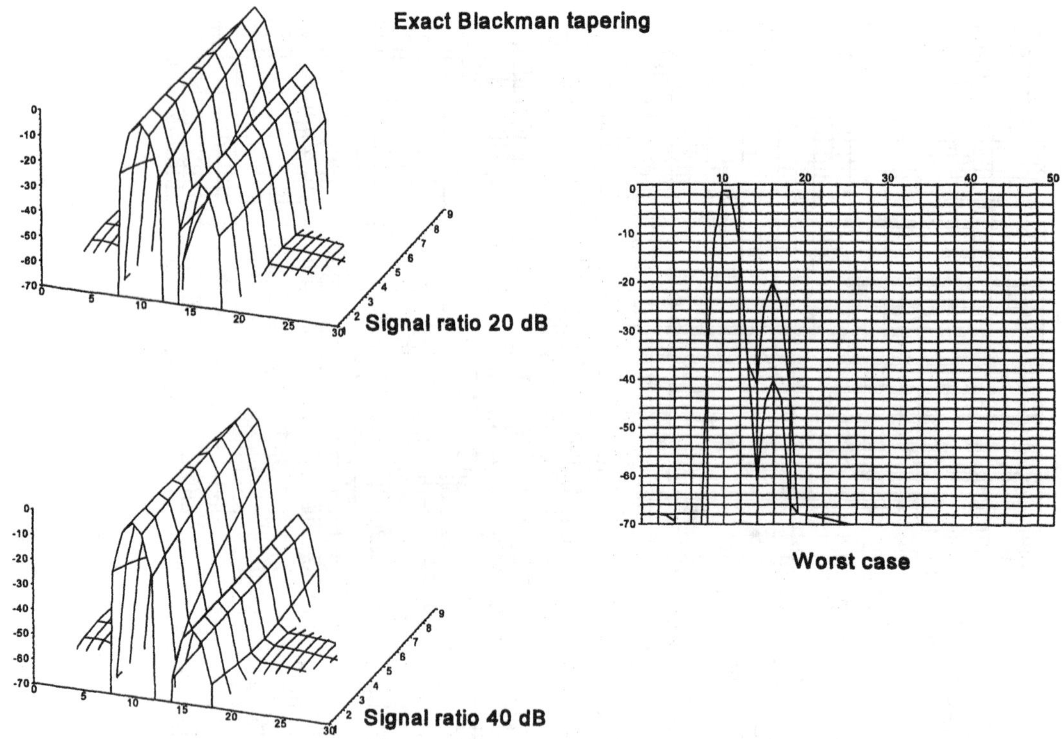

Figure B.30 Examples of the spectral leakage, or two-tone characteristics, for a 100 sample filter with exact Blackman tapering as the major signal frequency is varied from bin 10 to bin 11. The minor signal, 20 dB or 40 dB smaller, is in bin 16.

Closed forms for the Fourier transforms are given in Table B.13.

Table B.13
Closed forms of the Fourier transforms for Blackman and exact Blackman tapering

Name	Fourier transform
Blackman	$H(q') = \dfrac{-2\,(0.0286478897\,q'^2 - 0.2673803043)\,\sin(\pi q')}{q'\,(q'^2 - 4)\,(q'^2 - 1)}$
Exact Blackman	$H(q') = \dfrac{-2\left(-\dfrac{4q'^4}{1163} + \dfrac{130q'^2}{1163} - \dfrac{3969}{4652}\right)\sin(\pi q')}{\pi\,q'\,(q'^2 - 4)\,(q'^2 - 1)}$

B.2.5.3 Blackman-Harris tapering

The Blackman-Harris tapering functions for:

- Three term, 61 dB tapering function;
- Three term, 67 dB tapering function;
- Four term, 74 dB tapering function;
- Four term, 92 dB tapering function.

are plotted in Figure B.31 and their Fourier transforms in Figures B.32 and B.33.

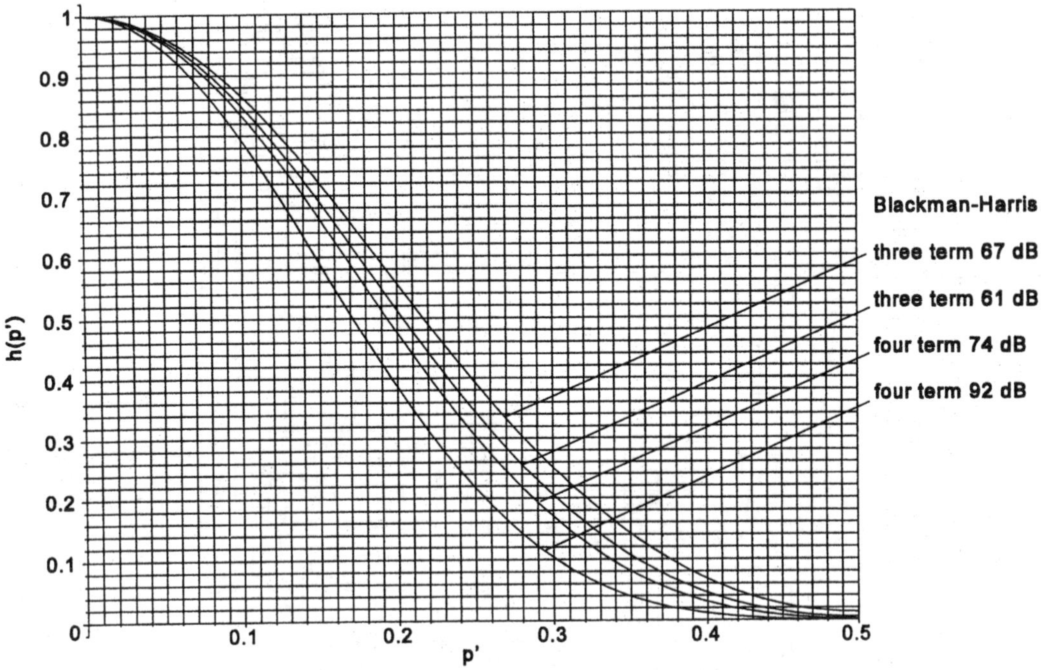

Figure B.31 Blackman-Harris tapering functions.

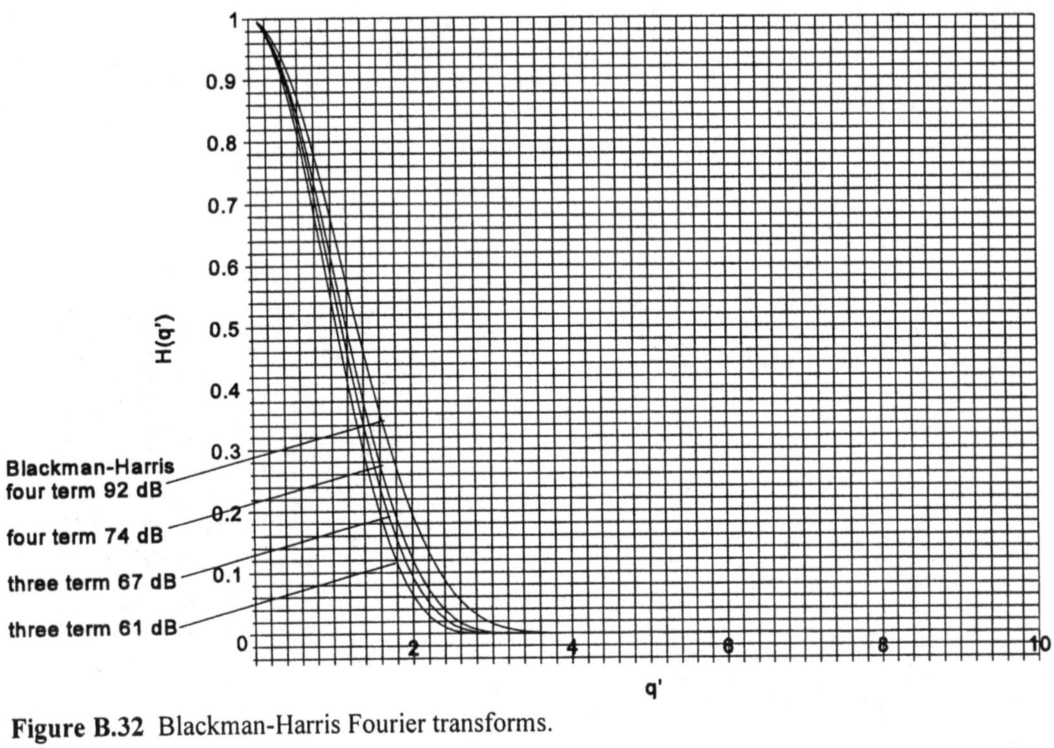

Figure B.32 Blackman-Harris Fourier transforms.

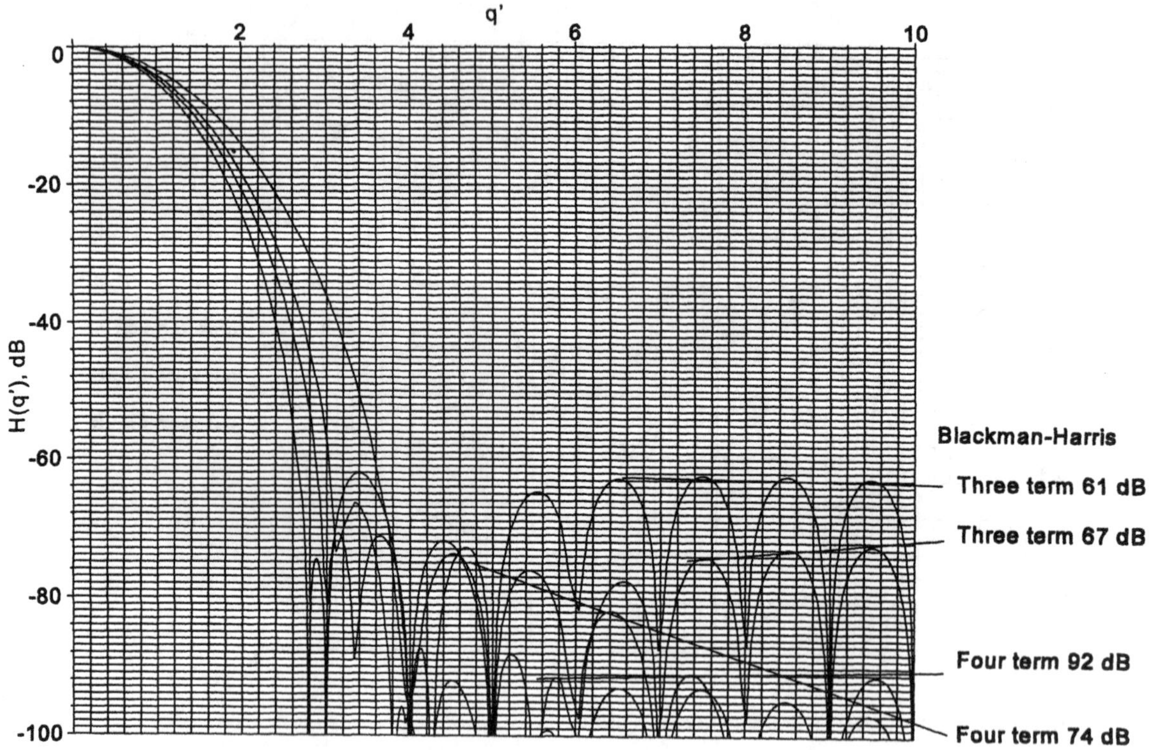

Figure B.33 Blackman-Harris Fourier transforms in decibels.

Figures B.34 and B.35 show examples of the effects of sampling on spectral leakage, or two-tone characteristics, when the discrete Fourier transform is used for 100 samples. The minor signal has a frequency corresponding to the 16th bin, and the major signal has its frequency varied between the 10th and 11th bins.

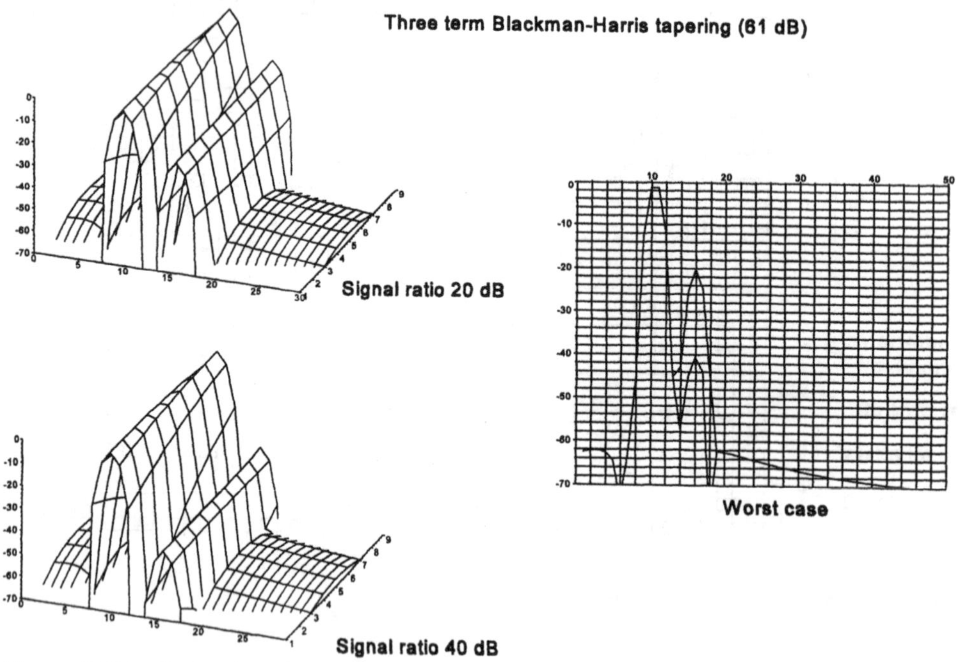

Figure B.34 Examples of the spectral leakage, or two-tone characteristics, for a 100 sample filter with three term Blackman-Harris 61 dB sidelobe tapering as the major signal frequency is varied from bin 10 to bin 11. The minor signal, 20 dB or 40 dB smaller, is in bin 16.

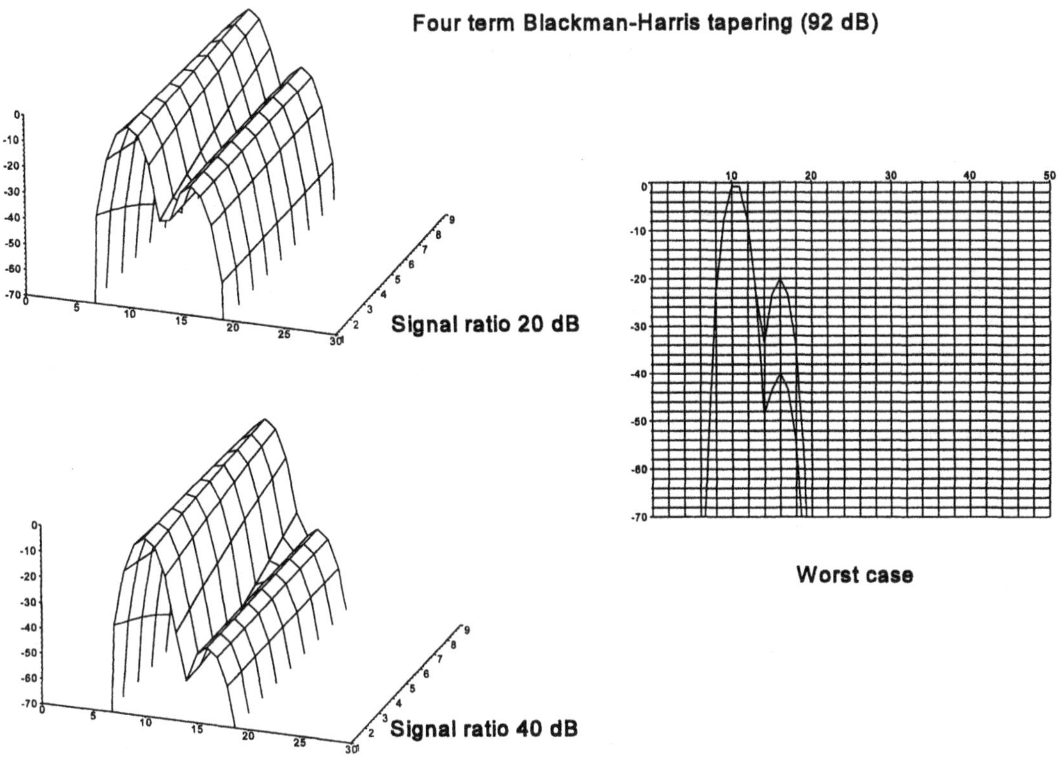

Figure B.35 Examples of the spectral leakage, or two tone-characteristics, for a 100 sample filter with four term Blackman-Harris 92 dB sidelobe tapering as the major signal frequency is varied from bin 10 to bin 11. The minor signal, 20 dB or 40 dB smaller, is in bin 16.

Closed forms for the Fourier transforms are shown in Table B.14.

Table B.14
Fourier transforms of examples of Blackman-Harris functions

Function name	Fourier transform
Blackman-Harris three term 67 dB	$H(q') = \dfrac{-2(-0.0007798592212q'^{4}+0.03265381968q'^{2}-.2694365863)\ \sin(\pi q')}{q'(q'^{2}-4)(q'^{2}-1)}$
Blackman-Harris three term 61 dB	$H(q') = \dfrac{-2\ (-0.002024450876q'^{4}+0.05254659600q'^{2}-0.2862178834)\ \sin(\pi q')}{q'(q'^{2}-4)(q'^{2}-1)}$
Blackman-Harris four term 92 dB	$H(q') = \dfrac{-2.(-0.000009549296584q'^{6}+0.004636183492q'^{4}-0.1949823123q'^{2}+2.055486090)\ \sin(\pi q')}{q'(q'^{2}-4)(q'^{2}-1)(q'^{2}-9)}$
Blackman-Harris four term 74 dB	$H(q') = \dfrac{-2(-0.0002355493157q'^{6}+0.02172464973q'^{4}-0.4248704867q'^{2}+2.305754054)\ \sin(\pi q')}{q'(q'^{2}-4)(q'^{2}-1)(q'^{2}-9)}$

B.2.6 Truncated Gaussian tapering

The tapering function is $\exp\left(-1.386\,(n\,p')^2\right)$ is plotted in Figure B.36 with the calculated characteristics in Table B.15.

Table B.15
Table of values for a number of truncated Gaussian tapering functions

	Value of *n*				
	1	1.7	2.4	2.8	3.2
Edge illumination, dB	-3.0103	-8.6998	-17.3393	-23.6008	-30.8255
A, effective length, *coherent gain*	0.8956	0.7465	0.5986	0.5270	0.4668
A, dB	-0.9581	-2.5392	-4.4576	-5.5634	-6.6174
C, effective power, *incoherent power gain*	0.8100	0.5978	0.4414	0.3798	0.3326
C, dB	-0.9150	-2.2346	-3.5513	-4.2045	-4.7809
D (see Section B.1)	0.4298	1.8544	3.3775	4.0804	4.7096
G (see Section B.1)	0.0559	0.0289	0.0132	0.0086	0.0058
H (see Section B.1)	0.0680	0.0473	0.0290	0.0212	0.0154
Efficiency η, *processing gain*	0.9901	0.9323	0.8117	0.7313	0.6552
Efficiency η, *processing gain*, dB	-0.0431	-0.3045	-0.9063	-1.3589	-1.8365
Noise beamwidth	1.0100	1.0726	1.2320	1.3674	1.5263
Half-power beamwidth	0.9309	1.0244	1.1822	1.3036	1.4453
RMS beamwidth	0.7284	1.7613	2.7661	3.2778	3.7630
RMS aperture	1.6507	1.3811	1.0883	0.9474	0.8328
First sidelobe, dB	-15.4306	-20.6991	-31.8631	-39.1452	—
Falloff, dB/octave	-6	-6	-6	-6	-6
Scalloping loss	0.6674	0.7194	0.7825	0.8174	0.8484
Scalloping loss, dB	-3.5129	-2.8609	-2.1301	-1.7517	-1.4283
Worst case loss	0.6641	0.6946	0.7050	0.6990	0.6867
Worst case loss, dB	-3.5560	-3.1654	-3.0364	-3.1106	-3.2648

The closed form for the Fourier transform is

$$H(q') = \frac{\sqrt{\pi}\,e^{-\frac{\pi^2 q'^2}{2\ln(2)n^2}}\left[\operatorname{erf}\left(\frac{\ln(2)n^2 + j\pi q'}{n\sqrt{2\ln(2)}}\right) + \operatorname{erf}\left(\frac{\ln(2)n^2 - j\pi q'}{n\sqrt{2\ln(2)}}\right)\right]}{2n\sqrt{2\ln(2)}}$$

where erf(x) denotes the Gaussian error function, $\int_x^{\infty} e^{-x^2}dx$.

The Fourier transforms for a number of truncated Gaussian tapers are shown in Figures B.37 and B.38.

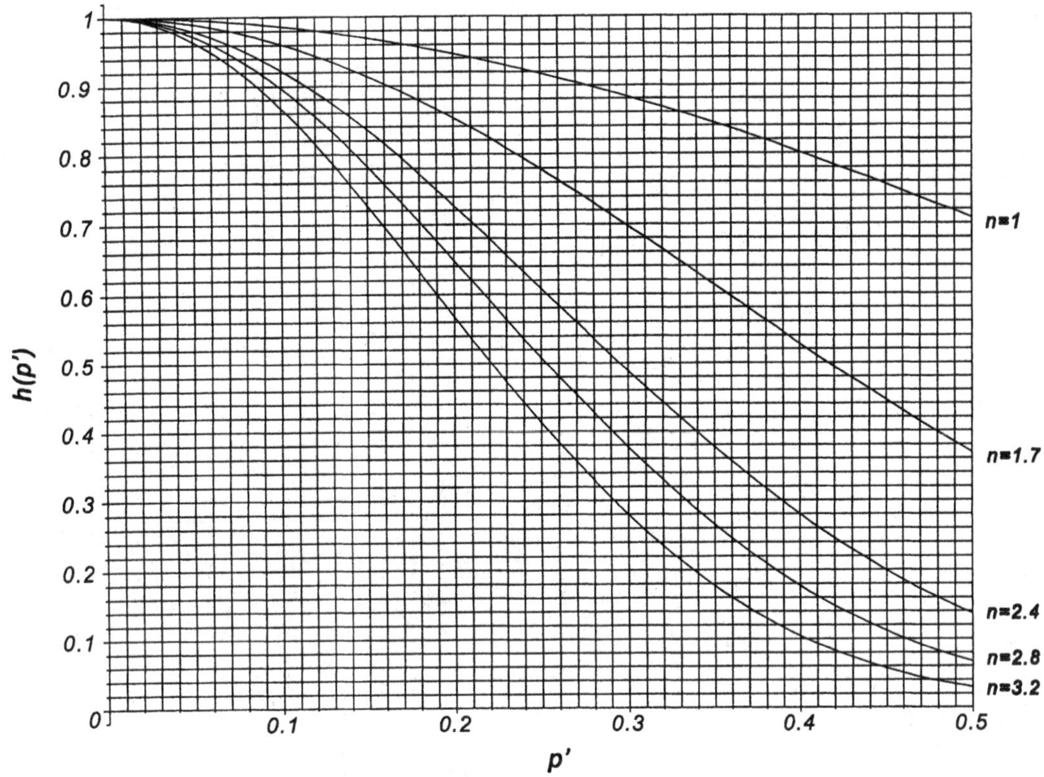

Figure B.36 Truncated Gaussian tapering function.

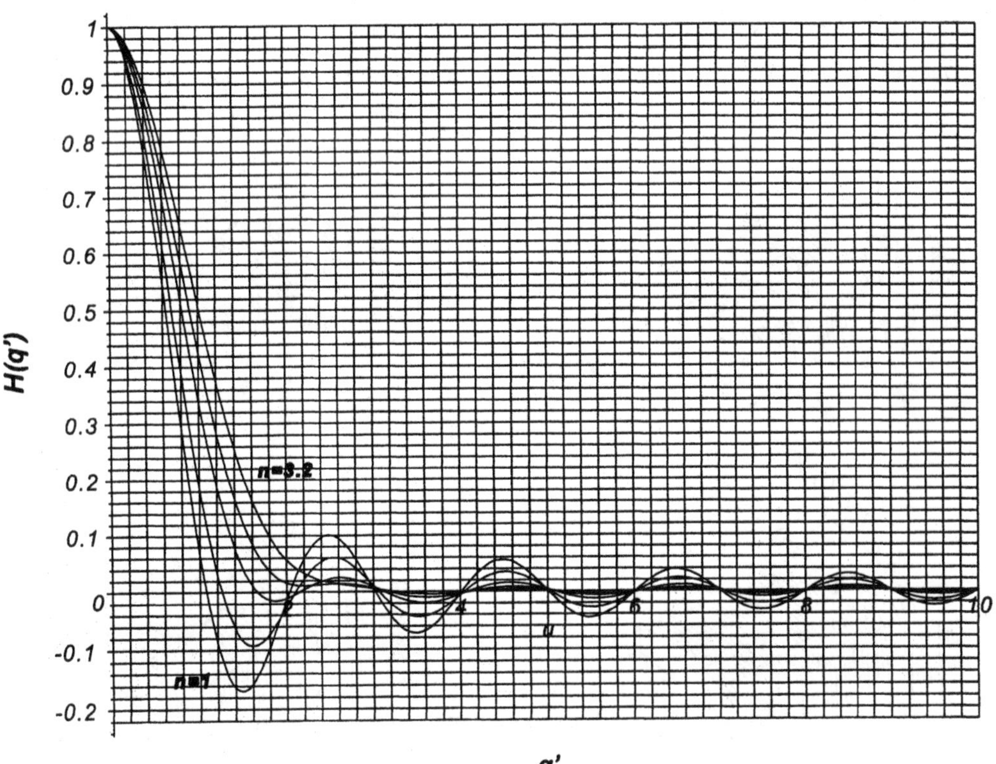

Figure B.37 Fourier transforms for truncated Gaussian tapering for $n = 1, 1.7, 2.4, 2.8, 3.2$.

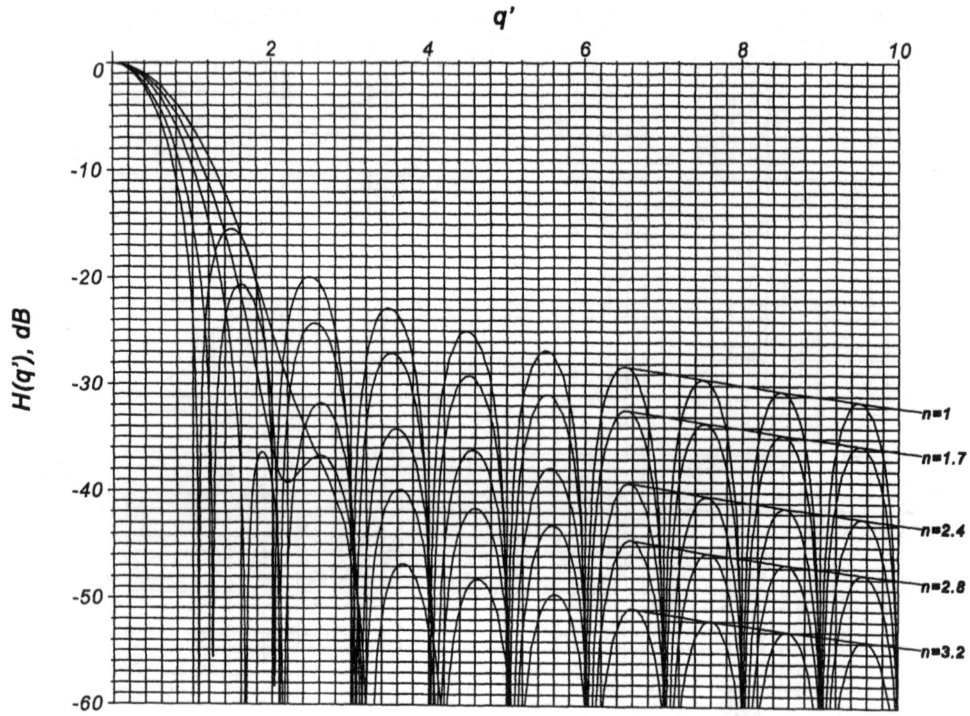

Figure B.38 Fourier transforms in decibels for truncated Gaussian tapering for $n = 1$, 1.7, 2.4, 2.8, 3.2.

Figure B.39 shows an example of the effects of sampling on spectral leakage, or two-tone characteristics, when the discrete Fourier transform is used for 100 samples. The minor signal has a frequency corresponding to the 16th bin and the major signal has its frequency varied between the 10th and 11th bins.

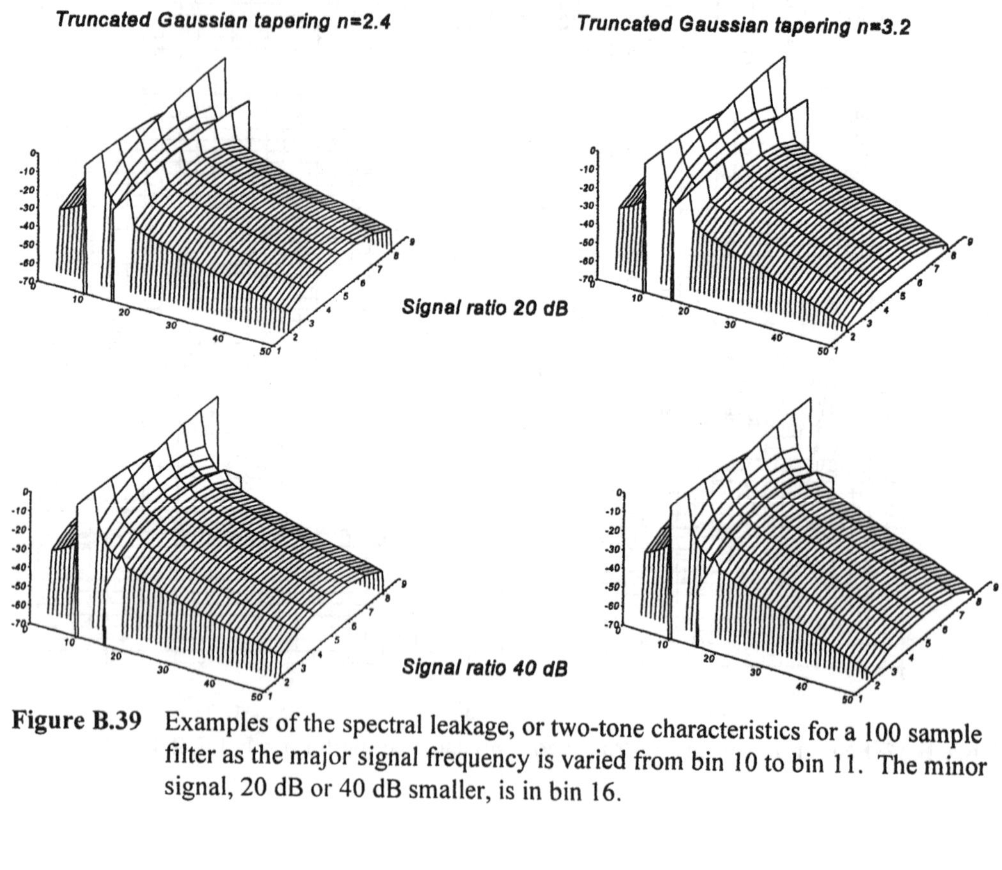

Figure B.39 Examples of the spectral leakage, or two-tone characteristics for a 100 sample filter as the major signal frequency is varied from bin 10 to bin 11. The minor signal, 20 dB or 40 dB smaller, is in bin 16.

Cuts through the worst cases in Figure B.39 are shown in Figure B.40.

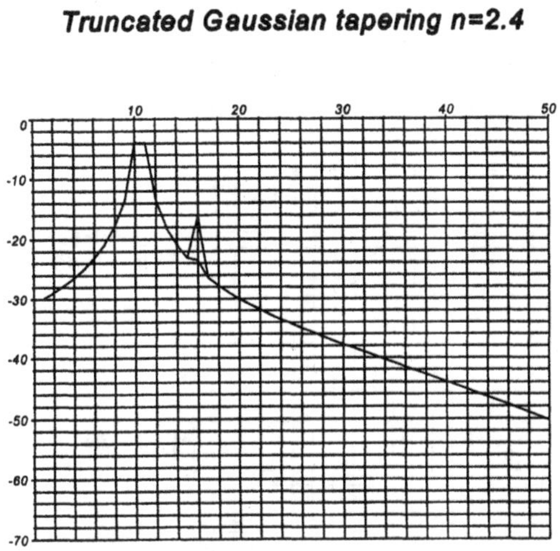

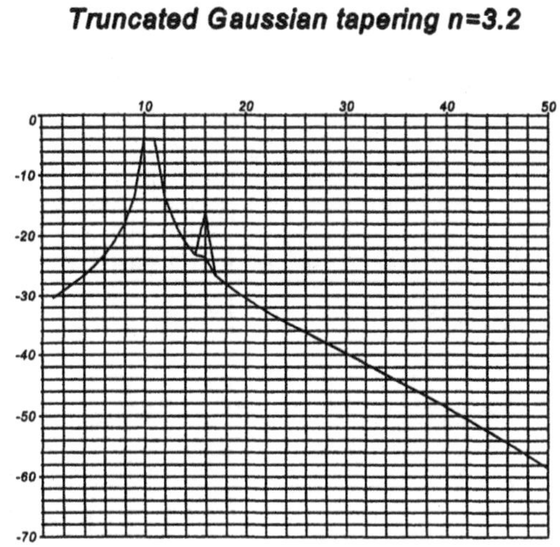

Figure B.40 Worst case masking of the minor signal with two cases of truncated Gaussian tapering.

B.2.7 Even Taylor tapering

The tapering functions and antenna pattern functions are given in the text in Section 5.2.2. The characteristics of even Taylor tapering are given in Table B.16, illustrated in Figures B.42 and B.43, and the tapering function plotted in Figure B.41.

Table B.16
Table of values for examples of Taylor tapering

	Sidelobe level, dB					
	20	25	30	35	40	45
$\bar{n}$	2	3	4	5	6	8
A, effective length, *coherent gain*	0.7586	0.6927	0.6418	0.6005	0.5661	0.5369
A, dB	-2.3992	-3.1893	-3.8519	-4.4304	-4.9426	-5.4028
C, effective power, *incoherent power gain*	0.6047	0.5317	0.4827	0.4462	0.4175	0.3939
C, dB	-2.1848	-2.7432	-3.1634	-3.5044	-3.7930	-4.0456
D (see Section B.1)	—	—	—	—	—	—
G (see Section B.1)	0.0322	0.0224	0.0168	0.0133	0.0109	0.0092
H (see Section B.1)	0.0510	0.0412	0.0342	0.0288	0.0247	0.0214
Efficiency η, *processing gain*	0.9518	0.9024	0.8534	0.8080	0.7674	0.7316
Efficiency η, *processing gain*, dB	-0.2144	-0.4460	-0.6885	-0.9260	-1.1496	-1.3572
Noise beamwidth	1.0506	1.1082	1.1718	1.2377	1.3031	1.3668
Half-power beamwidth	0.9943	1.0606	1.1247	1.1875	1.2484	1.3074
RMS beamwidth	—	—	—	—	—	—
RMS aperture	1.4501	1.2892	1.1718	1.0832	1.0141	0.9581
First sidelobe, dB	-20.6960	-25.4587	-30.3073	-35.2200	-40.1648	-45.1272
Falloff, dB/octave	—	—	—	—	—	—
Scalloping loss	0.7041	0.7361	0.7622	0.7842	0.8028	0.8186
Scalloping loss, dB	-3.0469	-2.6612	-2.3581	-2.1112	-1.9080	-1.7390
Worst-case loss	0.6870	0.6993	0.7042	0.7049	0.7033	0.7002
Worst-case loss dB	-3.2613	-3.1072	-3.0466	-3.0372	-3.0577	-3.0961

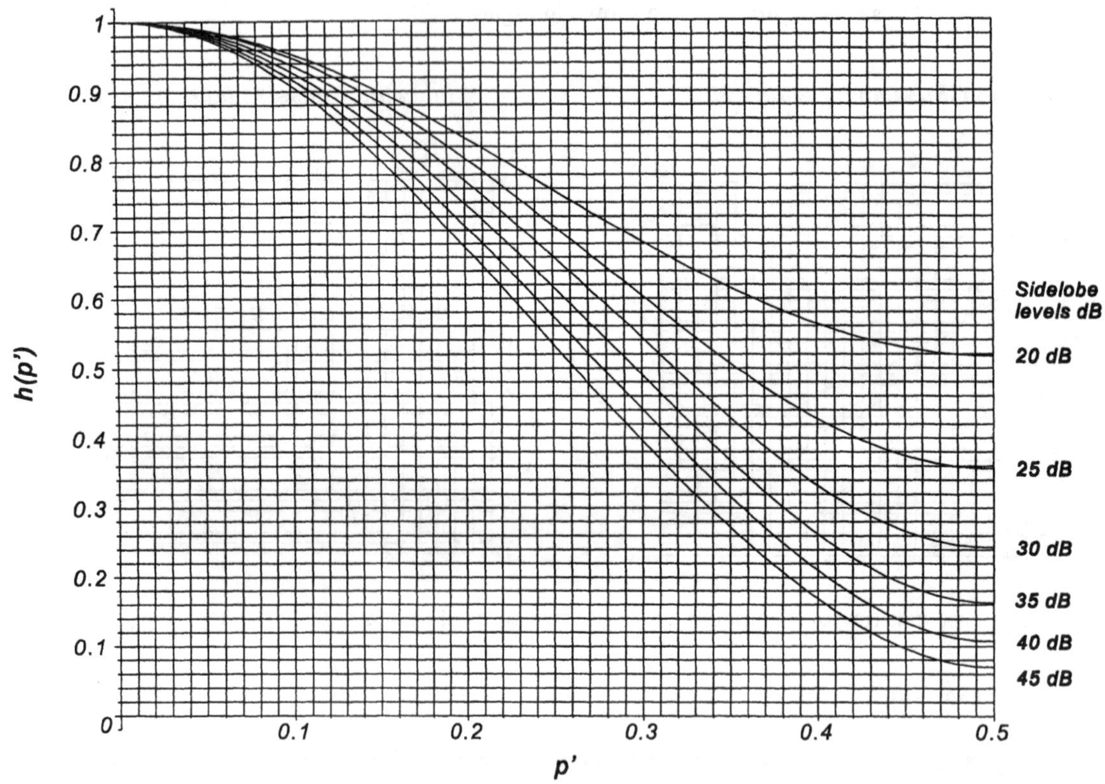

Figure B.41 Taylor tapering functions for sidelobe levels from 20 dB to 45 dB.

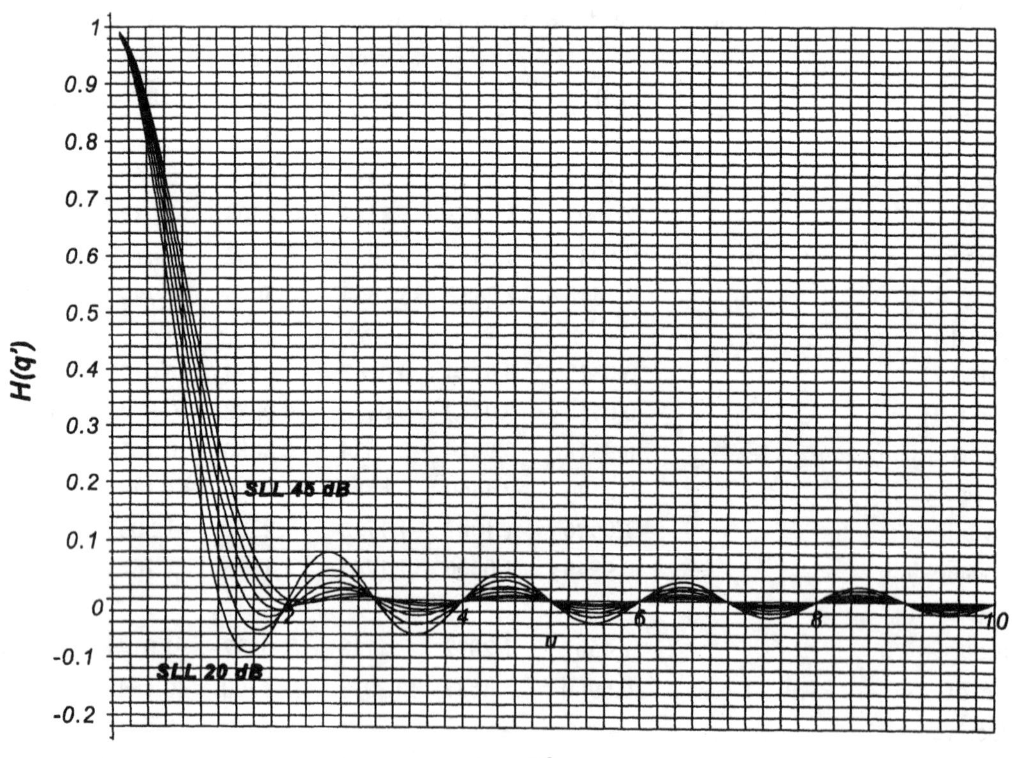

Figure B.42 Fourier transforms for Taylor tapering functions with sidelobe levels between 20 dB and 45 dB.

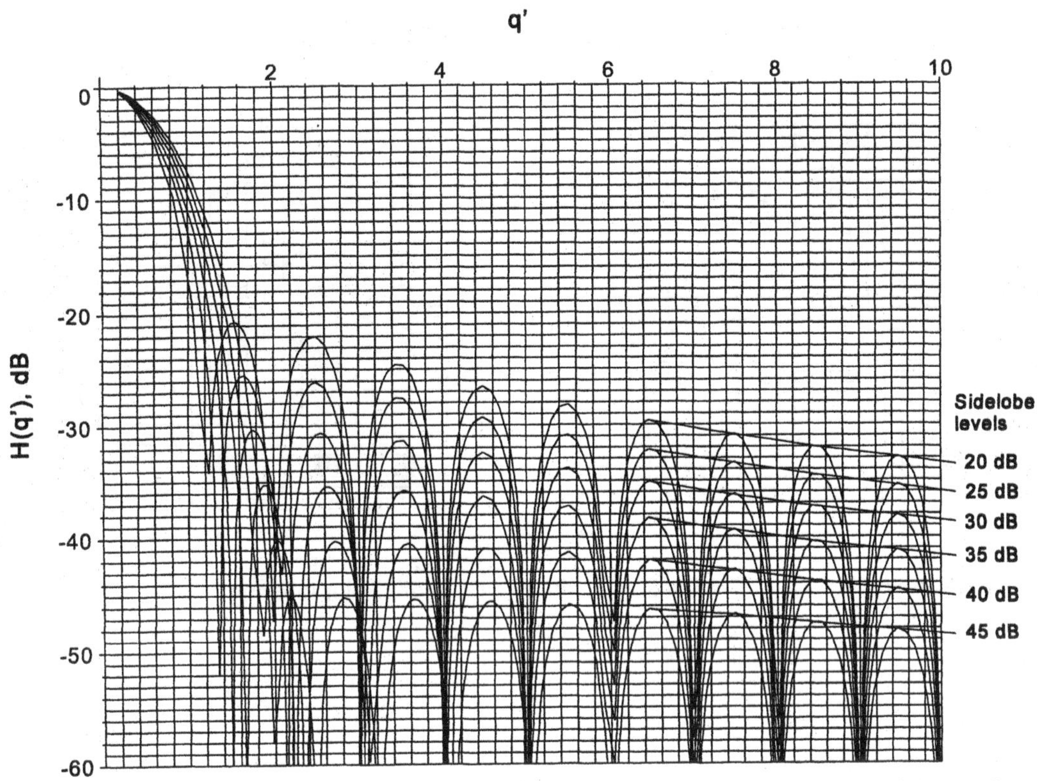

Figure B.43 Fourier transforms in decibels for Taylor tapering functions with sidelobe levels between 20 dB and 45 dB.

Figures B.44 and B.45 show an example of the effects of sampling on spectral leakage or two-tone characteristics when the discrete Fourier transform is used for 100 samples. The minor signal has a frequency corresponding to the 16th bin, and the major signal has its frequency varied between the 10th and 11th bins.

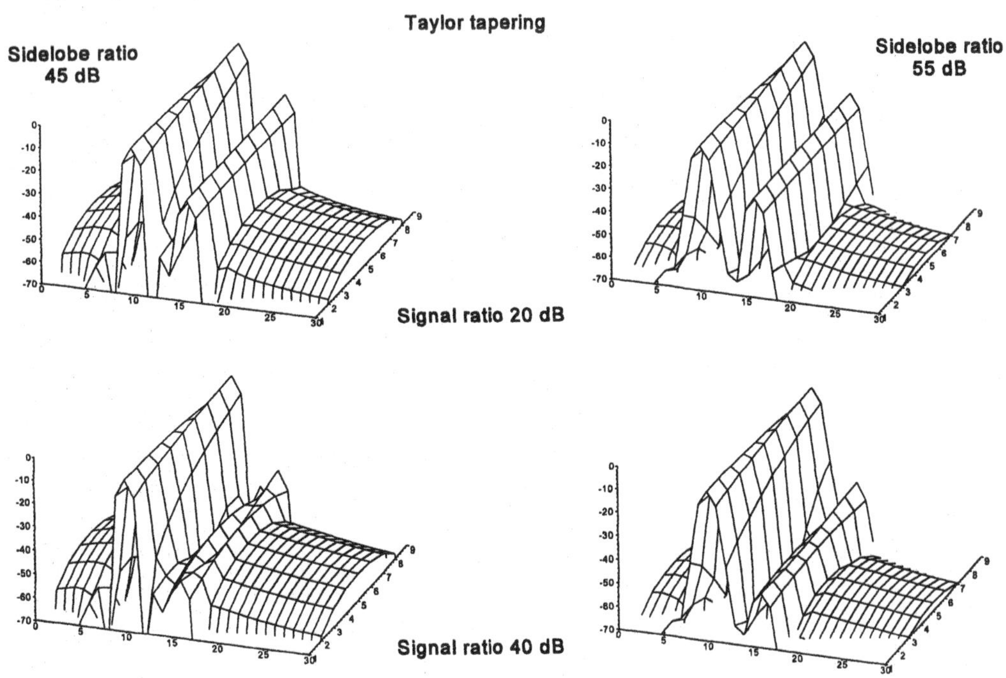

Figure B.44 Examples of the spectral leakage, or two-tone characteristics, for a 100 sample filter as the major signal frequency is varied from bin 10 to bin 11. The minor signal, 20 dB or 40 dB smaller, is in bin 16.

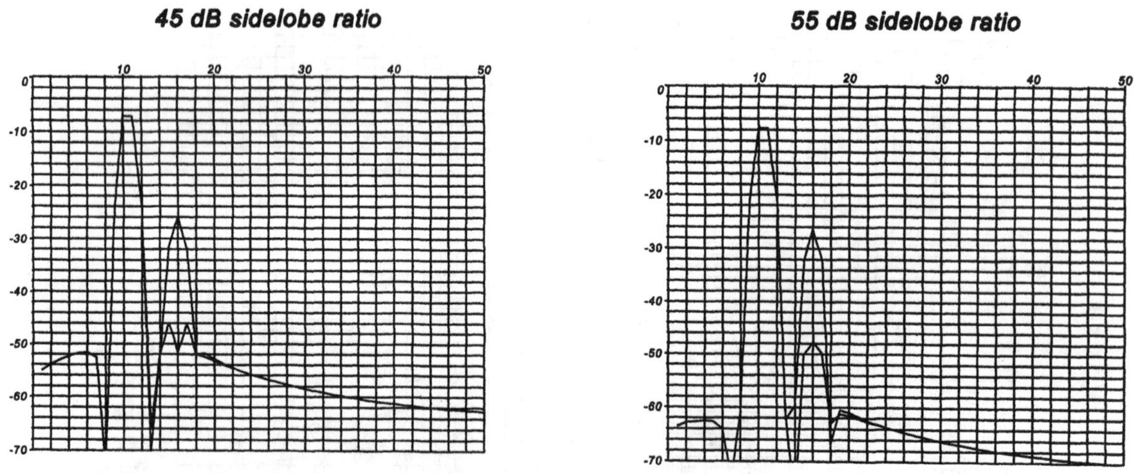

Figure B.45 Worst-case masking of the minor signal with two cases of Taylor tapering.

B.3 TAPERING WITH DISCRETE ELEMENTS

B.3.1 Dolph-Chebyshëv tapering for a discrete 10 element system

The tapering weights are given by [4]

$$weight[K] = \frac{N-1}{N-K} \sum_{S=0}^{K-2} \frac{(K-2)!\,(N-K)!\,\alpha^{S+1}}{S!\,(K-2-S)!\,(S+1)!\,(N-K-S-1)!} \qquad \alpha = \left[\tanh \frac{\operatorname{arccosh}(SLR)}{N-1}\right]^2$$

where SLR is the voltage ratio for the sidelobes;
N is the number of elements.

Examples are plotted in Figure B.46 with their characteristics in Table B.17.

Table B.17
Table of values for Chebyshëv tapering with 10 elements

	Sidelobe level, dB					
	20.0	25.0	30.0	35.0	40.0	45.0
A, effective length, *coherent gain*	0.7871	0.7043	0.6469	0.6046	0.5720	0.5461
A, dB	-2.0796	-3.0452	-3.7826	-4.3706	-4.8526	-5.2548
C, effective power, *incoherent power gain*	0.6438	0.5482	0.4940	0.4577	0.4311	0.4106
C, dB	-1.9122	-2.6108	-3.0627	-3.3939	-3.6538	-3.8654
D (see Section B.1)	—	—	—	—	—	—
G (see Section B.1)	—	—	—	—	—	—
H (see Section B.1)	—	—	—	—	—	—
Efficiency η, *processing gain*	0.9622	0.9048	0.8473	0.7986	0.7588	0.7262
Efficiency η, *processing gain*, dB	-0.1674	-0.4345	-0.7199	-0.9767	-1.1987	-1.3893
Noise beamwidth	1.0393	1.1052	1.1803	1.2522	1.3179	1.3770
Half-power beamwidth	0.9746	1.0594	1.1353	1.2030	1.2636	1.3178
RMS beamwidth	—	—	—	—	—	—
RMS aperture	—	—	—	—	—	—
First sidelobe, dB	19.7269	24.9131	29.9717	34.9907	39.9969	44.9990
Falloff, dB/octave	0	0	0	0	0	0
Scalloping loss	0.6934	0.7356	0.7663	0.7893	0.8071	0.8212
Scalloping loss, dB	-3.1798	-2.6667	-2.3122	-2.0550	-1.8611	-1.7108
Worst-case loss	0.6802	0.6998	0.7053	0.7054	0.7031	0.6998
Worst-case loss, dB	-3.3472	-3.1011	-3.0321	-3.0317	-3.0598	-3.1001

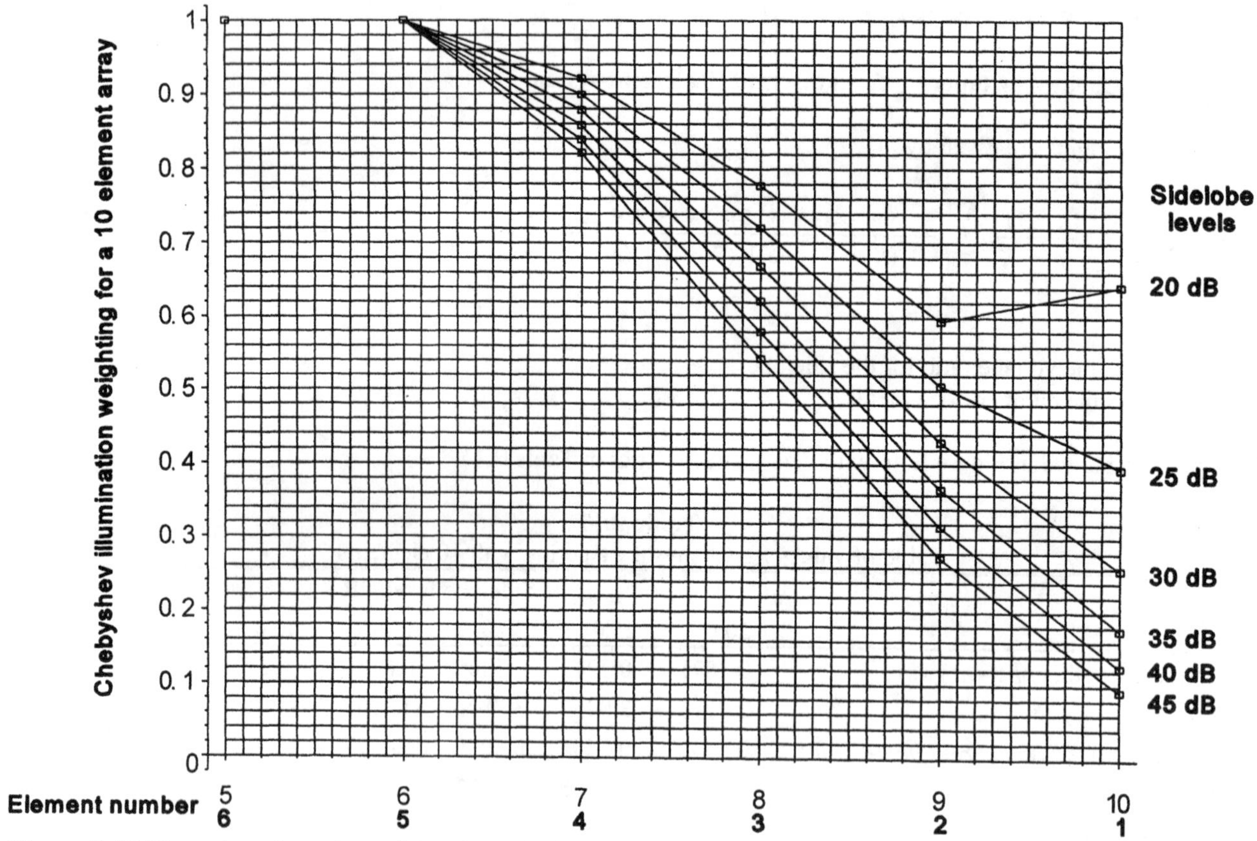

Figure B.46 The points show the weights for a 10 element Dolph-Chebyshëv array.

The closed form for the Fourier transform is given by

$$\begin{aligned} H(q') &= T_{N-1}(x_0 \cos \pi q') \\ &= \cos[(N-1) \arccos(x_0 \cos \pi q')] \end{aligned}$$

The order of the Chebyshëv T_{N-1} function determines the number of visible sidelobes, and x_0 defines the sidelobe level. The Chebyshëv characteristics for a number of examples are shown in Figures B.47 and B.48.

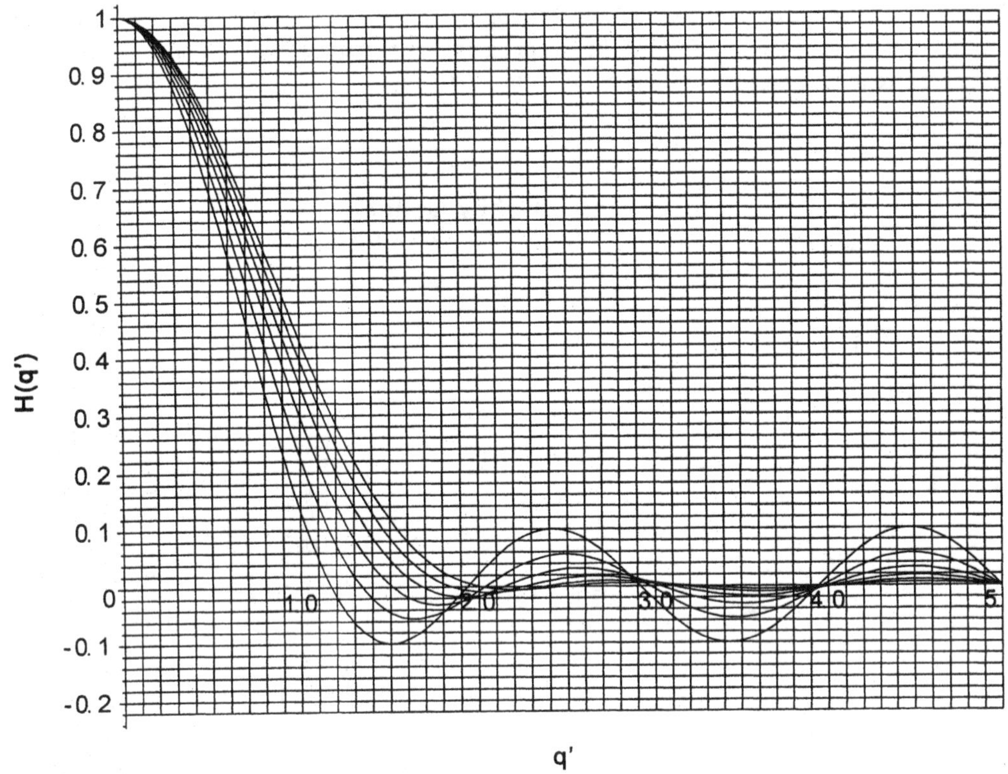

Figure B.47 Fourier transform for a discrete 10 element Dolph-Chebyshëv array.

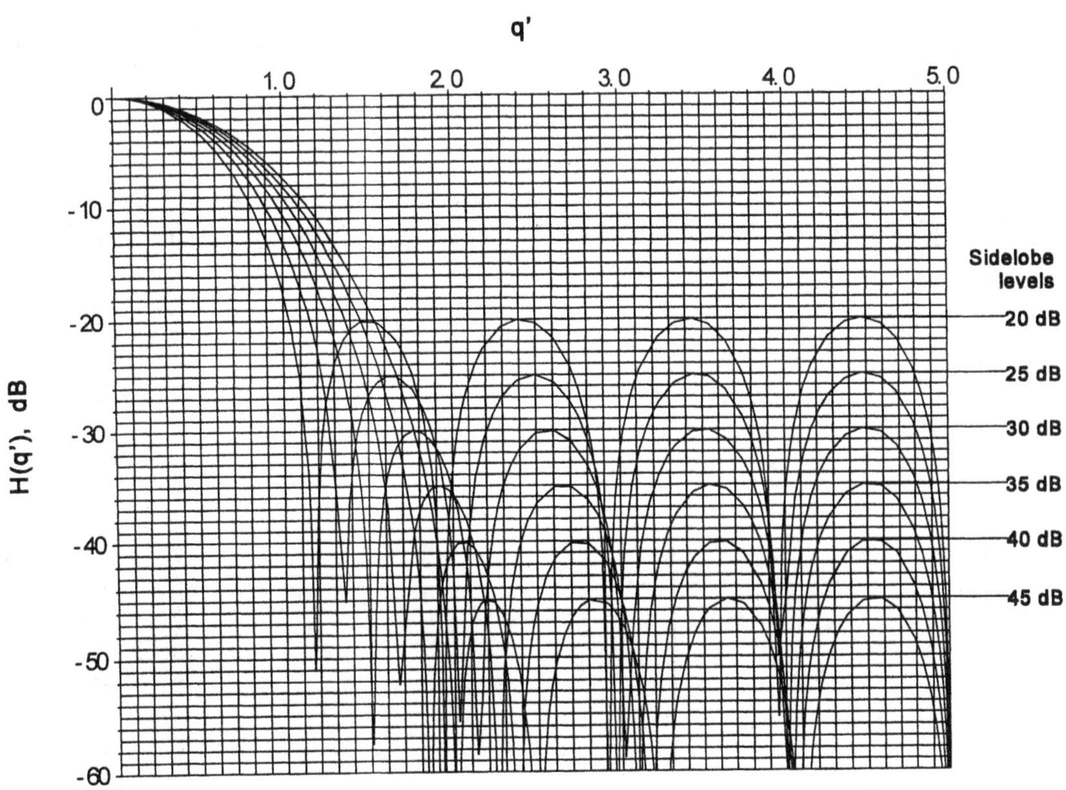

Figure B.48 Fourier transform decibels for a discrete 10 element Dolph-Chebyshëv array.

Figure B.49 shows an example of the effects of sampling on spectral leakage, or two-tone characteristics, when the discrete Fourier transform is used for 100 samples. The minor signal has a frequency corresponding to the 16th bin, and the major signal has its frequency varied between the 10th and 11th bins.

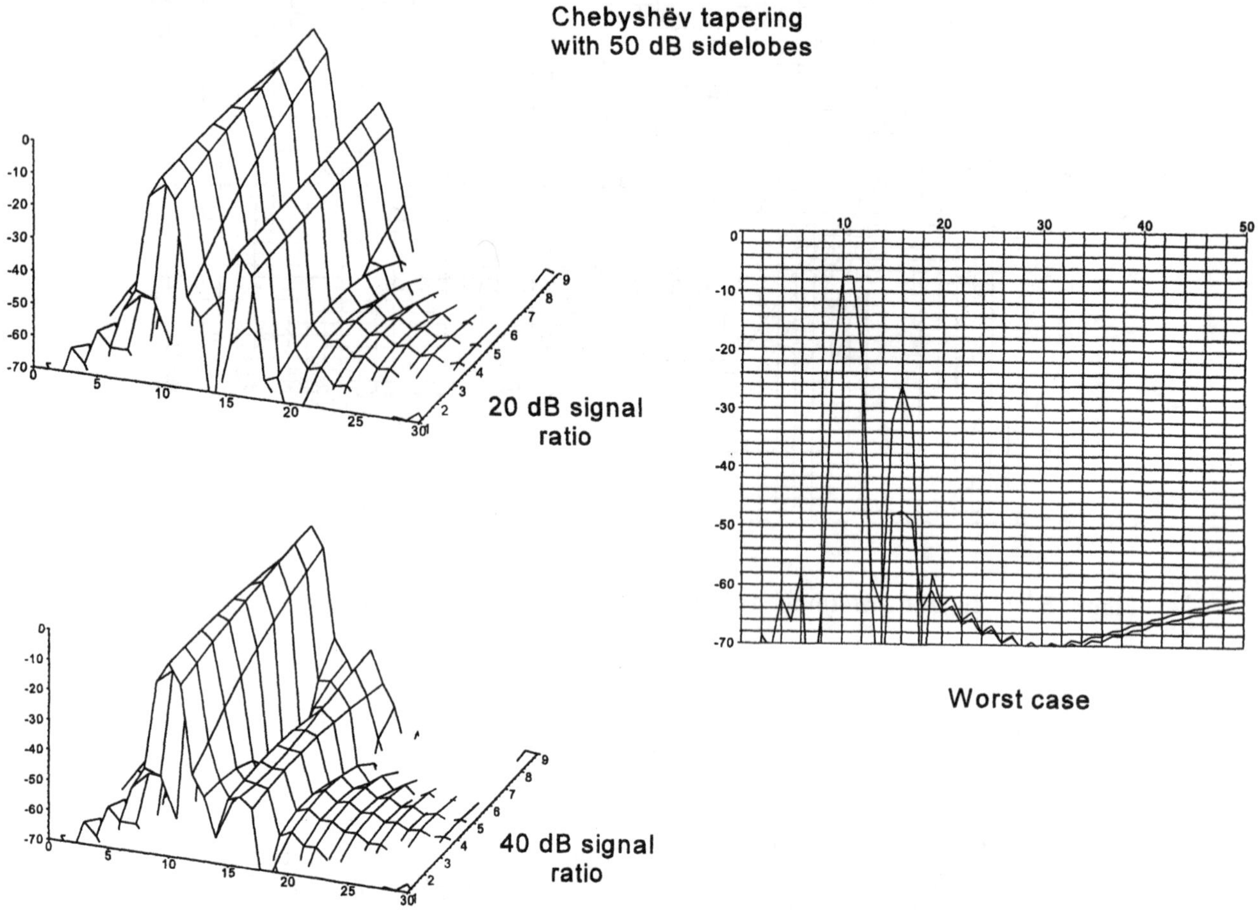

Figure B.49 Examples of the spectral leakage or two-tone characteristics for a 100 sample filter as the major signal frequency is varied from bin 10 to bin 11. The minor signal, 20 dB or 40 dB smaller, is in bin 16.

B.4 TAPERING OR ILLUMINATION FUNCTIONS FOR CIRCULAR ANTENNAS

B.4.1 Circular $(1 - 4r'^2)^n$ tapering

The tapering function is $(1 - 4r'^2)^n$, the characteristics of some examples are shown in Table B.18 and plotted in Figure B.50.

Table B.18

Table of values for circular $(1 - 4r'^2)^n$ tapering

	Power of n					
	0 Uniform	1 Parabolic	2	3	4	5
A, Effective aperture area	1.0000	0.5000	0.3333	0.2500	0.2000	0.1667
A, dB	0.0000	-6.0206	-9.5424	-12.0412	-13.9794	-15.5630
C, Effective radiated power	1.0000	0.3333	0.2000	0.1429	0.1111	0.0909
C, dB	0.0000	-4.7712	-6.9897	-8.4510	-9.5424	-10.4139
Aperture efficiency, η	1.0000	0.7500	0.5556	0.4375	0.3600	0.3056
Aperture efficiency η, dB	0.0000	-1.2494	-2.5527	-3.5902	-4.4370	-5.1491
Noise beamwidth	1.0000	1.3333	1.8000	2.2857	2.7778	3.2727
Half-power beamwidth	1.0290	1.2697	1.4727	1.6515	1.8130	1.9614
First sidelobe, dB	-17.5711	-24.6392	-30.6104	-35.9607	-40.9098	-45.5742

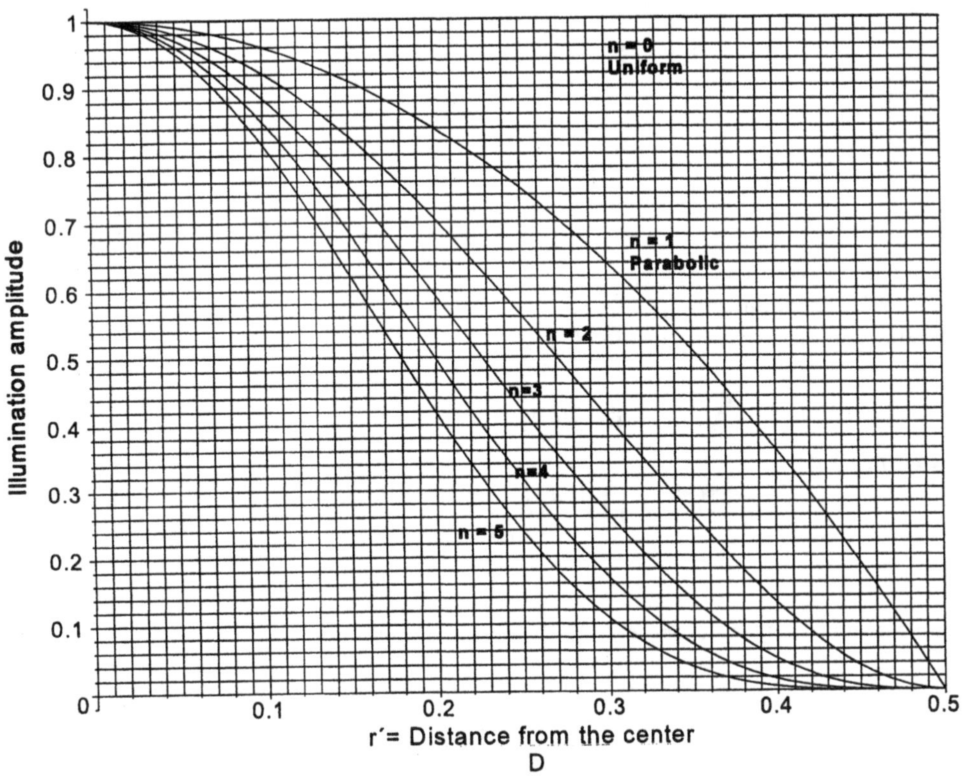

Figure B.50 The aperture functions for $(1–4r'^2)^n$ circular tapering.

Table B.19 gives the following closed expressions (MAPLE) in terms of the generalized hypergeometric function (or Barnes's extended hypergeometric function) [5].

Table B.19
Fourier transforms for $(1 - 4r'^2)^n$ tapering

Tapering function	Fourier transform
$(1-(2r')^2)^0$	$\frac{\pi}{4}\,\text{hypergeom}\left([],[2],\left(\frac{\pi q'}{2}\right)^2\right)$
$(1-(2r')^2)^1$	$2\pi\left(\frac{1}{8}\,\text{hypergeom}\left([],[2],\left(\frac{\pi q'}{2}\right)^2\right)-\frac{1}{16}\,\text{hypergeom}\left([2],[1,3],\left(\frac{\pi q'}{2}\right)^2\right)\right)$
$(1-(2r')^2)^2$	$\frac{\pi}{12}\,\text{hypergeom}\left([],[4],\left(\frac{\pi q'}{2}\right)^2\right)$
$(1-(2r')^2)^3$	$\frac{\pi}{16}\,\text{hypergeom}\left([],[5],\left(\frac{\pi q'}{2}\right)^2\right)$
$(1-(2r')^2)^4$	$\frac{\pi}{20}\,\text{hypergeom}\left([],[6],\left(\frac{\pi q'}{2}\right)^2\right)$
$(1-(2r')^2)^5$	$\frac{\pi}{24}\,\text{hypergeom}\left([],[7],\left(\frac{\pi q'}{2}\right)^2\right)$

where $n[i]$ is a list of j numerator coefficients;
$d[i]$ is a list of m denominator coefficients;
the generalized hypergeometric function [5] is

$$\text{hypergeom}(n,\ d,\ z) = \sum_{k=0}^{\infty} \frac{z^k \prod_{i=1}^{j} \frac{\Gamma(n[i] + k)}{\Gamma(n[i])}}{k! \prod_{i=1}^{m} \frac{\Gamma(d[i] + k)}{\Gamma(d[i])}}.$$

The Fourier transforms for the examples in Table B.18 are shown in Figures B.51 and B.52.

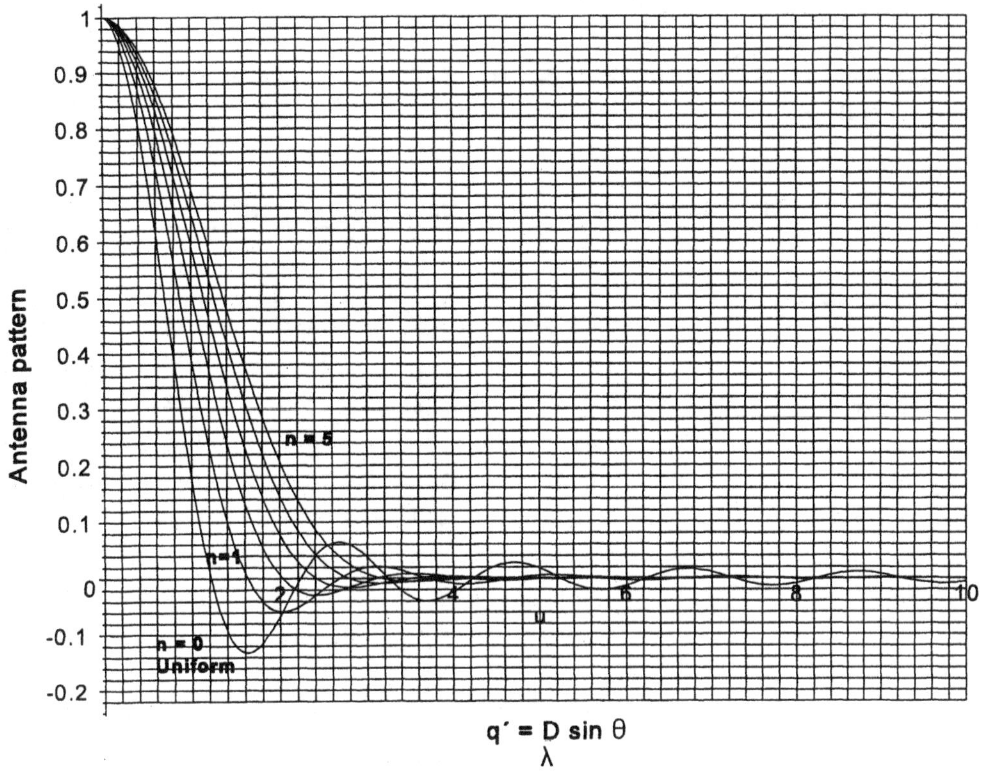

Figure B.51 Fourier transforms for $(1 - 4r'^2)^n$ circular illumination.

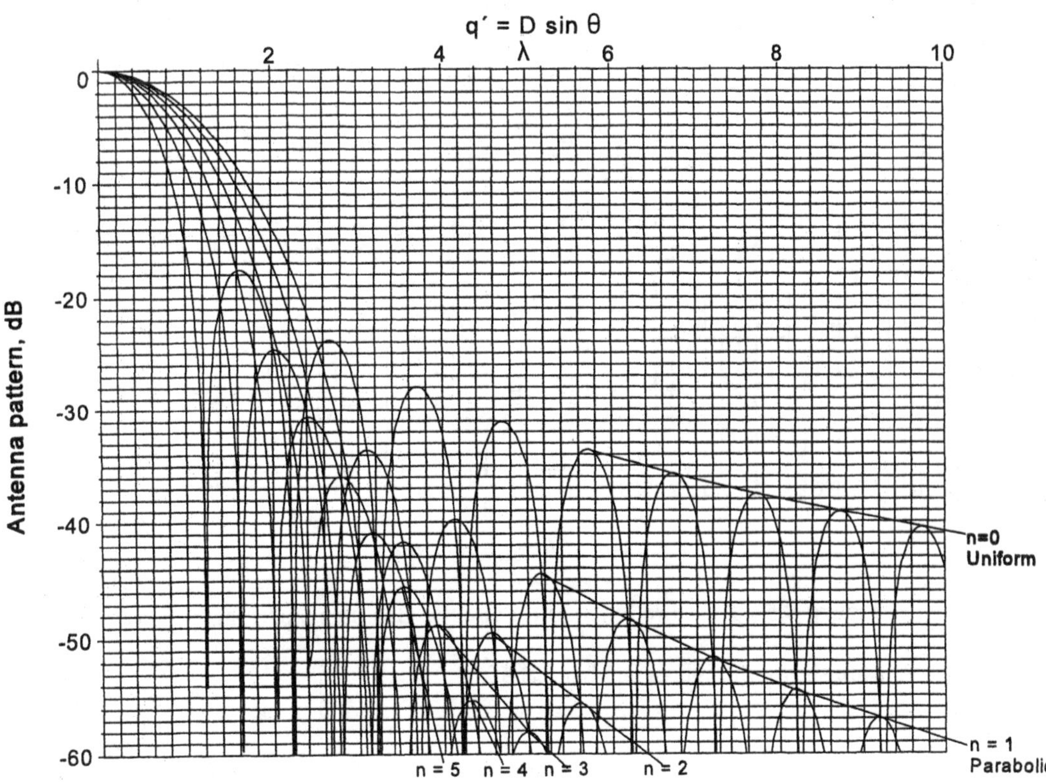

Figure B.52 Fourier transforms for $(1 - 4r'^2)^n$ circular tapering in decibels.

B.4.2 Circular truncated Gaussian tapering

The aperture function is $\exp(-1.386\, n\, r'^2)$ and the characteristics of a number of examples are shown in Table B.20. They are plotted in Figure B.53.

Table B.20
Table of values for circular truncated Gaussian tapering

	Values of *n*				
	1	1.7	2.4	2.8	3.2
A, Effective aperture area	0.8455	0.6325	0.4338	0.3446	0.2744
A, dB	-1.4573	-3.9785	-7.254	-9.2541	-11.2312
C, Effective radiated power	0.722	0.4328	0.2466	0.1838	0.1412
C, dB	-1.4144	-3.6373	-6.0809	-7.3573	-8.5015
Aperture efficiency, η	0.9902	0.9244	0.7633	0.6461	0.5334
Aperture efficiency, η, dB	-0.0429	-0.3412	-1.1731	-1.8969	-2.7298
Noise beamwidth	1.0099	1.0817	1.3101	1.5477	1.8749
Half-power beamwidth	1.0618	1.1313	1.2544	1.3541	1.4762
First sidelobe, dB	-19.2015	-23.1945	-33.9243	-54.8388	—

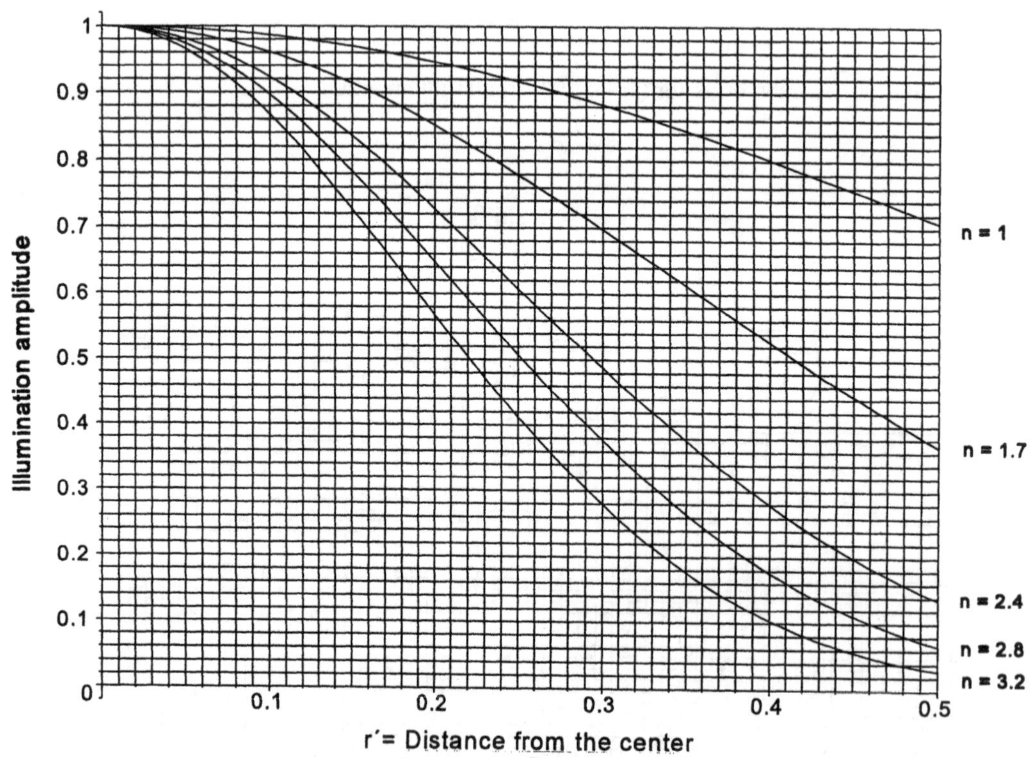

Figure B.53 Aperture function for circular Gaussian tapering.

The Fourier transforms are shown in Figures B.54 and B.55.

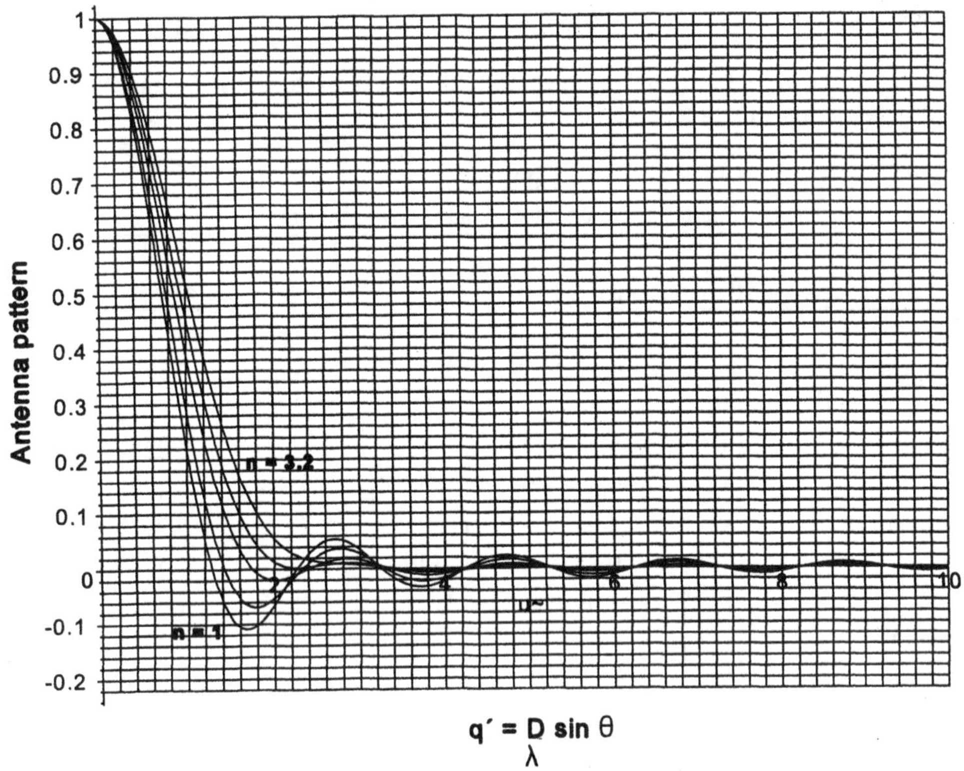

Figure B.54 Antenna pattern with circular Gaussian illumination.

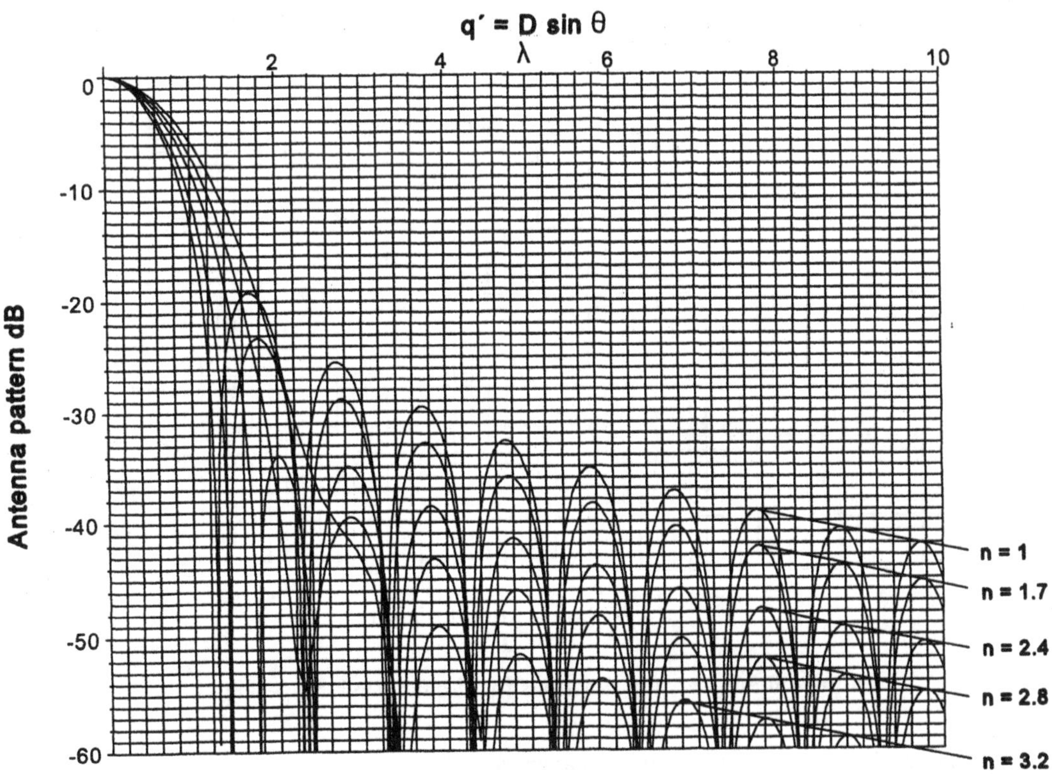

Figure B.55 Antenna pattern with circular Gaussian tapering, decibels.

B.4.3 Circular Taylor tapering

For the aperture and antenna pattern functions, see Section 5.2.2. The characteristics of some examples are shown in Table B.21, and the aperture functions are plotted in Figure B.56.

Table B.21
Table of values for circular Taylor tapering

	Sidelobe levels, dB					
	20	25	30	35	40	45
$\bar{n}$	2	3	4	5	6	8
A, Effective aperture area	0.6836	0.5619	0.5070	0.4429	0.3993	0.3633
A, dB	-3.3037	-5.0065	-5.8990	-7.0733	-7.9733	-8.7954
C, Effective radiated power	0.4836	0.3450	0.3031	0.2522	0.2240	0.2007
C, dB	-3.1554	-4.6213	-5.1841	-5.9826	-6.4972	-6.9740
Aperture efficiency, η	0.9664	0.9151	0.8482	0.7779	0.7119	0.6574
η, dB	-0.1483	-0.3852	-0.7149	-1.0907	-1.4760	-1.8214
Noise beamwidth	1.0347	1.0927	1.1789	1.2855	1.4048	1.5210
Half-power beamwidth	1.0900	1.1336	1.1848	1.2393	1.2945	1.3445
First sidelobe, dB	-21.9482	-26.0945	-30.7209	-35.5273	-41.8994	-45.2881

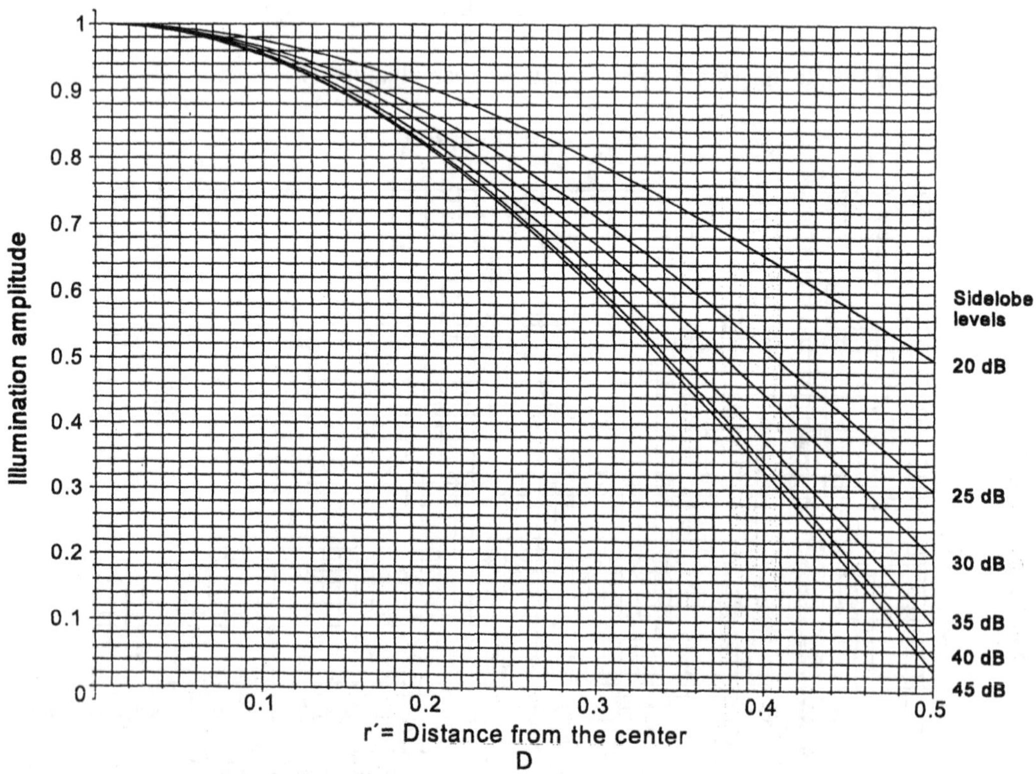

Figure B.56 Aperture functions for circular Taylor tapering.

Figures B.57 and B.58 show the circular Taylor characteristics.

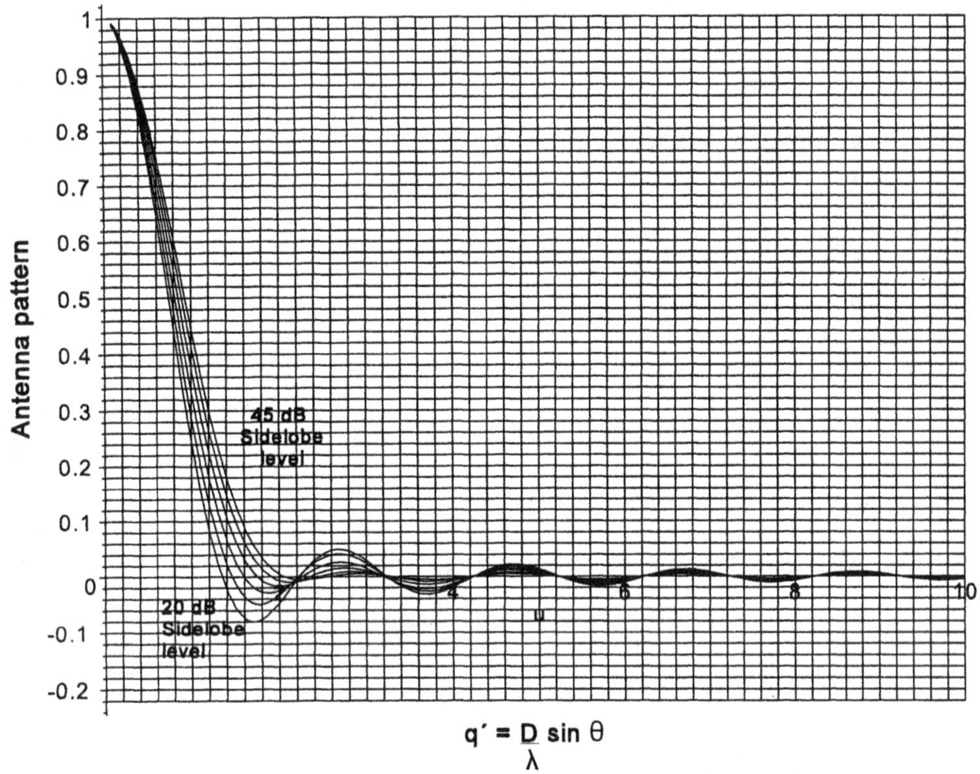

Figure B.57 Antenna patterns with circular Taylor illumination.

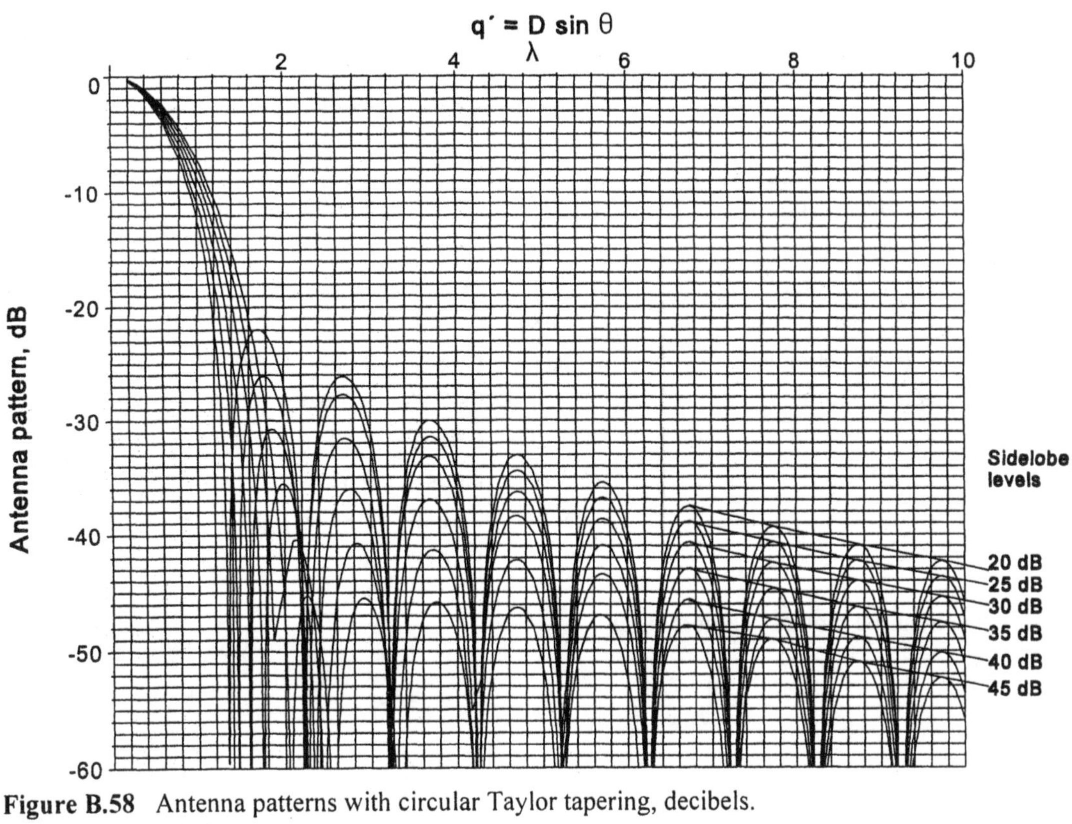

Figure B.58 Antenna patterns with circular Taylor tapering, decibels.

B.5 ODD TAPERING FUNCTIONS

Odd tapering functions are used to reduce the sidelobes in radar for monopulse antennas. This section gives a number of examples:

- Odd rectangular tapering, Section B.5.1;
- Odd triangular tapering, Section B.5.2;
- Odd cosine to the power *n* tapering, Section B.5.3;
- Odd truncated Rayleigh tapering, Section B.5.4;
- Odd Taylor derivative tapering, Section B.5.5;
- Bayliss tapering, Section B.5.6;
  - Linear Bayliss tapering, Section B.5.6.1;
  - Circular Bayliss tapering, Section B.5.6.2;
- Zolotarëv tapering, Section B.7.

B.5.1 Odd rectangular tapering

The tapering function is illustrated in Figure B.59, and its characteristics given in Table B.22.

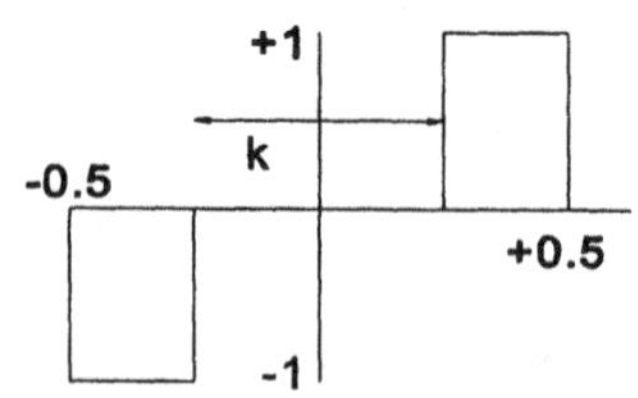

Figure B.59 Rectangular tapering.

Table B.22
Table of values for odd rectangular tapering

	Value of *k*				
	1.0000	0.8000	0.6000	0.4000	0.2000
C, effective power	1.0000	0.8000	0.6000	0.4000	0.2000
C, dB	0.0000	-0.9691	-2.2185	-3.9794	-6.9897
D (see Section B.1)	—	—	—	—	—
F (see Section B.1)	1.5708	1.5080	1.3195	1.0053	0.5655
G (see Section B.1)	0.0833	0.0827	0.0780	0.0653	0.0407
H (see Section B.1)	0.0833	0.0827	0.0780	0.0653	0.0407
K_r	0.8660	0.9295	0.9391	0.8764	0.6971
K_r, dB	-1.2494	-0.6349	-0.5453	-1.1464	-3.1336
G_r, dB	-2.7979	-2.3880	-2.8448	-4.1992	-7.0337
First sidelobe, dB	-13.4648	-18.1648	-8.5903	-6.0409	-7.3886

The aperture functions are plotted in Figure B.60.

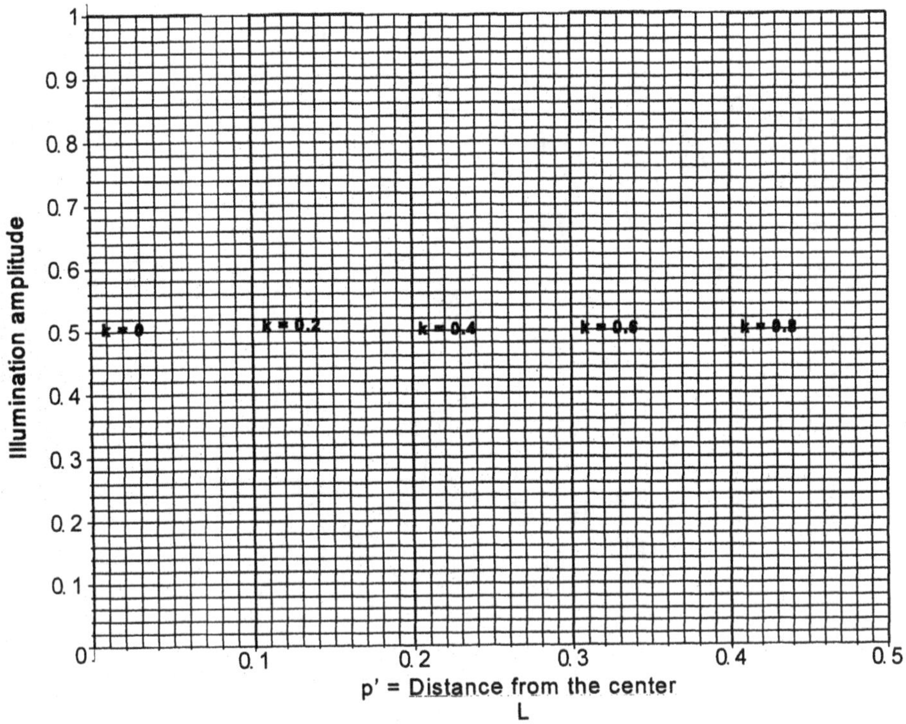

Figure B.60 Aperture function for odd rectangular tapering.

There is no closed form for the Fourier transform. The calculated Fourier transforms are given for the examples in Figures B.61 and B.62.

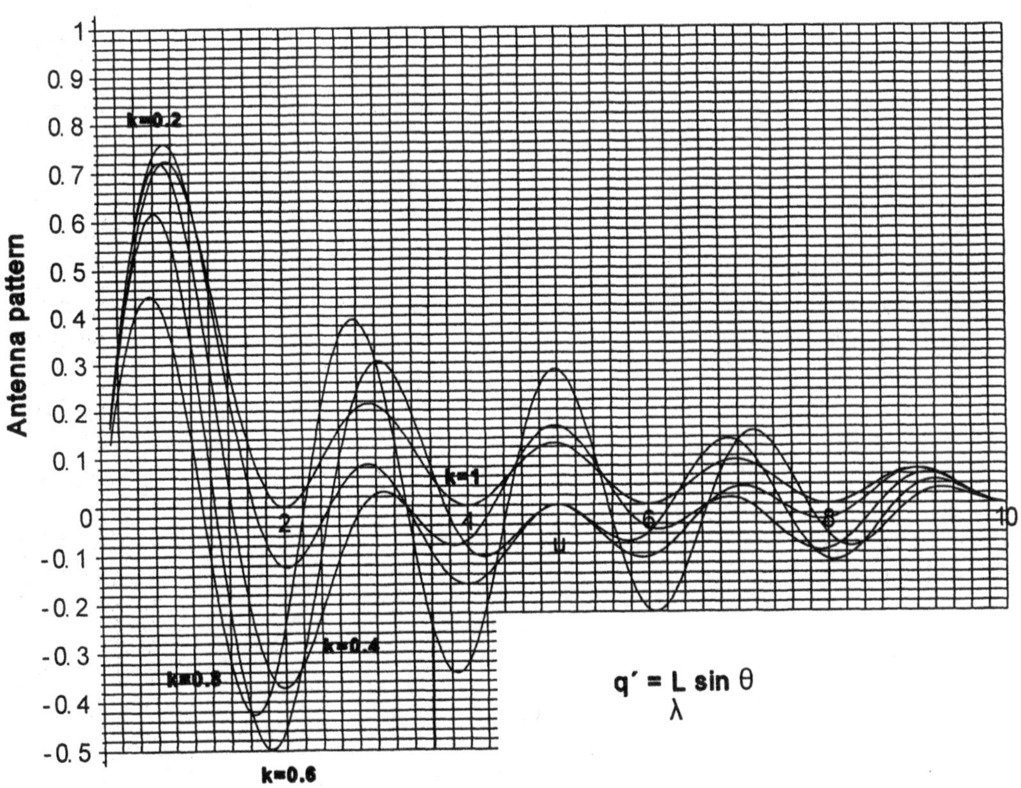

Figure B.61 Antenna diagram with odd rectangular illumination.

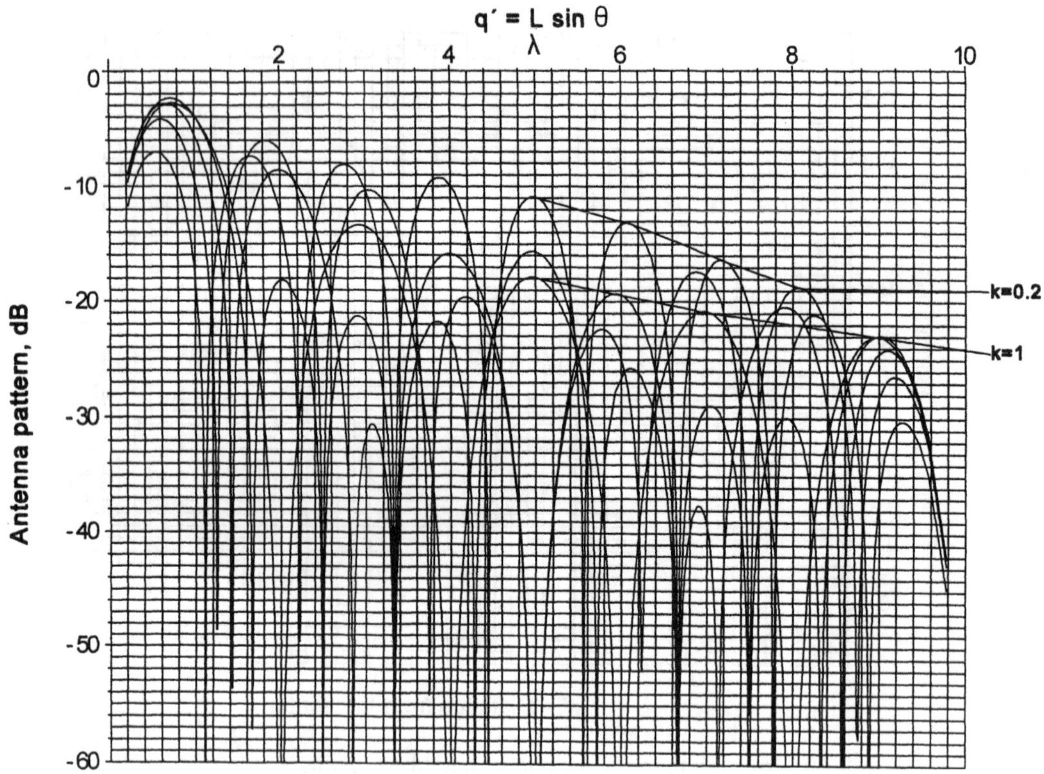

Figure B.62 Antenna diagram with odd rectangular tapering, decibels.

B.5.2 Odd triangular tapering

The aperture function is shown in Figure B.63, the characteristics given in Table B.23, and plotted in Figure B.64.

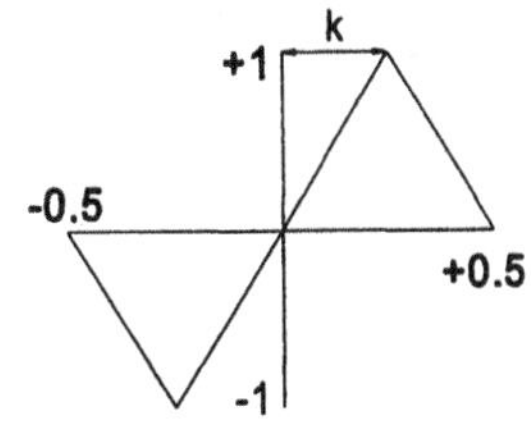

Figure B.63 Triangular tapering.

Table B.23
Table of values for odd triangular tapering

	Value of k					
	0.0000	0.1000	0.2000	0.3000	0.4000	0.5000 Linear
C, effective power	0.3333	0.3333	0.3333	0.3333	0.3333	0.3333
C, dB	-4.7712	-4.7712	-4.7712	-4.7712	-4.7712	-4.7712
D (see Section B.1)	4.0000	25.0000	16.6667	16.6667	25.0000	4.0000
F (see Section B.1)	0.5236	0.6283	0.7330	0.8378	0.9425	1.0472
G (see Section B.1)	0.0083	0.0127	0.0190	0.0273	0.0377	0.0500
H (see Section B.1)	0.0208	0.0258	0.0325	0.0408	0.0508	0.0625
Slope, K_r	0.5000	0.6000	0.7000	0.8000	0.9000	1.0000
K_r, dB	-6.0206	-4.4370	-3.0980	-1.9382	-0.9151	0.0000
G_r, dB	-5.1718	-3.8040	-3.0332	-2.6441	-2.4773	-2.4134
First sidelobe, dB	—	—	-24.8157	-14.0659	-11.3251	-10.7180

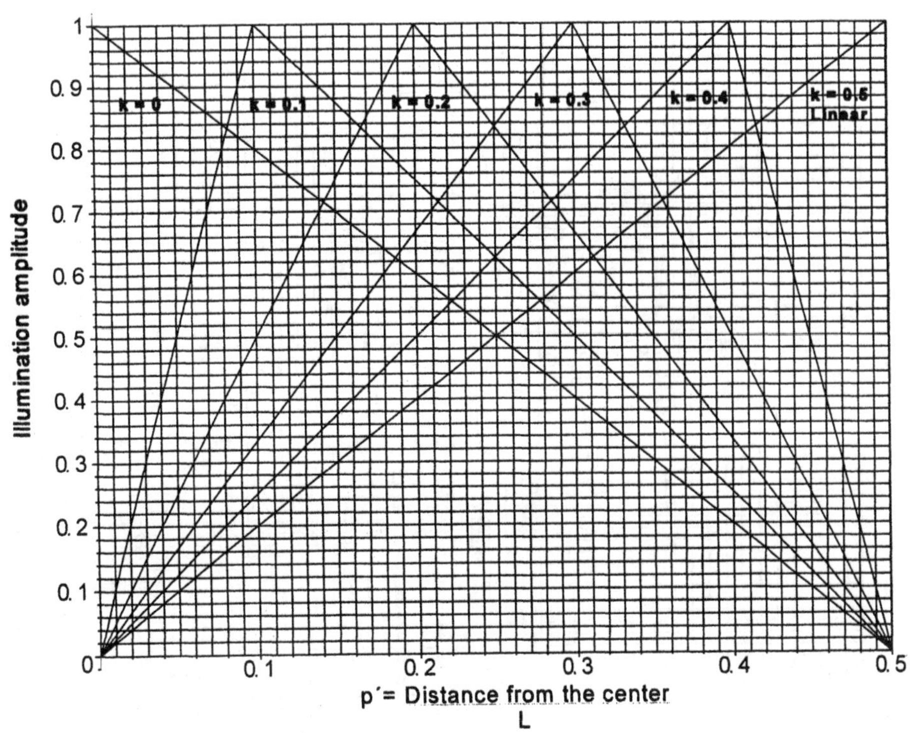

Figure B.64 Aperture functions for odd triangular tapering.

There is no closed form for the Fourier transform, and the calculated Fourier transforms for the examples are plotted in Figures B.65 and B.66.

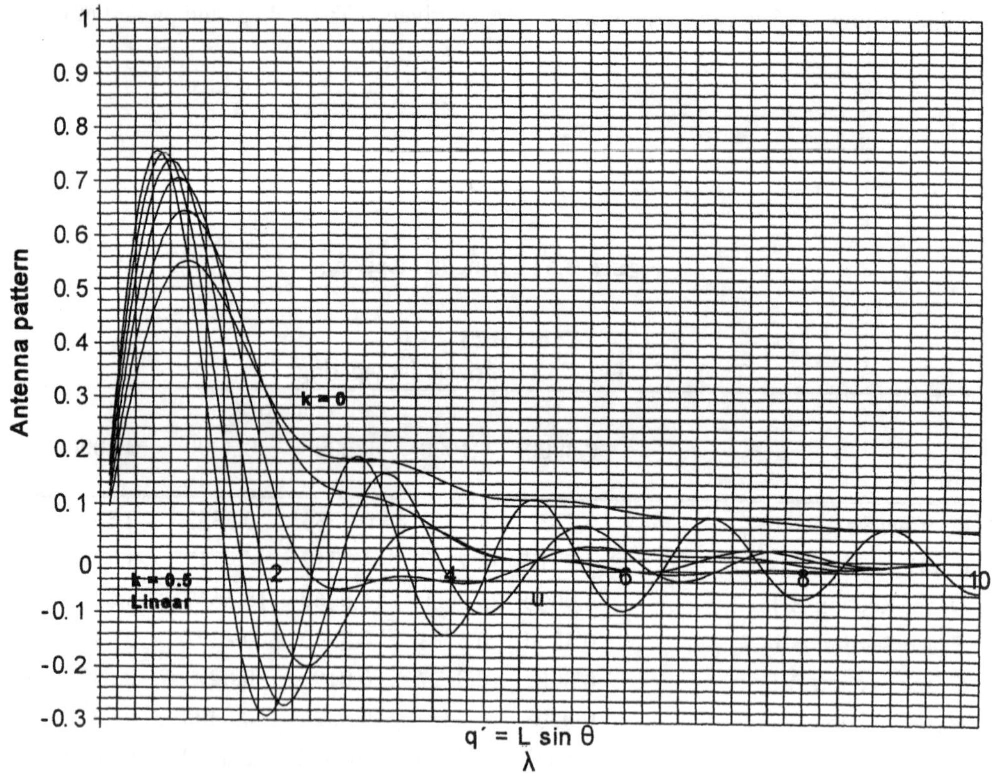

Figure B.65 Antenna diagrams with odd triangular illumination.

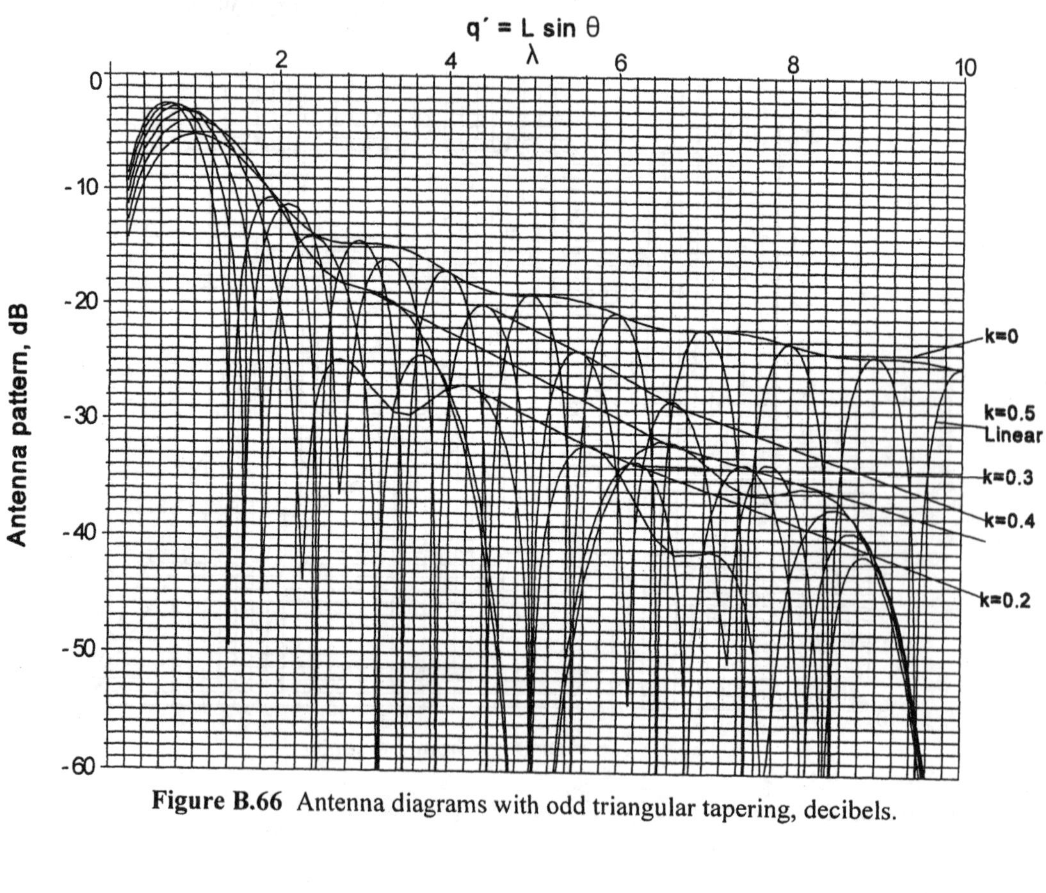

Figure B.66 Antenna diagrams with odd triangular tapering, decibels.

B.5.3 Odd cosine to the power *n* tapering

The tapering function is $p' \cos^n \pi p'$ and the characteristics of integer powers are shown in Table B.24 and plotted in Figure B.67.

Table B.24

Table of values for odd cosine to the power *n* tapering

	Power of *n*				
	0	1	2	3	4
	Linear	$x \cos \pi x$	$x \cos^2 \pi x$	$x \cos^3 \pi x$	$x \cos^4 \pi x$
C, effective power	0.3333	0.5121	0.4364	0.3818	0.3428
C, dB	-4.7712	-2.9062	-3.6016	-4.1813	-4.6500
D (see Section B.1)	1.0000	0.6612	0.3487	0.2673	0.2254
F (see Section B.1)	1.0472	1.0606	0.7828	0.6107	0.4985
G (see Section B.1)	0.0500	0.0403	0.0229	0.0149	0.0107
H (see Section B.1)	0.0625	0.0518	0.0330	0.0229	0.0170
Slope, K_r	1.0000	0.8171	0.6533	0.5449	0.4694
K_r, dB	0.0000	-1.7543	-3.6973	-5.2744	-6.5692
G_r, dB	-2.4356	-2.5176	-3.2	-3.7784	-4.2514
First sidelobe, dB	-10.7190	-18.3342	-26.0588	-33.3838	-40.4151

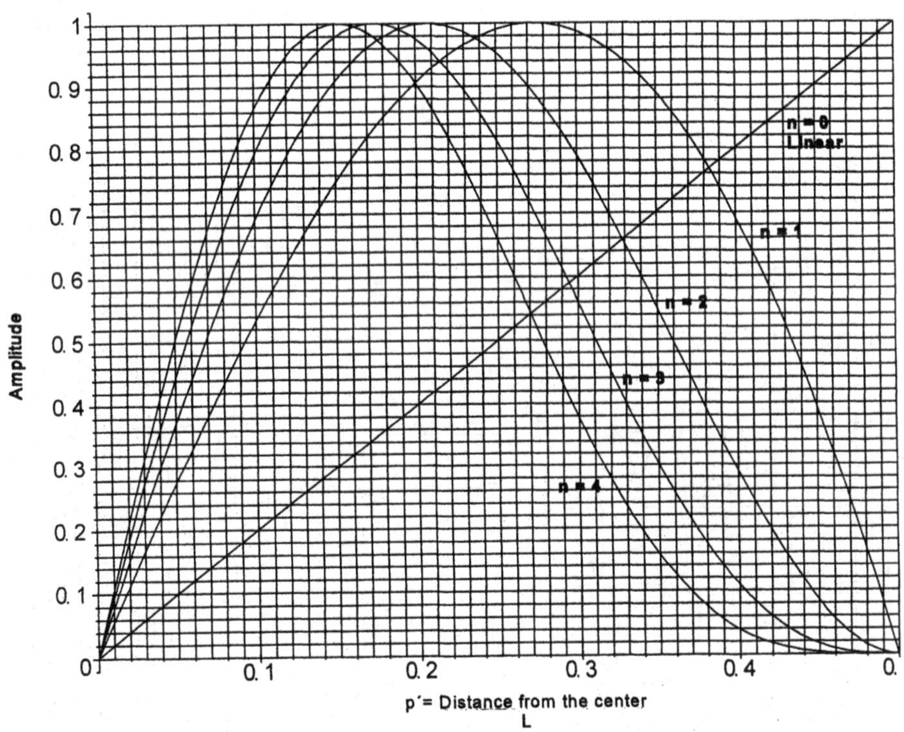

Figure B.67 Aperture function for odd cosine to the power *n* tapering.

The closed forms for the Fourier transforms are given in Table B.25 and plotted in Figures B.68 and B.69.

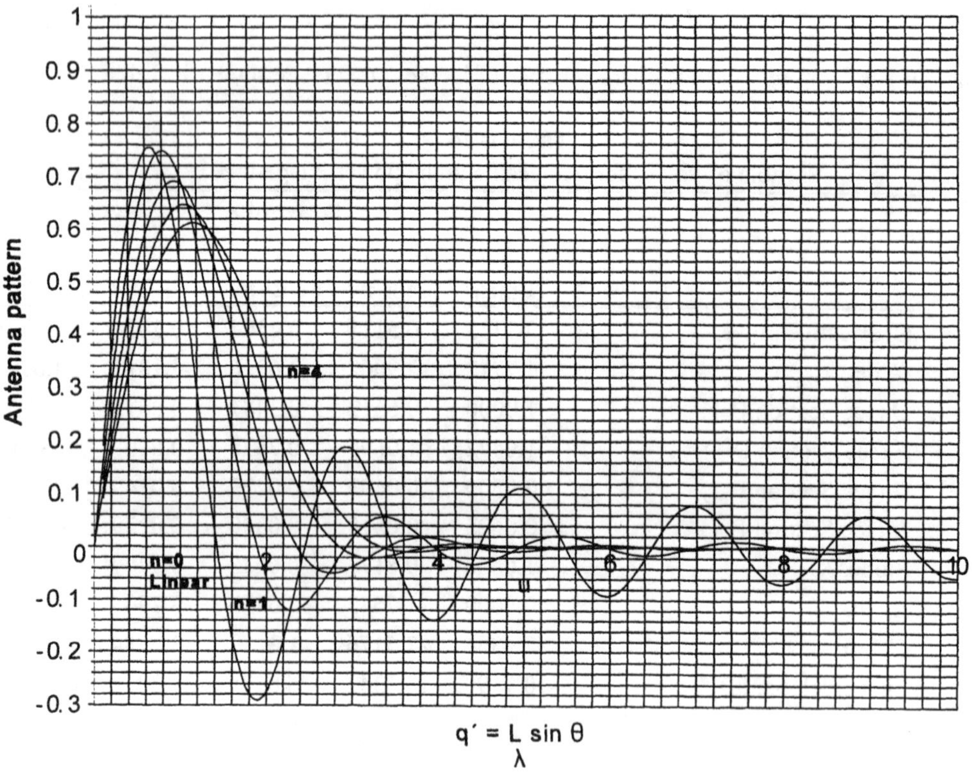

Figure B.68 Antenna pattern with odd cosine to the power n illumination.

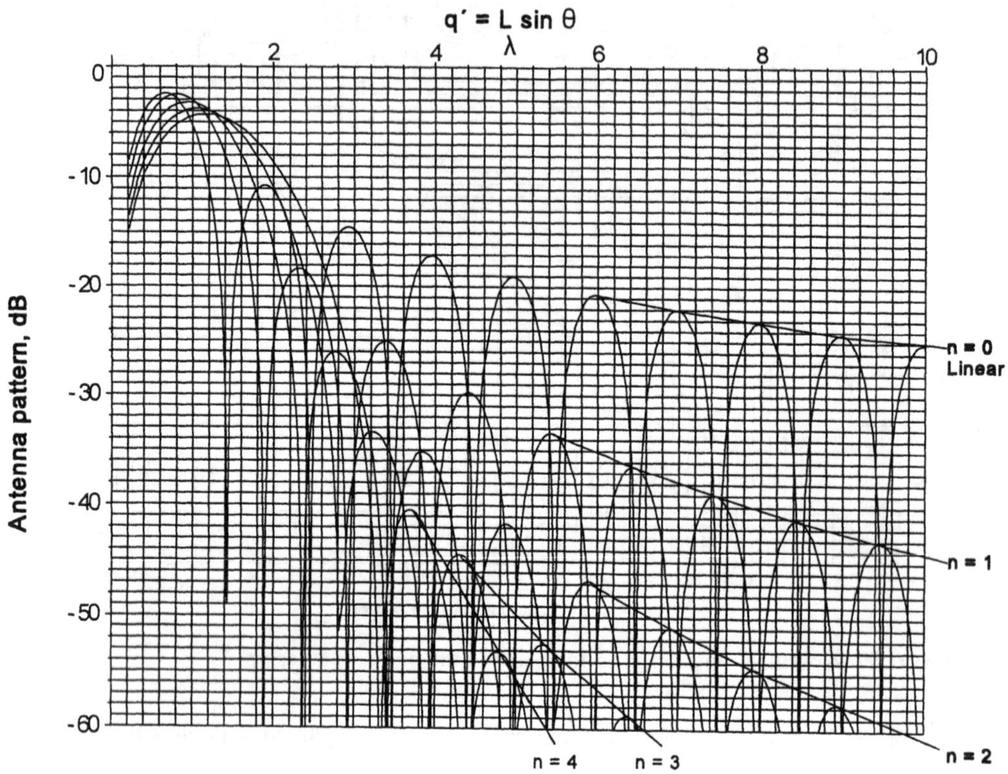

Figure B.69 Antenna pattern with odd cosine to the power n tapering, decibels.

Some closed forms for the Fourier transform for cosine to the power n tapering are given in Table B.25.

Table B.25
Closed forms for the Fourier transform for cosine to the power n tapering

Tapering function	Antenna pattern
$p'\cos^0 \pi p'$	$-\frac{\sqrt{3}(q'\cos(q')-\sin(q'))}{q'^2}$
$p'\cos^1 \pi p'$	$-\frac{2\sqrt{6}\pi^2((4q'^2-\pi^2)\sin q'+8q'\cos(q'))}{\sqrt{\pi^2-6}(4q'^2+\pi^2)^2}$
$p'\cos^2 \pi p'$	$-\frac{\pi^3(-q'^3\cos(q')+3q'^2\sin(q')+\pi^2 u\cos(q')-\pi^2\sin(q'))}{q'^2\sqrt{2\pi^2-15}((q')^4-2\pi^2 q'^2+\pi^4)}$

For higher powers, the closed form is too large.

B.5.4 Odd truncated Rayleigh tapering

The aperture function is $2.74\, n p' \exp(-1.386(np')^2)$ and the characteristics of some examples are given in Table B.26 and plotted in Figure B.70.

Table B.26
Table of values for odd Rayleigh tapering

	Value of n					
	1.2000	2.1750	2.5200	2.8800	3.3400	3.8700
C, effective power	0.5134	0.6070	0.5559	0.4980	0.4329	0.3741
C, dB	-2.8951	-2.1681	-2.5497	-3.0279	-3.6366	-4.2702
D (see Section B.1)	5.0737	11.0478	13.9299	16.7482	19.9128	23.2138
F (see Section B.1)	1.2883	1.2864	1.1498	0.9902	0.7977	0.6171
G (see Section B.1)	0.0680	0.0567	0.0433	0.0315	0.0209	0.0135
H (see Section B.1)	0.0743	0.0678	0.0577	0.0468	0.0345	0.0239
Slope, K_r	0.9913	0.9103	0.8502	0.7736	0.6684	0.5563
K_r, dB	-0.0762	-0.8162	-1.4097	-2.2295	-3.4986	-5.0942
G_r, dB	-2.2320	-2.2393	-2.4334	-2.7530	-3.2714	-3.8979
First sidelobe, dB	-12.4841	-19.4700	-25.8033	-49.9206	—	—

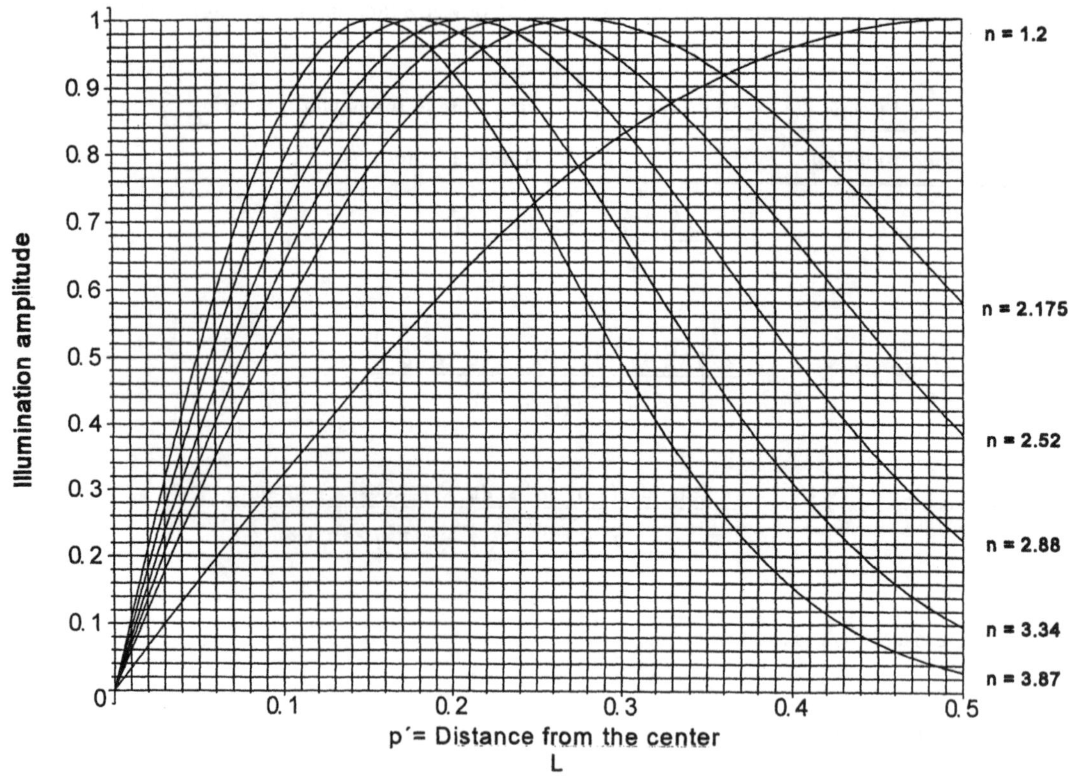

Figure B.70 Aperture function for odd truncated Rayleigh tapering.

There is no simple closed form for the Fourier transform, and calculated Fourier transforms are shown in Figures B.71 and B.72.

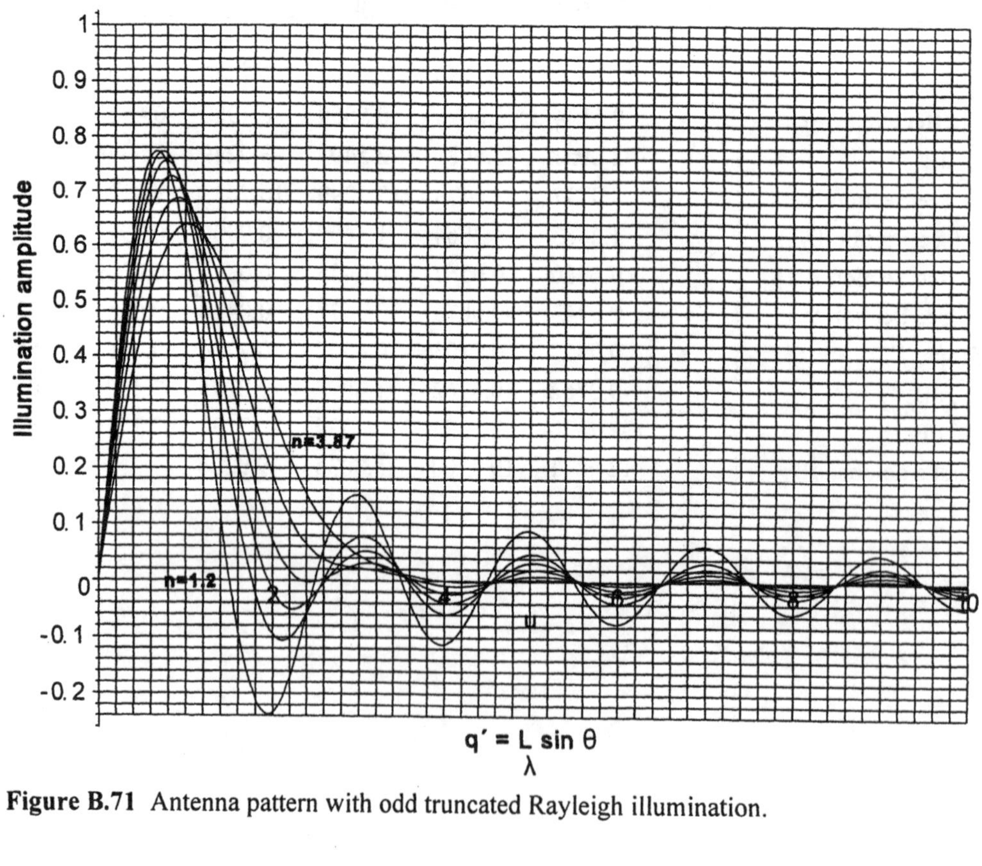

Figure B.71 Antenna pattern with odd truncated Rayleigh illumination.

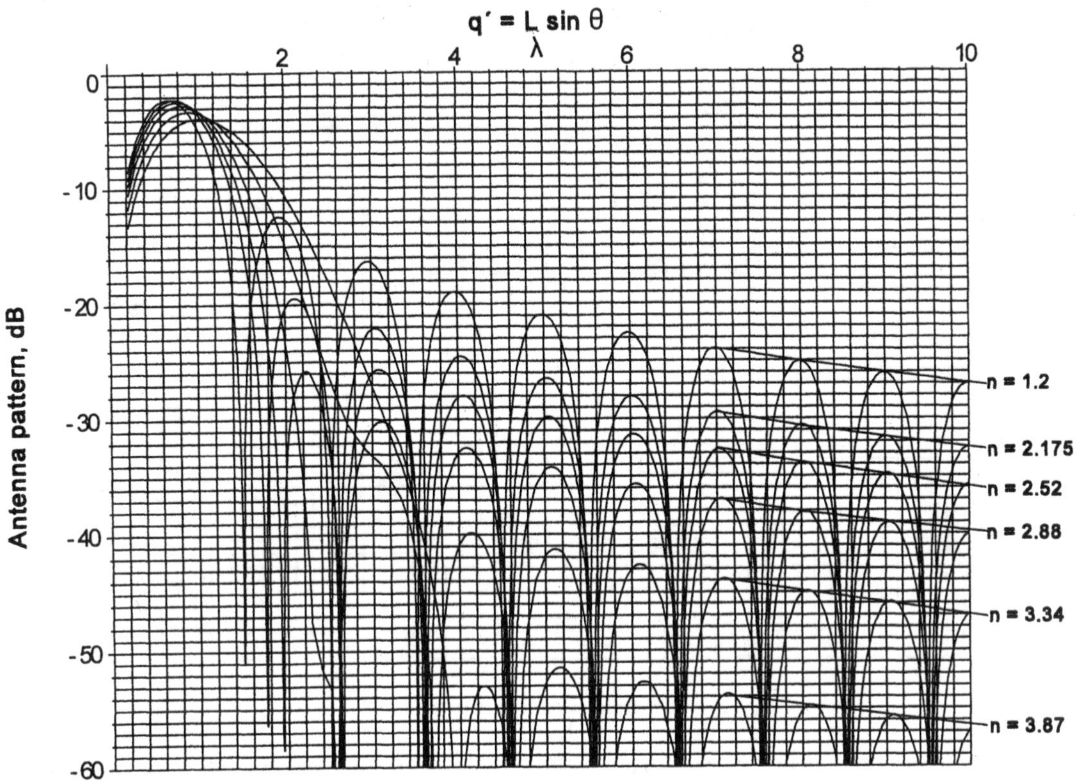

Figure B.72 Antenna pattern with odd truncated Rayleigh tapering, decibels.

B.5.5 Odd Taylor derivative tapering

The aperture function is given by

$$h(p') = p'\left(1 + 2\sum_{n=1}^{\bar{n}-1} F_n(A, \bar{n}) \cos(2\pi n p')\right)$$

where F_n is defined in (5.37). Characteristics of some examples are shown in Table B.27 and plotted in Figure B.73.

Table B.27
Table of values for odd Taylor derivative tapering

	Sidelobe level, dB					
	20.0	25.0	30.0	35.0	40.0	45.0
C, effective power	0.4814	0.6795	0.6078	0.5559	0.5207	0.4930
C, dB	-3.1748	-1.6778	-2.1623	-2.5504	-2.8342	-3.0716
D (see Section B.1)	—	—	—	—	—	—
F (see Section B.1)	1.2388	1.4275	1.2922	1.1726	1.0730	0.9870
G (see Section B.1)	0.0610	0.0743	0.0571	0.0453	0.0375	0.0318
H (see Section B.1)	0.0706	0.0784	0.0684	0.0597	0.0526	0.0467
Slope, K_r	0.9843	0.9547	0.9138	0.8671	0.8198	0.7750
K_r, dB	-0.1373	-0.4026	-0.7826	-1.2387	-1.7256	-2.2138
G_r, dB	-2.2081	-2.1791	-2.2473	-2.3818	-2.5500	-2.7289
First sidelobe, dB	-14.1560	-17.1622	-20.6054	-24.3524	-28.2916	-32.3483

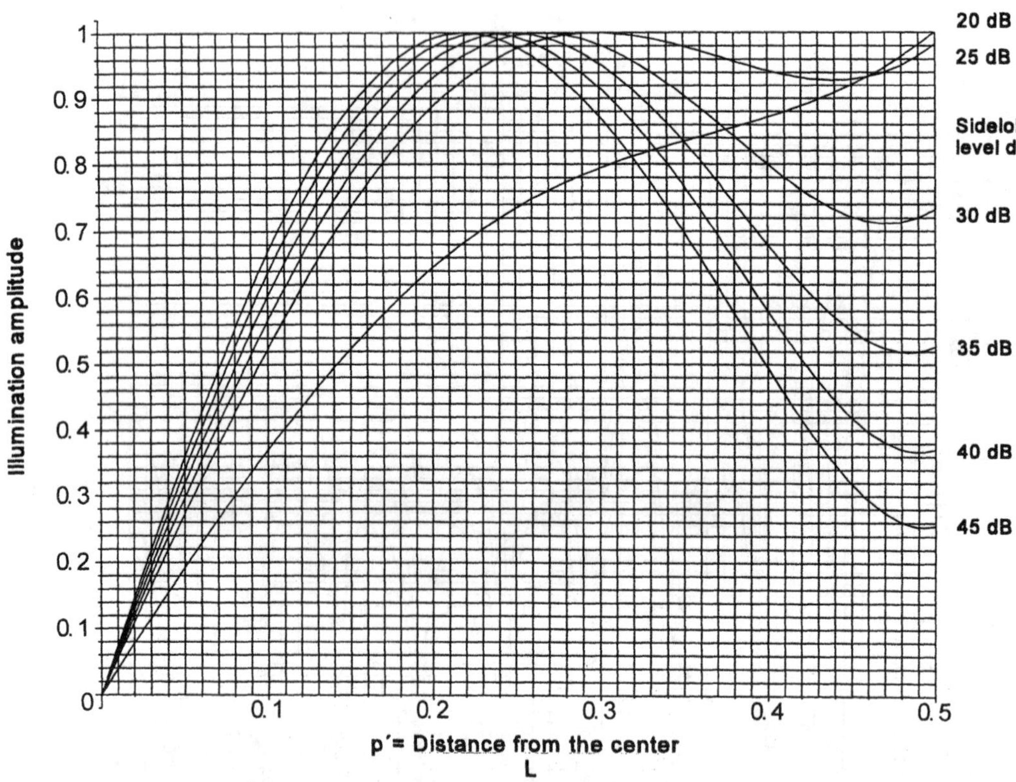

Figure B.73 Aperture functions for odd Taylor derivative antenna diagrams with different design sidelobe levels.

There are no simple closed term functions for the Fourier transforms shown in Figures B.74 and B.75.

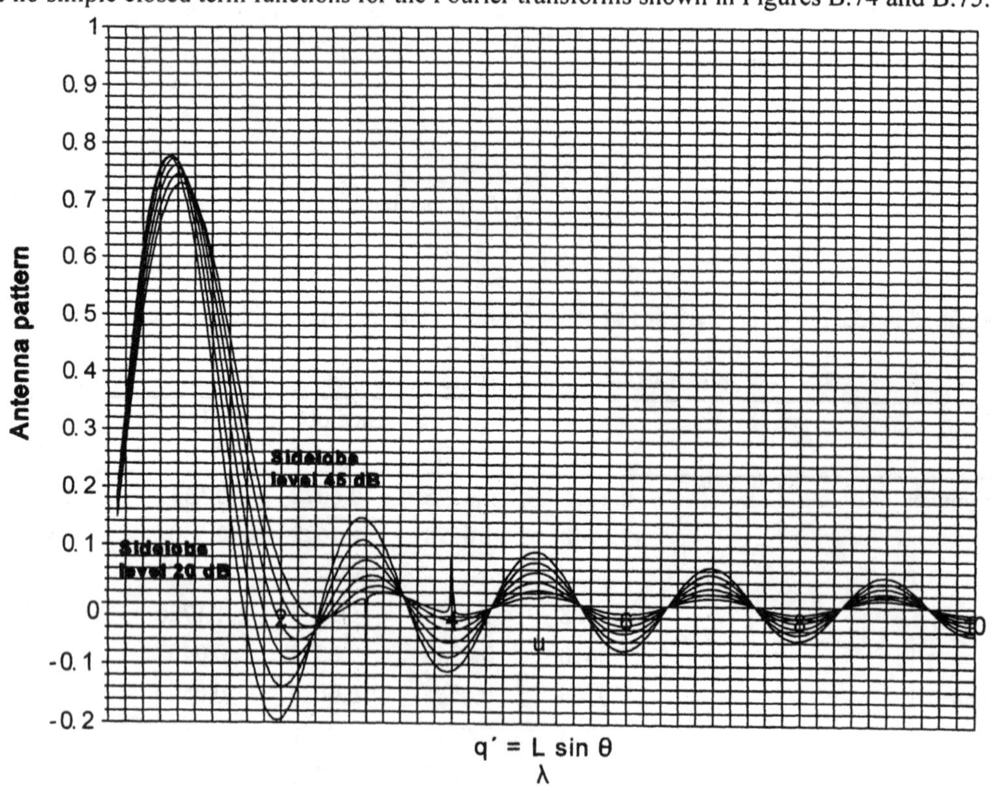

Figure B.74 Odd Taylor derivative antenna diagrams with different design sidelobe levels.

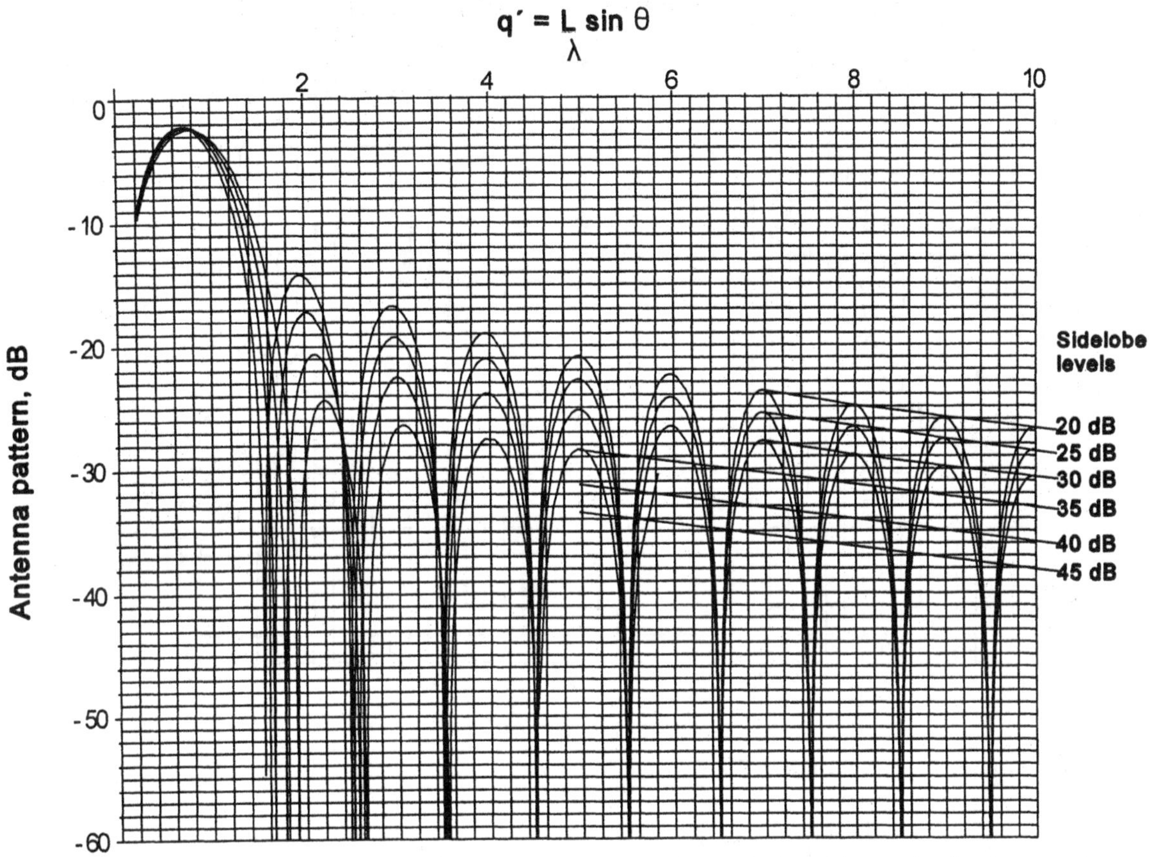

Figure B.75 Odd Taylor derivative antenna diagrams with different design sidelobe levels.

B.5.6 Bayliss tapering

The methods of calculating the Bayliss distribution are given in [6, 7], and Section 5A.3 in Appendix 5A. In his paper Bayliss started with the circular distribution, for consistency the next section describes the linear distribution first.

B.5.6.1 Linear Bayliss tapering

The diagrams that follow show examples of the Bayliss antenna pattern shapes and unfortunately further analysis is beyond the capability of a personal computer in reasonable time. In all the examples with sidelobe levels from –20 dB to –45 dB $\bar{n} = 5$.

The Bayliss tapering function is defined from the antenna pattern, and examples are shown in Figures B.77 and B.78. The aperture function is derived from the inverse Fourier transforms in Figure B.76, keeping the order of aperture functions followed by the Fourier transforms in the rest of this appendix..

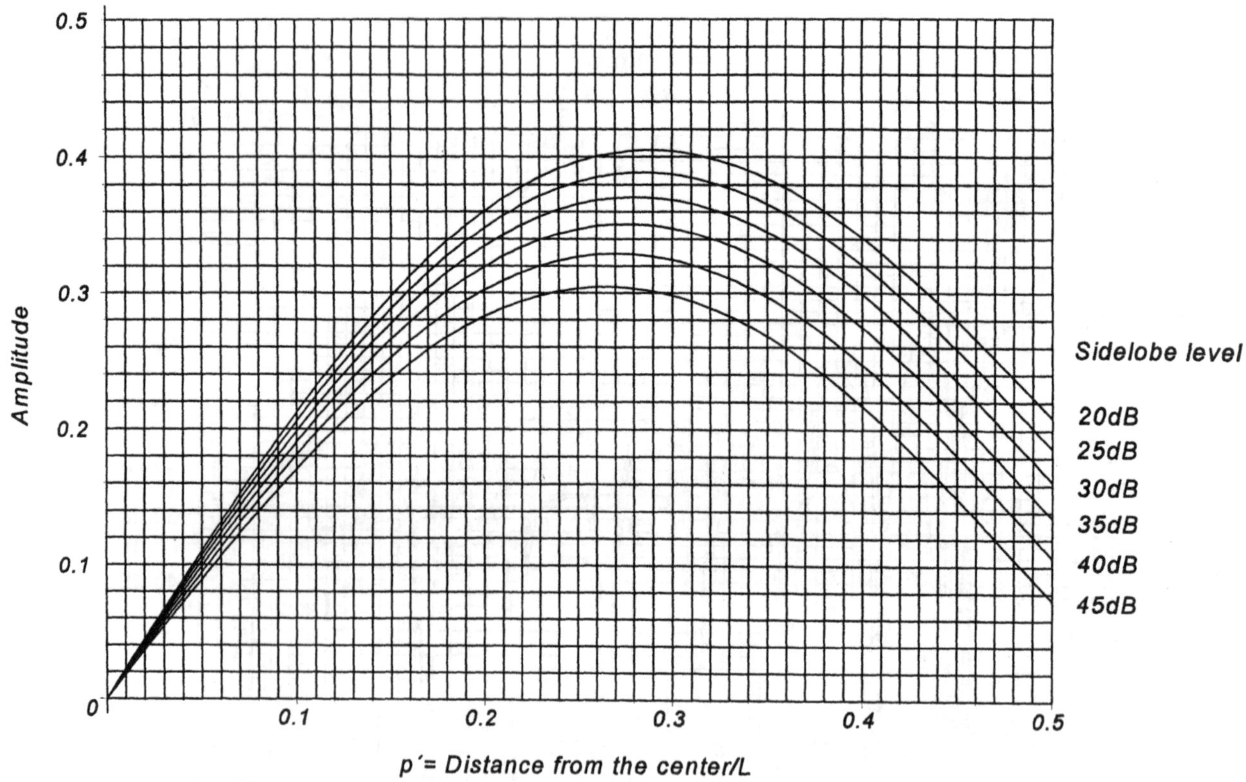

Figure B76 The aperture function for linear Bayliss tapering ($\bar{n} = 5$).

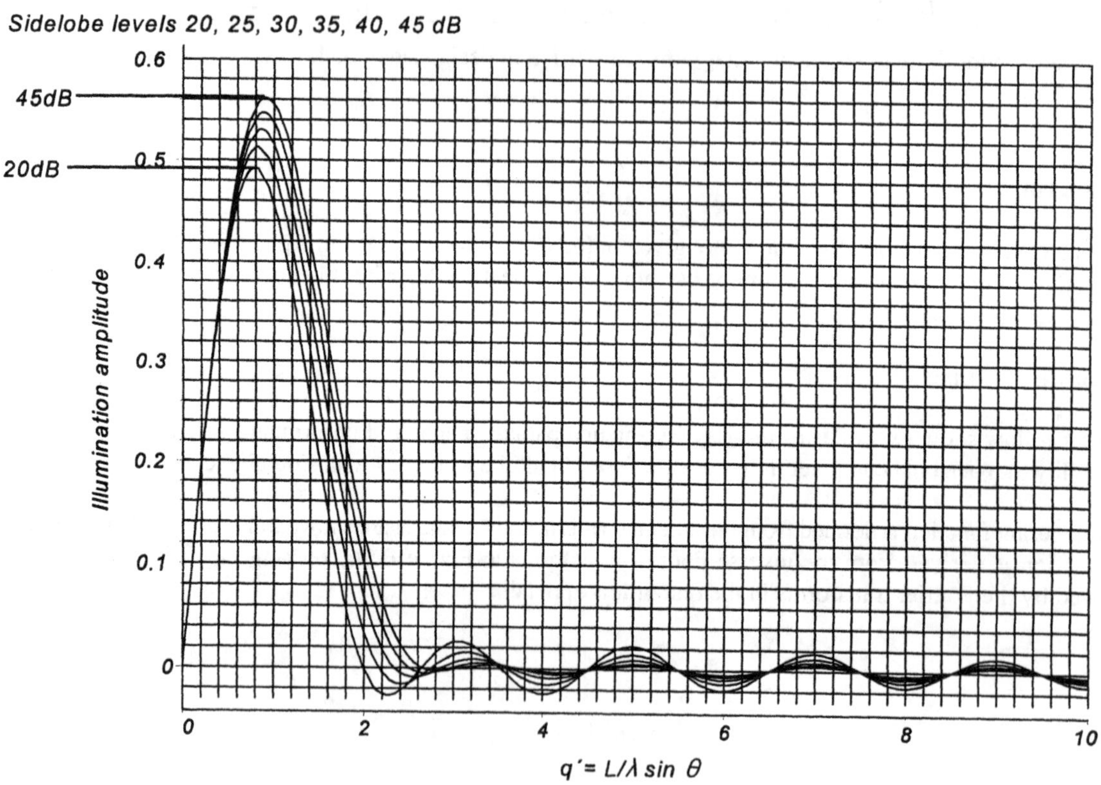

Figure B77 Antenna patterns for linear Bayliss tapering ($\bar{n} = 5$).

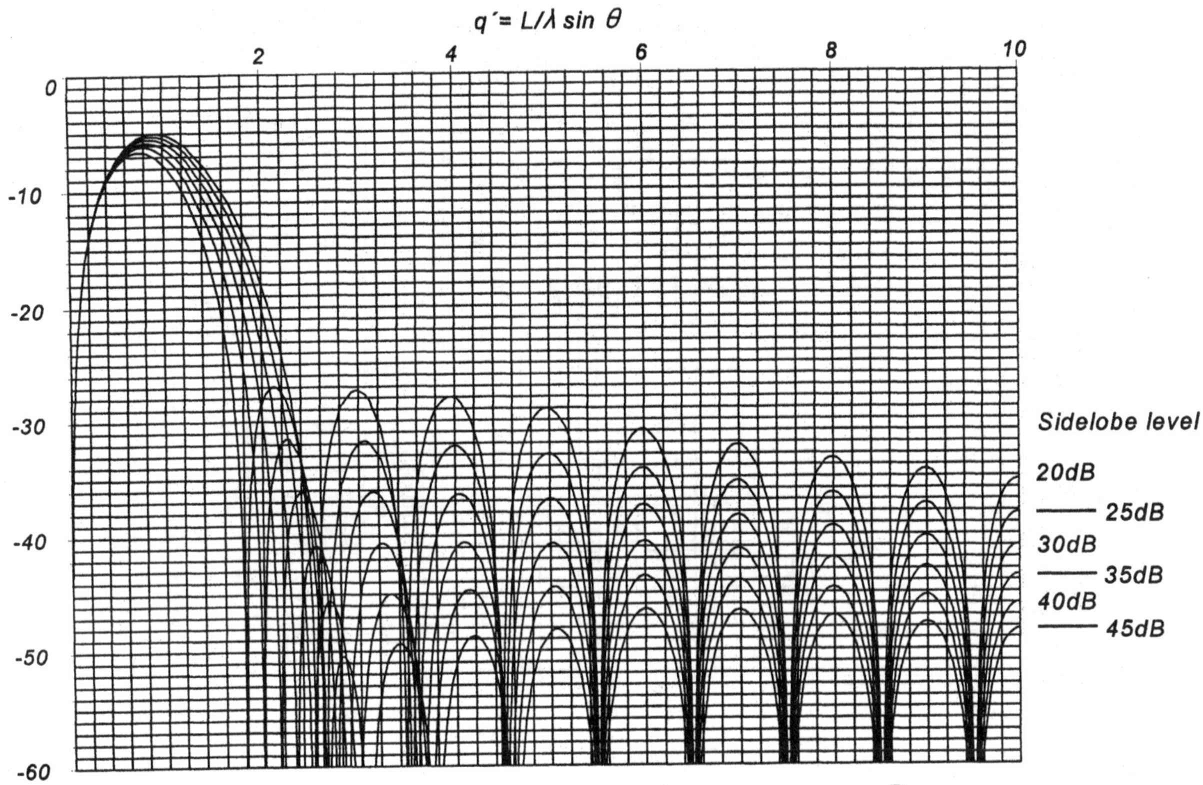

Figure B78 Antenna patterns in decibels for linear Bayliss tapering in decibels ($\bar{n}$ = 5).

B.5.6.2 Circular Bayliss tapering

The original reference [6] starts with the analysis for a tracking radar with a circular antenna. As with the linear distribution, further analysis takes too long on a normal personal computer. The example shown has, as with the linear case, $\bar{n}$ = 5.

As with the linear Bayliss tapering function, the circular Bayliss tapering function is defined from the antenna pattern and examples are shown in Figures B.80 and B.81. The aperture function is derived from the inverse Fourier transforms in Figure B.79, keeping the order of aperture functions followed by the Fourier transforms in the rest of this appendix.

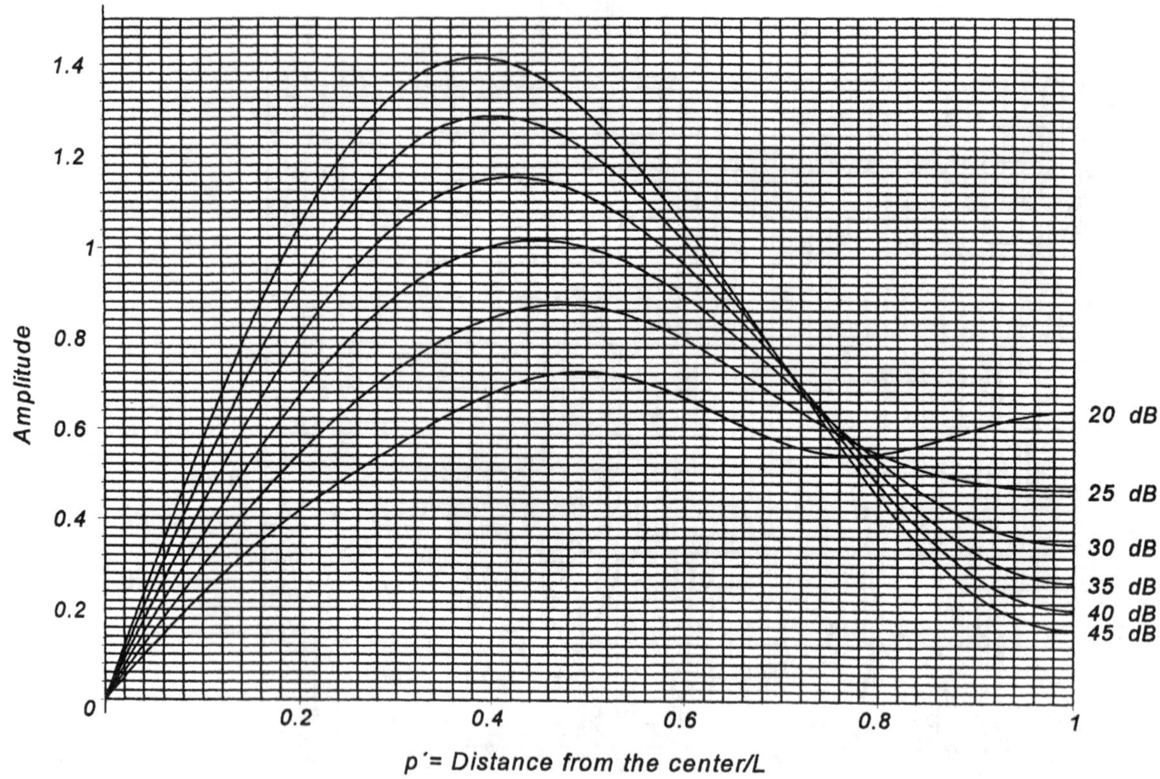

Figure B79 Aperture function for circular Bayliss tapering ($\bar{n}$ = 5).

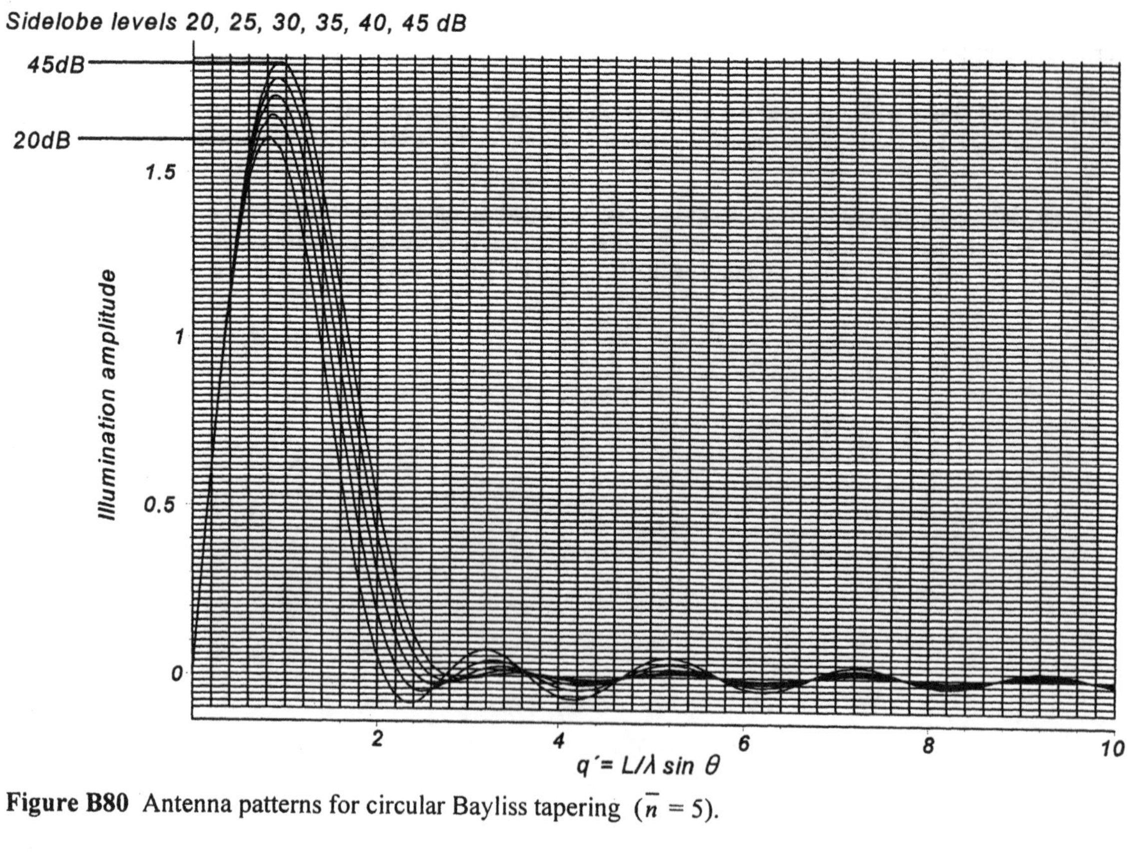

Figure B80 Antenna patterns for circular Bayliss tapering ($\bar{n}$ = 5).

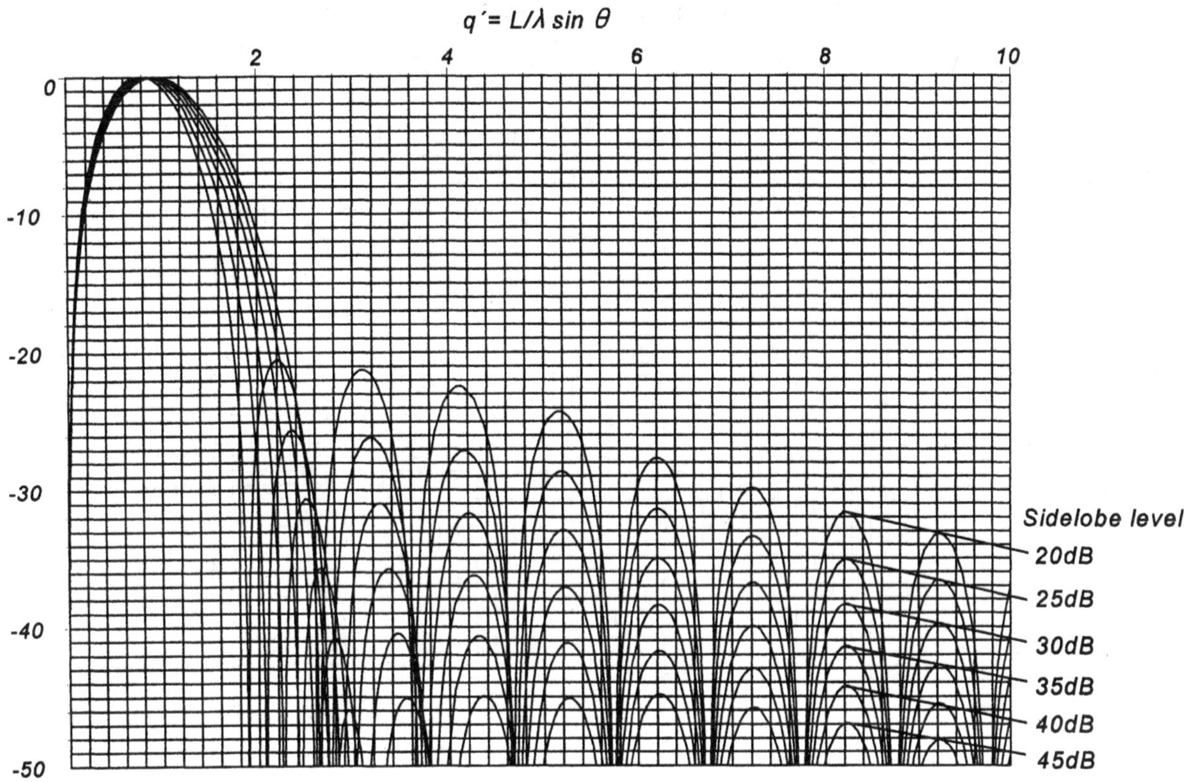

Figure B81 Antenna patterns for circular Bayliss tapering in decibels ($\bar{n} = 5$).

B.5.7. Zolotarëv tapering

Zolotarëv tapering [8, 9] (see Section 5.3.1.2) is the equivalent of Chebyshëv tapering (see Sections 5.1.4.1 and B.3.1) for monopulse arrays, and Appendix 5A describes the calculations required. For tables of performance values, the reader should refer to [9].

The examples shown are for a 21 element array with 10 elements on the left hand side, 10 elements on the right hand side driven in antiphase, with the center element not driven. The elements are spaced half a wavelength apart. Figure B.82 shows the discrete raw Zolotarëv pattern with its design sidelobe level oscillating between +1 and −1 and the aperture function derived from it.

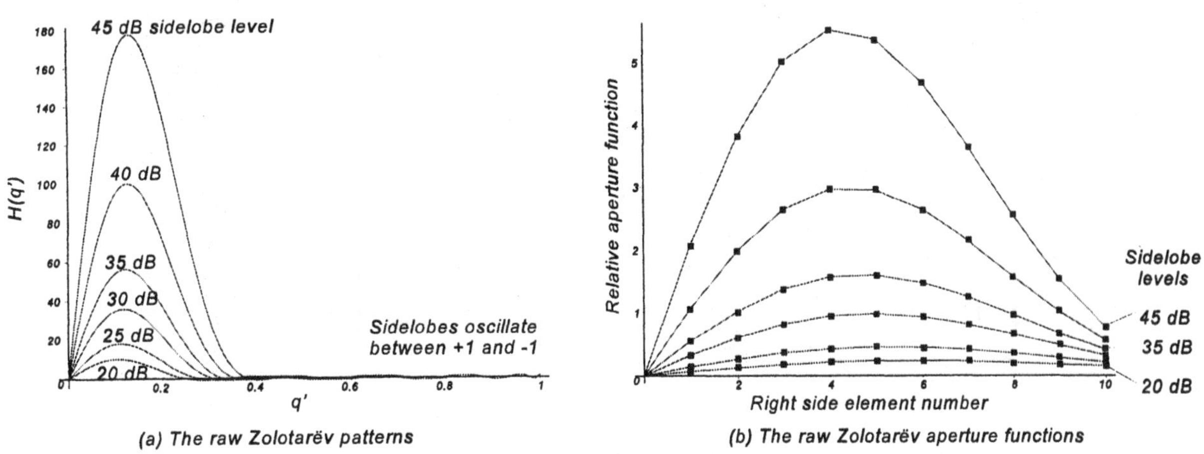

Figure B.82 Raw Zolotarëv patterns and aperture functions for a 21 element array.

The normalized aperture function is shown in Figure B.83.

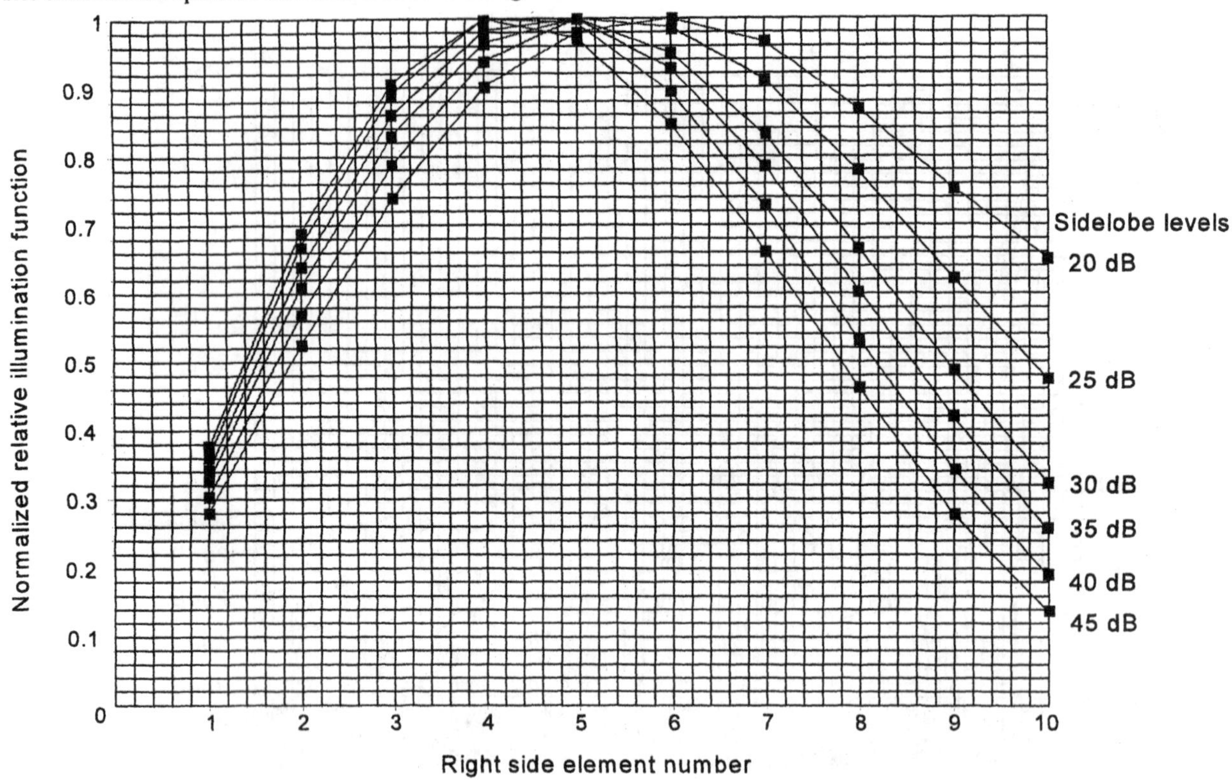

Figure B.83 Normalized Zolotarëv illumination function.

The normalized pattern is shown in Figures B.84 and B.85.

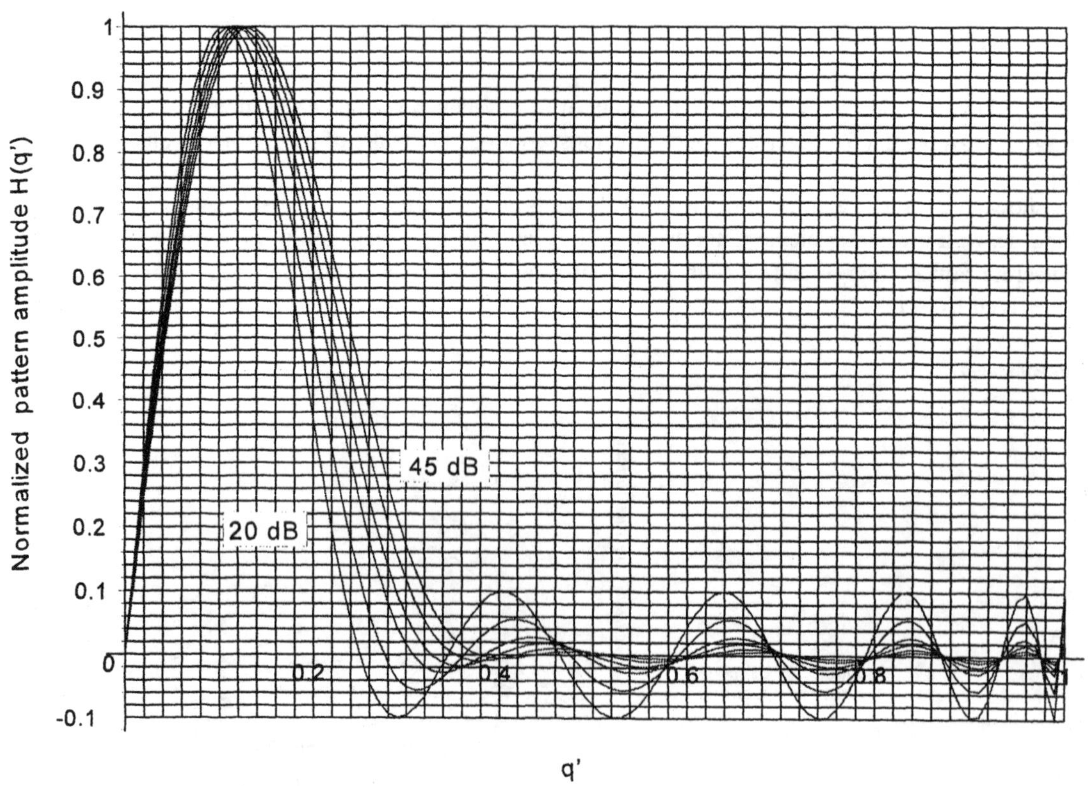

Figure B.84 Normalized Zolotarëv array pattern.

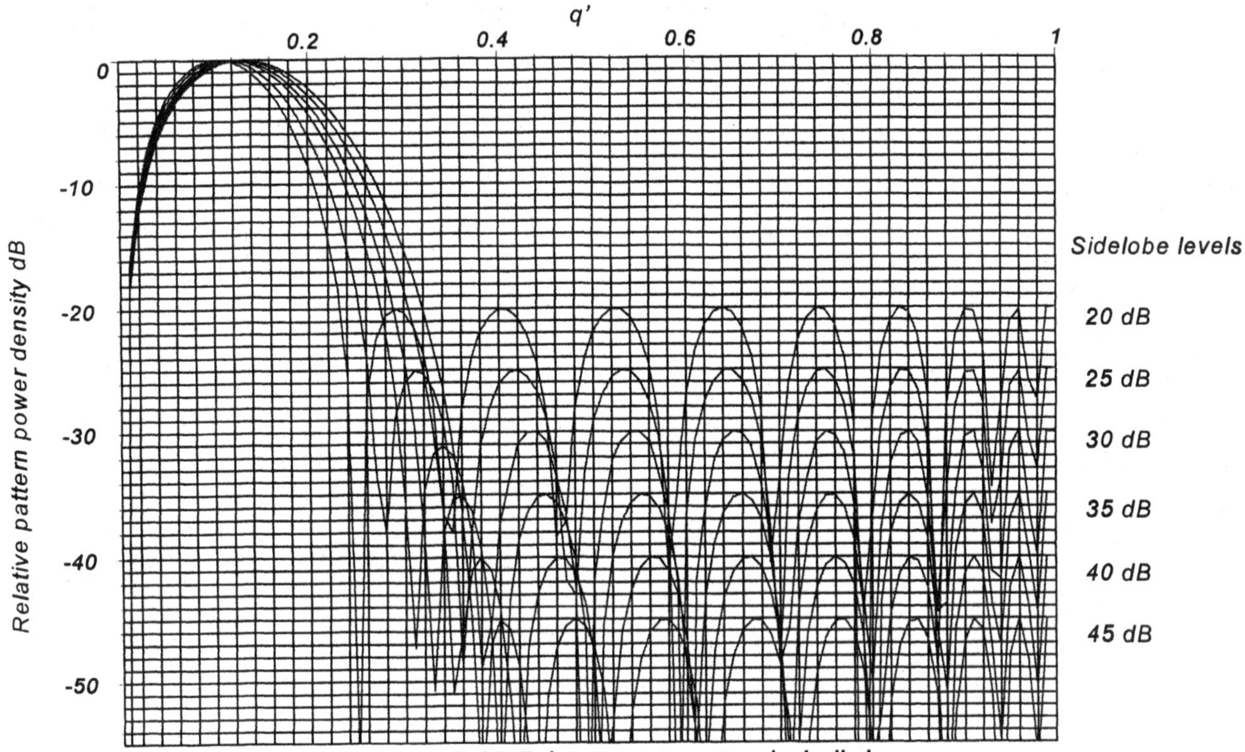

Figure B.85 Zolotarëv array pattern in decibels.

REFERENCES

1. Barton, D. K., and H. R. Ward, *Handbook of Radar Measurement*, Englewood Cliffs, New Jersey: Prentice-Hall, 1969.
2. Harris, F. J., "On the Use of Windows for Harmonic Analysis with the Discrete Fourier Transform", *Proceedings of the IEEE*, Vol. 86, No. 1, January 1978.
3. Elliott, D. E., and R. K. Rao, *Fast Transforms*, Orlando, Florida: Academic Press, 1982.
4. Skillman, W. A., *Radar Calculations Using the TI-59 Programmable Calculator*, Dedham, Massachusetts: Artech House, 1983.
5. Bronshtein, I. N., and K. A. Semendyayev, *Handbook of Mathematics*, Leipzig: B. G. Teubner Verlagsgesellschaft, 1985, p. 414.
6. Bayliss, E. T., "Design of Monopulse Antenna Difference Patterns with low Side Lobes", *Bell System Technical Journal*, Vol. 47, May-June 1968, pp. 623-650.
7. Hansen, R. C., *Phased Array Antennas*, New York: John Wiley, 1998.
8. McNamara, D. A., "Direct Synthesis of Optimum Difference Patterns for Discrete Linear Arrays Using Zolotarev Polynomials", *Proceedings of the IEE*, H Vol. 140, No. 6, December 1993.
9. McNamara, D. A., "Performance of Zolotarev and Modified-Zolotarev Difference Pattern Array Distributions", *IEE Proceedings on Microwave Antennas and Propagation*, Vol. 141, 1st February 1994.

Appendix C

Frequency band letters

Radar development started in the 1930s and was shrouded in secrecy, especially as to which frequencies were used. This secrecy led to the use of the code letters for frequency bands in Britain and the United States as shown in Table C.1.

Table C.1

British and United States radar frequency bands

British band letters [1]		United States band letters [2]	
Frequency, GHz	Band	Frequency, GHz	Band
		0.42 - 0.45	P
1 - 2	L	1 - 2	L
2 - 4	S	2 - 4	S
4 - 8	C	4 - 8	C
8 - 12	X	8 - 12	X
12 - 18	J	12 - 18	K_u
18 - 26	K	18 - 27	K
26 - 40	Q	27 - 40	K_a
40 - 60	V	40 - 75	V
60 - 90	O	75 - 110	W

The former arbitrary letter systems have been superseded by those for electromagnetic surveillance, as shown in Table C.2.

Table C.2

Letter frequency bands used for electromagnetic surveillance

Band	Frequency range, GHz
A	0 - 0.25
B	0.25 - 0.5
C	0.5 - 1.0
D	1.0 - 2.0
E	2.0 - 3.0
F	3.0 - 4.0
G	4.0 - 6.0
H	6.0 - 8.0
I	8.0 - 10.0
J	10.0 - 20.0
K	20.0 - 40.0
L	40.0 - 60.0
M	60.0 - 100.0

REFERENCES

1. Sword, S. S., *Technical History of the Beginnings of Radar*, London: IEE Press, 1968.
2. *IEEE Standard Radar Definitions, IEEE Standard 686-1990,* New York: The Institute of Electrical and Electronic Engineers, 1993.

List of symbols

The book covers many disciplines, each with its own library of symbols. Only the more general symbols are shown in the list and, when not generally used, are referenced by section, in brackets (X.YY) for equations, or table number.

Symbol	Definition	First or principally used in
$\bar{\sigma}$.	Average reflectivity	(6.76)
NF	Noise figure	(7.4)
α	Azimuth	Figure 1.7
α_{atm}	Integrated gain for atmospheric noise	(5.110)
α_{cosmic}	Noise gain (integrated over the solid noise beamwidth) of the antenna	(5.110)
α_{ground}	Integrated sidelobe gain over the solid angle subtended by the ground	(5.110)
α_{sun}	Integrated sidelobe gain over the solid angle subtended by the sun	(5.110)
Γ	Reflection coefficient	(6.52)
γ	Euler's constant (0.5772156649)	(11.5)
$\gamma(a, x)$	Incomplete gamma function	(12.26)
$\Gamma(z)$	Gamma function, argument z	(6.32)
Δ	Difference channel signal	(13.1)
$\Delta(\theta)$	Difference antenna characteristic	(11.106)
$\delta(t)$	Dirac function	(2.30)
ϵ	Eccentricity	(5.92)
ϵ_0	Permittivity of free space	(4.4)
ϵ_r	Relative permittivity of the reflecting surface	(6.59)
ζ	Zenith angle	Figure 1.7
η	Antenna efficiency	(5.14)
$H(v, k)$	Jacobi Eta function with modulus k	(5.53)
η_x	Efficiency of the linear array compared with a uniformly excited array	(5.17)
θ	Angle from the normal to an array	(5.3)
θ	Angle, general	

Symbol	Definition	First or principally used in
θ	Argand angle	Figure 1.7
θ_3	Half-power or 3 dB beamwidth	Table 5.3
θ_6	Half-voltage or 6 dB beamwidth	Table 5.3
θ_m	Pointing angle from the normal	(5.3)
θ_n	Noise beamwidth	Table 5.3 (5.18)
λ	Wavelength, m	
λ	exp(j $2\pi/3$)	(2.11)
$\lambda_{cut\text{-}off}$	Cutoff wavelength of a waveguide	(4.3)
λ_g	Guide wavelength	(5.91)
λ_g	Guide wavelength, m	(4.5)
λ_I	Interrogation wavelength	(1.12)
λ_R	Receiving wavelength	(1.12)
μ	Population mean	(15.2)
μ_0	Permeability of free space.	(4.4)
μ_n	The n th zero of $J_1'(\pi x)$	(5.58)
ξ	Variable in characteristic functions	(15.14)
$\Xi(\xi)$	Form of Fourier transform in statistics characteristic functions	(15.14)
ρ	Reflection coefficient	(4.2)
ρ	Pulse compression ratio	(3.11)
ρ	Collapsing loss	(11.16)
Σ	Sum channel signal	(13.1)
σ	Radar cross-section, m²	(1.7)
σ	Dilation factor	(5.34)
σ	Conductivity, S/m;	(4.4)
σ	Standard deviation, general	(1.17)
σ^2	Population variance	(15.5)
$\Sigma(\theta)$	Sum antenna characteristic	(11.106)
σ_θ	Root mean square aperture	Table 13.2
σ_B	Bistatic radar cross-section	(1.16)
σ_f	Root mean square frequency error	Table 13.2
σ_s	Conductivity of the surface layer, S/m	(6.59)

Symbol	Definition	First or principally used in
σ_t	Root mean square delay time error	Table 13.2
τ	Pulse width, s	(1.8)
τ_{flank}	Flank time, s	(3.5)
τ_j	Jitter	(1.2)
ϕ_c	Carrier phase	(2.9)
ϕ_m	Modulation phase	(2.9)
χ^2	Used for the chi-squared distribution	(6.34)
A	Carrier frequency amplitude	(2.9)
A	Effective antenna length (normalized)	(5.16)
A	Antenna area	(5.13)
A_l	Sidelobe constant	(5.31)
B	Receiver bandwidth, Hz	(1.8)
B	Modulation frequency amplitude	(2.9)
C	Effective radiated power (normalized)	(5.16)
C	Capacitance, F	(2.1)
C	Constant	(1.11)
c	Velocity of light	
c	Velocity of light	(1.16)
$C(\xi)$	Characteristic function	(12.8)
$C(\xi)$	Form of Fourier transform in statistics in this book	(15.16)
$C(\xi)$	Characteristic function	(2.32)
$cn(M, k)$	Jacobi elliptic function	(5.53)
CR	Cancellation ratio	(3.18)
D	Diameter of a circular array	(5.39)
D	Divergence factor	(6.66)
D	Directivity of an array	(5.12)
D	Length of the Fresnel region	(5.8)
d	Distance between two elements, m	
$dn(M, k)$	Jacobi elliptic function	(5.53)
D_S	Signal-to-noise ratio required for detection	(1.16)
D_s	Critical signal-to-noise ratio for detection	(12.7)

Symbol	Definition	First or principally used in
D_u	Duty cycle	(3.16)
e	Partial pressure of the water vapor, in hPa (millibars)	(6.18)
e	Base of natural logarithms, about 2.718..	
$E(x)$	Expected value of x	(15.3)
e_{ell}	Ellipticity	(5.92)
E_j	Effective radiated power density of the jammer, W/Hz	(14.10)
F	Propagation factor	(6.52)
f	Frequency, Hz	(2.1)
f	Frequency of the radar, Hz	(1.16)
$f^*(t)$	Complex conjugate of $f(t)$	(16.28)
$F(f)$	Fourier transform of $f(t)$	(11.51)
${}_1F_1(a, b, z)$	Confluent hypergeometric function	(12.43)
f_c	Carrier frequency, Hz	(2.9)
f_{curoff}	Cutoff frequency of a waveguide	(4.3)
$f_{Doppler.}$	Doppler frequency	(2.10)
f_m	Modulation frequency	(2.9)
F_n	Coefficients in a Taylor illumination	(5.38)
F_R	Propagation factor for reception	(1.16)
f_r	Pulse repetition frequency, Hz	(3.16)
F_T	Propagation factor for transmission	(1.16)
G	Antenna gain	(5.14)
$G(u)$	Antenna pattern	(5.32)
G_R	Gain of the receiving antenna	(1.6)
G_r	Antenna gain on receive	(1.9)
G_{rj}	Gain of the receiving antenna in the direction of the jammer	(14.10)
G_T	Gain of the transmitting antenna	(1.16)
G_t	Antenna gain (transmit)	(1.3)
h	Height above the earth's surface	(6.7)
$H(\omega)$	Fourier transform, ω convention	Table 16.1
$H(f)$	Fourier transform, f convention	Table 16.1
$h(k)$	Discrete function	(16.37)

Symbol	Definition	First or principally used in
$H(n)$	Discrete Fourier transform	(16.37)
$H(p)$	Laplace transform	Table 16.1 (16.65)
$h(r)$	Radius function	Table 16.1
$h(t)$	Waveform function in time	
$H(u)$	Hankel transform, $u' = r/\lambda \sin\theta$	Table 16.1 (16.68)
$H(z)$	z transform	Table 16.1 (16.71)
H_{max}	Maximum height for a cosecant squared beam	(5.81)
I	Used for in-phase signals	(10.3)
$I(u, p)$	Pearson's form of the incomplete gamma function	(15.35)
$I_0(x)$	Bessel function of zero order and imaginary argument	
j	$\sqrt{-1}$	
$J_1'(x)$	Differential of the Bessel function of the first kind, order one	(5.58)
$j_n(z)$	Spherical Bessel function of the first kind order n and argument z	(6.26)
$J_n(x)$	Bessel function of the first kind and order n	
k	Modulus for complete elliptic integral of the first kind	(5.53)
k	Used in constant k atmosphere	(6.19)
k	Boltzmann's constant	(1.8)
$K(k)$	Complete elliptic integral of the first kind, modulus k	(5.53)
K_l	Amplitude constant in (5.1)	(5.1)
k_1	Complementary modulus for complete elliptic integral of the first kind	(5.53)
K_B	Wind speed on the Beaufort	(6.93)
L	Losses	(1.9)
L	Product of the losses	(1.16)
l	Slat depth, m	(5.89)
L_{aj}	Atmospheric loss between the jammer and the radar	(14.10)
L_B	Baseline length	(1.13)
L_I	Total losses on the interrogation path	(1.12)
$\ln(x)$	Naperian or natural logarithm of x	
$\log_{10}(x)$	Common logarithm of x	
L_R	Total losses on the reply path	(1.12)

Symbol	Definition	First or principally used in
LSB	Least significant bit	(10.1)
M	Water content of the fog in g/m	(6.24)
m	Sample mean	(15.1)
$M_o(t)$	Moment generating function about the origin	(15.6)
N	Number of pulses	(12.9)
n	Complex index of reflection	(6.59)
n	Index of refraction	(6.18)
N_s	Refractivity at the earth's surface in N units	(6.7)
p	Atmospheric pressure, in hPa (millibars)	(6.18)
$P(x)$	Cumulative probability	
$p(x)$	Probability density of x	
$p(x)$	Probability density function	
P_d	Probability of detection	(11.17) (12.19)
P_{fa}	Probability of false alarm	(11.19) (12.3)
P_{IR}	Interrogator receiver sensitivity for the required probability of decoding a reply	(1.12)
P_{RR}	Transponder receiver sensitivity for replying to an interrogation.	(1.12)
P_t	Peak transmitter power, W	(3.16)
$\mathrm{P_t}$	Peak transmitter power	(1.3)
Q	Used for quadrature-phase signals	(10.3)
R	Voltage sidelobe ratio	(5.30)
R	Range	Figure 1.7
R	Resistance, ohm	(2.1)
R	Signal-to-noise ratio	(12.19)
R_I	Interrogation range	(1.12)
R_j	Range of the jammer, m;	(14.10)
R_R	Range to receiver	(1.12)
R_R	Distance from the scatterer to the receiver	(1.16)
R_R	Range scatterer to receiver	(1.13)
R_T	Range transmitter to scatterer	(1.13)
R_T	Distance from the transmitter to the scatterer	(1.16)

Symbol	Definition	First or principally used in
s^2	Sample variance	(15.4)
Si(x)	Sine integral	(8.2)
SLL	Sidelobe level	
SLL	Sidelobe level	(5.27)
$sn(M, k)$	Jacobi elliptic function	(5.53)
T	Temperature of the atmosphere, in K	(6.18)
T	Noise temperature, K	(1.8)
t	Time in seconds	(2.9)
T_0	Temperature of the input terminator	(7.5)
T_e	Effective source temperature, Kelvin	(5.108)
t_{far}	False alarm time	(12.4)
$t_{fa\text{-}old}$	Marcum's false alarm time	(12.3)
T_{out}	Output temperature, Kelvin	(5.109)
T_s	Source temperature, Kelvin	(5.108)
T_{sys}	System noise temperature	(7.7)
T_T	Loss temperature, Kelvin	(5.108)
T_t	Time between pulses	(3.2)
u'	Normalized angle from normal	(5.7)
V	Voltage, volt, general	
v	Velocity, m/s	
$VSWR$	Voltage standing wave ratio	(4.1)
W	Width of a synthetic aperture, m	(5.117)
w	Width of an array	(5.8)
W_K	Weighting function	(11.87)
Y_b	Threshold value	(11.10) , (12.13)
y_n (z)	Spherical Bessel function of the second kind order n and argument z	(6.26)
Y_T	Sample threshold	(11.11)
$z(n)$	Location of n th zero	(5.35)
$Z_n(x)$	Zolotarëv polynomial order n	(5.53)
$\otimes$	Convolution operator	(16.20)
★	Correlation operator	(16.28)

About the author

Hamish Meikle obtained a diploma in electrical engineering from Brighton Technical College in Sussex, England, in 1961. After a two-year graduate apprenticeship in Woolwich, England, with additional course work at Brunel University, he spent a short time working as a sales engineer for telephone transmission equipment, then went to Germany to work on radar and associated equipment. Since then, he has worked in the fields of operational research, radar systems, systems software, and associated fields as an independent systems consultant in western Europe.

Index

A-scope 38, 502, 606
Accuracy 22, 489, 494, 496, 497, 500-502, 540
Ament, W.S. 229
Amplitude modulation 28
Amplitude versus azimuth 507
Amplitude detector 282, 604
Amplitude modulation 326, 329
AMTI, see Area moving target indicator
Angle errors 498, 513
Anomalous propagation 191, 194, 276
Antenna 37, 489-492, 494, 495, 497-501, 505, 507, 509-511, 513, 515, 517-520, 522, 523, 622, 649
Antenna, Gain 104
Antenna, Leakage 109, 156
Antenna stability 518
Aperture efficiency 106
Area moving target indicator 365, 433-435
Array factor 99, 130
Arrays 97, 108, 109, 116, 120, 129, 131-133, 146, 147, 149, 153, 168, 178, 179, 494, 502
Attenuation in the atmosphere 195
Autocorrelation 31, 45, 305, 311, 321, 305, 311, 321, 569, 570, 577, 606
Azimuth 4-6, 417, 418, 428, 432, 437, 438, 444, 489-491, 494-497, 500-502, 504-507, 512-515, 517, 521, 522, 524

B-scope 4, 502, 606
Backlobe 110, 156, 246
Barker code 62, 63, 65, 315, 319-321
Barrage jamming 107, 168, 231, 519, 520
Bartlet tapering 621
Bayliss, E.T. 97, 121, 123, 125-128, 179, 183, 621, 664, 675-679, 681
Beam shape loss 156, 160, 177, 511, 523
Beam shape loss, Coherent integration 156
Beam steering 132, 133, 514

Beamwidth 16, 23, 101, 232, 234, 236, 245-247, 432, 433, 435, 441, 448, 496, 500, 517, 518, 524, 581, 582, 584, 622, 626, 629, 633, 637, 646, 649
Beaufort wind scale 244-246
Bias errors 507, 511, 512
Binary integration 385, 478
Binomial distribution 209, 551, 552
Bipolar video 337, 349, 387
Bistatic radar 17-21, 185
Bivariate Gaussian distribution 37-40, 210, 331
Blackman-Harris tapering 621, 636-639, 642-645
Blackman tapering 621, 636-645
Blade polarizer 147, 234, 490
Blanking maps 432, 436
Blocking loss 152, 153
Bochner tapering 621
Bottle-brush noise 338

C-scope 377, 504, 607
CACFAR, see Cell averaging constant false alarm rate
Canceler, double 392, 404-406
Cartesian coordinates 18, 27, 605
Cartesian detector 32, 33, 37, 338, 340, 515, 605
CCIR 188, 248
CFAR, see constant false alarm rate
Chaff 59, 107, 168, 231, 246, 248, 329, 370, 386, 408, 436-438, 509, 519, 520
Characteristic function 45, 531-533, 536, 541-543, 546, 549, 550, 555, 607
Chebychëv tapering 109-112, 114, 179, 605, 621, 653-655
Chi-squared distribution 546-548
Chi distribution 548
Circular Taylor pattern 118, 662, 663
Circular polarizer 147, 148
Circular polarization 523
Circular error probable 540
Circular polarization 147-151, 168, 201, 203, 204, 224, 251, 436
Circular aperture 115, 116, 128
Circular truncated Gaussian tapering 660
Circulators 84, 85, 90, 91, 95
Clutter 1, 13-16, 28, 58, 106, 185, 194, 201, 205, 206, 214, 217, 227, 231, 232, 235-248, 250, 252, 274-276, 282-284, 286, 287, 289, 304, 306, 307, 314, 319, 350-352, 361, 363-370, 373, 374, 378-380, 385-392, 394, 396, 402-407, 409-412, 418-420, 423, 425-429, 432, 433, 435, 436, 432-439, 441, 443, 444, 446, 447, 475, 502, 509, 512, 518, 519, 521, 523, 524, 538, 541, 544, 550, 552, 553, 584, 585, 600, 607, 608, 610, 611, 613, 614
Clutter improvement 423, 509
Clutter maps 179, 364, 404, 407
Coaxial cables 86, 89
Coherent detector 7, 11, 32, 62, 66, 73, 76, 287-289, 330, 334, 336, 329, 334, 336, 338-340, 386, 387, 605

Coded pulse anticlutter system 607
Coherent integration 425, 444, 453, 458, 462, 470, 475, 476
Coherent moving target indicator 387, 396, 600
Coherent memory filter 408-410, 442
Collapsing loss 10, 377, 378, 485, 607
Combining the videos 377
Complex sampling 343
Conical scanning 490
Constant false alarm rate 283, 286, 300, 364, 369, 370, 373, 374, 428, 436, 437, 444, 452, 511, 607, 608
Cell averaging constant false alarm rate 283, 366, 370, 373, 374, 386, 428, 432, 436, 437, 444, 523, 607
Constant k atmosphere 190, 494
Control of gain 274
Conventions 25, 50, 453, 555, 619
Conventions, Engineers' and physicists' 151
Convolution 45, 46, 359, 532, 565-567, 569, 577, 608
Coordinate systems 5
Corner reflectors 201, 204
Correction of errors 354
Correlation 28, 31, 45, 46, 77, 214-216, 302-307, 312, 316, 369, 384, 396, 397, 415, 444, 515, 501, 568, 569, 577, 599, 601, 602, 606, 608
Correlation of neighboring pulses 397
Cosine to the power n 629-631, 664, 669-671
Cosine on a pedestal 633
Cosine wave 35, 43, 355, 558, 559, 564, 592
Cosine edges 56
Cosine squared edges 56, 57
Cosmic noise 37, 162, 164, 165, 273, 280
Costas codes 72, 315, 326
CPACS, see Coded pulse anticlutter system 607
Croney, J. 366, 369, 444
Cross-gating 376
CRPL atmosphere 188, 192, 195, 251, 253, 266-271
Cumulative distribution function 529, 536, 540, 541, 544, 546, 549, 552

D-scope 504, 506, 608
Deception 613
Desert 220, 223, 226
Detection control maps 432, 436
Detector 30, 32, 33, 37, 38, 214, 276, 282, 285, 289, 291, 292, 301, 491, 492, 496, 511, 514, 515, 518, 524, 600, 603-605, 611, 613
Determination of position 9, 489
Dicke-fix receiver 282, 298, 300, 366
Dielectric 89, 148
Differentiation of the video 366
Diffraction 101, 109, 133, 156
Digital beam forming 162
Dilation factor 114, 119, 126, 127
Diplexers 91
Directional coupler 82, 93, 95
Directivity 94, 104-106, 108, 116, 131
Directrix 97, 98, 100, 115, 118, 146, 153
Dirichlet or rectangular tapering 621
Discrete phase shift modulated 312
Discrete radiators 98, 111, 129, 131, 133
Discrete distribution 551
Displaced phase center antenna 438, 439, 608
Displays 10, 274, 282, 365, 366, 369, 377, 402, 444, 502-507, 609-612, 614-617
Divergence factor 227-229
Dolph-Chebyshëv tapering 109-112, 114, 179, 621, 653-655
Doppler frequency 3, 5, 6, 19, 21-23, 33, 107, 178, 291, 298, 308, 309, 314-316, 319, 321-322, 329-331, 334, 337, 340, 343, 363, 364, 370, 374, 378, 381, 386, 391, 401-403, 407, 408, 409, 411, 412, 414, 417, 418, 425, 428-432, 436, 438-443, 494-496, 507, 514, 515, 518, 521, 584, 585, 610-612
Double canceler 392, 404-406
DPCA, Displaced phase center antenna
Driver 75, 76, 78, 81
Dry land 220, 226
Dummy loads 92, 93
Dwell time 447, 448, 453, 524
Dynamic range 8, 9, 14, 22, 27, 101, 273-276, 282, 283, 285-287, 291, 300, 306, 307, 333, 334, 337, 349, 350, 364, 366, 367, 396, 407, 490, 491, 418, 433, 518, 521, 613

E-scope 502, 504, 506, 608
Eccentricity 35, 151, 353
ECM, see Electronic countermeasures
Effective radiated power 121
Effective aperture 620
Effective area 186
Effective radiated power 519
Effects of the atmosphere 185, 187
Efficiency 57, 106-110, 118
Electronic counter measures 107, 168, 398, 399, 609
Elliptical coordinates 18, 19, 21
Ellipticity 18, 35, 151, 152, 353
Erlang distribution 546, 556
Escort jammer 107, 231, 520
Even Taylor tapering 649
Exact Blackman 636, 640-642
Expected value 530, 531
Exponential atmosphere 188, 195, 253, 260-271

F-scope 504, 506, 609
Fading 6, 201, 205, 384, 447, 475, 483, 608
False alarm control 432, 436
False alarm number 448, 609

False alarm probability 370, 372, 376, 449-452, 454, 456, 457, 462, 463, 465-467, 470, 471, 474, 475, 478, 480-482, 485, 609
False alarm rate 14, 283, 286, 300, 364, 366, 369, 370, 373, 374, 377, 378, 385, 428, 435-437, 443, 444 452, 484, 495, 511, 607, 608, 611
False alarm time 448, 609
Families of probability distributions 527, 534
Fast fluctuation 213, 214, 464, 477, 478
Fast time constant 366-369, 609
Finite impulse response 381, 393, 414, 568, 592, 597
Fire control radar 298, 489, 490, 493, 494, 505, 603
Flat earth 217, 218
Fluctuation 22, 185, 205, 206, 213-217, 378, 443, 447, 452, 453, 456-458, 460, 462-478, 500, 514, 608, 609
Focus 97, 120
Fog 176, 199, 519
Fourier transform 43-45, 48, 50, 531-533, 555-561, 563-566, 568, 570-572, 574, 576, 577, 579, 586, 590, 591, 593, 600, 620, 621
Frank code 66-68, 70, 315, 321-323, 609
Fraunhofer region 100, 101, 173, 512
Frejer tapering 621
Frequency band letters 683
Frequency diversity 6, 8, 21, 22, 82, 214, 285, 375, 376, 418, 429, 431, 478, 483-485, 488, 515, 604, 608
Fresnel region 100, 101, 231
FTC, see Fast time constant

G-scope 504, 506, 609
Gamma distribution 206-209, 213, 371, 447, 449, 452, 458, 464, 468, 544-547, 549, 556
Gamma distribution family 544
Gating the video 375
Gated time control 274
Gating 432

Gaussian distribution 37-40, 45, 49, 209, 210, 329, 331, 449, 477, 534, 537, 539, 545, 604
Gaussian filter 294, 295, 297
Gaussian noise 41, 300, 337, 345, 347, 453, 605
Gaussian pulse 25, 47, 48, 297, 563
Glint 215, 217, 501, 513, 610
Grating lobes 8, 130, 131, 133, 134, 136, 141, 146, 321, 323, 325
Ground reflections 185, 217, 221, 231, 251, 494, 518
GTC, see Gated time control 274

H-scope 504, 506, 610
Hamming, R.W. tapering 585, 587-589, 621, 636-639
Hankel transform 555, 581, 590, 591
Hann, J. von, tapering 585-589, 621, 629
Hanning, see Hann, J. von
Hardware losses 152
Harmonics 36, 52, 72, 342, 352
Hawk missile radars 1, 4
Headphones 234, 329, 337, 408
Histogram 529
Horn 91, 97, 98, 103, 109, 110, 133, 146-148, 153, 156, 162
Huygens principle 98
Hybrid coupler 94, 610-612

I-scope 504, 506, 610
Identification friend or foe 1, 610
IFF, see Identification friend or foe
IIR, see Infinite impulse response
Illumination function loss 152, 153
Incoherent detectors 330, 331
Incomplete gamma function 462, 485, 544, 545
Infinite impulse response 382, 385
Interference 8, 103, 107, 156, 178, 205, 215, 217, 227, 231, 247, 273, 280-283, 285, 342, 363, 364, 370, 374, 376, 384, 437, 447, 490, 506, 563, 603, 604, 609, 610
Interfering or jamming signals 519
Intermediate frequency 8, 9, 37, 273, 275-277, 279-283, 355
Inverse Fourier transform 566, 593
Inverse square law 274
Iris 147
Isolators 91
Isotropic antenna 105

J-scope 502, 506, 610
Jam strobe 610, 616
Jammer, 506, 616
Jamming 8, 103, 107, 119, 121, 153, 168, 169, 178, 231, 232, 283, 287, 298, 300, 348, 363, 370, 376, 437, 483, 490, 494, 506, 509, 519, 603, 608, 610, 613, 616, 617
Joint probability 209

K-scope 506, 611

L-scope 505, 507
Land clutter 21, 235, 239, 240, 242, 246, 370, 541
Land clutter spectrum 242
Laplace transform 555, 590, 591
Leakage 83, 94, 274, 423, 584
Lens 98, 133, 156, 170, 251, 511
Lens loss 187, 192, 193, 511
Limiters 91, 279, 282
Limiting 9, 14, 232, 273, 282-284, 286, 287, 291, 298-300, 311, 319, 320, 364, 366, 387, 402-405, 444, 517, 523, 524, 607, 608

Linear and rectangular radiators 98
Linear detector 37, 485, 349, 409
Linear frequency modulation 3, 59, 62, 66, 69, 72, 301, 302, 309, 314, 315, 321, 325
Linear polarization 147, 148, 151
Linear processing 28, 301, 364
LO, see Local oscillator
Lobing 49, 221, 227, 228, 489, 501, 505, 616
Local oscillator 62, 280, 281, 287-289, 316, 518, 613
Log fast time constant (FTC) 366
Log-normal 14, 15, 205, 206, 238, 239, 241, 245, 246, 374, 402, 426, 434, 536-538, 556
Log-normal distribution 206, 239, 426, 537, 538
Logarithmic video 23, 364, 366

M-scope 507
Magnetron 6, 10, 11, 51, 73, 76
Magnetron radars 73, 76, 289, 354, 438, 439, 583
Maps 188, 276, 286, 364, 365, 404, 407, 425, 428, 432, 433, 435-437, 490
Marcum, J.I. 9, 205, 377, 378, 442-444, 448, 452, 453, 455-457, 475-477, 485-487, 490, 609
Marsh land 220, 222, 225
Matched filter 9, 273, 282, 283, 285, 291, 293, 294, 296, 298-301, 316, 323, 327, 329, 341
Matching filter 23, 291, 292, 294, 295, 300, 327, 511, 523
Maximum range 6, 10, 12, 19, 22, 510, 521, 523
Mean value 37, 38, 40, 45, 47, 527-531, 533, 534, 536, 538, 540, 544, 545, 547, 548, 552, 558, 559, 568, 583, 601, 602, 606, 607, 611
Meander 148
Measurement of noise 344
Median 530, 536, 538, 540
Microstrip line 89
Mixer 168, 279-281, 289, 354
Mode 529, 536, 538, 540, 544, 547, 583, 613
Modulation 3-5, 25, 28, 29, 32, 33, 51, 53, 59, 62, 63, 65-67, 69, 70, 72, 282, 289, 290, 298, 299, 301, 302, 304, 305, 309-312, 314, 315, 321, 325, 326, 329, 330, 336, 338, 341, 357, 378, 408, 409, 512, 517, 562-564, 577, 586
Modulator power 57, 578
Moment generating function 530, 531
Monitoring 82, 90, 93, 95
Monopulse 19, 23, 119, 121, 123, 146, 178, 179, 274, 490, 492-494, 497, 500, 501, 513, 514, 521, 564, 578, 603, 608, 611, 619
Monostatic radar 7, 19-21
Moving target detector 21, 24, 364, 370, 374, 381, 407, 409, 410, 418, 420, 425, 428, 432, 435, 436, 441, 442, 444, 475, 511, 514, 515, 518, 579, 599, 603, 604, 611, 613
Moving target indicator 6, 21, 23, 33, 337, 365, 381, 387, 388, 390-394, 396-400, 402, 404-407, 409, 416, 418, 425, 427, 428, 432-436, 438, 441, 443, 444, 475, 490, 511, 524, 575, 597, 600, 603, 604, 607, 611, 613
MTD, see Moving target detector
MTI, see Moving target indicator
Multiple paths 217
Multiplier-accumulator 414

N-scope 507, 611
Negative exponential distribution 300, 370, 449, 458, 546, 549, 550
Negative phase sequence components or signals 33, 35-37, 43, 307, 312, 326, 340, 343, 344, 355, 357, 605
Nitzberg, R., 444
Noise 8-10, 13-16, 20-23, 25, 37-41, 43, 45, 83, 87, 95, 216, 232, 238, 240, 246, 248, 251, 252, 273-283, 285-287, 290, 291-294, 296-300, 304, 307, 311, 314, 320, 326, 328, 329-334, 337-339, 344-348, 363, 364, 366-373, 377, 378, 384, 387, 391, 397, 402, 404, 447-476, 478, 484-487, 489, 492, 495, 499, 505, 507, 509-515, 519, 523, 524, 536, 540, 541, 544, 546, 547, 549, 552, 605-610, 612, 620
Noise beamwidth 108, 167, 448, 581, 582, 622, 626, 629, 633, 637, 646, 649, 653
Noise clipping 369
Noise received from an antenna 164
Number of echoes 447, 495

Odd cosine to the power n 664, 669, 670
Odd rectangular 664-666
Odd Taylor derivative 664, 673-675
Odd triangular 664, 667, 668
Odd truncated Rayleigh 664, 671-673

Optimum filter 294, 296
Output stage 10, 51, 52, 54, 57, 76, 78, 81
P1 code 68-70, 315, 323-325
P2 code 69, 315, 324
P3 code 69-71, 325
P4 code 70, 71, 323, 325
Parabolic earth 191
Parabolic 621, 626, 629, 657
Parallel feed 132
Parallel plate 148
Parseval's theorem 577
Parzen, E. tapering 553, 621
Pearson's form of the incomplete gamma function 447, 485, 544
Phase detector 6, 33, 337, 339, 349, 387, 439, 491, 492, 604, 605

Phase modulation 5, 51, 59, 62, 63, 72, 289, 301, 304, 312, 315, 326, 378
Phase shift 26, 134, 135, 137, 312, 315, 316, 321, 330, 334
Phased array 132, 135, 149, 170, 179, 364, 397, 494, 497, 513, 514, 522
Plan position indicator 33, 200, 366, 378, 386, 502, 507, 612, 615, 616
Planar array 98, 109, 155, 156, 162
PLD, see Pulse length discriminator
Poisson distribution 527
Polar coordinates 5, 27, 28, 104, 115, 339, 568, 605
Polar detector 33, 38, 338, 339, 605
Polyphase codes 65, 66, 72, 321
Polyphase demodulation 28
Polyphase modulation 25, 28
Polyphase noise 38
Polyphase power 30-32, 36, 340, 351, 604
Population mean 528
Population variance 528
Positive phase sequence component 33, 36
Power dividers 152, 155
Power output stage 51
PPI see Plan position indicator
Pretrigger 10, 73
Primary radar 1, 17, 608, 612
Probability distribution 38, 40, 45, 527, 555, 565
Probability of detection 10, 14-16, 21, 41, 447, 451-457, 460-463, 465-467, 470, 471, 473-475, 478, 480-482, 484, 485, 487, 521, 610-612
Pulse compression 9, 23, 45, 52, 53, 58, 59, 62, 65, 67, 73, 292, 298, 301-304, 307, 309, 311, 312, 314-316, 319, 320, 326, 328, 340, 349, 350, 355, 359, 369, 370, 517, 523, 555, 562, 578, 609, 613
Pulse length (or width) discriminator 366, 409
Pulse modulator 51, 76, 78, 81

Quality 376, 385, 423, 432, 437, 485, 490, 495, 511, 565
Quantization 134, 138, 350, 351, 358-361, 508, 512, 513, 523, 524
Quantization noise loss 359

Radial velocity 1, 5, 6, 21, 23, 233, 246, 316, 337, 351, 494, 495, 515, 521, 612
Radio frequency amplifier 52, 277, 279, 280, 300
Radio frequency filter 273, 280, 281
Radio frequency (RF) modulator 75
Rain 1, 6, 22, 35, 152, 199, 200, 203, 209, 231, 246-251, 365, 366, 369, 370, 374, 386, 405, 406, 409, 428, 429, 432, 433, 436-438, 511, 519, 523, 524
Rain and chaff spectra 248
Rain maps 432, 433, 436
Raised cosine 621
Random errors 507, 512
Random numbers 205, 552, 553
Random walk 209, 210, 300
Range 1, 3-6, 8-10, 12, 14-19, 21-23, 513-515, 530, 532, 552, 509-513, 515, 517-521, 523, 525
Range errors 512
Range height indicator 507, 616
Range jitter loss 358, 359, 523
Range-height paper 8, 185, 188, 190, 192, 253-259
Rayleigh distribution 40, 41, 210, 211, 241, 242, 245-247, 332, 346, 369, 370, 464, 538-541, 549, 551
Rayleigh noise 40, 332, 344, 346, 347, 478, 546
Receiver 8-10, 28, 37, 291, 292, 298, 300, 311, 312, 323, 363, 364, 366, 368, 378, 404, 448, 483, 509-512, 518, 519, 523, 608-610, 613, 614
Rectangular pulse 41, 42, 64, 291-294, 296, 321, 327, 560-562, 572, 574, 575, 587, 588
Reflection 82, 84, 91, 110, 194, 195, 203-205, 217-227, 229, 230, 251, 513
Reflector 49, 97, 155, 246, 464, 468, 472
Refraction 8, 185, 188, 190, 192, 194, 195, 244, 251, 494, 504, 512
Rejection of echoes 385
Resolution 21-23, 38, 236, 246, 250, 301, 307, 316, 327, 328, 509, 515, 517, 518, 521, 524, 578, 585, 606, 610, 612, 616
Rice models 211, 213
Ricean distribution 40, 41, 433, 454, 541-543
Riesz tapering 621
Ring modulator 336
Rotary joints 91, 92
Roughness 229, 244

Sample mean 528, 529
Sampling waveforms and spectra 341
Saturation errors 350, 352
Sawtooth 506, 507
Scanning loss 160, 161, 177, 511, 523
Scattering 1, 8, 12, 19, 20, 233, 366, 438, 552, 614
Sea clutter 231, 244-246, 250, 541
Sea water 218-221, 224, 227, 228
Search radars 369, 381, 407, 513, 514
Second censoring 432, 436, 490
Secondary radar 2, 17, 21, 83, 84, 86, 92, 274, 276, 384, 494, 497, 513, 603, 607, 608, 610, 613, 614
Self-protection jammer 107, 519, 520
Sensitivity time control 8, 10, 14, 22, 139, 140, 178, 201, 274-279, 282, 284-287, 232, 248, 291, 300, 352, 364, 366, 377, 425, 432, 436, 613, 614
Series feed 132
Shaped beams 138, 227
Short time constant 9, 366
Sidelobes 58, 65, 72, 101-104, 106-112, 114-116, 118, 121, 123-126, 130, 131, 134, 154, 201, 276, 302, 304-308, 311, 312, 314, 316, 319, 321, 323, 325-328, 340, 506, 517, 520, 530, 577, 578, 584, 585, 587, 588, 600, 610, 613, 653, 654

Sidelobe blanking 364, 375
Sidelobe canceler 187, 364
Signal processor 8, 9, 23, 443, 491, 613
Sine wave 36, 63, 355, 559, 560, 572, 592, 607
Single canceler 403
Single phase sampling 341

Single canceler 168, 385, 390, 391, 393, 402-404, 418
SLB, see sidelobe blanking
SLC, see sidelobe canceler
Slow fluctuation 214, 458, 460, 468, 476, 477
Snidman, D.A. 488
Snow 200, 231, 246, 247, 250, 251, 432, 436, 511, 523, 524
Spectral leakage 423, 584, 621, 624, 628, 632, 635, 639, 641, 642, 644, 645, 648, 651, 656
Sphere 156, 185, 186, 202-204, 522
Spillover loss 152, 153
Square law detector 37, 332
Squint 132, 133, 178, 613
Stability 8, 22, 24, 76-78, 81, 86, 96, 178, 360, 509, 518, 521, 524
Staggering 21, 398-401, 405, 407, 431, 444, 604, 613
Standard deviation 11, 22, 23, 37, 40, 45, 47, 48, 497-500, 507, 508, 511, 515, 521, 527, 529, 531, 536, 538, 540, 544, 545, 547, 552, 569, 601, 602
Standard error of the mean 497, 536
STC, see Sensitivity time control
Step or quantization errors 350, 351
Straddling loss 358, 359, 523, 583
Stripline 89

Sun 164, 165, 167, 231, 251, 273, 437, 512
Suppressed carrier 330, 605
Surface tolerance losses 155, 177
Surveillance radars 14, 21, 236, 253, 606, 608, 614, 615
Swept gain 201, 274, 614
Swerling models 209, 211, 212, 465
Swerling case I 205, 372, 442, 452, 458, 460-464, 475, 476, 483, 486
Swerling case II 452, 464-467, 472, 475, 476, 483, 484, 486
Swerling case III 205, 452, 468-472, 475, 476, 483, 486
Swerling case IV 452, 472-475, 486
Switches 10, 91-93, 95
Symbols xxiii, 50, 617
Symmetrical components 25, 33, 351
Symmetry errors 350, 351
Synchronous detector 336, 337, 348, 349, 387, 604
Synthetic aperture 98, 170, 171, 173, 174, 176, 177, 180, 187, 443, 614
Synthetic array 170, 171, 173, 176, 187, 364, 612, 614

TACCAR, see Time averaged clutter coherent airborne radar
Tapered illumination 108, 109, 116, 131
Tapering 22, 23, 58, 59, 65, 66, 381, 498, 501, 502, 511, 657-663
Tapering functions 110, 115, 118, 121, 168, 302, 304, 306, 307, 312, 316, 321, 327, 420, 421, 423, 555, 578, 579, 583, 585, 587, 588, 619-621
Taylor antenna pattern 97, 109-111, 114, 115, 118, 121-123, 125, 126, 179, 181, 183, 290, 307, 649, 650, 652, 662, 663, 664, 673-675
Threshold 9, 12, 21, 216, 363, 364, 369-374, 376-378, 384, 385, 391, 397, 408, 425, 428, 429, 431, 433-437, 446, 447, 450-452, 454, 455, 458, 460, 462, 465, 470, 478, 483, 490, 495, 496, 509, 519, 527, 549, 555, 585, 611, 612
Time averaged clutter coherent airborne radar 438, 439, 614
Timing 7, 9, 11, 12, 22, 23, 309, 364, 388, 399, 512, 562, 611
Tracking radars 133, 216, 500, 513, 514, 518, 608
Transmission line 8, 22, 81, 278, 279
Transmit-receive tubes 91
Transmitter 8-10, 12-14, 16, 17, 19-23, 28, 51, 53, 72, 509, 511-514, 518, 523, 524, 619
Trapezoidal edges 54, 55
Trapezoidal aperture function 109, 110
Trapezoidal tapering 622-624
Triangular aperture function 109, 110, 121
Triangular tapering 291, 299, 321, 622-625, 664, 667, 668
Trigger generation 10-12, 24, 73, 76
Triplexer 91, 485
Truncated Gaussian aperture or tapering function 109, 110, 118, 646, 647, 649, 660
Tukey, J.W. 445, 574, 576, 600
Tuners 85, 95

Unambiguous Doppler frequency 22, 429, 431
Uncompressed pulses 292
Uniform aperture or tapering function 99, 104, 107-111, 115, 116, 118, 121, 131, 138, 141, 152, 533, 534, 552, 622-626, 629, 657

Variance 45, 528, 529, 531, 534, 536, 540, 544, 547, 549, 568, 602
Vector detector 9, 30, 301
Vector moving target indicator 388
Vector representation 27, 338
Video integration 384, 385, 496, 597
Voltage standing wave ratio 84-86, 88, 90-93, 95
Volume clutter 246
Voltage standing wave ratio 84-86, 88, 91, 95
von Hann, see Hann, J. von

VSWR, see Voltage standing wave ratio

Waveguide 8, 22, 51, 52, 72, 81-96, 518, 524, 603, 611, 612
Waveguide switch 8, 82, 92, 93
Weibull clutter 241, 246, 369
Weibull distribution 549-551
Weierstrass tapering 621
Weil, T.A. 193, 251
Weinstock models 208, 209
Wien bridge 25, 26
Woodward-Lawson 139, 140, 142

z transform 381, 555, 590-592
Zolotarëv tapering function 97, 121, 123-125, 181, 182, 605, 621, 664, 679-681

Recent Titles in the Artech House Radar Series

David K. Barton, Series Editor

Adaptive Antennas and Phased Arrays for Radar and Communications, Alan J. Fenn

Advanced Techniques for Digital Receivers, Phillip E. Pace

Advances in Direction-of-Arrival Estimation, Sathish Chandran, editor

Airborne Pulsed Doppler Radar, Second Edition, Guy V. Morris and Linda Harkness, editors

Bayesian Multiple Target Tracking, Lawrence D. Stone, Carl A. Barlow, and Thomas L. Corwin

Beyond the Kalman Filter: Particle Filters for Tracking Applications, Branko Ristic, Sanjeev Arulampalam, and Neil Gordon

Computer Simulation of Aerial Target Radar Scattering, Recognition, Detection, and Tracking, Yakov D. Shirman, editor

Design and Analysis of Modern Tracking Systems, Samuel Blackman and Robert Popoli

Detecting and Classifying Low Probability of Intercept Radar, Phillip E. Pace

Digital Techniques for Wideband Receivers, Second Edition, James Tsui

Electronic Intelligence: The Analysis of Radar Signals, Second Edition, Richard G. Wiley

Electronic Warfare in the Information Age, D. Curtis Schleher

ELINT: The Interception and Analysis of Radar Signals, Richard G. Wiley

EW 101: A First Course in Electronic Warfare, David Adamy

EW 102: A Second Course in Electronic Warfare, David L. Adamy

Fourier Transforms in Radar and Signal Processing, David Brandwood

Fundamentals of Electronic Warfare, Sergei A. Vakin, Lev N. Shustov, and Robert H. Dunwell

Fundamentals of Short-Range FM Radar, Igor V. Komarov and Sergey M. Smolskiy

Handbook of Computer Simulation in Radio Engineering, Communications, and Radar, Sergey A. Leonov and Alexander I. Leonov

High-Resolution Radar, Second Edition, Donald R. Wehner

Introduction to Electronic Defense Systems, Second Edition, Filippo Neri

Introduction to Electronic Warfare, D. Curtis Schleher

Introduction to Electronic Warfare Modeling and Simulation, David L. Adamy

Introduction to RF Equipment and System Design, Pekka Eskelinen

Microwave Radar: Imaging and Advanced Concepts, Roger J. Sullivan

Millimeter-Wave Radar Targets and Clutter, Gennadiy P. Kulemin

Modern Radar Systems, Second Edition, Hamish Meikle

Modern Radar System Analysis, David K. Barton

Modern Radar System Analysis Software and User's Manual, Version 3.0, David K. Barton

Multitarget-Multisensor Tracking: Applications and Advances Volume III, Yaakov Bar-Shalom and William Dale Blair, editors

Principles of High-Resolution Radar, August W. Rihaczek

Principles of Radar and Sonar Signal Processing, François Le Chevalier

Radar Cross Section, Second Edition, Eugene F. Knott et al.

Radar Evaluation Handbook, David K. Barton et al.

Radar Meteorology, Henri Sauvageot

Radar Reflectivity of Land and Sea, Third Edition, Maurice W. Long

Radar Resolution and Complex-Image Analysis, August W. Rihaczek and Stephen J. Hershkowitz

Radar Signal Processing and Adaptive Systems, Ramon Nitzberg

Radar System Analysis and Modeling, David K. Barton

Radar System Performance Modeling, Second Edition, G. Richard Curry

Radar Technology Encyclopedia, David K. Barton and Sergey A. Leonov, editors

Range-Doppler Radar Imaging and Motion Compensation, Jae Sok Son et al.

Signal Detection and Estimation, Second Edition, Mourad Barkat

Space-Time Adaptive Processing for Radar, J. R. Guerci

Theory and Practice of Radar Target Identification, August W. Rihaczek and Stephen J. Hershkowitz

Time-Frequency Transforms for Radar Imaging and Signal Analysis, Victor C. Chen and Hao Ling

For further information on these and other Artech House titles, including previously considered out-of-print books now available through our In-Print-Forever® (IPF®) program, contact:

Artech House
685 Canton Street
Norwood, MA 02062
Phone: 781-769-9750
Fax: 781-769-6334
e-mail: artech@artechhouse.com

Artech House
46 Gillingham Street
London SW1V 1AH UK
Phone: +44 (0)20 7596-8750
Fax: +44 (0)20 7630-0166
e-mail: artech-uk@artechhouse.com

Find us on the World Wide Web at: www.artechhouse.com

www.ingramcontent.com/pod-product-compliance
Lightning Source LLC
LaVergne TN
LVHW062133200926
848484LV00007B/1806